All-Russian Conference with International Participation on Modern Problems of Continuum Mechanics and Explosion Physics 2017

Dedicated to the 60th Anniversary of Lavrentyev Institute of Hydrodynamics SB RAS

Journal of Physics: Conference Series Volume 894

AA006762

Novosibirsk, Russia
4 – 8 September 2017

ISBN: 978-1-5108-5034-7
ISSN: 1742-6588

Printed from e-media with permission by:

Curran Associates, Inc.
57 Morehouse Lane
Red Hook, NY 12571

Some format issues inherent in the e-media version may also appear in this print version.

Copyright© (2017) by the Institute of Physics
All rights reserved. The material featured in this book is subject to
IOP copyright protection, unless otherwise indicated.

Printed by Curran Associates, Inc. (2017)

For permission requests, please contact the Institute of Physics
at the address below.

Institute of Physics
Dirac House, Temple Back
Bristol BS1 6BE UK

Phone: 44 1 17 929 7481
Fax: 44 1 17 920 0979

techtracking@iop.org

Additional copies of this publication are available from:

Curran Associates, Inc.
57 Morehouse Lane
Red Hook, NY 12571 USA
Phone: 845-758-0400
Fax: 845-758-2633
Email: curran@proceedings.com
Web: www.proceedings.com

Table of contents

Volume 894

All-Russian Conference with international participation "Modern problems of continuum mechanics and explosion physics" dedicated to the 60th anniversary of Lavrentyev Institute of Hydrodynamics SB RAS

4–8 September 2017, Novosibirsk, Russian Federation

Accepted papers received: 15 September 2017
Published online: 22 October 2017

Preface

Papers

IOP Conf. Series: Journal of Physics: Conf. Series **894** (2017) 011001 doi:10.1088/1742-6596/894/1/011001

Preface

The present volume of the *Journal of Physics: Conference Series* represents contributions from participants of the All-Russian Conference with international participation «Modern problems of continuum mechanics and explosion physics» dedicated to the 60th anniversary of Lavrentyev Institute of Hydrodynamics SB RAS held at the Technopark of Novosibirsk Akademgorodok, Russia, September, 4-8 2017, organised by the Lavrentyev Institute of Hydrodynamics.

The conference is traditional and attracts attention of mathematicians and physicists. Scientific activities of the Conference include a discussion of a wide range of problems: mathematical problems of continuum mechanics, physics and mechanics of high-energy processes, fluid mechanics and mechanics of solids. The program of conference included 18 plenary presentations (30 minutes), 170 oral presentations (15 minutes) and 50 posters given by more than 220 participants from 23 cities of Russia (Moscow, Novosibirsk, Vladivostok, Tomsk, Yakutsk, Perm, Novokuznetsk, Biisk, Lys'va, Kemerovo, Snezhinsk, Volgograd, Tyumen, Barnaul, Yekaterinburg, Krasnoyarsk, Irkutsk), Portugal (Lisbon), Kazakhstan (Almaty), Thailand (Nakhon Ratchasima), Denmark (Lyngby), Spain (Oviedo) and France (Marseille). A significant part of talks (about 25 %) were presented by young researchers and postgraduate students.

We would like to thank the Speakers for their significant contributions to the conference. We also would like to thank the members of the Organizing Committee. We cannot end without expressing our many thanks to our sponsors (Federal Agency for Scientific Organizations (FASO Russia), Russian Foundation for Basic Research and Technology Company Schlumberger, International Mathematical Center of Novosibirsk State University) for their financial support.

Novosibirsk, September 2017

Alexander Chesnokov, Eduard Pruuel, Vladimir Shelukhin

Content from this work may be used under the terms of the Creative Commons Attribution 3.0 licence. Any further distribution of this work must maintain attribution to the author(s) and the title of the work, journal citation and DOI.
Published under licence by IOP Publishing Ltd

IOP Conf. Series: Journal of Physics: Conf. Series **894** (2017) 011002 doi:10.1088/1742-6596/894/1/011002

Peer review statement

All papers published in this volume of *Journal of Physics: Conference Series* have been peer reviewed through processes administered by the proceedings Editors. Reviews were conducted by expert referees to the professional and scientific standards expected of a proceedings journal published by IOP Publishing.

Content from this work may be used under the terms of the Creative Commons Attribution 3.0 licence. Any further distribution of this work must maintain attribution to the author(s) and the title of the work, journal citation and DOI.
Published under licence by IOP Publishing Ltd

IOP Conf. Series: Journal of Physics: Conf. Series **894** (2017) 012001 doi:10.1088/1742-6596/894/1/012001

The energy method. Application to PDEs of hydrodynamics with nonstandard growth

S N Antontsev[1,2] **and S I Shmarev**[3]
[1]CMAF-CIO, University of Lisbon, Portugal
[2] Lavrentyev Institute of Hydrodynamics SB RAS, Novosibirsk, Russia
[3] University of Oviedo, Spain.

E-mail: antontsevsn@mail.ru, shmarev@uniovi.es

Abstract. We present a method for the study of localization properties of solutions of nonlinear equations and systems. An example of application of the method is furnished by results of analysis of a mathematical model for a class of non-Newtonian fluids.

1. Introduction

Numerous mathematical models of continuum mechanics involve combined systems of nonlinear PDEs where different components of the solution, such as velocity, density, pressure, temperature, satisfy equations of different type. These equations may degenerate or become singular at certain values of the solutions or their derivatives which causes localization of solutions in space or time. The typical localization effects are the finite speed of propagation of disturbances from the initial and boundary data, the waiting time phenomenon, extinction or blow-up of solutions in a finite time.

The standard approach to the study of localization properties consists in comparison with suitable explicit solutions or sub/super solutions of the same equation, see, e.g., [1]. However, this method ceases to be applicable if the explicit solutions are not available, or when the maximum principle fails. An alternative approach is based on analysis of the local energy functions which satisfy nonlinear ordinary differential inequalities stemming from the PDE or the system of PDEs being studied. We refer to the monographs [2, 3] for an insight into the method and to papers [4]-[12] for its applications to the study of localization properties of solutions of the mathematical models of hydrodynamics, including the flows of Non-Newtonian viscous fluids.

In this note, we make an emphasis on the possibility of application of the method to the models which include the so-called PDEs with nonstandard growth, i.e., the PDEs with variable nonlinearity. For such models, the method of local energies has turned out to be the most effective in the study of the localization properties of solutions - see [3, 13]. The method is illustrated by the proof of extinction in a finite time of solutions of the system describing the flows of incompressible non-homogeneous non-Newtonian fluids.

2. Flows of non-homogeneous non-Newtonian fluids

Let $\Omega \subset \mathbb{R}^N$ be a bounded domain, $N \geq 2$, and $\partial\Omega \in \text{Lip}$. We consider the incompressible flow governed by the laws of balance of mass and momentum, which lead to the following system of

Content from this work may be used under the terms of the Creative Commons Attribution 3.0 licence. Any further distribution of this work must maintain attribution to the author(s) and the title of the work, journal citation and DOI.
Published under licence by IOP Publishing Ltd

equations posed in the cylinder $Q_T = \Omega \times (0, T)$:

$$D_t\rho \equiv \frac{\partial \rho}{\partial t} + \mathbf{v} \cdot \nabla \rho = 0 \qquad \text{(the transport equation, hyperbolic)}, \tag{2.1}$$

$$\operatorname{div} \mathbf{v} = 0 \qquad \text{(the incompressibility equation, elliptic)}, \tag{2.2}$$

$$\rho D_t \mathbf{v} = -\nabla p + \operatorname{div} \mathbf{S} + \rho \mathbf{f} \qquad \text{(the balance of momentum, parabolic)}. \tag{2.3}$$

System (2.1)–(2.3) is completed by the initial and boundary conditions

$$\rho(x,0)\mathbf{v}(x,0) = \rho_0(x)\mathbf{v}_0(x), \quad \rho(x,0) = \rho_0(x) \text{ as } x \in \Omega, \tag{2.4}$$

$$\mathbf{v}(x,t) = 0 \text{ for } (x,t) \in \Gamma_T = \partial\Omega \times (0,T). \tag{2.5}$$

Here $\mathbf{v}(x,t)$, $\rho(x,t)$ and $p(x,t)$ stand for velocity, density and pressure in the fluid, $\mathbf{S}$ is the deviatoric part of the Cauchy stress tensor, $\mathbf{D}$ is the strain rate tensor and $\mathbf{f}(x,t)$ is the prescribed mass force. The tensor $\mathbf{S}$ is a (nonlinear) tensor function of the strain rate tensor $\mathbf{D}$ and the thermodynamic characteristics of the flow, for example, temperature θ, density ρ, or entropy. The tensor $\mathbf{S}$ is usually defined by the constitutive relation

$$\mathbf{S} = \mathbf{F}(\mathbf{D}, \rho, \theta) \equiv \nu(|\mathbf{D}|, \rho, \theta)\mathbf{D}, \tag{2.6}$$

see [14, 15] for further details and justifications. A few examples of fluids with different constitutive laws (2.6):

- *viscoplastic fluids* (Eu. Bingham, 1922)

$$\nu(|\mathbf{D}|) := \begin{cases} \infty \Rightarrow \mathbf{D} = \mathbf{0}, & |\mathbf{S}| \leq |\mathbf{S}_0|, \quad \mathbf{S}_0 \text{ is a yield stress} \\ \nu_0 + |\mathbf{S}_0||\mathbf{D}|^{-1}, & |\mathbf{S}| > |\mathbf{S}_0|, \quad \nu_0 = const. > 0; \end{cases}$$

- *generalized Newtonian fluids* (A. de Waele, 1923; W. Ostwald, 1925)

$$\nu(|\mathbf{D}|) = \nu_0 |\mathbf{D}|^{q-2}, \qquad \nu_0 = Const. > 0,$$

  which include viscoplastic fluids ($q = 1$), pseudoplastic or shear-thinning fluids ($1 < q < 2$), Newtonian fluids ($q = 2$), dilatant or shear-thickening fluids ($q > 2$);
- *a mixture of a Newtonian fluid with thickening agents* (A. W. Sisko, 1958)

$$\nu(|\mathbf{D}|) = \nu_\infty + \nu_0 |\mathbf{D}|^{q-2}, \quad \nu_\infty,\ \nu_0 = Const. > 0.$$

A special class of fluids is constituted by *electrorheological fluids*, which consist of solid particles dispersed in an insulating liquid. Their characteristic feature is the possibility of variation of the rheological properties under the influence of an exterior electric field. The mathematical models of motion of such fluids involve the constitutive equation (2.6) with variable exponents of nonlinearity, which depend on the electric field: $\nu \sim \nu_0 + \nu_1 |\mathbf{D}|^{q(\mathbf{E})}$ with constant ν_0, ν_1 - see [16]. The flow is described by a modified system of Navier-Stokes equations and the system of Maxwell equations for $\mathbf{E}$. Similar constitutive relations with $q \equiv q(\theta)$ are used in modelling of *thermorheological fluids* whose rheology changes together with the variation of the temperature $\theta(x,t)$ - see [17, 18] for an analysis of the pertinent mathematical models.

The study of well-posedness of the mathematical model and the analysis of the qualitative properties of solutions are often disconnected. For this reason in what follows we assume that the arguments of the function ν in (2.6) (the thermodynamic characteristics ρ, θ of the flow or the electric field $\mathbf{E}$) are known functions of (x,t). We will also assume that $\mathbf{F} = \mathbf{F}(\mathbf{D}, x, t)$ satisfies the coercivity condition: for every symmetric tensor $\mathbf{D} \in \mathbb{R}^{N \times N}$

$$\begin{aligned} &\forall (x,t)\in Q_T \quad \delta|\mathbf{D}|^{q(x,t)} \leq \mathbf{F}(\mathbf{D}):\mathbf{D} = F^{ij}D^{ij}, \quad \text{where } |\mathbf{D}|^2 = D^{ij}D^{ij}, \\ &1 < \underline{q} \leq q^-(t) \leq q(x,t) \leq q^+(t) \leq \overline{q} < \infty, \quad 0 < \delta \equiv \delta(\rho) < \infty, \end{aligned} \tag{2.7}$$

with given functions $q(x,t)$, $q^{\pm}(t)$, $\delta(\rho)$ and constants $\underline{q}$, $\overline{q}$.

System (2.1)-(2.6) is a system of combined type. Indeed: if we assume that in (2.1) the velocity $\mathbf{v}(x,t)$ is given, then (2.1) can be regarded as a hyperbolic equation for the density $\rho(x,t)$, and if in (2.3) the density $\rho(x,t)$ is known, then system (2.2)-(2.3) becomes a pseudo-parabolic system of equations for the velocity $\mathbf{v}$ and pressure p. The specific type of system (2.1)-(2.3) is unimportant for the applicability of the method.

3. Energy solutions

It is assumed that the initial data satisfy the conditions

$$E(0) = \frac{1}{2}\int_\Omega \rho(x,0)|\mathbf{v}(x,0)|^2 dx < \infty, \qquad \frac{1}{M} \leq \rho_0 \leq M \equiv const. \tag{3.1}$$

The solution $\{\mathbf{v},\rho\}$ of problem (2.1)-(2.5) is understood in the weak sense, i.e., as functions satisfying the integral identities which appear after multiplying equations (2.1), (2.3) by a smooth solenoidal test-function and integrating by parts. In the result, the pressure p is excluded from the weak formulation.

There exists an abundant literature devoted to the issues of existence and uniqueness of solutions of problem (2.1)-(2.4). In the incompressible homogeneous fluid ρ is constant and equation (2.1) fulfills automatically. For this class of fluids, solvability of the classical Navier-Stokes system with constant ρ and relation (2.7) with the parameters $\delta = const$ and $q = 2$ was studied in [19]-[21]. The flows of incompressible homogeneous trembling electrorheological fluids were considered in [10]. The authors of [10] studied solvability of the problem and the asymptotic behavior of solutions as $t \to \infty$. Global in time existence of weak solutions to problem (2.1)-(2.5) for nonhomogeneous fluids with $\rho(x,t) \not\equiv const$ was proved in [14, 15],[22]-[30] for various constitutive relations (2.7). In particular, for the fluids with the constitutive relation $\mathbf{S} = 2\mu\mathbf{D}$, $\mu = const$, the existence of weak solutions was proved in [14] in the class of functions

$$\{\mathbf{v},\rho\} \in W_q = \left\{ \begin{array}{l} \mathbf{v} \in L^\infty(0,T;L^2(\Omega)) \cap L^q\left(0,T;W_0^{1,q}(\Omega)\right) \\ \mathbf{D}(\mathbf{v}) \in L^q(Q_T), \ \operatorname{div}\mathbf{v} = 0, \ \dfrac{1}{M} \leq \rho \leq M \end{array} \right\}, \qquad q = const > 1.$$

Let us introduce *the energy function*

$$E(t) = \frac{1}{2}(\rho\mathbf{v},\mathbf{v})_\Omega = \frac{1}{2}\int_\Omega \rho(x,t)|\mathbf{v}(x,t)|^2\, dx$$

and consider the *energy solutions* which satisfy the relations

$$E'(t) + (\mathbf{F}:\mathbf{S},1)_\Omega = (\rho\mathbf{f},\mathbf{v})_\Omega, \qquad E'(t) = \frac{dE}{dt}(t). \tag{3.2}$$

Equality (3.2) can be derived from the integral identity for equation (2.3) with the test-function $\mathbf{v}$, equations (2.1)-(2.2) and the formulas of integration by parts (see, e.g., [10]):

$$\begin{aligned} &\left(\rho\frac{d\mathbf{v}}{dt},\mathbf{v}\right)_\Omega = \frac{1}{2}\frac{d}{dt}(\rho\mathbf{v},\mathbf{v})_\Omega \equiv \frac{dE}{dt}(t), \\ &(-\nabla p + \operatorname{div}\mathbf{S},\mathbf{v})_\Omega = (p,\operatorname{div}\mathbf{v})_\Omega - (\mathbf{F}:\mathbf{S},1)_\Omega = -(\mathbf{F}:\mathbf{S},1)_\Omega. \end{aligned}$$

IOP Conf. Series: Journal of Physics: Conf. Series **894** (2017) 012001 doi:10.1088/1742-6596/894/1/012001

4. Variable Lebesgue and Sobolev spaces

The analysis of the problem with variable nonlinearity in (2.6) is performed in the context of Lebesgue and Sobolev spaces with variable exponents [31]. Given a measurable function $p : \Omega \mapsto (1, \infty)$, we denote by $L^{p(\cdot)}(\Omega)$ the set of measurable functions $f(x)$ on Ω such that

$$A_{p(\cdot)}(f) = \int_{\Omega} |f(x)|^{p(x)}\, dx < \infty, \qquad p(x) \in [p^-, p^+] \subset (1, \infty) \tag{4.1}$$

with some constants $p^{\pm}$. The set $L^{p(x)}(\Omega)$ endowed with the Luxemburg norm,

$$\|f\|_{p(\cdot),\Omega} \equiv \|f\|_{L^{p(\cdot)}(\Omega)} = \inf\left\{\lambda > 0 : \ A_{p(\cdot)}\left(\frac{f}{\lambda}\right) \le 1\right\}, \tag{4.2}$$

becomes a Banach space. For every $f \in L^{p(\cdot)}(\Omega)$ the relations between the norm and the modular $A_p(f)$ is given by the inequalities

$$\min\left\{A_{p(\cdot)}^{\frac{1}{p^-}}(f), A_{p(\cdot)}^{\frac{1}{p^+}}(f)\right\} \le \|f\|_{p(\cdot),\Omega} \le \max\left\{A_{p(\cdot)}^{\frac{1}{p^-}}(f), A_{p(\cdot)}^{\frac{1}{p^+}}(f)\right\} \tag{4.3}$$

with constants $p^{\pm}$ from condition (4.1). We assume that for all $(x, y) \in \overline{\Omega}$ and every $t \in [0, T]$

$$1 < \underline{q} \le q^-(t) \le q(x,t) \le q^+(t) \le \overline{q} < \infty, \qquad |q(x,t) - q(y,t)| \le \omega(|x-y|), \tag{4.4}$$

where the modulus of continuity ω satisfies the condition $\overline{\lim}_{\tau \to 0^+} \omega(\tau) \ln \frac{1}{\tau} = C_\omega < \infty$. Let $W_0^{1,p(\cdot)}(\Omega) = \left\{u \in L^{p(\cdot)}(\Omega) \cap W_0^{1,1}(\Omega),\ |\nabla u|^{p(x)} \in L^1(\Omega)\right\}$ and

$$\mathcal{W}_{q(\cdot)}(Q_T) = \left\{ \mathbf{v} : Q_T \mapsto \mathbb{R}^n \ \middle|\ \begin{array}{l} \mathbf{v} \in \left(L^\infty(0,T; L^2(\Omega) \cap L^1(0,T; W_0^1(\Omega))\right)^n; \\ \operatorname{div} \mathbf{v} = 0, \quad \int_0^T \int_\Omega |\mathbf{D}(\mathbf{v})|^{q(x,t)}\, dxdt < \infty \end{array} \right\}.$$

The key technical tool used in the analysis is Korn's inequality for $\mathbf{v}(x,t) \in \mathcal{W}_{q(\cdot)}(Q_T)$.

Lemma 4.1. *If $q(x,t)$ satisfies (4.4), $\mathbf{v} = (v_1, \ldots, v_n)$ with $v_i(\cdot, t) \in W_0^{1,q(\cdot,t)}(\Omega)$ for a.e. $t \in [0,T]$, then*

$$\frac{1}{C}\|\nabla \mathbf{v}(\cdot,t)\|_{q(\cdot,t),\Omega} \le \|\mathbf{D}(\mathbf{v}(\cdot,t))\|_{q(\cdot,t),\Omega}, \qquad K\|\mathbf{v}(\cdot,t)\|_{r(\cdot,t),\Omega} \le \|\mathbf{D}(\mathbf{v}(\cdot,t))\|_{q(\cdot,t),\Omega} \tag{4.5}$$

where $q < N$, $r \le \dfrac{qN}{N-q} < \infty$, $K = K\left(C_\omega, \underline{q}, \overline{q}, N, \Omega\right)$, $C = C\left(C_\omega, \underline{q}, \overline{q}, N, \Omega\right)$.

We refer to [31, Th.14.3.21] for the proof of (4.5) with $\mathbf{v}$, q depending only on x and q satisfying the log-continuity condition (4.4) in Ω. For $q = const$ the proof of Korn's inequality can be found in [32]. Using (4.5) with $r = 2$, applying the embedding $W^{1,2}(\Omega) \subset L^2(\Omega)$ and inequalities (4.3) we derive the following chain of inequalities: for $\underline{q} \ge \frac{2N}{N+2}$

$$\frac{1}{K}\|\mathbf{v}(\cdot,t)\|_{2,\Omega} \le \|\mathbf{D}(\mathbf{v}(\cdot,t))\|_{q(\cdot),\Omega} \le \max\left\{A_{q(\cdot,t)}^{\frac{1}{q^-(t)}}(\mathbf{D}(\mathbf{v}(\cdot,t))), A_{q(\cdot,t)}^{\frac{1}{q^+(t)}}(\mathbf{D}(\mathbf{v}(\cdot,t)))\right\}. \tag{4.6}$$

5. Pseudoplastic fluids. Localization in time via analysis of the energy function

Let us consider the energy solutions of system (2.1)-(2.3) which satisfy identity (3.2). Assume that

$$\sup_{t\in(0,T)} E(t) \le \overline{E} < \infty, \qquad \int_0^T \int_\Omega |\mathbf{D}|^{q(x,t)}\,dxdt < \infty, \qquad \frac{1}{M} \le \rho(x,t) \le M < \infty \tag{5.1}$$

with some constants $\overline{E}$ and $M > 0$. The last two inequalities are byproducts of the maximum principle for the solutions of equation (2.1) and restriction (3.1) on $\rho(x,0)$. Without loss of generality we may assume that $\overline{E} \le 1$. It follows from (4.6), (5.1) that

$$C_0 E^{\frac{\overline{q}}{2}}(t) \le C_0 E^{\frac{q^+(t)}{2}}(t) \le \delta \int_\Omega |\mathbf{D}(\mathbf{v}(x,t))|^{q(x,t)}\,dx \le (\mathbf{F}:\mathbf{S},1)_\Omega, \quad C_0 = C_0(K,M,\overline{E}). \tag{5.2}$$

Theorem 5.1. *Let $\{\mathbf{v},\rho,p\}$, $\mathbf{v}\in \mathcal{W}_{q(\cdot)}(Q_T)$, be a weak solution of problem (2.1)–(2.5). Assume that condition (3.1) is fulfilled and, additionally,*

$$\text{(2.7) holds with } \delta = \delta_0 = const \text{ and } q(x,t) \in \left(\frac{2N}{N+2}, 2\right). \tag{5.3}$$

(1) If $\mathbf{f}\equiv 0$, then

$$E(t) = \frac{1}{2}\int_\Omega \rho(x,t)|\mathbf{v}(x,t)|^2 dx = 0 \qquad \text{for } t \ge t^* = \frac{2E^{(2-\overline{q})/2}(0)}{C_0(2-\overline{q})}.$$

In particular, $\mathbf{v}\,(x,t)\equiv 0$ in $Q_T \cap \{t\ge t^*\}$.

(2) Let us denote $u_+ = \max\{u,0\}$. Assume that $\mathbf{f}\not\equiv 0$, $\overline{q}<2$ and

$$\|\mathbf{f}(\cdot,t)\|_{2,\Omega} \le \epsilon\,(1-t/t_f)_+^{\overline{q}/(2-\overline{q})} \tag{5.4}$$

with $\varepsilon = const > 0$, $t_f > t^*$. If the data satisfy the condition

$$\frac{\overline{q}E(0)}{(2-\overline{q})t_f} + C_0 E^{\frac{\overline{q}}{2}}(0) = \varepsilon\sqrt{2M}, \tag{5.5}$$

then there exists a constant C such that $E(t) \le C\,[1-t/t_f]_+^{\overline{q}/(2-\overline{q})}$, that is, $E(t)=0$ for $t\ge t_f$ and $\mathbf{v}(x,t)\equiv 0$ in $Q\cap\{t\ge t_f\}$.

The mechanical meaning of Theorem 5.1 is that if the flow of a nonhomogeneous non–Newtonian pseudoplastic fluid is driven by the initial data, then the fluid becomes immobile in a finite time. If the flow is stirred by the source term $\mathbf{f}\ne 0$ that vanishes at the instant t_f, then the fluid is at rest for all $t\ge t_f$, provided that the intensity of the source is appropriately small.

Sketch of the proof. Applying (2.7), (3.2) and (5.2) we derive the nonlinear integral and differential inequalities

$$\begin{aligned} E(t) + C_0\int_0^t E^{\frac{\overline{q}}{2}}(\tau)d\tau &\le E(t) + C_0\int_0^t E^{\frac{q^+(\tau)}{2}}(\tau)d\tau \\ &\le E(t) + \int_0^t (\mathbf{F}:\mathbf{S},1)_\Omega d\tau = E(0) + \int_0^t (\rho\mathbf{f},\mathbf{v})_\Omega d\tau, \end{aligned}$$

$$E'(t) + C_0 E^{\frac{\overline{q}}{2}}(t) \le E'(t) + C_0 E^{\frac{q^+(\tau)}{2}}(t) \le E'(t) + \int_0^t (\mathbf{F} : \mathbf{S}, 1)_\Omega d\tau = (\rho \mathbf{f}, \mathbf{v})_\Omega. \tag{5.6}$$

If $\mathbf{f} \equiv 0$, we obtain the ordinary differential inequality $E'(t) + C_0 E^{\frac{\overline{q}}{2}}(t) \le 0$ and after a straightforward integration arrive at the estimate $E(t) \le \left[E^{\frac{2-\overline{q}}{2}}(0) - tC_0\frac{2-\overline{q}}{2}\right]$. Since $E(t) \ge 0$ by definition, it is necessary that $E(t) \equiv 0$ for all $t \ge t^*$. Under the assumptions of item (2), from (5.4), (5.6) we derive the following inequality:

$$E'(t) + C_0 E^{\frac{\overline{q}}{2}}(t) \le \sqrt{2M}(1 - t/t_f)_+^{\overline{q}/(2-\overline{q})}, \ (E \le 1).$$

Let $W(t)$ be a solution of the problem

$$W'(t) + C_0 W^{\frac{\overline{q}}{2}}(t) = \sqrt{2M}(1 - t/t_f)_+^{\overline{q}/(2-\overline{q})}, \quad W(0) = E(0).$$

Under condition (5.5) the problem for W has a unique solution $W(t) = E(0)\left[1 - t/t_f\right]_+^{\overline{q}/(2-\overline{q})}$, which is a majorant for $E(t)$, whence $E(t) \le W(t)$. □

The borderline case $q^+(t) \nearrow 2$ as $t \to \infty$ was considered in [10]. The analysis of behavior of the fluid as $t \to \infty$ reduces to the study of the ordinary differential inequality with the variable exponent $E'(t) + C_0 E^{\frac{q^+(t)}{2}}(t) \le (\rho \mathbf{f}, \mathbf{v})_\Omega$. Properties of the functions satisfying the inequalities of this type were studied in [3, Ch.6]. Analogous localization results were proved for pseudoplastic fluid with vanishing or unbounded initial density $\rho(x, 0) = \rho_0(x)$ under the assumptions

$$\|1/\rho_0(x)\|_{L^m(\Omega)} \le C_1, \quad \|\rho_0(x)\|_{L^M(\Omega)} \le C_2, \quad \min\{m, M\} > 1, \tag{5.7}$$

$$\sqrt{\rho_0}\mathbf{v}_0 \in L^2(\Omega), \quad E(0) = \frac{1}{2}\int_\Omega \rho_0 |\mathbf{v}_0|^2 dx < \infty; \quad \delta |\mathbf{D}(\mathbf{r})|^q \le \mathbf{F}(\mathbf{r}) : \mathbf{D}(\mathbf{r}) \ \forall \mathbf{r} \in \mathbb{R}^N$$

with $\delta = const > 0$, $q \in \left(\frac{2MN}{N(M-1)+2M}, 2\right)$, $M > \frac{N}{2}$. Condition (5.7) allows the initial density ρ_0 to vanish or become infinite on any set of zero measure.

Acknowledgments
This work of the first author was supported by the Research Grant Num. 15-11-20019 of Russian Science Foundation, Russia. The second author acknowledges the support of the Research Grant MTM2013-43671-P, MICINN, Spain.

References

[1] Samarskii A A, Galaktionov V A, Kurdyumov S P and Mikhailov A P 1995 *Blow-up in Quasilinear Parabolic Equations* (Berlin: Walter de Gruyter & Co.)
[2] Antontsev S N, Díaz J I and Shmarev S 2002 *Energy methods for free boundary problems. Applications to Nonlinear PDEs and Fluid Mechanics* Progress in Nonlinear Differential Equations and their Applications Vol 48 (Boston: Birkhäuser Boston, Inc.)
[3] Antontsev S and Shmarev S 2015 *Evolution PDEs with Nonstandard Growth Conditions. Existence, Uniqueness, Localization, Blow-up* (*Atlantis Studies in Differential Equations* vol 4) (Paris: Atlantis Press)
[4] Antontsev S N, Epikhov G P and Kashevarov A A 1986 *Mathematical System Modelling of Water Exchange Processes* (Novosibirsk: "Nauka" Sibirsk. Otdel.)
[5] Antontsev S N and de Oliveira H B 2007 *Kyoto Conf. on the Navier-Stokes Equations and their Applications* RIMS Kôkyûroku Bessatsu, B1 (Res. Inst. Math. Sci. (RIMS), Kyoto) pp 21–41

[6] Antontsev S N and Oliveira H B 2007 *Free Boundary Problems* (*Internat. Ser. Numer. Math.* vol 154) (Basel: Birkhäuser) pp 23–32
[7] Antontsev S N, Díaz J I and Oliveira H 2005 *Progress in Nonlinear Diferential Equations and Their Applications, (Proc. TPDE,Obidos, Portugal, June 7-10, 2003)* vol 61 (Obidos, Portugal, June 7-10)
[8] Antontsev S N and de Oliveira H B 2014 *Adv. Differential Equations* **19** 441–472
[9] Antontsev S N and de Oliveira H B 2011 *J. Math. Anal. Appl.* **379** 802–817
[10] Antontsev S N and de Oliveira H B 2014 *Nonlinear Anal. Real World Appl.* **19** 54–66
[11] Antontsev S N and de Oliveira H B 2016 *Rev. R. Acad. Cienc. Exactas Fís. Nat. Ser. A Math. RACSAM* **110** 729–754
[12] Antontsev S N and Khompysh K 2017 *J. Math. Anal. Appl.* **446** 1255–1273
[13] Antontsev S N and Shmarev S I 2014 *Nonlinear Anal.* **95** 483–498
[14] Böhm M 1985 *J. Differential Equations* **60** 259–284 ISSN 0022-0396
[15] Fernández-Cara E, Guillen F and Ortega R R 1997 *Nonlinear Anal., Theory Methods Appl.* **28** 1079–1100
[16] Růžička M 2000 *Electrorheological Fluids: Modeling and Mathematical Theory* Lecture Notes in Mathematics Vol 1748 (Berlin: Springer)
[17] Zhikov V V 2008 *Tr. Mat. Inst. Steklova* **261** 101–114
[18] Antontsev S N and Rodrigues J F 2006 *Ann. Univ. Ferrara, Sez.,VII. Sci. Mat.* **52** 19–36
[19] Ladyzhenskaya O A 1969 *The mathematical Theory of Viscous Incompressible Flow* (New York: Gordon and Breach Science Publishers) Mathematics and its Applications, Vol. 2
[20] Temam R 1977 *Navier-Stokes equations. Theory and Numerical Analysis* (Amsterdam: North-Holland Publishing Co.) Studies in Mathematics and its Applications, Vol. 2
[21] Galdi G P 1994 *An Introduction to the Mathematical Theory of the Navier-Stokes Equations* Springer Tracts in Natural Philosophy Vol 38 (New York, Berlin: Springer-Verlag)
[22] Antontsev S N, Kazhikhov A V and Monakhov V N 1990 *Boundary Value Problems in Mechanics of Nonhomogeneous Fluids* (Amsterdam: North-Holland Publishing Co.)
[23] Diening L, Růžička M and Wolf J 2010 *Ann. Sc. Norm. Super. Pisa Cl. Sci. (5)* **9** 1–46
[24] Berselli L C, Diening L and Rů˘ zička M 2010 *J. Math. Fluid Mech.* **12** 101–132
[25] Málek J, Nečas J, Rokyta M and Růžička M 1996 *Weak and measure-valued solutions to evolutionary PDEs* (London: Chapman Hall)
[26] Wolf J 2007 *J. Math. Fluid Mech.* **9** 104–138
[27] Frehse J and Rů˘ zička M 2008 *Math. Z.* **260** 355–375
[28] Zhikov V V and Pastukhova S E 2010 *Nonlinear partial differential equations and related topics* (*Amer. Math. Soc. Transl. Ser. 2* vol 229) (Providence, RI: Amer. Math. Soc.) pp 233–252
[29] Fernández-Cara E, Guillén F and Ortega R R 1998 *Ann. Scuola Norm. Sup. Pisa Cl. Sci. (4)* **26** 1–29
[30] Simon J 1990 *SIAM J. Math. Anal.* **21** 1093–1117
[31] Diening L, Harjulehto P, Hästö P and Růžička M 2011 *Lebesgue and Sobolev Spaces with Variable Exponents* Lecture Notes in Mathematics Vol. 2017 (Berlin: Springer)
[32] Mosolov P P and Mjasnikov V P 1971 *Dokl. Akad. Nauk SSSR* **201** 36–39

IOP Conf. Series: Journal of Physics: Conf. Series **894** (2017) 012002 doi:10.1088/1742-6596/894/1/012002

Influence of a surface plates inhomogeneity on a translation oscillations of a drop

A A Alabuznev

Institute of Continuous Media Mechanics UB RAS, Perm, Russia
Perm State University, Perm, Russia

E-mail: alabuzhev@icmm.ru

Abstract. Natural and forced oscillations of a sandwiched fluid drop are investigated. A drop is cylindrical in equilibrium, surrounded by another liquid and bounded axially by two parallel solid planes. Hocking's boundary conditions hold on the contact line: the velocity of the contact line motion is proportional to the deviation of the contact angle from its equilibrium value. The Hocking's parameter (so-called wetting parameter) is the proportionality coefficient in this case. This parameter is considered as a function of coordinates, i.e. solid plates have nonuniform surface. The translation vibration force is perpendicular to the symmetry axis of the drop. The solution of the boundary value problem is found using Fourier series of Laplace operator eigen functions. Both translation mode and other azimuthal modes are excited because energy passes from first azimuthal modes to others due to nonuniform surfaces.

1. Introduction

Motion of triple contact line is one of the most important and well-known in the multiphase hydrodynamics [1-5]. The most frequently used condition for contact line velocity is the one applied by Hocking [6] for investigation of standing waves between two vertical walls. This condition assumes a linear relationship between the velocity of the contact line motion and the contact angle

$$\frac{\partial \zeta^*}{\partial t^*} = \Lambda^* \vec{k} \cdot \vec{\nabla} \zeta^*, \tag{1}$$

where ζ^* is the deviation of the interface from the equilibrium position, $\vec{k}$ is the external normal to the solid surface, Λ^* is a phenomenological constant (the so-called wetting parameter or Hocking parameter) having the dimension of the velocity. There are two important limit of the boundary condition (1): (a) $\zeta^* = 0$ – the requirement of the fixed contact line (pinned-end edge condition), (b) $\vec{k} \cdot \vec{\nabla} \zeta^* = 0$ – the constant contact angle. The Hocking parameter was constant in all papers in which this condition was used [7-17]. Note that in the article [18] it was suggested to consider this parameter as complex.

In the present article, we consider the azimuthal oscillations of cylindrical fluid drop which surrounded by other ideal liquid. We assume that Hocking parameter is considered as a function of coordinates. The natural and forced translational oscillations of cylindrical drop for case of uniform plates were investigated in [9, 17].

Content from this work may be used under the terms of the Creative Commons Attribution 3.0 licence. Any further distribution of this work must maintain attribution to the author(s) and the title of the work, journal citation and DOI.
Published under licence by IOP Publishing Ltd

2. Problem formulation

Following [9], consider the oscillations of an incompressible fluid drop of density ρ_i^* and kinematic viscosity ν_i^* surrounded by other fluid of density ρ_e^* and kinematic viscosity ν_e^*. The system is bounded by two parallel solid surfaces which are separated by a distance h^* (see Figure 1). The vessel is closed at infinity. In equilibrium the drop has circle cylindrical form with radius R_0^*. The equilibrium contact angle θ_0 is equal to 0.5π. The system is affected by a vibration field with amplitude A^* and frequency ω^*. The vibration force is directed perpendicularly the symmetry axis of the drop. We assume that the frequency is high to neglect the dissipative effects caused by acoustic radiation and viscous dissipation, i.e. $\omega^* R_0^* \ll c$ and $\delta = (\nu^*/\omega^*)^{1/2} \ll R_0^*$, where c is the sound velocity, δ is the viscous boundary-layer thickness. The amplitude of external force is considered small in the sense that $A^* \ll R_0^*$.

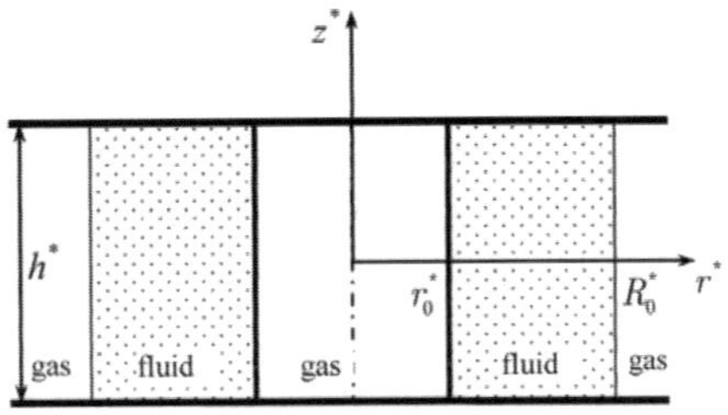

Figure 1. Problem geometry.

We work in cylindrical coordinates r^*, α and z^* because of the problem symmetry. Let the lateral surface of the drop be described by $r^* = r_0^* + \zeta^*(\alpha, z^*, t^*)$, where ζ^* is the surface deviation from equilibrium. The liquid motion is irrotational in the accepted approximations, which makes it convenient to introduce the velocity potential. Thus, the dynamics of the liquid is described by the Bernoulli and Laplace equations. We use the following quantities as the measurement units: R_0^* for length, h^* for height, $\left(\left(\rho_i^* + \rho_e^*\right) R_0^{*3} \sigma^{-1}\right)^{1/2}$ for time, $A^*\left(\left(\rho_i^* + \rho_e^*\right) R_0^{*3} \sigma^{-1}\right)^{-1/2}$ for velocity, $A^* R_0^*\left(\left(\rho_i^* + \rho_e^*\right) R_0^{*3} \sigma^{-1}\right)^{-1/2}$ for velocity potential, $A^*\sigma\left(R_0^*\right)^{-2}$ for pressure, A^* for surface deviation. The amplitude of oscillations is considered small $\varepsilon = A^*/R_0^*$, which allows us to linearize the governing equations and simplify the boundary conditions. Thus, the dimensionless linear boundary value problem is determined by

$$\Delta\varphi_j = 0,\; p_j = -\rho_j\left(\frac{\partial \varphi_j}{\partial t} - \omega^2 r e^{i\omega t}\cos\alpha\right),\; j = i, e, \tag{2}$$

$$r = 1:\; \left[\frac{\partial \varphi}{\partial r}\right] = 0,\; [p] = \zeta + \frac{\partial^2 \zeta}{\partial \alpha^2} + b^2\frac{\partial^2 \zeta}{\partial z^2},\; \frac{\partial \zeta}{\partial t} = \frac{\partial \varphi}{\partial r}, \tag{3}$$

$$z = \pm 1/2:\; \frac{\partial \varphi}{\partial z} = 0, \tag{4}$$

$$r = 1,\; z = \pm 1/2:\; \frac{\partial \zeta}{\partial t} = \mp\lambda(\alpha)\frac{\partial \zeta}{\partial z},\; \lambda(\alpha) = \lambda_0\left|\sin\left(k\cos(\alpha)\right)\right|, \tag{5}$$

Boundary value problem (2)–(5) contains five dimensionless parameters: aspect ratio – $b = R_0^*/h^*$, the Hocking constant (wetting parameter) – $\lambda_0 = \Lambda^* b\left(\left(\rho_e^* + \rho_i^*\right) R_0^* \sigma^{-1}\right)^{-1/2}$, the frequency –

$\omega=\omega^{*}\left(\left(\rho_e^{*}+\rho_i^{*}\right)R_0^{*}\sigma^{-1}\right)^{1/2}$, the density of the external liquid – $\rho_e=\rho_e^{*}\left(\rho_e^{*}+\rho_i^{*}\right)^{-1}$, the density of the drop liquid – $\rho_i=\rho_i^{*}\left(\rho_e^{*}+\rho_i^{*}\right)^{-1}$, $\rho_e+\rho_i=1$. The function $\lambda(\alpha)$ describes

3. Natural oscillations

To investigate the problem, it is convenient to begin with a consideration of the natural oscillations of a cylindrical drop. In further analysis our interest deals with the odd azimuthal oscillation modes due to the function $\lambda(\alpha)$ and even vertical modes due to the external force. The system of equations and boundary conditions (2)-(5) can be written without vibrational force. Representing the fields of the velocity potentials and surface deviation as

$$\varphi_i(r,\alpha,z,t)=\mathrm{Re}\left(i\Omega\sum_{m=0}^{\infty}\sum_{n=0}^{\infty}a_{2m+1,n}F_{2m+1,n}^{i}(r)\cos(2\pi nz)\cos((2m+1)\alpha)e^{i\Omega t}\right), \qquad (6)$$

$$\varphi_e(r,\alpha,z,t)=\mathrm{Re}\left(i\Omega\sum_{m=0}^{\infty}\sum_{n=0}^{\infty}b_{2m+1,n}F_{2m+1,n}^{e}(r)\cos(2\pi nz)\cos((2m+1)\alpha)e^{i\Omega t}\right), \qquad (7)$$

$$\zeta(z,\alpha,t)=\mathrm{Re}\left(\sum_{m=0}^{\infty}\left(\sum_{n=0}^{\infty}c_{2m+1,n}\cos(2\pi nz)+d_{2m+1}G_{2m+1}(z)\right)\cos((2m+1)\alpha)e^{i\Omega t}\right), \qquad (8)$$

where Ω is an eigen frequency, $F_{2m+1,0}^{i}(r)=r^{2m+1}$, $F_{2m+1n}^{i}(r)=\mathrm{I}_{2m+1}(2\pi nbr)$ for $n\geq 1$, $F_{2m+1,0}^{e}(r)=r^{-(2m+1)}$, $F_{2m+1,n}^{e}(r)=\mathrm{K}_{2m+1}(2\pi nbr)$ for $n\geq 1$, $G_1(z)=z^2$, $G_{2m+1}(z)=\mathrm{sh}\left(b^{-1}z\sqrt{(2m+1)^2-1}\right)$ for $m\geq 1$, $\mathrm{I}_m(r)$ and $\mathrm{K}_m(r)$ are modified Bessel functions of first and second kind, respectively.

Substituting solutions (6)–(8) into Eqs. (2)–(5), we obtain a spectral-amplitude problem of complex algebraic equations series which eigenvalues are the values of the natural oscillation frequency Ω. This problem is not given because of their complexity and bulkiness. The equations have complex solutions (were solved numerically with the usage of the two-dimensional secant method), this leads to oscillation damping due to the dissipation on the contact line.

Figures 2 and 3 show the real part of $\mathrm{Re}(\Omega)$ (oscillation frequency) and imaginary part $\mathrm{Im}(\Omega)$ (damping ratio) of the complex natural frequency Ω for the oscillation modes with azimuthal number 1 and 3 and zero vertical number, i.e. $\Omega_{1,0}$ and $\Omega_{3,0}$, respectively. The frequency decreases as the capillary parameter increases (see Figures 2a, 3a), and translation frequency $\mathrm{Re}(\Omega_{1,0})$ vanishes for a critical value λ_0 for any parameters (see [9, 17] for details). That is, such "bending" oscillations exist only for sufficiently small λ_0 and are due to strong interaction of the drop with the plate. At higher values λ_0, the restoring force acting on the drop on the side of the substrate is not sufficient for the occurrence of oscillatory motion. In the case of a fixed edge angle, the drop freely slides over the substrate and there aren't oscillations which accompanied by the motion of the drop center. The damping decrement $\mathrm{Im}(\Omega_{1,0})$ has a branching point of the solution, after which one of the branches of the solution increases very rapidly with growth of λ_0 (Figure 2b). This solution is responsible for the attenuation of free translational oscillations. In general case, the damping decrement $\mathrm{Im}(\Omega)$ has maximum for finite value λ_0 and tends to zero for limit cases $\lambda_0\to 0$ and $\lambda_0\to\infty$ (Figure 3b). The frequencies values are almost no different from theirs equilibrium values for finite and large wavenumber k (Figure 2,3). However the case of pinned contact line is realized for small k as $\lambda(\alpha)\to\lambda_0 k|\cos(\alpha)|$ for $k \ll 1$.

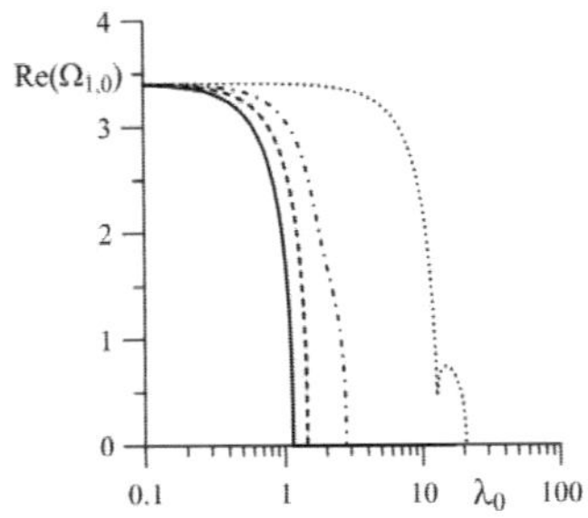

(a)

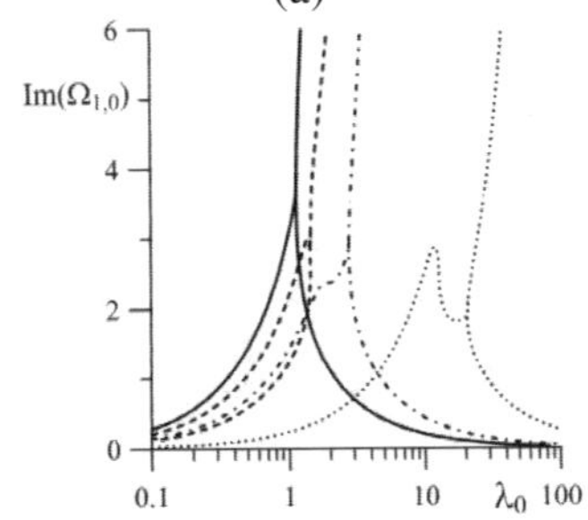

(b)

Figure 2. Frequency (a) and damping ratio (b) of volume natural oscillations vs wetting parameter λ_0 for $\Omega_{1,0}$ ($b=1$, $\rho_i=0.7$).

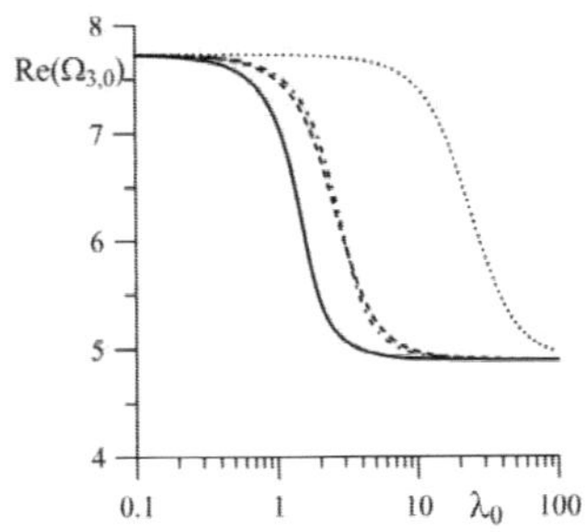

(a)

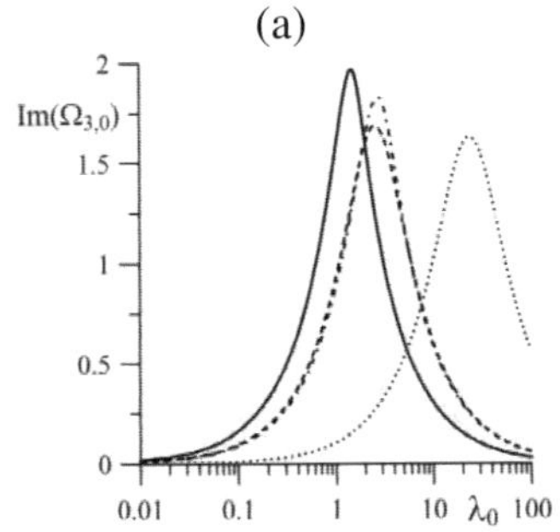

(b)

Figure 3. Frequency (a) and damping ratio (b) of volume natural oscillations vs wetting parameter λ_0 for $\Omega_{3,0}$ ($b=1$, $\rho_i=0.7$).

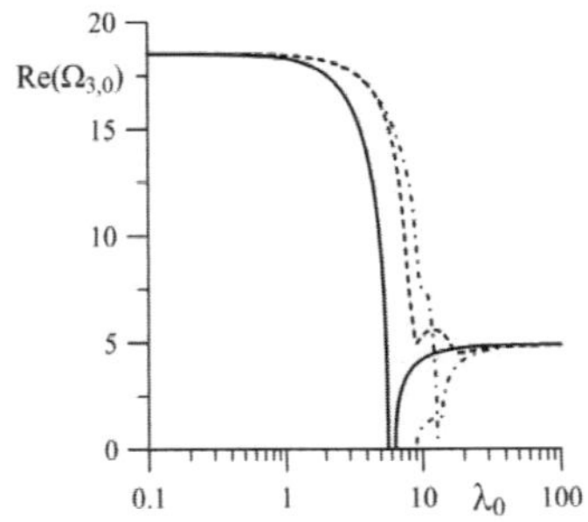

(a)

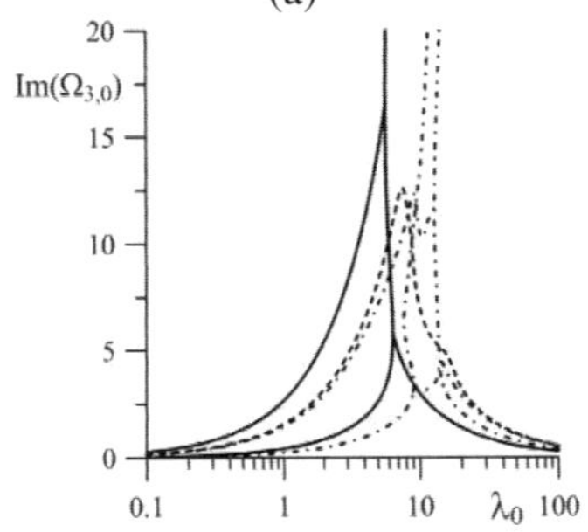

(b)

Figure 4. Frequency (a) and damping ratio (b) of volume natural oscillations vs wetting parameter λ_0 for $\Omega_{3,0}$ ($b=3$, $\rho_i=0.7$).

uniform plates – solid line, $k=0.1$ – dotted, $k=1$ – dashed, $k=10$ – dot-and-dashed.

For the azimuthal modes $\Omega_{m,0}$ ($m \geq 2$) there are interval of values λ_0 on which the frequencies $\mathrm{Re}(\Omega_{m,0})$ vanishes (Figure 4a). Length of this interval increases with growth of the parameter b (compare Figure 3a with Figure 4a). This effect is associated with dissipation during to the interaction of the contact line with the plates as the increase of parameter b corresponds to an increase in the drop radius so contact line length (see [9] for details). The similar effect was obtained for volume oscillations of cylindrical bubble [19]. However heterogeneity of plate surface changes the magnitude of dissipation and can lead to the disappearance of this effect. The frequency $\mathrm{Re}(\Omega_{3,0})$ does not go to zero for any values of parameter λ_0 at $k=1$ (Figure 4). The azimuthal mode $\Omega_{3,0}$ has very difficult behaviour at $k=10$ which needs a future research.

4. Forced oscillations

Here we consider the problem of forced oscillations. The solution in this case is similar to solution (6)–(8) with exception for the time dependence: oscillation frequency is the forced frequency ω.

The dependence of the surface oscillations amplitude in the layer centre ζ_0 and at the upper plate ζ_s, the form of contact line and the contact angle deviation on the upper plate on the forced frequency ω is given in Figures 5, 6 for different values of the Hocking parameter λ_0 and of the aspect ratio b. Note that the curves have a resonant shape in the limiting case $\lambda_0 \to 0$. In other limiting case $\lambda_0 \to \infty$ oscillations amplitude tends to $|\rho_i - \rho_e|$. The amplitude of the contact line oscillations is finite for any values of the parameter λ_0. Results for homogenous plates can be seen in [13, 17] more details.

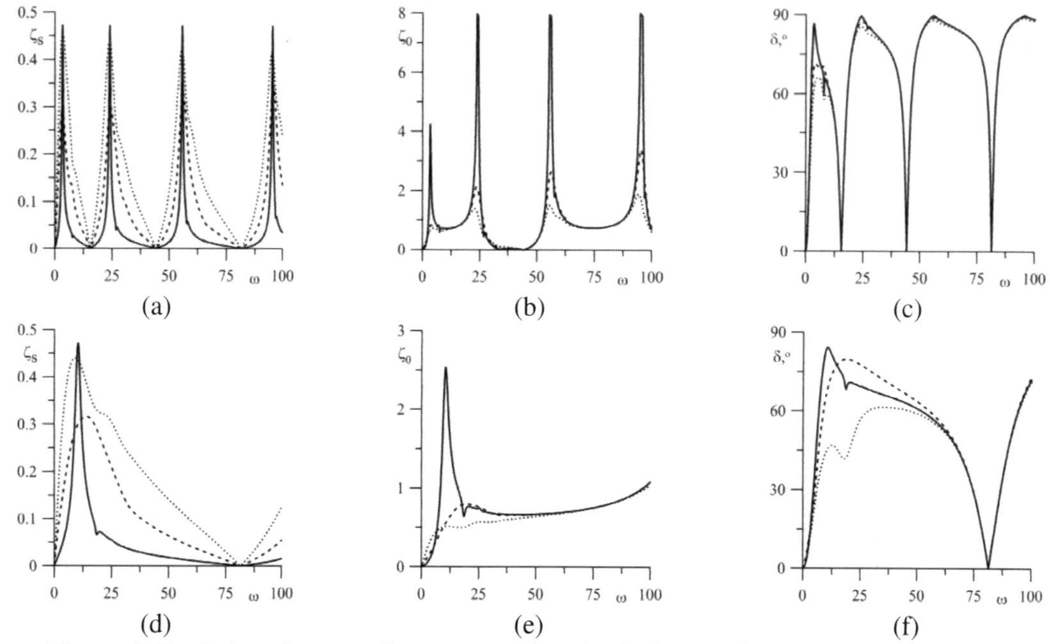

Figure 5. Deviation of contact line at upper plate (a, d), drop surface at center layer (b, e) and contact angle at upper plate vs frequency ($\rho_i = 0.7$), $k = 0.1$ – solid line, $k = 1$ – dotted, $k = 10$ – dashed.
(a)-(c) – $b = 1$, $\lambda_0 = 1$, (d)-(f) – $b = 3$, $\lambda_0 = 5$.

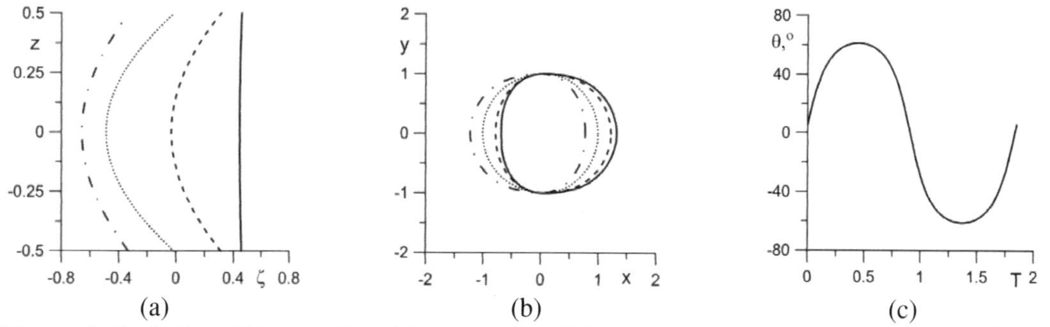

Figure 6. Evolution of drop surface (a), contact line (b) and contact angle (c) at upper plate ($b = 1$, $\rho_i = 0.7$, $\omega = 3.4$). $T = 2\pi\omega^{-1}$ – oscillation period,
(a)-(b) $t = 0$ – solid line, $t = 0.125T$ – dashed, $t = 0.25T$ – dotted, $t = 0.375T$ – dot-and-dashed.

The translation mode is excited by the first (Figures 5a, 6a,b) for any wavenumber k . This mode exists at $\lambda_0 = 1$ and $b = 1$ and the first resonance peak corresponds to main translation frequency (Figures 5a,b). Drop makes longitudinal periodic oscillations in this case (Figures 6a,b), the edge angle varies widely (Figure 6c). Note that there are “antiresonance” frequencies such that the contact line amplitude goes to zero. Contact angle also does not change at these points (Figure 5c). As was shown above, the first mode disappears with increasing parameter λ_0 and the third mode becomes the main resonance at $\lambda_0 = 5$ and $b = 3$ (Figures 5d,e). However, the third mode also can disappear at the aspect ratio b and wavenumber k are changed (Figures 3, 4). Figure 5d shown how the amplitude of the contact line changes with increasing of wavenumber k . Value of k plays the role of an effective wetting parameter as λ_0 . Thus the main frequency of third mode goes to zero as wavenumber k

growths (see Figures). As results the resonance peak vanishes and the amplitude of surface oscillations goes to $\rho_i - \rho_e$.

5. Conclusion

The behaviour of cylindrical drop between solid plates has been considered taking into account the dynamics of the contact angle under translational vibrations. The solid plates have nonuniform surfaces by function $\lambda(\alpha) = \lambda_0 \left|\sin(k\cos(\alpha))\right|$. The main purpose of this paper is to propose a method for studying the drop forced oscillation on an inhomogeneous substrates and determining λ. The investigation of natural oscillations has shown that wavenumber k plays the role of effective wetting parameter. Also the main frequency of azimuthal mode may not disappear for different values of k, in contrast to the homogeneous case.

For small values of the parameter λ_0, i.e. with a weak energy dissipation, the amplitude of the surface forced oscillations is large and tends to infinity in the limit $\lambda_0 \to 0$. The amplitude of the contact line oscillations is finite for any values of λ_0. There are "antiresonant" frequencies, i.e. such vibration frequencies, for which the contact line does not move for any values of λ_0, and the contact angle does not change. Moreover the change of a drop volume (i.e. aspect ratio b) can suppress resonances.

Acknowledgments
This work was supported by the Russian Science Foundation No. 14-21-00090.

6. References

[1] Voinov O V 1976 *Fluid Dyn* **11** 714
[2] De Gennes P G 1985 *Rev. Mod. Phys.* **57** 827
[3] Pukhnachev V V and Semenova I B 1999 *J. Appl. Mech. Tech. Phy.* **40** 594
[4] Zhang L and Thiessen D B 2013 *J. Fluid Mech.* **719** 295
[5] Snoeijer J H and Andreotti B 2013 *Annu. Rev. Fluid Mech.* **45** 269
[6] Hocking L M 1987 *J. Fluid Mech.* **179** 253
[7] Borkar A and Tsamopoulus J 1991 *Phys. Fluids A* **3** 2866
[8] Perlin M, Schultz W W and Liu Z 2004 *Wave Motion* **40** 41
[9] Alabuzhev A A and Lyubimov D V 2007 *J. Appl. Mech. Tech. Phys.* **48** 686
[10] Shklyaev S and Straube A V 2008 *Phys. Fluids* **20** 052102
[11] Fayzrakhmanova I S and Straube A V 2009 *Phys. Fluids* **21** 072104
[12] Fayzrakhmanova I S, Straube A V and Shklyaev S 2011 *Phys. Fluids* **23** 102105
[13] Alabuzhev A A and Lyubimov D V 2012 *J. Appl. Mech. Tech. Phys.* **53** 9
[14] Alabuzhev A A and Kaysina M I 2016 *J. Phys.: Conf. Ser.* **681** 012043
[15] Alabuzhev A A 2016 *Computational Continuum Mechanics* **9** 316 (In Russian)
[16] Alabuzhev A A 2016 *J. Appl. Mech. Tech. Phys.* **57** 1006
[17] Alabuzhev A A 2016 *Computational Continuum Mechanics* **9** 453 (In Russian)
[18] Miles J W 1991 *J. Fluid Mech.* **222** 197
[19] Alabuzhev A A 2014 *Computational Continuum Mechanics* **7** 151 (In Russian)

IOP Conf. Series: Journal of Physics: Conf. Series **894** (2017) 012003 doi:10.1088/1742-6596/894/1/012003

The dynamics of hemispherical drop under the influence of a alternating electric field

A A Alabuzhev[1,2] **and M A Kashina**[2]
[1]Institute of Continuous Media Mechanics UB RAS, Perm 614013, Russia
[2]Perm State University, Perm 614990, Russia

E-mail: alabuzhev@mail.ru

Abstract. The forced oscillations of incompressible fluid sessile drop under the alternating electric field are considered. Under equilibrium the drop has the hemispherical form with right contact angle. The drop is surrounded by an incompressible fluid with another density. The external uniform electric field acts as an external force that causes motion of the contact line. In order to describe this contact line motion the modified Hocking boundary condition is applied: the velocity of the contact line is proportional to the deviation of the contact angle and the speed of the fast relaxation processes, which frequency is proportional to twice the frequency of the electric field. It is shown that under uniform electric field the axisymmetric oscillation amplitude at the resonance frequency increases with the increase of the Hocking parameter.

1. Introduction

The behaviour of fluid drops under external electrical field continues to attract research efforts [1,2], in particular the dynamics and manipulation processes is electrowetting (EW) [2,3] and electrowetting-on-dielectric (EWOD) [4-6]. EWOD has been put into applications in various fields, such as digital (droplet) microfluidic devices for bioanalysis (lab-on-a-chip) [7, 8], variable-focus liquid lenses [9,10], electronic display technology [11,12] etc. The typical electrowetting devices are shown in Figure 1.

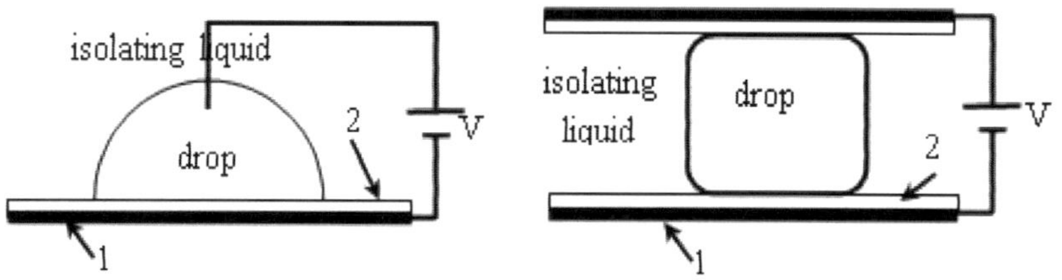

Figure 1. Typical schematic configurations.
1 – electrode, 2 – dielectric layer.

The Young–Lippmann equation is used most often to describe the varying contact angle θ of the drop under a direct current (DC) voltage [2-6]:

$$\cos\vartheta = \cos\vartheta_0 + E_w,\ E_w = 0.5CV^2\sigma_{ic}^{-1},\ C = \varepsilon\varepsilon_0 d^{-1},\ \cos\vartheta_0 = \sigma_{ic}^{-1}\left(\sigma_{ip} - \sigma_{cp}\right), \quad (1)$$

[1] To whom any correspondence should be addressed.

Content from this work may be used under the terms of the Creative Commons Attribution 3.0 licence. Any further distribution of this work must maintain attribution to the author(s) and the title of the work, journal citation and DOI.
Published under licence by IOP Publishing Ltd

where E_w is EW number (it represents the ratio of the electrostatic energy to the liquid – surrounding fluid interfacial energy), C is capacitance per unit area, V is the value of the applied DC voltage, ϑ_0 is the contact angle without applied voltage – equilibrium contact angle, which is defined by well-known Young's equation, σ is the interfacial tension between the drop of conducting fluid (c) and the isolating (surrounding) fluid (i) and the dielectric plate (p), d is the thickness of the dielectric film, ε_0 and ε are the permittivity of vacuum and the dielectric layer, respectively. For alternating current (AC) instead of V^2 in equation (1) the square value of the voltage, or effective voltage, U^2 is used [3]. Although Young–Lippmann equation (1) is widely accepted, different hypotheses have been advanced to derive it [2,6].

It should be possible to obtain zero contact angle (complete wetting and the contact angle does not change) by continuing to increase the applied voltage according to the equation (1), i.e. this equation is adequate for the voltage V which is smaller than the critical voltage V_* [2,3,6]. At $V \geq V_*$ the electrowetting contact angle does not increase with increasing of applied voltage, this effect is known as contact angle saturation. The mechanism of the contact angle saturation has not yet been elucidated and is the object of discuss [2].

We proposed in [13] another effective boundary condition on the basis of Hocking's equation [14] for cylindrical drop:

$$\frac{\partial \zeta^*}{\partial t^*} = \pm \Lambda^* \left(\frac{\partial \zeta^*}{\partial z^*} + A^* \cos\left(2\omega^* t^*\right) \right), \tag{2}$$

where ζ^* is the deviation of the drop interface from the equilibrium position, z^* is axial coordinate, Λ^* is a phenomenological constant (the so-called wetting parameter or Hocking parameter), having the dimension of velocity, A^* is effective amplitude, ω^* is AC frequency. The second term in boundary condition (2) is external action and written similarly to the Young–Lippmann equation (1): $\cos\vartheta = E_w = \tilde{E}_w V^2$ as $\cos\vartheta_0 = 0$ in our problem, therefore $\zeta_z \sim \operatorname{ctg}\vartheta = E_w / \sqrt{1 - E_w^2} = \tilde{E}_w V^2 / \sqrt{1 - \tilde{E}_w^2 V^4} \approx \tilde{E}_w V^2 + O\left(V^6\right) \approx \tilde{E}_w V^2 \sim V^2$. The special cases of boundary condition (2) are correspond to the fixed contact line $\zeta^* = 0$ (the so-called pinned-end edge condition) and infinitely fast relaxation of the contact angle $\partial\zeta^* / \partial z^* \sim \cos\left(2\omega^* t^*\right)$ (the contact angle quickly adjusts to the slowly changing force). For example, the Hocking's equation [14] was used in the studies of oscillations of a capillary bridge [15], sessile drop and bubble [16-18], "sandwiched" drop and bubble [19-21] and film interface [22].

In the present paper, we consider the behavior of a sessile drop under AC. In order to describe the motion of the contact line the modified boundary condition (2) is used: the velocity of the contact line is proportional to the deviation of the contact angle and the speed of the fast relaxation processes, whose frequency is proportional to twice of the electric field frequency. Once again note, that natural and power "mechanical" oscillations of sessile drop with taking into account the contact angle dynamics were examined in [17]. In future we plan to compare the results with experimental data (for example, [6]).

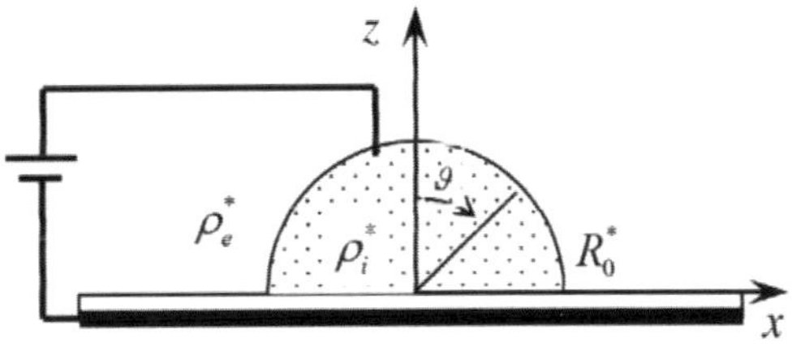

Figure 2. Problem geometry (1 – electrode, 2 – dielectric layer).

IOP Conf. Series: Journal of Physics: Conf. Series **894** (2017) 012003 doi:10.1088/1742-6596/894/1/012003

2. Problem formulation

The problem formulation largely coincides with the articles [13].Consider a hemispherical liquid drop of radius a sitting atop a solid substrate (Figure 2). The sessile drop and unbounded ambient fluid are characterized by densities ρ_j^* and kinematic viscosities ν_j^* ($j = i, e$, here and in the following, quantities with subscript i refer to the drop, and those with subscript e to the surrounding liquid). The equilibrium form of drop is hemispherical of radius R_0^* and the equilibrium contact angle ϑ_0 between the side drop surface and the solid plane is assumed to be right. We assume that, on the one hand, the AC frequency ω^* is large enough for the viscosity can be ignored – $\delta^2 = \nu^* \left(\omega^*\right)^{-1} \ll R_0^{*2}$, and, on the other hand, the oscillation frequency is small enough, so that we can use the incompressibility condition – $\omega^* R_0^* \ll c$ (δ is the boundary-layer thickness, c is the sound velocity). The characteristic amplitude of oscillations of the drop A^* is small compared to the equilibrium radius R_0^*: $\varepsilon = A^* \left(R_0^*\right)^{-1} \ll 1$.

We use $\left(\sigma^{-1}\rho_i^* R_0^{*3}\right)^{1/2}$, R_0^*, ρ_i^*, A^*, $A^*\sigma\left(R_0^*\right)^{-2}$, $A^*\sigma^{1/2}\left(\rho_i^* R_0^{*3}\right)^{-1/2}$ as the scales for the time, length, height, density, deviation of drop surface from its equilibrium position, pressure, and velocity potential, respectively. The axisymmetrical problem is described by the following equations and boundary conditions to spherical coordinates (r, ϑ):

$$\Delta\Phi_j = 0, \quad P_j = -\rho_j \frac{\partial \Phi_j}{\partial t}, \quad j = i, e, \tag{3}$$

$$r = 1: \left[\frac{\partial \Phi}{\partial r}\right] = 0, \quad [P] = -\left(\Delta_\vartheta + 2\right) Z, \quad \frac{\partial Z}{\partial t} = \frac{\partial \Phi_i}{\partial r}, \tag{4}$$

$$\vartheta = \frac{\pi}{2}: \quad \frac{\partial \Phi_j}{\partial \vartheta} = 0, \tag{5}$$

$$r = 1, \ \vartheta = \frac{\pi}{2}: \quad \frac{\partial Z}{\partial t} = -\lambda\left(\frac{\partial Z}{\partial \vartheta} + a\cos(2\omega t)\right), \tag{6}$$

where Φ is the potential of fluid velocity, P is the fluid pressure, Z is the interface deviation from equilibrium position, $\rho_i = 1$, $\rho_e = \rho$, the square brackets denote the jump in the quantity at the interface between the external liquid and the drop. The boundary-value problem (3)–(6) involves four parameters: the dimensionless density, the wetting parameter, the AC frequency and amplitude

$$\rho = \rho_e^*\left(\rho_i^*\right)^{-1}, \ \lambda = \Lambda^*\left(\sigma^{-1}\rho_i^* R_0^*\right)^{1/2}, \ \omega = \omega^*\left(\sigma^{-1}\rho_i^* R_0^{*3}\right)^{1/2}, \ a = 0.5A^*C\left(\sigma^{-3}\rho_i^* R_0^{*3}\right)^{1/2}.$$

3. Inviscid problem

In order to investigate the problem it is convenient to begin with a consideration of inviscid fluids, but there is dissipation in this case. This damping is due to the interaction the contact line with solid plate [14-22]. Because external driving is periodic, we proceed to complex amplitudes by representing all the fields as $F(\vec{r}, t) = \mathrm{Re}\left(f(\vec{r})e^{i2\omega t}\right)$. We expand the amplitudes in series with respect to the Legendre polynomials (similarly [17])

$$\varphi_i(r, \vartheta) = 2i\omega\sum_{n=0}^{\infty} A_n P_{2n}(\theta) r^{2n}, \ \varphi_e(r, \vartheta) = 2i\omega\sum_{n=0}^{\infty} B_n P_{2n}(\theta) r^{-2n-1}, \ \zeta(\theta) = \sum_{n=0}^{\infty} C_n P_{2n}(\theta) \tag{7}$$

where $\theta = \cos\vartheta$. These solutions are written down to automatically satisfy the requirements of being regular as $r \to 0$ (drop) and $r \to \infty$ (external fluid) and part of boundary conditions, as given first two relations in Eq. (4) and by Eq. (5). Substituting these relations into the dynamic boundary condition [see the final relation in 2 Eqs. (5)], we arrive at another representation for ζ

IOP Conf. Series: Journal of Physics: Conf. Series **894** (2017) 012003 doi:10.1088/1742-6596/894/1/012003

$$\zeta = \theta D - 4\omega^2 \sum_{n=0}^{\infty} 2n\Omega_n^{-2} A_n P_{2n}(\theta),\ \Omega_n^2 = 2n(2n-1)(2n+1)(2n+2)(2n\rho+2n+1)^{-1} \quad (8)$$

where D is a yet-to-be-determined amplitude of the contact angle oscillations and Ω_n has the meaning of eigen frequencies associated with the even eigen modes for a spherical drop (so-called Rayleigh's modes [16,17]). Note that the term with $n=0$ also contributes into this representation, as at $n=0$ we have $2n\Omega_n^{-2} = -0.5$. By comparing Eqs. (7) and (8) for ζ, we obtain the amplitudes

$$A_0 = \frac{D}{4\omega^2},\ A_n = \frac{\alpha_n \Omega_n^2 D}{2n(\Omega_n^2 - 4\omega^2)},\ B_n = -2n(2n+1)^{-1} A_n,\ C_n = 2nA_n \quad (9)$$

which are expressed in terms of coefficients α_n defined as

$$\alpha_n = -(4n+1)(2n-1)^{-1}(2n+2)^{-1} P_{2n}(0),\ \theta = \sum_{n=0}^{\infty} \alpha_n P_{2n}(\theta). \quad (10)$$

Finally, taking into account boundary condition (6), we figure out

$$D = \left(1 - \frac{i8\omega^3}{\lambda} \sum_{n=0}^{\infty} \frac{\alpha_n P_{2n}(0)}{\Omega_n^2 - 4\omega^2}\right)^{-1}. \quad (11)$$

We note that as all the constants A_n are proportional to the same complex factor D (see (7) and (11)), we conclude that all the Rayleigh modes oscillate in phase, though with a certain phase shift relative to the "external force" exerted at the contact line. Thus, the shape oscillations can be represented as a standing wave, and therefore no traveling waves propagate along the surface in the system under consideration.

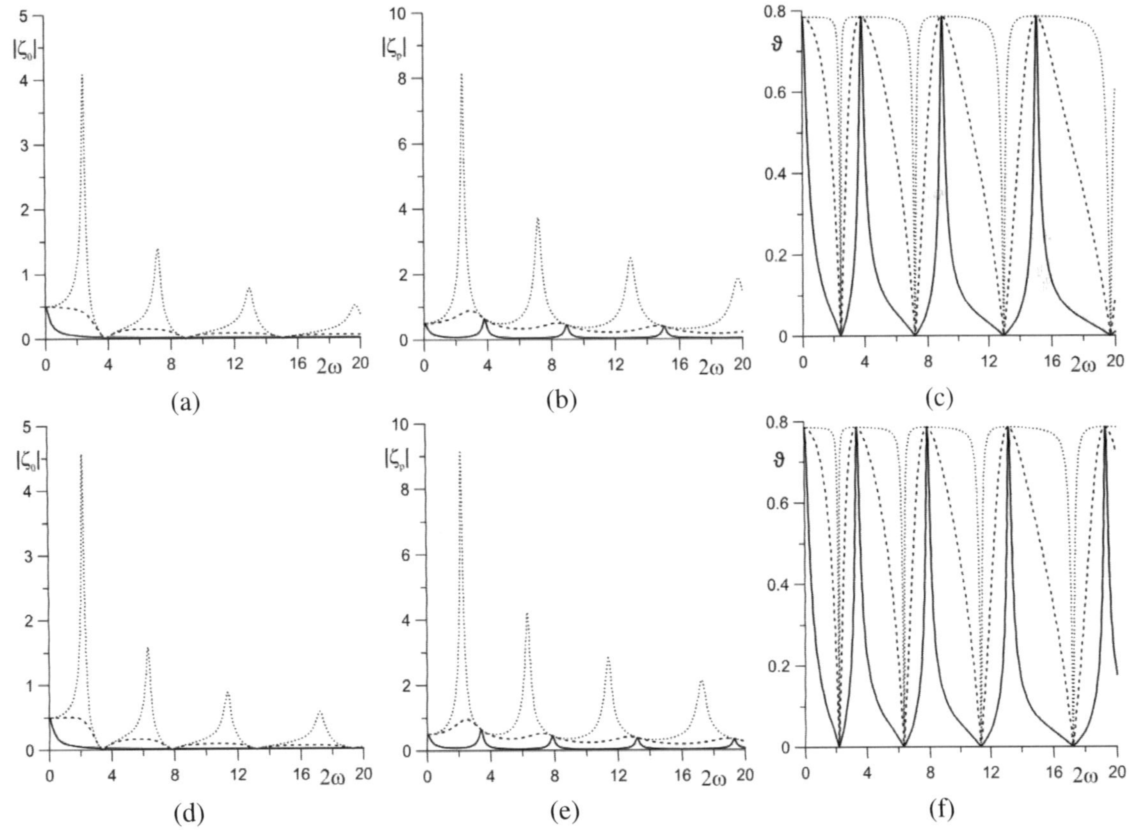

Figure 3. Response characteristics of drop oscillations. Amplitude of surface oscillation in the pole (a,d) and on the substrate (b,e) and contact line deviation (c,f) vs frequency ($a=1$).

IOP Conf. Series: Journal of Physics: Conf. Series **894** (2017) 012003 doi:10.1088/1742-6596/894/1/012003

$\lambda = 0.1$ (solid line), $\lambda = 1$ (dashed line), $\lambda = 10$ (dotted line), (a-c) $\rho = 0.5$, (d-f) $\rho = 1$.

The dependence of the surface oscillation amplitude and the contact angle vs the frequency of the driving force is given in Figure 3 for different values of the Hocking parameter and of the density. Obviously, the larger the wetting parameter λ, the lower the interaction of the contact line with the plate and larger the oscillation amplitude (Figure 3 a, d). Also, as it was mentioned in [16-21], at certain values of eigen frequencies the drop motion is independent of the wetting parameter: at any λ the contact line remains motionless. At such points the amplitude of the contact line goes to zero. This is so-called "antiresonance" frequencies. Values of these frequencies do not depend on λ and depend on the parameters of the drops (density ρ etc.). Note that the maximum deviation of the contact angle does not depend on the density (Figure 3 c, f).

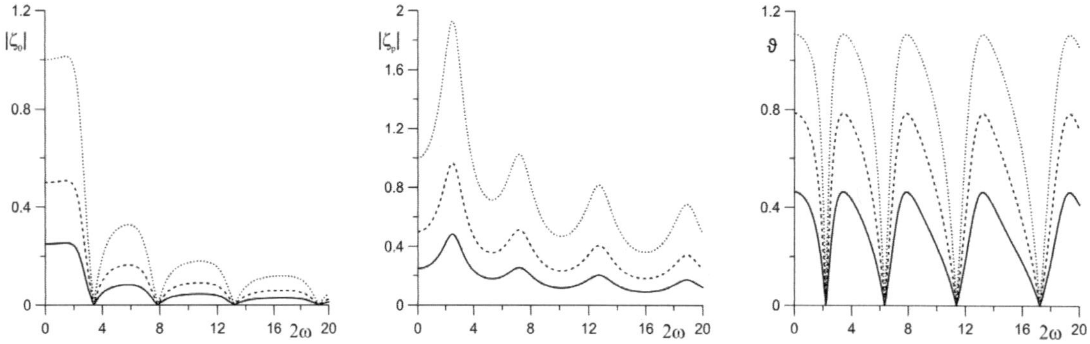

(a) (b) (c)

Figure 4. Response characteristics of drop oscillations. Amplitude of surface oscillation in the pole (a) and on the substrate (b) and contact line deviation (c) vs frequency ($\lambda = 1$, $\rho = 1$).
$a = 0.5$ (solid line), $a = 1$ (dashed line), $a = 2$ (dotted line).

Figure 4 shows response of the system for different values of the amplitude a for a given contrast of densities $\rho = 1$. The oscillations amplitude increases with growth of a (Figure 4 a, b) but the deviation of contact angle increases also (Figure 4 c).

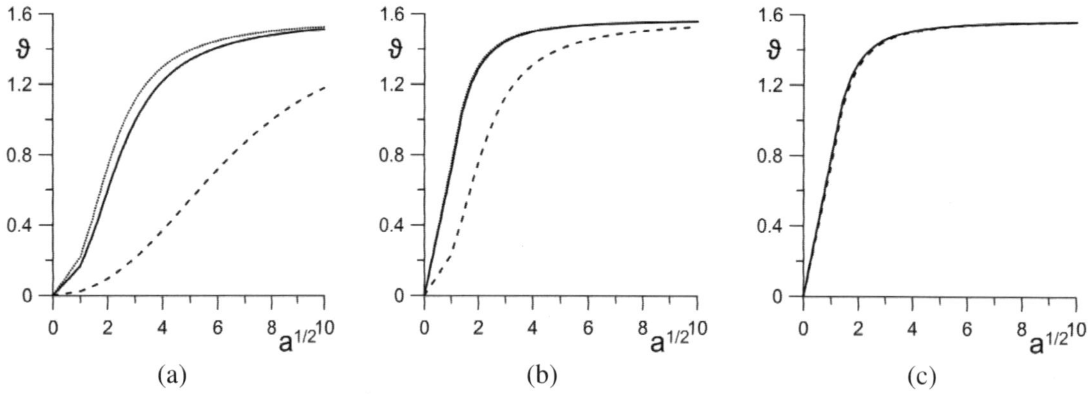

(a) (b) (c)

Figure 5. The contact angle ϑ vs square root of amplitude a for three different values of the Hocking parameter λ and of the oscillation frequency ω ($\rho = 1$).
(a) $\lambda = 0.1$, (b) $\lambda = 1$, (c) $\lambda = 10$,
$\omega = 0.5$ (solid line), $\omega = 1$ (dashed line), $\omega = 1.5$ (dotted line).

The contact angle versus square root of amplitude a (i.e. proportional to AC potential V) is given in Figure 5 for different values of the Hocking parameter λ and AC frequency ω. The responses obtained in qualitative agreement with the experimental data. However the maximum deviation of contact angle tends to π/2, i.e. $\vartheta \to 0$ or $\vartheta \to \pi$ whereas contact angle is finite in experiments.

Different wetting situations are addresses by changing the values of the parameter λ and range between two limiting cases: i) for $\lambda = 0$, the contact line motion is completely uncoupled from the external field – the contact line remains fixed and the droplet overall is motionless; ii) the opposite limit, $\lambda \gg 1$, corresponds to infinitely fast relaxation of the contact angle, when the latter quickly adjusts to the slowly changing force. In both these situations, there is no dissipation at the contact line. Apparently the fluid viscosity must be taken into account. It is giving a finite amplitude for both limit cases of Hocking boundary especially in limiting case $\lambda \gg 1$.

4. Conclusions

The dependences of surface amplitude and contact angle from oscillation amplitude and frequency are obtained for Hocking parameter and density. It is shown that the curve of contact angle ϑ vs square root of amplitude obtained in qualitative agreement with the experimental data. However the maximum deviation of contact angle tends to π/2, i.e. $\vartheta \to 0$ or $\vartheta \to \pi$ whereas contact angle is finite in experiments. Consequently, there are three ways of further development of problem solution: 1) fluid viscosity (see for example [23]), 2) contact angle hysteresis (see for example [24, 25]), 3) arbitrary equilibrium contact angle [20]. Both of these cases limit the amplitude of oscillation and may limit the change of the contact angle.

5. References

[1] Melcher J R and Taylor G I 1969 *Annual Review of Fluid Mechanics* **1** 111
[2] Mugele F and Baret J-C 2005 *J. Phys.: Condens. Matter.* **17** 705
[3] Chen L and Bonaccurso E 2014 *Adv. Colloid Interface Sci.* **210** 2
[4] Berge B 1993 *Comptes Rendus Acad. Sci. II* **317** 157
[5] Zhao Y-P and Wang Y 2013 *Rev. Adhesion Adhesives* **1** 114
[6] Chevalliot S, Kuiper S and Heikenfeld J 2012 *J. of Adhesion Sci. Tech.* **26** 1909–30
[7] Hua Z, Rouse J L, Eckhardt A E, Srinivasan V, Pamula V K, Schell W A, Benton J L, Mitchell T G and Pollack M G 2010 *Anal Chem* **82** 2310-16
[8] Li J, Wang Y,Chen H and Wan J 2014 *Lab Chip* **14** 4334-37
[9] Kuiper S and Hendriks B H W 2004 *Appl. Phys. Lett.* **85**, 1128–30
[10] Li C and Jiang H 2014 *Micromachines* **5** 432-41
[11] Hayes R A and Feenstra B J 2003 *Nature* **425** 383–5
[12] Roques-Carmes T, Hayes R A, Feenstra B J and Schlangen L J M 2004 *J. Appl. Phys* **95** 4389–96
[13] Alabuzhev A A and Kaysina M I 2016 *J. Phys.: Conf. Ser.* **681** 012043
[14] Hocking L M 1987 *J. Fluid Mech.* **179** 253-66
[15] Borkar A and Tsamopoulus J 1991 *Phys. Fluids A* **3** 2866–74
[16] Lyubimov D V, Lyubimova T P and Shklyaev S V 2004 *Fluid Dynamics* **39** 851–62
[17] Lyubimov D V, Lyubimova T P and Shklyaev S V 2006 *Phys. Fluids* **18** 012101
[18] Shklyaev S and Straube A V 2008 *Phys. Fluids* **20** 052102
[19] Alabuzhev A A and Lyubimov D V 2007 *J. Appl. Mech. Tech. Phys.* **48** 686–93
[20] Alabuzhev A A and Lyubimov D V 2012 *J. Appl. Mech. Tech. Phys.* **53** 9–19
[21] Alabuzhev A A 2016 *Appl. Mech. Tech. Phys.* **53** 9–19
[22] Perlin M, Schultz W W and Liu Z 2004 *Wave Motion* **40** 41–56
[23] Bostwick J B and Steen P H 2016 *Soft Matter* **12** 8919-26
[24] Fayzrakhmanova I S and Straube A V 2009 *Phys. Fluids* **21** 072104
[25] Fayzrakhmanova I S, Straube A V and Shklyaev S 2011 *Phys. Fluids* **23** 102105

IOP Conf. Series: Journal of Physics: Conf. Series **894** (2017) 012004 doi:10.1088/1742-6596/894/1/012004

Invisibility problem in acoustics, electromagnetism and heat transfer. Inverse design method

G Alekseev[1,2], A Tokhtina[2] and O Soboleva[1]
[1]Institute of Applied Mathematics, Vladivostok, Russia
[2]Far Eastern Federal University, Vladivostok, Russia
E-mail: alekseev@iam.dvo.ru

Abstract. Two approaches (direct design and inverse design methods) for solving problems of designing devices providing invisibility of material bodies of detection using different physical fields – electromagnetic, acoustic and static are discussed. The second method is applied for solving problems of designing cloaking devices for the 3D stationary thermal scattering model. Based on this method the design problems under study are reduced to respective control problems. The material parameters (radial and tangential heat conductivities) of the inhomogeneous anisotropic medium filling the thermal cloak and the density of auxiliary heat sources play the role of controls. A unique solvability of direct thermal scattering problem in the Sobolev space is proved and the new estimates of solutions are established. Using these results, the solvability of control problem is proved and the optimality system is derived. Based on analysis of optimality system, the stability estimates of optimal solutions are established and numerical algorithms for solving particular thermal cloaking problem are proposed.

1. Introduction

In recent years significant research has focused on design of invisibility cloaking devices. Beginning with pioneering papers [1, 2, 3, 4] the large number of papers was devoted to developing different schemes of cloaking material objects. These schemes include metamaterial cloaking based on transformation optics (TO) proposed by Pendry et al. [2], conformal method proposed by Leonhardt [3], plasmonic cloaking method based on scattering cancellation proposed by Alú and Engheta [4], mantle cloaking [5], etc.

The first works in this field were focused on the electromagnetic cloaking. Then the main results of the electromagnetic cloaking theory were expanded to an acoustic cloaking [6] and to cloaking static (magnetic, electric and thermal) fields (see, e.g., [7, 8, 9]). Papers [10, 11] are devoted to studying invisibility problems in X-ray tomography.

Development of the above-mentioned approaches have opened up the opportunities for creation the invisibility cloaking design strategies. They obtained the name of direct design strategies as they were based on solving the forward electromagnetic (acoustic or static) problems. It should be noted that the invisibility devices (hereafter, cloaks) designed on the basis of direct strategies possess serious drawbacks. The main one is the difficulty of their technical realization. For example, the design of the TO-based cloaks involves extreme values of constitutive parameters and spatially varying distributions of the permittivity and permeability tensors which are very difficult to implement [12].

Content from this work may be used under the terms of the Creative Commons Attribution 3.0 licence. Any further distribution of this work must maintain attribution to the author(s) and the title of the work, journal citation and DOI.
Published under licence by IOP Publishing Ltd

That is why the another cloak design strategy began develop recently. It obtained the name of inverse design as it is related with solving inverse electromagnetic (acoustic or static) problems (see [12, 13]). The optimization method forms the core of the inverse design methodology. This enables us to solve some substantial limitations of previous cloaking solutions. A growing number of papers is devoted to applying the inverse design methodology in various cloaking problems. Among them we mention [14, 15, 16] where numerical optimization algorithms are applied for finding the unknown material parameters of TO-based cloak and papers [17, 18, 19] devoted to theoretical analysis of cloaking problems using the optimization approach.

Optimization method is applied and in this paper for solving inverse problems for the 3D stationary thermal scattering model. These problems arise when designing thermal cloaking devices. We prove the solvability of direct and control problems for the thermal scattering model under study and derive the optimality system which describes necessary conditions of extremum. Based on it's analysis we establish the stability estimates of optimal solutions and propose a numerical algorithm for solving the particular control problem.

2. Statement of direct thermal scattering problem

In this Section we formulate and study direct problem of thermal scattering in bounded domain D. As in brief note [19] we begin with introducing externally applied thermal field T^e. Considering for the simplicity the case of rectangular parallelepiped $D = \{(x,y,z) : |x| < x_0, |y| < y_0, |z| < z_0\}$ (see Figure 1), we assume that external field T^e in D is created by two horizontal boundaries $\Gamma_1 : z = -z_0$ and $\Gamma_2 : z = z_0$ which are kept at temperatures T_1 and $T_2 < T_1$, respectively, while the lateral boundaries are thermally insulated. Then external field T^e satisfies equation $k_0 \Delta T^e = 0$ in D and the following boundary conditions: $T^e|_{z=-z_0} = T_1$, $T^e|_{z=z_0} = T_2$, $\partial T^e/\partial x|_{x=\pm x_0} = 0$, $\partial T^e/\partial y|_{y=\pm y_0} = 0$. Here k_0 is a constant thermal conductivity of homogeneous isotropic medium (background) filling D.

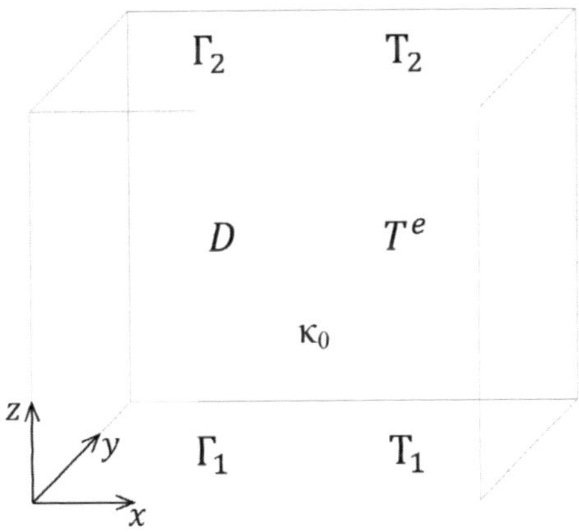

Figure 1. The geometry of the problem without cloak.

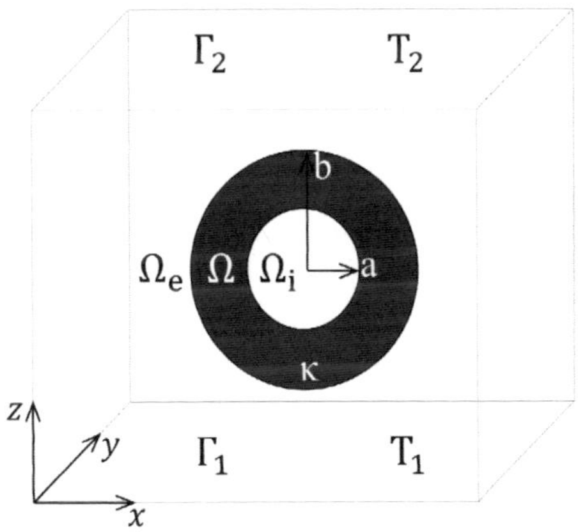

Figure 2. The geometry of the problem for the case when a cloak is placed in D.

We assume further that a material shell (Ω, κ) having the form of the spherical layer $a < r < b$ in spherical coordinates (r, θ, φ) which is filled by anisotropic medium characterized by heat conductivity tensor κ is placed into D (see Figure 2). Placing the shell (Ω, κ) into D leads to appearing the scattered thermal response T^{sc} in $D\backslash\Omega$ which can be determined by solving the direct thermal scattering problem. In order to formulate the latter problem we denote by Ω_i (or Ω_e) the interior (or exterior) of Ω in D. Then the mentioned thermal scattering problem consists of finding a triple of functions: T_i in the interior Ω_i of Ω, T_0 in Ω and T_e in the exterior Ω_e of Ω satisfying equations

$$k_0\Delta T_i = 0 \text{ in } \Omega_i, \;\; \mathrm{div}(\kappa \mathrm{grad} T_0) = 0 \text{ in } \Omega, \;\; k_0\Delta T_e = -f \text{ in } \Omega_e, \tag{1}$$

boundary conditions

$$T_e|_{z=-z_0} = T_1, \;\; T_e|_{z=z_0} = T_2, \;\; \partial T_e/\partial x|_{x=\pm x_0} = 0, \;\; \partial T_e/\partial y|_{y=\pm y_0} = 0 \tag{2}$$

and continuity conditions on internal and external components of boundary of Ω. They have the form [13, ch.4]

$$T_i = T_0, \; k_0\partial T_i/\partial n = (\kappa \nabla T)\cdot \mathbf{n} \text{ at } r = a, \; T_e = T_0, \; k_0\partial T_e/\partial n = (\kappa\nabla T)\cdot\mathbf{n} \text{ at } r = b. \tag{3}$$

Here $\mathbf{n}$ is the outward unit normal to the boundary of Ω.

Let us define a weak solution of problem (1)–(3). Preliminarily, we introduce a number of function spaces to be used in the subsequent analysis. We will use the space $H^1(\tilde{\Omega})$ where $\tilde{\Omega}$ is one of the domains $\Omega_i, \Omega, \Omega_e$, D and spaces $L^\infty(\Omega)$, $H^s(\Omega)$, $L^2(Q)$ where $Q \subset D$ is an arbitrary subset. The scalar products and norms in $H^1(D)$, $H^s(\Omega)$, $L^2(Q)$ will be denoted by $(\cdot,\cdot)_{1,D}$, $\|\cdot\|_{1,D}$, $(\cdot,\cdot)_{s,\Omega}$, $\|\cdot\|_{s,\Omega}$, $(\cdot,\cdot)_Q$, $\|\cdot\|_Q$. Set $L^\infty_{\mu_0} = \{\mu \in L^\infty(\Omega) : \mu(\mathbf{x}) \geq \mu_0\}$, $H^s_{\mu_0}(\Omega) = \{\mu \in H^s(\Omega) : \mu(\mathbf{x}) \geq \mu_0\}$, $\mu_0 = \text{const} > 0$. We note that continuous compact embedding $H^s(\Omega) \subset L^\infty(\Omega)$ for $s > 3/2$ takes place and the following estimate holds: $\|\mu\|_{L^\infty(\Omega)} \leq C_s\|\mu\|_{s,\Omega}$ $\forall \mu \in H^s(\Omega)$, $s > 3/2$. Here C_s is the constant dependent on s and Ω.

Now we introduce the following main assumption to the conductivity tensor κ describing the properties of the medium filling the shell Ω:

(i) tensor κ is diagonal in spherical coordinates (r, θ, φ) and its diagonal components (radial, polar and azimuthal conductivities) k_r, k_θ and k_φ do not depend on φ and satisfy

$$k_r \in L^\infty_{k_r^0}(\Omega), \; k_\theta \in L^\infty_{k_\theta^0}(\Omega), \; k_r^0 = \text{const} > 0, \; k_\theta^0 = \text{const} > 0, \; k_\varphi = k_\theta.$$

This assumption enables us to simplify substantially the study of the direct scattering problem (1)–(3) and to look for its solution in the subspace X of the space $H^1(D)$ consisting of functions $T := (T_i, T_0, T_e)$ independent of φ. Using the general formula

$$\mathrm{grad} S = \frac{\partial S}{\partial r}\mathbf{e}_r + \frac{1}{r}\frac{\partial S}{\partial \theta}\mathbf{e}_\theta + \frac{1}{r\sin\theta}\frac{\partial S}{\partial\varphi}\mathbf{e}_\varphi$$

for $\mathrm{grad}S$ in spherical coordinates where $\mathbf{e}_r$, $\mathbf{e}_\theta$ and $\mathbf{e}_\varphi$ are unit basis vectors and taking into account the independence of elements of X on angle φ we define the norm in X by

$$\|S\|_X^2 := \|S\|_D^2 + \|\nabla S\|^2_{\Omega_i\cup\Omega_e} + \left\|\frac{\partial S}{\partial r}\right\|^2_\Omega + \left\|\frac{1}{r}\frac{\partial S}{\partial\theta}\right\|^2_\Omega. \tag{4}$$

We introduce once more assumption concerning boundary functions T_1 and T_2:

(ii) $T_1 \in H^{1/2}(\Gamma_1), T_2 \in H^{1/2}(\Gamma_2)$ and there exists a function $T^0 \in X$ such that

$$T^0|_{\Gamma_1} = T_1, \; T^0|_{\Gamma_2} = T_2, \; \|T^0\|_X \leq C_T := C_D\left(\|T_1\|_{1/2,\Gamma_1} + \|T_2\|_{1/2,\Gamma_2}\right).$$

Here C_D is a constant which depends only on D.

Now we are able to define a weak solution of problem (1)–(3). Let $X_0 := S \in X : S|_{\Gamma_1} = S|_{\Gamma_2} = 0$. We multiply every equation in (1) by $S \in X_0$, integrate by part and add the results obtained. Using Green formulae, assumption (i) and boundary conditions (2), (3), we arrive at the following relations for finding a triple $T = (T_i, T_0, T_e) \in X$:

$$a(k_0, k; T, S) := a_0(k_0; T, S) + a_1(k_r; T, S) + a_2(k_\theta; T, S) = \langle F, S\rangle \ \ \forall S \in X_0,\ T|_{\Gamma_1} = T_1,\ T|_{\Gamma_2} = T_2. \tag{5}$$

Here, $k := (k_r, k_\theta)$, $a_0(k_0; ., .)$, $a_1(k_r; ., .)$, $a_2(k_\theta; ., .)$ or F are bilinear or linear forms defined by

$$a_0(k_0; T, S) := k_0 \int_{\Omega_i \cup \Omega_e} \nabla T \cdot \nabla S d\mathbf{x},\ a_1(k_r; T, S) := \int_\Omega \left(k_r \frac{\partial T}{\partial r} \frac{\partial S}{\partial r} \right) d\mathbf{x},$$

$$a_2(k_\theta; T, S) := \int_\Omega \left(\frac{k_\theta}{r^2} \frac{\partial T}{\partial \theta} \frac{\partial S}{\partial \theta} \right) d\mathbf{x},\ \langle F, S\rangle = (f, S)_{\Omega_e}.$$

We call a triple $T := (T_i, T_0, T_e) \in X$ satisfying (5) a weak solution of problem (1)–(3).

Using Hölder inequality, formula (4) for norm $\|\cdot\|_X$ and Poincaré inequality $\|\nabla T\|_D^2 \geq \delta \|T\|_X^2$ for $T \in X_0$ where the constant $\delta > 0$ depends on D one can easily derive the following estimates:

$$|a(k_0, k; T, S)| \leq \left(k_0 + \|k_r\|_{L^\infty(\Omega)} + \|k_\theta\|_{L^\infty(\Omega)}\right) \|T\|_X \|S\|_X,\ |\langle F, S\rangle| \leq \|f\|_{\Omega_e} \|S\|_X \ \forall T, S \in X,$$

$$a(k_0, k; T, T) \geq k^0 \|\nabla T\|_D^2 \geq \delta k^0 \|T\|_X^2 \ \ \forall T \in X_0,\ k^0 = \min(k_0, k_r^0, k_\theta^0).$$

These estimates mean that the linear form F is continuous on X while the bilinear form $a(k_0, k; \cdot, \cdot)$ is continuous on X and is coercive on X_0. Based on the Lax-Milgram theorem one can prove then that under conditions (i), (ii) a weak solution $T := (T_i, T_0, T_e) \in X$ of problem (1)–(3) exists and is unique. More precisely, the following theorem holds.

Theorem 2.1. *Let, under assumptions (i), (ii), $K_1 \subset L^\infty_{k_r^0}(\Omega)$, $K_2 \in L^\infty_{k_\theta^0}(\Omega)$ are nonempty bounded sets. Then for any pair $(k_r, k_\theta) \in K_1 \times K_2$ and for any $f \in L^2(\Omega_e)$ direct problem (1)–(3) has a unique weak solution $T = (T_i, T_0, T_e) \in X$ which satisfies estimate*

$$\|T\|_X \leq C_0 C_1 (C_T + \|f\|_{\Omega_e}),\ \ C_0 = \left(\delta k^0\right)^{-1}.$$

Here, constant C_1 depends on K_1 and K_2 but is independent of k_r, k_θ, k_φ.

3. Statement of inverse problem. Using optimization method. Main results

We recall that our purpose is analysis of inverse problems arising when developing technologies of designing thermal cloaking devices. These problems consist of finding conductivities k_r, k_θ and the density f of auxiliary heat sources placed in the domain Ω_e from the respective cloaking condition. For solving these problems we apply optimization method. This approach is based on introducing the cost functional to be minimized which adequately corresponds to the inverse problem of designing the device for approximate cloaking. As a result, the initial cloaking problem is reduced to study of the respective control problem using the well known methods of solving extremum problems. As a cost functional we choose one of the following:

$$I_1(T) = \|T - T_d\|_Q^2 = \int_Q (T - T_d)^2 d\mathbf{x},\ \ I_2(T) = \|T - T_d\|_{1,Q}^2.$$

Here, function $T_d \in L^2(Q)$ (or $T_d \in H^1(Q)$) models the thermal field measured in some subset $Q \subset D \setminus \bar{\Omega}$. As controls we choose variable conductivity coefficients $k_r(\mathbf{x})$ and $k_\theta(\mathbf{x})$ of the anisotropic inhomogeneous medium filling the shell Ω and the heat source density f in Ω_e. By definition, parameters k_r and k_θ have the sense of passive controls while f plays the role of active control. So our control problem corresponds to using mixed cloaking strategy (see [13]).

In the particular case when $Q \subset \Omega_e$ and $T_d = T^e$ functional $I_1(T)$ has the meaning of the $L^2(Q)$-norm of the external scattered thermal response T^{sc} in the set Q. Therefore parameter $k^{opt} = (k_r^{opt}, k_\theta^{opt})$ owing to which the minimum of functional $I_1(T)$ is achieved corresponds to the thermal cloaking problem. In the general case when $T_d = T^e + T_1^{sc}$ where T_1^{sc} is a thermal scattered response of a certain object, the minimizer k^{opt} of the cost functional I_1 describes an approximate solution of the respective thermal illusion problem [13, ch.4].

Let $K = K_1 \times K_2 \times K_3$, $u = (k_r, k_\theta, f)$. Define the operator $G := (G_0, G_1, G_2) : X \times K \to X_0^* \times H^{1/2}(\Gamma_1) \times H^{1/2}(\Gamma_2)$ where $\langle G_0(T,u), S\rangle = a(k; T, S) - \langle F, S\rangle$ for all $S \in X_0$, $G_1 T = T|_{\Gamma_1} - T_1$, $G_2 T = T|_{\Gamma_2} - T_2$ and rewrite weak formulation (5) of problem (1)–(3) as operator equation $G(T,u) = 0$. It is assumed that controls k_r, k_θ and f can change in sets K_1, K_2, K_3 and the following condition holds:

(j) $K_1 \subset H^s_{k_r^0}(\Omega)$, $K_2 \subset H^s_{k_\theta^0}(\Omega)$, $K_3 \subset L^2(\Omega_e)$ are nonempty convex closed sets, where $s > 3/2$, $k_r^0 = \text{const} > 0$, $k_\theta^0 = \text{const} > 0$; $\alpha_0 > 0$.

We consider the following control problem:

$$J(T,u) := \frac{\alpha_0}{2} I(T) + \frac{\alpha_1}{2}\|k_r\|^2_{s,\Omega} + \frac{\alpha_2}{2}\|k_\theta\|^2_{s,\Omega} + \frac{\alpha_3}{2}\|f\|^2_{\Omega_e} \to \inf, \quad G(T,u) = 0. \tag{6}$$

Here, $I(T)$ is a cost functional, α_0, α_1, α_2 and α_3 are nonnegative parameters which serve to regulate the relative importance of each of the terms in (6). Denote by $Z_{ad} = \{(T,u) \in X \times K : G(T,u) = 0, I(T) < \infty\}$ the set of admissible pairs for problem (6).

We apply the mathematical procedure developed in [20] for studying control problems arising when optimization method is applied when solving inverse problems for linear diffusion-convection models. Based on this procedure, one can prove the following results.

Theorem 3.1 *Let, under assumptions (i), (ii), (j), $I : X \to \mathbf{R}$ be a weakly lower semicontinuous functional and Z_{ad} be a nonempty set. Let $\alpha_1 \geq 0$, $\alpha_2 \geq 0$, $\alpha_3 \geq 0$ and K_1, K_2, K_3 be bounded sets or $\alpha_1 > 0$, $\alpha_2 > 0$, $\alpha_3 > 0$ and functional $I(T)$ is bounded below. Then, control problem (6) has at least one solution $(T,u) \equiv (T, k_r, k_\theta, f) \in X \times K_1 \times K_2 \times K_3$.*

Theorem 3.2 *Let, under assumptions (i), (ii), (j), $\alpha_1 > 0$, $\alpha_2 > 0$, $\alpha_3 > 0$ or $\alpha_1 \geq 0$, $\alpha_2 \geq 0$, $\alpha_3 \geq 0$ and K_1, K_2, K_3 be bounded sets. Then control problem (6) for $I = I_k(T)$, $k = 1, 2$, has at least one solution $(T, k_r, k_\theta, f) \in X \times K_1 \times K_2 \times K_3$.*

Theorem 3.3 *Let, under assumptions (i), (ii), (j), the pair $(\hat{T}, \hat{u}) \equiv (\hat{T}, \hat{k}_r, \hat{k}_\theta, \hat{f}) \in X \times K_1 \times K_2 \times K_3$ be a solution of problem (6) and let a functional $I(T)$ be continuously differentiable in the point $\hat{T}$. Then there exists a unique Lagrange multiplier $(\hat{R}, \hat{\zeta}_1, \hat{\zeta}_2) \in X_0 \times H^{1/2}(\Gamma_1)^* \times H^{1/2}(\Gamma_2)^*$ which is the solution of the adjoint problem in the form of the Euler-Lagrange equation:*

$$a(k_0, \hat{k}; \Psi, \hat{R}) + \langle \hat{\zeta}_1, \Psi\rangle_{\Gamma_1} + \langle \hat{\zeta}_2, \Psi\rangle_{\Gamma_2} = -(\alpha_0/2)\langle I_1'(\hat{T}), \Psi\rangle \ \forall \Psi \in X, \tag{7}$$

and the following variational inequalities hold:

$$\alpha_1(\hat{k}_r, k_r - \hat{k}_r)_{s,\Omega} + a_1((k_r - \hat{k}_r)\hat{T}, \hat{R}) \geq 0 \ \forall k_r \in K_1, \tag{8}$$

$$\alpha_2(\hat{k}_\theta, k_\theta - \hat{k}_\theta)_{s,\Omega} + a_2((k_\theta - \hat{k}_\theta)\hat{T}, \hat{R}) \geq 0 \ \forall k_\theta \in K_2, \tag{9}$$

$$\alpha_3(\hat{f}, f - \hat{f})_{\Omega_e} - (f - \hat{f}, \hat{R})_{\Omega_e} \geq 0 \ \forall f \in K_3. \tag{10}$$

Direct problem (5), the Euler-Lagrange equation (7) which has the meaning of the adjoint problem for the adjoint state $(\hat{R}, \hat{\zeta}_1, \hat{\zeta}_2)$ and variational inequalities (8)–(10) with respect to controls $\hat{k}_r, \hat{k}_\theta, \hat{f}$ constitute the optimality system for control problem (6). Based on analysis of the optimality system, one can establish sufficient conditions on the data which provide the uniqueness and stability of solutions of particular control problems. More precisely, the following theorem holds for the case of cost functional $I_1(T) = \|T - T_d\|^2_Q$.

Theorem 3.4 *Let, in addition to assumptions (i), (ii), (j), K_1, K_2 and K_3 be bounded sets and let the quadruples $(T^{(l)}, k_r^{(l)}, k_\theta^{(l)}, f^{(l)})$ be solutions of control problem (6) for $I = I_1(T)$ corresponding to given functions $T_d^{(l)} \in L^2(Q)$, $l = 1, 2$. Let the following conditions take place:*

$$\alpha_1(1-\varepsilon) > 5\alpha_0 C_0^2 C_s^2 M_T^0 M_T, \;\; \alpha_2(1-\varepsilon) > 5\alpha_0 C_0^2 C_s^2 M_T^0 M_T, \;\; \alpha_3(1-\varepsilon) > 5\alpha_0 C_0^2 M_T^0 M_T,$$

where $\varepsilon \in (0,1)$ is an arbitrary number, while constants M_T and M_T^0 are defined by

$$M_T := C_0 C_1 \sup_{f\in K_3} (C_T + \|f\|_{\Omega_e}), \;\; M_T^0 := M_T + \max\left(\|T_d^{(1)}\|_Q, \|T_d^{(2)}\|_Q\right).$$

Then the following stability estimates hold:

$$\|T_1 - T_2\|_Q \le \|T_d^{(1)} - T_d^{(2)}\|_Q, \;\; \|k_r^{(1)} - k_r^{(2)}\|_{s,\Omega} \le \sqrt{\alpha_0/\varepsilon\alpha_1}\|T_d^{(1)} - T_d^{(2)}\|_Q,$$

$$\|k_\theta^{(1)} - k_\theta^{(2)}\|_{s,\Omega} \le \sqrt{\alpha_0/\varepsilon\alpha_2}\|T_d^{(1)} - T_d^{(2)}\|_Q, \;\; \|f_1 - f_2\|_\Omega \le \sqrt{\alpha_0/\varepsilon\alpha_3}\|T_d^{(1)} - T_d^{(2)}\|_Q,$$

$$\|T_1 - T_2\|_X \le C_0\left[\left(C_s\sqrt{\alpha_0/\varepsilon\alpha_1} + C_s\sqrt{\alpha_0/\varepsilon\alpha_2}\right) + \sqrt{\alpha_0/\varepsilon\alpha_3}\right]\|T_d^{(1)} - T_d^{(2)}\|_Q.$$

4. Numerical algorithms

The optimality system (5), (7)–(10) derived above can be used to design efficient numerical algorithms for solving control problem (6). The simplest one for the functional $I_1(T)$ can be obtained by applying the fixed point iteration method for solving the optimality system. The m-th iteration of this algorithm consists of finding unknown values T^m, R^m, k_r^{m+1}, k_θ^{m+1} and f^{m+1} for given $k^m = (k_r^m, k_\theta^m)$, f^m, $m = 0, 1, 2...$, beginning with given initial values k_r^0, k_θ^0, f^0 by sequentially solving following problems:

$$a(k^m; T^m, S) + \langle\zeta_1^m, \Psi\rangle_{\Gamma_1} + \langle\zeta_2^m, \Psi\rangle_{\Gamma_2} = (f^m, S)_{\Omega_e} \; \forall S \in X_0, \;\; T^m|_{\Gamma_1} = T_1, \; T^m|_{\Gamma_2} = T_2,$$

$$a(k^m; \Psi, R^m) = -\alpha_0(T^m - T^d, \Psi)_Q \; \forall \Psi \in X, \tag{11}$$

$$\alpha_1(k_r^{m+1}, k_r - k_r^m)_{s,\Omega} + a_1((k_r - k_r^{m+1})T^m, R^m) \ge 0 \; \forall k_r \in K_1, \tag{12}$$

$$\alpha_2(k_\theta^{m+1}, k_\theta - k_\theta^m)_{s,\Omega} + a_2((k_\theta - k_\theta^{m+1})T^m, R^m) \ge 0 \; \forall k_\theta \in K_2,$$

$$\alpha_3(f^{m+1}, f - f^m)_{\Omega_e} - ((f - f^{m+1}), R^m)_{\Omega_e} \ge 0 \; \forall f \in K_3.$$

For discretization and solving problems (11), (12) one can use open source software freeFEM++ (www.freefem.org) based on finite element method. Along with this algorithm one can apply alternative algorithm of global optimization and in particular the particle-swarm method used in [21] for solving problems of designing 2D layered thermal cloaking shells.

5. Conclusion

In this paper, we studied control problems for the 3D thermal scattering model. These problems arise when optimization method is applied for solving cloaking problems for a respective thermal scattering model. Radial and polar heat conductivities k_r and k_θ of the inhomogeneous medium filling the cloaking shell and the density f of auxiliary heat sources placed in the domain Ω_e play the role of controls. We studied some new properties of solutions of direct thermal scattering problem, proved the solvability of direct and control problems and derived the optimality system describing the necessary conditions of extremum. Based on analysis of the optimality system, we established the uniqueness and stability estimates of optimal solutions for particular control problems. Besides, we proposed numerical algorithms for solving particular cloaking problem. We plan to devote a forthcoming paper to studying the properties of the algorithms and to analysis of results of numerical experiments performed using these algorithms.

IOP Conf. Series: Journal of Physics: Conf. Series **894** (2017) 012004 doi:10.1088/1742-6596/894/1/012004

Acknowledgments

The first author was supported by the Russian Science Foundation (project no. 14-11-00079), the second and third authors were supported by the Russian Foundation for Basic Research (project no.16-01-00365-a).

References

[1] Dolin L S 1961 On a possibility of comparison of three-dimensional electromagnetic systems with nonuniform anisotropic filling *Izu. Vuzov Radiofizika* **4** 964–7

[2] Pendry J B, Shurig D and Smith D R 2006 Controlling electromagnetic fields *Science* **312** 1780–82

[3] Leonhardt U 2006 Optical conformal mapping *Science* **312** 1777–80

[4] Alú A and Engheta N 2005 Achieving transparency with plasmonic and metamaterial coatings *Phys. Rev. E* **72** 016623

[5] Alú 2009 A Mantle cloak: invisibility induced by a surface *Phys. Rev. B* **80** 245115

[6] Cummer S A, Popa B I, Schurig D, Smith D R, Pendry J, Rahm M and Starr A 2008 Scattering theory derivation of a 3D acoustic cloaking shell *Phys. Rev. Lett.* **100** 024301

[7] Sanchez A, Navau C, Prat-Camps J and Chen D X 2011 Antimagnets: controlling magnetic fields with superconductormetamaterial hybrids *New J. Phys.* **13** 093034

[8] Yang F, Mei Z L and Jin T Y et al. 2012 DC electric invisibility cloak *Phys. Rev. Lett.* **109** 053902

[9] Han T and Qiu C W 2016 Transformation Laplacian metamaterials: recent advances in manipulating thermal and dc fields *J. Opt.* **18** 044003

[10] Anikonov D S, Nazarov V G and Prokhorov I V 1997 Visible and invisible media in tomography *Dokl. Math.* **56** 955–8

[11] Anikonov D S, Nazarov V G and Prokhorov I V 2002 *Poorly visible media in X-Ray Tomography* (Utrecht: VSP)

[12] Xu S, Wang Y, Zhang B and Chen H 2013 Invisibility cloaks from forward design to inverse design *Sci. China Inf. Sci.* **56** 120408

[13] Alekseev G V 2016 *The Invisibility Problem in Acoustics, Optics and Heat Transfer* (Vladivostok: Dalnauka)

[14] Xi S, Chen H, Zhang B, Wu B I and Kong J A 2009 Route to low-scattering cylindrical cloaks with finite permittivity and permeability *Phys. Rev. B* **79** 155122

[15] Popa B I and Cummer S A 2009 Cloaking with optimized anisotropic layers *Phys. Rev. A* **79** 023806

[16] Wang X and Semouchkinaa E 2013 A route for efficient non-resonance cloaking by using multilayer dielectric coating *Appl. Phys. Lett.* **102** 113506

[17] Alekseev G V 2013 Control of boundary impedance in two-dimensional material-body cloaking by the wave flow method *Comput. Math. and Math. Phys.* **53** 1853–69

[18] Alekseev G V 2013 Cloaking of material objects by controlling the impedance boundary condition for Maxwell's equations *Dokl. Phys.* **58** 482–6

[19] Alekseev G V and Levin V A 2016 An optimization method for the problems of thermal cloaking of material bodies *Dokl. Phys.* **61** 546–50

[20] Alekseev G V, Vakhitov I S and Soboleva O V 2012 Stability estimates in identification problems for the convection-diffusion-reaction equation *Comput. Math. and Math. Phys.* **52** 2190–205

[21] Alekseev G V, Levin V A and Tereshko D A 2017 Optimization analysis of the thermal-cloaking problem for a cylindrical body *Dokl. Phys.* **62** 71–5

IOP Conf. Series: Journal of Physics: Conf. Series **894** (2017) 012005 doi:10.1088/1742-6596/894/1/012005

Analysis of the in situ stress contrast impact on a planar hydraulic fracture propagation using the fully coupled poroelastic model

A N Baykin[1,2] and S V Golovin[1,2]

[1] Novosibirsk State University, Novosibirsk, Russia
[2] Lavrentyev Institute of Hydrodynamics SB RAS, Novosibirsk, Russia

E-mail: `alexey.baykin@gmail.com`, `svgolovin@gmail.com`

Abstract. In this paper, we use a fully coupled model for 3D planar fracture propagation within a poroelastic reservoir in order to analysis the non-symmetry of the fracture as it develops in a 3-layer reservoir with different compressive stresses in each layer. We demonstrate that the fracture predominantly develops in the layer with the lower compressive stress until it gets arrested due to the equality of the total leakoff to the volume of pumped fluid.

1. Introduction

Mathematical modeling of hydraulic fracturing attracts great attention in the scientific literature due to its importance in applied fields such as intensification of hydrocarbon production. Recent advances in the modeling are presented in the review papers [1, 2]. Roughly, the single-fracture models are classified as 2D and pseudo 3D ones that use various simplifying assumptions related to the geometry of the fracture, and also planar 3D and fully 3D models that use a solution of the complete 3D problem in order to model the spatial propagation of the fracture. On the other hand, most of the models are based on either numerical or analytical solutions of pure elastic equations coupled with lubrication equation for the fluid flow within the fracture, and the approximate Carter formula for estimation of the fluid leakoff from the fracture to the reservoir. More advanced approach consists of observing the poroelasticity model that has an essential advantage due to correct description of the fluid exchange between the fracture and the reservoir, and also due to account for the coupled influence of the pore fluid pressure on the stress state near the fracture, and the modification of filtration due to inhomogeneous stress.

In papers [3, 4], we demonstrated that inhomogeneity of the permeability may significantly influence the dynamics of the fracture's development. This analysis was based on the numerical model for the hydraulic fracture propagation in poroelastic medium such that in each section of the fracture by the vertical plane the width of the fracture remains constant (the same assumption as used in the classical KGD model). In present paper we generalize our numerical model to the 3D case by assuming the a planar fracture propagation without any limitations to the width of the fracture in all directions along the plane. At present, we do not model the fracture toughness by assuming that the fracture propagates in the viscosity-dominated regime. The fracture is growing due to the pressure action of the fluid pumped into the fracture through the horizontal well that intersects the fracture at a given point. As the reservoir itself is regarded

Content from this work may be used under the terms of the Creative Commons Attribution 3.0 licence. Any further distribution of this work must maintain attribution to the author(s) and the title of the work, journal citation and DOI.
Published under licence by IOP Publishing Ltd

IOP Conf. Series: Journal of Physics: Conf. Series **894** (2017) 012005 doi:10.1088/1742-6596/894/1/012005

as a porous medium, there is a natural fluid exchange between the reservoir and the fracture. During the fracture growth its area increases and, consequently, the volume of fluid that leaks off from the fracture to the reservoir increases as well. After some time, the fracture comes to the state when the volume of leaked fluid coincides with the volume of fluid pumped into the fracture. After this moment, the fracture is arrested [5] by this fluid balance and does not propagate further, provided the pumping rate of fracturing fluid does not grow.

As a simple analysis of the model's capabilities, we analyze the shape of the fracture at the time of fracture's arrest as it propagates in a 3-layered reservoir with different confining stresses in each layer. The confining stress is higher in the surrounding layers in comparison with the central layer, where the source of fracturing fluid is located. It is shown that even for the small contrast of the confining stresses, the fracture develops mostly in the central layer until it gets arrested. Note, that this effect might be distorted by the introduction of a finite fracture toughness, especially, if the toughness would be different in each layer.

2. Mathematical formulation of the problem

The mathematical model presented in this paper is a direct generalization to the 3D case of the hydraulic fracture model presented and analyzed in our earlier works [3, 6, 4].

We consider a planar hydraulic fracture propagating due to the pressure of fluid that is pumped into the fracture through the horizontal wellbore intersecting the fracture at a given point. We use a Cartesian system of coordinates such that Oxy plane coincides with the fracture's plane, Oy axis is directed upwards, and Oz is orthogonal to the fracture plane. The coordinates' origin coincides with the position of the well.

The reservoir is considered to be poroelastic meaning that it is represented as a two-phase continuum characterized by the displacement vector $\mathbf{u}$ of the elastic solid matrix (skeleton) and the pressure p of the fluid within the pore space. The elastic response to the applied stress is assumed to be linear with the moduli λ and μ as the skeleton's parameters. The filtration properties are characterized by the porosity ϕ and the permeability k_r. The speed $\mathbf{q}$ of the fluid flow in the porous medium is governed by the Darcy law $\mathbf{q} = -(k_r/\eta_r)\nabla p$ for a single-phase Newtonian fluid with the effective viscosity η_r.

It is supposed that the fluid filtrating from the fracture to the reservoir has the same viscosity as the pore fluid. However, the fluid injected into the fracture has different viscosity η_f. It corresponds to the normal situation in hydraulic fracturing when the fracturing fluid is a high-viscous gel and only its low-viscous base fluid is filtrated into the reservoir. The geological prestress state of the reservoir is taking into account by introducing the initial stress tensor $\boldsymbol{\tau}_0$.

Following [3, 6, 4] the mechanics of the poroelastic medium in quasi-static state is governed by the following system of equations:

$$\operatorname{div}\boldsymbol{\tau} = 0, \quad \boldsymbol{\tau} = \boldsymbol{\tau}_0 + \lambda \operatorname{div}\mathbf{u}\,\mathrm{I} + 2\mu\,\mathcal{E}(\mathbf{u}) - \alpha p\,\mathrm{I}, \tag{1}$$

$$S_\varepsilon \frac{\partial p}{\partial t} = \operatorname{div}\left(\frac{k_r}{\eta_r}\nabla p - \alpha\frac{\partial \mathbf{u}}{\partial t}\right), \tag{2}$$

where $\mathcal{E}(\mathbf{u})$ is the Cauchy's strain tensor $2\mathcal{E}(\mathbf{u})_{ij} = \partial u_i/\partial x_j + \partial u_j/\partial x_i$ $(i,j = 1,2)$, I is the identity tensor, α is the Biot coefficient responsible for the solid-fluid coupling. The storativity S_ε reflects the dependence of the Lagrangian porosity ϕ on $\epsilon = \operatorname{tr}\mathcal{E}$ and p as in [7]:

$$\frac{\partial\phi}{\partial t} = \alpha\frac{\partial\epsilon}{\partial t} + S_\varepsilon\frac{\partial p}{\partial t}, \quad S_\varepsilon = \frac{(\phi_0-\alpha)(1-\alpha)}{K}, \tag{3}$$

where $K = \lambda + \dfrac{2\mu}{3}$ is the bulk modulus, ϕ_0 is the initial porosity.

Symmetry of the problem with respect to Ox-axis makes it possible to solve equations (1)–(2) within domain $\Omega = [-L_x/2, L_x/2] \times [-L_y/2, L_y/2] \times [0, L_z/2]$ which cross-section is shown in

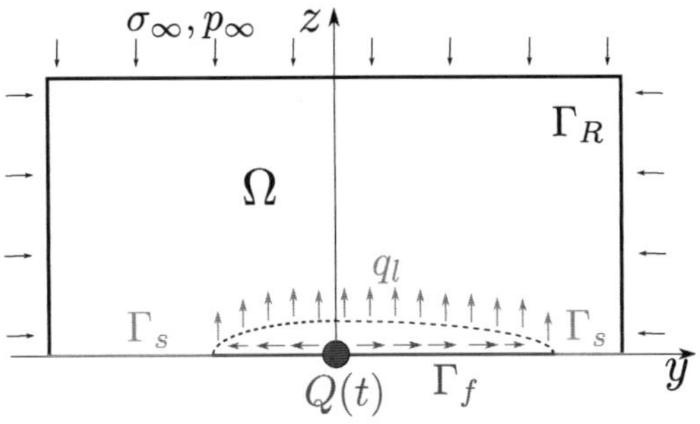

Figure 1. The cross-section of the fracture by plane $x = 0$.

Figure 1. Over the outer boundary $\Gamma_R = \partial\Omega \setminus \{z = 0\}$, the confining far-field stress $\boldsymbol{\sigma}_\infty$ and the constant pore pressure $p = p_\infty$ are applied:

$$\Gamma_R : \quad p = p_\infty, \quad \boldsymbol{\tau}\langle\mathbf{n}\rangle = \boldsymbol{\sigma}_\infty, \quad (\boldsymbol{\tau}\langle\mathbf{n}\rangle)_i = \tau_{ij} n_j. \tag{4}$$

Henceforth, $\mathbf{n}$ and $\mathbf{s}$ denote the outer normal and tangential unit vectors to the boundary of the domain Ω; the summation over the repeating index is implied. Passing to the limit $t \to 0$ implies that the presstress $\boldsymbol{\tau}_0$ satisfies the same boundary condition: $\boldsymbol{\tau}_0\langle\mathbf{n}\rangle|_{\Gamma_R} = \boldsymbol{\sigma}_\infty$. On the part of the boundary occupied by the fracture Γ_f it is set the condition that the fracture wall is subjected to the normal load generated by the fluid pressure p_f. The 2D fluid flow inside the fracture is governed by the mass conservation law in the lubrication theory approximation:

$$\frac{\partial w}{\partial t} = \frac{\partial}{\partial x}\left(\frac{w^3}{3\eta_f}\frac{\partial p_f}{\partial x}\right) + \frac{\partial}{\partial y}\left(\frac{w^3}{3\eta_f}\frac{\partial p_f}{\partial y}\right) + q_\ell + Q\delta(x, y, z), \tag{5}$$

where the leak-off rate q_ℓ is proportional to the pressure p gradient over the fracture's wall. The pumping rate Q is a multiplier of the Kronecker delta function concentrated at the origin. Over the remaining boundary part $\Gamma_s = \{z = 0\} \setminus \Gamma_f$ the symmetry conditions hold. The lag between the fracture front and fluid front inside the fracture is neglected. Also for the sake of simplicity we limit ourselves to the zero fracture toughness case so that the flow is assumed to be in the viscosity-dominated regime.

3. Numerical algorithm

For the numerical solution we apply the Finite Element Method. To this purpose, we use the weak formulation of the problem extended by the penalty integral term for prevention of the interpenetration of fracture walls. All the details of the numerical implementation of the algorithm are similar to the ones described in papers [3, 6, 4]. As in the cited algorithm, we resolve the nonlinear terms via the fixed-point iterations. The numerical code is implemented in the open-source software package FreeFem++ [8].

4. Case studies

The numerical model proposed in previous sections is employed to study the influence of the non-uniformness of closure stress on the fracture propagation. In our numerical experiment we consider the reservoir consists of three layers with different closure stress (see Figure 2). It corresponds to the case when the reservoir's pay zone, where the fracture is initiated, is confined by two symmetrical stress barriers with higher in situ stresses. This condition is realized by the special choice of the pre-stress τ_0 that satisfies the following condition over the fracture path [4]:

$$\mathbf{n} \cdot \tau_0 \langle \mathbf{n} \rangle |_{\Gamma_f} = \begin{cases} \sigma_{pay}, & |y| < H_{pay}/2, \\ \sigma_{barrier}, & |y| \geqslant H_{pay}/2. \end{cases} \tag{6}$$

In simulations the following set of parameters is used: the reservoir sizes $L_x = 80$ m, $L_y = 80$ m, $L_z = 60$ m, the pay zone height $H_{pay} = 10$ m, the in situ stress in the pay zone $\sigma_{pay} = 10$ MPa and in the confining layers $\sigma_{barrier} = 10$–12 MPa, Young's modulus $E = 17$ GPa, Poisson's ratio $\nu = 0.2$, initial porosity $\phi_0 = 0.2$, permeability $k_r = 10^{-14}$ m^2, Biot coefficient $\alpha = 0.75$, reservoir fluid viscosity $\eta_r = 0.001$ Pa·s, fracturing fluid viscosity $\eta_f = 0.1$ Pa·s, pumping rate $Q = 0.02$ m^3/s.

The calculations continue until the fracture is arrested by the equality of the total leakoff and the fluid inflow. The confining stress within the stress barrier is changed from 10 to 12 MPa. The purpose of the numerical experiment is to estimate the deviation of the fracture from the circular one for different values of the confining stress contrast $\Delta\sigma = \sigma_{barrier} - \sigma_{pay}$ between the pay zone and the barriers.

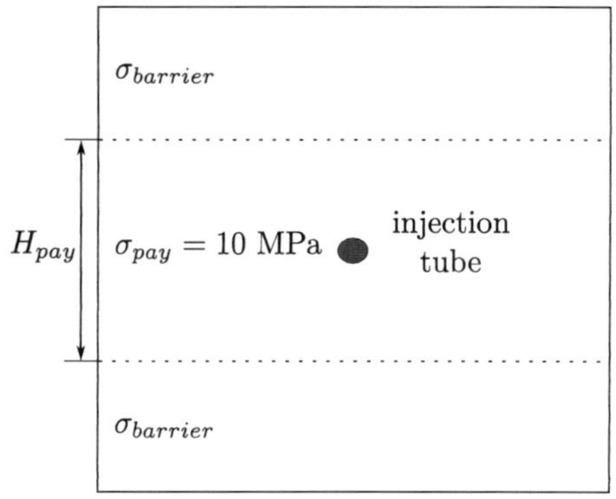

Figure 2. Schematics of the reservoir: the pay zone is surrounded by two barrier layers.

Results of calculations are demonstrated in Figures 3–6. One can see that the profile of the area occupied by the fracture barely depends on the confining stress above the limit of $\Delta\sigma = \sigma_{barrier} - \sigma_{pay} = 0.4$ MPa. For the higher values of the stress contrast, the fracture becomes shorter in the upper direction and longer along the horizontal axis.

5. Conclusion

A numerical model of a planar hydraulic fracture propagating in a three dimensional poroelastic medium based on the approach developed in in [3, 6] is presented. The model accounts for the influence of the pore fluid and the elastic rock deformation on the growth of the fracture. The fluid exchange between the reservoir and the fracture is naturally incorporated into the model. Also the model is able to take into account the inhomogeneity of reservoir properties and geological stresses [4].

The model is employed to analyze the influence of the in situ stress contrast in the confining layers on the final fracture geometry. Modeling of the three-layer reservoir with the stress contrast between the central and two surrounding layers demonstrates that even a small contrast of confining stresses leads to propagation of fracture only within the layer with a smaller confining stress. Increase of the value of stress contrast above 20% does not significantly change the fracture dynamics. It is also shown that in every case fracture comes to a stationary state as the total volume of the leakoff coincides with the volume of fluid pumped into the fracture.

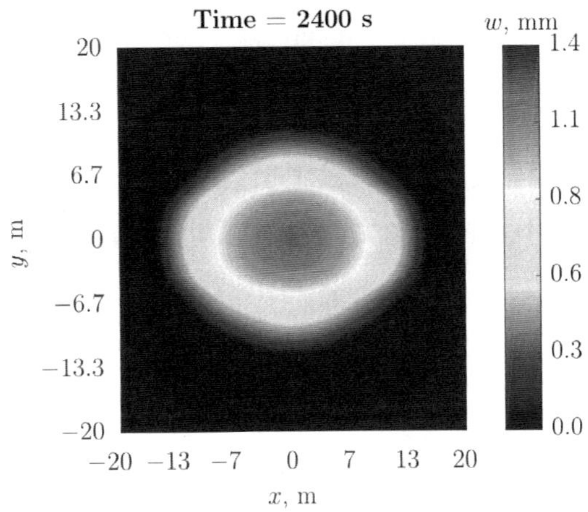

Figure 3. Hydraulic fracture half-width distribution at $t = 2400$ s. The in situ stress contrast is $\Delta\sigma = 0.8$ MPa.

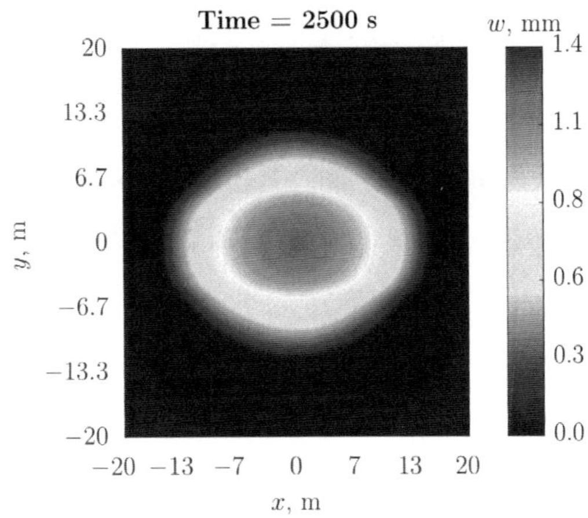

Figure 4. Hydraulic fracture half-width distribution at $t = 2500$ s. The in situ stress contrast is $\Delta\sigma = 0.8$ MPa.

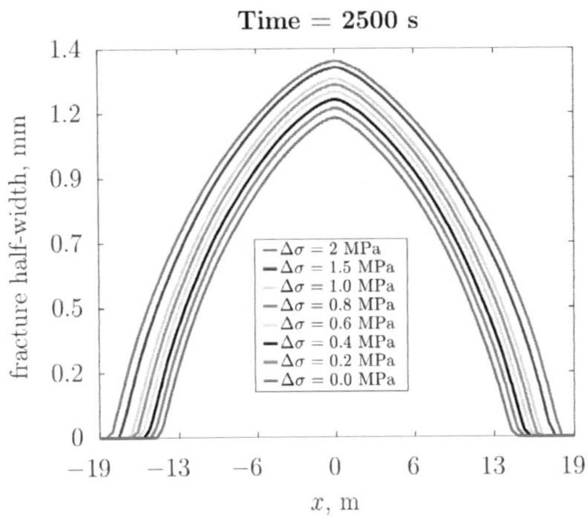

Figure 5. Fracture half-width profiles along line $\{z = 0\}$ at $t = 2500$ s for different in situ stress in confining layers.

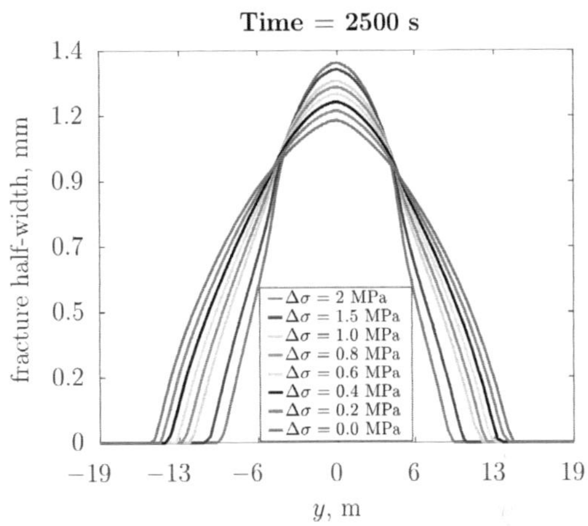

Figure 6. Fracture half-width profiles along line $\{y = 0\}$ at $t = 2500$ s for different in situ stress in confining layers.

Acknowledgments
This work was supported by the RFBR grant (Project No. 16-01-00610) and by President Grant for Leading Russian Scientific Schools of the Russian Federation (Grant N.Sh.-8146.2016.1).

References
[1] Adachi J, Siebrits E, Peirce A and Desroches J 2007 *Int. J. Rock Mech. Min. Sci.* **44** 739–757
[2] Detournay E 2016 *Ann. Rev. Fluid Mech* **48** 311–339
[3] Baykin A and Golovin S 2016 *J. Phys.: Conf. Ser.* **722** 012003
[4] Baykin A N and Golovin S V 2016 https://arxiv.org/abs/1610.09471
[5] Kovalyshen Y 2010 *Fluid-driven fracture in poroelastic medium* Ph.D. thesis University of Minnesota Minneapolis
[6] Golovin S V and Baykin A N 2016 https://arxiv.org/abs/1608.08001
[7] Coussy O 2004 *Poromechanics* (Chichester: John Wiley & Sons Ltd.)
[8] Hecht F 2012 *J. Numer. Math.* **20** 251–265 ISSN 1570–2820

Evaluation of the stress-strain state of a twisted rod made from an anisotropic material in the shear direction at creep

I A Banshchikova

Lavrentyev Institute of Hydrodynamics, Novosibirsk, Russia

E-mail: binna@ngs.ru

Abstract. Torsion of circular samples cut out in the longitudinal direction of a plate from a transversally-isotropic alloy with reduced resistance to creep in a direction at angle of 45° to the direction of the plate normal is simulated. To verify the calculations in the finite-element program, a lower and upper estimate of the torsion angle was obtained based on the principles of minimum of the total power and minimum of the additional dissipation power.

1. Introduction

Modern materials have the properties of anisotropy. That creates certain difficulties when processing and in predicting service life. The work [1] gives the results of experiments on creep at constant stresses on tension and compression of samples cut out in directions: longitudinal (along the rental), transverse plate normal and at an angle of 45° to the normal direction of the plate. These data indicate that the direction at an angle of 45° to the plate normal is the weakest direction for some alloys (V95pchT2, 1161, AK4-1).

The investigation of the problems of rods torsion in the conditions of anisotropic creep considered in [2, 3] is continued in the present paper. These works contain a description of the ANSYS Beam189 and Solid45 elements testing for calculation of circular rod. Since obtaining an analytical solution under anisotropic creep conditions even for a rod of circular cross-section is problematic, in order to estimate the reliability of finite element modeling results, the present work considers the simplest upper and lower bounds for the rate of torsion angle.

2. Theory

Considering torsion of a rod of arbitrary cross section with free ends, the creep strain rates are related to the stresses by the relations [2–4]:

$$\dot{\gamma}_{23} = \eta_{23} = \dot{W}_{,2} + \theta x_1 = 2T^{n-1} A_{23}\sigma_{23},$$
$$\dot{\gamma}_{13} = \eta_{13} = \dot{W}_{,1} - \theta x_2 = 2T^{n-1} A_{31}\sigma_{13}, \tag{1}$$

where $T = \left(2A_{23}\sigma_{23}^2 + 2A_{31}\sigma_{13}^2\right)^{0,5}$; $W(x_1, x_2)$ is function corresponding to the displacement of cross-section points (warping) in the x_3 direction along the rod and arising from the torsion; θ is rate of torsion angle per unit length, n is a creep constant of the material. The coefficients of the quadratic form A_{ij} are defined as follows [5]

$$2A_{23} = 4B_{23}^m - A_{22} - A_{33},\ 2A_{31} = 4B_{31}^m - A_{33} - A_{11},\ 2A_{12} = 4B_{12}^m - A_{11} - A_{22}$$

Content from this work may be used under the terms of the Creative Commons Attribution 3.0 licence. Any further distribution of this work must maintain attribution to the author(s) and the title of the work, journal citation and DOI.
Published under licence by IOP Publishing Ltd

IOP Conf. Series: Journal of Physics: Conf. Series **894** (2017) 012006 doi:10.1088/1742-6596/894/1/012006

$$2A_{11} = B_{22}^m + B_{33}^m - B_{11}^m,\quad 2A_{22} = B_{33}^m + B_{11}^m - B_{22}^m,\quad 2A_{33} = B_{11}^m + B_{22}^m - B_{33}^m, \tag{2}$$

Here B_{11}, B_{22}, B_{33} are constants of one-dimensional creep in three main directions (1 - longitudinal, 2 - along the normal to the plate, 3 - transverse); B_{12}, B_{23}, B_{31} are similar constants in three directions along the axes in the coordinate system obtained by rotating the original coordinate system by 45°; $m = 2/(n+1)$.

In the case of a transversally-isotropic alloy with a reduced resistance to creep at an angle of 45° to the direction of the plate normal, we have

$$B_{11} = B_{22} = B_{33} = B_{31} = B_O,\quad B_{12} = B_{23} = B_\Delta. \tag{3}$$

From (1), the stress components are

$$\sigma_{23} = \frac{\eta_{23}}{2A_{23}} G^{(1-n)/n},\quad \sigma_{13} = \frac{\eta_{13}}{2A_{31}} G^{(1-n)/n}, \tag{4}$$

where $G = \left(\eta_{13}^2/(2A_{31}) + \eta_{23}^2/(2A_{23})\right)^{0,5}$, $G = T^n$.

The equation of equilibrium for a rod with free ends [4]

$$\partial\sigma_{13}/\partial x_1 + \partial\sigma_{23}/\partial x_2 = 0. \tag{5}$$

Boundary conditions on the contour of the cross section are as follows

$$\sigma_{13}n_1 + \sigma_{23}n_2 = 0. \tag{6}$$

The compatibility condition for creep strain rates is transformed to the form

$$\frac{\partial}{\partial x_2}\left(2A_{31}T^{n-1}\sigma_{13}\right) - \frac{\partial}{\partial x_1}\left(2A_{23}T^{n-1}\sigma_{23}\right) = -2\theta. \tag{7}$$

Torque is

$$M = \iint_S (\sigma_{23}x_1 - \sigma_{13}x_2)dx_1dx_2. \tag{8}$$

Considering torsion of a solid circular rod of isotropic material, the warping of section is missing and the solution can be obtained analytically. In the case of an arbitrary cross-section rod of isotropic material, various approximate methods are considered in [4], including the method of combining solutions of the elastic and ideal-plastic problems. Finite-element discretization and finite-difference approximation methods for solving the torsion problems of isotropic rods of arbitrary cross section are considered in [6–8]. A detailed analysis of torsion of anisotropic elastic rods is given in [9]. In particular, an analytic solution of Saint-Venant for the elliptical cross-section, a method for finding a solution by means of a functions of a complex variable, as well as approximate methods of solving by means of energy methods are considered. A cyclic creep torsion of thick-walled tubular specimens was analyzed in [10] by taking the creep-induced anisotropy into account.

The solution of torsion problem can be obtained by converting permitting ratios to the differential equation with respect to function of warping $W(x_1,x_2)$ [6–7], or with respect to function of stresses $F(x_1,x_2)$, such that $\sigma_{13} = \partial F/\partial x_2$, $\sigma_{23} = -\partial F/\partial x_1$ [4, 9]. Then torque (8) is equal to

$$M = \iint_S \left(\frac{(W_{,2} + \theta x_1)x_1}{2A_{23}} - \frac{(W_{,1} - \theta x_2)x_2}{2A_{31}}\right)\left(\frac{(W_{,2} + \theta x_1)^2}{2A_{23}} + \frac{(W_{,1} - \theta x_2)^2}{2A_{31}}\right)^{\frac{k_1}{2}} dx_1dx_2 \tag{9}$$

or

$$M = 2\iint_S F dx_1dx_2, \tag{10}$$

where $k_1 = (1-n)/n$. In both cases, for the rods of an arbitrary cross-section, the solution of the system (1)–(8) can be obtained using finite element method or finite-difference approximation. Approximate estimates can be useful to verify these solutions.

For a lower estimate of the rate of torsion angle, we use the condition of minimum of the total power [4]:

$$I_1 = \iint_S \left(\frac{n}{n+1} G^{k_2} - \theta \left(\frac{x_1 \eta_{23}}{2A_{23}} - \frac{x_2 \eta_{13}}{2A_{31}} \right) G^{k_1} \right) dx_1 dx_2 = \min , \tag{11}$$

$k_2 = (n+1)/n$. Suppose that the function of warping similar to the case of anisotropic elasticity [9] approximately has the form

$$W(x_1, x_2) = \theta W_0(x_1, x_2) = \theta c_1 x_1 x_2 , \tag{12}$$

where c_1 is a constant. Function (12) satisfies the equation of compatibility (7). Taking into account (1), (12) expression (11) is transformed to

$$I_1 = \theta^{k_2} \iint_S \left(\frac{1}{k_2} \left(\frac{x_1^2 (c_1+1)^2}{2A_{23}} + \frac{x_2^2 (c_1-1)^2}{2A_{31}} \right)^{\frac{k_2}{2}} - \left(\frac{x_1^2 (c_1+1)}{2A_{23}} - \frac{x_2^2 (c_1-1)}{2A_{31}} \right) \left(\frac{x_1^2 (c_1+1)^2}{2A_{23}} + \frac{x_2^2 (c_1-1)^2}{2A_{31}} \right)^{\frac{k_1}{2}} \right) dx_1 dx_2 \tag{13}$$

Finding the constant c_1 at which the integral (13) achieves the minimum, we find the the rate of torsion angle from (9)

$$\theta = \left(\frac{M}{R_1} \right)^n , \tag{14}$$

where $R_1 = 4 \int_0^a \int_0^{f(x_2)} \left(\frac{x_1^2 (c_1+1)}{2A_{23}} - \frac{x_2^2 (c_1-1)}{2A_{31}} \right) \cdot \left(\frac{x_1^2 (c_1+1)^2}{2A_{23}} + \frac{x_2^2 (c_1-1)^2}{2A_{31}} \right)^{k_1/2} dx_1 dx_2$,

$f(x_2) = (a^2 - x_2^2)^{0.5}$, a is the radius of the rod cross section.

For the upper estimate of the rate of torsion angle, we use the condition of minimum of the additional dissipation power [4, 9]:

$$I_2 = \iint_S \left(\frac{1}{n+1} T^{n+1} - 2\theta F \right) dx_1 dx_2 = \min . \tag{15}$$

Suppose that the stresses function has the form

$$F(x_1, x_2) = c_2 \theta^{1/n} F_0(x_1, x_2) = c_2 \theta^{1/n} \left(1 - \left(\frac{x_1^2 + x_2^2}{a^2} \right)^{k_2/2} \right) , \tag{16}$$

where c_2 is a constant. Function (15) up to a multiplier is the solution of torsion problem of circular rod made of an isotropic material under conditions of steady-state creep. This function satisfies the equation of equilibrium and boundary conditions. Substituting (16) into (15), from the equation $dI_2 / dc_2 = 0$ we find the constant at which (15) reaches a minimum:

$$c_2 = \left(\iint_S 2F_0 dx_1 dx_2 \Big/ \iint_S \left(2A_{23} F_{0,1}^2 + 2A_{31} F_{0,2}^2 \right)^{(n+1)/2} dx_1 dx_2 \right)^{1/n} . \tag{17}$$

Then taking (16), (17) into account, we find the the rate of torsion angle from (10)

$$\theta = \left(\frac{M}{c_2 R_2} \right)^n , \tag{18}$$

where $R_2 = 8 \int_0^a \int_0^{f(x_2)} F_0 dx_1 dx_2$.

3. Results and discussion

Based on the experiments carried out for the V95pchT2 alloy (almost pure alloy V95 treated in mode T2) at T = 180°C, the weakest direction at an angle of 45° to the to the normal direction of the plate under creep conditions was determined and the following coefficients were obtained for the creep strain rates on tension [1]: $B_O = 6.3\cdot 10^{-31}$ MPa^{-n} s^{-1} for longitudinal, transverse and the plate normal directions; $B_\Delta = 3.9\cdot 10^{-30}$ MPa^{-n} s^{-1} for direction at the angle of 45 ° to the plate normal. The creep constant n = 10 is the same for all directions and does not depend on the sign of the applied load.

The calculations are performed with parameters: the radius of the rod a=0.02 м, the length L=0.1 m, torque M=3500 Nm, the Poisson's ratio ν=0.4, Young's modulus E=55000 MPa, the time of torsion t=600 s. The longitudinal direction of the plate coincides with the direction of the x_3 axis of the rod. The direction of the normal to the plate coincides with the direction of the x_2 axis.

Table 1 for comparison includes the full angle of torsion $\varphi = Lt\theta$ (radians):

- calculated in ANSYS using the Beam189 and Solid45 (angle estimated by the displacement) elements without taking into account the initial elastic component;
- the approximate estimates (14), (18);
- the analytical solutions for a rod cut out from an isotropic material ($B_{ij} = B_O$) [4]

$$\theta = \left(\sqrt{3}\right)^{n+1} \frac{B_O}{a}\left(\frac{3+1/n}{2\pi a^3} M\right)^n ; \qquad (19)$$

- the analytical solutions for a rod cut out in the normal direction of a plate of transversally-isotropic alloy (3) [2–3]

$$\theta = \frac{\left(2A_{12}\right)^{(n+1)/2}}{a}\left(\frac{3+1/n}{2\pi a^3} M\right)^n . \qquad (20)$$

Table 1. Full angle of torsion φ (radian).

type of symmetry	direction of rod cut	angle (formula)	Beam189	Solid45
isotropic	any	0.18 (19)	0.20	0.20
transversally-isotropic	longitudinal	0.54 (14)/ 0.8(18)	0.53	0.72
transversally-isotropic	normal	1.77 (20)	1.77	1.87

Work [3] gives the torsion angle for the mesh of Solid45 elements: 10 divisions along the radius, 40 along the boundary of the cross section (circle), 10 along the axis of the rod, a time step of 5 s in the steady-state creep. Work [2] gives the torsion angle for the mesh of Beam189 elements: 10; 32; 10; 5. Table 1 shows the angle values for the smaller mesh of Solid45 elements and time step: 16; 64; 20; 1. A mesh of Beam189 elements is the same, but the time step in the steady-state creep is reduced to 1 s. To apply torque load on one of the ends of the solid model, Shell181 elements were used, which creates a small edge effect and gives an uneven twist angle. Therefore, the angle was calculated from the displacements of the points near the end, where the shell elements are absent.

Table 2 shows the values of the maximum stresses calculated by different methods for a rod cut out longitudinally from a transversally-isotropic plate. If $c_1 = 0$ in (14), then there is no warping.

For a rod cut out from a material under the assumption of isotropy, the stresses are $\sigma_{13} = \sigma_{23} = 217.7$ MPa for Solid45 and $\sigma_{13} = \sigma_{23} = 215.6$ MPa for the Beam189. For a rod cut out in the normal direction of a plate of transversally-isotropic material, stresses are $\sigma_{13} = \sigma_{23} = 216.1$ MPa for Solid45 and $\sigma_{13} = \sigma_{23} = 215.4$ MPa for the Beam189.

Table 2. The maximum stresses σ_{13}, σ_{23} and full angle of torsion φ for a rod cut out longitudinally from a transversally-isotropic plate.

value	formula (14), $I_1 = \min$	formula (14), $c_1 = 0$	formula (18), $I_2 = \min$	Beam189	Solid45
σ_{13} (MPa)	238.2	240.1	215.9	239.6	238.3
σ_{23} (MPa)	194.3	190.4	215.9	190.2	201.9
φ (radian)	0.54	0.51	0.8	0.53	0.72

Figure 1*a* and Figure 1*b* show the isolines of the stresses σ_{13} and σ_{23} at the torsion of rod cut out in the longitudinal direction x_3 (coincide with Z) made of an alloy with the reduced resistance to creep strain in the direction 45° to the normal direction of plate x_2 (coincide with Y).

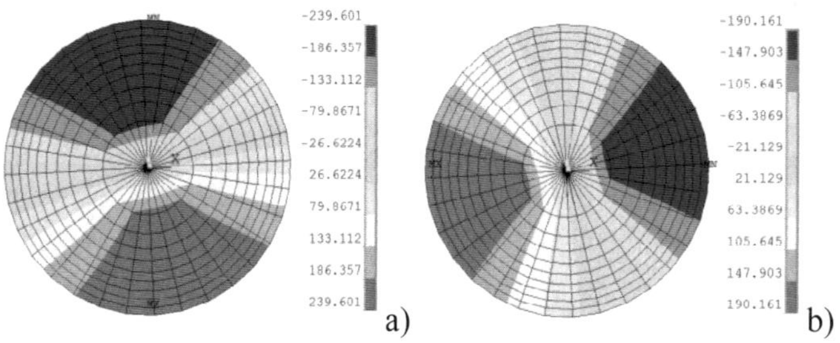

Figure 1. Stresses isolines (MPa) of a rod cut out in the longitudinal direction from a transversally-isotropic plate calculated by the Beam189 element σ_{13} (a), σ_{23} (b).

It should be noted that the calculations using Solid45 for small mesh takes considerable time. Given this, the methods proposed can be useful for a preliminary analysis of loads.

4. Conclusion

The torsion of circular rods cut out in the longitudinal direction of a plate from a transversally isotropic alloy with reduced resistance to creep deformation in the direction at an angle of 45 ° to the direction of the plate normal is considered. The obtained lower and upper estimates of the torsion angle rate based on the principles of minimum of the total power and minimum of the additional dissipation power describe the results of numerical simulation with a good accuracy. The developed technique can be applied to the problems of torsion of rods of arbitrary cross section for predicting loads and estimating of stresses.

Acknowledgments

This work was partially supported by the Russian Foundation for Basic Research (project codes 15-01-07631-a and 16-08-00713-a).

References

[1] Gorev B V and Masanov I 2009 *Tekhnologiya Mashinostroeniya* **7** 13–20

[2] Banshchikova I A, Tsvelodub I Yu and Petrov D M 2015 *Physics and Mathematics, Kazan. Gos. Univ. Uchen. Zap. Ser. Fiz.-Mat. Nauki* **157**(3) 34–41

[3] Banshchikova I A, Petrov D M and Tsvelodub I Yu 2016 *Journal of Physics: Conf. Series* **722**(1) 012004

[4] Kachanov L M 1960 *The Theory of Creep* (Moscow: Fizmatgiz) p 455

[5] Sosnin O 1965 *Appl. Mech. Tech. Phys.* **6**(6) 67–70
[6] Banshchikova I A, Gorev B V and Sukhorukov I V 2002 *Appl. Mech. Tech. Phys.* 43(3) 448–56
[7] Banshchikova I A and Sukhorukov I V 2003 *Computational Technologies - Regional Bulletin Of The Est* 8 104–10
[8] Banshchikova I A, Sukhorukov I V and Biao Yuan 2004 *Computational Technologies - The KazNU Journal, Mathematics, Mechanics and Computer Science Edition* 9 248–53
[9] Lekhnitskii S 1971 *Torsion of Anisotropic and Inhomogeneous Rods* [in Russian] (Moscow: Nauka) p 240
[10] Shutov A V, Larichkin A Yu and Shutov V A 2017 ZAMM DOI: 10.1002/zamm.201600286

Time of creep fracture of axisymmetrically loaded structures

I A Banshchikova[1], **I V Lubashevskaya**[1,2]

[1] Lavrentyev Institute of Hydrodynamics SB RAS, 15 Prospekt Lavrentyeva, Novosibirsk, 630090, Russia
[2] Novosibirsk State University, 1 Pirogova str., Novosibirsk, 630090, Russia

E-mail: binna@ngs.ru

Abstract. A stress-strain state and a time duration up to fracture are calculated with allowance for two-stage behavior of a rotating disk under creep conditions. The duration of the stages is investigated depending on the choice of the version of the creep kinetic theory and the geometric dimensions of the disk. The first stage is the accumulation of damage and the beginning of fracture in some area of the body, where the accumulated damage reaches a critical value. The second stage is the spread of the fracture front and the complete destruction of the body. A calculation method has been developed which reduces the solution of the unsteady-state creep problem to the solution of an analogous steady-state problem.

1. Introduction

Evaluation of durability of structural elements is an actual problem. Usually, normative life of structural elements is identified by the onset of their destruction. But the time of destruction onset does not determine durability, so the determination of the additional (residual) life of the structure is an important task [1-9].

In the present work, a study of the stress-strain state and the duration of deformation of a rotating disk accounting for two-stage behavior is carried out. The solution uses an approximate calculation method based on the kinetic creep theory, taking into account the accumulation of damage [7–9]. The time of fracture onset is the time when, in some area of the structure, the accumulated damage reaches a critical value. This is the first stage of latent destruction.

The second stage is the propagation of the creep fracture front and the complete destruction of the body. The calculation method consists in the fact that the solution of the unsteady-state creep problem reduces to solving an analogous steady-state creep problem. This technique was previously developed for a ring loaded axisymmetric plates under bending [10–12] and thick-walled vessels (pipes), loaded by internal pressure [13, 14]. For a rotating disk, the duration of the stages is investigated, depending on the choice of the version of the kinetic theory of creep and the geometric dimensions of the disk.

2. Theory

2.1. Structural equations and the time of fracture onset of the axisymmetrically loaded rotating disk

Creep of a uniformly heated disk with simultaneous consideration of damage accumulation in the material is considered from the phenomenological point of view. The internal, outer and current radiuses of the disk are a_0, b_0, r, $a_0 \le r \le b_0$, the thickness of the disk is $h = h(r)$, $h(r) = h_0 \cdot r^{-k}$, $k \ge 0$, h_0 is a constant (see Figure 1).

Content from this work may be used under the terms of the Creative Commons Attribution 3.0 licence. Any further distribution of this work must maintain attribution to the author(s) and the title of the work, journal citation and DOI.
Published under licence by IOP Publishing Ltd

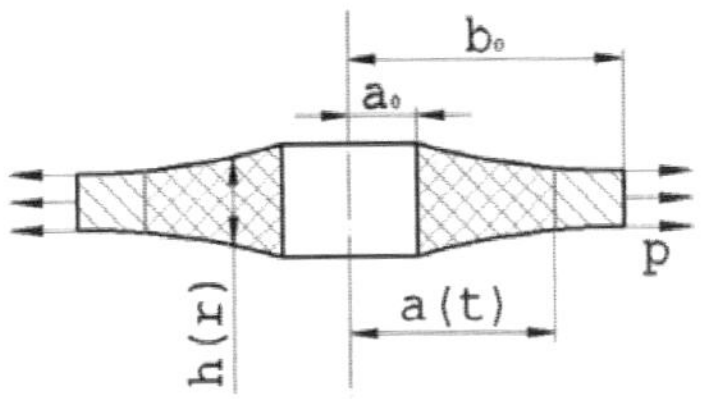

Figure 1. Diagram of a disk.

A plane stress state is assumed in the disk ($\sigma_z = 0$), the material is incompressible ($\eta_\varphi + \eta_r + \eta_z = 0$) and $\eta_r = dv/dr$, $\eta_\varphi = v/r$; here $v(r,t)$ is the displacement radial rate of an arbitrary point of the disk, $\eta_\varphi, \eta_r, \eta_z$ are tangential, radial and axial components of the tensor of creep strain rates. The z axis is perpendicular to the plane of the disk, and the axes r, φ are directed as usual [7, 8].

The stresses σ_φ, σ_r at any instant of time satisfy the equation of equilibrium

$$d(hr\sigma_r)/dr - h\sigma_\varphi + h\Omega^2 r^2 = 0 \qquad (1)$$

and the boundary conditions $\sigma_r(a_0) = 0$, $\sigma_r(b_0) = p$, where p is a uniform tensile pressure at the rim of the disk; $\Omega^2 = \gamma\omega^2$, γ is the density of the disk material, ω is the angular velocity of the disk. Integrating equation (1) prescribed boundary conditions and taking into account that $h = h_0 \cdot r^{-k}$, we obtain

$$\sigma_r(r,t) = \left(\frac{\Omega^2}{3-k} a_0^{3-k} + \int_{a_0}^{r} \sigma_\varphi r^{-k} dr \right) r^{k-1} + \frac{\Omega^2}{3-k} r^2, \qquad (2)$$

$$\int_{a_0}^{b_0} \sigma_\varphi r^{-k} dr = P, \qquad P = pb_0^{1-k} + \frac{\Omega^2}{3-k}(b_0^{3-k} - a_0^{3-k}). \qquad (3)$$

Physical relationships use Rabotnov's kinetic theory [7, 9]

$$\eta_{ij} = \frac{W}{\sigma_e} \frac{\partial \sigma_e}{\partial \sigma_{ij}}, \qquad W = \frac{B_1 \sigma_e^{n+1}}{\varphi_1(\mu)}, \qquad i, j = 1,2,3; \qquad (4)$$

$$\frac{d\mu}{dt} = -\frac{B_2 \sigma_{*e}^{g+1}}{\varphi(\mu)}, \qquad \mu(r^*, t_*) = 0, \qquad (5)$$

where η_{ij}, σ_{ij} are the components of the creep strain rates and stresses tensors; σ_e, σ_{*e} are first-order functions which are homogeneous with respect to stresses; W is the power of dissipated energy, $W = \sigma_{ij}\eta_{ij}$. B_1, n, B_2, g are the constants of creep and long-term strength; μ is a parameter describing the accumulation of damage during the creep of the material from the phenomenological point of view. The time of onset of body fracture t_* (the duration of the first stage) is calculated from the condition for achieving the parameter of damage μ its critical value of zero in some coordinates x_{k*}^*, $k = 1,2,3$.

If $\varphi_1(\mu) = \varphi(\mu)$, then the system (4), (5) represents the most widespread version of the kinetic theory, from which as the special cases follow the theory of short-term creep, the energy variant of creep and long-term strength, the hardening theory with its various modifications. If $\varphi_1(\mu) = 1$, then we get a variant of the kinetic theory of creep in the formulation of Kachanov [8]. In the following it is assumed that $\varphi_1(\mu) = \mu^{m_1}$, $\varphi(\mu) = \mu^m$, $0 \le m_1 \le m$.

In (4), (5) the maximum shear stress is chosen as σ_e и σ_{*e}. Consider the case when the condition $\sigma_\varphi > \sigma_r > \sigma_z = 0$ is satisfied at each point of the disk at any time up to destruction. Therefore $\sigma_e = \sigma_{*e} = \sigma_\varphi / 2$ and system (4), (5) reduced to:

$$\eta_\varphi = \bar{B}_1 \sigma_\varphi^n / \mu^{m_1}, \quad \eta_r = 0, \quad \eta_z = -\bar{B}_1 \sigma_\varphi^n / \mu^{m_1}, \qquad \bar{B}_1 = B_1 / 2^{n+1}, \tag{6}$$

$$\mu^m d\mu = -\bar{B}_2 \sigma_\varphi^{g+1}, \quad \mu(r,0) = 1, \quad \mu(r^*, t_*) = 0, \quad \bar{B}_2 = B_2 / 2^{g+1}. \tag{7}$$

Since $\eta_r = 0$, then $dv/dr = 0$ and it follows that $v(r,t)$ is a function of time only, i.e $v = C(t)$. Then

$$\eta_\varphi = C(t)/r, \quad \eta_r = 0, \quad \eta_z = -C(t)/r \tag{8}$$

and taking into account (3), (6):

$$\sigma_\varphi(r,t) = \frac{P}{J_1} \cdot \frac{[\mu(r,t)]^{m_1/n}}{X(t)} \cdot r^{-1/n}. \tag{9}$$

In (9) for convenience we introduce the notation:

$$J_1 = \int_{a_0}^{b_0} r^{-(k+1/n)} dr, \quad \left[\frac{C(t)}{\bar{B}_1}\right]^{1/n} = \frac{P}{J_1} \cdot [X(t)]^{-1}. \tag{10}$$

Substituting (9) into the equation for the parameter of damage (7) and (3), we obtain

$$\int_1^\mu [\mu(r,t)]^{(mn - m_1(g+1))/n} d\mu = -[(m+1) t^0(r)]^{-1} \int_0^t [X(\tau)]^{-(g+1)} d\tau, \tag{11}$$

$$\int_{a_0}^{b_0} [\mu(r,t)]^{m_1/n} r^{-(k+1/n)} dr = J_1 X(t), \tag{12}$$

where

$$t^0(r) = \left[(m+1) \bar{B}_2 (P/J_1)^{g+1} r^{-(g+1)/n}\right]^{-1} \tag{13}$$

Thus, the solution of the problem of the stress-strain state calculation of the disk and the time of its fracture has been reduced to solving the system of equations (11) and (12). Integrating (11), we obtain

$$\mu^{m_1/n} = \left[1 - \frac{\nu}{t^0(r)} \int_0^t X^{-(g+1)} d\tau\right]^\beta, \quad \beta = \frac{m_1}{n + mn - m_1(g+1)}, \quad \nu = \frac{n + mn - m_1(g+1)}{n(m+1)}. \tag{14}$$

Substituting (14) into (12):

$$\int_{a_0}^{b_0} \left[1 - \frac{\nu}{t^0(r)} \int_0^t X^{-(g+1)} d\tau\right]^\beta r^{-(k+1/n)} dr = J_1 X(t). \tag{15}$$

Solving (15), we find $X(t)$, then the function $\mu(r,t)$ is determined from (14). Knowing these functions, σ_φ and σ_r are computed from (9) and (2), η_φ and η_r are calculated from (8) using (10). The time of onset of disk fracture is determined by the condition $\mu(r^*, t_*) = 0$. It is easy to show that $r^* = a_0$, i.e. the fracture begins at the inner surface of the disk. In order that the integral on the left-hand side of equation (11) make sense in the case $\mu = 0$, restrictions on the constants of the material must be imposed, which are described in detail in [13].

Let $m_1 = 0$. Then, from (14) we get $\beta = 0$, $\nu = 1$, and system (4), (5) is a variant of the kinetic theory of creep in the formulation of Kachanov. From (15) we get $X(t) = 1$. From (2), (8), (9), taking into account the second equation (10), it follows that the stress-strain state is stationary. From (14)

$$\mu(r,t) = \left[1 - \left(t/t_*^0\right)\left(a_0/r\right)^{(g+1)/n}\right]^{1/(m+1)}, \quad t_*^0 = \left[(m+1) \bar{B}_2 (P/J_1)^{g+1} a_0^{-(g+1)/n}\right]^{-1}, \tag{16}$$

$t_*^0 = t^0(a_0)$ is the time of the beginning of the disk fracture according to (13) and $r^* = a_0$.

2.2. *Duration of the propagation of the fracture front*

The calculation of the duration of the fracture front propagation is one of the most important problems for practice. In the case of the considered problem: at the instant of time $t = t_*$, the inner surface of the disk is destroyed, and then the fracture front, whose boundary is a circle with a radius $r = a(t)$, moves toward the outer surface of the disk $r = b_0$ (Figure 1). It is obvious, that $a_0 < a(t) < b_0$, $a(t_*) = a_0$, $a(t_{*f}) = b_0$, t_{*f} is the time of disk fracture. Further relevant parameters and functions related to the fracture front will be marked by index 'f'. In particular, at any point in time $\tau > t_*$ will be $J_1(\tau) = J_{1f}$, $P(\tau) = P_f$ and they are calculated from (3), (10) and (13) with a_0 replaced by $a(\tau)$; stresses σ_r and σ_φ are still calculated from (2) and (9). Next, performing the same calculations as before, we get at the instant of time t instead of (11), (12):

$$\int_1^{\mu(t)} [\mu(r,t)]^{(mn-m_1(g+1))/n} d\mu = -\int_0^t [(m+1)t^0(r,\tau)]^{-1}[X(\tau)]^{-(g+1)} d\tau, \tag{17}$$

$$\int_{a(t)}^{b_0} [\mu(r,t)]^{m_1/n} r^{-(k+1/n)} dr = J_{1f}(t)X(t). \tag{18}$$

Integrating (17), we have instead of (14):

$$\mu(r,t)^{m_1/n} = \left[1 - \nu\int_0^t [t^0(r,\tau)]^{-1} X^{-(g+1)} d\tau\right]^\beta.$$

On the front line $r = a(t)$ we have $\mu(a(t),t) = 0$, i.e. taking into account (13)

$$1 - \nu\int_0^t \left[(m+1)\bar{B}_2\left[P_f(\tau)/J_{1f}(\tau)\right]^{g+1} a^{-(g+1)/n}\right]\cdot[X(\tau)]^{-(g+1)} d\tau = 0. \tag{19}$$

Using (19), for the damage parameter we obtain $\mu(r,t)^{m_1/n} = \left[1 - (r/a(t))^{-(g+1)/n}\right]^\beta$ and (18) reduces to the form

$$\int_{a(t)}^{b_0} \left[1 - (r/a(t))^{-(g+1)/n}\right]^\beta r^{-(k+1/n)} dr = J_{1f}(t)X(t). \tag{20}$$

To obtain the equation of motion of the fracture front, we differentiate expression (19) with respect to t, applying in general the rule of differentiation of the integral with respect to a parameter (here t acts as a parameter). After the simplest operations, we get:

$$a^{-1+(g+1)/n} da/dt = \frac{n\nu(m+1)\bar{B}_2}{g+1}\left(\frac{P_f(t)}{J_{1f}(t)X(t)}\right)^{g+1}. \tag{21}$$

Taking (20) into account, (21) can be reduced to the form

$$a^{-1+(g+1)/n}[P_f(a)]^{-(g+1)}\left(\int_{a(t)}^{b_0}\left[1 - (r/a(t))^{-(g+1)/n}\right]^\beta r^{-(k+1/n)} dr\right)^{g+1} da = \left(\frac{n\nu(m+1)\bar{B}_2}{g+1}\right)dt, \tag{22}$$

where $P_f(a) = pb_0^{1-k} + \dfrac{\Omega^2}{3-k}(b_0^{3-k} - a^{3-k})$. Introducing new variables $\rho = a/a_0$, $1 \le \rho \le \beta_1$, $\beta_1 = b_0/a_0$, $y = r/a$, $1 \le y \le \beta_1/\rho$, and integrating (22), we obtain $a(t)$ and the duration of the fracture front propagation $\Delta t_* = t_{*f} - t_*$

$$\Delta t_* = \frac{g+1}{nv\bar{B}_2(m+1)} \int_1^{\beta_1} \rho^d \cdot \left(p\beta_1^{1-k} + \varpi\left(\beta_1^{3-k} - \rho^{3-k}\right)\right)^{-(g+1)} \cdot \left(\int_1^{\beta_1/\rho} y^{-(k+1/n)} \left(1 - y^{-(g+1)/n}\right)^{\beta} dy \right)^{g+1} d\rho,$$

Where $d = -1-(k-1)(g+1)$, $\varpi = a_0^2\Omega^2/(3-k)$.

3. Results and discussion

An example the rotating disk loaded on the outer contour by a tensile pressure $p = 20\,\text{MPa}$ is considered. We estimate the influence of the choice of the variant of the creep kinetic theory and the geometric dimensions of the disk on the duration of the first and second stages. The following constant of the material corresponding to the D16 alloy are used for calculations at a temperature of 250°C $B_1 = 3.5172 \cdot 10^{-15}(\text{MPa}^{-n}/\text{h})$, $n = 6$, $m = 14$, $g = 4.75$, $B_2 = 2.7563 \cdot 10^{-15}(\text{MPa}^{-g-1}/\text{h})$ [9, 11]. Dimensions of the disk are $a_0 = 0.025\,\text{m}$, $b_0 = 0.1\,\text{m}$, $\Omega = 70\,(\text{MPa}^{1/2}/\text{m})$.

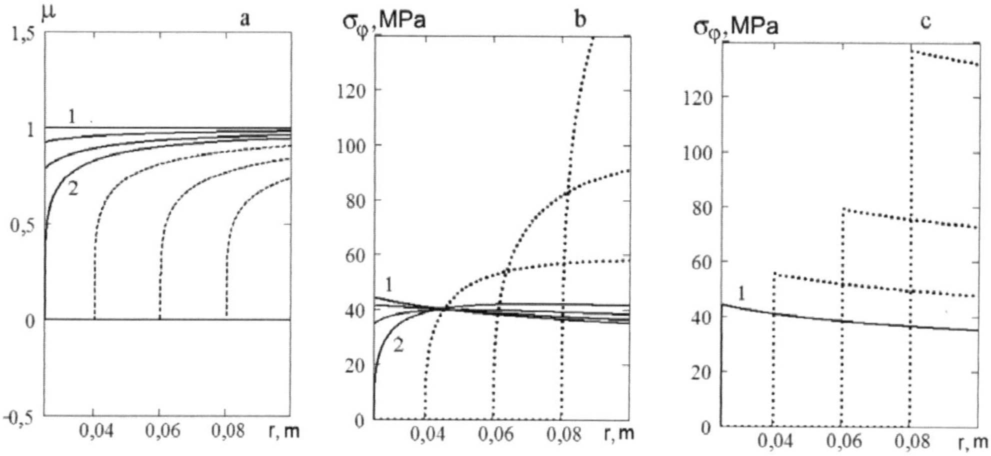

Figure 2. Dependence $\mu(r,t)$ – (a) and $\sigma_\varphi(r,t)$ – (b) for $m_1 = 10$ and $\sigma_\varphi(r,t)$ – (c) for $m_1 = 0$.

Lines in Figure 2 a, b show the dependence $\mu(r,t)$ and the diagrams $\sigma_\varphi(r,t)$ in the section $a_0 \le r \le b_0$ for some fixed time instants for $m_1 = 10$: solid lines correspond to the stage of latent destruction (lines 1 for $t = 0$, line 2 for $t = t_*$); dashed lines correspond to the stage of propagation of the fracture front.

Similar lines in Figure 2 c show the diagrams $\sigma_\varphi(r,t)$ for $m_1 = 0$ (a variant of the creep theory in Kachanov's formulation). In this case, redistribution of stresses up to the beginning of fracture is absent, i.e. the state is stationary (solid line 1 in figure 2 c). Here the function $\mu(r,t)$ for $m_1 = 0$ is analogous to the function $\mu(r,t)$ shown in Figure 2 a.

Table 1 presents the results of calculating of the first stage duration t_* and $\Delta t_*/t_* = (t_{*f} - t_*)/t_*$ for $m_1 = 0$ (Kachanov's creep theory) and $m_1 = 10$ (Rabotnov's kinetic theory) depending on a_0, k, Ω ($p = 20\,\text{MPa}$, $b_0 = 0.1\,\text{m}$).

Table 1. Duration t_* of the first stage and $\Delta t_*/t_*$ for $m_1 = 0$ and 10.

			$m_1 = 10$		$m_1 = 0$	
b_0/a_0	k	Ω, $(\text{MPa}^{1/2}/\text{m})$	t_*, $(10^3\ \text{h})$	$\Delta t_*/t_*$, (%)	t_*, $(10^3\ \text{h})$	$\Delta t_*/t_*$, (%)

5	0	0	351.9	17.9	195.1	50.2
5	0.5	70	49.1	12.8	28.6	39
4	1	70	119.4	6.3	76.7	24.4
4	0.5	70	30.9	9.1	19.0	32
4	0	70	8.9	12.7	5.3	42
4	0	30	120.8	12.1	71.0	39.3
4	0	0	266.7	11.9	156.8	38.4
2	0	0	31.8	1.9	23.5	13.5
2	0.5	30	35.8	1.6	26.9	12.4
2	0.5	70	2.6	1.8	1.99	14.5

4. Conclusion

The method of calculation of stress-strain state with account of damage accumulation which used to calculate such structures as plates under bending [10–12] and pipes under internal pressure [13, 14], is developed for the analysis of a rotating disk. On the basis of the obtained results it can be concluded that the duration of the stages of front propagation of the fracture, as in the case of pipes under internal pressure, can range from a few tenths to tens of percent of the time latent fracture, depending on the choice of the variant of creep theory, fracture criterion and geometrical dimensions of the structure.

Acknowledgments

This work was partially supported by the Russian Foundation for Basic Research (project codes 15-01-07631-a and 16-08-00713-a).

References

[1] Kachanov L M 1961 Time of Fracture under Creep Conditions *Problems of Continuum Mechanics* (Moscow: AN SSSR) 186–201 (In Russian)
[2] Hayhurst D R 1973 *Trans. ASME. J. Appl. Mech* **40**(4) 88-95
[3] Skrzypek J J, Ganczarski A W 1999 *Modeling of Material Damage and Failure of Structures. Theory and Applications* (Berlin: Springer- Verlag) p 326
[4] Lokoshchenko A M, Sokolov A V 2014 *Mechanics of Solids* **49(**1) 49–58
[5] Lokoshchenko A M, Fomin L V 2016 *J. Appl. Mech. Tech. Phy.* 57(5) 792–800
[6] Altenbach H, Altenbach J, Naumenko K 1997 *Mechanics of Time-Dependent Materials* **1**(2) 181–93
[7] Rabotnov Yu N 1969 *Creep Problems in Structural Members* (Amsterdam: North-Holland) p 836
[8] Kachanov L M 1960 *Teorija Polzuchesti [The Theory of Creep]* (Moscow, Fizmatgiz) p 455 (in Russian)
[9] Nikitenko A F 1997 *Polzuchest' i Dlitel'naya Prochnost' Metallicheskikh Materialov [Creep and Long-term Strength of Metal Materials*] (Novosibirsk: Novosibirsk State University of Architecture and Civil Engineering) p 278 (In Russian)
[10] Zaev V A, Nikitenko A F 1993 *J. Appl. Mech. Tech. Phy.* **34**(3) 423–6
[11] Banshchikova I A, Nikitenko A F 2006 *J. Appl. Mech. Tech. Phy.* **47**(5) 747–56
[12] Banshchikova I A 2015 *Computationl Continuum Mechanics* **8**(4) 359–68
[13] Nikitenko A F, Lyubashevskaya I V 2005 *Strength Mater.* **37**(5) 460–70
[14] Nikitenko A F, Lyubashevskaya I V 2007 *J. Appl. Mech. Tech. Phy.* **48**(5) 766–73

IOP Conf. Series: Journal of Physics: Conf. Series **894** (2017) 012008 doi:10.1088/1742-6596/894/1/012008

Mathematical modeling of the detonation wave and inert particles interaction at the macro and micro levels

I. A. Bedarev[1] and A. V. Fedorov[1, 2]

[1]Khristianovich Institute of Theoretical and Applied Mechanics Siberian Branch, Russian Academy of Sciences, Novosibirsk, 630090 Russia
[2]Novosibirsk State Technical University, Novosibirsk, 630073 Russia

E-mail: bedarev@itam.sc.ru, fedorov@itam.nsc.ru

Abstract. Calculation technology for modelling of two-dimensional detonation flows in a system of the reacting gas mixture-inert particles has been developed. The problems associated with the cellular detonation suppression are researched. The interaction of the detonation wave with inert particles is modeled using the macro and micro approaches. First, the interaction with Al_2O_3 particles with a diameter of 100 μm and 1 mm at the macro level was investigated using a continuum approach. Values of the volume concentration resulting in a change of the detonation cell size, weakening of the detonation wave, and suppression detonation are obtained. Secondly, a simulation of the detonation wave interaction with particles at the micro level has been carried out. A wave structure of the flow near the particles is obtained and the mechanism of detonation suppression was shown at volume concentrations of particles close to those obtained in modeling at the macro level.

1. Introduction

It is known that detonation waves in gas and heterogeneous combustible mixtures have a cellular structure [1-2]. It has also been established that the addition of inert particles to an explosive mixture contributes to the suppression of detonation [3-12]. The interaction between detonation waves and inert particles have been studied mainly in one dimension, resulting in estimates of the effect of volume fraction, particle diameter, and their thermophysical properties on the ability to attenuate and suppress detonation. However, the nonuniform structure of the detonation cell undoubtedly influences the parameters of mixtures of explosive gases with chemically inert particles and the limiting detonation characteristics. It can be noted that the interaction of cellular detonation waves in gases and mixtures of gas with inert particles has been the subject of a few studies, e.g., [13-14]. Thus, it is of both theoretical and practical interest to investigate the detonation-wave parameters and estimate the detonation cell size in mixtures of combustible gases and inert particles.
The interaction of a detonation wave with inert particles is investigated in the work by means of two approaches. On the one hand, the modeling of the problem at the macro level is carried out, when the cellular detonation wave is suppressed by a cloud of stationary particles. On the other hand, the direct simulation of the detonation wave interaction with fixed particles at the micro level has been studied.

2. Mathematical model and numerical implementation

In the initial stage of the study, we consider the interaction of a cellular detonation wave with a cloud of inert fixed isothermal particles or a porous filter. This assumption is possible in the case of

Content from this work may be used under the terms of the Creative Commons Attribution 3.0 licence. Any further distribution of this work must maintain attribution to the author(s) and the title of the work, journal citation and DOI.
Published under licence by IOP Publishing Ltd

IOP Conf. Series: Journal of Physics: Conf. Series **894** (2017) 012008 doi:10.1088/1742-6596/894/1/012008

interaction with relatively large particles, where the velocity and temperature of the particles do not vary significantly during the passage of the detonation wave. To test this assumption, we estimated the velocity and thermal relaxation times for a single particle of alumina (Al_2O_3), sand (SiO_2), and tungsten carbide (WC) of diameter 1 μm ÷1 mm in the gas flow behind a detonation wave. The calculations of the ratio of the velocity and thermal relaxation times of the particles showed that for alumina and sand particles, which have comparable densities and similar heat capacities, for diameters of 1-100 μm, the velocity relaxation time somewhat exceeds the thermal relaxation time. For tungsten carbide particles, having a high density and lower specific heat, the velocity relaxation time is about an order of magnitude higher than the thermal relaxation time over the whole range of diameters, with this difference decreasing somewhat with increasing diameter. The estimates allow us to judge the applicability of the process model used in this study. It can be seen from the tables that, for particles larger than 100 μm, the thermal and thermal relaxation time is an order of magnitude higher than the gas-dynamic time of transit of the detonation wave through the mixture. Thus, particles of this size can be considered fixed and isothermal.

The mathematical model includes the Favre-averaged Navier-Stokes equations for a multicomponent gas mixture with allowance for chemical reactions, supplemented by the SST modification of the k-ω model of turbulence. The interaction with fixed particles for the macro level problem was determined by means of a source term in the equations of momentum and energy. The solution was carried out in a two-dimensional formulation. As boundary conditions on the channel walls, standard no-slip conditions for velocities were setThe heat fluxes were equal to zero on the walls when solving the problem at the macro level in the continuum approach. The particle surfaces were assumed to be isothermal with a temperature of 300 K at the micro level.

As in any problem with chemical reactions, when modeling flows with detonation waves, the problem arises of choosing an appropriate kinetic scheme for description of the combustion process. A scheme with reduced or detailed kinetics can be chosen. The detailed kinetics allows more accurately reproducing parameters such that the ignition delay time and burning time of the mixture, especially at the limiting modes, but it requires a great deal of computer time. In [15], three models of the chemical kinetics of hydrogen combustion in oxygen and three gas-dynamic models of the flow of the reacting mixture behind the front of the initiating shock wave were analyzed. In [16], the use of certain kinetic models of the hydrogen-air mixture is justified to describe detonation. The verification performed on the experimental data on the ignition delay time and on the velocity of the detonation wave made it possible to determine the kinetic parameters of the model. Based on the detailed and reduced kinetic mechanisms of nonequilibrium chemical transformations within the framework of the ANSYS Fluent software, a mathematical technology was developed to describe the spread of cellular detonation in the channels. It is shown that the detonation cell size corresponds to the experimental data. Chemical reactions in the gas phase at this paper were modeled by a reduced one-stage kinetic model. In [16], this kinetic scheme was verified by the experimental data on the ignition delay times and the detonation wave velocity under various conditions.

The ANSYS Fluent software was used to simulate the two-dimensional problem of propagation and suppression of a detonation wave. An explicit fourth-order Runge-Kutta scheme and an implicit second-order scheme were used for approximation in time. The explicit scheme was found to be preferred for calculations of inviscid flows. The calculations results show that the implicit scheme has a somewhat larger numerical dissipation than the explicit fourth-order method, which leads to a greater smearing of shock-wave fronts. However, due to constraints on the stability conditions, the use of the explicit scheme for problems with viscosity significantly increases the computation time. The viscous formulation in this case will require a too small time step from the point of view of feasibility of the calculation on a computer. Convective terms are approximated in space using a second-order AUSM upwind scheme.

In the calculations, we used a moving quadrangular mesh that thickened towards the surface of the body and adapted, as to the density gradient, to the gas-dynamic features of the flow (shock waves, contact discontinuities, rarefaction waves). In the vicinity of particles moving in the gas under the action of the propagating shock wave the mesh moved together with the particles moving in the gas flow arising under the action of the shock wave. In constructing the mesh, the parameter y^+ on the particle surface was checked, which in all calculations did not exceed unity. Figure 1 shows an example of such a computational mesh adapted in the solution process for micro level problem. From this figure it is clearly seen how the computational mesh adapts itself to the features of the flow – the bow and hanging shock waves and the transmitted shock wave that arise in the flow in the vicinity of the particles system.

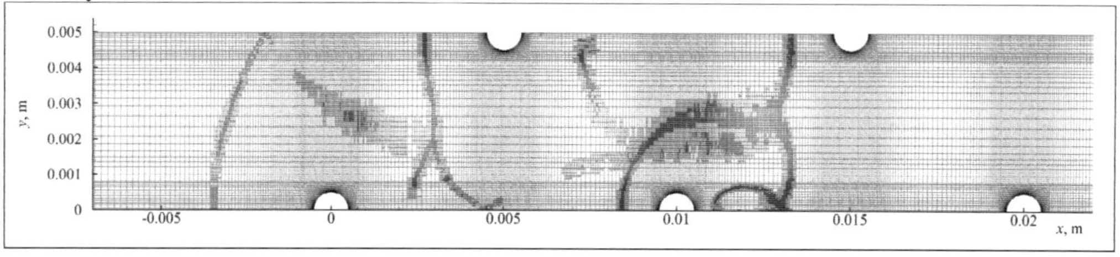

Figure 1. Example of the dynamically adaptive computational mesh.

3. Formulation of the problem

Plane overdriven detonation wave was used as the initial data to obtain a cellular detonation wave. Then cellular detonation wave was used at channel inlet, and a mixture of hydrogen and air and particles distributed in space were located in a low-pressure chamber (LPC) ($p_0 = 1$ atm, $T_0 = 296$ K), (Figure 3) for macro level problem. After the start of the calculation, the detonation wave moved from left to right and was attenuated by the presence of cold particles (Al_2O_3). At this stage of the study, the cloud of particles filled the entire LPC.

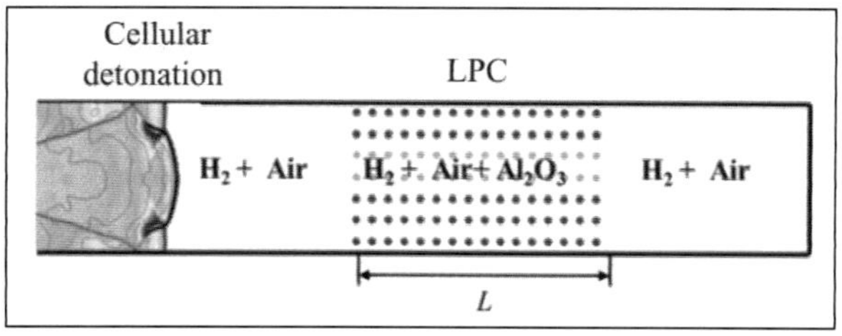

Figure 2. Diagram of the computational domain for the macro level problem.

The plane detonation wave with parameters close to the Chapman-Jouguet solution was set in the inlet section of the channel for solving the micro level problem. The flow diagram can be understood from Figure 1. From three to seven particles were arranged in a staggered order in the calculation area. The upper and lower boundaries of the computational domain were symmetry lines.

4. Calculation results

4.1. Macro level problem

Let us now consider the problem of the interaction of a cellular detonation wave with a cloud of inert fixed isothermal particles. The particles have a diameter of 100 μm and 1 mm. The left boundary of the cloud is at x =0.2 m. The cloud occupies all the space to the right boundary of the computational domain. The diameter of the detonation cell for the stoichiometric mixture of hydrogen and air is approximately 15 mm; accordingly, three cells are accommodated within the channel width at the initial time. Figure 3 shows the maximum pressure field in time for particles with diameter d = 100 μm and volume fraction $m_2 = 10^{-4}$. For this concentration, there is no significant change in the structure of the cellular wave. The number of cells is constant over the entire length of the cloud. Under these conditions, the deficit of the detonation wave velocity is $\eta = D/D_{CJ} = 0.93$, where D is the DW velocity and D_{CJ} is the Chapman-Jouguet velocity.

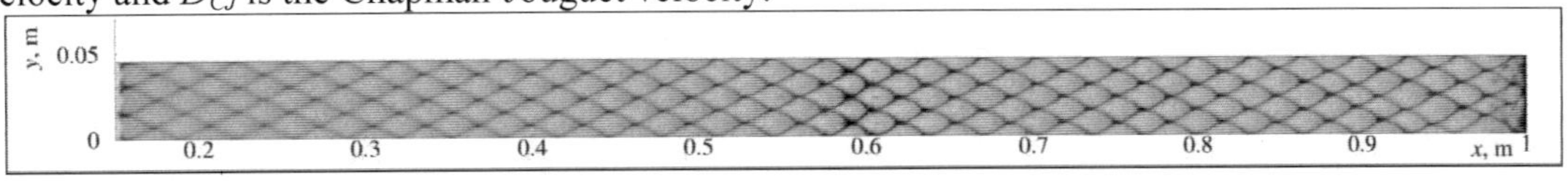

Figure 3. Time history of the maximum pressure field during interaction with a cloud of particles for $d = 100$ μm and $m_2 = 10^{-4}$.

Increasing the volume fraction of particles by an order of magnitude changes the structure of the cellular wave. Density isolines at different times are shown in Figure 4. At the time of 112 μsec, three more cells are accommodated within the channel width, but at the time of 228 μsec, the wave structure changes and the cell size increases.

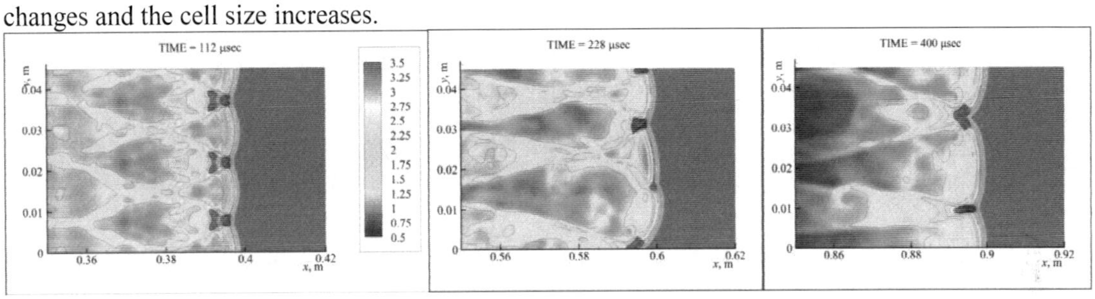

Figure 4. Density isolines at different times for particles with $d = 100$ μm and $m_2 = 10^{-3}$.

This is confirmed by the maximum pressure picture shown in Figure 5. In the channel cross- section at $x = 0.6$ m, an irregular cellular detonation propagation regime is observed, and at $x = 0.9$ m, one and one-half cell is accommodated within the channel width. The velocity deficit in this case is $\eta = 0.89$.

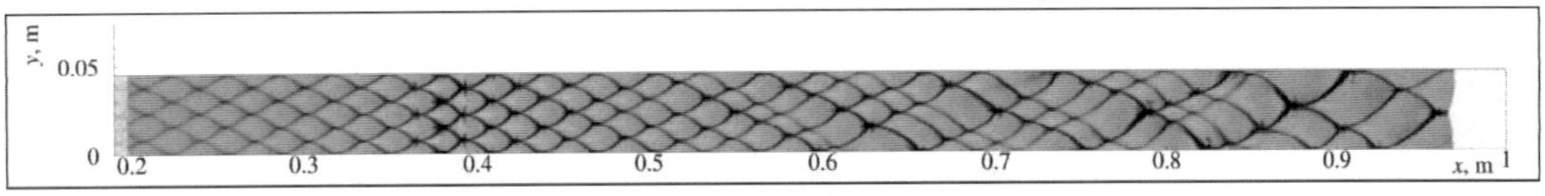

Figure 5. Time history of the maximum pressure field during interaction with a cloud of particles for $d = 100$ μm and $m_2 = 10^{-3}$.

Increasing the volume fraction of particles to 10^{-2} leads to rapid failure and quenching of detonation. The results of the calculations are shown in Figure 7 and 8. Detonation failure is observed already at a time of 70 μs and at $x = 0.27$ m. At a time of 260 μs and at $x = 0.41$ m, the cellular detonation degenerates into a shock wave without further re-initiation.

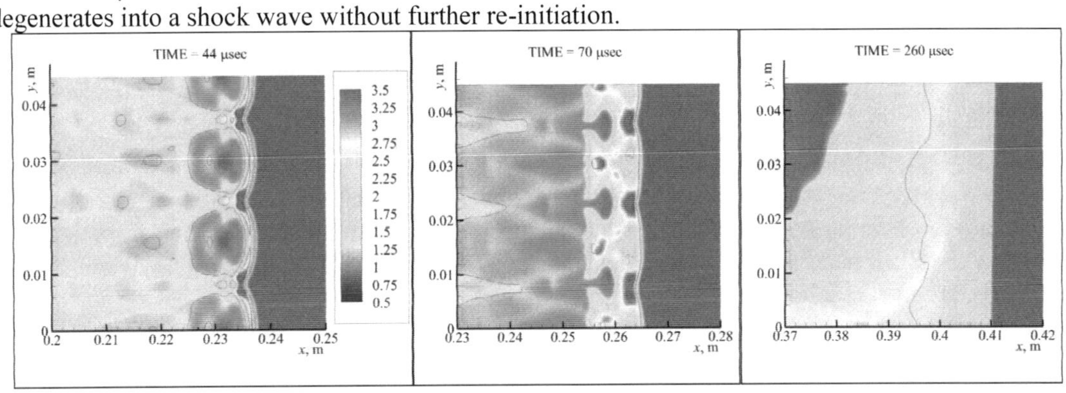

Figure 6. Density isolines at different times for particles with $d = 100$ μm and $m_2 = 10^{-2}$.

Interaction with a cloud of particles $d = 1$ mm is shown in Figure 8, 9. The maximum pressure field in time for volume fraction $m_2 = 5{\cdot}10^{-2}$ and $m_2 = 10^{-1}$ are presented. The volume fraction value for detonation suppression increases with growing of the particles diameter. Cell size increases and the deficit of the detonation wave velocity is $\eta = 0.75$ at volume fraction $m_2 = 5{\cdot}10^{-2}$ (Figure 8). Detonation suppression occurs at volume fraction value $m_2 = 10^{-1}$ (Figure 9).

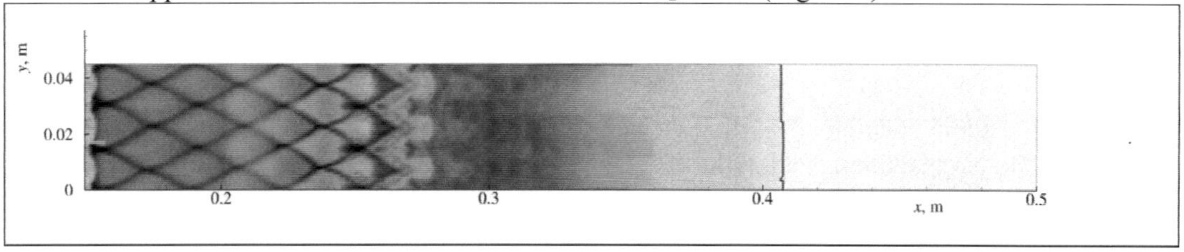

Figure 7. Time history of the maximum pressure field during interaction with a cloud of particles for $d = 100$ μm and $m_2 = 10^{-2}$.

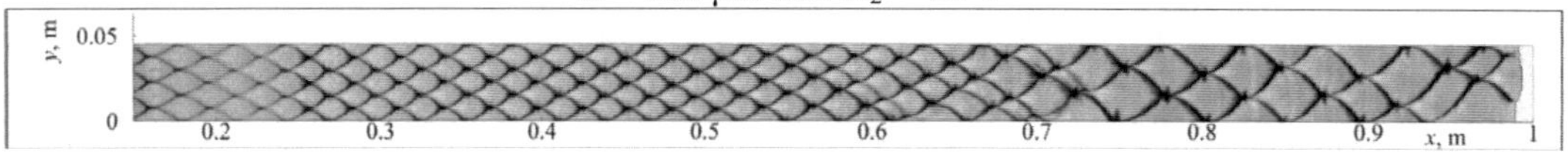

Figure 8. Time history of the maximum pressure field during interaction with a cloud of particles for $d = 1$ mm and $m_2 = 5{\cdot}10^{-2}$.

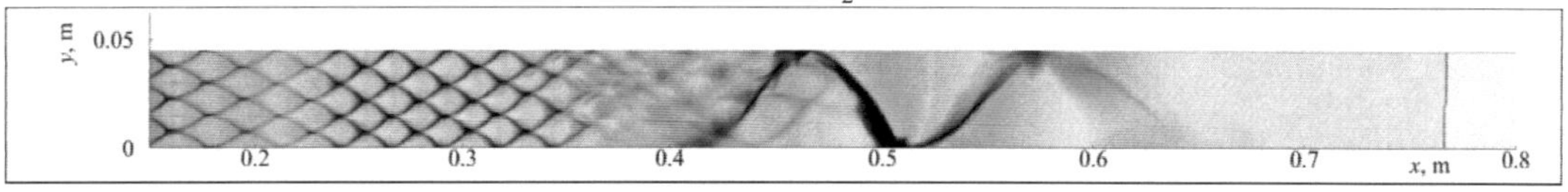

Figure 9. Time history of the maximum pressure field during interaction with a cloud of particles for $d = 1$ mm and $m_2 = 10^{-1}$.

4.2. Micro level problem

Simulation of the plane detonation wave passing through a grid of fixed particles with 1 mm diameter is performed. Calculations for three, five, and seven particles located in a staggered manner at different relative distances between the particles $k = l/d$ (l – distance between particles, d – particle diameter) are done. Analysis of shock-wave configurations in the flow behind detonation wave has shown that, depending on the transverse distance between the particles, different modes of shock waves interaction can arise. The collective configuration of shock waves is realized at small distances. As the distance between the bodies increases, a transition from a collective flow around the particles to an individual flow around each particle (first, with Mach interaction and then with regular interaction of shock waves) occurs. Figure 10 shows pressure flowfield for detonatuion wave interaction with three particles at $k = 5$ at different times. This distance between the particles corresponds to a volume fraction ~ 10^{-2}. Note that the upper and lower boundaries of the computational domain are symmetry lines. We can observe the formation of a collective shock wave in front of the particles in this figure. The collective shock wave passes through a regular (time 14 μs) and Mach (25 μs) interaction at the process of its formation. Particles are a source of perturbations for a plane detonation front. A plane detonation wave is transformed into a cellular-like wave at a time of 25 μs. Detonation suppression is not observed for this case.

When the distance between particles is reduced to $k = 3$ (corresponds to a volume fraction ~ 10^{-1}) and the length of the "cloud" is increased (up to 7 particles), a separation of the detonation wave front into a shock wave and a combustion wave is occurred. Figure 11 shows the temperature fields at different times. The separation can be observed at the time of 18 μs in Fig. 11. Than the combustion wave overtake the shock wave and the re-initiation of detonation takes place. With an increase of the "cloud" length it is possible to completely suppress detonation without re-initiation.

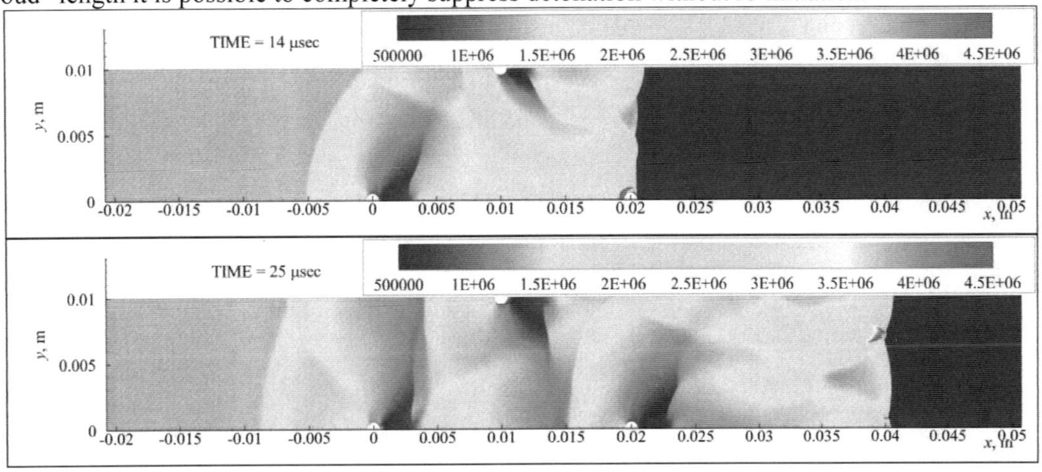

Figure 10. The pressure flowfield at different times for three particles at $k = 5$.

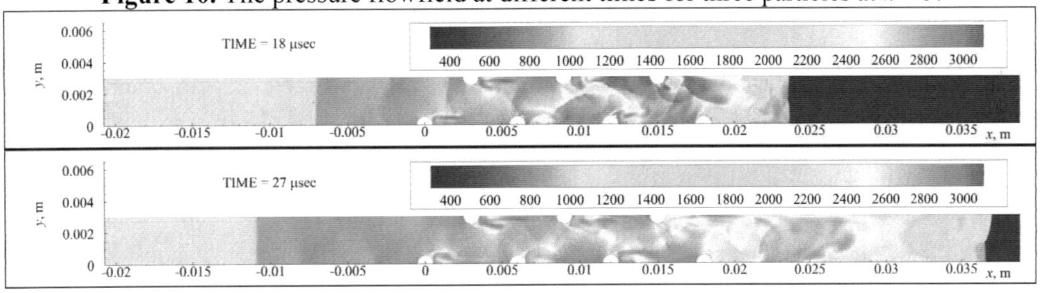

Figure 11. The temperature flowfield at different times for seven particles at $k = 3$.

5. Conclusions

- Within the framework of the ANSYS Fluent software, a mathematical technology was developed to describe 2-D cellular detonation wave interaction with inert particles.
- The values of the particles volume fraction are obtained resulting in a change of the detonation cell size, weakening of the detonation wave, and detonation suppression.
- A technology for modeling the interaction of a detonation wave with particles at a micro level has been developed. A wave structure of the flow near the particles is obtained. The mechanism of detonation suppression is shown at volume concentrations of particles close to those obtained in modeling at the macro level.

Acknowledgments
This work was partially supported by the Russian Foundation for Basic Research (Grant No. 17-08-00634) and the Russian Science Foundation (Grant No. 16-19-00010).

References

[1] Strehlow R A 1985 *Combustion Fundamentals* (New York: McGraw-Hill Book Company)
[2] Vasil'ev A A, Mitrofanov V V and M. E. Topchiyan 1987 Detonation Waves in Gases *Combust., Expl., Shock Waves* **23** 605–623
[3] Kazakov Yu V, Fedorov A V and Fomin V M 1989 Normal detonation regimes in relaxing media *Combust., Expl., Shock Waves* **25** 109–115
[4] Tropin D A, Fedorov A V and Bedarev I A 2010 Mathematical modeling of detonation

suppression in a hydrogen-oxygen mixture by inert particles *Combust., Expl., Shock Waves* **46** 332–343
[5] Fedorov A V and Tropin D A 2011 Determination of the critical size of a particle cloud necessary for suppression of gas detonation *Combust., Expl., Shock Waves*, **47** 464–472
[6] Tropin D A and Fedorov A V 2014 Physicomathematical modeling of detonation suppression by inert particles in methane-oxygen and methane-hydrogen-oxygen mixtures *Combust., Expl., Shock Waves* **50** 542–546
[7] Tropin D A and Fedorov A V 2014 Mathematical Modeling of Detonation Wave Suppression by Cloud of Chemically Inert Solid Particles *Combustion Science and Technology* **186** 1690–1698
[8] Fedorov A V, Fomin P A, Tropin D A and Chen J R 2012 Physicomathematical modeling of suppression of gaseous detonation by chemically inert particles *Journal of Engineering Physics and Thermophysics* **85** 359–367
[9] Fedorov A V, Fomin P A, Tropin D A and Chen J R 2012 Parameters, limits, attenuation, and suppression of detonation in mixtures of an explosive gas with chemically inert microparticles *Journal of Engineering Physics and Thermophysics* **85** 368–381
[10] Fedorov A V and Tropin D A 2013 Modeling of detonation wave propagation through a cloud of particles in a two-velocity two-temperature formulation *Combust., Expl., Shock Waves* **49** 178–187
[11] Fedorov A V, Fomin P A, Tropin D A and Chen J R 2012 Modeling of reflection of detonation and shock waves from a rigid wall in mixtures of a reactive gas and chemically inert particles *Journal of Engineering Physics and Thermophysics* **85** 614–619
[12] Kratova Yu V, Khmel T A and Fedorov A V 2012 Heterogeneous Detonation Propagation in Channels with Abrupt Area Expansion *Science and Technology of Energetic Materials*. **73** (6) 157–161.
[13] Fomin P A and Chen D R 2009 Effect of chemically inert particles on parameters and suppression of detonation in gases *Combust., Expl., Shock Waves* **45** 303–313
[14] Fedorov A V and Kratova Yu V 2013 Calculation of detonation wave propagation in a gas suspension of aluminum and inert particles *Combust., Expl., Shock Waves* **49** 335–347
[15] Bedarev I A and Fedorov A V 2006 Comparative Analysis of Three Mathematical Models of Hydrogen Ignition *Combust., Expl., Shock Waves* **42** 19–26
[16] Bedarev I A Rylova K V and Fedorov A V 2015 Application of detailed and reduced kinetic schemes for the description of detonation of diluted hydrogen–air mixtures *Combust., Expl., Shock Waves* **51** 528–539

IOP Conf. Series: Journal of Physics: Conf. Series **894** (2017) 012009 doi:10.1088/1742-6596/894/1/012009

Real gas flow about a round cone

E A Biberdorf[1,2] and A M Blokhin[1,2]

[1] Novosibirsk State University, Novosibirsk, Russia

[2] Sobolev Institute of Mathematics, Novosibirsk, Russia

E-mail: biberdorf@ngs.ru

Abstract. We propose a numerical method for a supersonic van der Waals gas flow about a round cone. This method enables one to calculate shock waves attached to the cone and flow parameters between the shock wave and the surface area of the cone. The advantages of the algorithm are demonstrated on some examples.

1. Introduction

We consider the supersonic van der Waals gas flow about a round cone. We assume that the upstream Mach number $M_\infty > 1$ is large enough and the shock wave is attached to the cone (see Figure 1).

The state equation of van der Waals gas has the form

$$p = \frac{RT}{V - V_0} - \frac{A}{V^2},$$

where p is the pressure, V is the molar volume of the gas, R is the universal gas constant, and T is the temperature. The constants V_0 and A provide the correct description of intermolecular forces. It is well known [1],[2] that the van der Waals gas can exist not only in the gaseous or liquid state but also in the two-phase state (Figure 2, see also [3]).

We take

$$e = c_V^* T\left(1 - \frac{A_1 T}{2}\right) - \frac{A}{V} \tag{1}$$

as the equation for the internal energy. Here e is the internal energy, c_V^* is the heat capacity, A_1 is a constant. Usually, the equation of the internal energy is used with $A_1 = 0$, but the method proposed below enables one to use this equation in the more complicate form (1).

2. Equations and their reduction to a dimensionless form

The problem of a supersonic van der Waals gas flow about a round cone can be described by the system of PDE's of gas dynamics. But since the problem is axisymmetric this system is transformed into a system of ODE's. Following [4], we introduce some new constants for its reduction to a dimensionless form:

$$\alpha = \frac{A}{\widehat{p}_\infty \widehat{V}_\infty^2}, \qquad \beta = \frac{V_0}{\widehat{V}_\infty}, \qquad 0 < \beta < 1,$$

Content from this work may be used under the terms of the Creative Commons Attribution 3.0 licence. Any further distribution of this work must maintain attribution to the author(s) and the title of the work, journal citation and DOI.
Published under licence by IOP Publishing Ltd

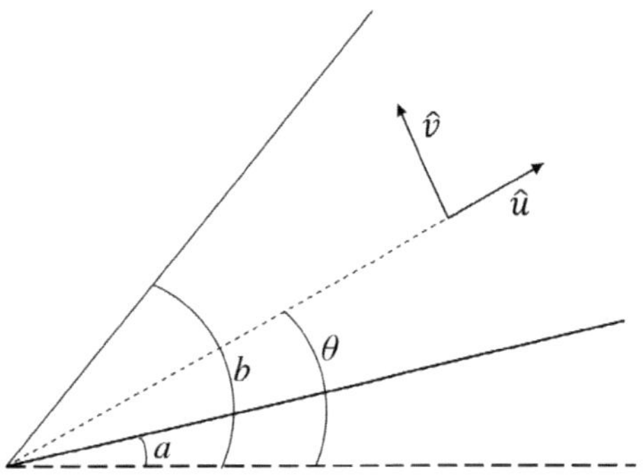

Figure 1. Supersonic gas flow about a round cone.

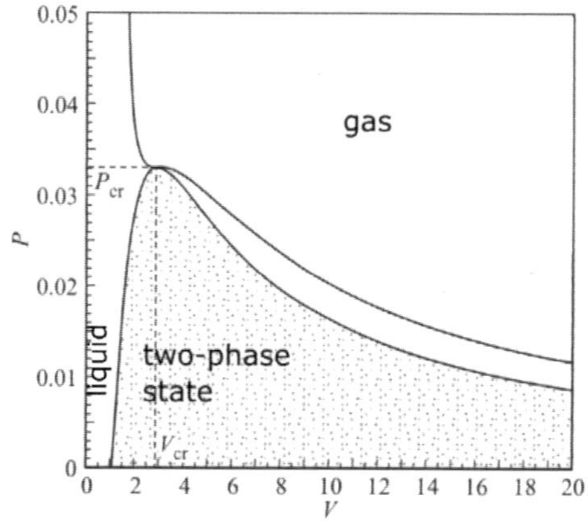

Figure 2. Regions of gaseous, liquid and two-phase states for Nitrogen.

$$\gamma_\infty = 1 + \frac{R}{c_V^*(1 - A_1\widehat{T}_\infty)}, \qquad \gamma^* = 1 + \frac{R}{c_V^*}, \qquad \gamma^* > \gamma_\infty > 0.$$

Then the system in a dimensionless form reads ($s = \theta - a$, $0 \le s \le b - a = \delta$, see Figure 1):

$$\widehat{u}'(s) = \widehat{v}(s), \tag{2}$$

$$\widehat{v}'(s) + \widehat{u}(s) = \frac{\widetilde{R}(s)}{M_\theta^2 - 1}, \tag{3}$$

$$\widehat{V}(s)\widehat{p}'(s) = -\widehat{v}(s)\frac{U_\infty^2}{\widehat{p}_\infty \widehat{V}_\infty}\frac{\widetilde{R}(s)}{M_\theta^2 - 1}, \tag{4}$$

$$\widehat{V}'(s) = \frac{\widehat{V}(s)\widehat{v}(s)U_\infty^2}{\widehat{c}^2(s)}\frac{\widetilde{R}(s)}{M_\theta^2 - 1}. \tag{5}$$

Here we use the following notations: $\widehat{u}(s)$, $\widehat{v}(s)$, $\widehat{p}(s)$, $\widehat{\rho}(s)$ are the components of the dimensionless velocity, the dimensionless pressure and the dimensionless density respectively;

$$\widetilde{R}(s) = \widehat{u}(s) + \widehat{v}(s)\mathrm{ctg}(s + a);$$

$$M_\theta^2(s) = \frac{\widetilde{M_\infty^2}\widehat{v}^2(s)}{\mathbf{C}(s)}, \qquad \widetilde{M_\infty^2} = M_\infty^2\left(\gamma_\infty\frac{1+\alpha}{1-\beta} - 2\alpha\right);$$

$$\widehat{c}^2(s) = \widehat{p}_\infty\widehat{V}_\infty\mathbf{C}(s), \;\; \mathbf{C}(s) = \frac{\widehat{p}(s) + \alpha\widehat{\rho}^2(s)}{\widehat{\rho}(s)(1 - \beta\widehat{\rho}(s))}\left(1 + \frac{\gamma^* - 1}{1 + \alpha_1\widehat{\Lambda}(s)}\right) - 2\alpha\widehat{\rho}(s);$$

$$\widehat{\Lambda}(s) = (1 - \beta\widehat{\rho}(s))(\widehat{p}(s) + \alpha\widehat{\rho}^2(s))/\widehat{\rho}(s), \qquad \widehat{\Lambda}(\delta) = \Lambda.$$

It is convenient to perform the additional transformation

$$W(s) = \sin(s + a)\widehat{u}(s) + \cos(s + a)\widehat{v}(s),$$
$$U(s) = \cos(s + a)\widehat{u}(s) - \sin(s + a)\widehat{v}(s) - 1.$$

In these terms system (2)–(5) becomes

$$W'(s) = \frac{\operatorname{ctg}(s+a)}{M_\theta^2 - 1} W(s), \tag{6}$$

$$U'(s) + \operatorname{tg}(s+a)W'(s) = 0, \tag{7}$$

$$\widehat{p}'(s) = -\widehat{\rho}(s)\widetilde{M}_\infty^2 \left[\frac{W^2(s)+U^2(s)}{2} + U(s)\right], \tag{8}$$

$$\widehat{\rho}'(s) = \frac{\widehat{p}'(s)}{\mathbf{C}(s)}. \tag{9}$$

This system is supplemented with the condition of impermeability

$$\cos(a)W(0) - \sin(a)U(0) = \sin(a) \tag{10}$$

on the surface area of cone and the four conditions

$$W(\delta) = \cos(b)\sin(b)k, \tag{11}$$

$$\widehat{p}(\delta) = 1 + \widetilde{M}_\infty^2 \sin^2(b)k, \tag{12}$$

$$\widehat{\rho}(\delta) = \frac{1}{1-k}, \tag{13}$$

$$\sin(b)W(\delta) + \cos(b)U(\delta) = 0 \tag{14}$$

on the shock front which follow from the Rankine–Hugoniot conditions. Here k is the root of the function $Q(k,b)$, $0 < k < 1-\beta$, $0 < \beta < 1$:

$$Q(k,b) = \Lambda\left(1 + \frac{\alpha_1}{2}\Lambda\right) - \Lambda_\infty\left(1 + \frac{\alpha_1}{2}\Lambda_\infty\right) - (\gamma^* - 1)k\left(\frac{\alpha}{1-k} + \frac{\widehat{p}(\delta)+1}{2}\right) = 0. \tag{15}$$

3. Numeric method

Let M_∞, a, α, β, γ_∞ and γ^* be given constants. Because of the nonlinearity of the problem the proposed algorithm has an iterative structure.

3.1. Iterations

If the parameters k and b are fixed, then the boundary value problem (6)–(13) has a usual form and can be solved in any convenient way. For this propose we use a special form of equations (6)–(7). Assuming that the function M_θ is known (in practice, we get M_θ from the previous iteration), the boundary value problem (6)–(7), (10)–(11) can be solved explicitly:

$$W(s) = \frac{\sin^2(a)\widehat{\mu}}{\sin(s+a)\widetilde{I}(s)},$$

$$U(s) = \frac{\widetilde{\mu}}{\cos(a)} + \sin^2(a)\widehat{\mu}R(s) - \operatorname{tg}(s+a)W(s),$$

where

$$\widetilde{I}(s) = \exp\left\{\int_0^s \frac{\operatorname{ctg}(\tau+a)M_\theta^2(\tau)}{1-M_\theta^2}d\tau\right\}, \qquad R(s) = \int_0^s \frac{d\tau}{\sin(\tau+a)\cos^2(\tau+a)\widetilde{I}(\tau)},$$

$$\widetilde{\mu} = -\frac{\cos^2(a)\sin^2(a)R(\delta)}{1+\cos(a)\sin^2(a)R(\delta)}, \qquad \widehat{\mu} = -\frac{\cos(a)}{1+\cos(a)\sin^2(a)R(\delta)}.$$

Note that in these terms the boundary condition (14) is reduced to the equation

$$k = \frac{\sin^2(a)\widehat{\mu}}{\cos(b)\sin^2(b)\widetilde{I}(\delta)}. \quad (16)$$

We then use the calculated functions $U(s)$ and $W(s)$ by solving numerically the Caushy plroblem (8)–(9), (12)–(13).

For calculating the values k and b for the next iteration we find the point of the intersection of the graphs of the function $\bar{k}(b)$ defined by (16) and the implicit function $\bar{\bar{k}}(b)$ defined by (15).

3.2. First approximation

To built the first approximation for the solution we use the same fact connected with the intersection of curves (15) and (16), but in this case we put $M_\theta(s) \equiv 0$. It follows that

$$\widetilde{I}(s) \equiv 1, \quad R(s) = \frac{1}{2}\ln\frac{1-\cos^2(a)}{1-\cos^2(a+s)} + \frac{1}{\cos(a+s)} - \frac{1}{\cos(a)},$$

$$\widetilde{\mu}(s) = -\frac{\cos^2(a)\sin^2(a)R(a+s)}{1+\cos(a)\sin^2(a)R(a+s)}, \quad \widehat{\mu}(s) = -\frac{\cos(a)}{1+\cos(a)\sin^2(a)(a+s)}.$$

The typical examples of such functions (15) and (16) are given in Figure 3. A lot of numerical

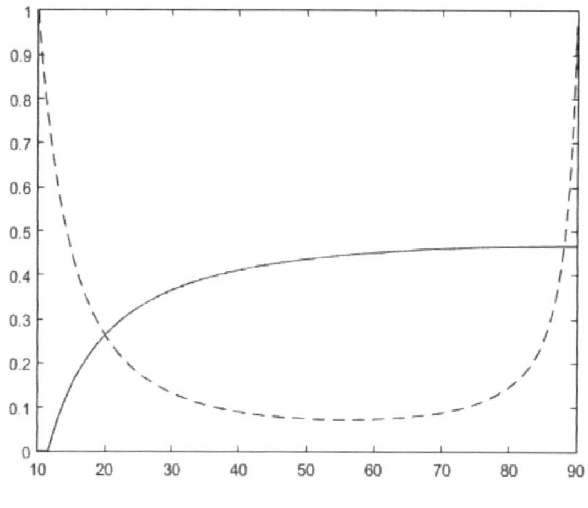

Figure 3. The functions $\bar{k}(b)$ (solid) and $\bar{\bar{k}}(b)$ (dashed) by $M_\theta \equiv 0$, $M_\infty = 5$, $\gamma_\infty = 1.8$, $\gamma^* = 1.9$, $\alpha = 4$, $\beta = 1/3$, $a = 10°$.

experiments show that this way of choosing the initial values of k and b gives the fast convergence of the algorithm.

3.3. Alternative version of the algorithm

If we fix the angle b instead of M_∞, then we can use the same idea of constructing the algorithm for finding the Mach number associated with the shock wave with the angle b.

4. Numerical examples

Let us put $M_\infty = 5$, $\gamma_\infty = 1.8$, $\gamma^* = 1.9$, $\alpha = 4$, $\beta = 1/3$, $a = 10°$. We perform the algorithm described in sections 3.1, 3.2 with the discretization step 0.01°. As the result we get the two angles for the shock wave $b_1 = 19.29°$ and $b_2 = 88.34°$. For the given precision the algorithm converges in 2–3 iterations.

Using the algorithm modification from section 3.3 we find pairs (b, M_∞) for several angles of the cone (see Figure 4). The minimal upstream Mach number for which the shock wave is attached to the cone can be defined by calculating of the minimums of the function $M_\infty(b)$ (see Figure 5).

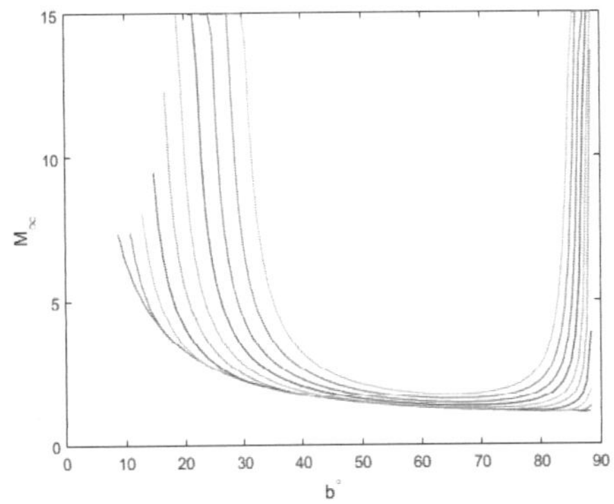

Figure 4. Dependence of Mach number on the shock wave angle for the cone angles $a = 3°, 5°, \ldots, 21°$.

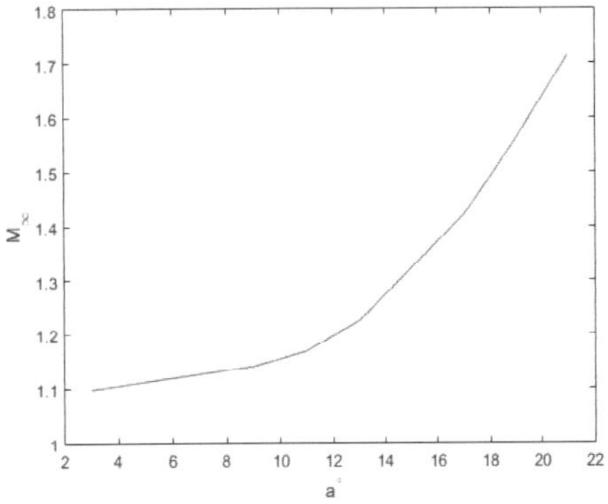

Figure 5. Dependence of the minimal upstream Mach number for which the shock wave is attached to the vertex of the cone on the cone angle.

4.1. Atmosphere

Since Nitrogen occupies the largest part of the Earth atmosphere (78%), for modeling the conic flow in the atmosphere we use the following data for this gas:

$$A = 1.348\,10^{-6}\text{atm.cm}^6/\text{moll}^2, \qquad V_0 = 3.86\,10^{-5}\text{m}^3/\text{moll}, \qquad c_V^* = 4.45\,10^2\text{J}\,\text{kg}^{-1}\text{K}^{-1}.$$

The parameter A_1 is not defined in the literature, and we thus put $A_1 = 0$. The influence of a nonzero value A_1 is demonstrated below as well.

The pressure and the density of the gas on the height 0 m are

$$\hat{p}_\infty = 1.01325\,10^5 \text{ Pa}, \qquad \hat{\rho}_\infty = 1.225 \text{ kg/m}^3.$$

We consider the case $a = 3°$, $M_\infty = 2$. The discretization step is $0.01°$. As the result of modeling we get two shock waves with the angles $b_1 = 30.0134°$ (weak shock wave) and $b_2 = 89.9240°$ (strong shock wave). The solutions are showed in Figure 6.

Other simulations show the following. The numerical solution corresponding to the height of 10000 m ($\hat{p}_\infty = 2.64999\,10^4$ Pa, $\hat{\rho}_\infty = 0.41351$ kg/m^3) is slightly different from the described above. The increase of the Mach number decreases the angle of the weak shock wave (to $b_1 = 7.6019$ by $M_\infty = 8$) and increases the jump of the pressure and the density corresponding to the strong shock wave. A positive nonzero value A_1 decreases the angle of the strong shock wave (to $b_2 = 75.9763°$ by $A_1 = 4 \cdot 10^{-6}$) and significantly changes the values of the pressure and the density (see Figure 7).

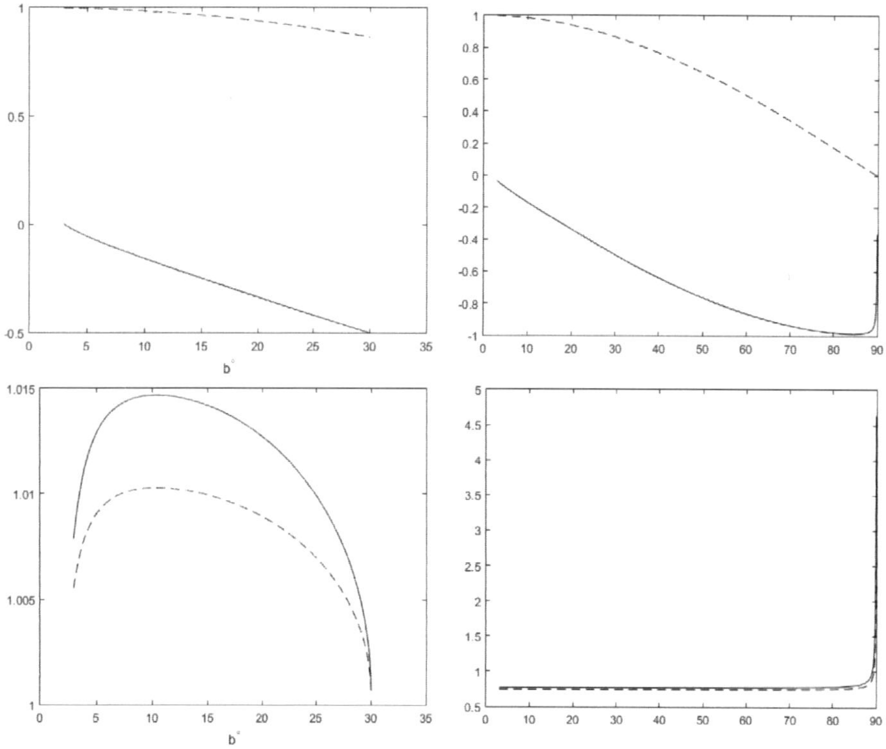

Figure 6. The values of the pressure $\widehat{p}$ (solid line) and the density $\widehat{\rho}$ (dashed line) are pictured at the top, and the velocity components $\widehat{v}$ (solid line), $\widehat{u}$ (dashed line) are pictured at the bottom. The solution corresponding to b_1 is pictured at the left, and the solution corresponding to b_2 at the right.

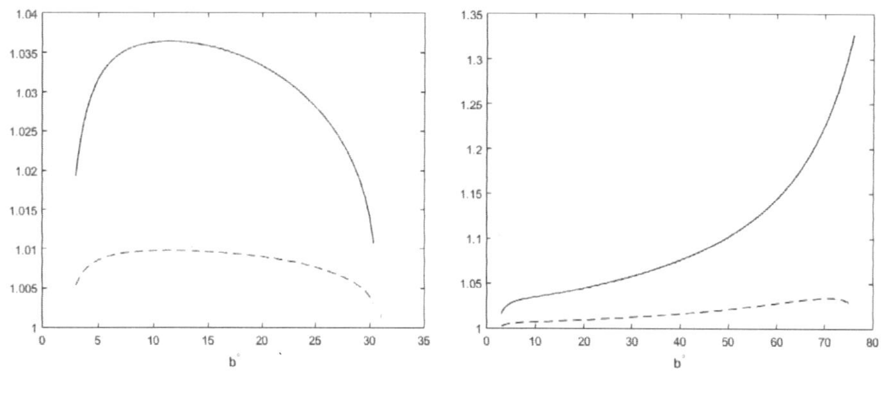

Figure 7. The values of the pressure $\widehat{p}$ (solid line), and the density $\widehat{\rho}$ (dashed line) corresponding to $b_1 = 30.3514°$ are pictured at the left, and ones corresponding to $b_2 = 75.9763°$ at the right. $A_1 = 4 \cdot 10^{-6}$.

Acknowledgments

This research was supported by RFBR grant 17-01-00791. The authors thank A. Goldin for his help in illustrations.

References

[1] Ter Haar D and Wergeland H 1960 *Elements of Thermodynamics* (NY: Addison-Wesley)
[2] Kubo R 1970 *Thermodynamics* (Amsterdam: North-Holland)
[3] Blokhin A M and Goldin A Yu 2016 *Tech. Phys.* **61** 1813-20
[4] Blokhin A M and Biberdorf E A 2015 *Vychisl. Tekhnol.* **20** 29-43

IOP Conf. Series: Journal of Physics: Conf. Series **894** (2017) 012010 doi:10.1088/1742-6596/894/1/012010

On General boundary value problem for an elliptic higher order equation in the plane with constant real coefficients

Yu A Bogan
Institute of Hydrodynamics, Novosibirsk, Russia
E-mail: bogan@hydro.nsc.ru

Abstract. By means of a new approach, the general boundary value problem for a higher order elliptic equation with two independent variables, and a normal set of boundary conditions and simple complex characteristics is reduced to the Fredholm system of integral equations in a bounded region with a smooth boundary.

1. Introduction

Consider a linear partial differential equation of elliptic type in two variables

$$L(u) = L(\frac{\partial}{\partial x_1}, \frac{\partial}{\partial x_2})u = 0,$$

where

$$L(\xi, \eta) = \sum_{k=0}^{2m} a_k \xi^k \eta^{2m-k}$$

is a homogeneous polynomial with constant real coefficients of degree $2m, m \geq 2$ and such that $L(\xi, \eta) > 0$ (ellipticity) for all real non-vanishing vectors (ξ, η). Reduction of boundary value problems to regular integral equations for higher order elliptic equations and systems of equations was discussed in a very general situation by Lopatinskii [1]. S. Agmon in his classical article [2] solved the Dirichlet problem for a higher order elliptic equation in the plane by means of the multiple layer potential, constructed by him. The technique developed by him is very complicated; even in the simplest case of a fourth-order equation it leads on to lengthy computations. Some boundary value problems connected with higher order equations, were discussed later in [3, 4, 5] by the help of the "simple layer" method, advanced by G. Fichera in [3]. It is necessary to recall that S. Agmon considered the very general situation of an elliptic equation with multiple complex characteristics. The considerably simplified situation, when an elliptic equation with real coefficients has simple complex characteristics, is studied in this article. Here necessary computations are nearly trivial. A particular case of a fourth order equation from theory of elasticity was studied by the author in [6]. The case of multiple characteristics can be considered as the result of the limiting transition from the case of the simple ones.

Content from this work may be used under the terms of the Creative Commons Attribution 3.0 licence. Any further distribution of this work must maintain attribution to the author(s) and the title of the work, journal citation and DOI.
Published under licence by IOP Publishing Ltd

2. Preliminaries

Let Q be a simply-connected bounded plane region with the Lyapunov boundary ∂Q; i.e., it is assumed, that its boundary ∂Q has the uniformly Hölder continuous inward normal field $\nu(z)$. Here $C^k(Q)$ and $C^k(\bar{Q})$ are spaces of real k-order continuously differentiable functions in Q and $\bar{Q}$, respectively. By $C^{0,\alpha}(\bar{Q})$ and $C^{0,\alpha}(\partial Q)$ are meant the spaces of real continuous functions satisfying in $\bar{Q}$ and ∂Q the uniform Hölder condition with the exponent $\alpha, 0<\alpha<1$, respectively. $C^{k,\alpha}(\bar{Q})$ is the subclass of $C^k(\bar{Q})$ such that $D^\nu u \in C^{0,\alpha}(\bar{Q}), |\nu| = k$. Here ν is multi-index $\nu = (k_1, k_2)$, and

$$D^\nu u = \frac{\partial^\nu u}{\partial x_1^{k_1} \partial x_2^{k_2}}, \quad k_1 + k_2 = \nu.$$

In a similar way the spaces $C^{l,\alpha}(\partial Q)$ are defined.

The aim of this and next sections is to show that for the general boundary problem

$$L(u) = L(\frac{\partial}{\partial x_1}, \frac{\partial}{\partial x_2})u = 0, \tag{1}$$

$$B_h(u) = \sum_{i=1}^{n_h} b_{ik} \frac{\partial^{n_h} u}{\partial x_1^{n_h - i} \partial x_2^i}|_{\partial Q} = g_h(s), g_h \in C^{2m-n_h-1,\alpha}(\partial Q), h = 1, 2, \ldots, m. \tag{2}$$

the Fredholm alternative holds. We assume, that the system of boundary conditions (2) is normal, that is, there are m boundary conditions, $n_i \neq n_j, i \neq j$, the order of any boundary condition is lesser $2m$, and

$$\sum_1^{n_h} b_{ih} \nu_1^{n_h - i} \nu_2^i \neq 0$$

where $\nu = (\nu_1, \nu_2)$ is the normal vector of the boundary. With the problem (1), (2) is related the next boundary problem for an ordinary differential equation: find the solution $v(t)$ of the equation

$$\sum_{k=0}^{2m} a_k i^k \frac{d^{2m-k} v}{dt^{m-k}} = 0, \tag{3}$$

equal to zero at infinity and satisfying the boundary condition

$$\sum_{i=1}^{n_h} a_{ik} i^k \frac{d^{n_h} v}{dt^{n_h - i}}|_{t=0} = g_h, h = 1, 2, \ldots, m, \tag{4}$$

We construct now the complete system of linearly independent solutions of equation (3), choose from them m equal to zero at infinity. Let $\omega(z,t)$ be the matrix, whose columns are chosen m solutions, $\delta(z)$ is the determinant of the matrix, obtained by substitution of $\omega(z,t)$ into the left-hand side of equation (4). Then the Lopatinsky condition means that $\delta(z) \neq 0$ for all $z \in \partial Q$. It was proved in [7], that the principle assertion pertaining to this circle of problems, consists in the following: the Lopatinsky condition is sufficient for the problem (1)-(2) to be Nöterian, and necessary for this problem to be solvable with the exactness up to finite dimensional sub-spaces. It is known about the problem (1)-(3), that it can be reduced to the system of regular integral equations, if the condition of Lopatinsky is satisfied. We consider below the simplest situation when the equation and boundary conditions have constant coefficients, and the solution belongs to Hölder class of functions. With respect to the problem it is known, that in can be reduced to the system of regular integral equations if the Lopatinsky condition is satisfied.

Our computations in this section are purely formal; their validity will be confirmed later. A general solution of equation (1) can be written as a sum of m quasi-harmonic functions,

$$u(x_1, x_2) = \sum_{k=1}^{m} w_k(x_1, x_2).$$

Since equation (1) is elliptic with real constant coefficients, the characteristic equation

$$L(1, \lambda) = \sum_{k=0}^{2m} a_k \lambda^{2m-k} = 0$$

has m pairs of (distinct) complex conjugate roots $\lambda_k, \overline{\lambda_k}, Im\lambda_k > 0, k = 1, 2, \ldots, m$. Let $\lambda_k = \alpha + i\beta_k, \beta_k > 0, k = 1 \ldots m$ be roots with a positive imaginary part. A quasi-harmonic function $w_k(x_1, x_2)$ is defined as a solution of the elliptic second order equation

$$(\beta_k^2 + \alpha_k^2)\frac{\partial^2 w_k}{\partial x_1^2} - 2\alpha_k\beta_k\frac{\partial^2 w_k}{\partial x_1 \partial x_2} + \frac{\partial^2 w_k}{\partial x_2^2} = 0,$$

which reduces to Laplacian under the change of independent variables $y_1 = x_1 + \alpha_k x_2, y_2 = \beta_k x_2$. In a simply connected region any quasi-harmonic function $w_k(x_1, x_2), k = 1, \ldots m$ can be represented as a real part of a holomorphic function of the (complex) argument $z_k = x_1 + \lambda_k x_2, k = 1, \ldots m$,

$$w_k(x_1, x_2) = Re\varphi_k(z_k), \quad z_k = x_1 + \lambda_k x_2, \quad k = 1, \ldots, m.$$

3. Construction of potentials

Put $z_k = x_1 + \lambda_k x_2, t_k(s) = x_1(s) + \lambda_k x_2(s), k = 1, 2, \ldots, m$. The solution of the problem (1),(2) is a sum of m analytic functions

$$u(x_1, x_2) = \sum_{n=1}^{m} \varphi_n(x_1 + \lambda_n x_2).$$

Then

$$\frac{\partial^{m_k} u}{\partial x_1^{m_k - p} \partial x_2^p} = Re \sum_{n=1}^{m} \lambda_n^p \varphi_n^{(m_k)}(z_k), k = 1, 2, \ldots, m, \tag{5}$$

and

$$B_k(u) = Re \sum_{p=0}^{m_k} \sum_{n=1}^{m} b_{pn} \lambda_n^p \varphi_n^{(m_k)}(z_n).$$

Put

$$\varphi_n^{(m_k)}(z_n) = \frac{1}{\pi i} \int_{\partial Q} \frac{b_n(s)\, dt_n}{t_n - z_n}, n = 1, 2, \ldots, m. \tag{6}$$

As result,

$$B_k(u) = Re \sum_{p=0}^{m_k} \sum_{n=1}^{m} a_{pn} \lambda_n^p \frac{1}{\pi i} \int_{\partial Q} \frac{b_n(s)\, dt_n}{t_n - z_n}, n = 1, 2, \ldots, m. \tag{7}$$

Recall, that by Plemelj's jump formula, when a point $z \in Q$ tends to a point $t_0 = x_1(s_0) + ix_2(s_0) \in \partial Q, s_0 \in (0, L)$ inside a region, then

$$\lim_{z\to t_0} \frac{1}{\pi i}\int\limits_{\partial Q} \frac{g_j\,dt_j}{t_j - z_j} = g_j(s_0) + \frac{1}{\pi i}\int\limits_{\partial Q} \frac{g_j\,dt_j}{t_j - t_{j0}}.$$

Here $t_{j0} = x_1(s_0) + \lambda_j x_2(s_0), j = 1, 2, \ldots, m.$

Assume, that densities $d_n \in C^{0,\alpha}(\partial Q)$ and use Plemelj jump formulas at the boundary for the Cauchy type integral. Choose now a set of m real functions $f_k(s), k = 1, 2, \ldots, m$. Determine densities $d_k(s), k = 1, 2, \ldots, m$ from the linear system of equations

$$\sum_{p=0}^{m_k}\sum_{n=1}^{m} b_{pk}\lambda_n^p d_n = f_k(s), \qquad k = r, \tag{8}$$

$$\sum_{p=0}^{m_k}\sum_{n=1}^{m} b_{pk}\lambda_n^p = 0, \qquad k \neq r. \tag{9}$$

Densities $d_n(s)$ and a potential $B_k(x_1, x_2)$ are uniquely defined from the system (10).

Denote by $C = (c_{ij}), i, j = 1, 2, \ldots, m$ the matrix of coefficients before $d_n, n = 1, 2, \ldots, m$. Let $f_k(s) \in C^{0,\alpha}(0, L), k = 1, 2, \ldots, m$ – is a vector of m real valued functions. Choose the densities as a solution of a linear system of equations

$$\sum_{k=1}^{m} a_{k,s} d_k(s) = f_k(s), \quad k = 1, 2, \ldots, m. \tag{10}$$

It is obvious, that system has a unique solution, if its determinant $\delta = det(a_{i,j})$ is different from zero. Here

$$a_{i,j} = \sum_{i=0}^{m_k} b_{j,i}\lambda_j^p, i, j = 1, \ldots m.$$

Denote by Δ_m the Vandermond determinant, built on the numbers $\lambda_i, i = 1, \ldots, m$. We assert, that the determinant δ can be written as

$$\delta = \Delta_m \delta_1,$$

where δ_1 is a symmetric polynome of roots of the characteristic equation. Indeed, columns of the matrix C differ only by numbers of the characteristic equation and δ becomes equal to zero when for some $i \neq j$ $\lambda_i = \lambda_j$; hence by the Bezout theorem any difference $\lambda_i - \lambda_j$ is a multiple of $det\ C$. Therefore, δ_1 depends only on the coefficients of equation (1).

Note, that it is the place, where we get a distinction from the standard approach. We see, that there is no necessity to use the fundamental solution of the equation, to substitute it into the corresponding Green's formula, reducing the boundary problem to the system of singular integral equations. In the standard approach densities $d_k k = 1, 2, \ldots, m$ are chosen to be real functions, and there is no necessity to solve the previous system of equations. By Cramer's formula we obtain, that

$$d_k = \delta^{-1}\sum_{j=1}^{m} f_j A_{j,k}, \quad k = 1, 2, \ldots, m.$$

Here $A_{j,k}$ is the cofactor of $a_{j,k}$. Therefore, the function $B_k(u)$ can be written as

$$B_k(u) = Re\sum_{n=1}^{m} a_{k,n}\frac{1}{\pi i\delta}\int\limits_{\partial Q}\sum_{j=1}^{m} g_j A_{j,n}\frac{dt_n}{t_n - z_n}. \tag{11}$$

Rewrite the previous equation as

$$B_k u = Re \sum_{n=1}^{m} a_{k,n} \frac{1}{\pi i \delta} \int_{\partial Q} \sum_{j=1}^{m} g_j A_{j,n} \{ \frac{d\,t_n}{t_n - z_n} - \frac{d\,t_k}{t_k - z_k} \} +$$

$$Re \sum_{n=1}^{m} a_{k,n} \frac{1}{\pi i \delta} \int_{\partial Q} \sum_{j=1}^{m} g_j A_{j,n} \frac{d\,t_k}{t_k - z_k}. \tag{12}$$

By known properties of determinants we have the relations

$$\sum_{n=1}^{m} a_{k,n} A_{k,n} = \delta, k = 1, 2, \ldots m, \sum_{n=1}^{m} a_{k,n} A_{l,n} = 0, k \neq l. \tag{13}$$

Then for any $k = 1, 2, \ldots m$, we can write $B_k(u)$ as

$$B_k(u) = Re \frac{1}{\pi i} \int_{\partial Q} \frac{f_k\, d\,t_k}{t_k - z_k} + Re \frac{1}{\pi i \delta} \sum_{n=1}^{m} a_{k,n} \int_{\partial Q} \sum_{j=1}^{m} f_j A_{j,n} (\frac{d\,t_n}{t_n - z_n} - \frac{d\,t_k}{t_k - z_k}). \tag{14}$$

As result, we obtain at the boundary the system of equations:

$$f_k(s_0) + Re \frac{1}{\pi i} \int_{\partial Q} \frac{g_k\, d\,t_k}{t_k - t_{k0}} +$$

$$Re \frac{1}{\pi i \delta} \sum_{n=1}^{m} a_{k,n} \int_{\partial Q} \sum_{j=1}^{m} f_j A_{j,n} (\frac{d\,t_n}{t_n - t_{n0}} - \frac{d\,t_k}{t_k - t_{k0}}) = g_k(s_0), k = 1, 2, \ldots, m. \tag{15}$$

It is obvious, that expression (14), specialized at the straight boundary, for example, half-plane $z = 0$, gives an explicit rapidly decaying solution of problem (1)-(2), in a half plane. As equation (1) has real coefficients and is strongly elliptic, all summands in (14) are compact operators in $C^{0,\alpha}(\partial Q)$ and the boundary value problem (1)-(2) is Fredholm with zero index. Summing up, we get

Theorem. Assume, that a bounded region Q has a Lyapunov boundary ∂Q, boundary data $g_k(s) \in C^{2m-m_k-1}, k = 1, 2, \ldots, m$. Then the boundary value problem (1)-(2) is Fredholm in the Hölder class of functions, if the determinant $\delta \neq 0$; It means, that

- a) The homogeneous boundary problem (1)-(2) has only finite number of linearly independent solutions;
- b) if the homogeneous boundary problem (1)-(2) does not have nonzero solutions, then the corresponding inhomogeneous problem always has a unique solution and
- c) if the number of linearly independent solutions of the homogeneous problem is equal to s, then for solvability of the inhomogeneous problem (1)-(2) it is necessary and sufficient, that the vector $g = (g_1(s), g_2(s), \ldots, g_m(s))$ of boundary data has to satisfy s conditions of orthogonality

$$\int_{\partial Q} f v^j\, d\,s = 0,$$

where v^j are linearly independent solutions of the homogeneous adjoint problem.

References

[1] Lopatinskii Ya B 1953 On a method of reducing boundary problems for a system differential equations of elliptic type to regular equations [in Russian] *Ukrain. Math. J.* **5** 123-51

[2] Agmon S 1957 Multiple layer potentials and the Dirichlet problem for the higher order elliptic equations in the plane *I. Comm. Pure Appl. Math.* **10** 179-240

[3] Fichera G 1961 Linear elliptic equations of higher order in two independent variables and singular integral equations, with applications to anisotropic inhomogeneous elasticity *Partial differential equations and continuum mechanics* ed R E Langer (Madison: Wisconsin press) pp 55-80

[4] Lanzara F 2007 On BVPS for strongly elliptic systems with higher order boundary conditions *Georgian Math. J.* **14** 145-67

[5] Bogan Yu A 2000 On the potential-theoretic method for an elliptic fourth-order equation from the anisotropic theory of elasticity [In Russian] *Sib. J. Indust. Math.* **3** 29-34

[6] Agmon S, Douglis A and Nirenberg L 1959 Estimates near the boundary for solutions of elliptic partial differential equations, satisfying general boundary conditions *I. Comm. Pure Appl. Math.* **12** 623-727

[7] Vekua I N 1962 *Generalized analytical functions* ed I N Sneddon (Oxford: Pergamon press)

[8] Agmon S 1957 Multiple layer potentials and the Dirichlet problem for higher order elliptic equations in the plane *Comm. Pure Appl. Math.* **10** 1957 179-239

[9] Ricci P E 1974 Sui potenziali di semplice strato per le equazione ellittiche di ordine superiore in due variabili *Rend. di Matematica* **7** 1-39

[10] Siradjudinov M M 1997 Boundary problems for general elliptic systems in the plane *Izv. Russian Ac. Sci. Ser. math.* **61** 37-176

IOP Conf. Series: Journal of Physics: Conf. Series **894** (2017) 012011 doi:10.1088/1742-6596/894/1/012011

On non-symmetric axial corner-layer flow

A V Boiko[1,2], S V Kirilovskiy[1], Y M Nechepurenko[1,3] and T V Poplavskaya[1]

[1]Khristianovich Institute of Theoretical and Applied Mechanics SB RAS, Novosibirsk, Russia
[2]University of Tyumen, Tyumen, Russia
[3]Institute of Numerical Mathematics RAS, Moscow, Russia

E-mail: boiko@itam.nsc.ru

Abstract. The problem of asymmetric incompressible axial flow in a corner formed of two intersecting plates at a right angle is considered. The asymptotic behaviour of the flow far away from the corner is analysed. Two types of asymptotic behaviour are found. It is shown that the flow is very sensitive to the asymmetry parameter. A comparison of the results with computations of full Navier–Stokes equations was performed.

1. Introduction

The axial corner-layer flows serve to mimic different flows at joints of aerodynamic bodies. The prevalence of such flows is the cause of their experimental and theoretical study for decades [1]. However, due to complexity of laboratory and numerical modelling, the relevant information, except some simple cases, is rather limited. For example, there are difficulties in implementing the symmetric flow with zero stream-wise pressure gradient in laboratory experiments [2]. Even relatively small inaccuracies in installing the model in a wind tunnel give rise to an asymmetry of the flow and to a stream-wise pressure gradient. The only theoretical study visiting the non-symmetric corner-layer flows [3] suggests using a flow asymmetry parameter to categorize the flows, but presents neither relevant boundary conditions, nor examples of such flows.

In this paper, self-similar symmetric and asymmetric axial laminar flows at right corner are considered. For the first time, a system of quite general asymptotic boundary conditions that may be useful for numerical solution to reduce the dimension of the problem in the asymmetric case is obtained. A numerical analysis of the boundary conditions is given. Computations of the asymmetric flows are presented.

2. Self-similar incompressible corner-layer flow with stream-wise pressure gradient

Let us consider in Cartesian coordinates (x^*, y^*, z^*) a semi-infinite right dihedral corner with faces at coordinates $(x^*, 0, z^*)$ and $(x^*, y^*, 0)$, with the borders at coordinates $(0, 0, z^*)$ and $(0, y^*, 0)$, so that the corner edge has coordinates $(x^*, 0, 0)$, where $x^* \geq 0$, $y^* \geq 0$, $z^* \geq 0$. The corner is exposed to an axial (along the edge) steady laminar flow of Newtonian fluid with kinematic viscosity ν and velocity $\mathbf{u}^*(x^*, y^*, z^*)$, so that $\mathbf{u}^*(x^*, kz^*, z^*) \to (U_\infty \cdot (x^*/L)^m, 0, 0)$ when $z^* \to \infty$ $(k > 0)$, where $U_\infty > 0$ and $L > 0$ are the characteristic velocity and the length, respectively. The value m describes the stream-wise pressure gradient (when $m = 0$, the gradient is equal to zero [3]).

Content from this work may be used under the terms of the Creative Commons Attribution 3.0 licence. Any further distribution of this work must maintain attribution to the author(s) and the title of the work, journal citation and DOI.
Published under licence by IOP Publishing Ltd

Let us define the Reynolds number as $\mathrm{Re} = U_\infty L/\nu$ and introduce new variables [4] with

$$\eta = c\frac{y^*}{x^*}, \quad \zeta = c\frac{z^*}{x^*}$$

and

$$u(\eta,\zeta) = \frac{u^*(x^*,y^*,z^*)}{U_\infty(x^*L)^m}, \quad v(\eta,\zeta) = 2c\frac{v^*(x^*,y^*,z^*)}{U_\infty(x^*L)^m}, \quad w(\eta,\zeta) = 2c\frac{w^*(x^*,y^*,z^*)}{U_\infty(x^*L)^m},$$

where $c = \sqrt{\mathrm{Re}(1+m)/2}$. In accordance with the theoretical analyses performed in previous studies cited above the dependent variables can be expressed as the velocity potentials

$$\phi = \frac{(1-m)\eta u - v}{1+m}, \quad \psi = \frac{(1-m)\zeta u - w}{1+m}$$

and the modified vorticity $\theta = \psi_\eta - \phi_\zeta$. The advantage of them in further analysis and computations is that two of four resulting similarity equations are linear:

$$u_{\eta\eta} + u_{\zeta\zeta} + \phi u_\eta + \psi u_\zeta + \frac{2m}{1+m}\left(1-u^2\right) = 0, \tag{1}$$

$$\phi_\eta + \psi_\zeta = \frac{2u}{1+m}, \tag{2}$$

$$\psi_\eta - \phi_\zeta = \theta, \tag{3}$$

$$\theta_{\eta\eta} + \theta_{\zeta\zeta} + \phi\theta_\eta + \psi\theta_\zeta + \frac{2u}{1+m}\left[\theta + (1-m)(\eta u_\zeta - \zeta u_\eta)\right] = 0, \tag{4}$$

with the boundary conditions at the walls

$$u = \phi = \psi = 0,$$

and at infinity

$$u(k\zeta,\zeta) \to 1, \quad \theta(k\zeta,\zeta) \to 0, \quad \zeta \to \infty.$$

3. Boundary conditions

The given boundary conditions are insufficient for solving equations (1)–(4) numerically. Instead, they can be replaced with asymptotic boundary conditions derived by analysing the behaviour of the flow far away from the corner edge [5]. Without loss of generality let us consider the limit as $\zeta \to \infty$ and $\eta = \text{const}$. Assuming the Reynolds number large, the boundary-layer approximation can be used in the limit yielding the following asymptotic solutions [5]:

$$u(\eta,\zeta) = U_0(\eta) + \frac{U_1(\eta)}{\zeta} + O\left(\frac{1}{\zeta^2}\right), \quad \phi(\eta,\zeta) = \Phi_0(\eta) + \frac{\Phi_1(\eta)}{\zeta} + O\left(\frac{1}{\zeta^2}\right),$$
$$\psi(\eta,\zeta) = \zeta\Psi_0(\eta) + \Psi_1(\eta) + O\left(\frac{1}{\zeta}\right), \quad \theta(\eta,\zeta) = \zeta\Theta_0(\eta) + \Theta_1(\eta) + O\left(\frac{1}{\zeta}\right).$$

Let us introduce the flow asymmetry parameter γ [3] such that the condition $\psi = (1+\gamma)/(1+m)\zeta + \lambda_1 + O(1/\zeta)$, where λ_1 is a constant to be defined, is satisfied at the bisecting line. Then,

IOP Conf. Series: Journal of Physics: Conf. Series **894** (2017) 012011 doi:10.1088/1742-6596/894/1/012011

the resulting system of governing equations can be reduced to the following system of three ordinary differential equations:

$$U_{0,\eta\eta} + \Phi_0 U_{0,\eta} + \frac{2m}{1+m}\left(1 - U_0^2\right) = 0, \tag{5}$$

$$\Phi_{0,\eta} + \Psi_0 = \frac{2U_0}{1+m}, \tag{6}$$

$$\Psi_{0,\eta\eta} + \Phi_0 \Psi_{0,\eta} + \left[\Psi_0^2 - \left(\frac{1}{1+m} + \gamma\right)^2\right] + \frac{1-m}{1+m}\left(1 - U_0^2\right) = 0, \tag{7}$$

with the boundary conditions

$$U_0(0) = \Phi_0(0) = \Psi_0(0) = 0, \tag{8}$$

$$U_0(\infty) = 1, \quad \Psi_0(\infty) = \frac{1}{1+m} + \gamma. \tag{9}$$

Leaving only the principal terms in the expansion makes the asymptotic boundary conditions for equations (1)–(4) incompatible when $\eta \to \infty$ and $\zeta \to \infty$, so the next terms must be taken into account. The corresponding equations are linear and after some manipulations can be reduced to the following system:

$$U_1 \equiv 0,$$
$$\Phi_1 \equiv 0$$
$$\Theta_1 = \Psi_{1,\eta},$$
$$\Psi_{1,\eta\eta} + \Phi_0 \Psi_{1,\eta} + \Psi_0 \Psi_1 = \left(\frac{1}{1+m} + \gamma\right)\lambda_1.$$

with the boundary conditions $\Psi_1(0) = 0$ and $\Psi_1(\infty) = \lambda_1$, where

$$\lambda_1 = \lim_{\eta\to\infty} \Psi_1^{(+\gamma)}(\eta) = \lim_{\zeta\to\infty}\left[\Phi_0^{(-\gamma)}(\zeta) - \left(\frac{1}{1+m} + \gamma\right)\zeta\right],$$

and the superscripts denote the asymmetry parameters.

The system of equations (5)–(9) for the principal terms in the expansion is non-linear. Therefore, an additional analysis of its solutions depending on m and γ is required. For comparison of the results with the data of some previous studies, it is convenient also to introduce the Hartree pressure gradient parameter $\beta = 2m/(1+m)$. The principal terms in the expansion of the boundary conditions were found on solving the system (5)–(9) as a boundary-value problem by setting $U_0'(0)$ as an additional boundary condition and finding β in the course of the solution [6]. To integrate the equations the `ode15s` solver of MATLAB was used [7].

Similar to the analysis of symmetric corner flow [8] it was found that the equations have two branches of solutions. The branch corresponding to the solutions with larger wall shear stress $\sim U'(0)$ is called the upper one, and the other branch is called the lower one. Examples of the obtained diagrams of $U_0'(0)$ and $\Psi_0'(0)$ vs. β for $\gamma = \pm 0.25$ and 0 are shown in Figure 1. It is seen that the change of γ leads to a change of the minimum of β. Since the boundary conditions for the asymmetric flows are defined by a pair $\pm\gamma$, then the self-similar flows in the corner are possible only for such β, for which there are asymptotic solutions in both cases.

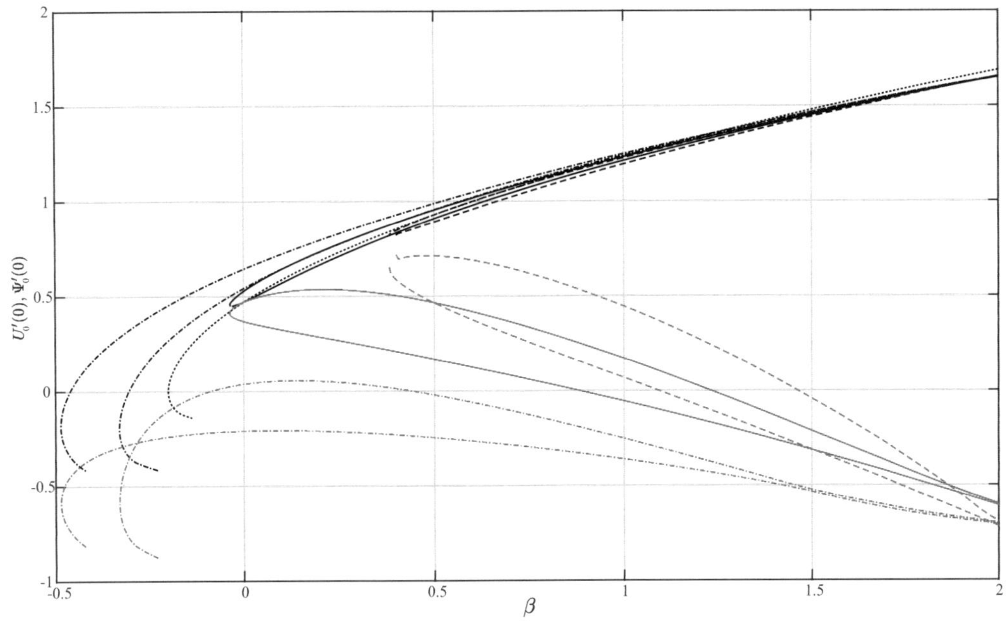

Figure 1. Dependencies of $U_0'(0)$ (*black*) and $\Psi_0'(0)$ (*red*) as functions of β for $\gamma = -0.25$ (*dashed*), $\gamma = 0$ (*solid*), and $\gamma = 0.25$ (*dash-dotted*).

4. Corner-layer flow

For the numerical solution of (1)–(4) with the corresponding boundary conditions, the equations (2) and (3) were pre-transformed by differentiation and further algebraic manipulations to yield

$$\begin{aligned}
u_{\eta\eta} + u_{\zeta\zeta} + \phi u_\eta + \psi u_\zeta + \frac{2m}{1+m}\left(1 - u^2\right) &= 0, \\
\phi_{\eta\eta} + \phi_{\zeta\zeta} &= \frac{2u_\eta}{1+m} - \theta_\zeta, \\
\psi_{\eta\eta} + \psi_{\zeta\zeta} &= \frac{2u_\zeta}{1+m} + \theta_\eta, \\
\theta_{\eta\eta} + \theta_{\zeta\zeta} + \phi\theta_\eta + \psi\theta_\zeta + \frac{2u}{1+m}\left[\theta + (1-m)(\eta u_\zeta - \zeta u_\eta)\right] &= 0,
\end{aligned}$$

that is more convenient for numerical integration. The NITSOL package [9] used widely for solving this kind of non-linear problems showed, however, a very slow convergence. Analysis of the problem indicates that it is related apparently to nearly degenerate Jacobian of the considered set of equations. This is in accordance with findings of previous investigators of the symmetric problem. Therefore, an original method of solution, which will be described in a separate publication, was developed.

The computations show that the flow is very sensitive to the value of asymmetry parameter γ. For example, Figure 2 demonstrates the solutions for both branches as absolute values of the normalized transverse velocity components $q = \sqrt{v^2 + w^2}$ at $\beta = 0$ with $\gamma = 0$ and 0.01 (the range of γ is 0 to ± 0.5). The results with $\gamma = 0$ is the same as, e.g., obtained in [3]. Meanwhile, when $\gamma = 0.01$, a significant distortion compared to the symmetric case takes place, the stream-wise velocity component experiencing much weaker change.

An important issue associated with the presence of multiple branches of asymptotic boundary conditions is the feasibility of resulting self-similar corner flows in practice. Its formal solution is

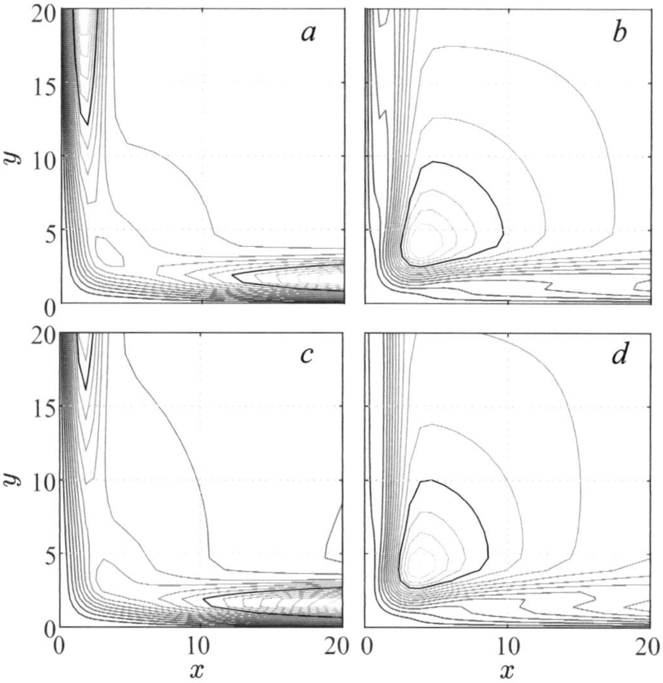

Figure 2. Contour lines of absolute values of spanwise velocity components q in the absence of stream-wise pressure gradient ($\beta = 0$) for the lower (*left*) and the upper branches of the boundary conditions (*right*) and the asymmetry parameter $\gamma = 0$ (*top*) and 0.01 (*bottom*). The contour line step is 0.2.

possible, e.g., by considering the transient problems by analogy with the symmetric case [8] that is beyond the scope of the present study. However, a qualitative comparison of the asymmetric self-similar flows with the results of computations of the full Navier–Stokes equations with ANSYS Fluent was performed. The details of both the numerical grid and the scheme will be described elsewhere. The obtained velocity distributions indicate in favour of the feasibility of the solution of the upper branch at least when β is close to 0 that is also true for the symmetric flow [8].

5. Conclusions

The symmetric and asymmetric self-similar flows of viscous incompressible fluid along a semi-infinite dihedral corner with stream-wise pressure gradients were considered. The geometry of the problem was reduced to the consideration of the flow in quadrant on the plane with the origin coinciding with the vertex of the corner. For the region of the flow bounded by a rectangle with one vertex coinciding with the origin, the system of equations for compatible asymptotic boundary conditions was obtained and analysed numerically. Two types of the asymptotic behaviour was found. It is shown that the flow is very sensitive to the value of the asymmetry parameter.

Acknowledgments

The study was performed with the financial support of the Russian Foundation for Basic Researches Grant No. 16–08–00354.

References

[1] Kornilov V 2013 *Three-Dimensional Turbulent Near-Wall Flows in Streamwise Corners* 2nd ed (Novosibirsk: Publishing House of Siberian Branch of RAS) (in Russian)
[2] Zamir M 1981 *Proc. Roy. Soc.* A **377** 269–88
[3] Dhanak M R and Duck P W 1997 *Proc. Roy. Soc.* A **453** 1793–815
[4] Vasiliev V I 1996 *AIAA J.* **34** 946–52
[5] Rubin S G 1966 *J. Fluid Mech.* **26** 97
[6] Keller H B and Cebeci T 1972 *J. Comp. Phys.* **10** 151–61
[7] Shampine L F and Reichelt M W 1997 *SIAM J. Sci. Comp.* **18** 1–22
[8] Ridha A 1992 *J. Eng. Math.* **26** 525–37
[9] Pernice M and Walker H F 1998 *SIAM J. Sci. Comp.* **19** 302–18

The analysis of solutions behaviour of Van der Pol Duffing equation describing local brain hemodynamics

A A Cherevko[1,2], E E Bord[2], A K Khe[1,2], V A Panarin[3] and K J Orlov[3]

[1] Lavrentyev Institute of Hydrodynamics of the Siberian Branch of the Russian Academy of Sciences, Novosibirsk, Russia
[2] Novosibirsk State University, Novosibirsk, Russia
[3] Siberian Federal Biomedical Research Center named after Academician E. N. Meshalkin, Novosibirsk, Russia

E-mail: cherevko@mail.ru, bord.elizaveta@yandex.ru

Abstract. This article proposes the generalized model of Van der Pol — Duffing equation for describing the relaxation oscillations in local brain hemodynamics. This equation connects the velocity and pressure of blood flow in cerebral vessels. The equation is individual for each patient, since the coefficients are unique. Each set of coefficients is built based on clinical data obtained during neurosurgical operation in Siberian Federal Biomedical Research Center named after Academician E. N. Meshalkin. The equation has solutions of different structure defined by the coefficients and right side. We investigate the equations for different patients considering peculiarities of their vessel systems. The properties of approximate analytical solutions are studied. Amplitude-frequency and phase-frequency characteristics are built for the small-dimensional solution approximations.

1. Introduction

Different pathologies of cerebral blood vessel network are very dangerous for human health and life. For example, arterial aneurysm [1, 2] rupture may lead to cerebral hemorrhage, blood stroke and very often death. That is why studying of bloodstream behavior in the cerebral blood vessel network is a problem of a great importance.

There are different approaches to modelling of bloodstream behavior in the surroundings of anomalies. One of them is to construct empirical model based on experimental data obtained during clinical measurements. This approach is described in this paper [3] and proved to be effective in different applications.

The equation of a nonlinear oscillator of Van der Pol — Duffing (VPD) is a simple model for description of bloodstream behaviour. While building the model velocity $u = u(t)$ is chosen as the control function, i.e. the right side of the equation, and pressure $q = q(t)$ is considered as sought quantity. Such equations have large and complex sets of solutions with the structure determined by different values of coefficients [4].

Being quite simple this model describes the behaviour of hemodynamic parameters near pathologies quite well.

Content from this work may be used under the terms of the Creative Commons Attribution 3.0 licence. Any further distribution of this work must maintain attribution to the author(s) and the title of the work, journal citation and DOI.
Published under licence by IOP Publishing Ltd

IOP Conf. Series: Journal of Physics: Conf. Series **894** (2017) 012012 doi:10.1088/1742-6596/894/1/012012

2. Choosing mathematical model

2.1. Obtaining clinical data

The hemodynamic parameters are measured during the neurosurgical operations with the help of ComboMap device with ComboWire sensor. Blood pressure and velocity are measured simultaneously inside the vessel. The measurements are made both in arterial and venous pool in different areas of vasculature. The obtained pressure and velocity are written into the file with ADC device [5, 6] at frequency $\approx 200\,\text{Hz}$. The obtained data are cleaned by using software noise wavelet filters to get rid of high-frequency noise, vibrations and the low-frequency oscillations corresponding to the respiratory rhythm. Gabor wavelet decomposition is used for this purpose [7].

2.2. Justification of the mathematical model choice

As the structure of arterial vessel is complex and multilayer, the model should consider how active vessel walls and vessel surroundings influence the bloodstream. Under such conditions the processes should be described by the multidimensional system. But we can measure and study only the projection of this system on the "velocity - pressure" plane.

So the question about such simplification correctness arises. The answer is positive in the following sense: pressure and velocity can be represented as the combination of fast and slow bloodstream changes (1)

$$p(t) = p_m(t) + a_p(t)q(t); v(t) = v_m(t) + a_v(t)u(t), \tag{1}$$

where $p_m(t), v_m(t)$ are mean values, $a_p(t), a_v(t)$ are amplitude values and $q(t), u(t)$ are dimensionless normalized values of pressure and velocity.

The values $p_m(t), v_m(t), a_p(t), a_v(t)$ are "slow variables" and $q(t), u(t)$ are "fast variables". Slow changes are caused by the reaction to the medical intervention, whereas fast occur during one cardiac cycle. The behaviour of slow variables is regulated by central nervous system, while the behaviour of fast dimensionless variables is regulated by local vessel reaction to the pulse wave. Thus, being quite simple for investigation, fast variables describe hemodynamic parameters quite good. Our model helps to predict the fast variables behaviour.

3. VDP model

The model of generalized Van der Pol–Duffing equation [8] was suggested to identify the characteristic behaviour of "fast" hemodynamic parameters in the surroundings of vascular pathologies

$$\varepsilon q''(t) + P_2(q)q' + P_3(q) = ku(t). \tag{2}$$

The functions $q = q(t)$ and $u = u(t)$ in (2) are dimensionless and normalized values of blood pressure and velocity, such that $|q| \leq 1$, $|u| \leq 1$. Moreover, in case of arteries velocity is the control function and pressure is found as the solution of equation. Polynomials $P_2(q) = a_0 + a_1 q + a_2 q^2$, $Q_3(q) = b_1 q + b_2 q^2 + b_3 q^3$ characterize friction and elasticity of the system. Real coefficients $\{a_i, b_j, k\}$ $(i = 0, 1, 2; j = 1, 2, 3)$ are built based on clinical data with the methods of inverse problem theory [9]. Finally, coefficient $\varepsilon = 10^{-3}$ corresponds to the relaxation oscillation character.

The model was confirmed experimentally with the large amount of clinical data. The pressure values obtained in the experiment coincide well with the pressure values predicted by the model.

Later the behaviour of model solutions is studied in case when the right side of the model is replaced by the harmonic function. The function has the form of $B\cos(\omega t)$ depending from two parameters which are frequency ω and amplitude B. These parameters are chosen from physiologically significant range of values. Such investigation allows to identify the important properties of the model. Though due to nonlinear character of the model full research of solution properties could not be conducted.

Numerical study of amplitude-phase-frequency solution characteristics shows the system having complex behaviour in concerned range of right side frequencies and amplitudes [10].

4. Existing theorem

Let apply Wong theorem [11] to the equation (2). Theorem satisfaction guarantees the existence of bounded solution and freedom from oscillation stop. The application of Wong theorem to the equation (2) gives the necessary conditions on the coefficients (3)

$$\begin{aligned} a_3 &> 0 \\ b_3 &> 0 \\ a_1 - \frac{a_2{}^2}{4a_3} &> 0. \end{aligned} \tag{3}$$

There is an example of theorem accomplishment (Figure 1: a - b), the form of solution is close to linear. The left picture on the figure is solution plot in polar coordinates and the right picture is the graph of solution in yellow and right side in blue. When the theorem is not fulfilled, the nonlinear effects appear on solution (Figure 1: c - d) and the oscillation stop occurs.

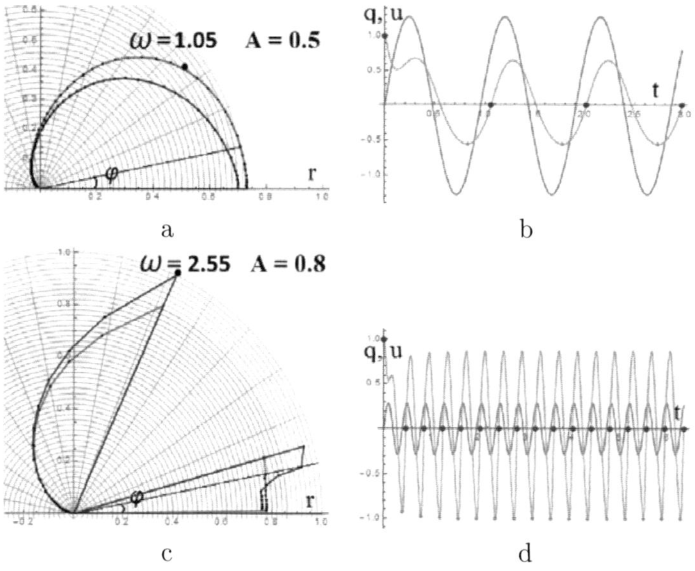

Figure 1. Examples of solution behaviour depending on Wong theorem. a - b: Theorem is fulfilled, the form of solution is close to linear. c - d: Theorem is not fulfilled, the form of solution is nonlinear.

5. Analytical study by Galerkin method

Galerkin method is often used to find approximate solutions of systems with essential nonlinearities [11]. For this sought solution is represented as the linear combination of linearly independent functions.

The solution is approximated with trigonometric polynomials. The approximate solution of initial equation can be represented in the following form (4) in case of units period $T = 2\pi$.

$$u_m(t) = c_0 + \sum_{i=1}^{m} (c_{2i-1} \sin it + c_{2i} \cos it), \tag{4}$$

where $c_0, c_1, ..., c_{2m}$ are constants.

Compute the residual projection on the solution expansion subspace $\{1,\ \sin it,\ \cos it,\ i = 1, \ldots, m\}$. And than equating residual to zero, we obtain the equations for identification of the coefficients $c_0, c_1, ..., c_{2m}$.

5.1. One-dimensional approximation

Consider the one-dimensional Galerkin approximation. The approximate solution is of the form $q = A\cos(\omega t)$. The equation (5) connects amplitude and frequency

$$3A_3 b_3 - 4Bk + 4A(b_1 - \varepsilon\omega^2) = 0. \tag{5}$$

Build the amplitude-frequency characteristic, where ω is the right side frequency, A is the amplitude of approximate solution and B is the right side amplitude. Different lines on the plot (Figure 2: a) correspond to different values of B. Coefficient a_i specifying damping and b_2 do not affect the behaviour of approximate solution.

5.2. Two-dimensional approximation

Consider the two-dimensional case. Solution representation is of the form $q = A\cos(\omega t - \phi)$. The equations for amplitude (6) and frequency (7) split:

$$\begin{aligned} &16B^2k^2 - A^2(4b_1 + 3A^2b_3)^2 + ((4a_1 + A^2a_3)^2 - \\ &-8(4b_1 + 3A^2b_3)\varepsilon)\omega^2 + 16\varepsilon^2\omega^4 = 0, \end{aligned} \tag{6}$$

$$\begin{aligned} &B^3k^3(a_3\omega\cos(\varphi) - 3b_3\sin(\varphi))^3 + \\ &+4Bk\omega^2(-a_3b_1 + 3a_1b_3 + a_3\varepsilon\omega^2)^2 * \\ &*(a_1\omega\cos\varphi + (-b_1 + \varepsilon\omega^2)\sin\varphi) = 0. \end{aligned} \tag{7}$$

Build the amplitude-frequency characteristic (Figure 2: b). Coefficients a_2 and b_2 do not influence the amplitude behaviour. Build the phase-frequency characteristic (Figure 2: c). Coefficients a_2, b_2 do not influence the phase behaviour.

5.3. Three-dimensional approximation

Consider three-dimensional approximation. Solution is of the form $q = A_0 + A_1 cos(\omega t - \varphi_1)$. Obtain equations for two amplitudes (8, 9)

$$2A_0b_1 + 2{A_0}^2b_2 + {A_1}^2b_2 + 2{A_0}^3b_3 + 3A_0{A_1}^2b_3 = 0, \tag{8}$$

$$
\begin{gathered}
16B^2k^2 - {A_1}^6(9{b_3}^2 + {a_3}^2\omega^2) + \\
+8{A_1}^4(-3b_1b_3 + (2b_2 + 3A_0b_3)^2 - a_1a_3\omega^2 - (A_0a3(a_2 + A_0a_3) - 3b_3\varepsilon)\omega^2) - \\
-16{A_1}^2({b_1}^2 + 2{A_0}^3b_2b_3 + 3{A_0}^4{b_3}^2 + ((a_1 + A_0(a_2 + A_0a_3))^2 - \\
-2(b_1 + A_0(2b_2 + 3A_0b_3))\varepsilon)\omega^2 + \epsilon^2\omega^4) = 0,
\end{gathered} \tag{9}
$$

where amplitudes do not depend on phase. Build the amplitude-frequency characteristic (Figure 2: d). Amplitudes A_0 and A_1 also do not depend from right side parameters and damping coefficients.

Phase equation (10) does not split off and phase depends on A_0 and A_1

$$
\tan\phi_1 = \frac{(4a_1 + 4A_0a_2 + 4{A_0}^2a_3 + {A_1}^2a_3)\omega}{4b_1 + 8A_0b_2 + 12{A_0}^2b_3 + 3{A_1}^2b_3 - 4\varepsilon\omega^2}. \tag{10}
$$

Build the phase-frequency characteristic (Figure 2: e). Phase does not depend explicitly from right side amplitude of initial equation.

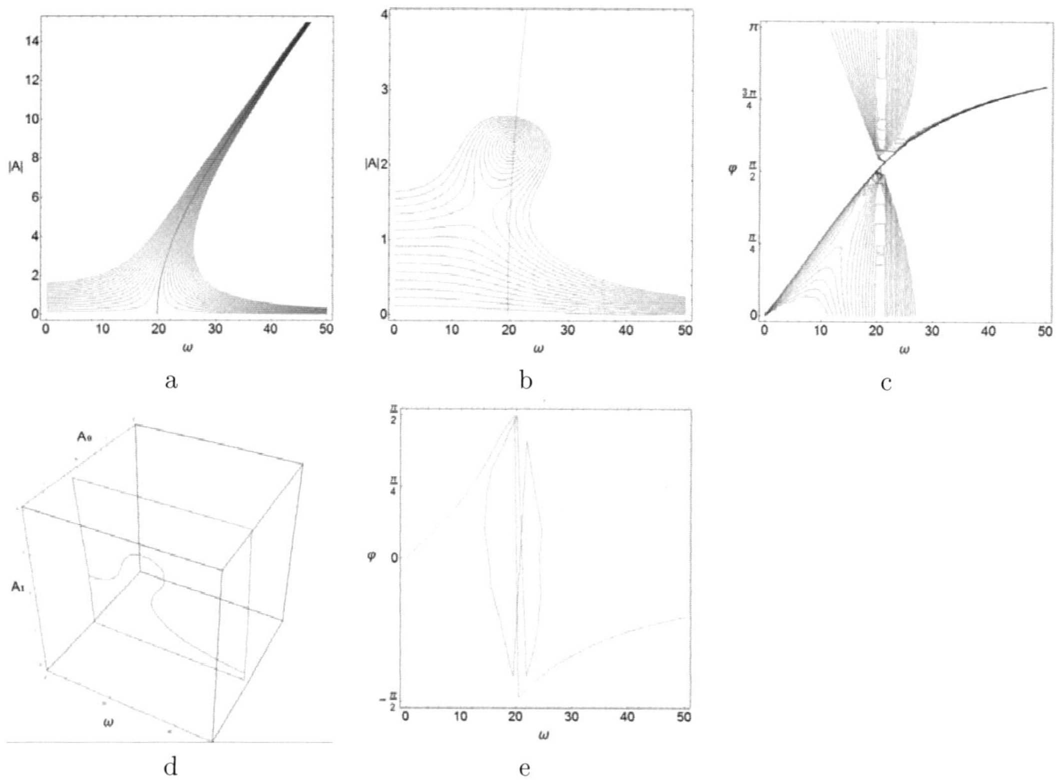

Figure 2. Examples of amplitude-frequency and phase-frequency characteristics for one patient. a: Amplitude-frequency characteristic in case of one-dimensional approximation. b-c: Amplitude-frequency and phase-frequency characteristics in case of two-dimensional approximation. d-e: Amplitude-frequency and phase-frequency characteristics in case of three-dimensional approximation.

For all considered amplitude- and phase-frequency characteristics, the ambiguous dependence of solution amplitude and phase on the right side frequency near a physiologically significant frequency range is characteristic. Under unfavourable circumstances this can result to a dangerous from the medical point of view switching of oscillatory regimes.

6. Conclusion

The study of generalized Van der Pol–Duffing equations built on the clinical data of real patients is presented in this paper. The equation has solutions of different structure defined by the coefficients and right side. Conditions on the coefficients guaranteing the absence of oscillation stop are found out. Amplitude-frequency and phase-frequency characteristics are built for Galerkin small-dimensional solution approximations in case of one patient. The dependencies of these characteristics from equation coefficients are found out.

Acknowledgments

The study was completed thanks to the support of Russian Science Foundation (project No. 17-11-01156).

References

[1] Klimov A B, Ryabuhin V E, Orlov K Yu and Krilov V V 2012 *Surgery of the Cerebral Aneurysms* **3** 231–68
[2] Krylov V V (ed) 2011 *Surgery of the Cerebral Aneurysms* vol 1 (Moscow: T. A. Alexeeva)
[3] Tikhonov A N and Arsenin V Ya 1979 *Methods of Solving Ill-Posed Problems* (Moscow: Nauka)
[4] 1951 Cartwright M L and Littlewood J C Some fixed point theorems *Ann. Math.* **54** 1–37
[5] Chupakhin A P, Cherevko A A, Khe A K, Telegina N J, Krivoshapkin A L, Orlov K J, Panarin V A and Baranov V I 2012 *Circulation Pathology and Cardiac Surgery* **16** 27–31
[6] Khe A K, Cherevko A A, Chupakhin A P, Krivoshapkin A L and Orlov K J 2016 Endovascular blood flow measurement system *J. Phys.: Conf. Series* **722** 012041
[7] Malla S 2005 *Wavelets in Signal Processing* (Moscow: Mir)
[8] Butenin N V, Nejmark Yu I and Fufaev N L (ed) 1987 *Introduction in Nonlinear Oscillation Theory* (Moscow: Nauka)
[9] Cherevko A A, Mikhaylova A V, Chupakhin A P, Ufimtseva I V, Krivoshapkin A L and Orlov K J 2016 Relaxation oscillation model of hemodynamic parameters in the cerebral vessels *J. Phys.: Conf. Series* **722** 012045
[10] Cherevko A A, Bord E E, Khe A K, Panarin V A, Orlov K J and Chupakhin A P 2016 Using non-linear analogue of Nyquist diagrams for analysis of the equation describing the hemodynamics in blood vessels near pathologies *J. Phys.: Conf. Series* **722** 012005
[11] Kryukov B. I. (ed) 1984 *Forced Oscillations of Essentially Nonlinear Systems* (Moscow: Mashinostoenie)

IOP Conf. Series: Journal of Physics: Conf. Series **894** (2017) 012013 doi:10.1088/1742-6596/894/1/012013

Numerical simulation of nonlinear dynamics of 1D pulsating detonations

S P Borisov[1,2] **and A N Kudryavtsev**[1,2]

[1] Khristianovich Institute of Theoretical and Applied Mechanics, Novosibirsk, Russia
[2] Novosibirsk State University, Novosibirsk, Russia

E-mail: borisov@itam.nsc.ru, alex@itam.nsc.ru

Abstract. The development of 1D instability of a detonation wave is numerically simulated for a two-stage chemical model. The shock-fitting approach is employed to track the leading detonation front. In order to determine its motion, the equation for the acceleration of the shock wave derived from the Rankine-Hugoniot conditions and the characteristic relations is integrated along with the reactive Euler equations. The fifth-order WENO scheme is used, time stepping is performed with the four-stage Runge-Kutta-Gill method. It is shown that in a certain range of parameters of the problem (the degree of overdrive f, the dissociation energy E_d and the activation energy E_a), the Zeldovich-Neumann-Döring stationary solution is unstable with respect to 1D disturbances. The evolution of disturbances at later nonlinear stages is studied. Nonlinear saturation of the growth of disturbances leads to the formation of a stable limit cycle. When changing the parameters of the problem, the period doubling bifurcation can occur leading to the appearance of pulsations with two different maxima of the amplitude.

1. Introduction

Linear stability of detonation waves (DWs) was studied almost exclusively using the simplest one-step chemical mechanism with one irreversible Arrhenius-type reaction [1]. Nonlinear dynamics of 1D unstable detonations in the framework of this model was also studied in detail [2, 3, 4, 5]. It was shown that, for weakly unstable detonations, the growth of disturbances leads to a periodic self-oscillating regime corresponding to a stable limit cycle in the phase plane. With increasing the activation energy, several successive period doubling bifurcations occur followed by transition to a chaotic regime.

The simplest one-step model correctly describes many features of the detonation phenomenon. However, it does not correspond to any specific combustible mixture and cannot claim quantitative agreement with experimental data. When solving practical problems connected with the propagation of detonations in gaseous mixtures, two-stage models, first proposed in [6], are most often used. In this paper we consider the nonlinear dynamics of 1D disturbances in a two-stage chemical model developed in [7, 8] for describing the detonation combustion of hydrogen and widely used later in numerical simulation of detonations.

2. Governing equations and numerical techniques

2.1. Two-stage chemical model

Instead of considering detailed kinetic schemes, which can include, depending on the composition of the combustible mixture, tens, hundreds, and even thousands of chemical reactions, in the

Content from this work may be used under the terms of the Creative Commons Attribution 3.0 licence. Any further distribution of this work must maintain attribution to the author(s) and the title of the work, journal citation and DOI.
Published under licence by IOP Publishing Ltd

IOP Conf. Series: Journal of Physics: Conf. Series **894** (2017) 012013 doi:10.1088/1742-6596/894/1/012013

two-stage model the chemical transformations that occur when a fuel burns are described by two variables, the fraction of the induction period Y, $0 \le Y \le 1$, and the current molar mass of the mixture μ. The flow is governed by the Euler equations (in our case, 1D ones) complemented with two additional equations:

$$\frac{\partial \mathbf{Q}}{\partial t} + \frac{\partial \mathbf{F}}{\partial x} = \mathbf{S}, \qquad \mathbf{Q} = [\rho, \rho u, \rho E, \rho Y, \rho\mu]^T, \tag{1}$$
$$\mathbf{F} = \left[\rho u, \rho u^2 + p, u(\rho E + p), \rho u Y, \rho u \mu,\right]^T, \qquad \mathbf{S} = (0, 0, 0, \rho W_Y, \rho W_\mu)^T.$$

The pressure and the total energy are

$$p = \frac{\rho R T}{\mu}, \qquad E = e(T, \mu) + \frac{1}{2}u^2, \tag{2}$$

where is the internal energy is given by

$$e(T,\mu) = E_d\left(\frac{1}{\mu} - \frac{1}{\mu_{min}}\right) + \left[\frac{\mu}{\mu_{min}} + \frac{1-\delta}{2} + \left(\frac{\mu}{\mu_{min}} + \delta - 1\right)\frac{\theta/T}{e^{\theta/T} - 1}\right]\frac{RT}{\mu}. \tag{3}$$

Here, for the stoichiometric hydrogen–oxygen mixture, $E_d = 4.6024 \cdot 10^8$ J/kmol is the mean dissociation energy of reaction products, $\mu_{min} = 6$ kg/kmol and $\mu_{max} = 18$ kg/kmol are the molar masses in completely dissociated and completely recombined states, respectively, $\delta \equiv \dfrac{\mu - \mu_{min}}{\mu_{max} - \mu_{min}}$, $\theta = 3500$ K is the effective temperature of excitation of vibrational degrees of freedom, and R is the universal gas constant.

In a two-stage model it is assumed that the entire process of chemical transformations can be divided into two stages, the induction stage and the heat release stage. During the induction stage, the molar mass of the gas does not change, there is no heat release, and the variable Y, initially equal to 1, changes according to the equation

$$\frac{dY}{dt} = W_Y = -\frac{1}{\tau_{ind}}, \qquad \tau_{ind} = \frac{K_a \mu_{O_2}}{\rho z}\exp\left(\frac{E_a}{RT}\right), \tag{4}$$

where $E_a = 7.17556 \cdot 10^7$ J/kmol is the activation energy, $K_a = 5.38 \cdot 10^{-11}$ kmol·s/m^3 is the pre-exponential factor, $\mu_{O_2} = 32$ kg/kmol is the molar mass of oxygen, and $z = 8/9$ is the mass fraction of oxygen.

When Y reaches zero the molar mass starts to change, that is accompanied by heat release. The molar mass changes according to the equation

$$\frac{d\mu}{dt} = W_\mu = W_1(\mu)\rho^2 - W_2(T,\mu)\rho, \qquad W_1(\mu) = \frac{4K_+}{\mu}\left(1 - \frac{\mu}{\mu_{max}}\right)^2, \tag{5}$$
$$W_2(T,\mu) = 4K_+K_-\left(\frac{\mu}{\mu_{min}} - 1\right)\left(\frac{T}{T_0}\right)^{\beta/2}\left[1 - \exp\left(-\frac{\theta}{T}\right)\right]^\beta \exp\left(-\frac{E_d}{RT}\right).$$

Here $K_+ = 6 \cdot 10^8$ m^6/(kmol^2·s) is recombination rate constant, $K_- = 1.769 \cdot 10^3$ kmol/m^3 is the equilibrium constant, T_0 is the initial temperature of the mixture, and $\beta \equiv \dfrac{\mu_{max}}{\mu_{max} - \mu_{min}}$.

The profile of the stationary solution (Zeldovich–von Neumann–Döring, ZND, solution) obtained with the two-stage model has a kink at the point where the induction period expires. This nonphysical singularity can lead to problems, in particular to the loss of accuracy in numerical simulations, and significantly affect the stability characteristics since the equations of the linear stability theory include derivatives of the basic flow quantities. To avoid the potential problems, we modified the two-stage model having replaced W_μ, W_Y by

$$\widetilde{W}_\mu = W_\mu \exp\left(-\eta Y\right), \qquad \widetilde{W}_Y = W_Y \left[1 - \exp\left(-\eta Y\right)\right], \tag{6}$$

Where η is a smoothing parameter. Thus, the two stages cease to be sharply separated. Figure 1 shows the temperature profiles of the stationary solution of the equations (1) for various values of the parameter η. Hereafter dimensionless variables are plotted with the induction length ℓ_{ind} as the reference length, the pre-shock speed of sound as the reference velocity and other flow properties measured in their pre-shock values.

In the subsequent calculations, $\eta = 10$ is taken. As can be seen from the figure, at this value the profile is very close to that obtained from the original model, with a well-defined induction zone in which heat release can be considered negligible.

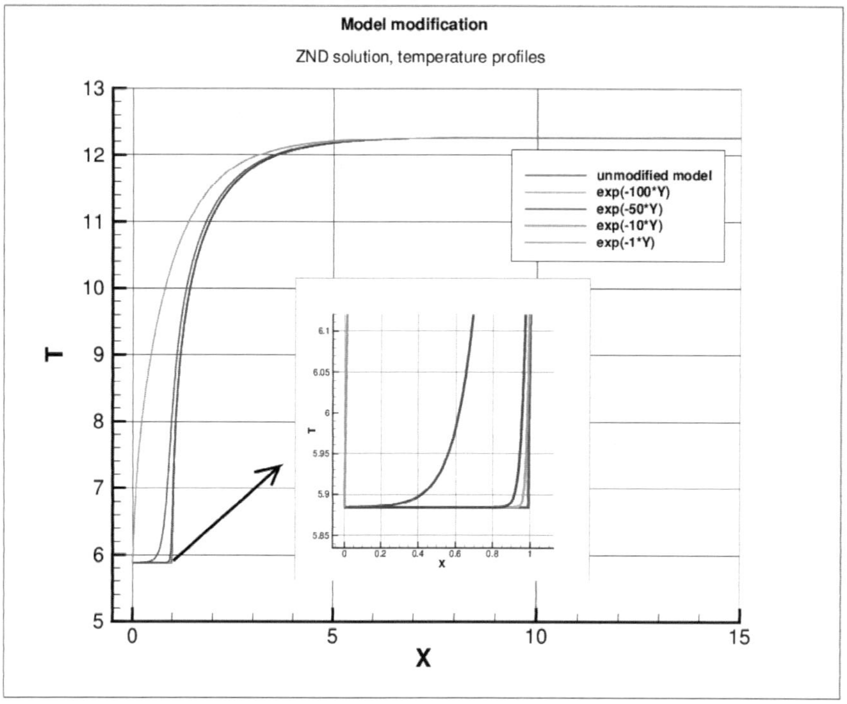

Figure 1. ZND temperature profiles at different values of the smoothing parameter.

2.2. Shock-fitting approach

Numerical simulations are performed using the shock-fitting approach. The leading detonation front is one of the boundaries of the computational domain. To determine its position and velocity, the equations of motion of the front are integrated along with the reactive Euler equations (1). By differentiating the Rankine-Hugoniot conditions with respect to time, the acceleration of the shock wave can be expressed through the time derivative of any of the flow quantities just behind the shock. The latter, in turn, can be replaced by spatial derivatives using the Euler equations. Since the spatial derivatives at the boundary have to be approximated by one-sided finite differences, it is most correct, from numerical point of view, to express the acceleration of DW in terms of $\mathcal{L}$, a linear combination of the derivatives of the flow quantities entering into the relation on the characteristic arriving at the shock wave [9]. For the two-stage model it is given by the following formula:

$$\mathcal{L} = \left(\frac{1}{\rho}\frac{\partial \rho}{\partial t} + \frac{\gamma}{c}\frac{\partial u}{\partial t} + \frac{1}{T}\frac{\partial T}{\partial t} - \frac{1}{\mu}\frac{\partial \mu}{\partial t}\right) = -(u+c)\left(\frac{1}{\rho}\frac{\partial \rho}{\partial x} + \frac{\gamma}{c}\frac{\partial u}{\partial x} + \frac{1}{T}\frac{\partial T}{\partial x} - \frac{1}{\mu}\frac{\partial \mu}{\partial x}\right) - \left(\frac{1}{T}\frac{e_\mu}{e_T} + \frac{1}{\mu}\right)W_\mu, \qquad \gamma \equiv 1 + \frac{R}{\mu e_T}, \qquad c^2 = \frac{\gamma R T}{\mu}. \quad (7)$$

Here e_T and e_μ are the partial derivatives of the internal energy with respect to relevant variables and c is the speed of sound.

The fifth-order WENO scheme [10] is used to evaluate the spatial derivatives. The length of the computational domain is $50\,\ell_{ind}$, the number of grid points is 5,000. Time integration is performed with the four-stage Runge–Kutta–Gill scheme [11]. Since the dependence of the internal energy on the temperature (3) is nonlinear, at each stage of the Runge–Kutta–Gill integration the temperature is determined with Newton iterations. The same iterative method is used in calculating the flow quantities just behind the shock wave from the Rankin-Hugoniot relations.

3. Results

The stationary ZND solution is used as initial data in numerical simulations. The influence of the degree of overdrive $f = (\mathrm{M}/\mathrm{M}_{CJ})^2$, the activation energy E_a, and the dissociation energy E_d was studied. Here M is the DW Mach number and M_{CJ} is the Chapman–Jouguet Mach number.

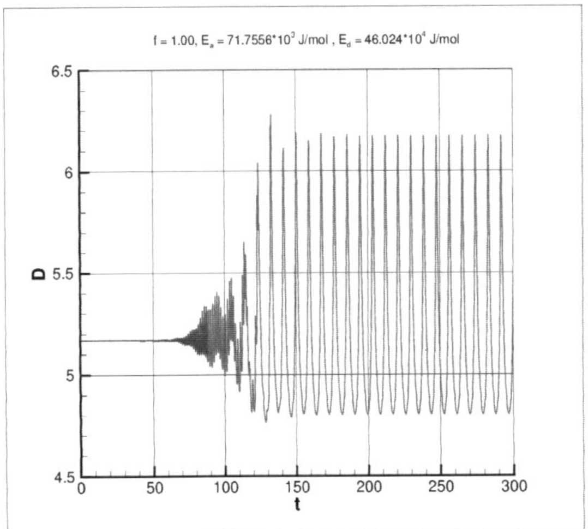

Figure 2. Evolution of DW velocity at $f = 1$, $E_a = 7.17556 \cdot 10^7$ J/kmol, and $E_d = 4.6024 \cdot 10^8$ J/kmol.

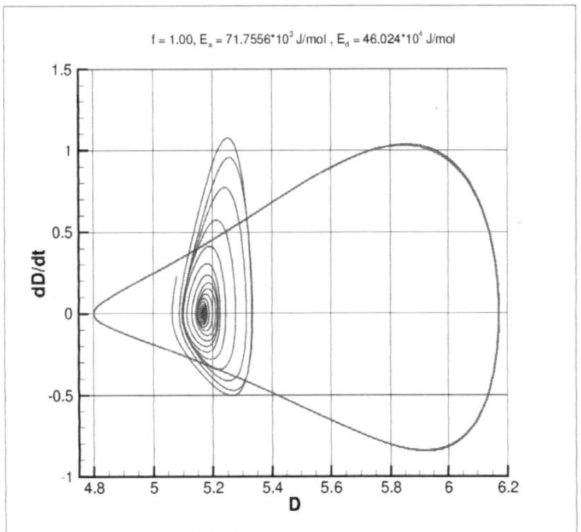

Figure 3. Phase portrait at $f = 1$, $E_a = 7.17556 \cdot 10^7$ J/kmol, and $E_d = 4.6024 \cdot 10^8$ J/kmol.

3.1. Influence of the degree of overdrive f

The numerical simulation of the propagation of the self-sustaining DW has shown that at $f = 1$, $E_a = 7.17556 \cdot 10^7$ J/kmol, and $E_d = 4.6 \cdot 10^8$ J/kmol the DW is unstable to 1D disturbances. The instability leads to the formation of a pulsating detonation regime. Figure 2 shows the evolution of the DW velocity. The initial stage of the instability development is singled out in blue. The oscillations are periodic and their amplitude is constant starting at about $t = 200$.

Figure 3 shows the phase plane $(D, dD/dt)$. Only the initial (blue) and final (red) parts of the phase trajectory are depicted. It is obvious that the initial point representing the ZND solution is an unstable focus under these parameters. Moving away from this point, the trajectory after $t \approx 200$ winds onto a stable limit cycle corresponding to periodic self-oscillations.

An increase in f to 1.05 does not lead to a qualitative change in the picture, although the instability develops more slowly. However, at $f = 1.1$ the flow becomes stable. The initial oscillations caused by perturbations originated from the limited accuracy of the calculation of the ZND solution dampen over time and, after that, the DW propagates at a constant speed.

3.2. Influence of the activation energy E_a

Increasing the activation energy leads to a more unstable state. Thus, at $f = 1$, $E_a = 7.42556 \cdot 10^7$ J/kmol, $E_d = 4.6024 \cdot 10^8$ J/kmol, the instability develops significantly faster and the amplitude of the DW velocity oscillations increases. Figure 4 shows that the character of the oscillations has changed significantly so that two maxima of different magnitude are observed during the period. The corresponding limit cycle, which the trajectory reaches at $t \approx 120$, has a self-intersection point, see figure 5. Thus, as a result of the cycle doubling bifurcation, the system enters a two-period regime.

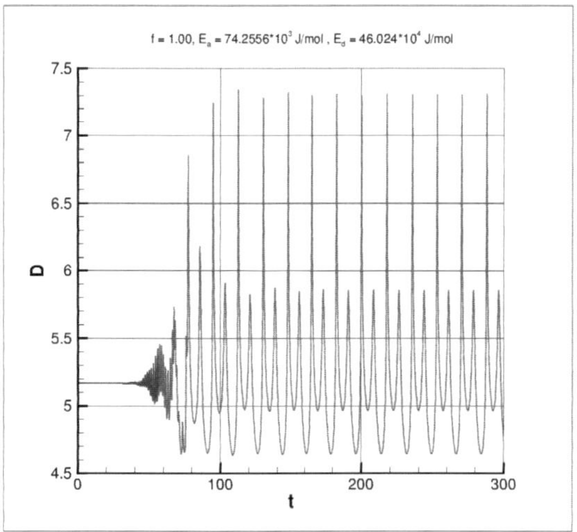

Figure 4. Evolution of DW velocity at $f = 1$, $E_a = 7.42556 \cdot 10^7$ J/kmol, and $E_d = 4.6024 \cdot 10^8$ J/kmol.

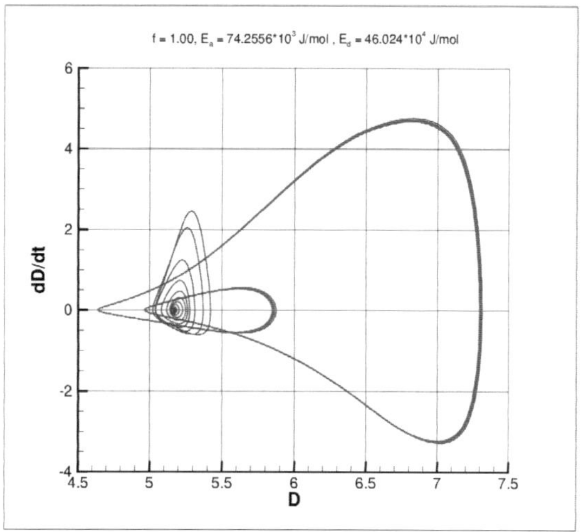

Figure 5. Phase portrait at $f = 1$, $E_a = 7.42556 \cdot 10^7$ J/kmol, and $E_d = 4.6024 \cdot 10^8$ J/kmol.

A further increase in E_a leads to an even stronger instability. In fact, even at $E_a = 7.67556 \cdot 10^7$ J/kmol, it is difficult to identify a stable cycle on the phase plane and the flow becomes chaotic.

A decrease in activation energy, on the contrary, has a stabilizing effect. At $E_a = 6.92556 \cdot 10^7$ J/kmol the system is still unstable and the self-oscillation regime occurs at $t \approx 700$. The decrease in the activation energy to $E_a = 6.67556 \cdot 10^7$ J/kmol leads to suppression of the instability.

3.3. Influence of the dissociation energy E_d

As can be expected, a decrease in E_d leads to an increase in instability, while an increase in E_d cause its weakening. Thus, as the dissociation energy is reduced to $E_d = 4.4024 \cdot 10^8$ J/kmol (as earlier, $f = 1$, $E_a = 7.17556 \cdot 10^7$ J/kmol), the picture is qualitatively similar to that shown in figures 4 and 5.

With the increase in E_d up to $4.8024 \cdot 10^8$ J/kmol the pulsations of the DW velocity are first visible only by the time moment $t \approx 150$, see figure 2. The corresponding picture in the phase plane, figure 7, shows very distinctively the instability development at the initial stage, a transient process, during which gradually damped oscillations around the limit cycle occur, and the final transition to the periodic regime.

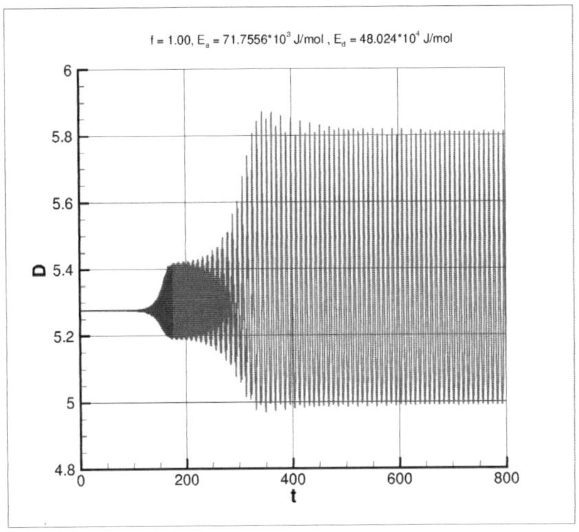

Figure 6. Evolution of DW velocity at $f = 1$, $E_a = 7.17556 \cdot 10^7$ J/kmol, and $E_d = 4.8024 \cdot 10^8$ J/kmol.

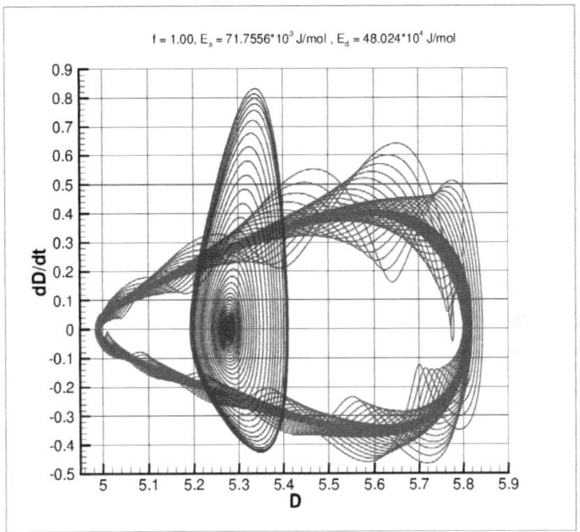

Figure 7. Phase portrait at $f = 1$, $E_a = 7.17556 \cdot 10^7$ J/kmol, and $E_d = 4.8024 \cdot 10^8$ J/kmol.

4. Conclusions

It is shown that a plane self-sustaining DW, which is a stationary solution of the Euler equations with a two-stage chemical model, is unstable with respect to 1D disturbances. The development of instability leads to the formation of a pulsating detonation regime with a periodic variation in the velocity of the leading DW front. Overdriven detonations are more stable; an increasing in the degree of overdrive f up to 1.1 completely suppresses the instability.

The effect of the activation energy E_a and the dissociation energy E_d on the 1D instability is studied. Detonations become more unstable with increasing E_a and, as a result of the cycle doubling bifurcation, two-period oscillations appear with two different maxima of the DW velocity during the period. A further increase in the activation energy leads to a highly unstable state with chaotic dynamics. On the contrary, a decrease in E_a can completely suppress the instability.

Variation of the dissociation energy E_d has the opposite effect; the DW becomes more stable as the E_d increases and more unstable as it decreases.

Acknowledgments

This work was supported by the Russian Foundation for Basic Research (joint Russia-India project No. 16-57-48007).

References

[1] J H S Lee 2008 *The Detonation Phenomenon* (Cambridge: Cambridge University Press)
[2] L He and J H S Lee 1995 The dynamical limit of one-dimensional detonations *Phys. Fluids* **7**(5) 1151-8
[3] Y Daimon and A Matsuo 2003 Detailed features of one-dimensional detonations *Phys. Fluids* **15**(1) 112-22
[4] A R Kasimov and D S Stewart 2004 On the dynamics of self-sustained one-dimensional detonations: A numerical study in the shock-attached frame *Phys. Fluids* **16**(10) 3566-78
[5] A K Henrick, T D Aslam and J M Powers 2006 Simulations of pulsating one-dimensional detonations with true fifth order accuracy *J. Comput. Phys.* **213**(1) 311-29
[6] V P Korobeinikov and V A Levin 1969 Strong explosion in a combustible gas mixture *Fluid Dyn.* **4**(6) 30-2
[7] Ya A Nikolaev and P A Fomin 1982 Analysis of equlibrium flows of chemically reacting gases *Combust. Explos. Shock Waves* **18**(1) 53-8
[8] Ya A Nikolaev and D V Zak 1988 Agreement of models of chemical reactions in gases with the second law of thermodynamics *Combust. Explos. Shock Waves* **24**(4) 461-4
[9] S P Borisov and A N Kudryavtsev 2016 Linear and nonlinear effects in detonation wave structure formation *J. Phys.: Conf. Series* **722** 012022
[10] G S Jiang and C-W Shu 1996 Efficient implementation of Weighted ENO schemes *J. Comp. Phys.* **126**(1) 202-28
[11] R J Thompson 1970 Improving round-off in Runge–Kutta computations with Gill's method *Comm. ACM* **13**(12) 739-40

Flow of viscoplastic suspensions in a hydraulic fracture: implications to overflush

S A Boronin[1], A A Osiptsov[1] and J Desroches[2]
[1] Skolkovo Institute of Science and Technology (Skoltech), Moscow, Russia
[2] Services Petroliers Schlumberger, Paris, France
E-mail: s.boronin@skoltech.ru

Abstract. The study is devoted to modeling of multiphase flows of immiscible viscoplastic fluids in a hydraulic fracture. In the framework of the lubrication approximation, three-dimensional Navier-Stokes equations are reduced to hyperbolic transport equations for the fluid tracers and a quasi-linear elliptic equation in terms of the fluid pressure. The governing equations are solved numerically using the finite-difference approach. A parametric study of the displacement of Bingham fluids in a Hele-Shaw cell is carried out. It is found that fingers developed through the pillar of a yield-stress suspension trigger the development of unyielded zones. An increase in the Bingham number leads to an increase in the so-called finger shielding effect, which manifests itself via an increase in the overall finger penetration zone and a decrease in the total number of fingers. The effect of flow parameters on the displacement of hydraulic fracturing proppant-laden suspension by a clean fluid in the vicinity of the perforation zone is carried out. This particular case is considered in application to overflush at the end of a stimulation treatment, when a small portion of a thin clean fluid is injected to wash out the particles from the wellbore into the fracture. It is found that an increase in the yield stress and the viscosity contrast between the fracturing and the overflush fluids typically reduces the area of the cavity thus mitigating the risk of loosing the conductive path between the wellbore and the fracture after the fracture closure.

1. Introduction

Proppant transport models incorporated into existing hydraulic fracturing simulators describe the flow of particle-laden suspension inside a hydraulic fracture in the framework of the lubrication approximation using the power-law rheological model (see [1]). Hydraulic fracturing suspensions with large concentration of solids or fibers show a yield-stress behavior in rheological experiments [2, 3], which is not taken into account in the proppant transport models implemented into commercial simulators of hydraulic fracturing.

A state of the art in the modeling of injection of particle-laden suspensions into hydraulic fractures is the family of 2D width-averaged models based on the lubrication approximation to Navier-Stokes equations [1, 4, 5, 6]. In the regime of non-inertial settling, the momentum conservation equation for particles is reduced to an algebraic relation for the particle velocity slip in the vertical direction given by the Stokes formula with a correction for hindered-settling effects due to a finite particle volume fraction. In the case of Newtonian suspension rheology, the total momentum conservation equation for the suspension is reduced to the linear expression of the fluid (or mixture) velocity through the pressure gradient (similar to the Darcy law, hence a well-known analogy between filtration and a flow in a Hele-Shaw cell). The dependence of the

Content from this work may be used under the terms of the Creative Commons Attribution 3.0 licence. Any further distribution of this work must maintain attribution to the author(s) and the title of the work, journal citation and DOI.
Published under licence by IOP Publishing Ltd

width-averaged fluid velocity on the pressure gradient is nonlinear, if the rheology of the carrier fluid or the suspension as a whole is non-Newtonian (see, e.g., [7, 8, 9]).

A large number of papers deal with the Saffman-Taylor (S-T) instability accompanying displacement of fluids in a narrow plane channel or annulus, starting from the pioneering study by Muskat [10]. The instability at the interface between the fluids is triggered when a high-viscosity fluid is displaced by a low-viscosity one. Note that in the absence of a certain "cut-off" mechanism (molecular diffusion or surface tension), the growth rate of disturbances at the interface between the fluids increases unboundedly with a decrease in the wavelength. Review of studies on the S-T instability in Newtonian fluids is presented in [11].

In the present study, we continue to analyze the displacement of viscoplastic fluids in a Hele-Shaw cell approximating a hydraulic fracture. This work was started in [12]. A parametric study of different injection scenarios of yield-stress fluids is carried out. In particular, we analyzed the effect of flow parameters on the shape and dimensions of the particle-free zone developed during overflush at the end of a hydraulic fracturing treatment. It is the final stage of a hydraulic fracturing operation, when the proppant-laden slurry is displaced by a low-viscosity particle-free fluid in order to clean the well and perforations.

2. Problem Formulation

We consider the flow of immiscible incompressible fluids with the Bingham rheology in a narrow Hele-Shaw cell approximating a single wing of a hydraulic fracture (we assume that the flow in the fracture wings is identical, so that there is a symmetry with respect to the perforation zone). A detailed derivation of 2D width-averaged governing equations is presented in [12], while here we present only the final system of equations in the non-dimensional form:

$$\frac{\partial w C_i}{\partial t} + \nabla \cdot (w C_i \mathbf{v}) = -2 C_i v_l, \; i = 0,\, 1,\, 2, \tag{1}$$

$$\nabla \cdot \left(\frac{w^3}{12\mu_m} G(\phi) \left[\nabla p + \mathrm{Bu} \rho_m \mathbf{e}_y \right] \right) = \frac{\partial w}{\partial t} + 2 v_l, \tag{2}$$

$$\mathbf{v} = -\frac{w^2}{12\mu_m} G(\phi) \nabla p, \; G(\phi) = 1 - 3\phi + 4\phi^3, \; \phi = \frac{\mathrm{Bn}\tau_m}{w|\nabla p|}, \tag{3}$$

$$\rho_m = C_0 + \zeta_1 C_1 + \zeta_2 C_2, \; \mu_m = C_0 + \xi_1 C_1 + \xi_2 C_2,$$

$$\zeta_i = \frac{\rho_i}{\rho_0}, \; \xi_i = \frac{\mu_i}{\mu_0}, \; \mathrm{Bu} = \frac{\rho_0 g d^2}{\mu_0 U}, \; \mathrm{Bn} = \frac{\tau_0 d}{U \mu_0}.$$

Here, Cartesian coordinate system Oxy is introduced in the cell plane, so that y-axis (with the basis vector $\mathbf{e}_y$) is vertical and origin O is located in the bottom left corner of the computational domain; C_i are the fluid tracer concentrations (with i being the number of fluid, so that $i = 0$ corresponds to the yield-stress fluid filling the slot initially); $w(x, y, t)$ is the width of the Hele-Shaw cell (currently it is a prescribed function of coordinates and time, while in the model describing the hydraulic fracture propagation, w is obtained via coupling the hydrodynamic equations describing the flow inside a hydraulic fracture with geomechanics equations describing the fracture growth [1]); $\mathbf{v}$ is the width-averaged fluid velocity; v_l is the velocity of fluid leak-off through the porous walls; G is the correction to fluid mobility due to the yield-stress rheology ($G = 1$ for Newtonian fluid); differential operator '∇' acts in the (x, y) plane as we applied the averaging procedure along the cell width. The flow scales are as follows: L is the cell length, U is the scale of the injection velocity, d is the cell width scale, ρ_0 is the fracturing fluid density, μ_0 and τ_0 are the fracturing fluid plastic viscosity and yield stress, respectively; g is the gravity acceleration. Non-dimensional parameters are as follows: Bu is the Buoyancy number, Bn is the

 IOP Publishing
doi:10.1088/1742-6596/894/1/012014

Bingham number; ζ_i is the fluid density ratio; ξ_i is the fluid viscosity ratio. The flow domain according to the scaling introduced above is $(x, y) \in [0,\, 1] \times [0,\, H/L]$, where H is the height of the Hele-Shaw cell.

For hyperbolic equations (1), we impose initial distribution of fluid tracers and boundary conditions for fluid tracers and velocity at the inlet:

$$x = 0 : \; C_i^{in} = C_i(y, t),\; i = 0,\, 1,\, 2; \tag{4}$$

$$t = 0 : \; C = C_{i0}(x, y),\;\; i = 0,\, 1,\, 2. \tag{5}$$

The quasi-elliptic equation for pressure (2) requires either Neumann or Dirichlet boundary conditions to be specified at the boundaries of the flow domain. At the top and bottom boundaries, we impose the no-flow condition. There is a specified velocity at the inlet segment of the left vertical boundary. We assume that the flux at the right vertical boundary is horizontal (which is the variant of a "soft" non-reflecting outlet boundary condition). The corresponding boundary conditions for pressure are formulated as follows:

$$x = 0 : \; \frac{\partial p}{\partial x} = -\frac{12\mu_m}{G(\phi)w^2}, \tag{6}$$

$$x = 1 : \; \frac{\partial p}{\partial y} = -\mathrm{Bu}\rho_m(1, y) \Leftrightarrow p(1, y) = -\mathrm{Bu}\int_0^y \rho_m(1, s)\mathrm{d}s, \tag{7}$$

$$y = 0 \text{ and } y = h : \; \frac{\partial p}{\partial y} = 0. \tag{8}$$

Here, $h = H/L$ is the channel height-to-length ratio, and we assumed that the velocity at the inlet is constant, so that in a dimensionless form it is unity. Note that for a shorter notation, the boundary conditions (4)–(8) are formulated for the flow configuration when the inlet zone occupies the entire height of the fracture.

3. Numerical Implementation and Validation

The governing equations (1)–(3) coupled with the boundary and initial conditions (4)–(8) are solved numerically using the finite-difference method and a rectangular staggered grid. The advection equations are solved using the second-order TVD flux-limiting scheme, while the quasi-linear elliptic pressure equation is solved using the iterative process with the multigrid solver applied to the solution of a linearized equation. The details of numerical algorithm and thorough validation of the model (1)–(3) are presented in [12]. In particular, the model was validated against the following experiments made in Hele-Shaw cells: (i) gravitational slumping of a heavy oil in a confined cell; (ii) Saffman-Taylor instability during the displacement of a water-glycerin solution by water; (iii) set of experiments with the channeling of Newtonian and power-law fluids through the cell filled initially with a yield-stress fluid. The simulations presented below are carried out using the 513×513 mesh.

4. Results and Discussion

A parametric study of the interaction between viscoplastic and viscous fluids in a Hele-Shaw cell is conducted using the research code based on the model and its implementation described above. It is found that when a yield-stress fluid is displaced by a high-viscosity Newtonian fluid (so that there is no Saffman-Taylor instability at the interface), the viscoplastic fluid behaves very similar to the viscous fluid (Figure 1*a*,*b*). The fingers of a low-viscosity fluid penetrating through the viscoplastic fluid trigger the development of unyielded zones (Figure 1*c*,*d*).

IOP Conf. Series: Journal of Physics: Conf. Series **894** (2017) 012014 doi:10.1088/1742-6596/894/1/012014

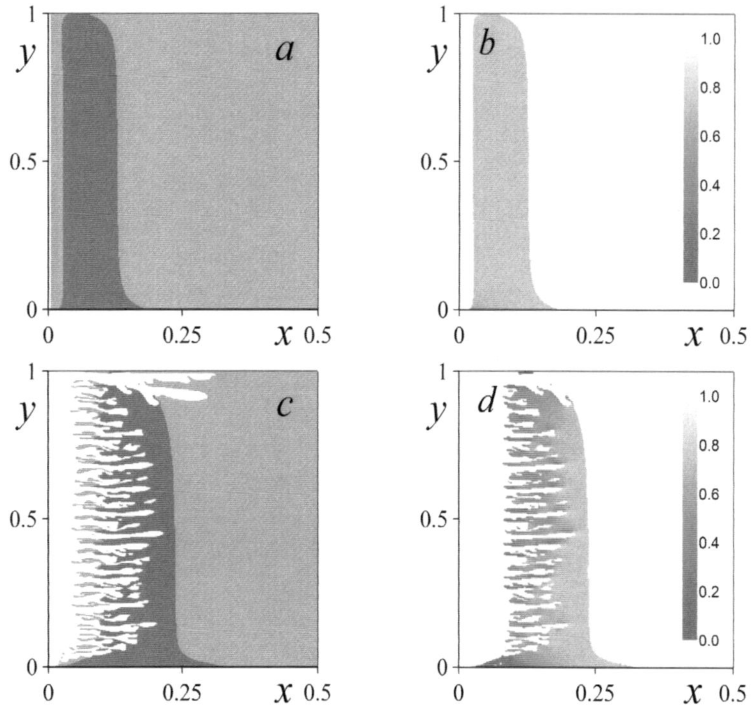

Figure 1. Distribution of fluids (a, c) and parameter G (3) describing the unyielding of the viscoplastic Fluid 0 (b, d) in a plane channel during a certain injection sequence. Viscoplastic Fluid 0 is red, high-viscosity Fluid 1 is blue and low-viscosity Fluid 2 is white. The inlet velocity is $4.76 \cdot 10^{-2} m/s$, $\xi_1 = 1.43$, $\xi_2 = 8.72 \cdot 10^{-4}$, $\zeta_1 = \zeta_2 = 0.5$, $\mathrm{Bu} = 12.9$, $\mathrm{Bn} = 1.43$, $t = 0.122$ (a, b) and $t = 0.235$(c, d).

The effect of Bingham number on the Saffman-Taylor instability between the yield-stress fluid and the viscous fluid is studied (see Figure 2). It is found that an increase in the Bingham number intensifies the finger shadowing effect: the growth rate of small fingers is damped, while the longer fingers grows faster and the total number of fingers is decreased. As a result, the finger penetration length is increased with an increase in the Bingham number.

A multistage hydraulic fracturing job in shales (typically in the U.S. Land) is usually followed by an injection of a small portion of particle-free low-viscosity fluid to clean up the well from the proppant. This stage is called an overflush. During this process, a portion of the clean fluid enters the hydraulic fracture and displaces the particle-laden suspension away from the perforations. Therefore, there is a risk of fracture closure in this proppant-free unsupported cavity, which would result in a dramatic decrease in the fracture conductivity. We carried out a sensitivity study of the shape and the area of the particle-free cavity developed during the displacement of a particle-laden suspension by a clean fluid in the vicinity of fracture inlet zone. We found that there are three qualitatively different displacement regimes, namely: (i) the slumping-dominated scenario Figure 3a ; (ii) an intermediate scenario Figure 3b; and (iii) the fingering-dominated scenario Figure 3c.

The size of the particle-free cavity is determined by the largest semi-circle with the diameter containing the perforation zone and containing only overflush fluid. The slumping-dominated regime provokes the development of a large particle-free zone at the lop of the hydraulic fracture, which would result in a significant decrease in the productive fracture area after closure. Among

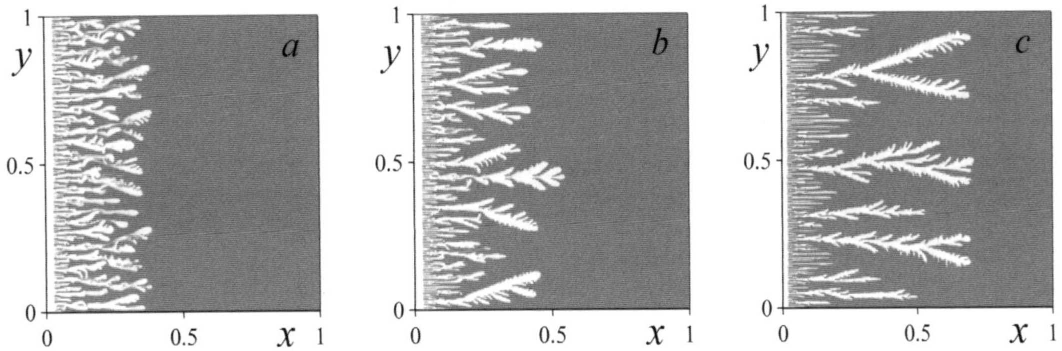

Figure 2. Effect of Bingham number on the fingering of a viscous fluid (white) through the viscoplastic fluid (red) for Bn $= 0.36$ (a), Bn $= 1.43$ (b) and Bn $= 5.72$ (c). The inlet velocity is $4.76 \cdot 10^{-2} m/s$, $\xi = 8.72 \cdot 10^{-4}$, $\zeta = 1$ and $t = 0.203$.

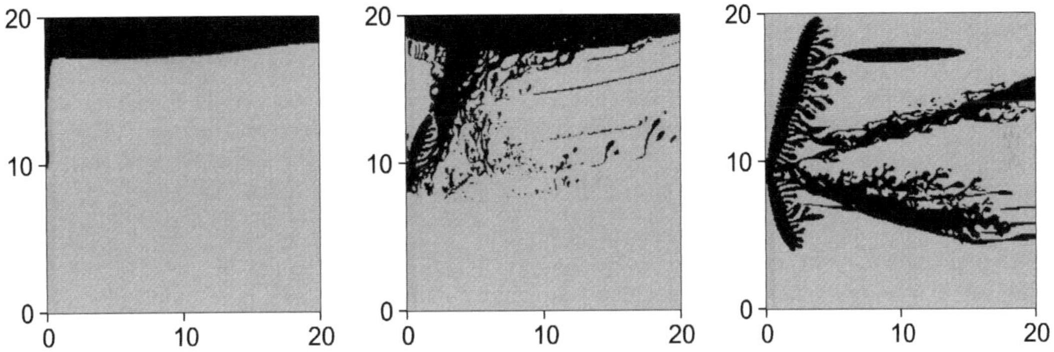

Figure 3. Typical regimes of displacement of a high-viscosity suspension by a low-viscosity fluid in the vicinity of perforations during hydraulic fracture overflush: slumping-dominated (left) at Bu $= 14.5$, Bn $= 9 \cdot 10^{-2}$, $\xi = 0.164$; intermediate (middle) at Bu $= 0.145$, Bn $= 9 \cdot 10^{-4}$, $\xi = 1.64 \cdot 10^{-3}$ and fingering-dominated (right) at Bu $= 0.36 \cdot 10^{-2}$, Bn $= 2.74 \cdot 10^{-4}$, $\xi = 5 \cdot 10^{-4}$.

the other two regimes, the smallest area of the particle-free cavity is found in the case of a finger-dominated regime, which is achieved by an decrease in the overflush-to-fracturing fluid viscosity ration, or an increase in the suspension yield stress (see Figure 4). Also numerical simulations showed that the size of particle-free cavity in the vicinity of perforations is not sensitive to flow parameters as soon as the fingering-dominated overflush regime is maintained. Therefore, both in terms of the operation control and the area of the particle-free zone, fingering-dominated overflush regime is preferable.

5. Conclusions

In the framework of a lubrication approximation, the displacement of a viscoplastic fluid by viscous fluids in a Hele-Shaw cell is studied. It is found that the yield-stress rheology of the fluid results in a modification to the fluid mobility, which is the coefficient of proportionality between the pressure gradient and the width-averaged fluid velocity. In contrast to the flow of Newtonian fluids in narrow channels, the pressure equation describing the flow of viscoplastic fluid is strongly non-linear.

Based on the numerical simulations, we carried out the parametric study of interactions

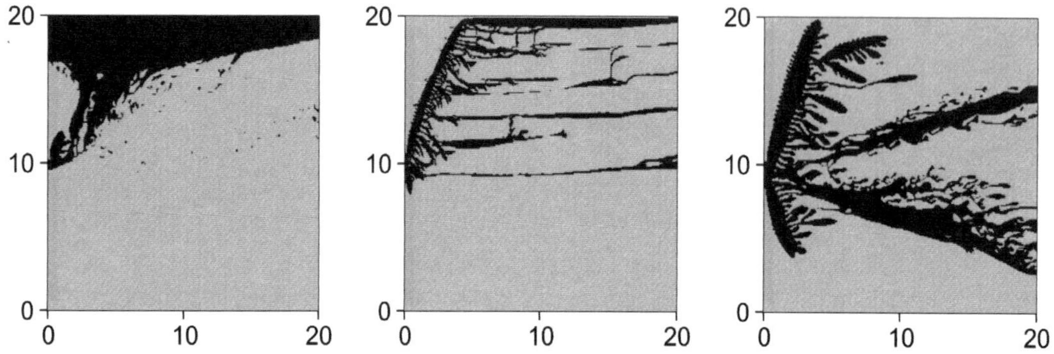

Figure 4. Change between the intermediate overflush regime (left, Bu = 0.36, Bn = 0, $\xi = 5 \cdot 10^{-3}$) and the fingering-dominated overflush regime by an increase in yield stress (middle, Bn = $2.74 \cdot 10^{-2}$) or a decrease in the overflush-to-fracturing fluid viscosity ratio (right, $\xi = 5 \cdot 10^{-4}$).

between viscous and viscoplastic fluids in a Hele-Shaw cell. It is found that when the yield-stress fluid is displaced by a viscous one and the interface is flat, the viscoplastic fluid behaves very similar to a viscous fluid. The instability triggered at the interface between the fluids leads to the development of unyielded zones in the viscoplastic fluid. During the displacement of a yield-stress fluid by a low-viscosity fluid, an increase in the Bingham number leads to an increase in the finger penetration length and a decrease in the total number of fingers.

The overflush stage at the end of a hydraulic fracturing treatment is studied with the aim to minimize the area of a particle-free zone in the vicinity of perforations inside a hydraulic fracture, which mitigates the risk of loosing well-to-fracture hydraulic connection. Numerical simulations of the displacement of a viscous or a viscoplastic fracturing fluid by a low-viscosity overflush fluid in the hydraulic fracture demonstrated that there are three qualitatively different overflush scenarios: (i) slumping-dominated; (ii) intermediate; (iii) fingering-dominated. The smallest area of particle-free zone in the vicinity of perforations is achieved in the fingering scenario, which occurs when either there is a large viscosity contrast between the fracturing and the overflush fluids, large injection rate or the fracturing fluid shows a strong yield-stress behavior.

Acknowledgments
Startup funds of Skolkovo Institute of Science and Technology (Skoltech) are gratefully acknowledged.

References
[1] Adachi J, Siebrits E, Peirce A and Desroches J 2007 *Int. J. Rock Mech. Mining Sci.* **44** 739–57
[2] Economides M and Nolte K 2000 *Reservoir Stimulation* 3rd ed (Wiley)
[3] Lecampion B and Garagash D 2014 *J. Fluid Mech.* **759** 197–235
[4] Pearson J 1994 *J. Non-Newtonian Fluid Mech.* **54** 503–13
[5] Hammond P 1996 *Int. J. Multiphase Flow* **22** 104
[6] Boronin S and Osiptsov A 2010 *Dokl. Phys.* **55** 192–202
[7] Bittleston S and Hassager O 1992 *J. Non-Newtonian Fluid Mech.* **42** 19–36
[8] Frigaard I, Howison S and Sobey I 1994 *J. Fluid Mech.* **263** 133–50
[9] Gorodtsov V and Yentov V M 1997 *J. Appl. Math. Mech.* **61** 111–26
[10] Muskat M 1934 *J. Appl. Phys.* **5** 250–64
[11] Homsy G 1987 *J. Non-Newtonian Fluid Mech.* **19** 271–311

[12] Boronin S, Osiptsov A and Desroches J 2015 *Int. J. Multiphase Flow* **76** 47–63

IOP Conf. Series: Journal of Physics: Conf. Series **894** (2017) 012015 doi:10.1088/1742-6596/894/1/012015

Study of control problems for the stationary MHD equations

R V Brizitskii[1,2]

[1] Institute of Applied Mathematics FEB RAS, Vladivostok, Russia
[2] Far Eastern Federal University, Vladivostok, Russia

E-mail: mlnwizard@mail.ru

Abstract. The optimal control problems for the stationary magnetohydrodynamic equations under inhomogeneous mixed boundary conditions for a magnetic field are considered. The role of control in control xs under study is played by normal component of the magnetic field on the part of the boundary. In the capacity of cost fucntionals quadratic tracking–type functionals for a velocity, magnetic field or pressure are taken.

1. Introduction. Statement of the boundary value problem

Great attention has been paid recently to the optimal control problems for the models of magnetohydrodynamics (MHD) for a viscous conducting incompressible fluid (see [1–6]).

In this paper we study control problems for the new MHD model in which stationary equations of magnetohydrodynamic are considered under mixed boundary conditions for a magnetic field and under Dirichlet boundary condition for a velocity with tracking–type functionals. The role of controls in control problems under study is played by a normal component of the magnetic field on the part of the boundary.

One of the features of this work consist in proving of the solvability of optimal control problem under minimum requirements on a normal component of the magnetic field given on the part of boundary and playing the role of control. This result, in contrast to [3], allows to use a simple L^2–norm for the normal component of the magnetic field instead of the $H^{1/2}$–norm on the part of the boundary by Tikhonov regularization of considered incorrect control problem. This regularization is needed to prove stability and uniqueness of the control problem's solution.

Let Ω be a bounded domain of space $\mathbf{R}^3$ with boundary $\partial\Omega$ consisting of two parts Σ_ν and Σ_τ. In this paper we study control problems for the stationary magnetohydrodynamic equations of viscous incompressible fluid

$$-\nu\Delta\mathbf{u} + (\mathbf{u}\cdot\nabla)\mathbf{u} + \nabla p - æ\,\mathrm{rot}\,\mathbf{H}\times\mathbf{H} = \mathbf{f},\quad \mathrm{div}\,\mathbf{u} = 0, \tag{1}$$

$$\nu_1\mathrm{rot}\,\mathbf{H} - \rho_0^{-1}\mathbf{E} + æ\mathbf{H}\times\mathbf{u} = \nu_1\mathbf{j},\quad \mathrm{div}\,\mathbf{H} = 0,\quad \mathrm{rot}\,\mathbf{E} = \mathbf{0}, \tag{2}$$

considering in domain Ω under following inhomogeneous boundary conditions:

$$\mathbf{u}|_{\partial\Omega} = \mathbf{g},\ \mathbf{H}\cdot\mathbf{n}|_{\Sigma_\tau} = q,\ \mathbf{H}\times\mathbf{n}|_{\Sigma_\nu} = \mathbf{0},\ \ \mathbf{E}\times\mathbf{n}|_{\Sigma_\tau} = \mathbf{k}. \tag{3}$$

Here $\mathbf{u}$ is the velocity vector, $\mathbf{H}$ and $\mathbf{E}$ are magnetic and electric fields, respectively, $p = P/\rho_0$, where P is the pressure, $\rho_0 = \mathrm{const}$ is a fluid density, $æ = \mu/\rho_0$, $\nu_1 = 1/\rho_0\sigma = æ\nu_m$, ν and

Content from this work may be used under the terms of the Creative Commons Attribution 3.0 licence. Any further distribution of this work must maintain attribution to the author(s) and the title of the work, journal citation and DOI.
Published under licence by IOP Publishing Ltd

ν_m are constant kinematic and magnetic viscousity coefficients, σ is a constant conductivity, μ is a constant magnetic permeability, $\mathbf{n}$ is the outer normal to $\partial\Omega$, $\mathbf{j}$ is the current density. In the remainder of the paper we will refer to problem (1)–(3) for given functions $\mathbf{f}, \mathbf{j}, \mathbf{g}, \mathbf{k}$ and q as Problem 1. We note that all the quantities in (1)–(3) are dimensional and their physical dimensions are defined in terms of SI units. Physically the boundary conditions for the electromagnetic field in (3) correspond to the situation when the part Σ_ν of the boundary $\partial\Omega$ is a perfect insulator.

For the first time a global solvability of Problem 1 under homogeneous boundary conditions for a velocity and electromagnetic field was proved in [11] without the requirement $\overline{\Sigma}_\tau \cap \overline{\Sigma}_\nu = \emptyset$. It is essential to use mathematical thools from [12–14] A global solvability of the inhomogeneous mixed Problem 1 was proved in [15], see also [16, 17]. The mixed boundary condition for a magentic field in (3) generalizes previously used two types of boundary conditions see [2, 9, 18, 19].

2. Function spaces. The preliminary results

Below we will use the Sobolev spaces $H^s(D)$, $s \in \mathbf{R}$, $H^0(D) \equiv L^2(D)$, where D denotes Ω or the boundary $\partial\Omega$. Corresponding spaces of vector–functions are denoted by $H^s(D)^3$ and $L^2(D)^3$. The inner products and norms in the spaces $H^s(\Omega)$ and $H^s(\Omega)^3$ are denoted by $(\cdot,\cdot)_{s,\Omega}$ and $\|\cdot\|_{s,\Omega}$. The inner products and norms in $L^2(\Omega)$ and $L^2(\Omega)^3$ are denoted by $(\cdot,\cdot)$ and $\|\cdot\|$. By $\|\cdot\|_{1,\Omega}$ and $|\cdot|_{1,\Omega}$ we denote norm and seminorm in $H^1(\Omega)$ or $H^1(\Omega)^3$. For arbitrary Hilbert space H by H^* we denote the dual space of H. As in [13, 14] we assume that the following conditions to Ω are satisfied:

(i) Ω is a bounded domain in $\mathbf{R}^3$ and the boundary $\partial\Omega$ is the union of a finite number of disjoint closed C^2 surfaces, each surface having finite surface area;

(ii) Σ_τ is nonempty open subset of $\partial\Omega$ with $M+1$ disjoint nonempty open components $\{\sigma_0, \sigma_1, ..., \sigma_M\}$ and there is a positive d_0 such that dist $d(\sigma_i, \sigma_j) \geq d_0 > 0$ when $i \neq j$ and $M \geq 1$. The boundary of each σ_i is either empty or $C^{1,1}$ curve. We set $\Sigma_\nu = \partial\Omega \setminus \overline{\Sigma}_\tau$.

Let $\mathcal{D}(\Omega)$ be the space of infinitely differentiable compactly supported functions in Ω, $H_0^1(\Omega)$ be the closure of $\mathcal{D}(\Omega)$ in $H^1(\Omega)$, $V = \{\mathbf{v} \in H_0^1(\Omega)^3 : \operatorname{div}\mathbf{v} = 0\}$, $H^{-1}(\Omega)^3 = (H_0^1(\Omega)^3)^*$, $L_0^2(\Omega) = \{p \in L^2(\Omega) : (p, 1) = 0\}$, $H^1(\Omega, \Sigma_\tau) = \{\varphi \in H^1(\Omega) : \varphi|_{\Sigma_\tau} = 0\}$, $C_{\Sigma_\tau 0}(\overline{\Omega})^3 := \{\mathbf{v} \in C^0(\overline{\Omega})^3 : \mathbf{v}\cdot\mathbf{n}|_{\Sigma_\tau} = 0,\ \mathbf{v}\times\mathbf{n}|_{\Sigma_\nu} = \mathbf{0}\}$. In addition to the spaces introduced above we will use the spaces $H(\operatorname{div},\Omega) = \{\mathbf{v} \in L^2(\Omega)^3 : \operatorname{div}\mathbf{v} \in L^2(\Omega)\}$, $H(\operatorname{curl},\Omega) = \{\mathbf{v} \in L^2(\Omega)^3 : \operatorname{curl}\mathbf{v} \in L^2(\Omega)^3\}$, $H^0(\operatorname{curl},\Omega) = \{\mathbf{v} \in H(\operatorname{curl},\Omega) : \operatorname{curl}\mathbf{v} = \mathbf{0}\}$ and the space $H_{DC}(\Omega) = H(\operatorname{div},\Omega) \cap H(\operatorname{curl},\Omega)$, equipped with the Hilbert norm $\|\mathbf{u}\|_{DC}^2 := \|\mathbf{u}\|_\Omega^2 + \|\operatorname{div}\mathbf{u}\|_\Omega^2 + \|\operatorname{curl}\mathbf{u}\|_\Omega^2$.

We will use the following Green's formulae [7]

$$\int_\Omega \mathbf{v}\cdot\operatorname{grad}\varphi\, dx + \int_\Omega \operatorname{div}\mathbf{v}\,\varphi\, dx = \int_{\partial\Omega} \mathbf{v}\cdot\mathbf{n}\,\varphi\, d\sigma \quad \forall\mathbf{v} \in H^1(\Omega)^3,\ \varphi \in H^1(\Omega), \tag{4}$$

$$\int_\Omega (\mathbf{v}\cdot\operatorname{curl}\mathbf{w} - \mathbf{w}\cdot\operatorname{curl}\mathbf{v})dx = \int_{\partial\Omega} (\mathbf{v}\times\mathbf{n})\cdot\mathbf{w}_T d\sigma \quad \forall\mathbf{v},\mathbf{w} \in H^1(\Omega)^3. \tag{5}$$

Let $H^1_{\Sigma_\tau 0}(\Omega)^3$ be the closure of the space $C_{\Sigma_\tau 0}(\overline{\Omega})^3 \cap H^1(\Omega)^3$ with respect to the norm $\|\cdot\|_{1,\Omega}$ and $H_{DC\Sigma_\tau}(\Omega)$ be the closure of $C_{\Sigma_\tau 0}(\overline{\Omega})^3 \cap H^1(\Omega)^3$ with respect to the norm $\|\cdot\|_{DC}$. Let

$$\mathcal{H}_{\Sigma_\tau}(\Omega) = \{\mathbf{h} \in L^2(\Omega)^3 : \operatorname{div}\mathbf{h} = 0,\ \operatorname{curl}\mathbf{h} = \mathbf{0} \text{ in } \Omega,\ \mathbf{h}\cdot\mathbf{n}|_{\Sigma_\tau} = 0,\ \mathbf{h}\times\mathbf{n}|_{\Sigma_\nu} = \mathbf{0}\},$$

$$\mathcal{H}_{\Sigma_\nu}(\Omega) = \{\mathbf{h} \in L^2(\Omega)^3 : \operatorname{div}\mathbf{h} = 0,\ \operatorname{curl}\mathbf{h} = \mathbf{0} \text{ in } \Omega,\ \mathbf{h}\cdot\mathbf{n}|_{\Sigma_\nu} = 0,\ \mathbf{h}\times\mathbf{n}|_{\Sigma_\tau} = \mathbf{0}\},$$

$$V_{\Sigma_\tau}(\Omega) = \{\mathbf{v} \in H_{DC\Sigma_\tau}(\Omega) : \operatorname{div}\mathbf{v} = 0 \text{ in } \Omega\} \cap \mathcal{H}_{\Sigma_\tau}(\Omega)^\perp. \tag{6}$$

A number of properties of the function spaces introduced above has been proved in [13, 14]. We formulate these properties as the following theorem.

Theorem 2.1. *We assume that conditions (i), (ii) hold. Then:*
1) the spaces $\mathcal{H}_{\Sigma_\tau}(\Omega)$ *and* $\mathcal{H}_{\Sigma_\nu}(\Omega)$ *are finite dimensional;*
2) $H_{DC\Sigma_\tau}(\Omega) = H^1_{\Sigma_\tau 0}(\Omega)^3$ *and the norm* $\|\cdot\|_{DC}$ *is equivalent to the norm* $\|\cdot\|_{1,\Omega}$ *;*
3) there is a constant δ_1 *dependent on* Ω *and* Σ_τ *such that the coercitivity inequality holds*

$$\|\mathrm{curl}\,\mathbf{h}\|^2 \geq \delta_1 \|\mathbf{h}\|^2_{1,\Omega} \quad \forall \mathbf{h} \in V_{\Sigma_\tau}(\Omega); \tag{7}$$

4) the following orthogonal decomposition holds:

$$L^2(\Omega)^3 = \nabla H^1(\Omega, \Sigma_\tau) \oplus \mathrm{curl}\, H_{DC\Sigma_\tau}(\Omega) \oplus \mathcal{H}_{\Sigma_\nu}(\Omega) \tag{8}$$

and $\mathrm{curl}\, H_{DC\Sigma_\tau}(\Omega) \equiv \mathrm{curl}\, V_{\Sigma_\tau}(\Omega)$.

Along with spaces $H_{DC}(\Omega)$ and $H^0(\mathrm{curl}, \Omega)$ we will use their subspaces

$$\mathcal{H}^{1/2}_{\mathrm{div}}(\Omega) := H^{1/2}(\Omega)^3 \cap \{\mathbf{h} \in H_{DC}(\Omega) : \mathrm{div}\mathbf{h} = 0 \text{ in } \Omega,\ \mathbf{h} \times \mathbf{n} = \mathbf{0} \text{ on } \Sigma_\nu\} \cap \mathcal{H}_{\Sigma_\tau}(\Omega)^\perp,$$

$$H^0_{\Sigma_\tau}(\mathrm{curl}, \Omega) := \{\mathbf{e} \in H^0(\mathrm{curl}, \Omega) : \mathbf{e} \times \mathbf{n}|_{\Sigma_\tau} \in L^2_T(\Sigma_\tau)\}$$

equipped natural norms $\|\mathbf{h}\|_{\mathcal{H}^{1/2}_{\mathrm{div}}(\Omega)} = l\|\mathbf{h}\|_{1/2,\Omega} + \|\mathrm{curl}\,\mathbf{h}\|_\Omega$, $\|\mathbf{e}\|_{H^0_{\Sigma_\tau}(\mathrm{curl},\Omega)} := \|\mathbf{e}\|_\Omega + \|\mathbf{e} \times \mathbf{n}\|_{\Sigma_\tau}$.

Here l is a demensional factor of the dimension $[l] = L_0$ and its value is equal to 1, L_0 denote the SI dimensions of the length. The spaces $\mathcal{H}^{1/2}_{\mathrm{div}}(\Omega)$ and $H^0_{\Sigma_\tau}(\mathrm{curl}, \Omega)$ will be used below for describing the magnetic and electric field, respectively.

Let $H^1_{\mathrm{div}}(\Omega) := \{\mathbf{v} \in H^1_T(\Omega) : \mathrm{div}\,\mathbf{v} = 0\}$. Along with Theorem 2.1 we will use also a number of properties of bilinear and trilinear forms related with linear and nonlinear terms in equations (1), (2). We formulate them as the following Lemma (see for details [13, 14]).

Lemma 2.1. *Under condition (i) there exist constants* $\delta_i = \delta_i(\Omega) > 0$ *and* $\gamma_i = \gamma_i(\Omega) > 0$, $i = 0, 1$, *depending on* Ω *such that*

$$(\nabla\mathbf{v}, \nabla\mathbf{v}) \geq \delta_0 \|\mathbf{v}\|^2_{1,\Omega} \ \ \forall \mathbf{v} \in H^1_0(\Omega)^3, \ (\mathrm{rot}\,\mathbf{\Psi}, \mathrm{rot}\,\mathbf{\Psi}) \geq \delta_1 \|\mathbf{\Psi}\|^2_{1,\Omega}\ \forall \mathbf{\Psi} \in V_{\Sigma_\tau}(\Omega),$$

$$|((\mathbf{u}\cdot\nabla)\mathbf{v}, \mathbf{w})| \leq \gamma_0 \|\mathbf{u}\|_{1,\Omega}\|\mathbf{v}\|_{1,\Omega}\|\mathbf{w}\|_{1,\Omega},\ |(\mathrm{rot}\mathbf{u} \times \mathbf{v}, \mathbf{w})| \leq \gamma_1 \|\mathbf{u}\|_{1,\Omega}\|\mathbf{v}\|_{1,\Omega}\|\mathbf{w}\|_{1,\Omega}\ \forall \mathbf{u}, \mathbf{v}, \mathbf{w} \in H^1(\Omega)^3,$$

$$|(\mathrm{rot}\mathbf{u} \times \mathbf{v}, \mathbf{w})| \leq \gamma_1 \|\mathbf{u}\|_{\mathcal{H}^{1/2}_{\mathrm{div}}(\Omega)} \|\mathbf{v}\|_{\mathcal{H}^{1/2}_{\mathrm{div}}(\Omega)} \|\mathbf{w}\|_{1,\Omega}\ \forall \mathbf{u}, \mathbf{w} \in \mathcal{H}^{1/2}_{\mathrm{div}}(\Omega), \mathbf{v} \in H^1(\Omega)^3. \tag{9}$$

Let the following conditions hold in addition to (i), (ii):
(iii) $\mathbf{f} \in H^{-1}(\Omega)^3$, $\mathbf{j} \in L^2(\Omega)^3$; (iv) $\mathbf{g} \in H^{1/2}_T(\partial\Omega)$, $q \in L^2(\Sigma_\tau)$, $\mathbf{k} \in (\gamma_\tau|_{\Sigma_\tau}) H^0_{\Sigma_\tau}(\mathrm{curl}, \Omega)$.

As usual, while studying control problems for MHD system we will deal with weak form of Problem 1. It consists of finding a triple $(\mathbf{u}, \mathbf{H}, p) \in H^1_{\mathrm{div}}(\Omega) \times \mathcal{H}^{1/2}_{\mathrm{div}}(\Omega) \times L^2_0(\Omega)$ satisfying

$$\nu(\nabla\mathbf{u}, \nabla\mathbf{v}) + \nu_1(\mathrm{rot}\mathbf{H}, \mathrm{rot}\mathbf{\Psi}) + ((\mathbf{u}\cdot\nabla)\mathbf{u}, \mathbf{v}) + æ[(\mathrm{rot}\mathbf{\Psi} \times \mathbf{H}, \mathbf{u}) - (\mathrm{rot}\mathbf{H} \times \mathbf{H}, \mathbf{v})] -$$

$$-(\mathrm{div}\mathbf{v}, p) = \langle \mathbf{f}, \mathbf{v}\rangle + (\nu_1 \mathbf{j}, \mathrm{rot}\,\mathbf{\Psi}) + \rho_0^{-1}(\mathbf{k}, \mathbf{\Psi})_{\Sigma_\tau}\ \forall (\mathbf{v}, \mathbf{\Psi}) \in H^1_0(\Omega)^3 \times V_{\Sigma_\tau}(\Omega), \tag{10}$$

$$\mathrm{div}\mathbf{u} = 0 \text{ in } \Omega,\ \ \mathbf{u} = \mathbf{g} \text{ on } \Sigma,\ \ \mathbf{H}\cdot\mathbf{n} = q \text{ on } \Sigma_\tau. \tag{11}$$

In order to obtain (10) one should multiply the first equation in (1) by a function $\mathbf{v} \in H^1_0(\Omega)^3$, the first equation in (2) by rot $\mathbf{\Psi}$ where $\mathbf{\Psi} \in V_{\Sigma_\tau}(\Omega)$, integrate over Ω, apply Green's formulas, add the obtained results and make use of the identity

$$(\mathbf{E}, \mathrm{curl}\,\mathbf{\Psi}) = \int_{\Sigma_\tau} (\mathbf{E} \times \mathbf{n}|_{\Sigma_\tau}) \cdot \mathbf{\Psi}_T d\sigma = (\mathbf{k}, \mathbf{\Psi}_T)_{\Sigma_\tau} = (\mathbf{E}_0, \mathrm{curl}\,\mathbf{\Psi})\ \forall \mathbf{\Psi} \in V_{\Sigma_T(\Omega)}. \tag{12}$$

The identity (10) does not contain electric field $\mathbf{E} \in H^0_{\Sigma_\tau}(\mathrm{curl}, \Omega)$ which was eliminated with the help of (12). However, using a condition on a boundary vector $\mathbf{k}$ in (iii) vector $\mathbf{E}$ can be reconstructed uniquely from triple $(\mathbf{u}, \mathbf{H}, p) \in H^1_{\text{div}}(\Omega) \times \mathcal{H}^{1/2}_{\text{div}}(\Omega) \times L^2_0(\Omega)$ satisfying (10) so that the first equation in (2) holds a.e. in Ω (see details in [15]). This allows us to refer below to mentioned triple $(\mathbf{u}, \mathbf{H}, p)$ as a weak solution to Problem 1.

Let in addition to (i)–(iv) the following condition holds

(v) $\overline{\Sigma}_\tau \cap \overline{\Sigma}_\nu = \emptyset$.

The following theorem from [15] provides the existence of a weak solution to Problem 1:

Theorem 2.2. *Under assumptions (i)–(v) there exists a weak solution $(\mathbf{u}, \mathbf{H}, p)$ to Problem 1 and the following estimates hold:*

$$\|\mathbf{u}\|_{1,\Omega} \le M_{\mathbf{u}}, \ \|\mathbf{H}\|_{\mathcal{H}^{1/2}_{\text{div}}(\Omega)} \le M_{\mathbf{H}}, \ \|p\|_\Omega \le M_p. \tag{13}$$

Here $M_{\mathbf{u}}, M_{\mathbf{H}}, M_p$ are continuous nondecreasing functions of $\|\mathbf{f}\|_{-1,\Omega}$, $\|\mathbf{j}\|_\Omega$, $\|\mathbf{k}\|_{\Sigma_\tau}$, $\|\mathbf{g}\|_{1/2,\Sigma}$, $\|q\|_{\Sigma_\tau}$.

To prove the Problem 1's local uniqueness we introduce the fucntional space

$$\tilde{V}_{\Sigma_\tau}(\Omega) = \{\mathbf{v} \in H(\text{rot}, \text{div}; \Omega) : \text{div}\,\mathbf{v} = 0, \mathbf{v}\cdot\mathbf{n}|_{\Sigma_\tau} = 0, \ \mathbf{v}\times\mathbf{n}|_{\Sigma_\nu} = \mathbf{0}\} \cap \mathcal{H}_{\Sigma_\tau}(\Omega)^\perp.$$

It's obvious that in a common case $V_{\Sigma_\tau}(\Omega) \subset \tilde{V}_{\Sigma_\tau}(\Omega)$.

Lemma 2.2. *Let conditions (i), (ii), (v) hold and function $\mathbf{f} \in L^2(\Omega)^3$ satisfies equality $(\mathbf{f}, \text{rot}\,\mathbf{v}) = 0$ for all $\mathbf{v} \in V_{\Sigma_\tau}(\Omega)$. Then $(\mathbf{f}, \text{rot}\,\mathbf{h}) = 0$ for all $\mathbf{h} \in \tilde{V}_{\Sigma_\tau}(\Omega)$.*

Proof. From relation $(\mathbf{f}, \text{rot}\,\mathbf{v}) = 0$ for all $\mathbf{v} \in V_{\Sigma_\tau}(\Omega)$ on the strength (8) results $\mathbf{f} = \nabla\varphi + \mathbf{e}$ a.e. in Ω, where $\varphi \in \nabla H^1(\Omega, \Sigma_\tau)$, $\mathbf{e} \in \mathcal{H}_{\Sigma_\nu}(\Omega)$.

Under condition (v) the embeding $\tilde{V}_{\Sigma_\tau}(\Omega) \subset H^1(\Omega)^3$ is true (see [20, p. 174]) which gives an opportunity to use Green's formula (5) to the inner product $(\mathbf{e}, \text{rot}\,\mathbf{\Psi})$:

$$(\mathbf{e}, \text{rot}\,\mathbf{\Psi}) = (\mathbf{e}\times\mathbf{n}, \mathbf{\Psi}_T)_\Sigma = (\mathbf{e}\times\mathbf{n}, \mathbf{\Psi}_T)_{\Sigma_\tau} + (\mathbf{e}\times\mathbf{n}, \mathbf{\Psi}_T)_{\Sigma_\nu} = 0 \ \ \forall\mathbf{\Psi} \in \tilde{V}_{\Sigma_\tau}(\Omega).$$

Let us introduce the space $C^\infty_{0,\Sigma_\tau}(\overline{\Omega}) = \{h \in C^\infty(\overline{\Omega}) : h = 0 \text{ on } \Sigma_\tau\}$. It's clear that $C^\infty_{0,\Sigma_\tau}(\overline{\Omega})$ embedded densely in $H^1(\Omega, \Sigma_\tau)$ by norm $\|\cdot\|_{1,\Omega}$ at condition (v). Let φ_m be a sequence of functions from the space $C^\infty_{0,\Sigma_\tau}(\overline{\Omega})$, converging to $\varphi \in H^1(\Omega, \Sigma_\tau)$ by the norm $\|\cdot\|_{1,\Omega}$.

Let us apply the Green's formula (5) to the inner product $(\nabla\varphi_m, \text{rot}\,\mathbf{\Psi})$:

$$(\nabla\varphi_m, \text{rot}\,\mathbf{\Psi}) = (\nabla\varphi_m\times\mathbf{n}, \mathbf{\Psi}_T)_\Sigma = (\nabla\varphi_m\times\mathbf{n}, \mathbf{\Psi}_T)_{\Sigma_\tau} + (\nabla\varphi_m\times\mathbf{n}, \mathbf{\Psi}_T)_{\Sigma_\nu} = 0 \ \ \forall\mathbf{\Psi} \in \tilde{V}_{\Sigma_\tau}(\Omega).$$

It's not comlicated to show further that

$$||(\nabla\varphi_m, \text{rot}\,\mathbf{\Psi})| - |(\nabla\varphi, \text{rot}\,\mathbf{\Psi})|| \le \|\varphi_m - \varphi\|_{1,\Omega}\|\mathbf{\Psi}\|_{1,\Omega} \to 0 \text{ at } m \to \infty \ \forall\mathbf{\Psi} \in \tilde{V}_{\Sigma_\tau}(\Omega). \ \blacksquare$$

From lemma 2.2 follows that the Problem 1's weak formulation is also true for all functions $(\mathbf{v}, \mathbf{\Psi}) \in V \times \tilde{V}_{\Sigma_\tau}(\Omega)$, but not only for functions $(\mathbf{v}, \mathbf{\Psi}) \in V \times V_{\Sigma_\tau}(\Omega)$.

From the corollary 7.4 in [12, p. 981] the coercive inequality follows $\|\text{rot}\,\mathbf{h}\|_\Omega \ge \alpha\|\mathbf{h}\|_\Omega$ for all $\mathbf{h} \in \tilde{V}_{\Sigma_\tau}(\Omega)$, according to which $\|\cdot\|_{\tilde{V}_{\Sigma_\tau}(\Omega)} \equiv \|\text{rot}\,\cdot\|_\Omega$.

As at (v) the embeding takes place $\tilde{V}_{\Sigma_\tau}(\Omega) \subset H^1(\Omega)^3$, then the inequality, illustrating the equivalence of two norms on the space $\tilde{V}_{\Sigma_\tau}(\Omega)$, is met:

$$\|\text{rot}\,\mathbf{h}\|_\Omega \ge \delta_1'\|\mathbf{h}\|_{1,\Omega} \ \ \forall\mathbf{h} \in \tilde{V}_{\Sigma_\tau}(\Omega). \tag{14}$$

Then from Theorem 2.2 and Lemma 2.2 follows

Theorem 2.3. *Under assumptions (i)–(v) there exists a weak solution $(\mathbf{u}, \mathbf{H}, p)$ of Problem 1 and a priori estimate (13) takes place. If, besides, functions $\mathbf{f}, \mathbf{j}, \mathbf{k}$, $\mathbf{g}$ and q are small (or "viscosity coefficients" ν, ν_m are large) in the sense*

$$\gamma_0 M_{\mathbf{u}} + \gamma_1(\sqrt{æ}/2)M_{\mathbf{H}} < \delta_0\nu, \ \ \gamma_1 M_{\mathbf{u}} + \gamma_1(\sqrt{æ}/2)M_{\mathbf{H}} < \delta_1'\nu_m, \tag{15}$$

where constants $\delta_0, \gamma_0, \gamma_1$ are introduced in Lemma 2.1, constant δ_1' is from (14), then the weak solution is unique.

3. Statement and solvability of control problem

The control problem consists in minimization of certain functionals depending on the state $(\mathbf{u}, \mathbf{H}, p)$ and other unknown functions (controls) satisfying the state equations (1)–(3).

As the cost functional we choose one of the following:

$$I_1(\mathbf{v}) = \|\mathbf{v} - \mathbf{v}_d\|_Q^2,\ I_2(\mathbf{H}) = \|\mathbf{H} - \mathbf{H}_d\|_Q^2,\ I_3(\mathbf{H}) = \|\mathbf{H} - \mathbf{H}_d\|^2_{\mathcal{H}^{1/2}_{\rm div}(Q)},\ I_4(p) = \|p - p_d\|_Q^2. \quad (16)$$

Here the function $\mathbf{v}_d \in L^2(Q)^3$ denotes some desired velocity field given in a subdomain $Q \subset \Omega$. Functions $\mathbf{H}_d \in L^2(Q)^3$ (or $\mathbf{H}_d \in \mathcal{H}^{1/2}_{\rm div}(Q)$) and $p_d \in L^2(Q)$ have similar sense for the magnetic field or pressure.

As controls we choose the boundary function q and assume that q is changed over set K satisfying the condition

(j) $K \subset L^2(\Sigma_\tau)$ is nonempty convex closed set.

Setting $\mathbf{x} = (\mathbf{u}, \mathbf{H}, p) \in X = H^1_{\rm div}(\Omega) \times \mathcal{H}^{1/2}_{\rm div}(\Omega) \times L^2_0(\Omega)$, $Y = H^{-1}(\Omega)^3 \times V_{\Sigma_\tau}(\Omega)^* \times L^2_0(\Omega) \times H_T^{1/2}(\Sigma) \times L^2(\Sigma_\tau)$, we introduce an operator $F \equiv (F_1, F_2, F_3, F_4) : X \times K \times H_T^{1/2}(\Sigma) \to Y$ by

$$\langle F_1(\mathbf{x}), (\mathbf{v}, \mathbf{\Psi})\rangle = \nu(\nabla\mathbf{u}, \nabla\mathbf{v}) + \nu_1({\rm rot}\mathbf{H}, {\rm rot}\mathbf{\Psi}) + ((\mathbf{u}\cdot\nabla)\mathbf{u}, \mathbf{v}) - ({\rm div}\mathbf{v}, p) +$$

$$+ æ[({\rm rot}\mathbf{\Psi}\times\mathbf{H}, \mathbf{u}) - ({\rm rot}\mathbf{H}\times\mathbf{H}, \mathbf{v})] - \langle\mathbf{f}, \mathbf{v}\rangle - (\nu_1\mathbf{j}, {\rm rot}\,\mathbf{\Psi}) - \rho_0^{-1}(\mathbf{k}, \mathbf{\Psi}_T)_{\Sigma_\tau}\ \forall(\mathbf{v}, \mathbf{\Psi}) \in H^1_0(\Omega)\times V_{\Sigma_\tau}(\Omega),$$

$\langle F_2(\mathbf{x}, r)\rangle = -({\rm div}\mathbf{u}, r)\ \forall r \in L^2_0(\Omega)$, $F_3(\mathbf{x}, \mathbf{g}) = \mathbf{u}|_\Gamma - \mathbf{g} \in H_T^{1/2}(\Sigma)$, $F_4(\mathbf{x}, q) = \mathbf{H}\cdot\mathbf{n} - q \in L^2(\Sigma_\tau)$.

and rewrite the weak form (10), (11) of Problem 1 in the form of the operator equation

$$F(\mathbf{x}, q) = F(\mathbf{u}, \mathbf{H}, p, q) = 0. \quad (17)$$

Let $I : X \to \mathbf{R}$ be a weakly lower semicontinuous cost functional. Consider the following optimal control problem:

$$J(\mathbf{x}, q) = (\mu_0/2)I(\mathbf{x}) + (\mu_1/2)\|q\|^2_{\Sigma_\tau} \to \inf,\ F(\mathbf{x}, q) = 0,\ (\mathbf{x}, q) \in X \times K. \quad (18)$$

Here $\mu_0 > 0$ and $\mu_1 \geq 0$ are nonnegative parameters.

Theorem 3.1. *Let under assumptions (i), (ii) and (j), $\mu_0 > 0$, $\mu_1 > 0$ or $\mu_0 > 0$, $\mu_1 \geq 0$ and K be the bounded set. Then problem (18) has at least one solution for $I = I_k$, $k = \overline{1,4}$.*

Proof. Let us denote by $(\mathbf{x}_m, q_m) \in Z_{ad}$, $m \in \mathbf{N}$ a minimizing sequence, for which $\lim_{m\to\infty} J(\mathbf{x}_m, q_m) = \inf_{(\mathbf{x},q)\in Z_{ad}} J(\mathbf{x}, q) \equiv J^*$.

On the strength of Theorem 3.1, the following estimates hold for the control q_m and corresponding components of Problem 1's solution: $\|q_m\|_{\Sigma_\tau} \leq c_1$, $\|\mathbf{u}_m\|_{1,\Omega} \leq c_2$, $\|\mathbf{H}_m\|_{\mathcal{H}^{1/2}_{\rm div}(\Omega)} \leq c_3$, $\|p_m\|_\Omega \leq c_4$, where $c_1, c_2, \ldots$ are some constants, not depending on m. From given estimetes follows that there are weak limits $q^* \in L^2(\Sigma_\tau)$, $\mathbf{u}^* \in H^1_{\rm div}(\Omega)$, $\mathbf{H}^* \in \mathcal{H}^{1/2}_{\rm div}(\Omega)$, $p^* \in L^2_0(\Omega)$ of some subsequences of sequences $\{q_m\}$, $\{\mathbf{u}_m\}$, $\{\mathbf{H}_m\}$, $\{p_m\}$, it can be considered that at $m \to \infty$

$$q_m \to q^* \text{ weakly in } L^2(\Sigma_\tau),\ \mathbf{u}_m \to \mathbf{u}^* \text{ weakly in } H^1(\Omega)^3 \text{ and strongly in } L^4(\Omega)^3,$$

$$p_m \to p^* \text{ weakly in } L^2(\Omega),\ \mathbf{H}_m \to \mathbf{H}^* \text{ weakly in } \mathcal{H}^{1/2}_{\rm div}(\Omega) \text{ and strongly in } L^2(\Omega)^3. \quad (19)$$

It is clear that $q^* \in K$, $F_2(\mathbf{x}^*, q^*) = 0$, $F_3(\mathbf{x}^*, q^*) = \mathbf{0}$, $F_4(\mathbf{x}^*, q^*) = 0$. Let us show that $F_1(\mathbf{x}^*, q^*) = 0$, i.e. that

$$\nu(\nabla\mathbf{u}^*, \nabla\mathbf{v}) + \nu_1({\rm rot}\,\mathbf{H}^*, {\rm rot}\,\mathbf{\Psi}) + ((\mathbf{u}^*\cdot\nabla)\mathbf{u}^*, \mathbf{v}) - ({\rm div}\,\mathbf{v}, p^*) + æ[({\rm rot}\,\mathbf{\Psi}\times\mathbf{H}^*, \mathbf{u}^*) - ({\rm rot}\,\mathbf{H}^*\times\mathbf{H}^*, \mathbf{v})] =$$

$$= \langle\mathbf{f}, \mathbf{v}\rangle + (\mathbf{j}, {\rm rot}\,\mathbf{\Psi}) + (\mathbf{k}, \mathbf{\Psi}_T)_{\Sigma_\tau}\ \forall(\mathbf{v}, \mathbf{\Psi}) \in H^1_0(\Omega)^3 \times V_{\Sigma_\tau}(\Omega). \quad (20)$$

For this purpose let us note that $\mathbf{u}_m, \mathbf{H}_m$, p_m satisfy the relation

$$\nu(\nabla\mathbf{u}_m,\nabla\mathbf{v})+\nu_1(\operatorname{rot}\mathbf{H}_m,\operatorname{rot}\mathbf{\Psi})+((\mathbf{u}_m\cdot\nabla)\mathbf{u}_m,\mathbf{v})+æ[(\operatorname{rot}\mathbf{\Psi}\times\mathbf{H}_m,\mathbf{u}_m)-(\operatorname{rot}\mathbf{H}_m\times\mathbf{H}_m,\mathbf{v})]=$$

$$-(\operatorname{div}\mathbf{v},p_m)=\langle\mathbf{f},\mathbf{v}\rangle+(\mathbf{j},\operatorname{rot}\mathbf{\Psi})+(\mathbf{k},\mathbf{\Psi}_T)_{\Sigma_\tau}\quad\forall(\mathbf{v},\mathbf{\Psi})\in H_0^1(\Omega)^3\times V_{\Sigma_\tau}(\Omega) \tag{21}$$

Let us pass to a limit in (21) at $m\to\infty$. From (19) it follows that all linear summands in (21) pass to corresponding linear summands in (20) at $m\to\infty$. Now the nonlinear summands starting with $(\operatorname{rot}\mathbf{H}_m\times\mathbf{H}_m,\mathbf{v})$ can be considered. It is clear that

$$(\operatorname{rot}\mathbf{H}_m\times\mathbf{H}_m,\mathbf{v})-(\operatorname{rot}\mathbf{H}^*\times\mathbf{H}^*,\mathbf{v})=(\operatorname{rot}\mathbf{H}_m\times(\mathbf{H}_m-\mathbf{H}^*),\mathbf{v})+(\operatorname{rot}(\mathbf{H}_m-\mathbf{H}^*)\times\mathbf{H}^*,\mathbf{v}).$$

Let us choose a random number $\varepsilon>0$ and fix a test function $\mathbf{v}\in H_0^1(\Omega)^3$. It can be proved that such number $M=M(\varepsilon,\mathbf{v})$ exists:

$$|(\operatorname{rot}\mathbf{H}_m\times(\mathbf{H}_m-\mathbf{H}^*),\mathbf{v})|\le\varepsilon\quad\forall m\ge M. \tag{22}$$

As $\mathcal{D}(\Omega)^3$ embeded densely in $H_0^1(\Omega)^3$ by norm $\|\cdot\|_{1,\Omega}$ that for any function $\mathbf{v}\in H_0^1(\Omega)^3$ there is a sequence $\mathbf{v}_n\in\mathcal{D}(\Omega)^3$, converging to $\mathbf{v}$ by norm $\|\cdot\|_{1,\Omega}$ at $n\to\infty$. Using $\mathbf{v}_n$, we have

$$|(\operatorname{rot}\mathbf{H}_m\times(\mathbf{H}_m-\mathbf{H}^*),\mathbf{v})|\le|(\operatorname{rot}\mathbf{H}_m\times(\mathbf{H}_m-\mathbf{H}^*),\mathbf{v}_n)|+|(\operatorname{rot}\mathbf{H}_m\times(\mathbf{H}_m-\mathbf{H}^*),\mathbf{v}_n-\mathbf{v})|\ \forall m,n\in\mathbf{N} \tag{23}$$

On the strength of uniform boundedness by m of values $\|\operatorname{rot}\mathbf{H}_m\|_\Omega$ and $\|\mathbf{H}_m-\mathbf{H}^*\|_{L^3(\Omega)^3}$ there is such number $N{=}N(\varepsilon,\mathbf{v})$ that for the second summand in (23) the estimate takes place

$$|(\operatorname{rot}\mathbf{H}_m\times(\mathbf{H}_m-\mathbf{H}^*),\mathbf{v}_n-\mathbf{v})|\le\varepsilon/2\ \ \forall n\ge N,\ \forall m\in\mathbf{N}. \tag{24}$$

Let us consider the first summand in (23). By Holder inequality for any function $\mathbf{v}_n\in\mathcal{D}(\Omega)^3$ the inequality is true $|(\operatorname{rot}\mathbf{H}_m\times(\mathbf{H}_m-\mathbf{H}^*),\mathbf{v}_n)|\le\|\operatorname{rot}\mathbf{H}_m\|_\Omega\,\|\mathbf{H}_m-\mathbf{H}^*\|_\Omega\,\|\mathbf{v}_n\|_{L^\infty(\Omega)^3}$.

As $\|\operatorname{rot}\mathbf{H}_m\|_\Omega\le\|\mathbf{H}_m\|_{\mathcal{H}^{1/2}_{\rm div}(\Omega)}\le c_4$ and $\mathbf{H}_m\to\mathbf{H}^*$ in $L^2(\Omega)^3$ at $m\to\infty$ by (19), then, starting with some number $M=M(\varepsilon,\mathbf{v},N)$, the estimate $|(\operatorname{rot}\mathbf{H}_m\times(\mathbf{H}_m-\mathbf{H}^*),\mathbf{v}_n)|\le\varepsilon/2$ holds for all $m\ge M$, $n=N$. From this and from (23), (24) the estimate (22) follows.

On the strength of randomness of ε it means that

$$|(\operatorname{rot}\mathbf{H}_m\times(\mathbf{H}_m-\mathbf{H}^*),\mathbf{v})|\to0 \text{ at } m\to\infty\ \forall\mathbf{v}\in H_0^1(\Omega)^3. \tag{25}$$

From the weak convergence of the sequence $\mathbf{H}_m$ in the space $\mathcal{H}^{1/2}_{\rm div}(\Omega)$ the weak convergence of $\operatorname{rot}\mathbf{H}_m$ in $L^2(\Omega)^3$ takes place. In such case

$$(\operatorname{rot}(\mathbf{H}_m-\mathbf{H}^*)\times\mathbf{H}^*,\mathbf{v})=(\operatorname{rot}(\mathbf{H}_m-\mathbf{H}^*),\mathbf{H}^*\times\mathbf{v})\to0 \text{ at } m\to0, \tag{26}$$

as $\mathbf{H}^*\times\mathbf{v}\in L^2(\Omega)^3$. From (22)–(26) it follows that $(\operatorname{rot}\mathbf{H}_m\times\mathbf{H}_m,\mathbf{v})\to(\operatorname{rot}\mathbf{H}^*\times\mathbf{H}^*,\mathbf{v})$ at $m\to\infty$ for all $\mathbf{v}\in H_0^1(\Omega)^3$.

Applying similar methods we show that $(\operatorname{rot}\mathbf{\Psi}\times\mathbf{H}_m,\mathbf{u}_m)\to(\operatorname{rot}\mathbf{\Psi}\times\mathbf{H}^*,\mathbf{u}^*)$ at $m\to\infty$ for all $\mathbf{\Psi}\in V_{\Sigma_\tau}(\Omega)$. (From the definition of the space $V_{\Sigma_\tau}(\Omega)$ follows that the subspace of the space of smooth function $C^\infty_{\Sigma_\tau}(\Omega)=\{\mathbf{h}\in C^\infty(\overline{\Omega}):\mathbf{h}\cdot\mathbf{n}|_{\Sigma_\tau}=0,\ \mathbf{h}\times\mathbf{n}|_{\Sigma_\nu}=\mathbf{0}\}$ is dense in $V_{\Sigma_\tau}(\Omega)$ by norm $\|\cdot\|_{1,\Omega}$.)

As $\mathbf{u}\in H^1(\Omega)^3$ without any problems we can conclude that $((\mathbf{u}_m\cdot\nabla)\mathbf{u}_m,\mathbf{v})\to((\mathbf{u}^*\cdot\nabla)\mathbf{u}^*,\mathbf{v})$ at $m\to\infty$ (see [7]).

Passing to the limit in (20) at $m\to\infty$, coming to (21), and from continuity of the functional J on $X\times L^2(\Sigma_T)$ we have that $J(\mathbf{x}^*,q^*)=J^*$. ■

Acknowledgments

This work was supported by the Russian Science Foundation (project No 14-11-00079) and the Russian Foundation for Basic Research (project No. 16-01-00365 A).

References

[1] Hou L and Meir A 1995 Boundary optimal control of MHD flows *Appl. Math. Optim.* **32** 143–62
[2] Alekseev G 2004 Solvability of control problems for stationary equations of magnetohydrodynamics of a viscous fluid *Sib. Math. J.* **45** 197–213
[3] Alekseev G and Brizitskii R 2012 Stability estimates of solutions of control problems for the stationary equations of magnetic hydrodynamics *Diff. Eq.* **48** 397–409
[4] Griesse R and Kunish K 2006 Optimal control for stationary MHD system in velocity–current formulation *SIAM J. Control. Optim.* **45** 1822–45
[5] Gunzburger M and Trenchea C 2007 Analysis of an optimal control problem for the three–dimensional coupled modified Navier–Stokes and Maxwell equations *J. Math. Anal. Appl.* **333** 295–310
[6] Ravindran S 2008 On the dynamics of controlled magnetohydrodynamic systems *Nonlinear Anal. Model. Control* **13** 351–77
[7] Alekseev G 2010 *Optimization in stationary problems of heat and mass transfer and magnetic hydrodynamics* (Moscow: Nauchniy Mir)
[8] Alekseev G and Brizitskii R 2010 On uniqueness and stability of solutions of extremum problems for stationary Navier–Stokes equation *Diff. Eq.* **46** 70–82
[9] Solonnikov V 1960 On some stationary boundary value problems in magnetohydrodynamics *Trudy Mat. Inst. Steklov.* **59** 174–87 (in Russian)
[10] Gunzburger M D, Meir A J and Peterson J S 1991 On the existence, uniqueness,and finite element approximation of solution of the equation of stationary, incompressible magnetohydrodynamics *Math. Comp.* **56** 523–63
[11] Alekseev G V and Brizitskii R V 2014 Solvability of the boundary value problem for stationary magnetohydrodynamic equations under mixed boundary conditions for the magnetic field *Appl. Math. Lett.* **32** 13–8
[12] Fernandes P and Gilardi G 1997 Magnetostatic and electrostatic problems in inhomogeneous anisotropic media with irregular boundary and mixed boundary conditions *Math. Models Methods Appl. Sci.* **7** 957–91
[13] Auchmuty G 2004 The main inequality of 3D vector analysis *Math. Models Methods Appl. Sci.* **14** 79–103
[14] Auchmuty G and Alexander J S 2006 Finite energy solutions of mixed 3d div-curl systems *Q. Appl. Math.* **64** 335–57
[15] Alekseev G 2016 Mixed boundary value problems for steady-state magnetohydrodynamic equations of viscous incompressible fluid *Comp. Math. Math. Phys.* **56** 1426–39
[16] Alekseev G V 2016 Solvability of an inhomogeneous boundary value problem for the stationary magnetohydrodynamic equations for a viscous incompressible fluid *Diff. Eq.* **52** 739–48
[17] Alekseev G 2016 Mixed Boundary Value Problems for Stationary Magnetohydrodynamic Equations of a Viscous Heat-Conducting Fluid *J. Math. Fluid Mech.* **18** 591–607
[18] Consiglieri L, Necasova S and Sokolowski J 2009 Incompressible Maxwell–Boussinesq approximation: existence, uniqueness and shape sensitivity *Control Cybern.* **38** 1193–1215
[19] Brizitskii R and Tereshko D 2007 On the solvability of boundary value problems for the stationary magnetohydrodynamic equations with inhomogeneous mixed boundary conditions *Diff. Eq.* **43** 246–58
[20] Alonso A and Valli A 1996 Some remarks on the characterization of the space of tangential traces of $\mathbf{H}(\mathrm{rot};\Omega)$ and the construction of the extension operator *Manuscr. Math.* **89** 159–78

IOP Conf. Series: Journal of Physics: Conf. Series **894** (2017) 012016 doi:10.1088/1742-6596/894/1/012016

The influence of the structure of ultrafine-grained aluminium alloys on their mechanical properties under dynamic compression and shock-wave loading

I G Brodova[1], A N Petrova[1, 2], O B Naimark[3], O A Plekhov[3], S V Razorenov[4] and E V Shorokhov[5]

[1] M.N. Miheev Institute of Metal Physics of Ural Branch of Russian Academy of Sciences, 620990, 18 S. Kovalevskaya Street, Ekaterinburg, Russia
[2] Ural Federal University, 620002, 19 Mira street, Ekaterinburg, Russia
[3] Institute of Continuous Media Mechanics of the Ural Branch of Russian Academy of Science, 614013, 1 Academician Korolev Street, Perm, Russia
[4] Institute of Problems of Chemical Physics of Russian Academy of Sciences, 142432, 1 Academician Semenov Avenue, Chernogolovka, Moscow region, Russian Federation
[5] Federal State Unitary Enterprise "Russian Federal Nuclear Center – Academician E.I. Zababakhin All-Russian Research Institute of Technical Physics", 456770, 13 Vasiliev Street, PO box 245, Snezhinsk, Chelyabinsk region, Russia

E-mail: brodova@imp.uran.ru

Abstract. Mechanical properties of commercial A3003 and A7075 aluminum alloys having ultrafine-grained structure formed under dynamic channel angular pressing were investigated. It was shown that yield strength of ultrafine-grained alloys at the wide range of strain rates from 2 10^{-3} to 1.5 10^{5} s^{-1} are higher than of coarse-grained alloys. The values of dissipated energy ratio of the ultrafine-grained alloys were compared concerning their different structures and strain rate of 3.5-6.6 10^{3} s^{-1}. Under shock-wave loading the decreasing of strength was observed for both investigated UFG alloys against quasi-static deformation at the strain rate of 10^{3} s^{-1}.

1. Introduction

Industrial aluminum alloys are current and extensively used materials for different applications. For secure operation of constructions, high requirements are imposed on strength and plasticity of aluminum alloys. It is common knowledge that mechanical properties of aluminum alloys are governed by structure and their phase composition. In recent years, the various methods of severe plastic deformation are widely used for structural refinement of the metallic materials. Materials with the grain size below 1μmwere so named ultrafine-grained (UFG) materials. It was shown that UFG materials have advanced static mechanical properties in comparison with conventional coarse-grained materials. However, at high strain-rate deformation and shock wave loading this benefit decays or sometimes even disappears [1-7]. Experimental and theoretical researches indicated that at specified deformation conditions under shock wave loading the initial temperature effects on elastic-plastic

Content from this work may be used under the terms of the Creative Commons Attribution 3.0 licence. Any further distribution of this work must maintain attribution to the author(s) and the title of the work, journal citation and DOI.
Published under licence by IOP Publishing Ltd

properties and a fracture behavior [2, 5, 6]. Thus, at the temperature about melting point the anomalous growth of an elastic precursor amplitude (approximately by four times) and sharp reduction of spall strength were demonstrated. Many researchers showed that deformation resistance and spall strength depend on grain size, type of the defects and their density [3, 5, 8, 9].Nevertheless, experimental data about strain rate dependence of mechanical properties of different aluminum alloys are quiet controversial and do not answer the questions about structural parameters contribution to deformation mechanisms at dynamic conditions.

The aim of the present study was systematical investigation of mechanical behavior of UFG aluminum alloys at the wide range of strain rates. The UFG alloys were produced by means of unique method of dynamic pressing. As it was shown [10-12] structure and mechanical properties of such alloys distinct from UFG alloys processed by static methods of severe plastic deformation. The studying of their behavior under shock wave conditions gives new knowledge about the character of phenomena in FCC metals.

2. Experimental

2.1. Material and processing

The rods of commercial A3003 (Al–1.5 Mn–0.1 Zn–0.05 Cu–0.6 Si–0.7 Fe in wt%) and A7075 (Al–7.0 Zn–2.3 Mg–1.8 Cu–0.5 Si–0.5 Fe–0.25 Cr in wt%) alloys having length of 75 mm and diameter of 16 mm were processed by the method of severe plastic deformation – Dynamic channel angular pressing (DCAP). Due to the action of powder gases pressure on the piston, the alloys were deformed with the strain rate of the order of 10^5 s^{-1}. The unique feature of the DCAP is the combination of a shock-wave loading and following high-strain rate shear deformation. The DCAP scheme is described in details in [10, 11]. DCAP provides grain refinement in the alloys to 200-500 nm. The A3003 alloy was deformed by 1 (DCAP-1) and 4 (DCAP-4) pressing cycles and the A7075 alloy was deformed by 2 pressing cycles (DCAP-2). Repeated deformation was performed using B_c route. Typical structures of DCAP alloys are shown in figure 1, 2.

After one pressing cycle the structure of A3003 alloy is composed by sells and subgrains with low-angle boundaries (figure 1a) and crystallites with high-angle boundaries (figure 1b). Inserted in fig. 1b micro-diffraction pattern argues for high angle boundaries formation in the area. After four DCAP cycles mix structure contains recrystallized grains with the absence of dislocations and the crystallites with high-angle grain boundaries (figure 1c).The average size of structural fragments is 500-600 nm in A3003 alloy after DCAP. In the A7075 alloy two cycles of DCAP leads to stronger structural refinement. Figure 2 demonstrates UFG structure with high dislocation density and non-equilibrium high-angle boundaries. The average grain size in this case is 200 nm.

2.2. Mechanical testing

For tensile testing, flat specimens with a gage length of 17 mm, thickness of 0.5 mm and width of 6 mm machined from both as received and DCAP processed bars were pulled to failure using a ZWICK/RoellZ050 testing machine at a strain rate of 2 10^{-3} s^{-1}. The specimens were prepared and tested according to ASTM 8/E-8M-08.

Cylindrical samples with lengths of 4 mm and diameters of 7 mm were machined from both the as received and DCAP processed bars. The longitudinal axis of the specimens coincide with the pressing direction of the bars. Than the specimens were compressed dynamically at split Hopkinson pressure bar at strain rate in the range of 3.5-6.6 10^3 s^{-1} using the procedure described in [13, 14]. At least three specimens were tested under the same experimental conditions. The stress -σ, and strain-ε, were determined in-situ during dynamic loading based on elastic pulses in input and output measuring bars using the relationships presented in [13].

During dynamic compression, the surface temperature of the specimens was detected with a FLIP SC 5000 infrared camera. The spatial resolution of the camera is 0.2 mm, the temperature sensitivity is 0.025 K, the recording frequency is 3 kHz. Under dynamic loading conditions, the heat losses due to

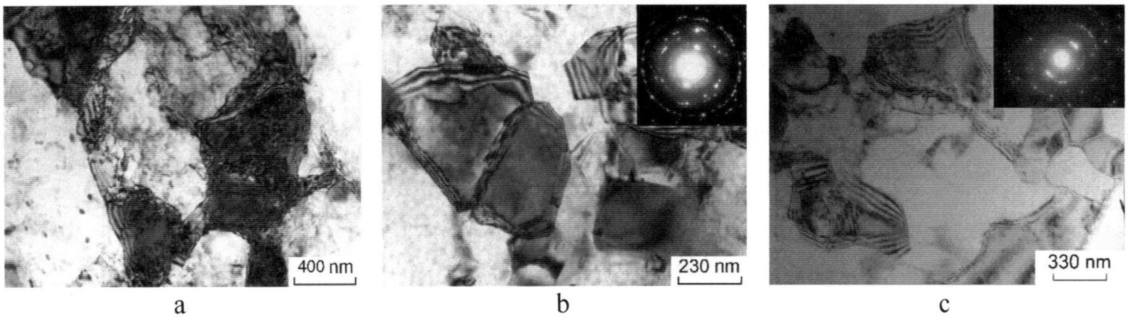

Figure 1. Structure of the DCAP-1 A3003 (a, b) and the DCAP-4 A3003 alloys (c).

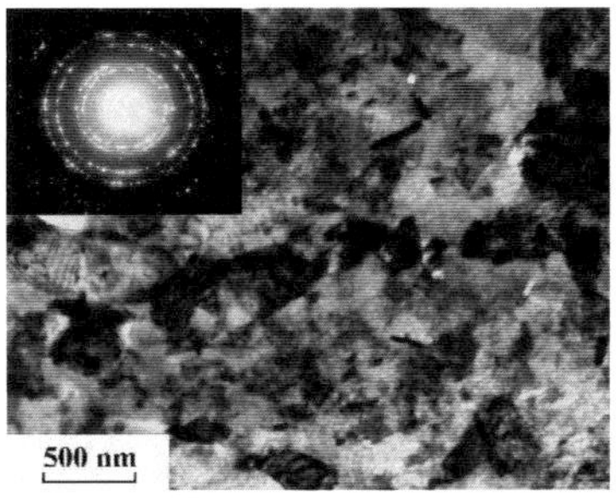

Figure 2. Structure of the DCAP-2 A7075 alloy.

exchange with the environment are negligible. Therefore, based on the temperature evolution data during compression, the fraction of dissipated energy was determined as

$$\beta = E_2/E_1,$$

where E_1 - is the energy consumed for specimen deformation, E_2 - is the energy converted into heat due to deformation [14].

For shock-wave experiments disc shaped specimens with the thickness of 2 mm were machined from the as received and DCAP processed bars in such a way that longitudinal axis of the specimens coincide with the pressing direction of the bars. The specimens were loaded by the stroke of a flat aluminum plate ~0.4 mm thick, which was accelerated to a velocity of 630±30 m/s using explosive devices [1, 2].The experimental scheme is shown in figure 3. Under the given loading conditions, the pressure of the impact compression was 4–5 GPa. The deformation rate varied from 1.2 10^5 to 1.6 10^5 s^{-1}. Velocity of the back free surface of the specimens - $u_{fs}(t)$ was recorded by means of VISAR laser Doppler velocimetry [15] with a time resolution of ~1 ns.

The Hugoniot elastic limit σ_{HEL} and the yield stress Y were calculated from the elastic precursor front amplitude u_{fsHEL} according to velocity profiles as follows:

IOP Conf. Series: Journal of Physics: Conf. Series **894** (2017) 012016 doi:10.1088/1742-6596/894/1/012016

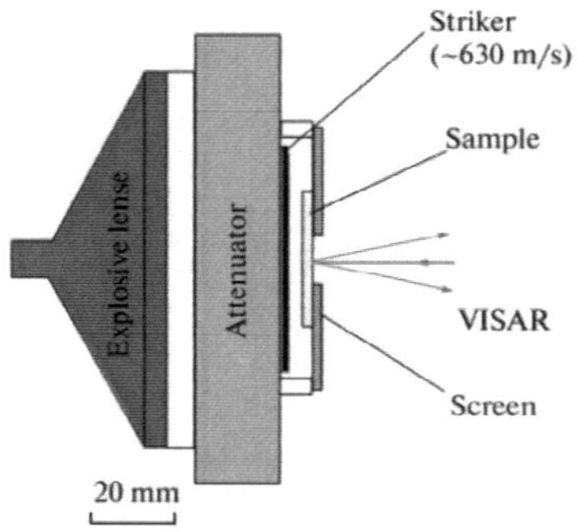

Figure 3. The scheme of the shock-wave loading experiment.

$$\sigma_{HEL} = \rho_0 c_l u_{fsHEL}/2$$

$$Y = \frac{1-2\nu}{1-\nu}\sigma_{HEL} = 2\sigma_{HEL}\left[\frac{c_s^2}{c_l^2}\right] \cong 1.5\sigma_{HEL}\left(1-\frac{c_b^2}{c_l^2}\right),$$

where ρ_0 is the initial density of the material, C_b, C_l, C_s– are the bulk, longitudinal, and shear sound velocities in material respectively, ν – Poisson's ratio.

Consider an allowance due to elastic–plastic behavior of the investigated materials -δ, the spall strength - σ_{sp} was determined from the following relation [1]:

$$\sigma_{sp} = 1/2\rho_0 c_b(\Delta u_{fs} + \delta),$$

Where Δu_{fs} is a drop of the velocity from the maximum to the first minimum due to the formation and propagation of reflection wave.

Material microhardness H was determined by means of indentation on PMT-3 facility with the load of 0.2 N. The measuring error did not exceed 10%. Brinell hardness test at a load of 250 kg with a 10 mm diameter ball was used to measure hardness of the alloys.

2.3. Microstructural characterization

The grain structure of the alloys was examined using Philips CM-30 transmission electron microscope (TEM) operating at an accelerating voltage of 200 kV. The average size of structural fragments was calculated based on dark-field images using the SIAMS image analysis software. The X-ray diffraction (XRD) analysis was performed on a DRON-3 diffractometer using CoK_α radiation. The Profile software was used to process the data by the approximation method and to calculate a root-mean square lattice microstrain $\langle\varepsilon^2\rangle^{1/2}$, crystal lattice parameter – a, and the size of coherent scattering region (D_{CSR}). Dislocation density was determined from the following relation:

$$\rho_d = \frac{2\sqrt{3}\langle\varepsilon^2\rangle^{1/2}}{(D_{CSR} \times b)},$$

where $b = a\sqrt{2}/2$ is Burgers vector for fcc metals.

3. Results

3.1. Dynamic compression at split Hopkinson pressure bar

Figures 4a, b shows stress-strain curves for DCAP-2 A7075 alloy and for DCAP-1 A3003 alloy obtained at different strain rates in a range of 3.5-6.6 10^3 s^{-1}. For the DCAP-2 A7075 alloy, the anomalous (inverse) dependence of yield stress on strain rate is established. There is the decreasing of yield stress on 27% when strain rate changes from 4 10^3 s^{-1} to 6 10^3 s^{-1}. Contrary, for DCAP-1 A3003 the increasing of yield stress is observed with increasing of strain rate, the same as for coarse-grained alloys (figure 4c). Comparing with coarse grained materials the yield stress of DCAP-2 A7075 alloy is by 20% lower and of DCAP-1 A3003 alloy is by 28% higher at the same strain rate about 6 10^3 s^{-1}. For the DCAP-1 A3003 alloy, the strengthening effect in reference to CG state also depends on strain rate. Thus, yield strength is increased by 47% at strain rate of 4 10^3 s^{-1}against 28 % at strain rate of 6 10^3 s^{-1}, when compare with coarse-grained alloy. The yield stress value reaches 260 MPa and 280 MPa at low strain rate, and reaches 190 MPa and 295 MPa at high strain rate for UFG alloys A7075 and A3003 respectively.

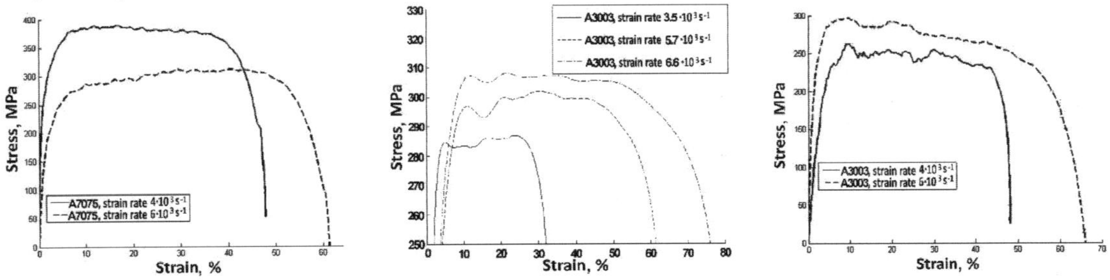

a b c

Figure 4. Stress-strain curves for UFG DCAP-2 A7075 alloy (a), UFG DCAP-1 A3003 alloy (b) and for CG A3003 alloy (c) at different strain rates.

To qualify dissipation capability of different alloys the values of dissipated energy ratio were compared at different strain rate [16]. It can be seen from figure 5, that the amount of dissipated energy is different for alloys in the UFG state having different structures and in the CG state. For instance, DCAP-2 A7075 and DCAP-1 A3003 alloys dissipate more elastic energy then the CG alloys respectively, but DCAP-4 A3003 alloy, conversely, dissipates less energy than the CG alloy at all investigated strain rates. The comparison of dissipated energy ratio of UFG A3003 alloy with two distinct types of structure shows that alloy with recrystalizated grains dissipates less and stores more energy. Energy dissipation capability of UFG DCAP-1 A3003 alloy does not depend on strain rate. However, for UFG DCAP-2 A7075 alloy the increasing of strain rate from 4 10^3 s^{-1} to 6 10^3 c^{-1} results in the growth of dissipated energy ratio from 0.62 to 0.8.

3.2. Shock wave loading

The strength properties of the DCAP-1, DCAP-2, and CG A3003 and A7075 alloys at a strain rate of (1.2-1.6) 10^5 s^{-1} are set out in a table 1, where σ_{HEL} is a Hugoniot elastic limit; Y is a dynamic yield strength; σ_{sp} is a spall strength.

Table 1. Dynamic strength properties of UFG and CG alloys.

Material	σ_{HEL}, GPa	Y, GPa	σ_{sp}, GPa
DCAP-2 A7075	0.42	0.21	1.32
CG A7075	0.28	0.14	1.38
DCAP-1 A3003	0.34	0.18	1.34
CG A3003	0.25	0.13	1.25

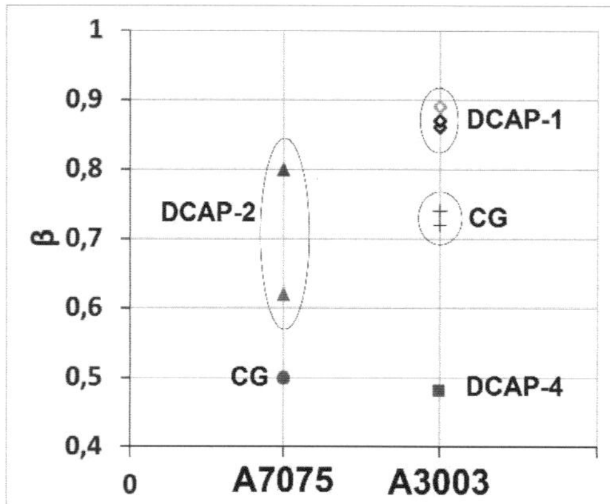

Figure 5. Dissipated energy ratio for A7075 and A3003 alloys at different strain rates.

It was shown that for DCAP-2 A7075 alloy Hugoniot elastic limit is by 50% greater than that for CG alloy, for DCAP-1 A3003 alloy σ_{HEL} is greater that for the CG one by 35%. The same changes are observed for dynamic yield strength. Spall strength is the same for DCAP-2 and CG alloy A7075, but for DCAP-1 A3003 alloy spall strength is increased by 7% in comparison with CG alloy.

Figure 6 demonstrates the dependence of yield strength on strain rate in a wide range of strain rates of 2 10^{-3}-1.6 10^{5} s^{-1}. Coarse-grained alloys strengthen up to strain rate in a range of 10^{3}-10^{4} s^{-1}. The same trend was observed for UFG A3003 alloy. Then alloys undergo softening at a strain rate about10^{5} s^{-1}. The dynamic yield strength, obtained at the strain rate about 10^{5} s^{-1}, is approximately the same as the yield strength at static conditions. The UFG A7075 alloy exhibits different and complicated behavior. Thus, the yield strength decreases at dynamic conditions compared with static conditions. The slight increasing of yield strength by 30 MPa is observed at strain rate 10^{5} s^{-1}. The comparison of the yield strength determined during static tensile tests and shock wave loading shows that the increment of the strength in regard to coarse grained alloy in the case of UFG A7075 alloy is 107% at static conditions and is 50 % in dynamics. In the case of UFG A3003 alloy the increment is 55% at static conditions and is 38% at dynamic conditions.

4. Discussion

The key attribute that distinguish DCAP from other static severe plastic deformation techniques is the superposition of several deformation modes of compression and tension due to the circulation of shockwaves and rarefaction waves and of the simple shear. Such complicated loading in combination with high strain rate promotes effective structural refinement upon 1-2 cycles of dynamic pressing and enhances significantly strength and hardness of alloys [10-12]. According to XRD analysis dislocation density increases by an order of magnitude and reaches the value more than 10^{15} m^{-2}. It is known, that evolution of ensembles of structural defects such as dislocation nucleation, sliding, interaction and annihilation is governed by strain rate. Therefore, the high level of internal stress can be explained by escalation of dislocation mobility as well as dislocation density due to activation of additional sliding systems [17]. Experimentally reviled peculiarities of structural formation during DCAP of aluminum alloys were confirmed completely by modeling of the process of high strain-rate deformation of metals using the equations of mechanics of continuum media containing defects [18].

It was established earlier that different structural states were formed depending on chemical composition of the alloy and DCAP mode [10-12, 19, 20]. Structures after DCAP of investigated alloys are shown in figure 1 and figure 2. The distinctive characteristics of the structural states are grain-subgrain size, dislocation density, proportion of high-angle and low-angle boundaries. It was determined that the non-equilibrium ultrafine-grained structure with high dislocation density of 2 10^{15} m^{-2} was formed in the DCAP-2 A7075 alloy.

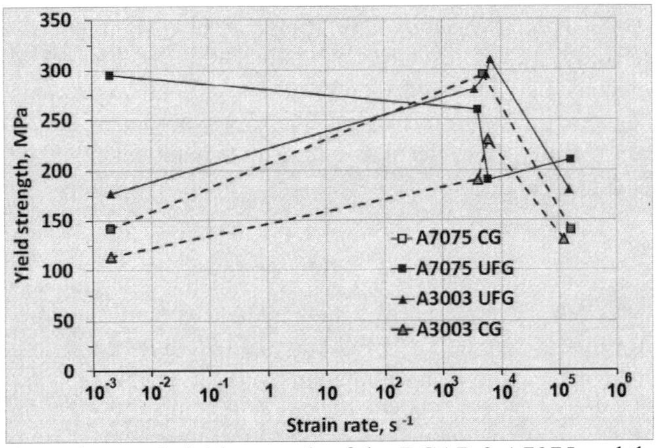

Figure 6. Strain rate dependence of yield strength of the DCAP-2 A7075 and the DCAP-1 A3003 and CG alloys.

The less defective mix UFG structure was formed in DCAP-4 A3003 alloy. The dislocation density in the DCAP-4 A3003 alloy is by an order of magnitude lower than in the DCAP-2 A7075 alloy. Since the main origin of internal stress is dislocation low-angle boundaries and disclinations in triple points [21], than the value of stored energy during DCAP will be different for the UFG A7075 and A3003 alloys. Structural evolution under further deformation is governed by the value of stored energy and connected with the relaxation of the internal stress. Thus, two different scenarios of the evolution of the structure during dynamic compression were observed for two investigated alloys at strain rate (3.5-6.6) 10^3 s^{-1}. The analysis of TEM images showed [16] that the grain-subgrain size maintains, well-developed dislocation substructure is formed and new low-angle boundaries appear in the DCAP-4 A3003 alloy after dynamic compression. Thus, the energy of dynamic compression was expend in the accumulation of the defects in the structure and therefore the increasing of stored energy was observed. Quite the opposite, non-equilibrium grain boundaries transform to more equilibrium state, the reduction of general dislocation density occurs in the DCAP-2 A7075 alloy. The evolution of the structure of the DCAP-2 A7075 alloy showed the activation of internal stresses relaxation processes. Transition electron microscopy data are complimented by XRD analysis. Due to dynamic compression of DCAP-2 alloy the root-mean square lattice microstrain was reduced from 0.17% to 0.12%, hardness was decreased by 300 MPa. The structure of DCAP-1 A3003 alloy responds the same way as DCAP-2 A7075 alloy. Consider that the less the proportion of stored energy than the more energy transforms in to heat, the ratio of dissipated energy of DCAP-2 A7075 alloy is 80 %, of DCAP-1 A3003 is 86%, and of DCAP-1 A3003 is 48% at strain rate 6 10^3 s^{-1}.

Irrespectively from chemical composition, a cell submicrocrystalline deformed structure was formed in the coarse grained alloys during dynamic compression. Furthermore, the root-mean square lattice microstrain increased and hardness grew by 200 MPa. Therefore, the part of the compression energy is consumed for defect formation due to that the proportion of stored energy increases.

DCAP UFG alloys possess higher hardness than CG alloys. Hardness increment is 700 MPa for DCAP-2 A7075 alloy and is 450 MPa for the DCAP-1 A3003 alloy. If there is no phase transformations than a strength and hardness increases due to grain boundaries length increasing and dislocation hardening. Comparison of the static mechanical properties have shown that yield strength and tensile strength of DCAP-2 alloy A7075 is 2 times higher and of the DCAP-1 A3003 alloy 1.5 times higher than of coarse grained alloys respectively. The difference of yield strength strongly depends on strain rate in the range of 10^3 s^{-1}. Mechanical behavior of UFG alloys under dynamic compression at strain rate of 10^3 s^{-1} is connected with structural evolution, which was described earlier. Thus, for CG alloys as well as for DCAP-1 A3003 alloy we observed normal (direct) yield strength-strain rate dependence, when for DCAP-2 A7075 alloy the inverse strain-rate dependence of the yield strength was established (figure 6). Such anomalous behavior testify, that the deformation mechanism changes from dislocation sliding in CG alloy for grain boundary sliding for UFG A7075 alloy.

Under shock-wave loading the decreasing of deformation resistance was observed for both investigated UFG alloys against the strain rate of 10^3 s^{-1}. However, according to the table 1 [20] and figure 6 the dynamic characteristic of elastic-plastic transition is higher for UFG alloys than for CG alloys under shock-wave loading condition.

5. Conclusions

It was established that UFG A3003 alloy exhibits higher yield strength against CG alloy at a wide strain rate range from 2 10^{-3} to 1.5 10^5 s^{-1}. The UFG alloy A7075 undergo softening as regards to CG alloy at strain rate of (4-6) 10^3 s^{-1}. At this strain rates the alloy have inverse yield strength-strain rate dependence. At shock wave loading condition spall strength of both A3003 and A7075 UFG alloys is not poorer than of their CG analogs. Energy dissipation capability of UFG DCAP aluminum alloys is higher than of coerce grained alloys in the event that nonequilibrium heavily deformed structure is formed.

Acknowledgments

The work was performed in the framework of the state assignment of Russian Federal Agency for Scientific Organizations (theme “Structure”, no. 01201463331).

The authors thank the administration of the UrFU for access to the scientific equipment of the UrFU, supported by program 211 of the Government of the Russian Federation (agreement no. 02.A03.21.0006).

The electron-microscopic studies were performed at the Center of Collaborative Access “Testing Center of Nanotechnology and Advanced Materials, M. N. Miheev Institute of Metal Physics, Ural Branch, Russian Academy of Sciences.

References

[1] Antoun T, Curran D R, Razorenov S V, Seaman L, Kanel G I and Utkin A V 2003 *Spall Fracture* (New York: Springer) p 404

[2] Kanel G I, Razorenov S V and Fortov V E 2004 *Shock-Wave Phenomena and Properties Condensed Matter* (New York: Springer) p 321

[3] Razorenov S V, Garkushin G V, Kanel G I and Ignatova O N 2012 *Physics of the Solid State* **54**(1) 790-7

[4] Garkushin G V, Razorenov S V and Kanel G I 2008 *Technical Physics. The Russian Journal of Applied Physics* Effect of structural factors on submicrosecond strength of D16T aluminum alloy **53** 11 pp 1441-1446

[5] Razorenov S V, Kanel G I and Fortov V E 2003 *The Physics of Metals and Metallography* **95**(1) 86-91

[6] Shtremel M A 2012 *Russian metallurgy (Metally)* **10** 843-52

[7] Evstifeev A D, Gruzdkov A A and Petrov Y V 2013 *Technical Physics. The Russian Journal of Applied Physics* **58**(7) 989-93

[8] Borodin E N and Mayer A E 2012 *Physics of the Solid State* **54**(4) 808-15

[9] Borodin E N, Petrov Y V, Mayer A E and Gruzdkov A A 2014 *Physics of the Solid State* **56**(12) 2470-9

[10] Brodova I G, Shirinkina I G, Antonova O A, Shorokhov E V and Zhgilev I I 2007 *Materials science and engineering* **503**(1-2) 103-5

[11] Petrova A N, Shirinkina I G, Shorokhov E V, Minaev I V, Zhgilev I N and Abramov A V 2010 *Reviews on Advanced Materials Science* **25**(2) 128-35

[12] Shirinkina I G, Petrova A N, Brodova I G, Pilyugin V P and Antonova O A 2012 *The Physics of Metals and Metallography* **113**(2) 170-5

[13] Kolsky H 1949 *Proc. Phys. Soc* **62B** 676-700

[14] Plekhov O, Naimark O, Chudinov V and Leontev V 2009 *Technical Physics Letters* **35**(1) 92-5

[15] Barker L M and Hollenbach R E 1972 *Journal of Applied Physics* **43** 4669–75

[16] Petrova A N, Brodova I G, Plekhov O A, Naimark O B and Shorokhov E V 2014 *Technical Physics. The Russian Journal of Applied Physics* **59**(7) 989-96

[17] Meyers M A and Murr L E 1981 *Shock Waves and High-Strain-Rate Phenomena in Metals* (NewYork: Plenum Press) p 1101

[18] Borodin E N and Mayer A E 2013 *Technical Physics. The Russian Journal of Applied Physics* **58**(8) 1159-1163

[19] Petrova A N, Brodova I G and Razorenov S V 2017 *Physics of Metals and Metallography* **118**(6) 601–7

[20] Brodova I G, Petrova A N, Razorenov S V and Shorokhov E V 2015 *The Physics of Metals and Metallography* **116**(5) 519-26

[21] Koneva E V, Popova N A, Ignatenko L N and Pekarskaya E E 2000 *Investigation and applications of severe plastic deformation* (New York: Kluwer Acad. Publ.) 121-6

IOP Conf. Series: Journal of Physics: Conf. Series **894** (2017) 012017 doi:10.1088/1742-6596/894/1/012017

Vortices appearance and diffusion

A M Gaifullin[1,2], D A Gadzhiev[1,2] and A V Zubtsov[1]

[1] Central Aerohydrodynamic Institute, Zhukovskii, Russia
[2] Moscow Institute of Physics and Technology, Dolgoprudnyi, Russia

E-mail: gaifullin@tsagi.ru

Abstract. The problem of a vortex flow generation by a rotating infinite cylinder is considered. The solution satisfying the two-dimensional unsteady equations describing a viscous gas motion is constructed. Asymptotic and numerical solutions of the problem are obtained. The comparison of the results is given.

1. Introduction

A variety of ways of the vortices formation is known: a vortex sheet shedding from a rigid surface, the flow interaction with the surface, shear layers interaction, etc. The correct description of the vortices diffusion process is important for the solution of a lot of problems. The examination of the point vortex diffusion in viscous incompressible fluid problem can be found in most of books on hydrodynamics. The problem of two opposite vortices diffusion has been solved in [1]. The case in which diffusive and convective processes change in time identically is interesting in a sense of the vortices appearance and diffusion distinctive features investigation. This is the case of the Navier - Stokes equations self-similar solution [2]. The flow interaction with a moving surface is considered in [3].

In this paper, a vortex generation and diffusion by the rotating cylinder for the cases of both incompressible and compressible flows are studied.

2. Rotating cylinder in incompressible fluid

The transverse velocity field

$$w = \frac{\Gamma_*}{r}, \tag{1}$$

where Γ_* is circulation divided by 2π and r is the distance from the vortex, is an either Euler equations and Navier - Stokes equations exact solution. However, in the second case the permanent power supply is required to maintain the velocity field (1).

A fluid moving with the velocity (1) has infinite kinetic energy, so to create the field (1) in the whole flow domain is impossible. However, there is a way to create the field (1) in the limited flow domain. Let an infinitely extended cylinder with the radius r_* start to rotate around its axis at the time $t = 0$ so that its surface speed is $w = \Gamma_* / r_*$. Because of the action of the viscous forces, a fluid near the cylinder starts moving.

We consider that axisymmetric flow with radial velocity equal to zero is implemented. That is, the flow in the absence of disturbances leading possibly to the flow instability is studied. Hence, the flow around the cylinder at $t \geq 0$ must satisfy the equation

Content from this work may be used under the terms of the Creative Commons Attribution 3.0 licence. Any further distribution of this work must maintain attribution to the author(s) and the title of the work, journal citation and DOI.
Published under licence by IOP Publishing Ltd

$$\frac{\partial\Gamma}{\partial t}=\nu\left(\frac{\partial^2\Gamma}{\partial r^2}-\frac{1}{r}\frac{\partial\Gamma}{\partial r}\right), \quad (2)$$

where $\Gamma(r,t)=rw$.

We write the boundary conditions at $t\geq 0$

$$\Gamma(r_*,t)=\Gamma_*, \quad \Gamma(\infty,t)=0 \quad (3)$$

and the initial condition at $r>r_*$

$$\Gamma(r,0)=0.$$

We must pay a special attention for the second condition from (3). This means that for any time t there is the distance r where the disturbances due to the cylinder rotation are small and in the limit $r\to\infty$ they may be considered to tend to zero and, consequently, the circulation $\Gamma(\infty,t)$ tends to zero too.

Subsequently, for the cases of either incompressible and compressible flows the solution is determined at large times when the scale of the domain in which $\Gamma(r,t)\sim\Gamma_*$ is much larger than r_*.

At first we determine the problem solution in the far domain $r>>r_*$. Because of the equation (2) and the boundary conditions (3), the solution must depend on the variables t, r and the value ν. Only one dimensionless variable $\eta=r/\sqrt{\nu t}$ can be composed from these three variables. This dimensionless variable determines the asymptotic domain $\Omega_1: r\sim\sqrt{\nu t}$ in which $\Gamma(r,t)=\Gamma(\eta)$. The solution is

$$\Gamma(r,t)=\Gamma_*\exp\left(-\eta^2/4\right)=\Gamma_*\exp\left(-r^2/4\nu t\right) \quad (4)$$

The solution (4) satisfies the second boundary condition from (3) but does not satisfy the first one. Consequently, the solution in the inner domain $\Omega_2: r\sim r_*$ must be considered. The inner limit of the outer expansion, that is, the solution (4) limit as $\eta\to 0$, is

$$\lim_{\eta\to 0}\Gamma(r,t)=\Gamma_*.$$

In the domain Ω_2, the leading-order behavior of Γ must satisfy the equation (2) and the boundary condition $\Gamma=\Gamma_*$ as $r/r_*\to\infty$. The solution is

$$\Gamma=\Gamma_*.$$

3. Rotating cylinder in compressible fluid

The two-dimensional axisymmetric flow characteristics must satisfy the equations

$$\frac{\partial(\rho v)}{\partial t}+\frac{\partial\left(\rho v^2+p\right)}{\partial r}+\frac{\rho}{r}\left(v^2-w^2\right)=\frac{2}{3}\frac{\partial}{\partial r}\left[\mu\left(2\frac{\partial v}{\partial r}-\frac{v}{r}\right)\right]+2\frac{\mu}{r}\left(\frac{\partial v}{\partial r}-\frac{v}{r}\right),$$

$$\frac{\partial(\rho w)}{\partial t}+\frac{\partial(\rho v w)}{\partial r}+2\frac{\rho v w}{r}=\frac{\partial}{\partial r}\left[\mu\left(\frac{\partial w}{\partial r}-\frac{w}{r}\right)\right]+2\frac{\mu}{r}\left(\frac{\partial w}{\partial r}-\frac{w}{r}\right),$$

$$\rho c_p\frac{\partial T}{\partial t}+\rho v c_p\frac{\partial T}{\partial r}-\frac{\partial p}{\partial t}-v\frac{\partial p}{\partial r}=$$

$$=\frac{c_p\mu}{\Pr}\left[\frac{1}{r}\frac{\partial}{\partial r}\left(r\frac{\partial T}{\partial r}\right)\right]+\frac{c_p}{\Pr}\frac{\partial\mu}{\partial r}\frac{\partial T}{\partial r}+\mu\left[2\left(\frac{\partial v}{\partial r}\right)^2+2\left(\frac{v}{r}\right)^2+\left(\frac{\partial w}{\partial r}-\frac{w}{r}\right)^2-\frac{2}{3}\left(\frac{\partial v}{\partial r}+\frac{v}{r}\right)^2\right], \quad (5)$$

$$\frac{\partial\rho}{\partial t}+\frac{\partial(\rho v)}{\partial r}+\frac{\rho v}{r}=0,$$

$$p=\rho RT.$$

Here, r, θ are polar coordinates, v, w are corresponding velocity components, t is time, ρ is density, p is pressure, c_p is specific heat at constant pressure, T is temperature, R is gas constant, $\Pr$ is the Prandtl number, μ is dynamic viscosity depending on temperature. In this paper, to be definite, we assume $\mu/T = \text{const}$, $c_p = \text{const}$ and $\Pr = \text{const}$.

The cylinder surface ($r = r_*$) boundary conditions associated with equations (5):

$$v = 0,\; w = \frac{\Gamma_*}{r_*},\; f\left(T, \frac{\partial T}{\partial r}\right) = 0. \tag{6}$$

The far-field boundary conditions ($r \to \infty$):

$$v = 0,\; w = 0,\; T = T_0,\; p = p_0,\; \rho = \rho_0,\; \mu = \mu_0.$$

The latter condition from (6) means that either temperature, temperature derivative or their combination can be applied on the cylinder surface. The significant (transverse velocity is not exponentially small) flow disturbances due to the cylinder rotation are propagated to the finite scale expanding in time. The disturbances out of this scale are small, so the flow characteristics as $r \to \infty$ must satisfy the unperturbed parameters.

As in the case of incompressible flow, we firstly determine the problem solution in the far domain $\Omega_1 : r >> r_*$. The study is limited by large Reynolds numbers $\mathrm{Re} = \Gamma_* \rho_0 / \mu_0 = \Gamma_* / \nu_0 >> 1$ and Mach numbers $\mathrm{M}_* = \Gamma_* / r_* a_0 \sim O(1)$. Here, $a_0 = \sqrt{\gamma R T_0}$ is the unperturbed stream sound speed, where γ is the ratio of specific heats, which is assumed to be constant. We note that the number M_* is introduced as the velocity on the cylinder surface divided by the unperturbed stream sound speed.

The scale of the far domain is $r \sim \sqrt{\nu_0 t}$ too. The gas flow characteristics in this domain can be divided into two types. The first is disturbances due to the cylinder rotation. The typical transverse velocity is estimated as $w \sim \Gamma_* / r$. The second is disturbances due to the temperature factor defined as the cylinder surface temperature divided by the far field temperature, which can exist even in the absence of the cylinder rotation. Both types of the temperature, pressure and density disturbances are small in the domain Ω_1 compared to corresponding magnitudes in the gas being at rest. Consequently, the problem linearization is possible and the disturbances of each type can be studied independently on another.

Firstly, we consider the disturbances due to the cylinder rotation. The typical Mach number M_1 in the domain Ω_1 is defined as the typical transverse velocity $w \sim \Gamma_* / \sqrt{\nu_0 t}$ divided by the unperturbed stream sound speed:

$$\mathrm{M}_1 = \frac{\Gamma_*}{a_0 \sqrt{\nu_0 t}} << 1.$$

The equations (5) are followed by the asymptotic form of the variables

$$w = \mathrm{M}_1 a_0 w_{1,1}(\eta, t),\; v = \frac{\mathrm{M}_1^3 a_0}{\mathrm{Re}} v_{1,1}(\eta, t),\; T = T_0\left(1 + \mathrm{M}_1^2 T_{1,1}(\eta, t)\right),$$

$$p = p_0\left(1 + \mathrm{M}_1^2 p_{1,1}(\eta, t)\right),\; \rho = \rho_0\left(1 + \mathrm{M}_1^2 \rho_{1,1}(\eta, t)\right),\; \mu = \mu_0\left(1 + \mathrm{M}_1^2 T_{1,1}(\eta, t)\right),$$

where $\eta = r / \sqrt{\nu_0 t}$.

For the dimensionless variables, the leading-order equations following from the equations (5) are

$$\frac{\partial p_{1,1}}{\partial \eta} = \gamma \frac{w_{1,1}^2}{\eta},$$

$$\frac{\partial^2 w_{1,1}}{\partial \eta^2}+\frac{\partial w_{1,1}}{\partial \eta}\left(\frac{1}{\eta}+\frac{\eta}{2}\right)+w_{1,1}\left(\frac{1}{2}-\frac{1}{\eta^2}\right)=t\frac{\partial w_{1,1}}{\partial t},$$

$$\frac{\gamma}{\gamma-1}\left[\frac{1}{\Pr}\left(\frac{\partial^2 T_{1,1}}{\partial \eta^2}+\frac{1}{\eta}\frac{\partial T_{1,1}}{\partial \eta}\right)+T_{1,1}+\frac{\eta}{2}\frac{\partial T_{1,1}}{\partial \eta}-t\frac{\partial T_{1,1}}{\partial t}\right]-\left(p_{1,1}+\frac{\eta}{2}\frac{\partial p_{1,1}}{\partial \eta}-t\frac{\partial p_{1,1}}{\partial t}\right)+\gamma\left(\frac{\partial w_{1,1}}{\partial \eta}-\frac{w_{1,1}}{\eta}\right)^2=0,$$

$$\frac{\partial v_{1,1}}{\partial \eta}+\frac{v_{1,1}}{\eta}=\rho_{1,1}+\frac{\eta}{2}\frac{\partial \rho_{1,1}}{\partial \eta}-t\frac{\partial \rho_{1,1}}{\partial t}, \tag{7}$$

$$p_{1,1}=\rho_{1,1}+T_{1,1}.$$

The self-similar solution of the second equation from (7) according to the incompressible flow solution is

$$w_{1,1}=\frac{A}{\eta}\exp\left(-\eta^2/4\right). \tag{8}$$

To require that the circulation as $\eta\to 0$ must be equal to the circulation on the cylinder surface is not appropriate because the solution is obtained in the domain Ω_1 whereas the cylinder surface boundary condition must be satisfied when the solution in the near-cylinder domain is constructed. So the value of the coefficient A is determined later. The equations (7) using the expression (8) result in

$$p_{1,1}=\gamma A^2\int_{\infty}^{\eta}\frac{\exp\left(-\eta^2/2\right)}{\eta^3}\mathrm{d}\eta.$$

The asymptotic behaviour of the equations solution as $\eta\to 0$ is

$$w_{1,1}\to\frac{A}{\eta},\ p_{1,1}\to-\frac{\gamma A^2}{2\eta^2},\ T_{1,1}\to-\frac{(\gamma-1)\Pr A^2}{\eta^2},\ \rho_{1,1}\to\frac{A^2}{\eta^2}\left[(\gamma-1)\Pr-\frac{\gamma}{2}\right],\ v_{1,1}\to 0.$$

The disturbances of the second type are determined in the presence of heat convection in the absence of the transverse velocity. In that case, the asymptotic expansion of the variables is written as follows:

$$T=T_0\left[1+\varepsilon T_{1,2}(\eta,t)\right],\ \rho=\rho_0\left[1+\varepsilon\rho_{1,2}(\eta,t)\right],v=\varepsilon\sqrt{\frac{v_0}{t}}v_{1,2}(\eta,t),\ p=p_0\left[1+\varepsilon\frac{\rho_0 v_0}{p_0 t}p_{1,2}(\eta,t)\right],$$

$$\mu=\mu_0\left[1+\varepsilon T_{1,2}(\eta,t)\right].$$

The value of $\varepsilon \ll 1$ is determined later. For the dimensionless temperature, density and viscosity disturbances, we obtain the following equations:

$$t\frac{\partial T_{1,2}}{\partial t}=\frac{\partial^2 T_{1,2}}{\partial \eta^2}+\frac{\partial T_{1,2}}{\partial \eta}\left(\frac{1}{\eta}+\frac{\eta \Pr}{2}\right), \tag{9}$$

$$\rho_{1,2}=-T_{1,2}.$$

The self-similar solution of the first equation from (9)

$$T_{1,2}=\int_{\infty}^{\eta}\frac{e^{-\eta^2\Pr/4}}{\eta}\mathrm{d}\eta.$$

The asymptotic behaviour of the solution as $\eta\to 0$

$$T_{1,2}\to\ln\eta+c_1,$$

where $c_1=\dfrac{C+\ln\Pr}{2}-\ln 2$, C is the Euler's constant.

Since the problem in the domain Ω_1 is linear, we can write the solution as $\eta\to 0$:

$$w\to\frac{\mathrm{M}_1 a_0 A}{\eta}=\frac{\Gamma_* A}{r},\ p\to p_0\left(1-\mathrm{M}_1^2\frac{\gamma A^2}{2\eta^2}\right)=p_0\left(1-\frac{\Gamma_*^2\gamma A^2}{2a_0^2 r^2}\right),$$

$$T \to T_0\left[1-\mathrm{M}_1^2\frac{(\gamma-1)\Pr A^2}{\eta^2}+\varepsilon\ln\eta+\varepsilon c_1\right]=T_0\left[1-\frac{\Gamma_*^2(\gamma-1)\Pr A^2}{a_0^2 r^2}+\varepsilon\ln\frac{r}{\sqrt{\nu_0 t}}+\varepsilon c_1\right]. \quad (10)$$

The expression (10) is followed that transverse velocity and pressure tend to functions independent on time whereas temperature has both the disturbance independent on time and the disturbance dependent on time.

When the temperature $T=T_*$ close to the temperature T_0 and the Mach number $\mathrm{M}_*=\Gamma_*/r_*a_0<<1$ are given on the cylinder surface, that is, disturbances are small, it is easily shown that the solution (10) can be extended to the cylinder surface. In this case

$$A=1,\ \varepsilon=\left[\frac{T_*-T_0}{T_0}+\mathrm{M}_*^2(\gamma-1)\Pr\right]\Big/\left(\ln\frac{r_*}{\sqrt{\nu_0 t}}+c_1\right).$$

In the case of the thermally-insulated cylinder surface ($\partial T/\partial r=0$ при $r=r_*$) for small disturbances we obtain

$$A=1,\ \varepsilon=-2\mathrm{M}_*^2(\gamma-1)\Pr,$$

$$T_*=T_0\left[1-2\mathrm{M}_*^2(\gamma-1)\Pr\left(\ln\frac{r_*}{\sqrt{\nu_0 t}}+c_1+\frac{1}{2}\right)\right]. \quad (11)$$

The expression (11) is followed that the gas temperature on the cylinder surface is larger than the unperturbed gas temperature and increases in time slowly (logarithmically).

In the $\mathrm{M}_*\sim O(1)$ case, the temperature disturbances reach the order of unity on the scale $r\sim \mathrm{M}_*r_*=\Gamma_*/a_0$, according to the expression (10). Then, we study the flow characteristics in the strong disturbances domain Ω_2 turning from the independent variables r, t to the new independent variables r_2, t, where $r_2=ra_0/\Gamma_*$.

The expression (10) determines the asymptotic form of the flow characteristics in the domain Ω_2 as follows:

$$w=a_0w_2(r_2,t),\ p=p_0p_2(r_2,t),\ T=T_0T_2(r_2,t),\ \rho=\frac{\rho_0 p_2}{T_2},\ \mu=\mu_0T_2.$$

The leading-order behaviour of the dimensionless variables w_2, p_2, T_2 is determined from the following equations:

$$\frac{\partial}{\partial r_2}\left[\mu\left(\frac{\partial w_2}{\partial r_2}-\frac{w_2}{r_2}\right)\right]+2\mu\left(\frac{\partial w_2}{\partial r_2}-\frac{w_2}{r_2}\right)=0,$$

$$\frac{\partial p_2}{\partial r_2}=\gamma\frac{p_2}{T_2}\frac{w_2^2}{r_2}, \quad (12)$$

$$\left(\frac{\partial^2 T_2}{\partial r_2^2}+\frac{1}{r_2}\frac{\partial T_2}{\partial r_2}\right)T_2+\left(\frac{\partial T_2}{\partial r_2}\right)^2+\Pr(\gamma-1)T_2\left(\frac{\partial w_2}{\partial r_2}-\frac{w_2}{r_2}\right)^2=0.$$

The first equation from (12) results in

$$\frac{\partial w_2}{\partial r_2}-\frac{w_2}{r_2}=\frac{\alpha}{\mu r_2^2}. \quad (13)$$

The constant α is determined from the condition of matching the solution in the domain Ω_1

$$w_2=\frac{A}{r_2}\ \text{as}\ r_2\to\infty. \quad (14)$$

Therefore, $\alpha=-2A\mu_0$.

The equation (13) substitution in the third equation from (12) and the definition $u = T_2^2$ give

$$\frac{\partial^2 u}{\partial r_2^2} + \frac{1}{r_2}\frac{\partial u}{\partial r_2} + \frac{8A^2 \Pr(\gamma - 1)}{r_2^4 \sqrt{u}} = 0 \quad (15)$$

with the corresponding condition as $r_2 \to \infty$

$$u = 1 - 2\frac{A^2 \Pr(\gamma-1)}{r_2^2} + 2\varepsilon\left(\ln\frac{r_2\Gamma_*}{a_0\sqrt{\nu_0 t}} + c_1\right). \quad (16)$$

Thus, the equations (13), (15) with the boundary conditions (14), (16) as $r_2 \to \infty$ and two boundary conditions on the cylinder surface (at $r_2 = 1/\mathrm{M}_*$) must be solved. The first cylinder surface boundary condition is the non-slip condition for the transverse velocity

$$w_2 = \mathrm{M}_*. \quad (17)$$

The second is the condition for the cylinder surface temperature

$$f_2\left(T_2, \frac{\partial T_2}{\partial r_2}\right) = 0. \quad (18)$$

The system of two ordinary differential equations (13), (15) with the boundary conditions (14), (16) - (18) has been solved numerically. Two unknown constants A и ε were fitted using iterative method so that the cylinder surface boundary conditions (17), (18) were fulfilled.

Using the solutions in the domains Ω_1 and Ω_2, the composite solution of the problem has been constructed.

In addition to the asymptotic study, the numerical solution of the cylinder rotating in a compressible fluid problem using the commercial program ANSYS Fluent 16.2.0 has been carried out. The comparison of asymptotic and numerical results is represented in the Figure 1 and Figure 2.

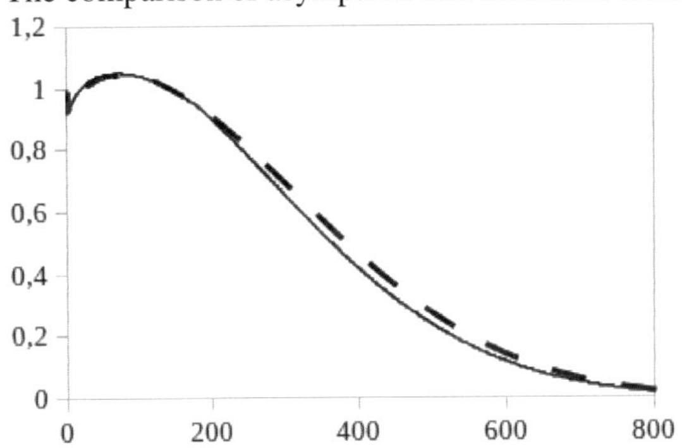

Figure 1. The circulation radial distribution $\frac{\Gamma(r_2)}{\Gamma_*}$ for the case $\mathrm{M}_* = 1$ at time $r_*/\sqrt{\nu t} = 5\cdot 10^{-3}$. The solid line is the composite asymptotic solution; the dashed line is the numerical solution using ANSYS Fluent.

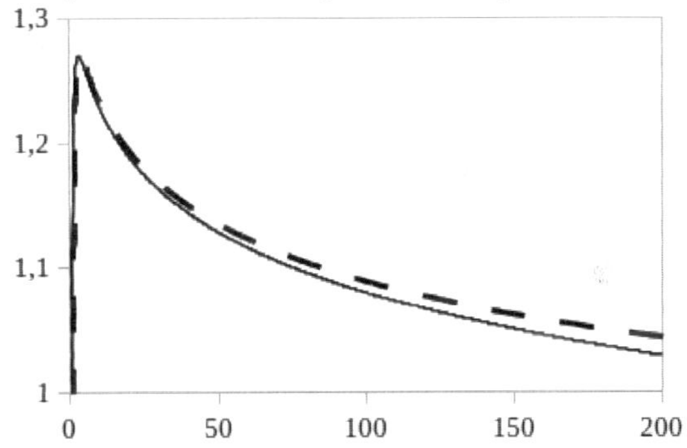

Figure 2. The temperature radial distribution $\frac{T(r_2)}{T_0}$ for the case $\mathrm{M}_* = 1$ at time $r_*/\sqrt{\nu t} = 5\cdot 10^{-3}$. The solid line is the composite asymptotic solution; the dashed line is the numerical solution using ANSYS Fluent.

Acknowledgements
The work was supported by RFBR (the project № 16-01-00128).

References
[1] Gaifullin A M and Zubtsov A V 2004 Diffusion of two vortices *J. Fluid Mech.* **39** 112
[2] Gaifullin A M 2005 Self-similar unsteady viscous flow *J. Fluid Mech.* **40** 526
[3] Gaifullin A M 2006 Flow past a plate with an upstream-moving surface *J. Fluid Mech.* **41** 375

IOP Conf. Series: Journal of Physics: Conf. Series **894** (2017) 012018 doi:10.1088/1742-6596/894/1/012018

Impact interaction of shells and structural elements of spacecrafts with the particles of space debris and micrometeoroids

A V Gerasimov[1], S V Pashkov[1] and Yu.F. Khristenko[1]

[1]Research Institute of Applied Mathematics and Mechanics by Tomsk State University, Tomsk, Russia

E-mail: ger@niipmm.tsu.ru

Abstract. Space debris formed during the launch and operation of spacecrafts in the circumterrestrial space, and the flows of micrometeoroids from the depths of space pose a real threat to manned and automatic vehicles. Providing the fracture resistance of aluminum, glass and ceramic spacecraft elements is an important practical task. These materials are widely used in spacecraft elements such as bodies, tanks, windows, glass in optical devices, heat shields, etc.

1.Introduction

The Lagrangian approach that has well-known advantages for multi-contact interactions of bodies during impact, especially when solving three-dimensional problems, can be successfully used for high-velocity impact problems considering the fragmentation of solids under intensive dynamic loading. One of the factors determining the fracture behavior of real materials is the natural heterogeneity of the material structure that influences the distribution of physical and mechanical material characteristics over the volume of the body under study. This factor can be considered in the equations for the mechanics of deformable solids by using the random distribution for the initial deviations of strength properties from the nominal value (simulation of initial defect material structures). Solving three-dimensional problems considers the natural heterogeneous real material structure that has an effect on the distribution of physical and mechanical material characteristics over the volume of structural elements and is one of the main factors determining the fracture behavior of real materials. Therefore, in order for the numerically simulated process to be in good agreement with experimental data, it is necessary to introduce disturbances into the physical and mechanical characteristics of the medium to be fractured, i.e. to specify a random distribution of the factors which determine the strength properties of the material.

2.Governing equations

The problems are solved using 3D formulation, considering a natural heterogeneous material structure that influences the distribution of physical and mechanical characteristics over the volume of structural elements and is one of the factors determining the fracture behavior.

The equations describing the spatial adiabatic motion of a solid compressible medium comprise the equation of mass conservation, the momentum equation and the energy equation [1-3].

The equations which consider the corresponding thermodynamic effects connected with adiabatic compression and the strength of a medium should be added to these equations. In the general case,

Content from this work may be used under the terms of the Creative Commons Attribution 3.0 licence. Any further distribution of this work must maintain attribution to the author(s) and the title of the work, journal citation and DOI.
Published under licence by IOP Publishing Ltd

when forces act on a deformed solid, there is a change in the volume (density) and the shape of the solid for different dependencies. Therefore, the stress tensor is the sum of the spherical tensor and deviator of the stress tensor.

To describe the shear strength of a body, the Prandtl - Reuss equations and the von Mises yield condition were used [1]. The state equation of the solid was chosen in the form of Mi-Grüneisen [1]. The criterion of ultimate equivalent plastic strain was used as a shear fracture criterion [1]. The fracture of glass and ceramics was considered to be the fracture of a brittle material without a plastic strain zone that is typical for metal strain. The Johnson-Cook relations were used to simulate the fracture of ductile materials (aluminum, steel) [4]. The Johnson-Holmquist (JH2) relations were used to simulate the fracture of brittle materials (glass, ceramics) [5].

Solving three-dimensional problems considers the natural heterogeneous real material structure that has an effect on the distribution of physical and mechanical material characteristics over the volume of structural elements and is one of the main factors determining the fracture behavior of real materials. Therefore, in order for the numerically simulated process to be in good agreement with experimental data, it is necessary to introduce disturbances into the physical and mechanical characteristics of the medium to be fractured, i.e. to specify a random distribution of the factors which determine the strength properties of the material. The natural fragmentation of projectiles and a target is simulated by introducing a probabilistic mechanism for the distribution of initial defects in the structure of the material to describe the formation of cracks.

The computation technique uses tetrahedral cells and is based on the combination of the Wilkins method for the calculation of internal body points and the Johnson method for the calculation of contact interactions [2,7-8]. The ideology and methodology of a probabilistic approach applied to the solid fracture problems is given in detail in the monograph [9].

A new three-stage light-gas gun (RF patent No. 2490580) has been developed, manufactured and mounted to study the processes of high-velocity interaction. It allows different studies of high-velocity impact processes to be conducted at velocities of up to 8 km/s.

3.Test computations

The problem of the ring fracture [10] and the problem of the two-layered target (fiberglass ST-HT + alloy D16) fracture by a ball (G52986) were numerically simulated for the normal and oblique impact with the target surface at an impact velocity of 900 m/s as test computations [11]. The computations were in good agreement with the experimental data.

4.Computations

The main tasks which should be solved during the high-velocity impact of space objects with particles are to preserve the integrity of different spacecraft elements (bodies, tanks, solar panels, etc.) subjected to the impact of sufficiently large fragments and reduce erosion for the surface of structural elements exposed to the flows of small and ultrafine particles.

The first problem considers the impact of a cluster of six spherical elements with one and two targets at a different velocity and a different angle to the surface of the target. The impact angle is counted from the normal line to the face of the plate in the clockwise direction. The size of the plates was 5x5 cm, the thickness was 0.35 cm, and the material was aluminum. The material of the ball 0.56 cm in diameter was steel. The group of balls was placed in a circle with a diameter of 1.5 cm, and all the balls simultaneously interacted with the target. The target was fractured and a distinctive fragment of the target flew at a velocity of 400 m/s, which was less compared to the velocity of the projectile fragments (~ 450 m/s, Figure 1) at an impact velocity of 800 m/s. It can be concluded that the secondary flow of fragments can cause significant fracture to the second target.

IOP Conf. Series: Journal of Physics: Conf. Series **894** (2017) 012018 doi:10.1088/1742-6596/894/1/012018

a b

Figure 1. Group impact of compact elements with a thin target at a velocity of 800 m/s: (a) - t = 10 μs; (b)-t = 75 μs.

The consequences of this impact are clearly observed with increasing the impact velocity. Figure 2 shows the interaction of the cluster of projectiles with a two-layered spaced target for the time t = 20, 50, 77 μs. The distance between the plates was 0.3 cm. The flow of the target fragments which are formed after the penetration of the first target differs in velocity. For the projectiles, the velocity is about 1100 m/s, and for the target fragments, the velocity is varied from 600 m/s to 430 m/s. After the penetration of the second target, the velocity of the projectile fragments is decreased to 650 m/s, and the velocity of the target fragments is a spectrum from 370 to 250 m/s. The formed powerful flow of fragments poses a great threat to the object protected by the target.

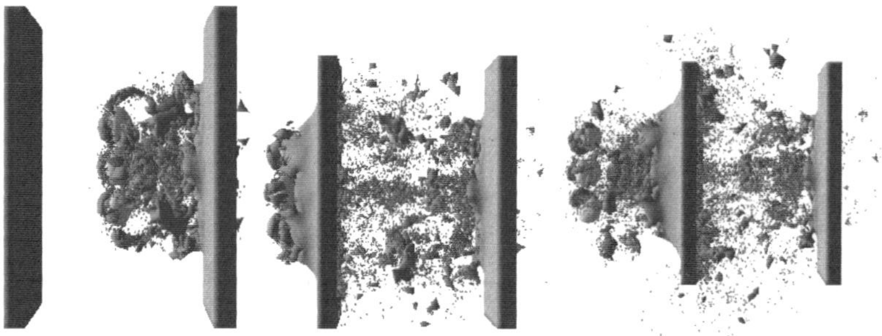

a b c

Figure 2. Group normal impact of compact elements with a thin target at a velocity of 1500 m/s: (a)-t = 20 μs; (b) - t = 50 μs; (c)-t = 77 μs.

The deviation of the impact angle by 60° does not significantly change the fracture behavior of the plate. For these and following computations, the size of the plates was 8x8 cm and the thickness was 0.5 cm. Here, as in the previous problems, a hole is formed, which is approximately equal to the diameter of the initial circle of the balls on the target, and also a multi-velocity flow of the projectile and target fragments is formed. Several stages of this process are shown in Figure 3 (time t = 10, 50, 130 μs).

a b c

Figure 3. Group impact of compact elements with a thin target at an angle of 60° and a velocity of 1500 m/s: (a)-t = 10 μs; (b)-t = 50 μs; (c)-t = 130 μs.

IOP Conf. Series: Journal of Physics: Conf. Series **894** (2017) 012018 doi:10.1088/1742-6596/894/1/012018

The impact of a group of 7 tungsten projectiles 5 mm in diameter at a velocity of 1000 m/s with a two-layered glass and ceramic target is shown in Fig. 4. The radius of the circle along which the elements of the group were located was 8 mm. The diameter of the target was 3 cm, and the thickness of the layers was 3 mm.

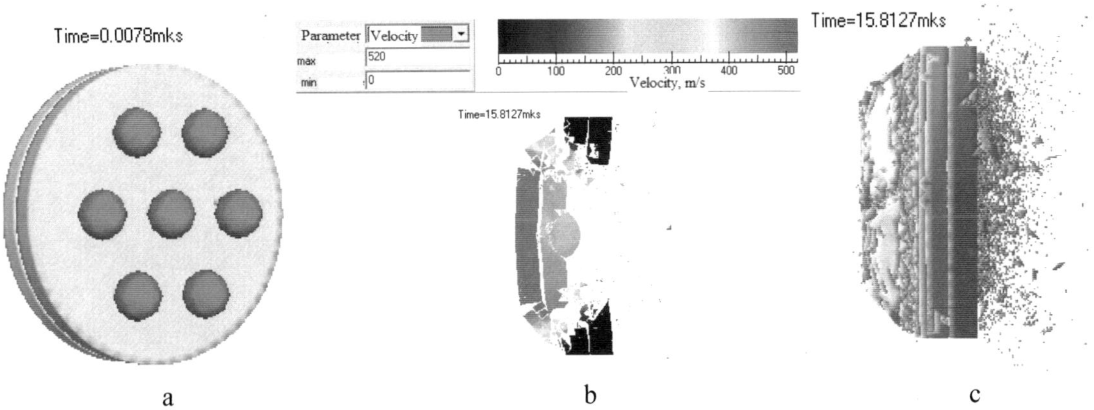

Figure 4. Group normal impact of 7 compact tungsten elements with a two-layer target (glass-ceramics) at a velocity of 1000 m/s: (a)-initial configuration; (b)-two-dimensional cross-section of the computational area; (c)-3D computations for the impact and fracture of the target; t = 15, 8127 μs.

During impact interaction, there is the intensive fragmentation of the first glass layer with its partial ejection from the front surface of the target, and a fragment in the form of a "plug" flies out of the second ceramic layer (Figure 4 (b)) at a residual velocity of 520 m/s.

Figure 5 shows the computations for the impact of the group of 7 steel, aluminum and ceramic particles 1 mm in diameter at a velocity of 3000 m/s with two-layered aluminum, glass and ceramic targets. The radius of the circle along which the elements of the group were located was 4 mm. The diameter of the target was 2 cm, and the thickness of the layers was 1 mm.

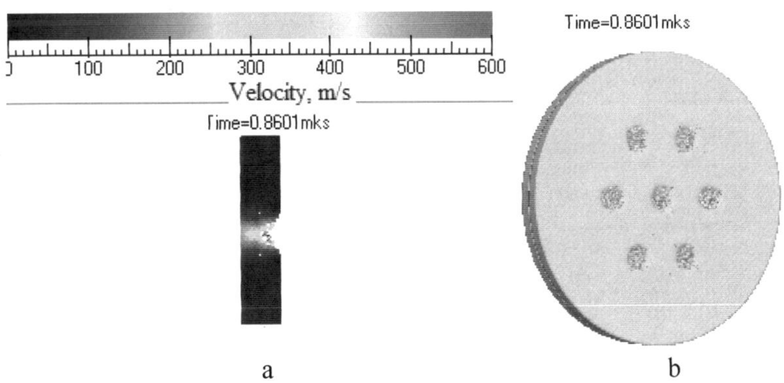

Figure 5. Group normal impact of 7 compact ceramic elements with a two-layer target (aluminum-aluminum) at a velocity of 3000 m/s: (a)-two-dimensional cross-section of the computational area; (b)-3D computations of the impact process; t = 0.86 μs.

During the impact of ceramics with aluminum plates, there is the intensive fragmentation of ceramic projectiles and the reduction in the velocity of their fragments up to zero. The penetration of the target is not observed, but we can track the formation of craters, their depth and size on the front surface.

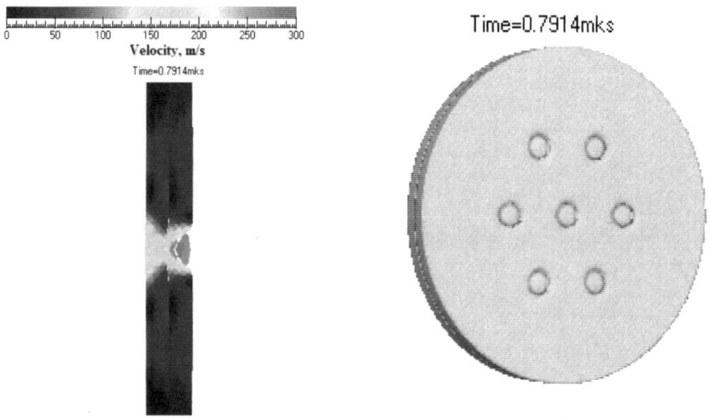

a b

Figure 6. Group normal impact of 7 compact aluminum elements with a two-layer target (aluminum-aluminum) at a velocity of 3000 m/s: (a)-two-dimensional cross-section of the computational area; (b)-3D computations of the impact process; t = 0.63 μs.

During the impact of an aluminum projectile with aluminum plates (Figure 6), there is the deformation of projectiles, an increase in the size of a crater and a reduction in the velocity of projectiles to zero. The penetration of the target is not observed, and there is no noticeable fracture of bodies. In contrast to the brittle fracture of ceramics, the plastic deformation of aluminum projectiles is observed.

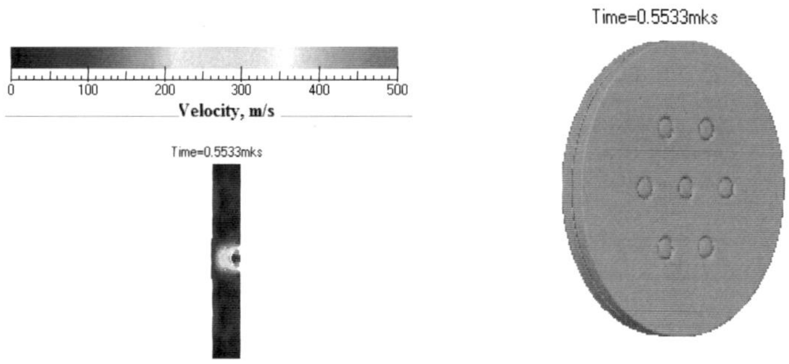

a b

Figure 7. Group normal impact of 7 compact aluminum elements with a two-layer target (ceramics-glass) at a velocity of 3000 m/s: (a)-two-dimensional cross-section of the computational area; (b) -3D computations of the impact process; t = 0.5533 μs.

At present, there is a need in ceramic coatings to protect the glass elements of spacecrafts. Figure 7 shows the computations for the glass-ceramic system subjected to impact loading by an aluminum projectile. The results obtained allow the deformation of the glass-ceramic system to be qualitatively estimated and the dimensions of protective layers to be quantitatively analyzed for the different impact velocities and sizes of projectiles.

The results obtained show that the group impact of a cluster of high-velocity elements leads to the significant fracture of targets and the formation of a hole, the size of which exceeds the total area of projectiles in the plane projection of the target. The formed fragmentation field behind the first target leads to the fracture of the second protective plate and can cause irreparable damage to the main equipment. A decrease in the angle of impact leads to the ricochet of projectiles and a change in the deformation of an aluminum plate. For the small sizes of projectiles, the penetration of the first layer is observed in the construction and the integrity of the second layer is preserved.

Acknowledgments
This work was partially funded by the Russian Foundation for Basic Research, the grants No. 16-08-00037 A, No. 15-08-06689 and No. 15-08-02265 A.

References

[1] Stanyukovich K *et al* 1975 *Physics of Explosion* (Moscow: Nauka)
[2] Wilkins M 1967 *Computational Methods in Hydrodynamics* (Moscow: Mir)
[3] Wilkins M 1999 *Computer simulation of dynamic phenomena* (Berlin-Heidelberg-New-York: Springer)
[4] Kuzkin V and Mikhalyuk D 2010 *Computational Continuous Mechanics* **3** 32
[5] Egorova M and Parshikov A 2015 *Physico-Chemical Kinetics in Gas Dynamics* **16** 1
[6] Kreyenhagen K *et al* 1970 *AIAA Journal* **8** 2147
[7] Johnson G, Colby D and Vavrick D 1979 *Int. J. Numer. Methods Engng.* **14** 1865
[8] Johnson G 1981 *Trans. ASME. J. of Appl. Mech.* **48** 30
[9] Gerasimov A *et al* 2016 *High-velocity impact. Simulation and Experiment* (Tomsk: STL)
[10] Gerasimov A, Alexeev N, Mikhailov V and Pashkov S 2005 *AIP Conference Proceedings* **849** 57
[11] Gerasimov A, Pashkov S and Surkov V 2007 *Russian Physics Journal* **50** 46

Modeling of thermomechanical processes in woven composite material at blow by the striking element

V O Kaledin[1], A Ye Gilyova[1], O N Budadin[2] and S O Kozelskaya[2]

[1] Department of Mathematics and Mathematical Modeling of Novokuznetsk Institute (Brunch) of FGBOU VO Kemerovo State University, Russia
[2] Central Research Institute for Special Machinery, JSC, Khotkovo, Moscow Region, Russia

The paper studies mechanical and thermomechanical interaction processes between striking element and multilayered woven barrier made of polymeric fabric. A simplified mathematical model is proposed. This model describes the inhibition of the damaging element due to energy dissipation, that is spent on irreversible strains of fabric threads stretching, friction during the mutual slippage of threads and barrier material heating. The defining equations parameters are estimated by identification according by the experimental data. The computational experiment results using the mathematical model can be used to diagnose armor protection according to the measurement of dynamic temperature fields.

1. Introduction

One of perspective methods of armored protection quality assessment is the analysis of the dynamic temperature field on a surface at interaction with the striking element [1, 2]. However, full-scale experimental research is an expensive action, and therefore the development of a theoretical model is relevant. The model for determining the energy absorption capacity of armored barriers and the dynamic temperature field on the surface allows for a comparative evaluation of materials and their optimal placement in armor structures without expensive experiments.

2. Thermomechanical model for interaction processes between armored protection and striking element

The squared layered woven sample is considered (Figure 1). Each layer is formed by two mutually perpendicular families of filaments: the weft and the base. The sample is on a malleable base. At the moment of impact, the central part of the sample contacts the striker, the mass M and its initial velocity V0 are known.

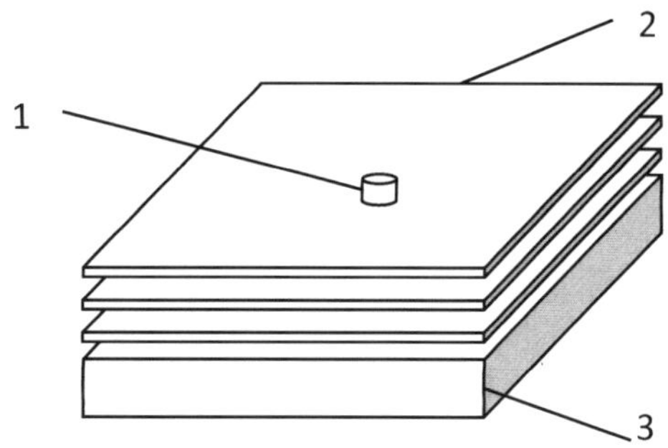

Figure 1. Layered woven pattern with a striking element: 1 - impactor, 2 - layer of tissue, 3 - base.

Content from this work may be used under the terms of the Creative Commons Attribution 3.0 licence. Any further distribution of this work must maintain attribution to the author(s) and the title of the work, journal citation and DOI.
Published under licence by IOP Publishing Ltd

The speed of the impactor is slowed down in the interacting process; kinetic energy turns into frictional energy threads, work of structural changes and destruction of threads material, irreversible stress energy in the basis and heating of a sample, the impactor and threads. At the same time variable conditions of model change in time: movements U, speed $\dot{U}$, internal stress Σ and temperature T. It is required to construct mathematical model which displays variables impacts Ξ={M, V0} on variable states H={U, $\dot{U}$, Σ, T}:

$$H = \Psi(\Xi). \tag{1}$$

We will enter Lagrangian coordinates for each fabric layer (Figure 2) where axes α and β are sent to along threads, and the axis n is perpendicular the layer planes. These coordinates are time-independent in the process of layer movement, and Euler coordinates *(x, y, z)* change over time. We will demand that axes n of all Lagrangian systems remained parallel to an axis z of Euler system.

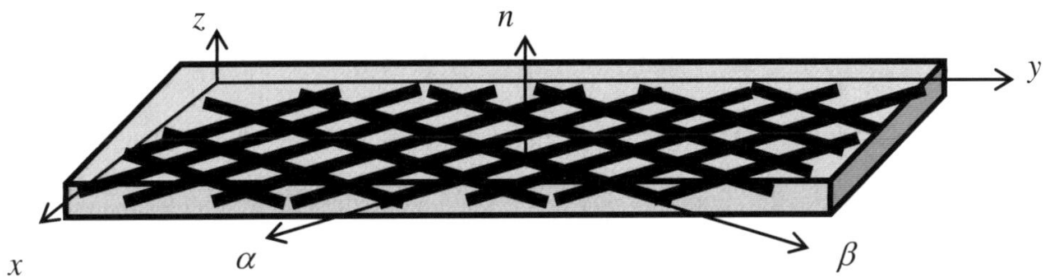

Figure 2. Euler coordinates (x, y, z) and Lagrangian coordinates (α, β, n)

The provision of a layer of fabric is defined by the current provision of a coordinate surface (α, β) – a median surface of a layer. Each point of the surface moves along an axis z. Full points movement of threads consists of the middle surface's motion and relative movement of thread within a layer. Thus, the figurative movement is defined by shifts along an axis z, and relative – along Lagrangian coordinate lines. Simplifying model, we will accept that the relative shift of threads points occurs along the direction of the same thread. The movement components along the reference vectors that are tangent to lines α and β refer to different between (base and weft) of the same layer

Taking into account the made simplification, knowing figurative movements, relative lengthening along threads of a basis and a weft will be equal

$$\lambda_\alpha = \sqrt{1+\left(\frac{\partial u_z}{\partial \alpha}\right)^2}, \quad \lambda_\beta = \sqrt{1+\left(\frac{\partial u_z}{\partial \beta}\right)^2}, \tag{2}$$

and relative rapprochement of layers

$$\lambda_n = \frac{\dfrac{\partial u_z}{\partial z}}{\sqrt{1+\left(\dfrac{\partial u_z}{\partial \alpha}\right)^2+\left(\dfrac{\partial u_z}{\partial \beta}\right)^2}} \tag{3}$$

The accounting dissipative forces is the most complex challenge of the model creation. Friction forces are the most essential, they arise at the mutual shift of layers and when slipping threads in woven layers. We will allocate two components of dissipative force – "viscous" "rigid". The "rigid" component of dissipative force limits growth of mutual shifts at a small tension, and the "viscous" component limits the speed of threads shift basis relatively a weft.

The movement of the modelled object is unambiguously described by functions of two coordinates: figurative movements of layers u_z, relative movements of basis threads of each fabric layer u_α and relative movements of threads weft u_β We will apply the variation principle of Lagrange according to which the variation of work external forces is equal to work internal forces and inertial forces on variations the generalized movements:

$$\int_V \rho \dot{V}_i \delta u_i dV + \int_V \sigma_k \delta\varepsilon_k dV + \int_V \tau_{nk} \delta\gamma_{nk} dV + \int_S f_k \delta u_k dS = P_i \delta u_i . \tag{4}$$

Here the index *i* belongs to Euler basis, an index *k* – to Lagrangian. Integration on the volume *V* is carried out at calculation of a inertia force variation works , tension in threads and the conditional tension of interlayered shift. The works variation of dissipative forces at relative slipping of threads is calculated as integration on layers surfaces *S*.

The formulated model is closed, but the equations can't be solved analytically. For integration of the equation of the movement (4) it is necessary to use a numerical method.

3. The numerical scheme for the allowing system of the equations

The multilayered sample, rectangular in the plan is a set of final elements – parallelepipeds. Knots of elements settle down on a median surface of a layer. The pliable basis is allocated as a separate layer which moves together with the lower layer of fabric. Thus, the knots layer of a grid is connected with each fabric layer.

Initial knots coordinates is Lagrangian coordinates which don't change in the course of the movement. The Lagrangian basis is knotted with everyone. In this basis the axis α is directed along a fabric basis, β axis - lengthways a weft, and the axis z is directed down and doesn't change the situation at deformation.

Movements of all knots unambiguously define the current configuration of model, and they are the generalized movements. We will designate the generalized movement corresponding to degree of freedom Δ, through u_Δ (numbers of degrees of freedom are designated by capital Greek letters). The number of freedom degrees of discrete model is final, therefore movements variations δu can't be any coordinates functions in (5), the only variation of the movements field answers each freedom degree.

$$\delta u^{(\Delta)}(\alpha,\beta) = \psi_\Delta(\alpha,\beta) \cdot \delta u_\Delta . \tag{5}$$

Here δu_Δ - variation of the generalized movement u_Δ,

$\psi_\Delta(\alpha,\beta)$ - basic function of coordinates,

$\delta u^{(\Delta)}(\alpha,\beta)$ - variation of the movements field at a variation of the generalized movement u_Δ.

For receiving a discrete analog of the inertial member (first composed in equality (4)) it is necessary to consider that the generalized movements are referred to Lagrangian basis, and the vector of acceleration has to be expressed in Euler coordinates. It is necessary to cover separately the movement of threads basics and threads a weft as their movements and speeds are various.

Speed of each of families of threads consists of the figurative speed directed along an Euler axis *z*, and the relative speed directed on a tangent to the deformed layer surface. Resulting relative speed in Euler basis, we will receive:

$$\dot{u}_x = \dot{u}_\alpha \cos(x,\alpha) + \dot{u}_\beta \cos(x,\beta) ,$$
$$\dot{u}_y = \dot{u}_\alpha \cos(y,\alpha) + \dot{u}_\beta \cos(y,\beta) , \tag{6}$$

$$\dot{u}_z = \dot{u}_{ez} + \dot{u}_\alpha \cos(z,\alpha) + \dot{u}_\beta \cos(z,\beta),$$

where $\dot{u}_{ez}$ – transport velocity.

Further, we interpolate figurative and relative speeds, using basis (6), and we will find kinetic energy by integration on volume in Lagrangian coordinates. Lowering intermediate calculations, we will receive that kinetic energy T has an appearance of a square form of rather figurative and relative speeds of final element model knots $\dot{u}$, and square form matrix depends on the current movements:

$$T = \frac{1}{2}\dot{u}_\Delta M_{\Delta\Gamma}\dot{u}_\Gamma. \quad (7)$$

We will receive inertial composed discrete analog from Lagrange's equation of the second sort:

$$\int_V \rho\ddot{u}_i\delta u_i dV = \delta u_\Delta\left[M_{\Delta\Gamma}\ddot{u}_\Gamma + \dot{M}_{\Delta\Gamma}\dot{u}_\Gamma\right]. \quad (8)$$

Further we will consider third and fourth composed in the equation (8) which, contain two components – "rigid", linearly depending from the generalized movements, and "viscous", proportional to the generalized speeds of discrete model. The variation of work of these forces can be presented in shape:

$$\int_V \tau_k \delta u_k dV = \delta u_\Delta\left(R_{f\,\Delta} + K_{f\,\Delta\Gamma}u_\Gamma + C_{f\,\Delta\Gamma}\dot{u}_\Gamma\right), \quad (9)$$

where R_f - constant composed,

K_f - slipping rigidity matrix,

C_f - slipping viscosity matrix.

Finally, the discrete analog of the equation (4) has an appearance system of the linear differential equations with float factors

$$M_{\Delta\Gamma}\ddot{u}_\Gamma + C_{\Delta\Gamma}\dot{u}_\Gamma + K_{\Delta\Gamma}u_\Gamma = R_\Delta, \quad (10)$$

where M – masses matrix,

C – total viscosity matrix with $\dot{M}_{\Delta\Gamma}$,

K – total rigidity matrix,

R – total vector of constants (on a small time step) the generalized forces.

The system of the equations (10) decides under an entry condition:

$$u_\Delta(0) = 0, \quad \dot{u}_\Delta(0) = v_{\Delta 0}, \quad (11)$$

where $v_{\Delta 0}$ - initial velocity.

For numerical integration of the differential equation with matrix coefficients (10) under entry conditions (12) we will enter a small step on time τ and we approximate derivatives the differential relations (the top index designates number of timepoint):

$$Y^{t+1} = \frac{X^{t+1} - X^t}{\tau}, \quad \dot{Y}^{t+1} = \frac{Y^{t+1} - Y^t}{\tau}, \quad (12)$$

where X – generalized movements column matrix,

$Y = \dot{X}$ - generalized speeds column matrix.

Substitution (12) in (10) gives the following differential equation:

$$\begin{cases} \left(M^t + \tau C^t + \tau^2 K^t\right) Y^{t+1} = M^t Y^t - \tau\left(K^t X^t + R^t\right) \\ X^{t+1} = X^t + \tau Y^{t+1}. \end{cases} \tag{13}$$

The spectral analysis of the scheme (13) shows her absolute stability [3]. At the same time, this scheme has positive viscosity (isn't conservative) that demands adjustment of a step depending on the current speed and a posteriori assessment of an error of integration.

4. Computing program implementation technique

The algorithm used for formation system (13), its numerical integration on time and armored barrier material condition parameters calculations on each temporary step consists of a large number of the interconnected. This has led to complexity of the computing program and debugging difficulties. Essential simplification of computer program development is reached by use for a functional and object paradigm [4], which the algorithm is built on functional objects – program structural units. The main component of program realization is the functional and object scheme which reflects functional objects dependence. Algorithm elementary steps are calculation of value, which a functional object is responsible. Functional and object schemes and specialized classes of objects in language C ++ have been developed for solution realization.

Calculation results (movement, speeds, tension and temperature) are unloaded to file; the dynamic graphics creation tool program is developed for their viewing and schedules creation.

5. Power analysis for interaction processes between striking element and armored protection

Kinetic energy of the striking element in the course of interaction with a textile armored package is spent on: work of deformation of threads at their stretching, work of friction forces when slipping (prodyorgivaniye) of threads, the message of kinetic energy to threads of fabric which dissipates further in the form of heat.

Change of kinetic energy of the armored package consisting of 12 layers of fabric of a twill weave with a density of material of 1450 kg/m^3 was investigated. The armored package had area density of 2 kg/m^2 and settled down on the plasticine basis. The steel ball weighing 1,05 g acted as the striking element.

We will show kinetic energy integrated size distribution in the course of impact of the striking element with textile armor at an initial speed 200 m/s and 400 m/s, kinetic energy of the striker is equal 20 J and 80 J respectively.

In the Figure 3 the curve of distribution of kinetic energy is presented, on a horizontal axis time of contact of the striking element with fabric, and on vertical – kinetic energy is postponed.

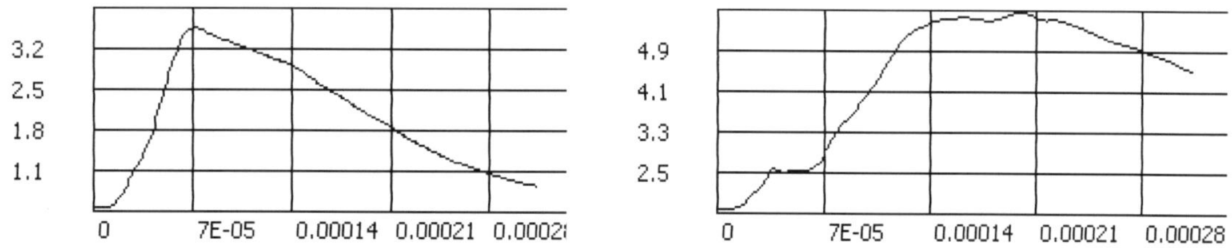

a б

Figure 3. A curve of kinetic energy distribution at a striker initial speed: a – 200 m/s, b – 400 m/s.

In the course of impact of the striker with armored fabric growth of kinetic energy of a woven package before achievement of value 3,5 J in a case with initial speed in 200 m/s and 5,6 J is observed at a speed 40 m/s. Further kinetic energy begins to decrease, turning into irreversible work of deformations of threads (Figure 4) and work of internal friction of threads (Figure 5).

At an initial speed of the striker in 400 m/s the saved-up irreversible work of destruction makes 14% of initial kinetic energy of the striking element whereas at speed of 200 m/s only less than 2%. Such distinction is explained by existence of destructions of threads at speed of 400 m/s.

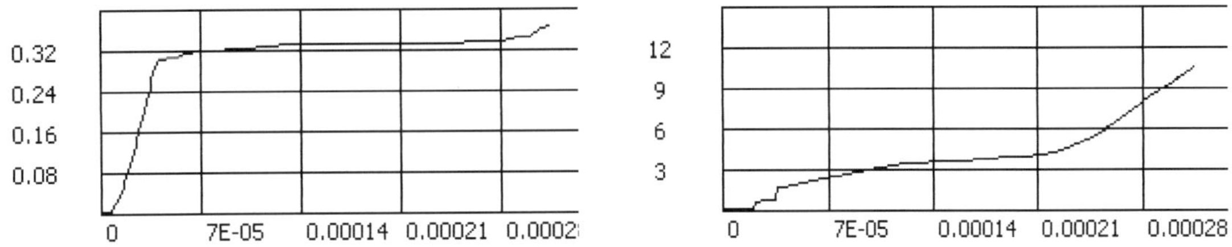

a б

Figure 4. A curve of irreversible deformations work of distribution at a striker initial speed: a – 200 m/s, b – 400 m/s

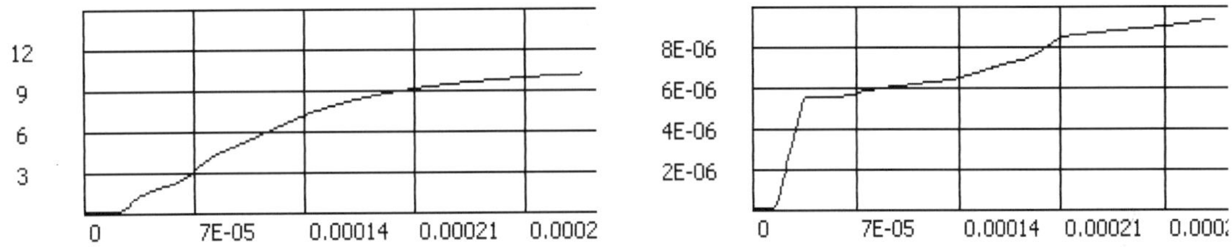

a б

Figure 5. A curve of internal friction work distribution at a striker initial speed: a – 200 m/s, b – 400 m/s

In the Figure 5, opposite result are observed. At speed of 200 m/s about 50% fall to the share of work of internal friction because of slipping of threads, and at an initial speed of 400 m/s only 10^{-7} %.

6. Conclusions

The mathematical model is developed. This model adequately describes the physical processes happening at interaction between the striking element and multilayered armored composite material: the joint movement of armored material and the striking element, irreversible deformation of threads, slipping of threads, losses of kinetic energy on friction and irreversible deformation, and also heating. The quality control possibility of an armored composite material an is shown on the basis of the dynamic temperature fields analysis. Results of theoretical researches qualitatively correspond to the known experimental data. Individual control of model by identification of physicomechanical parameters is necessary for receiving quantitative coordination of calculation and an experiment. Calculations allow to assess the striker kinetic energy, which is spent for friction forces and irreversible work of the stretching tension.

References

[1] Kharchenko E F and Ermolenko A F 2013 *Composite, textile and composite armor material: in 3 volumes. Vol. 1. Mechanisms of interaction with ballistic striking elements* (Moscow) p 294

[2] Barynin V A, Budadin O N and Kul'kov A A 2013 *Modern technologies of non-destructive testing of structures from polymeric composite materials* (Moscow: Izdatel'skii dom "Spektr") p 242

[3] Samarskiy A A 1989 *Theory of differential schemes* (Moscow: Nauka) p 616

[4] Kaledin V O, Kryukova Y S, Nagaytseva N V and Ravkovskaya E V 2014 *Software Systems for Algorithmization of Numerical Solution of Continuum Mechanics Problems* (News of Altai state University vol 1 №1) 161–164

IOP Conf. Series: Journal of Physics: Conf. Series **894** (2017) 012020 doi:10.1088/1742-6596/894/1/012020

Molecular-dynamics investigation of the nano-bar stability

E I Golovneva[1a], I F Golovnev[1] and L A Merzhievsky[2]

[1] Institute of Theoretical and Applied Mechanics SB RAS, Novosibirsk, 630090, Russia

[2] Institute of Hydrodynamics SB RAS, Lavrentiev str., 15, Novosibirsk, 630090, Russia

E-mail: elena@itam.nsc.ru

Abstract. This paper presents the analysis of the Eulerian instability of the nano-bar by the molecular dynamics method. The results shows that the same critical value of the external compressive stress is the bifurcation point both for the quasi-static structure contraction (alternation of the modes of instability and further fracture) and for the impact action (alternation of the energy absorption modes and occurring wave of transformation defects).

1. Introduction

The series of works [1 – 3] are address to the analysis of the wave processes in nano-structures under the action of the external compressive stress within the range up to 4 GPa.

Performed analysis gave three intervals of the external compressive stresses. In the first interval, below 2.6 GPa, the total energy absorbed by the system is described by the quadratic dependency in the applied stress. In the second interval, from 2.6 to 2.8 GPa, the process of linear wave propagation incrementally transits to the mode of formation of transformation (or amorphization) waves. And in the third interval from 2.8 GPa to 4 GPa, this dependency was close to the linear one.

It stimulated the research of the Eulerian stability of nano-structures, but on the atomic scale, to avoid initial preset of any phenomenological models aside for the interatomic interaction potentials. By now, there are considerable amount of works addressed to the stability of various nano-size structure.

Note that this term means for the authors the structure stability in respect to its chemical composition, shape, size [4 – 8]. In some papers, the molecular dynamics (MD) method was used to study the chemical structure stability, like [9]. Moreover, under study is the thermodynamic stability [10 – 11] and the non-linear response of the structure to the external action [12]. The authors [13] research the plastic deformation in stressed crystallites under irradiation by MD simulation. MD method got broad areas of applications in last decades, so in the work [14], scattering processes of argon beam impinging on tungsten surface are investigated numerically by applying MD simulations.

However, almost no works are devoted to the analysis of the Eulerian stability of the bar. And especially few works are addressed to the solution of this task by the molecular dynamics method. Quite frequently the continuum methods are used when considering nano-objects and stating the task of Eulerian stability definition, like [15 – 18]. In [15], under consideration are spinning nanotubes under the action of the tangential stress. The motion equations have been obtained on the base of the Hamilton principle in the nanotube energy terms, regarding the rotational speed and compressive

Content from this work may be used under the terms of the Creative Commons Attribution 3.0 licence. Any further distribution of this work must maintain attribution to the author(s) and the title of the work, journal citation and DOI.
Published under licence by IOP Publishing Ltd

tangential stress. The authors compare the results obtained within the Euler theory with the ones obtained within the Timoshenko theory (the Euler–Bernoulli beam theory (EBT) and Timoshenko beam theory (TBT)). Thus this task has been solved here in the framework of the continuum mechanics. In [16], they applied the Galerkin finite-elements method. Multi-layer carbon nanotubes are the object of analysis in [17]. There, the nonlocal Euler Bernoulli beam theory is proposed for the free vibration and stability analyses of a multiple-nanobeam system (MNBS) using the Eringen nonlocal continuum theory.

The work [19] is the particular case. Here, the molecular dynamics method is used to analyze the nano-mechanical behavior of the β-SiC nano-wire at the extension, compression, spinning, combined extension and rotation, and combined compression and rotation. The Tersoff potential is used, the crystal orientation is chosen [111]. It is discovered that the defect depends on the size and temperature. At the uniaxial compression, the SiC nano-bar is crumpled by the column settling or bending regarding the length and diameter of the nano-bar, which agrees with the analysis of the equivalent continuum structures, on the base of the Euler theory of bar bending. The nano-bars are crumpled via the phase transition from the crystal structure to amorphous one in several atomic layers at the uniaxial spinning extension. Under the combined load, the spinning with the low spinning speed does not influence the defect and bending modes, but the critical stress decreases as the spinning speed rises. The plastic deformation occurs in the bending zone at the further rise of the combined load.

The nano-bars of various cross section shapes were considered: the hexagonal, triangular, rectangular, rhombohedric, octagonal, circular ones. But the length of the supercell was always similar. The authors found that the bars with the hexagonal cross section had the minimal potential energy, i.e. the most stable structure. It agrees with the experimental data wherein it is shown that the structures with the natural shape of the cross section are most stable. Moreover, σ-ε diagrams were plotted for the nano-bars of various diameters and at various temperatures.

The criticism of [19] includes the fact that the supercell length (supercell, 3.02, 4.53 nm) is too small to avoid the cross of the disturbance wave and the disturbance wave reflected from the opposite side of the nano-bar. The other difference of the results presented in this paper is that under consideration is metal, not ceramics. What is especially important, the aim of the work is to study not only the Eulerian stability of the bar, but the effect of the loss of bar stability on the transformation of the sample crystal structure.

This work is devoted to the analysis of the relation between the synergetic transition in the intensive dynamic process with the stability loss in the quasi-static process under the action of the compressive stress.

To do this, the specific task of compression of metal nano-structures shaped as rectangular parallelepipeds and getting of uniform states along the nano-structure length was solved.

The characteristic scale of the studied phenomena in the space and time is about 10^{-9}cm – 10^{-12}s, respectively. Hence, experimental investigation of such processes is impossible now. It caused the application of the molecular dynamics method based on the first principles. The method is based on the calculation of atom trajectories in the phase space on the base of the classic motion equations, but in the potential fields found in the framework of the quantum theory regarding electronic states. The calculations gave the fullest information on the system: the coordinates and impulses of every atom at every time instant. It permits performing detained investigations of the process.

2. Physical system and mathematical model

The technique and software complex for the MD analysis of the disturbances propagation in the nano-structures were adjusted on the example of a copper cluster shaped as the rectangular parallelepiped, its sizes were 50*5*5 and 200*5*5 crystal cells along the coordinate axes XYZ, respectively. The crystal orientation was chosen (1,0,0). To describe the interatomic interaction, the widely approved A. Voter EAM potential [20] for metals was chosen.

Initially, the atoms were in the nodes of the perfect crystal FCC lattice. Then, the system coordinates and impulses were found in the minimal potential energy state by the artificial viscosity method. Then the found values of the coordinates and impulses were used as initial data.

The calculation of macro- and meso- characteristics of the crystal is described in detail in [1].

The well-known velocity Verlet modification of the second accuracy order, with the time step 10^{-16}s was used as the numerical scheme [21]. It should be just stresses that in the case of the isolated system, the energy error is no more than 10^{-5} % within the time gap of 50 ps.

3. Calculation results

3.1. Adiabatic compression of the nano-structure

The value of the external compressing stress acting on the left side edge lying at the initial time instant in the $X = 0$ plane rose linearly within a certain time interval T_σ up to σ_0. Since the forces $\vec{f}_a$ acting on individual atoms are specified in the MD model as the initial value, they were calculated in the program by the simple formula

$$f_{ax} = \sigma_0 S_1 / n_1$$

where: S_1 is the area of the crystal left edge, and n_1 is the number of atoms on this edge. Since the edge area may vary somehow during the compression, it was re-calculated at each time step, similarly to the external forces acting on the atoms, providing that $\sigma_0 = const$. In other aspects, the movable clamp atoms were free.

To simulate the structure stop over the wall, the right edge (immovable clamp) was fixed in the space by means of the generalized potential

$$V(\vec{r}_i) = \frac{k}{2}\left(x_i - x_i^0\right)^4$$

where $\left(x_i^0, y_i^0, z_i^0\right)$ are the coordinates of the ith atom after cooling. Such a fixation enables to atoms to move freely in the YZ plane, but limits their motion along the X axis.

Mass calculations are carried out for the external compressing stress σ_0 values within the range up to 4 GPa with the step 0.1 GPa.

The preliminary method proposed by the authors was the quasi-static loading of the system resulting in the stationary and spatially uniform state of the system [22]. The monotony rise of the external load (under the condition of smallness of its time derivative) and artificial viscosity are the necessary conditions of such an action.

For the structure compression, the external stress σ_{ext} linearly rising within 10 ps up to the preset value σ_0 was supplied to the crystal left edge.

The viscous force model was used at the growing external loads

$$\vec{f}_i = -\nu \vec{p}_i,$$

acting on each i-th atom ($\vec{p}_i$ is the impulse of the i-th atom). The value of the viscosity coefficient is $\nu = 0.1$ in the program entities. It enabled to simulate the quasi-static growth of the external load and remove the wave processes in the nano-structure.

At the same time, the dependency of the potential energy on time shows that the time of system relaxation is above 100 ps for the chosen parameters of the external stress relaxation and viscosity value.

All these factors caused the check of the interval duration of the system relaxation into the equilibrium state. Figure 1 shows the dependency of external forces deflection on the clamps

$$errF = \lg\left(\frac{|F_{ex1}|-|F_{ex2}|}{|F_{ex1}|}\cdot 100\%\right)$$

on the time within the gap of 500 ps.

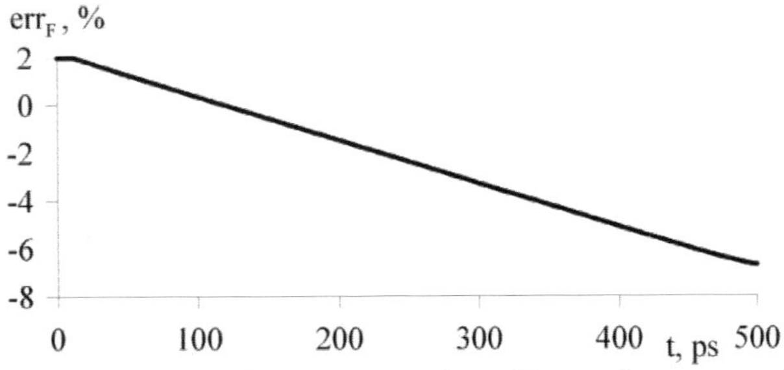

Figure 1. External forces on clamps versus time. External stress amplitude 1 GPa.

Evident that the deflection linearly decreases within this interval from 100% to 10^{-6} %. It means that the sum of the external forces acting on the system equals to zero accurate to the eighth place.

3.2. Analysis of instability characteristics at the adiabatic compression of the x nano-structure
Since the major task of this stage is the sought of the critical compressive stress at which the structure does not break, the numerical experiment was performed with the external load interval from 1 GPa to 4 GPa. It was preconditioned that the system should survive the time gas of 500 ps without fracture. The pre-check showed that by this time, the parameter distribution over the structure was quite uniform. As a result is was shown that the structure did not break up to 2.3 GPa within this time interval. Figure 2 presents the illustration of the sought of this utmost stress.

To illustrate the behavior of the system stability within the interval of the external stresses causing the propagation of the transformation wave in the structure [1], figure 3 shows the dependency of the potential energy on time for the external stress above 2.7 GPa.

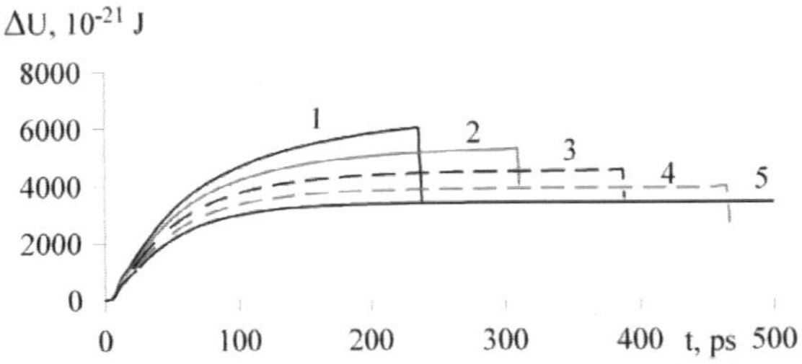

Figure 2. Potential energy versus time. 1 – 2.69 GPa, 2 – 2.6 GPa, 3 – 2.5 GPa, 4 – 2.4 GPa, 5 – 2.3 GPa

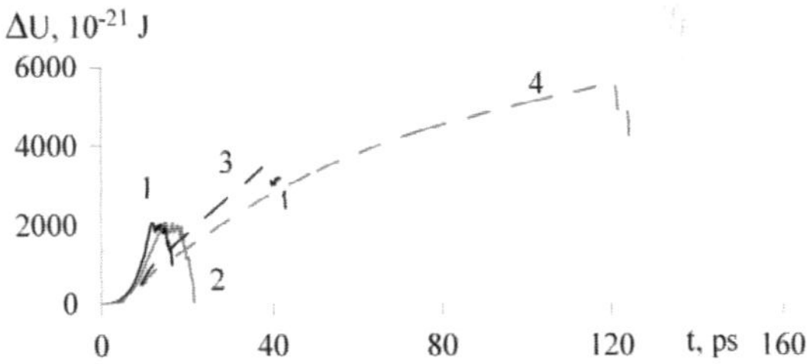

Figure 3. Potential energy versus time. 1 –4 GPa, 2 – 3.5 GPa, 3 – 3.0 GPa, 4 – 2.75 GPa

Figures 4–5 give the results of such detailed processing: the critical time instants of the fracture and accumulated potential energy versus the external applied stress.

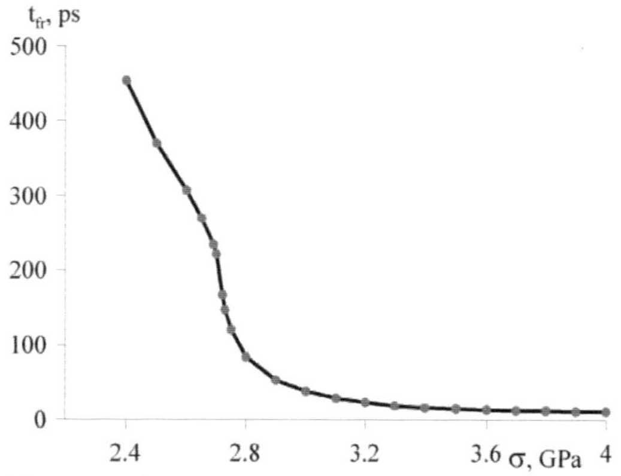

Figure 4. Fracture time instant versus external stress.

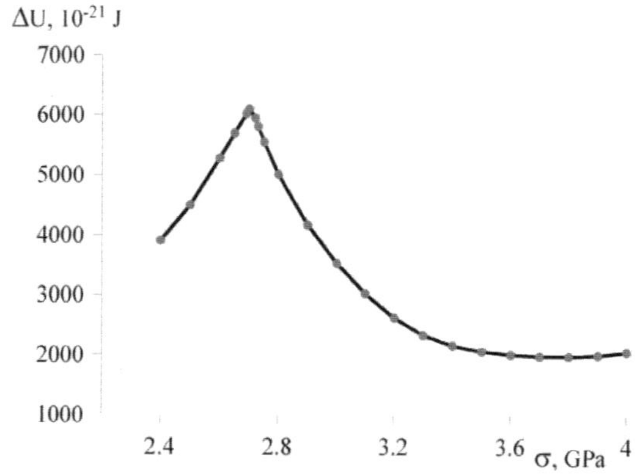

Figure 5. Variation of the potential energy versus external stress.

For convenience, the calculated points are charted in the graphs. Evident that this technique of data processing enables to determine the critical value of the external stress (the bifurcation point), at which the modes of energy absorption alternate at the impact action [1] accurate to 0.01 GPa. Its value is 2.70 GPa. For the stresses below this value, the linear section of time reduction and linear growth of the potential energy variation are observed as the stress grows. Then both the time and critical potential energy decrease as the applied stress rises in rough accordance with the hyperbolic law.

4. Conclusions

Thus, the same critical value of the compressive external stress is the bifurcation point both for the quasi-static structure compression (alternation of the instability and further fracture modes), and for the impact action (alternation of the energy absorption modes and occurring wave of transformation defects).

Acknowledgments

The work is supported by RFBR, grant No 16-01-00468 «Multilevel models of deformation and fracture of solids».

References

[1] Golovnev I F, Golovneva E I, Merzhievsky L A and Fomin V M 2013 *Phys. Mesomech.* **16** (4) 294–302
[2] Golovnev I F, Golovneva E I, Merzhievsky L A et al 2014 *International Conference on Physiacal Mesomechanics of Multilevel Systems Book Series: AIP Conference Proceedings* **1623** 171–174
[3] Golovnev I F, Golovneva E I and Merzhievsky L A 2016 *Phys. Mesomech.* **19** (5) 66–73 (for the present in Russian only)
[4] Luo C L, Zhou Y H and Zhang Y 2000 *Acta Physica Sinica* **49** (1) 54–56
[5] Sun D Y and Gong X G 2000 *Surface Science* **445** (1) 41–48
[6] Mitas L, Grossman J C, Stich I, et al 2000 *Phys. Rev. B.* **84** (7) 1479–1482
[7] Bastug T, Erkoc S, Hirata M, et al 2000 *Physica E* **8**(3) 223–229
[8] Kabir M, Mookerjee A and Bhattacharya A K 2004 *Phys. Rev. A.* **69** (4) 043203
[9] Bakaev V A, Steele W A and Pantano C G 2001 *Journal of Chem. Phys.* **114** (21) 9599–9607
[10] Kadau K and Entel P 2002 *Phase transition* **75** (1-2) 59–65
[11] Shim J H, Lee B J and Cho Y W 2002 *Surface Science* **512** (3) 262–268
[12] Psakhe S G, Zolnikov K P and Korostelev S Y 1995 *Pisma v Zhurnal Tekhnicheskoi Fiziki* **21** (3) 1–5

[13] Korchuganov A V, Zolnikov K P, Kryzhevich D S, Chernov V M and Psakhie S G 2016 *Physics of Atomic Nuclei* **79** (7), 1193–1198
[14] Ozhgibesov M S, Leu T S, Cheng C H and Utkin A V 2012 *Journal of Molecular Graphics and Modelling* **38** 375–381
[15] Hosseini-Hashemi Sh and Ilkhani M R 2017*Meccanica* **52** (4-5) 1107–1121
[16] Eltaher M A, Hamed M A, Sadoun A M and Mansour A 2014 *Applied Mathematics and Computation* **229** 260–272
[17] Karlicic D, Kozic P and Pavlovic R 2016 *Applied Mathematical Modeling* **40** (2) 1599–1614
[18] Karlicic D**,** Ozvat S, Cajic M, Kozic P and Pavlovic R 2016 *Fasta Universitatis (Series Mechanical Engineering)* **14** (1) 75–88
[19] Wang Z G, Zu X T, Gao F and Weber W J 2008 *Physical Review B.* **77** (22) 224113
[20] Voter A F 1993 *Los Alamos Unclassified Technical Report* (# LA-UR-93-3901)
[21] Allen M P and Tildesley D J 1987 Computer Simulation of Liquids (O New York: Oxford University Press) p 385
[22] Golovnev I F, Golovneva E I and Fomin V.M. 2003 *Phys. Mesomech.* **6** (5-6) 41–45

IOP Conf. Series: Journal of Physics: Conf. Series **894** (2017) 012021 doi:10.1088/1742-6596/894/1/012021

Mathematical modeling and numerical calculation of composite structures

S K Golushko[1,2]

[1] Novosibirsk State University, Novosibirsk, Russia,
[2] Institute of Computational Technologies, SB RAS, Novosibirsk, Russia

E-mail: s.k.golushko@gmail.com

Abstract. The report is devoted to modeling the properties of composite materials. Two major approaches are considered: phenomenological and structural [1]. Within the framework of the first approach reinforced materials are modeled as homogeneous anisotropic medium with efficient physical and mechanical properties. In this case mechanical parameters of the material are determined basing on experimental data. In a structural approach, physical and mechanical parameters of the composite are expressed in terms of the parameters of its components and design of reinforcement that open up opportunities for improvement of the properties of composite structures. The mathematical relations describing the nonlinear elastic three-point bending of isotropic and reinforced beams with account of different strength and stiffness behavior in tension and compression are obtained. An algorithm for numerical solution of corresponding boundary-value problems is proposed and implemented. Results of numerical modeling have been compared to acquired data for polymer matrix and structural carbon fiber reinforced plastics.

1. Introduction

Thin plates and shells are common elements of many modern structures within the aircraft and shipbuilding, mechanical engineering, the petroleum, gas and chemical industry. Opportunities of their using have considerably extended with occurrence of composite materials (CM). Because of CM lightness, strength, rigidity they are essentially better compared to traditional metals and alloys in specific characteristics. Having an opportunities for turning the stress-strain state (SSS) of structures, thus providing the best of those structures and corresponding CM usage. Increase of requirements to durability and reliability of modern constructions results is necessity of using the nonlinear and non-classical theories of shells and plates alongside with the classical and linear one. There are two levels of nonhomogeneity characterizing the fibrous composite materials: the micrononhomogeneity corresponded to the presence of two materials phases (a matrix or binding and a reinforcement or filling material), and the macrononhomogeneity, corresponded to the presence in a material of focused micrononhomogeneous layers with different space orientation. The main goal of the micromechanics of composites is the definition of the effective modules of elasticity.

Content from this work may be used under the terms of the Creative Commons Attribution 3.0 licence. Any further distribution of this work must maintain attribution to the author(s) and the title of the work, journal citation and DOI.
Published under licence by IOP Publishing Ltd

2. Structural Models of Composite Materials

For most of composite material models and all that considered here we can write the relations between average stresses $\sigma_{\alpha\beta}$, $\tau_{\alpha 3}$ and deformations $e_{\alpha\beta}$, $\gamma_{\alpha 3}$ (generalized Hooke's Law):

$$\sigma_{\alpha\alpha} = a_{\alpha\alpha}e_{\alpha\alpha} + a_{\alpha\beta}e_{\beta\beta} + a_{\alpha 3}\cdot 2e_{\alpha\beta} - a_{\alpha\Theta}\Theta,$$

$$\sigma_{\alpha\beta} = a_{\alpha 3}e_{\alpha\alpha} + a_{\beta 3}e_{\beta\beta} + a_{33}\cdot 2e_{\alpha\beta} - a_{3\Theta}\Theta, \quad (1)$$

$$\gamma_{\alpha 3} = q_{\alpha\alpha}\tau_{\alpha 3} + q_{\alpha\beta}\tau_{\beta 3},$$

where Θ is the increase of temperature. Relations (1) are called the thermoelasticity relations, or, when no temperature influence is considered, simply elasticity relations.

The structural model of fiber reinforced composite described in [2, 3, 4, 5] has become a foundation of large amount of current researches. Now it is widely used while simulating the behavior of composite structures. The model is based on the following assumptions: the stress-strain state into isotropic elastic fibres and into entire volume of isotropic ideally elastic matrix is homogeneous; fibers and matrix are deformed jointly along the direction of reinforcement; stresses in fibers and in matrix corresponding to other directions are equal.

For computing the effective elastic modulus of unidirectional fiber reinforced composite the Reuss-Voigt average was used giving the following formulae

$$E_1 = \omega_f E_f + \omega_m E_m, \quad E_2 = \frac{E_f E_m}{\omega_f E_m + \omega_m E_f},$$

$$\nu_{12} = \omega_f \nu_f + \omega_m \nu_m, \quad G = \frac{G_f G_m}{\omega_f G_m + \omega_m G_f}, \quad (2)$$

where all the terms having squared Poisson coefficients are neglected.

Here E_1, E_2 are effective moduli along and across the direction of reinforcement, G is effective share modulus, ν_{12} is effective Poisson coefficient in the plain of layer; E, ν, ω with "f" and "m" indices are elastic moduli, Poisson coefficients and volume fractions of matrix and fibres correspondingly, herewith $\omega_m + \omega_f = 1$.

On the ground of symmetry of compliance tensor one has

$$\nu_{21} = \nu_{12}E_2E_1^{-1}.$$

Formulae for effective coefficients of thermal expansion have the form

$$\alpha_1 = \omega_f\alpha_f + \omega_m\alpha_m, \quad \alpha_2 = \frac{\omega_f\alpha_f E_f + \omega_m\alpha_m E_m}{\omega_f E_f + \omega_m E_m}.$$

In description of model it is noted that among formulae for effective moduli, those obtained using Reuss averaging (in particular formulae for G) lead to the most inaccurate results. Estimations for G obtained using variational method are also obtained, and it is shown that lower bound

$$G = \frac{(1+\omega_f)G_f + \omega_m G_m}{(1+\omega_f)G_m + \omega_m G_f}G_m$$

gives more accurate approximation than (2) does. Hereinafter share moduli of matrix and fibers are

$$G_m = \frac{E_m}{2(1+\nu_m)}, \quad G_f = \frac{E_f}{2(1+\nu_f)}.$$

Components of effective stiffness tensor for unidirectionally reinforced layer in case of state of plane stress have the following form:

$$A_{\alpha\alpha\alpha\alpha} = \frac{E_\alpha}{1-\nu_{12}\nu_{21}}, \quad A_{1122} = \frac{\nu_{21}E_1}{1-\nu_{12}\nu_{21}}, \quad A_{1212} = G.$$

Unwritten expressions can be obtained using symmetry rule or vanish. Hereinafter we assume α, $\beta = 1, 2$ and $\alpha \neq \beta$.

The coefficients in the relations (1) are for example defined by the formulas from [1, 4, 5].

3. Bending deformation model with account for different strength and stiffness behavior in tension and compression

Three-point bending flexural test has been one of the standard techniques to determine physical and mechanical characteristics of materials. Figure 1 shows a scheme of physical model of three-point bending of a beam with the rectangular cross section $b \times 2h$, and the span l between the supports. The left edge of the beam is hinged, while the right one is supported freely. The force P is applied to the center of the beam. The model neglects the shape of the supports and assumes the occurring load P and support responses R_A and R_B to be concentrated. In addition, the model neglects the possible heterogeneity of the deformations in the direction normal both to the longitudinal direction and to the load direction.

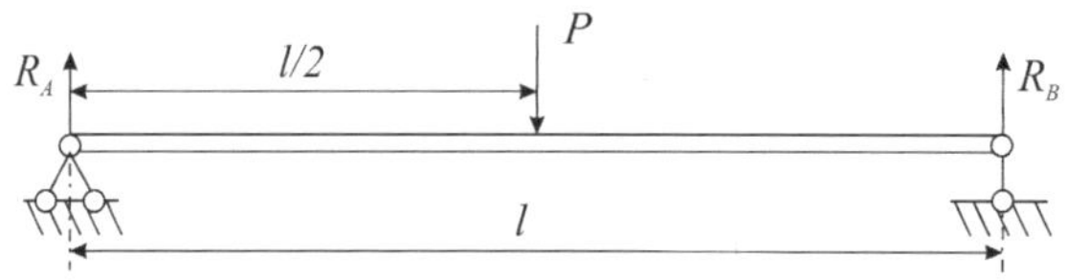

Figure 1. Three-point bending of a rectangular-sectioned beam.

In this case the beam's upper part undergoes compression strain in the longitudinal direction, bottom part – tension strain. VSE-1212 polymer matrix and VKU-28 structured CFRP react differently to tension and compression. The effect of accounting for this factor on the computational results is essential. Further, these results are compared to acquired ones.

Due to very low deformation rates, the classical theory of beam bending can be regarded as satisfactory for description of the equilibrium state. To this end it is convenient to consider the beam's median surface as a reference one.

The beam's stress-strain state is characterized by the following values determined on the reference surface: the shear force $Q(x)$, the bending moment $M(x)$, the longitudinal force $N(x)$, and by the longitudinal displacement and bend ($u(x)$, $w(x)$ respectively). The corresponding equilibrium equations are written as:

$$\frac{dN}{dx} = 0, \qquad \frac{dQ}{dx} = 0, \qquad \frac{dM}{dx} = Q. \tag{3}$$

The responses R_A and R_B are determined by the ratio $R_A = R_B = P/2$. The bending moments at the support points are equal to zero: $M_A = M_B = 0$. The solution of the equation system (3) can be expressed as:

$$N = 0, \qquad Q(x) = \begin{cases} P/2, & 0 \leq x \leq l/2, \\ -P/2, & l/2 \leq x \leq l, \end{cases}$$
$$M(x) = \begin{cases} Px/2, & 0 \leq x \leq l/2, \\ -P(x-l)/2, & l/2 \leq x \leq l, \end{cases} \tag{4}$$

Strain distribution for the beam's thickness can be obtained from the Kirchhoff-Love kinematic hypotheses:

$$\varepsilon(x, z) = e(x) + z\kappa(x), \tag{5}$$

$$e(x) = \frac{du}{dx}, \qquad \kappa(x) = -\frac{d^2w}{dx^2}, \tag{6}$$

where $\varepsilon(x, z)$ is the strain in the beam; $e(x)$ is the median surface strain; and $\kappa(x)$ denotes changes in the median surface curvature. As mentioned above, the beam undergoes tension and compression strain, whose interface will be marked as z_1. In this case for the section area $-h \leq z \leq z_1$ the strain will be negative, and for $z_1 \leq z \leq h$ positive. At the interface of these two states the strains ε vanish, so the interface itself is determined as:

$$z_1 = -\frac{e}{\kappa}, \qquad -h \leq z_1 \leq h. \tag{7}$$

The constitutive equation can be expressed as:

$$\sigma^{\pm}(x, z) = f_i^{\pm}(\varepsilon), \tag{8}$$

where the superscript "+" refers to the areas with positive strains and "—" – to the area with negative ones; $f(\varepsilon)$ denotes the approximation selected for the stress-strain curve (a linear function, a polynomial, or a combination of linear and power-law functions).

The longitudinal force N and the bending moment M in the beam cross section are determined by the equations:

$$\begin{aligned} N &= b\left(\int_{-h}^{z_1} \sigma^- dz + \int_{z_1}^{h} \sigma^+ dz\right), \\ M &= b\left(\int_{-h}^{z_1} \sigma^- z dz + \int_{z_1}^{h} \sigma^+ z dz\right). \end{aligned} \tag{9}$$

Having substituted (9) with the relations (5), (7), (8) into and integrated it over the beam thickness one obtains a system of equations to determine κ, e:
at $0 \leq x \leq l/2$

$$\begin{cases} N(\kappa, e, x) = 0, \\ M(\kappa, e, x) = Px/2, \end{cases} \tag{10}$$

at $l/2 < x \leq l$

$$\begin{cases} N(\kappa, e, x) = 0, \\ M(\kappa, e, x) = -P(x - l)/2. \end{cases} \tag{11}$$

The system of equations (10)–(11) in general case is nonlinear, but in the case of piecewise linear constitutive equations which take into account different strength and stiffness behavior in tension and compression expressed as

$$\sigma^{\pm}(x, z) = E^{\pm}\varepsilon, \tag{12}$$

it can be solved analytically. In the nonlinear case, the Newton method is applied to solve equations (10)-(11), and then, the linearized system

$$N(\varepsilon_0, \kappa_0) + \frac{\partial N(\varepsilon_0, \kappa_0)}{\partial \varepsilon}(\varepsilon - \varepsilon_0) + \frac{\partial N(\varepsilon_0, \kappa_0)}{\partial \kappa}(\kappa - \kappa_0) = 0,$$

$$M(\varepsilon_0, \kappa_0) + \frac{\partial M(\varepsilon_0, \kappa_0)}{\partial \varepsilon}(\varepsilon - \varepsilon_0) + \frac{\partial M(\varepsilon_0, \kappa_0)}{\partial \kappa}(\kappa - \kappa_0) = M(x)$$

can be solved for unknown values

$$\kappa = F(\varepsilon_0, \kappa_0, M(x)), \quad \varepsilon = G(\varepsilon_0, \kappa_0),$$

where ε_0 and κ_0 are the initial approximations, and $M(x)$ is determined from (4).

As the initial approximation at small values of the load P, the solutions obtained for the linear constitutive equations (12) were used. Since the computation is performed with a relatively small increment of P, in case of big values of P one can use the computation results acquired at a previous step as the initial approximation for current step.

Having determined the change of median surface curvature from the equations (10)-(11)

$$\kappa(x) = \begin{cases} \kappa_1(x), & \text{at } x \in [0, l/2), \\ \kappa_2(x), & \text{at } x \in [l/2, l], \end{cases}$$

one can write a differential equation to determine the beam bend. For that purpose the bend function is expressed as:

$$w(x) = \begin{cases} w_1(x), & \text{at } x \in [0, l/2), \\ w_2(x), & \text{at } x \in [l/2, l]. \end{cases}$$

Using equation (6) and the beam's fixing conditions, a system of equations can be derived:

$$\frac{d^2 w_1}{dx^2} = -\kappa_1, \quad \frac{d^2 w_2}{dx^2} = -\kappa_2,$$

$$w_1(0) = w_2(l) = 0, \quad w_1(l/2) = w_2(l/2),$$

$$\frac{dw_1(l/2)}{dx} = \frac{dw_2(l/2)}{dx}.$$

The solution of these equations can be obtained using the methods of solving boundary-value problems for systems of ordinary differential equations. For that purpose the modified collocation and least-residuals method [6, 7] was applied.

4. Numerical analysis of beam deformation. Comparative analysis of the simulation and experimental data

Within the framework of the presented study, numerical analysis of deformation processes in VSE-1212 polymer matrix and VKU-28 structured carbon fiber is proposed. It is based on approximation of stress-strain curves and the three-point bending model.

Further three different specimens with the geometrical sizes $l \times 2h \times b$ are considered:

1) specimen 1 – VSE-1212 polymer matrix, $75 \times 4.78 \times 10.05$ mm,
2) specimen 2 – VKU-28 structured carbon fiber (the specimen was cut out along the reinforcements), $90 \times 3.45 \times 9.85$ mm;
3) specimen 3 – VKU-28 structured carbon fiber (the specimen was cut out perpendicular to the reinforcements), $90 \times 3.40 \times 9.95$ mm

In Figure 2, one can see the simulation results for beam three-point bending, obtained through different approaches to approximation of the constitutive equations, and their comparison with the experimental data.

Applying the linear dependencies to tension and compression has not resulted in adequate approximation even for the 30% of curve. Using more complex than quadratic approximation

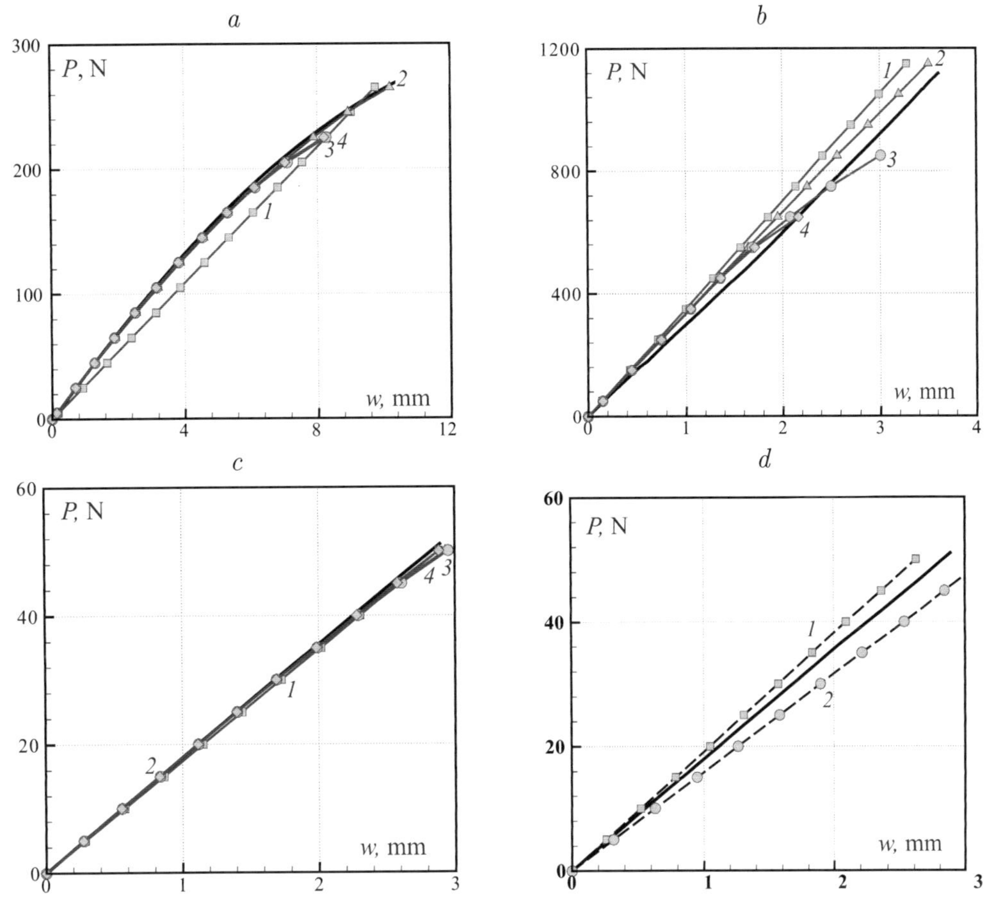

Figure 2. Experimental (solid curves) dependencies of beam-deflection and load obtained in simulation: linear approximation (1); quadratic approximation by a polynomial of the second degree (2); cubic approximation (3); linear and power-law approximation (4); a–c are specimens 1–3 respectively; d – the solution to a three-point bending problem without account for the different strength and stiffness behavior in tension (curve 1) and compression (curve 2). The solid line shows the results of mechanical tests.

laws at first led to a significant deviation from the experimental curve, and then to divergence of the Newton method iteration process. This is explained by the fact that in tension tests, due to their fragility, the strain range for the polymer matrix specimens was limited to 2%, while in the bending tests the strains in tension zone reached 4-5%.

Thus, to solve the bending problem the tension curve was extrapolated into the domain of high strains. The extrapolations obtained using a polynomial of the third degree, and by linear and power-law function reached the maximum too quickly and then started to decrease, which is against the physics behind the deformation process. A similar effect was observed when calculating the bending of the carbon-fiber specimens cut out along direction of reinforcement filler.

The calculations using quadratic approximation and extrapolation of tension curves and approximation of compression curves within a short (up to 6%) segment have turned out to be best for qualitative and quantitative description of the nonlinear character of VSE-1212 polymer matrix bending. In the case of the specimen cut out perpendicular to direction of its reinforcement all the approximations have shown the results close to experiment. At the same time for the test with maximum load the best option has still been application of quadratic approximations.

Taking different strength and stiffness behavior in tension and compression into account has an essential effect. As it was demonstrated above, the tension tests of VKU-28 specimens produced nonlinear stress-strain curves, while the difference of characteristics between tension and compression reached 5-7% for the longitudinal reinforcements and 12-15% – for the transverse ones. For the polymer matrix this difference exceeded 15%. However if bending tests have been performed to determine an elasticity module of CFRPs, different strength and stiffness behavior in tension and compression is compensated and one obtains some averaged characteristic. It is useful to consider the effect of the way for determining and setting of the mechanical characteristics on modeling of three-point bending of the carbon-fiber beam cutout perpendicular to its reinforcements. Figure 2 *d* shows the solutions obtained while using a linear approximation of the constitutive equations with equal elastic moduli for tension and compression: for curve 1 the modulus was obtained from tension experiments, for curve 2 – from compression ones. As one can see the calculated linear results without account for the different strength and stiffness behavior in tension and compression have differed from the results of mechanical tests (the solid curve) by more than 15 %.

Most of real CFRP structures under day-to-day service conditions bear complex loads that result in formation of tension, compression and bending zones as well as their combinations in the structures. Applying the traditional methods for determination of material characteristics in combination with linear deformation models (in particular those that do not account for the different strength and stiffness behavior in tension and compression) for calculation of such structures, one risks to distort the deformation and stress pattern significantly, which, in its turn, results in either underestimation or overestimation of the structure's strength and rigidity. Keeping in mind that carbon fibers are used for manufacturing of high-duty structures, their computation demands different strength and stiffness behavior in tension and compression to be taken into account.

Acknowledgments

The study was supported by project NSh-7214.2016.9.

References

[1] Golushko S K and Nemirovskiy Yu V 2008 *Direct and Inverse Problems of Mechanics of Composite Plates and Shells* (Moscow: Fizmatlit) (in Russian)

[2] Bolotin V V and Novitchkov Y N 1980 *The Mechanics of Multilayered Constructions* (Moscow: Mashinostroeniye) (in Russian)

[3] Nemirovsky Y V and Reznikov B S 1986 *Strength of Elements of Composite Structures* (Novosibirsk: Nauka) (in Russian)

[4] Golushko S K 2005 *Direct and inverse problems in the mechanics of composite plates and shells Computational Science and High Performance Computing II, Notes on Numerical Fluid Mechanics and Multidisciplinary Design* vol 88 (Berlin: Springer) pp 205-27

[5] Golushko K S, Golushko S K and Yurchenko A V 2011 *On modeling of mechanical properties of fibrous composites Computational Science and High Performance Computing IV, Notes on Numerical Fluid Mechanics and Multidisciplinary Design* vol 115 (Berlin: Springer) pp 107-20

[6] Golushko S K, Idimeshev S V and Shapeev V P 2013 *Application of collocations and least residuals method to problems of the isotropic plates theory Computational Technologies* **18**(6) 31–43 (in Russian)

[7] Golushko S K, Idimeshev S V and Shapeev V P 2014 *Development and application of collocations and least residuals method to the solution of problems in mechanics of anisotropic laminated plates Computational Technologies* **19**(5) 24–36 (in Russian)

Calculation and design of lattice cylindrical shells manufactured of unidirectional CFRPs

S Golushko[1] **and B Semisalov**[2]

[1] Institute of Computational Technologies SB RAS, Novosibirsk, Russia
[2] Novosibirsk State University, Novosibirsk, Russia

E-mail: s.k.golushko@gmail.com, ViBiS@ngs.ru

Abstract. The work is devoted to numerical simulation of the stationary stress-strain state of lattice cylindrical shells made of unidirectional carbon fiber reinforced plastics and to their optimal design. After averaging stiffness characteristics of lattice structure, the displacement and stress formulations for continuum anisotropic equivalents of lattice shells are given on the basis of classical relations of 3D theory of elasticity. While using a special fast pseudo-spectral algorithm, the solutions to problems of axial compression of anisogrid shells are obtained and compared one to another. The algorithm provides exponential decrease of error of approximation and requires small computational resources. Using this algorithm we have found the optimal geometrical parameters of structure ensuring its bearing capacity under given non-uniform loads.

1. Introduction

Anisogrid structures are grid shells manufactured of unidirectional carbon fiber reinforced plastics (CFRPs) by automated method of continuous filament winding. These structures demonstrate enhanced properties of strength and stiffness and have significant potential in the field of construction of rockets and spacecrafts, aircrafts and deep-diving vehicles, buildings and high-duty structures [1].

Anisogrid shells of zero Gaussian curvature including conical payload adapters of launch vehicles and cylindrical elements of their airframes (see Figure 1) are efficient technological solution for space industry. Tendency to minimize the weight of such structures leads to decrease of safety factors during their design that requires the description of stress-strain state (SSS) to be **reliable** and **accurate**, and the numerical simulation to have high **time efficiency**. This work aims at numerical simulation of stationary SSS of anisogrid cylindrical shells accounting for three mentioned principles.

To this end we applied the continuum approach verified in [2], used the classical equations of elasticity theory without any additional hypotheses [3], derived two formulations of problems and designed a special pseudo-spectral method to solve them based on Chebyshev and Fourier approximations of unknown functions [4] that allows for estimation of approximation and round-off errors [5].

2. 3D formulations of problem describing deformation of anisogrid shells

Assume unidirectional CFRP to be orthotropic and quasi-homogeneous material with tension-compression (E, E_T) and share (G, G_T) moduli describing stiffness along and across the direction

Content from this work may be used under the terms of the Creative Commons Attribution 3.0 licence. Any further distribution of this work must maintain attribution to the author(s) and the title of the work, journal citation and DOI.
Published under licence by IOP Publishing Ltd

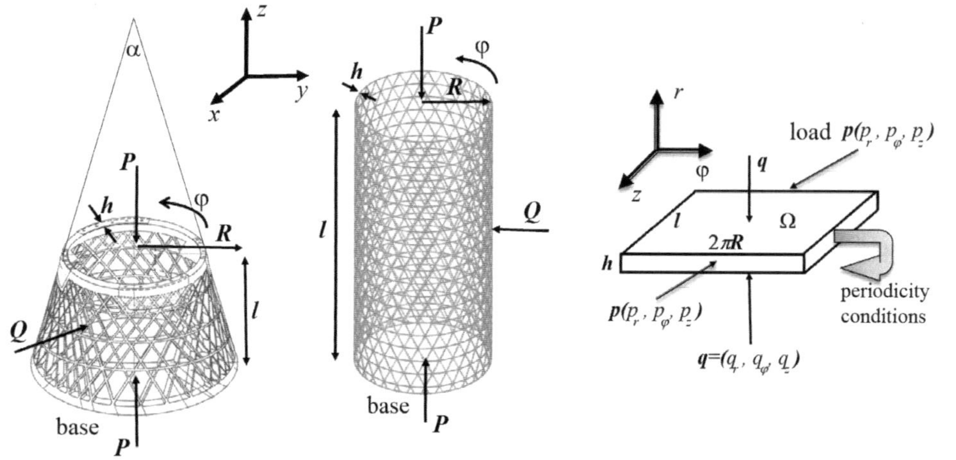

Figure 1. Conical and cylindrical anisogrid shells represented in Cartesian and cylindrical coordinate systems.

of its fibers and also with Poisson's coefficients ν, ν_T, ν_L, [6]– [8]. In this case the relation between stresses and strains in ribs of anisogrid shell can be described by generalized Hooke's law. Further, the continuum model is used for averaging the stiffness characteristics of lattice shells and for passing to its solid anisotropic equivalent (see [2], [9]).

Anisogrid shell consists of several sets of unidirectional ribs. Each set has its own number j ($j = 1, ..., N_s$) and can be defined by angle of rib's slope ϕ^j specified in coordinate system associated with shell, by thickness of ribs δ^j and by distance between ribs a_j. In accordance with the concept of "smearing" [2], each set of ribs should be replaced with a solid surface with characteristics of stiffness obtained by averaging those of ribs by volume. After summing the characteristics over all sets of ribs in [6], [7] a relation between stresses and strains occurring in the continuum equivalent of lattice shell was derived. It is as follows

$$\overline{\sigma} = \mathcal{A}\overline{\varepsilon}. \tag{1}$$

Relations (1) are written in Cartesian coordinate system (x, y, z) (see Figure 1), associated with shell for components of stress and strain tensors $\overline{\sigma} = (\sigma_z, \sigma_y, \sigma_x, \tau_{yz}, \tau_{xz}, \tau_{xy})$, $\overline{\varepsilon} = (\varepsilon_z, \varepsilon_y, \varepsilon_z, \gamma_{yz}, \gamma_{xz}, \gamma_{xy})$. Matrix of stiffness coefficients $\mathcal{A}$ has block diagonal form: $A_{ij} \neq 0$ as $i, j = 1, ..., 4$ and $i, j = 5, 6$, their values are expressed by stiffness characteristics of CFRPs and geometrical parameters of sets of ribs. Other values $A_{ij} = 0$.

2.1. Displacement-based formulation of the problem

Now it is convenient to pass from (x, y, z) to the cylindrical coordinate system (r, φ, z). Let l be height of shell, h be its constant thickness, R be radius of middle surface of shell (Figure 1). In this section problems of deformation of anisogrid cylindrical shells are posed in the domain

$$\Omega = \{(r, \varphi, z) : 0 \leq z \leq l,\ 0 \leq \varphi \leq 2\pi,\ R - h/2 \leq r \leq R + h/2\}.$$

Constitutive equations (1) in system (r, φ, z) remain valid. Lets supplement them with equilibrium equations

$$
\begin{aligned}
L_r &= \frac{\partial}{\partial r}(r\sigma_r) - \sigma_\varphi + \frac{\partial \tau_{\varphi r}}{\partial \varphi} + \frac{\partial}{\partial z}(r\tau_{zr}) + F_r r = 0,\\
L_\varphi &= \frac{\partial \sigma_\varphi}{\partial \varphi} + \frac{\partial}{\partial z}(r\tau_{z\varphi}) + \frac{\partial}{\partial r}(r\tau_{r\varphi}) + \tau_{r\varphi} + F_\varphi r = 0,\\
L_z &= \frac{\partial}{\partial z}(r\sigma_z) + \frac{\partial}{\partial r}(r\tau_{rz}) + \frac{\partial \tau_{\varphi z}}{\partial \varphi} + \tau_{r\varphi} + F_z r = 0,
\end{aligned} \tag{2}
$$

and compatibility equations

$$
\begin{aligned}
&\varepsilon_r \approx \frac{\partial u}{\partial r}, \quad \varepsilon_\varphi \approx \frac{1}{r}\frac{\partial v}{\partial \varphi} + \frac{u}{r}, \quad \varepsilon_z \approx \frac{\partial w}{\partial z},\\
&\gamma_{r\varphi} \approx \frac{1}{r}\frac{\partial u}{\partial \varphi} + r\frac{\partial}{\partial r}\left(\frac{v}{r}\right), \quad \gamma_{\varphi z} \approx \frac{\partial v}{\partial z} + \frac{1}{r}\frac{\partial w}{\partial \varphi}, \quad \gamma_{rz} \approx \frac{\partial w}{\partial r} + \frac{\partial u}{\partial z},
\end{aligned} \tag{3}
$$

where u, v, w are displacements of points of shell depending on r, φ, z; F_r, F_φ, F_z are components of field of strength in shell. Substituting (3) into (1) and (1) into (2) one obtains resolving system of differential equations with unknown displacements in Ω:

$$
\triangle_1 u = f_1(u, v, w, A_{ij}), \quad \triangle_2 v = f_2(u, v, w, A_{ij}), \quad \triangle_3 w = f_3(u, v, w, A_{ij}), \tag{4}
$$

where $\triangle_{1,2,3}$ are differential operators, $f_{1,2,3}$ are linear but rather cumbersome functions of the displacements, of their first and second derivatives and of stiffness coefficients A_{ij}. Solution to the system of equations (4) is assumed to be smooth enough and 2π-periodic with respect to φ coordinate. Two types of boundary conditions can be specified for the system. They are Dirichlet conditions (displacements are specified on border) and Neumann conditions (loads $p = (p_x, p_y, p_z)$ and $q = (q_x, q_y, q_z)$ are specified on border), see Figure 1. In last case the values of normal derivatives of u, v, w are specified
on $\Gamma = \{(r, \varphi, z) : z = 0, l, \ 0 \le \varphi \le 2\pi, \ R - h/2 \le r \le R + h/2\}$ and
on $\Upsilon = \{(r, \varphi, z) : 0 \le z \le l, \ 0 \le \varphi \le 2\pi, \ r = R - h/2, R + h/2\}$.
The complete expressions for all of equations and boundary conditions are given in [6].

2.2. *Stress-based formulation of the problem*

Stress-based formulation of theory of elasticity should take into account Saint-Venant's compatibility condition. This condition means that the medium remains solid while deforming. Discontinuities and voids are absent. It can be expressed in mathematical sense as continuity of functions u, v, w that is provided by vanishing of six independent components of Riemann-Christoffel tensor, [3]:

$$
R_{mnpq} = 0, \quad mnpq \in \{1212, 1313, 2323, 1213, 2123, 3132\}.
$$

In case of infinitely small values of relative displacements in curvilinear coordinate system, these conditions have the following form, see [14]:

$$
\begin{aligned}
&\frac{\partial^2 \varepsilon_{mq}}{\partial x^n \partial x^p} + \frac{\partial^2 \varepsilon_{np}}{\partial x^m \partial x^q} - \frac{\partial^2 \varepsilon_{nq}}{\partial x^m \partial x^p} - \frac{\partial^2 \varepsilon_{mp}}{\partial x^n \partial x^q} - 2\varepsilon_{rs}(\Gamma^r_{qm}\Gamma^s_{pn} - \Gamma^r_{mp}\Gamma^s_{qn}) +\\
&+ 2\Gamma^r_{np}\varepsilon_{mqr} + 2\Gamma^r_{qm}\varepsilon_{npr} - 2\Gamma^r_{nq}\varepsilon_{mpr} - 2\Gamma^r_{mp}\varepsilon_{nqr} = 0,
\end{aligned} \tag{5}
$$

where (x^1, x^2, x^3) are curvilinear coordinates, presence of index r implies summation over all values of $r = 1, 2, 3$, $\varepsilon_{npr} = \dfrac{1}{2}\left(\dfrac{\partial \varepsilon_{pr}}{\partial x^n} + \dfrac{\partial \varepsilon_{rn}}{\partial x^p} - \dfrac{\partial \varepsilon_{pn}}{\partial x^r}\right)$, Γ^k_{ij} are Christoffel symbols of the second kind, ε_{ij} are physical projections of components of strain tensor $\overline{\varepsilon}$ on curvilinear axes. In case of cylindrical coordinates $\varepsilon_{ij} = H_i H_j \overline{\varepsilon}_{ij}$, $i, j = 1, 2, 3$, where $H_1 = 1$, $H_2 = r$, $H_3 = 1$ are Lame coefficients, $\Gamma^1_{22} = -r$, $\Gamma^2_{21} = \Gamma^2_{12} = 1/r$, other $\Gamma^k_{ij} = 0$. From (5) one obtains

$$\begin{aligned}
U_r(\overline{\varepsilon}) &= \frac{1}{r}\frac{\partial^2 \varepsilon_r}{\partial \varphi^2} + \frac{1}{r}\frac{\partial}{\partial r}\left(r^2 \frac{\partial \varepsilon_\varphi}{\partial r}\right) - \frac{\partial \varepsilon_r}{\partial r} - \frac{2}{r}\frac{\partial^2 (r\gamma_{r\varphi})}{\partial r \partial \varphi} = 0,\\
U_\varphi(\overline{\varepsilon}) &= \frac{\partial^2 \varepsilon_r}{\partial z^2} + \frac{\partial^2 \varepsilon_z}{\partial r^2} - 2\frac{\partial^2 \gamma_{rz}}{\partial r \partial z} = 0,\\
U_z(\overline{\varepsilon}) &= \frac{\partial^2 \varepsilon_\varphi}{\partial z^2} + \frac{1}{r^2}\frac{\partial^2 \varepsilon_z}{\partial \varphi^2} + \frac{1}{r}\frac{\partial \varepsilon_r}{\partial r} - \frac{2}{r}\frac{\partial}{\partial z}\left(\frac{\partial \gamma_{\varphi z}}{\partial \varphi} + \gamma_{rz}\right) = 0,\\
U_{r\varphi}(\overline{\varepsilon}) &= -\frac{1}{r}\frac{\partial^2 \varepsilon_r}{\partial \varphi \partial z} + \frac{\partial}{\partial r}\left(\frac{1}{r}\frac{\partial (r\gamma_{\varphi z})}{\partial r}\right) - \frac{1}{r^2}\frac{\partial^2 (r^2 \gamma_{r\varphi})}{\partial r \partial z} - \frac{\partial^2}{\partial r \partial \varphi}\left(\frac{\gamma_{rz}}{r}\right) = 0,\\
U_{rz}(\overline{\varepsilon}) &= r\frac{\partial}{\partial z}\left(\varepsilon_r - \frac{\partial (r\varepsilon_\varphi)}{\partial r}\right) - \frac{\partial^2 \gamma_{rz}}{\partial \varphi^2} + \frac{\partial^2 (r\gamma_{\varphi z})}{\partial r \partial \varphi} + \frac{\partial^2 (r\gamma_{r\varphi})}{\partial \varphi \partial z} = 0,\\
U_{\varphi z}(\overline{\varepsilon}) &= \frac{\partial^2}{\partial r \partial \varphi}\left(\frac{\varepsilon_z}{r}\right) + \frac{\partial^2 \gamma_{r\varphi}}{\partial z^2} - r\frac{\partial^2}{\partial r \partial z}\left(\frac{\gamma_{\varphi z}}{r}\right) - \frac{1}{r}\frac{\partial^2 \gamma_{rz}}{\partial \varphi\, \partial z} = 0.
\end{aligned} \tag{6}$$

To derive resolving system, the following transformations were performed.

1. System (1) is resolved with regard to $\overline{\varepsilon}$: $\overline{\varepsilon} = \mathfrak{B}\overline{\sigma}$, the elements B_{ij} of matrix $\mathfrak{B} = \mathcal{A}^{-1}$ were computed, $i, j = 1, ..., 6$.

2. Substitution of components of $\overline{\varepsilon}$ into (6) was performed. As a result, the expressions $U_\alpha(\overline{\sigma})$ $U_{\alpha\beta}(\overline{\sigma})$ were obtained with $\alpha, \beta \in \{r, \varphi, z\}$.

3. The relations for $\dfrac{\partial^2 \tau_{r\varphi}}{\partial r \partial \varphi}$, $\dfrac{\partial^2 \tau_{\varphi z}}{\partial \varphi \partial z}$, $\dfrac{\partial^2 \tau_{rz}}{\partial r \partial z}$ were obtained using differentiation of linear transformations of (2). The results of its substitution into formulae for $U_\alpha(\overline{\sigma})$ $U_{\alpha\beta}(\overline{\sigma})$ was denoted by $U^*_\alpha(\overline{\sigma})$ $U^*_{\alpha\beta}(\overline{\sigma})$.

4. The resolving system of differential equations for stresses consists of six relations
$\dfrac{U^*_\varphi + U^*_z}{B_{11}} = 0$, $\dfrac{U^*_r + U^*_z}{B_{22}} = 0$, $\dfrac{U^*_r + U^*_\varphi}{B_{33}} = 0$, $\dfrac{U^*_{r\varphi}}{B_{44}} = 0$, $\dfrac{U^*_{rz}}{B_{55}} = 0$, $\dfrac{U^*_{\varphi z}}{B_{66}} = 0$.

Thus, a system of elliptic type equations for unknown components of stress tensor $\overline{\sigma}$ was obtained in the form

$$\begin{aligned}
&\Lambda_1 \sigma_1 = g_1(\overline{\sigma}, B_{ij}),\ \triangle^2 \Lambda_2 = g_2(\overline{\sigma}, B_{ij}),\ \triangle^3 \Lambda_3 = g_3(\overline{\sigma}, B_{ij}),\\
&\Lambda_4 \tau_{12} = g_{12}(\overline{\sigma}, B_{ij}),\ \Lambda_5 \tau_{13} = g_{13}(\overline{\sigma}, B_{ij}),\ \Lambda_6 \tau_{23} = g_{23}(\overline{\sigma}, B_{ij}),
\end{aligned} \tag{7}$$

where $\Lambda_1 - \Lambda_6$ are differential operators, $g_{1,2,3}$, $g_{12,13,23}$ are linear but rather cumbersome functions of stresses, of their first and second derivatives and of coefficients B_{ij}.

The system should be supplemented with the values of stresses on the border of domain Ω. The derivation of resolving equations is computerized and implemented using symbolic computations in Wolfram Mathematica environment.

3. Fast computational algorithm and numerical simulations

3.1. Pseudo-spectral method without saturation

The 3D boundary-value problems of displacement and stress formulations (4), (7) have a high computational complexity due to presence in them of small parameters. They are physical parameters, like relations of elastic moduli E_T/E, G_T/G, and geometrical parameters, like the relation a_j/δ^j and h^2 (square of the relative shell's thickness) that will appear as a scale factor at the highest order derivatives in (4), (7) after approximating them.

The solutions to these problems require application of fast and highly accurate algorithms providing the possibility to control the errors. For this purpose a new pseudo-spectral algorithm was developed on the basis of iterational stabilization method, collocation method and the so-called approximations without saturation. Particulary, for approximation of unknown displacements or stresses the Kronecker products of interpolation polynomials with Chebyshev nodes (with respect to coordinates r, z) and with Dirichlet kernel (with respect to φ coordinate) were used, see [6]– [8]. Following the works by K. I. Babenko [11] we can conclude, that the accuracy of such interpolations strictly corresponds to that of best polynomial approximations for any finite order of smoothness of sought-for function, while for infinitely differentiable functions an exponential (or geometrical) convergence can be achieved (this case was described in detail in [13]).

The resolving systems of equations (4), (7) have elliptic type and the applied load can be approximated using smooth function with steep gradients. Thus, the solutions are of high order of smoothness. This ensures the efficiency of applied approximations and the minimization of computational efforts.

Another way for reducing these efforts is a new approach for passage from differential problems to linear algebra. In this work we use the collocation method to derive the tensors approximations of derivatives that yields the Sylvester tensor equations. To solve these equations we use a special fast method proposed in [4]. The essential advantage of this method is the slow growth of condition numbers of matrices forming the tensors with grows of number of interpolation nodes. Detailed description of the algorithm together with estimates of its approximation and round-off errors can be found in [5].

3.2. Simulation of non-uniform axial compression of a cylindrical shell

Under assumption that the lower base of structure is rigidly fixed, the proposed algorithm was used for computing displacements of the upper one with respect to z.

$$w^* = \max_{j,k} w(r_j, \varphi_k, l), \quad w^{**} = \max_{j,k} \int_0^l \sigma_z(r_j, \varphi_k, z)dz,$$

where w is numerical solution obtained for displacement formulation of problem (4) (**problem 1**); σ_z is solution obtained for stress formulation of problem (see (7)) – **problem 2**; w^{**} is defined in accordance with Cesaro's formulae. Let N, K, M be the numbers of interpolation nodes of algorithm with respect to r, φ, z axis correspondingly. Here we assume $N = K = M$ and denote $\lambda = |w^* - w^{**}|/w^*$. Further, T is running time of algorithm in seconds, values $æ_N$ and ε_R are estimations of approximation and round-off errors of algorithm, they were computed while solving both problems for w and σ_z. All these values were obtained using one core of Intel Core i3-3217U 1.8 Ghz CPU with DDR3 6Gb 1600 Mhz memory.

Here a solution to the problem of non-uniform compression is described. This problem may occur in many applications, for example, while describing the deformation of first stage adapter of Proton rocket. The first stage has six engines placed round a circle under the adapter (see https://en.wikipedia.org/wiki/Proton_(rocket_family)). Engine thrust is $F_E = 10.02$MN, load from the top in the moment of rocket's start can be estimated as $F_L = 2.94$MN. Let us specify the following values of parameters: $R = 2$m, $l = 2$m, $h = 16$mm, $a_h = 125.47$mm, $a_c = 139.6$mm, $\delta_h = 7.5$mm, $\delta_c - 4.2$mm. These values for lattice structure were taken from [9], [10]. The non-uniform load of six engines on the bottom of shell should be specified in accordance with the mechanical experiments with structure, but these data in open literature is absent. Here we consider the test example with load of the form

$$P(r,\varphi,0) = \frac{F_E}{c_L}[\sin 3\varphi]^{10}\ \text{MN},\quad c_l = \int\limits_{R-h/2}^{R+h/2}\int\limits_{0}^{2\pi} r(\sin 3\varphi)^{10} d\varphi dr = \frac{63h\pi r}{128}.$$

The results of solution to the problems (4), (7) with the described loads are given in Table 3.2 (here we take different values of N and $K = 41$, $M = 7$).

Table 1. Results obtained for non-uniform axial compression of cylindrical shell

N	Problem	$æ_N$,%	ε_R,%	T, s	$w^{*,**}$, mm	λ,%
7	1	0.1398	2.51e-3	526.1	13.12	1.8
	2	0.4218	1.31e-5	19.3	13.36	
11	1	0.222	3.08e-2	768	13.24	
	2	0.4344	1.35e-4	35.5	13.372	
23	1	1.7458	1.1	1675.9	13.124	
	2	0.4346	97.6	280.8	13.372	

Note that for a given non-uniform load the applied algorithm allows to obtain the solution of problem 1 only for small values of M ($M = 5 - 13$) and it requires more time than the solution of problem 2. The reason is that the computational stability of boundary-value problem with Neumann conditions is much lower then that with Dirichlet conditions; that affects the values of round-off errors and the convergence of applied iteration process. But in any case, as it is seen from Table 3.2, the round-off error for this test does not exceed 1.1%.

Figure 2 contains the solutions of problem 2 with $N = 21$, $K = 71$, $M = 21$. They are values of σ_z in section $r = 2$ (Figure 2, a) and of u in section $z = 0$ (Figure 2, b). For this case we found that $\varepsilon_R = 0.0342\%$ and $æ_N = 1.94 \times 10^{-3}\%$. Thus, we may say, that the solutions have five accurate significant digits.

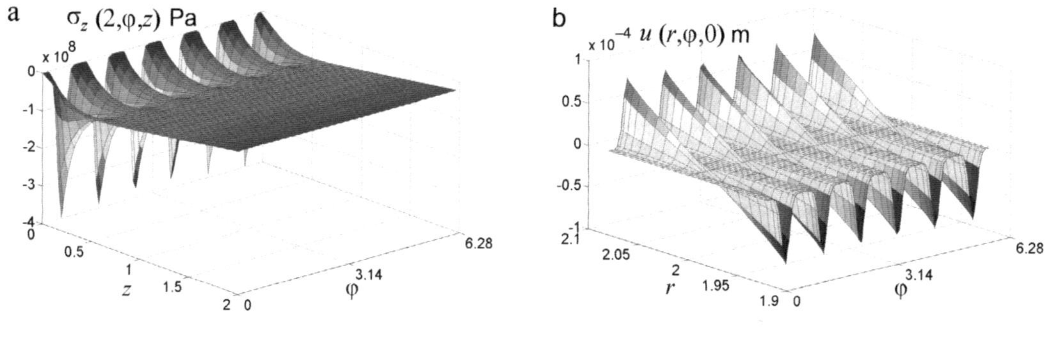

Figure 2. Numerical solutions to the problem of non-uniform axial compression.

3.3. Design of a cylindrical shell bearing non-uniform axial loads

Here by varying the values of angles of spiral ribs slope ϕ^2 ($\phi^2 = -\phi^3$) and the thicknesses of ribs h we shall design the cylindrical shell of minimal mass bearing the axial loads described above. As the criterions of loss of bearing capacity the following expressions have been considered

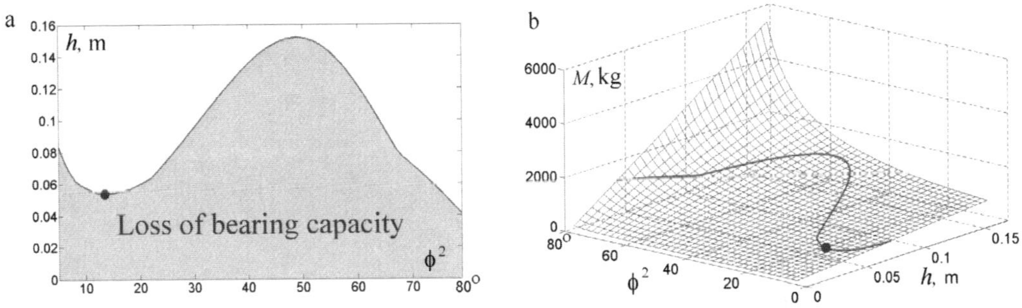

Figure 3. Analysis of the bearing capacity of a shell – (a) values of ϕ^2, h ensuring the bearing capacity, (b) dependence of M on ϕ^2, h. The points on graphs show optimal values.

- Exceeding the CFRPs ultimate stress while tension or compression along (s_{TL}, s_{CL}) or across (s_{TT}, s_{CT}) the direction of fibers, or while applying share loads (s_S).
- The loss of stability of shell's rib. For a rib of length l with cross section $b \times h$ and with E and G elasticity and share moduli measured in tests with loads applied along the direction of fibers, the ultimate load is (see [12])

$$T_L = \frac{\pi^2 K D}{K l^2 + \pi^2 D}, \; K = bhG, \; D = EJ, \; J = \frac{bh^3}{12}. \tag{8}$$

Using (8) the ultimate loads on structure's ribs were computed taking the following values from [9]: $s_{TL} = 2\times10^9$Pa, $s_{CL} = 1.2\times10^9$Pa, $s_{TT} = 5\times10^7$Pa, $s_{CT} = 1.7\times10^8$Pa, $s_S = 7\times10^7$Pa.

Performing the calculations using the described models and methods and also the bisection method we separated on the plane (ϕ^2, h) the domain of parameters ensuring the bearing capacity of structure for a given load (see Fig 3, a). The mass of shell was computed as a sum of masses of all ribs assuming that the density of CFRP is $\rho \approx 1500$kg/m^3 (see Fig 3, b). Taking into account the obtained bounds on the values of ϕ^2, h, the values ensuring the minimal weight of structure were computed. They are $\phi^2 = 12.78°$, $h = 53.84$mm. The minimal mass of shell is $M = 249.176$kg. Let us note, that in the case of uniform load in [10], [12] it was recommended to take $\phi^2 = 26.6°$ as optimal angle. In the case of considered non-uniform load it leads to increase of structure's mass up to $M = 445.5$kg, that is 79 percent more than in optimal case.

References

[1] Vasiliev V V, Barynin V A, Razin A F, Petrokovsky S A and Halimanovich V I 2009 Anisogrid composite grid structures – development and application in space technology *Composites and nanostructures* **3** 38–50 (in Russian)

[2] Obraztsov J F, Rybakov L S and Mishustin I V 1996 Deformation-analysis methods for elastic spaceframe systems of regular structure *J. on composite mechanics and design* **2** 3–14

[3] Sedov L I 1997 *Mechanics of Continuous Media* (Singapore: World Scientific) Vol. 1

[4] Semisalov B V 2014 Nonlocal algorithm for solving the Poisson equation and it's applications *Computational Mathematics and Mathematical Physics* **54** 1110–35 (in Russian)

[5] Semisalov B V 2016 A fast nonlocal algorithm for solving Neumann-Dirichlet boundary value problems with error control *Vychisl. Metody Programm.* **17** 500–22 (in Russian)

[6] Golushko S K and Semisalov B V 2015 Numerical modeling of anisogrid structures deformation using schemes of high accuracy without saturation *Mathematical modeling and Numerical Methods* **2** 23–45 (in Russian)

[7] Golushko S K, Idimeshev S V and Semisalov B V 2014 *Methods for solving boundary-value problems of mechanics of composite plates and shells: school-book of the curse "Direct and inverse problems of composite mechanics"* (Novosibirsk: DTI DT SB RAS, electronic resource ISBN 978-5-9905791-0-1) (in Russian)

[8] Golushko S K and Semisalov B V 2016 On the modeling and simulation of anisogrid shells of zero gaussian curvature with error control *Proc. of int. conf. "Mathematical and Information Technologies, MIT-2016"* (Vrnjacka Banja, Serbia - Budva, Montenegro) 28 August – 5 September, pp. 84–5

[9] Vasiliev V V and Morozov E V 2007 *Advanced Mechanics of Composite Materials* (Great Britain: Elsevier)
[10] Azarov A V 2012 Continuum and discrete models of lattice composite cylindrical shells *J. on Composite Mechanics and Design* **18** 121–30
[11] Babenko K I 2002 *Fundamentals of numerical analysis* (Moscow-Izhevsk: Regular and chaotic dynamics) (in Russian)
[12] Vasiliev V V 1988 *Mechanics of Composite Material Structures* (M: "Mechanical engineering") (in Russian).
[13] Boyd J 2000 *Chebyshev and Fourier Spectral Methods* (Ann Arbor, Michigan: University of Michigan) Second Edition
[14] Amenzade Yu A 1979 *Theory of Elasticity* (Moskow: Mir Publishers)

IOP Conf. Series: Journal of Physics: Conf. Series **894** (2017) 012023 doi:10.1088/1742-6596/894/1/012023

Three-dimensional thermocapillary flow regimes with evaporation

V B Bekezhanova[1] **and O N Goncharova**[2,3]

[1] Institute of Computational Modeling SB RAS, Krasnoyarsk, Russia
[2] Institute of Thermophysics SB RAS, Novosibirsk, Russia
[3] Altai State University, Barnaul, Russia

E-mail: vbek@icm.krasn.ru, gon@math.asu.ru

Abstract.
A three-dimensional problem of evaporative convection in a system of the immiscible media with a common thermocapillary interface is studied. New exact solution, which is a generalization of the Ostroumov – Birikh solution of the Navier – Stokes equations in the Oberbeck – Boussinesq approximation, is presented in order to describe the joint flows of the liquid and gas – vapor mixture in an infinite channel with a rectangular cross-section. The motion occurs in the bulk force field under action of a constant longitudinal temperature gradient. The velocity components depend only on the transverse coordinates. The functions of pressure, temperature and concentration of vapor in the gas are characterized by the linear dependence on the longitudinal coordinate. In the framework of the problem statement, which takes into account diffusive mass flux through the interface and zero vapor flux at the upper boundary of the channel, the influence of the gravity and intensity of the thermal action on flow structure is studied. The original three-dimensional problem is reduced to a chain of two-dimensional problems which are solved numerically with help of modification of the method of alternating directions. Arising flows can be characterized as a translational-rotational motion, under that the symmetrical double, quadruple or sextuple vortex structures are formed. Quantity, shape and structure of the vortexes also depend on properties of the working media.

1. Introduction

Comprehensive study of convection accompanied by evaporation/condensation is motivated by widespread application of the evaporating media in many manufacturing processes and experimental engineering systems. Among them the advanced mini- and microscale cooling technologies, are emphasized. The technologies with use of joint flows of evaporating liquid and co-current gas flux are applied in heat exchangers and heat pipes, jet-spray evaporative cooling systems, electronic setups of planar geometry with multibranch microreactors and microchips, shear-driven film evaporators, etc. For a detail review of analytical, numerical and experimental investigations of the convective fluid flows with mass transfer at interface due to evaporation or condensation we refer to [1]. Preliminary theoretical study of convective processes and evaporation/condensation dynamics can help us to clarify physical aspects of evaporative convection, to define the driving mechanisms and to examine the phenomena in the liquid media caused by the gas flow. Obtained analytical results allow one to predict the outcome of physical experiments and to specify ways of improving the technologies.

Content from this work may be used under the terms of the Creative Commons Attribution 3.0 licence. Any further distribution of this work must maintain attribution to the author(s) and the title of the work, journal citation and DOI.
Published under licence by IOP Publishing Ltd

The structure of the joint liquid and gas – vapor flows can be described on the basis of the Boussinesq approximation of the Navier – Stokes equations. Seeing the problems of evaporative convection are the multiparameter and very difficult ones, various methods of investigations should be applied. One of the most important and effective methods is obtaining the exact solutions of the governing equations. Exact solutions allow one to model the convective fluid flows rather rapidly and to obtain the qualitative and quantitative flow characteristics (see, for instance, [2–5]). The exact solutions give a possibility to evaluate the degrees of influence of various factors on flow characteristics and to validate the mathematical models.

In the paper we present a generalization of the well-known Ostroumov – Birikh solution [6, 7] for three dimensional non-axis-symmetrical case. We carry out the mathematical modeling of the three-dimensional convective fluid flows with evaporation to analyze the influence of intensity of the gravity force and size of the liquid layer thickness on the flow patterns and temperature distribution. The study is performed with respect to the Soret and Dufour effects in the vapor – gas phase.

2. Governing equations

Let two viscous incompressible fluids (liquid and gas-vapor mixture) fulfill an infinite horizontal channel. Liquid layer $\Omega_1 = \{(x,y,z) : -x_0 < x < 0,\, 0 < y < 1,\, -\infty < z < \infty\}$ and gas-vapor layer $\Omega_2 = \{(x,y,z) : 0 < x < x^0,\, 0 < y < 1,\, -\infty < z < \infty\}$ are separated by the thermocapillary interface Γ given here by the equation $x = 0$. The boundaries of the domains Ω_1, Ω_2 defined by $x = -x_0$, $x = x^0$, $y = 0$ $y = 1$ are the fixed impermeable walls. Note, that the Cartesian coordinate system is chosen so that the gravity acceleration vector $\mathbf{g}$ is directed opposite to the Ox axis ($\mathbf{g} = -g\mathbf{i}$, $\mathbf{i}$ is the unit vector of Ox).

Let the linear size of the flow domains in the y-direction h be the characteristic length. We introduce the characteristic values of parameters for the coupled problem on the basis of the characteristic values of the liquid parameters in Ω_1, namely: u_*, T_*, $p_* = \rho_1 u_*^2$ are the characteristic velocity, temperature and pressure, respectively. The stationary three-dimensional convective flows of j-th medium (here and subsequently $j = 1, 2$ relate to the liquid and gas – vapor mixture, respectively) will be described by the Oberbeck – Boussinesq approximation of the Navier – Stokes equations [8] in the non-dimensional form:

$$(\mathbf{v}_j \cdot \nabla)\mathbf{v}_j = -\eta_j^p \nabla p_j + \frac{\eta_j^{\mathbf{v}}}{Re}\Delta \mathbf{v}_j + \mathbf{G}_j, \quad \operatorname{div} \mathbf{v}_j = 0, \quad \mathbf{v}_j \cdot \nabla T_j = \frac{\eta_j^T}{RePr}(\Delta T_j + \underline{\bar{\delta}\Delta C}), \quad (1)$$

$$\mathbf{v}_2 \cdot \nabla C = \frac{1}{Pe}(\Delta C + \underline{\bar{\alpha}\Delta T_2}). \quad (2)$$

Here $\mathbf{v}_j$ is the velocity vector, T_j is the temperature, p_j is the pressure (deviation of pressure from the hydrostatic one), C is the vapor concentration function, $\mathbf{G}_1 = \mathbf{i}(Gr/Re^2)T_1$, $\mathbf{G}_2 = \mathbf{i}(\bar{\beta}(Gr/Re^2)T_2 + \gamma(Ga/Re^2)C)$. The equations (1), (2) with marked terms are considered to model the gas-vapor flows in the upper layer. The effects of diffusive thermal conductivity and thermodiffusion (or the Dufour and Soret effects) are taken into account in the gas phase. The equations (1) without marked terms are used to describe the liquid flows in the lower layer. We supposed here that the vapor is a passive admixture, the vapor diffusion in the gas phase is described by the diffusion equation (2). For more details with respect to the problem statement and substantiation of the used assumptions we refer to [5].

The problem is characterized by the following dimensionless parameters: $Re = u_* h/\nu_1$ is the Reynolds number, $Pr = \nu_1/\chi_1$ is the Prandtl number, $Gr = \beta_1 T_* g h^3/\nu_1^2$ is the Grashof number, $Ga = gh^3/\nu_1^2$ is the Galilei number, $g = |\mathbf{g}|$, $Pe = u_* h/D$ is the Peclet number (diffusive Peclet number), $\eta_1^p = 1$, $\eta_2^p = 1/\bar{\rho}$, $\eta_1^{\mathbf{v}} = 1$, $\eta_2^{\mathbf{v}} = \bar{\nu}$, $\eta_1^T = 1$, $\eta_2^T = \bar{\chi}$, $\bar{\rho} = \rho_2/\rho_1$, $\bar{\nu} = \nu_2/\nu_1$, $\bar{\chi} = \chi_2/\chi_1$, $\bar{\beta} = \beta_2/\beta_1$ are the ratios of the densities ρ_j and coefficients of kinematic viscosity ν_j, thermal diffusivity χ_j and thermal expansion β_j of the gas and liquid, respectively; γ is the concentration coefficient of the gas density; $\bar{\alpha} = \alpha T_*$, $\bar{\delta} = \delta/T_*$, the coefficients α and δ characterize the Soret and Dufour effects in the gas-vapor layer, D is the coefficient diffusion of vapor in the gas.

3. Exact solution of the Oberbeck – Boussinesq equations

3.1. Form of exact solution

We construct the exact solution of the equations (1), (2), which is characterized by dependence of the components of the liquid and gas velocity vectors $\mathbf{v}_j = (u_j, v_j, w_j)$ on the transverse coordinates (x, y). The temperature and pressure functions T_j, p_j and the vapor concentration C have the terms Θ_j, q_j, Φ, which also depend on the transverse coordinates:

$$\begin{gathered} u_j = u_j(x,y), v_j = v_j(x,y), w_j = w_j(x,y), \quad T_j = -Az + \Theta_j(x,y), \quad C = Bz + \Phi(x,y), \\ p_1 = -A\frac{Gr}{Re^2}xz + q_1(x,y),\; p_2 = -A\bar{\rho}\bar{\beta}\frac{Gr}{Re^2}xz + B\bar{\rho}\gamma\frac{Ga}{Re^2}xz + q_2(x,y). \end{gathered} \tag{3}$$

The coefficients $A = A_* h/T_*$ and $B = B_* h$ determine the constant longitudinal temperature and concentration gradients along the interface; A_* and B_* are the corresponding dimensional gradients.

The presented solution is an analogue of the Ostroumov – Birikh solution [6, 7]. Group-theoretical nature of class of the Ostroumov – Birikh type solutions was proved in [9].

3.2. Boundary conditions

In order to find the functions u_j, v_j, w_j, q_j, Θ_j, Φ_j we formulate the boundary conditions on the interface and fixed impermeable walls of the channel. The interface conditions are formulated on the basis of the conservation laws and some additional assumptions [8, 10, 11]. Conditions of continuity of tangential velocities and temperature are assumed to be fulfilled on the thermocapillary interface $x = 0$: $v_1 = v_2$, $w_1 = w_2$, $T_1 = T_2$.

The kinematic and dynamic conditions are to be valid on the interface $x = 0$. For considered exact solution the first condition is written as $u_1 = u_2 = 0$. Projections of the dynamic boundary condition on the tangential vectors to the interface are

$$(u_{1y} + v_{1x}) - \bar{\rho}\bar{\nu}(u_{2y} + v_{2x}) = -\frac{Ma}{RePr}\Theta_{1y}, \qquad w_{3x} - \bar{\rho}\bar{\nu}w_{2x} = A\frac{Ma}{RePr}.$$

Projection on the normal vector $\mathbf{n}$, that coincides with the unit vector of the Ox axis, is [10]:

$$-Re(q_1 - q_2) + 2(u_{1x} - \bar{\rho}\bar{\nu}u_{2x}) = \frac{1}{Ca}2H\sigma.$$

Here $Ma = \sigma_T T_* h/(\rho_1\nu_1\chi_1)$ is the Marangoni number, $Ca = \rho_1\nu_1 u_*/\sigma_0$ is the capillary number, H is the mean curvature of the interface $x = 0$ (in the case of non-deformed interface $H = 0$), σ is the surface tension ($\sigma = 1 - MaCa/(RePr)\,(T - T_0)$); the positive constants σ_0 and σ_T are some relative values of the surface tension and temperature coefficient of surface tension, respectively.

At $x = 0$ the heat transfer condition with respect to the diffusive mass flux due to evaporation and the mass balance equation are formulated: $T_{1x} - \bar{\kappa}T_{2x} - \bar{\delta}\bar{\kappa}C_x = -\bar{\lambda}M$, $M = -(C_x + \bar{\alpha}T_{2x})$. Here $\bar{\kappa} = \kappa_2/\kappa_1$ is the ratio of the thermal conductivity coefficients κ_j. The evaporative mass flow rate M is a function of y defined on the interface $x = 0$. The characteristic value of the evaporative mass flow rate is chosen equal to $M_* = D\rho_2/h$. The linearized form of an equation for saturated vapor concentration at the interface is used to define the vapor concentration function [3, 5]

$$C|_{x=0} = C_*(1 + \bar{\varepsilon}(T_2 - T_0)). \tag{4}$$

Here C_* is the saturated vapor concentration at $T_2 = T_0$ (T_0 is equal to 20° in [2,3]), $\bar{\varepsilon} = \varepsilon T_*$, $\varepsilon = \lambda\mu/(R^* T_0^2)$, λ is the latent heat of evaporation ($\bar{\lambda} = \lambda D\rho_2/(\kappa_1 T_*)$), μ is the molar mass of the evaporating liquid, R^* is the universal gas constant. The relation (4) is a consequence of the Clapeyron – Clausius equation and the Mendeleev – Clapeyron equation for an ideal gas (for details we refer to [5]). In virtue of (4) the following relation between gradients of temperature and concentration (A and B) will take place $B = -C_*\bar{\varepsilon}A$.

The no-slip conditions for velocity fields and the conditions of thermal insulating of the lateral walls are set on the fixed impermeable walls of the channel; the condition of absence of vapor flux on the upper and lateral rigid boundaries is used in the present statement:

$$\begin{aligned} &x = -x_0: \ \mathbf{v}_j = 0,\ T_{1x} = 0; \qquad x = x^0: \ \mathbf{v}_j = 0,\ T_{2x} = 0,\ C_x = 0; \\ &y = 0: \ \mathbf{v}_j = 0,\ T_{jy} = 0,\ C_y = 0; \qquad y = 1: \ \mathbf{v}_j = 0,\ T_{jy} = 0,\ C_y = 0. \end{aligned} \tag{5}$$

Conditions for the temperature and vapor concentration function on the fixed walls provide a fulfillment of the full heat flux condition with respect to the Dufour effect and of full mass flux condition with respect to the Soret effect [12]: $(\partial T_2/\partial n) + \bar{\delta}(\partial C/\partial n) = 0$, $(\partial C/\partial n) + \bar{\alpha}(\partial T_2/\partial n) = 0$.

4. Numerical results of investigations of the fluid flow regimes with evaporation

4.1. Scheme of solution of the coupled problems

Instead of the transverse components u_j, v_j of the velocity vectors the stream functions $\psi_j(x,y)$ and vorticity $\omega_j(x,y)$ are introduced according to usual rules $u_j = (\psi_j)_y$, $v_j = -(\psi_j)_x$, $\omega_j = v_{j_x} - u_{j_y}$. We reformulate the boundary conditions in terms of these new unknown functions. Construction of the stationary solution (3) for the two-layer convective fluid flows is reduced to a sequential solving the corresponding two-dimensional problems for the functions $w_1(x,y)$ and $w_2(x,y)$, $\Theta_1(x,y)$ and $\Theta_2(x,y)$, $\Phi(x,y)$, $\psi_j(x,y)$ and $\omega_j(x,y)$ in the working domains $\widetilde{\Omega}_j$ ($\widetilde{\Omega}_1 = \{(x,y) : -x_0 < x < 0,\ 0 < y < y^0\}$, $\widetilde{\Omega}_2 = \{(x,y) : 0 < x < x^0,\ 0 < y < y^0\}$).

Numerical algorithm based on the longitudinal transverse finite difference scheme known as the method of alternating directions [8, 10, 13] was constructed to perform the fluid flow simulations. A general scheme of solution of the coupled problem is described in [11] and [14] (see also [10]).

4.2. Regimes of flows

Numerical investigations are carried out in order to describe the possible flow topology for the liquid – gas system like ethanol – nitrogen and HFE-7100 – nitrogen with following physicochemical properties [15, 16] (all the values are presented in the order "ethanol, HFE-7100 and nitrogen" or only "ethanol and HFE-7100", respectively): $\rho = \{0.79 \cdot 10^3,\ 1.5 \cdot 10^3,\ 1.2\}$ (kg/m^3); $\nu = \{0.15 \cdot 10^{-5},\ 0.38 \cdot 10^{-6},\ 0.15 \cdot 10^{-4}\}$ (m^2/s); $\beta = \{0.108 \cdot 10^{-2},\ 1.8 \cdot 10^{-3},\ 3.67 \cdot 10^{-3}\}$ (K^{-1}); $\sigma_T = \{0.8 \cdot 10^{-4},\ 1.14 \cdot 10^{-4}\}$ (N/(m·K)); $D = \{0.135 \cdot 10^{-4},\ 0.7 \cdot 10^{-5}\}$ (m^2/s); $L = \{8.55 \cdot 10^5,\ 1.11 \cdot 10^5\}$ (W·s/kg); $k = \{0.1672,\ 0.07,\ 0.02717\}$ (W/(m·K)); $\chi = \{0.89 \cdot 10^{-7},\ 0.4 \cdot 10^{-7},\ 0.3 \cdot 10^{-4}\}$ (m^2/s); $\gamma = \{-0.62,\ -0.5\}$; $C_* = \{0.1,\ 0.6\}$ (at $T_0 = 10^o C$, relation (4)); $\bar{\varepsilon} = \{0.1, 0.04\}$ (relation (4)); Dufour coefficient $\delta = \{10^{-4}, 10^{-4}\}$ (K); Soret coefficient $\alpha = \{10^{-4}, 10^{-4}\}$ (K^{-1}). Computations are performed with the values of the Grashof number equal to $4.7 \cdot 10^4$ and 470 for the ethanol – nitrogen system and to $1.22 \cdot 10^6$ and $1.22 \cdot 10^4$ for HFE-7100 – nitrogen system in the case of normal ($g = g_0 = 9.81$ m/s^2) and low gravity ($10^{-2} g_0$ m/s^2), respectively. The values of the characteristic temperature drop T_* and characteristic length h are equal to $T_* = 10$ K and $h = 10^{-2}$ m. The Reynolds number was chosen equal to 1 so that the characteristic velocity u_* is equal to $0.15 \cdot 10^{-5}$ (m/s) for ethanol and $0.38 \cdot 10^{-4}$ (m/s) for HFE-7100.

IOP Conf. Series: Journal of Physics: Conf. Series **894** (2017) 012023 doi:10.1088/1742-6596/894/1/012023

The examples of the fluid flows (stream lines and trajectories) and corresponding temperature distributions are presented in Figure 1 – 4 for ethanol – nitrogen and Figure 5 – 7 for HFE-7100 – nitrogen systems in the three-dimensional channels, when the thickness h_l of the liquid layer is $h_l = 0.25$ or $h_l = 0.1$, the thickness h_g of the gas-vapor layer is $h_g = 0.5$ (corresponding values of the dimensional thicknesses are equal to $h_l^* = 0.25$ cm or $h_l^* = 0.1$ cm, $h_g^* = 0.5$ cm). In the numerical experiments the non-dimensional longitudinal temperature gradient A has the values $A = 0.1$, $A = 0.3$, $A = 0.4$ (the values of the dimensional longitudinal temperature gradient are equal to $A_* = 100$ K/m, $A_* = 300$ K/m, and $A_* = 400$ K/m, respectively).

For better visualization (Figure 1, 3, 5, 6) the first and second components of the liquid and gas velocities have been multiplied by factors 100 and 10, respectively. The flows of the liquids (lower layers) can be characterized as a rotational motion of a smaller intensity than in the gas layers (upper layers).

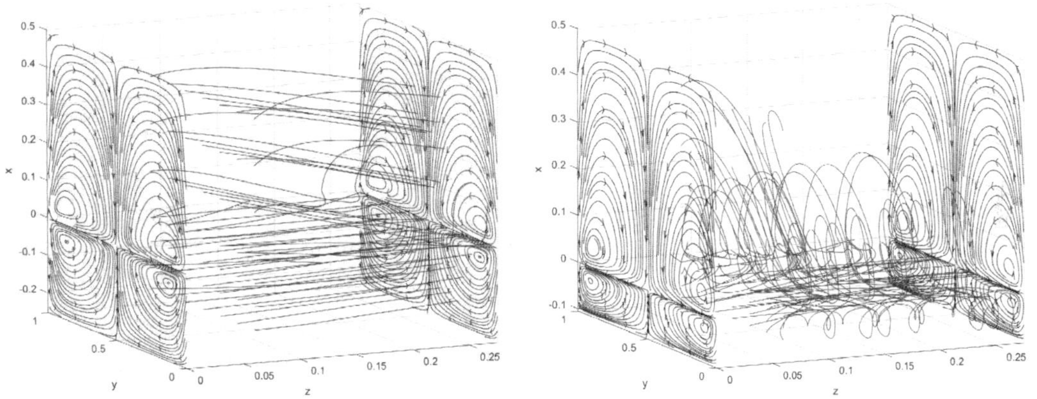

Figure 1. Stream lines and trajectories in ethanol – nitrogen system at $A = 0.1$, $Gr = 470$: $h_l = 0.25$ (left); $h_l = 0.1$ (right).

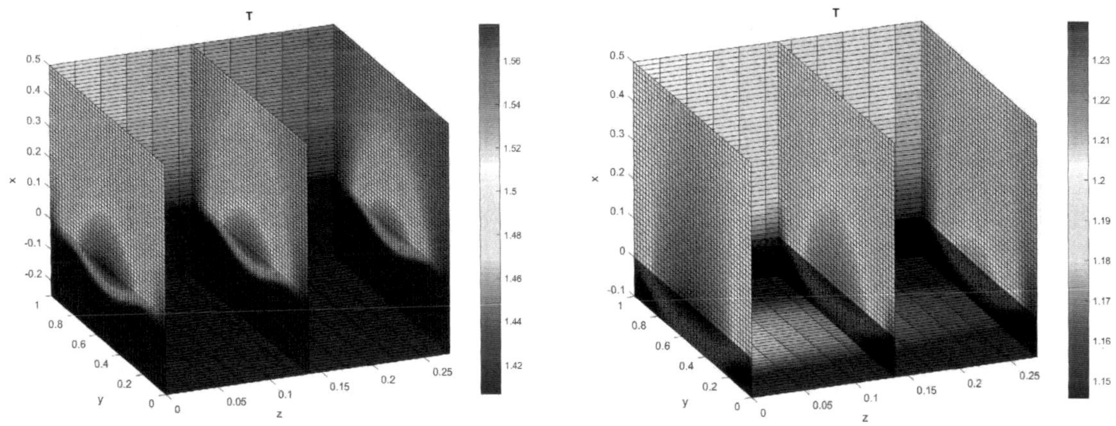

Figure 2. Distribution of the temperature in ethanol – nitrogen system at $A = 0.1$; $Gr = 470$: $h_l = 0.25$ (left); $h_l = 0.1$ (right).

Under low gravity the flows in both phases are characterized by clearly marked symmetric two-vortex structure at $A = 0.1$ both for $h_l = 0.1$ and $h_l = 0.25$ (Figure 1). Vortexes in the liquid have the thermocapillary nature and vortex centers are located close to the interface in the "corners". Such topology of the flow arises due to liquid spreading on the interface from hot central thermal roll to the lateral walls (see thermal pattern in Figure 2). Two longitudinal shafts with the opposite circulation of the liquid are generated by the Marangoni effect. With changing the gravitational influence the qualitative structure of the velocity and temperature fields is not modified; the two-vortex structure is kept in the media and the thermal billow is formed. Only the quantitative variations of hydrodynamical, temperature and concentration characteristics are observed. But in the thin layer the rotational motion component becomes more significant due to action of the thermocapillary effect, which generates in the thin layer more intensive liquid spreading from the center of the thermal billow to the periphery (to walls).

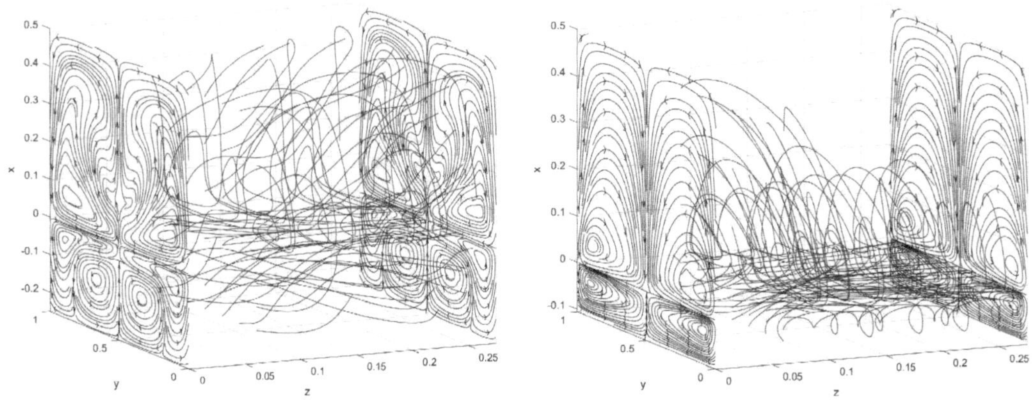

Figure 3. Stream lines and trajectories in ethanol – nitrogen system at $A = 0.3$: $Gr = 470$, $h_l = 0.25$ (left); $Gr = 47000$, $h_l = 0.1$ (right).

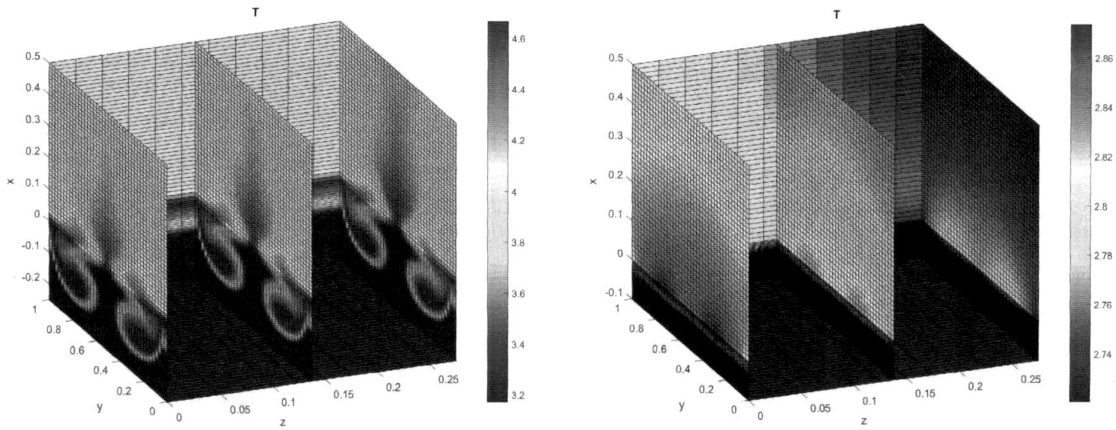

Figure 4. Distribution of the temperature in ethanol – nitrogen system at $A = 0.3$: $Gr = 470$, $h_l = 0.25$ (left); $Gr = 47000$, $h_l = 0.1$ (right).

At the same time the bright qualitative differences are found by simulation of the HFE-7100 – nitrogen flows (Figure 5). At $A = 0.1$ we observe the two-vortex structure in the gas layer and "quadruple" flow in the liquid layer under microgravity (such flow topology with formation of four longitudinal thermocapillary rolls is typical for both considered geometrical configuration).

IOP Conf. Series: Journal of Physics: Conf. Series **894** (2017) 012023 doi:10.1088/1742-6596/894/1/012023

In the system with $h_l = 0.25$ two additional small corner vortices appear in the liquid layer in the case of normal gravity. Formation of the extra vortexes is explained by weaker viscous effects in HFE-7100 in comparison with ethanol. Furthermore, in the terrestrial condition near-bottom-wall reverse flows in HFE-7100 appear (see trajectories in right picture of Figure 5). Thermal pattern is similar to structure, presented in Figure 2. But transversal size of arising thermal rolls is more than in ethanol - nitrogen system; cold "spots" arise near the lateral walls.

More complicated vortex patterns in fluids are observed in the case with more intensive thermal load (at $A = 0.3$, see Figure 3, 6). Splitting two big vortex into four asymmetrical swirls in the liquid layer occurs in the ethanol – nitrogen system in microgravity conditions. Two of the vortices are deformed into "angled" vortexes, their cores are located in "corners" near the lateral walls and stay close to the interface (left picture of Figure 3). Liquid particles move on the interface from the hot spots formed in "corners" (Figure 4) to center zone due to the Marangoni effect and go down because of unstable thermal stratification. Thus, the upper part of the "angled" vortex is generated by thermocapillary effect. In microgravity conditions in ethanol the thermal "horns" appears (left picture in Figure 4). Formation of such inhomogeneities of the temperature field is induced by condensation supplying the heat inflow in the liquid layer. Under weak gravity the thermal stratification in the system is less prominent. Thermoclines (in the form of the thermal "horns" in the liquid layer) arise due to condensation and are supported by the thermal properties of the working liquid (see Table 1). In the terrestrial conditions vortex structures in the liquid layer are weakly distorted and keep "dipolar" pattern (right picture in Figure 3) for both values of h_l, but a localization of the vortex cores is changed, they moves from the surface into the interior of liquid layer. In the layer with $h_l = 0.25$ the second core appears inside the vortex. The temperature regime differs sufficiently with changing thickness of the liquid layer: at smaller h_l a sufficiently homogeneous hot near-surface layer is formed (a surface hot film, see right picture in Figure 4), in the gas-vapor layer the thermal billow falls apart two small "shafts".

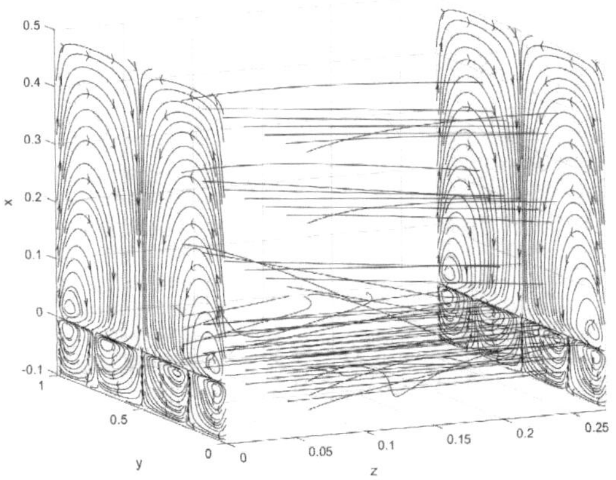

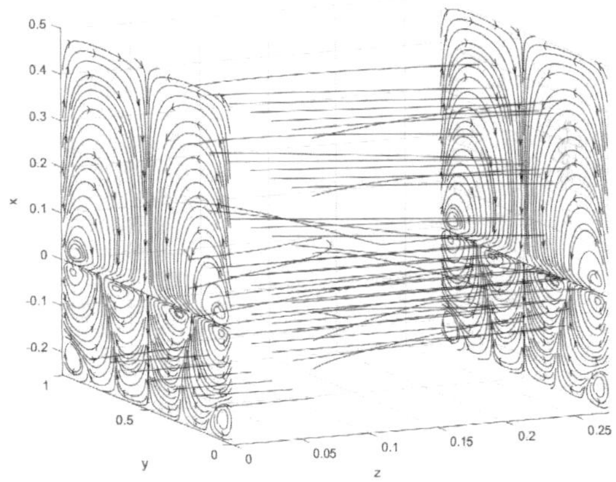

Figure 5. Stream lines and trajectories in HFE-7100 – nitrogen system at $A = 0.1$: $Gr = 12200$, $h_l = 0.1$ (left); $Gr = 1220000$, $h_l = 0.25$ (right).

In HFE-7100 – nitrogen system in the case with more intensive longitudinal temperature regime (Figure 6) the liquid flows are characterized by four separate vortices under microgravity and six swirls in the terrestrial conditions for both values of h_l. Note, that the transverse size of the "inner" vortexes, which appear in the central part of the channel, decreases. At $g = g_0$ and $h_l = 0.25$ near-bottom-wall reverse flow is observed and the additional vortexes in the corners are generated (see right picture in Figure 6). In the thin layer the similar flow picture is not observed. For the system the flow regimes differ only in the liquid layers (compare Figure 5

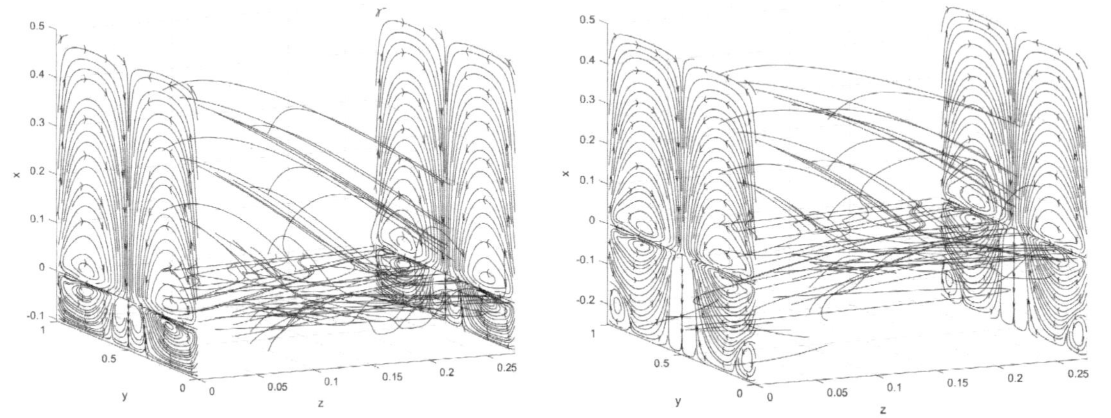

Figure 6. Stream lines and trajectories in HFE-7100 – nitrogen system; $A = 0.3$: $Gr = 12200$, $h_l = 0.1$ (left); $Gr = 1220000$, $h_l = 0.25$ (right).

and 6), whereas the fluid flow structures are rebuilt in both upper and lower layers in ethanol – nitrogen system (Figure 1 and 3). In the case of more intensive thermal load the temperature field in HFE-7100 – nitrogen system is not changed, we observe only variation of the transversal size of the thermal roll formed in the central part of the channel (like to pattern in Figure 3).

Further increasing the temperature gradient leads to formation the thermal shafts in the liquid layer and to growth of number of the longitudinal thermocapillary rolls in the thin layer of HFE-7100 liquid (Figure 7).

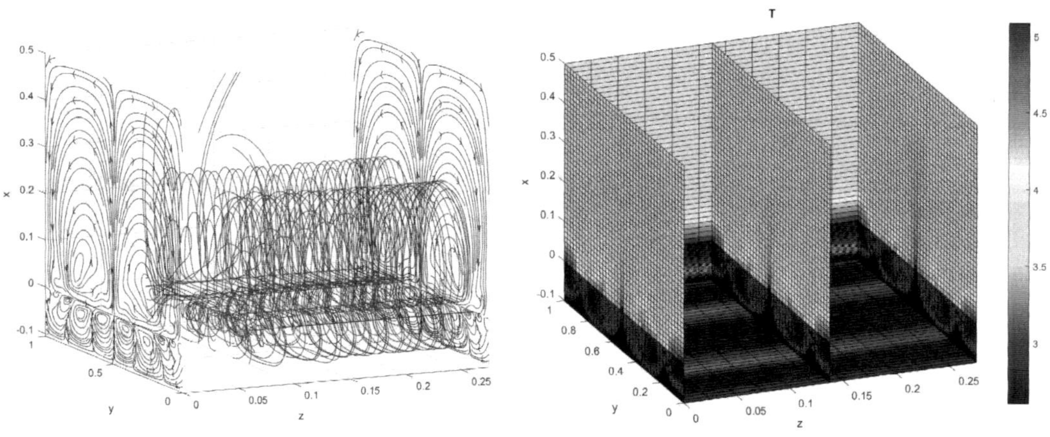

Figure 7. Stream lines and trajectories (left) and distribution of the temperature (right) in HFE-7100 – nitrogen system at $A = 0.4$, $h_l = 0.1$, $Gr = 12200$.

In all the cases flow patterns are symmetric with respect to the plane $y = 0.5$ in both phases. Thus, we have different flow topology and essentially varied temperature patterns for the various gas – liquid systems. More intensive rotational flow has been observed under normal gravity. It is explained by the more significant influence of the temperature stratification.

5. Conclusions

The stationary coupled problem of the gravitational and thermocapillary convection with respect to evaporation is formulated in the paper. The exact solution is constructed in the three-dimensional case to describe the convective flows in an infinite channel of a rectilinear cross section, it means without assumption of axial symmetry of the flow domains. This solution is the analogue of the Ostroumov – Birikh solution of the convection equations which additionally include the Dufour and Soret effects in the gas phase. The flows of both fluids (of a liquid and gas – vapor mixture) modeled with the help of the exact solution can be characterized as a translational motion and progressively rotational flows and be realized in the various forms. The qualitative and quantitative differences are confirmed for the flows of various working fluids (ethanol – nitrogen and HFE-7100 – nitrogen systems). The numerical investigations allow one to analyze the possible flow structure with respect to the intensity of the gravitation field and longitudinal temperature gradients created on the interface. Intensity and structure of the liquid flows depends on intensity of the gravitation field and on interface temperature regime. Topologically various structures of flows (double, quadruple or sextuple vortex patterns) are formed due to combined influence of the thermocapillary and convective mechanisms and evaporation/condensation process, affecting the thermal pattern of the flows.

Acknowledgments

This work is supported by the Russian Science Foundation (project RSF 15-19-20049).

References

[1] Bekezhanova V B and Goncharova O N 2017 Problems of the evaporative convection (Review) *Journal of Applied Mathematics and Mechanics* (accepted for publication)

[2] Shliomis M I and Yakushin V I 1972 Convection in a two-layers binary system with an evaporation, *Collected papers: Uchenye zapiski Permskogo Gosuniversiteta, seriya Gidrodinamika* **4** 129-40

[3] Goncharova O N and Rezanova E V 2014 Example of an exact solution of the stationary problem of two-layer flows with evaporation at the interface *Journal of Applied Mechanics and Technical Physics* **55**(2) 247-57

[4] Goncharova O N, Rezanova E V, Lyulin Yu V and Kabov O A 2015 Modeling of two-layer liquid-gas flow with account for evaporation *Thermophysics and Aeromechanics* **22**(5) 655-61

[5] Bekezhanova V B and Goncharova O N 2016 Stability of the exact solutions describing the two-layer flows with evaporation at interface *Fluid Dynamics Research Journal* **48**(6) 061408

[6] Ostroumov G A 1952 *Free convection under the conditions of an internal problem* (Moscow-Leningrad: Gostekhizdat Press)

[7] Birikh R V 1969 Thermocapillary convection in a horizontal layer of liquid *Journal of Applied Mechanics and Technical Physics* **3** 43-5

[8] Andreev V K, Gaponenko Yu A, Goncharova O N and Pukhnachov V V 2012 *Mathematical models of convection (de Gruyter Studies in Mathematical Physics)* (Berlin/Boston: De Gruyter)

[9] Pukhnachov V V 2000 *Group-theoretical nature of the Birikh's solution and its generalizations* (Krasnoyarsk: Book of Proc. Symmetry and differential equations) pp 180-3

[10] Goncharova O N, Kabov O A and Pukhnachov V V 2012 Solutions of special type describing the three dimensional thermocapillary flows with an interface *Int. J. Heat Mass Transfer* **55** 715-25

[11] Goncharova O N and Kabov O A 2016 Investigation of the two-layer fluid flows with evaporation at interface on the basis of the exact solutions of the 3D problems of convection *Journal of Physics: Conference Series* **754** 032008(1-6)

[12] Landau L D and Lifshitz E M 1987 *Fluid Mechanics* vol 6 (Butterworth – Heinemann)

[13] Roache P J 1976 *Computational Fluid Dynamics* (Albuquerque: Hermosa Publishers)

[14] Bekezhanova V B, Goncharova O N and Shefer I A 2017 Problems of the evaporative convection (Review). Part II. Three-dimensional flows *Journal of Siberian Federal University. Mathematics & Physics* (submitted for publication)

[15] Weast C R C 1979 *Handbook of Chemistry and Physics* (Boca Raton: CRC Press Inc)

[16] Lyulin Y, Kabov O, Iorio C S, Chikov S, Glushchuk A, Marchuk I and Queeckers P 2009 *Liquids – candidates for CIMEX-1 experiments on ISS* (Manuscript: CIMEX Meeting. Bruxelles, May 15)

IOP Conf. Series: Journal of Physics: Conf. Series **894** (2017) 012024 doi:10.1088/1742-6596/894/1/012024

An analytical solution of the ill–posed Cauchy problem for Lame equation in a rectangle

Yu Grigor'ev

North-Eastern Federal University, Yakutsk, Russia

E-mail: grigyum@yandex.ru

Abstract. In this paper we present an analytical method for the inverse Cauchy problem of Lame equation of the elasticity theory. A rectangular domain is frequently used in engineering structures and we only consider the analytical solution in a two-dimensional rectangle, wherein a missing boundary condition is recovered from a full measurement of the stresses and displacements on an accessible boundary. The essence of the method consists in solving three independent Cauchy problems for the Laplace and Poisson equations. For each of them the Fourier series is used to formulate a first-kind Fredholm integral equation for the unknown function of data. Then, we use a Lavrentiev regularization method and the termwise separable property of kernel function allows us to obtain a closed-form regularized solution. As a result, for the displacement components, we obtain solutions in the form of a sum of series with three regularization parameters. The uniform convergence and error estimation of the regularized solutions are proved.

1. Introduction

Let us have an element of construction with a boundary $S = S_1 \cup S_2$ consisting of two parts S_1 and S_2. The part S_1 is accessible for all measurements and the part S_2 is not accessible. In engineering practice it is often necessary to reconstruct the stress-strain state of such structural element by means of measured data on S_1. In this case within an elastic model we have the Cauchy problem for the Lame equation of theory of elasticity. The Cauchy problem for elliptic equations is a so-called ill-posed problem. The main difficulty for solving such problems is numerical instability. For the Cauchy problem in linear elasticity there exist numerous ways for numerical solving. Many of these methods can be classified as the Tikhonov type regularization methods, the finite element method, the boundary element method, etc. (sf. [1, 2, 3, 4]). With regard to analytical solutions it is known only usage of Carleman's method in the elasticity theory [5, 6, 7]. In this paper we present an analytical method for the Cauchy problem solution of Lame equation in a rectangle. This method generalizes the Liu method [8] for an analytical solution of the Cauchy problem of Laplace equation in the rectangle.

2. Formulation of a problem

In numerous publications under the name of Cauchy problem for the Laplace equation there exist two types of problems. In them it is necessary to find a harmonic function in $\Omega \in \mathbb{R}^3$. At first one has a cylindrical domain $\Omega = S \times (0, T)$ with a lower base $S \in \mathbb{R}^2$ and a lateral surface S_1. On S the Cauchy conditions are given: $u = f$, $\partial u/\partial n = g$, and some boundary conditions

Content from this work may be used under the terms of the Creative Commons Attribution 3.0 licence. Any further distribution of this work must maintain attribution to the author(s) and the title of the work, journal citation and DOI.
Published under licence by IOP Publishing Ltd

are given on the lateral surface S_1, for example, $u = 0$, on an upper base there are no given boundary conditions. Such problems can be named as an initial boundary value problem for the Laplace equation (see Ch.9 in [1]). For another type we have a bounded domain $\Omega \in \mathbb{R}^3$ with a smooth boundary $S = S_1 \cup S_2$ and the Cauchy conditions are given on the part S_1 of a boundary S. We will deal with the first type of the Cauchy problem and it's analogous in the elasticity theory. It is necessary to note that thc first type of Cauchy problem is not considered for analytical solutions in the elasticity theory. All analytical results in the elasticity theory concern the second type problem and use Carleman's method.

Let x, y, z be the Cartesian coordinates, for the convenience we also will simultaneously use index notations: x_i $(i = 1, 2, 3)$. An equilibrium equation in the theory of elasticity is the Lame equation

$$L\mathbf{u} \equiv \mu\triangle\mathbf{u} + (\lambda + \mu)\nabla(\nabla \cdot \mathbf{u}) = 0. \tag{1}$$

Hooke's Law expresses connections between components σ_{ij} of stress tensor and components ε_{ij} of deformation tensor and for the linear elasticity has the form

$$\sigma_{ij} = (\lambda + 2\mu)(\nabla \cdot \mathbf{u})\delta_{ij} + 2\mu\varepsilon_{ij}, \quad \varepsilon_{ij} = \frac{1}{2}\left(u_{i,j} + u_{j,i}\right). \quad (i, j = 1, 2, 3) \tag{2}$$

Here and below, the comma and an index (or a variable) denote the corresponding partial derivative. For simplicity we will consider the plane deformation when components of displacement $\mathbf{u}$ have the form $u_x = u(x, y)$, $u_y = v(x, y)$, $u_z = 0$. In such case equilibrium equations in stresses components have the form

$$\begin{aligned} \sigma_{xx,x} + \sigma_{xy,y} &= 0, \\ \sigma_{xy,x} + \sigma_{yy,y} &= 0. \end{aligned} \tag{3}$$

Let Ω be the rectangle $\mathbf{r} \in (0, a) \times (0, b) \in \mathbb{R}^2$. The problem to be solved is the next:

$$\begin{aligned} &L\mathbf{u}(\mathbf{r}) = 0, \quad \mathbf{r} \in \Omega; \\ &\sigma_{yy} = u_x = u_y = 0, \quad \sigma_{xy} = \sigma(x), \quad \text{on } y = b; \\ &\sigma_{xx} = u_y = 0 \quad \text{on } x = 0 \text{ and } x = a. \end{aligned} \tag{4}$$

Thus, we have the Cauchy problem for the Lame equation with the Cauchy data on the upper base, with zero boundary conditions on the lateral sides of rectangle and without any boundary conditions on the lower base of a rectangle. On the upper base only one boundary function $\sigma(x)$ from four is taken not equal to zero only for the simplicity.

3. Method of solution

3.1. Cauchy problem for the auxiliary harmonic function

Let us introduce an auxiliary function proportional to the divergence of a displacement vector

$$f = (\lambda + 2\mu)\nabla \cdot \mathbf{u}. \tag{5}$$

It is well known that this function is harmonic. From Hook's Low (2) we have

$$\begin{aligned} \sigma_{xx} &= \lambda\nabla \cdot \mathbf{u} + 2\mu\varepsilon_{xx} = \lambda\nabla \cdot \mathbf{u} + 2\mu u_{x,x} = \lambda\nabla \cdot \mathbf{u} + 2\mu(\nabla \cdot \mathbf{u} - u_{y,y}), \\ \sigma_{yy} &= \lambda\nabla \cdot \mathbf{u} + 2\mu\varepsilon_{yy} = \lambda\nabla \cdot \mathbf{u} + 2\mu u_{y,y} = \lambda\nabla \cdot \mathbf{u} + 2\mu(\nabla \cdot \mathbf{u} - u_{x,x}), \end{aligned}$$

and we have the next expressions for our auxiliary function (5)

$$f = \sigma_{xx} + 2\mu u_{y,y},$$
$$f = \sigma_{yy} + 2\mu u_{x,x}. \tag{6}$$

Then, after differentiating of these expressions and using equilibrium equations (3) and (1) we obtain

$$f_{,y} = \sigma_{yy,y} + 2\mu u_{x,xy} = -\sigma_{xy,x} + 2\mu(\nabla\cdot\mathbf{u} - u_{y,y})_{,y} = -\sigma_{xy,x} + 2\mu\nabla\cdot\mathbf{u}_{,y} - 2\mu u_{y,yy}$$
$$= -\sigma_{xy,x} + 2\mu\nabla\cdot\mathbf{u}_{,y} - 2[-\mu u_{y,xx} - (\lambda+\mu)\nabla\cdot\mathbf{u}_{,y}] = -\sigma_{xy,x} + 2\mu u_{y,xx} + 2f_{,y},$$

from here we have

$$f_{,y} = \sigma_{xy,x} - 2\mu u_{y,xx}. \tag{7}$$

Analogically one can obtain the next similar formula

$$f_{,x} = \sigma_{xy,y} - 2\mu u_{x,yy}. \tag{8}$$

Let the boundary conditions of the Cauchy problem (4) provide of a solution smoothness $\mathbf{u} \in C^2(\overline{\Omega})$. In this case all formulas (6)–(8) are valid up to the boundary of a rectangle. Then from the formulas (6)–(8) and boundary conditions from (4) we obtain boundary conditions for the function f and we have Cauchy problem for the harmonic function f:

$$\begin{aligned} &\Delta f = 0, \quad \mathbf{r} \in \Omega, \\ &f = 0, \quad f_{,y} = -h(x) \quad \text{on } y = b, \\ &f = 0 \quad \text{on } x = 0 \text{ and } x = a, \end{aligned} \tag{9}$$

where the function $h(x)$ is expressed in terms of the initial boundary functions from (4)

$$h(x) = -\sigma'(x).$$

The regularized solution f^α of this problem is obtained by using the Liu [8] method in the next form:

$$\begin{aligned} &f^\alpha(x,y) = \sum_{k=1}^{\infty} a_k^\alpha \frac{\sinh\dfrac{k\pi(b-y)}{a}}{\sinh\dfrac{k\pi b}{a}} \sin\frac{k\pi x}{a}, \\ &a_k^\alpha = \frac{k\pi\sigma_k \sinh\dfrac{k\pi b}{a}}{k\pi + \alpha a \sinh\dfrac{k\pi b}{a}}, \\ &\sigma_k = \frac{2}{a}\int_0^a \sigma(\xi)\cos\frac{k\pi\xi}{a}d\xi, \end{aligned} \tag{10}$$

where α is a parameter of regularization. In this method the integral equation of the first kind is solved by Lavrent'ev's regularization method using the Fourier method. The kernel in this integral equation has a termwise separable property. This property is the main reason for a solution to be obtained in a closed form.

For the solution (10) are valid the next results presented in [8], including an error estimation:

Theorem 1. If the Neumann datum $h(x)$ is bounded in the interval $x \in [0, a]$, then for any $\alpha > 0$ and $y_0 > 0$ for all $x \in [0, a]$ and $y \in [y_0, b]$ the regularized solution (10) converges uniformly and this series can be termwise differentiable any number of times.

Theorem 2. Assume that the Neumann datum $h(x) \in L^2(0, a)$. Then the sufficient and necessary condition that the Cauchy problem (9) has a solution is that

$$\sum_{k=1}^{\infty} \frac{\sinh^2 \frac{k\pi b}{a}}{k^2\pi^2} \left(\int_0^a h(\xi) \sin \frac{k\pi\xi}{a} d\xi \right)^2 < \infty. \tag{11}$$

Theorem 3. If the Neumann datum $h(x)$ satisfies condition (11) and there exists an $\varepsilon \in (0, 1)$, such that

$$4\sum_{k=1}^{\infty} \left[\frac{a^2 \sinh^2 \frac{k\pi b}{a}}{k^2\pi^2} \right]^{1+\varepsilon} \left(\int_0^a h(\xi) \sin \frac{k\pi\xi}{a} d\xi \right)^2 := M^2(\varepsilon) < \infty, \tag{12}$$

then for any $\alpha > 0$ the regularized solution $f^\alpha(x, y)$ satisfies the following error estimation:

$$\|f^\alpha(x, y) - f(x, y)\|_{L^2(0,a)} \le \alpha^\varepsilon M(\varepsilon).$$

3.2. Cauchy problems for the components of displacements

After substitution $f^\alpha(x, y)$ in place of the $f(x, y)$ in the equilibrium equations (1) for the components of a displacement we have Poisson equations. It can be shown that for the each components $u_x(x, y)$ and $u_y(x, y)$ we can find Cauchy boundary conditions. Thus for the $u_x(x, y)$ we have the next Cauchy problem for the Poisson equations:

$$\begin{aligned}
\Delta u_x &= -\frac{\lambda + \mu}{\mu(\lambda + 2\mu)} f^\alpha_{,x}, \quad \mathbf{r} \in \Omega, \\
u_x &= 0, \quad u_{x,y} = \frac{1}{\mu}\sigma(x) \quad \text{on } y = b, \\
u_{x,x} &= 0 \quad \text{on } x = 0 \text{ and } x = a.
\end{aligned}$$

For the component $u_y(x, y)$ we have another Cauchy problem:

$$\begin{aligned}
\Delta u_y &= -\frac{\lambda + \mu}{\mu(\lambda + 2\mu)} f^\alpha_{,y}, \quad \mathbf{r} \in \Omega, \\
u_y &= u_{y,y} = 0 \quad \text{on } y = b, \\
u_y &= 0 \quad \text{on } x = 0 \text{ and } x = a.
\end{aligned}$$

These problems are solved by the method similar to Liu. Thus, for the solution u_x and u_y of the Cauchy problem (4) we have regularized expressions similar to (10), but with additional regularization parameters β and γ. Now we give the final formulas for these solutions:

$$\begin{aligned}
u_x^{\alpha,\beta}(x, y) &= w_x^\alpha(x, y) + v_x^{\alpha,\beta}(x, y), \\
u_y^{\alpha,\gamma}(x, y) &= w_y^\alpha(x, y) + v_y^{\alpha,\gamma}(x, y);
\end{aligned}$$

$$w_x^\alpha(x,y) = \sum_{k=1}^{\infty} w_{xk}^\alpha(y)\cos\frac{k\pi x}{a},$$
$$w_{xk}^\alpha(x,y) = \frac{\lambda+\mu}{\mu(\lambda+2\mu)}\frac{a_k^\alpha}{\sinh(\frac{k\pi b}{a})}\left[y\cosh\frac{k\pi(b-y)}{a} - b\frac{\sinh\frac{k\pi y}{a}}{\sinh\frac{k\pi b}{a}}\right];$$
$$v_x^{\alpha,\beta}(x,y) = b_0(b-y) + \sum_{k=1}^{\infty} b_k^{\alpha,\beta}\frac{\sinh\frac{k\pi(b-y)}{a}}{\sinh\frac{k\pi b}{a}}\cos\frac{k\pi x}{a},$$

$$b_k^{\alpha,\beta} = \frac{a\sinh\frac{k\pi b}{a}}{\mu\left(k\pi+\beta\sinh\frac{k\pi b}{a}\right)} \left[\frac{\lambda+\mu}{2(\lambda+2\mu)}\frac{a_k^\alpha}{\sinh\frac{k\pi b}{a}}\left(1-\frac{k\pi b}{a}\coth\frac{k\pi b}{a}\right) - \sigma_k\right];$$

$$w_y^\alpha(x,y) = -\frac{\lambda+\mu}{\mu(\lambda+2\mu)}\sum_{k=1}^{\infty} a_k^\alpha y\frac{\sinh\frac{k\pi(b-y)}{a}}{\sinh\frac{k\pi b}{a}}\sin\frac{k\pi x}{a};$$

$$v_y^{\alpha,\gamma}(x,y) = \sum_{k=1}^{\infty} c_k^{\alpha,\gamma}\frac{\sinh\frac{k\pi(b-y)}{a}}{\sinh\frac{k\pi b}{a}}\sin\frac{k\pi x}{a},$$
$$c_k^{\alpha,\gamma} = \frac{\lambda+\mu}{2\mu(\lambda+2\mu)}\cdot\frac{k\pi b}{k\pi+\gamma\sinh\frac{k\pi b}{a}}a_k^\alpha,$$
$$b_0 = -\frac{1}{\mu}\sigma_0 = -\frac{1}{a\mu}\int_0^a \sigma(\xi)d\xi.$$

If the boundary function $\sigma(x)$ in (4) is bounded in closed interval, then all of these regularized solutions converge uniformly and can be differentiated termwise any number of times. When the parameters $\alpha,\beta,\gamma\to 0$ then the regularized solutions $u_x^{\alpha,\beta}(x,y), u_y^{\alpha,\gamma}(x,y)$ tend to the exact solutions.

4. Error estimation

For the regularized auxiliary function $f^\alpha(x,y)$ we have error estimation according of the Theorem 3. For the component of a displacement $u_x^{\alpha,\beta}(x,y)$ the next error estimation theorem can be proved:

Theorem 4. If the Neumann datum $h(x)$ satisfies condition (11) and there exist $\varepsilon\in(0,1)$ and $\delta\in(0,1)$, such that conditions (12) and the next

$$4\sum_{k=1}^{\infty}\left[\frac{a^2\sinh^2\frac{k\pi b}{a}}{k^2\pi^2}\right]^{1+\delta}\left(\int_0^a q(\xi)\sin\frac{k\pi\xi}{a}d\xi\right)^2 := N^2(\delta) < \infty,$$
$$q(x) = -\frac{1}{\mu}\sigma(x) + w_{,y}^\alpha(x,b)$$

IOP Publishing
doi:10.1088/1742-6596/894/1/012024

are hold, then for any $\alpha > 0$, $\beta > 0$ the regularized solution $u_x^{\alpha,\beta}(x,y)$ satisfies the following error estimation:

$$\|u_x(x,y) - u_x^{\alpha,\beta}(x,y))\|_{L^2(0,a)} \leq \sqrt{\frac{b}{2}}\alpha^{\varepsilon}M(\varepsilon) + \beta^{\delta}N(\delta).$$

For the component of a displacement $u_y^{\alpha,\gamma}(x,y)$ an analogical error estimation theorem also can be formulated. In the paper of C.S. Liu [8] the examples of effective numerical calculations by means of presented formula for the $f^{\alpha}(x,y)$ are presented.

5. Conclusion

The Liu method for solving the Cauchy problem for the Laplace equation is generalized for the Cauchy problem of the linear elasticity. This allows us to obtain the regularized solution of the problem in the closed form by using the Fourier and Lavrent'ev methods. The obtained regularized solution converges uniformly and the error estimation of regularized solution is proved. For simplicity we only consider a two–dimensional rectangular domain. The method can be used in the cases of other boundary conditions, it can be generalized for a three–dimensional case. Also it is interesting to note that the used auxiliary function $f = (\lambda + 2\mu)\nabla \cdot \mathbf{u}$ is a scalar part of a regular quaternion function connected with the solution $\mathbf{u}$ of the Lame equation (1) [9, 10, 11]. Thus it may be possible to use presented method in the problem of the regular extension into a domain of the quaternion function defined on a part of its boundary.

Acknowledgments

This work was supported by grant 15-41-05081 from the Russian Foundation for Basic Research.

References

[1] Kabanikhin S I 2012 *Inverse and Ill-posed Problems : Theory and Applications* (Berlin-Boston: Walter de Gruyter GmbH & Co. KG)
[2] Kozlov V A, Maz'ya V G and Fomin A V 1991 An iterative method for solving the Cauchy problem for elliptic equations *Comput. Math. Math. Phys.* **31** 45–52
[3] Marin L and Johansson B T 2010 A relaxation method of an alternating iterative algorithm for the Cauchy problem in linear isotropic elasticity *Comput. Methods Appl. Mech. Engrg.* **199** 3179–96
[4] Durand B, Delvare F and Bailly P 2011 Numerical solution of Cauchy problems in linear elasticity in axisymmetric situations *Int. J. Sol. Str.* **48** 3041–53
[5] Tarkhanov N 1995 *The Cauchy Problem for Solutions of Elliptic Equations* (Berlin: Akademie Verlag)
[6] Shlapunov A 1996 On the Cauchy problem for the Lame system *ZAMM* **76** 215–21
[7] Makhmudov O I and Niezov I E 2000 A Cauchy problem for the system of elasticity equations *Diff. Equat.* **36** 749–54
[8] Liu C S 2011 An analytical method for the inverse Cauchy problem of Laplace equation in a rectangular plate *J. Mech.* **27** 575–84
[9] Grigor'ev Yu M 2014 Three-dimensional quaternionic analogue of the Kolosov-Muskhelishvili formulae hypercomplex analysis *New perspectives and applications, Trends in Mathematics (Eds. S. Bernstein, U. Kaehler, I. Sabadini, F. Sommen)* (Basel: Birkhauser) 145–66
[10] Grigor'ev Yu 2016 Three-dimensional analogue of Kolosov-Muskhelishvili formulae *Modern Trends in Hypercomplex Analysis, Trends in Mathematics (Eds. S. Bernstein, U. Kaehler, I. Sabadini, F. Sommen)* (Basel: Birkhauser) 203-15
[11] Grigor'ev Yu M 2017 Regular quaternionic polynomials and their properties *Complex Variables and Elliptic Equations* **62** 1343–63

IOP Conf. Series: Journal of Physics: Conf. Series **894** (2017) 012025 doi:10.1088/1742-6596/894/1/012025

Supercritical stratified flow over an uneven bottom

D S Denisenko[1,2] and N I Makarenko[1,2]

[1] Lavrentyev Institute of Hydrodynamics, Novosibirsk, Russia

[2] Novosibirsk State University, Novosibirsk, Russia

E-mail: d.denisenko@g.nsu.ru

Abstract. The solvability of nonlinear problem on steady internal waves in supercritical flow of a heavy stratified fluid over rough topography is investigated. Existence of solution in the Hardy-type function classes is proven by the perturbation technique using small parameter which characterizes typical height of obstacles. It is shown that the solution is analytic with respect to this parameter.

1. Introduction

In this paper, we study the nonlinear problem on stationary flows of stratified fluid affected by the gravity over an uneven bottom. Such a flows are of interest due to their importance in meteorology and oceanology [1]. There are two different regimes of steady-state flow which are called sub- and supercritical flows. In the case of subcritical flow over the obstacle, the most known structure is stationary configuration with attached wave train having periodic asymptotics in the downstream direction. Respectively, the supercritical regime is characterized by that the wave structures form in the vicinity of the obstacle, but these disturbances are damped far downstream. Mathematical problem on stationary free surface flow of homogeneous ideal fluid over an uneven bottom was studied in [2,3]. The boundary value problem for the Dubreil–Jacotin –Long (DJL) equation [4, 5], which models the motion of stratified fluid over the obstacles, was considered in [6, 7]. These papers contain the proof of existence of solutions for sufficiently fast upstream flow (the Froude numbers should be much greater than the critical value). In present work, we show the solvability of the nonlinear problem for the DJL equation in the whole range of supercritical values of the Froude numbers.

2. Statement of the problem

We consider steady two–dimensional flow of an inhomogeneous incompressible fluid in a finite–depth layer over an uneven bottom. Basic dimensionless constants are the Boussinesq parameter σ and the squared inverse densimetric Froude number λ defined by the formulae

$$\sigma = \frac{N^2 H}{g}, \qquad \lambda = \frac{\sigma g H}{U^2}.$$

Here N is a constant buoyancy frequency, H is the thickness of undisturbed fluid layer (Fig. 1), g is the gravity acceleration, and U is the velocity of upstream flow. Constance of the frequency N means that the fluid density depends exponentially on height at rest. The flow over the obstacles

Content from this work may be used under the terms of the Creative Commons Attribution 3.0 licence. Any further distribution of this work must maintain attribution to the author(s) and the title of the work, journal citation and DOI.
Published under licence by IOP Publishing Ltd

is modelled by a boundary value problem for the DJL equation which has the following form in dimensionless variables:

$$\psi_{xx} + \psi_{yy} + \lambda(\psi - y) = \frac{\sigma}{2}\,(\psi_x^2 + \psi_y^2 - 1), \tag{1}$$

$$\psi|_{y=ah(x)} = 0, \qquad \psi|_{y=1} = 1, \tag{2}$$

$$\psi(x,y) \to y \quad (x \to -\infty), \tag{3}$$

where $\psi(x,y)$ is unknown stream function. The parameter σ defines the slope of the density profile, and the parameter λ indicates whether the upstream flow is sub- or supercritical. Namely, the flow is supercritical if the parameters λ and σ satisfy the inequality

$$\lambda < \frac{\sigma^2}{4} + \pi^2, \tag{4}$$

and the flow is subcritical in opposite case. The value $\lambda = \sigma^2/4 + \pi^2$ is noted as the critical value of the Froude number λ. In what follows, the condition (4) is assumed to be valid.

The equation (1) is equivalent to the system of Euler equations of inhomogeneous fluid flow [4, 5]. It is assumed that the flow domain is bounded from above by the rigid lid, and the bottom shape is defined by smooth function $h(x)$ which decays as $|x| \to \infty$. Kinematic boundary conditions at the bottom and at the lid have the form (2). In this work, we use the dimensionless height of the obstacle $a = d/H$ as a natural small parameter, where d is typical height of the obstacles. The radiation condition (3) means that there are no waves in the upstream flow.

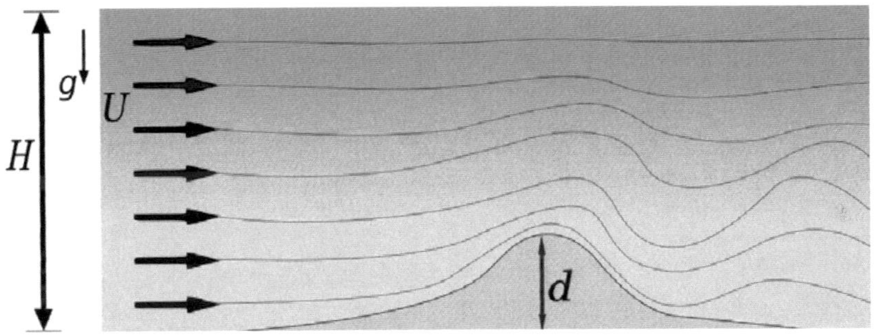

Figure 1. The motion of a stratified flow over obstacles.

It is assumed that all the streamlines $\psi = \text{const}$ have one-to-one projection on the x-axis, so the flow has no overturns. By that, we can use semi-Lagrangian independent variables (x, ψ) known as the von Mises variables. In this case, the shape of streamlines can be found in the form $y = \psi + w(x,\psi)$, and the boundary value problem (1)–(3) can be rewritten with new unknown function w as follows:

$$\begin{gathered} w_{xx} + w_{\psi\psi} - \sigma w_\psi + \lambda w = f(w), \\ w(x,0) = ah(x), \qquad w(x,1) = 0, \\ w(x,\psi) \to 0 \qquad (x \to -\infty), \end{gathered} \tag{5}$$

where the nonlinear operator $f(w)$ has the form

$$f(w) = -\frac{\partial}{\partial x}\frac{w_x w_\psi}{1 + w_\psi} + \frac{1}{2}\frac{\partial}{\partial \psi}\frac{w_x^2 + 3w_\psi^2 + w_\psi^3}{(1 + w_\psi)^2} - \frac{\sigma}{2}\frac{w_x^2 + 3w_\psi^2 + w_\psi^3}{(1 + w_\psi)^2}.$$

Thus, the original boundary value problem (1)–(3), formulated in the flow region $ah(x) < y < 1$, reduces to the problem (5) in the rectilinear strip $(x,\psi) \in \Pi = \mathbb{R} \times [0,1]$.

IOP Conf. Series: Journal of Physics: Conf. Series **894** (2017) 012025 doi:10.1088/1742-6596/894/1/012025

3. The analysis of linearized problem

We consider here nonhomogeneous boundary value problem arising from linearized equation (5) with a given right-hand side function $f(x,\psi)$, and with fixed parameters σ, λ, a:

$$\begin{gathered} w_{xx} + w_{\psi\psi} - \sigma w_\psi + \lambda w = f(x,\psi), \\ w(x,0) = ah(x), \qquad w(x,1) = 0, \\ w(x,\psi) \to 0 \qquad (x \to -\infty). \end{gathered} \tag{6}$$

Using partial Fourier transform of w with respect to horizontal variable x,

$$\widehat{w}(\xi,\psi) = \mathbf{F}[w](\xi,\psi) = \frac{1}{2\pi}\int\limits_{-\infty}^{\infty} e^{i\xi x} w(x,\psi)dx,$$

we represent the solution $w(x,\psi)$ to the problem (6) by explicit formula

$$w = \mathbf{F}^{-1}\left[\mathbf{G}\langle\widehat{f}\rangle + \mathbf{K}\langle\widehat{h}\rangle\right], \tag{7}$$

where $\mathbf{F}^{-1}$ denotes inverse Fourier transformation in x. Integral operator $\mathbf{G}$ has the form

$$\mathbf{G}\langle\widehat{f}\rangle = \int\limits_0^1 G(\xi,\psi,s)\widehat{f}(\xi,s)ds, \tag{8}$$

where the Green function $G(\xi,\psi,s)$ is given by the formula

$$G(\xi,\psi,s) = \frac{e^{\frac{\sigma}{2}(\psi-s)}}{\sqrt{\xi^2-\nu^2}\,\mathrm{sh}\sqrt{\xi^2-\nu^2}}\begin{cases} \mathrm{sh}\sqrt{\xi^2-\nu^2}(s-1)\,\mathrm{sh}\,(\sqrt{\xi^2-\nu^2}\psi), & \psi \leqslant s, \\ \mathrm{sh}\,(\sqrt{\xi^2-\nu^2}s)\,\mathrm{sh}\sqrt{\xi^2-\nu^2}(\psi-1), & \psi > s, \end{cases}$$

as $\xi^2 > \nu^2 = \lambda - \frac{\sigma^2}{4}$, and

$$G(\xi,\psi,s) = \frac{e^{\frac{\sigma}{2}(\psi-s)}}{\sqrt{\nu^2-\xi^2}\sin\sqrt{\nu^2-\xi^2}}\begin{cases} \sin\sqrt{\nu^2-\xi^2}(s-1)\sin(\sqrt{\nu^2-\xi^2}\psi), & \psi \leqslant s, \\ \sin(\sqrt{\nu^2-\xi^2}s)\sin\sqrt{\nu^2-\xi^2}(\psi-1), & \psi > s. \end{cases}$$

as $\xi^2 < \nu^2$. The operator $\mathbf{K}$ is defined by the relation

$$\mathbf{K}\langle\widehat{h}\rangle = a\widehat{h}(\xi)e^{\frac{\sigma}{2}\psi}K(\xi,\psi), \tag{9}$$

where the multiplier $K(\xi,\psi)$ has the form

$$K(\xi,\psi) = \begin{cases} \dfrac{\mathrm{sh}\sqrt{\xi^2-\nu^2}(1-\psi)}{\mathrm{sh}\sqrt{\xi^2-\nu^2}}, & \xi^2 > \nu^2, \\[2ex] \dfrac{\sin\sqrt{\nu^2-\xi^2}(1-\psi)}{\sin\sqrt{\nu^2-\xi^2}}, & \xi^2 < \nu^2. \end{cases}$$

The function G and K are defined at $\xi^2 = \nu^2$ by continuity.

The formulae (7)–(9) provide a formal solution, so we need to specify it in more detailes. For instance, an analytic extension of the Fourier transform $\widehat{w}(\xi,\psi)$ on the complex plane $\zeta = \xi + i\eta$ is important here. The analyticity strip $|Im\,\zeta| < \alpha$ of the function $\widehat{w}$ is evidently restricted by zeros of the functions sh $\sqrt{\xi^2-\nu^2}$, sin $\sqrt{\nu^2-\xi^2}$. These square roots possess branching points, but the map $\widehat{w}$ is not multivalent since their representation involves only even analytic functions of the variable $\kappa = \sqrt{\xi^2-\nu^2}$. In accordance with the condition (4), we are interested in the solution which satisfies inequality $\nu^2 < \pi^2$. In this case, zeros of functions sh $\sqrt{\xi^2-\nu^2}$, sin $\sqrt{\nu^2-\xi^2}$ lie on the imaginary axis and form the set $\{\pm\sqrt{|k^2\pi^2-\nu^2|}i : k \in \mathbb{Z}\}$. Hence, the quantity

$$\alpha_* = \sqrt{\pi^2-\nu^2} = \sqrt{\pi^2+\frac{\sigma^2}{4}-\lambda} \tag{10}$$

presents natural bound for decay exponent $\alpha < \alpha_*$ of the solution $w(x,\psi)$ as $|x| \to \infty$.

Let us introduce the function spaces where we are looking for a solution to the boundary value problem (6). Suppose that real numbers $\rho > 0$, $\alpha > 0$ are given, and real interval $[a,b]$ is chosen. We denote by $\Pi_{[a,b]}$ a strip $\mathbb{R}\times[a,b]$. The class $E_\rho^\alpha\left(\Pi_{[a,b]}\right)$ consists of the functions $w(x,\psi)$ that are defined in the strip $\Pi_{[a,b]}$, and these functions admit the Fourier transform $\widehat{w}(\zeta,\psi)$ which is analytic with respect to $\zeta = \xi + i\eta$ in the complex domain $|Im\,\zeta| < \alpha$, and $\widehat{w}(\zeta,\psi)$ is jointly continuous in closed set $\{(\zeta,\psi) : |\eta| \leqslant \alpha_1, a \leqslant \psi \leqslant b\}$ for any $\alpha_1 < \alpha$. The norm of the function $w(x,\psi)$ is defined as follows:

$$\|w\|_{\rho,\alpha} = \sup_{|\eta|<\alpha}\int_{-\infty}^{\infty} e^{\rho|\xi|}\sup_{\psi\in[a,b]}|\widehat{w}(\xi+i\eta,\psi)|d\xi. \tag{11}$$

Since this norm is finite, the function $w(z,\psi)$ is analytic with respect to $z = x+iy$ in the domain $|y| < \rho$, and their Fourier transform $\widehat{w}$ is jointly continuous for $|y| \leqslant \rho$. By the definition of the norm (11), both functions $\widehat{w}_\pm(\zeta,\psi) = \exp(\pm\rho\zeta)\widehat{w}(\zeta,\psi)$ belong to the Hardy class H^1 for every fixed $\psi \in [a,b]$ in the strip $|Im\,\zeta| < \alpha$. Therefore, in accordance with properties of the Hardy class H^1 [8], the function $w \in E_\rho^\alpha\left(\Pi_{[a,b]}\right)$ decays exponentially as $|x| \to \infty$ in such a way that

$$|w(x+iy,\psi)| \leqslant \|w\|_{\rho,\alpha}e^{-\alpha|x|}$$

for $x \in \mathbb{R}$, $|y| \leqslant \rho$.

The class of the Fourier transforms $\widehat{w}(\zeta,\psi)$ mentioned above is a special case of generalized Hardy classes [9]. In addition to the Banach space $E_\rho^\alpha\left(\Pi_{[a,b]}\right)$, we consider also the spaces $E_{\rho,2}^\alpha\left(\Pi_{[a,b]}\right)$ of functions that belong to $E_\rho^\alpha\left(\Pi_{[a,b]}\right)$ together with their derivatives up to n-th order, equipped with the norm

$$\|w\|_{E_{\rho,2}^\alpha} = \sum_{0\leqslant m+k\leqslant 2}\|D_x^m D_\psi^k w\|_{\rho,\alpha}.$$

Also, we define the class $E_\rho^\alpha(\mathbb{R})$ of the function of a single variable x, equipped with the norm (11) (the spaces $E_{\rho,2}^\alpha(\mathbb{R})$ are defined in a similar way). In accordance with the notation for the strip $\Pi = \Pi_{[0,1]}$, we find the solutions of the problem (6) in the class $E_{\rho,2}^\alpha(\Pi)$ with exponent $\alpha < \alpha_*$, where α_* is determined in (10). The condition that the solution derivatives should belong to the spaces $E_\rho^\alpha(\Pi)$ is caused by the fact that the nonlinear boundary value problem (5) has the right-hand side operator f depending on the second-order derivatives $D_x^m D_\psi^n w$ $(m+n \leqslant 2)$.

In this section we apply the techniques of estimates developed in the paper [10]. The following Lemma gives us preliminary estimates of the Green function G. For this purpose, we introduce the cut-off functions $M_j : \mathbb{R}_+ \to \mathbb{R}_+ (j = 0,1,2)$ by setting

$$M_j(t) = 1 \quad (0 \leqslant t \leqslant 1), \quad M_j(t) = t^{-j} \quad (t \geqslant 1).$$

The subscript j indicates here the decay rate of the Fourier transforms as $|Re\,\zeta| \to \infty$.

Lemma 1. *The following estimates hold in the strip* $|Im\,\zeta| \leqslant \alpha < \alpha_*$ *where* α_* *is given by* (10):

$$\int_0^1 |D_\psi^j G(\zeta,\psi,s)|ds \leqslant C_1(\alpha,\alpha_*,\nu)M_{2-j}(Re\,|\zeta|) \quad (j=0,1).$$

Proof. At first, we choose a real number $L > 0$, and consider separately the cases $|Re\,\zeta| \leqslant L$ and $|Re\,\zeta| > L$. Using elementary inequalities for hyperbolic functions of complex- and real variable

$$|\operatorname{sh}\xi| \leqslant |\operatorname{sh}\zeta| \leqslant \operatorname{ch}\xi, \tag{12}$$

$$\left|\frac{\operatorname{sh}\zeta}{\zeta}\right| \leqslant \frac{\operatorname{sh}\xi}{\xi} \leqslant \frac{2\beta}{\sin\beta}\left|\frac{sh\zeta}{\zeta}\right| \quad (|Im\,\zeta| \leqslant \beta < \pi), \tag{13}$$

we obtain that continuous function $G(\zeta,\psi,s)$ is uniformly bounded on the compact set $|Re\,\zeta| \leqslant L$. Hence we have the estimate

$$\int_0^1 |G(\zeta,\psi,s)|ds \leqslant \frac{\sqrt{\pi^2+\alpha^2-\alpha_*^2}}{\sin\sqrt{\pi^2+\alpha^2-\alpha_*^2}}C_2(L).$$

Further, applying inequalities (12) and (13) by $|Re\,\zeta| > L$, we find that the source integral has the second-order decay with respect to the quantity $|\gamma| = |Re\sqrt{\zeta^2-\nu^2}|$:

$$\int_0^1 |G(\zeta,\psi,s)|ds \leqslant \frac{C_3(L)}{|\gamma|^2}.$$

Finally, using trigonometric representation $\zeta^2-\nu^2 = \tau e^{i\theta}$, which implies $|\gamma| = \sqrt{\tau}|\cos\frac{\theta}{2}|$ and $\theta = \arg(\zeta^2-\nu^2) = \arg(\zeta-\nu)+\arg(\zeta+\nu)$, we estimate the value of $|\gamma|$ by $L > 2|\nu|$ as follows:

$$\frac{1}{|\gamma|} \leqslant \frac{C_4(\alpha,\nu)}{|\xi|}.$$

The estimate of integral, mentioned in the Lemma for $j=1$, is obtained in a similar way. □

The following Theorem gives the sufficient condition for existence of the solution to linear problem (6) in the space $E^\alpha_{\rho,2}(\Pi)$.

Theorem 1. *Suppose that the upstream flow is supercritical in the sense of* (4), *and the bottom shape is such that* $h(x) \in E^\alpha_{\rho,2}(\mathbb{R})$, $f(x,\psi) \in E^\alpha_\rho(\Pi)$ *with* $\alpha < \alpha_*$, *where* α_* *is given by* (10). *Then the formula (7)–(9) represents the solution to the problem (6) belonging to the class* $E^\alpha_{\rho,2}(\Pi)$. *Furthermore, for* $0 \leqslant m+n \leqslant 2$ *the following estimates hold*:

$$\|D_x^m D_\psi^n w\|_{E^\alpha_\rho(\Pi)} \leqslant C_5(\alpha,\alpha_*,\nu)\left(a\|h\|_{E^\alpha_{\rho,2}(\mathbb{R})} + \|f\|_{\rho,\alpha}\right).$$

Proof. According to the formulae (7)–(9), we have:

$$\|D_x^m D_\psi^n w\|_{E^\alpha_\rho(\Pi)} = \sup_{|\eta|<\alpha}\int_{-\infty}^{\infty} e^{\rho|\xi|}\sup_{\psi\in[0,1]}\left|\zeta^m D_\psi^n \widehat{w}(\xi+i\eta,\psi)\right|d\xi \leqslant J_1+J_2,$$

where is denoted

$$J_1 = \sup_{|\eta|<\alpha} \int_{-\infty}^{\infty} e^{\rho|\xi|} \sup_{\psi\in[0,1]} \left| \zeta^m \int_0^1 D_\psi^n G(\zeta,\psi,s)\widehat{f}(\zeta,s)ds \right| d\xi$$

and

$$J_2 = \sup_{|\eta|<\alpha} \int_{-\infty}^{\infty} e^{\rho|\xi|} \sup_{\psi\in[0,1]} \left| \zeta^m a\widehat{h}(\zeta) D_\psi^n (e^{\frac{\sigma}{2}\psi} K(\zeta,\psi)) \right| d\xi.$$

Let us consider the case $n \neq 2$. Using Lemma 1, we obtain for J_1 the estimate

$$J_1 \leqslant \frac{\sqrt{\pi^2+\alpha^2-\alpha_*^2}}{\sin\sqrt{\pi^2+\alpha^2-\alpha_*^2}} C_6 \|f\|_{\rho,\alpha}.$$

Since the derivatives $h'(x)$ and $h''(x)$ belong to the class $E_\rho^\alpha(\mathbb{R})$, we have for J_2 the estimate

$$J_2 \leqslant \frac{\sqrt{\pi^2+\alpha^2-\alpha_*^2}}{\sin\sqrt{\pi^2+\alpha^2-\alpha_*^2}} C_7 \|h\|_{E_{\rho,2}^\alpha}.$$

In the case $n = 2$, the estimate of the Theorem 1 follows, due to the equations (6), from above-obtained solution estimates. □

4. The analysis of the nonlinear problem

According to results of the Section 3, the nonlinear problem (5) is equivalent to the nonlinear operator equation

$$w = \Phi(w,a,h,\lambda,\sigma), \tag{14}$$

where is denoted

$$\Phi(w,a,h,\lambda,\sigma) = \mathbf{F}^{-1}\left[\mathbf{G}\langle\widehat{f}(w)\rangle + \mathbf{K}\langle\widehat{h}\rangle\right].$$

Thus, the problem (5) reduces to the finding a fixed point of the operator Φ. Note that the norm in $E_\rho^\alpha\left(\Pi_{[a,b]}\right)$ is multiplicative due to the convolution theorem for the Fourier transform:

$$\|wv\|_{\rho,\alpha+\beta} \leqslant \|w\|_{\rho,\alpha}\|v\|_{\rho,\beta} \tag{15}$$

where $w \in E_\rho^\alpha$, $v \in E_\rho^\beta$. The nonlinear operator f can be presented by convergent Taylor series in powers w, similarly to the geometric progression series, so the map $f(w)$ is analytic at w in a neighborhood of the element $w = 0$. Using this fact, we obtain by (15) the estimate

$$\|f(w)-f(v)\|_{\rho,\alpha} \leqslant (28+6\sigma)\frac{\|w\|_{E_{\rho,2}^\alpha}+\|v\|_{E_{\rho,2}^\alpha}}{(1-\|w\|_{E_{\rho,2}^\alpha}-\|v\|_{E_{\rho,2}^\alpha})^4}\|w-v\|_{E_{\rho,2}^\alpha}. \tag{16}$$

Fixe the real parameters λ, σ, a and the function h, we obtain from (16) under the estimates of Theorem 1 the following inequality which is valid for $\|w\|_{E_{\rho,2}^\alpha} \leqslant r$, $\|v\|_{E_{\rho,2}^\alpha} \leqslant r$:

$$\|\Phi(w,a,h,\lambda,\sigma)-\Phi(v,a,h,\lambda,\sigma)\|_{E_{\rho,2}^\alpha} \leqslant \frac{C_8(\alpha,\alpha_*,\nu)r}{(1-2r)^4}\|w-v\|_{E_{\rho,2}^\alpha}.$$

Therefore, the mapping $\Phi(w,a,h,\lambda,\sigma)$ is the contraction in the ball $B_r = \{w \in E_{\rho,2}^\alpha(\Pi) : \|w\|_{E_{\rho,2}^\alpha} < r\}$, where the radius $r > 0$ is limited as follows:

$$r < C_9\frac{\sin\sqrt{\pi^2+\alpha^2-\alpha_*^2}}{\sqrt{\pi^2+\alpha^2-\alpha_*^2}}. \tag{17}$$

For $a = 0$, the problem (14) has the trivial solution $w = 0$. Then, using that mapping Φ depends continuously on a, we can select the radius r such that the mapping Φ is contractor on the closed ball $\bar{B}_r$. According to the contraction mapping principle, there is a unique fixed point $w = w(a, h, \lambda, \sigma) \in \bar{B}_r$ which gives unknown solution. Due to the point uniqueness we have $w(0, h, \lambda, \sigma) = 0$. As $\Phi(w, a, h, \lambda, \sigma)$ is jointly continuous of the λ, σ, a, h, the constructed solution is continuous of this parameters. Moreover, since Φ is analytic with respect to w and a, the solution is analytic in a. Thus, we have proved the following theorem.

Theorem 2. *Let the upstream flow be supercritical in the sense of* (4). *Suppose that the function $h(x)$, which defines the bottom shape, is in $E^{\alpha}_{\rho,2}(\mathbb{R})$ with $\alpha < \alpha_*$, where α_* is given by* (10). *Then the boundary value problem* (5) *has the solution $w \in E^{\alpha}_{\rho,2}(\Pi)$ which is analytic with respect to the parameter a.*

Note that the restrictions, formulated for the parameters λ, σ, a and the function h, have clear physical meaning. The conditions (4) means that the velocity U of upstream flow should be not small. The fact that the derivatives of the function h belong to the spaces $E^{\alpha}_{\rho}(\mathbb{R})$ suggests that the bottom shape is nearly flat. In particular, the quantity $a\|h\|_{E^{\alpha}_{\rho,2}}$ is limited due to (17). It means that the height of obstacles, as well as their steepness, should not exceed a certain value. Otherwise, as it is known from the theory of stratified fluid [5], the flow may be blocked in the front of the obstacle. Constructed solution has limited derivatives by small a, so the streamlines have bijective projection on the x-axis. In addition, analyticity in a justifies, at least in the supercritical case, the formal perturbation procedure used in [11] by the construction of approximate solutions describing non-linear lee wave structures. According to (4) and (10), we obtain that the parameters α, $\alpha_* \to 0$ as the Froude number λ tends to the critical value $\lambda = \sigma^2/4 + \pi^2$. Therefore, the ball, which contains the solution w, vanishes in this limit case due to the estimate (17). Possible reason may be the change of the flow regime by the transition to the subcritical flow, so the nonlinear wave structures can bifurcate at critical value of λ.

5. Conclusion

The nonlinear boundary value problem on steady 2D flows of an inhomogeneous incompressible fluid in a finite–thickness layer over an uneven bottom is studied. The existence of exact solution is proved under the assumption that upstream flow satisfies the super-criticality condition $\lambda < \sigma^2/4 + \pi^2$, and the bottom shape is nearly flat. The solution of inhomogeneous linearized problem is constructed in explicit form (with the corresponding estimates). It is shown that the nonlinear solution is analytic with respect to the parameter of the obstacle height.

Acknowledgements

This work was supported by the Russian Foundation for Basic Research (grant No 15-01-03942).

References

[1] Grimshaw R (Ed.) 2001 *Environmental Stratified Flows* (Boston: Kluwer)
[2] Nalimov V I 1982 *Dinamika Sploshn. Sredy* **58** 108–55 (in Russian)
[3] Iguchi T 2003 *Wave Motion* **37** 219–39
[4] Dubreil-Jacotin M L 1935 *Atti Acad. Lincei Rend. Cl. Sci. Fis. Mat. Nat.* **21(6)** 344–46
[5] Yih C S 1980 *Stratified flows* (N.-Y.: Acad. Press)
[6] Krutitskii P A 1996 *Appl. Math. Lett.* **9** 41–6
[7] Krutitskii P A 1997 *Int. J. Non-Linear Mech.* **32** 483–8
[8] Koosis P 1980 *Introduction to Hp Spaces* (N.-Y.: Cambridge University Press)
[9] Hille E, Phillips R S 1968 *Functional Analysis and Semi-Groups* (Providence: AMS)
[10] Golosov K V and Makarenko N I 2013 *J. Math. Sci.* **195** 805–14
[11] Makarenko N I and Maltseva J L 2014 *J. Appl. Mech. Tech. Phys.* **55** 311–17

IOP Conf. Series: Journal of Physics: Conf. Series **894** (2017) 012026 doi:10.1088/1742-6596/894/1/012026

The choice of the energy embedding law in the design of heavy ionic fusion cylindrical targets

GV Dolgoleva[1] and A I Zykova[1]

[1] National Research Tomsk State University, Tomsk, Russia

E-mail: Arven2022@mail.ru

Abstract. The paper considers the numerical design of heavy ion fusion (FIHIF) targets, which is one of the branches of controlled thermonuclear fusion (CTF). One of the important tasks in the targets design for controlled thermonuclear fusion is the energy embedding selection whereby it is possible to obtain "burning" (the presence of thermonuclear reactions) of the working DT region. The work is devoted to the rapid ignition of FIHIF targets by means of an additional short-term energy contribution to the DT substance already compressed by massively more longer by energy embedding. This problem has been fairly well studied for laser targets, but this problem is new for heavy ion fusion targets. Maximum momentum increasing is very technically difficult and expensive on modern FIHIF installations. The work shows that the additional energy embedding ("igniting" impulse) reduces the requirements to the maximum impulse. The purpose of this work is to research the ignition impulse effect on the FIHIF target parameters.

1. Introduction

A microtarget is a layered system wherein one layer, called a "working" layer, consists of a deuterium-tritium blend (DT) or another blend wherein thermonuclear reactions occur. The target is either irradiated from the outside, or the energy is embedded in the inner layer. Under the action of this energy, the "working" region is compressed and "burning" begins. So thermonuclear reactions occur, as a result of which energy is released.

The main microtargets design tasks for controlled thermonuclear fusion (CTF) are a selection of the geometry (size and composition of layers) and the law of energy embedding wherein it is possible to obtain "burning" (the presence of thermonuclear reactions) of the working DT region. At the same time, the energy output as a result of thermonuclear reactions should be greater than the embedded energy (the thermonuclear gain coefficient is greater than one). And an important question is the amount of energy that is being embedded.

The work is devoted to the rapid ignition of FIHIF targets by means of an additional short-term embedding of energy to the DT substance already compressed by massively more longer by energy embedding. This problem has been fairly well studied for laser targets [1,2], but this problem is new for heavy ion fusion targets.

The maximum impulse increase is very difficult technically and expensively on modern FIHIF installations. The paper shows that additional energy embedding ("igniting" impulse) reduces the requirements to the maximum impulse (driver). The purpose of this work is to research the ignition impulse effect on the FIHIF target parameters.

Content from this work may be used under the terms of the Creative Commons Attribution 3.0 licence. Any further distribution of this work must maintain attribution to the author(s) and the title of the work, journal citation and DOI.
Published under licence by IOP Publishing Ltd

IOP Conf. Series: Journal of Physics: Conf. Series **894** (2017) 012026 doi:10.1088/1742-6596/894/1/012026

2. Research objective

There is considered the cylindrical layered system (target) of the form:

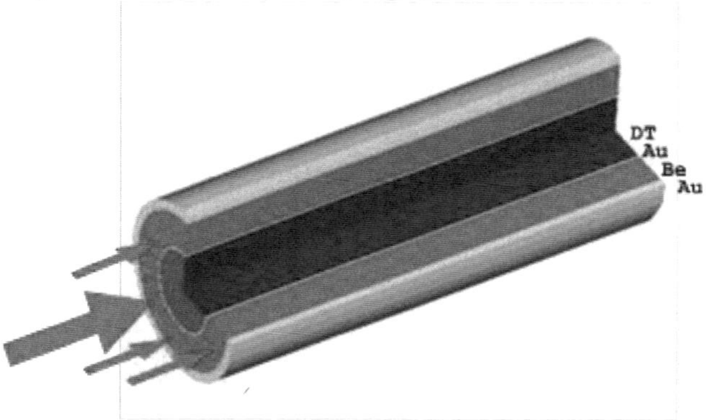

Figure 1. Appearance of the target

The target is an axisymmetric cylindrical structure (of a certain length) consisting of nested layers made of different substances.

Functionality of the system layers:

- a central layer (rod) of DT is energy source in the implementation of thermonuclear fusion reactions;
- a shell of a heavy substance (Au) is a pusher which compresses the DT layer during energy cumulation and at the same time is a shield for thermal insulation of DT fuel from a heated energy embedding region;
- the energy embedding region is intended for high-speed heavy ion beams deceleration and accumulating energy from an external source;
- an outer heavy shell (Au) is designed to reduce the radial expansion of the layers outward, which increases the amount of cumulative energy.

The target ends remain free. Through them heavy ion fluxes of high energy are injected left and right into the light layer (Be) from an external source. The energy embedding mechanism is realized by means of focused beams directed parallel to the system axis. To realize axial symmetry of motion, the beams rotate in circular orbits with a high frequency. The ion flux deceleration is carried out in the light layer almost evenly along the length. Such energy embedding keeps the motion axisymmetric.

The ions deceleration in the target layers matter causes intense heating and subsequent expansion. In this case, the motion in the axis direction leads to the energy cumulation in its vicinity. Expansion from the open ends, as calculations in the complete (two-dimensional) formulation [3] show, cover only limited parts of the target, while retaining its large part unaffected. Therefore, in a significant part of the target, outside the vicinity of the ends, the motion remains only radial (one-dimensional). The movement in this part of the target is studied, i.e. there is one-dimensional target of the form:

m_0 (DT) m_1(Au) m_2(Be) m_3(Au)

|--------------|--------------------|///////////////|------------------|

0 L_0 L_1 L_2 L_3

Figure 2. Geometry of the target section perpendicular to its axis.

A unshocked compression idea of the working DT region is used when designing targets.

IOP Conf. Series: Journal of Physics: Conf. Series **894** (2017) 012026 doi:10.1088/1742-6596/894/1/012026

What does unshocked compression give?

Large compressions and large densities can be obtained even endless if one has sufficient energy to compress.

To reproduce unshocked compression at the right boundary of the working (DT) region, the velocity and pressure in the case of an ideal gas must be respectively equal (these dependences were obtained by Stanyukovich K.P. [4]):

$$u_l = \frac{2}{\gamma-1}c_0\left[1-\left(1-\frac{c_0}{L_0}t\right)^{-\frac{\gamma-1}{\gamma+1}}\right] \quad (1)$$

$$P_l = \frac{\rho_0}{\gamma}c_0^2\left(1-\frac{c_0}{L_0}t\right)^{-\frac{2\gamma}{\gamma+1}} = P_0\left(1-\frac{c_0}{L_0}t\right)^{-\frac{2\gamma}{\gamma+1}} \quad (2)$$

where L_0 - length of compressible layer, ρ_0, c_0 - respectively initial density and sound velocity, γ - adiabatic index, t - current time.

Within the framework of the shell system (Figure 2.), the energy embedding law $Q(t)$ ($dE_2 / dt = Q(t)$) in the region of mass m_2 can be selected such that on the right boundary of the region with mass m_0 we the velocity and pressure values $u_l(t)$, $P_l(t)$, necessary for carrying out the unshocked compression of this region, i.e. blend of DT gas with mass m_0 [7], are obtained.

$$Q(t) = 2\gamma c_0^3 G L_0^{\alpha} \Big/ \left((\gamma-1)^2 m_2\right) \cdot$$

$$\left\{-(1-\xi^{-(\gamma-1)/(\gamma+1)})^2 + \frac{\left[\frac{V_2(0)}{aL_0^{\alpha}} + (\gamma-1) + 2\xi - (\gamma+1)\xi^{2\gamma/(\gamma+1)}\right]}{\gamma+1}\xi^{-1}\right\}\xi^{2\gamma/(\gamma+1)}L_0^{\alpha-1},$$

$$\xi = 1 - c_0 t/L_0 \; (0 \le \xi \le 1), \quad G = m_0(1-2k_2)/\gamma + 2k_1/(\gamma+1) + (\gamma+1)m_0^2 k_3/\gamma^2, \quad (3)$$

$$a = \left(\left(m_1+m_2+m_3+(\gamma+1)\rho_0 L_0^{\alpha}/(2\gamma)\right)\right)\Big/\left((\gamma-1)\left(m_2/2+m_3\right)\right),$$

$$k_1 = m_1 + \frac{m_2}{3} + \left(m_1+\frac{m_2}{2}\right)\left(\frac{m_1 m_2}{3} + m_1 m_3 + \frac{m_2 m_3}{6}\right)\Big/(m_3+\frac{m_2}{2})^2,$$

$$k_3 = 3\left(m_3+m_2/3\right)\Big/\left(2(m_3+m_2/2)^2\right).$$

where m_i- layers mass, α- parameter that depends on the geometry type.

Figure 3 shows the function type $Q(t)$ is for a specific target with c_0=0.01 and L_0=0.1 (the length is measured in cm, the time is 10-7 seconds, the power $Q(t)$ is 0.01 TW/g).

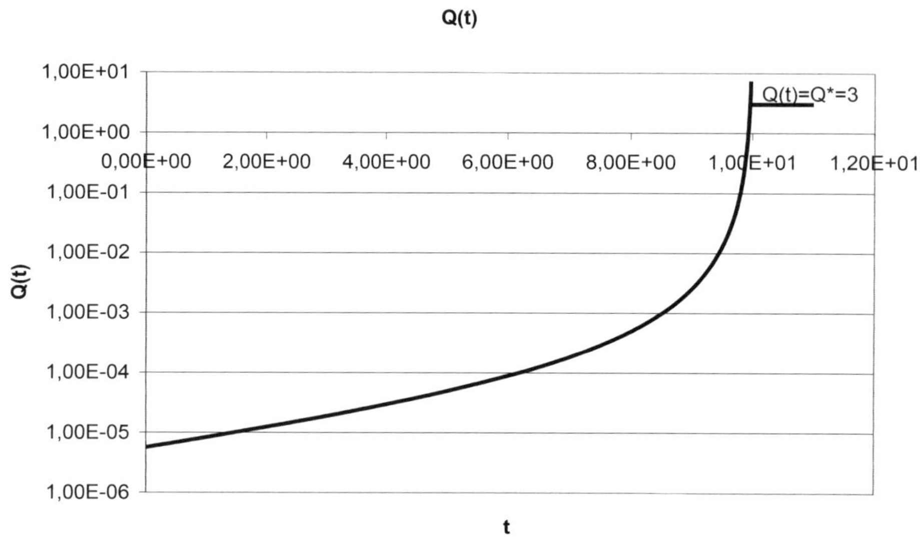

Figure 3. Function type $Q(t)$.

$Q(t)$ quantitatively differs in different calculations depending on the geometry (m_i), the maximum value of Q^*, the values of c_0 and L_0, but the curves form is the same (with peaking).

It stands to mention a number of the moments arising in the unshocked compression.

Firstly, with unshocked compression, it is possible to obtain high parameters of the compressible gas (DT gas density).

Secondly, the main shortcoming in the immediate implementation of unshocked compression is the need for unlimited energy embedding growth and it can be eliminated. This is achieved by the kinetic energy accumulation in the heavy shell with the mass m_1 in the energy embedding course. The energy embedding termination is offset by the subsequent release of the stored kinetic energy gas in the region m_1 (Figure 2).

Thirdly, under classical unshocked (cold) compression in the fuel for burning, the temperature of its heating is not sufficient, although the "ignition" criterion "$\rho \int r \sim \int_{DT} \rho dr$" is fulfilled with a large margin due to high densities under the unshocked compression. The final temperature can be raised by increasing the initial heating (by increasing the parameter c_0), as well as by breaking the unshocked compression mode at its final stage.

During microtargets development, it is also necessary to take into account the design characteristics of the installation where experiments with this target are planned. The proposed "plan" of energy embedding in the shell microtarget design, the variant of which is shown in Fig. 2, will be described

According to the found law $Q(t)$ the energy embedding is determined by the time t^* ($0 \le t \le t^*$), wherein the aggravation of the impulse $Q(t)$ reaches the maximum value for setting $Q^* = Q(t^*)$. Then the energy embedding occurs up at a constant Q^* to the time t^{**}. The moment t^{**} is determined by the achievement of an average region density with a mass m_0 of a value that would exceed the ignition criterion ρR. However, the set values ρR for unshocked compression do not yet ensure the ignition of the DT gas blend. Its temperature is insufficient. Then it can be do two things. Firstly, it can be to invest energy with the maximum exacerbation magnitude into the region with the mass m_2 up to a certain value, until the target ignites (one-impulse energy embedding). And secondly, it is possible to make additional energy embedding in the form of a impulse with intensity F_{DT} ("igniting" impulse) at the moment t^{**} when the necessary density criterion for ignition in the DT gas region is reached. Due to this, it is possible to effect active burning of DT gas in the DT region (two-impulse energy embedding).

IOP Conf. Series: Journal of Physics: Conf. Series **894** (2017) 012026 doi:10.1088/1742-6596/894/1/012026

The one impulse regime of compression and burning of FIHIF targets is similarly described in [6, 7].

3. Calculations results

Figure 4 shows a computational comparison of the two-impulse and one-impulse compression and burning scenarios is given in the example of the cylindrical target. Calculations were made using models that adequately describe the plasma physics [8].

		DT		Au		Be		Au		
r	0		0.1		0.105		0.195		0.222	(см)
ρ_0		0.05		19.3		1.8		19.3		(г/см^3)

Figure 4. Specific type of target.

The motion of the system in the two-temperature approximation, the energy transfer by electrons and ions, the kinetics of thermonuclear reactions, the spectral energy transfer by photons and their interaction with matter are calculated. The external energy embedding $Q(t)$ is given in the region *Be*. Figure 3 shows the form of the function $Q(t)$.

The system of units wherein the results are given: the time in 110-7 sec, the length in cm, the energy in 0.1 MJ.

There is introduced the following notation:

t^*- an instant of time before which the energy embedding is carried out according to the scheme of unshocked compression;

$Q^*=Q(t^*)$ – maximum exacerbation of energy embedding;

$F(t^*)$ – full energy embedding (according to the scheme of unshocked compression);

t^{**} - a moment of the external energy embedding end and a moment of the second impulse of the direct embedding in DT;

$F(t^{**})$ – energy embedding in the time interval $0<t<t^{**}$;

F_{DT} – an instantaneous energy embedding in the region DT at two-impulse energy;

F_{emb} – energy, nested in the target for two inpulses;

F_{ign}- avalue released in the DT thermonuclear energy;

K - coefficient of thermonuclear gain (ratio F_{ign}/F_{emb}).

The table 1 shows the results of target calculations for two-impulse and single-impulse energy embedding.

Table 1. Results of target calculations

№ calculation	I	II	III	IV
t^*	9.9786	9.9786	9.96476	9.96994
$F(t^*)$	0.02625	0.02625	0.0246	0.0309
Q^*	30	30	50	75
t^{**}	10.2126	10.2126	10.25	10.157
$F(t^{**})$	0.38384	0.38384	0.2679	0.3895
F_{DT}	0	0.03	0	0
F_{emb}	0.38384	0.4138	0.2679	0.3895
F_{ign}	0.0002	15.72	0.996	3.417
K	0.0012	21.4	3.72	8.77

The first column is the result of calculating $Q^* = 3$, without additional energy embedding in the DT layer. In this case, about 0.38 energy units are invested in the target and it does not light up (F_{ign} = 0.0002) The second column shows the result of calculations for additional energy embedding in the DT gas region (two-impulse energy embedding). The second impulse is F_{DT} = 0.03, i.e. about 1% of the energy is invested from the nested one at the first impulse. The additional impulse value is chosen to equalize the energy embedding values for the single-impulse and two-impulse energy embedding, although calculations show that the target can be ignited at a lower energy of the second impulse.

The third and fourth columns are the calculations results with the single-impulse energy embedding, but an increased value of Q^*.

The third column is the result of calculation with $Q^* = 50$, and the fourth is the results of calculations with $Q^* = 75$.

The approach wherein the ignition is carried out by means of the additional energy directly into the DT fuel by the second impulse, is attractive. It is possible to ignite the target with a smaller exacerbation amount, and the energy output for an equally enclosed energy is 5.8 times larger (look at columns 2 and 4).

4. Conclusion

The ignition impulse performs the same role as increase of the main energy embedding power aggravation from $Q^* = 30$ to $Q^* = 75$.

The igniting impulse presence activates the "burning" of the target. With the same embedded energy, the thermonuclear gain coefficient increases by 5.5 times.

Acknowledgments

The results were obtained during the implementation of the project No. 9.9036.2017/8.9 within the state task framework of the Ministry of Education and Science of Russia.

IOP Conf. Series: Journal of Physics: Conf. Series **894** (2017) 012026 doi:10.1088/1742-6596/894/1/012026

References

[1] Gus'kov S Yu Thermonuclear gain and parameters of fast ignitions ICF- targets 2005 *Laser and Particle Beams* **23** 255

[2] Caruso A and Strangio S. Ignition of high gain targets de entropy injection 2000 *Laser and Particle Beams* **18** 25

[3] Zabrodin A B and Prokopov G P Numerical simulation methods of two-dimensional unsteady flows of the heat-conducting gas in the three-temperature 1998 *Problems of atomic science and technology, series: Mathematical modeling of physical processes* **3** 3

[4] Stanukovich K P Unsteady motion of a continuous medium. 1971 (Moscow: Sience)

[5] Dolgoleva G V and Zabrodin A V Cumulation of energy in layered systems and realization of unshocked compression 2004 (Moscow: Fizmatlit)

[6] Dolgoleva G V and Zabrodin A V Development of thermonuclear targets based on the realization of the unshocked compression concept 2002 *Aeromechanics and gas dynamics* **2** 48

[7] Dolgoleva G V, Zabrodin A V and Khabibulin O B Computational design of microtargets based on the implementation of the unshocked compression concept 2003 *Proceedings of the Institute of Mathematics and Mechanics, Ural Department of the Russian Academy of Sciences* **9** 1

[8] Dolgoleva G V Method for calculating of the two-temperature emitting gas motion (SND) 1983 *Problems of atomic science and technology, series: Methods and programs for the numerical solution of mathematical physics problems* **2**(13) 29

IOP Conf. Series: Journal of Physics: Conf. Series **894** (2017) 012027 doi:10.1088/1742-6596/894/1/012027

Pressure driven laminar flow of a power-law fluid in a T-channel

O A Dyakova and O Yu Frolov

Tomsk State University, 36 "Lenina" prospect, Tomsk, 634050, Russia

E-mail: dyakova_o@ftf.tsu.ru

Abstract. Planar flow of a non-Newtonian fluid in a T-channel is investigated. The viscosity is determined by the Ostwald–de Waele power law. Motion of the fluid is caused by pressure drop given in boundary sections of the T-channel. On the solid walls, the no slip boundary condition is used. The problem is numerically solved with using a finite difference method based on the SIMPLE procedure. As a result of this study, characteristic flow regimes have been found. Influence of main parameters on the flow pattern has been demonstrated. Criteria dependences describing basic characteristics of the flow under conditions of the present work have been shown.

1. Introduction

A T-channel is a very common element of pipelines networks for transportation of fluids and gases. This element has a variety of engineering applications in irrigation systems, wastewater treatment, biomechanics, pharmaceuticals, as well as, oil and gas, polymer industries, and in many other areas. The fluid flow in a T-channel is characterized by separation of the flow into two parts, and recirculation zones can be formed in the region of the separation. In engineering practice, it is essential to understand the main characteristics of the flow in separating and reattaching flows.

The flow of Newtonian fluids in a T-channel has been studied both experimentally and numerically to obtain basic information of the flow structure and give quantitative estimation. The results of experimental research of a Newtonian fluid flow in a T-channel for various Reynolds number and geometric sizes of the channel are presented in [1-3]. The results of numerical simulation of a turbulence Newtonian flow obtained with using the software package are demonstrated in [4]. Authors of the paper [5] investigated both laminar and turbulence flows of Newtonian and non-Newtonian fluids for various Reynolds number.

With regard to flows of non-Newtonian fluids, steady and unsteady non-Newtonian fluid flows for a range of the Reynolds number Re=10-1000 were investigated in [6]. The data obtained in the study [6] are in close agreement with experimental and numerical results of [1, 7-9]. Authors of [10] solved the problem of the flow of the non-Newtonian fluid describing by the Carreau–Yasuda model in a T-channel for a range of the Reynolds number Re=50-1000. Characteristics of laminar power-law flow in a T-channel for a range of the Reynolds number Re=5-200 and the power-law index n=0.2-1 are presented in [11].

In the studies listed above, the flow rate was given in boundary sections of a T-channel. In the research [12], a numerical simulation of the flow of a Newtonian incompressible fluid in channels of complex geometry including fluid flow in a T-channel was performed when the pressure drop was

Content from this work may be used under the terms of the Creative Commons Attribution 3.0 licence. Any further distribution of this work must maintain attribution to the author(s) and the title of the work, journal citation and DOI.
Published under licence by IOP Publishing Ltd

given in boundary sections. Also, the authors of [12] proved existence and uniqueness of the solution of such problems.

The objective of the present work is to study of characteristics of the flow of a power-law fluid moving in a T-channel under given pressure drop in boundary sections for a range the Reynolds number Re=5-350 and the power-law index n=0.6-1.4.

2. Problem statement and mathematical formulation

Planar steady-state flow of a non-Newtonian incompressible fluid in a T-channel is considered. The flow region is limited by solid walls *MKF*, *EDC*, *AB* (Fig.1). Motion of the fluid is caused by pressure drop given in boundary sections *AM*, *FE*, *BC* of the T-channel. Mathematical problem statement includes momentum and continuity equations which in dimensionless vector form are written as follows:

$$\begin{cases} \mathrm{Re}(\mathbf{U}\cdot\nabla)\mathbf{U} = -\mathrm{Re}\nabla p + \nabla\cdot(2\mathrm{B}\mathbf{E}), & (1) \\ \nabla\cdot\mathbf{U} = 0, & (2) \end{cases}$$

where $\mathbf{U}$ – velocity vector with components (U,V) in the Cartesian coordinate system, p – the pressure, $\mathrm{Re} = \dfrac{\rho U_0^{2-n} L^n}{\mu_0}$ – the Reynolds number, ρ – the fluid density, μ_0 – the power-law consistency index, $\mathbf{E}$ – the rate of deformation tensor. To scale the length and the velocity, L (the width of the boundary section *AM*) and $U_0 = \sqrt{\dfrac{\Delta p}{\rho}}$ are used, respectively. Here $\Delta p = p_{AM} - p_{BC}$ – the pressure drop between boundary sections *AM* and *BC*. The dimensionless pressure is defined as $p = \dfrac{p^* - p^*_{BC}}{\Delta p}$ where the symbol «*» is used to denote dimension values of the pressure.

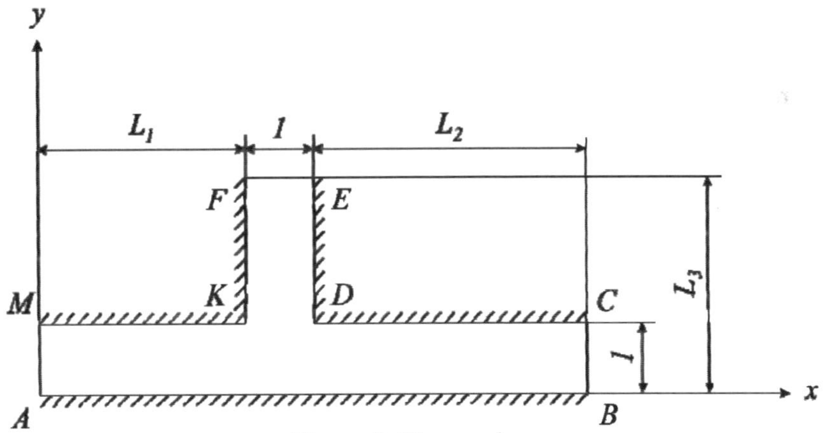

Figure 1. Flow region.

The viscosity of non-Newtonian fluid is determined by the Ostwald–de Waele power law [13]:

$$\mathrm{B} = (A)^{n-1},$$

where $A = \sqrt{2\left(\dfrac{\partial U}{\partial x}\right)^2 + \left(\dfrac{\partial U}{\partial y} + \dfrac{\partial V}{\partial x}\right)^2 + 2\left(\dfrac{\partial V}{\partial y}\right)^2}$ – dimensionless intensity of the rate of deformation tensor, n – the power-law index.

For pressure, boundary conditions are following:

$$\begin{aligned} p_{AM} &= 1, \quad x = 0,\ 0 \le y \le 1 \\ p_{BC} &= 0, \quad x = L_1 + L_2 + 1,\ 0 \le y \le 1 \\ p_{FE} &= p_1,\ L_1 \le x \le L_1 + 1,\ y = L_3 \end{aligned} \tag{3}$$

On solid walls, no slip boundary condition implying vanishing velocity vector on the solid walls is used:

$$\mathbf{U} = 0 \tag{4}$$

The problem solution is reduced to finding both the velocity and pressure fields which satisfy the equations (1),(2) with the given boundary conditions (3),(4).

3. Numerical method

The problem is numerically solved. To find steady-state velocity and pressure fields, the relaxation method is used [14]. This method assumes the addition of time derivative of the function **U** in the equation (1).The obtained system is discretized by the finite difference method based on the SIMPLE procedure [15]; rectangular staggered grid is used.

For pseudo-plastic fluid, there is a singularity of «infinite» viscosity as $A \to 0$. To ensure the stability and accuracy of calculations in the regions of small values of A, additional regularization of the rheological model is necessary. In our study, a modified model of the rheological equation is used. According to this equation, the viscosity is determined by expression $B = (A + \varepsilon)^{n-1}$. The proposed modification allowing the passage to the power-law model as $\varepsilon \to 0$ extends the range of variation of n for a stable calculation [16]. The procedure of regularization is justified both physically and from point of view of the computational fluid dynamics. On the one hand, regularization provides a high but finite value of the viscosity B at low shear rates which corresponds to the physical content of the concept of the viscosity. And on the other hand, the modification of the expression for the viscosity B facilitates the convergence of the computational algorithm.

To verify the numerical algorithm developed for this study, test calculations were performed. Computations for a sequence of grids were carried out to check the approximate convergence. Calculated value of the velocity in the boundary section BC was compared with analytical solution of the problem of the steady-state non-Newtonian fluid flow in the planar channel with given pressure drop per length unit δp

$$U_0(y) = \left|\mathrm{Re}\ \delta p\ (y - 0.5)\right|^{\frac{n-1}{n}},\ 0 \le y \le 1 \tag{5}$$

The value δp was numerically determined in vicinity of the boundary BC where one-dimensional fluid flow is realized. In Tables 1 and 2, dependences of the error E on the spatial grid step h and regularization parameter ε, respectively, are presented. The value E was calculated by the equation $E = \max\limits_i \left\{ \dfrac{|U_0(y_i) - U_i|}{U_0(y_i)} \right\}$, where $U_0(y_i)$ –the analytic value, U_i – the calculated value of the velocity in grid node «i» on the boundary BC.

Data of Table 1 indicate the approximate convergence. In further calculations, the spatial grid step h=1/40 was chosen. It can be seen from Table 2, the error E decreases with decreasing ε. However, starting from a certain value of ε, the error E increases with decreasing ε that indicates a weakening of the role of regularization. Further calculations were carried out for the value $\varepsilon = 0.01$.

Table 1. Dependence of the error E on the spatial grid step h (Re = 40, n =0.6).

H	E, %
1/10	7.8
1/20	3.1
1/40	0.34

Table 2. Dependence of the error E on the regularization parameter ε (Re = 40, n=0.6, h=1/40).

E	E, %
0.1	2.2
0.01	0.27
0.001	0.34

4. Numerical results

The analysis of the problem statement indicates the characteristics of the flow are depending on geometric sizes of the channel and three parameters: the Reynolds number (Re), the power-law index (n), and the pressure given on the boundary FE (p_1). In this research, sizes of the T-channel were chosen to avoid the flow influence in vicinity of the corner points K and D on the character of the flow near the boundary sections AM, FE, BC (L_1=3, L_2=4, L_3=3).

According to numerical calculations, the steady-state flow is formed for the problem investigated. Distribution of the flow characteristics for Re=40 is presented in Fig.2. Close to the boundaries AM, FE, and BC, there is the planar-parallel non-Newtonian flow with parabolic velocity profile (5), while the pressure gradient is constant and directed along the flow direction. Transient flow regions appear in vicinity of the sections with corner points.

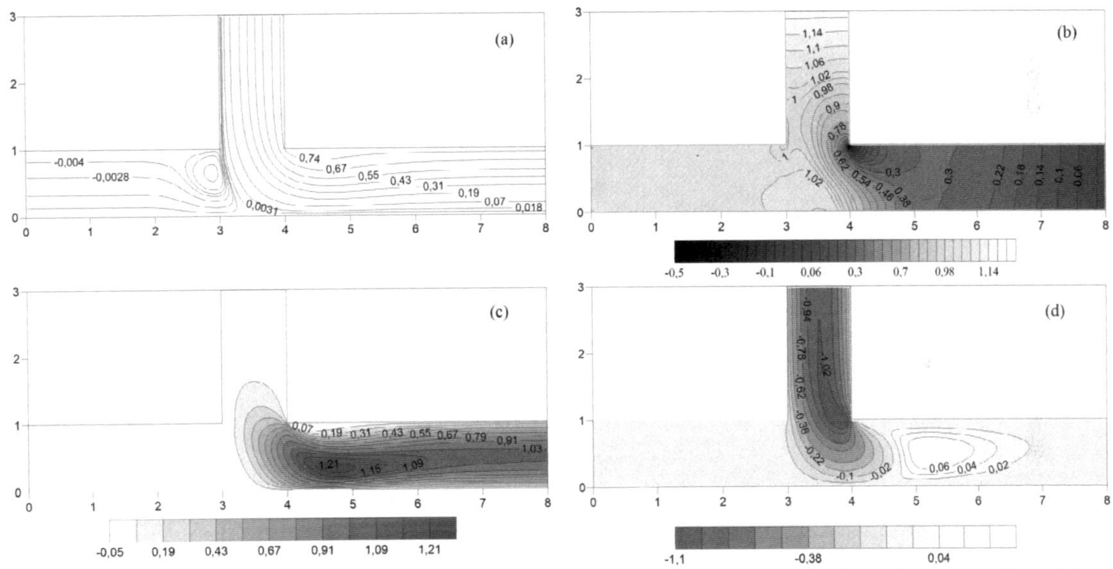

Figure 2. Distribution of the flow characteristics for Re=40 (n=0.6, p_1=1.2, (a) – the streamline, (b) – the pressure field, (c) –the velocity field U, (d)–the velocity field V).

Fig. 3 demonstrates influence of the Reynolds number on the steady-state flow pattern, and two characteristic flow regimes can be distinguished for the considered range of variation of this parameter.The first flow regime ($\mathrm{Re} \le 30$): the flow without recirculation zones is realized; the fluid flows into the boundary sections AM and FE and flows out into the boundary section BC. Growth of the inertial forces leads to the fluid flow moving from the border FE begins to press on the flow moving from the border AM harder. As a result, the recirculation zone in vicinity of the solid wall AB is formed. Further behavior of the fluid flow is characterized by the growth of recirculation zone sizes with increasing Re. At the critical value of the Reynolds number Re=38.45 (Fig. 3, c), the recirculation zone closes the cross-section in the direction of the border AM. The second flow regime is realized at Re>38.45. New flow regime is described by division of the flow moving from the border FE into two parts. Consequently, the fluid flows into the boundary section FE and flows out into the boundary sections *AM* and *BC*. The recirculation zone is formed near the border *MN*, and sizes of it decrease with increasing values of the Reynolds number.

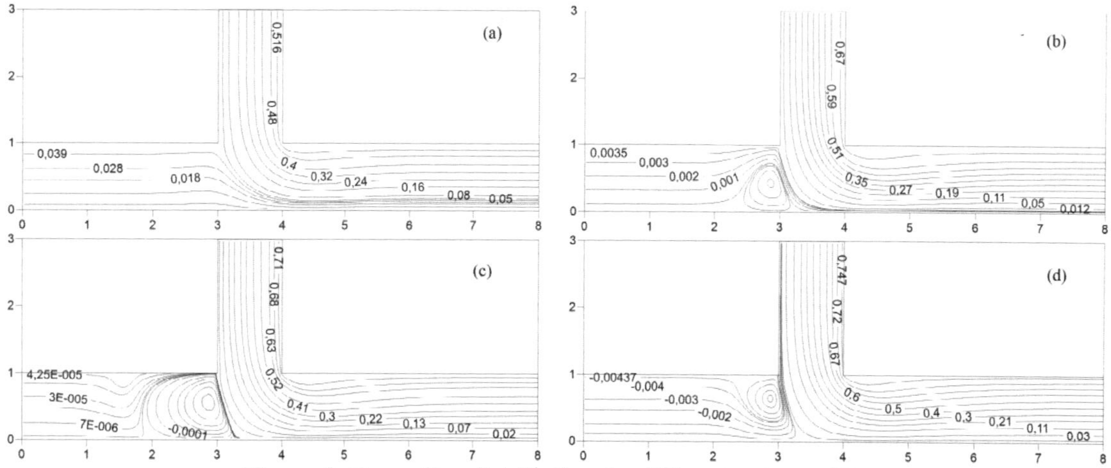

Figure 3. Streamline distribution for different values of Re (n=0.6, p_1=1.2, (a) – Re=30, (b) – Re=37, (c) – Re=38.45, (d) – Re=40).

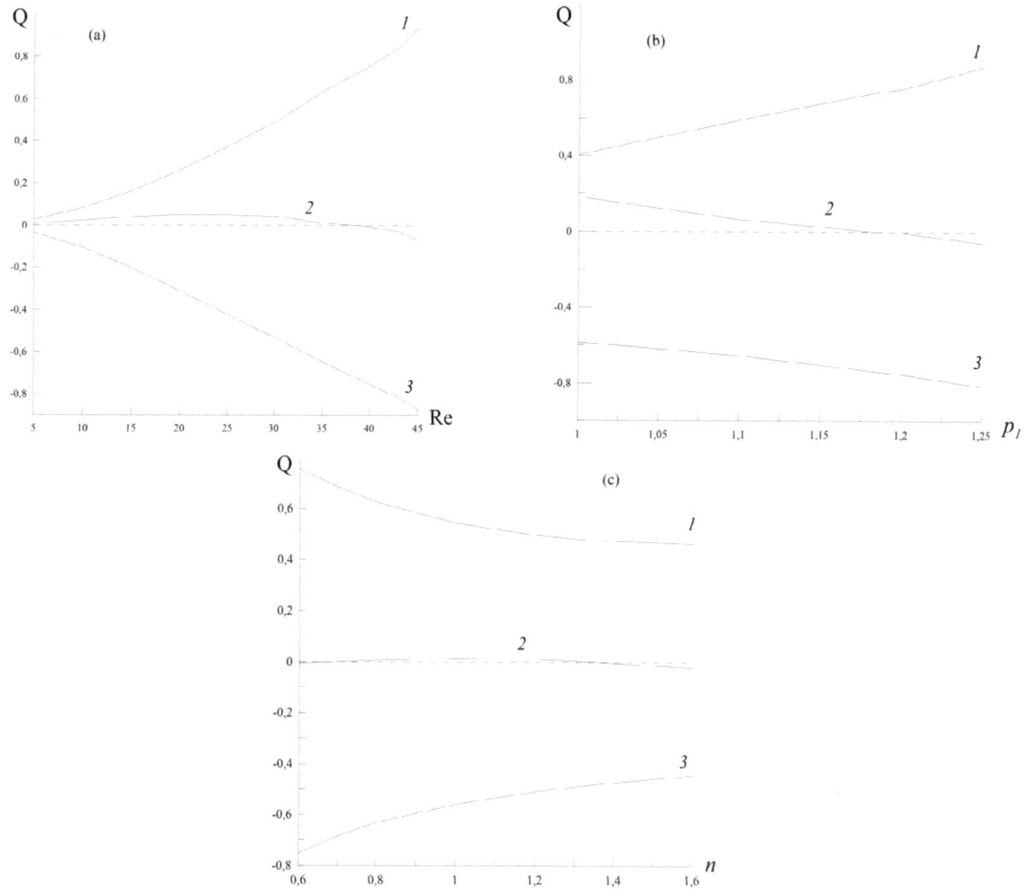

Figure 4. Flow rate through the sections *AM*, *FE* and *BC* as a function of (a) – Re (n=0.6, p_1=1.2); (b) – p_1 (Re=40, n=0.6); (c) – n (Re=40, p_1=1.2).

Fig. 4 (a) demonstrates change of the volume flow rate through the sections *AM*, *FE*, and *BC* with increasing Re. Positive values of the volume flow rate Q correspond to the case when the fluid flows into the T-channel, and negative values of it – when the fluid flows out of the T-channel. It is seen that the absolute volume flow rate through the boundary sections *BC* and *FE* increases for the considered range of the Reynolds number. The volume flow rate through the boundary section *AM* smoothly increase at the beginning, and, after reaching its maximum value, the volume flow rate decreases up to negative values passing through the zero value at Re = 38.45 that corresponds to a change of the flow regime.

Results of research of the head-capacity characteristics depending on the value of the pressure p_1 are presented in Fig. 4 (b). Increase of the pressure given on the border *FE* leads to growth of the absolute value of the flow rate through the boundary sections *BC* and *FE*. The flow rate through the border *AM* decreases and reaches zero at the pressure p_{FE}=1.185 indicating the second flow regime which describes by separation of the flow moving from the border *FE*. In this case, character of dependence of the flow pattern on the pressure p_1 is similar to that observed in the study of influence of the Reynolds number on the flow pattern, as presented above.

Fig. 4 (c) shows change of the flow fate through the boundary sections *AM*, *FE*, and *BC* with the power-law index *n*. It can be seen that the flow rate curve through the zero value two times that indicates the change of the flow regime.

5. Conclusion

The planar flow of the power-law incompressible fluid in the T-channel has been studied. The motion of the fluid has been caused by pressure drop given in the boundary sections of the T-channel. On the solid walls, the no slip boundary condition has been used.

Investigation of the flow characteristics depending on the Reynolds number (Re=5-350), the pressure given in the border *FE* (p_1=1-1.25), and the power-law index (*n*=0.6-1.4) has been carried out. The range of change of these parameters has been chosen so that the planar-parallel non-Newtonian flow with parabolic velocity profile was realized in vicinity of the boundary sections *AM*, *FE*, and *BC*.

As a result of this study, it has been found that the change of the flow regime occurs with change of these parameters, and the recirculation zone is formed in the stream. Sizes and location of the recirculation zone changes with change of these parameters too. Estimation of the influence of the main parameters on the flow pattern has been performed. Criteria dependences describing characteristics of the flow under conditions of the present work have been presented.

Acknowledgements

This research was supported by the Grant of the President of the Russian Federation for the Young Russian Scientists (MK-710.2017.1).

References

[1] Liepsch D, Moravec S, Rastogi A K and Vlachos N S 1982 Measurements and calculation of laminar flow in a ninety degree bifurcation *J. Biomech.* **15** 473-85

[2] Khodadadi J M, Nguyen T M and Vlachos N S 1986 Laminar forced convective heat transfer in a two-dimensional 90° bifurcation *Numer. Heat Transfer* **9** 677-95

[3] Grace J L and Priest M S 1958 *Division of Flow in Open Channel Junctions* (Auburn, Ala.: Engineering Experiment Station, Alabama Polytechnic Institute)

[4] Shamloo H and Pirzadeh B 2008 Investigation of characteristics of separation zones in T-Junctions *WSEAS Transactions on Mathematics* **7** 303-12

[5] Benes L, Louda P, Kozel K, Keslerova R and Stigler J 2013 Numerical simulations of flow through channels with T-junction *Applied Mathematics and Computation* **219** 7225-35

[6] Miranda A I P, Oliveira P J and Pinho F T 2008 Steady and unsteady laminar flows of Newtonian and generalized Newtonian fluids in a planar T-junction *Int. J. Numerical Methods*

in Fluids **57** 295-328

[7] Khodadadi J M, Vlachos N S, Liepsch D and Moravec S 1988 LDA measurements and numerical prediction of pulsatile laminar flow in a plane 90-degree bifurcation *J. Biomech. Eng.* **110** 129-36

[8] Khodadadi J M 1991 Wall pressure and shear stress variations in a 90-deg bifurcation during pulsatile laminar flow *J. Fluids Eng* **113** 111-5

[9] Moshkin N and Yambangwi D 2009 Steady viscous incompressible flow driven by a pressure difference in a planar T-junction channel *Int. J. Comput. Fluid Dyn.* **23** 259-70

[10] Matos H M and Oliveira P J 2013 Steady and unsteady non-Newtonian inelastic flows in a planar T-junction *Int. J. Heat and Fluid Flow* **39** 102-26

[11] Khandelwal V, Dhiman A and Baranyi L 2015 Laminar flow of non-Newtonian shear-thinning fluids in a T-channel *Computers & Fluids* **108** 79-91

[12] Kuznetsov B G, Moshkin N P and Smagulov Sh 1983 *Numerical investigation of the viscous incompressible fluid flow in channels of the complex geometry with given pressure* (Proceedings of the Institute of the Theoretical and Applied Mechanics of the Siberian Branch of the Academy of Sciences of the USSR Vol. 14) (in Russian)

[13] Shul'man Z P 1975 *Convective Heat and Mass Transfer of Rheologically Complex Liquids* (Moscow: Énergiya) (in Russian)

[14] Godunov S K and Ryabenkii V S 1987 *Difference Schemes* (North-Holland: Elsevier Science Ltd)

[15] Patankar S 1980 *Numerical Heat Transfer and Fluid Flow* (New York: Taylor & Francis)

[16] Shrager G R, Kozlobrodov A N and Yakutenok V A 1999 *Simulation of Hydridynamic Processes in the Processing Technology of Polymer Materials* (Tomsk: Tomsk State University Publishing) (in Russian)

IOP Conf. Series: Journal of Physics: Conf. Series **894** (2017) 012028 doi:10.1088/1742-6596/894/1/012028

Vragov's boundary value problem for an implicit equation of mixed type

I E Egorov

Northeastern Federal University, Yakutsk, Russia

E-mail: IvanEgorov51@mail.ru

Abstract. We study a Vragov boundary value problem for a third-order implicit equation of mixed type with an arbitrary manifold of type switch. These Sobolev-type equations arise in many important applied problems. Given certain constraints on the coefficients and the right-hand side of the equation, we demonstrate, using nonstationary Galerkin method and regularization method, the unique regular solvability of the boundary value problem. We also obtain an error estimate for approximate solutions of the boundary value problem in terms of the regularization parameter and the eigenvalues of the Dirichlet spectral problem for the Laplace operator.

1. Introduction

Due to their importance in applications to gas dynamics, boundary value problems for nonclassical equations of mixed type were studied by F. Tricomi, S. Gellerstedt, M. A. Lavrentyev, A. V. Bitsadze, and others [1–11]. One can find examples [7–9] of implicit PDEs that arise in mathematical models of viscoelasticity, electrodynamics, semiconductor physics, polymer mechanics and other processes of modern physics. The equations of this kind are said to be of Sobolev type.

Note that the regular solvability of the first boundary value problem for equations of Sobolev type was studied [12–14].

Here, a Vragov [10] boundary value problem for a third-order implicit equation of mixed type will be studied. Given certain constraints on the coefficients and the right-hand side of the equation, the unique regular solvability of the boundary value problem will be demonstrated, using nonstationary Galerkin method and regularization method. Moreover, an error estimate for approximate solutions of the boundary value problem in terms of the regularization parameter and the eigenvalues of the Dirichlet spectral problem for the Laplace operator will be obtained.

2. Statement of the boundary value problem

Let Ω be a bounded domain in R^n with a smooth boundary γ. Denote $Q = \Omega \times (0,T)$, $\Gamma = \gamma \times (0,T)$, $T > 0$, $\Omega_t = \Omega \times \{t\}$, $t \in [0,T]$.

In the cylindrical domain Q consider the equation

$$Lu \equiv k(x,t)u_{tt} - \triangle u_t + a(x,t)u_t + c(x)u = f(x,t), \quad (x,t) \in Q. \tag{1}$$

The coefficients here are assumed to be smooth functions.

Content from this work may be used under the terms of the Creative Commons Attribution 3.0 licence. Any further distribution of this work must maintain attribution to the author(s) and the title of the work, journal citation and DOI.
Published under licence by IOP Publishing Ltd

Note that the equation (1) belongs to the class of equations of mixed type with a non-fixed manifold where the type changes. Introduce the set

$$P_0^{\pm} = \{(x,0) : k(x,0) \gtrless 0, x \in \Omega\}, \quad P_T^{\pm} = \{(x,T) : k(x,T) \gtrless 0, x \in \Omega\}.$$

Boundary value problem. Find a solution of equation (1) in Q, such that

$$u|_{\Gamma} = 0, \;\; u|_{t=0} = 0, \quad u_t|_{\overline{P_0^+}} = 0, \quad u_t|_{\overline{P_T^-}} = 0. \tag{2}$$

3. Derivation of the a priori bound

In an anisotropic Sobolev space $W_2^{m,s}(Q)$ introduce the norm

$$\|u\|_{m,s}^2 = \int\limits_Q \left[\sum_{|\alpha| \le m} (D^\alpha u)^2 + (D_t^s u)^2 \right] dQ, \quad u \in W_2^{m,s}(Q),$$

with $\|u\|_{m,m} = \|u\|_m$ for $u \in W_2^{m,m}(Q) = W_2^m(Q)$.

Denote

$$(u,v)_0 = \int\limits_\Omega u(x)v(x)dx, \; \forall\, u,v \in L_2(\Omega)$$

the scalar product in the space $L_2(\Omega)$ and $(u,v) = \int\limits_0^T (u,v)_0 dt$ for $u,v \in L_2(Q)$, $\|u\|^2 = (u,u)$.

Introduce the class of functions

$$C_L = \left\{ u(x,t) : \; u \in W_2^2(Q), \; u_{x_i x_j t} \in L_2(Q), \; i,j = \overline{1,n} \text{ and conditions (2) hold} \right\}.$$

Lemma 1 *Assume that* $c(x) \ge 0, \;\; a - \frac{1}{2}k_t \ge \delta > 0.$

Then the following inequality holds:

$$(Lu, u_t) \ge C_1 \left[\|u\|_1^2 + \int_Q \sum_{i=1}^n u_{tx_i}^2 dQ \right], \quad C_1 > 0$$

for all functions $u \in C_L$.

Proof. For a function $u(x,t) \in C_L$

$$(Lu, u_t) = \int_Q \left[\left(a - \frac{1}{2}k_t \right) u_t^2 + \sum_{i=1}^n u_{tx_i}^2 \right] dQ + I, \tag{3}$$

where

$$I \equiv \frac{1}{2} \int_{P_T^+} k u_t^2 dx - \frac{1}{2} \int_{P_0^-} k u_t^2 dx + \frac{1}{2} \int_{\Omega_T} c u^2 dx \ge 0.$$

Due to inequalities

$$\int_Q u^2 dQ \le T^2 \int_Q u_t^2 dQ,$$

$$\int_Q \sum_{i=1}^n u_{x_i}^2 dQ \le T^2 \int_Q \sum_{i=1}^n u_{tx_i}^2 dQ,$$

the statement of Lemma 1 follows from (3).

For $\varepsilon > 0$ define $L_\varepsilon u = -\varepsilon u_{ttt} + Lu$. Assume the functions $\varphi_k(x)$ are the solutions of a spectral problem

$$-\Delta\varphi = \lambda\varphi,\ x \in \Omega,\ \varphi|_\gamma = 0.$$

The functions $\varphi_k(x)$ form an orthonormal basis in $L_2(\Omega)$, and their corresponding eigenvalues satisfy the conditions $0 < \lambda_1 \le \lambda_2 \le ...$ and $\lambda_k \to +\infty$ for $k \to \infty$.

Construct an approximate solution $u^{N,\varepsilon}(x,t)$ of the boundary value problem (1)–(2) of the form

$$u^{N,\varepsilon}(x,t) \equiv v(x,t) \equiv \sum_{k=1}^{N} c_k^{N,\varepsilon}(t)\varphi_k(x), \quad N \ge 1, \quad \varepsilon > 0,$$

where $c_k^{N,\varepsilon}(t)$ are defined as the solution of the following boundary value problem for the third-order ordinary differential equation system:

$$(L_\varepsilon u^{N,\varepsilon}, \varphi_l)_0 = (f, \varphi_l)_0, \tag{4}$$

$$c_l^{N,\varepsilon}|_{t=0} = 0,\ D_t c_l^{N,\varepsilon}|_{t=0} = 0,\ D_t^2 c_l^{N,\varepsilon}|_{t=T} = 0,\ l = \overline{1,N}, \tag{5^1}$$

when $k(x,0) > 0,\ \ k(x,T) \ge 0$,

$$c_l^{N,\varepsilon}|_{t=0} = 0,\ D_t c_l^{N,\varepsilon}|_{t=0} = 0,\ D_t c_l^{N,\varepsilon}|_{t=T} = 0,\ l = \overline{1,N}, \tag{5^2}$$

when $k(x,0) > 0,\ \ k(x,T) < 0$,

$$c_l^{N,\varepsilon}|_{t=0} = 0,\ D_t^2 c_l^{N,\varepsilon}|_{t=0} = 0,\ D_t c_l^{N,\varepsilon}|_{t=T} = 0,\ l = \overline{1,N}, \tag{5^3}$$

when $k(x,0) \le 0,\ \ k(x,T) < 0$,

$$c_l^{N,\varepsilon}|_{t=0} = 0,\ D_t^2 c_l^{N,\varepsilon}|_{t=0} = 0,\ D_t^2 c_l^{N,\varepsilon}|_{t=T} = 0,\ l = \overline{1,N}, \tag{5^4}$$

when $k(x,0) \le 0,\ \ k(x,T) \ge 0$.

From Lemma 1 immediately follows

Lemma 2 *Assume the conditions of Lemma 1 hold, and $f \in L_2(Q)$ satisfies either of the conditions $k(x,0) > 0,\ \ k(x,T) \ge 0$, or $k(x,0) > 0,\ \ k(x,T) < 0$, or $k(x,0) \le 0,\ \ k(x,T) < 0$, or $k(x,0) \le 0,\ \ k(x,T) \ge 0$. Then for an approximate solution of the boundary value problem (1)–(2) the following bound holds:*

$$\varepsilon\|v_{tt}^2\| + \|v\|_1^2 + \sum_{i=1}^{n} \|v_{tx_i}\|^2 \le C_2\|f\|^2,\ C_2 > 0.$$

Lemma 3 *Assume that $c(x) \ge 0$, $a - \frac{1}{2}|k_t| \ge \delta > 0$, $f \in W_2^{0,1}(Q)$ and either of the four conditions $k(x,0) > 0,\ \ k(x,T) \ge 0,\ \ f(x,0) = 0$, or $k(x,0) > 0,\ \ k(x,T) < 0,\ f(x,0) = 0,\ f(x,T) = 0$, or $k(x,0) \le 0,\ \ k(x,T) < 0,\ f(x,T) = 0$, or $k(x,0) \le 0,\ \ k(x,T) \ge 0$ holds. Then for an approximate solution of the boundary value problem (1)–(2) the following bound holds:*

$$\varepsilon\|v_{ttt}^2\| + \int_Q \left[v_{tt}^2 + \sum_{i=1}^{n} v_{ttx_i}^2\right] dQ \le C_3\|f\|_{0,1}^2,\ C_3 > 0.$$

Proof. From (4) it is not hard to obtain an equality

$$\varepsilon\|v_{ttt}^2\| + \int_Q \left[\left(a + \frac{1}{2}k_t\right) v_{tt}^2 + \sum_{i=1}^{n} v_{ttx_i}^2 + (a_t + c)v_t v_{tt}\right] dQ + J = (f_t, v_{tt}), \qquad (6)$$

where

$$J \equiv -\frac{1}{2}\int_{\Omega_T} k v_{tt}^2 dx + \frac{1}{2}\int_{\Omega_0} k v_{tt}^2 dx - \int_{\Omega_T} c v v_{tt} dx.$$

Note that if either of boundary value conditions (5^1) or (5^4) holds then $J \geq 0$. If the boundary value condition (5^2) or (5^3) is satisfied then $-k(x,T) \geq k_0 > 0$, and the second term in the expression above for J is nonnegative. The well-known trace theorem then yields the inequality

$$\int_{\Omega_T} v^2(x,T)dx \leq C_4\|v\|_1^2, \ C_4 > 0.$$

Using this inequality, together with the bound of Lemma 2, and applying the Cauchy inequality to the equality (6), yields the required a priori bound of Lemma 3.

Lemma 4 *Under the conditions of Lemma 3, the approximate solutions of the boundary value problem* (1)–(2) *satisfy the bound*

$$\|\Delta v_t\| \leq C_5\|f\|_{0,1}, \ C_5 > 0.$$

Proof. Due to the boundary conditions (5^p), $p = \overline{1,4}$, the following equality holds:

$$\varepsilon \int_Q \sum_{i=1}^{n} v_{ttx_i}^2 dQ + \int_Q \left[(\Delta v_t)^2 - (k v_{tt} + a v_t + c v)\Delta v_t\right] dQ = -(f, \Delta v),$$

which, together with the Cauchy inequality and Lemmas 2, 3, yields the bound of Lemma 4.

4. Regular solvability and the error estimate

Note that Lemma 4 implies the a priori bound

$$\|\Delta v\| \leq C_5 T\|f\|_{0,1}.$$

Then from Lemmas 2, 3, 4 follows the a priori bound

$$\|u^{N,\varepsilon}\|_2 + \|\Delta u_t^{N,\varepsilon}\| \leq C_6\|f\|_{0,1}, \ C_6 > 0. \qquad (7)$$

The bound (7) provides the basis for the proof of the following theorem.

Theorem 1 *Assume all conditions of Lemma 3 are satisfied. Then the boundary value problem* (1)–(2) *has a unique solution* $u(x,t)$ *in* C_L*, and the following inequality holds:*

$$\|u\|_2 + \|\Delta u_t\|_2 \leq C_7\|f\|_{0,1}. \ C_7 > 0. \qquad (8)$$

Using inequalities (7)–(8), as well as Lemma 1, in a fashion analogous to that of the works [15, 16], the error estimate for nonstationary Galerkin method is obtained.

Theorem 2 *Assume all conditions of Lemma 3 are satisfied. Then the following error estimate for nonstationary Galerkin method holds:*

$$\|u - u^{N,\varepsilon}\|_1 + \|u_t - u_t^{N,\varepsilon}\|_{1,0} \leq C_8\|f\|_{0,1}(\sqrt{\varepsilon} + \lambda_{N+1}^{-1/2}), \ C_8 > 0,$$

where $u(x,t)$ *is the exact solution of the boundary value problem* (1)–(2).

Acknowledgments
This research was supported by the Ministry of Education and Science of Russia through the basic part of state order (project No 6069).

References
[1] Smirnov M M 1978 *Equations of Mixed Type* (Providence, R.I.: AMS)
[2] Djuraev T D, Sopuev A and Mamajonov M 1986 *Boundary Value Problems for the Parabolic-hyperbolic Type Equations* (Tashkent: FAN)
[3] Yakubov S Y 1985 *Linear Differential-operator Equations and Their Applications* (Baku: Elm)
[4] Moiseev E 1988 *Mixed-type Equations with Spectral Parameter* (Moscow: Moscow State University)
[5] Kuzmin A 1990 *Non-classical Equations of Mixed Type and the Application in the Gas Dynamics* (Leningrad: LGU)
[6] Egorov I and Fedorov V 1995 *Nonclassical Higher-order Equations of Mathematical Physics* (Novosibirsk: Computing Centre of SB RAS)
[7] Demidenko G V and Upsenskii S V 2003 *Partial Differential Equations and Systems Not Solvable with Respect to the Highest-order Derivative* (CRC Press)
[8] Egorov I, Pyatkov S and Popov S 2000 *Nonclassical Operator Differential Equations* (Novosibirsk: Nauka)
[9] Sveshnikov A G, Al'shin A B, Korpusov M O and Pletner Yu D 2007 *Linear and Nonlinear Equations of Sobolev Type* (Moscow: Fizmatlit)
[10] Vragov V 1977 On the theory of boundary value problems for mixed-type equations in space *Differentsial'nye Uravneniya* **13** 1098–105
[11] Kozhanov A I 1994 Boundary value problems for some classes of higher-order equations that are unsolved with respect to the highest derivative *Sib. Math. J.* **35** 324–40
[12] Kozhanov A I 1997 Existence of regular solutions for a certain class of equations of Sobolev type of variable direction *Mat. Zametki Yakutsk. Gos. Univ.* **4** 39–48
[13] Potapova S 2012 Boundary value problems for pseudohyperbolic equations with a variable time direction *TWMS J. Pure Appl. Math.* **3** 73–91
[14] Kozhanov A I and Potapova S V 2014 The Dirichlet problem for a class of composite type equations with a discontinuous coefficient of the highest derivative *Dal'nevost. Mat. Zh.* **14** 48–65
[15] Egorov I E and Tikhonova I M 2015 A modified galerkin method for vragov problem *Sib. Élektron. Mat. Izv.* **12** 732–42
[16] Egorov I E, Fedorov V E and Tikhonova I M 2016 Modified galerkin method for the second order equation of mixed type and estimate of its error *Vestnik YuUrGU. Ser. Mat. Model. Progr.* **9** 30–9

IOP Conf. Series: Journal of Physics: Conf. Series **894** (2017) 012029 doi:10.1088/1742-6596/894/1/012029

Non-symmetrical separated flow along the parabolic wing

A M Gaifullin[1,2] and V V Zhvick[1,2]

[1]Moscow Institute of Physics and Technology, Dolgoprudnyi, Russia
[2]Central Aerohydrodynamic Institute, Zhukovskii,, Russia

E-mail: gaifullin@tsagi.ru, vladzhvick@yandex.ru

Abstract. Inviscid incompressible separated flow along the parabolic wing is explored. The partition is placed in the symmetry plane. The problem is considered without an assumption of flow symmetry. Self-similar solution is constructed. Both symmetrical and non-symmetrical solutions are observed. Fields of self-similar trajectories and stability are analysed.

1. Introduction

A flow along a body with a small aspect ratio is known to be non-symmetrical at high angles of attack, even so the shape of the body and its position relative to the flow are symmetrical.

The phenomenon can be explained [1] by the symmetrical flow instability at high angles of attack, so one of the stable non-symmetrical solutions is realized.

In the present study a non-symmetrical separated flow along a parabolic wing is explored. The wing is assumed to be curved on the parabolic law and has the parabolic shape in plan. Also the wing has a partition in the symmetry plane and the partition height grows on the parabolic law. Because of a small aspect ratio the unsteady analogy can be used. It allows us to reduce a steady three-dimensional problem to an unsteady two-dimensional one.

The symmetrical self-similar solution of this problem was obtained [2] and experimentally studied [3]. The solution is two discrete vortex filaments escaped from the wing vertex.

2. Statement of the problem

The parabolic wing geometry (Figure 1) is described by the relationships:

$$l(Z) = 2\varepsilon\alpha Z^{1/2}, \; Y(Z) = -\varepsilon\beta Z^{1/2}, \; h(Z) = \varepsilon\gamma Z^{1/2} \; . \tag{2.1}$$

Here, $l(Z)$ is the dependence of the wing width on the longitudinal coordinate Z, $Y(Z)$ is the intersection of the wing and the partition, $h(Z)$ is the partition height, ε is the dimensional parameter, α, β, γ are dimensionless constants, $0 \le Z < +\infty$.

Inviscid incompressible steady separated flow along the parabolic wing is explored. Far from the wing the flow is uniform with the velocity V_∞ along the Z axis. A solution depends on two dimensionless parameters: the relative partition height b (the quotient of the partition height to the wing half-width) and the wing flexion a_0 (the quotient of the wing flexure depth to the wing half-width)

$$b = \gamma / \alpha, \quad a_0 = \beta / \alpha \; . \tag{2.2}$$

Content from this work may be used under the terms of the Creative Commons Attribution 3.0 licence. Any further distribution of this work must maintain attribution to the author(s) and the title of the work, journal citation and DOI.
Published under licence by IOP Publishing Ltd

A solution should satisfy the Chaplygin–Zhukovsky condition on the sharp edges. The flow is required to be unperturbed far from the wing. The normal component of the velocity to the boundary is to be zero.

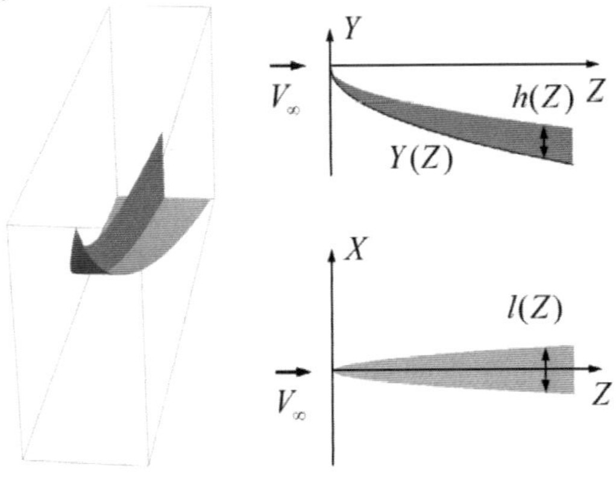

Figure 1. Wing geometry ($a_0 = 5$, $b = 1.5$).

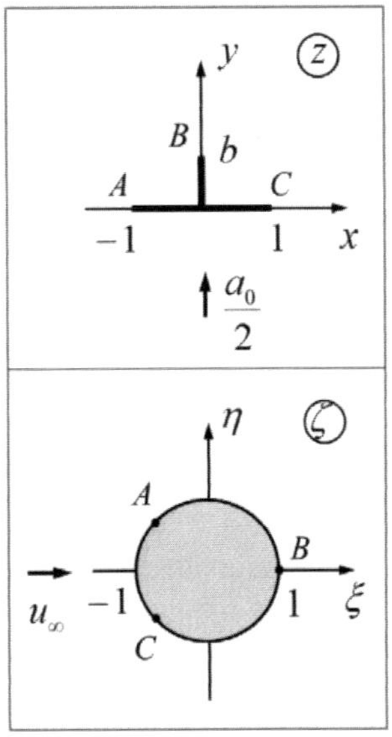

Figure 2. Geometry in self-similar and conformal planes.

3. Derivation of the equations

The unsteady analogy is valid for $Z >> \varepsilon Z^{1/2}$ or $Z >> \varepsilon^2$, when the wing aspect ratio is assumed to be small. Due to the unsteady analogy the problem is reduced to an expanding plate motion. Meanwhile the law, describing the expansion of the plate and its motion along the Y axis, is determined by substitution the expression $Z = V_\infty t$ in equalities (2.1) taking (2.2) into account:

$$l(t) = 2at^{1/2}, \quad h(t) = abt^{1/2}, \quad Y(t) = aa_0 t^{1/2}, \quad a = \varepsilon \alpha V_\infty^{\ 1/2}.$$

Choose the coordinate system with the origin in the point of intersection of the partition and the plate. The axis x_1 is directed along the plate and the axis y_1 is directed along the partition. The free-stream flow velocity along the axis y_1 in this coordinate system is

$$-\frac{\mathrm{d}Y}{\mathrm{d}t} = \frac{1}{2} aa_0 t^{-1/2}.$$

This problem is self-similar because there are not some characteristic scales of space and time. The circulation of vortices escaped from the sharp edges of the plate and the partition does not depend on time. Therefore, a continuous descent of a vortex sheet from the edges is not realized. There are three discrete vortices escaped from the edges at the time origin.

Introduce the complex coordinate in the physical plane $z_1 = x_1 + \mathrm{i}y_1$ and in the self-similar plane $z = x + \mathrm{i}y$, where $z_1 = at^{1/2} z$. The velocity and circulation of the vortices in the physical and self-similar variables are connected by the relationships

$$\mathbf{u}_1 = at^{-1/2}\mathbf{u}, \quad \Gamma = a^2 G.$$

In the self-similar variables the plate has unity half-width and the free-stream flow velocity is $a_0 / 2$.

Transform the plate with the partition in the plane z to the cylinder of unity radius in the plane ζ (Figure 2) using the next conformal transformations:

$$\sigma(z)=\sqrt{z^2-1}, \qquad \mu(\sigma)=p\sigma+q, \qquad \zeta(\mu)=\mu+\sqrt{\mu^2-1}, \tag{3.1}$$

where

$$p=\frac{-2\mathrm{i}}{1+\sqrt{1+b^2}}, \qquad q=\frac{1-\sqrt{1+b^2}}{1+\sqrt{1+b^2}}.$$

The regular branches of the roots are chosen by the conditions: $\sigma(+\infty)=+\infty$, $\zeta(+\infty)=+\infty$.

In the conformal plane $\zeta=\xi+\mathrm{i}\eta$ the cylinder is in the free-stream flow with the velocity

$$u_\infty=\frac{1+\sqrt{1+b^2}}{8}a_0$$

directed on the ξ axis. The edges of the plate with the partition

$$z_{A,C}=\mp 1, \qquad z_B=ib$$

come to the points on the cylinder surface

$$\zeta_{A,C}=q\pm\mathrm{i}\sqrt{1-q^2}, \qquad \zeta_B=1.$$

A flow complex potential w in the conformal plane is a superposition of the free-stream flow, the dipole in the center of the cylinder, three vortices escaped from three sharp edges and three adjoint vortices in the inverse points inside the cylinder. The complex adjoint velocity is

$$\frac{\mathrm{d}w}{\mathrm{d}\zeta}=u_\infty\left(1-\frac{1}{\zeta^2}\right)+\frac{1}{2\pi\mathrm{i}}\sum_{j=1}^{3}\left(\frac{G_j}{\zeta-\zeta_j}-\frac{G_j}{\zeta-1/\overline{\zeta}_j}\right). \tag{3.2}$$

Here, j = 1, 2, 3 is the vortex number, G_j is the dimensionless intensity (circulation) of the vortex with the number j, ζ_j is the complex coordinate of the vortex with the number j.

The vortex moves in the physical plane with the velocity

$$\frac{\mathrm{d}\overline{z}_{1j}}{\mathrm{d}t}=\lim_{z_1\to z_{1j}}\left(\frac{\mathrm{d}w_1}{\mathrm{d}z_1}-\frac{1}{2\pi\mathrm{i}}\frac{\Gamma_j}{z_1-z_{1j}}\right), \tag{3.3}$$

where Γ_j is the dimensional intensity (circulation) of the vortex with the number j, w_1 is the flow complex potential in the physical plane. Taking into account the relationships

$$z_1=at^{1/2}z, \qquad w_1=a^2w, \qquad \Gamma_j=a^2G_j.$$

we obtain the equation of the vortex motion in the self-similar plane from the expression (3.3):

$$t\frac{\mathrm{d}\overline{z}_j}{\mathrm{d}t}+\frac{\overline{z}_j}{2}=\lim_{z_1\to z_{1j}}\left(\frac{\mathrm{d}w}{\mathrm{d}z}-\frac{1}{2\pi\mathrm{i}}\frac{G_j}{z-z_j}\right). \tag{3.4}$$

Hence we obtain the condition of the vortex rest in the self-similar plane

$$\frac{\overline{z}_j}{2}=\lim_{z\to z_j}\left(\frac{\mathrm{d}w}{\mathrm{d}z}-\frac{1}{2\pi\mathrm{i}}\frac{G_j}{z-z_j}\right).$$

Substitute in this condition the expression (3.2) and take into account the transformations (3.1):

$$\frac{\overline{z}_j}{2}\frac{\mathrm{d}z_j}{\mathrm{d}\zeta_j}=u_\infty\left(1-\frac{1}{\zeta_j^2}\right)+\frac{1}{2\pi\mathrm{i}}\sum_{\substack{k=1\\k\neq j}}^{3}\left(\frac{G_k}{\zeta_j-\zeta_k}-\frac{G_k}{\zeta_j-1/\overline{\zeta}_k}\right)-\frac{1}{2\pi\mathrm{i}}\frac{G_j}{\zeta_j-1/\overline{\zeta}_j}+\frac{G_j}{2\pi\mathrm{i}}A_j \tag{3.5}$$

where

$$\frac{\mathrm{d}z_j}{\mathrm{d}\zeta_j}=\frac{1-1/\zeta_j^2}{2p}\frac{\sigma_j}{z_j}, \qquad A_j=\lim_{\zeta\to\zeta_j}\left(\frac{1}{\zeta-\zeta_j}-\frac{1}{z-z_j}\frac{\mathrm{d}z}{\mathrm{d}\zeta}\right)=-\frac{1-1/\zeta_j^2}{4p\sigma_j z_j^2}-\frac{1}{\zeta_j(\zeta_j^2-1)}.$$

According to the Chaplygin–Zhukovsky conditions the velocities at the edges A, B, C are finite

$$\left.\frac{\mathrm{d}w_1}{\mathrm{d}z_1}\right|_{A,B,C} < \infty .$$

In conformal plane

$$\left.\frac{\mathrm{d}w}{\mathrm{d}\zeta}\right|_{A,B,C} = 0$$

as

$$\left.\frac{\mathrm{d}\zeta}{\mathrm{d}z}\right|_{A,B,C} = \infty .$$

These relationships are reduced to three real equations

$$4\pi u_\infty \operatorname{Im}\zeta_m + \sum_{j=1}^{3} G_j \frac{\left|\zeta_j\right|^2 - 1}{\left|\zeta_j - \zeta_m\right|^2} = 0 \quad (3.6)$$

where $m = A, B, C$ is an edge number.

The system of three complex (3.5) and three real (3.6) equations is reduced by extraction of real and imaginary parts to the system of nine real equations with nine unknown quantities. They are the vortex coordinates in conformal plane and their intensities. We can express the vortex intensities through the coordinates because the equations (3.6) are linear relative to the former. After substitution of these expressions into the equations (3.5) we will have the system of six real equations that can be solved by Newton method.

4. Self-similar trajectories

A self-similar trajectory is a fixed liquid particle trajectory in the self-similar plane z.

Liquid particle velocities in the physical and the self-similar planes are connected by the relation

$$\frac{\mathrm{d}z_1}{\mathrm{d}t} = \frac{1}{2}at^{-1/2}z + at^{1/2}\frac{\mathrm{d}z}{\mathrm{d}t} .$$

Express these velocities through the complex potential

$$\frac{\mathrm{d}w_1}{\mathrm{d}z_1} = \frac{1}{2}at^{-1/2}\overline{z} + at^{1/2}\frac{\mathrm{d}\overline{z}}{\mathrm{d}t}$$

or

$$t\frac{\mathrm{d}\overline{z}}{\mathrm{d}t} = \frac{\mathrm{d}w}{\mathrm{d}z} - \frac{\overline{z}}{2} .$$

Multiplied the right and the left parts of the latter equation by an element of self-similar trajectory $\mathrm{d}z$, we obtain the complex equation with real left part. Therefore, the self-similar trajectory equation [2] is

$$\operatorname{Im}\left[\left(\frac{\mathrm{d}w}{\mathrm{d}z} - \frac{\overline{z}}{2}\right)\mathrm{d}z\right] = 0 .$$

It enables to determine in every point the unity vector $\mathrm{d}z / \left|\mathrm{d}z\right|$ that is tangential to the trajectory.

In Figure 3 we have reproduced the self-similar trajectories for the wing with the flexion $a_0 = 4.6$. It was observed the non-symmetrical solution continuously turns into the symmetrical one if the partition height decreases. Such transition occurs at the critical partition height $b_* \approx 1.10666$. Figure 3 (a) shows a weakly non-symmetrical solution at the partition height slightly more than b_*.

The non-symmetrical solution is proved to tend to a non-symmetrical limit as the partition height tends to infinity. This limit is reached at $b \approx 4$. Figure 3 (c) shows the solution at the extremely high partition $b = 150$.

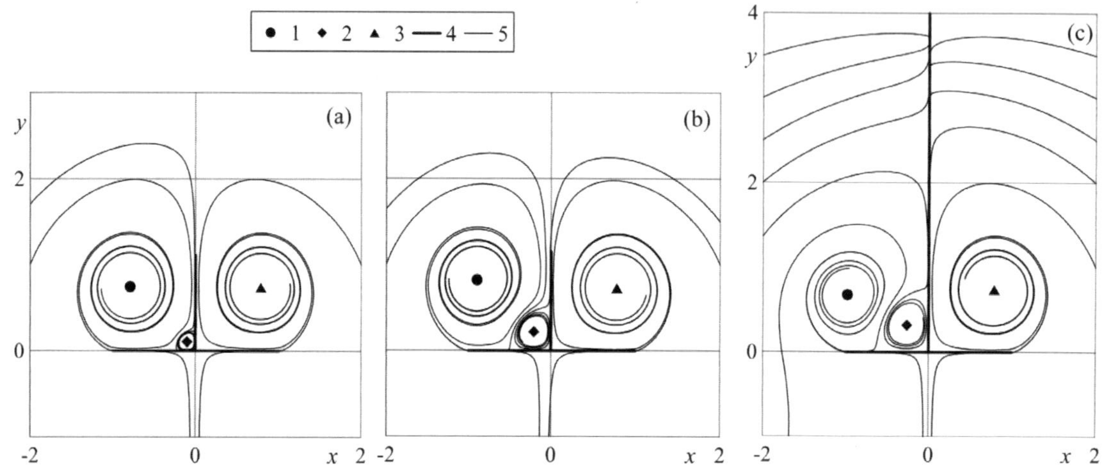

Figure 3. Self-similar trajectories, $a_0 = 4.6$; $b = 1.110$ (a), $b = 1.150$ (b), $b = 150$ (c). 1, 2, 3 – vortices position, 4 – rigid body, 5 – self-similar trajectories.

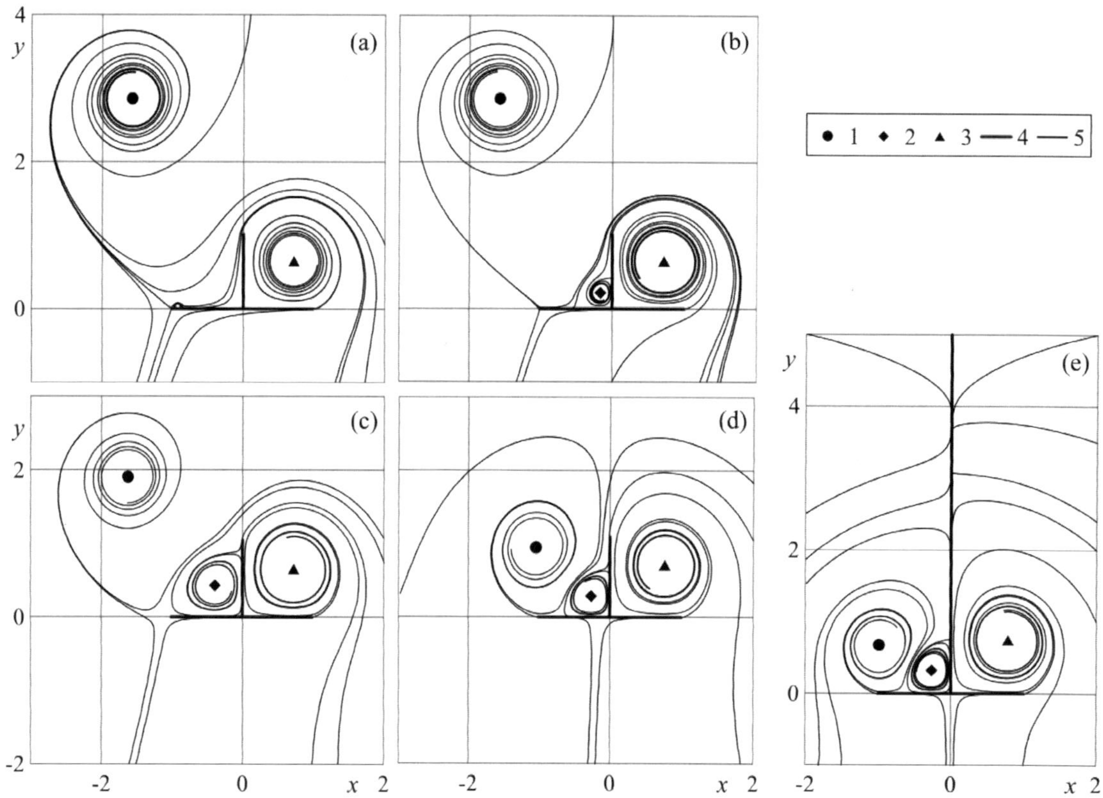

Figure 4. Self-similar trajectories, $a_0 = 4.7$; $b = 1.010$ (a), $b = 1.015$ (b), $b = 1.050$ (c), $b = 1.100$ (d), $b = 150$ (e). 1, 2, 3 – vortices position, 4 – rigid body, 5 – self-similar trajectories.

The critical partition height increases as the flexion decreases. It was obtained $b_* \approx 150$, if $a_0 = 2.816$, and $b_* \approx 1.5$, if $a_0 = 3.787$. Non-symmetrical solutions at $a_0 \leq 2.8$ are not observed.

A continuous transition of a non-symmetrical solution to a symmetrical one is observed at $2.8 < a_0 \leq 4.6$. If $a_0 \geq 4.7$ a non-symmetrical solution does not tend to a symmetrical one if the partition height decreases. In Figure 4 (a) the vortex escaped from the partition is close to the left edge and its intensity is very small relative to the side vortices intensities. In Figure 4 (e) we have reproduced the limiting non-symmetrical solution at the extremely high partition $b = 150$.

In Table 1 and Table 2 we have presented the intensities of the vortices in Figure 3 and Figure 4.

Table 1. Vortices intensities at $a_0 = 4.6$.

b	G_1	G_2	G_3
1.110	-20.22	2.44	20.37
1.150	-19.46	7.80	20.89
150	-16.82	11.12	20.34

Table 2. Vortices intensities at $a_0 = 4.7$.

b	G_1	G_2	G_3
1.010	-24.27	0.18	33.89
1.015	-24.22	1.40	33.85
1.050	-25.43	11.22	30.48
1.100	-20.05	10.08	22.47
150	-17.17	11.49	20.85

5. Stability analysis

A vortex sheet generation on the expanding plate is considered to check a stability of the solutions under discussion. As an initial condition at $t_0 = 1$ we take the self-similar solution, i.e. the flow is assumed to be developed without any perturbation till $t_0 = 1$. The vortex sheet is simulated by the discrete vortices method. During the time step Δt three vortices are escaped from the edges and move on the position determined by the complex quantity $\dot{z}_m \Delta t$, where $\dot{z}_m$ is an initial vortex velocity that can be calculated from the physical fluid velocity at the edge number m ($m = A, B, C$). The vortex intensity is determined from the Chaplygin–Zhukovsky conditions.

During the time step the vortices in the calculation area move on the quantity $\dot{z}_j \Delta t$, where $\dot{z}_j$ is calculated using the formula (3.4). Note, the vortex velocity in self-similar plane is inversely proportional to the time.

The vortex sheet is developed by the algorithm till the state shown in Figure 5 (a) (the vortex sheets escaped from the side edges have three quarters of a coil and the vortex sheet escaped from the partition have a quarter of a coil). A further vortex sheet evolution is described by the "vortex-section" approach. If a vortex crosses the horizontal section in Figure 5 (a), it "drops" on a large vortex so the large vortex intensity is incremented on the "dropped" vortex intensity and the position moves in their vorticity center. It is necessary to "drop" vortices because a lot of vortices in a confined area lead to the calculation process instability.

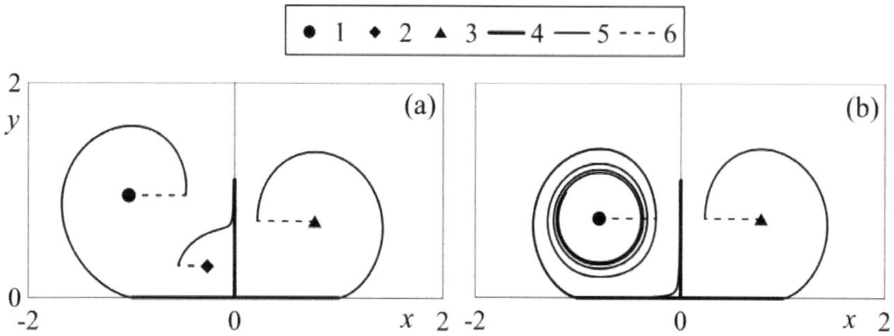

Figure 5. Vortex sheet evolution, a_0 = 4.7, b = 1.1; calculation with non-symmetrical initial condition (a), calculation with symmetrical initial condition (b). 1, 2, 3 – vortices position, 4 – rigid body, 5 – vortex sheet, 6 – section.

For a symmetrical solution stability analysis a vortex sheet is assumed to escape from the partition due to small perturbations. The solution is understood as stable if the vortex sheet weakly changes intensities and positions of the large vortices. The solution is understood as unstable if third large vortex is generated and a solution will tend to a non-symmetrical one.

The algorithm was used for stability analysis of symmetrical and non-symmetrical solutions shown in Figure 3 (b) and Figure 4 (d), and weakly non-symmetrical solution shown in Figure 3 (a). The vortex sheet was stable at all these cases during long time period and its intensity decreased. While testing the symmetrical solution the vortex sheet escaped from the partition makes many coils (Figure 5 (b)) so this solution is stable in accordance with the criterion mentioned above. The algorithm cannot be used for the stability analysis in cases shown in Figure 4 (a), (c) because of more complicated flow pattern. The stability in these cases is not tested.

Acknowledgments

The work was carried out with support from the Russian Foundation for Basic Research (project No. 16-01-00128A).

References

[1] Goman M G, Zakharov S B and Hrabrov A N 1985 Symmetrical and non-symmetrical separated flow along a small aspect ratio wing with a fuselage *Uchenye zapiski TsAGI* **16 (6)** 1–8

[2] Nikol'skii A A, Betyaev S K and Malyshev I P 1971 On the limiting separated self-similar flow of an inviscid fluid *Problems in Applied Mathematics and Mechanics* (Moscow: Nauka) pp 262–268

[3] Bakulin V L and Gayfullin A M 1989 Experimental study of the flow in the cores of a vortex structure *Fluid mechanics. Soviet research.* **18 (1)** 42–46

IOP Conf. Series: Journal of Physics: Conf. Series **894** (2017) 012030 doi:10.1088/1742-6596/894/1/012030

Self-Exciting Oscillations of Elastic Tube Conveying Fluid at Laminar and Turbulent Flow Regimes

J S Zayko[1] **and V V Vedeneev**[1,2]

[1]Institute of Mechanics, Lomonosov Moscow State University, Moscow, Russia
[2]Faculty of Mechanics and Mathematics, Lomonosov Moscow State University, Moscow, Russia

E-mail: juliazaiko@yandex.ru

Abstract. Behavior of elastic tubes conveying fluids has been extensively studied during last 50 years in the context of biological and physiological applications. A lot of experimental works are conducted in a large range of Reynolds numbers, and mostly at turbulent flow regimes. But it is known that biofluids often circulate at laminar regimes. The unit based on the "Starling resistor" is commonly used to study experimentally the stability of the tube with fluid flowing into it. The behavior of the tube as the part of such unit has become a separate issue, and is also widely studied experimentally, analytically and numerically. In this work we experimentally investigate the influence of the flow regime on the stability boundary and limit cycle oscillations of latex thin-walled tube. It is obtained that there are two main differences. At first, the oscillation frequency more essentially depends on a pressure drop in the tube for a fixed flow rate at laminar, than at turbulent regimes. Secondly, the oscillation amplitude is larger for the turbulent regimes, than for the laminar, when the other parameters are the same.

1. Introduction

An extensive review of advances in experimental, analytical and computational investigations of the flow into an elastic tube and the stability of such system is given in [1]. In numerous experiments researches observe such phenomena as pressure drop/flow rate relations, wave propagation and generation of instabilities. Various theoretical models are proposed to define the behavior of the elastic tube conveying fluid [2 – 4]; [3] outlines several models that have been developed to describe standard experiments in Starling resistor. There are a lot of numerical simulations of the behavior of the elastic tubes conveying fluids [2, 3, 5 – 7].

In [8, 9] detailed description of the behavior of thick-walled silicone rubber tubes conveying aqueous fluid is given. Pressure drop/flow rate curves for steady flow and equilibrium characteristics of the collapsed tube are obtained, various types of oscillations of the collapsible tube are exemplified, and little detail of the wave-forms of pressure, flow rate and area are revealed. It is shown, that the frequency of self-excited oscillations is not strongly dependent on the tube length, and the effect of the tube length is to predispose the system to a particular mode of oscillation. The investigations of the flow through thin-walled axisymmetric tubes are presented in [2]. Cross-sectional area from transmural pressure and pressure drop/flow rate relations, obtained in experiments, are used to verify the proposed lumped parameter system model. The validity of the model is confirmed by the comparison of experimental and simulation results.

Content from this work may be used under the terms of the Creative Commons Attribution 3.0 licence. Any further distribution of this work must maintain attribution to the author(s) and the title of the work, journal citation and DOI.
Published under licence by IOP Publishing Ltd

Although biofluid flows are generally laminar [7], most experimental studies span a wide range of the Reynolds numbers (from hundred to several thousand) and mostly deal with turbulent flows. In [10] it is shown that for thick-walled tubes oscillations are not possible at laminar regimes. In [5] the laminar flows of various viscosities through thin-walled tubes are investigated to facilitate numerical simulations of the flow in the "Starling resistor". Results of the experiments show that the flow rate and Reynolds number at the oscillation onset do not strongly depend on downstream rigid pipe length, the oscillation frequency depends on this parameter significantly; the oscillation amplitude decreases as fluid viscosity increases.

In this paper we find the stability boundaries of the tube when the fluid flows at laminar and turbulent regimes, analyze the effect of the flow regime (laminar vs turbulent) on the limit cycle properties, the flow rate limitation and the influence of the fluid viscosity on the tube behavior.

2. Experimental Unit and Conditions

An experimental apparatus for the recirculation of liquid through the elastic tube is shown in Figure 1. Working fluids are water and glycerin mixtures of various concentrations. The working fluid flows in a closed loop and is isolated from the ambient air to protect the glycerin mixture from aeration. The fluid is carried to the base tank by the pump and enters to the elastic tube from there. Latex Penrose tubes of 8mm diameter are attached at each end to the rigid tubes of the same diameters. Elastic tubes with a length-to-diameter ratio from 50:1 to 35:1 are used. The results presented do not depend on the elastic tube length.

An external pressure p_e in the chamber is constant. The flow through the elastic tube occurs under the pressure drop $\Delta p = p_1 - p_2$. p_1 and p_2 are upstream and downstream pressures. The pressure drop in the tube is changed by the flow rate Q or downstream pressure p_2. The flow rate Q is controlled by an adjustable resistance and is measured by the flowmeter. p_2 is controlled by the position of a draining hose. Absolute and differential pressure sensors and microphone are used to measure the frequency and the amplitude of the elastic tube oscillations.

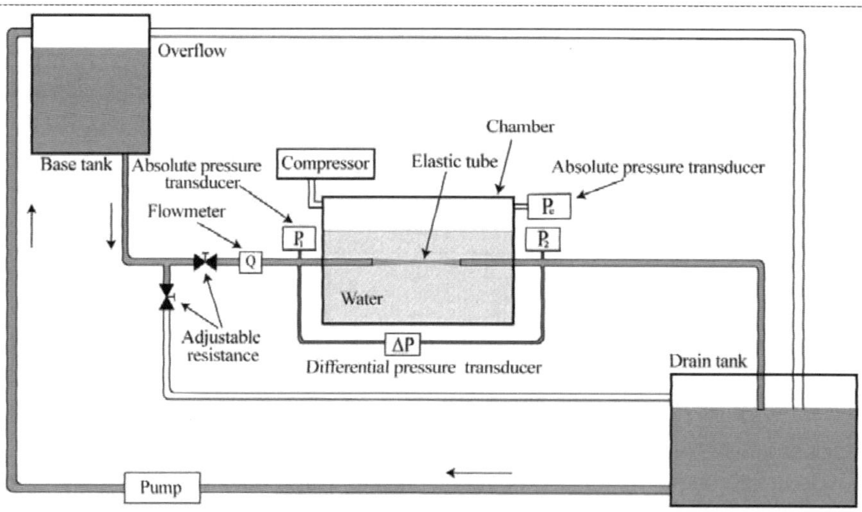

Figure 1. Apparatus for investigations of self-exciting oscillations of elastic tubes conveying fluid.

At first, we conduct the experiments at turbulent flow regimes in the flow rate range Q = 1.5-9 l/min with the step ~ 1 l/min. The working fluid is water in this case. For each flow rate we change the pressure drop Δp in the tube, find the stability boundary and the pressure drop Δp_s, which corresponds to the last cylindrical form of the tube before it begins to collapse. After it, for each fixed flow rate and corresponding Δp_s, we use the formula for the resistance coefficient at laminar regimes in the tube $\lambda = 64/Re$, and calculate an appropriate viscosity (and concentration) of the glycerin mixture, which provides for each flow rate the same pressure drop in the tube at laminar regime. So we can compare the tube behavior at turbulent and laminar regimes, when the other factors (the geometry, the flow rate and the pressure drop) are equal. This is particularly significant, when we compare the stability boundaries at laminar and turbulent regimes.

3. Results of Experiments

At first, we study the influence of the fluid viscosity on the stability boundary (in terms of Q and p_2) and limit cycle for both turbulent and laminar regimes. Reynolds number based on the Penrose tube diameter are varied in the range 3000<Re<18000 for turbulent, and 400<Re<2000 for laminar regime. The experiments reveal that the stability boundary and the character of limit cycle oscillations do not significantly depend on the fluid viscosity for both turbulent and laminar regimes. When the stability is lost while keeping $p_1 - p_e$ constant and increasing Δp, the tube first oscillates in the following manner (Figure 2): two collapses followed by a delay in the stable state, then again two collapses, etc. For higher Δp, the tube collapses three times followed by a delay; then four, five, and up to nine times; for higher Δp single-frequency oscillations are finally established and lock the tube. The limit cycle of this type occurs for laminar and turbulent regimes, but the dependence on Δp is more pronounced at turbulent regimes. For laminar regimes single-frequency oscillations lock the tube a little quicker. Stability boundaries are closely adjacent for laminar and turbulent regimes and fluid viscosity does not affect them (Figure 3).

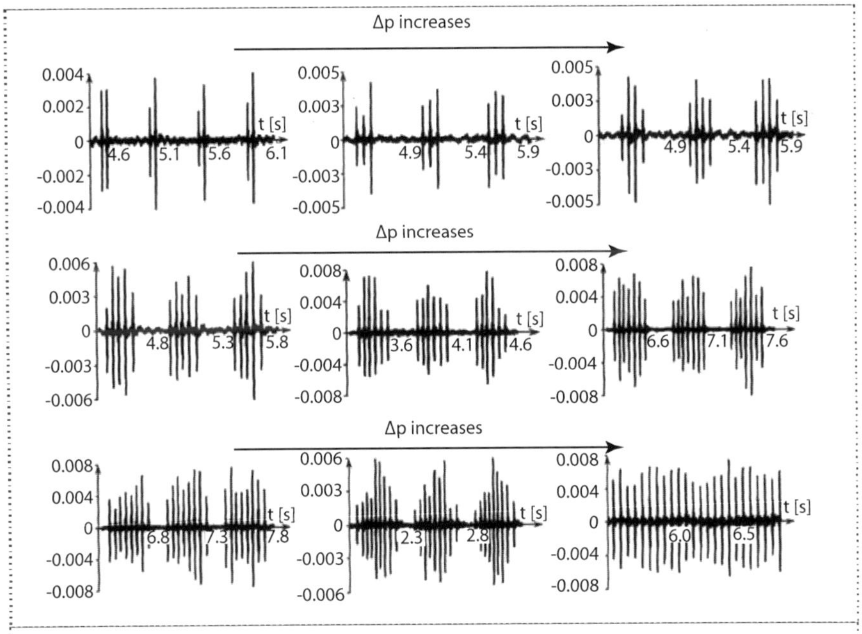

Figure 2. Various types of oscillations (corresponding to various numbers of consecutive tube collapses followed by a delay). Microphone readings. Turbulent regime.

It is significant, that after the tube loses the stability an acquainted phenomena, called "flow rate limitation", is observed for both laminar and turbulent flow regimes. It is that the flow rate does not change, when the pressure drop increases.

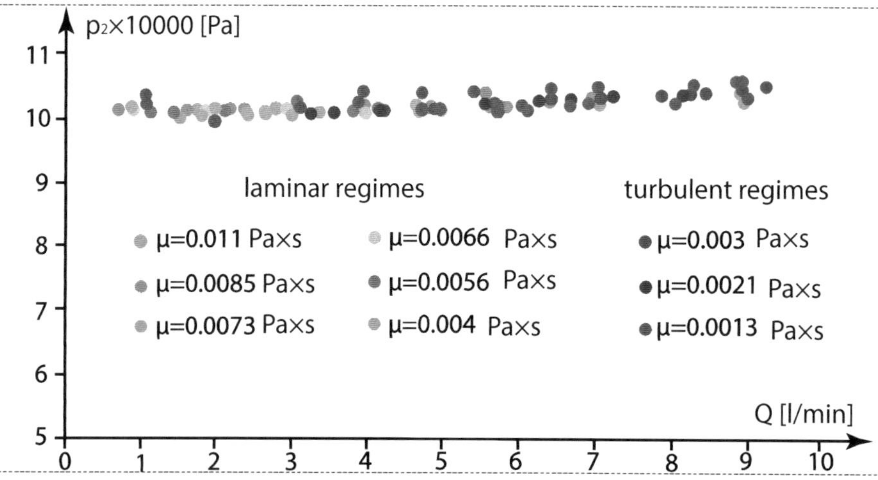

Figure 3. Stability boundaries. Dependence on fluid viscosity and flow regime.

The first difference between turbulent and laminar regime is that the oscillation frequency after the establishing of single-frequency oscillation is more significantly affected by the pressure drop at laminar regimes. In Figure 4 the deepest color corresponds to the highest frequency, white to the stability. So we see that the color changes faster for fixed Q with Δp increase at laminar, than at turbulent regimes.

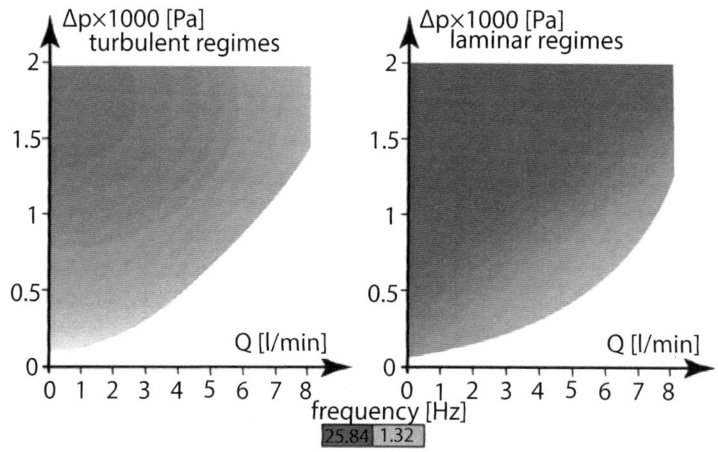

Figure 4. The oscillation frequency vs the flow rate and pressure drop relations for turbulent and laminar regimes.

The second difference is that the oscillation amplitude at laminar regimes is essentially lower than at turbulent, so that sometimes the oscillations do not fully block the tube at laminar regimes (Figure 5).

IOP Conf. Series: Journal of Physics: Conf. Series **894** (2017) 012030 doi:10.1088/1742-6596/894/1/012030

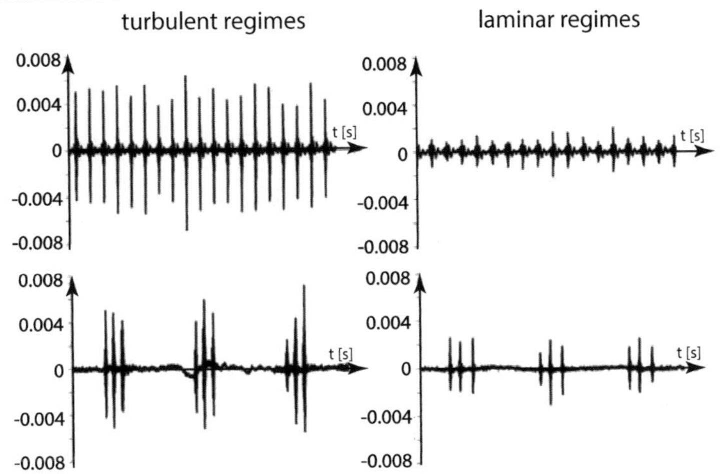

Figure 5. Oscillation amplitudes. Dependence on the flow regime.

References

[1] James B Gr and Oliver E J 2004 *Annu. Rev. Fluid Mech.* **36** 121-47
[2] Katz A I, Chen Yu and Moreno A H 1969 *Biophysical Journal* **9** 1261-79
[3] Pedley T J and Luo X Y 1998 *Theoret. Comput. Fluid Dynamics* **10** 277-94
[4] Jensen O E 1990 *J. Fluid Mech.* **220** 623-59
[5] Bertram C D and Tscherry J 2006 *J. of Fluids and Structures* **22** 1029-45
[6] Heil M and Hazel A 2011 *Annu. Rev. Fluid Mech.* **43** 141-62
[7] Hazel A and Heil M 2003 *J. Fluid Mech.* **486** 79-103
[8] Bertram C D 1986 *J. Biomechanics* **19** 61-9
[9] Bertram C D, Raymond C J and Pedley T J 1990 *J. of Fluids and Structures* **4** 125-53
[10] Bertram C D and Elliot N S J 2001 *Proceedings of the ASME Summer Bioengineering Conference 27 June – 1 July, Snowbird* **50** 383-4

IOP Conf. Series: Journal of Physics: Conf. Series **894** (2017) 012031 doi:10.1088/1742-6596/894/1/012031

Experimental studies of laminar-turbulent transition on a body of revolution

A V Dovgal, B Yu Zanin and A M Sorokin

Khristianovich Institute of Theoretical and Applied Mechanics, Siberian Branch of the Russian Academy of Sciences, Novosibirsk 630090, Russian Federation

Abstract. The focus of the present paper is hydrodynamic stability and transition to turbulence on an axisymmetric body. The objective is to trace the evolution of perturbed flow close to the surface of experimental model with increase of the angle of attack starting with zero incidence. In what follows, we briefly summarize our wind-tunnel data on this topic which were obtained at low subsonic velocities through hot-wire measurements and flow visualization. As is found, in conditions of laminar boundary-layer separation and flow instability, even small variations of the body incidence may have a profound effect on the flow pattern.

1. Motivation

A subject of long-term studies in subsonic fluid dynamics is the spatio-temporal structure of unstable flow on bodies of revolution. An axisymmetric configuration at zero angle-of-attack of the body represents a canonical one which was examined in recent decades. As a result, a lot of research data have been obtained which clarify the mean and nonstationary flow characteristics in conditions of the hydrodynamic instability initiating the growth of velocity perturbations, vortex formation, and transition to turbulence, see e.g. [1−10].

Indeed, an idea about the axial symmetry is appropriate and really helpful at investigation of the instability phenomena. However, at a non-zero angle of attack the flow becomes three-dimensional [11] so that the determination of its quantitative characteristics both in experiments and calculations becomes much more complicated. In this context, an issue is how large the deviation of the body position from zero incidence should be to make the axisymmetric approximation to the unstable flow no longer valid. This is the point which we are interested in here most of all. Note that in the present experimental configuration we deal with laminar boundary layer separation from the body surface. One expects, similar flow behavior on bodies of revolution is typical of quiet environmental conditions and fairly low Reynolds numbers.

2. Experimental arrangement

The experimental runs were performed in the subsonic T-324 wind tunnel of ITAM SB RAS which is a low-turbulent closed-circuit facility. The closed test section of the tunnel is of 4-m length with a square cross section of 1 × 1 m^2, the free-stream turbulence level making $Tu \leq 0.04\%$. The test model sketched in figure 1 represented an axisymmetric body, 1140-mm in length and of 100-mm midsection radius. During the experiments the model was installed at different angles of attack α ranging from 0 to 20° with an accuracy of 0.1°.

Quantitative data on the flow pattern over the body of revolution were obtained through hot-wire measurements using an AN 1003 constant-temperature anemometer from A.A. Lab Systems Ltd and a

Content from this work may be used under the terms of the Creative Commons Attribution 3.0 licence. Any further distribution of this work must maintain attribution to the author(s) and the title of the work, journal citation and DOI.
Published under licence by IOP Publishing Ltd

one-wire probe which was calibrated in the free stream against a Pitot-static tube. The probe was positioned in the measurement region by an automated 3D traversing mechanism. The hot-wire readings were digitized by a 14-bit analog-to-digital converter and further processed on a PC in MATLAB environment. Besides, visualization of the mean near-wall flow was carried out using a mixture of a titanium dioxide powder with kerosene which was applied onto the axisymmetric body.

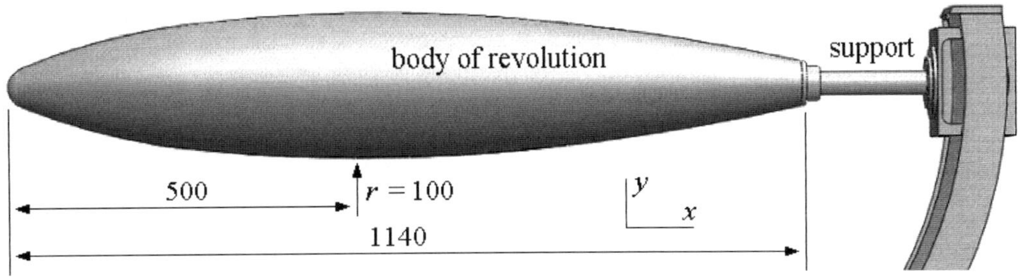

Figure 1. Experimental model, dimensions are in millimeters.

The data were acquired at the Reynolds number $\mathrm{Re}_r = U_0 r/\nu$ ranging from 63000 to 126000 where U_0 is the oncoming flow velocity varying from 10 to 20 m/s and r is the midsection radius of the experimental model. In what follows, x is the streamwise coordinate measured from the nose of axisymmetric body and y is the radial distance from the wall.

3. Wind-tunnel results

Naturally, at zero angle of attack an axisymmetric flow over the model is observed. In the present experimental conditions the laminar boundary-layer separates in the aft part of the body. The mean-velocity contours plotted in figure 2(a) demonstrate a local region of flow separation which is prone to instability. The streamwise growth of velocity perturbations in the separated shear layer results in the laminar-turbulent transition. High amplitudes of the disturbances making about 10 to 15 percent of U_0 in the downstream sections of the measurement domain are typical of the near-wall turbulence, see figure 2(b). In more detail, the stability characteristics of the present axisymmetric flow were reported in [12]. In particular, the behavior of small-amplitude oscillations was found to be in a quantitative agreement with that of the instability waves at laminar boundary-layer separation in plane configurations.

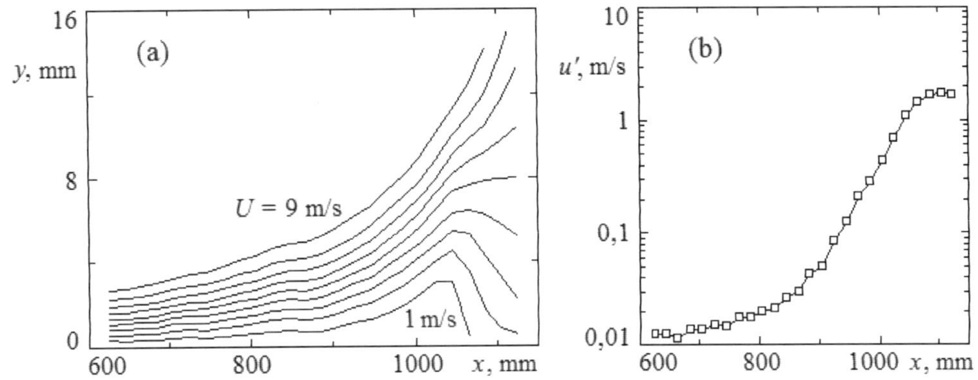

Figure 2. Mean-velocity contours with the step of 1 m/s (a) and the maximum rms amplitude of velocity perturbations integrated over the frequency spectrum up to 500 Hz (b); $\mathrm{Re}_r = 63000$, $\alpha = 0$.

The axial symmetry of the mean flow breaks down when passing from zero angle of attack to a small incidence of several degrees [13]. In this case, laminar flow separation still occurs on the windward side of the body while the boundary layer is attached to the surface on the leeward side, see figure 3. Because of much different stability characteristics of the attached flow and of the separated boundary layer, a pronounced azimuthal nonuniformity of the amplifying velocity perturbations is generated. One can observe in figure 4 that at the angles of attack $\alpha = 2^\circ$ and 4° the amplitudes of laminar flow disturbances on the windward and leeward sides of the body in some cross sections differ by an order of magnitude.

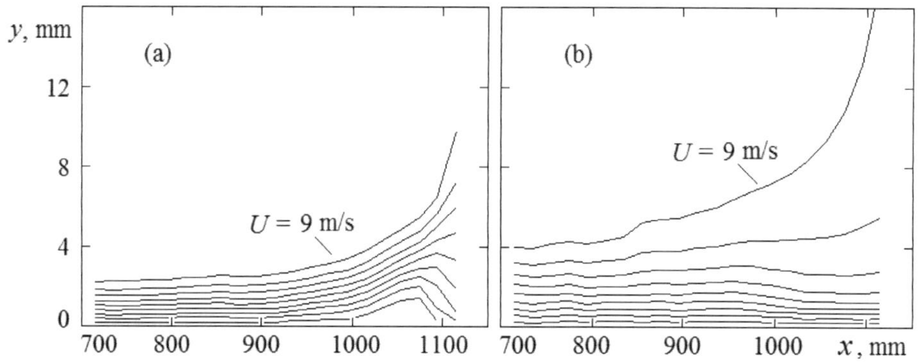

Figure 3. Mean-velocity contours with the step of 1 m/s at the windward (a) and the leeward (b) sides of axisymmetric body; $Re_r = 63000$, $\alpha = 4^\circ$.

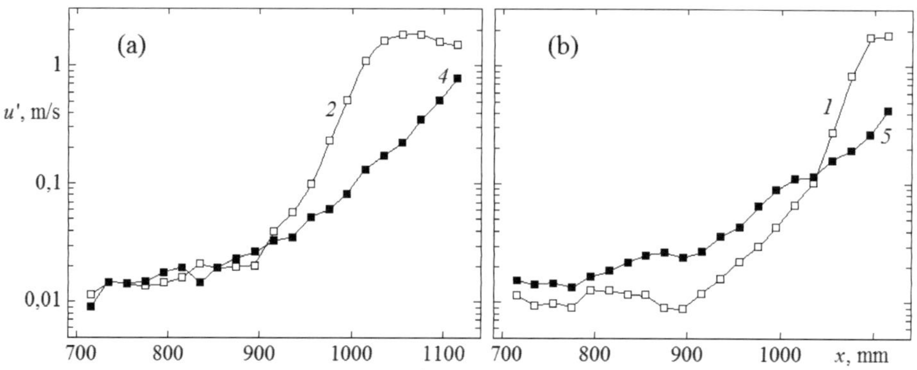

Figure 4. Maximum rms amplitude of velocity perturbations integrated over the frequency spectrum up to 500 Hz at $\alpha = 2^\circ$ (a) and 4° (b), $Re_r = 63000$; windward (□) and leeward (■) sides of the experimental model.

Even more distinct transformation of the unstable flow around the model occurs when the angle of attack is further increased and the local boundary layer separation is replaced with the global one. Then, in a fully 3D separated flow the reliability of hot-wire measurements with one-wire probes goes down. Instead, visualization by surface coatings, although providing qualitative data only, is helpful to obtain a panoramic pattern. An example is given in figure 5. At the angles of attack $\alpha = 10^\circ$ and 20° the separated flow is featured with a pair of large-scale streamwise vortices originating in the mid-chord (figure 5(a)) and in the upstream (figure 5(b)) sections of the model. Similar vortex behavior of the flow past axisymmetric bodies at incidence was observed in a number of studies, for graphic illustrations see [14].

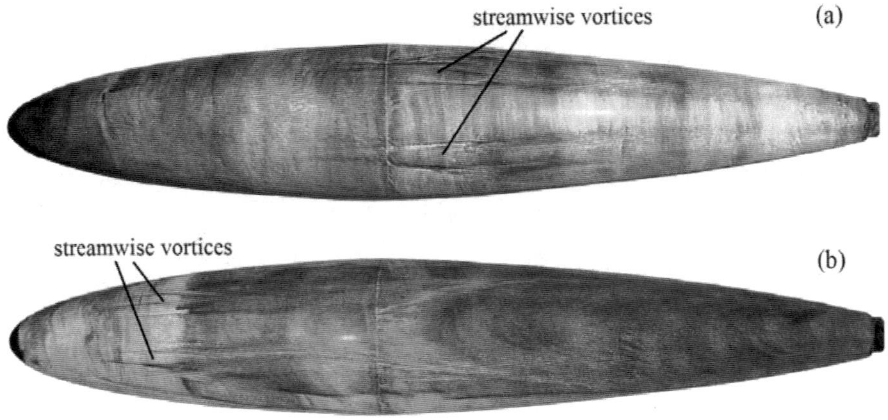

Figure 5. Mean-flow patterns on the leeward side of experimental model at α = 10° (a) and 20° (b); Re_r = 126000, flow is from left to right.

4. Concluding remarks

The above wind-tunnel results illustrate a variety of flow regimes on a body of revolution at low subsonic velocities. In the present configuration of experimental model, the hydrodynamic events at small angles of attack are dominated by the shear-layer instability of a locally separated boundary layer, while at increasing the body incidence, a globally separated flow with the prevalence of large-scale vortex motion is observed.

The key point of the present consideration is the profound response of the flow pattern to small variations of the angle of attack at its close-to-zero values. The effect is basically due to different stability characteristics of the separated and attached boundary layers in the aft part of the body. As a result, minor deviations of its angular position from zero incidence generate a nonuniformity of the mean flow and, hence, strong variations of the nonstationary velocity component in the azimuthal direction. In this respect, the unstable flow over the body of revolution with laminar boundary-layer separation becomes far from being axisymmetric.

References

[1] Gates E M 1980 Observations of transition on some axisymmetric bodies *Laminar-Turbulent Transition* ed Eppler R and Fasel H (Berlin: Springer) pp 351-63

[2] Kiya M 1987 Structure of flow in leading-edge separation bubbles *Boundary-Layer Separation* ed Smith F T and Brown S N (Berlin: Springer) pp 57-71

[3] Sigurdson L W and Roshko A 1988 The structure and control of a turbulent reattaching flow *Turbulence Management and Relaminarization* ed Liepmann H W and Narasimha R (Berlin: Springer) pp 497-514

[4] Kiya M, Mochizuki O, Tamura H, Nozawa T, Ishikawa R and Kushioka K 1991 Turbulence properties of an axisymmetric separation-and-reattaching flow *AIAA J.* **29** 936-41

[5] Kiya M, Shimizu M, Mochizuki O, Ido Y and Tezuka H 1993 Active forcing of an axisymmetric leading-edge turbulent separation bubble *AIAA Paper*/93–3245

[6] Michalke A, Kozlov V V and Dovgal A V 1995 Contribution to the instability of laminar separating flows along axisymmetric bodies. Part I: Theory *Eur. J. Mech., B/Fluids* **14** 333-50

[7] Dovgal A V., Kozlov V V and Michalke A 1995 Contribution to the instability of laminar separating flows along axisymmetric bodies. Part II: Experiment and comparison with theory *Eur. J. Mech., B/Fluids* **14** 351-65

[8] Michalke A 1995 Receptivity of axisymmetric boundary layers to excitation by a Dirac point source at the wall *Eur. J Mech., B/Fluids* **14** 373-93

[9] Hammache M, Browand F K and Blackwelder R F 2002 Whole-field velocity measurements around an axisymmetric body with a Stratford-Smith pressure recovery *J. Fluid Mech.* **461** 1-24
[10] Hudy L M, Naguib A and Humphreys W M 2007 Stochastic estimation of a separated-flow field using wall-pressure-array measurements *Phys. Fluids* **19** 024103 1-18
[11] Tobak M and Peake D J 1982 Topology of three-dimensional separated flows *Ann. Rev. Fluid Mech.* **14** 61-85
[12] Dovgal A V, Zanin B Yu and Sorokin A M 2013 Flow instability in the zone of laminar axisymmetric boundary layer separation *Fluid Dynamics* **48** 747-53
[13] Dovgal A V, Zanin B Yu and Sorokin A M 2014 Stability of the laminar flow on the body of revolution at incidence *Thermophysics and Aeromechanics* **21** 401-6
[14] Van Dyke M 1982 *An Album of Fluid Motion* (Stanford, California: The Parabolic Press)

IOP Conf. Series: Journal of Physics: Conf. Series **894** (2017) 012032 doi:10.1088/1742-6596/894/1/012032

The influence of coating technologies on stress-strain characteristics of the sample at periodic loading

K V Zakharchenko[1, 2], V P Zubkov[3], V I Kapustin[2], E A Maksimovski[4] and A V Talanin [3]

[1]Lavrentyev Institute of Hydrodynamics, Novosibirsk, Russia
[2]Novosibirsk State Technical University,Novosibirsk, Russia
[3]LLC Science and Technology Initiative, Novosibirsk, Russia
[4]Nikolaev Institute of Inorganic Chemistry, Novosibirsk, Russia.

E-mail: zaharchenkok@mail.ru

Abstract. The article is devoted to the research on influence of coating technologies on stress-strain characteristics of a heterogeneous sample (the substrate-coating system) at periodic stress-controlled loading. The comparison of stress-strain characteristics of samples with three types of surface layer showed that the coatings lead to the change in stress at which inelastic phenomena appear in the material. Apart stress-strain characteristics of samples, microrelief on the samples' surface and formation of a slipband in the grain structure of the coatings were studied in the experiment. It is stated that cold dynamic spraying, which is performed by centrifugal acceleration of particles in vacuum, makes it possible to obtain a coating with better strength and stress-strain characteristics in comparison with cladding.

1. Introduction

There are well-known modern innovative technologies that allow applying metal particles on a detail and creating coatings not only on flat surfaces, but also on surfaces of complex geometrical shapes. Such technologies include, for example, dynamic spraying. This method uses kinetic energy to create a firm joint of the coating to the substrate and has been already used in aircraft engineering, machine manufacturing and production of oil and gas refining equipment [1]. The current trends in the development and improvement of the technology are focused on reducing the temperature and increasing velocity of spraying particles [1-3]. It allows avoiding oxidation of particles, obtaining a better adherence of coating to substrate and a hardened layer of the substrate from cold work hardening after particles hit.

If aluminium powder is used as the sprayed material and applied it to the details' surface made of aluminium alloys as an electrochemical protection, it is possible to obtain a coating identical to cladding layer by chemical composition.

The cladding layer obtained by the classical method is highly corrosion-resistant and widely used in industry. The thickness of the cladding layer usually varies from 1.5 up to 10% of the semi finished product's thickness. However, such a coating makes the fatigue-resistance lower which is confirmed by experimental studies [4]. It is stated that the fatigue resistance of the cladding sheets made by classical method can decrease by 35% [5].

Content from this work may be used under the terms of the Creative Commons Attribution 3.0 licence. Any further distribution of this work must maintain attribution to the author(s) and the title of the work, journal citation and DOI.
Published under licence by IOP Publishing Ltd

IOP Conf. Series: Journal of Physics: Conf. Series **894** (2017) 012032 doi:10.1088/1742-6596/894/1/012032

For the quantitative analysis of the coating technology influence on the mechanical properties of a sample a comparative analysis is needed. Standard samples with different types of coatings can be analyzed that allow defining the stress-strain and fatigue-resistance characteristics of a sample.

The purpose of the research is an experimental study of the influence of aluminium coating technology on the regularities of inelastic stress of samples made of the D16T (1163) alloy.

2. Samples, equipment, methods

To perform a comparative analysis of the technology effect on the mechanical properties of the semi-finished product, three parties of smooth samples were made of the D16AT aluminium alloy. The samples had a rectangular cross section of 12x10 mm^2 and a gage length of 50 mm, which made it possible to fix extensometers for measuring longitudinal and transverse stresses.

The first sample party had a cladding layer made of a commercially pure aluminium brand AD1. The thickness of a cladding layer was determined by TU 1-92-161-90 and for this alloy was no more than 4% (Fig. 1a).

From the surface of the samples of the second party the cladding layer was removed by dimensional chemical etching. The chemical etching was carried out in sodium alkali (15% NaOH) at the temperature of 80℃ followed by neutralization in a solution of nitric acid (30% HNO3), washing in cold running water and drying in a warm air steam. The cladding layer removal rate was 50 micron/min. This made it possible to remove the coating without deforming underlying basic material and without creating residual stresses or hardening the surface (Fig. 1c).

The samples of the third party had a coating made by cold dynamic spraying (Fig. 1b). To fulfill the conditions of the comparative experiment, the cladding layer was etched off the surface of the third party, and the coating of the required thickness was made by cold dynamic spraying. The protective coating was made by the cold dynamic spraying machine [6].

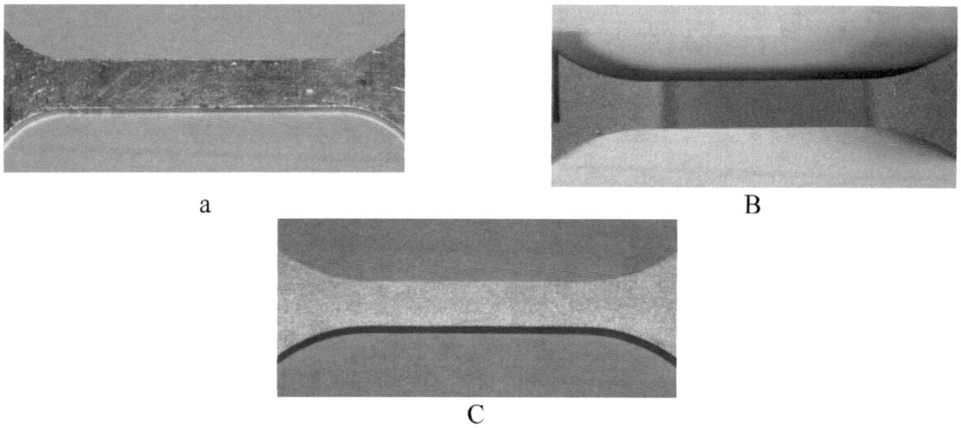

Figure 1. Samples with a surface layer represented by: a) cladding, b) coating by cold dynamic spraying c) with the removed coating.

The modes under which the metal particles were sprayed were selected experimentally and were as follows: the rotational speed of the gun - 27600 rpm, depth of the vacuum in the chamber - $2x10^{-1}$ Mmhg. The coating time - 8 minutes. Aluminum powder ASD-1 was used as the sprayed material. The fraction of the sprayed powder - 20 ... 30 microns.

The thickness and coating homogeneity obtained by cold dynamic spraying was controlled by transverse microsections using a scanning electron microscopy equipment Hitachi S-3400N (shown in Fig. 2a), b). With the help of Back Scattered Electron Detector the image was obtained in the composite contrast which makes it possible to separate visually the substrate and coating. This method allows defining the elements in the surface layer and equitability of particles over the entire substrate surface.

IOP Conf. Series: Journal of Physics: Conf. Series **894** (2017) 012032 doi:10.1088/1742-6596/894/1/012032

The performed analysis of the cross-section microstructure shows that the coating thickness for the D16T alloy is uniform and lies in the range from 95 to 120 microns.

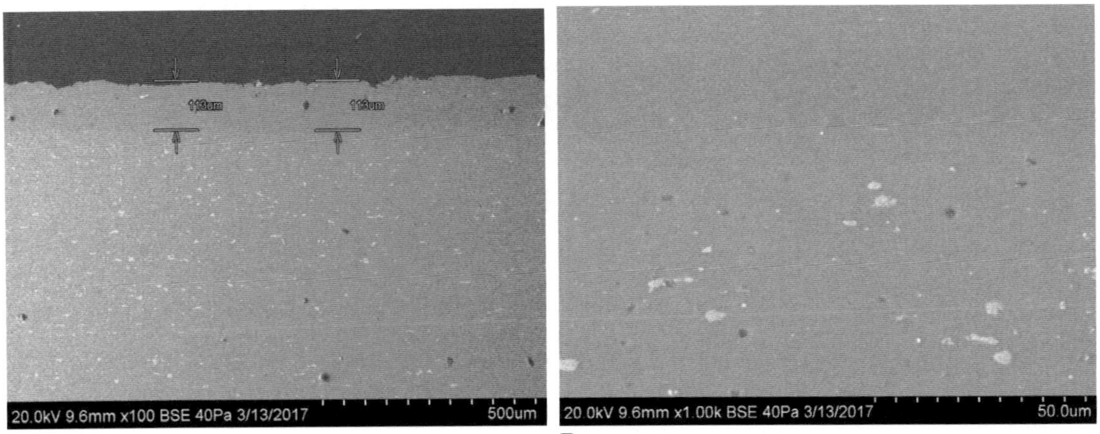

a B

Figure 2. The microstructure of the aluminum powder coating on the surface of the D16T alloy with an increase in 100 (a) and 1000 times (b), respectively, made by cold dynamic spraying.

To determine the stress-strain characteristics of samples from three parties, they were tested on the universal servo-hydraulic test machine Instron 8801 according to a program reported in [4]. The loading program was a sequence of blocks with stepwise increasing stress amplitude. Each block contained 100 cycles of harmonic loading with a frequency of 1 Hz of constant amplitude and zero average stress in the cycle. The amplitude of the loading for each stage increased by 1000 N. The increase of total stress tensor-component was measured by standard extensometers: No. 2620-601 Dynamic Extensometer, Transverse / Diametral Extensometer No. W-E-404-F.

The methods of defining stress-strain characteristics described in [7] made it possible to determine the stress-strain state at which irreversible changes in the sample occur preceded the destruction.

3. The main results

Figure 3 in the coordinates: $\varepsilon_{x\max}$, $\varepsilon_{y\max}$, $\varepsilon_{x\min}$, $\varepsilon_{y\min}$ versus $\sigma_{x\max}$, $\sigma_{x\min}$, represents the experimental diagrams of the D16T samples stress: 1 with a cladding layer, 2 without a cladding layer, 3 coated by cold dynamic spraying under a symmetrical stress cycle, allowing determining the characteristics of the limiting stress cycle. Here: $\varepsilon_{x\max}$ and $\varepsilon_{y\min}$ - are the longitudinal and transverse stresses, respectively, at the maximum stress in the cycle $\sigma_{x\max}$; $\varepsilon_{x\min}$ and $\varepsilon_{y\max}$ are longitudinal and transverse stress, respectively, at the minimum stress in the cycle $\sigma_{x\min}$.

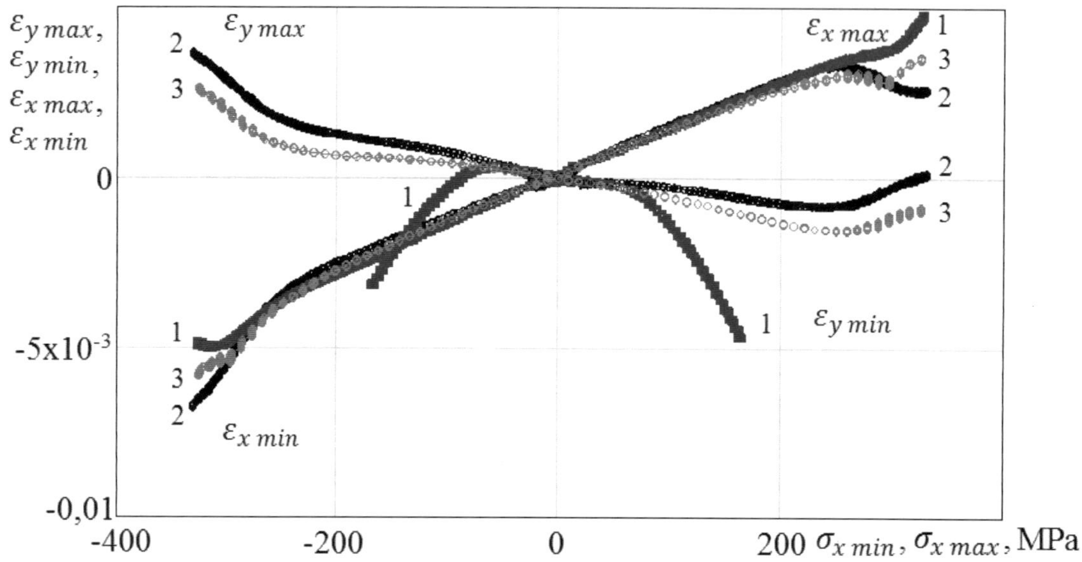

Figure 3. Influence of coating technology on the stress regularities during periodic loading: 1 - sample with a cladding layer; 2 - uncoated sample; 3 - a sample with a coating made by cold dynamic spraying.

Figure 3 shows that on the diagram of transverse stress $\varepsilon_{y\max}$, $\varepsilon_{y\min}$ for the sample with a cladding layer 1 at amplitude 80 MPa, nonequilibrium stress processes are activated.

Figure 3 shows that the stress diagrams for 2 and 3 are close in the entire loading interval. The sample coated by cold dynamic spraying (curve 3) has identical stress-strain characteristics as the uncoated sample 2, and tends to unilaterally accumulate transverse and longitudinal stress.

The comparative analysis of the tensor components ratios of the samples stress with cladding samples and samples coated by cold dynamic spraying (1 and 3) reveals that when the ultimate stress (80 and 250 MPa, respectively) is reached, nonequilibrium stress processes are activated in the material, which are followed by a change in the shape and volume of the sample surface.

As follows from the diagrams, the inelastic behavior of a sample without a cladding layer and a sample coated by cold dynamic spraying comes out at a stress amplitude in the 250 MPa cycle; a sample with a cladding layer - 80 MPa. It shows that the D16T alloy surface treatment by cold dynamic spraying with aluminum powder ASD-1 does not impair the ability to resist stress during periodic loading in comparison with the uncoated D16T sample.

The research of the samples' surface using optical microscopy before and after loading shows that the localization of the microrelief and formation of slipbands in the grain structure of the coating (Fig. 4) take place in the surface layer of the sample with cladding [8]. That does not happen in the surface layer of samples from other parties (uncoated and coated by cold dynamic spraying).

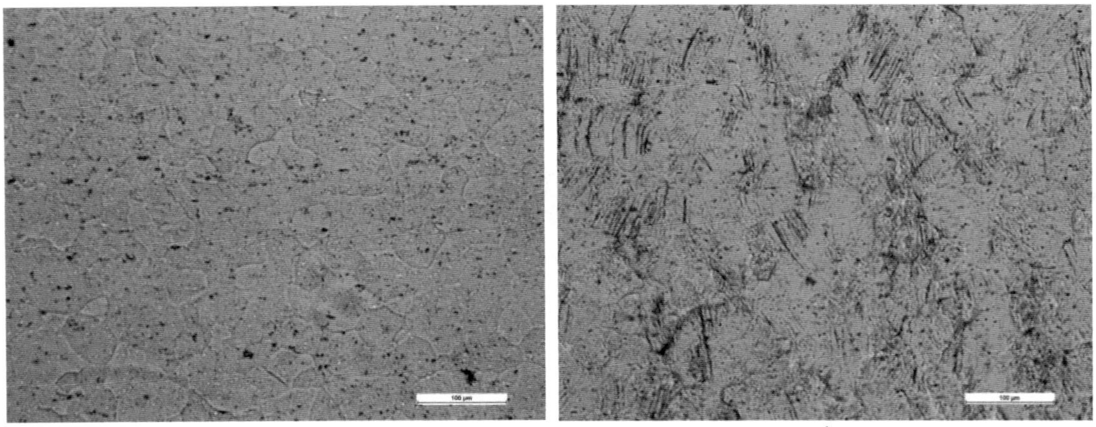

a b

Figure 4. Microstructure of the D16T alloy surface with a cladding layer before (a) and after (b) periodic loading.

4. Conclusions

It has been experimentally proved that cold dynamic spraying makes it possible to create a coating that, in comparison with the classical method of protecting the samples surface with a cladding layer, increases the stress-strain characteristics of the material under symmetrical periodic loading. It allows predicting that the fatigue-resistance of samples coated by cold dynamic spraying does not decrease, but increases in relation to samples with cladding.

References

[1] Alkhimov A P, Klinkov S V, Kosarev V F and Fomin V M 2010 *Cold gas-dynamic spraying. Theory and practice (*Publishing house: FIZMATLIT) p 538

[2] Karthikeyan J 2007 *The advantages and disadvantages of the cold spray coating process. The Cold Spray Materials Deposition Process. Fundamentals and Applications. A volume in Woodhead Publishing Series in Metals and Surface Engineering* p 62-71

[3] Xiao-Tao L, Cheng-Xin L, Fu-Lin S and Guan-Jun Y 2014 *High velocity impact induced microstructure evolution during deposition of cold spray coatings: A review // Surface and Coatings Technology* **254** p 11-20

[4] Zakharchenko K V and Kapustin V I 2015 C*omparative Analysis of the Surface Layer Effect on the Stress-Strain Characteristics of D16AT Alloy. Industrial Laboratory. Materials Diagnostics* **81(7)** p 51-56

[5] Tumanova A T, Kvasova F I and Friedlyder I N (ed.) 1972 *Aluminum* (Moskow: Metallurgy)

[6] Zubkov P I and Zubkov V P 2009 *Method and device for coatings* RU Patent no.2369445

[7] Zakharchenko K V, Kapustin V I and Larichkin A Y 2014 *About the influence of ceramic coatings on the stress-strain characteristics of the alloy D16AT (Al-Cu-Mg). Obrabotka metallov (Metal working and material science)* **3** p 37-44

[8] Kapustin V I, Gileta V P and Zakharchenko K V 2011 *The experimental study of regularities of aluminum alloys deformation in case of regular stresses. Obrabotka metallov (Metal working and material science)* **4** p 40-43

IOP Conf. Series: Journal of Physics: Conf. Series **894** (2017) 012033 doi:10.1088/1742-6596/894/1/012033

Numerical simulation for the propagation and action of shock waves during explosive synthesis

S A Zelepugin[1,2], O V Ivanova[2], A S Yunoshev[3] and A S Zelepugin[1,2]

[1] National Research Tomsk State University, 36 Lenin Ave., 634050, Tomsk, Russia
[2] Tomsk Scientific Center of the Siberian Branch of the Russian Academy of Sciences, 10/3 Akademichesky Ave., 634055, Tomsk, Russia
[3] Lavrentyev Institute of Hydrodynamics of the Siberian Branch of the Russian Academy of Sciences, 15 Lavrentyev Ave., 630090, Novosibirsk, Russia

Abstract. The paper presents numerical and experimental studying the behavior of porous reactive mixtures (Al/S, Al/Tf and Al/Tf/C) under explosive loading using a multicomponent medium model. The initial dispersion of reactive mixtures is varied to evaluate its contribution to the development of chemical reactions. The results have shown that the shock wave, reflected from the bottom of the ampoule in the form of a compression wave, encounters the shock wave propagating in the sample, which leads to a sharp increase in the pressure and the rate of chemical reactions at the bottom of the sample. The high rate of heat release and the sharp increase in pressure lead to the ampoule fracture. It is found that the higher the initial dispersion of the mixture components is, the more intensive the development of chemical reactions and the ampoule fracture are.

1. Introduction

The obtaining and use of new advanced materials is often connected with extreme conditions such as high-speed processes, high pressures and temperatures. At present, explosive technologies are used mainly in the metalworking production for molding, welding, cutting, hardening and compacting. Many of these technologies have been already introduced into production, while the potentials and perspectives of explosive loading have not been studied thoroughly yet, and so far this direction has not become a technology due to the lack of experimental data and numerical multicomponent medium models which can consider the cumulative effect of mechanical, chemical and physical processes and estimate the role of each factor. During explosive loading, there is a significant energy release due to exothermic reactions in reacting mixtures, which, on the one hand, can lead to a self-sustaining propagation of chemical reactions in mixtures and, on the other hand, to the failure of ampoules, as well as the complete fracture [1-3]. It is necessary to mention that the initiation and development of shock-induced reactions are determined not only by the temperatures reached but also by the dispersion of initial components [4]. In addition, the initial dispersion of components can contribute to the fracture of the cylindrical ampoule. For the analysis of these processes, it is necessary to develop mathematical models which in combination with available experimental data will provide reliable information on the behavior of materials, including the mechanisms and kinetics of physical and chemical transformations and ways to form the new states of substances.
In the work, the aluminum-sulfur, aluminum-fluoroplast and aluminum-fluoroplast-carbon mixtures placed in a cylindrical ampoule were studied under explosive loading on the basis of a mathematical multicomponent medium model considering the dispersion of the initial components.

Content from this work may be used under the terms of the Creative Commons Attribution 3.0 licence. Any further distribution of this work must maintain attribution to the author(s) and the title of the work, journal citation and DOI.
Published under licence by IOP Publishing Ltd

2. Formulation of the problem

The governing equations of continuity, momentum, and energy for the nonstationary adiabatic motion of each component in some fixed volume V of a compressible mixture bound by surface S can be written in the following form [5]:

$$\frac{\partial}{\partial t}(\alpha_i \rho_i) + \nabla \alpha_i \rho_i \upsilon_i = 0, \ (i = 1, 2, \ldots, N),$$

$$\alpha_i \rho_i \frac{\mathrm{d}_i \upsilon_i}{\mathrm{d}t} = \nabla \sigma_i + \alpha_i \sum_{j=1}^{N} \alpha_j \mathbf{R}_{ji}, \ (i = 1, 2, \ldots, N),$$

$$\alpha_i \rho_i \frac{\mathrm{d}_i E_i}{\mathrm{d}t} = \sigma_i \varepsilon_i + \alpha_i \sum_{j=1}^{N} \alpha_j F_{ji}, \ (i = 1, 2, \ldots, N),$$

$$\text{where } \frac{\mathrm{d}_i}{\mathrm{d}t} \equiv \frac{\partial}{\partial t} + \upsilon_i^k \frac{\partial}{\partial x^k}.$$

Here, t is the time, ρ_i is the density of the i-th component equal to the mass of the i-th component per unit volume of the i-th component, υ_i is the velocity vector, E_i is the internal specific energy, ε_i is the strain rate tensor, $\sigma_i = -P_i \delta_i + S_i$ is the stress tensor, P_i is pressure, S_i is the stress deviator, R_{ji} is the intensity of the momentum exchange between the j-th and i-th components, F_{ji} is the intensity of the energy exchange between the j-th and i-th components, N is the number of components.

Volume fractions of the mixture occupied by each component are given by $\alpha_1 + \alpha_2 + \ldots + \alpha_N = 1$, ($\alpha_i \geq 0$), $\alpha_i = \rho_i^* / \rho_i$, where ρ_i^* is the reduced density (mass of the i-th component per unit volume).

In modeling chemical reactions under shock wave loading, we use the zeroth-order kinetic relation characterized by a constant rate of chemical transformations [1, 2, 6]:

$$J_{ji} = \frac{\mathrm{d}\eta}{\mathrm{d}t} = \begin{cases} 0, & \textit{if } \eta = 1 \textit{ or } (T_i < T_\eta \textit{ and } P < P_\eta) \\ f(P_\eta), & \textit{if } \eta < 1 \textit{ and } (T_i \geq T_\eta \textit{ or } P \geq P_\eta) \end{cases},$$

$$f(P_\eta) = \begin{cases} K_0, & \textit{if } P < P_\eta \\ K_p K_0, & \textit{if } P \geq P_\eta, \end{cases}$$

where T_i is the temperature, P is the matched pressure of components, and T_η, P_η, K_p, K_0 are the constants, η is the conversion degree.

Evolution of porosity in the material (compression and growth of pores) is simulated using a kinetic model of the active type, which determines changes in the specific volume of pores influencing on the material properties and causing stress relaxation [7]:

$$\frac{dV_{fi}}{dt} = \begin{cases} 0 \ \text{ for } |P_{si}| \leq P_i^* \ \text{ or } \ (P_{si} > P_i^* \text{ and } V_{fi} = 0) \\ -\text{sign}(P_{si}) K_{fi} (|P_{si}| - P_i^*)(V_{2i} + V_{fi}) \\ \text{for } P_{si} < -P_i^* \text{ or } \ (P_{si} > P_i^* \text{ and } V_{fi} > 0) \end{cases}$$

where $P_i^* = P_{ki} V_{1i} / (V_{fi} + V_{1i})$, P_{si} is the pressure in the solid (undamaged) part of the i-th component in the mixture, V_{1i}, V_{2i}, P_{ki}, K_{fi} are the experimentally determined constants of the material.

The equal pressures of the components are considered to be a condition for joint deformation of components during the interaction in the mixture, which determines the volume concentrations of the components:

$$P = P_i(V_i, E_i) = P_j(V_j, E_j) = \ldots = P_N(V_N, E_N).$$

3. Experimental results

3.1. Explosive loading of the aluminium-sulfur mixture

To study experimentally the explosive synthesis, the powder mixture of aluminum and sulfur (Al/S) was used as a loaded material that was placed in a cylindrical ampoule. The powders were mixed in an AGO-2U planetary mill. The mass fractions of components were 65/35 (Al/S), which was close to the stoichiometry of the aluminum sulfide (Al_2S_3) formation. The initial porosity of the mixture components to be loaded was 0.4 in the experiments. The pressed mixture was placed in a steel cylindrical ampoule. The height of the cylindrical ampoule was 95 mm, the external diameter was 20 mm. The thickness of the lateral wall of the ampoule was 3 mm. The ampoule was closed with lids 10 and 20 mm in sickness. The ampoule was loaded by an explosive (ammonite 6GV) with addition of NaCl in mass proportion 1/1. The external diameter of the explosive was 50 mm. The density of the explosive was 1.0 g/cm^3, and the detonation velocity was 2.8 km/s.

During loading the ampoule was broken (Figure 1). The ampoule was first broken at the bottom part, and then it was broken along the full length. After the experiment many crystallized drops with sizes up to 4 mm were found on the ampoule internal surface. X-ray phase analysis of the material collected from the ampoule showed that this was aluminum sulfide.

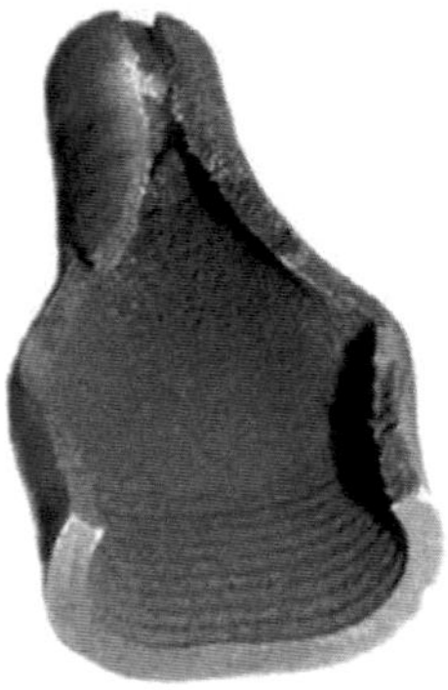

Figure 1. The ampoule after the experiment.

3.2. Explosive loading of the aluminium- fluoroplastic mixture

In the following experiment (Figure 2), the powder mixture of aluminum and fluoroplastic (Al/Tf) was used as a loaded material. The powders were mixed in an AGO-2U planetary mill. The mass fractions of components were 30/70 (Al/Tf), which was close to the stoichiometry of the aluminum fluoride (AlF_3) formation.

Figure 2. Fragments of the ampoule after explosive loading.

The initial porosity of the loaded mixture was 0.5 (the ratio between the volume of pores and the total volume). The pressed mixture was placed in a steel cylindrical ampoule. The height of the cylindrical ampoule was 95 mm, the external diameter was 20 mm. The thickness of the lateral wall of the ampoule was 3 mm. The ampoule was closed with lids 10 mm in sickness. The ampoule was loaded by an explosive (ammonite 6GV). The external diameter of the explosive was 45 mm. The density of the explosive was 1.0 g/cm^3, and the detonation velocity was 3.85 km/s. The ampoule was broken into large fragments during the experiment (Figure 2).

3.3. Explosive loading of the aluminium-fluoroplastic-carbon mixture

In the following experiment, a cylindrical ampoule was loaded, which contained a three-component mixture of aluminum, fluoroplastic and carbon. Carbon was added to the mixture of aluminum and fluoroplastic in a proportion of 2/3, where graphite had two mass fractions and the mixture of Al/Tf had one mass fraction. The components were mixed in an AGO-2U planetary mill, and then the pressed mixture was placed in a steel cylindrical ampoule. The height of the cylindrical ampoule was 81 mm, the external diameter was 20 mm. The thickness of the lateral wall of the ampoule was 3 mm. The ampoule was closed with lids 21 and 13 mm in sickness. The measured porosity was 0.25±0.005. The ampoule was loaded by an explosive (ammonite 6GV). The external diameter of the explosive was 45 mm. The density of the explosive was 1.0 g/cm3, and the detonation velocity was 3.85 km/s. After experiment the ampoule was uniformly compressed from top to bottom. To determine the phase composition of the sample, the ampoule was cut into 5 parts (Figure 3). Detonation wave propagated from the left sample to the right one. The difference is clearly observed in the central and peripheral part of the sample. The reaction was complete only at the center of the sample. Carbon and AlF_3 are present among the products of the reaction.

Figure 3. Ampoule cut into 5 parts after explosive loading.

4. Numerical results

To study the explosive synthesis, the numerical computations were carried out by using the model of a multicomponent medium and the finite element method [7-9]. We numerically solved an axisymmetric problem of a cylindrical steel ampoule containing a porous Al/S mixture under explosive loading. The mass fractions of the components were as follows: Al – 0.35; S – 0.65, which corresponded to the stoichiometry of the aluminum sulfide (Al_2S_3) formation after initiation of a chemical reaction.
The height and the diameter of the cylindrical mixture sample were 65 mm and 14 mm, respectively. The thickness of the lateral wall of the ampoule was $\Delta h = 3$ mm; the thickness of the lids was 20 mm. The external diameter of the ampoule was 20 mm. In the computations, the action of the explosive surrounding of the ampoule was modeled as the pressure of the explosion products acting on the horizontal and vertical layers of the ampoule during propagation of the detonation front. The detonation velocity was 3.3 km/s and corresponded to the experimental evaluations.
In the work, we assume that a chemical reaction initiation criterion (pressure criterion) and a reaction rate depend on the initial value of the mixture dispersion. For the aluminum and sulfur powders with low dispersion, we chose the reaction initiation criteria in terms of temperature, $T_\eta = 933$ K (aluminum melting temperature), and pressure, $P_\eta = 1.6$ GPa. Reaction rate K_0 was taken to be 301.6 GJ/(kg s), and

$K_p = 2.0$. For the aluminum and sulfur powders with high dispersion, we chose $T_\eta = 933$ K and $P_\eta = 1.0$ GPa.

Figure 4 demonstrates the distribution of pressure profiles in the Al/S high dispersion mixture placed in a cylindrical ampoule at different moments of time, describing the dynamics of the shock wave process during the explosive synthesis of aluminum sulfide.

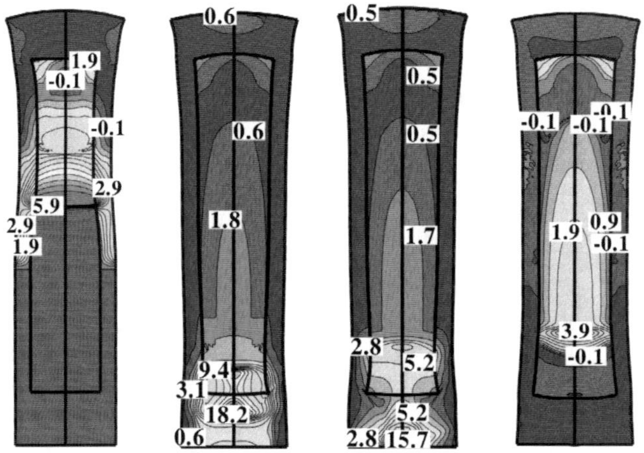

a b c d

Figure 4. Distribution of pressure profiles (GPa) for the high dispersion reactive mixture (Al-S) in an axial section of the ampoule at different moments of time: (a) 15 μs, (b) 27 μs, (c) 28 μs, and (d) 30 μs.

A shock wave caused by the explosion products propagates along the sample. When the shock wave is reflected from the bottom of the ampoule in the form of a compression wave, the pressure at the bottom of the sample increases by several times (Figure 4b) and is greater than 18 GPa. At the same time, high pressers are observed in the sample for some time (Figure 4d), while the shock wave propagated along the ampoule already stopped acting.

Figure 5 demonstrates the explosive synthesis of aluminum sulfide in the Al/S high dispersion mixture placed in a cylindrical ampoule at different moments of time.

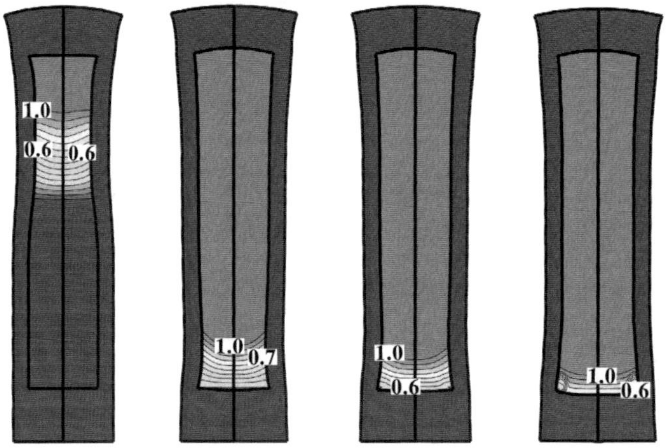

a b c d

Figure 5. Explosive synthesis of aluminum sulfide in the high dispersion reactive mixture (Al/S) in an axial section of the ampoule at different moments of time: (a) 15 μs, (b) 27 μs, (c) 28 μs, and (d) 30 μs.

The results show (Figure 5) that the reaction in the high dispersion Al/S mixture was complete with the formation of aluminium sulfide (100%).

IOP Conf. Series: Journal of Physics: Conf. Series **894** (2017) 012033 doi:10.1088/1742-6596/894/1/012033

Figure 6 demonstrates the distribution of pressure profiles in the Al/S low dispersion mixture.

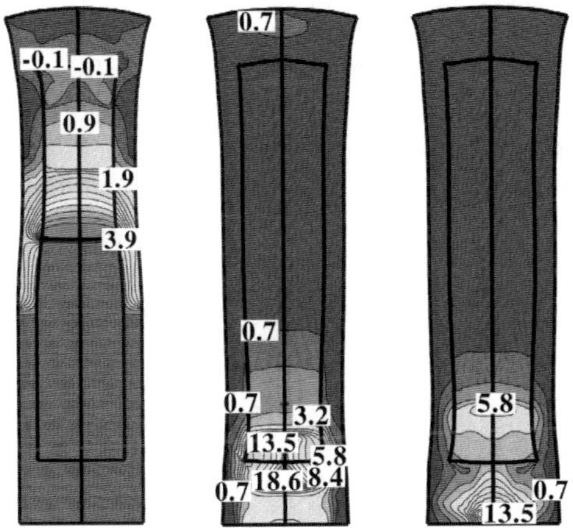

a b c d

Figure 6. Distribution of pressure profiles (GPa) for the low dispersion reactive mixture (Al/S) in an axial section of the ampoule at different moments of time: (a) 15 μs, (b) 27 μs, (c) 28 μs, and (d) 30 μs.

During the propagation of the shock wave, the higher pressure is maintained for a longer time in the high dispersion sample compared to the low dispersion one, which indicates the development of a faster chemical reaction and the contribution of porosity to the increase in pressure and temperature. The shock wave reflected from the bottom of the ampoule encounters the shock compression wave propagating in the sample, which leads to a sharp increase in the pressure at the bottom of the sample (Figure 6b) and is greater than 18 GPa. After the encounter, the shock waves start propagating in opposite directions (Figure 6c). By 30 μs, high residual pressures (3.9 GPa) are observed in the high dispersion mixture (Figure 4d), while only negative residual pressures (-0.1 GPa) are observed in the low dispersion mixture (Figure 6d).

Figure 7 demonstrates the explosive synthesis of aluminum sulfide in the Al/S low dispersion mixture placed in a cylindrical ampoule at different moments of time.

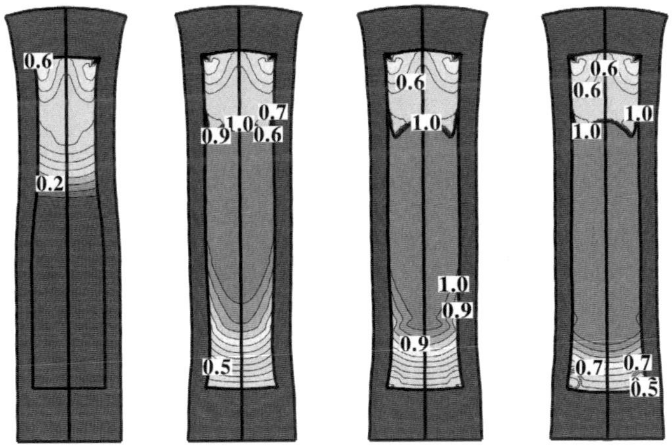

a b c d

Figure 7. Explosive synthesis of aluminum sulfide in the low dispersion reactive mixture (Al/S) in an axial section of the ampoule at different moments of time: (a) 15 μs, (b) 27 μs, (c) 28 μs, and (d) 30 μs.

The results show (Figure 7) that not all the low dispersion mixture components reacted to form aluminium sulfide.

5. Conclusion

The analysis of numerical and experimental results allows us to conclude about the contribution of the porosity and dispersion of mixtures to the dynamics of the shock wave propagation and the development of chemical reactions during the solid-phase synthesis in cylindrical ampoules subjected to explosive loading. During the propagation the shock wave, reflected from the bottom of the ampoule in the form of a compression wave, encounters the shock wave propagating in the sample, which leads to a sharp increase in the pressure and the rate of chemical reactions in the low and middle part of the sample. It should be noted that the reaction in the high dispersion Al/S mixture is complete with the formation of aluminium sulfide (100%). Despite the sharp increase in pressure and temperature, not all the low dispersion mixture components react to form aluminium sulfide. The contribution of high pressures and temperatures is not sufficient for the complete reaction in the low dispersion mixture.

The experimental and numerical results allowed us to find that the higher the initial dispersion and porosity of the mixture components, the more intensive the ampoule fracture. In some experiments, there was a complete fracture of the ampoule.

References

[1] Zelepugin S, Ivanova O, Yunoshev A and Zelepugin A 2015 *Letters on materials* **5(4)** 468-72
[2] Zelepugin S A, Ivanova O V, Yunoshev A S and Zelepugin A S 2016 *IOP Conf. Series: Mat. Sci. and Eng.* **127(1)** 012057
[3] Batsanov S S 1994 *Effects of explosions on materials* (New York: Springer-Verlag)
[4] Zelepugin S A, Ivanova O V, Yunoshev A S, Zelepugin A S 2017 *IOP Conf. Series: Mat. Sci. and Eng.* **177(1)** 012071-1
[5] Nigmatulin R I 1991 *Dynamics of Multiphase Media* vol 1 (New York: Hemisphere) p 507
[6] Shteinberg A S, Berlin A A, Denisaev A A and Mukasyan A S 2011 *Int. J. of Self-Propagating High-Temperature Synthesis* **20(4)** 259-65
[7] Gorel'skii V A, Zelepugin S A and Sidorov V N 1994 *Int. Appl. Mech.* **30(3)** 193-98
[8] Johnson G R 2011 *Int. J. Impact Eng.* **38(6)** 456-72
[9] Kanel G I, Fortov V E, Razorenov S V 2007 *Physics-Uspekhi* **50(8)** 771

Impact compaction of ultra high molecular weight polyethylene

B S Zlobin, A A Shtertser, V V Kiselev and S D Shemelin

Design and Technology Branch, Lavrentyev Institute of Hydrodynamics, Novosibirsk, 630090, Russia

E-mail: asterzer@mail.ru

Abstract. Ultra high molecular weight polyethylene (UHMWPE) is a partially crystalline polymer with unique physical and mechanical characteristics. Therefore it finds a wide use in present-day mechanical engineering. Experiments on impact compaction of UHMWPE granules were performed on the special facility designed for consolidation of powders. Bulk samples were made out of GUR 4150 powder with molecular weight of $9.2 \cdot 10^6$ g/mol. The method of impact compaction permits to make not only pure UHMWPE compacts, but material with micro- and nano-sized additions as well, and produce composites with reinforcing metal structures.

1. Introduction

Ultra high molecular weight polyethylene (UHMWPE) is unique material with high toughness and wear resistance, high shock resistance, low friction, and many other remarkable properties. Bulk material is made out of UHMWPE powder by different methods, mostly by the hot molding technique. Having a great molecular weight (millions of g/mol), this material does not pass into the viscous-flow state even at high temperatures up to its melting point, besides, UHMWPE has high viscosity in the liquid state. These properties hinder producing of bulk material by forming operation. That's why the existing industrial technologies (hot molding, ram extrusion, sintering, etc.) are expensive and complicated. In this connection, dynamic loading methods are of interest, as they possibly can simplify the process of powder consolidation. The first encouraging results were obtained in [1] where the method of cyclic high-velocity forming was used. In present studies the impact compaction technique, consisting in sequence of blows affecting the powder portion, was employed. Bulk samples of pure UHMWPE and of UHMWPE with micro- and nanosized additives were made. Besides, composite samples with reinforcing metal structure were produced by the said technique.

2. Experiments on UHMWPE compaction

The first stage experiments on the impulse compaction of the UHMWPE powder performed in Lavrentyev Institute of Hydrodynamics are described in [2]. In present study we have used the same experimental facility (hydro-pneumatic apparatus) providing the impact rate up to 9 blows per second. The supposed method was designated in [2] as cyclic impact compaction (CIC). Bulk samples were made out of the commercial GUR 4150 powder with molar mass of $9.2 \cdot 10^6$ g/mol. A peculiarity of CIC method is that material remains in the solid state during the treatment, hence the starting content of crystal phase (about 66%) remains in the compact. The tensile strength of the produced material

Content from this work may be used under the terms of the Creative Commons Attribution 3.0 licence. Any further distribution of this work must maintain attribution to the author(s) and the title of the work, journal citation and DOI.
Published under licence by IOP Publishing Ltd

reaches 45.3 MPa and elongation can be as much as 370%. The CIC method is the promising one for the production of small articles, similar to those shown in figure 1. The obtained compacts require minimal after-treatment, like those produced by closed-die forming technique. But in contrast to this technology CIC permits to make parts directly from the starting powder material.

Figure 1. Examples of parts made by CIC.

Figure 2. UHMWPE compacts with additives (from left to right) of TiO_2, SiO_2 and SiC.

Presently, there are the works addressed to the production and characterization of polymer composites on the base of UHMWPE with different additives [3, 4]. This direction in development of new materials with enhanced properties is considered as having of good prospects. Positive results of CIC application for pure UHMWPE powder enabled us to make experiments with UHMWPE powder modified by micro- and nano-sized TiO_2, SiO_2 and SiC additives at the amount of 2 wt.%. The procedure of powder modification is described in detail in [2]. The obtained compacts with the density of 0.94 – 0.95 g/cm^3 are shown in figure 1.

3. Properties of compacts made by CIC

Cylindrical samples for tensile tests were made out of the compacts containing micro- and nano-additives (figure 3*a*). The sample deformable length was 10 mm, diameter 6 mm. The tests were carried out on the testing machine ZDM-5 at the elongation velocity of 6.5 mm/min. The tested samples were fixed in the special holders, since the design of ZDM-5 does not permit testing of small-size specimens. The outlook of samples after the tests is presented in figure 3*b*. It should be noted that after unloading specimens shrink so that their length decreases by about 30%. The test results are the following: tensile strength σ = 23.1, 31.4 and 27.7 MPa for specimens with SiC, TiO_2 and SiO_2 additives, respectively; elongation δ = 250, 290 and 300% for specimens with SiC, TiO_2 and SiO_2 additives, respectively.

The presence of fine ceramic particles in the polymer volume usually increases its wear resistance. To compare the wear resistance of samples obtained by CIC and those made of PTFE and commercial UHMWPE, the experiments were made on dry friction of said polymers against the hardened steel counterpart with the surface ground to the roughness of grade 8 (0.438 μm). Four cylindrical samples were prepared for the tests: the first one was made of pure UHMWPE by CIC, the second one was made of UHMWPE + 2 wt% SiO_2 by CIC, the third one was made of commercial UHMWPE, and the forth one was made of PTFE (figure 4). Friction of polymer samples against steel counter face was organized in the forth – and – back movement with an amplitude of 100 mm. Total friction path amounted to 4,000 m for samples made of UHMWPE and 200 m for that made of PTFE, the pressing force was 151.8 N. The wear of samples was characterized by their mass loss after the tests. The precision laboratory balance LV 210-A was used for sample weighing. The tests were made on BiSS UTM servo-hydraulic testing machine in the Laboratory of Mechanical and Climatic Testing in the

Industrial Park of Novosibirsk Academic Township (Akademgorodok). Because of the limited test time we failed to compare the wear resistance of UHMWPE samples, since their mass loss was close to the balance limit of accuracy. Tests have shown the mass loss of 0.5, 0.5, 0.3 and 15 mg for the first, second, third and forth sample, respectively. So, taking into account the mass loss and friction path, we can conclude that UHMWPE material made by CIC has the wear resistance at least 600 times as much as that of PTFE.

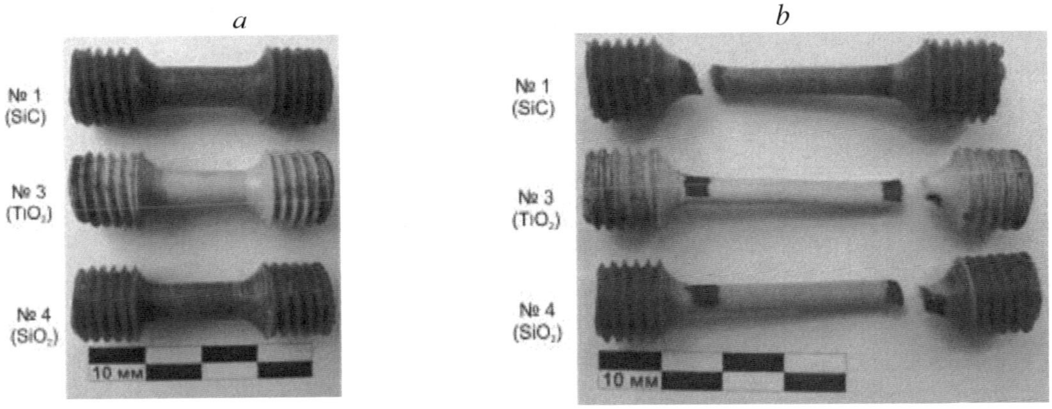

Figure 3. Samples from the UHMWPE modified by additives of SiC, TiO_2, and SiO_2 (2 mass %): *a* – before the tensile tests; *b* – after the tests.

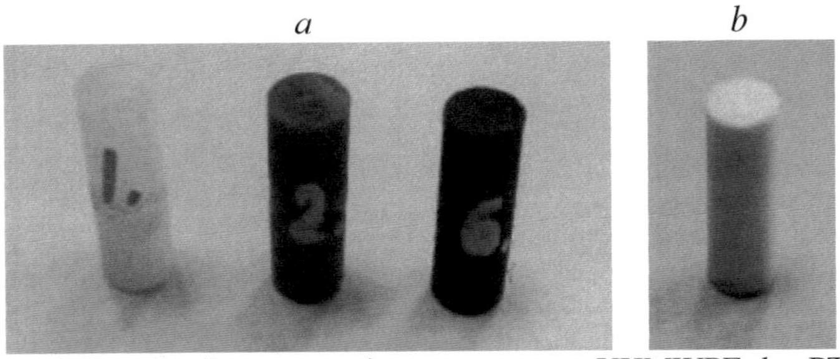

Figure 4. Samples for wear-resistance tests: *a* – UHMWPE, *b* – PTFE.

Composite materials on the base of UHMWPE with metal reinforcement are of great interest. Since polymers chemically not interact with different other materials, including metals, there are difficulties in manufacturing of such composites. In engineering practice the polymer-metal joining is provided by different design solutions based on mechanical locking of composite components. To make polymer – metal composite, on the first stage the discs with the thickness of 2 – 4 mm were produced by CIC, then, on the second stage CIC was used for stamping the multilayered item consisting of alternate layers of polymer and perforated titanium bands with the thickness of 0.5 mm. Figure 5*a* presents the 5-layered compact produced by the described technological route. Analysis of the compact macrostructure shows that it is dense all over the volume and there are no cavities in the polymer – metal interface and in perforation areas (figure 5*b*). Figure 5*c* shows the specimen for tensile test cut out of the composite compact.

Figure 6*a* presents the tensile diagram of pure UHNWPE bulk material made by CIC, this is a typical polymer elongation diagram [2]. Figure 6*b* presents the tensile diagram for the specimen shown in figure 5*c*. This diagram was obtained on the Zwick Z100 testing machine at elongation velocity of 1 mm/min.

a *b* *c*

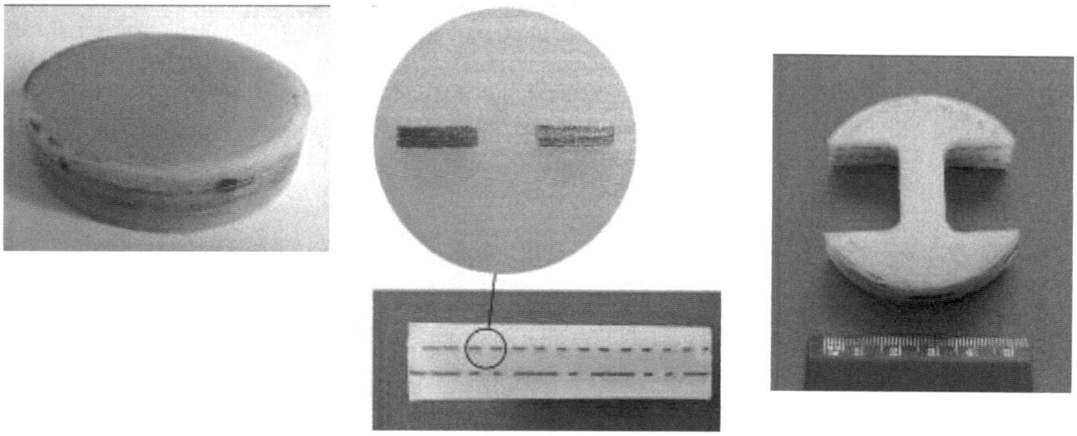

Figure 5. Polymer-metal composite: *a* – five-layer compact; *b* – compact structure; *c* – specimen for the tensile test.

a *b*

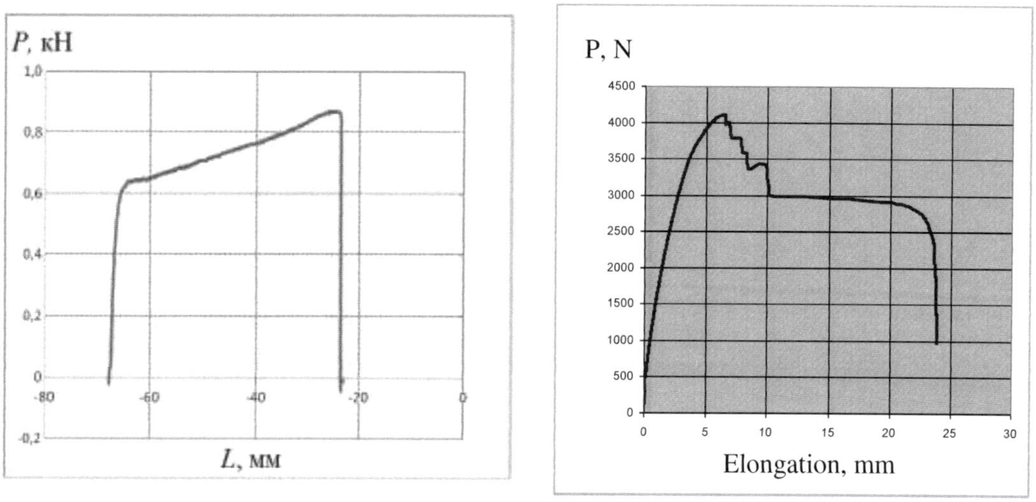

Figure 6. Tensile diagrams of specimens: *a* – pure UHMWPE [2]; *b* – five-layer composite.

It is evident that diagrams shown in figures 6*a* and 6*b* differ dramatically. Deformation curve of the composite specimen is somehow similar by character to tensile curves of metals. The maximal load value in figure 6*b* indicates that metal and polymer layers behave as a single whole.

The impact toughness of composite material was detected on the ram impact machine KM-30. The sample sizes followed the recommendations of the Russian standard GOST 4647-80 (figure 7*a*). The tests demonstrated that composite samples, made by CIC method, do not fail completely. This is also typical for the samples made of pure UHMWPE [2]. Therefore it is difficult to estimate the impact toughness of obtained compacts quantitatively. However, we can suppose that UHMWPE compacts, especially reinforced with metal layers, can successfully sustain the impact loading and possibly the action of piercing projectiles.

a *b*

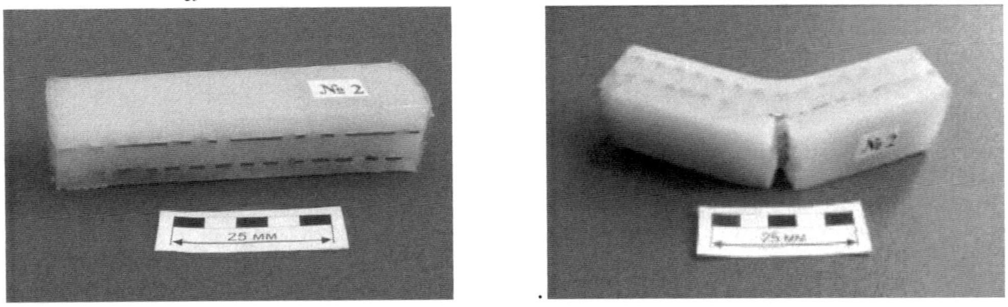

Figure7. Composite sample: *a* – before the impact toughness test, *b* – after the test.

Besides the above-mentioned composite samples with flat metal reinforcing inserts, the experiments were carried out on CIC treatment of UHMWPE powder with the cylindrical steel element inserted into the powder volume. After compaction the steel element inside the compact was joined to the steel rod as is shown in figure 8*a*. Then the steel rod and UHMWPE compact were fixed in holders of the ZDM-5 testing machine (UHMWPE via special transition device) and tensile test was done. Figure 8*b* shows the sample after the test. It is evident, that steel rod with diameter of 5 mm was broken at the force of 8212 N and cylindrical element was not pulled out of the compact. This is another demonstration of CIC capability to induce the strong joining between composite components.

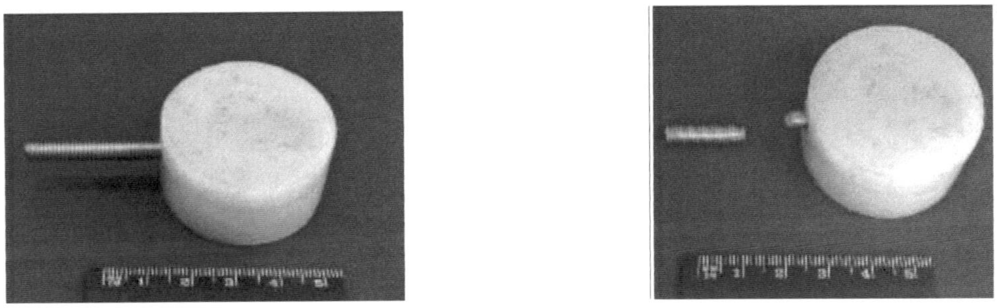

Figure 8. Metal - polymer sample: *a* – before the tensile test; *b* – after the test

4. Conclusion

The results of present studies show that the method of cyclic impact compaction has a significant potential in manufacturing of composites on the base of ultra high molecular weight polyethylene. These composites can contain either micro- and nano-additives or metal reinforcing structures.

References

[1] Jaufres D, Lame O, Vigier G and Dore F 2007 *Polymer* **48** 6374-6383

[2] Zlobin B S, Shtertser A A, Kiselev V V, Shemelin S D, Poluboyarov V A and Zhdanok A A 2017 *J. Appl. Mech. Tech. Phys.* **58** 435-442

[3] Selyutin G E, Gavrilov V V, Voskresenskaya E N, Zakharov V A, Nikitin V E and Poluboyarov V A 2010 *Chem. Sustain. Develop.* **18** 301-314

[4] Gogoleva O V, Petrova P N, Popov S N and Okhlopkova A A 2015 *J. Frict. Wear* **36** 301-305

IOP Conf. Series: Journal of Physics: Conf. Series **894** (2017) 012035 doi:10.1088/1742-6596/894/1/012035

Numerical simulation of the drop spreading on a horizontal plane

N A Zyuzina[1] **and V V Ostapenko**[1,2]

[1]Lavrentyev Institute of Hydrodynamics, Novosibirsk, Russia
[2]Novosibirsk State University, Novosibirsk, Russia

E-mail: nzyuzina1992@gmail.com

Abstract. A shallow water model of film flow is used to describe the process of a drop spreading on a horizontal plane, taking into account the liquid viscosity, heat-mass transfer and surface tension forces. To simulate this flow numerically in polar coordinates an unconditionally stable implicit difference scheme has been developed, which (in the case of evaporation) can calculate the drop spreading over a dry surface without special allocation of the drop boundary. A region of dimensionless parameters is singled out under which the evaporating droplet, as a result of surface tension forces, transforms into a circular ring before complete evaporation.

1. Introduction

The process of a liquid drop spreading over a horizontal plane is of classical interest and have been studied by various authors, since many industrial applications (lubrication, coating, painting) require the spreading of a liquid on a solid. The independent directions of investigations are a migration of liquid droplets on the solid surfaces [1], spreading of evaporating (volatile) drops [2–5], evaporation of sessile drops [6, 7], dynamic contact angles and moving contact lines [8, 9], gravity effect [10–12]. Such researches are useful in understanding of wetting phenomena [13] and for some technological and biological applications (coating, drying, spray cooling, tissue spreading, cell sorting).

In this paper, using the hydrodynamic approach [9,14,15], we apply a shallow water model of film flow to describe the process of a drop spreading on a horizontal plane, taking into account the liquid viscosity, heat-mass transfer and surface tension forces. To simulate this flow numerically in polar coordinates an unconditionally stable implicit difference scheme has been developed, which (in the case of evaporation) can calculate the drop spreading over a dry surface without special allocation of the drop boundary. A region of dimensionless parameters is singled out under which the evaporating droplet, as a result of surface tension forces, transforms into a circular ring before complete evaporation.

2. Mathematical Model

Let us consider a viscous incompressible fluid flowing at a velocity $\boldsymbol{u} = (u, v, w)$ along the horizontal plane α, which in the Cartesian coordinate system $\{x, y, z\}$ is given by the equation $z = 0$. The fluid is subjected to the gravity force, which generates an acceleration $\boldsymbol{g} = (0, 0, -g)$ directed vertically downward. We assume that in the cylindrical coordinate system (r, α, z) : $x = r\cos\alpha$, $y = r\sin\alpha$ the horizontal velocity components are determined by

Content from this work may be used under the terms of the Creative Commons Attribution 3.0 licence. Any further distribution of this work must maintain attribution to the author(s) and the title of the work, journal citation and DOI.
Published under licence by IOP Publishing Ltd

formulas $u = q\cos\alpha$, $v = q\sin\alpha$, where the modulus $q = \sqrt{u^2+v^2}$ of the horizontal velocity vector $\boldsymbol{v} = (u, v)$, as well as other scalar flow parameters, do not depend on the angle α. In this case, the fluid flow in the cylindrical coordinate system is described by the following Navier-Stokes, continuity and thermal conductivity equations:

$$q_t + qq_r + wq_z + p_r = \mu\left(q_{rr} + q_{zz} + q_r/r - q/r^2\right), \tag{1}$$

$$w_t + qw_r + ww_z + p_z + g = \mu\left(w_{rr} + w_{zz} + w_r/r\right), \tag{2}$$

$$q_r + w_z + q/r = 0, \tag{3}$$

$$\theta_t + q\theta_r + w\theta_z = a\left(\theta_{rr} + \theta_{zz} + \theta_r/r\right), \tag{4}$$

where p is the specific pressure, θ is the temperature, μ is kinematic viscosity and a is thermal diffusivity.

As the boundary conditions for the liquid, we set the adhesion condition on the plane α:

$$q(r,0,t) = w(r,0,t) = 0, \tag{5}$$

the flow potentiality, the kinematic and dynamic conditions on the free surface $z = h$:

$$\left(q_z - w_r\right)|_{z=h} = 0, \tag{6}$$

$$\gamma\left(h_t + qh_r - w\right) = \lambda\left(\theta_z - h_r\theta_r\right)/\varphi, \quad \varphi = \sqrt{1+(h_r)^2}, \quad z = h, \tag{7}$$

$$p = 2\mu\left((h_r)^2 q_r + w_z - h_r(w_r + q_z)\right)/\varphi^2 - \sigma\left(h_{rr}/\varphi^3 + h_r/(r\varphi^2)\right), \quad z = h, \tag{8}$$

where $h = h(r,t)$ is the liquid depth, γ is the latent heat, λ is the thermal conductivity, σ is the surface tension coefficient. The boundary values for the temperature are given by the formulas

$$\theta(r,0,t) = \theta_0, \quad \theta(r,h,t) = \theta_1, \quad \theta_0, \theta_1 = const. \tag{9}$$

Let us introduce the following dimensionless variables

$$r_* = \frac{r}{L}, \quad z_* = \frac{z}{H}, \quad t_* = \frac{t}{T}, \quad h_* = \frac{h}{H}, \quad q_* = \frac{q}{Q}, \quad w_* = \frac{w}{W}, \quad p_* = \frac{p}{P}, \quad \theta_* = \frac{\theta}{\Theta}, \tag{10}$$

$$g_* = \frac{g}{G}, \quad \mu_* = \frac{\mu}{M}, \quad a_* = \frac{a}{A}, \quad \lambda_* = \frac{\lambda}{\Lambda}, \quad \gamma_* = \frac{\gamma}{\Gamma}, \quad \sigma_* = \frac{\sigma}{S},$$

where L, H, T, $Q = L/T$, $W = H/T$, P, Θ, G, M, A, Γ, Λ and S are characteristic wave length, depth, time, horizontal and vertical velocity, pressure, temperature, gravity acceleration, viscosity, thermal diffusivity and conductivity, latent heat and surface tension respectively. Eq. (3) written in dimensionless variables (9), (10) remains unchanged; Eq. (1), (2), (4) and boundary condition (6)–(8) become

$$\varepsilon T\left[Q^2(q_t + qq_r + wq_z) + Pp_r\right] = M\mu\left[q_{zz} + \varepsilon(q_{rr} + q_r/r - q/r^2)\right], \tag{11}$$

$$\varepsilon T(w_t + qw_r + ww_z) + Pp_z + GHg = M\mu\left[w_{zz} + \varepsilon(w_{rr} + w_r/r)\right], \tag{12}$$

$$\varepsilon L^2(\theta_t + q\theta_r + w\theta_z) = TAa\left[\theta_{zz} + \varepsilon(\theta_{rr} + \theta_r/r)\right], \tag{13}$$

$$\left(q_z - \varepsilon w_r\right)|_{z=h} = 0, \tag{14}$$

$$\varepsilon L^2\Gamma\gamma\left(h_t + qh_r - w\right) = T\Theta\Lambda\lambda\left(\theta_z - \varepsilon h_r\theta_r\right)/\varphi, \quad \varphi = \sqrt{1+\varepsilon(h_r)^2}, \quad z = h, \tag{15}$$

$$Pp = \frac{2M\mu}{T\varphi^2}\left(w_z + \varepsilon(h_r)^2 q_r - h_r(q_z + \varepsilon w_r)\right) - \frac{SH\sigma}{L^2\varphi^3}\left(h_{rr} + \frac{\varphi h_r}{r}\right), \quad z = h, \tag{16}$$

where $\varepsilon = H^2/L^2 \ll 1$ is the small parameter of long-wave approximation [16]. Here and elsewhere the asterisk at the dimensionless variables is omitted for brevity.

Assuming that the characteristic quantities satisfy conditions

$$M = O(\varepsilon),\ \varepsilon Q^2 = o(\varepsilon),\quad P = ML^2/(TH^2) = GH = SH/(L^2),\quad \Gamma H^2 = T\Theta\Lambda \quad \Rightarrow$$
$$L = \sqrt{S/H},\quad T = ML^2/(GH^3) = MS/(GH^4) = \Gamma H^2/(\Theta\Lambda),$$

from Eq. (11)–(16) in the long-wave approximation, we obtain

$$p_r = \mu q_{zz},\quad p_z + g = 0,\quad \theta_{zz} = 0, \tag{17}$$

$$q_z|_{z=h} = 0,\quad (h_t + qh_r - w - \lambda\theta_z/\gamma)\,|_{z=h} = 0,\quad (p + \sigma(rh_r)_r/r)\,|_{z=h} = 0. \tag{18}$$

Integrating the second equation (17) with allowance for the third boundary condition (18), we have

$$p = g(h - z) + p(h) = g(h - z) - \sigma(rh_r)_r/r,\quad z \in [0, h]. \tag{19}$$

Integrating the first equation (17) with taking into account Eq.(19), boundary condition (5) and first boundary condition (18), we obtain

$$q = (p_r/\mu)\left(z^2/2 - hz\right),\quad p_r = gh_r - \sigma\left((rh_r)_r/r\right)_r. \tag{20}$$

Integrating the third equation (17) with allowance for the boundary condition (9), we have

$$\theta_z = b,\quad \theta = bz + \theta_0,\quad b = (\theta_1 - \theta_0)/h, \tag{21}$$

Integrating the continuity equation (3) with respect to z from $z = 0$ to $z = h$ taking into account Eq. (20), (21) and second boundary condition (18), we obtain differential equation

$$h_t = \frac{1}{3r}\left[\alpha_1(rh^3h_r)_r - \alpha_2\left(rh^3\left(\frac{(rh_r)_r}{r}\right)_r\right)_r\right] + \frac{\beta}{h} \tag{22}$$

for calculating the film thickness $h(r,t)$, where $\alpha_1 = g/\mu$, $\alpha_2 = \sigma/\mu$ and $\beta = \lambda(\theta_1 - \theta_0)/\gamma$. We note that Eq. (22) can be regarded as a particular case of the equation derived in [17]. Multiplying Eq. (23) by $2h$, introducing the notation $\eta = h^2$ and taking into account that $h_r = \eta_r/(2h)$, we have

$$\eta_t = \frac{h}{3r}\left[\alpha_1(rh^2\eta_r)_r - \alpha_2\left(rh^3\left(\frac{1}{r}\left(\frac{r\eta_r}{h}\right)_r\right)_r\right)_r\right] + 2\beta. \tag{23}$$

Performing differentiation in the second term in square brackets, we find

$$\eta_t = \frac{h}{3r}\left[\alpha_1(rh^2\eta_r)_r - \alpha_2\left(k_1\eta_r - k_2\eta_{rr} + k_3\eta_{rrr} + k_4\eta_{rrrr}\right)\right] + 2\beta,\quad \eta = h^2, \tag{24}$$

where

$$k_1 = h^2/r^2 + (h_r - 2h/r)h_r + (3rh_r - 2h)h_{rr} - rhh_{rrr},$$
$$k_2 = h^2/r + hh_r + rhh_{rr},\quad k_3 = 2h^2,\quad k_4 = rh^2.$$

The advantage of Eq. (24), in comparison with Eq. (22), is that it lacks a singularity at $h = 0$, which makes it possible to simulate the propagation of a droplet over a dry surface without specifying its boundary. For Eq. (24), we set the following initial-boundary value problem

$$h(r,0) = \begin{cases} \sqrt{9 - r^2}, & 0 \le r \le r_0, \\ h_0, & r_0 \le r \le r_1, \end{cases}\qquad r_0 = \sqrt{9 - h_0^2},\ r_1 = 20, \tag{25}$$

$$h_r(0,t) = h_{rrr}(0,t) = h_r(r_1,t) = h_{rrr}(r_1,t) = 0, \tag{26}$$

which simulates the spreading of a drop over the residual layer of the initial thickness $h_0 \le 1$.

3. Numerical simulation

Figures 1–4 show the results of numerical simulation of the problem (24)–(26) by an implicit, unconditionally stable finite-difference scheme, in the implementation of which a five-point sweep and nonlinearity iterations were applied. The calculations were performed on a rectangular finite-difference grid with parameter $\alpha_1 = 1$. The dashed line in figures shows the initial position of the drop and solid lines show its free surface at four subsequent times. Figure 1 corresponds to the condensation at $\beta = 1$, parameter $\alpha_2 = 1$ and the initial thickness of the residual layer $h_0 = 0.5$. Figure 2–4 correspond to the evaporation at $\beta = -1$ and $h_0 = 1$. The calculations shown in Figure 2 were carried out with the parameter $\alpha_2 = 0$, in Figure 3 – with $\alpha_2 = 1$ and in Figure 4 – with $\alpha_2 = 5$.

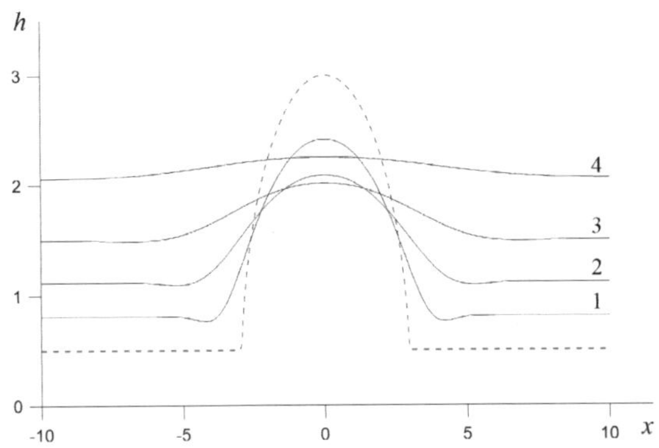

Figure 1. Numerical simulation of a drop spreading in the case of condensation at $\beta = 1$ with $\alpha_1 = \alpha_2 = 1$ at the following times: $t = 0.2$ (line 1), $t = 0.5$ (line 2), $t = 1$ (line 3), $t = 2$ (line 4).

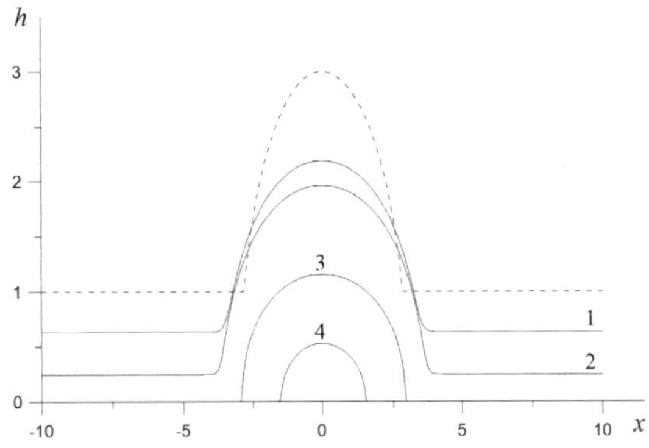

Figure 2. Numerical simulation of a drop spreading in the case of evaporation at $\beta = -1$ with $\alpha_1 = 1$, $\alpha_2 = 0$ at the following times: $t = 0.3$ (line 1), $t = 0.47$ (line 2), $t = 1.3$ (line 3), $t = 1.8$ (line 4).

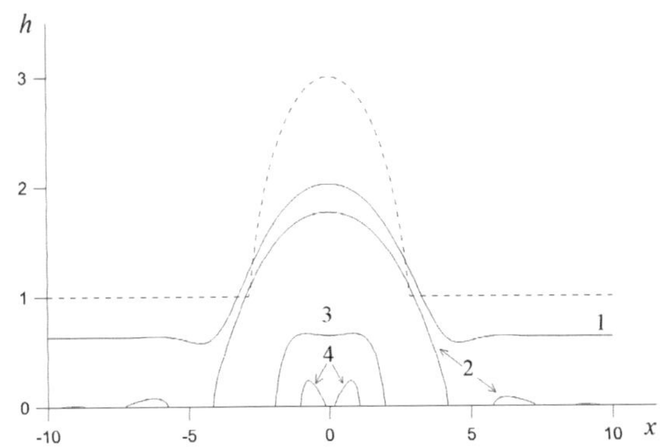

Figure 3. Numerical simulation of a drop spreading in the case of evaporation at $\beta = -1$ with $\alpha_1 = \alpha_2 = 1$ at the following times: $t = 0.3$ (line 1), $t = 0.5$ (line 2), $t = 1.7$ (line 3), $t = 1.93$ (line 4).

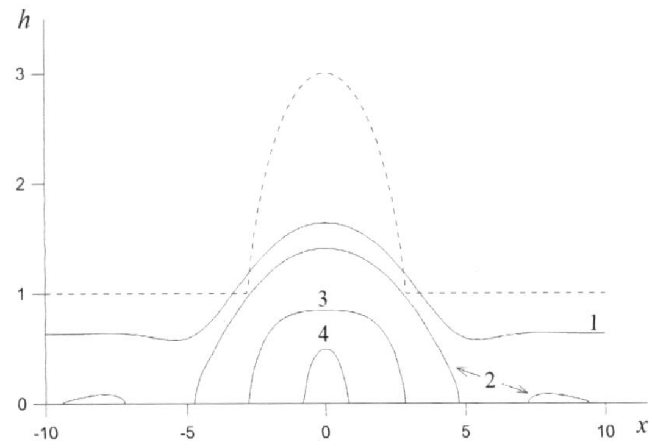

Figure 4. Numerical simulation of a drop spreading in the case of evaporation at $\beta = -1$ with $\alpha_1 = 1$, $\alpha_2 = 5$ at the following times: $t = 0.3$ (line 1), $t = 0.5$ (line 2), $t = 1.1$ (line 3), $t = 1.6$ (line 4).

It follows from these calculations that for a finite value of the parameter α_2 responsible for the surface tension, a sequence of damped waves is formed on the surface of the residual layer over which the drop spreads (only the first downward wave is clearly visible in Figures 1,3,4). In the case of condensation, the increasing residual layer gradually absorbs these waves and at the time $t \approx 1$ (line 3 in Figure 1) the surface of the residual layer becomes monotonous. In the case of evaporation, the amplitude of these waves gradually increases, which leads to the appearance of a sequence of decreasing annular film layers before the complete evaporation of the residual layer. Only the first annular film layer from this sequence is clearly visible in Figures 3,4 (line 2 in these figures). The calculations shown in the Figure 3 also revealed the following interesting fact: the surface tension at $\alpha_2 = 1$ forms a hollow at the drop top which increases in time (line 3) and leads to the transformation of the drop into an annular film layer (line 4) before its complete evaporation.

4. Conclusion

This article proposes a shallow water model of the film flow for describing the process of a drop spreading on a horizontal plane, taking into account the gravity acceleration, liquid viscosity, heat-mass transfer and surface tension forces. The main advantage of this model is that the resulting differential equation (24) describing the evolution of the thickness h of a liquid film has no singularity at $h = 0$ which allows to simulate the propagation of a droplet over a dry surface without specifying its boundary. This, in particular, makes it possible to numerically determine the values of the triple angle at the gas-liquid-solid interface without additional heuristic assumptions.

Acknowledgments

The authors are grateful to Prof. V.V. Kuznetsov for useful comments made during the discussion of this work. The work was supported by the Russian Science Foundation (grant No. 15-19-10025).

References

[1] Smith M K 1995 *J. Fluid Mech.* **294** 209
[2] Potash M and Wayner P C 1972 *J. Heat Mass Transfer* **15** 1851
[3] Ruiz O E and Black W Z 2002 *J. Heat Transfer* **124** 855
[4] Ajaev V S 2005 *J. Fluid Mech.* **528** 279
[5] Strotos G, Gavaises M, Theodorakakos A and Bergeles G 2008 *Int. J. Heat Mass Transfer* **51** 1516
[6] Hu H and Larson R G 2005 *Langmuir* **21** 3963
[7] David S, Sefiane K and Tadrist L 2007 *Colloids Surf. A Physicochem. Eng. Asp.* **298** 108
[8] Greenspan H P 1978 *J. Fluid Mech.* **84** 125
[9] Dussan V 1979 *Annu. Rev. Fluid Mech.* **11** 371
[10] Shikhmurzaev Y D 1997 *Phys. Fluids* **9** 266
[11] Reznik S N and Yarin A L 2002 *Int. J. Multiph. Flow* **28** 1437
[12] Bartashevich M V, Kuznetsov V V and Kabov O A 2010 *Micrograv. Sci. Technol.* **22** 107
[13] Gennes P G 1985 *Rev. Mod. Phys.* **57** 827
[14] Tanner L H 1979 *J. Phys. D Appl. Phys.* **12** 1473
[15] Voinov V V 1995 *Int. J. Multiph. Flow* **21** 801
[16] Friedrichs K O 1948 *Comm. Pure Appl. Math.* **1** 109
[17] Kabova Yu, Kuznetsov V V and Kabov O 2014 *Interfacial Phenom. Heat Transfer* **2** 85

IOP Conf. Series: Journal of Physics: Conf. Series **894** (2017) 012036 doi:10.1088/1742-6596/894/1/012036

The influence of compliant coatings on skin friction in the turbulent boundary layer

O O Ivanov[1], V V Vedeneev[1], V M Kulik[2] and A V Boiko[3]

[1]Institute of Mechanics, Lomonosov Moscow State University, Moscow, Russia
[2]Institute of Thermophysics, Russian Academy of Sciences, Novosibirsk, Russia
[3]Institute of Theoretical and Applied Mechanics, Russian Academy of Sciences, Novosibirsk, Russia

E-mail: ololivan@gmail.com

Abstract. The results of an experimental study of the effect of single-layer monolithic compliant coatings on friction in a turbulent boundary layer are presented. The techniques of manufacturing the coating and measuring their properties are described. A special test model (a section of a wing section of an infinite span) to be used in experiment was designed. The model has four replaceable plates: the solid (metal) plates and panels with compliant coatings 4,6,8,and 10 mm thick can be installed. The direct measurements of the total drag of the model and the local skin friction obtained by the Clauser chart method showed consistent results: drag increased up to 6.5% and 4%, respectively. The relationship between the results and theoretical predictions is discussed.

1. Introduction

The problem of drag reduction of bodies moving in fluids is one of the most important and practically significant in aerohydrodynamics. The use of compliant coatings in liquids is one of the most promising ways to reduce friction, turbulent noise, and to delay the laminar-turbulent transition passively (without additional energy supply).

The idea of drag reduction by compliant coatings was first proposed and studied by M.O. Kramer [1–4]. He obtained experimentally a significant reduction in resistance when using special coatings on a towed body similar to the skin of a dolphin. Followers observed both positive and negative results in the sense of the drag reduction with the Kramer-type or self-made coatings. Details of the studies on compliant coatings can be found in the thorough reviews [5,6,7]. In a large number of experiments, where "soft" coatings (with a porous or gel-like filler covered by a thin film) were used, the λ-shaped folds increased the drag. However, it is difficult to compare and generalize the results as in the majority of the studies the properties of coatings were not measured either at all, or carefully.

A certain success was achieved in experiments with a compliant surface to delay of the laminar-turbulent transition [8] that was also verified theoretically.

For practical use, stiff monolithic coatings are the most promising. The study [9] of such coatings conducted at speed of 10–20 m/s with at owed model in an open reservoir showed a drag reduction up to 20%. An attempt to repeat these results in laboratory conditions was undertaken in a cavitation tunnel of the University of Newcastle at speeds of 1-7 m/s [10]. In these experiments, the coatings of

Content from this work may be used under the terms of the Creative Commons Attribution 3.0 licence. Any further distribution of this work must maintain attribution to the author(s) and the title of the work, journal citation and DOI.
Published under licence by IOP Publishing Ltd

study [9] were used. The drag reduction up to 7% was obtained, as well as a decrease of the level of pressure pulsations at the wall, the longitudinal velocity pulsations, and the turbulence intensity.

The recent appearance of reliable technique for measuring the dynamic viscoelastic properties of coatings over a wide range of frequencies [11, 12], stimulates new studies of compliant coatings. Now it is possible to attribute quantitatively the properties of the coating with its effectiveness in drag reduction or the laminar-turbulent transition delay.

2. Experimental Setup

The experiments were conducted in the water tunnel of Institute of Mechanics, Lomonosov Moscow State University. The operating speed range was 5–16 m/s, the level of turbulence of water tunnel was ~0.2%, the static pressure was maintained at the same level during all runs. For the study, a special model with four replaceable plates (Fig. 1) was designed. The model has a shape of section of an infinite-span symmetric wing. The length of the model is 1190 mm, width is 100 mm, height is 117 mm. Test plates have dimensions of 300 x 20 x 117 mm (LxWxH). The model was installed in a two-dimensional test section of water tunnel with dimensions of 2000 x 1000 x 120 mm. The geometry of the model was optimized by numerical simulations in ANSYS CFX to minimize the pressure gradient along the test plates of the model. The velocity field distribution along the model for the operating speed 20m/s is shown in Fig. 2.

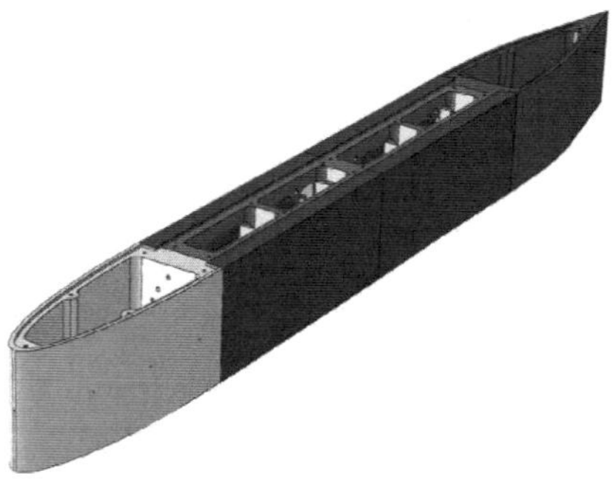

Figure 1. The test model. The leading section is grey, test panels are blue, the trailing section is red.

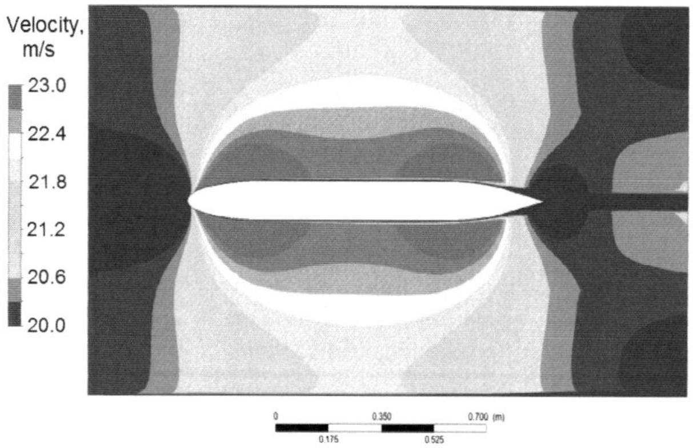

Figure 2. The velocity distribution around the test model at 20 m/s.

The measurements were conducted in two stages. In the first stage, the total drag of the model was measured using one-component drag balance. To determine the contribution of friction force over the test plates into the total drag, a numerical simulation of the flow, including settling chamber, nozzle, operating section, and a part of diffuser was conducted, and this contribution was found to be nearly 20%for the velocities of 5–25 m/s. Therefore, a change in the skin friction on the plates of 5% will result in about 1% corresponding change in the total drag.

In the second stage, the longitudinal mean velocity profiles of the turbulent boundary layer over second test plate were measured by laser Doppler anemometer (LDA). From these profiles, the local skin friction coefficients were determined using the modified Clauser method [13]. The tripping wire with a diameter of 1 mm was mounted on the leading section of the model at the distance of 29 mm from the leading edge to obtain downstream a fully developed turbulent boundary layer. Measurements were conducted at the positions x = 629, 729, 829 mm downstream of the leading edge.

3. Coating manufacturing and measurement of their properties.

Viscoelastic coatings were produced from silicone rubber Mold Max 10 (Smooth-On firm), by polymerizing at room temperature and normal pressure with added catalyst. Before mixing the components, it was degassed during 2 hours in an extruder with the under pressure of –0.85 atm. The resulting mixture was then mixed during 10 minutes in a low-speed mixer to exclude appearance of gas bubbles. Two compliant coatings and samples for measuring the viscoelastic properties were produced simultaneously of the mixture. A special mould (Fig. 3) was used to manufacture compliant coatings and to avoid air bubble formation during mixture pouring and shrinkage.

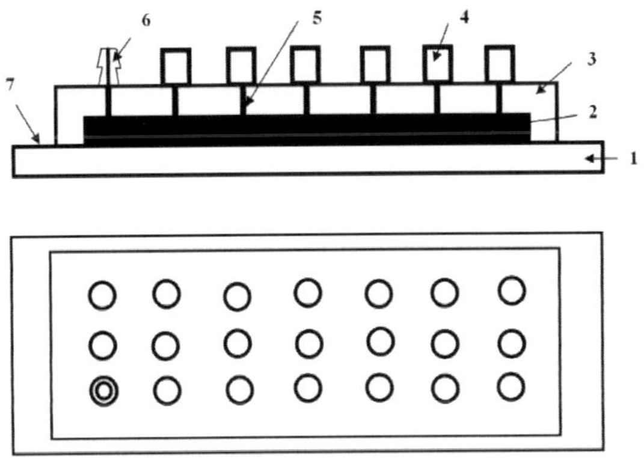

Figure 3. The scheme of the mould for coating manufacturing: flat base (1), compliant layer (2), test plate body (3), anti-shrinkage channels (4) and cylinders (5), fitting (6), film (7).

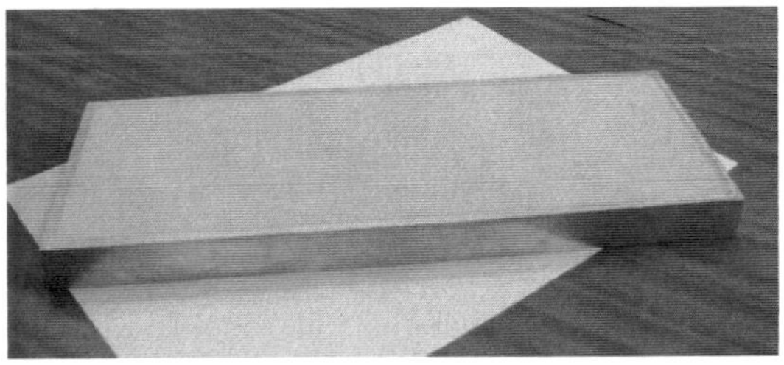

Figure 4. The test plate with compliant layer.

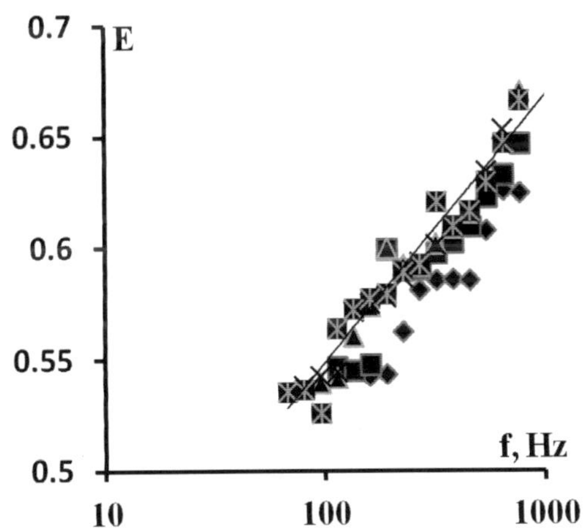

Figure 5. Relation between dynamic modulus of elasticity of Mold Max 10 and load frequency.

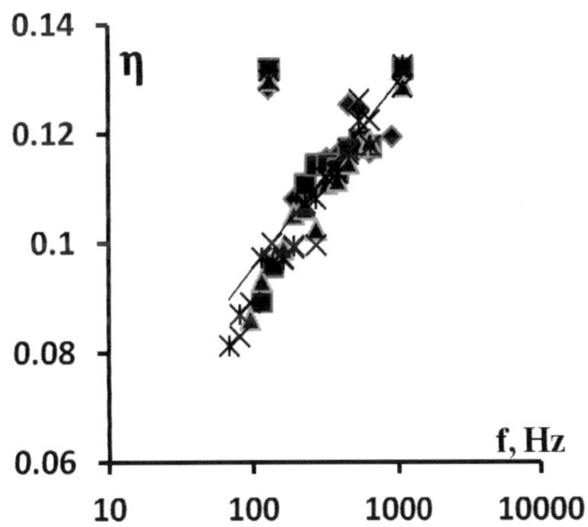

Figure 6. Relation between loss tangent of viscoelastic material and load frequency.

The method for measuring viscoelastic properties of the coatings is described in detail in [12,13]. After the measurements, the following dependencies the dynamic modulus of elasticity and the loss tangent versus oscillation frequency of deformation were obtained (see also Figs. 5 and 6). The dynamic modulus of elasticity (MPa) $E = 0.0521\,ln(f) + 0.3087$, the loss tangent $\eta = 0.017\,ln(f) + 0.0141$.

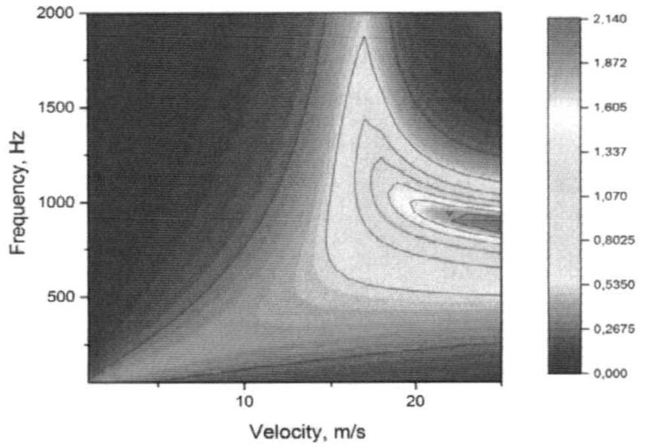

Figure 7. Non-dimensional normal component of compliance vs. pressure fluctuation frequency and flow velocity (8 mm coating).

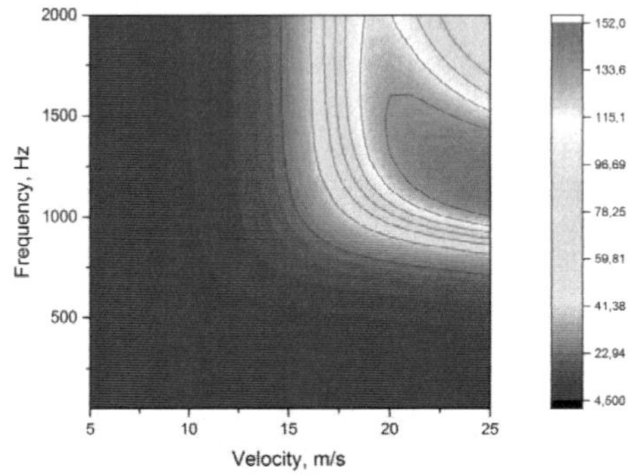

Figure 8. Variation of the phase angle (in degrees) of compliance versus pressure fluctuation frequency and flow velocity (8 mm coating).

4. Prediction of drag reduction

The response of the compliant coating to an actual level of external pressure was theoretically investigated in [14]. The amplitude of deformation of the surface of "stiff" compliant coatings is less than the thickness of the viscous sublayer [14] so that the coating always remains hydraulically smooth. However, in a region of coating–flow interaction frequencies (in the vicinity of the resonance frequency of the coating), the velocity of its surface motion is comparable with the turbulent velocity fluctuations near the wall.

For the current coatings, the complex compliance (the ratio of the strain deformation of a coating surface to the applied pressure) and the phase angle on the stream velocity and the pulsation frequency were obtained as functions of flow velocity and pressure fluctuation frequency. A compliant coating will effectively interact with coherent structures of turbulent flow only at those speeds and frequencies where its compliance is maximum. In all other regions of parameters, it will only slightly differ from the solid wall. Analyzing Figs. 6 and 7, two scenarios of interaction can be distinguished:

- broadband, when the flow velocity coincides with the elongated crest, and the compliance does not reach a maximum, passing along the low-speed slope of the peak (flow velocity 10–20 m/s);
- resonant, at higher frequencies (flow velocity >20 m/s).

5. Results and discussion

5.1. Measurements of the total drag

In the experiments conducted 5–6 months after the viscoelastic coatings manufacturing, the total drag of the model was measured. Experiments with metal and coated plates conducted at the speeds of 7–15 m/s, the Reynolds number based on the model length is equal to $(7.3-15.7)\times10^6$. The coefficients of total drag $C_x = F / \rho V_\infty^2 S$ (*S* is the area of the midsection of the model) vs. the speed of incoming flow are compared for the solid wall (empty squares) and the compliant coating (filled markers) in Fig. 8.

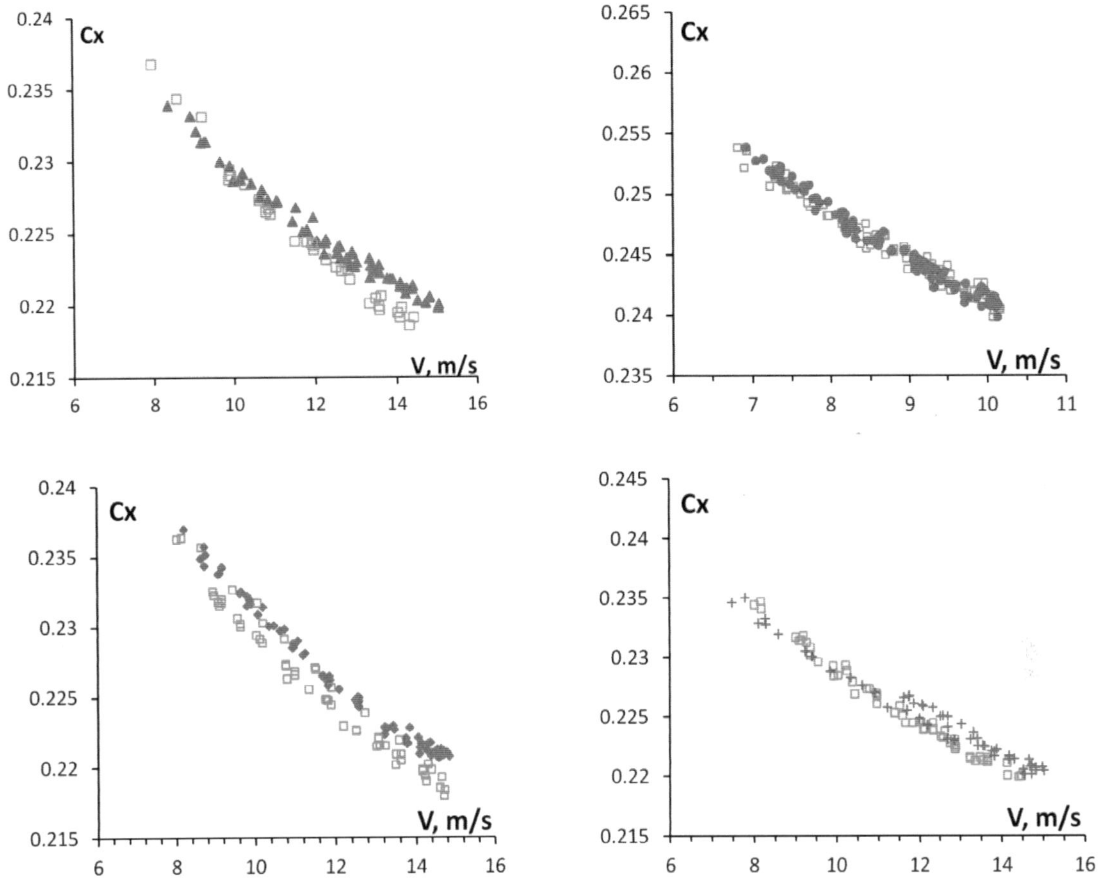

Figure 9. Total drag coefficient versus flow velocity. From top left to bottom right:4 mm (triangles), 6 mm (circles), 8 mm (diamonds), 10 mm (crosses).

The compliant coatings increase the total drag coefficient at the end of the operating speed range. On average, the difference between the coefficients for the solid wall and coatings is 0.001–0.0025, which in relative terms is 0.4–1.3%. In terms of the change in friction, drag over coatings increases up to 2–6.5%. For the 6-mm coating, the model was installed at a small angle of attack, so that the absolute values of C_x are greater than those for other thicknesses.

5.2. Local skin friction measurements
The second stage was conducted 12–15 months after the manufacture of the coatings. To find the local friction, the modified Clauser chart method [13] was used; namely, the error between the Musker profile [16] and the experimental points was minimized (Fig. 9).

The local skin friction coefficients vs. momentum thickness Reynolds number $Re_2 = V_\infty \theta / \nu$ are shown in Fig.10. The lower solid line corresponds to relation $\Sigma_x = 0.0256 \cdot Re_2^{-0.25}$ [15], the upper solid line is $\Sigma_x = 0.0131 \cdot Re_2^{-1/6}$ [15]. As seen, in this series of tests the coatings also increase the drag up to 4%.

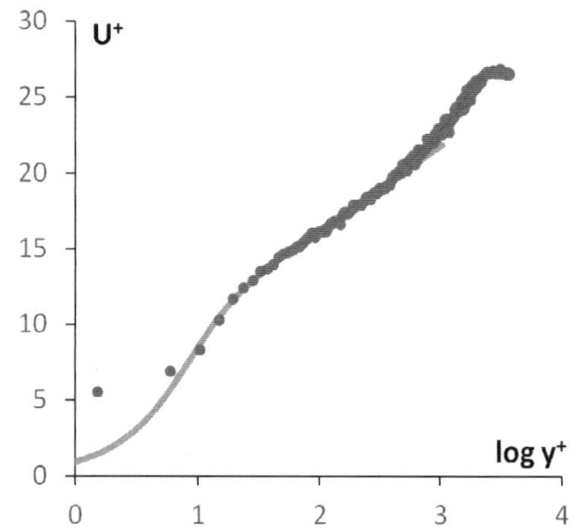

Figure 10. The example of turbulent boundary layer velocity profile in law of the wall scale. Solid line is the Musker profile [16].

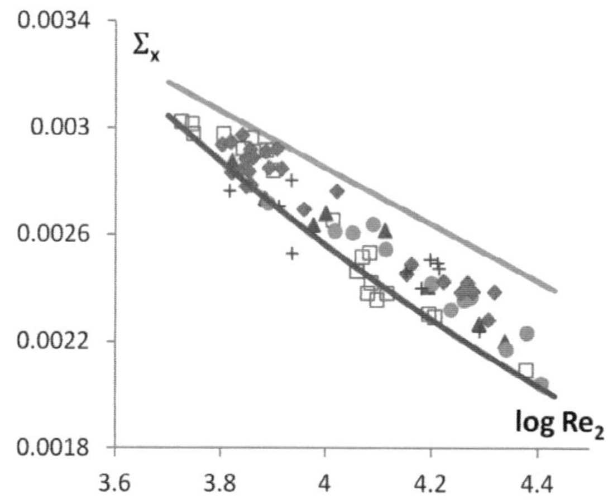

Figure 11. Local skin friction coefficient vs. the momentum thickness Reynolds number. The upper and lower lines are the empirical relations [15].

Note that the results of measurements at the first and second stages are in satisfactory agreement with each other. The theory predicts the maximum interaction region after 15 m/s (17.5 m/sis the peak of the wide-frequency-band interaction). The total drag coefficients of the model with coatings start to differ from those for metal plates at a speed of ~12 m/s, which also indicates the consistency of the theory and experiment, and the predictive capabilities of this theory.

The disadvantage of the theory [12, 14] is that it can show the region of most intensive interaction between the boundary layer and compliant coating, but cannot answer whether it will yield the drag reduction or drag increase. This is also seen from the results of the present study. Given that there is no reliable theory predicting drag reduction in turbulent flow, experimental methods are still the only (besides DNS) way to study compliant coatings. From this point of view, results of the present research can be considered as the contribution into the database of coating properties and their effect.

6. Conclusions

The results of the laboratory measurements of single-layer coating effect on skin friction in turbulent boundary layer are presented. Two stages of direct and indirect measurements are conducted, and the drag increase up to 6.5% and 4% for two stages, respectively, is obtained. Results are compared with theoretical prediction, and satisfactory agreement is obtained.

References

[1] Kramer M O 1957 *J. Aeronaut. Sci.* **24** 459-60
[2] Kramer M O 1960 *J. Am. Soc. Navel Engrs.* **72** 25-34
[3] Kramer M O 1961 *J. Am. Soc. Navel Engrs.* **73** 103-7
[4] Kramer M O 1962 *J. Am. Soc. Navel Engrs.* **74** 341-8
[5] Bushnell D M, Hefner J N and Ash R L 1977 *Phys. Fluids* **20** S31-S48
[6] Gad-el-Hak M, Blackwelder R F and Rilly J J 1984 *J. Fluid Mech.* **140** 257-80
[7] Gad-el-Hak M 1996 *Appl. Mech. Rev.* **49** S147-60
[8] Carpenter P W 1990 *Viscous drag reduction in boundary layers* Progress in Astronautics and Aeronautics (AIAA) **123** 79-113
[9] Kulik V M, Poguda I S and Semenov B N 1991 *Recent Developments in Turbulence Management* ed Choi K-S (Dordrecht: Kluwer) pp 263-89

[10] Choi K S, Yang X, Clayton B R, Glover E J, Atlar M, Semenov B N and Kulik V M 1997 *Proc. Royal Soc.* A **453** 2229-40
[11] Kulik V M, Semenov B N, Boiko A V, Seoudi B, Chun H H and Lee I 2009 *Exp. Mechanics* **49** 417-25
[12] Kulik V M, Semenov B N, Boiko A V, Seoudi B, Chun H H and Lee I 2010 *Int. J. of Solids and Structures* **45** 374-82
[13] Rodríguez-López E, Bruce P J and Buxton O R 2015 *Exp. Fluids* **56** 68
[14] Kulik V M 2012 *Int. J. Heet Fluid Flow* **33** 232-41
[15] Schlichting H, Gersten K, Krause E and Oertel H 1960 *Boundary-Layer Theory* (New York: McGraw-Hill)
[16] Musker A 1979 *AIAA J* **17** 655-7

IOP Conf. Series: Journal of Physics: Conf. Series **894** (2017) 012037 doi:10.1088/1742-6596/894/1/012037

On a criterion of incipient motion and entrainment into suspension of a particle from cuttings bed in shear flow of non-Newtonian fluid

Yaroslav Ignatenko[1], Oleg Bocharov[1] and Roland May[2]

[1] Baker Hughes, 630090 Novosibirsk, Russia

[2] Baker Hughes, 29221 Celle, Germany

E-mail: `yaroslav.ignatenko@bakerhughes.com`

Abstract. Solids transport is a major issue in high angle wells. Bed-load forms by sediment while transport and accompanied by intermittent contact with stream-bed by rolling, sliding and bouncing. The study presents the results of a numerical simulation of a laminar steady-state flow around a particle at rest and in free motion in a shear flow of Herschel–Bulkley fluid. The simulation was performed using the OpenFOAM open-source CFD package. A criterion for particle incipient motion and entrainment into suspension from cuttings bed (Shields criteria) based on forces and torques balance is discussed. Deflection of the fluid parameters from the ones of Newtonian fluid leads to decreasing of the drag and lift forces and the hydrodynamic moment. Thus, the critical shear stress (Shields parameter) for the considered non-Newtonian fluid must be greater than the one for a Newtonian fluid.

1. Introduction

Moving of spherical particles along a plane in a shear flow or fluid at rest is a common phenomenon in nature as well as in technical solutions. It involves river drifts and drilling mud transporting cuttings out of a well. Both processes can be described using this simplified model, which requires determination of hydrodynamic forces and moments affecting a particle near the wall in a viscous fluid.

The flows heterogeneity has a significant effect on the hydrodynamic processes around a moving particle, such as the drag force. At the same time, it creates a force perpendicular to the flow known as lift force. Its value is usually small if compared to the drag force, but it still plays an important role both in sedimentation and sediment spreading. When a particle approaches a wall, the lift force increases dramatically and becomes comparable to the drag force [1]. It occurs due to three main mechanisms such as a shear flow, sphere rotation, and presence of the wall [2].

One of the first publications devoted to flow around a sphere near the wall was written by Bretherton [3] & Goldman et al. [4, 5]. Both studies relied on the Stokes approximation. Goldman et al. obtained an analytical solution for the forces and moment affecting a spherical particle near the wall. The authors considered two cases: a sphere at rest in a moving shear flow and a moving rotating sphere in a fluid flow at rest. For the last case, it was shown that the drag force and moment grew logarithmically as the distance to wall reduced. Due to the Stokes approximation, the lift force value was equal to zero. Krishnan and Leighton [2] demonstrated

Content from this work may be used under the terms of the Creative Commons Attribution 3.0 licence. Any further distribution of this work must maintain attribution to the author(s) and the title of the work, journal citation and DOI.
Published under licence by IOP Publishing Ltd

that at small Reynolds numbers, the forces affecting a rotating sphere moving along the wall could be presented as a sum of six components. The first three are the forces caused by the shear flow, sphere motion, and its rotation. The remaining three are the forces produced by the interaction of the components listed above. This theory, including the lift-off criterion, was experimentally confirmed by King and Leighton [6]. Accounting for the particle roughness allowed avoiding a logarithmic singularity when compared against the theoretical results.

However, the previously mentioned studies concerned small Reynolds numbers. If moderate Reynolds numbers are considered, the problem of particle and bubble motions at a finite distance along the wall in a fluid at rest was investigated by Takemura [7] and Zeng et al. [8] respectively. Zeng et al. [9] studied the forces and moment affecting a sphere in a shear flow near the wall up to the instant the sphere touching the wall.

Lee and Balachandar [10] developed correlation ratios to describe both drag and lift forces at moderate Reynolds numbers for a sphere moving along a plane. They based the correlations on the theoretical study by Krishnan and Leighton [2], summing the six components, as well as on the results by Zeng [8, 9] for a particle moving along the wall.

In one of the most recent publications Lee and Balachandar [11] generalized the results obtained in the previous studies devoted to the forces affecting a spherical particle, and modeled the free steady motion of a particle on a plane in a shear flow. The steady flow in this study was obtained through modeling of an unsteady flow under the assumption of no friction force between the sphere and the plane while the drag force and moment affecting the particle were equal to zero. Apart from the lift force, the authors obtained the particles free-motion velocity, its rotation rate and the time its motion becomes steady. In their further publications [12] the authors concentrated on a spherical particle moving along a rough surface in a turbulent flow. Thus, for Newtonian fluid at moderate Reynolds numbers, the problem of a particle moving on a plane or near to it has been studied sufficiently.

Bocharov and Kushnir [13] investigated the forces affecting a spherical particle near the wall in a shear flow of a power-law fluid. It was demonstrated that at a distance to the wall exceeding three diameters of the sphere and the Reynolds numbers less than 50 (determined from flow velocities opposite the particles), the lift force comprised less than 5% of the drag force. In case, the distance to the wall was less than one diameter, the lift force, trying to push the particle away from the wall, increased as the distance got smaller at constant Reynolds numbers and reached 25-30% of the drag force.

The current publication presents the results of parametric study of the motion of a spherical particle moving along a plane with and without slip in a shear flow of the Herschel–Bulkley fluid.

2. Problem statement

The study considered the steady rolling of a spherical particle with the diameter d and the velocity vp moving along a plane the (Figure 1) in a shear flow of non-Newtonian (Herschel–Bulkley) fluid. It was assumed that the sphere moved in the same direction as the flow. There was no friction between the sphere and the wall. The origin of Cartesian coordinates matched the spheres center, where the OX axis correlated with the flow, the OY one- with the planes normal, and ω denoted the rotational rate of the particle along the OZ axis. The fluid flow was described by a steady system of the Navier-Stokes equations:

$$\rho u \cdot \nabla u = -\nabla p + \nabla \cdot \tau, \nabla u = 0,$$

where the viscous stress tensor τ depended on the local shear rate $\dot{\gamma}$:

$$\tau = \begin{cases} \frac{k\dot{\gamma}+\tau_0}{\dot{\gamma}}\dot{\gamma} & at\ \dot{\gamma} > 0 \\ 0 & at\ \dot{\gamma} = 0 \end{cases}, \text{ thus the effective viscosity was } \mu = \begin{cases} \frac{k\dot{\gamma}+\tau_0}{\dot{\gamma}} & at\ \dot{\gamma} > 0 \\ \infty & at\ \dot{\gamma} = 0 \end{cases}.$$

Here $\dot{\gamma} = 0.5\left(\nabla u + \nabla u^T\right)$ denoted the strain velocity tensor, - the second invariant of the strain velocity tensor of shear rate, n the power index, τ_0 the yield stress of fluid. The boundary conditions are:

$$u = \begin{cases} r \times \omega - v_p e_x & at \ r = d/2 \\ G\left(y + d/2\right) e_x & at \ x \to -\infty \end{cases}$$

where r denoted a radius vector from the center of the sphere, $\omega = (0, 0, -\omega)$ – the vector of rotational particle velocity, G – the flow shear rate. The characteristic length was the diameter

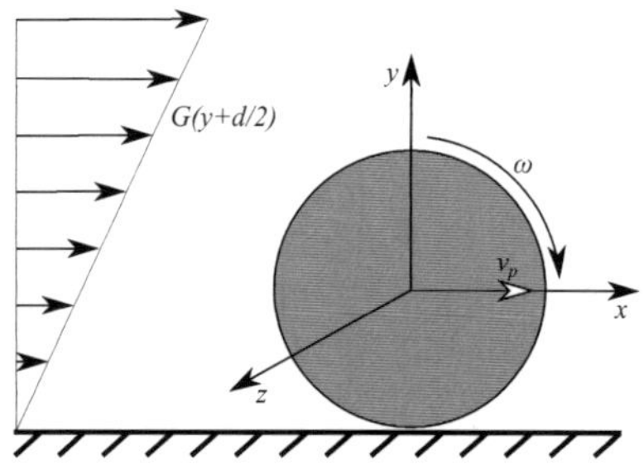

Figure 1. Mobile particle schematic.

d, and the characteristic velocity – the face velocity $v = Gd/2$. Hence, the flow could be characterized by the following set of dimensionless parameters such as $Re_s = G^2 d^2 \rho / 2(kG^n + \tau_0)$ that denoted shear –Reynolds number with the characteristic Herschel–Bulkley viscosity or $ResPL = G2 - nd^2\rho/2k$ – with the characteristic power–law viscosity, $V_p = v_p/v$ that is the spheres dimensionless velocity, n – the power index, Bn=?0/kGn – the Bingham number, and $\Omega = \omega/G$ – the spheres dimensionless rotational rate.

In the calculations, the Reynolds number varied in the range $0.1 \leq Re_s \leq 200$. Two cases were considered: particle at rest ($V_p = 0, \Omega = 0$) and free motion when the terminal velocity V_{pss} and the terminal rotating rate Ω_{pss} were found iteratively to satisfy the equalities $F_d = 0, M = 0$. The power index n varied in the range $0.4 \leq n \leq 1$ and the Bingham number – $0 \leq Bn \leq 10$. The drag force F_d, the lift force F_L and the hydrodynamic torque M were calculated by integration of the friction and pressure over the sphere surface area S.

For forces and torque analyze dimensionless parameters were used: $C_d = 4F_d/\pi\rho G^2 d^2$ to denote the drag force coefficient, $Y = C_d Re_s/24$ the normalized drag force coefficient, $C_L = 4F_L/\pi\rho G^2 d^2$ the lift force coefficient, $C_M = 8M/\pi\rho G^2 d^3$ the hydrodynamic torque coefficient.

3. Numerical Method and Verification

To avoid computational problems, as recommended in [6], a gap equal to $d/100$ between the sphere and the wall, was considered for in all the further studies modeling similar problems. Since we modeled a laminar flow with no symmetry plane ($OXOY$), only half of the flow was considered. The Navier–Stokes equations were solved by shifting the particles reference system and using a stationary computational mesh.

The computational volume was a rectangular 3D parallelepiped, whose surface was isomorphically reflected onto a hemisphere that transferred onto the spherical surface of the particle, forming a structured hexahedral computational grid. The distance from the sphere to the distant boundaries of the computational domain was equal to twenty-five diameters ($25d$) of the sphere. The maximum grid size on the sphere in the radial direction was $\pi d/100$. The size of the computation cell on the spheres surface did not exceed . The cell sizes had been selected

Table 1. Comparison of force characteristics against those in [9].

Re_s	value	Zeng, 2009 [9]	present	discrepancy, %
2	C_d	$2.349 \cdot 10$	$2.295 \cdot 10$	2.29
2	C_L	2.653	2.571	3.09
2	C_M	8.005	7.813	2.40
10	C_d	6.445	6.421	0.37
10	C_L	1.305	1.293	0.90
10	C_M	1.778	1.769	0.52
200	C_d	1.381	1.343	2.79
200	C_L	$3.384 \cdot 10^{-1}$	$3.244 \cdot 10^{-1}$	4.13
200	C_M	$1.044 \cdot 10^{-1}$	$1.021 \cdot 10^{-1}$	2.23

to make it possible for the grid to condense by a factor of two in all the directions without significant changes in the forces and moment ($< 1\%$) affecting the sphere.

The numerical solution of the Navie –Stokes equations was found using the OpenFOAM 3.0.1 open-source hydrodynamic package [14], whose solver (simpleFoam) was equipped with the SIMPLE-C algorithm. For discretization of the convective terms, a linear upwind scheme was applied. The coefficient of the lift force, drag force and moment of forces for $V_p = 0$ published by Zeng et al. [9] was used to test the computational mesh and the numerical method. For modeling, Zeng et al. [9] had used NEK5000, an open–source hydrodynamic package based on the spectral method providing high level of computational accuracy. The table below contains the values and their discrepancy, which does not exceed 5%.

4. Results and Discussion

4.1. Forces and Torque acting on a particle

Numerical simulation showed that the growing power law index n led to decreasing of drag force coefficient Y (Figure 2a). This, evidently, occurred because of the apparent viscosity decrease. Quite the reverse, the increasing Bingham number Bn led to increasing of Y, because of the apparent viscosity increase.

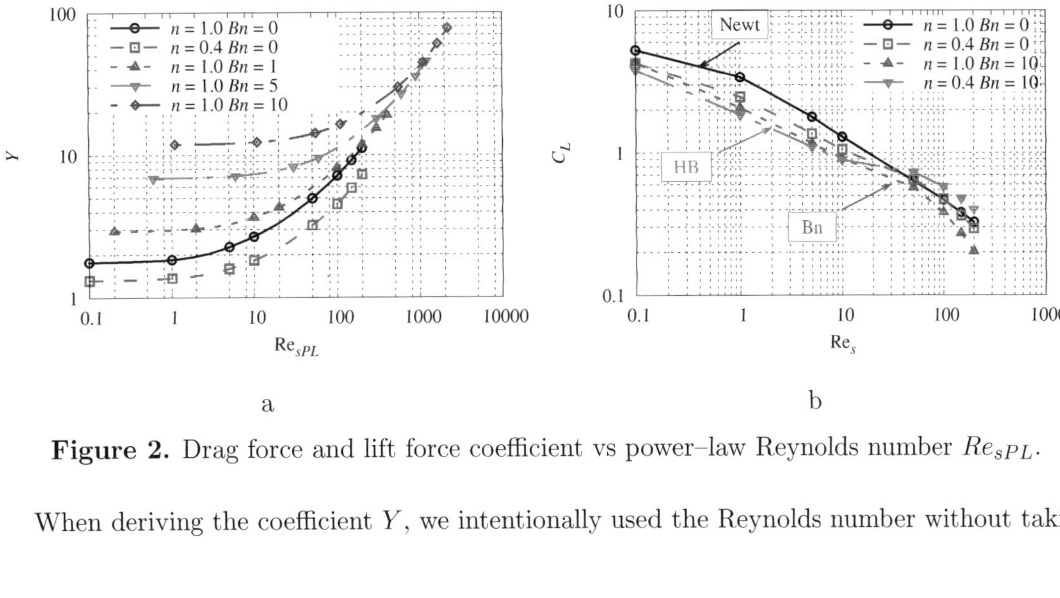

Figure 2. Drag force and lift force coefficient vs power–law Reynolds number Re_{sPL}.

When deriving the coefficient Y, we intentionally used the Reynolds number without taking into account the yield stress. The graph demonstrates that all the coefficient curves show the same asymptote as the Reynolds number increases independent of the Bingham exponent (Figure 2a). This holds true for the Herschel–Bulkley fluid, whose asymptotic inclination depends on the power index n. At the same time, despite the significant dependence of the drag coefficient on the fluid rheology, the lift force coefficient changes insignificantly (Figure 2b).

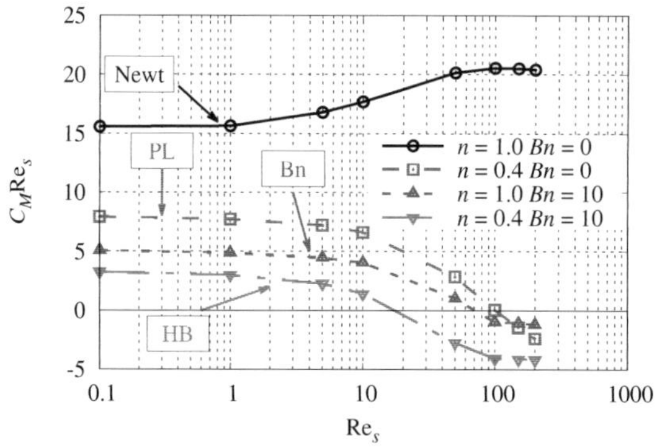

Figure 3. Torque coefficient behavior vs Reynolds number Re_s.

It was convenient to consider the moment of hydrodynamic forces as one multiplied by the Reynolds number Re_s. This expression does not change much for a wide range of the Reynolds numbers. Figure 3 shows that as the Reynolds number increases, the moment of forces may reduce significantly, becoming negative for non-Newtonian fluids. In such a way, the moment at the large Reynolds numbers may help the particle stay at rest. A very interesting effect observed here is that the values of $C_M Re_s$ are constant for both low and high values of the Reynolds numbers.

4.2. Free rolling particle

Lee et al. [11] consider the equilibrium rolling with a slip of a particle moved by a shear flow. The equilibrium condition here is the drag force F_d and moment equal M to zero. In our study, a similar problem statement was considered by making changes in the numerical algorithm. Calculating the solution we found the spheres equilibrium (terminal) velocity V_{pss} ($C_d = 0$) and rotation rate Ω_{pss} ($M = 0$). In the case of free rotation, the equilibrium velocity V_{pss} reduces, but it still may exceed one Figure 4a). As for the case without slip, the equilibrium velocity grows monotonically as the Reynolds number, n and Bn deviate from Newtonian values.

The particles free rotation rate does not change much in the Newtonian fluid, which is impossible to say about the power–law fluid where the increased limiting stress leads to a reduced dependence between the rotation rate and the Reynolds number (Figure 4b).

The lift force coefficient monotonously decreases as the rheological parameters deviate from Newtonian ones (Figure 5). However, lines for different fluid types are placed not as close as in the case in Figure 2b. For a free rolling particle in a Newtonian fluid, the lift force coefficient deviates by 20% in comparison to the particle at rest (Figure 5b). When fluid properties deviate from the Newtonian values a significant reduction of this ratio is observed. There are two opposite factors that determine the lift force in a shear flow of a Newtonian fluid. Increase in the velocity leads to decrease of the lift force and the rotation leads to its increase. But for a non-Newtonian fluid increase of rotation leads to a decrease of apparent viscosity and to significant losses in the lift force.

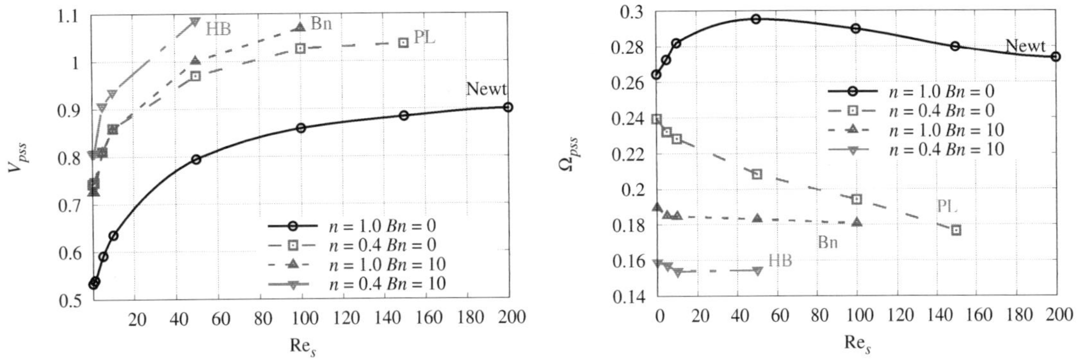

a b

Figure 4. Terminal velocity and rotation rate vs the Reynolds number Re_s and fluid properties.

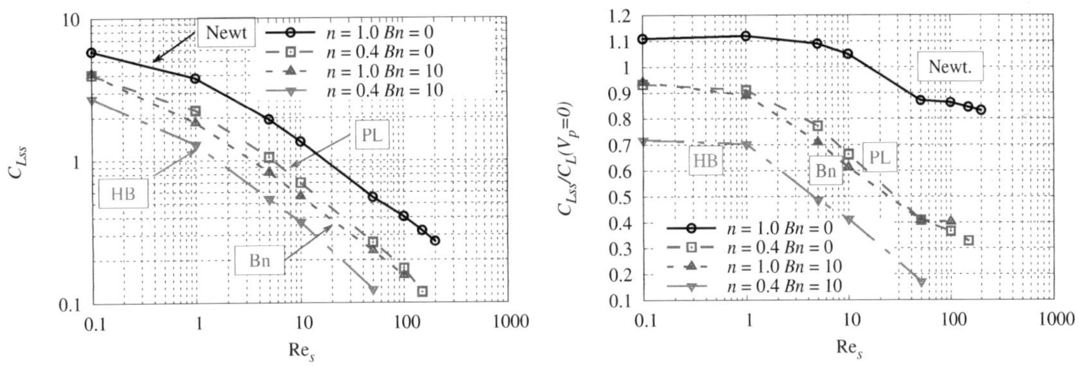

a b

Figure 5. Terminal lift force coefficient vs the Reynolds number Re_s and fluid properties.

4.3. Critical shear stress

Using forces and torque we could formulate a criterion for incipient motion and particle lift-off. Lets consider it for a case with particles of the same size. In Figure 6 you can see the scheme

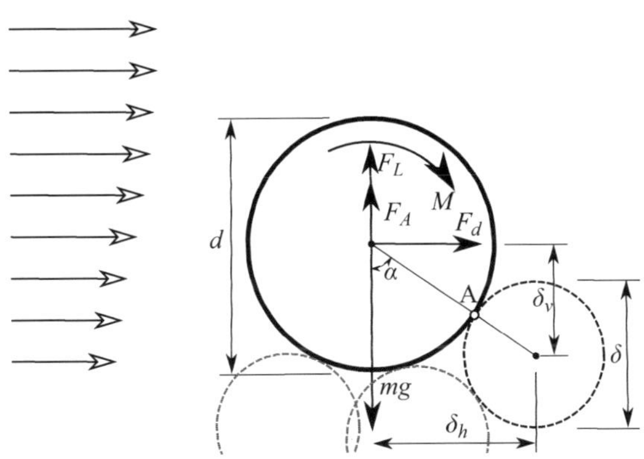

Figure 6. Sphere on a cuttings bed.

that we used to obtain force and torques balance equations for a spherical particle on a sediment bed. Consider the equilibrium of force moments relative to point A:

$$M + F_d\delta_v + F_L\delta_h - F_A\delta_h - mg\delta_h \geq 0$$

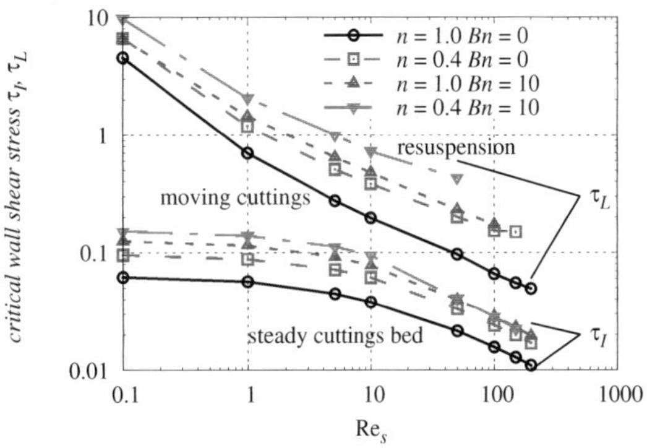

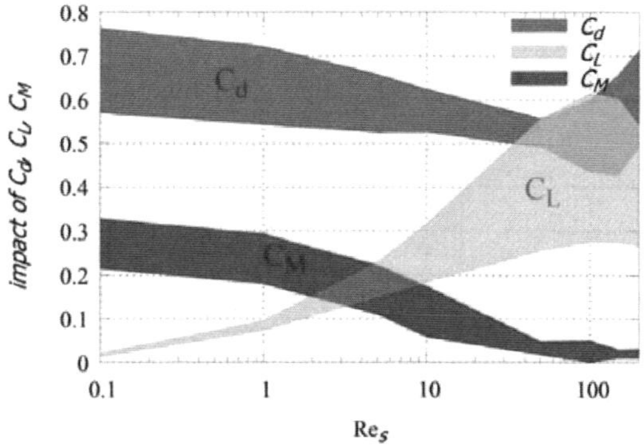

Figure 7. The critical Shields parameter vs the Reynolds number Re_s and fluid properties.

Figure 8. Relative contribution of different factors in incipient motion criterion.

For the spheres of the same size $\alpha \approx 60°$. Now, an equation for the Shields number can be obtained, which is derived as a ratio of the shear stress at the top of the bed to the lift force acting on the particle $\tau_I = \widetilde{\tau}_{wall}/(\rho_f - \rho_p)gd$ and previous parameters C_d, C_L, C_M and Re_s:

$$\tau_I = \frac{8}{\sqrt{3}Re_s}\left(C_d + \sqrt{3}C_L + 2C_M\right)^{-1}.$$

Let us consider the contribution of each factor (C_d, C_L and C_M) in as a function of τ_I. From Figure 8 we can say that for all types of fluids and a wide range of the Reynolds numbers the main input is given by the drag force C_d. The input of the lift force is negligible for small Re_s, but with growing Re_s the impact of C_L stands to be compatible with C_d. The torque C_M decreases as Re_s increases and its input to the criterion becomes negligible at the moderate Reynolds numbers.

The next phase of particle motion is its lift-off from the sediment bed. Neglecting the interaction with other particles, lets consider the lift force for a single particle while free motion. From Newtons law, we derive:

$$F_L - mg \geq 0$$

In its dimensionless form this inequality produces a criterion for contact loss in terms of Shields parameter τ_L:

$$\tau_L = \frac{8}{3C_L Re_s}$$

The effects of suggested criteria are shown on Figure 7. The domain of shear stresses τ on the wall bed now divided into three sub-domains: stationary sludge settlings ($\tau < \tau_I$), moving sludge settlings ($\tau_I \leq \tau < \tau_L$) and stirring-up (roiling) ($\tau \geq \tau_L$).

References

[1] Muthanna C, Nieuwstadt F T M and Hunt J C R 2005 *Experiments in Fluids* **39**(2)
[2] Krishnan G P and Leighton D T 1995 *Physics of Fluids* **7**(11)
[3] Bretherton F P 1962 *Journal of Fluid Mechanics* **14**(02)
[4] Goldman A, Cox R and Brenner H 1967 *Chemical Engineering Science* **22**(4)
[5] Goldman A, Cox R and Brenner H 1967 *Chemical Engineering Science* **22**(4)
[6] King M R, Leighton J and T D 1997 *Physics of Fluids* **9**(5)
[7] Takemura F and Magnaudet J 2003 *Journal of Fluid Mechanics* **495**
[8] Zeng L, Balachandar S and Fischer P 2005 *Journal of Fluid Mechanics* **536**
[9] Zeng L, Najjar F, Balachandar S and Fischer P 2009 *Physics of Fluids* **21**(3)
[10] Lee H and Balachandar S 2010 *Journal of Fluid Mechanics* **657**
[11] Lee H, Ha M Y and Balachandar S 2011 *International Journal of Multiphase Flow* **37**(2)
[12] Lee H and Balachandar S 2012 *Journal of Geophysical Research* **117**(F1)
[13] Bocharov O B and Kushnir D Y 2016 *Thermophysics and Aeromechanics* **23**(1)

IOP Conf. Series: Journal of Physics: Conf. Series **894** (2017) 012038 doi:10.1088/1742-6596/894/1/012038

On exact solutions of a heat-wave type with logarithmic front for the porous medium equation

A L Kazakov[1], A A Lempert[1], S S Orlov[2] and S S Orlov[1]

[1]Matrosov Institute for System Dynamics and Control Theory SB RAS, Irkutsk, Russia
[2]Irkutsk State University, Irkutsk, Russia

E-mail: kazakov@icc.ru, lempert@icc.ru

Abstract. The paper deals with a nonlinear second-order parabolic equation with partial derivatives, which is usually called "the porous medium equation". It describes the processes of heat and mass transfer as well as filtration of liquids and gases in porous media. In addition, it is used for mathematical modeling of growth and migration of population. Usually this equation is studied numerically like most other nonlinear equations of mathematical physics. So, the construction of exact solution in an explicit form is important to verify the numerical algorithms. The authors deal with a special solutions which are usually called "heat waves". A new class of heat-wave type solutions of one-dimensional (plane-symmetric) porous medium equation is proposed and analyzed. A logarithmic heat wave front is studied in details. Considered equation has a singularity at the heat wave front, because the factor of the highest (second) derivative vanishes. The construction of these exact solutions reduces to the integration of a nonlinear second-order ordinary differential equation (ODE). Moreover, the Cauchy conditions lead us to the fact that this equation has a singularity at the initial point. In other words, the ODE inherits the singularity of the original problem. The qualitative analysis of the solutions of the ODE is carried out. The obtained results are interpreted from the point of view of the corresponding heat waves' behavior. The most interesting is a damped solitary wave, the length of which is constant, and the amplitude decreases.

1. Introduction

We consider the nonlinear porous medium equation [1, 2] having the form

$$T_t = \mathrm{div}(k\nabla T), \quad k = k(T)$$

in the case of $k = k_0 T^\sigma$, $\sigma \in \mathbb{R}^{>0}$

$$T_t = k_0 \mathrm{div}(T^\sigma \nabla T). \tag{1}$$

Here $T = T(t, \mathbf{x})$ is an unknown function; t is time, $\mathbf{x}$ is a vector of spatial variables. Operators div and ∇ act on $\mathbf{x}$.

The existence of plane symmetry allows us to reduce (1) to one-dimensional equation

$$u_t = uu_{\rho\rho} + \frac{u_\rho^2}{\sigma}, \tag{2}$$

where $u = u(t, \rho)$ is a unknown function, depending on the time variable t and space variable ρ.

Content from this work may be used under the terms of the Creative Commons Attribution 3.0 licence. Any further distribution of this work must maintain attribution to the author(s) and the title of the work, journal citation and DOI.
Published under licence by IOP Publishing Ltd

$$u|_{\rho=\mathcal{F}(t)} = 0, \tag{3}$$

where $\rho - \mathcal{F}(t) = 0$ is a front of the heat wave, defined in the plane of the variables (t, ρ).

One of the classes of solutions of the porous medium equation is the heat waves (wave filtering) propagating on cold (zero temperature) background with a finite velocity. Geometrically the heat-wave type solution consists of two surfaces (perturbed solution $u(t, x) \geq 0$ and zero temperature background $u \equiv 0$), which are continuously joined along some sufficiently smooth line $x = b(t)$, called the front. In the linear case such solutions are apparently, known, since the time of J. Fourier.

It was Ya.B. Zeldovich, who first obtained solutions of (1), in the form of a heat wave for nonlinear heat conductivity problem [3]. In [4] similar results were obtained for filtration problems by G.I. Barenblatt. Heat waves in the class of piecewise-analytic functions were first considered by A.F. Sidorov [5, 6]. Using the methodology [6] the existence and uniqueness theorems are proved for solutions of one-dimensional [7, 8] and two-dimensional [9] problems. The method of power series [10] for the porous medium equation with special boundary conditions is applied in [11].

To solve (2)-(3) we reduce it to the Cauchy problem for nonlinear ordinary differential equations of second order with a singularity at the highest derivative. Such solutions of nonlinear partial differential equations in the literature are called "the exact solutions" [12, 13].

2. Construction of Exact Solutions

We assume that the solution of (2), (3) has the following form

$$u(t, \rho) = \psi(t, \rho) w(\xi), \quad \xi = \xi(t, \rho), \tag{4}$$

where $w(\xi)$ is an one-variable function, $\psi(t, \rho)$ is an elementary function, $\xi_t \xi_\rho \neq 0$.

Substituting (4) to (2) we have

$$\begin{aligned} &ww'' + \frac{1}{\sigma}(w')^2 + \left[2\left(\frac{1}{\sigma} + 1\right)\frac{\psi_\rho}{\psi\xi_\rho} + \frac{\xi_{\rho\rho}}{\xi_\rho^2}\right] ww' + \\ &+ \left(\frac{1}{\sigma}\frac{\psi_\rho^2}{\psi^2\xi_\rho^2} + \frac{\psi_{\rho\rho}}{\psi\xi_\rho^2}\right) w^2 - \frac{\xi_t}{\psi\xi_\rho^2} w' - \frac{\psi_t}{\psi^2\xi_\rho^2} w = 0. \end{aligned} \tag{5}$$

In order that (5) becomes an ODE we should solve an overdetermined system of partial differential equations

$$\frac{\psi_\rho}{\psi\xi_\rho} = a_1, \; \frac{\xi_{\rho\rho}}{\xi_\rho^2} = a_2, \; \frac{\psi_{\rho\rho}}{\psi\xi_\rho^2} = a_3, \; \frac{\xi_t}{\psi\xi_\rho^2} = a_4, \; \frac{\psi_t}{\psi^2\xi_\rho^2} = a_5, \tag{6}$$

where $a_l \in \mathbb{R}$, $l = \overline{1, 5}$.

Lemma 1. Let $a_1 = -2a_2$, $a_3 = 2a_2^2$, then

(i) system (6) is solvable;

(ii) Eq. (2) has an exact solution having form

$$u(t, \rho) = -\alpha^2 f'(t) v(\xi), \quad \xi = \frac{\rho}{\alpha} + f(t), \quad f(t) = \left[\begin{array}{l} c_1 t + c_2, \\ \ln(c_1 t + c_2)^{-\omega}, \end{array}\right. \tag{7}$$

where $v(\xi) = -w(\xi)/a_4, a_4 \neq 0$;

(iii) $v(\xi)$ satisfies the following ODE

$$vv'' + \frac{(v')^2}{\sigma} + v' + kv = 0. \tag{8}$$

Here $k = 0$, if $f(t) = c_1 t + c_2$ or $k = 1/\omega$, if $f(t) = \ln(c_1 t + c_2)^{-\omega}$.

The validity of Lemma 1 follows from the results presented in [14].

In order the solution of Eq. (2) in the form (7) satisfies the boundary-value condition (3) we have to append condition $v|_{\xi=0} = 0$. Therefore $\mathcal{F}(t) = c_1 t + c_2$ or $\mathcal{F}(t) = \ln(c_1 t + c_2)^{\alpha\omega}$. In the first case we have a linear solution $u(t, \rho) = -\sigma c_1(\rho - c_1 t - c_2)$.

Now we consider the second case in more detail. Eq. (8) becomes degenerate, because the highest derivative coefficient vanishes. To ensure it's solvability it is necessary to satisfy the condition $v'|_{\xi=0} = v_1 \in \{-\sigma, 0\}$.

So, Cauchy conditions for Eq. (8) are

$$v|_{\xi=0} = 0, \quad v'|_{\xi=0} = v_1 \in \{-\sigma, 0\}. \tag{9}$$

The solutions of (8)-(9) are concave functions (see Figure. 1). It is easy to see that the global maximum is $v = 0$ in the case $v_1 = 0$ (Figure 1b).

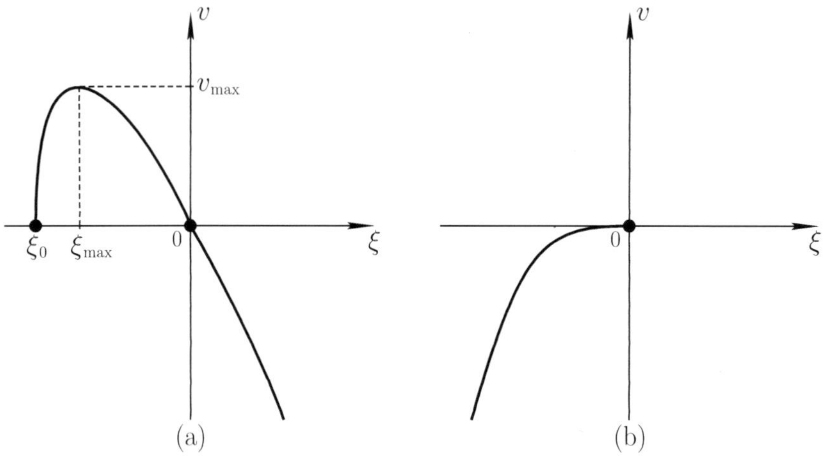

Figure 1. The behavior of the solution of (8)-(9): (a) for $v_1 = -\sigma$; (b) for $v_1 = 0$.

In the case $v_1 = -\sigma$ we have following Lemma.

Lemma 2. The following inequalities hold

$$-[\sqrt{\sigma(\sigma+1)} + \sigma + 1]\omega \leq \xi_0 \leq -\sigma\omega,$$
$$-(\sigma+1)\omega \leq \xi_{\max} \leq -\sigma\omega, \quad \frac{\sigma^2\omega}{2} \leq v_{\max} \leq \frac{\sigma(\sigma+1)\omega}{2}.$$

3. On a heat wave with logarithmic front

Here we investigate the behavior of heat-wave type solutions with a logarithmic front. Let the heat wave moves from the origin to the right (to the first octant). This means that for function $\mathcal{F}(t) = \ln(c_1 t + c_2)^{\alpha\omega}$ (see (7)), which determines the law of the front movement following conditions are held

$$\mathcal{F}|_{t=0} = 0, \mathcal{F}'|_{t=0} \geq 0,$$

and $c_2 = 1$, $\alpha\omega c_1 > 0$. Then, in accordance with (7) we have

$$u(t,\rho) = \frac{\alpha^2\omega}{t+1/c_1} v(\xi), \quad \xi = \frac{\rho}{\alpha} + \ln(c_1 t + 1)^{-\omega}. \tag{10}$$

Assuming (without loss of generality) that $\omega > 0$, we consider the two cases: $\omega > 0$, $\alpha > 0$, $c_1 > 0$ and $\omega > 0$, $\alpha < 0$, $c_1 < 0$.

A. Let $\omega > 0, \alpha > 0$, $c_1 > 0$. Obviously, in this case, if $v(\xi) \geq 0$, then $u(t,\rho) \geq 0$, therefore it is sufficient to consider the non-negative part of solution of (8) with respect to $v_1 = -\sigma$ (see Figure 1a). Because of $\xi_0 \leq \xi \leq 0$, then from (10) we obtain the field of heat wave propagation $\ln(c_1 t + 1)^{\alpha\omega} + \alpha\xi_0 \leq \rho \leq \ln(c_1 t + 1)^{\alpha\omega}$. Here we can see, that the heat wave has the back front $\rho = \ln(c_1 t + 1)^{\alpha\omega} + \alpha\xi_0$.

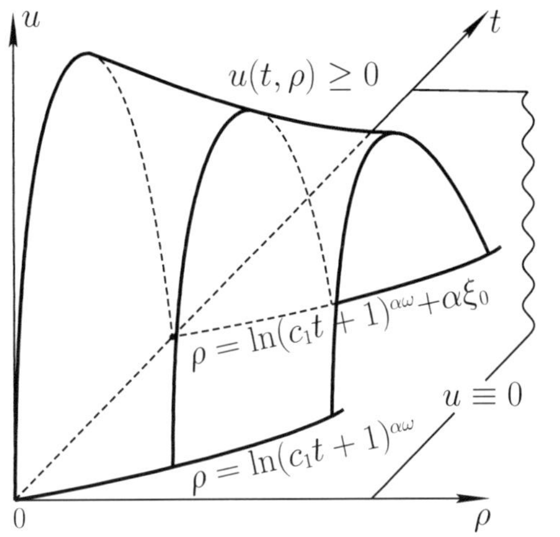

Figure 2. The heat wave with front $\rho = \ln(c_1 t + 1)^{\alpha\omega}$, $\alpha\omega > 0$, $c_1 > 0$.

Figure 2 shows schematically the behavior of a heat wave with a logarithmic front. The velocity of the main and back fronts decreases with time and tends to zero. From (10) and in view of the boundedness of $v(\xi)$ in $[\xi_0, 0]$ we obtain that $\lim_{t\to+\infty} u(t,\rho) = 0$ for each ρ in the field of heat wave propagation. So, the complete cooling-down of the heated half-space occurs in an infinite time.

Using methods of power geometry [15, 16] we obtain the following relation

$$u(t,\rho) \underset{\xi\to 0}{\sim} -\frac{\alpha^2\sigma\omega}{t+1/c_1}\left[\frac{\rho}{\alpha} + \ln(c_1 t + 1)^{-\omega}\right]. \tag{11}$$

It characterizes the behavior of the heat wave near the main front.

For any fixed ρ the solution (10) has a maximum $u_{\max}$ at time $t_{\max}$. The inequalities below follow from (10) and Lemma 2.

$$0 \leq u_{\max} \leq \frac{\alpha^2\sigma(\sigma+1)\omega^2 c_1}{2} e^{-\frac{\rho}{\alpha\omega}}, \quad \frac{e^{\frac{\rho}{\alpha\omega}} - 1}{c_1} \leq t_{\max} \leq \frac{e^{\frac{\rho}{\alpha\omega}+\sigma+1} - 1}{c_1}.$$

B. Let $\omega > 0$, $\alpha < 0$, $c_1 < 0$. Then $0 \leq t < -1/c_1$ and $t = -1/c_1$ is the vertical asymptote of the front. According to (10) the inequality $u(t,\rho) \geq 0$ holds, if $v(\xi) \leq 0$. Consequently, we have the following subcases.

B1. We consider the nonpositive part of the solution $v(\xi)$ of Cauchy problem (8) for $v_1 = -\sigma$ (see Figure 1a). Here $\xi \geq 0$, so from (10) we obtain the domain $\rho \leq \ln(c_1 t + 1)^{\alpha\omega}$ where heat wave exists. The heating of considered half-space to an arbitrary high temperature is carried out in finite time and there is no localization of heat. A similar situation corresponds to the HS blow-up regime [17], where $0 < -1/c_1 < +\infty$ is blow-up time (see Figure 3a). For any fixed ρ the following relation holds

$$u(t,\rho) \underset{t\to -1/c_1-0}{\sim} -\frac{\alpha^2\sigma \ln^2(c_1 t+1)^{\omega}}{2(\sigma+2)(t+1/c_1)}.$$

The asymptotic formula (11) describes the heat wave behavior in the neighborhood of the front for $\xi \to 0$.

B2. Here we consider the nontrivial solution $v(\xi)$ of Cauchy problem (8) for $v_1 = 0$ (see Figure 1b). Here $\xi \leq 0$, so from (10) we obtain the field of heat wave propagation $\rho \geq \ln(c_1 t+1)^{\alpha\omega}$. Such heat wave can not be generated by any boundary regime if $\rho = 0$ and it is not a solution of the Sakharov problem (Figure 3b). However, it is easy to see that the constructed heat wave is a solution of some initial problem for Eq. (2). In this case, the solution tends to a cold background faster than any surface having form

$$u_n(t,\rho) = \frac{\alpha^2\omega}{t+1/c_1}\left[\frac{\rho}{\alpha} + \ln(c_1 t+1)^{-\omega}\right]^n, \; n \in \mathbb{N}.$$

For each fixed $t \in [0, -1/c_1)$, the following relation holds

$$u(t,\rho) \underset{\rho\to +\infty}{\sim} -\frac{\sigma\rho^2}{2(\sigma+2)(t+1/c_1)}.$$

A detailed proof is carried out by the methods of power geometry [15, 16] as well. It is quite cumbersome and will be the subject of a separate article.

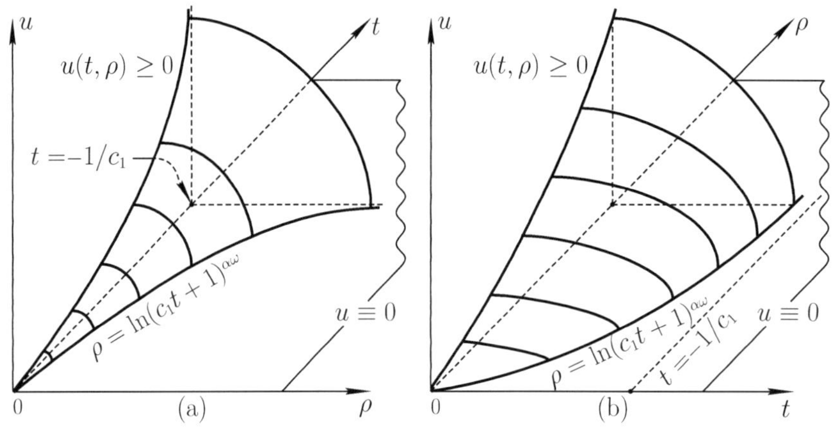

Figure 3. The heat wave with front $\rho = \ln(c_1 t+1)^{\alpha\omega}$ for $\alpha\omega < 0$, $c_1 < 0$: (a) blow up regime; (b) without boundary regime.

4. Conclusion

Summarizing the research, it can be noted that the main purpose of the study is to construct and analyze special exact solutions of the one-dimensional porous medium equation. Constructing of exact solutions of nonlinear equations of mathematical physics is one of the most important problem in modern continuum mechanics. In particular, comparison of the results of calculations with exact solutions is one of the most effective ways of verifying numerical algorithms.

The proposed method for exact solutions constructing is based on the transformation to new dependent and independent variables. It usually aims to reduce the number of independent variables. Thus, in the case of two independent variables, the construction of exact solutions reduces to the integration of ordinary differential equations.

In this paper we address the case when the heat wave front moves in the plane of the variables (ρ, t) according to the logarithmic law. The construction of the exact solution is reduced to the Cauchy problem for the second-order ordinary differential equation (ODE). This problem inherits a singularity of the original one, and therefore, it is solvable not for all values of the derivative at the initial point. The qualitative analysis of the solutions of the ODE is carried out. The obtained results are interpreted from the point of view the heat waves behavior. It is shown that the logarithmic law of the front moving corresponds to three possible configurations of the heat waves. The most interesting is a damped solitary wave, the length of which is constant, and the amplitude decreases.

Further research can be related both to the study of other laws of heat front movement (for example power or exponential), and to the consideration of cylindrical and spherical symmetries.

Acknowledgments
The reported study was particulary funded by RFBR according to the research projects No. 16-01-00608 and No. 16-31-00291.

References

[1] Vazquez J 2007 *The Porous Medium Equation: Mathematical Theory* (Oxford: Clarendon Press)
[2] Antontsev S and Shmarev S 2015 *Evolution PDEs with Nonstandard Growth Conditions* (Amsterdam: Atlantis Press)
[3] Zeldovich Y and Raizer Y 1966 *Physics of Shock Waves and High-Temperature Hydrodynamic Phenomena* vol 2 (New York: Academic Press)
[4] Barenblatt G, Entov V and Ryzhik V 1990 *Theory of Fluid Flow of Fluids Through Natural Rocks* (Dordrecht/Boston/London: Kluwer Academic Publ.)
[5] Vasin V and Sidorov A *Izvestiya Vysshikh Uchebnykh Zavedenii. Matematika.* **7** 13
[6] Sidorov A *Dokl. Akad. Nauk SSSR.* **280** 47
[7] Kazakov A and Lempert A *J. Appl. Mech. Tech. Phys.* **54** 97
[8] Kazakov A and Lempert A 2016 *J. Phys.: Conf. Series* **722** 012016
[9] Kazakov A and Kuznetsov P *J. Appl. Ind. Math.* **8** 227
[10] Teshukov V *Journal of Applied Mechanics and Technical Physics* **21** 261
[11] Furtak-Cole E, Telyakovskiy A and Cooper C 2013 *AGU Fall Meeting Abstracts*
[12] Galaktionov V and Svirshchevskii S 2006 *Exact Solutions and Invariant Subspaces of Nonlinear Partial Differential Equations in Mechanics and Physics* (New-York: Chapman and Hall/CRC)
[13] Meleshko S 2005 *Methods for Constructing Exact Solutions of Partial Differential Equations* (New-York: Springer)
[14] Kazakov A, Lempert A, Orlov S and Orlov S *CEUR Workshop Proceedings* **1839** 344
[15] Bruno A 2000 *Power Geometry in Algebraic and Differential Equations* (Amsterdam: Elsevier Science)
[16] Bruno A *Russian Math. Surveys* **59** 429
[17] Samarskii A, Galaktionov V, Kurdyumov S and Mikhailov A 1995 *Blow-up in Quasilinear Parabolic Equations* (Berlin; New-York: de Gruyter)

IOP Conf. Series: Journal of Physics: Conf. Series **894** (2017) 012039 doi:10.1088/1742-6596/894/1/012039

Peculiarity of she shaped-charge liner collapse concerning the unevenness in its cross-section

K A Karnaukhov, V D Baskakov, V V Korenkov and **O V Zarubina**
Department of Physics, Bauman Moscow State Technical University, Moscow, Russia

e-mail: karnaukhov93@mail.ru

Abstract. For harmonical components of the unevenness, simulating the misalignment of the inner and the outer surfaces of the shaped-charge liner the functional connections to estimate the angular deviation of the shaped-charge jets from symmetry axis were developed. Numerical calculations in CAE system ANSYS AUTODYN for the process of collapse of the ring with the consideration of alteration of its thickness in accordance with harmonical components of the unevenness were ran to estimate the distortion of cross-sectional shape of the shaped-charge liner throughout the process of collapse. Obtained results can be used in quantitative estimations of the influence of the inaccuracies of the shaped-charge liner on geometrical and kinematical characteristics of the shaped-charge jets.

Inaccuracies in the configuration of the shaped-charge (SC) have a great influence on the shaped-charge jet (SCJ) behavior: it deviates, the cross-sectional shape distorts (in case of axial symmetry) and the penetration depth decreases.

The majority of well-known mathematical models describing the influence of inaccuracies in the configuration of the SC on the behavior of SCJ, focuses on estimation of deviation in plane case (i.e., [3-6]), but pays little attention to the distortion of the cross-sectional shape.

Works [1,7] show that the inaccuracies in configuration of SC have complex structure, which can be described by the trigonometrical Fourier's series. For example thickness of the shaped-charge liner (SCL) in the plane of the cross-section can be represented by the following functional connection

$$\delta(\varphi) \approx \delta_0 + \sum_{i=1}^{I} \Delta\delta_i cos\,(i\varphi + \psi_i) \quad (1)$$

in which δ_0 – the nominal thickness of SCL, $\Delta\delta_i$ – the value of an amplitude of the $i-$ th harmonic of the inaccuracy of the SCL's thickness, , $i = 1,2 \dots I$ – the number of the harmonic, I – the number of the last valuable harmonic, $0 \le \varphi \le 2\pi$ – angular coordinate in the plane of a cross-section of the SCL, $0 \le \psi_i \le 2\pi$ – phase of the $i-$ th harmonic.

In addition, it has been discovered that the faceting of the SCJ caused by the harmonics with $i = 2, 3$ may have a valid influence on the ultimate elongation [1].

The purpose of this work is the development of the simulative representation of the means of an influence of the specific harmonical component (1) of the inaccuracy of SCL's thickness on the behavior of the SCJ in axially symmetrical case.

The collapse of the SCL in case of the only one harmonical component $i = 1$ in the structure of the inaccuracy of the SCL's thickness may be represented by the convergent movement of conical "sheet", that has misalignment $\Delta\delta_1$ of inner and outer surfaces, moreover, $\Delta\overline{\delta_1} = \frac{\Delta\delta_1}{\delta_0} = const$ for every cross-section of the conical "sheet" (Figure 1).

Content from this work may be used under the terms of the Creative Commons Attribution 3.0 licence. Any further distribution of this work must maintain attribution to the author(s) and the title of the work, journal citation and DOI.
Published under licence by IOP Publishing Ltd

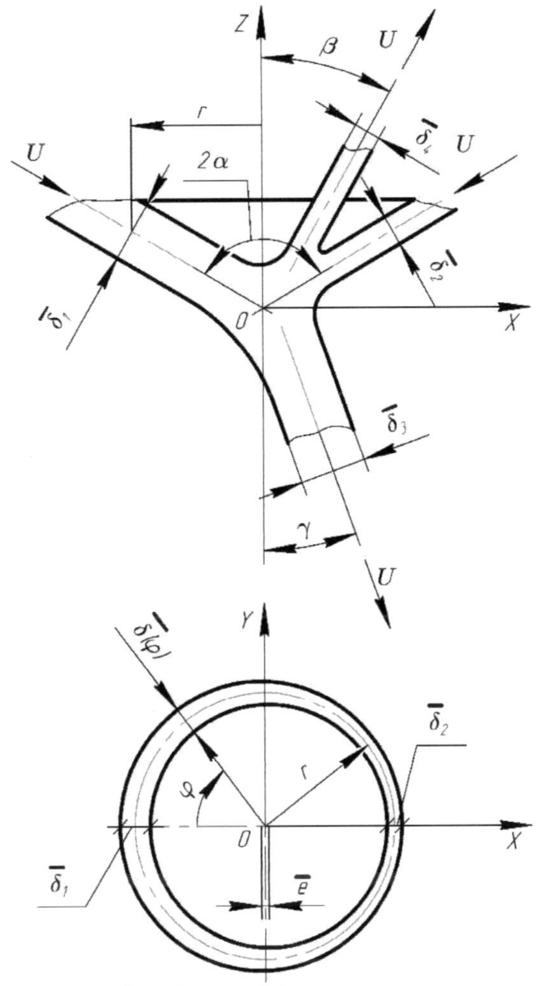

Figure 1. The scheme of the convergent movement of the SCL's conical "sheet"

The legend of the Figure 1 is as follows: U – the velocity of convergent movement of the "sheet"; $\overline{\delta_1} = \frac{\delta_1}{\delta_0} = 1 + \Delta\overline{\delta_1}, \overline{\delta_2} = \frac{\delta_2}{\delta_0} = 1 - \Delta\overline{\delta_1}$ – crest and narrow thickness of the SCL in the cross-section; $\overline{\delta_3} = \frac{\delta_3}{\delta_0}$ – slug's diameter (the outer jet); $\overline{\delta_4} = \frac{\delta_4}{\delta_0}$ – jet's diameter (the inner jet); $\bar{e} = \Delta\overline{\delta_1}$ – eccentricity of the outer and the inner and the outer surfaces; $\bar{\delta}(\varphi) = \frac{\delta(\varphi)}{\delta_0}$; 2α – an angle of the conical "sheet"; β, γ – the angular deviation of the jet and the slug; r – the radius of a random cross-section of the conical sheet.

The solution to a problem of the determination of the angular deviation β, γ of the jet and the slug is similar to the solution of the problem of impingement of two asymmetric plane jets of an ideal fluid [3].

Write down the law of conservation of momentum along the OX and OY axes:

$$2\bar{r}cos\alpha = (\frac{\overline{\delta_3}}{2})^2 cos\gamma - (\frac{\overline{\delta_4}}{2})^2 cos\beta, \quad (2)$$

$$\bar{r}\Delta\overline{\delta_1}sin\alpha = (\frac{\overline{\delta_3}}{2})^2 sin\gamma + (\frac{\overline{\delta_4}}{2})^2 sin\beta.$$

IOP Conf. Series: Journal of Physics: Conf. Series **894** (2017) 012039 doi:10.1088/1742-6596/894/1/012039

The system of equations (2) consists of 4 unknown variables $\overline{\delta_3}$, $\overline{\delta_4}$, β, γ. To find its solution we define the independence of the diameters $\overline{\delta_3}$ of the jet and $\overline{\delta_4}$ the slug from eccentricity $\bar{e}$ are determined by well-know law in case when $\Delta\overline{\delta_1} = 0$ [2, 3].

$$\overline{\delta_3} = 2\sqrt{2\bar{r}}cos\left(\frac{\alpha}{2}\right), \qquad (3)$$

$$\overline{\delta_4} = 2\sqrt{2\bar{r}}sin\left(\frac{\alpha}{2}\right).$$

Using (3) in (2) and taking into consideration the smallness of the angles β and γ (that leads to $sin\beta \approx \beta$, $sin\gamma \approx \gamma$, $cos\beta \approx 1 - \beta^2/2$, $cos\gamma \approx 1 - \gamma^2/2$) we obtain

$$\beta = \Delta\overline{\delta_1}f_1(\alpha)\,,\ \ \gamma = \Delta\overline{\delta_1}f_2(\alpha) \qquad (4)$$

in which $f_1(\alpha) = \frac{1}{1+tg\frac{\alpha}{2}}$, $f_2(\alpha) = \frac{tg\frac{\alpha}{2}}{1+tg\frac{\alpha}{2}}$.

Comparison of the results for axially symmetrical and plane cases [3] shows the fact that if the inaccuracies $\Delta\overline{\delta_1}$ in plane and axially symmetrical cases are equal the angular deviation in the plane case is twice as big as in the axially symmetrical case, which means that axially symmetrical jets are less sensitive to the inaccuracies of the thickness of the SCL, than plane.

Considering the identity of the approach of the mathematical modeling to the estimation of angular deviation of the jet and the slug in plane and axially symmetrical cases we can use the functional connection that approximates the results of numerical simulations, that was obtained in the work [3], to increase the accuracy of the estimation.

$$\beta = \Delta\overline{\delta_1}f(\alpha) \qquad (5)$$

where $f(\alpha) = 3(2\alpha)^{10}e^{-4,43(2\alpha)}$

For the estimation of the influence of harmonical components $i = 1, 2, 3$ etc on the deformation in the cross-section of the SCJ the numerical simulation of the collapse of the plane ring of a compressible fluid (water) in Lagrangian coordinates was analyzed by the means of CAE system ANSYS AUTODYN. The inner diameter of plane ring was 31 mm, the thickness of the ring's wall – 1.8 mm. The amount of cells on the thickness and the length of circular element was 4x32. The outer surface of the ring was under pressure of $P = 30\ GPa$. Values of the inaccuracies of the thickness were 0.15, 0.25, 0.33, 0.5.

Figure 2, 3 represent the form of ring elements of numerical simulations, circular elements are shown on the field of pressure. Analysis of these results shows that the distortion of the shape of the inner surface of the ring, that is responsible for the SCJ's cross-sectional shape occurs at any number of harmonic i. It was learned that at the great values of harmonical amplitudes the inner contour distorts tremendously, ensuring the contribution into the establishment of the number of centers of the SCJ formation.

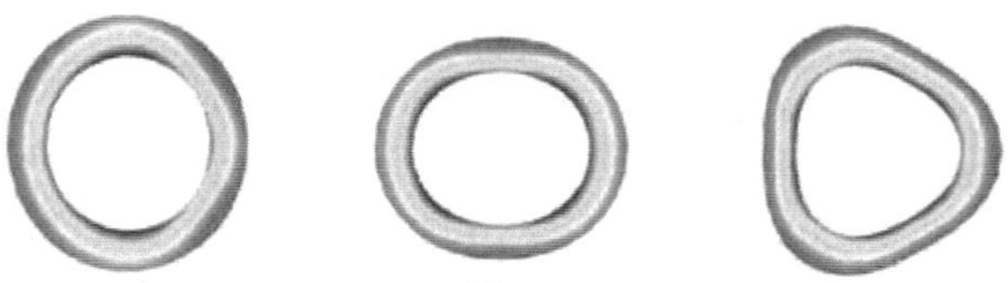

a) b) c)

Figure 2. Shape of the circular elements at the moment of collapse $t \approx 10\ \mu s$, inaccuracy of the thickness $\overline{\Delta\delta_i} = 0.15$: a) – $i = 1$, b) – $i = 2$, c) – $i = 3$

IOP Conf. Series: Journal of Physics: Conf. Series **894** (2017) 012039 doi:10.1088/1742-6596/894/1/012039

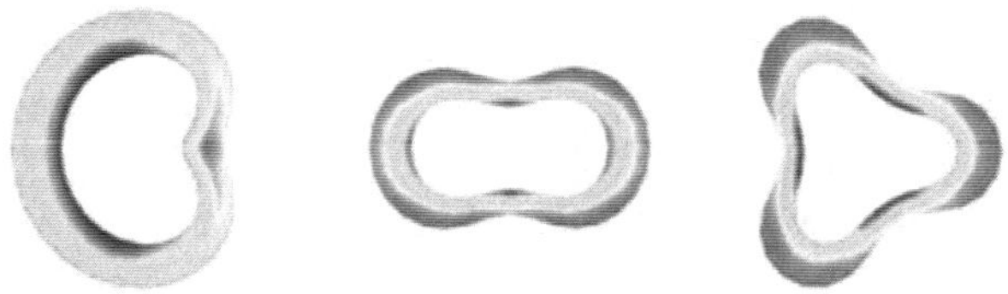

a) b) c)

Figure 3. Shape of the circular elements at the moment of collapse $t \approx 10\ \mu s$, inaccuracy of the thickness $\overline{\Delta\delta_i} = 0.5$: a) – $i = 1$, b) – $i = 2$, c) – $i = 3$

The $\bar{a}$ as the ratio of the diameter a of the circle, inscribed into the inner contour of the ring to diameter b of the circle, described in the inner contour of the ring. Figure 4 shows the dependency graphs of $\bar{a} = \bar{a}(a)$ for harmonics $i = 1, 2, 3$ in the course of a ring collapse.

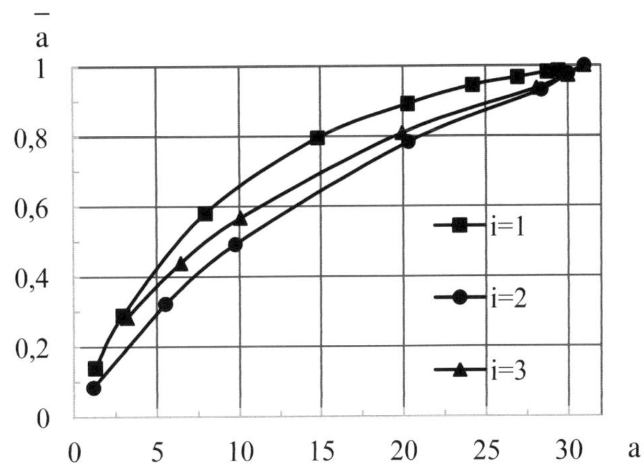

a)

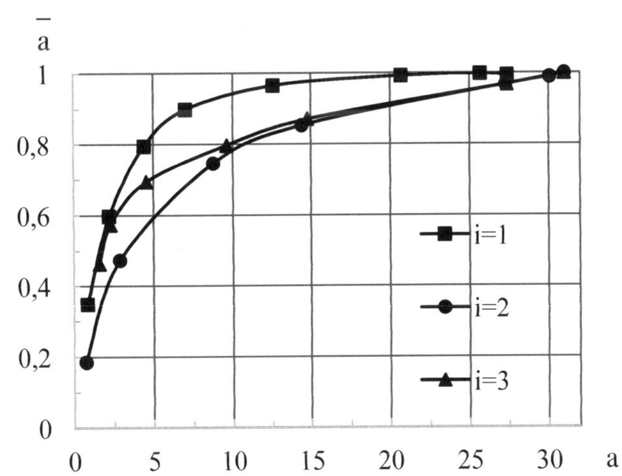

b)

Figure 4. Dependency graphs of $\bar{a} = \bar{a}(a)$ for harmonics $i = 1, 2, 3$ when the intolerance of thickness is: a) $\overline{\Delta\delta_i} = 0.5$, b) $\overline{\Delta\delta_i} = 0.15$

Analysis of Figure 4 shows that in case of equal values of the amplitudes of intolerances $\Delta\delta_i$ the most distortion comes from the harmonic with $i = 2$.

Numerical simulations not for the compressible fluid but for copper also took place. In that case the plots of $\bar{a} = \bar{a}(a)$ for fluid and copper were completely aligned. It indicates that the density of the material has no influence on the formation of the SCJ.

Conclusion

1. Mathematical model for the estimation of the influence of the axially symmetrical SCL's intolerances on the angular deviation of SCJ was developed.
2. The fact that the harmonic $i = 1$ defining the misalignment of the inner and the outer surfaces of SCL leads not to the angular deviation of SCJ only, but to the distortion of its cross-sectional shape was established.
3. The fact that the great values of the amplitude of the inaccuracy $\overline{\Delta\delta_i}$ of the SCL may cause the establishment of a number of SCJ's formation centers was shown.
4. It shows that the greatest input in the distortion of SCJ's cross-sectional shape was made by the harmonic $i = 2$ of trigonomeetrical partition law of thickness of the lining in circumferential direction.

References

[1] Tarasov V A, Baskakov V D and Dubovskoy M A 1995 *Defensive equipment* **4** 54–59
[2] Orlenko L P 2002 Physics of explosion – Ed. **3** p 656
[3] Baskakov V D, Zarubina O V, Karnaukhov K A and Tarasov V A 2016 *Bauman Moscow State Technical University bulletin of Natural sciences series* **2** 79–90
[4] Baskakov V D and Karnaukhov K A 2016 *J. Phys.: Conf. Series* **731** 012002
[5] Curtis J P 1998 *Proceedings of the 17th International Symposium on Ballistics, Midrand, South Africa* **2** 405–412
[6] Dr. Simcha Miller and Mr. Roy Ceder 2001 *Proceedings of the 19th International Symposium on Ballistics, Interlaken, Switzerland* 843–850

Hypersonic boundary layer stabilization by using a wavy surface

S V Kirilovskiy[1,2] and T V Poplavskaya[1,2]

[1]Khristianovich Institute of Theoretical and Applied Mechanics SB RAS, 630090, Novosibirsk, Russia
[2]Novosibirsk State University, 630090, Novosibirsk, Russia

Abstract. Numerical simulation of hypersonic (M_∞=6) flow and evolution of disturbances on a smooth plate and a shallow grooved plate was performed by solving two-dimensional Navier–Stokes equations. Computational soft-ware verification was conducted by comparison with existing data of pressure pulsations on plates surface. It was showed that wavy surface significantly decrease pressure pulsations on plate surface and does not increase the value of mean heat fluxes. Data about effect of wavy surfaces with different form on the disturbances intensity in hypersonic boundary layer was obtained.

1. Introduction

Laminar-turbulent transition in the compressible boundary layers was the subject of research for many years [1]. It is known that the laminar–turbulent transition in hypersonic flows depends on the parameters of the mean flow and may be initiated by two instability modes. The first mode is associated with Tolmin–Schlichting waves, which are vortex perturbations, and the second one is associated with inviscid instability and it is acoustic in nature. The second or Mack [2] mode dominates when the Mach number is fairly high. In this case, acoustic perturbations are trapped by a boundary layer, grow in amplitude and eventually may cause nonlinear processes [3–5] and turbulization of the flow (laminar–turbulent transition).

In the past decade gas permeable porous materials has been actively used for suppression of acoustic disturbance in order to prolong laminar flow regime [6–9]. These phenomena are based on dissipation of the energy of acoustic disturbances, which are a dominating mode of instability of a hypersonic boundary layer, owing to friction in pores of the coating.

On the other hand, the second-mode instability can also be affected by local shaping of the body surface. It is well known that increasing Mach number produces a stabilization effect on the flow in free shear layers and wakes (e.g., [10, 11]). It is natural to assume that a relatively long free shear layer formed near a streamlined surface may decrease the growth rates and damp the second-mode disturbances with a short wave length (of the order of the layer thickness). This could be achieved with the help of a shallow, grooved, wavy surface, which produces a relatively stable free shear layer bridging neighboring cavities. This wavy-wall stabilization (WWS) concept was confirmed by experiment and 2-D DNS [12].

This method is new and the integral effect of wavy surface is not been sufficiently investigated. The main objective of this work is to examine the stabilization of boundary layer by wavy surfaces of different type. These studies are useful for prediction of laminar-turbulent transition on wavy walls which are arising naturally in the swelling of heat-shielding panels of aircraft.

Content from this work may be used under the terms of the Creative Commons Attribution 3.0 licence. Any further distribution of this work must maintain attribution to the author(s) and the title of the work, journal citation and DOI.
Published under licence by IOP Publishing Ltd

2. Numerical simulation of supersonic flow over smooth plate and plate with wavy surface

Numerical simulation of supersonic flow was carried out by solving unsteady 2D Navier–Stokes equations using the ANSYS Fluent program package. Computations were performed using parameters according to free stream conditions in experiments performed in the wind tunnel Transit-M ITAM SB RAS[12]: Mach number M_∞=6.0, free stream velocity U_∞= 790.38 m/s, unit Reynolds number $Re_1 = 10.5 \times 10^6$ m^{-1}, total pressure $P_0 = 7 \times 10^5$ Pa, free stream static pressure P_∞ = 445 Pa, free stream static temperature T_∞ = 43,08 K, total temperature T_0 = 354,06K, wall temperature T_w = 293 K. Air was considered as ideal gas with heat capacity ratio γ=1.4 and Prandtl number Pr = 0.72, molecular viscosity μ were set by Sutherland's law. The left (input) boundary is located at 10 computational cells upstream from the leading edge of plate, the right (output) boundary was shifted to 40 computational cells from the trailing edge of the plate so that the flow in the exit section was completely supersonic. The computational domain height was chosen under the condition that the bow shock wave did not interact with the upper boundary of the computational domain (36mm). Plate length was taken equal to 176mm which is smaller than experimental model (350mm) in aim to create more precise mesh in wavy surface region.

Three type of wavy surface were considered (Figure 1). In all cases the beginning of wave surface region *xb*=52mm, the end of wave surface region *xe* = 160mm. Wavy surface 1 has 9 grooves in arc form and were defined by the equation *y*=h(cos(pi(*x*-*xb*)/*l*-1), where h=1.8mm – height of wavy element, *l*=12ᴍᴍ – length of wavy element. This shape of wavy surface corresponds to the experimental model [12] and is shown in Figure 1a. Wavy surface 2 has 9 grooves in sawtooth form with height and length as in wavy surface 1 (Figure 1b). Wavy surface 3 was a sequence of 6 triangular cavities, the same as on the wave surface 2.

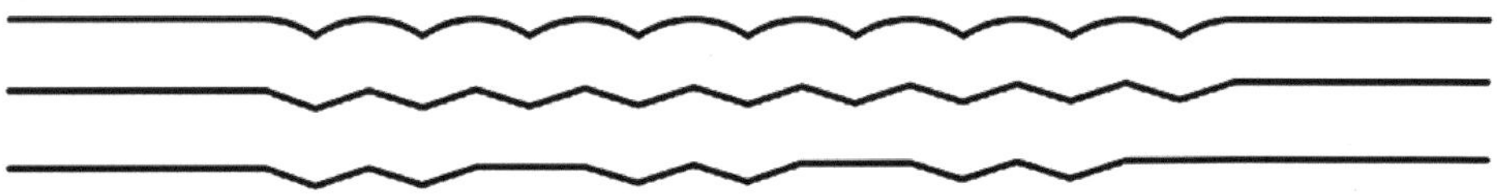

Figure 1. Plate with wavy surface of different shapes.

The drawing of model geometry and computational domain and creating of a regular computational grid were performed using ANSYS ICEM. The allocation of subdomain with a height of 0.003m was made near the surface of the plate, which corresponded to the thickness of the boundary layer (according to preliminary calculations) for detailed resolution of the boundary layer. Computational domain was divided into two subdomains: a subdomain with a height of 0.003 m, including boundary layer, with uniform grid and a subdomain outside this zone with a grid refinement toward the plate surface. The total number of meshes was 568000.

The computations were performed using the density-solver with explicit/implicit scheme of the second-order accuracy in space and the Roe-FDS method of convective flux splitting and explicit Runge–Kutta method. The problem of numerical simulation of a supersonic flow of air was solved in two stages. First, numerical simulation of steady flow was performed using implicit scheme in space. After that, unsteady disturbances of the periodic suction-blowing type are introduced locally (on the surface area from x_1 = 0.01m to x_2 = 0.015m). These disturbances were introduced by the periodic boundary condition on the transverse velocity component: $\upsilon_w(x,t) = A\sin\left(2\pi\frac{x-x_1}{x_2-x_1}\right)\sin(2\pi ft)$,

where f – frequency, $A = \varepsilon\rho_\infty U_\infty / \rho_w$ – amplitude of introduced disturbance. To ensure the linear evolution of introduced disturbances in front of the wavy region a small amplitude was chosen $\varepsilon = 2\times10^{-3}$. Boundary condition of the transverse velocity component was set at the region x_1x_2 on the wall surface by using UDF (User Defined Function). Up to ten processors of the Information-

Computational Center of the Novosibirsk State University and the Siberian Supercomputer Center were used in the computations.

3. Results

Primarily in test reason of computational method numerical simulation of flow over wavy plate 1 was performed to make comparison with computational data from [12]. The computed steady-flow Mach-number field over smooth plate and plate with wavy surface 1 are shown on figure 2a,b. In both cases viscous- inviscid interaction leads to formation of a weak shock wave near the leading edge region the slopes of which are the same. The cavities of the wavy surface induce oblique shocklets emanating

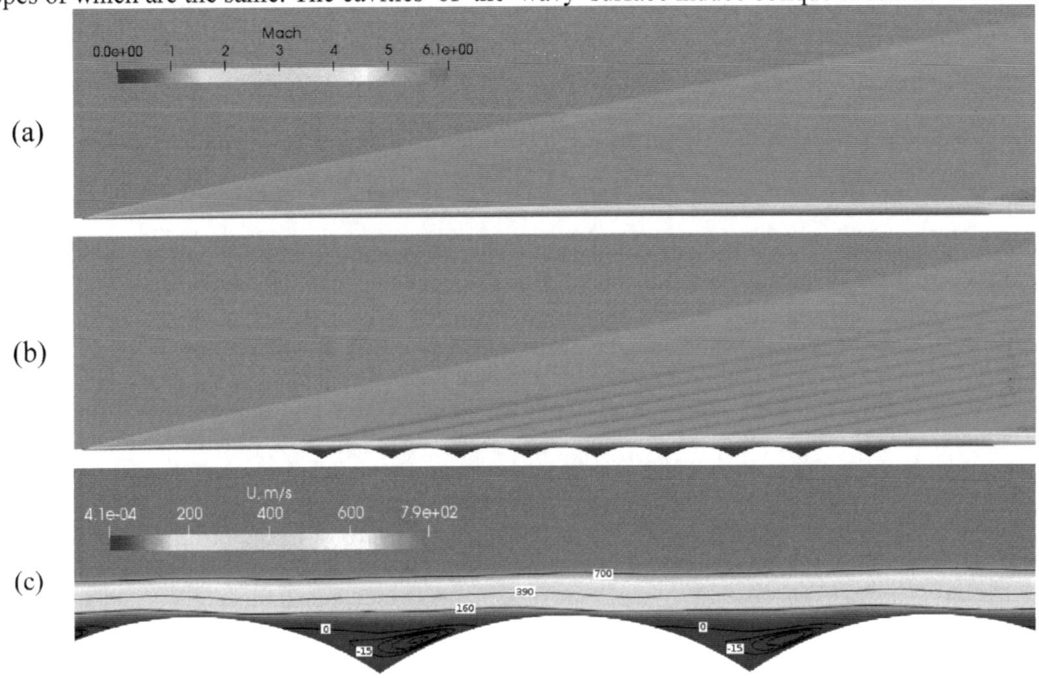

Figure 2. Mach-number field of steady flow (a,b) and velocity counters near the region of wavy surface with velocity isolines (c): (a) smooth plate, (b,c) plate with wavy surface 1.

from the reattachment regions on the tops of wavy surface the interaction of which with the flow in the boundary layer causes recirculation inside the cavities (with a negative horizontal velocity, as shown on Figure 2c). Near the separation and attachment points, a compression wave is clearly observed. The upper boundary of each separation zone and the entire layer above the wavy surface is almost a straight line, which is typical for supersonic flows. As a result, the upper boundary of the whole mixing layer over the wavy plate remains almost unchanged compared to the smooth plate case. The wavy surface has little effect on the global flow field over the boundary layer. At the same time, this surface generates a mixing layer that connects the cavities and resembles a free shear layer with almost parallel edges.

IOP Conf. Series: Journal of Physics: Conf. Series **894** (2017) 012040 doi:10.1088/1742-6596/894/1/012040

(a)

(b)

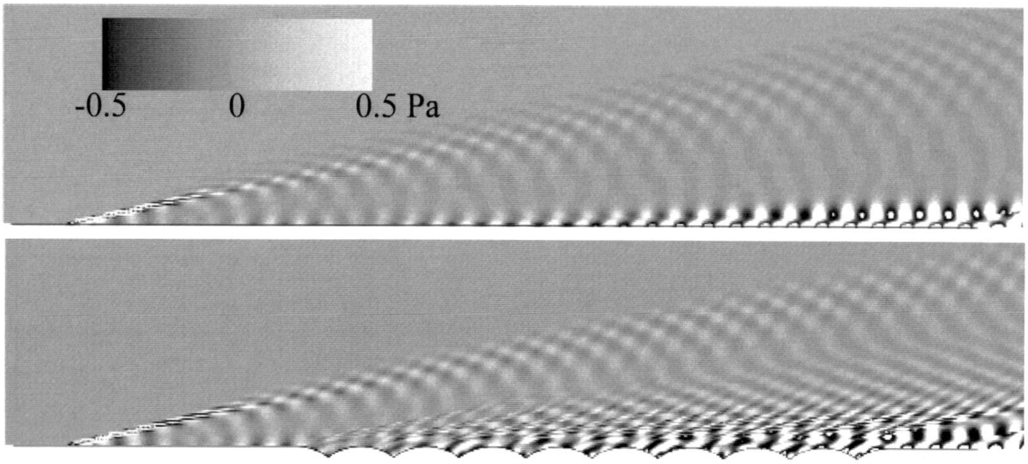

Figure 3. Instantaneous pressure pulsation field due to suction-blowing disturbances with amplitude A=8.76 and frequency f=138.74 kHz: a – smooth plate; b – plate with wavy surface 1.

Figure 3 shows the appearance of cellular structures in the flow, divided into two fronts, behind a source of blowing-suction disturbances for a smooth plate and a plate with wavy surface 1. In the case of a wavy plate, the disturbances in the shock layer (above the mixing layer and behind the bow shock) have similar behavior to the case of a smooth plate, because the upper boundary of the mixing layer is weakly affected by the separation bubbles within the cavities. The evolution of disturbances on the wavy plate in the near-wall region is completely other: disturbances interact with shocklets induced by tops of wavy surface and with the cavity flows. While the structure of the disturbances field is rather complicated, there is no evidence of local amplification of the disturbance amplitude, which could be caused by these interactions.

Figure 5 shows the longitudinal distributions of normalized pressure pulsations at the plate surface $\Delta c_p = (p - p_\infty)/(\rho_\infty U_\infty^2 / 2)$. In the case of a smooth plate, the pulsations increase at $x > 80$ mm and reach maximum amplitude at $x \sim 155$ mm. On the wavy surface the pressure pulsations decrease in amplitude in comparison with a smooth plate, and it is seen their variations along x which depend on the period of wall waviness. The wavy surface decrease the growth of disturbances because the free shear layer formed on the wavy surface is more stable than the boundary layer. The reduction of pulsations on a wavy surface confirms the concept of stabilization of a hypersonic boundary layer by a wavy wall.

(a) (b)

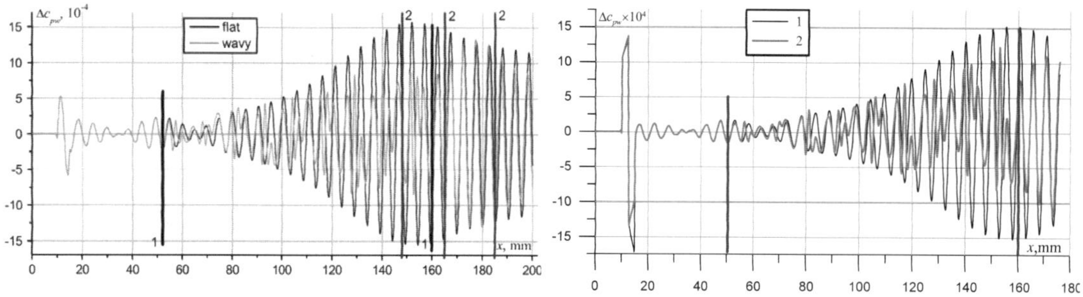

Figure 4. Longitudinal distributions of normalized pressure pulsations at the plate surface: a – figure from [12], b – numerical data from present work, A=8.76, f=138.74kHz: 1 – smooth plate; 2 – plate with wavy surface 1.

Figure 5a shows that the decrease of pulsation amplitude on a wavy surface is not instantaneous, since it persists with time. The values of the heat flux on the wavy surface are close to the values of the heat flux on the smooth plate. As shown in Figure 5b, the cavities cause relatively small periodic perturbations of the heat flux without an average overheating.

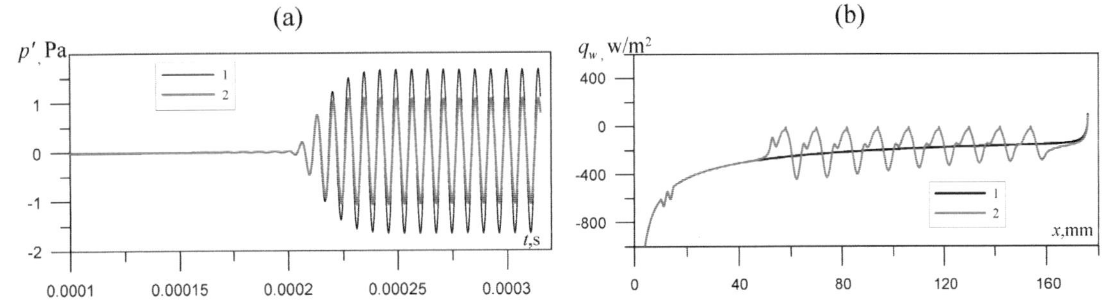

Figure 5. Time dependency of the calculated disturbance amplitudes of pressure coefficient on the wall at x = 165 mm (a) and heat flux distribution (b). A=8.76, f=138.74kHz: 1 – smooth plate; 2 – plate with wavy surface 1.

Figure 6 shows the Mach-number field of steady flow (a) and the velocity field at the wavy surface region with velocity isolines (b) in the flow over the plate with wave surface 2. One can see that in the case of wavy surface 2 (a sawtooth surface), the size of vortices appear in the region of cavities is somewhat larger than in case 1 (Figure 2c) because of the larger cavity size. It is also seen that the boundary of the mixing layer becomes more curved in the region of the waviness peaks. Despite some differences in the flow in the region of a wavy surface, the general behavior of the flow is analogous to case 1, i.e. the formation of a mixing layer which is more stable than the boundary layer.

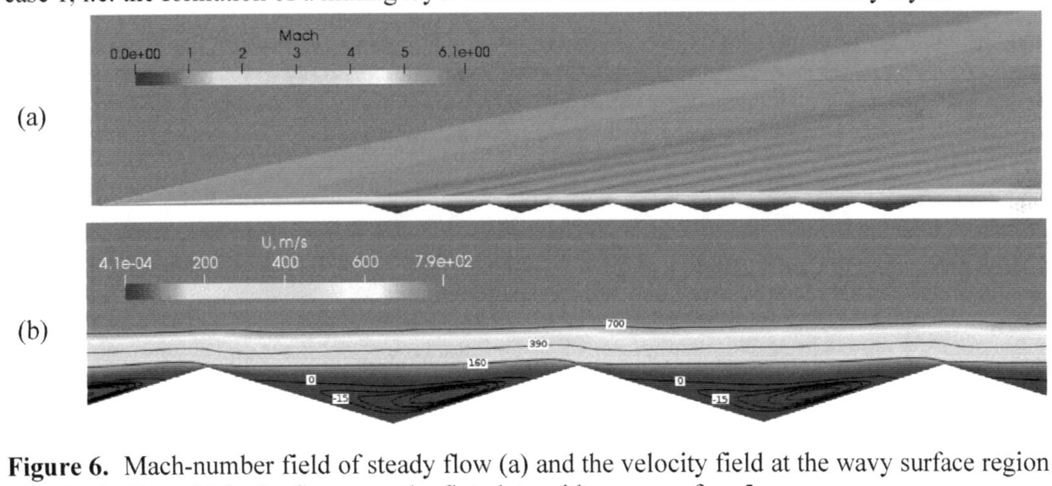

Figure 6. Mach-number field of steady flow (a) and the velocity field at the wavy surface region with velocity isolines (b) in the flow over the flat plate with wave surface 2.

Figure 7 shows the normalized pressure pulsations on the plates with different wavy surface (a,c) and the distribution of heat fluxes (b,d). Comparison of these data for three different types of wavy surface shows their similarity. Some differences in the distribution of heat flux are seen at the tops of wavy surface. It is seen that in comparison with the wavy surface 1 (Figure 5b), sawtooths surfaces 2 and 3 with sharp tops give peaks on the distributions of heat fluxes (Figure 7b,d).

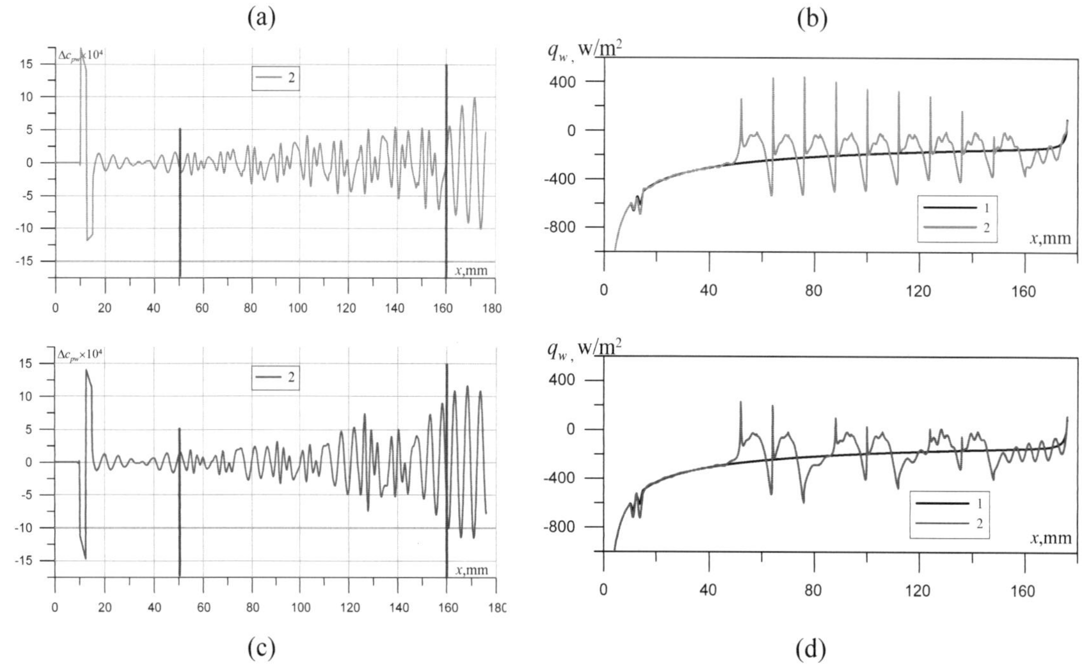

Figure 7. Normalized pressure pulsations on the plates with different wavy surface (a,c) and the distribution of heat fluxes (b,d): A=8.76, f=138.74kHz: 1 – smooth plate; 2 – wavy plate with wavy surface, (a,b) – waviness 2, (c,d) – waviness 3, blue vertical lines indicate the waviness position.

4. Conclusions

Numerical simulation of supersonic (M_∞ = 6) flow over the smooth plate and the plate with a wavy surface of different types was performed by solving the two-dimensional Navier–Stokes equations with using the ANSYS Fluent package. The program modules for introduction of disturbances of blowing-suction type into the boundary layer were created. The verification of the computational program was performed by comparison with the data available in the literature.

Numerical studies have shown that wavy surfaces form an almost parallel layer of mixing, connecting adjacent cavities with soft separations and reattachments. Such waviness slightly perturbs the external flow, and at the same time, weakens the growth of perturbations in the boundary layer and does not increase the values of the average heat fluxes.

It is shown that the flow behavior does not change for different types of wavy surface: a mixing layer is formed, which effectively prevents the growth of perturbations in the boundary layer.

Acknowledgments
This work is supported by the Russian Foundation for Basic Research (project No. 016-08-00782).

References

[1] Fedorov A V 2011 *Annual Review of Fluid Mechanics.* **43** 79–95
[2] Mack L M 1975 *AIAA J.* **13** 278–89
[3] Maslov A A and Mironov S G 2000 *J. Fluid Mech.* **412** 259–77
[4] Bountin D, Shiplyuk A and Maslov A 2008 *J. Fluid Mech.* **611** 427–42
[5] Maslov A A, Poplavskaya T V and Tsyryulnikov I S 2010 *Fluid Dynamics.* **45**(4) 552–8
[6] Fedorov AV, Shiplyuk A N, Maslov A A, Burov E Vand Malmuth N D 2003 *J. Fluid Mech.* **479** 99–124
[7] Egorov I V, Fedorov A V and Soudakov V G 2008 *J. Fluid Mech.* **601** 165-87

[8] Maslov A A, Mironov S G, Poplavskaya N V, Tsyryulnikov I S and Kirilovskiy S V 2012 *J. Appl. Mech. Tech. Phys*. **53**(2) 162-72
[9] Tsyryulnikov I S, Maslov A A, Mironov S G, Poplavskaya T V and Kirilovskiy S V 2015 *Technical Physics Letters*. **41**(2) 184–6
[10] Lees L and Gold H 1966 *Proceedings of the International Symposium on Fundamental Phenomena in Hypersonic Flows*, Cornell University Press, Ithaca
[11] Lysenko V I 1999 *J. Fluid Mech*. **392** 1–26
[12] Bountin D, Chimitov T, Maslov A, Novikov A, Egorov I, Fedorov A and Utyuzhnikov S 2013 *AIAA J*. **51**(5) 1201-10

Solution of internal ballistic problem for SRM with grain of complex shape during main firing phase

A E Kiryushkin and L L Minkov

Faculty of Physics and Engineering, Tomsk State University, Tomsk, Russia

E-mail: sashakir94@mail.ru

Abstract. Solid rocket motor (SRM) internal ballistics problems are related to the problems with moving boundaries. The algorithm able to solve similar problems in axisymmetric formulation on Cartesian mesh with an arbitrary order of accuracy is considered in this paper. The base of this algorithm is the ghost point extrapolation using inverse Lax-Wendroff procedure. Level set method is used as an implicit representation of the domain boundary. As an example, the internal ballistics problem for SRM with umbrella type grain was solved during the main firing phase. In addition, flow parameters distribution in the combustion chamber was obtained for different time moments.

1. Introduction

Solid rocket motors (SRM) have a relative simplicity of design and reliability in compare with other types of rocket motors. Despite this fact, they have a number of disadvantages, such as thrust control. Due to certain difficulties in active control of the SRM thrust, the main interests of the designers lie in various ways of "passive" thrust control. These ways include the choice of propellant, as well as the form of charge. Depending on the purpose of the SRM, an acceptable curve of the thrust versus time is determined. Then the shape of propellant grain, which provides the required dependence of the burning surface on time, is designed. In recent years, numerical modeling of processes in combustion chambers of rocket engines, as one of the tools for designing new SRMs, has been widely used. The ability to model internal chamber processes of SRM over firing period without the cost of production or using an engine test stand provides the opportunity for testing SRMs varied in form and allows one to obtain, ultimately, the optimal design of the rocket engine. As a rule, modern SRMs have grains of complex shape, characterized by 3-D or axisymmetric geometry. For some grains, the flow of combustion products in the chamber free volume of SRM does not significantly change in space. In this case, numerical models taking into account the complex shape of the grain with either a zero-dimensional or quasi-one-dimensional formulation of the problem for describing the flow of combustion products are applicable [1, 2].

However, in some cases it is necessary to consider a three-dimensional (axisymmetric) flow in conjunction with a change of the grain surface geometry, which greatly complicates the numerical implementation. This class of problems relate to problems with moving boundaries. For the numerical solution of this type of problem, unstructured moving computational grids can be used. Despite the simplicity of discretization of equations on such grids, this approach has a number of drawbacks. Firstly, the construction of a "qualitative" computing grid can occupy

Content from this work may be used under the terms of the Creative Commons Attribution 3.0 licence. Any further distribution of this work must maintain attribution to the author(s) and the title of the work, journal citation and DOI.
Published under licence by IOP Publishing Ltd

a considerable part of the total time of the problem solution. Secondly, this approach requires permanent rebuilding of the computational grid, which can lead to loss of accuracy. Thirdly, with time, various topological changes in the surface can occur, which leads to a significant complication of the algorithm. Fourthly, such schemes have an order of approximation not higher than the second.

In this paper, an algorithm that makes it possible to solve such problems with moving boundaries on a fixed Cartesian grid with an arbitrary order of accuracy in space and time has been developed. The peculiarities of the proposed algorithm are the following: 1) boundaries cross the computational grid in an arbitrary way, which complicates the specification of boundary conditions; 2) the realization of a numerical scheme in the vicinity of the boundaries requires the determination of the parameter values at the "ghost" points; 3) it is necessary to track the moving surface.

For the geometric representation of the surface and tracking its evolution over time, the level set method [3] is used. To take into account the boundary conditions and find the values of the flow parameters at points outside the computational domain ("ghost" points), the inverse Lax-Wendroff method developed by Shu [4, 5] is used.

2. Description of the numerical scheme

Let us consider the unsteady equations of gas dynamics with two spatial coordinates for the vector-function of conservative variables $U(x, y, t)$:

$$\begin{cases} U_t + F(U)_x + G(U)_y = S(U), \\ U(x, y, 0) = U_0(x, y), \end{cases} \tag{1}$$

in a bounded region $\Omega(t)$ with the relevant conditions specified on the boundary $\Gamma(t) = \partial\Omega(t)$ at the time t. The region $\Omega(t)$ is covered by a homogeneous Cartesian grid with steps along the x and y coordinates, equal, respectively $\Delta x = h_x$ and $\Delta y = h_y$, but the boundary of the region does not coincide with any of the coordinate lines. Schematically region $\Omega(t)$ and computational grid are shown in Figure 1. Then the semi-discrete approximation of equations (1) can be written in the form:

$$\frac{d}{dt}U_{i,j}(t) = -\frac{1}{h_x}(F_{i+\frac{1}{2},j} - F_{i-\frac{1}{2},j}) - \frac{1}{h_y}(G_{i,j+\frac{1}{2}} - F_{i,j-\frac{1}{2}}) + S_{i,j}, \tag{2}$$

where $F_{i+\frac{1}{2},j}$ and $G_{i,j+\frac{1}{2}}$ are numerical fluxes.

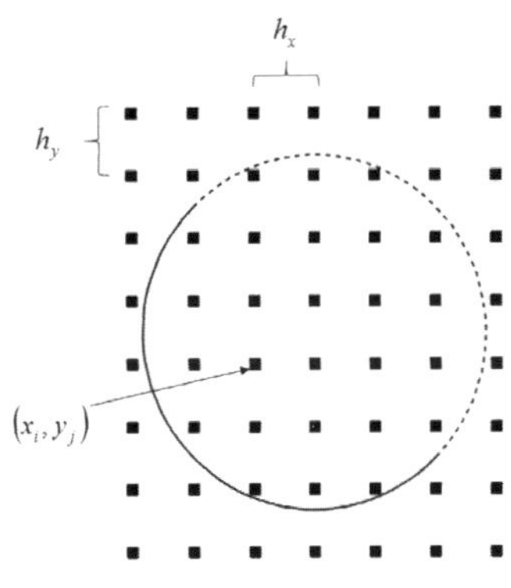

Figure 1. Region boundary and computational grid.

To integrate the system of ordinary differential equations (2) with respect to time, we use the Runge-Kutta method. To obtain the solution of a high order of accuracy over space either

IOP Conf. Series: Journal of Physics: Conf. Series **894** (2017) 012041 doi:10.1088/1742-6596/894/1/012041

an ENO scheme of the third order of accuracy or a WENO scheme of the fifth order of accuracy are used. Those schemes have a stencil consisting of seven points, three of which can be ghost points that are outside the solution region.

3. Inverse Lax-Wendroff procedure for fixed boundary

We write down the boundary conditions for the burning propellant surface, which will be assumed to be stationary. Those boundary conditions have the form of equations (3)-(5) , where ρ, p, u_n, u_τ are the density, the pressure, the normal and tangential components of velocity at the boundary, respectively, and m_t, H_0, ν are constants characterizing the propellant.

$$\rho u_n = m_t p^\nu, \tag{3}$$

$$\frac{k}{k-1}\frac{p}{\rho} + \frac{u_n^2 + u_\tau^2}{2} = H_0, \tag{4}$$

$$u_\tau = 0. \tag{5}$$

As noted in [4], to obtain values at ghost points it is more convenient to use the system of equations of gas dynamics written in a non-divergent form for primitive variables

$$W_t + A\left(W\right)W_x + B\left(W\right)W_y = 0,$$

where $W = (\rho, u, v, p)^T$, $A(W) = \begin{pmatrix} u & \rho & 0 & 0 \\ 0 & u & 0 & 1/\rho \\ 0 & 0 & u & 0 \\ 0 & \rho c^2 & 0 & u \end{pmatrix}$, $B(W) = \begin{pmatrix} v & \rho & 0 & 0 \\ 0 & v & 0 & 1/\rho \\ 0 & 0 & v & 0 \\ 0 & 0 & \rho c^2 & v \end{pmatrix}$ and c is the sound speed.

The system of equations of gas dynamics written in this form has the following left eigenvectors for the matrix $A(W)$.

$$L(W) = \begin{pmatrix} 0 & -1/2c & 0 & 1/2\rho c \\ -1/2\rho & 0 & 1/2 & 1/2\rho c \\ 1/2\rho & 0 & 1/2 & -1/2\rho c \\ 0 & 1/2c & 0 & -1/2\rho c \end{pmatrix}$$

When applying the inverse Lax-Wendroff procedure [4], the fourth extrapolation equation should be added to the three boundary conditions, which is written taking into account the zero components of the fourth eigenvector as:

$$l_{4,2}u_n + l_{4,2}p = V_4. \tag{6}$$

To solve the system of equations (3)-(5), (6) we rewrite it in another form:

$$u_n = a + bp, \tag{7}$$

$$\rho = \frac{m_t}{u_n}p^\nu, \tag{8}$$

$$u_\tau = 0, \tag{9}$$

$$\frac{b^2}{2}p^2 + bcp^{2-\nu} + abp + acp^{1-\nu} + \frac{a^2}{2} - H_0 = 0, \tag{10}$$

where $a = \frac{V_4}{l_{4,2}}$, $b = -\frac{l_{4,4}}{l_{4,2}}$, $c = \frac{k}{k-1}\frac{1}{m_t}$.

Thus, it is required to solve numerically a single nonlinear equation (10) with respect to pressure, and the remaining parameters are found from the corresponding equations (7) - (9). To solve Eq. (10), one can use the Newton method.

In order to obtain the first derivatives with respect to space, it is necessary to take the time derivatives of the boundary conditions (3)-(5), convert these derivatives to spatial ones using gas dynamic equations and add one extrapolation equation. First order scheme is used for calculations. Thus, there is no need in calculating first derivatives and it is omitted in this paper.

4. The level set method for representing the region boundary

As can be seen from the description of the inverse Lax-Wendroff procedure algorithm for determining the parameter values at ghost points, it is necessary to know a nearest point on the boundary and the normal to it. In the case of moving boundaries, it is also necessary to know the algorithm for changing the position of the boundary with time. Relatively simple regions, for the boundaries of which there exist analytic expressions, are considered in [4, 5]. However, such an analytical description of moving boundaries of burning propellant for the problems of SRM internal ballistics is possible only in rare cases. Thus, to keep track the burning surface, a numerical technique is needed. In this paper we used the level set method, which seems to be the most effective way to determine the position of a moving surface on a Cartesian grid. The level set method is widely used for various problems with interface boundaries and the necessary information can be found in the book [3].

5. Algorithm for calculation of flow parameters for SRM internal ballistics problems with account for propellant burn-out

The main firing phase of SRM is characterized by relatively small changes in parameters over time as compared with their change in space, which suggests that the combustion of propellant occurs in a quasi-stationary mode, and the pressure field has time to adjust to the change in the burning surface. In addition, the characteristic times for the flow of combustion products in a rocket engine and the motion of a burning surface differ by approximately ten times. Furthermore, the characteristic times for the flow of combustion products in SRM and for the motion of a burning surface differ by approximately 10^6 times.

The following algorithm is proposed:

1) The propellant surface is assumed to be fixed at each time step. The system of gas dynamics equations is solved on a fixed Cartesian grid according to the scheme (2) until a stationary solution is obtained. Values in ghost points are given by the inverse Lax-Wendroff method; 2) For each point near the burning surface, the speed of surface motion is calculated to integrate the level function in time [3]; 3) The new position of the burning propellant surface is determined; 4) The results are written to a file. This procedure is repeated until all the propellant is burnt out.

6. Physical statement of the problem

As an example, the calculations were carried out in the region depicted in Figure 2. The region is the motor housing in which the propellant grain and the nozzle block are located. The geometric parameters are as follows. The grain length is $L = 2.5$ m, the initial radius of channel is $R_0 = 0.2$ m, the final radius of channel is $R_k = 0.9$ m, the length of umbrella is $H = 0.523$ m, the width of umbrella is $h = 0.2$ m, the distance from left edge of the grain to the begin of umbrella is $l = 0.585$ m, the throat radius is $r_{cr} = 0.1$ m. The angle of inclination of the "umbrella" α is variable.

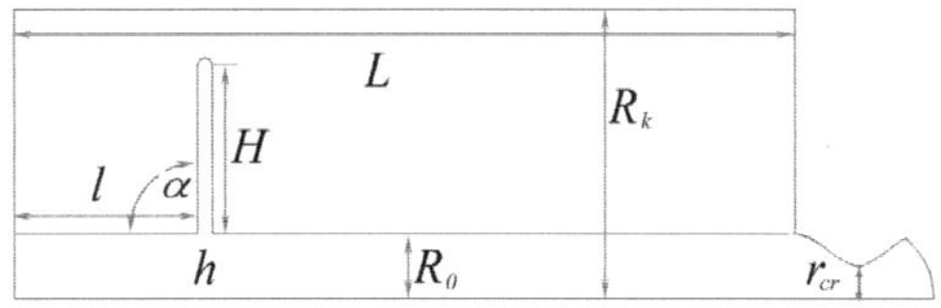

Figure 2. Geometry of region.

In solving this problem, the following assumptions were used: flow field is axisymmetric, combustion products are inviscid and ideal gas, propellant is homogeneous and the burning rate depends on the pressure according to the law: $\rho u = m_t p^\nu$.

7. Mathematical statement of the problem

The Euler equations written in the form (11)–(14) and the ideal gas law (15) are used to describe internal chamber processes for an axisymmetric region.

$$\frac{\partial \rho}{\partial t} + \frac{\partial \rho u}{\partial z} + \frac{\partial \rho v}{\partial r} = -\frac{1}{r}\rho v, \tag{11}$$

$$\frac{\partial \rho u}{\partial t} + \frac{\partial (\rho u^2 + p)}{\partial z} + \frac{\partial \rho v u}{\partial r} = -\frac{1}{r}\rho v u, \tag{12}$$

$$\frac{\partial \rho v}{\partial t} + \frac{\partial \rho u v}{\partial z} + \frac{\partial (\rho v^2 + p)}{\partial r} = -\frac{1}{r}\rho v^2, \tag{13}$$

$$\frac{\partial \rho E}{\partial t} + \frac{\partial \rho u H}{\partial z} + \frac{\partial \rho v H}{\partial r} = -\frac{1}{r}\rho v H, \tag{14}$$

$$p = \rho R T \tag{15}$$

The system of equations (11) - (14) is supplemented by initial and boundary conditions. There are four types of boundaries for the problem. The impermeability condition $u_n = 0$ is specified at solid walls. Condition $\frac{\partial}{\partial n} = 0$ is specified at the axis of symmetry. Eqs (3)-(5) are specified at the inflow boundary. Constant pressure $p = p^*$ is specified at the outflow boundary when the flow speed is subsonic.

Initial conditions are written as: $p = p_0$, $\rho = \rho_0$, $u = 0$, $v = 0$.

8. Values of parameters and constants

Numerical results were obtained with the following values of parameters and constants: $\alpha = 45°, 60°$; $m_t = 5.34 \cdot 10^{-3}\ \frac{kg}{m^2 s \cdot Pa^\nu}$; $H_0 = 10251150\ \frac{J}{kg}$; $\nu = 0.5$; $\rho^T = 1700\ \frac{kg}{m^3}$; $p_0 = p^* = 101325\ Pa$; $\rho_0 = 1.3\ \frac{kg}{m^3}$; $h_x = h_y = 0.2\ m$.

9. Numerical results

A scheme of the first order of accuracy with respect to space and time was used for solving the equations of gas dynamics. A scheme of the fifth-order accuracy in space and the third-order accuracy with respect to time was used to determine the level function. In Figures 3 and 4, one can see the evolution of the combustion surface over time for angles $\alpha = 45°$ and $\alpha = 60°$, respectively. The time interval through which isosurfaces are represented is equal to $\Delta t = 5\ s$.

It can be seen from the figures that for a given type of charge, combustion occurs in practically parallel layers, that is, the burning velocity along the surface varies insignificantly.

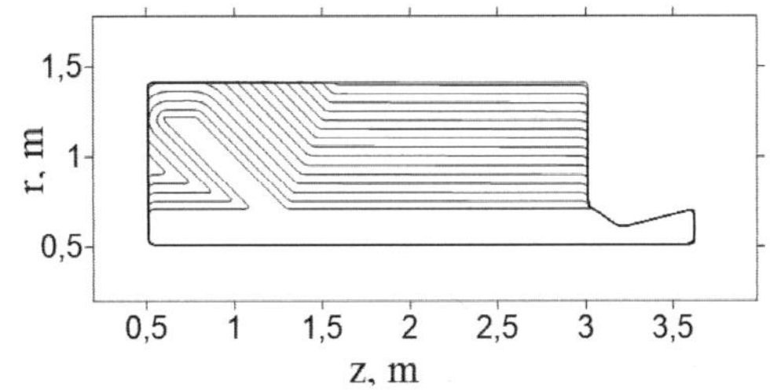

Figure 3. The shape of the propellant grain surface as a function of time, $\alpha = 45°$.

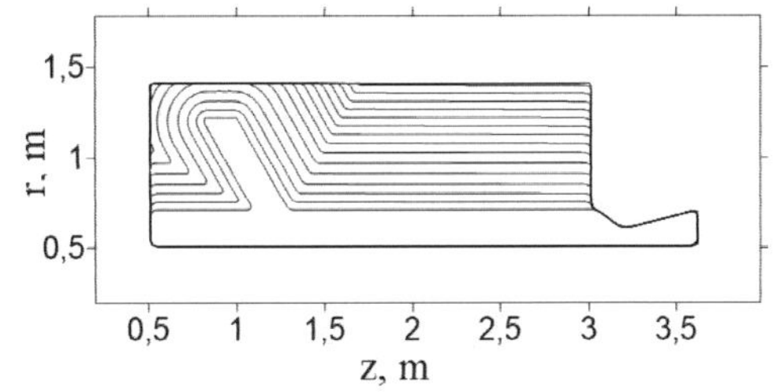

Figure 4. The shape of the propellant grain surface as a function of time, $\alpha = 60°$.

Figure 5 shows the dependencies of pressure averaged over the combustion chamber volume on time. This figure shows that the pressure in the combustion chamber increases to the point of time $t \approx 20\ s$ that corresponds to the complete combustion of propellant near the left wall of the housing. Then the combustion area begins to decrease, which leads to a drop in pressure. At the time $t \approx 70\ s$, the propellant is completely burned out and the gas inflow becomes zero, after which the pressure falls monotonically.

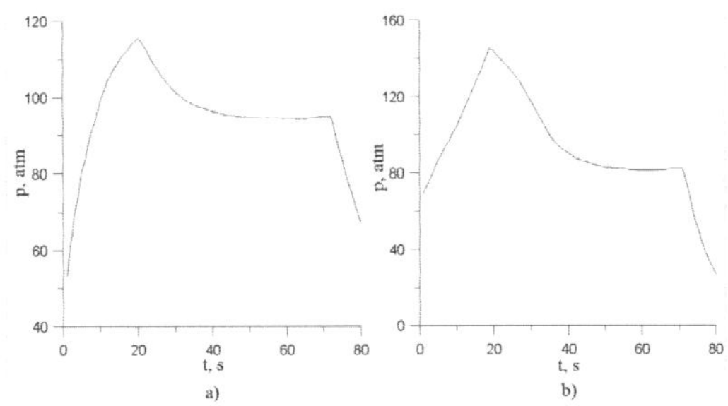

Figure 5. The mean chamber pressure as a function of time, a) $\alpha = 45°$, b) $\alpha = 60°$.

Figures 6, 7 show the distribution of pressure in the combustion chamber for the times $t = 5\ s$ and $t = 40\ s$, respectively. It can be seen from the figures that until the fuel burned out near the left wall, there is a pressure drop along the channel of about five atmospheres. For a period of time $40s < t < 70s$, the pressure in the combustion chamber can be considered constant.

IOP Conf. Series: Journal of Physics: Conf. Series **894** (2017) 012041 doi:10.1088/1742-6596/894/1/012041

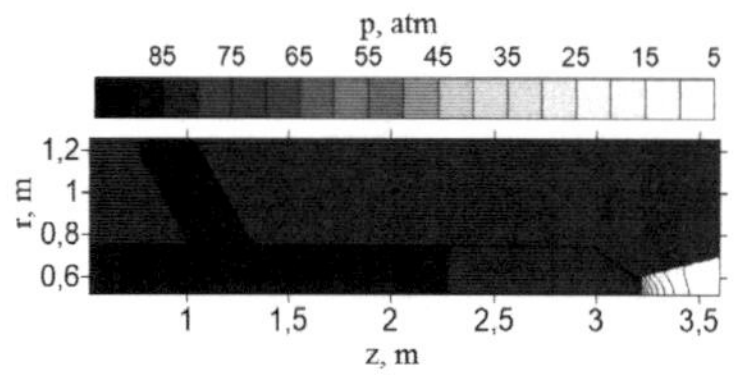

Figure 6. Distribution of pressure in the SRM chamber, $t = 5$ s.

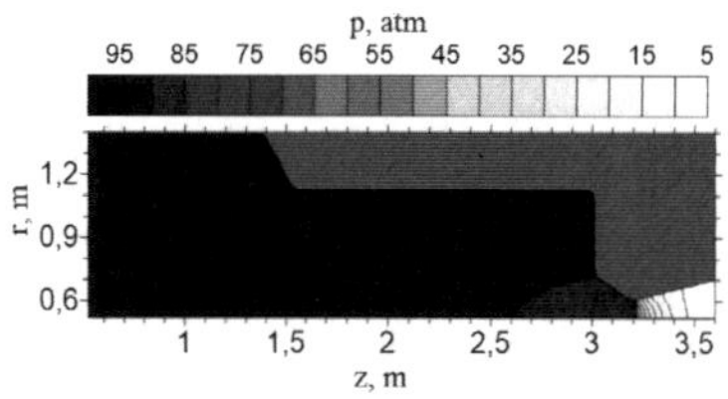

Figure 7. Distribution of pressure in the SRM chamber, $t = 40$ s.

10. Conclusion

In this paper, an algorithm was developed to solve the internal ballistics problem for solid-rocket motors with the umbrella type grain during main firing phase. The moving surface of the burning propellant was tracked using a level set method, the essence of which lies in the implicit representation of the burning surface as a zero level of some function. To obtain parameter values at ghost points the inverse Lax-Wendroff method, which provides the necessary order of the equations approximation, was used.

The coefficients of extrapolation polynomials that allow one to determine the values of the gas parameters at the gas-inflow boundary taking into account the non-linearity of the boundary conditions are obtained. In the case of the motion of the burning surface, a number of simplifying assumptions have been made, which make it possible to reduce the calculation time. As an example the dependencies of the pressure on time and the distributions of the flow parameters at different moments of time in the combustion chamber were obtained for a grain with an "umbrella" oriented at different angles of inclination to the channel generatrix.

Acknowledgments

This work was carried out with financial support from the Ministry of Education and Science of the Russian Federation (State assignment No. 9.9625.2017/8.9).

References

[1] Cavallini E. 2010 *Modeling and Numerical Simulation of Solid Rocket Motors Internal Ballistics* (PhD thesis)
[2] Sullwald W 2014 *Grain Regression Analysis* (Master's thesis) University of Stellenbosch
[3] Osher S and Fedkiw R 2003 *Level Set Methods and Dynamic Implicit Surfaces* vol 153 ed S S Antman, J E Marsden and L Sirovich (New York: Springer) p 273
[4] Tan S Wang C Shu C W and Ning J 2012 *J. Comput. Phys.* **231** 2510–27
[5] Tan S and Shu C W 2011 *J. Comput. Phys.* **230** 6023–36

Modeling of gas flows in radial micro-nozzles

S P Kiselev[1,2], V P Kiselev[1], V Yu Liapidevskii[3,4] and V N Zaikovskii[1]

[1]Khristianovich Institute of Theoretical and Applied Mechanics, Russian Academy of Sciences, Siberian Branch, 630090 Novosibirsk, Russia
[2]Novosibirsk State Technical University, 630092 Novosibirsk, Russia
[3]Lavrent'ev Institute of Hydrodynamics, Russian Academy of Sciences, Siberian Branch, 630090 Novosibirsk, Russia
[4]Novosibirsk State University, 630090 Novosibirsk, Russia

E-mails: kiselev@itam.nsc.ru, kiselevvp@itam.nsc.ru, zaikovskii@itam.nsc.ru, liapid@hydro.nsc.ru

Abstract. In the present paper, results of an experimental and numerical study of gas flows in radial micro-nozzles are reported. The radial micro-nozzle consists of two disks spaced apart by a less than one millimeter distance. To the inlet of the radial nozzle, a gas under high pressure is supplied; through the nozzle, the pressurized gas is ejected into ambient space. In the present study, we analyzed the flows in which the gas at the inlet to the micro-nozzle had a supersonic velocity. It was shown that, on the condition that the micro-nozzle width was smaller than some critical value, then, due to the wall friction force, a pseudo-shock arose in the micro-nozzle. In the pseudo-shock, the gas underwent deceleration to a subsonic velocity. A satisfactory agreement between the calculated and experimental data was achieved.

1. Introduction

Until recently, little attention was given to the study of supersonic gas flows in radial micro-nozzles. Exceptions here is the publication [1], in which supersonic gas flows in wider nozzles and at lower pre-chamber pressures in comparison with the present study were dealt with. However, by now novel technologies in which radial nozzles play an important role have emerged. S.P. Kiselev *et al.* [2] have reported experimental and numerical data proving the possibility of using radial micro-nozzles for microparticle acceleration and cold-spray application of coatings onto inner pipe surfaces. In [3], self-sustained oscillations that arose during the outflow of a supersonic jet into ambient space were studied. In the present paper, we report detailed data gained in experiments and numerical simulations in which the structure of supersonic gas flows in radial micro-nozzles was investigated.

2. Experimental procedure

The radial micro-nozzle was formed by two disks spaced a part by a distance smaller than one millimeter (this distance was the nozzle width). The experiments were carried out on a facility schematically shown in Figure 1. The experimental arrangement comprised an external disk 1 with radius $r_e = 36$ mm and an internal disk 2 of the same radius. The space between the disks was connected to a pre-chamber 3. The pre-chamber was a cylindrical tube with internal radius $r_1 = 5$ mm and external radius $r_2 = 9$ mm. The gas was supplied to the pre-chamber under pressure p_0 through

Content from this work may be used under the terms of the Creative Commons Attribution 3.0 licence. Any further distribution of this work must maintain attribution to the author(s) and the title of the work, journal citation and DOI.
Published under licence by IOP Publishing Ltd

twelve holes that were made on the external surface of the tube (in Figure 1, those holes are conventionally indicated with three arrows)

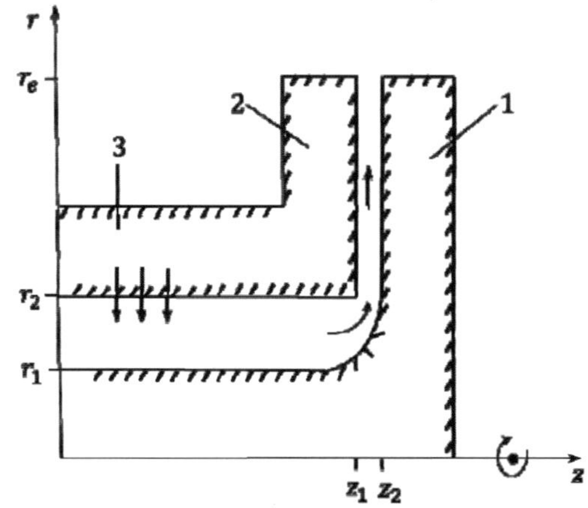

Figure 1. Gas flow in a radial micro-nozzle

For making the flow at the inlet to the nozzle more uniform, a rounding was provided at the joint between the pre-chamber and the external disk. The nozzle width $h = z_2 - z_1$ in different experiments was varied via displacement of the external disk.

As the working gas, cold air with temperature $T_0 = 300$ K was used. In different experiments, the gas pressure in the pre-chamber was varied within the interval $0.9 \text{ MPa} \le p_0 \le 1.5 \text{ MPa}$. The gas was discharged from the radial nozzle into ambient space, which was filled with air under normal conditions (pressure $p_\infty = 0.1$ MPa, temperature $T_\infty = 300$ K).In the experiments, the pressures on the surface of the external disk were measured. For performing such measurements, the external disk was provided with eleven pressure taps of diameter $d_i = 0.8$ mm that stretched along the disk radius. The pressure taps were connected to KIM tensometric pressure sensors, developed at ITAM SB RAS, with pressure tubes. In this configuration, the measurement accuracy of static pressure was 0.5 %.

3. Numerical simulation

The numerical modeling of the gas flow was carried out in the radial-nozzle flow region $r_2 < r < r_e$ in quasi-one-dimensional approximation. Two models for calculating the gas flow were used: a simplified model 1, in which the vorticity of the radial gas flow was neglected, and a complicated model 2, in which this vorticity, arising due to the non-uniform inflow of gas from the pre-chamber to the radial nozzle, was taken into account. Consider in succession the governing equations and the statement of the problem about the gas flow in a radial nozzle in both models.

3.1. Mathematical model 1

The equations of quasi-one-dimensional model 1 result from the averaging performed over the radial-nozzle cross section. The continuity, flow, and gas-energy equations for the mean flow quantities are

$$\frac{\partial \rho r}{\partial t} + \frac{\partial \rho v r}{\partial r} = 0, \quad \frac{\partial v}{\partial t} + v\frac{\partial v}{\partial r} + \frac{1}{\rho}\frac{\partial p}{\partial r} = -\frac{2\tau_W}{\rho h}, \quad p = \rho R T,$$
$$\frac{\partial}{\partial t}\rho r\left(C_V T + \frac{v^2}{2}\right) + \frac{\partial}{\partial r}\rho v r\left(H + \frac{v^2}{2}\right) = 0, \quad H = C_p T. \qquad (1)$$

In equations (1), v is the radial component of gas velocity, p is the pressure, ρ is the gas density, T is the gas temperature, C_V and C_p are the specific heats of the gas at constant volume and pressure, R is the universal gas constant, and τ_W is the wall friction force. Like in [1], the friction force τ_W was calculated by the formulas

$$\tau_W = \frac{C_f \rho v^2}{2}, \quad C_f = \frac{0.06}{(2\,\mathrm{Re})^{1/4}}, \quad \mathrm{Re} = \frac{vh}{\nu}, \tag{2}$$

where $\nu = \mu/\rho$ is the kinematic viscosity and μ is the dynamic viscosity, which for the case of air was calculated by the Sutherland formula [4]. The dependence of drag coefficient C_f on the Reynolds number Re was chosen as the one in a turbulent flow of incompressible fluid. As it was shown in [1], the mentioned formula can be used for calculating supersonic gas flows. In the latter case, the Reynolds number in formulas (2) is to be calculated at a temperature equal to half the sum of the static gas temperature and the nozzle wall temperature, $T' = (T + T_W)/2$. In the present study, the static temperature T was determined from the solution of system (1), whereas the wall temperature was assumed equal to the flow stagnation temperature $T_W = T_0$. The empirical constant 0.06 in the formula for the drag coefficient (see the second formula in (2)) was adjusted so that a best agreement between the calculated and experimental data could be achieved in one of the experiments for a nozzle with $h = 0.15$ mm. Subsequent numerical calculations for the flows at other nozzle-width values and the comparisons of calculated with experimental data were carried out at the chosen value of the constant. Note that in the study of [1] a value of 0.08 was adopted for the empirical constant.

System (1) was solved numerically by the explicit Lax–Wendroff difference scheme [5]. In calculations of flows with shock waves by this scheme, shock-wave smearing by scheme viscosity occurs. In the region $r_2 < r < r_e$, the problem about the decay of a pressure discontinuity specified at the initial time at some section $r = r_d$ of the nozzle was solved:

$$p = p_0,\ T = T_0 = T_\infty,\ v = 0,\ r_2 < r < r_d;\ p = p_\infty,\ T = T_\infty,\ v = 0,\ r_d < r < r_e, \tag{3}$$

where p_0 is the pressure in the pre-chamber, $p_0 > p_\infty$. After the decay of the discontinuity, a rarefaction wave propagates to the left; behind this wave, the gas undergoes acceleration as it moves toward the right-hand boundary of the nozzle. Initially, at the left boundary of the calculation domain, $r = r_2$, a condition of symmetry is posed. This condition holds till the time $t = t_*$, unless at this boundary a critical condition $\mathrm{M} = 1$ is achieved. Afterwards, all the three characteristics penetrate into the nozzle and, therefore, nozzle choking conditions, $p = p_*$, $T = T_*$, $v = v_* = c_*$, are to be posed at the left boundary (here, $\mathrm{M} = v/c$ is the Mach number, $c = \sqrt{\gamma RT}$ is the speed of sound, and $\gamma = 1.4$). Marked with the asterisks are the values of flow quantities at the critical point $r = r_2$; these values can be expressed in terms of the corresponding parameter values in pre-chamber using the well-known isentropic formulas [4]: $p_* = \beta^{\gamma/(\gamma-1)} p_0$, $T_* = \beta T_0$, $v_* = c_* = \beta^{1/2} c_0$, $\beta = 2/(\gamma+1)$. At the right boundary of the calculation domain, $r = r_e$, at a subsonic velocity $\mathrm{M}_e = v_e / c_e < 1$ the pressure in ambient volume, $p = p_\infty$, was posed, whereas at a supersonic speed $\mathrm{M}_e > 1$, conditions of symmetry were used. The change of boundary conditions was due to the fact that at subsonic velocity only one, left characteristic penetrated into the calculation domain, whereas at supersonic velocity all the three characteristics were coming out from this domain.

Consider a problem about the decay of a discontinuity in a nozzle with $h = 0.15$ mm which occurs at the point $r_d = 23$ mm. Initially, a pressure drop in a gas at rest with temperature $T_0 = 300$ K is specified. At the left of the point $r_d = 23$ mm, a uniform pressure $p_0 = 1.43$ MPa is assumed, and at the right of this point, we have $p_\infty = 0.1$ MPa. After the decay of the discontinuity, in the nozzle there arises an unsteady gas flow that reaches a steady state in a time $t_1 = 3\ \mu s$. The establishment of the steady nozzle flow is illustrated in Figure 2.

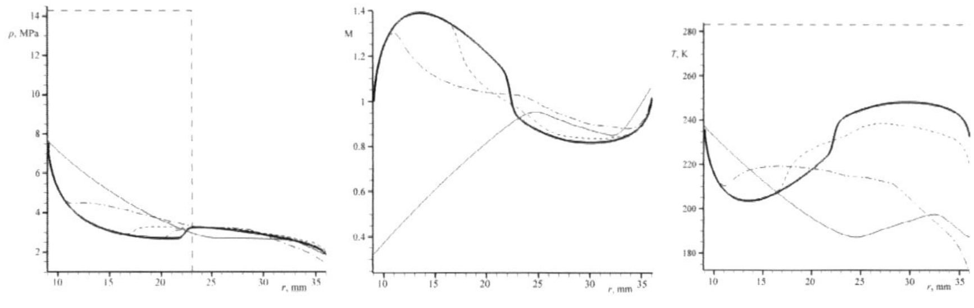

(a) (b) (c)

Figure 2. The calculated distributions of flow quantities in the nozzle at times t_i (0; 0.6; 1.2; 1.8; 2.4; 3.0, μs): (a) – $p(r,t_i)$; (b) – $M(r,t_i)$; (c) – $T(r,t_i)$

The dashed line in Figure 2 shows the initial distribution of flow quantities. The thin solid line and the dash-and-dot lines show the flow quantities in the unsteady flow, whereas the thick solid line shows the flow quantities in the established flow. Evidently, the choking condition $M = 1$ is realized in the established flow at the nozzle inlet. Behind the inlet, the flow accelerates to a supersonic velocity $M = 1.4$ ($r \approx 13.5$ mm) and, afterwards, the flow undergoes deceleration. The supersonic-flow deceleration region ends in a shock wave ($r \approx 22$ mm), in which the supersonic flow with $M > 1$ passes over into a subsonic flow with $M < 1$. Unlike in pipe flows [4], here the gas flow behind the shock wave first undergoes deceleration and, then, acceleration to the sonic velocity, $M = 1$. The gas temperature grows in value behind the shock wave to attain a maximum at the point at which the gas deceleration gives way to gas acceleration. At the nozzle outlet, there establishes a sonic flow with $M = 1$, like it also occurs in uniform-section pipes [4].

The variation of pressure with radius in three radial nozzles of width $h = 0.075$; 0.15; 0.5 mm as revealed in numerical calculations and experiments (respectively, lines and symbols) are shown in Figure 3. The data calculated by model 1 are shown with thick lines, and the data calculated by model 2 (this model is described below in Sec. 3.2), with thin lines.

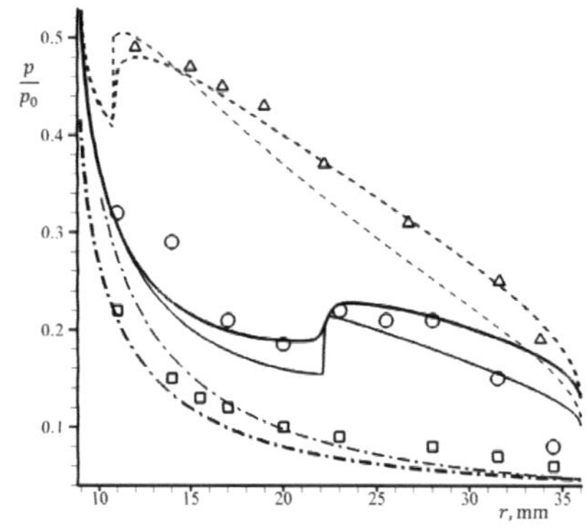

Figure 3. Normalized pressure versus radius for various micro-nozzles; the triangles and dashed lines referto $h = 0.075$ mm and $p_0 = 1.4$ MPa ; the circles and solid lines, to $h = 0.15$ mm and $p_0 = 1.4$ MPa ; and the squares and dash-and-dot lines, to $h = 0.5$ mm and $p_0 = 1.15$ MPa

As it is evident from Figure 3, the calculated data well agree with the experimental data. In the nozzle with $h = 0.5$ mm, a near-isentropic flow in which the pressure monotonically decreases with radius is realized. In the nozzle with $h = 0.15$ mm, we have a flow with a shock wave located in the section $r \approx 22$ mm. On decreasing the nozzle width to $h = 0.075$ mm, a displacement of the shock wave to the inlet section is observed. Behind the shock wave, there establishes a constant gradient of pressure, which compensates the action of wall friction force on the gas flow in the nozzle.

3.2. Mathematical model 2

A mathematical unsteady pseudo-shock model for a barotropic gas flow in a long plane channel capable of providing a description to the transition from supersonic to subsonic flow was developed in [6]. The model is based on a two-layer flow scheme with mass transfer that includes a potential supersonic flow core and a turbulent boundary-layer flow. As the boundary layers join together, the model transforms into a single-layer model for barotropic vortex gas flow. In microchannels analyzed in the present study, the two-layer flow region of a length not exceeding ten channel widths can be neglected. In the latter case, the pseudo-shock can be represented in long-wave approximation with a weak shock wave and with a subsequent region of intense vortex flow.

In radial nozzles, the equations of swirling barotropic gas flow assume the form (cp. with equations (5) in [6]):

$$\frac{\partial \rho r}{\partial t} + \frac{\partial \rho v r}{\partial r} = 0, \quad \frac{\partial \rho r v}{\partial t} + \frac{\partial(\rho r(v^2 + q^2) + rp)}{\partial r} = p - \frac{2r\tau_W}{h},$$

$$\frac{\partial}{\partial t}\left(\rho r\left(\frac{v^2}{2} + \frac{q^2}{2} + \varepsilon\right)\right) + \frac{\partial}{\partial r}\rho v r\left(\frac{v^2}{2} + \frac{3q^2}{2} + (\varepsilon + \frac{p}{\rho})v\right) = -\kappa \rho q^3 - \frac{2r\tau_W v}{h}, \quad (4)$$

$$p = p_0\left(\frac{\rho}{\rho_0}\right)^{\gamma}, \quad \varepsilon = \frac{p}{(\gamma - 1)\rho}.$$

Here, the quantity $\omega = q/h$ describes the mean vorticity of the flow, and the constant quantity κ defines the rate of energy dissipation. Note that the sub-class of continuous solutions (4) with $q \equiv 0$ coincides with the class of continuous solutions of the classical equations of barotropic gas flow. However, in the discontinuous solutions of equations (4), behind the shock-wave front there forms a vorticity leading to a smooth rise of pressure in the pseudo-shock.

Equations (4) are hyperbolic equations that have one contact characteristic and two sonic characteristics:

$$\frac{dr}{dt} = v, \quad \frac{dr}{dt} = v \pm \sqrt{c^2 + 3q^2}, \quad c^2 = \frac{\gamma p}{\rho}.$$

Thus, equations (4) are equations of the gas-dynamic type, and the numerical solutions of the problem about the structure of the transonic flow in a radial nozzle can therefore be obtained by standard methods. In the present study, for performing a comparison between the solutions of equations (4), on the one hand, and the experimental data and numerical solutions obtained above within model 1, on the other hand, we used a method developed by S.K. Godunov. As the initial data, the exact supersonic solution of equations (4) with $q \equiv 0$ and $C_f = 0$ was used. The boundary conditions were the same as those for model 1. At the left boundary of the calculation domain, the flow was assumed irrotational. The value of κ in our calculations weakly affected the solutions, and it therefore was put equal to zero. In determination of the dissipative term of (2), the air kinematic viscosity was assumed constant ($\nu = 1.2 \cdot 10^{-5}\ \mathrm{m/s^2}$). The calculated distributions of pressure for three designs of radial nozzle are shown in Figure 3 with thin lines together with their respective experimental data. It is seen that the data calculated by model 2 are close to the data calculated by model 1, and both calculated datasets satisfactorily agree with the experiment.

4. Conclusion

In sufficiently wide plane and radial nozzles, the pseudo-shock region occupies a substantial portion of the channel. Under such conditions, during the transition from supersonic to subsonic flow, a predominant fraction of energy dissipation (up to 90%) is spent for generation of the mechanical energy of vortex flow in the turbulent boundary layer. In the microchannels, the pseudo-shock width decreases markedly, allowing the use of quasi-one-dimensional gas flow models for pseudo-shock modeling. For making a comparison between the dissipative processes related with the growth of entropy and the generation of vorticity in microchannels, in the present study we have considered two alternative approaches, the classical one-dimensional gas-dynamics equations and the vortex-flow equations for barotropic gas. The performed numerical calculations of gas flows in microchannels have shown that both models yield close results. This circumstance is related with the fact that, in transonic gas flows in microchannels, the boundary-layer perturbation or separation regions behind pseudo-shocks extend to a distance of order ten microchannel widths. The ratio between the length of the perturbed boundary layer behind pseudo-shock and the microchannel radius is low, and both regions contribute little to the deceleration of the gas flow. The results of numerical calculations by both models fairly well agree with experimental data.

Acknowledgements. This work was partially supported by the Russian Foundation for Basic Research under Grant No. 16-01-00156-a.

References

[1] Moller P 1966 *J. Basic Eng. Series D* **88** 153-4
[2] Kiselev S P, Kiselev V P and Zaikovskii V N 2016 *J. of Appl. Mech. and Tech. Phys.* **57** 237-46
[3] Kiselev S P, Kiselev V P, Klinkov S V, Kosarev V F and Zaikovskii V N 2017 *Surface & Coating Technology* **313** 24-30
[4] Abramovich G N 1976 *Applied Gas Dynamics* (Moscow: Nauka) (in Russian)
[5] Anderson D A, Tannehill J C and Pletcher R H 1984 *Computational Fluid Mechanics and Heat Transfer* (Hemisphere Publishing Corporation)
[6] Lipatov I I, Liapidevskii V Yu and Chesnokov A A 2016 Doklady Academii Nauk **466** 545-9 (in Russian)

IOP Conf. Series: Journal of Physics: Conf. Series **894** (2017) 012043 doi:10.1088/1742-6596/894/1/012043

Influence of the solute immobilization on linear stability within the solute analogy of Horton–Rogers–Lapwood problem

L S Klimenko and B S Maryshev

Institute of Continuous Media Mechanics, Perm, Russia

E-mail: lyudmilaklimenko@gmail.com

Abstract. Linear stability analysis within the solute analogy of Horton–Rogers–Lapwood (HRL) problem has been investigated. Solute immobilization (solute sorption) of nanoparicles by the porous medium is taken into account within the fractal model of MIM approach. The solute concentration difference between the layer boundaries and the horizontal external filtration flux are assumed as constant. The system of equations that determine the frequency of neutral oscillations and the critical value of the Rayleigh-Darcy number is derived. Neutral curves of the critical parameters on the governing parameters are plotted. Stability maps are obtained numerically in a wide range of parameters of the system. It was found that taking immobilization into account leads to an increase in the critical value of the Rayleigh-Darcy number with an increase in the intensity of the external filtration flux. The case of weak time-dependent external flux is investigated analytically. It was shown that the modulated external flux leads to an increase in the critical value of the Rayleigh–Darcy number and a decrease in the critical wavenumber.

1. Introduction

The present work is devoted to the study of the linear stability problem for solutal convection in a horizontal layer of a porous medium at a given vertical concentration gradient with an imposed external horizontal filtration flux, taking into account the immobilization of the solute.

The problem in a similar formulation, but without immobilization and in the absence of an external flux, was solved in [1] for the first time. It was shown that the convective regime has a form of a set of convective cells, which width is equal to the thickness of the layer. The influence of an external horizontal flux on convective regimes was investigated in [2]. It was shown that, although the presence of external flux leads to generation of an oscillatory mode of the instability, but the intensity of the external flux affects only the frequency of the oscillations. The wavelength of critical perturbations and the critical value of the Rayleigh-Darcy number do not changed. The effect immobilization on the problem under consideration was investigated in [3]. The map of the observed convection regimes was obtained numerically. It was found that taking into account the immobilization the critical values of the parameters become dependent on the external flux intensity. In [3] the immobilization was considered in the framework of the MIM (mobile-immobile medium) model with linear sorption kinetics [4]. However, as shown in [5], the modeling of the immobilization will be more realistic if we use the fractal linear model of the mobile/non-mobile medium (fMIM) [6]. The present paper is devoted to such modeling.

Content from this work may be used under the terms of the Creative Commons Attribution 3.0 licence. Any further distribution of this work must maintain attribution to the author(s) and the title of the work, journal citation and DOI.
Published under licence by IOP Publishing Ltd

2. Problem statement

We study the solutal convection of a mixture of solid nanoparticles and ambient fluid in a horizontal layer of a porous medium under external filtration flux (Figure 1). The system is subjected to the gravity field. The solid nanoparticles are considered as a solute within the continuous approach. We took into account the effect of solute immobilization when the solute particles can stick to the solid matrix of the porous medium and do not move for a sufficient time. According to MIM approach [4], we consider that the total solute concentration is divided into two phases: the mobile phase (with volume concentration C) and the immobile phase (with Q). Solute concentrations at the upper and lower boundaries of the layer are kept constant. The solute is assumed to be heavier than the carrier fluid, thus the difference of densities leads to a solutal convection.

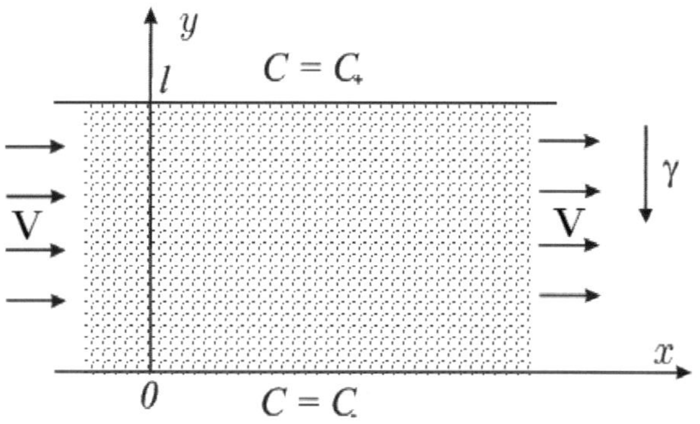

Figure 1. Problem configuration. The layer of thickness l is infinite in the x-direction. The flow through the layer with velocity $\mathbf{V}$ is imposed. The unit vector $\boldsymbol{\gamma}$ shows the direction of gravity field.

The equation of the solutal convection in porous media within the framework of the Boussinesq approximation in dimensionless form can be written as

$$\mathbf{V} = -\nabla p + R_p C \boldsymbol{\gamma}, \tag{1a}$$

$$\nabla \cdot \mathbf{V} = 0, \tag{1b}$$

$$\partial_t (C + Q) = \nabla^2 C - \mathbf{V} \cdot \nabla C, \tag{1c}$$

$$Q = \frac{\lambda}{\Gamma(1-\alpha)} \int_0^t \frac{C(\mathbf{r}, t')}{(t-t')^\alpha} dt'. \tag{1d}$$

The boundary conditions on horizontal and vertical plane read

$$y = 0: \quad C = 0, \quad V_y = 0, \tag{2a}$$

$$y = 1: \quad C = 1, \quad V_y = 0, \tag{2b}$$

$$x = \pm\infty: \quad V_x = \mathrm{Pe}\, f(t), \tag{2c}$$

where ∂_t denotes the time derivative, $\mathbf{V}$ is the two-dimensional flow velocity, p is pressure, $\boldsymbol{\gamma}$ is the unit vertical vector, $\Gamma(x)$ is the Euler gamma function. The kinetics of immobilization (equation (1d)) is described by the fractal MIM model [6] with two parameters: $0 < \alpha \le 1$ is the exponent of the Levy stable law and λ is the mobility parameter. The mobility parameter denotes the portion of the immobile concentration in the total concentration, meanwhile the exponent of the Levy stable law describes the relaxation of the system to the state of dynamical equilibrium between phases. For $\alpha = 0$ the relaxation time is infinite, the process of mobilization is blocked and equilibrium cannot be achieved; whereas the opposite case $\alpha = 1$ corresponds to the instantaneous relaxation ($Q = \lambda C$).

We choose l^2/D, l, $C_0 = C_+ - C_-$, D/l, $D\eta\phi/\kappa$ as the scales for time, length, concentration, velocity and pressure respectively. Here l is the layer thickness, η is dynamic viscosity and D, ϕ, κ are the effective diffusivity, porosity and permeability of porous media. In chosen scales the problem (equations (1a)-(2c)) is characterized by two governing parameters Rayleigh–Darcy number and Péclet number

$$R_p = \frac{\beta_c C_0 g l \kappa \rho_f}{\phi \eta D}, \; \mathrm{Pe} = \frac{Vl}{D}.$$

The problem (equations (1a)-(2c)) admitted uniform seepage solution, namely

$$C_h = y, \quad \mathbf{V}_h = \left(\mathrm{Pe}\, f(t), 0\right), \tag{3}$$

Let us consider the perturbation of the solution (equation (3)) in the following form: $c = C - y$, $\mathbf{v} = (u, w) = \left(V_x - \mathrm{Pe}\, f(t), V_y\right)$, $p = \tilde{p} - p_h$. Thus, in terms of the stream function ($u = -\partial_y \psi$, $w = \partial_x \psi$) the governing system (equations (1a)-(1d)) can be rewritten in linear form as following

$$\partial_t (c + q) + \mathrm{Pe}\, f(t) \partial_x c + \partial_x \psi = \nabla^2 c, \tag{4a}$$

$$\nabla^2 \psi = -R_p \partial_x c, \tag{4b}$$

$$q = \frac{\lambda}{\Gamma(1-\alpha)} \int_0^t \frac{c(\mathbf{r}, t')}{(t - t')^\alpha} dt', \tag{4c}$$

where ∂_x and ∂_y denote corresponding derivatives, while $\nabla^2 = \partial_x^2 + \partial_y^2$ is the two-dimensional Laplace operator. With boundary condition

$$c, q, \psi\big|_{y=0;1} = 0 \tag{5}$$

the system (equations (4),(5)) represents the linear stability problem.

3. Steady external flux

Let us consider the time-independent external filtration flux or $f(t) = 1$. In order to find the solution we applied the Laplace-Fourier transformation method [7] to the governing equation system (equations (4a)-(4c)). Let $\tilde{F}(\mathbf{r}, s) = \int_0^\infty F(\mathbf{r}, t) \exp\{-st\} dt$ is the Laplace transform of some function $F(\mathbf{r}, t)$ varying in time and $\hat{F}(\mathbf{k}, s) = \iiint_\infty F(\mathbf{r}, t) \exp\{-i\mathbf{k} \cdot \mathbf{r}\} d\mathbf{k}$ is the Fourier transform of $F(\mathbf{r}, t)$ varying in space. Thus in Laplace-Fourier space the system (Equations 4a-c) has the form

$$\hat{\tilde{c}}(k, s)\left(s + \lambda s^\alpha\right) - \hat{\tilde{c}}(k, 0) = -k^2 \hat{\tilde{c}} - \mathrm{Pe}\, i k_x \hat{\tilde{c}} - i k_x \hat{\tilde{\psi}}, \tag{6a}$$

$$-k^2 \hat{\tilde{\psi}} = -i k_x R_p \hat{\tilde{c}}, \tag{6b}$$

Substitution Equation 6b into Equation 6a defines the expression for mobile volume concentration in Laplace-Fourier space via

$$\hat{\tilde{c}}(k, s) = \hat{\tilde{c}}(k, 0)\left(s + \lambda s^\alpha + k^2 + i k_x \mathrm{Pe} - k_x^2 R_p k^{-2}\right)^{-1}, \tag{7}$$

which coincide with [5].

The inverse Laplace-Fourier transformation can be expressed as a sum of residues: $2\pi i \sum Res\left(\exp(st)/f(s)\right)$. According to Cauchy's residue theorem we can define the contour integral in a such way that $\exp(st) \to 0$. Thereby in order to analyze the stability of the solution of equation (7) its denominator has to be considered.

Let us find the neutral perturbation (for which $s = -i\omega$). Substitution it into denominator of equation (7) and extraction of real and imaginary parts gives the system of equation, which define the frequency of critical perturbations and the critical value of the Rayleigh–Darcy number as

$$\omega - \lambda \operatorname{Im}\left(i^{3\alpha}\right) \omega^\alpha - k_x \mathrm{Pe} = 0, \tag{8a}$$

$$R_p = \frac{\left(\pi^2 n^2 + k_x^2\right)^2}{k_x^2} + \frac{\left(\pi^2 n^2 + k_x^2\right)}{k_x^2} \lambda \operatorname{Re}\left(i^{3\alpha}\right)\omega^\alpha, \tag{8b}$$

It should be noted that without immobilization ($\lambda = 0$) these expressions (equation (8)) match the well-known solution, which has been first obtained in [2] for the case when the external flux has no influence on the convection threshold. Indeed substituting $\lambda = 0$ one can get

$$\omega = k_x \mathrm{Pe}, \tag{9a}$$

$$R_p = \frac{\left(\pi^2 n^2 + k_x^2\right)^2}{k_x^2}. \tag{9b}$$

This convection threshold (equation (9b)) is the same for the case without external filtration flux. As it can be seen from equation (8a), if external flux is absent ($\mathrm{Pe} = 0$) the oscillatory convection does not occur and $\omega = 0$. Thus the critical value of the Rayleigh–Darcy number is the same as described by equation (9b).

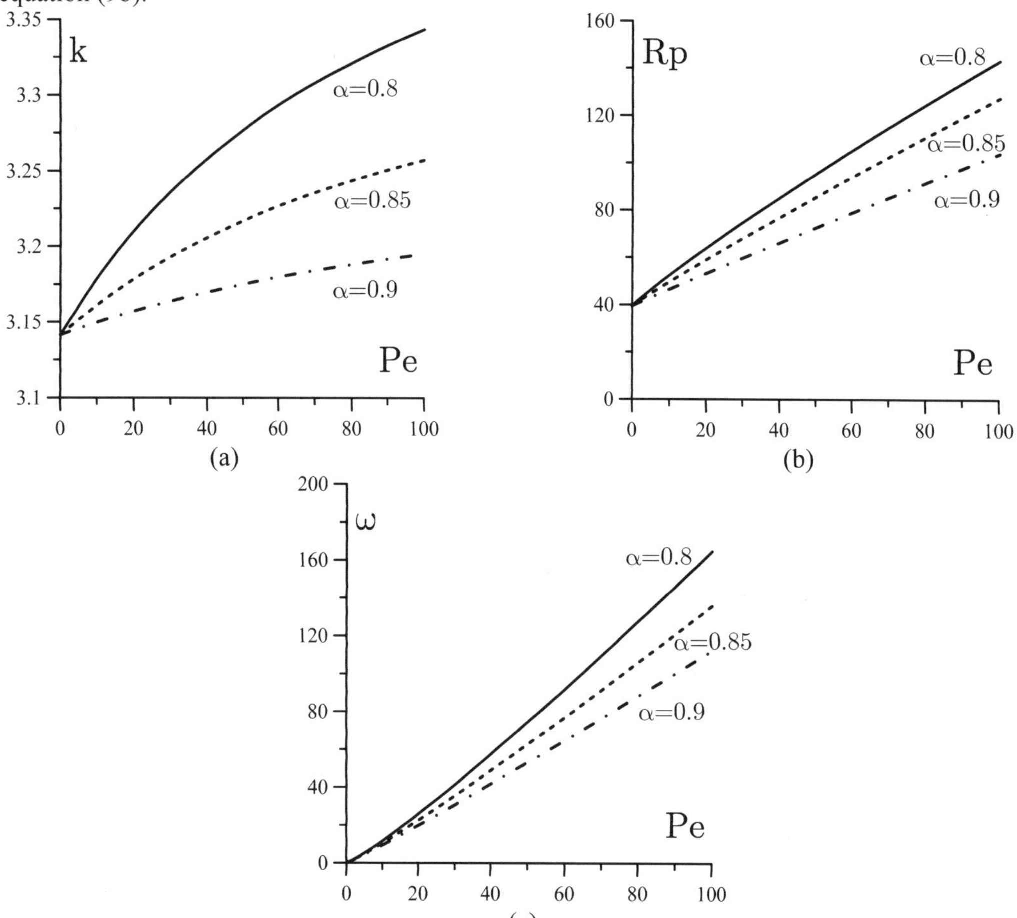

Figure 2. Dependences of the critical wave number (a), the Rayleigh–Darcy number (b) and the oscillation frequency (c) on the Péclet number for $\alpha = 0.8$ (solid line), $\alpha = 0.85$ (dash line) and $\alpha = 0.9$ (dash dot line). $\lambda = 2$.

3.1. Limiting case of instantaneous relaxation. Natural results can be obtained easily for limiting case $\alpha = 1$. From equation (8a) it can be seen that oscillation frequency has the form

$$\omega = \frac{k_x \mathrm{Pe}}{1+\lambda}. \tag{10}$$

Meanwhile equation (8b) for $\alpha = 1$ will give the same result for the critical Rayleigh–Darcy number as equation (9b). These results are reasonable and proof that equations (8a), (8b) are consistent. Indeed, this limiting case corresponds to the linear relation between immobile and mobile concentrations; $Q = \lambda C$, which means that instead of Equation 1c1 one can solve the ordinary diffusion equation with additional factor for time derivative $(1+\lambda)$. Thus, the frequency has to be reduced by this factor and convection threshold does not change.

3.2. General case. The numerical computations of equations (8a), (8b) are presented on Figure 2, for the values of the exponent of the Levy stable law widely used in experiments.

As it is shown on Figure 2, the dependencies become nonlinear with sufficient values of Peclet number. Hence, in case of steady external filtration flux the account of immobilization leads to increase of convection threshold with growth of filtration intensity.

4. Weak non-steady external flux

Let us consider the influence of time-dependent filtration flux on the convection threshold, so in this case $f = A\cos\Omega t$. We assume that the flux is weak, it allows to find the solution as a series of $\varepsilon = k_x Pe << 1$:

$$c = c^{(0)} + \varepsilon c^{(1)} + \varepsilon^2 c^{(2)} + \dots$$
$$q = q^{(0)} + \varepsilon q^{(1)} + \varepsilon^2 c^{(2)} + \dots$$
$$\gamma = \varepsilon \gamma^{(1)} + \varepsilon^2 \gamma^{(2)} + \dots$$

where $\gamma = k_x^2 R_p k^{-2} - k^2$ is the critical Rayleigh–Darcy number without immobilization (see equation (9b)).

In Laplace-Fourier space the system (equations (4a)-(4c)) yields

$$\hat{\tilde{c}}(k,s)(s+\lambda s^\alpha) - \hat{\tilde{c}}(k,0) = \gamma \hat{\tilde{c}}(k,s) - i\varepsilon \widehat{A\cos\Omega t\, c(r,t)}. \tag{11}$$

In the first-order the mobile volume concentration can be written as

$$\hat{\tilde{c}}^{(1)}(k,s) = -\frac{\hat{\tilde{c}}(k,0)}{(s+\lambda s^\alpha)}\left(\frac{\gamma^1}{(s+\lambda s^\alpha)} + iAf(k,s)\right), \tag{12}$$

where

$$f = \frac{1}{2\pi i}\int_0^\infty \cos\Omega t\, e^{-st} \int_{-i\infty}^{i\infty} (p+\lambda p^\alpha)^{-1} e^{pt}\, dt dp = \frac{1}{2}\left(\frac{1}{s+i\Omega+\lambda(s+i\Omega)^\alpha} + \frac{1}{s-i\Omega+\lambda(s-i\Omega)^\alpha}\right).$$

The inverse transformations of equation (12) give a linear growth due the term with γ^1, which corresponds to the pole of order 2. Thus, in the first order the perturbations of the Rayleigh–Darcy number vanishes: $\gamma^1 = 0$; meanwhile the solution for mobile concentration becomes

$$\hat{\tilde{c}}^{(1)}(k,s) = \frac{\hat{\tilde{c}}(k,0)}{s+\lambda s^\alpha}\left(-\frac{iA}{2}\left(\frac{1}{s+i\Omega+\lambda(s+i\Omega)^\alpha} + \frac{1}{s-i\Omega+\lambda(s-i\Omega)^\alpha}\right)\right). \tag{13}$$

In the second-order by analogy the solution of equation (11) can be obtained as

$$\hat{\tilde{c}}^{(2)}(k,s) = \frac{\hat{\tilde{c}}(k,0)}{s+\lambda s^\alpha}\left(-\frac{A^2}{4(s+\lambda s^\alpha)}\left(\frac{1}{(s+i\Omega+\lambda(s+i\Omega)^\alpha)} + \frac{1}{(s-i\Omega+\lambda(s-i\Omega)^\alpha)}\right) + \frac{\gamma^2}{s+\lambda s^\alpha} + \dots\right).$$

By vanishing the terms corresponding to the pole of order 2 the rectification of γ gives

$$\gamma^{(2)} = \frac{A^2}{4}\frac{2\lambda(\Omega)^\alpha \cos(\alpha\pi/2)}{\Omega^2 + \lambda^2\Omega^{2\alpha} + 2\lambda(\Omega)^\alpha \cos(\alpha\pi/2)}. \tag{14}$$

Thus, the expected correction to the critical Rayleigh–Darcy number is

$$R_p = \frac{\left(\pi^2 n^2 + k_x^2\right)^2}{k_x^2} + \left(\pi^2 n^2 + k_x^2\right)^2 \frac{A^2 \mathrm{Pe}^2}{4} \frac{\lambda\Omega^\alpha \cos(\pi\alpha/2)}{\Omega^2 + \lambda^2\Omega^{2\alpha} + \lambda\Omega^\alpha \cos(\pi\alpha/2)}. \tag{15}$$

Note, that in the absence of immobilization ($\lambda = 0$) or in limiting case with $\alpha = 1$, the correction (equation (14)) vanishes, which coincide with known results [8]. The dependence of the correction on external flux frequency is plotted on Figure 3. As it is expected the correction decrease dramatically with growth of external frequency.

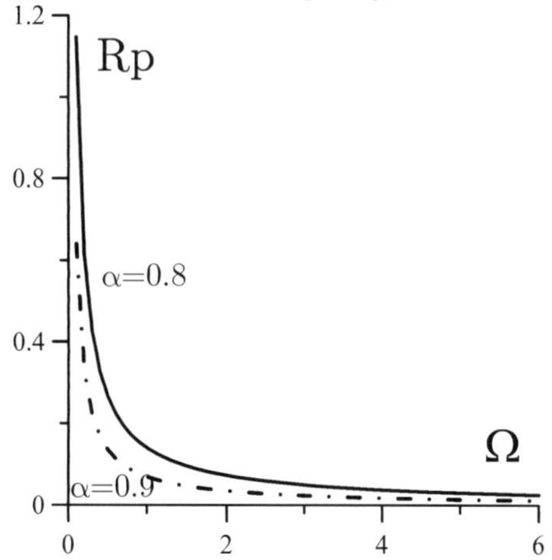

Figure 3. Dependence of correction to the critical Rayleigh–Darcy ($\gamma^{(2)}$ from equation (14)) on the external frequency Ω for $\alpha = 0.8$ (solid line) and $\alpha = 0.9$ (dash dot line). $\lambda = 2$; $A = 2$.

5. Conclusion

The linear stability problem for solutal convection in a horizontal layer of a porous medium at a given vertical concentration gradient with imposed external horizontal filtration flux is studied. The effect of solute immobilization when the solute particles can stick to the solid matrix of the porous medium is taken into account in the framework of fractal linear MIM model. The solute concentration difference between the layer boundaries and the horizontal external filtration flux are assumed as constant. Two cases are considered: steady and weak time-dependent external filtration flux.

In the case of steady flux the system of equations that determine the frequency of neutral oscillations and the critical value of the Rayleigh-Darcy number is derived (equations (8a),(8b)). Neutral curves of the critical parameters on the governing parameters are calculated. Stability maps are obtained numerically in a wide range of parameters of the system. It was found that taking immobilization into account leads to an increase in the critical value of the Rayleigh-Darcy number with an increase in the intensity of the external filtration flux.

The case of weak time-dependent external flux is investigated analytically. The results show that at low flow velocity the modulation makes additional positive contribution to the critical value of the Rayleigh–Darcy number. As a result, the formula, which determines the threshold value, consists of two positive terms: the first term is the main critical value without external flow, and the second term is related to the effect of the external flow modulation (equation (15)). Thus the introduction of modulation leads to stabilization.

The work was supported by the Russian Science Foundation (grant 14-21-0090).

References

IOP Conf. Series: Journal of Physics: Conf. Series **894** (2017) 012043 doi:10.1088/1742-6596/894/1/012043

[1] Horton C W and Rogers F T 1945 Convection currents in a porous medium *J. Appl. Phys* **16** 367-70
[2] Prats M 1963 The effect of horizontal fluid flow on thermally induced convection currents in porous mediums *J. Geophys. Res.* **71** 4835-8
[3] Maryshev B S 2015 The effect of sorption on linear stability for the solutal Horton–Rogers–Lapwood problem *Transport Porous Med.* **109** 747-64
[4] Van Genuchten M T and Wierenga P J 1976 Mass transfer studies in sorbing porous media I. Analytical solutions *Soil Sci. Soc. Am. J.* **40** 473-80
[5] Maryshev B 2016 A non-linear model for solute transport, accounting for sub-diffusive concentration decline and sorption saturation *Math. Model. Nat. Phenom.* **11** 179-90
[6] Schumer R, Benson D A, Meerschaert M M and Bauemer B 2003 Fractal mobile/immobile solute transport *Water Resour. Res.* **39** 1-10
[7] Metzler R and Klafter J 2000 The random walk's guide to anomalous diffusion: a fractional dynamics approach *Phys. Rep.* **339** 1-77
[8] Lyubimov D V and Teplov V S 1998 Stability of the modulated liquid seepage in a porous medium heated from below *Hydrodynamics* **11** 176-90

IOP Conf. Series: Journal of Physics: Conf. Series **894** (2017) 012044 doi:10.1088/1742-6596/894/1/012044

Stability of a horizontal shear flow

P V Kovtunenko[1,2]

[1] Lavrentyev Institute of Hydrodynamics of SB RAS, 630090 Novosibirsk, Russia
[2] Novosibirsk State University, 630090 Novosibirsk, Russia

E-mail: pkovtunenko@gmail.com

Abstract. In this work we study the stability of horizontal shear flows of an ideal fluid in an open channel. Stability conditions are derived in terms of the theory of generalized hyperbolicity of motion equations. We show that flows with monotonic convex profile are always stable, whereas flows with an inflexion point in the velocity profile might become unstable. To illustrate the criteria we give simple examples for stable and unstable flows. Then we derive a multilayered model that is an approximation of the original model and features a continuous piecewise linear velocity profile. We also formulate sufficient hyperbolicity conditions for the multilayered model.

1. Introduction

When modelling flows of fluid in channels and riverbeds it is sometimes necessary to investigate non-linear integro-differential systems of equations. This is caused mainly by a heterogeneity of the flow due to the flow geometry and friction effects. Here we study the horizontal-shear flow model of an ideal fluid in an open channel proposed in [1] that allows to generalize the concept of subcritical and supercritical flow and to study a flow around local constrictions and expansions [2]. Classes of exact solutions of this model are built and investigated in [3]. The current work is focused on the stability analysis of the flow. The stability criteria are derived in terms of generalized hyperbolicity — a theoretical approach developed by V. M. Teshukov [4, 5]. First we derive criteria similar to the Rayleigh and Fjortoft theorem [6]. Then we consider a multilayered approximation of the motion equations, characterized by a piecewise constant potential vorticity, which has proven to be efficient when describing shear flows [7, 8]. We also derive sufficient hyperbolicity conditions for the multilayered model using the same theoretical approach. A similar analysis has been performed in [9] for Benney equations [10] but for a wider class of flows that also features slippage between the layers of the multilayered model.

2. Mathematical model

We consider a system of equations that describes flow of a fluid in an open narrow channel [1, 2]:

$$\begin{gathered} u_t + uu_x + vu_y + gh_x = 0, \quad h_y = 0, \\ h_t + (uh)_x + (vh)_y = 0, \quad v\big|_{y=0} = v\big|_{y=Y} = 0, \end{gathered} \tag{1}$$

here t, x, y are time and Cartesian coordinates; $u(t,x,y), v(t,x,y)$ are horizontal velocity vector components; $h(t,x)$ is the flow depth; g is the gravitational constant; Y is the channel width.

Content from this work may be used under the terms of the Creative Commons Attribution 3.0 licence. Any further distribution of this work must maintain attribution to the author(s) and the title of the work, journal citation and DOI.
Published under licence by IOP Publishing Ltd

The spatial configuration of the flow is depicted in Figure 1. From equations (1) follows the conservation of potential vorticity $\Omega = u_y/h$ along the trajectories:

$$\Omega_t + u\Omega_x + v\Omega_y = 0. \tag{2}$$

From (2) it follows that system (1) allows a class of solutions with piecewise constant vorticity.

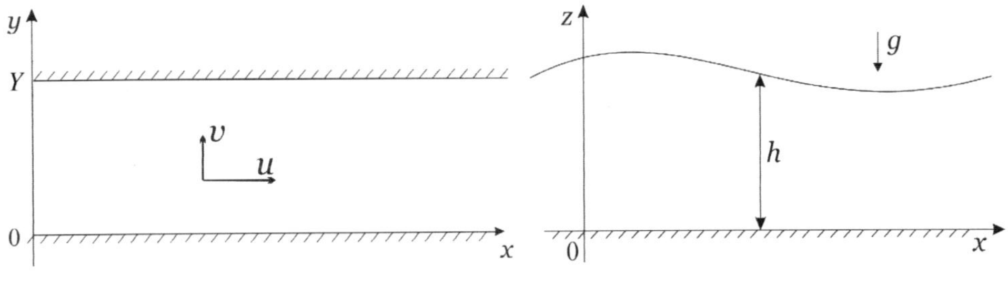

Figure 1. The spatial configuration of the flow.

The stability criteria are derived using the generalization of hyperbolicity theory [4, 5]. Because the hyperbolicity conditions for the system (1) are derived and studied in [1], here we only present them briefly. The characteristic function $\chi(k)$ of the system (1) has the form:

$$\chi(k) = 1 - \frac{gh}{Y}\int_0^Y \frac{dy}{(u-k)^2}. \tag{3}$$

And the system of equations (1) is hyperbolic in terms of [4], if

$$\Delta\arg\frac{\chi^+(u)}{\chi^-(u)} = 0, \quad \chi^\pm \neq 0. \tag{4}$$

Here $\chi^+(u)$ and $\chi^-(u)$ are the limit values of function $\chi(u)$ from the upper χ^+ and lower χ^- complex half-plane on segment $[u_0, u_1]$. The functions $\chi^+(u)$ and $\chi^-(u)$ are derived from integrating (3) by parts and applying Sokhotski–Plemelj theorem [11]:

$$\chi^\pm(u) = 1 + \frac{g}{Y}\left[\frac{1}{\Omega_1(u_1-u)} - \frac{1}{\Omega_0(u_0-u)} - \int_0^Y \frac{\partial}{\partial y'}\left(\frac{1}{\Omega(y')}\right)\frac{dy'}{u(y')-u} \mp \frac{\pi i}{u_y}\left(\frac{1}{\Omega}\right)_y\right].$$

The indices "0" and "1" correspond to the values on channel walls $y = 0$ and $y = Y$ respectively. So if the conditions (4) hold, then the characteristic equation $\chi(k) = 0$ has real roots only [1, 5].

3. Stability condition

It is convenient to verify hyperbolicity conditions for a flow with a monotonic velocity profile $u = U(y)$, $U'(y) > 0$ using the following function:

$$\Psi^\pm(U) = m(U)\chi^\pm(U), \quad m(U) = (U_1 - U)(U - U_0) > 0.$$

Significant here is that functions

$$\Psi^{\pm}(U) = m(U) + \frac{gh}{Y}\Bigg(\frac{U - U_0}{U_1'} + \frac{U_1 - U}{U_0'} +$$
$$+m(U)\int_0^Y \frac{U(\nu)''d\nu}{(U'(\nu))^2(U(\nu) - U(y))} \pm \pi i m(U)\frac{U''}{(U')^3}\Bigg)$$

do not have any singularities at $U = U_0, U = U_1$ due to the choice of $m(U)$ which, in it's turn, obviously has no effect on the conditions (4). On the complex plane (z_1, z_2) let us construct a closed contour C, which consists of contours C^+ and C^-, which are set parametrically:

$$C^+ : \quad z_1 = \mathrm{Re}[\Psi^+(U)], \quad z_2 = \mathrm{Im}[\Psi^+(U)];$$
$$C^- : \quad z_1 = \mathrm{Re}[\Psi^-(U)], \quad z_2 = \mathrm{Im}[\Psi^-(U)];$$

Note that C^+ and C^- are symmetric with respect to the z_1 axis. If the zero point $(0, 0)$ of the complex plane lies inside the contour C, then the argument of $\Psi^{\pm}$ will increase which indicates existence of complex characteristics. Stability conditions for flows that are described by Benney equations are formulated in [9]. Here we formulate similar stability criteria for the model (1). These models are mathematically equivalent, so only short explanations are given.

Statement 1 *Any flow with a monotonic convex velocity profile is stable (Rayleighs inflexion point theorem analogue).*

Indeed, let $U' > 0$, $U'' \neq 0$, then $\Psi^{\pm}(U_0) > 0$, $\Psi^{\pm}(U_1) > 0$, and $\mathrm{Im}\Psi^{\pm}$ does not change its sign on the interval $[U_0, U_1]$. Thus, given the inequality $\Psi^{\pm}(U_c) > 0$, where U_c is any point from $[U_0, U_1]$, the argument of Ψ does not grow.

To illustrate this, consider a flow with a monotonic convex velocity profile given by equation:

$$U(y) = \frac{3y}{2y + 1}, \quad y \in [0, 1]. \tag{5}$$

For the sake of simplicity let us consider $h = g = Y = 1$. Then the velocity profile of the flow is shown in Figure 2, and the corresponding contour C^- in Figure 3.

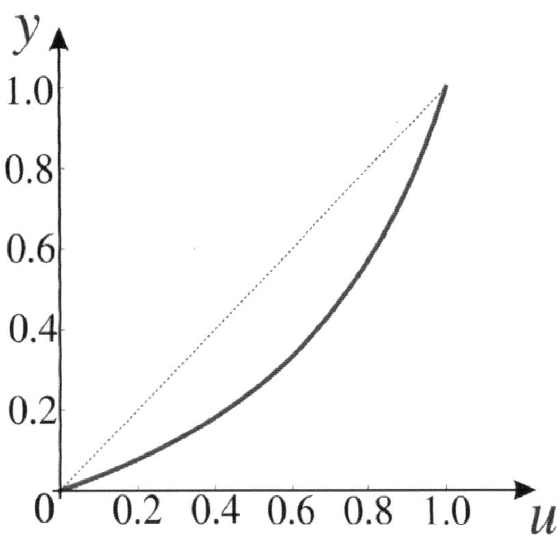

Figure 2. Velocity profile given by (5)

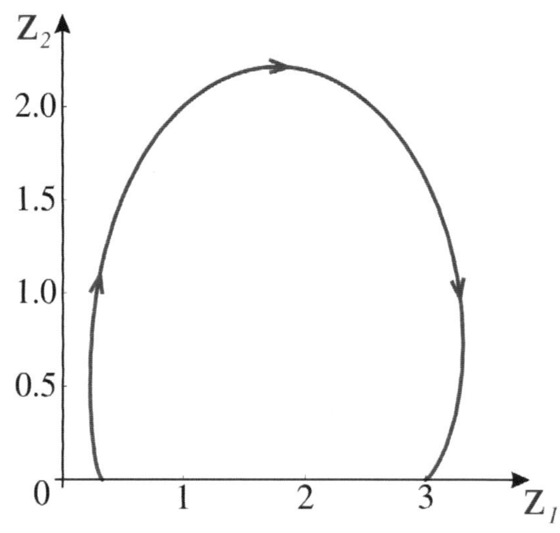

Figure 3. Contour C^- corresponding to (5)

Statement 2 *If $U''(U-U_c) \geq 0$, where $U_c = U(y_c)$, and y_c is the inflexion point $U''(y_c) = 0$, then the flow is stable (Fjortofts theorem analogue).*

Similar to the previous statement, $\Psi^{\pm}(U_0) > 0$, $\Psi^{\pm}(U_1) > 0$, and the sign of $\text{Im}\Psi^{\pm}$ is the same as of U''. Therefore, verification of stability reduces to determination of the sign of $\Psi^{\pm}(U_c)$.

$$\Psi^{\pm}(U_c) = m(U_c) + \frac{gh}{Y}\left(\frac{U_c - U_0}{U_1'} + \frac{U_1 - U_c}{U_0'} + m(U_c)\int_0^Y \frac{U(\nu)''(U(\nu) - U_c)d\nu}{(U'(\nu))^2(U(\nu) - U(y))^2}\right).$$

Considering that $m(U_c) \geq 0$ and $U''(U-U_c) \geq 0$, the following inequality holds $\Psi^{\pm}(U_c) > 0$. This means that the flow is stable.

To illustrate the second statement, consider a flow with a velocity profile given by equation:

$$U(y) = y^3 - y^2 + y, \quad y \in [0,1]. \tag{6}$$

For the sake of simplicity let us consider $h = g = Y = 1$. Then the velocity profile of the flow is shown in Figure 4, and the corresponding contour C^- in Figure 5.

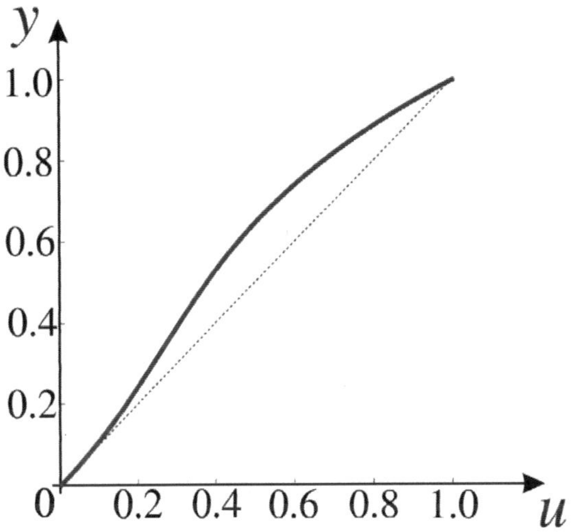

Figure 4. Velocity profile given by (6)

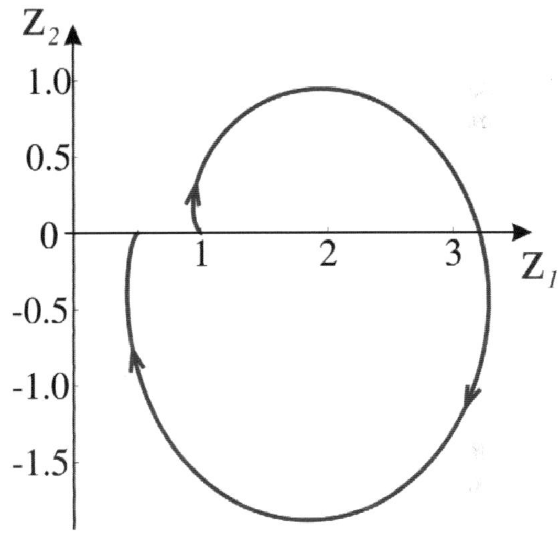

Figure 5. Contour C^- corresponding to (6)

Statement 3 *If $U''(U-U_c) \leq 0$, where $U_c = U(y_c)$, and y_c is the inflexion point $U''(y_c) = 0$ then the flow might be unstable.*

Though in this case all the same conditions as in Statement 2 hold, here we cannot guarantee that the following inequality holds: $\Psi^{\pm}(U_c) > 0$. Let us construct an example where a non-zero argument increase is possible, and hence the flow might become unstable. This will prove Statement 3. Consider a flow with a velocity profile given by equation:

$$U(y) = \frac{\tanh((y-y_c)a) + \tanh(ay_c)}{\tanh((1-y_c)a) + \tanh(ay_c)}, \quad y \in [0,1]. \tag{7}$$

Here we also consider $h = g = Y = 1$. The second derivative $U''(y)$ will be equal zero at the inflection point $y_c = 0.53$. Consider two different cases: $a = 2.4$ and $a = 3.6$. The velocity profiles of the flow are shown in Figure 6 and the corresponding contour C^- in Figure 7.

The dotted lines correspond to $a = 2.4$. In this case the point $(0,0)$ does not lie inside the contour C^-, which means that the argument increase of the function χ is equal zero and the

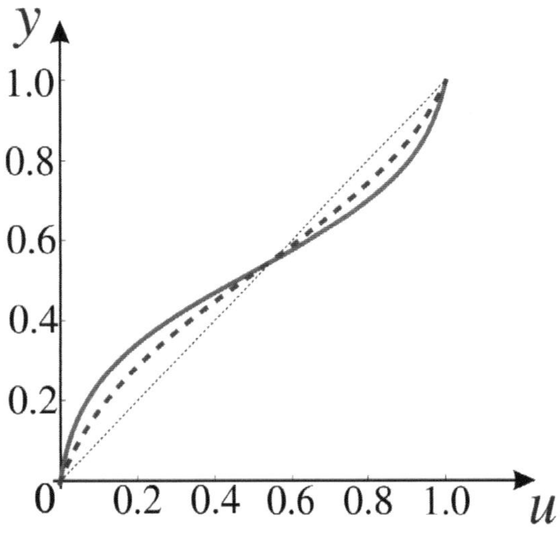

Figure 6. Velocity profile given by (7) for $a = 2.4$ (dotted) and $a = 3.6$ (solid)

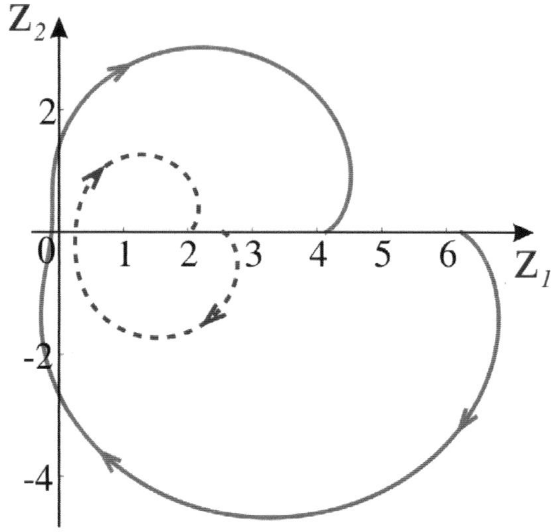

Figure 7. Contour C^- corresponding to (7) for $a = 2.4$ (dotted) and $a = 3.6$ (solid)

flow is stable. Solid lines represent the velocity profile and the contour C^- when $a = 3.6$. In this case the point $(0, 0)$ lies inside the contour C^-, so the flow is unstable.

In this way, stability conditions similar to Rayleigh's (Statement 1) and Fjortoft's (Statement 2) theorems [6] are formulated for the model (1).

4. Multilayered model

Now let us consider a class of flows with a piecewise linear velocity profile:

$$u = (y - y_{i-1})\Omega_i h + u_{i-1}, \quad y \in (y_{i-1}, y_i), \quad i = 1, ..., N. \tag{8}$$

Here the vorticities Ω_i in every single layer $\eta_i = y_i - y_{i-1}$ are constant; the values u_i, v_i are the velocity vector components at $y = y_i$ $(y_0 = 0, y_N = Y)$. At the border between layers the following kinematic conditions hold:

$$\frac{\partial y_i}{\partial t} + u_i \frac{\partial y_i}{\partial x} = v_i, \quad i = 1, ..., N + 1. \tag{9}$$

The substitution of the profile (8) into equations (1) result in a system of $N + 1$ equations:

$$\frac{\partial u_0}{\partial t} + u_0 \frac{\partial u_0}{\partial x} + g \frac{\partial h}{\partial x} = 0, \qquad \frac{\partial m_i}{\partial t} + \frac{\partial(\bar{u}_i m_i)}{\partial x} = 0, \tag{10}$$

that is used to determine the mass $m_i = h\eta_i$ of the fluid in each layer and velocity u_0 at the channel border $y = 0$. The depth of the fluid h and the average velocities in each layer $\bar{u}_i = (u_i + u_{i-1})/2$ are given by formulae:

$$h = \frac{1}{Y}\sum_{j=1}^{N} m_j, \quad \bar{u}_i = u_0 - \frac{1}{2}\Omega_i m_i + \sum_{j=1}^{i} \Omega_j m_j.$$

Horizontal velocities at the channel borders are calculated using the formula $u_i = y_i h \Omega_i + u_{i-1}$.

To analyse the stability of a flow in the framework of the model (10), we also use the more general result presented in [1]. Considering the piecewise linear velocity profile (8), the characteristic function (3) is rewritten as follows:

$$\chi(k) = 1 - \frac{g}{Y\Omega_1}\frac{1}{u_0 - k} + \frac{g}{Y\Omega_N}\frac{1}{u_N - k} - \frac{g}{Y}\sum_{i=1}^{N-1}\left(\frac{1}{\Omega_{i+1}} - \frac{1}{\Omega_i}\right)\frac{1}{u_i - k}. \tag{11}$$

Let us rewrite motion equations (10) in the form $U_t + A(U)U_x = 0$, where U is the vector of unknown functions and A is a square matrix of size $N + 1$. Then the equation that determines eigenvalues of A is connected to the characteristic function $\chi(k)$ in the following way:

$$\det(A - kI) = \chi(k)\prod_{i=0}^{N}(u_i - k).$$

Statement 4. *If all the layer vorticities* Ω_i *are ordered*

$$\Omega_N < \Omega_{N-1} < ... < \Omega_1 \quad \text{or} \quad \Omega_N > \Omega_{N-1} > ... > \Omega_1 \tag{12}$$

then equations (10) *are hyperbolic.*

To prove this statement one has to analyse the behaviour of the function $\chi(k)$ given by equation (11) when the conditions (12) hold. The function $\chi(k)$ has one root inside each interval $k \in (-\infty, u_0)$ and $k \in (u_N, \infty)$, because $\chi(k) \to 1$ when $k \to \pm\infty$ and $\chi(k) \to -\infty$ when $k \to u_0 - 0$, and $\chi(k) \to \infty$ when $k \to u_N + 0$. Furthermore, $\chi(k)' < 0$ on the interval $(-\infty, u_0)$, and $\chi(k)' > 0$ on (u_N, ∞). A similar analysis shows that $\chi(k)$ $\chi(k)$ has $N - 1$ roots on the intervals (u_i, u_{i+1}), $i = 0, ..., N$. A common form of $\chi(k)$ is shown in Figure 8.

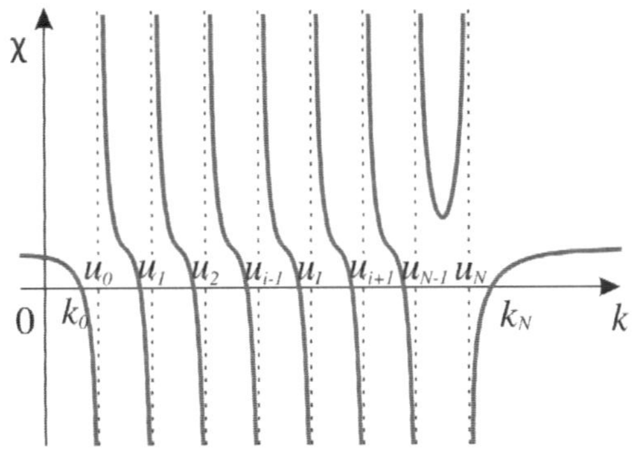

Figure 8. Common form of $\chi(k)$ when $\Omega_1 > ... > \Omega_N > 0$.

As in Section 3, conditions (12) resemble the Rayleigh's inflexion point theorem. Note that a similar result has recently been obtained for flows within the multilayered approximation of Benney equations [9], where possible slippage between the layers is also taken into account.

As for an example let us consider a case where the conditions (12) do not hold and the flow might become unstable, like in Statement 3. We consider $N = 3$, because in case of $N = 2$ all the velocity profiles are convex and thus the flow is always stable. In case of $N = 3$ the characteristic equation (11) is rewritten as follows:

$$\chi(k) = 1 - \frac{g}{Y}\left(\frac{u_1 - u_0}{\Omega_1(u_1 - k)(u_0 - k)} + \frac{u_2 - u_1}{\Omega_2(u_2 - k)(u_1 - k)} + \frac{u_3 - u_2}{\Omega_3(u_3 - k)(u_2 - k)}\right).$$

For the sake of simplicity here we take $g = Y = 1$, $\eta_1 = \eta_3 = 0.35$, $\eta_2 = 0.3$ and $u_0 = 0$.

Now, for the first example let us consider the following distribution of vorticities $\Omega_1 = 1.2$, $\Omega_2 = 0.2$ $\Omega_3 = 0.6$. This vorticity distribution does not satisfy the condition (12). The velocity profile and the characteristic function of this flow are shown in Figure 9. Here we see that the characteristic function $\chi(k)$ has four real roots and we can conclude that the flow is stable, although the condition (12) does not hold.

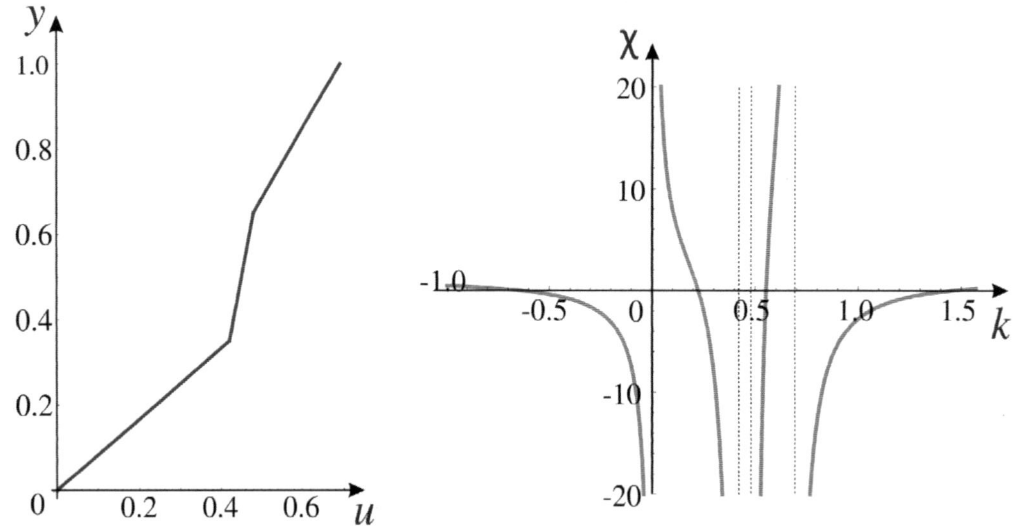

Figure 9. Velocity profile (left) and $\chi(k)$ (right) when $\Omega_1 = 1.2$, $\Omega_2 = 0.2$ $\Omega_3 = 0.6$.

For the second example let $\Omega_1 = 0.6$, $\Omega_2 = 1.2$ $\Omega_3 = 0.2$. Here the condition (12) also does not hold. The velocity profile and the characteristic function are shown in Figure 10. Here we see that the characteristic function $\chi(k)$ has only two real roots, and thus this flow is unstable.

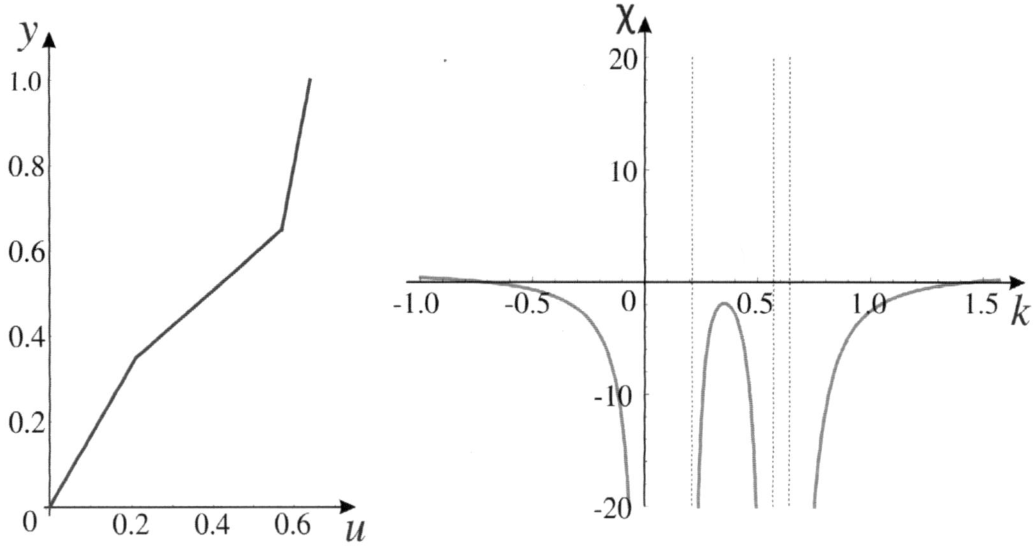

Figure 10. Velocity profile (left) and $\chi(k)$ (right) when $\Omega_1 = 0.6$, $\Omega_2 = 1.2$ $\Omega_3 = 0.2$.

IOP Conf. Series: Journal of Physics: Conf. Series **894** (2017) 012044 doi:10.1088/1742-6596/894/1/012044

5. Conclusion

For horizontal shear flows of ideal fluid in an open channel we obtain stability conditions, which are derived in terms of the theory of generalized hyperbolicity. These conditions are similar to Rayleigh and Fjortoft theorems. In particular we show that flows with a monotonic convex profile are always stable, whereas flows with an inflexion point in the velocity profile might become unstable. Illustrative examples for both stable and unstable flows are given.

We also study a class of solutions that is used in a number of works [7, 8] and is described by a piecewise linear velocity profile. For this class we derive a multilayered model that is an approximation of the original model. We also formulate sufficient hyperbolicity conditions for this model, verification of which reduces to analysis of the characteristic function. Illustrative examples that feature both stable and unstable flows are shown.

Acknowledgments

This work was supported by the Russian Foundation for Basic Research (Grant 16-01-00156) and by the President Grant for Leading Russian Scientific Schools of the Russian Federation (Grant N.Sh.-8146.2016.1). The author is grateful to A.A. Chesnokov for fruitful discussions and helpful tips.

References

[1] Chesnokov A A and Liapidevskii V Yu 2009 *J. Appl. Mech. Tech. Phys.* **50** 220-8
[2] Liapidevskii V Yu and Chesnokov A A 2009 *Fluid Dyn.* **44** 903-16
[3] Kovtunenko P V and Chesnokov A A 2011 *Sib. J. Ind. Math.* [in Russian] **14** 50-7
[4] Teshukov V M 1985 *Dokl. Akad. Nauk.* **284** 555-9
[5] Teshukov V M 1994 *J. Appl. Mech. Tech. Phys.* **35** 823-31
[6] Drazin P G 2002 *Introduction to Hydrodynamic Stability* (Cambridge: Cambridge University Press) p 278
[7] Teshukov V M, Russo G and Chesnokov A A 2004 *Math. Models Methods Appl. Sci.* **14** 1451-79
[8] Chesnokov A A and Kovtunenko P V 2014 *Stud. Appl. Math.* **132** 50-64
[9] Chesnokov A A, El G A, Gavrilyuk S L and Pavlov M V 2017 *SIAM J. Appl. Math.* **77** 1068-87
[10] Benney D J 1973 *Stud. Appl. Math.* **52** 45-50
[11] Gakhov F D 1990 *Boundary Value Problems* (New York: Dover) p 561

On catastrophic fracture of steel structures at temperatures lower than cold brittleness threshold

V M Kornev

Lavrentyev Institute of Hydrodynamics SB RAS, 15 Prospect Lavrentyeva, Novosibirsk, 630090, Russia

E-mail: kornev@hydro.nsc.ru

Abstract. The paper considers crack propagation in elements of homogeneous steel structures and those with welded joints. For analysis of failure of the structures, diagrams of quasi-brittle fracture have been plotted. When constructing quasi-brittle fracture diagrams, the model of elastic-plastic material having an ultimate strain was used. The data report for quasi-brittle fracture diagrams of common elements of structures has been given. Analysis of parameters used in the proposed model was carried out for temperatures near or lower the brittleness threshold. Parameters of the model are selected from two laboratory experiments (critical stress intensity factor and classical stress-strain diagram) performed at appropriate temperatures. It has been established that weld structures with cracks in the vicinity of a welded joint exhibit decreased crack toughness. The effect of structure break under monotonic loading conditions is clearly visible inasmuch as ultimate loads essentially decrease with increasing a crack length. The attention is given to the parameter characterizing plastic material deformation and exhaustion of plasticity resource under preliminary plastic material deformation. After the plasticity resource is exhausted, the temperature of brittleness threshold approaches a room temperature.

1. Introduction

Uzhik [1] describes (in the second chapter of his book) breakdowns of bridges, large petroleum reservoirs, gas holders, and ships at low temperatures. Given in Fig 6 from [1] are typical points of stress concentrations and nucleation cracks caused by them, which can lead to failure of a structure with decreasing temperature. Crack nucleation (points of stress concentrations) provoke [1, p 20]: "...1) welded joints or sites adjacent to shuts; 2) points of stress concentrations caused by the form of structural elements; 3) local defects (flaws, hollows, etc.) arising during production of some structural elements or due to poor quality treatment".When studying catastrophic breakdowns, it is worthwhile to pay attention to the loading that, as a rule, is a force controlled. Break of a structure occurs under action of loadings due to finite sizes of structural element with a crack.

Except for indicated points of fracture, which are mechanical in nature, attention should be given to degradation of properties of structural steels associated with the structure [2, 3].

A scheme of loading (the case of a force control) of a quasi-brittle material [4-6], which describes catastrophic breakdowns of homogeneous and weld structures due to decrease of a temperature below the cold brittleness threshold is considered below.

2. Diagrams of quasi-brittle fracture of compact specimens

Content from this work may be used under the terms of the Creative Commons Attribution 3.0 licence. Any further distribution of this work must maintain attribution to the author(s) and the title of the work, journal citation and DOI.
Published under licence by IOP Publishing Ltd

Judging from catastrophic breakdowns [1], all structure have, in one form or another, nucleus defects such as cracks. If crack lengths didn't reach a critical value at room temperatures, the structures continued to be used. Mechanical characteristics of the material (steel) change with decreasing temperature. The proposed model [4-6] uses a non-classic scheme of material fracture.

Suppose that when a macro-specimen was tested in laboratory experiment, a $\sigma-\varepsilon$ diagram of quasi-brittle material was obtained for various temperatures T, i.e., $\sigma=\sigma(T)$, $\varepsilon=\varepsilon(T)$. The simplest approximation of a real $\sigma-\varepsilon$ diagram of a tested material is a two-link polygonal line. During approximation, the original material is changed by and elastic ideally-plastic material, which has an ultimate strain. Parameters of this approximation $E,\sigma_Y,\varepsilon_0,\varepsilon_1,\mu$ are as follows: $E=const$ is the modulus of elasticity, $\sigma_Y(T)$ is the yield tensile strength and constant stresses acting in modified fashion of the Leonov-Panasyuk-Dagdale model, $\varepsilon_0(T)$ is the maximum elastic material strain ($\sigma_Y=E\varepsilon_0$), $\varepsilon_1(T)$ is the maximum material strain, $\mu=const$ is the Poisson coefficient. The $\sigma-\varepsilon$ approximation of the diagram within the range $\varepsilon_0<\varepsilon<\varepsilon_1$ may be interpreted as a perfect plasticity. It has been accepted in the study proposed below that the modulus of elasticity and Poisson coefficient are independent of temperature, that approximately corresponds to behavior of steels at low temperatures.

Let r be the grain diameter, more exactly, effective diameter of fracture structure for material with a regular structure [6]. The Neuber-Novozhilov approach permits solutions having a singular component with integrable singularity to be used for structural media. An opening mode crack is under consideration. Let this plane opening mode crack be propagating rectilinearly. Apart from a length l_0 of a real crack-cut, introduce into consideration a length of a model crack-cut. The lengths of model cracks are $l=l_0+\Delta$, the pre-fracture zone with the length Δ being located on the continuation of a real crack. The fracture problem has two linear scales: if the grain diameter r is defined by a material structure, then the second linear scale is developed by the system itself. It is the pre-fracture zone length that is the second linear scale Δ, which varies in accordance with the fact how the following parameters are changed *1)* a real crack length, and *2)* load intensity. It should be emphasized that the critical pre-fracture zone length Δ^* is a well-defined parameter ($l^*=l_0+\Delta^*$ is the critical macro-crack length) under single loading.

When plotting diagrams of quasi-brittle fracture, sufficient fracture criteria [4-6] are used when short, long, and medium length macro-cracks are considered. The sufficient criterion may be represented in the form of two relations for short macro-cracks

$$\frac{1}{r}\int_0^r \sigma_y(x,0)dx=\sigma_Y\,, \tag{1}$$

$$2v(-\Delta^*,0)=\delta^*\,. \tag{2}$$

Here $\sigma_y(x,0)$ are normal stresses on the crack continuation; Oxy is the rectangular coordinate system, the coordinate origin being coincident with a model crack tip in the modified Leonov-Panasyuk-Dagdale model, where the x-axis is directed along the crack plane and the y-axis is directed along the normal to the crack plain; $2v=2v(x,0)$ is the model macro-crack opening ($x<0$); δ^* is the critical opening of this crack, Δ^* is the critical length of a pre-fracture zone (critical values derived via sufficient and necessary fracture criteria are labeled by superscripts *and ^{0}, respectively). Attention is given to the fact that the proposed criterion (1) and (2) is two parametric.

A field of normal stresses $\sigma_y(x,0)$ on the continuation of model cracks $x>0$ can be represented as a sum of two summands

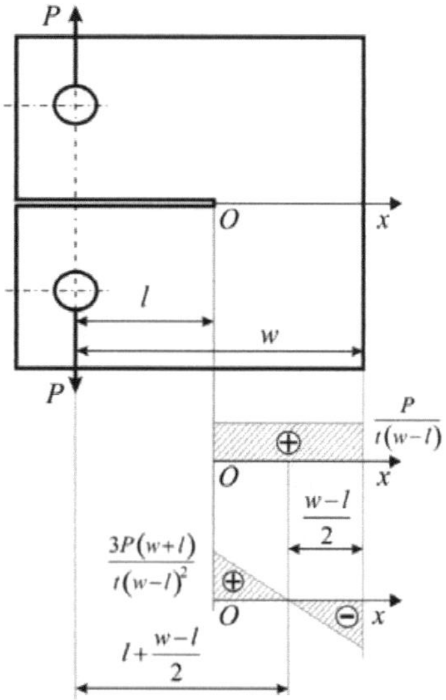

Figure 1. The compact specimen with a sharp crack

$\sigma_y(x,0) \cong K_{\mathrm{I}} / (2\pi x)^{1/2} + \sigma_{nom}$, $K_{\mathrm{I}} = K_{\mathrm{I}\infty} + K_{\mathrm{I}\Delta} > 0$,
$K_{\mathrm{I}\infty} > 0$, $K_{\mathrm{I}\Delta} < 0$, (3)

where σ_{nom} are nominal stresses, in other words, estimates of regular members of solutions in the vicinity of model crack tips,and these members have no singularities; $K_{\mathrm{I}\infty}$ is the stress intensity factor (SIF) generated by specified test conditions; $K_{\mathrm{I}\Delta}$ is the SIF generated by constant stresses $-\sigma_Y$ acting in a pre-fracture zone. The first and second summands in relation (3) are singular and regular parts of the solution, respectively. The total SIF K_{I} at the model crack tip is positive inasmuch as the low-scale yielding is considered.

The specimen with a crack is shown in Fig. 1 (P are forces applied during tests, w and t are the width and thickness of the specimen, respectively). The proposed approximation of normal stresses at tips of model cracks for compact specimens with a sharp crack has the following form at tips of model cracks for compact specimens

$$\sigma_{nom} = \sigma_{1nom} + \sigma_{2nom},\ \sigma_{1nom} = P\left[t(w-l)\right]^{-1},\ \sigma_{2nom} = 3P(w+l)\left[t(w-l)^2\right]^{-1}. \quad (4)$$

Here σ_{1nom} and σ_{2nom} are nominal stresses under tension and bending, respectively. The singularity of nominal stresses σ_{nom} for $l \to w$ reflects the break process of a structure at catastrophic failure of steel structures under conditions of low temperatures.

Under conditions of low-scale yielding for compact specimens with sharp cracks, we have for the total SIF

$$K_{\mathrm{I}} = K_{\mathrm{I}\infty}(P/tw, l, l/w) + K_{\mathrm{I}\Delta}(l, \Delta, \sigma_Y) > 0,\ K_{\mathrm{I}\infty} = (P/tw)Y(l/w)\sqrt{\pi l},\ K_{\mathrm{I}\Delta} \approx -2\sigma_Y\sqrt{2\Delta/\pi}, \quad (5)$$

$$Y(l/w) = 16{,}7 - 104(l/w) + 370(l/w)^2 - 574(l/w)^3 + 361(l/w)^4.$$

Opening of model cracks $2v$ in a compact specimen with a sharp crack is represented in the form

$$2v(-x,0) \approx \left[(\eta+1)/G\right]K_{\mathrm{I}}(l,\Delta,l/w)\left[(-x)/2\pi\right]^{1/2}, K_{\mathrm{I}} > 0, \eta_d = 3-4\mu, \eta_s = (3-\mu)/(1+\mu). \quad (6)$$

Coefficients η_d and η_s are given for the plane strain and plane stress states, $G = E/2(1+\mu)$ is shear modulus, and μ is the Poisson ratio.

The critical opening of model cracks δ^* in relation (2) is calculated as

$$\delta^* = (\varepsilon_1 - \varepsilon_0)a, \quad (7)$$

$$a = \left\{\left[K_{\mathrm{I}\infty 0}\right]^2 / \left[2\pi(\sigma_Y)^2\right]\right\}\left[3/2 + (1-2\mu)^2\right],\ a_w = \left\{\left[K_{\mathrm{I}\infty 0}\right]^2 / \left[\pi(\sigma_Y)^2\right]\right\}\left[3/2 + (1-2\mu)^2\right].$$

Identify the pre-fracture zone width in relation (7) with the width of plasticity zones at the real crack tip for compact specimens with sharp cracks made of a homogeneous material a and material with weld joint a_w (the joint plane is located in the tip plane [9-12]). The plasticity zone width a_w in the material with weld joint is half as large as the plasticity zone width in the homogeneous material, i.e., $a_w = a/2$.

After appropriate transformations, the initial equalities of criterion (1) – (2) go into approximate equalities for compact specimens with sharp cracks, if we take into account relations (3)-(7)

$$\frac{\sigma_\infty^*}{\sigma_Y} \approx \left[\frac{1}{1-l^*/w} + \frac{3\left(1+l^*/w\right)}{\left(1-l^*/w\right)^2} + \left(1 - \frac{3+2\left(1-2\mu\right)^2}{8\pi\left(1-\mu^2\right)} \frac{\varepsilon_1-\varepsilon_0}{\varepsilon_0} \right) Y\left(\frac{l^*}{w}\right)\sqrt{2\frac{l^*}{r}} \right]^{-1} . \quad (8)$$

$$\frac{\Delta^*}{l^*} \approx \frac{\left[3+2\left(1-2\mu\right)^2\right]^2}{2^9\left(1-\mu^2\right)^2} \left(\frac{\varepsilon_1-\varepsilon_0}{\varepsilon_0}\right)^2 \left[\frac{\sigma_\infty^*}{\sigma_Y} Y\left(\frac{l^*}{w}\right)\right]^2$$

$$\left\{\left[3+2\left(1-2\mu\right)^2\right]/8\pi\left(1-\mu^2\right)\right\}\left(\varepsilon_1-\varepsilon_0\right)/\varepsilon_0 < 1. \quad (9)$$

The last inequality (9) is a restriction at which quasi-brittle fracture occurs under conditions of low scale yielding of homogeneous material in a pre-fracture zone.

Under conditions of critical stresses σ_∞^0/σ_Y corresponding to the necessary fracture criterion, we have the following relation

$$\frac{\sigma_\infty^0}{\sigma_Y} \approx \left[\frac{1}{1-l_0/w} + \frac{3\left(1+l_0/w\right)}{\left(1-l_0/w\right)^2} + Y\left(\frac{l_0}{w}\right)\sqrt{2\frac{l_0}{r}} \right]^{-1}, \; l_0 < l^*. \quad (10)$$

If in two laboratory experiments, the critical SIF $K_{\mathrm{I}c}$ and classical $\sigma-\varepsilon$ diagram (more precise, its approximation), then, by three parameters $r, \sigma_Y, \left(\varepsilon_1-\varepsilon_0\right)/\varepsilon_0$, with allowance made for the Poisson ratio μ, it is possible to construct two critical curves σ_∞^*/σ_Y, σ_∞^0/σ_Y in the plane "crack length – stress" within the wide range of variation of crack lengths.

Plotted in Fig. 2 are dimensionless critical stresses σ_∞^0/σ_Y (curves 1 and 4), σ_∞^*/σ_Y (curves 2, 3, 5, and 6) of compact specimens with sharp cracks, solid and dashed lines reflecting behavior of specimens made of homogeneous material and material with weld joint ($\left(\left(\varepsilon_1-\varepsilon_0\right)/\varepsilon_0 = 3\right)$). For a specific implementation of calculations, parameters $w/r = 500$ and $w/r = 2000$ were chosen for curves 1 – 3 and for curves 4 – 6, respectively. In calculations, $\mu = 0$ was taken, inasmuch as the Poisson ration has weak influence on critical parameters. Pairs of curves 1 – 3 and 4 – 6 represent quasi-brittle fracture diagrams for studied specimen made of homogenous material. Pairs of curves 1 – 2 and 4 – 5 represent diagrams for studied specimen made of material with weld joint (the weld joint is located in the plane of a crack).

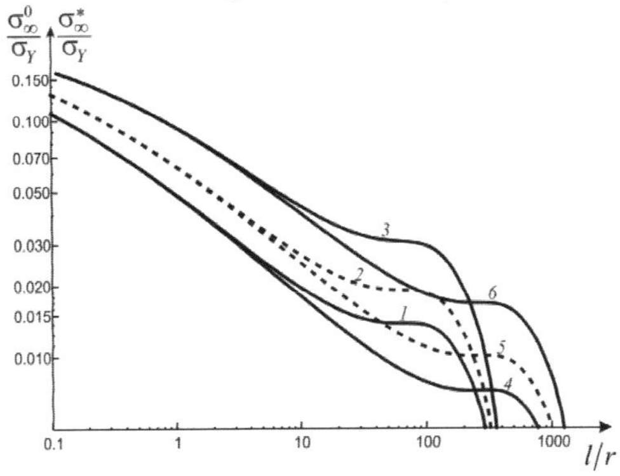

Figure 2. Critical stresses σ_∞^0/σ_Y and σ_∞^*/σ_Y of compact specimens

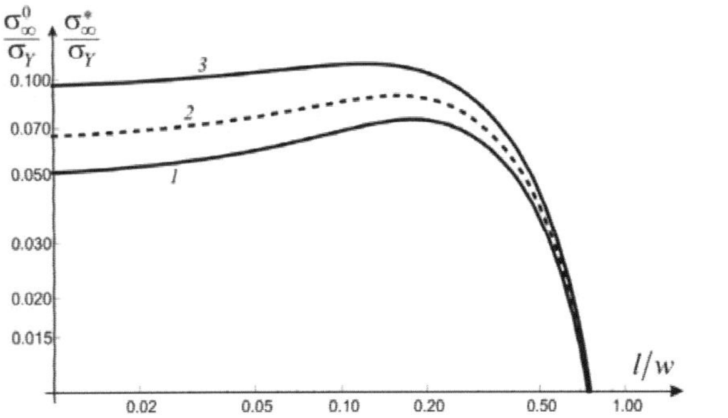

Figure 3. Critical stresses σ_∞^0/σ_Y, σ_∞^*/σ_Y versus l/w

The conditional critical stresses $\sigma_{\infty}^{0} / \sigma_Y$ (curve 1), $\sigma_{\infty}^{*} / \sigma_Y$ (curves 2 and 3) of compact specimen are plotted in Fig.3 versus the l / w ratio. When constructing the curves, relations (8) and (10) were used, the calculations being made for $2l / r = 1$ and $(\varepsilon_1 - \varepsilon_0) / \varepsilon_0 = 3$. Solid and dashed lines correspond to homogeneous and welded structures. The relatively weak dependence of critical stresses on growth of short racks l / w from 0.01 to 0.2 has been revealed.

3. Quasi-brittle fracture diagrams of specimens-counterparts of constructions

Apart from relations (8) and (10) for critical stresses of compact specimens, we give analogous relations for critical stresses (relations (11)-(13)) and representations of fracture diagrams (Figs. 4 and 5) of tensile band with cracks and critical moments of a bending specimens. Solid lines were used for homogeneous structures in Figs. 4 and 5 and dashed lines were used for structures with weld joints in the same figures. All the calculations were made for $(\varepsilon_1 - \varepsilon_0) / \varepsilon_0 = 3$.

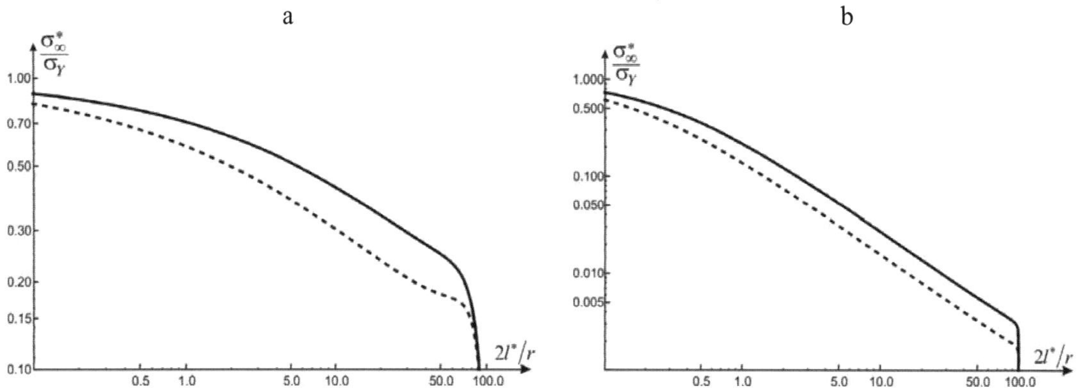

Figure 4. Critical stresses $\sigma_{\infty}^{*} / \sigma_Y$ of tensile band a) with internal crack of $2l$ in length b) with two edge cracks of length l

Relations of critical stresses $\sigma_{\infty}^{*} / \sigma_Y$ of the tensile band of width w with the internal crack of length $2l$ are as follows

$$\frac{\sigma_{\infty}^{*}}{\sigma_Y} = \left[\frac{1}{1-2\lambda} + \left(1 - \frac{3+2(1-2\mu)^2}{8\pi(1-\mu^2)} \frac{\varepsilon_1 - \varepsilon_0}{\varepsilon_0} \right) Y(\lambda) \sqrt{\frac{2l^*}{r}} \right]^{-1}, \tag{11}$$

$$Y(\lambda) = \cos^{-1/2} \pi\lambda, \ \lambda = l^* / w = (l^* / r) / (w / r).$$

Critical stresses $\sigma_{\infty}^{*} / \sigma_Y$ of a tensile band of the width w with two edge cracks of the length l have the form

$$\frac{\sigma_{\infty}^{*}}{\sigma_Y} = \left[\frac{1}{1-\lambda} + \left(1 - \frac{3+2(1-2\mu)^2}{8\pi(1-\mu^2)} \frac{\varepsilon_1 - \varepsilon_0}{\varepsilon_0} \right) Y(\lambda) \sqrt{\frac{2l^*}{r}} \right]^{-1}, \tag{12}$$

$$Y(\lambda) = \left[1 + 0{,}122\cos^2(\pi\lambda/2) \right] \sqrt{(2/\pi\lambda) tg(\pi\lambda/2)}, \ \lambda = 2l^* / w = (2l^* / r) / (w / r).$$

The bending moments M for symmetrical three-point bending can be represented as $M = PL / 4$, $L = 4b$, where L is the length of a bending beam span, b is the height of a bending specimen. The bending moments M^* of specimens with a crack of length l under bending (t is the beam thickness) are as follows

$$\widetilde{M}^* = \frac{6M^*}{tb^2\sigma_Y} = \left\{ \left(\frac{1}{1-\lambda}\right)^2 \left(1-\frac{r/b}{1-\lambda}\right) + \sqrt{\frac{2}{\pi}}\left[1-\frac{3+2(1-2\mu)^2}{8\pi(1-\mu^2)}\frac{\varepsilon_1-\varepsilon_0}{\varepsilon_0}\right] Y(\lambda)\sqrt{\frac{l^*}{b}\frac{b}{r}} \right\}^{-1},$$

$$Y(\lambda) = \frac{1{,}99-\lambda(1-\lambda)(2{,}15-3{,}93\lambda+2{,}7\lambda^2)}{(1+2\lambda)(1-\lambda)^{3/2}},\ \lambda = l^*/b = (l^*/r)/(b/r) \quad . \quad (13)$$

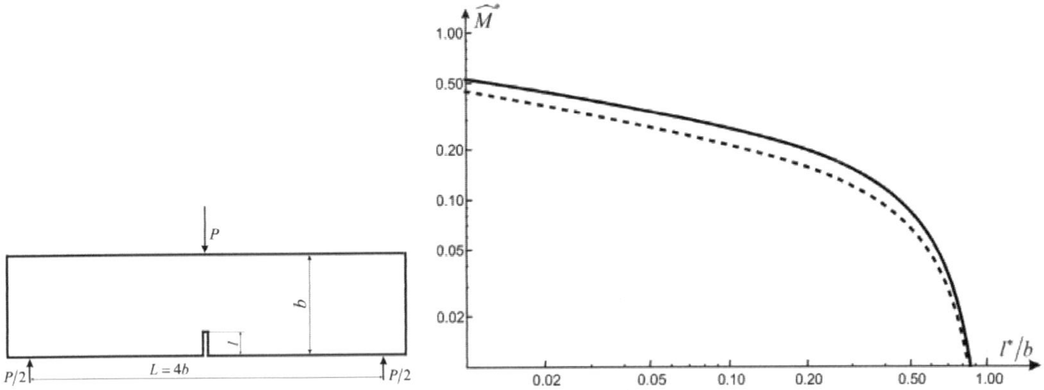

Figure 5. The scheme of loading and critical moments $\widetilde{M}^*$ of bending specimens.

4. Using quasi-brittle fracture diagrams in analyzing catastrophic failure

Consider the diagrams of quasi-brittle fracture of structural elements for analysis of catastrophic failures. The visual appearances of these diagrams do not change for various loading conditions. In order to select parameters of a proposed fracture model, the data derived from a critical SIF and from a $\sigma-\varepsilon$ diagram at corresponding temperatures are needed. All elements with weld joints could be capable to withstand lesser loads as compared to homogeneous structures. The effect of break of a structure under monotonous loading is pronounced inasmuch as the critical loadings are essentially decreased with increasing a crack length.

Attention should be paid to the special role of the $(\varepsilon_1-\varepsilon_0)/\varepsilon_0$ parameter when a response of the system to applied loading is formed. This parameter characterizes plastic properties not only of the original material, but also of material subject to cold plastic deformation in the pre-fracture zone under low-cycle loading [13]. The impact of cold plastic deformation of mild steels on an ultimate loading has been shown earlier. In some cases, the brittleness threshold T_0 was shifted into the region of ambient temperatures after essential plastic deformation. When cold-formed steel profiles are formed, they are usually hardened. When plasticity reserve of material is already partly consumed in locations where the profile is bent, use of such profiles provokes crack initiation when works are carried out under conditions of low temperatures.

5. Conclusion

The data report for quasi-brittle fracture diagrams of common elements of structures of homogeneous structures produced by welding has been given. For plotting such diagrams, one of versions of the quasi-linear fracture mechanics is used. It has been established that welded structures with cracks in the vicinity of a weld joint exhibit decreased resistance to cracking. When analyzing parameters entering the proposed model, the data on deformation of steels at temperatures in the vicinity of the cold brittleness threshold or lower are used. Diagrams of quasi-brittle fracture are proposed to be used when analyzing catastrophic failures of structural elements operating at temperatures lower than the cold brittleness threshold. The attention is given to the fact that steel structures after plastic deformation have decreased operational integrity.

IOP Conf. Series: Journal of Physics: Conf. Series **894** (2017) 012045 doi:10.1088/1742-6596/894/1/012045

References

[1] Uzhik GV 1957 Durability and Plasticity of Metals at Low Temperatures (Moscow: Publisher AS USSR) (in Russian)
[2] Syromyatnikova AS 2014 *Physical Mesomech* **17**(2) 85 (in Russian)
[3] Panin VE, Derevyagina LS, Lebedev MP, Syromyatnikova AS, Surikova NS, Pochivalov YaI and Ovechkin BB 2016 *Physical Mesomech* **19**(2) 5 (in Russian)
[4] Kornev VM 2010 *Physical Mesomech* **13**(1) 47 (in Russian)
[5] Kornev VM Demeshkin AG 2011 *J of Applied Mech Technical Physics.* **52** 6 975
[6] Kornev VM 2013 *Physical Mesomech* **16**(5) 25 (in Russian)
[7] Kornev VM 2013 *Engin failure analysis* **35** 533
[8] Kornev VM 2015 *J. Machinery Manufacture and Reliability* **44**(2) 140
[9] Kornev VM 2013 *Mekh Komp Mater i Konstr* **19**(4) 568 (in Russian)
[10] Kornev VM Kurguzov VD Astapov NS 2013 *Applied Composite Mater.* **20**(2) 129
[11] Kornev VM 2014 *Advances in Mater Science and Applic* **3**(4) 164
[12] Kornev VM 2015 *Applied Mechanics and Materials* **784** 179
[13] Kornev VM 2016 *Physical Mesomech* **19**(2) 80 (in Russian)

Quasi-solutions of genuinely nonlinear forward-backward ultra-parabolic equations

I V Kuznetsov[1,2] **and S A Sazhenkov**[1,2]

[1] Novosibirsk State University, Pirogova str. 2, 630090 Novosibirsk, Russia
[2] Lavrentyev Institute of Hydrodynamics, ac. Lavrentyev av. 15, 630090 Novosibirsk, Russia

E-mail: kuznetsov_i@hydro.nsc.ru

Abstract. In the present paper we have proved the existence of quasi-solutions of genuinely nonlinear forward-backward ultra-parabolic equations. Quasi-solutions are obtained with the help of the vanishing anisotropic temporal diffusion method. Moreover, at the present stage of our research we assume that various choices of temporal artificial diffusion coefficients lead to entropy solutions or to quasi-solutions. The latter assumption is the subject of our further scientific research.

Key words: entropy solution, forward-backward ultra-parabolic equation, genuine nonlinearity condition, kinetic solution

1. Introduction

Ultra-parabolic equations are used in the theory of boundary layers and in the mathematical models of Brownian motion [1–3]. Entropy solutions of nonlinear ultra-parabolic equations were studied in [4–7], see references therein. It is important to note that entropy solutions were firstly obtained in [8] for hyperbolic differential equations and later extended to various types of partial differential equations.

The research on the well-posedness of nonlinear forward-backward parabolic equations was started in [9–14]. Moreover, the history of the research on the linear case was described in [15].

Here we apply the vanishing anisotropic temporal diffusion (viscosity) method [16]. The technique of elliptic regularization was invented in [17] during the study of degenerate parabolic equations, and subsequently adapted for the Navier-Stokes equations [18], and for forward-backward parabolic equations [9, 11]. Independently, the elliptic regularization of hyperbolic nonlinear equations was studied in [19, 20]. Furthermore, singular limits of anisotropic elliptic perturbations were obtained in [21, 22]. Also, for hyperbolic equations the vanishing viscosity method with gradient dependent viscosity was applied in [23, 24].

In this paper we have obtained quasi-solutions of forward-backward ultra-parabolic equations. In comparison with [25] ($p_1 = p_2 = 2$), we have not obtained entropy solutions when $|p_1-2|+|p_2-2| \neq 0$. It is important to mention that quasi-solutions for hyperbolic equations were studied in [26–30] and, correspondingly, for ultra-parabolic equations [31]. In [27] it was shown that several types of regularization can lead to quasi-solutions but not to entropy solutions.

We have obtained a quasi-solution with the help of a corresponding sequence of weak solutions $\{u_\varepsilon\}_{\varepsilon>0}$ to problem $\mathbf{\Pi}_\varepsilon$ as $\varepsilon \to 0+$. The main difference from previous results on quasi-solutions is that only the right-hand sides of kinetic boundary conditions (4.7b)–(4.7e) are not

Content from this work may be used under the terms of the Creative Commons Attribution 3.0 licence. Any further distribution of this work must maintain attribution to the author(s) and the title of the work, journal citation and DOI.
Published under licence by IOP Publishing Ltd

sign-defined. A similar type of quasi-solutions for forward-backward p-parabolic equations was obtained in [32]. In the present paper we have used the vanishing diffusion method with the help of anisotropic $\boldsymbol{p}$-Laplacian, $\boldsymbol{p} = (2, \dots, 2, p_1, p_2) \in \mathbb{R}^{d+2}$, $p_1, p_2 > 1$. The physical meaning of the vanishing anisotropic diffusion method is that we take into account fast and slow diffusive regimes when $|p_1 - 2| + |p_2 - 2| \neq 0$, see [33, Chapter 5].

This paper is organized as follows. In section 2 we formulate the non-homogeneous Dirichlet problem $\mathbf{\Pi}_0$. Since this problem is ill-posed, in section 3 we formulate problem $\mathbf{\Pi}_\varepsilon$. Quasi-solutions to problem $\mathbf{\Pi}_0$ are obtained as singular limits of weak solutions to problem $\mathbf{\Pi}_\varepsilon$ as $\varepsilon \to 0+$, see sections 4 and 5. We have not proved yet that singular limits of weak solutions to problem $\mathbf{\Pi}_\varepsilon$ are also entropy solutions to problem $\mathbf{\Pi}_0$ even if $|p_1 - 2| + |p_2 - 2| \neq 0$. So, it is still an open question (see Remark 4).

2. Genuinely nonlinear forward-backward ultra-parabolic equation

Let scalar functions $a(z)$, $b(z)$ and vector function $\boldsymbol{\varphi}(z)$ satisfy the following conditions.

Conditions on $a, b\&\boldsymbol{\varphi}$. *Let $a, b \in C^2(\mathbb{R})$, $a(0) = b(0) = 0$, $\boldsymbol{\varphi}(z) = (\varphi_1(z), \dots, \varphi_d(z))$, $z \in \mathbb{R}$, $\varphi_j \in C^2(\mathbb{R})$, $j = 1, \dots, d$, $\boldsymbol{\varphi}(0) = \mathbf{0}$. Function a is non-monotonic. Moreover, a' and b' satisfy the genuine nonlinearity condition*

$$\operatorname{mes}\left\{\lambda \in \mathbb{R} : \ a'(\lambda)\theta + b'(\lambda)\vartheta = 0\right\} = 0 \tag{2.1}$$

for every $(\theta, \vartheta) \in \mathbb{S}^1$.

Under Conditions on $a, b\&\boldsymbol{\varphi}$, we are going to formulate boundary value problem $\mathbf{\Pi}_0$. Let $G_{T,S} = \Omega \times (0, T) \times (0, S)$, $\Gamma_0 = \overline{\Omega} \times \{t = 0\} \times [0, S]$, $\Gamma_T = \overline{\Omega} \times \{t = T\} \times [0, S]$, $\Xi_0 = \overline{\Omega} \times [0, T] \times \{s = 0\}$, $\Xi_S = \overline{\Omega} \times [0, T] \times \{s = S\}$, $\Gamma_l = \partial G_{T,S} \backslash (\Gamma_0 \cup \Gamma_T \cup \Xi_0 \cup \Xi_S)$, a bounded domain $\Omega \subset \mathbb{R}^d$ ($\operatorname{mes}\Omega < \infty$) has C^1-smooth boundary $\partial\Omega$. Here we deal with anisotropic Sobolev spaces, see, for example, [34]. The anisotropic Sobolev space $W_0^{1,\boldsymbol{p}}(G_{T,S})$ is equipped with the norm

$$\|u\|_{W_0^{1,\boldsymbol{p}}(G_{T,S})} = \sum_{i=1}^{d} \|\partial_{x_i} u\|_{L^2(G_{T,S})} + \|\partial_t u\|_{L^{p_1}(G_{T,S})} + \|\partial_s u\|_{L^{p_2}(G_{T,S})} + \|u\|_{L^1(G_{T,S})},$$

where $\boldsymbol{p} = (2, \dots, 2, p_1, p_2) \in \mathbb{R}^{d+2}$, $p_1, p_2 > 1$.

Problem $\mathbf{\Pi}_0$. *For arbitrary initial and final conditions $u_0^\Gamma, u_T^\Gamma \in C_0^{1,\alpha}(\Omega \times (0, S))$, $u_0^\Xi, u_S^\Xi \in C_0^{1,\alpha}(\Omega \times (0, T))$, $\alpha \in (0, 1)$, the unknown function $u : G_{T,S} \to \mathbb{R}$ satisfies*

$$\partial_t a(u) + \partial_s b(u) + \operatorname{div}_x \boldsymbol{\varphi}(u) = \Delta_x u, \quad (\boldsymbol{x}, t, s) \in G_{T,S}, \tag{2.2a}$$

$$u|_{\Gamma_0} \approx u_0^\Gamma(\boldsymbol{x}, s), \quad (\boldsymbol{x}, s) \in \Omega \times (0, S), \tag{2.2b}$$

$$u|_{\Gamma_T} \approx u_T^\Gamma(\boldsymbol{x}, s), \quad (\boldsymbol{x}, s) \in \Omega \times (0, S), \tag{2.2c}$$

$$u|_{\Xi_0} \approx u_0^\Xi(\boldsymbol{x}, t), \quad (\boldsymbol{x}, t) \in \Omega \times (0, T), \tag{2.2d}$$

$$u|_{\Xi_S} \approx u_S^\Xi(\boldsymbol{x}, t), \quad (\boldsymbol{x}, t) \in \Omega \times (0, T), \tag{2.2e}$$

$$u|_{\Gamma_l} = 0, \tag{2.2f}$$

in the form given in Definition 3 from section 4. The sign $\approx$ means the equality only on a part of the boundary.

Remark 1. *We formulate equation* (2.2a) *in the sense of distributions. Since function $a(z)$ is non-monotonic on $\mathbb{R}$, equation* (2.2a) *is a forward-backward ultra-parabolic equation. Moreover, a quasi-solution u can deviate from initial and final data u_0^Γ, u_T^Γ, u_0^Ξ and u_S^Ξ, see Remark 5. Therefore, the difficulty of problem $\mathbf{\Pi}_0$ is that equation* (2.2a) *and initial and final conditions* (2.2b)–(2.2e) *should be reformulated in the form of kinetic equalities given in Definition 3.*

3. Anisotropic elliptic regularization

We are going to construct a quasi-solution as a singular limit of weak solutions u_ε to the non-homogeneous Dirichlet problem $\mathbf{\Pi}_\varepsilon$ as $\varepsilon \to 0+$.

Problem $\mathbf{\Pi}_\varepsilon$. *For arbitrary initial and final conditions $u_0^\Gamma, u_T^\Gamma \in C_0^{1,\alpha}(\Omega \times (0,S))$, $u_0^\Xi, u_S^\Xi \in C_0^{1,\alpha}(\Omega \times (0,T))$, $\alpha \in (0,1)$, the unknown function u_ε satisfies the boundary value problem*

$$\partial_t a(u_\varepsilon) + \partial_s b(u_\varepsilon) + \operatorname{div}_x \boldsymbol{\varphi}(u_\varepsilon) = \Delta_x u_\varepsilon + \varepsilon \partial_t(|\partial_t u_\varepsilon|^{p_1-2}\partial_t u_\varepsilon) + \varepsilon \partial_s(|\partial_s u_\varepsilon|^{p_2-2}\partial_s u_\varepsilon),\ (\boldsymbol{x}, t, s) \in G_{T,S}, \tag{3.3a}$$

$$u_\varepsilon|_{\Gamma_0} = u_0^\Gamma, \quad u_\varepsilon|_{\Gamma_T} = u_T^\Gamma, \quad u_\varepsilon|_{\Xi_0} = u_0^\Xi, \quad u_\varepsilon|_{\Xi_S} = u_S^\Xi, \quad u_\varepsilon|_{\Gamma_l} = 0, \tag{3.3b}$$

in a weak sense, see Definition 1.

We assume here that $\varepsilon \in (0,1]$, $p_1, p_2 > 1$. Let $V^{1,\boldsymbol{p}}(G_{T,S}) = \{v \in W^{1,\boldsymbol{p}}(G_{T,S}) : \ v|_{\Gamma_l} = 0\}$.

Definition 1. *Function $u_\varepsilon \in L^\infty(G_{T,S}) \cap V^{1,\boldsymbol{p}}(G_{T,S})$ is called a weak solution to problem $\mathbf{\Pi}_\varepsilon$ if the following demands hold:*

1. *Let $\widehat{u} \in L^\infty(G_{T,S}) \cap V^{1,\boldsymbol{p}}(G_{T,S})$ be an extension of functions $u_0^\Gamma, u_T^\Gamma, u_0^\Xi$ and u_S^Ξ into $G_{T,S}$ such that $u_\varepsilon - \widehat{u} \in L^\infty(G_{T,S}) \cap W_0^{1,\boldsymbol{p}}(G_{T,S})$.*
2. *The following equality holds*

$$\int_{G_{T,S}} \big(-a(u_\varepsilon)\partial_t\phi - b(u_\varepsilon)\partial_s\phi - \boldsymbol{\varphi}(u_\varepsilon)\cdot\nabla_x\phi + \nabla_x u_\varepsilon \cdot \nabla_x \phi + \varepsilon|\partial_t u_\varepsilon|^{p_1-2}\partial_t u_\varepsilon \partial_t\phi + \varepsilon|\partial_s u_\varepsilon|^{p_2-2}\partial_s u_\varepsilon \partial_s\phi\big)\, d\boldsymbol{x}dtds = 0 \tag{3.4a}$$

for every $\phi \in L^\infty(G_{T,S}) \cap W_0^{1,\boldsymbol{p}}(G_{T,S})$.

Remark 2. *We can reformulate (3.4a) in the equivalent way:*

$$\int_{G_{T,S}} \big(\partial_t a(u_\varepsilon)\phi + \partial_s b(u_\varepsilon)\phi + \operatorname{div}_x\boldsymbol{\varphi}(u_\varepsilon)\phi + \nabla_x u_\varepsilon \cdot \nabla_x\phi + \varepsilon|\partial_t u_\varepsilon|^{p_1-2}\partial_t u_\varepsilon\partial_t\phi + \varepsilon|\partial_s u_\varepsilon|^{p_2-2}\partial_s u_\varepsilon \partial_s\phi\big)\, d\boldsymbol{x}dtds = 0. \tag{3.4b}$$

Remark 3. *We assume that extension $\widehat{u}$ into $G_{T,S}$ exists if u_0^Γ, u_T^Γ, u_0^Ξ and u_S^Ξ are from $C_0^{1,\alpha}$, $0 < \alpha < 1$. In the case $p_1 = p_2 = 2$ we deal with entropy solutions (see Remark 4) and with the help of [25, Theorem 1] we can decrease the smoothness of initial and final data: $u_0^\Gamma, u_T^\Gamma \in L^\infty(\Omega \times (0,S))$, $u_0^\Xi, u_S^\Xi \in L^\infty(\Omega \times (0,T))$.*

Proposition 1. *Under Conditions on $a, b \& \boldsymbol{\varphi}$, problem $\mathbf{\Pi}_\varepsilon$ has at least one weak solution u_ε for all $u_0^\Gamma, u_T^\Gamma \in C_0^{1,\alpha}(\Omega \times (0,S))$, $u_0^\Xi, u_S^\Xi \in C_0^{1,\alpha}(\Omega \times (0,T))$, $\alpha \in (0,1)$. Moreover, the maximum principle*

$$\|u_\varepsilon\|_{L^\infty(G_{T,S})} \le M = \max\Big(\|u_0^\Gamma\|_{L^\infty(\Omega\times(0,S))}, \|u_T^\Gamma\|_{L^\infty(\Omega\times(0,S))}, \|u_0^\Xi\|_{L^\infty(\Omega\times(0,T))}, \|u_S^\Xi\|_{L^\infty(\Omega\times(0,T))}\Big), \tag{3.5}$$

and the energy estimate

$$\int_{G_{T,S}} \Big(|\nabla_x u_\varepsilon|^2 + \varepsilon\,|\partial_t u_\varepsilon|^{p_1} + \varepsilon\,|\partial_s u_\varepsilon|^{p_2}\Big)\, d\boldsymbol{x}dtds < C \tag{3.6}$$

hold. The constant C does not depend on $\varepsilon \in (0,1]$.

This proposition can be proved with the help of results formulated in [25,35].

4. Kinetic formulation of forward-backward ultra-parabolic equations

In this section we deal with the kinetic formulation of forward-backward ultra-parabolic equations. Here we use methods developed in [25–27, 31, 36–41].

Consider the function χ which is defined in the following way

$$\chi(\lambda; \upsilon) = \begin{cases} +1, & \text{if } 0 < \lambda < \upsilon, \\ -1, & \text{if } \upsilon < \lambda < 0, \\ 0, & \text{elsewhere.} \end{cases}$$

Definition 2. *Let N be a positive integer, $\mathcal{O}$ be an open set of $\mathbb{R}^N$ and the function $h \in L^\infty(\mathcal{O} \times (-L, L))$ satisfying $0 \le h(\boldsymbol{z}, \lambda)\mathrm{sgn}(\lambda) \le 1$ for almost every $(\boldsymbol{z}, \lambda) \in \mathbb{R}^{N+1}$. It is said that h is a χ-function if there exists a function $\upsilon \in L^\infty(\mathcal{O})$ such that*

$$h(\boldsymbol{z}, \lambda) = \chi(\lambda; \upsilon(\boldsymbol{z}))$$

for a.e. $\boldsymbol{z} \in \mathcal{O}$. Note that $\upsilon(\boldsymbol{z}) = \int_{-L}^{L} h(\boldsymbol{z}, \lambda)\, d\lambda = \int_{-L}^{L} \chi(\lambda; \upsilon(\boldsymbol{z}))\, d\lambda$.

The following lemma formulated and proved in [40] guarantees the link between sequences of χ-functions and their limits.

Lemma 1. *Let $\mathcal{O}$ be an open set of $\mathbb{R}^N$ and $h_n \in L^\infty(\mathcal{O} \times (-L, L))$ be a sequence of χ-functions converging weakly to $h \in L^\infty(\mathcal{O} \times (-L, L))$. We define $\upsilon_n(\cdot) = \int_{-L}^{L} h_n(\cdot, \lambda)\, d\lambda$ and $\upsilon(\cdot) = \int_{-L}^{L} h(\cdot, \lambda)\, d\lambda$. Then the three assertions are equivalent:*

- *h_n converges strongly to h in $L^1_{\mathrm{loc}}(\mathcal{O} \times (-L, L))$,*
- *υ_n converges strongly to υ in $L^1_{\mathrm{loc}}(\mathcal{O})$,*
- *h is a χ-function.*

Definition 3. *Function $u : G_{T,S} \to \mathbb{R}$ is called a quasi-solution of problem $\mathbf{\Pi}_0$ if it satisfies the following equations:*

1. *(Kinetic equation)*

$$a'(\lambda)\partial_t \chi(\lambda; u) + b'(\lambda)\partial_s \chi(\lambda; u) + \boldsymbol{\varphi}'(\lambda) \cdot \nabla_x \chi(\lambda; u) = \Delta_x \chi(\lambda; u) + \partial_\lambda (m(\boldsymbol{x}, t, s, \lambda) + n(\boldsymbol{x}, t, s, \lambda)), \quad (4.7a)$$

2. *(Kinetic boundary conditions)*

$$a'(\lambda)\big(\chi(\lambda; u_0^{\tau,\Gamma}(\boldsymbol{x}, s)) - \chi(\lambda; u_0^{\Gamma}(\boldsymbol{x}, s))\big) - \delta_{(\lambda = u_0^{\Gamma}(\boldsymbol{x}, s))}\big(a(u_0^{\tau,\Gamma}(\boldsymbol{x}, s)) - a(u_0^{\Gamma}(\boldsymbol{x}, s))\big) = \partial_\lambda \mu_0^{\Gamma}(\boldsymbol{x}, s, \lambda), \quad (4.7b)$$

$$a'(\lambda)\big(\chi(\lambda; u_T^{\tau,\Gamma}(\boldsymbol{x}, s)) - \chi(\lambda; u_T^{\Gamma}(\boldsymbol{x}, s))\big) - \delta_{(\lambda = u_T^{\Gamma}(\boldsymbol{x}, s))}\big(a(u_T^{\tau,\Gamma}(\boldsymbol{x}, s)) - a(u_T^{\Gamma}(\boldsymbol{x}, s))\big) = -\partial_\lambda \mu_T^{\Gamma}(\boldsymbol{x}, s, \lambda), \quad (4.7c)$$

$$b'(\lambda)\big(\chi(\lambda; u_0^{\tau,\Xi}(\boldsymbol{x}, t)) - \chi(\lambda; u_0^{\Xi}(\boldsymbol{x}, t))\big) - \delta_{(\lambda = u_0^{\Xi}(\boldsymbol{x}, t))}\big(b(u_0^{\tau,\Xi}(\boldsymbol{x}, t)) - b(u_0^{\Xi}(\boldsymbol{x}, t))\big) = \partial_\lambda \mu_0^{\Xi}(\boldsymbol{x}, t, \lambda), \quad (4.7d)$$

$$b'(\lambda)\big(\chi(\lambda; u_S^{\tau,\Xi}(\boldsymbol{x}, t)) - \chi(\lambda; u_S^{\Xi}(\boldsymbol{x}, t))\big) - \delta_{(\lambda = u_S^{\Xi}(\boldsymbol{x}, t))}\big(b(u_S^{\tau,\Xi}(\boldsymbol{x}, t)) - b(u_S^{\Xi}(\boldsymbol{x}, t))\big) = -\partial_\lambda \mu_S^{\Xi}(\boldsymbol{x}, t, \lambda), \quad (4.7e)$$

where $\mu_0^\Gamma, \mu_T^\Gamma \in \mathcal{M}(\Omega \times (0,S) \times (-M,M))$, $\mu_0^\Xi, \mu_S^\Xi \in \mathcal{M}(\Omega \times (0,T) \times (-M,M))$, n, $m \in \mathcal{M}^+(G_{T,S} \times (-M,M))$, $n = \delta_{(\lambda=u)} \, |\nabla_x u|^2$

Remark 4. *Kinetic boundary conditions* (4.7b)–(4.7e) *are deduced with help of methods proposed in [36, 38]. When* $\mu_0^\Gamma, \mu_T^\Gamma$, μ_0^Ξ *and* μ_S^Ξ *are positive measures, Definition 3 corresponds to an entropy solution to problem* $\mathbf{\Pi}_0$*, see [25]. Moreover, in the case* $p_1 = p_2 = 2$ *an entropy solution is a singular limit of weak solutions to problem* $\mathbf{\Pi}_\varepsilon$ *as* $\varepsilon \to 0+$*. When* $|p_1 - 2| + |p_2 - 2| \neq 0$*, we can only prove that* $u = \lim\limits_{\varepsilon \to 0+} u_\varepsilon$ *is a quasi-solution to problem* $\mathbf{\Pi}_0$*.*

Remark 5. *In Definition 3 we have used the fact that a quasi-solution to problem* $\mathbf{\Pi}_0$ *has traces* $u_0^{\tau,\Gamma}$, $u_T^{\tau,\Gamma}$, $u_0^{\tau,\Xi}$ *and* $u_S^{\tau,\Xi}$ *in the* L^1 *sense, see [25, 28, 29, 31].*

The main result of the present paper is the following theorem.

Theorem 1. *Under Conditions* $a, b\&\boldsymbol{\varphi}$*, problem* $\mathbf{\Pi}_0$ *has at least one quasi-solution.*

When $p_1 = p_2 = 2$, it was proved in [25] that u_ε converges in $L^1(G_{T,S})$ to the unique entropy solution of problem $\mathbf{\Pi}_0$. When $|p_1 - 2| + |p_2 - 2| \neq 0$, the form of a quasi-solution to problem $\mathbf{\Pi}_0$ implies its nonuniqueness.

5. Brief scheme of the proof of Theorem 1

In order to prove Theorem 1, we partly repeat results from [25,37]. Using kinetic formulation of (2.2a), the genuine nonlinearity condition (2.1), the precomactness of $\int_{-M}^{M} \chi(\lambda; u_\varepsilon)\, d\lambda$ (see [41]) and Lemma 1, we prove that $\{u_\varepsilon\}_{\varepsilon>0}$ is a precompact set in $L^1(G_{T,S})$.

We need to introduce the convex entropy flux pair $(\eta, \boldsymbol{q})$:

$$\boldsymbol{q}'_\varphi(z) = \boldsymbol{\varphi}'(z)\eta'(z), \quad q'_a(z) = a'(z)\eta'(z), \quad q'_b(z) = b'(z)\eta'(z), \quad \eta''(z) \geq 0,$$
$$\boldsymbol{q}(z) = (\boldsymbol{q}_\varphi(z), q_a(z), q_b(z)), \quad z \in \mathbb{R}.$$

The entropy inequality

$$\partial_t q_a(u) + \partial_s q_b(u) + \mathrm{div}_x \boldsymbol{q}_\varphi(u) - \Delta_x \eta(u) \leq -\eta''(u)|\nabla_x u|^2 \tag{5.8}$$

is valid for every convex entropy flux pair $(\eta, \boldsymbol{q})$. We get this inequality from

$$\partial_t q_a(u_\varepsilon) + \partial_s q_b(u_\varepsilon) + \mathrm{div}_x \boldsymbol{q}_\varphi(u_\varepsilon) = \Delta_x \eta(u_\varepsilon) + \varepsilon\partial_t(|\partial_t u_\varepsilon|^{p_1-2}\partial_t \eta(u_\varepsilon))$$
$$+ \varepsilon\partial_s(|\partial_s u_\varepsilon|^{p_2-2}\partial_s \eta(u_\varepsilon)) - \eta''(u_\varepsilon)(|\nabla_x u_\varepsilon|^2 + \varepsilon|\partial_t u_\varepsilon|^{p_1} + \varepsilon|\partial_s u_\varepsilon|^{p_2}),$$

which is deduced by putting $\phi = \eta'(u_\varepsilon)\gamma$ in (3.4b), where γ is an arbitrary nonnegative finite test function in $G_{T,S}$.

Using the kinetic formulation, from (5.8) we obtain

$$\int_{-M}^{M} \eta'(\lambda)\big(a'(\lambda)\partial_t \chi(\lambda; u) + b'(\lambda)\partial_s \chi(\lambda; u) + \boldsymbol{\varphi}'(\lambda) \cdot \nabla_x \chi(\lambda; u) - \Delta_x \chi(\lambda; u)\big)\, d\lambda =$$
$$\int_{-M}^{M} \eta'(\lambda)\partial_\lambda(m+n)\, d\lambda = -\int_{-M}^{M} \eta''(\lambda)(m+n)\, d\lambda \leq 0. \tag{5.9}$$

Kinetic equation (4.7e) follows from (5.9) due to arbitrariness of η'.

Furthermore, kinetic boundary conditions (4.7b)–(4.7e) are deduced from the following equalities

$$q_a(u_0^{\tau,\Gamma}(\boldsymbol{x},s)) - q_a(u_0^{\Gamma}(\boldsymbol{x},s)) - \eta'(u_0^{\Gamma}(\boldsymbol{x},s))(a(u_0^{\tau,\Gamma}(\boldsymbol{x},s)) - a(u_0^{\Gamma}(\boldsymbol{x},s))) =$$
$$\int_{-M}^{M} \eta'(\lambda)\big(a'(\lambda)(\chi(\lambda;u_0^{\tau,\Gamma}(\boldsymbol{x},s)) - \chi(\lambda;u_0^{\Gamma}(\boldsymbol{x},s))) - (a(u_0^{\tau,\Gamma}(\boldsymbol{x},s)) - a(u_0^{\Gamma}(\boldsymbol{x},s)))\delta_{(\lambda=u_0^{\Gamma}(\boldsymbol{x},s))}\big)\,d\lambda =$$
$$\int_{-M}^{M} \eta'(\lambda)\partial_\lambda \mu_0^{\Gamma}(\boldsymbol{x},s,\lambda)\,d\lambda = -\int_{-M}^{M} \eta''(\lambda)\mu_0^{\Gamma}(\boldsymbol{x},s,\lambda)\,d\lambda,$$

$$q_a(u_T^{\tau,\Gamma}(\boldsymbol{x},s)) - q_a(u_T^{\Gamma}(\boldsymbol{x},s)) - \eta'(u_T^{\Gamma}(\boldsymbol{x},s))(a(u_T^{\tau,\Gamma}(\boldsymbol{x},s)) - a(u_T^{\Gamma}(\boldsymbol{x},s))) =$$
$$\int_{-M}^{M} \eta'(\lambda)\big(a'(\lambda)(\chi(\lambda;u_T^{\tau,\Gamma}(\boldsymbol{x},s)) - \chi(\lambda;u_T^{\Gamma}(\boldsymbol{x},s))) - (a(u_T^{\tau,\Gamma}(\boldsymbol{x},s)) - a(u_T^{\Gamma}(\boldsymbol{x},s)))\delta_{(\lambda=u_T^{\Gamma}(\boldsymbol{x},s))}\big)\,d\lambda =$$
$$-\int_{-M}^{M} \eta'(\lambda)\partial_\lambda \mu_T^{\Gamma}(\boldsymbol{x},s,\lambda)\,d\lambda = \int_{-M}^{M} \eta''(\lambda)\mu_T^{\Gamma}(\boldsymbol{x},s,\lambda)\,d\lambda,$$

$$q_b(u_0^{\tau,\Xi}(\boldsymbol{x},t)) - q_b(u_0^{\Xi}(\boldsymbol{x},t)) - \eta'(u_0^{\Xi}(\boldsymbol{x},t))(b(u_0^{\tau,\Xi}(\boldsymbol{x},t)) - b(u_0^{\Xi}(\boldsymbol{x},t))) =$$
$$\int_{-M}^{M} \eta'(\lambda)\big(b'(\lambda)(\chi(\lambda;u_0^{\tau,\Xi}(\boldsymbol{x},t)) - \chi(\lambda;u_0^{\Xi}(\boldsymbol{x},t))) - (b(u_0^{\tau,\Xi}(\boldsymbol{x},t)) - b(u_0^{\Xi}(\boldsymbol{x},t)))\delta_{(\lambda=u_0^{\Xi}(\boldsymbol{x},t))}\big)\,d\lambda =$$
$$\int_{-M}^{M} \eta'(\lambda)\partial_\lambda \mu_0^{\Xi}(\boldsymbol{x},t,\lambda)\,d\lambda = -\int_{-M}^{M} \eta''(\lambda)\mu_0^{\Xi}(\boldsymbol{x},t,\lambda)\,d\lambda,$$

$$q_b(u_S^{\tau,\Xi}(\boldsymbol{x},t)) - q_b(u_S^{\Xi}(\boldsymbol{x},t)) - \eta'(u_S^{\Xi}(\boldsymbol{x},t))(b(u_S^{\tau,\Xi}(\boldsymbol{x},t)) - b(u_S^{\Xi}(\boldsymbol{x},t))) =$$
$$\int_{-M}^{M} \eta'(\lambda)\big(b'(\lambda)(\chi(\lambda;u_S^{\tau,\Xi}(\boldsymbol{x},t)) - \chi(\lambda;u_S^{\Xi}(\boldsymbol{x},t))) - (b(u_S^{\tau,\Xi}(\boldsymbol{x},t)) - b(u_S^{\Xi}(\boldsymbol{x},t)))\delta_{(\lambda=u_S^{\Xi}(\boldsymbol{x},t))}\big)\,d\lambda =$$
$$-\int_{-M}^{M} \eta'(\lambda)\partial_\lambda \mu_S^{\Xi}(\boldsymbol{x},t,\lambda)\,d\lambda = \int_{-M}^{M} \eta''(\lambda)\mu_S^{\Xi}(\boldsymbol{x},t,\lambda)\,d\lambda,$$

which are valid for every convex entropy flux pair $(\eta, \boldsymbol{q})$, where $\mu_0^{\Gamma}, \mu_T^{\Gamma} \in \mathcal{M}(\Omega \times (0,S) \times (-M,M))$, $\mu_0^{\Xi}, \mu_S^{\Xi} \in \mathcal{M}(\Omega \times (0,T) \times (-M,M))$. When $p_1 = p_2 = 2$, it was shown in [25] that μ_0^{Γ}, μ_T^{Γ}, μ_0^{Ξ} and μ_S^{Ξ} are positive measures: $\mu_0^{\Gamma}, \mu_T^{\Gamma} \in \mathcal{M}^+(\Omega \times (0,S) \times (-M,M))$, $\mu_0^{\Xi}, \mu_S^{\Xi} \in \mathcal{M}^+(\Omega \times (0,T) \times (-M,M))$

There is still an open question about the convergence of u_ε to entropy solution u of problem $\mathbf{\Pi}_0$ as $\varepsilon \to +0$ when $|p_1 - 2| + |p_2 - 2| \neq 0$, see Remark 4.

6. Conclusion

In the present paper we have enriched results presented in [25] in the case when temporal artificial diffusion coefficients depend on partial derivatives of u_ε in t and s variables. Namely, we have shown that various choices of temporal artificial diffusion coefficients lead to quasi-solutions or to entropy solutions of problem $\mathbf{\Pi}_0$. The later assumption is still under discussion.

Acknowledgments

The work was supported by the Russian Science Foundation (grant no. 15-11-20019).

IOP Conf. Series: Journal of Physics: Conf. Series **894** (2017) 012046 doi:10.1088/1742-6596/894/1/012046

References

[1] Genčev T G 1963 *Doklady Akademii Nauk SSSR* **151** 265–8 (in Russian)
[2] Kolmogorov A N 1933 *Mathematische Annalen* **108** 149–60
[3] Kolmogorov A N 1934 *Annals of Mathematics* **35** 116–7
[4] Panov E Yu 2009 *Journal of Mathematical Sciences* **159** 180–228
[5] Panov E Yu 2011 *Annales de l'Institut Henri Poincaré (C) Non Linear Analysis* **28** 47–62
[6] Plotnikov P I and Sazhenkov S A 2005 *Journal of Mathematical Analysis and Applications* **304** 703–24
[7] Sazhenkov S A 2006 *Siberian Mathematical Journal* **47** 355–75
[8] Kruzhkov S N 1970 *Mathematics of the USSR-Sbornik* **10** 217–43
[9] Bocharov O B 1978 *Dynamika Sploshnoy Sredy* **37** 27–39 (in Russian)
[10] Höllig K 1983 *Transactions of the American Mathematical Society* **278** 299–316
[11] Monakhov V N 1998 *Dynamika Sploshnoy Sredy* **113** 107–13 (in Russian)
[12] Plotnikov P I 1999 *Journal of Mathematical Sciences* **93** 747–66
[13] Larkin N A, Novikov V A and Yanenko N N 1983 *Nonlinear Equations of a Variable Type* (Novosibirsk: Nauka) (in Russian)
[14] Yanenko N N 1983 *Journal of Soviet Mathematics* **21** 868–75
[15] Tersenov S A 1985 *Parabolic Equations with Changing Time Direction* (Nauka: Novosibirsk) (in Russian)
[16] Hughes T J R and Stewart J R 1996 *Journal of Computational and Applied Mathematics* **74** 217–29
[17] Lions J L 1973 *Perturbations Singuliéres dans les Problémes aux Limites et en Contrôl Optimal* (Berlin: Springer)
[18] Salvi R 1994 *Acta Applicandae Mathematicae* **37** 169–79
[19] Howes F A 1983 *Communications in Partial Differential Equations* **8** 847–74
[20] Kelley W G 1981 *Journal of Differential Equations* **40** 37–52
[21] Chipot M, Guesmia S and Sengouga A 2011 *Journal of Mathematical Sciences* **176** 828–43
[22] Chipot M M 2016 *Asymptotic Issues for Some Partial Differential Equations* (Zurich: Imperial College Press)
[23] Marcati P and Natalini R 1994 *Nonlinear Analysis: Theory, Methods & Applications* **23** 621–8
[24] Matas A and Merker J 2014 *Electronic Journal of Qualitative Theory of Differential Equations* **8** 1–14
[25] Kuznetsov I V 2017 *Siberian Electronic Mathematical Reports* **14** 710–31
[26] De Lellis C, Otto F and Westdickenberg M 2003 *Archive for Rational Mechanics and Analysis* **170** 137–84
[27] Hwang S and Tzavaras A E 2002 *Journal Communications in Partial Differential Equations* **27** 1229–54
[28] Panov E Yu 2005 *Journal of Hyperbolic Differential Equations* **2** 885–908
[29] Panov E Yu 2007 *Journal of Hyperbolic Differential Equations* **4** 729–70
[30] Panov E Yu 2010 *Archive for Rational Mechanics and Analysis* **195** 643–73
[31] Aleksic J and Mitrovic D 2013 *Journal of Hyperbolic Differential Equations* **10** 659–76
[32] Antontsev S N and Kuznetsov I V 2016 *Journal of Elliptic and Parabolic Equations* **2** 357–70
[33] Antontsev S and Shmarev S 2015 *Evolution PDEs with Nonstandard Growth Conditions: Existence, Uniqueness, Localization, Blow-up* (Paris: Atlantis Press)
[34] Kruzhkov S N and Kolodii I M 1983 *Russian Mathematical Surveys* **38** 188–9
[35] Kovalevsky A A 2015 *Journal of Mathematical Analysis and Applications* **432** 820–43
[36] Coclite G M, Karlsen K H and Kwon Y-S 2009 *Journal of Functional Analysis* **257** 3823–57
[37] Kuznetsov I V 2016 *Siberian Electronic Mathematical Reports* **13** 930–49
[38] Kwon Y-S 2008 *Journal of Mathematical Analysis and Applications* **340** 543–9
[39] Perthame B 2002 *Kinetic Formulation of Conservation Laws* (Oxford: Oxford Univ. Press.)
[40] Vasseur A 2001 *Archive for Rational Mechanics and Analysis* **160** 181–93
[41] Lazar M and Mitrovic D 2012 *Dynamics of Partial Differential Equations* **9** 239–60

IOP Conf. Series: Journal of Physics: Conf. Series **894** (2017) 012047 doi:10.1088/1742-6596/894/1/012047

Plastic deformation and wave formation on the interface of metals welded by ultrasound-assisted explosive welding

E V Kuz'min, S V Kuz'min, V I Lysak and A N Lata

Volgograd State Technical University, 28 Lenin Avenue, Volgograd, 400005, Russian Federation

E-mail: e.v.kuzmin@yandex.ru

Abstract. This paper presents the influence of the effect of ultrasound on the wave formation and plastic deformation in the metals welded by ultrasound-assisted explosive welding. It has been established that the influence of high-frequency acoustic waves on the metal leads to a reduction of the dynamic yield stress, which in turn leads to changes in the properties of the surface layers of metal and in the conditions of bonding between the collided plates upon explosive welding. It has been shown that the changes in the length and amplitude of waves that arise in the weld joint upon the explosive welding with the simultaneous action of ultrasonic vibrations is connected with a decrease in the magnitude of the deforming pulse and time of action of the compressive stresses that exceed the dynamic yield stress behind the point of contact.

1. Introduction

The high-speed collision of bodies upon the explosive welding is accompanied by a number of unique effects and phenomena, including wave formation, intense strongly-localized plastic deformation of the metal in the heat-affected zone, cumulation, and seizure of metals [1, 2]. At the same time, in a number of cases upon the explosive welding local regions of fused metal are formed in the weld joint, which decrease the strength and service properties of the welded composite. These regions are especially dangerous in the composites in which the base layers may interact and form intermetallic compounds, such as Cu + Al, Al + Fe, etc. It was shown that the formation of alloys is a consequence of shear plastic deformation, which leads to a nonuniform the uneven heat release in the welding joint [3–5].

Earlier [6], we carried out original studies of the metal bonding upon the explosive welding under the action of ultrasonic vibrations. It has been established that the action of ultrasonic vibrations on the zone of the formation of the weld joint in the process of explosive welding exerts a substantial influence on the structure and properties of the obtained joints, which manifests in a significant decrease in the parameters of the arising waves and in an increase in the microhardness of the metal of the heat-affected zone near the weld.

Changes in the microstructure were found (on the example of copper), which manifest in a decrease in the size of the blocks of the mosaic and an increase in the textural (second kind) stresses compared with the standard sample with an equilibrium structure. It has been shown that a change in the size and structure of the deformed zone upon the explosive welding with the application of ultrasonic vibrations is the result of the simultaneous action of these processes at the stage of the formation of the joint and it is almost independent of the original fine structure of the material.

Content from this work may be used under the terms of the Creative Commons Attribution 3.0 licence. Any further distribution of this work must maintain attribution to the author(s) and the title of the work, journal citation and DOI.
Published under licence by IOP Publishing Ltd

However, the problem of the influence of the amplitude and frequency of ultrasonic vibrations on the structure and properties of weld-joint zone remains open. The study of this problem is the purpose of this work.

2. Materials and methods

In experiments, we used as the objects of investigation aluminum AA1135 4 mm thick and as-annealed copper C11000 3.5 mm thick.

Explosive welding was carried out simultaneously with ultrasound vibrations supplied to a based plate in an explosion chamber according to the propagation of ultrasonic vibrations to the welding direction (Figure 1) [7]. A piezoceramic transducer was attached to a fixed plate with a metal rod welded to it. The distance between the based plate and the piezoceramic transducer in all experiments was kept constant (no more than 100 mm), and an ultrasonic generator was located outside the explosive chamber where it was turned on 30-60 seconds before detonating an explosive charge. For the sake of comparison, the same metal pairs were explosively welded under the same conditions without ultrasound processing (reference samples) (Figure 1, (4)).

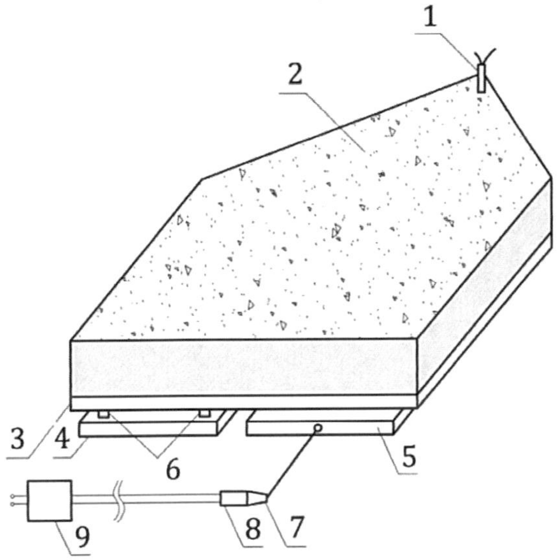

Figure. 1. Schematic illustration of ultrasound-assisted explosive welding: (1) electric blasting detonator, (2) explosive charge, (3) flyer plate, (4) still (reference) base plate, (5) vibrating base plate, (6) supporting feet, (7) waveguide, (8) piezoceramic transducer and (9) ultrasonic generator.

Parameters of explosive welding regimes (V*c* is the velocity of the collision point and V*i* is the velocity of the impact) were calculated using an EW Calc application software package [8]. The detonation velocity of an explosive substance was controlled by electrocontact method. Time registration was conducted with ChZ-63 electronic frequency meters.

As the source of ultrasonic waves, we used a UZGI-2 setup with a piezoceramic converter and with a waveguide in the form of conical and stepped concentrators, which are characterized by a minimum coefficient of transformation and by a wide range of the amplitudes of vibrations.

An OLYMPUS BX61 optical microscope was used for metallographic studies.

To study the features of plastic deformation of near-weld volumes of metal under the conditions of high-speed glancing collision, we used the procedure [7] based on the use of the model of layered samples with a thickness of the layers of 50 μm. Explosive welding with the simultaneous action of an ultrasound on the fixed plate was implemented using regimes that ensured the formation of a sinusoidal profile of the welding-joint line. For a comparison of the obtained results, the explosive welding of the same pairs of metals was also performed using identical regimes, but without the action of ultrasound (control samples).

3. Results and discussion

The research showed that the ultrasound-assisted explosive welding has a significant effect on the structure and properties of the heat-affected zone of the formed joints. The greatest changes are observed in the case of the oppositely directed ultrasonic vibrations relative to the welding direction, which will be further analyzed (hereinafter referred to as test samples). Under the action of oppositely directed longitudinal ultrasonic vibrations, parameters of the wavy interface of the welded aluminum joint (Figures 3 b, 3 c) decreased almost three times to yield an average wave height $2a \approx 0.18$ mm and wavelength $\lambda \approx 0.36$ mm, whereas for the reference samples $2a \approx 0.56$ mm and $\lambda \approx 0.88$ mm (Figures 3 b, 3 c). A similar effect was observed in the case of the explosive welding of the copper samples (Figures 2 b, 2 c).

The study of the plastic deformation of the metal in the heat-affected zone upon the explosive welding of uniform aluminum and copper plates under the action of ultrasound revealed the following features.

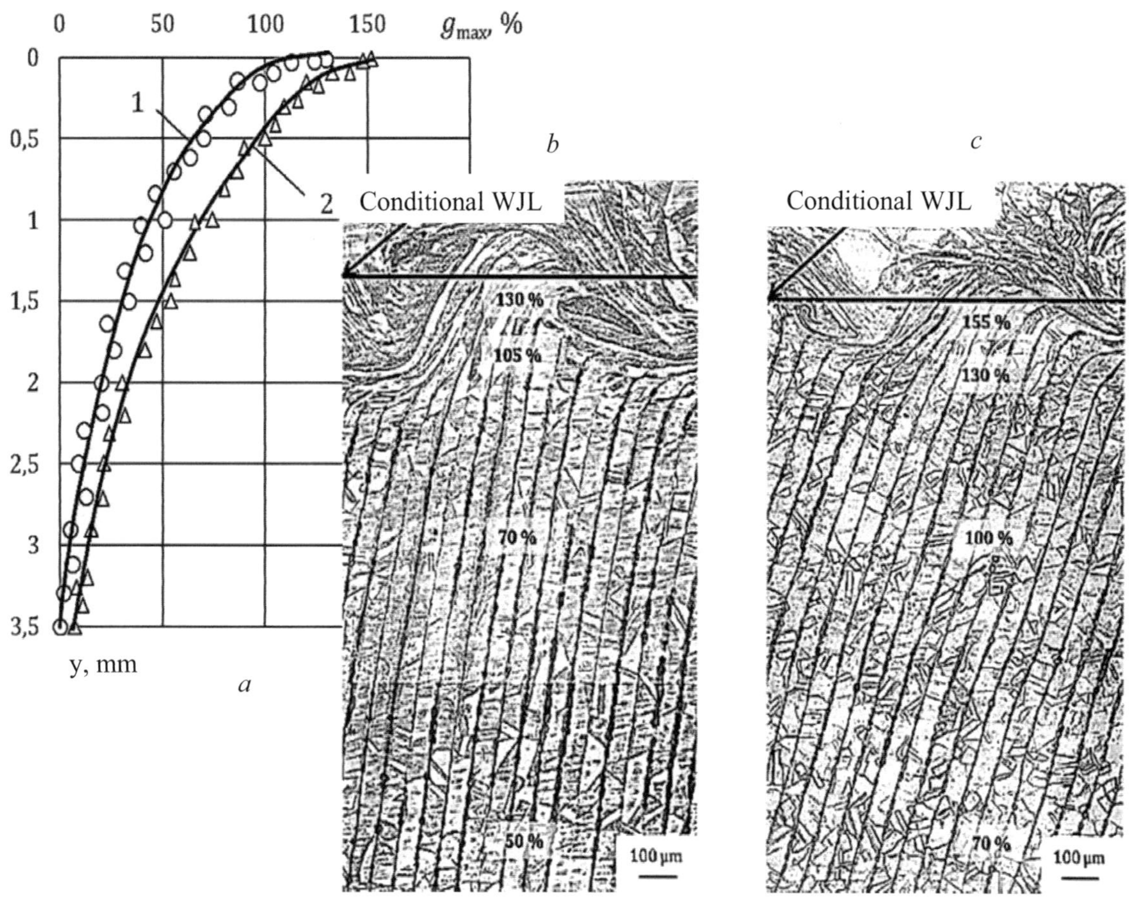

Figure 2. Diagrams of the maximum shears g_{max} (a) and the deformation of the transverse layered model of a copper joint in (b) control sample without the application of ultrasound and (c) in a sample with the action of ultrasound.

Upon the explosive welding with the simultaneous application of ultrasound, the maximum residual shearing deformation g_{max} measured in the close proximity to the conditional line of joining of the layers is about 155%. At the same time, in the control sample obtained under identical conditions of explosive welding but without the application of ultrasound, the shearing strain is smaller, i.e., g_{max} ~ 130%. With moving from the line of joining, g_{max} decreases sharply with a slightly different gradient (Figure 2 a). A similar distribution of the residual shearing deformation g_{max} is also observed in the case of the welding of model aluminum plates (Figure 3). Thus, following the explosive welding with the action of ultrasound the maximum residual shearing deformation g_{max} measured in the close proximity to the conditional welding joint line (WJL) in the aluminum samples is about 175%, whereas in the control sample without the action of the ultrasound, the values of g_{max} are ~155%.

The analysis of the distribution diagrams of g_{max} and of the microstructure of copper and aluminum samples show that, together with an increase in the maximum values of g_{max} and a decrease in the

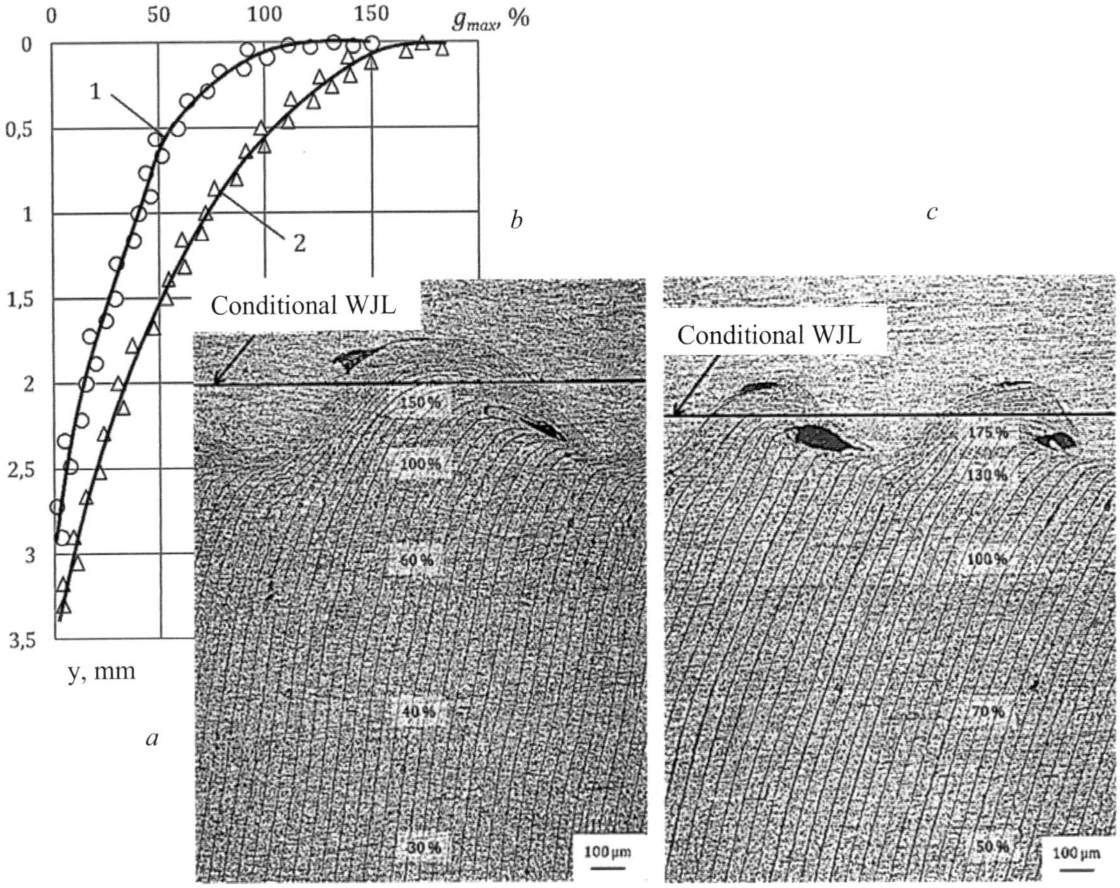

Figure 3. Diagrams of maximum shears g_{max} (a) and deformation of transverse layered model of an aluminum joint in (b) control sample without the application of ultrasound and (c) in sample with the application of ultrasound.

parameters of the interface between the layers, a larger volume of metal is involved into the plastic deformation upon the explosive welding under the action of ultrasound (Figures 2, 3). It is also worthwhile to note that the values of the residual shearing strain g_{max} measured at the identical distances from the conditional WJL are different for the samples with the application and without the application of the ultrasound. Taking into account the results obtained in [6, 7], the following assumption can be made. Upon the action of ultrasonic vibrations on the metal, an acoustoplastic effect takes place in it, which consists of a reduction in the yield stress of the material at a constant rate of deformation.

As a result, smaller power expenditures are required for the dislocations and vacancies to overcome barriers; thus, they can easier move and interact with each other. In turn, this leads to changes in the plastic properties of the surface layers of metal and to changes in the conditions for the formation of the bond between the colliding plates.

Let us try to explain the change of the parameters of the wave in the weld joint zone taking into account the results of the works [5, 9–11], in which it has been shown that the area limited by the diagram of g_{max} is nothing but the work of deformation determined by the following dependence [5, 8]:

$$A = S_k \cdot \int_0^{\delta} \varepsilon(y)dy \tag{1}$$

where $\varepsilon(y)$ is the current magnitude of the degree of deformation; δ is the thickness of the plate investigated; and S_k is the resistance of the material to deformation (or the maximum strength of the material, which depends on the ultimate tensile strength and relative reduction in area), which is numerically equal to the dynamic yield stress (640 MPa for copper; and 108 MPa for aluminum).

Note that the experiments were carried out under the identical conditions, where the energy spent for the plastic deformation (W_2) of the metal of the heat-affected zone upon the explosive welding was constant.

However, in all series of experiments with the application of high-frequency acoustic vibrations, we observed an increase in the area (25–35% on average depending on the grade of the material in the g_{max}-y coordinates) limited by the g_{max} diagram. An increase in this area indicates an increase in W_2, which is impossible, since the energy introduced by the high-frequency acoustic wave into the system under consideration is less by several orders of magnitude than that that comes from the explosion welding.

It is logical to assume that an increase in the area of the diagram of g_{max} upon the explosive welding with the application of ultrasound is connected with the change in the dynamic yield stress due to the development of the acoustoplastic effect. Based on the obtained diagrams of g_{max}, it is not difficult to estimate the values of the dynamic yield stress under the action of ultrasound for copper we obtained 458 MPa and 71.9 MPa for aluminum, respectively.

It has been shown in [5, 9–11] that the sizes of waves that rise in the weld joint zone are caused by the magnitude of the deforming pressure impulse I_d calculated through the following dependences:

at $\tau_p < \tau_c$

$$I_\partial = \int_{\tau_0}^{\tau_p} P_{max} e^{-\tau/\theta} d\tau = P_{max}\theta(1 - e^{-\tau_p/\theta}), \tag{2}$$

in which the rate of the pressure release θ after the transformations (under the assumption that in the process of interaction of the collided plates the contact pressure after the time τ_p will decrease from P_{max} to σ_τ^d takes on the form

$$\theta = \frac{\tau_p}{\ln \dfrac{P_{max}}{\sigma_\tau^d}}, \tag{3}$$

The peak pressure P_{max} of the pulse is determined by the dependence [12]

$$P_{max} = \frac{\rho_1 c_{o1} \rho_2 c_{o2}}{\rho_1 c_{o1} + \rho_2 c_{o2}} V_c, \tag{4}$$

(where ρ_1, ρ_2 and c_{o1}, c_{o2} are, correspondingly, the densities and the speeds of sound in the flyer plate and in the fixed plate, respectively) and changes exponentially; according to [13], the time of the interaction of the metals

$$\tau_c = \frac{2\delta_{min}}{c_o}, \tag{5}$$

where δ_{min} is the minimum thickness of the plate and c_o is the speed of sound.

The calculations according to the dependences (2–5) show that the decrease in the time constant θ reduces the value of the deforming pulse I_d (for copper with 4 kN·s/m^2 to 3.35 kN·s/m^2 and for aluminum from 0.9 kN·s/m^2 and 0.8 kN·s/m^2) and the time τ of the action of the compressive stresses that exceed after the point of contact at which the plastic flow of metal can occur. This is apparently one of the main factors that are responsible for a change in the length and amplitude of the wave in the weld joint zone upon explosive welding under the action of high-frequency acoustic vibrations.

4. Conclusions

The action of vibrations exerts a substantial influence on the structure and properties of the weld joint zone of copper and aluminum plates bonded using explosive welding, the parameters of the waves decrease by 2.0–2.5 times, there are hardly any regions of fused metal, and the strength of the tearing-off of the layers increases by 15–20% compared with samples obtained by explosive welding without the application of ultrasound.

It has been established that the influence of high-frequency acoustic waves on the metal leads to a reduction of the dynamic yield stress, which in turn leads to changes in the properties of the surface layers of metal and in the conditions of bonding between the collided plates upon explosive welding.

It has been shown that the changes in the length and amplitude of the wave in the weld joint zone upon the explosive welding under the action of high-frequency acoustic waves is connected with a decrease in the magnitude of the deforming pulse and the time of action of the compressive stresses that exceed the dynamic yield stress beyond the contact point at which the plastic flow of the metal is possible.

Acknowledgments

This work is executed at financial support of the Ministry of education and science of the Russian Federation in the framework of the "Council on grants of the President of the Russian Federation" (SP-1940.2016.1) and partially supported by the Russian Foundation for Basic Research (project 15-08-09438 A).

References

[1] Crossland B and Bahrani A S 1968 *Contemp. Phys.* **9** 71-87
[2] Carpenter S H and Wittman R H 1975 *Annu. Rev. Mater. Sci.* **5** 177-99
[3] Hoseini Athar M M and Tolaminejad B L 2015 *Mater. Design* **86** 516–25
[4] Gulenc B 2008 *Mater. Design* **29** 275-278
[5] Lysak V I and Kuz'min S V 2012 *J. Mater. Process. Techn.* **212** 150-6
[6] Peev A P, Kuz'min S V, Lysak V I, Kuz'min E V and Dorodnikov A N 2015 *Phys. Met. Metallogr.* **116** 817–22
[7] Peev A P, Kuz'min S V, Lysak V I, Kuz'min E V and Dorodnikov A N 2017 *Phys. Met. Metallogr.* **118** 485–93
[8] Lysak V I and Kuz'min S V 2015 *J. Mater. Process. Techn.* **222** 356-64
[9] Bondar' M P and Ogolikhin V M 1988 *Combust., Expl., Shock Waves* **24** 113-7
[10] Gordopolov Yu A, Dremin A N and Mikhailov A N 1978 *Combust., Expl., Shock Waves* **14** 472-80
[11] Godunov S K, Deribas A A and Kozin N S 1973 *J. Appl. Mech. Tech. Phys.* **12** 398-406
[12] Kachan M S and Trishin Yu A 1975 *Combust., Expl., Shock Waves* **11** 816-9
[13] Kolesnikov S Yu, Shelekhov E V and Epshtein G N 1983 *Izv. Vys. Ucheb. Zaved. Cher. Metall.* **7** 84-8

IOP Conf. Series: Journal of Physics: Conf. Series **894** (2017) 012048 doi:10.1088/1742-6596/894/1/012048

On the existence of global solutions to equations for mixtures of compressible viscous fluids

N A Kucher[1] and A A Zhalnina[1]

[1] Kemerovo State University, 650043 Kemerovo, Russia

E-mail: nakycher@rambler.ru, qwert1776@yandex.ru

Abstract. We prove existence of global generalized solutions for initial boundary value problem which describes the three-dimensional unsteady barotropic motions of binary mixtures of compressible viscous fluids in a bounded domain with impermeable boundary.

1. Introduction

Assume that a binary mixture of viscous gases occupies a bounded domain $\Omega \subset R^3$ in the Euclidean space of points $x = (x_1, x_2, x_3)$ with boundary of class $C^{2+\nu}$. A motion of the mixture is characterized by the velocity fields $\vec{u}^{(i)}(t,x)$, the densities $\rho_i(t,x)$ and also by pressures $p_i(t,x)$ of the mixture components ($i = 1,2$), which satisfy the following equations [1]

$$\partial_t(\rho_i \vec{u}^{(i)}) + div(\rho_i \vec{u}^{(i)} \otimes \vec{u}^{(i)}) + \nabla p_i = div\sigma^{(i)} + \vec{J}^{(i)} \; in \; Q_T = (0,T) \times \Omega, \tag{1}$$

$$\partial_t(\rho_i) + div(\rho_i \vec{u}^{(i)}) = 0 \; in \; Q_T, \; i = 1,2. \tag{2}$$

Here viscous stress tensors $\sigma^{(i)}$ are defined by the equalities

$$\sigma^{(i)}(\vec{u}^{(1)}, \vec{u}^{(2)}) = \sum_{j=1}^{2} \left[2\mu_{ij} \mathbf{D}(\vec{u}^{(j)}) + \lambda_{ij} div \vec{u}^{(j)} \cdot \mathbf{I} \right], \tag{3}$$

where $\mathbf{D}(\vec{u}) = \frac{1}{2}\left(\nabla \vec{u} + (\nabla \vec{u})^T\right)$, $\mathbf{I}$ is the identity tensor, constant viscosity coefficients μ_{ij} and λ_{ij}, $i,j = 1,2$, satisfy the conditions

$$\begin{gathered} \mu_{11} > 0, \;\; 4\mu_{11}\mu_{22} - (\mu_{12} + \mu_{21})^2 > 0, \;\; \nu_{11} > 0, \;\; \nu_{22} > 0, 4\nu_{11}\nu_{22} - (\nu_{12} + \nu_{21})^2 > 0, \\ \nu_{ij} = \lambda_{ij} + 2\mu_{ij}, \; i,j = 1,2. \end{gathered} \tag{4}$$

It is assumed the following relationship between the pressures p_i and the density ρ_i for the i-th component of the mixture $p_i = \rho_i^{\gamma_i}$, where $\gamma_i > 1, i = 1,2$, are adiabatic exponents. The summands $\vec{J}^{(i)} = (-1)^{i+1} \cdot a(\vec{u}^{(2)} - \vec{u}^{(1)})$, $a = const > 0$, $i = 1,2$. characterize the intensity of the momentum exchange between the mixture components [2, 3]. Equations (1) and (2) are supplemented with initial conditions

$$\rho_i \left|_{t=0} = \rho_i^0, \;\; \rho_i \vec{u}^{(i)} \right|_{t=0} = \vec{q}^{(i)} \Big|_{t=0} = \vec{q}_0^{(i)} \; in \; \Omega, \; i = 1,2, \tag{5}$$

Content from this work may be used under the terms of the Creative Commons Attribution 3.0 licence. Any further distribution of this work must maintain attribution to the author(s) and the title of the work, journal citation and DOI.
Published under licence by IOP Publishing Ltd

IOP Conf. Series: Journal of Physics: Conf. Series **894** (2017) 012048 doi:10.1088/1742-6596/894/1/012048

and with boundary conditions

$$\vec{u}^{(i)} = 0 \; on \; (0,T) \times \partial\Omega, \; i = 1,2. \tag{6}$$

It means, that boundary of flow region is motionless and impermeable.

Global theorems of existence and results of stabilisation of solutions for the unsteady equations for multivelocity continuum similar to (1)-(4) are received now only in case of one-dimensional movement with flat waves when the solution depends only on one spatial variable [3-5]. The first results for Stoks-like mixture model in case of more than one spatial variable are received in [6-7].

In [8] research of so-called quasistationary mixture model of compressible fluids was conducted in the bounded region with special boundary conditions. In [9] the stationary solutions of the Dirichlet problem for the full equations (1)-(4) in case of three spatial variables was constructed, and in [10-13] existence of weak solutions for equations describing the three-dimensional steady motions of binary mixtures of heat-conductive compressible viscous fluids was proved.

The reader is referred to [10, 15] for standard notation of functional spaces and terminology.

2. Statement of the problem

Motion of continuum is characterised by potential energy

$$E = E[\rho](t) = \sum_{i=1}^{2} \int_{\Omega} P_i(\rho_i(t,x))dx, \quad \rho = (\rho_1, \rho_2),$$

where nonnegative functions P_i are defined from the equations $sP_i'(s) - P_i(s) = p_i(s)$ up to a nonessential linear function (in particular, if $p_i = \rho_i^{\gamma_i}$, $\gamma_i > 1$, then $P_i(\rho_i) = (\gamma_i - 1)^{-1}\rho_i^{\gamma_i}$) and by energy dissipation rate

$$D = D[\vec{u}](t) = \sum_{i=1}^{2} \int_{\Omega} |\nabla \vec{u}^{(i)}|^2 dx, \quad \vec{u} = (\vec{u}^{(1)}, \vec{u}^{(2)}).$$

Multiplying the momentum balance equations (1) by $\vec{u}^{(i)}$ and integrating by parts, we arrive (using (6)) at the identity

$$\frac{d}{dt}\left\{\sum_{i=1}^{2}\left(\frac{1}{2}\int_{\Omega} \rho_i|\vec{u}^{(i)}|^2 dx + \int_{\Omega} \frac{\rho_i^{\gamma_i}}{\gamma_i - 1} dx\right)\right\} + \sum_{i=1}^{2}\int_{\Omega} \sigma^{(i)} : \nabla \vec{u}^{(i)} dx + a\int_{\Omega} |\vec{u}^{(2)} - \vec{u}^{(1)}|^2 dx = 0, \tag{7}$$

which gives the mathematical relationship for the total energy balance. From identity (7) we obtain estimates of the energy dissipation rate $D(t)$ and total energy $\mathcal{E}(t)$

$$\mathcal{E}(t) = \mathcal{E}[\rho, \vec{u}](t) = K[\rho, \vec{u}](t) + E[\rho](t) = \sum_{i=1}^{2} \int_{\Omega} \left[\frac{1}{2}\rho_i|\vec{u}^{(i)}|^2 + \frac{\rho_i^{\gamma_i}}{\gamma_i - 1}\right] dx. \tag{8}$$

Note that for any continuously differentiable function $a_i : R \to R$, every smooth solution of the continuity equations (2) satisfies the equalities

$$\partial_t a_i(\rho_i) + div\,(a_i(\rho_i)\vec{u}^{(i)}) + (\rho_i a_i'(\rho_i) - a_i(\rho_i)) div\,\vec{u}^{(i)} = 0, \quad i = 1,2, \tag{9}$$

which are called renormalized form of the equation (2). The functions ρ_i, satisfing this system are called renormalized solutions of equations (2). Relations (7)-(9) determine the specific definition of a generalized solution of the initial-boundary value problem (1)-(6).

Definition A. A generalized solution to the initial-boundary value problem (1)-(6) are non-negative functions $\rho_i \in L^\infty(0,T; L^{\gamma_i}(\Omega))$, $i = 1,2$, and vector fields $\vec{u}^{(i)} \in L^2\left(0,T; W_0^{1,2}(\Omega)\right)$, $i = 1,2$, such that:

(A1) The total energy $\mathcal{E}[\rho,\vec{u}](t)$ is locally integrable on an interval $(0,T)$, and the total energy equation (7) is fulfilled in the space of distributions $D'(0,T)$;

(A2) The momentum balance equations (1) are fulfilled in the space $D'((0,T)\times\Omega)$;

(A3) The continuity equations (2) are satisfied in the sense of renormalized solutions, i.e., for any differentiable functions a_i such that, $a_i(z)\equiv 0$ for all sufficiently large $z\in R$, for example $z\geq M$ (constants M are different for each of the functions a_i), equations (9) are fulfilled in $D'((0,T)\times\Omega)$. In addition, equations (2) hold true in the space of distributions $D'((0,T)\times R^3)$ if the functions ρ_i, $\vec{u}^{(i)}$ extended by zero over the whole space $R^3\backslash\Omega$.

Remark 1. It follows from equations (1)-(2) that the generalized solution of problem (1)-(6) belongs to the class

$$\rho_i\in C([0,T];L^{\gamma_i}_{weak}(\Omega)),\;\;\rho_i\vec{u}^{(i)}\in C([0,T];L^{\frac{2\gamma_i}{\gamma_i+1}}_{weak}(\Omega)), i=1,2,$$

and, therefore, the initial conditions (5) make sense. This means that it is required that the initial date ρ_i^0, $\vec{q}_0^{\,(i)}, i=1,2$, satisfy the conditions:

$$\rho_i^0\in L^{\gamma_i}(\Omega),\; \rho_i^0\geq 0\; a.e.\; in\; \Omega,\; \vec{q}_0^{\,(i)}\cdot 1_{\{\rho_i^0=0\}}=0\; a.e.\; in\; \Omega,\; \frac{1}{\rho_i^0}|\vec{q}_0^{\,(i)}|^2\cdot 1_{\{\rho_i^0>0\}}\in L^1(\Omega). \tag{10}$$

The main result of the article is given by the following theorem.

Theorem 2. Let $\Omega\subset R^3$ be a bounded domain with $C^{2+\nu}$, $\nu>0$, boundary. Let the initial data ρ_i^0, $\vec{q}_0^{\,(i)}$, $i=1,2$, satisfy conditions (10), $\gamma_i>3/2$ and total viscosity matrix $H=\{\nu_{ij}\}_{i,j=1}^2$ is triangular. Then for an arbitrary $T>0$, there exists at least one generalized solution of the problem (1)-(6).

Let us characterize briefly the main stages of the proof of this theorem.

Along with problem (1)-(6) we consider its regularization

$$\begin{gathered}\partial_t(\rho_i\vec{u}^{\,(i)})+div(\rho_i\vec{u}^{\,(i)}\otimes\vec{u}^{\,(i)})+\nabla(\rho_i^{\gamma_i})+\delta\nabla(\rho_i^{\beta_i})+\\ +\varepsilon\nabla\vec{u}^{(i)}\cdot\nabla\rho_i=div\sigma^{(i)}+\vec{J}^{(i)}\; in\; Q_T=(0,T)\times\Omega, i=1,2,\end{gathered}\tag{11}$$

$$\partial_t(\rho_i)+div(\rho_i\vec{u}^{(i)})=\varepsilon\triangle\rho_i\; in\; Q_T, i=1,2, \tag{12}$$

$$\rho_i\,|_{t=0}=\rho_i^0,\;\; \rho_i\vec{u}^{(i)}\Big|_{t=0}=\vec{q}_0^{\,(i)}\; in\;\Omega, i=1,2, \tag{13}$$

$$\vec{u}^{(i)}=0\; on\; (0,T)\times\partial\Omega, i=1,2, \tag{14}$$

$$\nabla\rho_i\cdot\vec{n}=0\; on\; (0,T)\times\partial\Omega, i=1,2, \tag{15}$$

where $\varepsilon>0$, $\delta>0$ are small parameters and values $\beta_i, i=1,2$ should be chosen sufficiently large. Equations (12) supplemented with Neumann boundary conditions (15).

A generalized solution of problem (1)-(6) will be obtained as the limit of a sequence of solutions to the regularized problem (11)-(15). Therefore, the first step in the proof of theorem 2 will be to construct solutions of the regularized problem by the method of finite-dimensional approximation. The next stage of the proof will be to justify the limit passages when the dissipation parameter ε go to zero. The difficulty of this procedure is due to insufficiency of estimates for the densities $\rho_i, i=1,2$. Here we use the technique developed in [13-16] for the classical model of Navier-Stokes, based on the weak regularity properties of the effective viscous fluxes. In this work also introduced analogues of the effective viscous fluxes $\tau_i\left(\rho_i,\vec{u}^{(1)},\vec{u}^{(2)}\right)=\rho_i^{\gamma_i}+\delta\rho_i^{\beta_i}-\sum\limits_{j=1}^{2}(\lambda_{ij}+2\mu_{ij})div\vec{u}^{(j)}$ and the following properties are proved

$$\overline{\tau_i\left(\rho_i,\vec{u}^{(1)},\vec{u}^{(2)}\right)\cdot\rho_k}=\overline{\tau_i\left(\rho_i,\vec{u}^{(1)},\vec{u}^{(2)}\right)}\cdot\overline{\rho_k},\;\; i,k=1,2. \tag{16}$$

Here $\overline{b(\rho)}$ denotes the weak limit of a sequence $b(\rho_\varepsilon)$. Properties (16) allow us to establish strong convergence of the densities and complete justification of the limit passages when $\varepsilon \to 0$. The final stage in the proof of theorem 2 consist in exclusion of the artificial pressures $\delta\rho_i^{\beta_i}, i = 1, 2$. To this end we apply the so-called truncation operators introduced in [17,18].

3. The existence of a strong solution of the auxiliary problem (11)-(15)

The following theorem gives the main result on the well-posedness of the regularized problem.

Theorem 3. Let the coefficients μ_{ij} and λ_{ij} satisfy the conditions (4) and adiabatic exponents $\gamma_i > \frac{3}{2}, i = 1, 2$. Let the parameters ε, δ, β_i are chosen so that $\varepsilon > 0$, $\delta > 0$, $\beta_i \geq 15, i = 1, 2$. Suppose that Ω be a bounded domain in R^3 with $C^{2,\theta}$ boundary, $\theta \in (0, 1]$, and

$$0 < \underline{\rho} \leq \rho_i^0 \leq \overline{\rho} < \infty, \ \rho_i^0 \in W^{1,\infty}(\Omega), \ \vec{q}_0^{(i)} \in L^2(\Omega).$$

Then there exist a couple $(\rho_{i,\varepsilon}, \vec{u}_\varepsilon^{(i)})$, $\rho_{i,\varepsilon} = \rho_{i,\varepsilon,\delta}$, $\vec{u}_\varepsilon^{(i)} = \vec{u}_{\varepsilon,\delta}^{(i)}, i = 1, 2$, with the following properties:

(i) $\rho_{i,\varepsilon} \in C^0(\bar{I}, L^{\beta_i}_{weak}(\Omega)), \ \rho_{i,\varepsilon} \in C^0(\bar{I}, L^p(\Omega)) \cap L^{\beta_i+1}(Q_T), \ 1 \leq p < \beta_i, \ I = (0, T).$

$$\rho_{i,\varepsilon} \geq 0 \ a.e. \ in \ Q_T, \ \rho_{i,\varepsilon}^{\frac{1}{2}\beta_i} \in L^2(I, W^{1,2}(\Omega)),$$

$$\partial_t\rho_{i,\varepsilon} \in L^{\frac{5\beta_i-3}{4\beta_i}}(Q_T), \ D^\alpha\rho_{i,\varepsilon} \in \left(L^{\frac{5\beta_i-3}{4\beta_i}}(Q_T)\right)^{3\times 3}, |\alpha| = 2.$$

$$\vec{u}_\varepsilon^{(i)} \in L^2(I, W_0^{1,2}(\Omega)), \ \vec{q}_\varepsilon^{(i)} = \rho_{i,\varepsilon}\vec{u}_\varepsilon^{(i)} \in C^0\left(\bar{I}, L_{weak}^{\frac{2\beta_i}{\beta_i+1}}(\Omega)\right), \ \vec{q}_\varepsilon^{(i)} \in L^2\left(I, L^{\frac{6\beta_i}{\beta_i+6}}(\Omega)\right),$$

$$\rho_{i,\varepsilon}|\vec{u}_\varepsilon^{(i)}|^2 \in L^\infty(I, L^1(\Omega)) \cap L^2\left(I, L^{\frac{6\beta_i}{4\beta_i+3}}(\Omega)\right), \ \nabla\vec{u}_\varepsilon^{(i)} \cdot \nabla\rho_{i,\varepsilon} \in L^{\frac{5\beta_i-3}{4\beta_i}}(Q_T),$$

$$\nabla\rho_{i,\varepsilon}, \ \rho_{i,\varepsilon}\vec{u}_\varepsilon^{(i)} \in L^{\frac{5\beta_i-3}{4\beta_i}}\left(I, E_0^{\frac{10\beta_i-6}{3\beta_i+3}, \frac{5\beta_i-3}{4\beta_i}}(\Omega)\right), \ \int_\Omega \rho_{i,\varepsilon}dx = \int_\Omega \rho_i^0 dx.$$

(ii) Equations (11) are fulfilled in the space of distributions $D'(Q_T)$

(iii)Equations

$$\frac{d}{dt}\int_\Omega \rho_{i,\varepsilon}\eta_i dx - \int_\Omega \rho_{i,\varepsilon}\vec{u}_\varepsilon^{(i)}\nabla\eta_i dx + \varepsilon\int_\Omega \nabla\rho_{i,\varepsilon} \cdot \nabla\eta_i dxdt = 0, \ \eta_i \in C^\infty(R^3), \ i = 1, 2,$$

are fulfilled in $D'(I)$.

(iv) The functions $\rho_{i,\varepsilon}, \vec{u}_\varepsilon^{(i)}$ satisfy the energy inequalities

$$\frac{d}{dt}\mathcal{E}_\delta(\rho_{i,\varepsilon}\vec{u}_\varepsilon^{(i)}) + c_0\int_\Omega (|\nabla\vec{u}_\varepsilon^{(1)}|^2 + |\nabla\vec{u}_\varepsilon^{(2)}|^2)dx +$$

$$+\varepsilon\int_\Omega \sum_{i=1}^2 (\gamma_i\rho_{i,\varepsilon}^{\gamma_i-2} + \delta\beta_i\rho_{i,\varepsilon}^{\beta_i-2})|\nabla\rho_{i,\varepsilon}|^2 dx + a\int_\Omega |\vec{u}_\varepsilon^{(2)} - \vec{u}_\varepsilon^{(1)}|^2 dx \leq 0, \ in \ D'(I),$$

where $\mathcal{E}_\delta(\rho_i, \vec{u}^{(i)}) = \sum_{i=1}^2 \int_\Omega \left\{\frac{1}{2}\rho_i|\vec{u}^{(i)}|^2 + \frac{\rho_i^{\gamma_i}}{\gamma_i - 1} + \delta\frac{\rho_i^{\beta_i}}{\beta_i - 1}\right\} dx.$

(v) For any $\eta_i \in C_0^\infty(\Omega)$ and $\vec{\varphi}^{(i)} \in C_0^\infty(\Omega)$ the following relations hold true

$$\lim_{t\to 0+} \int\limits_\Omega \rho_{i,\varepsilon}(t)\eta_i dx = \int\limits_\Omega \rho_i^0 \eta_i dx, \quad \lim_{t\to 0+} \int\limits_\Omega \rho_{i,\varepsilon}\vec{u}_\varepsilon^{(i)}(t)\vec{\varphi}^{(i)} dx = \int\limits_\Omega \vec{q}_0^{(i)} \cdot \vec{\varphi}^{(i)} dx.$$

(vi) If $\delta \in (0,1)$, then we have the following estimates, uniform with respect to the parameter ε:

$$\sum_{i=1}^2 \left(\|\vec{u}_\varepsilon^{(i)}\|_{L^2(I,W^{1,2}(\Omega))} + \|\rho_{i,\varepsilon}\|_{L^\infty(I,L^{\gamma_i}(\Omega))} + \delta^{1/\beta_i}\|\rho_{i,\varepsilon}\|_{L^\infty\left(I,L^{\beta_i}(\Omega)\right)} \right) \le L$$

$$\sum_{i=1}^2 \left(\sqrt{\varepsilon}\|\nabla\rho_{i,\varepsilon}\|_{L^2(Q_T)} + \varepsilon\|\nabla\rho_{i,\varepsilon}\|_{L^{\frac{10\beta_i-6}{3\beta_i+3}}(Q_T)} \right) \le L$$

$$\sum_{i=1}^2 \left(\|\rho_{i,\varepsilon}|\vec{u}_\varepsilon^{(i)}|^2\|_{L^\infty(I,L^1(\Omega))} + \|\rho_{i,\varepsilon}|\vec{u}_\varepsilon^{(i)}|^2\|_{L^2\left(I,L^{\frac{6\beta_i}{4\beta_i+3}}(\Omega)\right)} \right) \le L$$

$$\sum_{i=1}^2 \left(\|\rho_{i,\varepsilon}\vec{u}_\varepsilon^{(i)}\|_{L^\infty\left(I,L^{\frac{2\beta_i}{\beta_i+1}}(\Omega)\right)} + \|\rho_{i,\varepsilon}\vec{u}_\varepsilon^{(i)}\|_{L^2\left(I,L^{\frac{6\beta_i}{\beta_i+6}}(\Omega)\right)} \right) \le L,$$

$$\sum_{i=1}^2 \left(\|\rho_{i,\varepsilon}\vec{u}_\varepsilon^{(i)}\|_{L^{\frac{10\beta_i-6}{3\beta_i+3}}(Q_T)} + \varepsilon\|\nabla\rho_{i,\varepsilon}\cdot\nabla\vec{u}_\varepsilon^{(i)}\|_{L^{\frac{5\beta_i-3}{4\beta_i}}(Q_T)} \right) \le L,$$

Here L – positive constant independent of ε and δ.

4. The study of equations with artificial pressure

In this section we present the results of limit passages when the dissipation parameter ε go to zero.

Theorem 4 Let the coefficients μ_{ij}, λ_{ij} and the parameters γ_i, δ, $\beta_i, i = 1,2$ meet all requirements of theorem 2 and theorem 3. Suppose that Ω be a bounded domain in R^3 with boundary of class $C^{2,\theta}$, $\theta \in (0,1)$, and

$$\rho_i^0 \in L^{\beta_i}(\Omega), \ \ \rho_i^0 \ge 0 \ a.e. \ in \ \Omega,$$
$$\vec{q}_0^{(i)} \in L^{\frac{2\beta_i}{\beta_i+1}}(\Omega), \ \ \vec{q}_0^{(i)} \cdot 1_{\{\rho_i^0=0\}} = 0 \ a.e. \ in \ \Omega, \ \ \frac{|\vec{q}_0^{(i)}|^2}{\rho_i^0} \cdot 1_{\{\rho_i^0>0\}} \in L^1(\Omega).$$

Then there exists generalized solution $(\rho_{i,\delta}, \vec{u}_\delta^{(i)})$, $i = 1,2$, of problem

$$\partial_t(\rho_i\vec{u}^{(i)}) + div(\rho_i\vec{u}^{(i)} \otimes \vec{u}^{(i)}) + \nabla\rho_i^{\gamma_i} + \delta\nabla\rho_i^{\beta_i} = div\,\sigma^{(i)} + \vec{J}^{(i)} \ in \ Q_T, \tag{17}$$

$$\partial_t\rho_i + div(\rho_i\vec{u}^{(i)}) = 0 \ in \ Q_T, \tag{18}$$

$$\vec{u}^{(i)} = 0 \ on \ \partial\Omega\times(0,T), \ \ \rho_i|_{t=0} = \rho_i^0, \ \ \vec{q}^{(i)}\Big|_{t=0} = \rho_i\vec{u}^{(i)}\Big|_{t=0} = \vec{q}_0^{(i)}, i = 1,2, \tag{19}$$

with the properties:

(i)
$$\rho_{i,\delta} \in L^{\beta_i+1}(Q_T), \ \rho_{i,\delta} \in C^0(\bar{I}, L^{\beta_i}_{weak}(\Omega)) \cap C^0(\bar{I}, L^p(\Omega)), 1 \le p < \beta_i,$$
$$\rho_{i,\delta} \ge 0 \ in \ Q_T, \ \rho_{i,\delta} = 0 \ in \ (R^3\backslash\Omega)\times I,$$
$$\vec{u}_\delta^{(i)} \in L^2(I, W_0^{1,2}(\Omega)), \ \vec{u}_\delta^{(i)} = 0 \ in \ (R^3\backslash\Omega)\times I, \ \rho_{i,\delta}\vec{u}_\delta^{(i)} \in L^2(I, L^{\frac{6\beta_i}{\beta_i+6}}(\Omega)) \cap C^0(\bar{I}, L^{\frac{2\beta_i}{\beta_i+1}}(\Omega)),$$
$$\rho_{i,\delta}|\vec{u}_\delta^{(i)}|^2 \in L^\infty(I, L^1(R^3)) \cap L^2(I, L^{\frac{6\beta_i}{4\beta_i+3}}(\Omega)) \cap L^1(I, L^{\frac{3\beta_i}{\beta_i+3}}(\Omega)).$$

(ii) Equations (17) are fulfilled in $D'(Q_T)$

(iii) Equations (18) are fulfilled in the sense $\partial_t \rho_{i,\delta} + div(\rho_{i,\delta}\vec{u}^{(i)}_\delta) = 0$ in $D'(R^3 \times \Omega)$.
(iv) The initial conditions in (19) understood in the sense

$$\lim_{t\to 0+}\int_\Omega \rho_{i,\delta}(t)\eta_i dx = \int_\Omega \rho_i^0\eta_i\, dx,\ \eta_i \in C_0^\infty(\Omega),\ \lim_{t\to 0+}\int_\Omega \rho_{i,\delta}\vec{u}^{(i)}_\delta(t)\vec{\varphi}\, dx = \int_\Omega \vec{q}^{(i)}_0\vec{\varphi} dx,\ \vec{\varphi}\in C_0^\infty(\Omega).$$

(v) For $\delta \in (0,1)$ we have the following uniform estimates

$$\left\{\|\vec{u}^{(i)}_\delta\|_{L^2(I,W_0^{1,2}(\Omega))}, \|\rho_{i,\delta}\|_{L^\infty(I,L^{\gamma_i}(\Omega))}, \delta^{\frac{1}{\beta_i}}\|\rho_{i,\delta}\|_{L^\infty(I,L^{\beta_i}(\Omega))}, \|\rho_{i,\delta}|\vec{u}^{(i)}_\delta|^2\|_{L^\infty(I,L^1(\Omega))}\right\} \le L, \quad (20)$$

$$\begin{gathered}\|\rho_{i,\delta}\|_{L^{s_i}(Q_T)} \le L,\quad s_i = \gamma_i + \theta_i,\ \ \theta_i = \frac{2}{3}\gamma_i - 1,\\ \delta^{\frac{1}{\beta_i+\theta_i}}\|\rho_{i,\delta}\|_{L^{\beta_i+\theta_i}(Q_T)} + \|\rho_{i,\delta}\vec{u}^{(i)}_\delta\|_{L^\infty(I,L^{\frac{2\gamma_i}{\gamma_i+1}}(\Omega))} + \|\rho_{i,\delta}\vec{u}^{(i)}_\delta\|_{L^2(I,L^{\frac{6\gamma_i}{\gamma_i+6}}(\Omega))} \le L,\\ \|\rho_{i,\delta}|\vec{u}^{(i)}_\delta|^2\|_{L^1(I,L^{\frac{3\gamma_i}{\gamma_i+3}}(\Omega))} + \|\rho_{i,\delta}|\vec{u}^{(i)}_\delta|^2\|_{L^2(I,L^{\frac{6\gamma_i}{4\gamma_i+3}}(\Omega))} \le L.\end{gathered} \quad (21)$$

Where the constant L depends on the problem date, namely on

$$\mathcal{E}_{1,0} = \sum_{i=1}^{2}\int_\Omega \left\{\frac{1}{2}\rho_i^0|\vec{u}^{(i)}_0|^2 + \frac{(\rho_i^0)^{\gamma_i}}{\gamma_i - 1} + \frac{(\rho_i^0)^{\beta_i}}{\beta_i - 1}\right\} dx.$$

5. The limit passage when $\delta \to 0$.

In this section as a result of limit passage when $\delta \to 0$ we exclude artificial pressures and completes the proof of theorem 2.

Together with the inequalities (20)-(21) there are other estimates of the densities independent of the parameter δ.

Lemma 5. Let $\rho_{i,\delta}$, $\vec{u}^{(i)}_\delta$ is solution of problem (17)-(19) constructed in theorem 4. Then we have the following uniform with respect to $\delta \in (0,1)$ estimates

$$\begin{gathered}\|\rho_{i,\delta}\|_{L^{s_i}(Q_T)} \le L(\mathcal{E}_{1,0}),\ \ s_i = \gamma_i + \theta_i,\ \theta_i = \frac{2}{3}\gamma_i - 1,\ (\frac{3}{2} < \gamma_i < 6),\ \theta_i < \frac{1}{2}\gamma_i,\ \ (\gamma_i \ge 6),\\ \delta^{\frac{1}{\beta_i+\theta_i}}\|\rho_{i,\delta}\|_{L^{\beta_i+\theta_i}(Q_T)} \le L(\mathcal{E}_{1,0}),\end{gathered} \quad (22)$$

with constant L independent of the δ.

It follows from estimates (20), (22) that there are subsequences, such that

$$\begin{gathered}\vec{u}^{(i)}_\delta \to \vec{u}^{(i)}\ weakly\ in\ L^2(I,W_0^{1,2}(\Omega))\ and\ in\ L^2(I,W^{1,2}(R^3)),\ \vec{u}^{(i)} = 0\ in\ (R^3\backslash\Omega)\times I.\\ \rho_{i,\delta} \to \rho_i\ weakly\ in\ L^{\gamma_i+\theta_i}(R^3\times I),\\ \theta_i = \frac{2}{3}\gamma_i - 1\ (if\ \frac{3}{2} < \gamma_i < 6)\ and\ 0 < \theta_i < \frac{1}{2}\gamma_i\ (if\ \gamma_i \ge 6).\\ \rho_{i,\delta} \to \rho_i\ *-weakly\ in\ L^\infty(I,L^{\gamma_i}(R^3)),\ \rho_i \ge 0\ a.e.\ in\ Q_T,\ \rho_i = 0\ in\ (R^3\backslash\Omega)\times I.\\ \rho_{i,\delta}^{\gamma_i} \to \overline{\rho_i^{\gamma_i}}\ weakly\ in\ L^{\frac{\gamma_i+\theta_i}{\gamma_i}}(Q_T), \overline{\rho_i^{\gamma_i}} \ge 0\ a.e.\ in\ R^3\times I,\ \overline{\rho_i^{\gamma_i}} = 0\ in\ (R^3\backslash\Omega)\times I.\\ \delta\rho_{i,\delta}^{\beta_i} \to 0\ weakly\ in\ L^{\frac{\beta_i+\theta_i}{\beta_i}}(Q_T).\end{gathered} \quad (23)$$

In addition, we note the following properties

1. If $\rho_{i,\delta}$, $\vec{u}^{(i)}_\delta$, $i = 1,2$, is solution of problem (17)-(19) then $\rho_{i,\delta}$ are renormalized solutions of continuity equations(18).

2. The functions ρ_i, $\vec{u}^{(i)}$, $\rho_i^{\gamma_i}$ defined by the formulas (23) satisfy the continuity equations (18) in $D'(I\times R^3)$, the balance momentum equations (17) in $D'(Q_T)$ up to terms $\nabla\rho_i^{\gamma_i}$ (these terms

will be replaced by the term $\nabla\overline{\rho_i^{\gamma_i}}$). The limit function $\rho_i, i = 1, 2$, are renormalized solutions of continuity equations, i.e. for any functions $b_i \in C^0[0,\infty) \cap C^1(0,\infty)$ satisfying the conditions $|b_i'(t)| \le c \cdot t^{-\lambda_0}, t \in (0,1],\ \lambda_0 < 1,\ |b_i'(t)| \le c \cdot t^{\lambda_i}, t \ge 1,\ 1 < \lambda_0 < \frac{\gamma_i}{2} - 1, c > 0$, the equations $\partial_t\overline{b_i(\rho_i)} + div(\overline{b_i(\rho_i)} \cdot \vec{u}^{(i)}) + \overline{[\rho_i b_i'(\rho_i) - b_i(\rho_i)]div\vec{u}^{(i)}} = 0$ are fulfilled. We have the estimates

$$\|T_k(\rho_{i,\delta}) - T_k(\rho_i)\|_{L^{\gamma_i+1}(Q_T)} \le c,\ k \ge 1,\ T_k(s) = \begin{cases} s,\ if\ s \in [0,k), \\ k\ if\ s \in [k,\infty). \end{cases}$$

The above facts allow us to prove that under the requirements of theorems 2 and 3, there exist a subsequence $\rho_{i,\delta}$, such that $\rho_{i,\delta} \to \rho_i$ strongly in $L_p(Q_T), i = 1, 2,\ 1 \le p < \gamma_i + \theta_i(\gamma_i)$

From this result we finally obtain the equalities $\overline{\rho_i^{\gamma_i}} = \rho_i^{\gamma_i}, i = 1, 2$. This completes the proof of theorem 2.

References

[1] Rajagopal K R and Tao L 1995 *Mechanics of mixtures* (Singapore: World Sci)
[2] Kraiko A N and Nigmatulin R I 1972 Mechanics of multiphase media *Itogi Nauki. Gidromechanika* **6** 93–174 (in Russian)
[3] Kazhikov A V and Petrov A N 2008 Well-posedness of the initial-boundary value problem for a model system of equations of a multicomponent mixture *Selected works. Mathematical hydrodynamics* (Novosibirsk: IG SB RAS) 157-167 (in Russian)
[4] Petrov A N 1982 Well-posedness of initial-boundary value problems for onedimensional equations of mutually penetrating flows of ideal gases *Din. Splosh. Sredy* **56** 105–21 (in Russian)
[5] Zlotnik A A 1995 Uniform estimates and stabilization of solutions to equations of one-dimensional motion of a multicomponent barotropic mixture *Mat. Zametki* **58** 307–12 (in Russian)
[6] Frehse J, Goj S and Malek J 2005 On a Stokes-like system for mixtures of fluids *SIAM J. Math. Anal.* **36** 1259–81
[7] Frehse J, Goj S and Malek J 2005 A uniqueness result for a model for mixtures in the absence of external forces and interaction momentum *Appl. Math.* **50** 527–41
[8] Frehse J and Weigant W 2008 On quasi-stationary models of mixtures of compressible fluids *Appl. Math.* **53** 319–45
[9] Kucher N A and Prokudin D A 2009 Stationary solutions to the equations of a mixture of compressible viscous fluids *Sibirsk. Zh. Industr. Mat.* **12** 52–65
[10] Kucher N A and Prokudin D A 2011 *Stationary problems in viscous compressible continuum mechanics I* (Kemerovo: Kemerovo State University) (in Russian)
[11] Kucher N A and Prokudin D A 2012 *Stationary problems in viscous compressible continuum mechanics II* (Kemerovo: Kemerovo State University) (in Russian)
[12] Kucher N A, Mamontov A E and Prokudin D A 2012 Stationary solutions to the equations of dynamics of mixtures of heat-conductive compressible viscous fluids *Siberian Math. J.* **53** 1338–53
[13] Lions P L 1998 *Mathematical topics in fluid mechanics. Compressible models* (New York: Oxford University Press)
[14] Feireisl E and Novotny A and Petzeltova H 2001 On the existence of globally defined weak solutions to the Navier-Stokes equations *Math. Fluid Mech.* **3** 358–92
[15] Novotny A and Straskraba I 2004 *Introduction to mathematical theory of compressible flow* (New York: Oxford University Press)
[16] Plotnikov P I and Sokolowski J 2007 Stationary solutions of Navier–Stokes equations for diatomic gases *Russian Math. Surveys* **62** 561–93 (in Russian)
[17] Feireisl E, Matusu-Necasova S and Petzeltova H 1999 On the motion of a viscous compressible fluid driven by a time-periodic external force *Arch. Rational Mech. Anal.* **149** 69-96
[18] Feireisl E and Petzeltova H 1999 On compactness of solutions to the Navier–Stokes equations of compressible flow *J. Diff. Equ.* **163** 57–75

IOP Conf. Series: Journal of Physics: Conf. Series **894** (2017) 012049 doi:10.1088/1742-6596/894/1/012049

Modeling of detonation propagation in expansion channel in bidisperse mixture

Fedorov A V, Khmel` T A and Lavruk S A

Khristianovich Institute of Theoretical and Applied Mechanics SB RAS, Novosibirsk, Russia

E-mail: khmel@itam.nsc.ru

Abstract. The problem of detonation propagrtion in channels with linear expansion filled with two-fractional alumunium mixture dispersed in oxigen was investigated. Fractional composition of mixture with particle diameters 3.5 and 1 μm was varied. Transition from diffusion to kinetic regime of combustion was taken into account in the description of the chemical reactions for microsized aluminum particles. It was established that fractional composition of the mixture inflience on propagation regimes and formation of transverse waves of cellular detonation.

1. Introduction

The combustion of nano-dispersed aluminum powders, including in the detonation mode, attracts great interest in view of the wide prospects for their use. The problems of detonations in aluminum particles gas suspensions have been extensively studied experimentally and theoretically for the micro-dimensional particle size range.

Propagation of shock waves behind a rectangular edge in a heterogeneous monodisperse mixture and influence of particle concentrations and sizes on flow patterns was analyzed in [1]. It was noted that main features that distinguish wave patterns of flow from flows in gas media are caused by presence of zones of velocity and thermal relaxation of the phases. Similar problems for detonation waves in gas – particles mixtures of aluminum in oxygen were considered in [2-6]. Two types of diffraction problem were analyzed for different initial conditions: for a flat detonation wave (the Chapman-Jouguet flow) and for cellular detonation. It was demonstrated in [2] that even with a flat detonation wave as the initial conditions, a flow with transverse waves develops in the wide part of the channel behind the edge that characterize cellular detonation.

In [3] description of wave patterns in different diffraction regimes on the backward step of a flat detonation wave is given. The channel width and particle size were varied from 0.01 to 0.15 m for channel and from 1 μm to 3.5 μm for particles. The results allow us to classify the diffraction regimes similar to those observed in gas detonation: supercritical, critical, and subcritical. The works [4, 5] presented results of numerical simulation data on propagation of cellular detonation in a half-space behind the backward ledge in a gas suspension of aluminum particles with sizes 1.5 - 3.5 μm where channel width varied from 0.5 to 4 detonation cells. Wave patterns were analyzed in the process of reorganization of cellular structures in different propagation modes.

Real aluminum powders are characterized by heterogeneity and certain dispersion in particle sizes. In [5-7] processes of initiation and propagation of plane and cellular detonation waves in bidisperse and polydisperse mixtures of aluminum were studied. Two-dimensional flows in flat channel and

Content from this work may be used under the terms of the Creative Commons Attribution 3.0 licence. Any further distribution of this work must maintain attribution to the author(s) and the title of the work, journal citation and DOI.
Published under licence by IOP Publishing Ltd

IOP Conf. Series: Journal of Physics: Conf. Series **894** (2017) 012049 doi:10.1088/1742-6596/894/1/012049

channel with sudden expansion were analyzed in [5-6]. It was found that in mixtures with large dispersion in particle sizes or in bidisperse suspensions with equal partition of fractions the cellular detonations transform into plane detonation waves which propagate sustainably without transverse waves. It was established in [7] that the structures of plane waves refer to nonideal detonation and depend on the fractional composition of the mixture. Small additions of fine fraction lead to significant decreases in the initiation energy.

In the present work processes of propagation of detonation in a channel with section of linear expansion in bi-fractional aluminum suspensions of particles 3.5 μm and 1 μm are studied. The main goal of this work is to analyze effect of bidisperse fractional composition on propagation regimes and detonation flow characteristics.

2. Physical and mathematical model.

The Euler equations for two-dimensional flow in dilute gas particle suspensions follow from conservation laws for mass, momentum, and energy (the subscripts 1, 2, 3 indicate gas and reactive particles, respectively) [8]:

$$\frac{\partial \boldsymbol{W}}{\partial t}+\frac{\partial \boldsymbol{F}}{\partial x}+\frac{\partial \boldsymbol{G}}{\partial y}=\boldsymbol{\Gamma},\ \boldsymbol{W}=\begin{pmatrix}\boldsymbol{W}_1\\ \boldsymbol{W}_2\\ \boldsymbol{W}_3\end{pmatrix},\ \boldsymbol{F}=\begin{pmatrix}\boldsymbol{F}_1\\ \boldsymbol{F}_2\\ F_3\end{pmatrix},\ \boldsymbol{G}=\begin{pmatrix}\boldsymbol{G}_1\\ \boldsymbol{G}_2\\ G_3\end{pmatrix},\ \boldsymbol{\Gamma}=\begin{pmatrix}-\boldsymbol{\Gamma}_2-\boldsymbol{\Gamma}_3\\ \boldsymbol{\Gamma}_2\\ \boldsymbol{\Gamma}_3\end{pmatrix},$$

$$\boldsymbol{W}_i=\begin{pmatrix}\rho_i\\ \rho_i u_i\\ \rho_i v_i\\ \rho_i E_i\end{pmatrix},i=1,2,3,\ \boldsymbol{F}_1=\begin{pmatrix}\rho_1 u_1\\ p+\rho_1 u_1^2\\ \rho_1 u_1 v_1\\ \rho_1 u_1 E_1+p u_1\end{pmatrix},\ \boldsymbol{G}_1=\begin{pmatrix}\rho_1 v_1\\ \rho_1 u_1 v_1\\ p+\rho_1 v_1^2\\ \rho_1 v_1 E_1+p v_1\end{pmatrix}. \quad (1)$$

$$\boldsymbol{F}_i=\begin{pmatrix}\rho_i u_i\\ \rho_i u_i^2\\ \rho_i u_i v_i\\ \rho_i u_i E_i\end{pmatrix},\ \boldsymbol{G}_i=\begin{pmatrix}\rho_i v_i\\ \rho_i u_i v_i\\ \rho_i v_i^2\\ \rho_i v_i E_i\end{pmatrix},\ \boldsymbol{\Gamma}_i=\begin{pmatrix}-J_i\\ f_{xi}-J_i u_i\\ f_{yi}-J_i v_i\\ q_i+f_{xi}u_i+f_{yi}v_i-J_i E_i\end{pmatrix},i=2,3.$$

The system is enclosed by the equations of state and relationships for mass, momentum, and heat exchange between the gas and the particles

$$\vec{f}=\frac{\rho_i}{\tau_{ui}}(\vec{u}_1-\vec{u}_i),\ q_i=\frac{\rho_i c_{v,2}}{\tau_{Ti}}(T_1-T_i),\ J=\frac{\rho_i}{\tau_{\xi i}}(\xi_i-\xi_k)$$

$$p=\rho_1 R T_1,\qquad E_1=\frac{u_1^2}{2}+c_{v,1}T_1,\ E_2=c_{v2}T_i+Q+\frac{u_i^2}{2}, \quad (2)$$

Here indexes 1, 2 and 3 refer to gas and particles, respectively. We use the following notations: p is pressure; $\rho_i=m_i\rho_{ii}, u_i, E_i, T_i, c_{v,i}, m_i$, are mean densities, velocities, total energy per unit mass, temperatures, heat capacities, and volume concentrations of the components, $\xi_i=\rho_i/\rho$ is mass particle concentration, ξ_k is fraction of unburnt particles, ρ_{ii} are true densities, R is the reduced gas constant, Q is heat effect of particle combustion in the frame of reduced chemical kinetics.

The interphase interactions are presented by parameters J (the mass transfer), f (the interphase interaction force), and q (the heat transfer):

$$\tau_{ui} = 4d_i\rho_{22} / 3c_{Di}\rho_{11}\left|\vec{u}_1 - \vec{u}_i\right|, c_{Di}(\mathrm{Re}_i, M_{1i}) = \left(1+\exp(-\frac{0.43}{\mathrm{M}_{1i}^{4.67}})\right)\left(0.38+\frac{24}{\mathrm{Re}_i}+\frac{4}{\sqrt{\mathrm{Re}_i}}\right),$$

$$\tau_{Ti} = d_i^2\rho_{22}c_{v2} / 6\lambda_1\mathrm{Nu}_i,\ \mathrm{Nu}_i = 2+0.6\mathrm{Re}_i^{1/2}\mathrm{Pr}^{1/3}, \tag{3}$$

$$\mathrm{Re}_i = \frac{\rho_{11}d_i\left|\vec{u}_1 - \vec{u}_i\right|}{\mu_1},\ \mathrm{M}_i = \frac{\left|\vec{u}_1 - \vec{u}_i\right|\sqrt{\rho_{11}}}{\sqrt{\gamma_1 p}}.$$

where i=2, 3, μ_1 is the gas dynamic viscosity and γ_1 is the adiabatic exponent.

To describe the burning processes for micro-sized aluminum particles with diameter less than 3.5 μm the improved semi-empirical combustion model [9] is used which takes into account the transition from diffusion to kinetic regime of aluminum particle combustion. The equation of reduced Arrhenius-type kinetics is used with the activation energy value 32 kJ/mol. The dependence between the characteristic time of particle combustion and the particle diameter in accordance with the data [10] is taken in form $\tau_\xi = \tau_0(d/d_0)^{0.3}\exp(E_a/RT_2)$ with d_0 =3.5 μm, τ_0=0.294 μs.

The total density of particles is taken to correspond to stoichiometric composition $\rho_{20}+\rho_{30} = 1.34\,\mathrm{kg/m^3}$. The fractional composition of the discrete phase is characterize by the saturation parameter $\eta = \rho_{30}/(\rho_{20}+\rho_{30})$. The saturation parameter value is varied in range from 0 to 1. The particle diameters of bidisperse mixture are accepted d_2=3.5 μm and d_3=1 μm.

The initial- and boundary-value problem is posed as a problem of the plane CJ detonation wave propagation in a plane channel with a section of linear expansion. A linear expansion of the flat channel is located at a distance L_2 from the beginning of the channel with the angle of inclination of the wall α. We consider the half of the channel and the bottom dashed line in Fig. 1 is the line of symmetry. The initial position of the detonation front is L_1. We study the detonation propagation and exit to the expanding part of the channel. The calculations were performed for the channel width value H_1=0.03 m and the inclination angle α = 45° and 30°.

The numerical technique is based on the conservative flux-splitting schemes: the TVD scheme by Harten for gas and the Gentry-Martin-Daly scheme for particles. The numerical method has been tested earlier and applied for 2-D numerical simulations of the shock wave and detonation flows in [1-8].

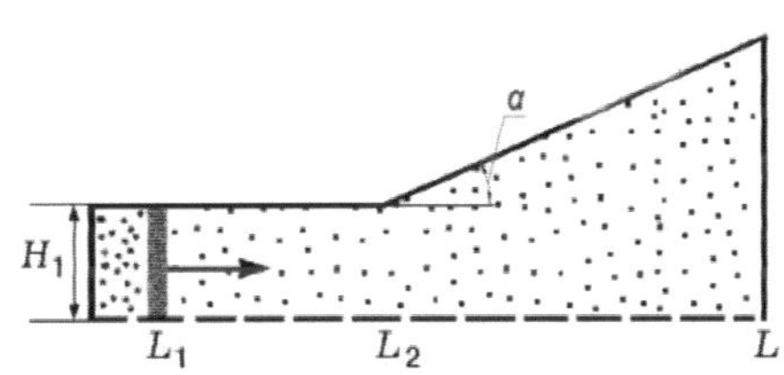

Figure 1. Detonation propagation in a channel with linear expansion: the computation domain and initial position of the DW.

3. Numerical results

Figure 2 shows the results of calculations for channel with inclination angle of the wall of 45°. In monodisperse suspension of 3.5 μm particles at given channel width the detonation failure regime is realized (Figure 2a). In the Schlieren image (Figure 2a) the separation of the leading front of the shock wave (SW) and the lagging combustion front (CW) can be seen, the distance between the fronts at the time of 0.26 ms is 30 mm.

When 5% large (3.5 μm) particles are replaced by smaller ones (1 μm) and the saturation parameter η=0.05, a transition to the critical mode of detonation propagation is observed (Figure 2b). Here, near the plane of symmetry there are separation of the shock wave front and combustion front (Figure 2b).

At the same time in the most part of channel detonation continues to propagate without a noticeable separation of the fronts. Transverse wave propagates along the front of the detonation wave which is formed as a result of re-initiation of detonation on the inclined wall. Pressure in the transverse wave reaches values of about 70 atm.

Further increase in the saturation parameter to η = 0.1 leads to a supercritical detonation propagation regime. Here the leading front propagates with stretching and without breakdown (Figure 2c). It also shows formation of transverse wave that moves along the leading front.

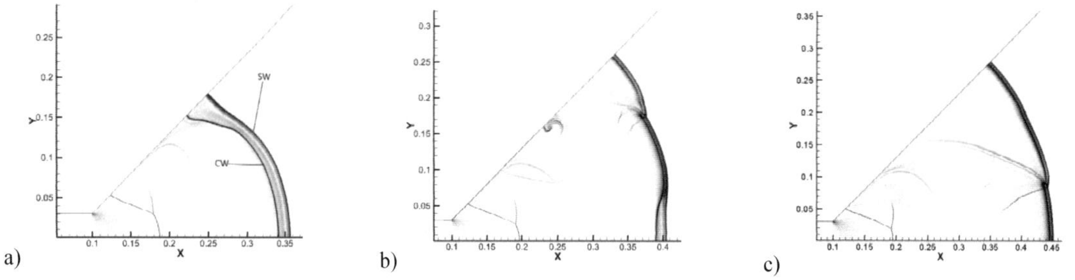

Figure 2. Detonation propagation modes depending on the saturation parameter t=0.24 ms: subcritical (η=0) a), critical (η=0.05) b), supercritical (η=0.1) c).

For the inclination angle 30° in Figure 3 shows the maximum pressure fields in the channel for different saturation parameters are presented. A critical regime is realized in a monodisperse mixture of 3.5 μm aluminum particles (Figure 3a). Partial detonation failure in Figure 3a can be observed as pressure drop of up to 20 atm in the region $0 < y < 0.04$ m on the leading edge near the plane of symmetry.

With increase in saturation parameter η from 0 to 0.1, the transition to supercritical detonation propagation regime without failing is observed in the maximal pressure fields (Figure 3b). With further increase in the saturation parameter a supercritical detonation propagation regime takes place (Figure 3c η = 0.3). It is also worth noting that with increase in the saturation parameter the maximal pressure in collisions of the transverse waves which propagate along the leading edge decreases. For η=0 and η=0.1 the maximal pressure reaches the values up to 70 atm, while for η = 0.3 it does not exceed 55 atm. At the same time inhomogeneities are practically absent on the detonation front (Figure 3c).

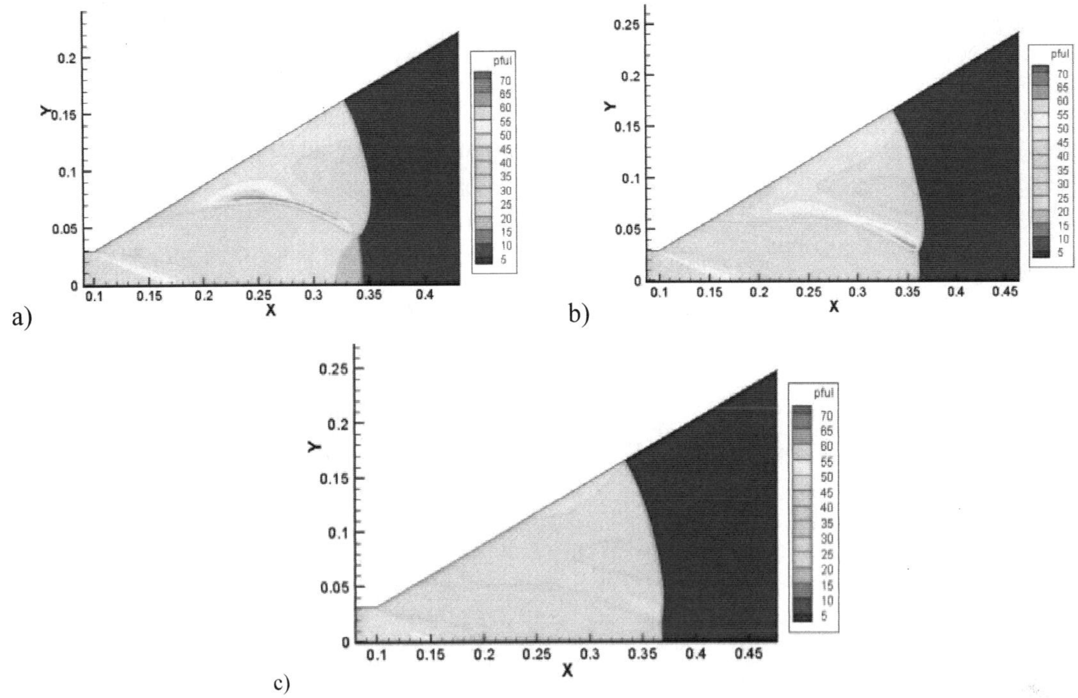

Figure 3. The effect of the fractional composition on the cellular detonation formation: maximal pressure histories; t=0.24 ms, η=0 (a); η=0.1 (b); η=0.3 (c).

Further increase in concentration of small particles leads to formation of the regime of detonation propagation in which the surface of the leading front is almost completely smoothened. Figure 4 shows the pressure fields in area near the leading front of detonation wave. It can be seen that as the fraction of fine particles in the mixture increases, the pressure in the transverse wave decreases from 65 atm with η=0.1 to 45 atm with η=0.5. With predominance of the fraction of fine particles in the mixture the transverse waves are again produced (η=0.7). It is observed that the pressure at the front increases up to 65 atm with η=0.7. Also formation of inhomogeneities at the front of the detonation wave again occurs.

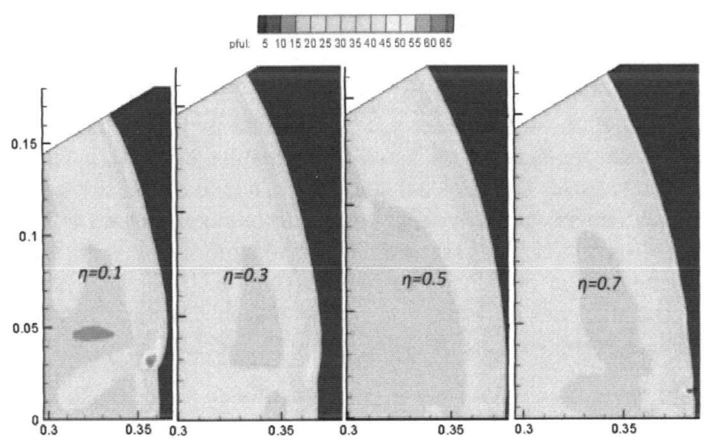

Figure 4. Pressure fields at the front of the detonation wave as a function of the saturation parameter.

With η=0.7, 0.9, 1.0, one can see the formation of cellular structure in the channel (Figure 5). In fields of maximum pressure (Figure 5), it can be seen that increase in the fraction of fine particles results in increase in the number of triple points and increase in maximum pressure in these regions. At η=0.9, 1.0 irregular cellular structures with relatively equal pressure values at triple points are formed in the channel. In general the structures of transverse waves in Figures 5b and 5c are similar.

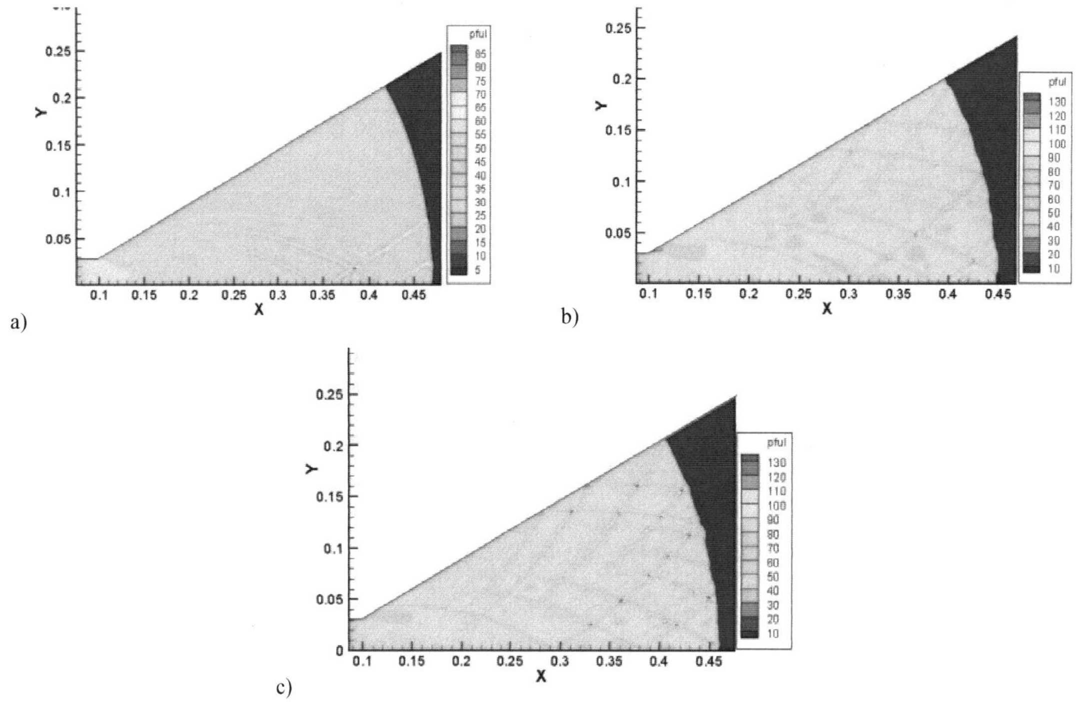

Figure 5. Maximal pressure histories at t=0.24 ms, η=0.7 (a); η=0.9 (b); η=*1* (c).

The problem of weakening of transverse waves in cellular detonation in bidisperse and polydisperse suspensions was discussed in [5-7] in analysis of the propagation of detonation in plane channels and channels with a discontinuity of the cross section. Applying similar arguments we can explain the weakening of transverse waves with mean values of the saturation parameter and in problem of the detonation wave exit into a channel with linear expansion. According to [5-7] growth of small perturbations at front of the detonation wave and formation of secondary transverse waves agrees with the acoustic theory proposed by Bartel in [11] and developed in [7] for heterogeneous mixtures. The specific feature of bidisperse mixtures is that they form peculiar flow structure in the relaxation zone behind the leading shock wave front. In this flow the acoustic perturbations being formed by primary transverse waves either not come back to the leading front (and secondary transverse waves do not appear) or partly come back with overcompressed or weakened sections of the front. It leads to the dispersion of transvers waves (weakening) and the reduction of peak pressures until the steady propagation in the Chapman-Jouguet regime. It can be assumed that weakening of transverse waves in the processes of detonation propagation in a channel with a linear expansion filled with bidispersed particle suspension in gas is due to similar mechanisms.

IOP Conf. Series: Journal of Physics: Conf. Series **894** (2017) 012049 doi:10.1088/1742-6596/894/1/012049

4. Conclusions

The problem of propagation of detonation in bidispersed mixtures in a channel with linear expansion was studied on the base of numerical simulations within the semiempirical model of detonation of microdispersed suspension of aluminum particles. The model takes into account the transition regime from diffusion to kinetic in the description of aluminum particle combustion. The flow patterns in the main propagation regimes in channels with inclination angle of the wall of 45 ° and 30 ° at $H_1 = 0.03$ m and variation of the fractional composition of particles with sizes of 3.5 and 1 μm are analyzed.

It is established that an increase in part of fine particles in the mixture leads to a change in the mode of propagation of detonation in the expanding channel. The transition from subcritical to critical regime occurs at η=0.05 and from critical to supercritical at η=0.1.

At close values of part of each fraction of bidisperse suspension the detonation propagation in channel is characterized by a weakening of transverse waves and a decrease in pressures peak in transverse waves. With the saturation parameter η=0.4-0.5, the transverse waves are almost completely absent as in propagation of detonation in bidisperse suspensions in flat channels. With the subsequent increase in fraction of fine particles the transverse waves form again and with predominance of small particles similar pattern of transverse waves as in the monodisperse fine particle suspension is observed.

5. Acknowledgements

The work was supported by the Russian Scientific Foundation (project No. 16-19-00010).

References

[1] Kratova Yu, Fedorov A and Khmel T 2008 *Combustion, Explosion, and Shock Waves* **44** 85
[2] Fedorov A, Khmel T and Kratova Y 2008 *Shock Waves* **18** 280
[3] Kratova Yu, Fedorov A and Khmel T 2009 *Combustion, Explosion, and Shock Waves* **45** 280
[4] Fedorov A, Khmel T and Kratova Y 2010 *Shock Waves* **20** 509
[5] Fedorov A and Khmel T 2008 *Comb. Exp. and Shock Waves* **44** 343
[6] Kratova Yu, Fedorov A and Khmel T 2011 *Comb. Exp. and Shock Waves* **47** 572
[7] Fedorov A and Khmel T 2008 *Comb. Exp. and Shock Waves* **44** 163
[8] Fedorov A 1992 *Combustion, Explosion, and Shock Waves* **28** 277
[9] Fedorov A and Khmel T 2016 *Nonequilibrium Processes in Physics and Chemistry II* (Ed. by A.M. Starik and S.M. Frolov) 341
[10] Sundaram D, Yang V and Zarko V 2015 *Combustion, Explosion, and Shock Waves* **51** 173
[11] Barthel H 1974 Physics of Fluids **17** 1547

IOP Conf. Series: Journal of Physics: Conf. Series **894** (2017) 012050 doi:10.1088/1742-6596/894/1/012050

Influence of the creep ageing process on the fatigue properties of components from V95pchT2 (analog 7175T76) and V95ochT2 (analog 7475) aluminium alloys

A Larichkin[1], K Zakharchenko[1, 3], B Gorev[1], V Kapustin[3] and E Maksimovskiy[4]

[1] Lavrentyev Institute of Hydrodynamics of SB RAS, Novosibirsk, Russia
[2] Novosibirsk State University, Novosibirsk, Russia
[3] Novosibirsk State Technical University, Novosibirsk, Russia
[4] Nikolaev Institute of Inorganic Chemistry of SB RAS, Novosibirsk, Russia

E-mail: larichking@gmail.com

Abstract. Influence of conditions of technological process of forming thick panels of a given geometry on fatigue limit of flat specimens from the V95ochT2 and V95pchT2 (analogues 7475 and 7175T76) alloys (Al-Mg-Cu-Zn) has been analysed. The process has been simulated experimentally on flat samples for temperatures 20, 165 and 420°C. The process includes: non-elastic strain in the range 10^{-5} - 10^{-2} s^{-1} up to 2% of total strain, followed by heat treatment according to T2 mode (quenching and aging). Fatigue life tests were carried out both on solid samples and on samples with a hole. It has been shown that resistance to fatigue of the observed alloys after forming at the annealing temperature (420°C) is comparable to the basic material resistance to fatigue. Meso-structure analysis showed absence of stress in grains. It is established that, on average, the shape of the grains is the same for a series of samples for different temperatures and loading rates. The results of testing samples with a hole showed that fatigue limit slightly decreases in samples which were previously deformed at (420°C), with respect to the durability of samples from the material in basic state. With an increase in rate of pre-strain, the relative number of cycles before destruction occurs increased.

1. Introduction

Strain ageing combined with CAF (creep age forming) is an interesting and strongly needed technology in aviation. This production method by forming of 7XXX aluminum alloys thin plates (not more than 15-20 mm) has proved to have the required accuracy in production of panels with double curvature without finishing works and preserve their long service life. These plates are used as wing skins and other elements of the airframe. Existing technologies require forming of panels at temperatures not exceeding the ageing temperature of the material. Obtaining of the target form of thick plates (more than 40 mm) by CAF is more complicated for technical reasons since it requires heavy presses in addition to heating.

Content from this work may be used under the terms of the Creative Commons Attribution 3.0 licence. Any further distribution of this work must maintain attribution to the author(s) and the title of the work, journal citation and DOI.
Published under licence by IOP Publishing Ltd

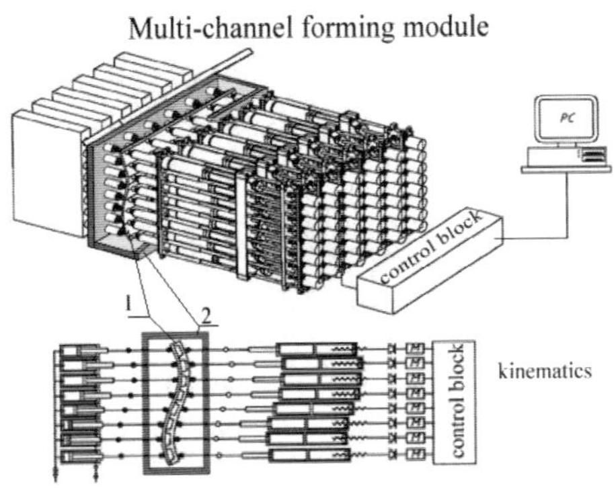

Figure 1. Kinematic scheme of multy-channel forming module: 1 – plate; 2 – oven.

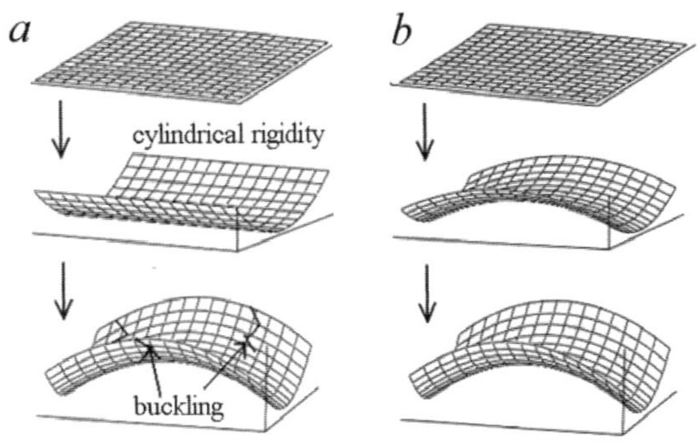

Figure 2. Forming a rectangular plate into the saddle surface: at the same constant rate of rod movement (a), with the individual rates of rod movement (b).

2. Idea and method

A well- known CAF method (creep age forming) [1] includes two processes: creep and heat treatment (ageing), which depends on time and temperature. This method is actively used in the aerospace industry to form big panels with a complex curvature. For example, the B1B and Hawk bombers, the executive jet Gufstream G-IV and the commercial Airbus jets A330 and A340, as well as the A380 [2, 3]. However, there are no examples of the reinforced panels made from the thick plates. Compared with other processes, CAF results in lower residual stresses leading to higher fatigue characteristics [2].

This research was conducted to address the problem of forming solid or ribbed plates of the V95ochT2 alloy (Al-Mg-Cu-Zn) (analog 7475) V95pchT2 alloy (analog 7175T76) for the jet industry. It is possible to make a complex geometry of the initially rectangular plate by using a multi-channel forming module (Figure 1).

The final product is made from the initial flat plate deformed in the plug-in equipment. Figure 1a shows the form and kinematics of the experimental multi-channel forming module. Here, 1 – plate, 2 - oven.

The module is designed for the accurate forming of smooth and ribbed large plates of single and double curvature in creep or close to superplasticity. The plates are deformed by the movement of two discretely set rod systems arranged oppositely to the detail's surface.

For reinforced panels, rods of the hydraulic puncheons are set against intersection of edges, for smooth – in rows within the panels' width. Swivel heads are placed at the end of the rods to form the workpiece without dents and corrugating. The movement of each rod is set individually by a computer program using a stepper motor that allows forming the details of double curvature in one technological step. Combined forming and heat treatment of the material in the slow strain modes due to irreversible creep strain significantly reduce forming forces and make the technological process cost effective.

Most of the existing technological solutions use load distribution with the constant rate of rod or snap movement. It should be noted that strain at the constant rate of rods movement and at individual rates of each rod movement leads to different results. Figure 2 shows the scheme of strain of a rectangular plate into the saddle surface. Figure 2a shows how the workpiece is first bent in the cylindrical surface. Then geometric buckling occurs at further bending along the element of cylinder to form corrugations.

Figure 2b shows bending of a plate from the initial to the final shape in two main directions simultaneously throughout the entire strain process. This changes the shape of a workpiece without buckling and may be either continuous or divided into a series of intermediate stages depending on the complexity of the product.

The technology of forming is based on the solution of the inverse problem of inelastic strain of the body in creep, followed by elastic springback. The essence of the problem lies in defining the force and shape of the equipment to form the workpiece, which will provide the specified curvature after removal of the load. The mathematical apparatus for solving such problems is given in [4]. The complexity of these problems is connected with the substantial anisotropy of the creep properties of the workpiece, as well as with the strength difference (SD) effect (tension-compression asymmetry) of the material.

In [5] creep phenomenological models are presented which were introduced in FEM to calculate the springback of the constructions after CAF. The plates with a constant thickness are considered. In [6] constitutive equations were established for AA7010 which take into account the creep effect and precipitation hardening. These equations are implemented in FEM to simulate the forming of thick plates of double curvature [7]. The authors report that it is possible to estimate the influence of the thickness of the plate and the process time on the springback [8].

In [9], the numerical solution of the problem of forming an element of double curvature in creep is given, taking into account the strength difference (SD) effect of the material. Leaving it aside leads to a difference between the calculated bending of the structure and that which is observed experimentally. In [10] the solution to the problem of torsion of the plate by permanent moment is given taking into account anisotropic properties of the material. It is mentioned that direction at the angle of 45 ° to the normal of the anisotropic plate is the least resisting to the creep. In [11] an anisotropic finite strain creep model was proposed for CAF applications.

The answer to the question about the influence of the characteristics of the process - rate and temperature of the workpiece forming - on the fatigue life of the formed structures is important for a modern aircraft manufacturing. The forming parameters are chosen in connection with the requirements to the final physical and mechanical properties of the material.

This leads to the question: how long is the fatigue life of a product following this process?

To answer this question we reconstructed the technological conditions of forming in the laboratory with samples of V95ochT2 (plate h = 15 mm) and V95pchT2 (plate h = 36 mm) (analogues 7475 and 7175T76) alloys (Al-Mg-Cu-Zn). Flat samples from V95ochT2 (4 mm thick, 10 mm wide) were cut longitudinally and transversely to plate rolling direction, and flat samples with hole from V95pchT2 (4 mm thick, 36 mm wide) were cut longitudinally to plate rolling direction.

Test algorithm for samples from V95ochT2 is as follows: the samples divided into three series I,II and III were pre-strained up to 2% of tensile deformation at three different temperatures (T = 20 °C, 165°C, 420°C), respectively. The temperature for the II series is the temperature of the second stage of artificial ageing (T2 mode), and the temperature for the III series corresponds to the annealing temperature [12, 13]. The tests were conducted at Zwick Roell Z100, and for each temperature the strain rate changed (from 10^{-5} to $8\cdot10^{-2}$ s^{-1}). The samples were then exposed to heat treatment at the T2 mode (quenching in water at 470 °C and ageing at 115 and 165° C during the day). Test algorithm for samples from V95pchT2 was the same but only for T = 420°C, and holes (6 mm diameter of the hole) waere drilled after heat treatment in the middle of samples.

The flat samples treated in this way were tested for fatigue life. Stress for fatigue tests was determined by methods [12, 13], including:

- in-situ measurement of strain characteristics allowing to describe the laws of inelastic strain of the material;
- in-situ measurement of the dissipative heating temperature of the sample surface, indicating the beginning of irreversible effects in the material.

A fatigue life test was performed at room temperature (frequency 10 Hz, amplitude of cycle σ_{max} = 250 MPa (for V95ochT2) and σ_{max} = 157 MPa (for V95pchT2), σ_{min} = 0 MPa) on an Instron 8801 fatigue testing machine.

The microstructure of the material was studied by scanning electron microscopy (on a Hitachi S-3400N instrument).

3. Results

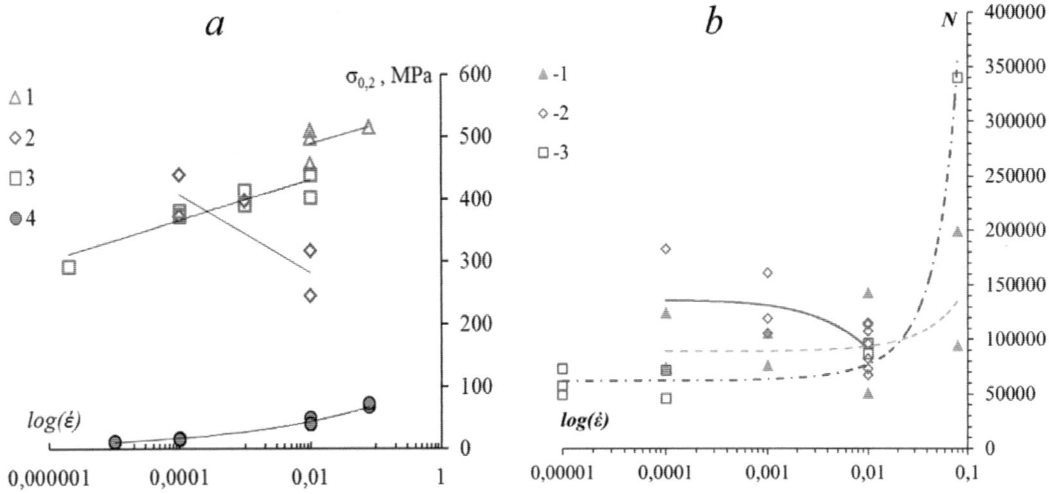

Figure 3. Results for V95ochT2: (a) - Dependence of $\sigma_{0,2}$ on the pre-strain rate; (b) - dependence of number of cycles before destruction from the pre-strain rate for different series of tests.

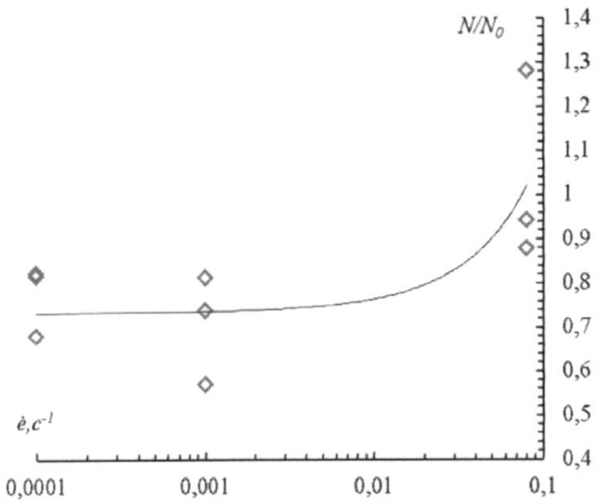

Figure 4. Results for V95pchT2 samples with hole. Dependence of relative fatigue limit from pre-strain rate.

The experiments have shown that the yield strength at higher temperature is almost 5 times smaller than at room temperature, and it rises with increasing pre-strain rate. Figure 3(a) shows the dependence of $\sigma_{0,2}$ on the pre-strain rate at different temperatures. Figure 3(a) shows the results for V95ochT2 samples: dependence of $\sigma_{0,2}$ on the pre-strain rate for different temperatures: 1 – Series I at T=24°C; 2 – Series II first direction at T=165°C; 3 – Series II orthogonal direction at T=165°C; 4 – Series III orthogonal direction at T=420°C. Figure 3(b) shows the number of cycles of the pre-strain for different temperatures. We can see that the fracture life for series I (20°C) and III (420°C) increases with increasing pre-strain rate, and for series II (165°C) this is decreasing.

Figure 4 shows values of relative fatigue limit as a function of the pre-strain rate for samples of the V95pchT2 alloy. Here, N is the number of cycles before fracture of the samples previously deformed at 420 °C, N_0 is the fatigue limit for the samples in the base state. Fatigue test of V95pchT2 alloy were on samples with hole.

Figure 5 shows the microstructure of the samples from V95ochT2 after stretching up to 2% strain at different temperatures and heat treatment. On Figure 5 (a, c, e, g) the images of grains are presented, and on (b, d, f, h) the images of microstresses are presented. Pre-strain temperatures are: (a), (b) T=24°C; (c), (d) T=165°C; (e), (f) T=420°C; (g), (h) no pre-strain – the base material;

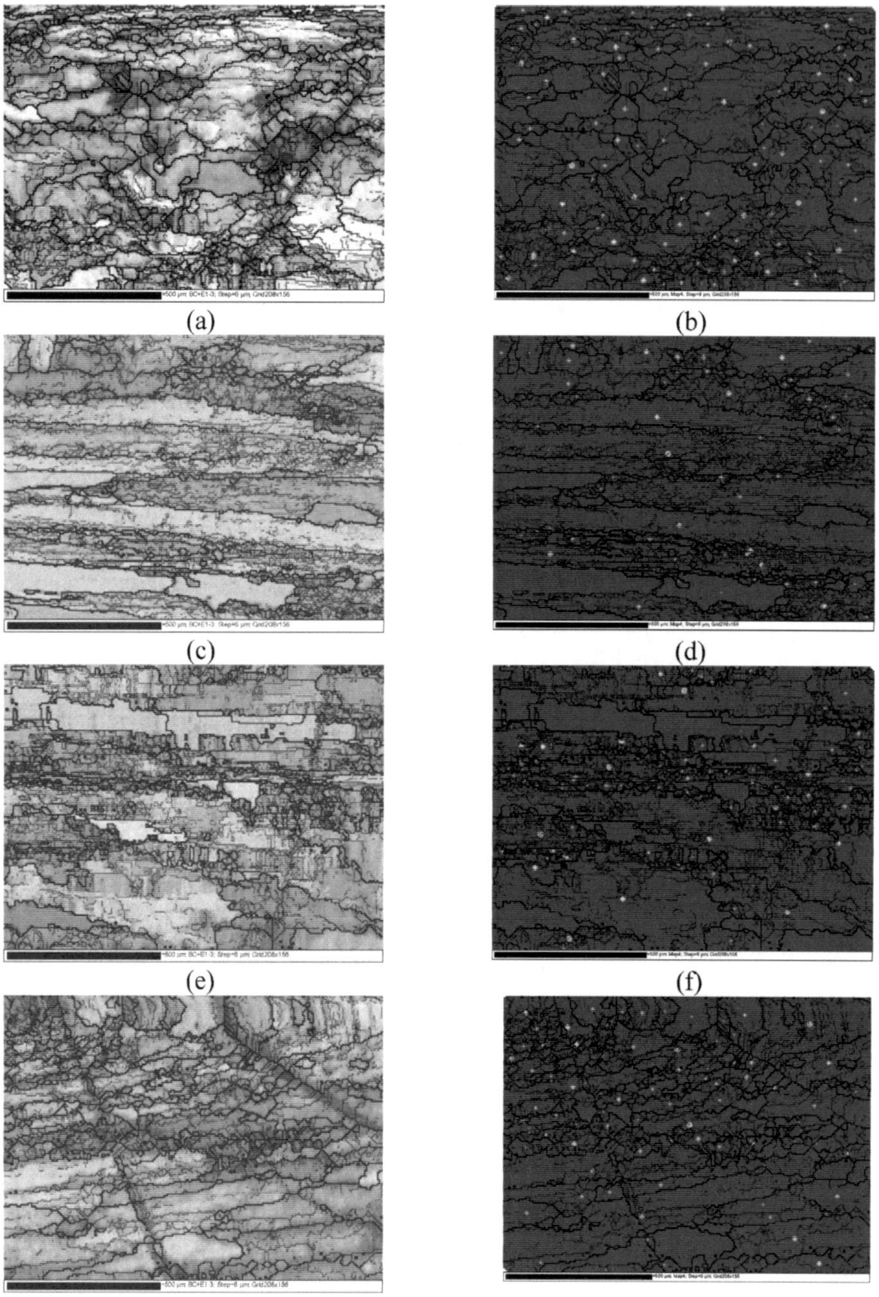

(a) (b) (c) (d) (e) (f) (g) (h)

Figure 5. Microstructure of V95ochT2 samples after pre-strain up to 2% of total strain at different temperatures. Pre-strain temperatures are: (a), (b) *T*=24°C; (c), (d) *T*=165°C; (e), (f) *T*=420°C; (g), (h) no pre-strain – the base material.

Analysis of microstructure images showed that the change in grain structure of the material after pre-strain and heat treatment (series I, II, III) did not occur, grain size did not change in comparison with the base material; residual stresses of the second kind in deformed samples (I, II, and III series) after heat treatment are absent.

4. Conclusion

These results allow determining optimal temperature and strain rates for the technological forming of the parts with complex geometry produced within a single cycle. The optimality criterion is a high fatigue limit of the product.

It is shown that the fatigue limit of V95ochT2 after pre-strain at the ageing temperature of 165°C with subsequent heat treatment becomes higher with the decreasing pre-strain rate (Figure 3b).

It is shown experimentally that at increasing strain rate in series III the trend to increase the number of cycles before destruction has been observed. In series I there is a trend to reduce the number of cycles before destruction, which correlates with the results of the forming method proposed in [14].

Pre-strain of the samples at annealing temperature (420°C) and a strain rate of 10^{-2} s^{-1}, followed by heat treatment in the T2 mode does not reduce the resistance to fatigue as compared with the samples deformed at 20°C.

5. Discussion

A number of modern works devoted to CAF at forming products from the constructional 7XXX aluminum alloys is worth mentioning.

The research on influence of pre-strain of the sheets made of the 7475 alloy on CAF is close to the research given in this article [15]. The authors say that the strength property varies by a low-to-peak-to-low manner as the increase of the deformation degree, and the radius-thickness ratio of the peak strength maintains a constant.

Some foreign authors justified the method of forming plates made of the Al-Zn-Mg-Cu alloy by CAF methods at the ageing temperature (at 165°C for 18 h) [16]. The authors show that the relationship between initial temper and CAF formability is investigated. Three tempers are selected as the initial tempers in CAF, viz., solution, retrogression and re-solution. The CAF formability of this alloy with initial temper of retrogression is the best, and the creep strain of the retrogression tempered specimen after creep aging of 18 h is about 1.21 and 1.34 times than that of the solution and the re-solution tempered specimens, respectively.

After the analysis of the relevant works it can be concluded that forming at the ageing temperature is commonly used for thin plates made from the V95ochT2 and V95pchT2 alloys since in this case forming and ageing are combined in one process.

In [17] forming and defining of the AA7475 ribbed wing panel springback (T73) in a vacuum bag and an autoclave at the temperature of ageing are considered (the recommended overageing procedure consists in a two step cycle of 6 to 8 hours at 121°C followed by 24 to 30 hours at 163 °C). However, the authors refer to the results of research on the influence of the effect of Retrogression and re-ageing treatments on the resistance to cracks for the materials of 7XXX series (AA7050 and AA7150), but different from V95ochT2 (7475).

The authors consider forming of a thin panel at the temperature of ageing, and they say that the magnitude of the springback can reach about 40 up to 70% of the imposed initial curvature. Such a high level of the springback can be reduced by forming at the annealing temperature. Also forming at the annealing temperature can significantly reduce the efforts and enables almost complete relaxation of the internal stresses; however, it adds two technological procedures: water quenching and ageing of the material. This article answers the question about the optimal strain mode, which allows forming thick panels and preserving their long service life.

Acknowledgements

The work is performed at the support of the grants of the Russian Foundation for Fundamental Research: No. 16-08-00713, 16-31-00460

References

[1] Lihua Zh, Jianguo L and Deanc T 2011 *International Journal of Machine Tools and Manufacture* **51**(1) 1

[2] Holman M 1989 *Journal of Mechanical Working Technology* 477
[3] Brewer H 1989 *In Proceedings of the Aircraft Design, Systems and Operations Conference* (Seattle: Edimet Spa) 6
[4] Tsvelodub I 1991 *The postulate of sustainability and its application to the creep theory of metallic materials* (Novosibirsk: Lavrentev institute of hydrodynamics) p 201
[5] Narimetla S, Peddieston J and Buchanan G 2000 *Materials Research Communications* 27 pp 631–636
[6] Ho K, Lin J and Dean T 2004 T *Journal of Materials Processing Technology* 153-154 pp 122–127
[7] Jeunechamps P, Ho K, Lin J, Ponthot J and Dean T 2006 *International Journal of Mechanical Sciences* 48 pp 621–629
[8] Aaron C. Lama, Zhusheng Sh, Haoliang Y, Li W and Catrin M 2015 *Journal of Materials Processing Technology* **219** 155
[9] Korobeinikov S, Oleinikov A, Gorev B and Bormotin K 2008 *Computational methods and programming: new computing technologies* **9**(1) 346
[10] Banshchikova I, Tsvelodub I and Petrov D 2015 *Scientific notes of the Kazan University. Physics and mathematics* **157**(3) 34 (In Russian)
[11] Shutov A V, Larichkin A Yu and Shutov V A 2017 *Journal of Applied Mathematics and Mechanics* **97**(8)
[12] Zakharchenko K and Kapustin V 2015 *Factory laboratory. Diagnosis materials.* **81**(7) 51
[13] Kapustin V, Gileta V and Tereshin E 2010 *Journal of Applied Mechanics and Technical Physics* **51**(3) 393
[14] Senatorova O 2006 *Method of producing a high-strength aluminum alloy*. Patent RF, no 2396367, MPK C22F 1/053 (2006.01) p 7
[15] Jin Zh, Yu W, Yunlai D and Xinming Zh 2016 *Materials Science and Engineering: A* **664** 126
[16] Chao L, He Y, Heng L, Nian Sh, Jin F and Lihua Zh 2016 *Chinese Journal of Aeronautics* **29**(5) 1445
[17] Inforzato D, Costa P, Fernandez F and Travessa D 2012 *Mat. Res.* **15**(4) (São Carlos) 596–602

IOP Conf. Series: Journal of Physics: Conf. Series **894** (2017) 012051 doi:10.1088/1742-6596/894/1/012051

Formation of annular cracks in glass during contact interaction

M A Legan[1,2], A N Novoselov[1] and N V Fedorova[1,2]

[1] Lavrentyev Institute of Hydrodynamics SB RAS, Novosibirsk, 630090, Russia
[2] Novosibirsk State Technical University, Novosibirsk, 630073, Russia

E-mail: legan@ngs.ru

Abstract. Experimental results are presented on the indentation of steel balls of various diameters into glass samples having a rectangular parallelepiped shape. The ultimate load in the formation of an annular crack in the vicinity of the contact region and the radius of this crack were determined experimentally. The annular crack appeared outside the contact area almost in all tests. Due to the fact that the diameter of the contact area was much smaller than the dimensions of the samples, the glass samples were considered as an elastic half-space. Huber's solution for Hertz's problem of pressing a ball into an elastic half-space was used to determine the field of contact stresses in the fracture zone. Local criterion for maximum stresses and nonlocal failure criteria (average stress criterion, the Nuismer criterion, and the gradient criterion) were used to model fracture in a contact interaction. The ultimate tensile stress and the critical stress intensity factor of the glass were experimentally determined on beams without a notch and with a notch for calculating a parameter having a dimension of length and entering nonlocal failure criteria. It is shown that the estimates of the radius of the annular crack that are closest to the experimental data give a gradient criterion among the applied criteria. Estimates of the ultimate tensile stress according to this criterion exceed the values obtained when the beams are bent, which can be explained by the scale factor.

1. Introduction

According to the solution of the Hertz problem on the ball pressing into the elastic half-space [1-3], the maximum tensile stress is reached on the surface of the half-space on the boundary of the circular contact area and acts on the section perpendicular to the radius from the point of initial contact, that is, the radial stress. Consequently, according to the local criterion of maximum stresses, failure must begin at the boundary of the contact area. However A.S. Argon et al. [4] experimentally established that when the balls are pressed into glass samples, the radius of the annular crack on the surface of the samples is greater than the radius of the contact region. This circumstance is a motivation for the experimental investigation of brittle fracture in the vicinity of the contact region and the modeling of this failure using nonlocal failure criteria, taking into account the inhomogeneity of the contact stress field. It is necessary to know the stress field not only on the surface, but also inside the half-space for the application of these criteria. The stress distribution for the Hertz problem of pressing a ball into an elastic half-space was obtained by M.T. Huber [2].

The values of the limiting load on the ball in the formation of an annular crack and the radius of this crack will be determined in an experimental study. Estimates of the ultimate load in the formation of an annular crack and the radius of this crack will be obtained by modeling the failure in the zone of contact stresses. A comparison of the obtained estimates with the experimental data will be carried out

Content from this work may be used under the terms of the Creative Commons Attribution 3.0 licence. Any further distribution of this work must maintain attribution to the author(s) and the title of the work, journal citation and DOI.
Published under licence by IOP Publishing Ltd

and a conclusion will be made about the consistency of the fracture criteria with the experimental results. The main attention will be paid to the correspondence of the theoretical and experimental values of the radius of annular cracks. It is known that annular cracks appear outside the contact area from experiments [4-8]. Therefore, the theoretical estimation of the radius of cracks can also be checked by comparison with the radius of the contact area.

2. Experiments and methods

The real-time video method was used to determine the force at which a crack is formed. The tests were carried out on a Zwick / Roell Z100 machine at the Lavrentyev Institute of Hydrodynamics SB RAS. USB-microscope was installed under steel protection. The objective of the microscope is directed to the center of the through hole Ø10 mm in the cover of protection. The glass sample is placed on top of this 10 mm thick protection. The steel ball is pressed into this glass sample. The microscope transmits the image to the computer in real time, and the installed software allows to videotape and take photographs. The operator notices the formation of a crack on the computer monitor, stops the experiment and fixes the ultimate cracking force. Glass samples were in the form of rectangular parallelepipeds, having a height of 10 mm and sides of 20 mm. Results of experiments are discussed in detail in [9, 10].

The radius of ring-shaped cracks r_c was determined using a measuring microscope. The radius of the contact region a was calculated from the formula [3]:

$$a = \sqrt[3]{\frac{3RP}{4E^*}} . \quad (1)$$

Here R is the radius of the ball; P is the force at which a crack is formed; E^* is the effective modulus of elasticity upon contact of a steel ball with a glass sample, which is calculated from the known values of the elasticity modulus and Poisson's ratio of the steel E_{st}, ν_{st} and glasses E_{gl}, ν_{gl}:

$$E^* = \left(\frac{1-\nu_{gl}^2}{E_{gl}} + \frac{1-\nu_{st}^2}{E_{st}} \right)^{-1} . \quad (2)$$

For steel $E_{st} = 2.11 \cdot 10^{11}$ Pa, $\nu_{st} = 2.28$. For glass $E_{gl} = 7 \cdot 10^{10}$ Pa, $\nu_{gl} = 0.2$.

3. Results and discussion

The local criterion of maximum tensile stress applied to simulate the fracture. When the ball is pressed into an elastic half-space, the tensile radial forces take place in the contact vicinity. The maximum value of the radial stress on the surface of the half-space is concentrated at the points of the boundary circular contact region. Therefore, according to the local criterion of maximum stresses, failure must begin at the boundary contact area. The radial stress depends on the radius r on the surface of the half-space outside the contact region in accordance with the formula [2, 5, 6]

$$\sigma_r = p_m \frac{1-2\nu}{2} \frac{a^2}{r^2} . \quad (3)$$

Here $p_m = \frac{P}{\pi a^2}$ is average pressure in the contact area; ν is Poisson's ratio.

The averaged experimental data obtained by indentation of balls with different diameters and the results of calculation from these data are summarized in Table 1. Figure 1 shows the dependence of the maximum tensile stresses at the surface of a half-space on the radius. Figure 1 depicts points whose values of coordinates are equal to the value of the limiting tensile stress of the glass obtained during the bending of the beams and the radii of annular cracks, respectively, for each of the ball's diameters. The numbers of the curves and points in Figure 1 correspond to the numbering of the data in Table 1. Table 1 and Figure 1 show that, firstly, the maximum tensile stresses with ball pressing are an order of magnitude higher than the limiting tensile stress for glass obtained when bending the beams, and secondly, the radii of annular cracks are larger than the radii of the contact area.

Table 1. Averaged experimental data and calculation results.

No	Diameter of the ball Mm	Breaking Load *P*, N	Radius of the annular crack r_c, mm	Radius of the contact area *a*, mm	Average pressure p_m, GPa	Maximum radial stress $\sigma_r^{\max}$, MPa
1	5.5	692	0.376	0.281	2.79	837
2	10	717	0.456	0.355	1.81	543
3	17	1115	0.670	0.498	1.43	429

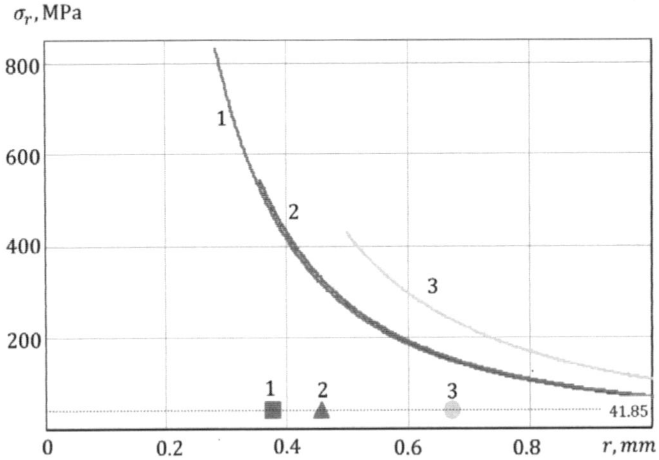

Figure 1. Dependencies of the maximum tensile stresses on the radius for different diameters of balls.

Taking into account the poor description of the experimental data by the criterion of maximum stresses, nonlocal failure criteria were used to estimate the ultimate tensile stress and the crack radius when the ball was pressed into glass samples. It is necessary to know the distribution of stresses not only on the surface, but also within the half-space for the application of these criteria. Formulas that can be used to calculate stresses at any point in a half-space for Hertz's problem of ball indentation were obtained by Huber [2]. The radial stress in Huber's solution is calculated by formula

$$\sigma_r = \frac{3}{2} p_m \left[\frac{1-2\nu}{3} \frac{a^2}{r^2} \left(1 - \frac{z^3}{\sqrt{u}^3} \right) + \frac{a^2 z^3}{(u^2 + a^2 z^2)\sqrt{u}} + \frac{z}{\sqrt{u}} \left(\frac{(1-\nu)u}{a^2+u} + (1+\nu)\frac{\sqrt{u}}{a} arctg\left(\frac{a}{\sqrt{u}} \right) - 2 \right) \right]. \quad (4)$$

Here z is the coordinate, measured from the surface of the half-space along its normal;
$u = \frac{1}{2}\left(r^2 + z^2 - a^2 + \sqrt{(r^2 + z^2 - a^2)^2 + 4a^2 z^2} \right)$.

3.1. Estimation by the criterion of average stresses. According to this criterion, the ultimate strength of the material σ_b is compared not with the maximum positive value of the first principal stress σ_1, but with the average normal stress

$$\sigma_{av} = \frac{1}{L_*} \cdot \int_0^{L_*} \sigma_n \, dz \quad (5)$$

which is calculated on area of size L_*, including an infinitesimal area σ_I at the considered point of the body, where both areas have a common normal n. The stress along the other side of the area is assumed to be constant. At the moment of the beginning of destruction $\sigma_{av} = \sigma_b$. The size of the averaging area is given by formula [11, 12]

$$L_* = \frac{2K_{Ic}^2}{\pi \sigma_b^2}. \quad (6)$$

IOP Conf. Series: Journal of Physics: Conf. Series **894** (2017) 012051 doi:10.1088/1742-6596/894/1/012051

Here K_{Ic} is the critical stress intensity factor. For sodium-calcium-silicate glass of grade M1, after annealing, the following was obtained:[9, 10] $K_{Ic} = 1{,}271\,\text{MPa}\,\text{m}^{1/2}$, $\sigma_b = 41{,}85\,\text{MPa}$. Then $L_* = 0{,}587\,\text{mm}$.

According to this criterion, the dependence of the average tensile stress on the radius is obtained upon indentation of balls of three diameters. In this problem, these are the dependences of the radial stress averaged over the segment L_* on the radius, which are shown in Figure 2. In calculating the average radial stress, integration is performed over z from 0 to L_*. The obtained dependencies have maxima, the coordinates of which along the r axis give estimates of the radius of annular cracks, and the values of the maxima give estimates of the ultimate tensile stress of the glass. The estimates obtained are shown in Table 2. Figure 2 shows the points which values of coordinates are equal to the value of the ultimate tensile stress of the glass obtained during the bending of the beams and the radii of annular cracks, respectively, for each of the diameters of the balls. The numbers of curves and points on the graph correspond to the numbering of the data in Table 2.

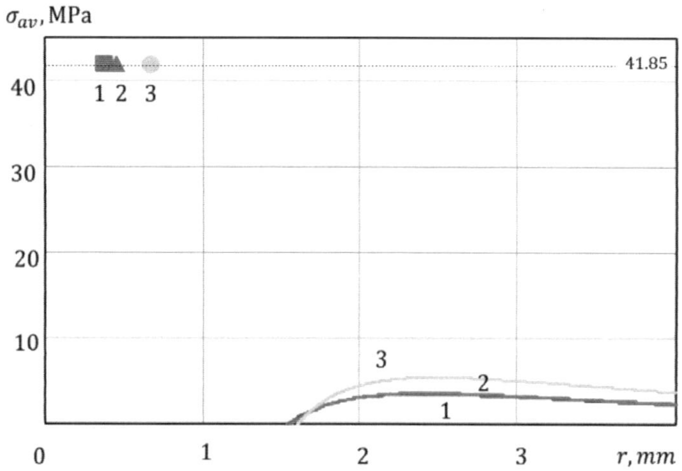

Figure 2. Dependencies of the average radial stresses on the radius for different diameters of balls.

Table 2. Results predicted by the criterion of average stresses

No	Diameter of the ball	Estimate of the radius of an annular crack along the coordinate of the maximum σ_{av}, mm	Maximum average radial stress σ_{av}^{max}, MPa
1	5.5	2.441	3.5
2	10	2.430	3.6
3	17	2.482	5.5

Table 2 and Figure 2 show that, firstly, the estimates of the ultimate tensile stress by the criterion of average stresses are an order of magnitude smaller than the ultimate tensile stress of the glass obtained when the beams are bent, and secondly, estimates of the radius of annular cracks are many times larger than the experimental values. Consequently, the criterion of average stresses poorly describes the experimental data on the destruction of glass when indenting steel balls.

3.2. Estimation by the Nuismer failure criterion for stresses at a distant point. In this criterion [11], the ultimate strength of material is not compared with the maximum tensile stress at the considered point of the body, but with the stress at the point considered at a distance C_*:

$$\sigma_N = \sigma_r(r, z = C_*), \quad (7)$$

$$C_* = L_*/4. \quad (8)$$

According to this criterion, the dependence of the distant stress on the radius is obtained by indenting balls of three diameters. In this problem, this is the dependence of the radial stress distant from the surface of a half-space on the distance C_* on the radius, which are presented in Figure 3. The obtained dependencies have maxima, the coordinates of which along the r axis give estimates of the radius of annular cracks, and the values of the maxima give estimates of the ultimate tensile stress of the glass. The obtained estimates are shown in Table 3. Figure 3 shows the points which coordinates are equal to the value of the ultimate tensile stress of the glass obtained when the beams are bent and the radii of annular cracks for each of the ball diameters, respectively. The numbers of curves and points on the graph correspond to the numbering of the data in Table 3.

Table 3 and Fig. 3 show that, firstly, the estimates of the ultimate tensile stress by Nuismer's failure criterion are lower than the ultimate tensile stresses of the glass obtained when the beams are bent, and secondly, estimates of the radius of annular cracks are larger than the experimental values. Consequently, the Nuismer criterion does not satisfactorily describe the experimental data on the destruction of glass when pressing steel balls.

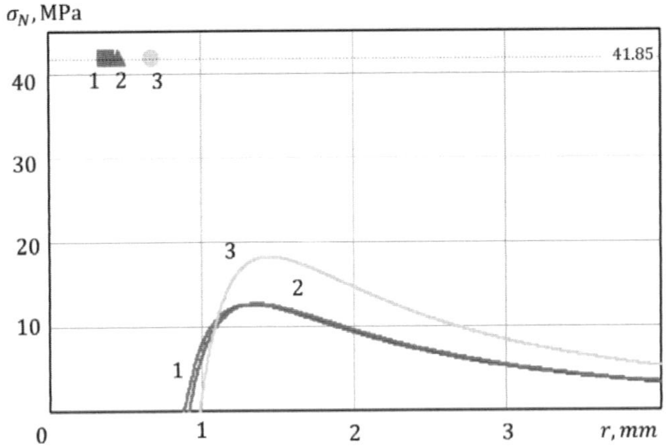

Figure 3. Dependencies of the distance radial stresses on the radius for different diameters of balls.

Table 3. Results predicted by the Nuismer's failure criterion

No	Diameter of the ball	Estimate of the radius of an annular crack along the coordinate of the maximum σ_N, mm	Maximum average radial stress σ_N^{max}, MPa
1	5.5	1.321	12.6
2	10	1.357	12.6
3	17	1.442	18.2

3.3. Estimation by the gradient failure criterion. In the gradient criterion [12-14], the strength of the material σ_b is compared not with the maximum stress, but with an effective stress σ_e for determining the start of failure. The effective stress is proportional to the first principal stress σ_l at the considered point of the body and depends on the relative gradient g_n of the positive normal stress σ_n acting perpendicular to the plane including the area of the first principal stress at the considered point of the body where the plane and the area have a common normal n:

$$g_n = \frac{|\operatorname{grad} \sigma_n|}{\sigma_n}. \tag{9}$$

The relative gradient is found using the solution of the corresponding elasticity problem. The expression for the effective stress is written in the form:

$$\sigma_e = \frac{\sigma_1}{1-\beta+\sqrt{\beta^2+L_* g_n}} \tag{10}$$

where L_* is a parameter having the dimension of length, which is expressed in terms of the known characteristics of the material σ_b and K_{Ic} by formula (6); β is a non-negative dimensionless parameter ($\beta \geq 0$), which can be considered as an approximation parameter. It is believed that the destruction in the vicinity of the consideration point begins when the effective stress is reaches the material ultimate strength and initially spreads over the action area of the first principal stress, and the failure condition is written in the form:

$$\sigma_e = \sigma_b . \tag{11}$$

For $z = 0$, on the surface of the half-space in the vicinity of the contact region, the first principal stress is the radial stress σ_r, which is used in the gradient criterion. We use formulas (3) and (4) to calculate the radial stress and its derivatives with respect to the coordinates r and z. Next, determine the relative gradient of the radial stress and the effective stress. We find on the surface of the half-space from (3):

$$\frac{\partial \sigma_r}{\partial r} = -p_m (1-2\nu)\frac{a^2}{r^3} .$$

Find the derivative $\frac{\partial \sigma_r}{\partial z}$, using the expression (4):

$$\frac{\partial \sigma_r}{\partial z} = \frac{3}{2} p_m \left[\frac{1-2\nu}{3}\frac{a^2}{r^2}\left(-\frac{3z^2}{u^{3/2}} + \frac{3}{2}\frac{z^3}{u^{5/2}}\frac{\partial u}{\partial z} \right) + \right.$$

$$+\frac{3a^2z^2\left(u^2+a^2z^2\right)u^{1/2} - a^2z^3\left[\frac{5}{2}u^{3/2}\frac{\partial u}{\partial z} + 2a^2zu^{1/2} + \frac{a^2z^2}{2u^{1/2}}\frac{\partial u}{\partial z}\right]}{\left(\left(u^2+a^2z^2\right)u^{1/2}\right)^2} + (1-\nu)\frac{\left(u^{1/2} + \frac{z}{2u^{1/2}}\frac{\partial u}{\partial z}\right)\left(a^2+u\right) - zu^{1/2}\frac{\partial u}{\partial z}}{\left(a^2+u\right)^2} +$$

$$\left. +(1+\nu)\left(\frac{1}{a} arctg\left(\frac{a}{u^{1/2}}\right) - \frac{\frac{z}{2u^{3/2}}}{1+\frac{a^2}{u}}\frac{\partial u}{\partial z} \right) - \frac{2u^{1/2} - z\frac{\frac{\partial u}{\partial z}}{u^{1/2}}}{u} \right]; \qquad \frac{\partial u}{\partial z} = \left(1 + \frac{r^2+z^2+a^2}{\sqrt{\left(r^2+z^2-a^2\right)^2 + 4a^2z^2}} \right) z$$

When z=0 and $r \neq a$ then $\frac{\partial u}{\partial z} = 0$ and $\frac{\partial \sigma_r}{\partial z} = \frac{3}{2} p_m \left(\frac{(1-\nu)\sqrt{u}}{a^2+u} + \frac{1+\nu}{a} arctg\frac{a}{\sqrt{u}} - \frac{2}{\sqrt{u}} \right)$;

$$u(r) = \frac{1}{2}\left(r^2 - a^2 + \sqrt{\left(r^2-a^2\right)^2} \right) .$$

The relative gradient is calculated from formula:

$$g_r = \frac{|\operatorname{grad}\sigma_r|}{\sigma_r}; \quad |\operatorname{grad}\sigma_r| = \sqrt{\left(\frac{\partial \sigma_r}{\partial r}\right)^2 + \left(\frac{\partial \sigma_r}{\partial z}\right)^2} . \tag{12}$$

Substituting formula (12) into (10), we find the effective stress at β=0. In accordance with the gradient criterion, the dependencies of the effective stress on the radius are plotted for indentation of balls of three diameters in Figure 4. These dependencies have maxima, the coordinates of which along the r axis give estimates of the radius of annular cracks, and the values of the maxima give estimates of the ultimate tensile stress of the glass. The obtained estimates are shown in Table 4. Figure 4 shows the points whose coordinates are equal to the value of the ultimate tensile stress of the glass obtained during the bending of the beams and the radii of annular cracks, respectively, for each of the diameters of the balls. The numbers of curves and points on the graph correspond to the numbering of the data in Table 4.

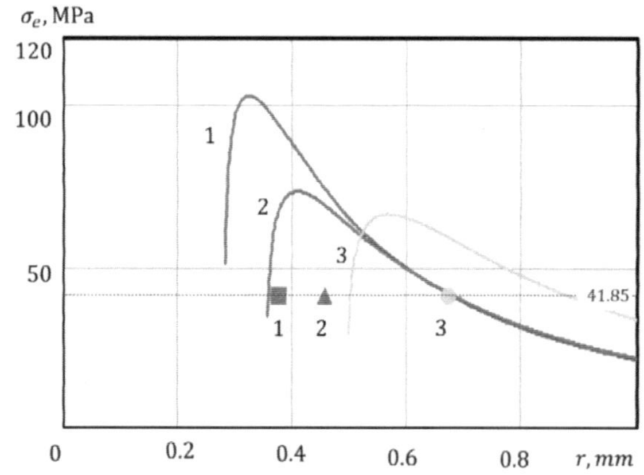

Figure 4. Dependencies of the effective stresses on the radius for different diameters of balls.

IOP Conf. Series: Journal of Physics: Conf. Series **894** (2017) 012051 doi:10.1088/1742-6596/894/1/012051

Table 4. Results predicted by the gradient failure criterion

No	Diameter of the ball	Estimate of the radius of an annular crack along the coordinate of the maximum σ_e, mm	Maximum average radial stress σ_e^{max}, MPa
1	5.5	0.322	102.6
2	10	0.407	73.4
3	17	0.568	66.6

Table 4 and Figure 4 show that, firstly, estimates of the ultimate tensile stress with respect to the gradient criterion are higher than the ultimate tensile stress of glass obtained from the bending of beams, which can be explained by the scale factor, and secondly, estimates of the radius of annular cracks are close to experimental data, but less than them. Thus, the gradient criterion is better describing experimental data on the destruction of glass when indenting steel balls than the other considered criteria.

4. Conclusion

The series of experiments were carried out to indentation of steel balls of different diameters into glass samples in the form of a rectangular parallelepiped. The ultimate forces in the formation of annular cracks in the vicinity of the contact region and the radii of these cracks were determined experimentally. Almost in all tests, annular cracks appeared outside the contact area. Since the diameter of the contact region was much smaller than the dimensions of the samples, the glass samples were considered as an elastic half-space. Huber's solution of Hertz's problem of indentation ball into an elastic half-space was used to determine the field of contact stresses in the fracture zone. Local maximum stress criterion and non-local failure criteria: the criterion of average stresses, the Nuismer criterion and the gradient criterion were used to simulate contact fracture of the samples. It is shown that the gradient criterion is most close to the experimental data from the all considered failure criteria. Estimates of the ultimate tensile stress according to this criterion exceed the values obtained in the bending of beams, which can be explained by the scale factor.

Acknowledgments

This work was partially supported by the Russian Foundation for Basic Research (project codes 15-01-07631 and 16-08-00713) and Grant of Government of Russian Federation No 14.W03.31.0002.

References

[1] Hertz H 1881 *J. für die reine und angewandte Mathematik* **92** 156–71
[2] Huber M T 1904 *Annalen der Physik,* **14**(4) 153–63
[3] Johnson K L 1985 *Contact mechanics* (Cambridge University Press) p 452
[4] Argon A S, Hori Y and Orowan E 1960 *J. American Ceramic Soc. Ceramic Abstr.* **43**(2) 86–96
[5] Wilshaw T R 1971 *J. Phys. D: Appl. Phys.* **4** 1567–81
[6] Kolesnikov Yu V and Morozov E M 2007 Mechanica contactnogo razrusheniya [*Mechanics of contact destruction*] (In Russian) (Moscow: Publishing house LCI) p 224
[7] Fischer-Cripps A C 2007 *Introduction to Contact Mechanics* (New York: Springer) p 221
[8] Batanova O A, Matvienko Yu. G. 2015 *Physical mesomechanics* (In Russian) **18**(2) 22–36
[9] Novoselov A N, Legan M A 2016 *Science. Industry. Defense: tr. 17 All-Russia. Scientific-techn. Conf.* (Novosibirsk: Publishing House of the NSTU) (In Russian) **1** 151–6
[10] Novoselov A N, Legan M A 2017 *Problems of Optimal Structures Design: tr. 4 All-Russia Conf.,* (Novosibirsk: NSASU) (In Russian) 201–9
[11] Whitney J M and Nuismer R J 1974 *J. Compos. Mater.* **8**(3) 253–65
[12] Legan M A 2013 *Vestnik NSU. Mathematic and Mechanics* (In Russian) **13**(3) 70–6
[13] Legan M A 1993 *J. Appl. Mech. Tech. Phys.* **34**(4) 585–92
[14] Legan M A 1994 *J. Appl. Mech. Tech. Phys.* **35**(5) 750–6

IOP Conf. Series: Journal of Physics: Conf. Series **894** (2017) 012052 doi:10.1088/1742-6596/894/1/012052

Application of the boundary elements method for modeling of the fracture of cylindrical bodies by hydraulic fracturing

M A Legan[1,2], V A Blinov[1], A Yu Larichkin[1] and A N Novoselov[1]

[1] Lavrentyev Institute of Hydrodynamics SB RAS, 15 Prospekt Lavrentyeva, Novosibirsk, 630090, Russia
[2] Novosibirsk State Technical University, 20 Prospekt K. Marksa, Novosibirsk, 630073, Russia

E-mail: legan@ngs.ru

Abstract. Experimental study of hydraulic fracturing of thick-walled cylinders with a central circular hole was carried out using the machine that creates a high oil pressure. Experiments on the compression fracture of the solid cylinders by diameter and rectangular parallelepipeds perpendicular to the ends were carried out with a multipurpose test machine Zwick / Roell Z100. Samples were made of GF-177 material based on cement. Ultimate stresses in the material under study were determined for three types of stress state: under compression, with a pure shear on the surface of the hole under frecking conditions and under a compound stress state under conditions of diametral compression of a solid cylinder. The value of the critical stress intensity factor of GF-177 material was obtained. The modeling of the fracturing process taking into account the inhomogeneity of the stress state near the hole was carried out using the boundary elements method (in the variant of the fictitious load method) and the gradient fracture criterion. Calculation results of the ultimate pressure were compared with values obtained analytically on the basis of the Lame solution and with experimental data.

1. Introduction

Lavrentyev Institute of Hydrodynamics SB RAS has an experimental machine that produces a high pulsating oil pressure, which can be used to experimentally investigate the fracturing process of samples made of brittle materials, in particular rocks and materials based on cement or gypsum. The maximum pressure can reach 40 MPa. A distinctive feature of the machine is the ability to create pulsating pressure with a frequency of up to 100 Hz. Thus, it is possible to investigate not only the formation of hydraulic fracturing cracks, but also their growth under cyclic loading.

In laboratory conditions, experiments can only be carried out with models of wells in rocks. Real wells have a significantly larger diameter than laboratory models. Therefore, it is important to investigate the scale factor, which, according to the literature, is relevant for hydraulic fracturing. Haimson [1] conducted laboratory tests on hydraulic fracturing of two rocks: limestone and granite. The diameter of the cylindrical hole imitating the borehole varied from 3 to 32 mm. With this increase in diameter, the critical pressure for limestone fell by a factor of 7, and for granite by a factor of 1.5. The results of similar experiments carried out on the same rocks are presented in [2, 3]. The diameter of the hole was changed over a wider range: in limestone – up to 50 mm, and in granite – up to

Content from this work may be used under the terms of the Creative Commons Attribution 3.0 licence. Any further distribution of this work must maintain attribution to the author(s) and the title of the work, journal citation and DOI.
Published under licence by IOP Publishing Ltd

100 mm. With increasing diameter, the critical pressure decreased, asymptotically approaching a constant value.

It is known that in the presence of stress concentration, the scale factor manifests itself at a much greater degree than under a homogenous stress state. Since the borehole in the mountain massif is a stress concentrator, the inhomogeneity of the stress state near the well has a significant effect on the strength of the geomaterials. Therefore, in order to estimate the ultimate value of pressure during fracturing of rocks, nonlocal fracture criteria can be used to take into account the inhomogeneity of the stress state. It is possible to apply the criterion of average stresses, gradient criterion of fracture, etc. A distinctive feature of these criteria is the presence of a parameter having the dimension of length and characterizing the heterogeneity of the structure of the material. It is shown [4 -7] that this parameter is expressed through two standard characteristics of the material – the tensile strength σ_b at uniaxial tension and the critical stress intensity factor K_{Ic}. The values characterizing the strength, crack resistance and heterogeneity of the structure of some brittle materials and rocks can be found in [8].

In the present study, the simulation of the fracturing process, taking into account the inhomogeneity of the stress state near the hole, was carried out using the boundary element method (in the variant of the fictitious load method) and the gradient fracture criterion.

2. Hydraulic fracturing experiments on cylindrical samples

Experimental research of hydraulic fracturing of thick-walled cylinders with a central circular hole was carried out using a machine that creates a high oil pressure. Cylinders were made of GF-177 material on the cement basis with polymer additives. The presence of polymer additives ensures the absence of shrinkage and cracking of the material upon solidification. The preparation of the solution was conducted by mixing the compound with water in a ratio recommended by the manufacturer – in the proportion of 0.29 liters of water per kilogram of the compound. The solution was poured into cylindrical molds located on the shaker, which used to remove air bubbles from the solution. Cylindrical samples had a diameter of 105 mm and a height of 100 mm. A central hole 10.5 mm in diameter was drilled along the axis of the sample. Thus, the sample was a thick-walled cylinder with an orifice, whose diameter is 10 times smaller than the diameter of the cylinder. With the help of such a sample, it is possible to simulate a circular hole in an infinite medium.

Figure 1. Destroyed sample and the loading system.

Loading of the sample by internal pressure was carried out under conditions of plane strain. The sample loading system consists of a metal tube 10 mm in diameter with radial holes. The tube was inserted into the sample hole. After installing the end rubber seals, a compressor oil was supplied to the tube, which proceeded via radial holes into the internal volume of the sample. The pressure determined using a sensor that transmits electrical signals to the oscilloscope, and a pressure gauge of the experimental machine. The experimental data recorded by the oscilloscope was transferred to the computer for further processing. The oil consumption was monitored by the oil flow sensor integrated into the unit.

After installing the loading system in the sample and connecting to the high-pressure oil line, the process of pressure growth started. The moment of fracture of the sample was observed by the pressure drop on the manometer, and also with the help of a video camera that observed the sample (Fig. 1)

IOP Conf. Series: Journal of Physics: Conf. Series **894** (2017) 012052 doi:10.1088/1742-6596/894/1/012052

The ultimate value of the pressure at the time of fracture was determined from a file recorded on a computer using a pressure and oscilloscope sensor. The readings of the pressure gauge and the pressure sensor almost coincided. Three samples were tested. The following values of the ultimate pressure were obtained: $p_1 = 3,3\,\text{MPa}$, $p_2 = 4,2\,\text{MPa}$, $p_3 = 3,6\ \text{MPa}$. The average value of the ultimate pressure for the three tests was $p_* = 3,7\,\text{MPa}$.

According to Lame's solution, under pressure p on the surface of a circular hole in an infinite medium, two stresses are acting at points on this surface: the circumferential tensile stress $\sigma_\varphi = p$ and the radial compressive stress $\sigma_r = -p$. Then the ultimate tensile stress during hydraulic fracturing is the circumferential stress at the surface of the hole $\sigma_\varphi = p_* = 3,7\,\text{MPa}$. Should be noted, that according to Lame's solution there is a stress state of pure shear on the surface of the hole and at other points of the infinite medium.

3. Fracture of solid cylinders by diametral compression

Smaller cylinders: 61 mm in diameter and 31 mm in thickness were produced from the same GF-177 material. When a solid cylinder is compressed along a diameter between rigid plates (Brazilian test), a constant tensile stress acts in the section passing through the load application points acting perpendicular to this section. With the exception of the vicinity of the load application points, the magnitude of this stress is practically constant and equals $\sigma = \dfrac{P}{\pi R t}$, where P is the compressive force, R and t are the radius and thickness of the cylinder. In addition, at the points of this section a compressive stress is applied along the cross-section, i.e. perpendicular to the tensile stress. Thus, in the section passing through the load application points, compound stress state is taking place. The compressive stress in the section is not constant, in the center of the cylinder it reaches a minimum in absolute value and is equal $\sigma_c = -\dfrac{3P}{\pi R t}$, i.e. three times higher than the tensile stress. Nevertheless, cylinders made of brittle materials, in which the tensile strength is much less than compression strength, are fractured by splitting the disk along the plane connecting the load application points. Thus, the fracture occurs mainly due to the action of tensile stresses, although the presence of a compressive stress affects the magnitude of the tensile stress upon fracture.

Three solid cylindrical samples were tested by diametral compression. Following values of the ultimate tensile stress were obtained $\sigma_1 = 1{,}91\,\text{MPa}$, $\sigma_2 = 1{,}72\,\text{MPa}$, $\sigma_3 = 1{,}82\,\text{MPa}$. The average value of the ultimate tensile stress for the three tests is $\sigma_* = 1{,}82\,\text{MPa}$. This value is 2 times smaller than the ultimate tensile stress during hydraulic fracturing.

4. Compression fracture and a limit state diagram scheme

Samples from GF-177 were made in the form of rectangular parallelepipeds for compression fracture tests. The average value of the compressive strength for the three tests is $\sigma_c = 10{,}2\,\text{MPa}$.
Based on the test results for two types of stress state, a limit state diagram was constructed using the Mohr- Coulomb strength criterion (Figure 2) and an estimation of the tensile strength of the material for uniaxial tension was obtained $\sigma_b = 3{,}92\,\text{MPa}$. A light square in the limit state diagram shows the result obtained during hydraulic fracturing. The experimental value of the ultimate tensile stress during hydraulic fracturing exceeds the estimation that can be obtained by the Mohr-Coulomb test, which can be explained by the influence of the inhomogeneity of the stress state during hydraulic fracturing.

5. Determination of the critical stress intensity factor

Two cylindrical samples with a central cut were made to determine the critical stress intensity factor of

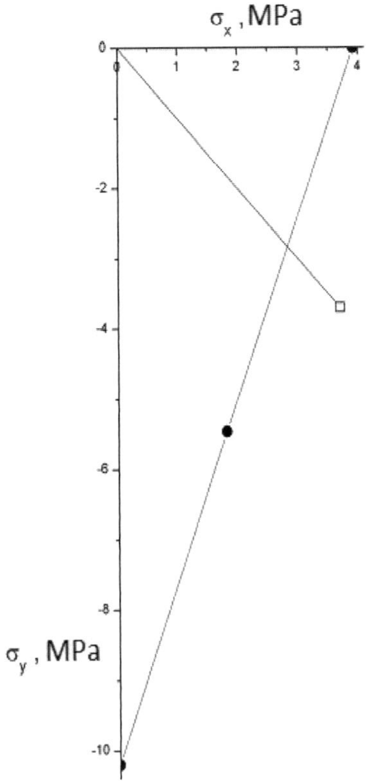

Figure 2. The limit state diagram.

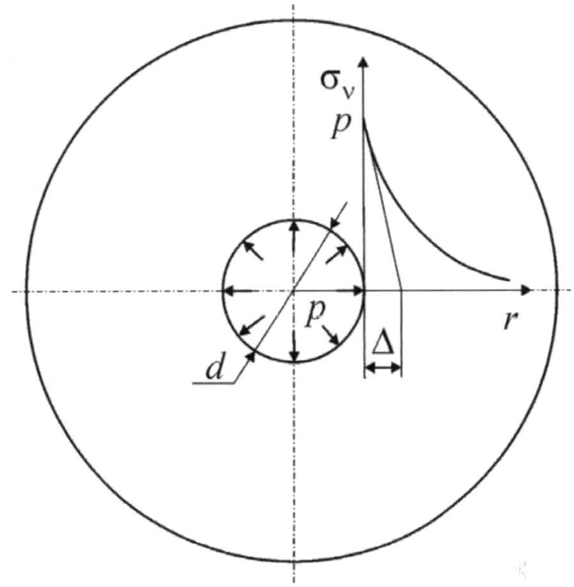

Figure 3. The sample loading scheme and the plot of the tensile stress.

GF-177 material. Samples were tested on a Zwick / Roell Z100 test machine by compressing between plates along the cutting line. According to two experiments, the average value of the critical stress intensity factor was obtained $K_{Ic} = 0{,}142\,\text{MPa}\cdot\text{m}^{1/2}$. Using this value of K_{Ic} and an estimation of the ultimate strength of the material $\sigma_b = 3{,}92\,\text{MPa}$, the parameter that enters into the average stress criterion and the gradient criterion was calculated $L_1 = (2/\pi)K_{Ic}^2/\sigma_b^2 = 0{,}835\,\text{mm}$.

6. Hydraulic fracture modeling using a gradient fracture criterion

A gradient criterion can be used to model fracture during frecking [5-7]. According to this criterion, with an inhomogenous stress distribution to determine the onset of fracture, the ultimate tensile stress of the material should be compared with the maximum value of the effective stress σ_e instead of maximum value of the first principal stress σ_1 (assumed as equivalent), that is higher than the equivalent stress. The effective stress is proportional to σ_1 and, in addition, depends on the local inhomogeneity of the stress field in the vicinity of the point under consideration and the representative size of the material's inhomogeneity. Local inhomogeneity of stress distribution is characterized by a relative gradient of positive normal stress acting on the plane including the normal plane of the first principal stress at the considered point of the body, where the plane and the platform have a common normal. Calculating the value $|\text{grad}\,\sigma_v|$, in some tasks is easier than the previously used value

$|\mathrm{grad}\,\sigma_1|$ [5, 6]. The relative gradient g_v is inversely proportional to the conditional size of the stress concentration zone Δ, which is shown in Fig. 3, i.e. $g_v = 1/\Delta$. The relative gradient is found using the solution of the corresponding problem of elasticity theory. The expression for the effective stress is written in the form

$$\sigma_e = \sigma_1 \Big/ \left(1 - \beta + \sqrt{\beta^2 + L_1 g_v}\right)$$

Here, L_1 is a parameter having the dimension of length and characterizing the heterogeneity (defectiveness) of the material; β is a non-negative dimensionless parameter, which can be considered as an approximation parameter.

The parameter L_1 is found in [5] from the condition for matching the gradient criterion with linear fracture mechanics and is expressed in terms of known material characteristics – the ultimate strength σ_b and the critical stress intensity factor K_{Ic} by the formula $L_1 = (2/\pi) K_{Ic}^2 / \sigma_b^2$.

We will assume that the fracture in the vicinity of the point under consideration begins when the effective tensile stress reaches the ultimate tensile stress and initially spreads through the action area of the maximum tensile stress. In the case of hydraulic fracturing, this is the area of action of the circumferential stress, which was confirmed experimentally (Figure 1).

According to Lame's solution, when the pressure p is acting on the surface of a circular hole in an infinite medium, then the circumferential and radial stresses are determined by the formulas $\sigma_\varphi = p\frac{a^2}{r^2}$ and $\sigma_r = -p\frac{a^2}{r^2}$. In this case, the tensile stress $\sigma_v = \sigma_\varphi$. The distribution of this stress along the radius is shown in Fig. 3. Note that in the absence of axial forces and stresses $\sigma_z = 0$, a plane strain occurs with $\varepsilon_z = 0$, since $\varepsilon_z = \frac{1}{E}\left[\sigma_z - \nu\left(\sigma_r + \sigma_\varphi\right)\right]$.

Calculation of the relative gradient yields $g_v = 4/d$, i.e. $\Delta = d/4$. Thus, the diameter of the hole enters the gradient fracture condition, which allows one to describe the scale effect upon fracture under conditions of a non-uniform stress state. Note that in the classical conditions of fracture, the size of the hole is not taken into account.

7. Combined use of the boundary element method and gradient fracture criterion

By using the modification of the program [9] in [10-12] the algorithm for joint application of the method of boundary elements (in the variant of the method of fictitious loads) and the gradient fracture criterion was implemented. For the problem of hydraulic fracturing of a thick cylinder, this algorithm was verified by comparing the numerical results with the results obtained by solving the Lame problem for a thick-walled tube. The outer and inner contours of the cylinder were divided into 360 elements. Thus the value $\sigma_* = 2{,}83\,\mathrm{MPa}$ was used as the limiting tensile stress in pure shear. This value was obtained by intersecting a ray passing from the origin in Fig. 2 at an angle of 45 degrees and corresponding to the loading during hydraulic fracturing, with a straight line drawn through two experimental points (compressive strength and Brazilian test ultimate strength). To determine the pressure at which fracture starts during a frecking, the effective stress in the gradient criterion was compared with σ_*. The inhomogeneity of the stress state in the vicinity of the hole leads to a decrease of σ_e in comparison with the circumferential stress σ_φ and, consequently, an increase in the limiting pressure p_* in comparison with σ_*. Thus, the joint application of the boundary element method and the gradient fracture criterion gives the ultimate pressure $p_* = 3{,}55\,\text{MPa}$, which is close to the mean experimental value $p_* = 3{,}7\,\text{MPa}$.

8. Conclusion

Experimental analysis of hydraulic fracturing of thick-walled cylinders with a central circular hole was carried out using a machine that creates a high oil pressure. Cylinders were made of GF-177 material on the cement basis with polymer additives. The value of the limiting tensile circumferential stress of 3.7 MPa is obtained. Also, limiting stresses in the material under study were determined under two more types of stress state: under compression and under a complex stress state under conditions of compression of a solid cylinder by diameter. The value of the critical stress intensity factor of material GF-177 was obtained. Using the method of boundary elements (in the variant of the fictitious load method) and the gradient fracture criterion, the fracturing process is modeled taking into account the heterogeneity of the stress state near the hole. A comparison of the results of calculations with the values of limit pressure obtained analytically based on Lame's solutions and experimental data was made.

Acknowledgments

This work was partially supported by the Russian Foundation for Basic Research (project codes 15-01-07631 and 16-08-00713) and Grant of Government of Russian Federation No 14.W03.31.0002.

References

[1] Haimson B C 1990 *Scale effects in rock masses: Proc. 1st Int. Workshop on scale effects in rock masses* (Rotterdam; Brookfield: Balkema) 89-101

[2] Haimson B C and Zhao Z 1991 *Rock Mechanics as a Multidisciplinary Science: Proc. 32nd US Symp.* (Rotterdam: Balkema) 191-9

[3] Cuisiat F D, Haimson B C 1992 *Int. J. Rock Mech. Min. Sci. & Geomech. Abstr.* **29**(2) 99-117.

[4] Whitney J M and Nuismer R J 1974 *J. Compos. Mater.* **8**(3) 253-65

[5] Legan M A 1993 *J. Appl. Mech. Tech. Phys.* **34**(4) 585–92

[6] Legan M A 1994 *J. Appl. Mech. Tech. Phys.* **35**(5) 750–6

[7] Legan M A 2013 *Vestnik NSU. Mathematic and Mechanics* (In Russian) **13**(3) 70–6

[8] Efimov V P 2013 *J. Appl. Mech. Tech. Phys.* **54**(5) 857-65

[9] Crouch S L and Starfield A M 1983 *Boundary element methods in solid mechanics* (London: Jeorge Allen & Unwin) p 322

[10] Sheremet A S and Legan M A 1999 *J. Appl. Mech. Tech. Phys.* **40**(4) 744-50

[11] Legan M A and Blinov V A 2016 *Science. Industry. Defense: tr. 17 All-Russia. Scientific-techn. Conf.* (Novosibirsk: Publishing House of the NSTU) (In Russian) **1** 129–34

[12] Legan M A and Blinov V A 2015 *Omskiy nauchniy vestnik* (Omsk: Publishing House of the OmSTU) (In Russian) 3(143) 349-52

On numerical model of one-dimensional time-dependent gas flows through bed of encapsulated phase change material

N A Lutsenko[1,2] and S S Fetsov[1,2]

[1] Institute of Automation and Control Processes FEB RAS, Vladivostok, 690041, Russia
[2] Far Eastern Federal University, Vladivostok, 690950, Russia

E-mail: nickl@inbox.ru, fetc95@mail.ru

Abstract. Mathematical model and numerical method are proposed for investigating the one-dimensional time-dependent gas flows through a packed bed of encapsulated Phase Change Material (PCM). The model is based on the assumption of interacting interpenetrating continua and includes equations of state, continuity, momentum conservation and energy for PCM and gas. The advantage of the method is that it does not require predicting the location of phase transition zone and can define it automatically as in a usual shock-capturing method. One of the applications of the developed numerical model is the simulation of novel Adiabatic Compressed Air Energy Storage system (A-CAES) with Thermal Energy Storage subsystem (TES) based on using the encapsulated PCM in packed bed. Preliminary test calculations give hope that the method can be effectively applied in the future for modelling the charge and discharge processes in such TES with PCM.

1. Introduction

Thermal Energy Storage (TES) can be considered as a solution of energy conservation problem which arises when the availability for energy does not coincide with demand. TES are applied in solar energy, central air cooling, air heating, waste heat recovery, etc. Phase Change Materials (PCM) are used in these devices to increase storage density. TES with PCM have been achieving lately a high penetration in the market in the areas of thermal protection and thermal inertia [1].

One of the effective methods to compose TES with phase change is packed bed of PCM enclosed in spherical particles, which are heated (cooled) by the Heat Transfer Fluid (HTF), flowing through it [2]. Last time this construction is adopted for novel Adiabatic Compressed Air Energy Storage (A-CAES). CAES differs from other storage methods, because it is available for high power ranges and energy capacities [3]; CAES is promising technology for energy storage with high reliability, economic feasibility, and low environmental impact [4]. The advantage of A-CAES is that the processes of gas compression and expansion occur adiabatically, so the heat of the compressed air is not lost and remains in the process for using in power generation. These properties make the A-CAES to be the most perspective technology for energy storage. Experimental analyzes and numerical modeling of CAES with PCM in the packed bed are performed in [5-7]; empirical investigations of the fixed and fluidized bed of the PCM blown by the heated air is presented in [8]. Modeling the processes of heat transfer in packed bed of the encapsulated PCM with air as the HTF and validation with experimental data are considered in [9,10].

Content from this work may be used under the terms of the Creative Commons Attribution 3.0 licence. Any further distribution of this work must maintain attribution to the author(s) and the title of the work, journal citation and DOI.
Published under licence by IOP Publishing Ltd

Since the problems with phase transitions have analytical solution in limited class of conditions, the numerical methods are developed for these problems [11]. These methods can be separated in two groups: methods with selection of the phase interface and shock-capturing methods. When flow of HTF in the packed bed of the encapsulated PCM is considered, the zone of the phase change is some area, where the transition takes place in every point simultaneously. So, the explicit phase interface is absent, and assumption of the instant phase transition is not actual. It does not allow to apply the known methods of calculating the phase transitions in similar problems.

When typical particle size in the packed bed thermal energy storage is relative small, TES can be considered as a porous medium. Mathematical model and numerical method were developed in [12] and tested in [12-14] for calculating the time-dependent one-dimensional gas flows through the porous heat-evolutional objects with known pressure at the boundaries and unknown gas flow rate. These model and method were modified and applied for simulating the time-dependent cooling processes of two-dimensional heat-evolutional objects [15] and heterogeneous combustion in porous media with unknown oxidant flow rate [16,17].

In the present work the numerical model based on previous results [12] is proposed for modeling the one-dimensional time-dependent gas flow through the packed bed of the encapsulated PCM; some test calculations are carried out.

2. Mathematical model

Let a motionless porous object consist of small balls made from phase change (solid – liquid) material (PCM) enclosed in envelope. Suppose that the porous object is bounded of impermeable non-heat-conductive side walls and is opened at the top and at the bottom. The PCM has its melting (crystallization) temperature T_m. When fraction of the PCM is much larger than fraction of the envelope material the last one can be neglected. Specific heat capacity, heat conductivity, density of the PCM are assumed to be constant during and after phase change. A gas is conveyed with known temperature and pressure into the inlet of the object (at the bottom); the gas flows up, changes its temperature as a result of the heat exchange with the particles, and flows out through the object top to the ambient air with known pressure.

Mathematical model is very similar to the previously developed model described in detail in [12]. It is constructed on the model of two interactive interpenetrated continua [18] and includes the equations of continuity, momentum conservation, state and energy for each medium:

$$\begin{gathered}
(1-a)\rho_c c\frac{\partial T}{\partial t} = -\alpha\left(T-T_g\right)+(1-a)\lambda\frac{\partial^2 T}{\partial x^2} \pm (1-a)\rho_c L\frac{\partial f}{\partial t},\\
\rho c_p\left(a\frac{\partial T_g}{\partial t}+u\frac{\partial T_g}{\partial x}\right) = \alpha\left(T-T_g\right)+a\frac{\partial P}{\partial t}+u\frac{\partial P}{\partial x}+\frac{\mu}{k}u^2,\\
\frac{1-(1-a)\chi}{a^2}\rho\left(a\frac{\partial u}{\partial t}+u\frac{\partial u}{\partial x}\right) = -\frac{\partial P}{\partial x}-\rho g-\frac{\mu}{k}u,\\
a\frac{\partial \rho}{\partial t}+\frac{\partial(\rho u)}{\partial x}=0,\quad P=\rho R T_g,\quad \mu = c_{s1}\frac{T_g^{3/2}}{c_{s1}+T_g},
\end{gathered} \tag{1}$$

where a is the porosity, c is the specific heat capacity of the PCM, c_p is the specific heat capacity of gas under constant pressure, c_{s1} and c_{s2} are the constants in Sutherland formula, f is the degree of phase change, g is the acceleration of gravity, k is the coefficient of permeability of the porous medium, L is the latent heat of PCM, P is the gas pressure, R is the gas constant, T is the temperature of the PCM, T_g is the temperature of the gas, t is the time, u is the gas filtration velocity ($u=av$, where v is the gas velocity), x the is spatial coordinate, α is the constant determining the intensity of the interphase heat exchange, λ is the coefficient of the heat-conductivity of PCM, μ is the coefficient of dynamic viscosity of the gas, ρ is the density of the gas, ρ_c is the density of the PCM, χ is the coefficient taking into account the inertial interaction of the phases in their relative motion [18].

It should be noted that the equations of motion and continuity of the PCM degenerate and are not included in the system since the PCM is assumed to be motionless with constant density.

The phase change is taken into account in the energy equation of the PCM by addition of the term with function *f*, which means the degree of phase change. This function is equal to 0 and 1 before and after the phase change respectively, and is unknown a priori when the phase change takes place. On the contrary, the PCM temperature is equal to T_m during the phase change and is unknown in other cases. Sign at the last term in the equation of energy for PCM is determined by the direction of the phase changing: the upper sign is selected for solidification; the lower sign is selected for melting.

The temperature and the pressure of the gas are known at the inlet of the object, the gas pressure at the outlet is known. Conditions of heat exchange are known on the both open boundaries (bottom and top) of the object. So, the boundary conditions for system (1) can be written as follows:

$$T_g\big|_{x=0}=T_{g0},\quad P\big|_{x=0}=P_0,\quad \lambda\frac{\partial T}{\partial x}\bigg|_{x=0}=\beta\left(T\big|_{x=0}-T_{g0}\right),$$
$$P\big|_{x=1}=P_a,\quad \lambda\frac{\partial T}{\partial x}\bigg|_{x=1}=\beta\left(T_g\big|_{x=H}-T\big|_{x=H}\right), \tag{2}$$

where P_0, P_a, T_{g0} are known values of the pressure and gas temperature respectively, β is the heat transfer coefficient.

To solve system (1) with boundary conditions (2) it is necessary to specify the values of sought parameters at the initial moment of time.

3. Numerical method

System (1) with boundary conditions (2) is solved by finite-difference method on the uniform grid with mesh size *h* and time step $\tau=rh^2$, where $r=const$. The algorithm is based on the previously proposed and tested in [12-14] numerical procedure for computing one-dimensional time-dependent gas flow through the porous object with internal heat-evolutional sources when the gas pressure at the open boundaries of the object is known. The equations of energy and motion are transformed into explicit finite-difference schemes; the continuity equation is written in implicit form and is solved using the Thomas algorithm [19]. The temperatures and the velocity of the gas are determined from the energy and motion equations respectively; the gas pressure is determined from the continuity equation, then the density is determined from the equation of perfect gas state. The boundary conditions for grid functions are defined from (2) trivially. This finite-difference scheme approximates system (1) with the first-order time accuracy and the second-order accuracy in space; the convergence takes place under a certain limitation on *r*. The main difference between the present method and the one described in [12] is the first equation in (1) (energy equation for PCM), which includes two unknown functions: *f* and *T*. Denoting the variation of time coordinate by superscript and the variation of space coordinate by subscript, let's write the finite-difference equation of energy for PCM (other equations are the same as in [12]):

$$\rho_c c\left(T_m^{n+1}-T_m^n\right)=-\frac{\tau\alpha}{1\text{-}a}\left(T_m^n-T_{gm}^n\right)+r\lambda\left(T_{m+1}^n-2T_m^n+T_{m-1}^n\right)\pm L\rho_c\left(f_m^{n+1}-f_m^n\right) \tag{3}$$

Explain the feature of the algorithm on an example of the following problem. Before starting moment, the temperature inside the porous object is equal to ambient temperature, which is lower than T_m; the gas does not flow. At the initial time moment, the gas begins to flow into the object inlet at a fixed pressure and temperature, which is higher than T_m. The gas flows up, heats the PCM to the melting temperature, and the phase transition begins after some time. Further heating of solid particles continues after the ending of phase transition until the stabilization of the PCM and gas temperatures. So, the whole process can be divided into three stages. The first stage is the heating of the PCM from initial temperature to the melting temperature; the second stage is the phase transition; the last stage is the further heating of the PCM until the temperature stabilization. Consider the algorithm for calculating *T* and *f* in every stage. At the first stage *f* is equal to 0 according to the definition. So, the last term in (3) is absent, and (3) is transformed into the following explicit scheme for determining *T*:

$$T_m^{n+1} = \left(1 - \frac{\tau\alpha}{(1\text{-}a)\rho_c c}\right) T_m^n + \frac{\tau\alpha}{(1\text{-}a)\rho_c c} T_{gm}^n + \frac{r\lambda}{\rho_c c}\left(T_{m+1}^n - 2T_m^n + T_{m-1}^n\right) \quad (4)$$

Computations using (4) continues until the melting temperature of the PCM is reached. Than the second stage (the phase transition) begins. The PCM temperature is equal to T_m, and left part of (3) is equal to 0. Since the melting is considered, lower sign at the last term in (3) is selected, and f is explicitly defined from (3):

$$f_m^{n+1} = f_m^n + \frac{1}{\rho_c L}\left(-\frac{\tau\alpha}{1\text{-}a}\left(T_m^n - T_{gm}^n\right) + r\lambda\left(T_{m+1}^n - 2T_m^n + T_{m-1}^n\right)\right). \quad (5)$$

Computations using (5) continues until f=1; than the phase change is completed, and the third stage (further heating of the PCM until the temperature stabilization) begins. The last term in (3) is equal to 0, and scheme (3) is transformed into (4) again.

When solidification is considered, the algorithm is similar except the selection of the upper sign at the last term in (3).

The advantage of the proposed method is that it does not require predicting the location of phase transition zone and can define it automatically as in a usual shock-capturing method. Due to this, the proposed method can be easily extended to 2D and 3D cases.

4. Calculations

The proposed numerical method was implemented and tested on simulation of the gas flow through the packed bed of the encapsulated PCM. The results were compared with results in [20], where simulation of the HTF flow through the bed of microfine particles of the PCM is considered. Air was used as HTF, and paraffin was used as PCM. Comparison between the results of the present numerical simulation and data in [20] is demonstrated in figure 1, which shows the PCM temperature at the outlet (x=2 m) of the porous object. As can be seen from the Figure 1, times of the beginning and the ending of the phase change coincide; times of the stabilization of the PCM temperature are close.

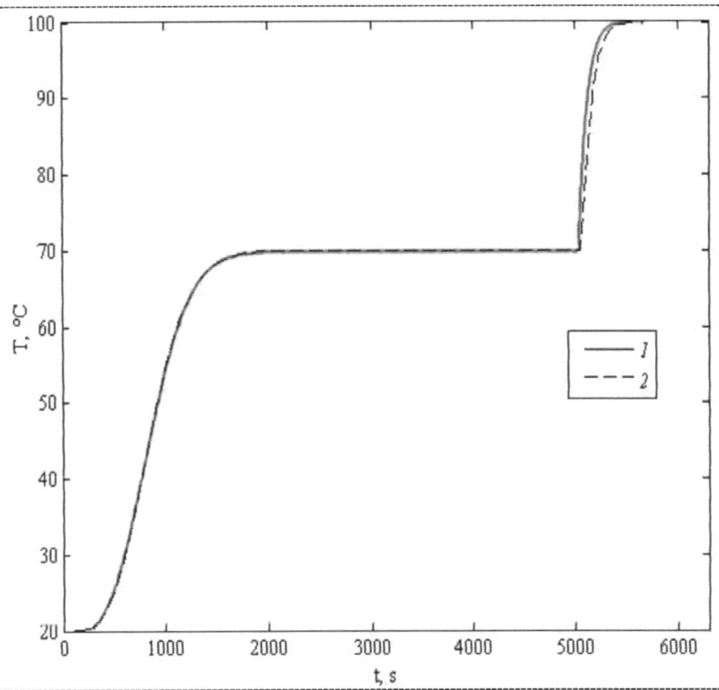

Figure 1. PCM temperature at the outlet (x=2 m) of the bed versus time for data in [20] (*1*) and for the present numerical simulation (2).

Figure 2 depicts the PCM temperature within the porous object in fixed time for data in [20] and for the present numerical simulation. As seen from the figure 2, both results are rather close. The difference between the results is due to the fact that the applied mathematical model has a number of differences from the model used in [20].The system of equations (1) takes into account more factors than the model in [20] like heat conductivity in PCM, power of internal forces in the gas etc.

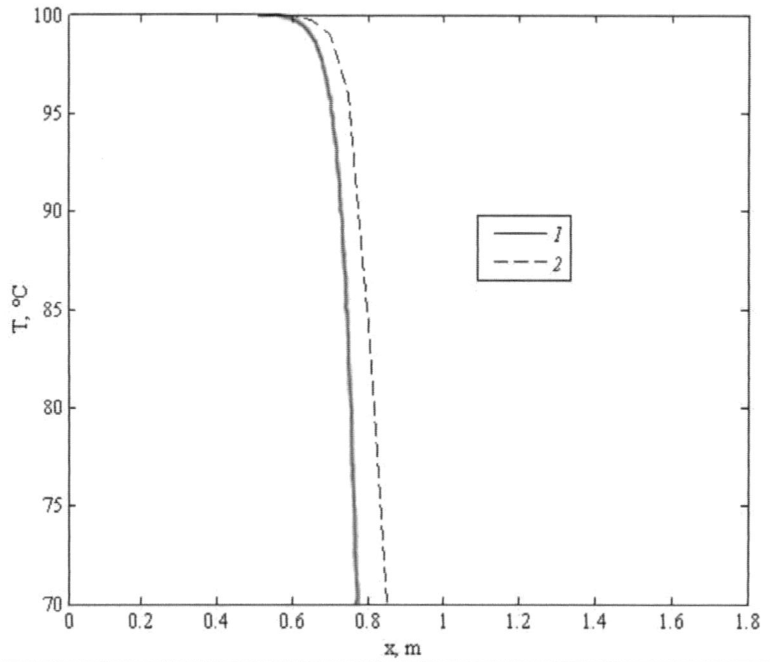

Figure 2. PCM temperature in time t=2314 s versus the height of the bed for data in [20] (*1*) and for the present numerical simulation (*2*).

In the near future, it is planned to modify the proposed numerical model for possibility to calculate the process when the gas velocity at the object inlet is fixed, and then to compare calculated result with experimental data under conditions of fixed gas velocity at the object inlet.

5. Conclusions

On the basis of the mathematical model and numerical method for simulating the time-dependent processes of gas flows through the heat-evolutional porous media with known pressure at the open boundaries and unknown gas flow rate, the mathematical model and numerical method are proposed for investigating the time-dependent one-dimensional gas flows through the packed bed of the encapsulated phase change material. The developed numerical model can be applied for the simulation of novel adiabatic compressed air energy storage system with thermal energy storage subsystem based on using the encapsulated PCM in packed bed. The advantage of the method is that it does not require predicting the location of phase transition zone and can define it automatically as in a usual shock-capturing method.

Acknowledgements

The work was supported financially by the Russian Foundation for Basic Research (project 16-01-00103-a) and the Far-Eastern Branch of the Russian Academy of Sciences (project 15-I-4-021).

IOP Conf. Series: Journal of Physics: Conf. Series **894** (2017) 012053 doi:10.1088/1742-6596/894/1/012053

References

[1] Zalba S, Marin J M, Gabeza L F and Mehling H 2003 *Appl. Therm. Energy* **23** 251
[2] Bedecarrats J P, Strub F, Falcon B and Dumas J P 1996 *Int J. Refrig.* **19** 187
[3] Castellani B, Presciutti A, Filliponi M, Nicolini A and Rossi F 2015 *Sustainability* **7** 9773
[4] Venkataramani G, Parankustan P, Ramalingam V and Wang J 2016 *Renuable and Sustainable energy reviews* **62** 895
[5] Peng H, Dong H and Ling X 2014 *Energy Conservation and Management* **81** 420
[6] Peng H, Li R, Ling X and Dong H 2015 *Applied Energy* **160** 1
[7] Peng H, Yang Yu, Li R and Ling X 2016 *Applied Energy* **183** 1361
[8] Izquierdo-Barrientos M A, Sobrino C and Almendros-Ibanez J A 2013 *Chemical Engineering Journal* **230** 573
[9] Nagano K, Takeda S, Mochida T and Shimakura K 2004 *Applied Thermal Engineering* **24** 2131
[10] Rady M 2009 *Applied Thermal Engineering* **29** 3149
[11] Samarskiy A A and Vaischevich P N 2003 *Numerical Heat Transfer* (Moscow: Editorial URSS) p 784 (in Russian)
[12] Lutsenko N A 2005 *Matematicheskoe Modelirovanie* **17** 120 (in Russian)
[13] Levin V A and Lutsenko N A 2005 *Combustion and Plasmochemistry* **3** 81 (in Russian)
[14] Levin V A and Lutsenko N A 2006 *Journal of Engineering Physics and Thermophysics* **79** 33
[15] Lutsenko N A 2014 *International Journal of Heat and Mass Transfer* **72** 602
[16] Lutsenko N A 2013 *Proceedings of the Combustion Institute* **34** 2289
[17] Lutsenko N A and Levin V A 2014 *Combustion Science and Technology* **186** 1410
[18] Nigmatulin R I 1978 *Principies of Mechanics of Heterogenious Media* (Moskow: Nauka) p 336 (in Russian)
[19] Tannehill J C, Anderson D A and Pletcher R H 1997 *Computational Fluid Mechanics and Heat Transfer, 2nd ed* (Washington, DC: Taylor&Francis)
[20] Teplitskii Yu S and Roslik A R 2015 *Journal of Engineering Physics and Thermophysics* **88** 1341

Smoldering of porous media: numerical model and comparison of calculations with experiment

N A Lutsenko[1, 2] and V A Levin[1, 3]

[1]Institute of Automation and Control Processes FEB RAS, Vladivostok, 690041, Russia
[2]Far Eastern Federal University, Vladivostok, 690950, Russia
[3]Institute of Mechanics, Lomonosov Moscow State University, Moscow, 119991, Russia

E-mail: NickL@inbox.ru

Abstract. Numerical modelling of smoldering in porous media under natural convection is considered. Smoldering can be defined as a flameless exothermic surface reaction; it is a type of heterogeneous combustion which can propagate in porous media. Peatbogs, landfills and other natural or man-made porous objects can sustain smoldering under natural (or free) convection, when the flow rate of gas passed through the porous object is unknown a priori. In the present work a numerical model is proposed for investigating smoldering in porous media under natural convection. The model is based on the assumption of interacting interpenetrating continua using classical approaches of the theory of filtration combustion and includes equations of state, continuity, momentum conservation and energy for solid and gas phases. Computational results obtained by means of the numerical model in one-dimensional case are compared with the experimental data of the smoldering combustion in polyurethane foam under free convection in the gravity field, which were described in literature. Calculations shows that when simulating both co-current combustion (when the smoldering wave moves upward) and counter-current combustion (when the smoldering wave moves downward), the numerical model can provide a good quantitative agreement with experiment if the parameters of the model are well defined.

1. Introduction

Combustion of solid porous media is one of the types of filtration combustion; it is quite common in nature and besides is used in some technological processes. From the point of view of mechanics, various objects can be modeled as porous media: for example, peatlands, coal dumps, landfills, grain elevators. When a porous medium burns, the reaction between oxygen and the condensed fuel often occurs directly on the surface of the fuel, that is, the fuel and oxidant are in different phases, so this combustion is heterogeneous. A widespread type of heterogeneous combustion in porous media is smoldering, which can be defined as a flameless exothermic surface reaction propagated with low velocity [1]. Burning of various porous objects during natural or man-made disasters (like peat fire) occurs under free (or natural) convection. In this case the flow rate of gas passed through the porous object is unknown a priori, and only the gas pressure at the object boundaries is known. Smoldering leads to peat megafires, which are the largest and some of the longest burning fires on Earth, cause huge environmental and economic damage, and are responsible for 15% of annual global greenhouse gas emissions [2]. To prevent and eliminate fires in porous objects, it is necessary to study the

Content from this work may be used under the terms of the Creative Commons Attribution 3.0 licence. Any further distribution of this work must maintain attribution to the author(s) and the title of the work, journal citation and DOI.
Published under licence by IOP Publishing Ltd

dynamics of gas flowing through them, since it can allow to understand how it is possible to limit the flow of oxidant (air) to the combustion zones and, thus, to extinguish the fire.

There are a lot of publications devoted to smoldering and another heterogeneous combustion of solid porous media under free filtration. Analytical investigations of solid porous burning under natural convection were carried out in [3] for counter-flow regime, when the gas and the combustion wave in porous object move in opposite directions, and in [4] for co-flow regime, when the gas moves in the same direction as the combustion wave. Interesting experimental results of smoldering combustion in some porous materials (cellulose, polyurethane, wood char powder) under free convection were obtained in [5–8]. In [9–14] numerical studies of smoldering combustion under free convection were carried out for various porous materials. At the same time, some important aspects of the smoldering in porous objects under natural conditions remain unstudied. In particular, the detailed gas dynamics in such porous objects has not yet been described due to the complexity of the problem – on the open borders of the simulated object not only the gas flow rate is unknown but also the directions of gas flow.

In the present work a numerical model is proposed for investigating smoldering in porous media under natural convection. The model is the modification of previously developed model [15–16] and can describe more complex conversion of solid porous fuel substance. Computational results obtained by means of the numerical model in one-dimensional case are compared with the experimental data of the smoldering combustion in polyurethane foam under free convection in the gravity field, which were described in [6].

2. Mathematical model and numerical method

Consider a motionless porous object, which has some impermeable non-heat-conducting borders and some boundaries opened to the atmosphere. In the porous medium the self-sustained wave of heterogeneous combustion (smoldering) can propagate. Cold gas can flow into the open walls of the porous object; the gas can flow through porous medium and flow out. Suppose that a solid porous substance consist of combustible component (or fuel), inert component and solid product of reaction, and at the same time the solid combustible material transforms into gaseous and solid products during the reaction with gaseous oxidizer, so we have the following expression:

$$\text{Solid fuel} + (\mu_g)\text{Oxidizer} \rightarrow (1+\mu_g-\mu_p)\text{Gaseous product} + (\mu_p)\text{Solid product}, \tag{1}$$

where μ_g and μ_p are the mass stoichiometric coefficients for oxidizer and solid product respectively.

The model is based on the assumption of interacting interpenetrating continua [17] using classical approaches of the theory of filtration combustion [18] and includes the equations of state, continuity, momentum conservation and energy for each phase (solid and gas):

$$
\begin{gathered}
\left(\rho_{cf}c_{cf}+\rho_{ci}c_{ci}+\rho_{cp}c_{cp}\right)\frac{\partial T_c}{\partial t}=-\alpha\left(T_c-T_g\right)+Q\rho_{cf0}W+\left(1-a_g\right)\lambda_c\Delta T_c,\\
\rho_g c_g\left(\frac{\partial T_g}{\partial t}+\left(\boldsymbol{v}_g\cdot\nabla\right)T_g\right)=\alpha\left(T_c-T_g\right)+a_g\lambda_g\Delta T_g,\\
\rho_g\left(1+\chi\left(1-a_g\right)\right)\left(\frac{\partial \boldsymbol{v}_g}{\partial t}+\left(\boldsymbol{v}_g\cdot\nabla\right)\boldsymbol{v}_g\right)=-a_g\nabla p+\rho_g\boldsymbol{g}-a_g^2\frac{\mu}{k_1}\boldsymbol{v}_g-\left(1-\mu_p\right)\rho_{cf0}W\boldsymbol{v}_g,\\
\rho_g\left(\frac{\partial C}{\partial t}+\left(\boldsymbol{v}_g\cdot\nabla\right)C\right)=\nabla\cdot\left(\rho_g D_g\nabla C\right)-\mu_g\rho_{cf0}W-\left(1-\mu_p\right)\rho_{cf0}WC,\\
\frac{\partial \rho_g}{\partial t}+\nabla\cdot\left(\rho_g\boldsymbol{v}_g\right)=\left(1-\mu_p\right)\rho_{cf0}W,\qquad p=\frac{\rho_g RT_g}{a_g M},\qquad \frac{\partial\eta}{\partial t}=W,
\end{gathered}
\tag{2}
$$

$$W=(1-\eta)C\rho_g k\exp\left(-\frac{E}{RT_c}\right),\qquad \rho_{cf}=(1-\eta)\rho_{cf0},\qquad \rho_{cp}=\mu_p\rho_{cf0}\eta,$$

$$a_g=a_{g0}+(a_{cf0}-a_{cpEnd})\eta,\qquad D_g=D_{g0}\left(\frac{T_g}{273}\right)^b,\qquad \mu=c_{s1}\frac{T_g^{1.5}}{c_{s2}+T_g}.$$

Here a is the volume concentration, b is the exponent in the expression for diffusion coefficient, C is the mass concentration of oxidizer, c is the specific heat capacity, c_{s1} and c_{s2} are the constants in Sutherland's formula, D_g is the diffusion coefficient of gas, E is the activation energy, $\boldsymbol{g}$ is the gravity acceleration, k is the pre-exponential factor in the expression for the rate of reaction, k_1 is the permeability coefficient, M is the molar mass of gas, p is the gas pressure, Q is the heat release of reaction, R is the universal gas constant, t is the time, T is the temperature, $\boldsymbol{v}_g$ is the gas velocity, W is the rate of chemical reaction, α is the constant determining the interphase heat transfer intensity, η is the degree of conversion of the combustible component of solid medium, λ is the thermal conductivity (including radiation heat transfer), μ is the dynamic viscosity of gas, ρ is the effective (or bulk) density (i.e., the product of phase density and volumetric concentration), χ is the coefficient, taking into account the inertial interaction of the phases in their relative motion [17], ∇ is the del (nabla) operator, Δ is the Laplace operator; subscripts: "0" denotes the initial moment, "c" denotes the condensed phase (solid medium), "End" denotes the end point, "i" denotes the inert component, "f" denotes the combustible component (fuel), "g" denotes the gas, "p" denotes the product.

A distinctive feature of the model is that the equation of momentum conservation for porous media is used, which can be considered as generalization of the classical Darcy's equation and is convenient for numerical implementation. Another feature of the model is that the dynamic viscosity of gas is temperature dependent by Sutherland's formula because the allowance for the temperature dependence of gas viscosity in its motion through a porous heat-evolutional medium can change the solution both quantitatively and qualitatively [19–20]. But the main essential feature of the model is the boundary conditions: the gas flow rate and gas velocity at the inlet to the porous object are unknown and have to be found from the solution of the problem. So, we assume that at the object boundaries, which are opened to the atmosphere, we know gas pressure and the conditions of heat exchange. Also at these boundaries we know either gas temperature and the mass concentration for the oxidizer or conditions for gas temperature and oxidizer concentration depending on the directions of the gas flow. At the impermeable boundaries of the object we use the conditions of non-heat-conductivity and impermeability. Thus the boundary conditions for system of equations (2) are as follows:

$$\begin{gathered} p\big|_{x\in G_1}=p_0(x),\qquad \lambda_c\partial T_c/\partial n\big|_{x\in G_1}=\beta\left(T_{g0}-T_c\big|_{x\in G_1}\right),\\ T_g\big|_{x\in G_1}=T_{g0}\ \text{and}\ C\big|_{x\in G_1}=C_0,\ \text{if}\ \boldsymbol{v}_g\big|_{x\in G_1}\cdot\boldsymbol{n}\big|_{x\in G_1}\le 0,\\ \partial T_g/\partial n\big|_{x\in G_1}=0\ \text{and}\ \partial C/\partial n\big|_{x\in G_1}=0,\ \text{if}\ \boldsymbol{v}_g\big|_{x\in G_1}\cdot\boldsymbol{n}\big|_{x\in G_1}>0,\\ \partial T_c/\partial n\big|_{x\in G_2}=0,\qquad \partial T_g/\partial n\big|_{x\in G_2}=0,\qquad \boldsymbol{v}_g\big|_{x\in G_2}\cdot\boldsymbol{n}\big|_{x\in G_2}=0, \end{gathered}\tag{3}$$

where G_1 is the object boundary opened to the atmosphere, G_2 is the impermeable boundary of the object, $\boldsymbol{n}$ is the outward vector directed normally to G_1 or to G_2, β is the heat removal coefficient.

It should be noted, that previously developed mathematical model, which is described in [15–16], is the special case of model (1–3) when μ_p is equal to 0. In the present work we investigate the only one-dimensional processes, as we consider the porous object with height *H*, which is bounded of impermeable non-heat-conducting side walls and is opened at the top and at the bottom, and at the same time the pressure at the object bottom and top corresponds to the atmospheric pressure at the assigned heights. So, the numerical method for proposed mathematical model (1–3) is very similar to the one used in [15–16] and described in detail in [21]. According to the method, the energy equations, momentum conservation equation and equation for oxidizer concentration are transformed into the explicit finite difference equations. The gas temperature, solid phase temperature, gas velocity and oxidizer concentration are determined from these equations. The continuity equation is transformed into the implicit finite difference equation. From this equation taking into account the perfect gas equation of state the gas pressure is determined using Thomas algorithm [22]. The effective gas density and the remaining unknown quantities are determined trivially from the perfect gas equation of state and other closure equations.

3. Comparison of calculations with experiment

Besides of gaining new knowledge, numerous calculations, which were carried out using the mathematical model (1–3) when μ_p is equal to 0, showed good qualitative coincidence between obtained computational results and characteristics of filtration combustion described in literature. However, the quantitative coincidence of the computational results with the experimental ones was not previously investigated. There are a lot of parameters in the mathematical model (1–3), some of them have a significant effect on the solution. So, for obtained precise results we should use precise parameter values. But analysis of papers with experimental data showed that the parameters specified in each paper do not allow us to accurately define all the parameter values of the mathematical model (1–3). Therefore, for defining the parameter values of the model, we should use information from different papers. But material used in different experiments on combustion may have different properties. For example, properties of different samples of peat and solid waste can be very different.

However, we assumed that the properties of the polyurethane foam used in various experiments on smoldering can be quite close. And in this way, we can compare the experimental data with the results of our calculations. To define the parameter values of polyurethane foam, we had analyzed a lot of papers and have used data mainly from [6, 9]. Kinetics parameters are obtained from [9] by reducing the two-step mechanism of reactions to the one-step mechanism. The value of the heat release of reaction incorporates the heat loss effect into the heat of combustion. So, we use the following parameters values:

$$H=0.15\,\text{m},\ \mu_g=0.536,\ \mu_p=0.1485,\ a_{g0}=0.975,\ a_{cpEnd}=a_{cf0}\cdot 0.1485,$$

$$Q=2.1\cdot 10^6\ \text{J/kg},\ k=3.92\cdot 10^{11}\ \text{1/s},\ E=1.4\cdot 10^5\ \text{J/mole},$$

$$\lambda_c=0.047/0.025=1.88\,\text{J/(m K s)},\ \lambda_g=0.034\ \text{J/(m K s)},\ \alpha=10^4\ \text{J/(m}^3\ \text{K s)},$$

$$\rho_{cf0}=a_{cf0}\cdot 1.06\cdot 10^3\,\text{kg/m}^3,\ \rho_{ci}=\left(0.025-a_{cf0}\right)\cdot 1.06\cdot 10^3\ \text{kg/m}^3,$$

$$c_{cf}=1.7\cdot 10^3\ \text{J/(kg K)},\ c_{ci}=1.7\cdot 10^3\ \text{J/(kg K)},\ c_{cp}=1.7\cdot 10^3\ \text{J/(kg K)},\ c_g=10^3\ \text{J/(kg K)},$$

$$c_{s1}=1.458\cdot 10^{-6}\ \text{kg/(m s K}^{1/2}),\ c_{s2}=110.4\,\text{K},\ R=8.31441\,\text{J/(mole K)},\ g=9.81\,\text{m/s}^2,$$

$$M=2.993\cdot 10^{-2}\ \text{kg/mole},\ D_{g0}=1.82\cdot 10^{-5}\ \text{m}^2\text{/s},\ b=1.724,\ \chi=0.5,\ \beta=20\,\text{J/(m}^2\ \text{K s)},$$

$$T_{g0}=300\,\text{K},\ C_0=0.23,\ p_0(0)=1.00001766\cdot 10^5\,\text{Pa},\ p_0(H)=10^5\,\text{Pa}.$$

Some of these parameter values are given in the above mentioned papers; some parameters do not have a strong effect on the solution hypothetically and can be specified approximately (and our calculations confirm this assumption). The main difficulties have arisen in defining the initial volume concentration of the solid combustible component (fuel) a_{cf0} and the change in permeability coefficient k_1. Both these parameters have a significant effect on the solution. We suppose, that permeability varies according to the formula $k_1 = k_{10} + (k_{1End} - k_{10})\eta$, where $k_{10} = 2.76 \cdot 10^{-9}$ m^2 [6], $k_{1End} = 5 \cdot 10^{-8}$ m^2 [6]. Taking into account the variations in heat release of polyurethane foam described in [9], we assume that a_{cf0} may be from $0.4 \cdot (1 - a_{g0}) = 0.01$ to $0.6 \cdot (1 - a_{g0}) = 0.015$.

Computational results obtained by means of the numerical model in 1D case are compared with the experimental data of the smoldering combustion in polyurethane foam under free convection in the gravity field, which were described in [6]. The comparison is carried out in the central part of the porous object, since it was revealed in [6] that smoldering in this zone is free from external effects and is the most representative. Without additional experimental measuring we cannot define very accurate heat losses at the ends of the porous object, so we cannot obtain precise numerical solutions at these zones.

According to the experiment [6], for downward propagation of the smolder wave (counter-current combustion, when the gas and the combustion wave in porous object move in opposite directions) the maximum smolder reaction temperature T_{sm} in the central zone under natural convection is from 370°C to 380°C, the smolder velocity v_{sm} is from $3 \cdot 10^{-5}$ m/s to $4 \cdot 10^{-5}$ m/s. Figure 1 depicts the maximum smolder reaction temperature and smolder velocity in the central zone of the porous object for different concentration of the solid fuel when the counter-current combustion (downward smoldering) takes place. It should be noted that graphs in this and next figures are obtained when mesh size h is equal to 0.01. As seen from the figure, the numerical solution depends weakly on the concentration of the solid combustible component and very close to the experimental data.

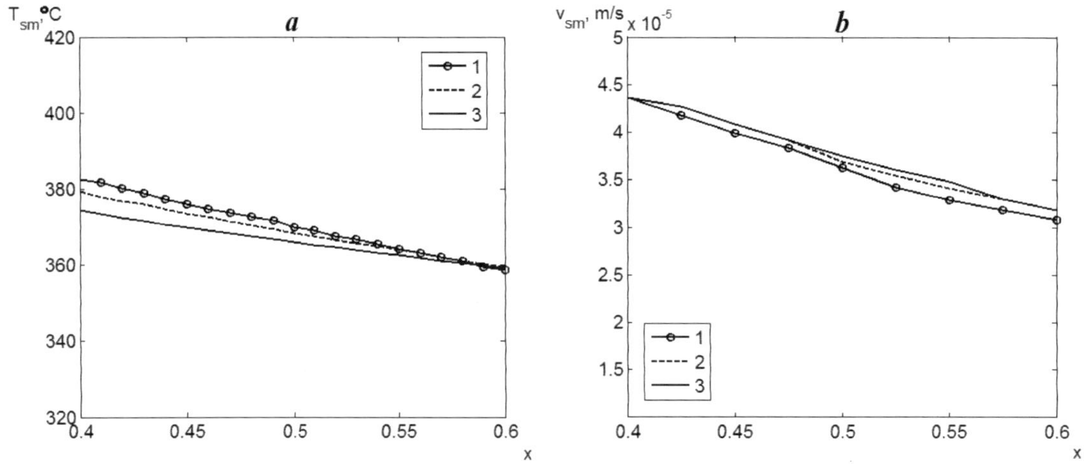

Figure 1. Maximum smolder reaction temperature T_{sm} (*a*) and smolder velocity v_{sm} (*b*) versus dimensionless distance from the bottom of the porous object x when concentration of the solid combustible component $a_{cf0} = 0.01$ (*1*), $a_{cf0} = 0.0125$ (*2*), $a_{cf0} = 0.015$ (*3*) for counter-current combustion (downward propagation of the smolder wave).

According to the experiment [6], for upward propagation of the smolder wave (co-current combustion, when the gas moves in the same direction as the combustion wave) the maximum smolder reaction temperature T_{sm} in the central zone under natural convection is from 360°C to

390°C, the smolder velocity v_{sm} is from $2.5 \cdot 10^{-5}$ m/s to $3.5 \cdot 10^{-5}$ m/s. Figure 2 shows the maximum smolder reaction temperature and smolder velocity in the central zone of the porous object for different concentration of the solid fuel when the co-current combustion (upward smoldering) takes place. As can be seen from the figure, the numerical solution depends essentially on the concentration of the solid combustible component and rather close to the experimental data. The differences between counter-current and co-current combustion obtained in the calculations may seem strange, however, they agree well with experimental observations. For example, it is said in [8] that upward smoldering is affected by many factors, and operational conditions must be rigidly controlled to achieve repeatability, while downward smoldering is stable. Our numerical calculations demonstrate the same.

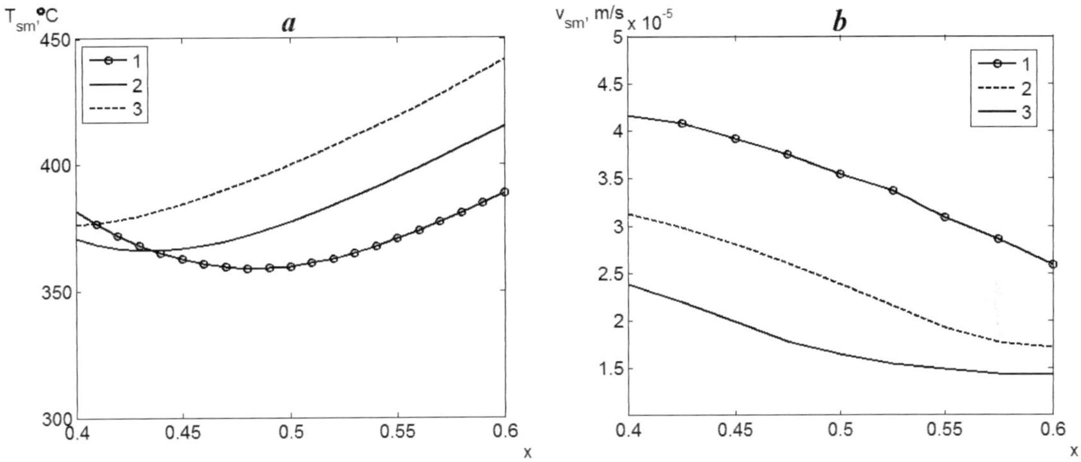

Figure 2. Maximum smolder reaction temperature T_{sm} (*a*) and smolder velocity v_{sm} (*b*) versus dimensionless distance from the bottom of the porous object x when concentration of the solid combustible component $a_{cf0} = 0.01$ (*1*), $a_{cf0} = 0.0125$ (*2*), $a_{cf0} = 0.015$ (*3*) for co-current combustion (upward propagation of the smolder wave).

The calculation accuracy of the used numerical method is demonstrated in figure 3 and figure 4, which show the results of computations with different mesh size. We can conclude that calculations even on rather a coarse grid (h = 0.025) give good results. Especially when investigating the smoldering of materials with unstable properties (like peat or solid waste), the use of fine grids may not lead to much more accurate results, since errors in defining the properties of these materials can lead to errors exceeding the accuracy of the calculations.

Thus, the proposed numerical model can provide a good quantitative agreement with experiment if the parameters of the model are well defined. Calculations of the counter-current combustion can be carried out and can give good results without a very accurate definition of some parameters, in contrast to the calculations of the co-current combustion.

Acknowledgements
The work was supported financially by the Russian Foundation for Basic Research (project 16-01-00103-a) and the Far-Eastern Branch of the Russian Academy of Sciences (project 15-I-4-021). The results were obtained using the equipment of Shared Resource Center "Far Eastern Computing Resource" IACP FEB RAS.

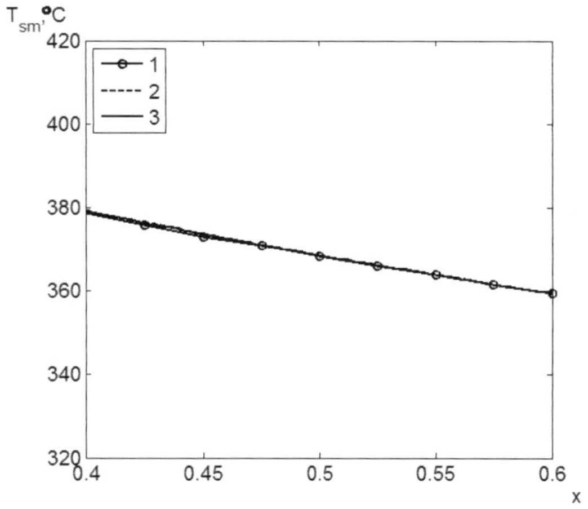

Figure 3. Maximum smolder reaction temperature T_{sm} versus dimensionless distance from the object bottom x when the concentration of the solid fuel $a_{cf0} = 0.0125$ and mesh size $h = 0.025$ (*1*), $h = 0.01$ (*2*), $h = 0.0025$ (*3*) for counter-current combustion.

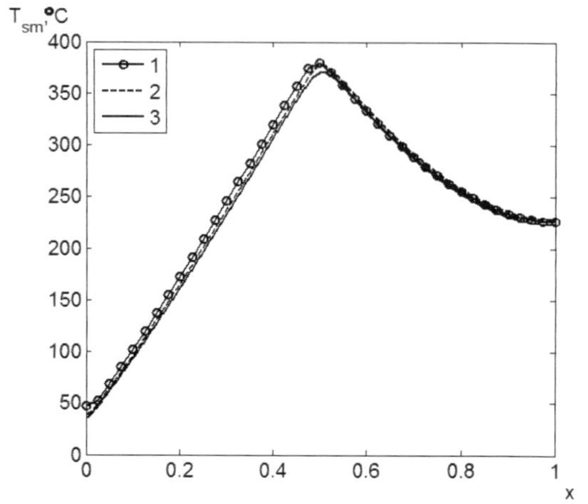

Figure 4. Temperature of the solid phase T_c versus dimensionless distance from the object bottom x when the concentration of the solid fuel $a_{cf0} = 0.0125$ and mesh size $h = 0.025$ (*1*), $h = 0.01$ (*2*), $h = 0.0025$ (*3*) for co-current combustion.

References

[1] Ohlemiller T J 1985 *Prog. Energy Combust. Sci.* **11** 277–310
[2] Rein G 2015 Smoldering-Peat Megafires: The Largest Fires on Earth *Coal and Peat Fires: a Global Perspective. Volume 4: Peat - Geology, Combustion, and Case Studies* ed G B Stracher, A Prakash and G Rein (Boston: Elsevier) pp 1–11
[3] Aldushin A P, Matkowsky B J and Schult D A 1996 *Combust. Flame* **107** 151–75
[4] Aldushin A P, Matkowsky B J and Schult D A 1997 *J. Eng. Math.* **31** 205–34
[5] Dosanjh S, Peterson J, Fernandez-Pello A C and Pagni P J 1986 *Acta Astronaut.* **13** 689–96
[6] Torero J L, Fernandez-Pello A C and Kitano M 1993 *Combust. Sci. and Tech.* **91** 95–117
[7] Torero J L and Fernandez-Pello A C 1995 *Fire Saf. J.* **24** 35–52
[8] He F and Behrendt F 2011 *Fire Saf. J.* **46** 406–13
[9] Ohlemiller T J, Bellan J and Rogers F 1979 *Combust. Flame* **36** 197–215
[10] Di Blasi C 1995 *Combust. Sci. and Tech.* **106** 103–24
[11] Rein G, Fernandez-Pello A C and Urban D L 2007 *Proc. Combust. Inst.* **31** 2677–84
[12] Huang X, Rein G and Chen H 2015 *Proc. Combust. Inst.* **35** 2673–81
[13] Huang X and Rein G 2015 *Int. J. Wildland Fire* **24** 798–808
[14] Yang J, Chen H and Liu N 2016 *Energy Fuels* **30** 8765–75
[15] Lutsenko N A 2013 *Proc. Combust. Inst.* **34** 2289–94
[16] Lutsenko N A and Levin V A 2014 *Combust. Sci. Technol.* **186** 1410–21
[17] Nigmatulin R I 1978 *Principles of the Mechanics of Heterogeneous Media* (Moscow: Nauka Publishers) (in Russian)
[18] Aldushin A P and Merzhanov A G 1988 The theory of filtration combustion: concept and state of research *Heat Wave Propagation in Heterogeneous Media* ed Y S Matros (Novosibirsk: Nauka Publishers) pp 9–52 (in Russian)
[19] Lutsenko N A and Levin V A 2006 *J. Eng. Phys. Thermophys.* **79** 33–9
[20] Levin V A and Lutsenko N A 2015 *Doklady Physics* **60** 255–8
[21] Lutsenko N A and Sorokova S N 2014 *Advanced Materials Research* **880** 115–20
[22] Tannehill J C, Anderson D A and Pletcher R H 1997 *Computational Fluid Mechanics and Heat Transfer, 2nd ed* (Washington, DC: Taylor&Francis)

IOP Conf. Series: Journal of Physics: Conf. Series **894** (2017) 012055 doi:10.1088/1742-6596/894/1/012055

Modeling of the kinetics of diffusion processes in the formation of a junction glass-steel

O N Lyubimova, A V Morkovin and V V Andreev

Far Eastern Federal University, Vladivostok, Russia

E-mail: berms@mail.ru

Abstract. Basic types of structure of the contact zone of glass and steel as a result of long-term temperature processing, and mathematical models describing diffusion processes are explored in this article. The diffusion models of iron ions proposed in this paper agree with the experimental results. The effectiveness of the application of the diffusion coefficient in the Stokes-Einstein approximation is proved.

1. Relevance of research

Technological features of manufacturing of structural composites in which the metal layers alternate with the layers of glass imply the reduction of the glass layer by the metal shells, that additionally protect glass and prevent the formation of microdefects on the glass, and increase the static and dynamic strength of the glass layer and composite as a whole 1,2. The temperature regime for the production of a glass-metal composite has special features. For example, under different technological conditions, the quality of glass and metal contact zone can vary from a "taut" connection to a vacuum-tight junction with a developed diffusion zone at the contact boundary. The modeling of concentration profiles in the diffusion layer will allow to predict the development of its structure and geometry and optimize the technological parameters of production. Most of studies of the diffusion of iron in glass are difficult to compare with the problem considered in this paper, with the exception of [2]–[6] works. A huge amount of work has been devoted to the mathematical modeling of diffusion processes, however, for each technological task, the problem of selecting the appropriate model and its correction remains topical in optimizing experimental studies. The purpose of this article is mathematical modeling of diffusion processes in the junction of glass and steel.

2. Research objectives

The obtaining of a glass-steel connection includes the formation of a complete physical contact and diffusion processes coming from the dissolution of the oxide layer on the steel in the glass and diffusion of iron in the glass. During the experimental studies, it was noticed that the reason and direct participant of the diffusion processes is the oxide layer on the steel surface [2]. Preliminary chemical treatment of the steel surface during further heating allows forming a wustite - FeO layer on its surface, and then the technology divides into two main directions: further joining without using solder and with solder based on sodium tetraborate ($Na_2B_4O_7 \times 10H_2O$). There are two main types of the bonding zone. The main feature of the zone (I) is the presence of two

Content from this work may be used under the terms of the Creative Commons Attribution 3.0 licence. Any further distribution of this work must maintain attribution to the author(s) and the title of the work, journal citation and DOI.
Published under licence by IOP Publishing Ltd

crystalline and amorphous phases different in concentration and structure: a small thickness (up to 5 micrometers) of crystalline phase with a high iron content, possibly this is a layer of fayalite $2FeO \cdot SiO_2$ and iron metasilicate $FeO \cdot SiO_2$, and a longer (up to 92 microns) amorphous phase with a lower iron content (Figure 1a). In case (II) only the amorphous phase (up to 252 μm) is observed, in this phase the iron concentration decreases uniformly in the direction from the joint zone (Figure 1 b).

In some samples with the zone (I), conglomerates of insoluble iron oxides (most likely hematite Fe_2O_3) are seen (Figure 1a). All samples have a zone supersaturated with iron, which crystallizes upon further cooling. After the analytical processing of the results of spectral analysis this zone contains up to 40% of iron (point 3, Figure 1a), while the iron content of the nearby zone does not exceed 20% (point 4, Figure 1a). Samples with the zone (II) do not have similar crystalline phase, which is the result of preliminary treatment of the surface of steel with sodium tetraborate, which prevents the oxidation of the steel surface upon heating and accelerates the solubility of wustite in glass.

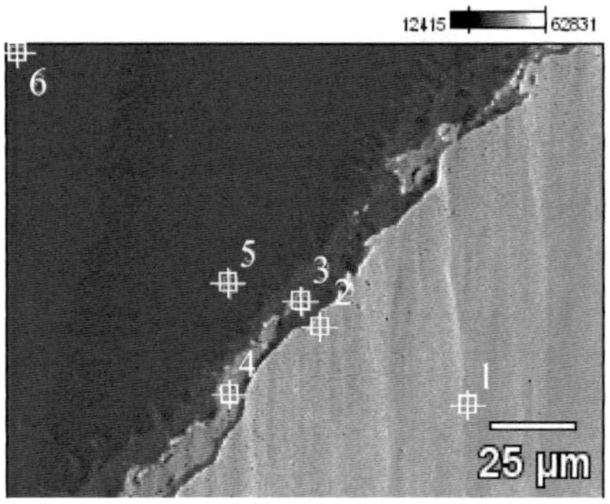

a)

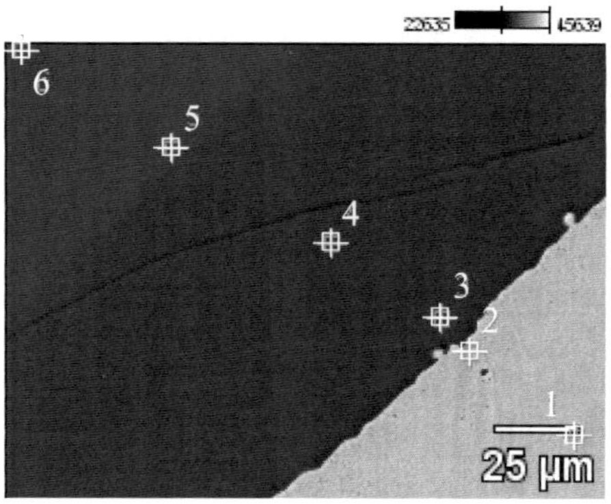

b)

Figure 1. Microstructure of two principal types of connection zone: (a) -I and (b) -II (Hitachi S-3400N), soldering temperature 800 ^{o}C, dwell time 100 minutes.

Analytical processing of intensity profiles with linear scanning (Figure 2) allows to distinguish typical dependencies for different methods of surface treatment: type I-cold chemical treatment and type II- cold chemical treatment with application of sodium teraborate, to note their similarity and differences. Despite the similar nature of the concentration profiles and the mechanism of iron diffusion, the mathematical description of types I and II will be different. Samples of the I type has step-like zone due to the presence of an initially wider oxide layer than in type II, as a consequence, the diffusion mobility of iron ions in this layer is limited and, with insufficient dwell time, forms a saturated layer, which subsequently crystallizes upon slow cooling.

3. Mathematical modeling

In general, the diffusion process at the glass and steel connection zone in a glass-metall composite resembles the creation of p-n junctions with two isolated stages: by doping for a short time into the surface layer of the semiconductor with the formation of a thin surface layer enriched with dopants and distillation-the distribution of the dopants along the thickness of the semiconductor during long time (sometimes dozens of hours) [7]. The use of additional treatment with sodium tetraborate prevents the growth of the oxide on the steel and its phase change, the saturation of the glass melt occurs from a thin layer of wustite dissolving first in the solder and then

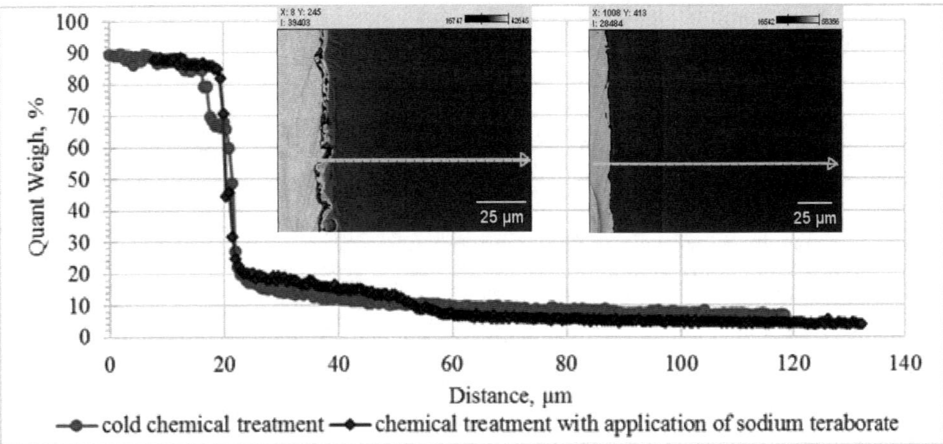

Figure 2. Analytical data of spectral analysis and schemes of linear scanning (Hitachi S-3400N).

diffusing into the glass melt. The iron diffusion model for type II can be described by models, accepted in the theory of semiconductors, for example, as diffusion from an infinitely thin layer into a semibounded body with a reflecting boundary, which corresponds to the condition of the presence of an excess concentration of iron in a thin surface layer, which is redistributed throughout the volume, decreasing on the surface and continuously increasing in volume [8], the initial and boundary conditions of the given problem have the form

$$t = 0 : \delta \to 0, C = C_0 \quad \text{for} \quad 0 \leq x \leq \delta \qquad \text{and} \qquad C = 0 \quad \text{for} \quad \delta < x < \infty, \tag{1}$$

$$x = 0 : t > 0, \frac{\partial C}{\partial x} = 0, \tag{2}$$

where C is the mass concentration of iron, C_0 is the concentration of diffusing iron in the oxide layer at the initial time, δ is the thickness of the oxide layer, D is the diffusion coefficient of iron in the glass melt. If the diffusion coefficient does not depend on x ($\frac{\partial D}{\partial x} = 0$), then equation

$$\frac{\partial C}{\partial t} = \frac{\partial}{\partial x}\left(D(x)\frac{\partial C}{\partial x}\right), \quad 0 < x < \infty, \quad t > 0, \tag{3}$$

taking into account the boundary conditions (1) and (2), has an analytic solution

$$C(x,t) = \frac{q}{\sqrt{\pi D t}} e^{-\frac{x^2}{Dt}}, \tag{4}$$

where $q = C_0\delta$ is the amount of substance concentrated at the initial time in a layer of thickness $\delta \to 0$. Formula (4) can be used to estimate the diffusion coefficient D from the available experimental data from the solution of the nonlinear equation

$$\frac{C(x_{exp}, t_{exp})}{q}\sqrt{\pi D t_{exp}} = e^{-\frac{x_{exp}^2}{D t_{exp}}}, \tag{5}$$

where $t_{exp}, x_{exp}, C(x_{exp}, t_{exp})$ and q are experimentally defined parameters.

The diffusion process for the I type zone can be described by the diffusion of iron oxide into a viscous glass melt, from the boundary δ, which characterizes the motion of a viscous melt along the oxide layer. Then the mathematical model represents a boundary value problem that includes the diffusion equations in two layers:

$$\frac{\partial C}{\partial t} = D_1 \frac{\partial^2 C}{\partial x^2}, \quad 0 < x < \delta(t), \quad t > 0, \tag{6}$$

$$\frac{\partial C}{\partial t} = \frac{\partial}{\partial x}\left(D_2 \frac{\partial C}{\partial x}\right), \quad \delta(t) < x < \infty, \quad t > 0, \tag{7}$$

with initial and boundary conditions:

$$t = 0 : \delta(0) = \delta_0, C = C_0, \quad \text{for} \quad 0 \le x \le \delta_0 \qquad \text{and} \qquad C = 0 \quad \text{for} \quad \delta_0 < x < \infty, \tag{8}$$

$$x = 0 : \quad t > 0, \quad \frac{\partial C}{\partial x} = 0, \tag{9}$$

$$x = \delta(t) : \quad t > 0, \quad D_1 \frac{\partial C}{\partial x} \mid_- = D_2 \frac{\partial C}{\partial x} \mid_+ \quad and \quad C \mid_- = KC \mid_+, \tag{10}$$

$$x \to \infty : \quad t > 0, C \to 0, \tag{11}$$

here D_1 and D_2 are the diffusion coefficients in the oxide layer and the glass, respectively, K is the solubility constant (determined experimentally), and $\delta(t)$ is the moving boundary of the viscous glass melt.

The diffusion process involves the dissolution of the oxide layer in the glass melt and its diffusion deep into the glass, the diffusion coefficients will have different approximations: in the oxide layer D_1, the mutual diffusion coefficient will be assumed constant until some time at which the oxide layer completely "dopes" into the glass melt, and D_2 depends on the viscosity of the glass and related with the Stokes-Einstein relation [4]

$$D_2 = \frac{k_B T}{6\pi r \eta(C)}, \tag{12}$$

where k_B is the Boltzmann constant, T is the temperature, r is the radius of the particle, $\eta(C)$ is the viscosity of the glass, which depends on the glass composition and temperature.

To determine $\delta(t)$ it is proposed to consider a mechanical model of the motion of the boundary of a viscous glass melt in the oxide layer, by analogy with the model for determining the time of formation of physical contact in a cylindrical junction of glass with a metal proposed in work [9]. It is necessary to solve the problem of Reynolds on the deformation of a viscous (glass) layer enclosed between two cylindrical plates, one of plates is fixed, force Q acts on the other plate.

Using the axial symmetry, the neglect of the acceleration of the particles in the viscous layer, the thickness of the metal plates, and the condition of constant motion of the upper plate make possible to rewrite the Navier–Stokes equations and the continuity condition as

$$\eta \frac{\partial^2 \nu_r}{\partial z^2} = \frac{\partial p}{\partial r}, \quad \frac{\partial p}{\partial z} = 0,$$

$$\frac{1}{r}\frac{\partial (r\nu_r)}{\partial r} + \frac{\partial \nu_z}{\partial z} = 0.$$

With boundary conditions

$$z = 0 : \quad \nu_z = 0, \quad \nu_r = 0,$$

$$z = \delta : \quad \nu_z = \frac{\partial h}{\partial t} = \nu(t), \quad \nu_r = 0,$$

$$r = R: \quad P = 0,$$

where η is the coefficient of dynamic viscosity, r and z are the radial and axial coordinates of the layer particle, respectively, ν_r is the velocity of the viscous layer in the radial direction; P is the pressure in the layer; ν_z is the velocity of the viscous layer in the axial direction; δ is the current thickness of the layer; $\nu(t)$ is the velocity of the upper cylinder in the axial direction, R is the radius of the metal plates. The change in the thickness of a viscous layer is due to a change in its radius as $\delta = \frac{R^2\delta_0}{r^2}$. The solution of the boundary value problem leads to a nonlinear integral equation

$$\int_{\delta_0}^{\delta} \frac{\eta(\tilde{\delta})}{\tilde{\delta}^3} d\tilde{\delta} = \frac{2Qt}{3\pi R^4}. \qquad (13)$$

The dynamic viscosity depends on the concentration, i.e. from the thickness of the saturation of the oxide layer by the melt. Depending on the analytical task η, the equation (13) can be solved analytically in the case of an integrand integrable in elementary functions, or numerically.

4. Discussion

The results of determining the experimental diffusion coefficients from formula (5) are shown in Figure 3 on the right. In Figure 3 on the left there are the experimental concentration profiles of iron for different technological regimes for samples of the II type, soldering temperature 850 ^{o}C.

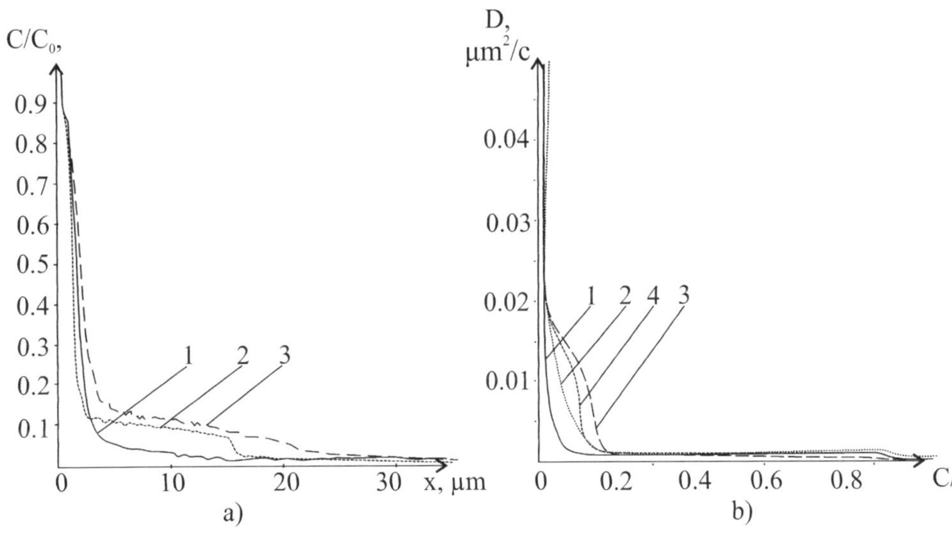

Figure 3. Experimental curves: a) the concentration distribution of iron for different dwell time: 1 - t=30 minutes, 2 - t=60 minutes, 3 - t=100 minutes; b) the diffusion coefficient: 1, 2, 3 - the experimental data by formula (5) correspondingly with experimental curves of concentration 1, 2 and 3 (a), 4- diffusion coefficient with approximation of the dependence of viscosity on the concentration of iron oxides by formula (12).

The numerical solution of problems (1)–(3), (6)–(13) has no special features, in this paper the numerical implementation was carried out using the implicit difference scheme, the sweep method and the quasilinear approximation of the diffusion coefficients from the time step [10]. There are a few studies about iron-containing glasses, in the calculations we use the approximation of the dependence of viscosity on the concentration of iron oxides proposed in work [4]. Comparing the diffusion coefficients calculated by formula (5) with the aid of experimental iron concentration profiles and using of formula (12) where the dependence of the viscosity of glass depends on the concentration of iron in it [4] for samples of type II (Figure 3b), it can be stated that the diffusion coefficient calculated from the Stokes–Einstein relation is not only qualitatively, but also quantitatively close to the experimental curves, so the usage of it for calculating the iron concentration in samples of type II is justified. Figure 4 presents the results of numerical calculations of the concentration profiles of iron from the models: (1)–(3) for type I with the approximation of the diffusion coefficients according to the formula (5) (Figure 4a) and (6) - (13) (Figure 4b).

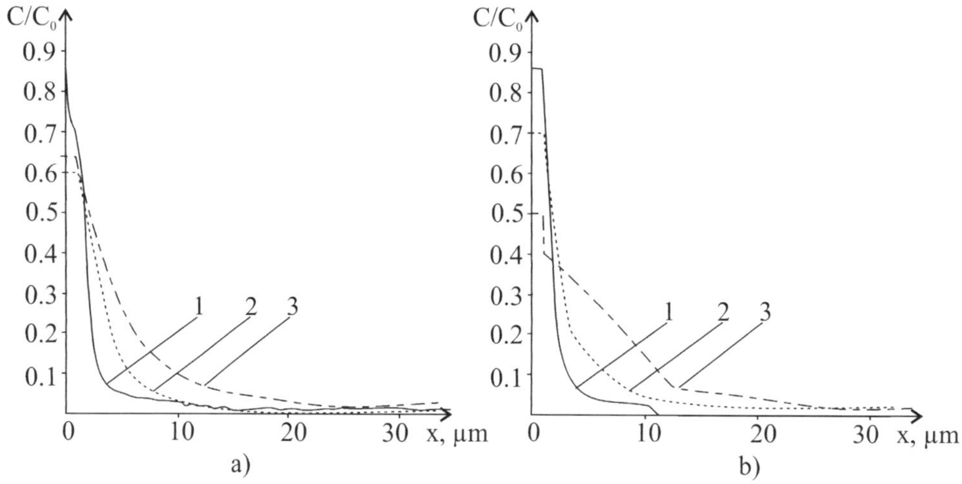

Figure 4. Concentration distribution of iron in samples: a) type II: 1- D=const [3], 2 - D by formula (12), 3 - D by formila (5); b) type I, D by formula (12) for different dweld time: 1 - t=40 minutes, 2 - t=60 minutes, 3 - t=100 minutes.

5. Conclusion

It is noted that the reason and direct participant of the diffusion processes in the welding of glass and steel is the oxide layer on the steel surface. The kinetics of the diffusion zone formation significantly depends of its characteristics and dwell time at the welding temperature. Depending on the technological regime, there are two main types of the connection zone structure. The results of calculations for the proposed models for each type are consistent with the experimental data. The model (6)–(13) is interesting and new. This model proposes a mechanical description of the motion of the boundaries of a viscous melt into an oxide layer. It is shown that the approximation of the diffusion coefficient by the Stokes–Einstein relation agrees with the experimental data, it isbpossible to state the relevance of studies about the dependence of the viscosity of glass on the concentration of metals in it for describing the diffusion processes in glasses.

References

[1] Pikul' V V 1999 Prospects for creating a layered composite based on glass materials *Prospective materials* 34–42
[2] Lyubimova O N, Lyubimov E V, Solonenko E P, Morkovin A V and Dryuk S A 2016 Research of structure, mechanical and operation properties of glass-metal composites. AIP Conf. Proc. **1783** 020141
[3] Makarova N O 1991 Coefficients of diffusion of nickel, cobalt and iron ions in borosilicate melts (Dissertation of Candidate of chemical sciences. Sverdlovsk, 143 p)

[4] Tatarintseva et al O S 2007 Effect of iron oxides on the viscosity and wetting capacity of silicate melts *The herald of Polzunov* 144–9
[5] Lyubimova O N, Morkovin A V and Dryuk S A 2017 The characteristics of the structure of the glass-steel junction zone in the technology of obtaining glass-metal composite *Materials Science* 3–7
[6] Gridasova E A 2013 *Increase of strength properties of glass as a result of metallization by the method of diffusion welding* (Dissertation of Candidate of technical sciences. Komsomolsk-on-Amur State technical university, Komsomolsk-on-Amur, 128 p)
[7] Raskin A A and Prokof'eva V K 2010 *Technology of Materials of Micro-, Opto- and Nanoelectronics: Educational Book. Part 1* (Moscow: BINOM. Knowledge lab) (in Russian)
[8] Bekman I N 2016 *Diffusion Mathematics* (Moscow: Publishing House "OntoPrint") (in Russian)
[9] Lyubimova O N and Dryuk S A 2017 Modeling the parameters of the temperature regime in the producing of a glass-metal composite *Thermophysics and Aeromechanics* **24** 127–135
[10] Samarskii A A and Nikolaev E S 1978 *Methods for Solving Grid Equations* (Moscow: Science) (in Russian)

Asymptotic behavior of solutions to perturbed superstable wave equations

I Y Kmit[1,2] **and N A Lyulko**[3,4]

[1] Department of Mathematics of Humboldt University of Berlin, Berlin, Germany
[2] Institute for Applied Problems of Mechanics and Mathematics of Ukrainian National Academy of Sciences, Kiev, Ukraine
[3] Sobolev Institute of Mathematics of Russian Academy of Sciences, Novosibirsk, Russia
[4] Novosibirsk State University, Novosibirsk, Russia

E-mail: kmit@mathematik.hu-berlin.de, natlyl@mail.ru

Abstract. The paper deals with initial-boundary value problems for the linear wave equation whose solutions stabilize to zero in a finite time. We prove that problems in this class remain exponentially stable in L^2 as well as in C^2 under small bounded perturbations of the wave operator. To show this for C^2, we prove a smoothing result implying that the solutions to the perturbed problems become eventually C^2-smooth for any $H^1 \times L^2$-initial data.

1. Motivation and our results

A linear system

$$\frac{d}{dt}x(t) = A(t)x(t), \quad x(t) \in X \quad (0 \le t \le \infty), \tag{1}$$

on a Banach space X is called *exponentially stable* if there exist positive reals γ and $M = M(\gamma)$ such that every solution $x(t)$ satisfies the estimate

$$\|x(t)\| \le Me^{-\gamma t}\|x(0)\|, \quad t \ge 0, \tag{2}$$

where $\|\cdot\|$ denotes the norm in X.

The papers [2, 3] address a stronger property of exponentially stable systems, known as superstability. They consider the Cauchy problem for the *autonomous* version of (1). Moreover, $A : X \to X$ is supposed to be the infinitesimal generator of a strongly continuous semigroup $T(t)$; see [4, 13]. A semigroup $T(t)$ is called *superstable* [2] if its stability index is $-\infty$, that is

$$\lim_{t\to\infty} \frac{\log \|T(t)\|}{t} = -\infty.$$

The superstability property implies that the system is exponentially stable and, moreover, the estimate (2) holds for every $\gamma > 0$.

The notion of superstability can be appropriately extended to second-order equations, studied in this paper.

An important subclass of superstable systems consists of the systems whose solutions stabilize to zero after some time. The time of the stabilization is called a *finite time extinction*. The

Content from this work may be used under the terms of the Creative Commons Attribution 3.0 licence. Any further distribution of this work must maintain attribution to the author(s) and the title of the work, journal citation and DOI.
Published under licence by IOP Publishing Ltd

following simple example shows which type of boundary conditions cause the superstability property for the wave equation in the one-dimensional case, see [2, 14]:

$$w_{tt} - a^2 w_{xx} = 0, \quad (x,t) \in \Pi, \tag{3}$$

$$\begin{aligned} &w(0,t) = 0, \quad t \in (0,\infty),\\ &(w_t + aw_x)(1,t) = 0, \quad t \in (0,\infty), \end{aligned} \tag{4}$$

$$\begin{aligned} &w(x,0) = w_0(x), \quad x \in [0,1],\\ &w_t(x,0) = w_1(x), \quad x \in [0,1], \end{aligned} \tag{5}$$

where a is arbitrary positive constant and $\Pi = (0,1) \times (0,\infty)$. Constructing solutions by the method of characteristics shows that all solutions stabilize to zero for $t > 2/a$. A class of boundary conditions ensuring the superstability property for the wave operator in the multidimensional case is described in [11]. Such boundary conditions naturally appear, for example, in the scattering theory [8, 9] and in the control theory [12].

A comprehensive review of the available results on asymptotic behavior of solutions to linear and quasi-linear first order hyperbolic problems can be found in [1, 7]. The present paper concerns superstable initial-boundary value problems for the one-dimensional wave equation. Specifically, we consider the problem (3), (5) with the boundary conditions either

$$\begin{aligned} &w(0,t) = p(w_t + aw_x)(0,t), \quad t \in (0,\infty),\\ &(w_t + aw_x)(1,t) = 0, \quad t \in (0,\infty), \end{aligned}$$

or

$$\begin{aligned} &w(1,t) = q(w_t - aw_x)(1,t), \quad t \in (0,\infty),\\ &(w_t - aw_x)(0,t) = 0, \quad t \in (0,\infty), \end{aligned}$$

where p and q are arbitrary constants. Note that the finite time extinction for the problems under consideration equals $2/a$ (see Section 3.1). In this paper we will investigate the problem (3), (6), (5) (the same result is true for the problem (3), (6), (5) as well).

Along with the equation (3) we will consider its perturbed version, namely

$$w_{tt} - a^2 w_{xx} + c(x,t)w = 0, \quad (x,t) \in \Pi, \tag{6}$$

where c is a two times continuously differentiable function such that c itself and its first order and second order derivatives are bounded on $\overline{\Pi}$.

Given $T > 0$, write $\Pi_T = (0,1) \times (0,T)$. By $H^1(0,1)$ we denote the space of functions $u : (0,1) \to R$ such that $u \in L^2(0,1)$ and its weak derivative $u' \in L^2(0,1)$. From [7] it follows that for any initial functions $w_0 \in H^1(0,1)$ and $w_1 \in L^2(0,1)$ the problem (6), (6), (5) has a unique L^2-generalized solution $w \in C([0,\infty), L^2(0,1))$ (see Section 2 for the definition). For this solution we prove the following perturbation theorem.

Theorem 1. Let $w_0 \in H^1(0,1)$ and $w_1 \in L^2(0,1)$. Then for any $\gamma > 0$ there exist $\varepsilon > 0$ and $M = M(\gamma)$ such that, whenever $\sup_{(x,t)\in\Pi} |c| < \varepsilon$, the L^2-generalized solution $w(x,t)$ to the problem (6), (6), (5) fulfills the bound

$$\|w(\cdot,t)\|_{L^2(0,1)} \le Me^{-\gamma t} \max(\|w_0\|_{H^1(0,1)}, \|w_1\|_{L^2(0,1)}), \quad t > 0. \tag{7}$$

This means that sufficiently small bounded perturbations of the zero order part of the wave equation (3) presurve the exponential stability in $L^2(0,1)$ of the unperturbed problem (3), (6), (5).

Definition 2. The problem (6), (6), (5) is called smoothing from $L^2(0,1)$ to $C^2([0,1])$ if there is $T > 0$ such that, given $w_0 \in H^1(0,1)$ and $w_1 \in L^2(0,1)$, the L^2-generalized solution $w(x,t)$ to this problem belongs to $C^2([0,1])$ whenever $t \ge T$.

We are prepared to formulate the following smoothing result.

Theorem 3. The problem (6), (6), (5) is smoothing from $L^2(0,1)$ to $C^2([0,1])$ with $T = 6/a$. Moreover, given $w_0 \in H^1(0,1)$ and $w_1 \in L^2(0,1)$, the L^2-generalized solution $w(x,t)$ fulfills the bound

$$\|\partial_{x,t}^{\alpha,\beta} w(\cdot,t)\|_{C([0,1])} \le M_1 e^{\omega t} \max(\|w_0\|_{H^1(0,1)}, \|w_1\|_{L^2(0,1)}), \qquad t \ge T, \tag{8}$$

where $\alpha + \beta \le 2$ and M_1, ω are constants not depending on t, w_0, and w_1.

Corollary 4. Let $w_0 \in H^1(0,1)$ and $w_1 \in L^2(0,1)$. Then for any $\gamma > 0$ there exist $\varepsilon > 0$ and $M_2 = M_2(\gamma) \ge M$ such that, whenever $\sup_{T>0}(\|c\|_{C^2(\overline{\Pi_T})}) < \varepsilon$, the L^2-generalized solution $w(x,t)$ to the problem (6), (6), (5) fulfills the bound

$$\|\partial_{x,t}^{\alpha,\beta} w(\cdot,t)\|_{C([0,1])} \le M_2 e^{-\gamma t} \max(\|w_0\|_{H^1(0,1)}, \|w_1\|_{L^2(0,1)}), \qquad t > T, \tag{9}$$

where $T = 6/a$, $\alpha + \beta \le 2$, and M_2, γ are constants not depending on t, w_0, and w_1.

2. Reduction to first order systems and L^2-generalized solutions

For proving Theorems 1 and 3, we will apply our results from [7], where we study the exponential stability of the solutions to the initial-boundary value problems for the general first order nonautonomous hyperbolic systems, that are small bounded perturbations of the corresponding superstable systems. To this end, let us introduce new variable $u = \partial_t w + a\partial_x w$ and rewrite the perturbed problem as the following 2×2-first order system:

$$\partial_t w + a\partial_x w - u = 0, \qquad \partial_t u - a\partial_x u + c(x,t)w = 0, \tag{10}$$

$$w(0,t) = pu(0,t), \qquad u(1,t) = 0, \tag{11}$$

$$w(x,0) = \varphi_1(x), \quad u(x,0) = \varphi_2(x), \tag{12}$$

where

$$\varphi_1(x) = w_0(x), \quad \varphi_2(x) = w_1(x) + a\partial_x w_0(x).$$

It is evident that the C^2-function w is a classical solution to the problem (6), (6), (5) if and only if the C^2-vector function $(w,u) = (w, \partial_t w + a\partial_x w)$ is a classical solution to the problem (10)–(12).

As it follows from [5], for $w_0 \in C^3[0,1]$ and $w_1 \in C^2[0,1]$ satisfying the zero order, the first order and the second order compatibility conditions between (11) and (12) the problem (10)–(12) has a unique classical solution $(w,u) \in C^2([0,1]) \times C^2([0,1])$, where the first component w is a classical solution to the problem (6), (6), (5). Note that then the compatibility conditions up to the second order between (6) and (5), see e.g. [10], follow immediately from the equality $u = \partial_t w + a\partial_x w$.

Let us introduce the notion of an L^2-generalized solution. Fix arbitrary $w_0 \in H^1(0,1)$ and $w_1 \in L^2(0,1)$ and sequences $\varphi_1^l \in C_0^\infty([0,1])^n$, $\varphi_2^l \in C_0^\infty([0,1])^n$ such that $\varphi_i^l \to \varphi_i$ in $L^2(0,1)$, $i = 1,2$. Note that, due to the fact that φ_i^l are compactly supported for all $l \in N$, they satisfy the compatibility conditions up to the second order between (11) and (12). By [5], given $l \in N$, the problem (10)–(12) has a unique classical solution, say (w^l, u^l), belonging to $C^2([0,1]) \times C^2([0,1])$. In [7] it is proved that these solutions satisfy the following estimate:

$$\max(\|w^l(\cdot,t)\|_{L^2(0,1)}, \|u^l(\cdot,t)\|_{L^2(0,1)}) \le M_3 e^{At} \max_{i=1,2}(\|\varphi_i\|_{L^2(0,1)}), \qquad \text{for all } t > 0, \tag{13}$$

for some constants M_3, A, not depending on t, φ_1, φ_2, and $l \in N$. Hence, there exist unique functions $w, u \in C([0,\infty), L^2(0,1))$ such that

$$\|w(\cdot,\theta) - w^l(\cdot,\theta)\|_{L^2(0,1)} \to 0, \ \|u(\cdot,\theta) - u^l(\cdot,\theta)\|_{L^2(0,1)} \to 0 \quad \text{as } l \to \infty,$$

IOP Conf. Series: Journal of Physics: Conf. Series **894** (2017) 012056 doi:10.1088/1742-6596/894/1/012056

uniformly in θ varying in the range $0 \le \theta \le t$, for every $t > 0$. The vector-function (w, u) is called an L^2-*generalized solution* to the problem (10)–(12), while the function w is called an L^2-*generalized solution* to the problem (6), (6), (5). Furthermore, the following estimate is true:

$$\|w(\cdot,t)\|_{L^2(0,1)} \le M_3 e^{At} \max(\|w_0\|_{H^1(0,1)}, \|w_1\|_{L^2(0,1)}) \quad \text{for all } t > 0, \tag{14}$$

what follows from (13).

3. Proofs of the main results

3.1. Proof of Theorem 1

Consider the problem (10)–(12). Due to the method of characteristics, the classical solution to this problem fulfills the following system of integral equations:

$$w(x,t) = \int_{t-x/a}^{t} u(a(\tau - t) + x, \tau)\, d\tau + pu\,(0, t - x/a)\,, \quad t > \frac{x}{a}, \tag{15}$$

$$u(x,t) = -\int_{t+(x-1)/a}^{t} [cw](x + a(t-\tau), \tau)\, d\tau, \quad t + \frac{x}{a} > \frac{1}{a}. \tag{16}$$

Note that if $c = 0$ then the finite time extinction for this problem equals $2/a$.

Letting $t > 4/a$ and substituting (16) into (15), we get an integral equation for w, namely

$$w(x,t) = -\int_{t-x/a}^{t}\int_{2\tau-t+(x-1)/a}^{\tau} [cw](a(\tau-\xi)+x,\xi)\, d\xi d\tau - p\int_{t-(x+1)/a}^{t-x/a} [cw](a(t-\tau)-x,\tau)\, d\tau.$$

Set $W(t) = \|w(\cdot,t)\|_{L^2(0,1)}$. Then the last equality entails

$$W(t) \le K \sup_{(x,t)\in\Pi} |c(x,t)| \int_{t-\frac{4}{a}}^{t} W(\tau)\, d\tau, \quad t > \frac{4}{a}, \tag{17}$$

the constant K being independent on t, w but only on p. On the other side, the estimate (14) causes the bound

$$W(t) \le K_1 \max(\|w_0\|_{H^1(0,1)}, \|w_1\|_{L^2(0,1)}), \quad 0 \le t \le \frac{4}{a}, \tag{18}$$

with the constant $K_1 = M_3 e^{\frac{4A}{a}}$. On the account of (17), (18) and [7, Lemma 5.1], for any $\gamma > 0$ there exist $\varepsilon > 0$ and $M = M(\gamma)$ such that, whenever $\sup_{(x,t)\in\Pi} |c(x,t)| < \varepsilon$, for the classical solution w to the perturbed problem (6), (6), (5) the following bound is true:

$$W(t) \le M e^{-\gamma t} \max(\|w_0\|_{H^1(0,1)}, \|w_1\|_{L^2(0,1)}), \quad t > 0.$$

Finally, the statement of the theorem follows from the definition of an L^2-*generalized solution* to the problem (6), (6), (5).

3.2. Proof of Theorem 3

First consider the problem (10)–(12) where the system (10) is replaced by the corresponding decoupled system, namely

$$\begin{aligned} &\partial_t w + a\partial_x w = 0, \quad \partial_t u - a\partial_x u = 0, \\ &w(0,t) = pu(0,t), \quad u(1,t) = 0, \\ &w(x,0) = \varphi_1(x), \quad u(x,0) = \varphi_2(x). \end{aligned} \tag{19}$$

IOP Conf. Series: Journal of Physics: Conf. Series **894** (2017) 012056 doi:10.1088/1742-6596/894/1/012056

This system is easily checked to be superstable. Indeed, for the classical solution (w, u) to (19) the method of characteristics gives the formulas

$$w(x,t) = w(0, t - \frac{x}{a}) = pu\left(0, t - \frac{x}{a}\right), \quad t > \frac{x}{a},$$

$$u(x,t) = u\left(1, t + \frac{x-1}{a}\right), \quad t + \frac{x}{a} > \frac{1}{a}.$$

Since $u(1,t) = 0$, we conclude that the finite time extinction for this problem is $2/a$.

Then the smoothing property of the problem (10)–(12) immeadiately follows from [6, Theorem 12] and [7, Theorem 2.5]. More specifically, there exists a positive real T such that for any $\varphi_1, \varphi_2 \in L^2(0,1)$ and any C^2-function c the L^2-generalized solution (w, u) to the problem (10)–(12) is two times continuously differentiable for $t > T$. Moreover, it fulfills the following estimate:

$$\max(\|\partial_{x,t}^{\alpha,\beta} w(\cdot,t)\|_{C[0,1]}, \|\partial_{x,t}^{\alpha,\beta} u(\cdot,t)\|_{C([0,1])}) \le M_1 e^{At} \max_{i=1,2}(\|\varphi_i\|_{L^2(0,1)}), \quad t > T,$$

where $\alpha + \beta \le 2$ and the constants M_1, A are independent of t and φ_1, φ_2. This entails that the first component w of this solution, which is the L^2-generalized solution to the original problem (6), (6), (5), fulfills Theorem 3, as desired.

Finally, to prove Corollary 4, we follow the argument used to prove [7, Theorem 2.7(ii)]. The estimate (9) will then follow from the estimates (7) and (8).

3.3. Generalization

Our results given by Theorems 1 and 3 can be easily extended to the second order equations involving first order terms, of the following type:

$$(\partial_t - a\partial_x + a_1(x,t))(\partial_t + a\partial_x)w + c(x,t)w = 0, \tag{20}$$

where the coefficient a_1 is a two times continuously differentiable function such that a_1 itself and its first order and second order derivatives are bounded on $\overline{\Pi}$. If the coefficient a is positive, then in the domain Π we consider the initial-boundary value problem (20), (6), (5)); if a is negative, then we consider the problem (20), (6), (5).

For the problem (20), (6), (5), in the proofs we encounter the following minor changes. Similarly to the above, the non-perturbed problem (20), (6), (5) (with $c \equiv 0$) has the finite time extinction $2/a$, while the first order system corresponding to (20) now reads as

$$\partial_t w + a\partial_x w = u, \qquad \partial_t u - a\partial_x u + a_1(x,t)u + c(x,t)w = 0 \tag{21}$$

and the boundary and the initial conditions (6) and (5) read as (11), (12). The solution to the first order problem (21), (11), (12) satisfies the following integral system:

$$w(x,t) = \int_{t-\frac{x}{a}}^{t} u(a(\tau - t) + x, \tau)\, d\tau + pu(0, t - \frac{x}{a}), \quad t > \frac{x}{a},$$

$$u(x,t) = e^{\int_1^x \frac{a_1(\xi, t + \frac{x-\xi}{a})}{a} d\xi} \int_{t+\frac{x-1}{a}}^{t} [-cwe^{\int_\eta^1 \frac{a_1(\xi, \tau + \frac{\eta-\xi}{a})\, d\xi}{a}}](\eta,\tau)|_{\eta = x + a(t-\tau)} d\tau, \quad t + \frac{x}{a} > \frac{1}{a}.$$

Acknowledgments

I. Kmit was supported by the VolkswagenStiftung Project "Modeling, Analysis, and Approximation Theory toward Applications in Tomography and Inverse Problems". N. Lyulko was supported by the Presidium of the Russian Academy of Sciences (basic research program, number I.5P).

IOP Conf. Series: Journal of Physics: Conf. Series **894** (2017) 012056 doi:10.1088/1742-6596/894/1/012056

References

[1] Bastin G and Coron J-M 2016 *Progress in Nonlinear Differential Equations and Their Applications* vol 88 (Basel: Birkhäuser)
[2] Balakrishnan A V 2005 *J. Appl. Math. and Comput.* **164** 321-6
[3] Creutz D, Mazo M Jr and Preda C 2013 Superstability and finite time extinction for C_0-semigroups *ArXiv e-prints* (*Preprint* 0907.4812)
[4] Hille E, Phillips R 1957 *Functional analysis and semi-groups* (Providence: American Math. Society)
[5] Kmit I 2008 *Intern. J. Dynamic Syst. Differ. Equat.* **1** 191-5
[6] Kmit I 2011 *Appl. Anal.* **90** 1609-34
[7] Kmit I and Lyulko N 2017 Perturbations of superstable linear hyperbolic systems *ArXiv e-prints* (*Preprint* 1605.04703)
[8] Lax P and Phillips R 1967 *Scattering theory* (New York: Academic Press)
[9] Lax P and Phillips R 1975 *Indiana Univ. Math. J.* **24** 1119-38
[10] Lyulko N 2004 *J. Math. Phys., Anal., Geom.* **11** 169-76
[11] Majda A 1975 *Indiana Univ. Math. J.* **24** 1119-33
[12] Moulay E and Perruquetti W 2006 *Advances in variable structure and sliding mode control* **334** (Berlin: Springer-Verlag) pp 23-41
[13] Pazy A 1983 *Semigroups of operators and applications to partial differential equations* (Berlin: Springer-Verlag)
[14] Perrollaz V and Rosier L 2014 *SIAM J. on Control and Optimiz.* **52** 143-63

IOP Conf. Series: Journal of Physics: Conf. Series **894** (2017) 012057 doi:10.1088/1742-6596/894/1/012057

Numerical investigations of oxides and silicates under shock-wave loading

K K Maevskii

Lavrent'ev Institute of Hydrodynamics of SB RAS, Novosibirsk, Russia

E-mail: konstantinm@hydro.nsc.ru

Abstract. By using thermodynamic equilibrium model TEC (thermodynamic equilibrium components) results of numerical experiments of modeling of shock wave loading of oxides, silicates and mixtures are presented. The model allows to consider the presence of a polymorphic phase transition under shock wave loading of the materials, which is important for accurate modeling of thermodynamic parameters of pure substances and mixtures. The significant change of volume of the region of phase transition components included in the mixtures allows us to expand the range of variation of thermodynamic parameters of the mixtures under shock wave loading. The results allow us to create purposefully the necessary conditions for the synthesis of new materials in future. Interest in the research of compressibility of such mixtures is associated with the possibility of creating materials with desired properties and with the properties of the materials themselves. The good agreement of these model calculations with the experimental data of different authors is obtained. The model allows to describe dynamic loading of solid and porous mixtures comprising oxides, silicates as components quartz SiO_2, Periclase MgO, Corundum Al_2O_3, Spinel $MgAl_2O_4$, Enstatite $Mg_2[Si_2O_6]$.

1. Introduction

It is necessary to change deliberately thermodynamic parameters for the implementation of the necessary conditions for the creation of materials with desired properties during the development of the method of shock-wave synthesis. The researches of compressibility of multicomponent mixtures are performed for this. Oxides and silicates, and also mixtures based on them are of great interest.

Quartz SiO_2, zirconium dioxide ZrO_2, periclase MgO, corundum Al_2O_3, and others are of great interest in themselves and in mixes [1–4]. For example, quartz SiO_2 is widely used in optical and electronic devices, in the manufacture of silica refractories and quartz glass. Periclase MgO is an important component of magnesite and dolomite refractory. Additional interest from the point of view of the researches of oxides is associated in particular with MgO, for which the phase transition at high pressure was determined [3]. Such researches parameters of the oxides at high dynamic loads not only allow to purposefully change the parameters of shock-wave loading of samples, but more plausible to describe the structure of the earth's crust.

Abrasive materials are produce from corundum Al_2O_3. Spinel $MgAl_2O_4$ is rare mineral of cubic crystal structure. It is mixed oxide of magnesium and aluminum (volumetric fraction MgO 28.2%; Al_2O_3 — 71.8%). The name of mineral Enstatite $Mg_2[Si_2O_6]$ is derived from the Greek words ἐνστάτης (estates) "the enemy", "the resistance", which is associated with resistance to the melting of enstatite. Modeling the thermodynamic parameters for mixtures with such components is performed for the development of promising areas to obtain materials with desired properties.

Content from this work may be used under the terms of the Creative Commons Attribution 3.0 licence. Any further distribution of this work must maintain attribution to the author(s) and the title of the work, journal citation and DOI.
Published under licence by IOP Publishing Ltd

2. Calculation model

In order to calculate thermodynamic parameters of porous mixtures under dynamic loadings, a model of shock-wave loading with allowance for gas in the pores is used [5–7]. The model was able to describe the parameters in a wide range of pressures under shock wave loading, with the possibility of taking into account polymorphic phase transition of the components of the mixtures. The model is based on the assumption of thermodynamic equilibrium of all the components of a material under shock-wave loading, which implies the equality of the velocities, pressures and temperatures. In order to describe the thermodynamic parameters of condensed phases, the equations of state of the Mie-Grüneisen type are used. The equations that determine the state of the condensed component are written as follows:

$$P(\rho,T)=P_C(\rho)+P_T(T)\text{, and } E(\rho,T)=E_C(\rho)+E_T(T) \quad (1)$$

$$P_T(\rho,T)=\Gamma*\rho*E_T(T),\; E_T(T)=c_V*(T-T_0) \quad (2)$$

Here P_C, E_C, P_T, and E_T , are the potential and thermal components of pressure and specific energy, respectively; c_V is the specific heat capacity; T_0 is the initial temperature. By taking into account the pressure region of applicability of this model > 5GPa that is of interest to us, initial energy E_0 of a substance under normal conditions is considered to be equal to zero. Cold pressure component P_C is described by a Tait-type equation. Therefore, the equation of state for a condensed component with current density ρ and initial density ρ_0 are as follows:

$$P(\rho,T)=A\left(\left(\frac{\rho}{\rho_0}\right)^n-1\right)+\Gamma c_V(T-T_0)\rho, \quad (3)$$

The ideal gas equation of state is taken for a gas. The conditions of dynamic compatibility at the wave front are written, namely, the conditions of conservation of the mass flux for each component of the material and the conditions of conservation of momentum and energy fluxes for the media considered as a whole. The following expression can be obtained for a material with n condensed components (μ_{n0} is the volume fraction of the i-th phase of the substance):

$$P=\frac{\sum_{i=1}^{n}A_i\frac{\mu_{i0}}{\sigma_i}\left[\left(h_i-\frac{k_i+1}{k_i-1}\right)\sigma_i^{k_i}+\frac{2k_i\sigma_i}{k_i-1}-h_i-1\right]}{\sum_{i=1}^{n}\frac{\mu_{i0}}{\sigma_i}h_i+\left(\frac{h_g}{\sigma_g}\right)\left(1-\sum_{i=1}^{n}\mu_{i0}\right)-1}, \quad (4)$$

Here $h_i=2/\Gamma_i+1$, $i=1\ldots n$, $h_g=2/(\gamma-1)+1$. $\sigma_i=\rho_i/\rho_{i0}$, $\sigma_g=\rho_g/\rho_{g0}$ is the compression ratio of components, ρ_g, ρ_{g0} current and initial gas densities, $\gamma = 1.41$ is the ratio of specific heat. The obtained equations, together with the equations of state of each component, are sufficient to find dependences $P(U)$ or $D(U)$ (P, U, and D are the pressure, mass velocity, and wave velocity, respectively; A, k the coefficients in the equations of state of condensed component. These dependences can be treated as shock adiabat of multicomponent material.

The function $\Gamma=\frac{P_C V}{E_C}$ that determines the contribution of the thermal component depends explicitly only on the temperature $\Gamma(T)$ in the model [5–7].

The model TEC may describe the thermodynamic parameters of the materials including the phase transition region [8]. The material is considered as the mixture of low-pressure phase and high-

IOP Conf. Series: Journal of Physics: Conf. Series **894** (2017) 012057 doi:10.1088/1742-6596/894/1/012057

pressure phase in this area. The simulation results allow us to describe the thermodynamic parameters of the mixture including the components experiencing a phase transition at high dynamic loads. The parameters of the equation of state only their components are used to calculate the shock-wave effects in such mixtures.

3. Numerical results

Calculations of thermodynamic parameters of Corundum Al_2O_3, Spinel $MgAl_2O_4$, Enstatite $Mg_2[Si_2O_6]$ and data obtaining on the basis of the experiments of [9, 10] are shown in Figure 1. The accurate descriptions of the available data are obtained. The model parameters allowing reliable description of the thermodynamic parameters of pure oxides and silicates under shock wave loading are used to simulate the parameters of the mixture of them as components [6].

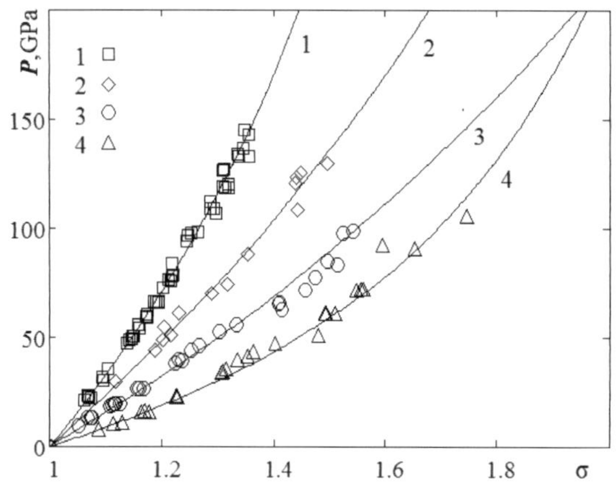

Figure 1. . Shock adiabata in variable – pressure compression. Calculation – line: 1 – corundum Al_2O_3 2 – spinel $MgAl_2O_4$, 3 – forsterite ceramic, $Mg_2[SiO_4]$, 4 – enstatite ceramic $Mg_2[Si_2O_6]$. The data of experiments 1–4 [9, 10].

Model TEC allows us to describe the shock adiabats in the region of phase transition. The reliable description of thermodynamic parameters of quartz SiO_2 taking into account polymorphic phase transition for samples of different porosity is obtained [8]. The similar modeling is performed for periclase MgO with a density of ρ_0=3.584 g/cm^3. The data for shock-wave loading of MgO is given in [3] up to value of pressure of 1000 GPa. In addition to the new field of dynamic parameters, the region of phase transition high pressure was discovered. The modeling results of thermodynamic parameters and the data from [3, 10] are shown in Figure 2. Thermodynamic parameters successfully modeling for MgO in the whole range of available experimental data are shown in Figure 2. The presence of a phase transition high pressure oxides, if they exist for other oxides, holds great promise for the analysis of the internal structure of the Earth given the large content of oxides in its composition [2].

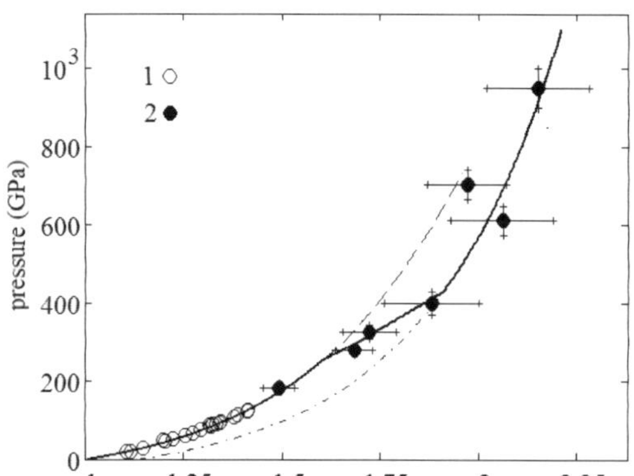

Figure 2. Shock adiabats of MgO in the coordinates pressure – compression. Line – calculation for MgO with phase transition, dotted line represents shock adiabata for low-pressure phase, dash-dotted line represents the calculation for the phase of high-pressure. Experimental data – 1 [10], 2 [3].

IOP Conf. Series: Journal of Physics: Conf. Series **894** (2017) 012057 doi:10.1088/1742-6596/894/1/012057

The accuracy of the description of thermodynamic parameters of the mixtures, taking into account phase transitions of the components under shock wave loading, was determined by calculations of shock wave loading of mixture with quartz. In Figure 3 the results are shown obtained for the mixtures of epoxy and quartz mixture, volume fraction vol% Epoxy(60)SiO_2(40) ρ_0=1.66 g/cm^3 consequently. The calculation corresponds well to the data of the experiments until the value of pressure 50 GPa. It is assumed that the phase transition of the components of the mixture starts under the same conditions as for the pure substances. Due to the fact that the calculation was carried out for the mixtures with low porosity, the estimated the pressure value to the beginning of the phase transition can be considered justified. This assumption has been confirmed by the calculations of the mixtures with two components undergo phase transition [11].

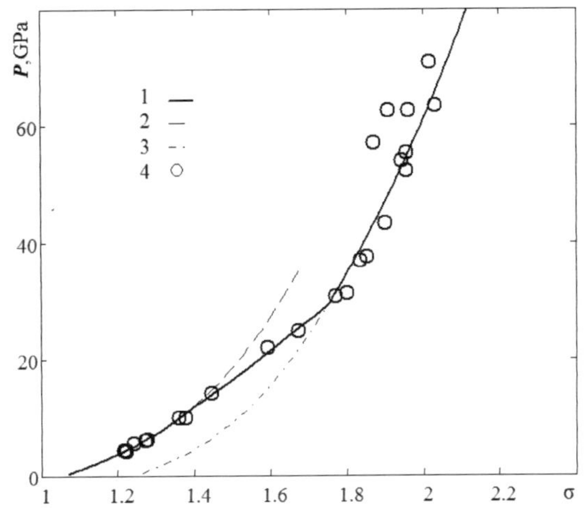

Figure 3. Shock adiabats in the coordinates of pressure – compression for Epoxy-quartz mixture.

The calculation for the mixture taking into account phase transition – solid line 1, the mixture with quartz low-pressure phase – dotted line 2, the mixture with quartz high-pressure phase (stishovite) – dash-dotted line 3. Data experiments: 4 [10].

The parameters that gave the opportunity to reliably describe the thermodynamic parameters of epoxy resins in a wide range of dynamic parameters, has allowed to describe with the precision of the experiment the thermodynamic parameters of the mixtures with other materials, which are characterized by high strength and heat resistance. The calculated and experimental data for these mixtures in the coordinates (pressure – compression) are shown in Figures 4–7. It is necessary to know the composition and density of the mixture for the description of thermodynamic parameters for shock wave loading. The component parameters are taken the same as in the description of pure substances. Thermodynamic parameters for the mixture of epoxy resin with Corundum, volume fraction, vol% Epoxy (60) Al_2O_3 (40), respectively, ρ_0 = 2.307 g/cm^3, the porosity, determined on the basis of data [10] m=0.93 are shown in Figure 4. The porosity m is defined as the ratio of the density of monolithic material to the density of the sample.

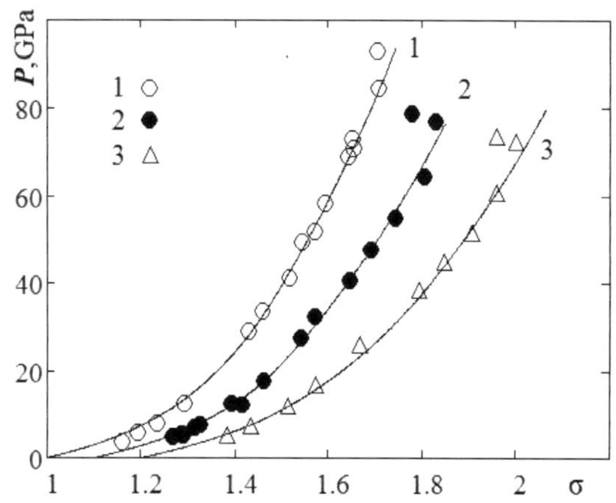

Figure 4. Shock adiabats in the coordinates of pressure – compression for the epoxy–corundum mixture 1, epoxy–spinel mixture 2, epoxy–enstatite mixture 3. Calculation – line. Data experiments: [10].

The modeling results of the thermodynamic parameters for the mixture of epoxy resin with Spinel, volume fraction, vol % Epoxy (60) $MgAl_2O_4$ (40), ρ_0 = 2.171 g/cm³, *m*=0.97 are shown here. The experiments data from [10] and modeling results of thermodynamic parameters, which are defined by the model TEC for mixture of epoxy resin with Enstatite, volume fraction, vol% Epoxy (60) Mg2[Si2O6] (40), ρ_0 = 2.017 g/cm3, *m*=0.93 are shown too.

The good correspondence of the calculations to data obtained on the basis of experiments for mixtures with epoxy resin are shown. The deviation of the calculated points for the mixture of quartz at pressure above 60 GPa and for the other mixtures above 80 GPa are probably due, in part, to the possible existence of polymorphic phase transition including for epoxy resin. In the presence of experimental data that allow to accurately, determining the region of phase transition, the model TEC can be used in this case. The possibility of calculating the thermodynamic parameters of a mixture with multiple components that are experiencing a phase transition under shock wave loading was shown in [11].

4. Conclusion

The data of the model TEC calculations correspond well to the experimental data by various authors. The model parameters, that allow us to reliably describe the behavior of pure materials under shock wave loading, were used to simulate thermodynamic parameters of oxides, silicates and mixtures based on them mixtures including components experiencing a phase transition under shock wave loading. A significant change in volume in the region of the polymorphic phase transition components included in the mixtures allows expanding the range of changes of thermodynamic parameters of the mixtures under shock wave loading. The results obtained allow creating purposefully the necessary conditions for the synthesis of new materials. This model allows for specific selection of compositions of mixtures and the ratios of their components with the aim of obtaining the specified characteristics under shock-wave loading of solid and porous materials and mixtures.

References

[1] Medvedev A B and Trunin R F 2012 *Phys. Usp.* **55** 773
[2] Milyavskii V V, Akopov F A, Anan'ev S Yu, Borovkova L B, Borodina T I, Valuev A V, Val'yano G E, Ziborov V S, Lukin E S, Popova N A and Savinykh A S 2013 *High Temperature* **51**(5) 640
[3] Miyanishi K, Tange Y, Ozaki N, Kimura T, Sano T, Sakawa Y, Tsuchiya T and Kodama R 2015 *Phys. Rev. E* **92** 023103
[4] Bukaemskii A A, Avramenko S S, and Tarasova L S 2002 *Combustion, Explosion and Shock Waves* **38**(4) 478
[5] Kinelovskii S A and Maevskii K K *J. Appl. Mech. Tech. Phys.* 2013 **54**(4) 524

[6] Kinelovskii S A and Maevskii K K 2016 *High Temperature* **54**(5) 675
[7] Kinelovskii S A and Maevskii K K 2016 *Technical Physics* **61**(8) 1244
[8] Maevskii K K and Kinelovskii S A *AIP Conf. Proc.* 2014 **1623** 391
[9] Levashov PR, Khishchenko K V, Lomonosov I V and Fortov V E *AIP Conf. Proc.* 2004 **706** 87
[10] S. P. Marsh (Ed.), *California University Press*, Berkeley, 1980
[11] Maevskii K K and Kinelovskii S A *AIP Conf. Proc.* 2016 **1783** 020143

IOP Conf. Series: Journal of Physics: Conf. Series **894** (2017) 012058 doi:10.1088/1742-6596/894/1/012058

Modeling viscous compressible barotropic multi-fluid flows

Alexander Mamontov[1,2] **and Dmitriy Prokudin**[1,2]
[1] Lavrentyev Institute of Hydrodynamics, 630090 Novosibirsk, Russia
[2] Novosibirsk State University, 630090 Novosibirsk, Russia

E-mail: aemamont@hydro.nsc.ru, prokudin@hydro.nsc.ru

Abstract. We study the system of equations which describes barotropic (isentropic) flows of viscous compressible multi-fluids (mixtures of fluids). We study the relations between pressure, densities, concentrations, viscosities and other parameters of the flow and of the medium.

1. Introduction

The flow of a mixture which consists of N viscous compressible fluids (gases) with a common pressure of the constituents (phases) may be described [1], [2] via the system of equations

$$\frac{\partial \rho_i}{\partial t} + \operatorname{div}(\rho_i \boldsymbol{v}) = 0, \quad i = 1, \ldots, N, \tag{1}$$

$$\frac{\partial(\rho_i \boldsymbol{u}_i)}{\partial t} + \operatorname{div}(\rho_i \boldsymbol{v} \otimes \boldsymbol{u}_i) + \alpha_i \nabla p = \operatorname{div} \mathbb{S}_i + \rho_i \boldsymbol{f}_i, \quad i = 1, \ldots, N. \tag{2}$$

Here ρ_i is the density of the i-th component (constituent) of the multi-fluid, $\boldsymbol{u}_i$ is the velocity field, and p is the pressure. The concentration $\xi_i = \rho_i/\rho$ of each component is defined (they compose the vector $\boldsymbol{\xi} = \{\xi_i\}_{i=1}^N$), where $\rho = \sum_{i=1}^N \rho_i$ is the total density of the multi-fluid. Then, $\boldsymbol{v} = \sum_{i=1}^N \alpha_i \boldsymbol{u}_i$ is the average velocity of the multi-fluid, and the coefficients $\alpha_i(\boldsymbol{\xi}) > 0$ are such that $\sum_{i=1}^N \alpha_i = 1$ (e. g., $\alpha_i = \xi_i$). The viscous stress tensors $\mathbb{S}_i$ are defined by the equalities

$$\mathbb{S}_i = \sum_{j=1}^N \widehat{\mathbb{S}}_{ij}, \quad \widehat{\mathbb{S}}_{ij} = \Big(2\mu_{ij}\mathbb{D}(\boldsymbol{u}_j) + \lambda_{ij}(\operatorname{div} \boldsymbol{u}_j)\mathbb{I}\Big), \qquad i, j = 1, \ldots, N,$$

where $\mathbb{D}(\boldsymbol{w}) = ((\nabla \otimes \boldsymbol{w}) + (\nabla \otimes \boldsymbol{w})^*)/2$ is the rate of deformation tensor of the vector field $\boldsymbol{w}$, $\mathbb{I}$ is the identity tensor, and the viscosity coefficients compose the matrices

$$\mathbf{M} = \{\mu_{ij}\}_{i,j=1}^N > 0, \quad \mathbf{H} = \mathbf{\Lambda} + \frac{2}{3}\mathbf{M} \geqslant 0, \quad \mathbf{\Lambda} = \{\lambda_{ij}\}_{i,j=1}^N.$$

Content from this work may be used under the terms of the Creative Commons Attribution 3.0 licence. Any further distribution of this work must maintain attribution to the author(s) and the title of the work, journal citation and DOI.
Published under licence by IOP Publishing Ltd

IOP Conf. Series: Journal of Physics: Conf. Series **894** (2017) 012058 doi:10.1088/1742-6596/894/1/012058

Finally, $\boldsymbol{f}_i$ are known fields of external body forces.

Generally speaking, the pressure p depends on the densities and on other thermodynamical variables (say, temperature(s)), so that the system (1), (2) must be complemented by the equation(s) for these variables (temperatures, energies, entropies etc.). However, the paper is limited by barotropic flows, so we concentrate on the justification of the state (constitutive) equations which characterize the relations between the following values: $\{\rho_i\}_{i=1}^N$, p, $\{\alpha_i\}_{i=1}^N$, $\{\lambda_{ij}\}_{i,j=1}^N$ and $\{\mu_{ij}\}_{i,j=1}^N$.

2. Pressure as a function of the densities

From (1) we have

$$\frac{\partial \rho}{\partial t} + \operatorname{div}(\rho \boldsymbol{v}) = 0, \qquad \frac{\partial \xi_i}{\partial t} + \boldsymbol{v} \cdot \nabla \xi_i = 0. \tag{3}$$

2.1. Motion of a material volume

Let the material volume V be occupied by a mixture of N ideal gases, and this multi-fluid volume is the subject of a quasistatic process, in which the thermodynamical variables are supposed to be constant over the whole volume. Concerning the system (1), (2), this means that the volume is rather small and consists of the constant ensemble of particles, i. e. it moves along the trajectories of the transport equations (1).

Each constituent of the multi-fluid is characterized by the following variables (parameters): density ρ_i, temperature θ (which is common for all constituents), entropy S_i, mass $m_i = \rho_i V$, molar mass M_i, partial pressure $p_i = \dfrac{\rho_i}{M_i} R\theta$ (R is the gas constant), degree of freedom of molecules ν_i, adiabatic index (heat capacity ratio, Poisson constant) $\gamma_i = 1 + \dfrac{2}{\nu_i}$ (i. e. $\nu_i = \dfrac{2}{\gamma_i - 1}$), and internal energy $U_i = \dfrac{\nu_i R}{2M_i} m_i \theta$, or specific internal energy $e_i = \dfrac{\nu_i R}{2M_i}\theta$.

The heat transfer into each constituent (from other constituents and from outside of the volume) equals

$$\theta dS_i = dU_i + p_i dV,$$

and the total heat transfer into the volume equals (here S is the total entropy of the volume)

$$\delta Q = \theta dS = \sum_{i=1}^N dU_i + \sum_{i=1}^N p_i dV.$$

2.2. Adiabatic process

If the process is adiabatic (isentropic), then $\delta Q = 0$, and we obtain

$$\sum_{i=1}^N dU_i + \sum_{i=1}^N p_i dV = 0. \tag{4}$$

Observe that

$$dU_i + p_i dV = R\theta \left(\frac{m_i \nu_i}{2M_i} d\ln\theta + \frac{m_i}{M_i} d\ln V \right). \tag{5}$$

Let us denote

$$\alpha = \sum_{i=1}^N \frac{m_i \nu_i}{2M_i}, \qquad \beta = \sum_{i=1}^N \frac{m_i}{M_i}.$$

We find from (4) and (5) that $\theta = C_1 V^{-\beta/\alpha}$. Taking into account the relation $m = \rho V$ (m is the total mass of the multi-fluid), we finally obtain $\theta = C_2 \rho^{\beta/\alpha}$.

IOP Conf. Series: Journal of Physics: Conf. Series **894** (2017) 012058 doi:10.1088/1742-6596/894/1/012058

2.3. Adiabatic index

Let us note that the value

$$\frac{\beta}{\alpha}=\frac{\sum\limits_{i=1}^{N}\frac{\rho_i}{M_i}}{\sum\limits_{j=1}^{N}\frac{\rho_j\nu_j}{2M_j}}=\frac{\sum\limits_{i=1}^{N}\frac{\xi_i}{M_i}}{\sum\limits_{j=1}^{N}\frac{\xi_j\nu_j}{2M_j}}$$

is not constant in the flow, however, in the partial case of multi-fluid with equal degree of freedom of molecules $\nu_i=\nu$ in all constituents (so that $\gamma_i=\gamma$) we obtain

$$\frac{\beta}{\alpha}=\frac{2}{\nu}=\gamma-1.$$

In a general case, β/α depends on the concentrations in a spacial point and can take any values from the interval $[\gamma_{\min}-1,\gamma_{\max}-1]$ (where $\gamma_{\min}$ and $\gamma_{\max}$ are correspondingly the minimal and maximal adiabatic indices between all constituents of the multi-fluid). Let us denote

$$\gamma=\gamma(\boldsymbol{\xi})=\frac{\beta}{\alpha}+1=1+\frac{\sum\limits_{i=1}^{N}\frac{\xi_i}{M_i}}{\sum\limits_{j=1}^{N}\frac{\xi_j}{M_j(\gamma_j-1)}}. \tag{6}$$

This quantity is a function of the concentrations, and hence (see $(3)_2$) is transported along the trajectories, i. e. it satisfies the equation

$$\frac{\partial\gamma}{\partial t}+\boldsymbol{v}\cdot\nabla\gamma=0, \tag{7}$$

but γ can be supposed to be constant in two situations: either if all γ_i coincide (then γ coincides with it), or if $\gamma=\text{const}$ in the initial state of the multi-fluid.

2.4. Completion of the constitutive equation

Thus, we consider an adiabatic movement of a material volume of a multi-fluid (which consists of the constant ensemble of particles), in which the distribution of the intensive variables is supposed to be uniform. As a result, the partial pressures have taken the form $p_i=C_3\frac{\rho_i}{M_i}\rho^{\beta/\alpha}$, and hence the total pressure

$$p=\sum_{i=1}^{N}p_i=C_3\rho^{\beta/\alpha}\sum_{i=1}^{N}\frac{\rho_i}{M_i}.$$

Thus,

$$p=K_1\rho^{\gamma-1}\widetilde{\rho}=K_1\rho^{\gamma}\widetilde{\xi},\qquad \widetilde{\rho}=\sum_{i=1}^{N}\frac{\rho_i}{M_i},\qquad \widetilde{\xi}=\sum_{i=1}^{N}\frac{\xi_i}{M_i}. \tag{8}$$

Observe that $C_1=V_0^{\beta_0/\alpha_0}\theta_0$, $C_2=\rho_0^{-\beta_0/\alpha_0}\theta_0$, $C_3=K_1=R\rho_0^{-\beta_0/\alpha_0}\theta_0$.

2.5. Variant of writing the constitutive equation

Only one of the densities is independent, since they are connected through the relations

$$\rho_i = \frac{m_i}{V} = \frac{m_i}{m}\rho = \frac{\rho_{0i}V_0}{\rho_0 V_0}\rho = \frac{\rho_{0i}}{\rho_0}\rho,$$

and we can choose either one of them as an unknown value, or some their combination, e. g., the total density ρ. As a result, we obtain

$$p_i = \frac{R\theta_0}{M_i}\rho_0^{-\gamma_0}\rho_{0i}\rho^{\gamma} \quad \Longrightarrow \quad p = K\rho^{\gamma}, \quad K = R\theta_0\rho_0^{-\gamma_0}\sum_{i=1}^{N}\frac{\rho_{0i}}{M_i}.$$

In this way, we see that, in a real flow, K is a function of the Lagrangian spacial coordinates, i. e. it is transported along the trajectories and satisfies the equation

$$\frac{\partial K}{\partial t} + \boldsymbol{v}\cdot\nabla K = 0, \tag{9}$$

but, as the simplest version, K can be accepted to be constant (this means a property of the initial distribution of the parameters of the multi-fluid).

2.6. Choice of the form of the constitutive equation

The relations (see Section 2.5)

$$p = K\rho^{\gamma} \tag{10}$$

and (see (8))

$$p = K_1\rho^{\gamma}\widetilde{\xi}, \tag{11}$$

are, of course, equivalent (since they are obtained from the same assumptions), as one can easily make sure via the direct verification of the identity $K = K_1\widetilde{\xi}$. Those relations are just the same, written in two different forms, as long as we discuss an adiabatic movement of a material volume. However, when we use the constitutive equation for the pressure in order to complete the system (1), (2), the choice of the appropriate form may become crucial (see Section 5.4).

2.7. Coefficients α_i

The quantities α_i which enter as factors at the pressures in (2), and also in the relations between $\boldsymbol{v}$ and $\boldsymbol{u}_i$, depend on the concentrations, and hence satisfy the equations

$$\frac{\partial \alpha_i}{\partial t} + \boldsymbol{v}\cdot\nabla\alpha_i = 0. \tag{12}$$

In the simplest case, α_i can be accepted to be constant (this means a property of the initial state of the multi-fluid).

3. Viscosities

The viscosity matrices in real multi-fluids take non-trivial forms. For instance, there is the following information [3], [4], [5] about the matrix $\mathbf{M}$. As $i \neq j$, the following representation holds

$$\mu_{ij} = \mu_{ij}^0(\widehat{\mu}_i, \widehat{\mu}_j)\xi_i\xi_j \exp\left(\frac{\alpha_{ij}\xi_i + \beta_{ij}\xi_j}{\xi_i + \xi_j}\right),$$

where $\widehat{\mu}_i$ are the "pure" viscosities of the constituents. The most frequent variant is

$$\mu_{ij}^0(\widehat{\mu}_i, \widehat{\mu}_j) = \sqrt{\widehat{\mu}_i\widehat{\mu}_j}, \qquad \alpha_{ij} = \beta_{ij} = 0, \tag{13}$$

but, generally speaking, α_{ij} and β_{ij} are the empiric constants. The diagonal entries of the matrix $\mathbf{M}$ are accepted as

$$\mu_{ii} = \widehat{\mu}_i \xi_i^2 + \sum_{j \neq i} \mu_{ij}.$$

Thus, in the simplest case (13), we can write

$$\mu_{ij} = \widehat{\nu}_i \widehat{\nu}_j + \widehat{\nu}_i (\widehat{\nu} - \widehat{\nu}_i) \delta_{ij}, \qquad \widehat{\nu}_i = \sqrt{\widehat{\mu}_i} \xi_i, \qquad \widehat{\nu} = \sum_{i=1}^{N} \widehat{\nu}_i. \tag{14}$$

The matrix $\mathbf{M}$, on the one hand, possesses an important diagonal predominance property, but on the other hand, it essentially depends on the concentrations.

Summarizing, we see that the viscosity matrices, in general, depend on the concentrations, and hence (see $(3)_2$) are transported along the trajectories and satisfy the equations

$$\frac{\partial \mathbf{M}}{\partial t} + (\boldsymbol{v} \cdot \nabla)\mathbf{M} = 0, \qquad \frac{\partial \boldsymbol{\Lambda}}{\partial t} + (\boldsymbol{v} \cdot \nabla)\boldsymbol{\Lambda} = 0, \tag{15}$$

but they can be accepted to be constant as a simplifying assumption which is justified at least in the case of the corresponding initial configuration.

The dependence of the viscosity matrices on the concentrations constitutes serious difficulties for the effective viscous flux (EVF) technique, which underlies the present existence theory for the equations of compressible viscous fluids. That is why, in the present theory, the simplifying assumption of constant viscosities is always accepted, but with the preservation of necessary properties of positiveness.

4. Energy balance

Regardless of the variant of the constitutive relation for the pressure (in the framework described in Section 2.6) and of the specification of the values $\{\alpha_i\}_{i=1}^N$, $\{\lambda_{ij}\}_{i,j=1}^N$ and $\{\mu_{ij}\}_{i,j=1}^N$, the total energy balance can be expressed as an integral relation. Indeed, let (10) hold, where K and γ are arbitrary constants satisfying (7) and (9). Observe that p in (11) satisfies these assumptions as well, as long as K_1 satisfies (9), and $\widetilde{\xi}$ satisfies $(3)_2$. Then, for any smooth function φ we have

$$d(\varphi(\gamma)p) = p\varphi'(\gamma)d\gamma + \varphi(\gamma)\rho^\gamma dK + \varphi(\gamma)K\rho^\gamma(\ln\rho)d\gamma + \varphi(\gamma)K\gamma\rho^{\gamma-1}d\rho.$$

Dividing by dt, using the notation $\dfrac{d}{dt} = \dfrac{\partial}{\partial t} + \boldsymbol{v} \cdot \nabla$ and the equation $(3)_1$, we obtain

$$\frac{\partial}{\partial t}(\varphi(\gamma)p) + \operatorname{div}(\varphi(\gamma)p\boldsymbol{v}) + \varphi(\gamma)(\gamma-1)p\operatorname{div}\boldsymbol{v} = 0. \tag{16}$$

Multiplying (1) by $-|\boldsymbol{u}_i|^2/2$, (2) (in the sense of the inner product) by $\boldsymbol{u}_i$ and summing, we deduce

$$\frac{\partial}{\partial t}\frac{\rho_i|\boldsymbol{u}_i|^2}{2} + \operatorname{div}\left(\frac{\rho_i|\boldsymbol{u}_i|^2}{2}\boldsymbol{v}\right) + \operatorname{div}(\alpha_i p\boldsymbol{u}_i) - p\operatorname{div}(\alpha_i\boldsymbol{u}_i) = \operatorname{div}(\mathbb{S}_i\boldsymbol{u}_i) - \mathbb{S}_i : \mathbb{D}(\boldsymbol{u}_i) + \rho_i\boldsymbol{u}_i \cdot \boldsymbol{f}_i.$$

Making the summation of these equalities and the equation (16) with $\varphi(\gamma) = 1/(\gamma-1)$, we obtain the desired balance

$$\begin{gathered}\frac{\partial}{\partial t}\left(\sum_{i=1}^N \frac{\rho_i|\boldsymbol{u}_i|^2}{2} + \frac{p}{\gamma-1}\right) + \operatorname{div}\left(\left[\sum_{i=1}^N \frac{\rho_i|\boldsymbol{u}_i|^2}{2} + \frac{\gamma p}{\gamma-1}\right]\boldsymbol{v}\right) = \\ = \operatorname{div}\left(\sum_{i=1}^N \mathbb{S}_i\boldsymbol{u}_i\right) - \sum_{i=1}^N \mathbb{S}_i : \mathbb{D}(\boldsymbol{u}_i) + \sum_{i=1}^N \rho_i\boldsymbol{u}_i \cdot \boldsymbol{f}_i.\end{gathered} \tag{17}$$

IOP Conf. Series: Journal of Physics: Conf. Series **894** (2017) 012058 doi:10.1088/1742-6596/894/1/012058

5. Examples of non-monotone pressure

During further development of the theory, the relation (17) is insufficient, and the monotonicity of the pressure with respect to the densities is necessary. Let us give examples which show that the equations of the type (11) are inconvenient in this sense. To do so, we will stay in the framework of the assumptions that $\gamma > 1$ and $K_1 = 1$ are constants.

5.1. Non-monotonicity with respect to $\widetilde{\rho}$

Let $N = 2$, $M_1 > M_2 > 0$. Set

$$\widetilde{\rho}^1 = 1, \qquad \widetilde{\rho}^2 = \left(\frac{M_1}{M_2}\right)^{\frac{\gamma-1}{2\gamma}} > 1, \qquad \rho^1 = M_1 - \varepsilon, \qquad \rho^2 = (M_2 + \varepsilon)\widetilde{\rho}^2, \tag{18}$$

where

$$\varepsilon = \frac{\sqrt{M_1 M_2}}{3} \cdot \frac{M_1 - M_2}{M_1 + M_2} \in (0, M_1 - M_2) \quad \Longrightarrow \quad \rho^1 \in (M_2, M_1) \quad \rho^2 \in (M_2\widetilde{\rho}^2, M_1\widetilde{\rho}^2).$$

Then all components of the corresponding "reconstructed" densities

$$\boldsymbol{\rho} = \frac{1}{M_1 - M_2}\begin{bmatrix} M_1(\rho - M_2\widetilde{\rho}) \\ M_2(M_1\widetilde{\rho} - \rho) \end{bmatrix}$$

are positive. However, the corresponding pressures

$$p^1 = (\rho^1)^{\gamma-1}\widetilde{\rho}^1 = (M_1 - \varepsilon)^{\gamma-1}, \qquad p^2 = (\rho^2)^{\gamma-1}\widetilde{\rho}^2 = (M_2 + \varepsilon)^{\gamma-1}(\widetilde{\rho}^2)^{\gamma},$$

as one can easily see, possess the property $p^2 < p^1$, and hence

$$(p^2 - p^1)(\widetilde{\rho}^2 - \widetilde{\rho}^1) < 0. \tag{19}$$

5.2. Non-monotonicity with respect to ρ

Under the assumptions of the preceding example, we set (18) again, but let us choose $\varepsilon \in (0, M_1 - M_2)$ so that the inequalities

$$\left(\frac{M_1}{M_2}\right)^{\frac{\gamma-1}{2\gamma}} < \frac{M_1 - \varepsilon}{M_2 + \varepsilon} < \left(\frac{M_1}{M_2}\right)^{\frac{1}{2}}$$

hold. This gives positive components of the densities again, but now $p^2 > p^1$, $\rho^2 < \rho^1$, and hence

$$(p^2 - p^1)(\rho^2 - \rho^1) < 0. \tag{20}$$

5.3. Integral non-monotonicity

The "pointwise" examples above let us construct the corresponding examples of functional density distributions in any domain which satisfy the conservation of the mass, i. e. $\int \boldsymbol{\rho}^1 = \int \boldsymbol{\rho}^2$, but with the loss of the monotonicity, i. e. with the integral analogs of the inequalities (19) and (20). To do so, it suffices to split the domain into two parts of equal measure, and then to take the values from the preceding examples on one part of the domain, and to swap densities on the other part. Similarly one can make sure that inequalities of the type

$$\int (p(\boldsymbol{\rho}^1) - p(\boldsymbol{\rho}^2))(\rho_j^1 - \rho_j^2) \geqslant 0$$

are impossible, as well as their analogs with any linear combinations of the densities instead of ρ_j.

IOP Conf. Series: Journal of Physics: Conf. Series **894** (2017) 012058 doi:10.1088/1742-6596/894/1/012058

5.4. Conclusions concerning admissible constitutive equations

Hence, if we are based on the constitutive equations of the type (11), i. e. if the unique (common) pressure depends on all densities but not on their special combination, then we have no hope to obtain inequalities of the form $\overline{\rho p} \geqslant \rho \overline{p}$ (bar stands for the weak limit) which play the crucial role in the EVF technique. In addition, such a situation (one pressure depends on several combinations of the densities) is unlikely for the thermodynamics.

On the other hand, the dependence of the pressures on separate densities would possibly be found reasonable if the momentum equations (2) contain partial pressures instead of the common one. However, in this case, the present EVF technique works only with diagonal or triangular viscosity matrices, which is of little interest for both physics and mathematics (see details in [2]).

6. Further development of the theory

One can formulate the initial boundary value problem for the system (1), (2) in a space-time cylinder, or the steady analog of this problem, and after that, one can construct approximate solution via an appropriate method. Basing on (17), after additional efforts, the estimates can be obtained which allow to select weakly convergent sequence of approximate solutions, and the focus is shifted to the justification of the equality $\overline{p} \sim p$ (here $\sim$ means equal action on the test functions) and/or strong convergence of the densities, for which the standard (in the viscous compressible flow theory) EVF technique is applied.

It is important to note that there exist two versions of the final step of the EVF argument. The first one is based on [6], P. 339 (it is applied in [7] and [8]), and the second one is based on [9], P. 188, or on more general arguments as on P. 187 (it is applied in [10]). In [2], a general scheme (in the first version) of applying this technique is written for the systems like (1), (2). This scheme is completely applicable to the system (1), (2), (10) provided that all quantities K, γ, $\{\alpha_i\}_{i=1}^N$, $\{\lambda_{ij}\}_{i,j=1}^N$ and $\{\mu_{ij}\}_{i,j=1}^N$ are constant. In the general case, we can state only the following properties of these quantities:

- for K: (9),
- for γ: (6) (possibly, it is sufficient to claim (7)),
- for $\{\alpha_i\}_{i=1}^N$: some explicit form of $\alpha_i(\boldsymbol{\xi})$ (possibly, it is sufficient to claim (12)),
- for $\{\lambda_{ij}\}_{i,j=1}^N$ and $\{\mu_{ij}\}_{i,j=1}^N$: some explicit form like (14) or any other described in Section 3 (possibly, it is sufficient to claim (15)).

Functioning of the EVF technique in the situation when at least one of these quantities is not constant, remains an open problem.

7. Bibliographic remarks

As a starting point for the multi-fluid model under consideration, we can cite [11] (Chapter 1) and [12] (Chapter 7). During the development of the existence theory, it became clear that the model admits and requires definite corrections, and we came, as a result, to several variants of the model, one of which is the system (1), (2). The main solvability results for this variant of the model, and for other ones, are obtained in [7], [8], [10], [13], [14], [15] and [16]. A detailed review of the results and the discussion of the model can be found in [1] and [2].

Acknowledgments

The authors were supported by the Russian Science Foundation (grant 15–11–20019).

References

[1] Mamontov A E and Prokudin D A 2013 *Methods and Applications of Analysis* **20**(2) 179
[2] Mamontov A E and Prokudin D A 2017 *Sib. Elect. Math. Reports* **14** 388
[3] Al-Sharif A, Chamniprasart K, Rajagopal K R and Szeri A Z 1993 *J. Tribol.* **115**(1) 46
[4] Benner J J, Sadeghi F, Hoeprich M R and Frank M C 2006 *J. Tribol.* **128**(2) 296
[5] Lee R T, Yang K T and Chiou Y C 2013 *Tribol. Int.* **66** 241
[6] Feireisl E and Novotný A 2009 *Singular Limits in Thermodynamics of Viscous Fluids* (Basel: Birkhäuser)
[7] Mamontov A E and Prokudin D A 2016 *Sib. Elect. Math. Reports* **13** 664 (in Russian)
[8] Mamontov A E and Prokudin D A 2017 *Sib. Math. J.* **58**(1) 113
[9] Novotný A and Straškraba I 2004 *Introduction to the Mathematical Theory of Compressible Flow* (Oxford: Oxford University Press)
[10] Mamontov A E and Prokudin D A 2016 *Sib. Elect. Math. Reports* **13** 541 (in Russian)
[11] Nigmatulin R I 1990 *Dynamics of Multiphase Media* vol 1 (New York: Hemisphere)
[12] Rajagopal K L and Tao L 1995 *Mechanics of Mixtures* (Singapore: World Scientific Publishing)
[13] Kucher N A, Mamontov A E and Prokudin D A 2012 *Sib. Math. J.* **53**(6) 1075
[14] Mamontov A E and Prokudin D A 2014 *Izv. Math.* **78**(3) 554
[15] Mamontov A E and Prokudin D A 2016 *Sib. Math. J.* **57**(6) 1044
[16] Prokudin D A and Krayushkina M V 2016 *J. Appl. Industr. Math.* **10**(3) 417

IOP Conf. Series: Journal of Physics: Conf. Series **894** (2017) 012059 doi:10.1088/1742-6596/894/1/012059

Global solvability of 1D equations of viscous compressible multi-fluids

A Mamontov[1] **and D Prokudin**[1,2]
[1]Lavrentyev Institute of Hydrodynamics, Novosibirsk, Russia
[2]Voronezh State University, Voronezh, Russia

E-mail: aemamont@hydro.nsc.ru, prokudin@hydro.nsc.ru

Abstract. We consider the model of viscous compressible multi-fluids with multiple velocities. We review different formulations of the model and the existence results for boundary value problems. We analyze crucial mathematical difficulties which arise during the proof of the global existence theorems in 1D case.

1. Introduction

The description of the motion of multi-component media is an interesting and rather little-studied problem both in physics/mechanics and in mathematics. There is no standard approach to simulating these motions, nor is there any developed mathematical theory concerning the existence, uniqueness and properties of solutions of initial-boundary value problems arising in this simulation.

In the present paper, we choose one of the numerous versions of simulating the motion of multi-component fluid mixtures, namely, a homogeneous mixture of viscous compressible fluids and a multi-velocity model. This means that all components (constituents) of the mixture are present at any point of the space and with the same phase, and each of them has its own local velocity. The interaction between the components occurs via the viscous friction and the exchange between momenta, and also using the heat exchange (in heat-conducting models). This kind of mixtures is also called a multi-fluid, see [1] for the details.

2. Mathematical model of multi-fluids

The mathematical model of a multi-fluid which consists of $N \geqslant 2$ components, includes (see [2] and [3]) the continuity equations for each constituent

$$\partial_t \rho_i + \operatorname{div}(\rho_i \boldsymbol{u}_i) = 0, \quad i = 1, \dots, N, \tag{1}$$

the momentum equations for each constituent

$$\partial_t(\rho_i \boldsymbol{u}_i) + \operatorname{div}(\rho_i \boldsymbol{u}_i \otimes \boldsymbol{u}_i) + \nabla p_i - \operatorname{div} \mathbb{S}_i = \rho_i \boldsymbol{f}_i + \boldsymbol{J}_i, \quad i = 1, \dots, N, \tag{2}$$

and the energy equations. Here ρ_i is the density of the i-th constituent of the multi-fluid, $\boldsymbol{u}_i$ is the velocity field, p_i is the pressure, $\mathbb{S}_i$ is the viscous part of the stress tensor $\mathbb{P}_i = -p_i \mathbb{I} + \mathbb{S}_i$:

$$\mathbb{S}_i = \sum_{j=1}^{N} \left(\lambda_{ij} (\operatorname{div} \boldsymbol{u}_j) \mathbb{I} + 2\mu_{ij} \mathbb{D}(\boldsymbol{u}_j) \right), \quad i = 1, \dots, N,$$

Content from this work may be used under the terms of the Creative Commons Attribution 3.0 licence. Any further distribution of this work must maintain attribution to the author(s) and the title of the work, journal citation and DOI.
Published under licence by IOP Publishing Ltd

where λ_{ij} and μ_{ij} are the viscosity coefficients, $\mathbb{D}$ is the rate of deformation tensor, and $\mathbb{I}$ is the identity tensor. The viscosity coefficients λ_{ij} and μ_{ij} compose the matrices $\mathbf{\Lambda} = \{\lambda_{ij}\}_{i,j=1}^{N}$ and $\mathbf{M} = \{\mu_{ij}\}_{i,j=1}^{N}$. Finally, $\boldsymbol{f}_i = (f_{i1}, \dots, f_{in})$ denotes the external body force (n is the dimension of the flow), and

$$\boldsymbol{J}_i = \sum_{j=1}^{N} a_{ij}(\boldsymbol{u}_j - \boldsymbol{u}_i), \quad i = 1, \dots, N, \quad a_{ij} = a_{ji}, \quad i, j = 1, \dots, N,$$

stands for the momentum exchange between the constituents (the momentum supply).

The viscosity matrices in real multi-fluids take non-trivial forms and, generally speaking (see [4] for the details), they depend on the concentrations $\xi_i = \dfrac{\rho_i}{\rho}$, where $\rho = \sum\limits_{i=1}^{N} \rho_i$ is the total density of the multi-fluid. The dependence of the viscosity matrices on the concentrations constitutes serious difficulties. We accept the simplifying assumption of constant viscosities, but with the preservation of necessary properties of positiveness. The thing is that it is very important (physically and mathematically) to validate Second law of thermodynamics which means

$$\sum_{i=1}^{N} \mathbb{S}_i : \mathbb{D}(\boldsymbol{u}_i) \geqslant 0. \tag{3}$$

Besides, in order to provide important mathematical property of ellipticity, it is necessary to validate the condition

$$\sum_{i=1}^{N} \int_{\Omega} \mathbb{S}_i : \mathbb{D}(\boldsymbol{u}_i) d\boldsymbol{x} \geqslant C \sum_{i=1}^{N} \int_{\Omega} |\nabla \otimes \boldsymbol{u}_i|^2 d\boldsymbol{x}, \tag{4}$$

where Ω is the flow domain, and $\boldsymbol{u}_i|_{\partial\Omega} = 0$, $i = 1, \dots, N$. The formulated positiveness or coercivity can be provided by the following properties of viscosity matrices: the properties $n\mathbf{\Lambda} + 2\mathbf{M} \geqslant 0$, $\mathbf{M} \geqslant 0$ provide (3), and the properties $\mathbf{\Lambda} + 2\mathbf{M} > 0$, $\mathbf{M} > 0$ provide (4).

A very important observation is that the viscosity matrices should not be diagonal. Momentum supply $\boldsymbol{J}_i$ in the momentum equations gives lower order terms (physically important, but mathematically causing no difficulties), and if the matrices are diagonal then $\boldsymbol{J}_i$ is the only connection between the constituents, so we have N Navier–Stokes systems connected only via lower order terms. Earlier such problems were relevant (even in 1D) (see, e. g., [5], [6] and [7]), but nowadays such results almost automatically come from the theory of mono-fluid systems governed by the compressible Navier–Stokes equations. If viscosity matrices are "complete" (general) then we have interesting mathematical problems (see the reviews in [1], [8] and the citations there).

The crucial problem in the multi-D flows is that an automatic extension of the theory of compressible Navier–Stokes equations to the theory of multi-fluids requires

$$\operatorname{div}\operatorname{div}\mathbb{S}_i = \mathrm{const}_i \cdot \Delta \operatorname{div} \boldsymbol{u}_i, \quad i = 1, \dots, N,$$

but we have

$$\begin{pmatrix} \operatorname{div}\operatorname{div}\mathbb{S}_1 \\ \cdots \\ \operatorname{div}\operatorname{div}\mathbb{S}_N \end{pmatrix} = \mathbf{N} \begin{pmatrix} \Delta \operatorname{div} \boldsymbol{u}_1 \\ \cdots \\ \Delta \operatorname{div} \boldsymbol{u}_N \end{pmatrix},$$

where $\mathbf{N} = \{\nu_{ij}\}_{i,j=1}^{N} = \{\lambda_{ij} + 2\mu_{ij}\}_{i,j=1}^{N}$ is the matrix of total viscosities. It is possible to obtain results in the case of triangular matrices $\mathbf{N}$ (see, e. g., [9] and [10]). However, for a general $\mathbf{N}$, it is important to radically review the existing techniques developed for the compressible Navier–Stokes system. If $n = 1$ then interesting problems also appear (see below).

The authors found a way to overcome the crucial problem of total viscosity matrix structure, that is to introduce the assumptions that

- the pressures in all constituents are equal to each other,
- the velocities of the constituents in the material derivative operators are replaced by the average velocity.

Both assumptions are physically realistic in many situations (see [4] for the details), and the main mathematical effect is that the model does not lose richness inherent to the multi-fluid models (different densities and velocities of the constituent are preserved), moreover, the variety of the model grows because this trick allows to eliminate the restrictions of the viscosity matrix, and hence to admit all possible viscous terms.

As a result, we come (see [4] and [8]) to the following system of equations:

$$\partial_t \rho_i + \operatorname{div}(\rho_i \boldsymbol{v}) = 0, \quad i = 1, \ldots, N, \tag{5}$$

$$\partial_t(\rho_i \boldsymbol{u}_i) + \operatorname{div}(\rho_i \boldsymbol{v} \otimes \boldsymbol{u}_i) + \alpha_i \nabla p = \operatorname{div} \mathbb{S}_i + \rho_i \boldsymbol{f}_i, \quad i = 1, \ldots, N. \tag{6}$$

Here $\boldsymbol{v} = \sum\limits_{i=1}^{N} \alpha_i \boldsymbol{u}_i$ is the average velocity of the multi-fluid, and the coefficients $\alpha_i \left(\{\xi_j\}_{j=1}^{N} \right) > 0$ are such that $\sum\limits_{i=1}^{N} \alpha_i = 1$ (for instance, $\alpha_i = \xi_i$). Thus, the values α_i, due to their dependence on the concentrations, satisfy the equations

$$\frac{\partial \alpha_i}{\partial t} + \boldsymbol{v} \cdot \nabla \alpha_i = 0, \quad i = 1, \ldots, N.$$

In the simplest version, α_i may be regarded as constants. The main existence results for the model (5), (6) are obtained in [11], [12], [13], [14] and [15].

To conclude the Section, let us note that the multidimensional existence theorems concern only weak solutions, whose regularity is not sufficient even for the uniqueness; smoothness increase is hindered by serious obstacles. Moreover, the difficulty of multidimensional problems eclipses the study of many qualitative properties of solutions, as well as related problems including modeling; this gives researchers the right to consider the corresponding questions first in the one-dimensional case. Hence, whereas in the solvability theory for the main boundary value problems of the viscous gas there was a shift of emphasis from the one-dimensional case to the multidimensional one already two decades ago, in many other domains of the theory, the one-dimensional problems stay at the forefront.

On the other hand, as well as in the multidimensional case, the classical one-dimensional results for mono-fluids cannot be reproduced for multi-fluids automatically, in particular, due to essentially different structure of the viscous terms, namely, the presence of non-diagonal viscosity matrices; this difference in difficulty *does not depend on the dimension of the flow.*

3. Statement of the 1D problem, formulation of the result, Lagrangian coordinates

Consider the rectangular Q_T (here and below $Q_t = (0,1) \times (0,t)$) with an arbitrary finite height $T > 0$, and the system of equations ($i = 1, \ldots, N$)

$$\partial_t \rho_i + \partial_x(\rho_i v) = 0, \quad v = \frac{1}{N} \sum_{i=1}^{N} u_i, \tag{7}$$

$$\rho_i\left(\partial_t u_i + v\partial_x u_i\right) + K\partial_x \rho^\gamma = \sum_{j=1}^{N} \mu_{ij}\partial_{xx}u_j, \quad \rho = \sum_{i=1}^{N}\rho_i, \tag{8}$$

with the following initial and boundary conditions ($i = 1, \ldots, N$):

$$\rho_i|_{t=0} = \rho_{0i}(x), \quad u_i|_{t=0} = u_{0i}(x), \quad x \in [0,1], \tag{9}$$

$$u_i|_{x=0} = u_i|_{x=1} = 0, \quad t \in [0,T]. \tag{10}$$

Definition 1. Refer as a *strong solution* to the problem (7)–(10) a collection of $2N$ functions $(\rho_1, \ldots, \rho_N, u_1, \ldots, u_N)$ such that the equations (7), (8) are valid a. e. in Q_T, the initial conditions (9) are satisfied for a. a. $x \in (0,1)$, the boundary conditions (10) hold for a. a. $t \in (0,T)$, and the following inequalities and inclusions hold ($i = 1, \ldots, N$)

$$\rho_i > 0, \quad \rho_i \in L_\infty\big(0,T; W_2^1(0,1)\big), \quad \partial_t\rho_i \in L_\infty\big(0,T; L_2(0,1)\big),$$

$$u_i \in L_\infty\big(0,T; W_2^1(0,1)\big) \bigcap L_2\big(0,T; W_2^2(0,1)\big), \quad \partial_t u_i \in L_2(Q_T).$$

The result is formulated as the following Theorem.

Theorem 2. *Let the initial conditions in* (9) *satisfy the assumptions*

$$\rho_{0i} \in W_2^1(0,1), \quad \rho_{0i} > 0, \quad u_{0i} \in \overset{\circ}{W}{}_2^1(0,1), \quad i = 1, \ldots, N,$$

the symmetric viscosity matrix $\boldsymbol{M}$ *be positive, the adiabatic index* $\gamma > 1$, *the constants* $K, T > 0$. *Then there exists the unique strong solution to the problem* (7)–(10) *in the sense of Definition 1.*

In the model case (when all densities are equal to each other), the proof of Theorem 2 is given in [16] and [17]. Let us comment the scheme of the proof of Theorem 2. First of all, we prove the local in time solvability of the initial boundary value problem which is obtained from (7)–(10) via the Galerkin method (with respect to the spacial variable x) in the momentum equations (8). The proof is made via the Leray–Schauder Fixed Point Theorem. Then, basing on the uniform estimates, we pass to the limit and establish the local solvability of the problem (7)–(10), i. e. on a small time interval $[0, t_0]$. In order to continue the solution over the whole interval $[0, T]$, we prove estimates in which the constants depend on the input data of the problem and on T, but not on t_0. In conclusion, we prove the uniqueness in a standard way. The key problem in the proof of Theorem 2 is the verification of the strict positiveness and boundedness of the densities (note that this problem is the key one for the complete model (1), (2) as well). We consider this problem in more detail in Section 4.

During the study of the problem (7)–(10), the parallel use of the Lagrangian coordinates is convenient. Let us accept $y(x,t) = \int_0^x \rho(s,t)\, ds$ and t as new independent variables. Then the system (7), (8) takes the form

$$\partial_t\rho_i + \rho\rho_i\partial_y v = 0, \quad v = \frac{1}{N}\sum_{i=1}^{N} u_i,$$

$$\frac{\rho_i}{\rho}\partial_t u_i + K\partial_y\rho^\gamma = \sum_{j=1}^{N}\mu_{ij}\partial_y(\rho\partial_y u_j), \quad i = 1, \ldots, N, \quad \rho = \sum_{i=1}^{N}\rho_i.$$

The domain Q_T is transformed into the rectangular $\Pi_T = (0,d)\times(0,T)$, where $d = \int\limits_0^1 \rho_0\,dx > 0$, $\rho_0 = \sum\limits_{i=1}^N \rho_{0i}$, and the initial and boundary conditions take the form $(i = 1,\dots,N)$

$$\rho_i|_{t=0} = \widetilde{\rho}_{0i}(y), \quad u_i|_{t=0} = \widetilde{u}_{0i}(y), \quad y \in [0,d],$$
$$u_i|_{y=0} = u_i|_{y=d} = 0, \quad t \in [0,T].$$

4. Review of 1D viscous gas theory, strict positiveness and boundedness of the densities in the complete model (1), (2)

Consider the initial boundary value problem for the 1D Navier–Stokes equations (in the Lagrangian coordinates):

$$\partial_t\rho + \rho^2\partial_y v = 0, \tag{11}$$
$$\partial_t v + K\partial_y\rho^\gamma = \mu\partial_y(\rho\partial_y v), \tag{12}$$
$$\rho|_{t=0} = \widetilde{\rho}_0, \quad v|_{t=0} = \widetilde{v}_0, \quad y\in[0,d], \tag{13}$$
$$v|_{y=0} = v|_{y=d} = 0, \quad t\in[0,T]. \tag{14}$$

The first a priori estimate is obtained via the standard arguments and takes the form:

$$\|v\|_{L_\infty\left(0,T;L_2(0,d)\right)} + \|\sqrt{\rho}\partial_y v\|_{L_2(\Pi_T)} + \|\rho\|_{L_\infty\left(0,T;L_{\gamma-1}(0,d)\right)} \leqslant C. \tag{15}$$

The equation (11) allows to express $\rho\partial_y v = -\partial_t \ln\rho$ and to substitute it into (12):

$$\partial_{ty}\ln\rho + K\partial_y\rho^\gamma = -\partial_t v. \tag{16}$$

Let us multiply this equality by $\partial_y \ln\rho =: w$ and integrate over $y\in(0,d)$, then we obtain

$$\frac{1}{2}\frac{d}{dt}\left(\int\limits_0^d w^2\,dy\right) + K\gamma\int\limits_0^d \rho^\gamma w^2\,dy = -\int\limits_0^d (\partial_t v)\,w\,dy. \tag{17}$$

The right-hand side may be transformed via integration by parts due to (16):

$$-\int\limits_0^d (\partial_t v)\,w\,dy = -\frac{d}{dt}\left(\int\limits_0^d vw\,dy\right) + \int\limits_0^d \rho|\partial_y v|^2\,dy. \tag{18}$$

Thus, after integration of (17) in t, using (18), we find that

$$\|w\|^2_{L_2(0,d)} + 2K\gamma\int\limits_0^t\int\limits_0^d \rho^\gamma w^2\,dyd\tau \leqslant$$

$$\leqslant \|w_0\|^2_{L_2(0,d)} - 2\int\limits_0^d vw\,dy + 2\int\limits_0^d v_0w_0\,dy + 2\int\limits_0^t \|\sqrt{\rho}\partial_y v\|_{L_2(0,d)}\,d\tau,$$

where $w_0 = w(0,t)$. Using Cauchy's inequality and the estimate (15), we derive

$$\|w(t)\|_{L_2(0,d)} \leqslant C \quad \forall t\in[0,T],$$

i. e. the norm of the derivative $\partial_y \ln \rho$ in $L_2(0,d)$ is bounded uniformly in $t \in [0,T]$.

The equation (11) and the conditions (13), (14) obviously imply that for every $t \in [0,T]$ the equality $\rho(z(t),t) = d$ holds with some $z(t) \in [0,d]$. Hence, we can use the presentation

$$\ln \rho(y,t) = \ln \rho(z(t),t) + \int_{z(t)}^{y} \partial_s \ln \rho(s,t)\, ds.$$

This leads to the fact that $|\ln \rho(y,t)| \leqslant |\ln d| + \sqrt{d}\|w\|_{L_2(0,d)} \leqslant C$, and hence

$$0 < C^{-1} \leqslant \rho(y,t) \leqslant C.$$

For the complete multi-fluid model (1), (2), difficulties appear even on the stage of introducing the Lagrangian coordinates. In fact, if the Lagrangian coordinates are related to the velocity of one constituent (as it was made in [5]), say, with the number m, then we come to the following system of equations:

$$\partial_t \rho_i + \rho_m (u_i - u_m) \partial_y \rho_i + \rho_i \rho_m \partial_y u_i = 0,$$

$$\frac{\rho_i}{\rho_m} \partial_t u_i + \rho_i (u_i - u_m) \partial_y u_i + K \partial_y \rho^\gamma = \sum_{j=1}^{N} \mu_{ij} \partial_y (\rho_m \partial_y u_j) + \frac{1}{\rho_m} \sum_{j=1}^{N} a_{ij}(u_j - u_i), \quad i = 1, \ldots, N.$$

The analysis of this system via 1D viscous gas technique is possible only in the case of the diagonal viscosity matrix.

For the modified multi-fluid model (7), (8), the direct application of the technique also fails. However, as we have seen, it is possible to overcome all difficulties and to obtain all necessary estimates.

5. Open problems

Let us identify several open problems in the 1D theory of viscous compressible multi-fluids:

- General model with complete viscosity matrix.
- Heat-conductive models.
- Asymptotics as $t \to +\infty$.
- Other boundary value problems.

Acknowledgments

The research was supported by the Ministry of Education and Science of the Russian Federation (grant 14.Z50.31.0037).

References

[1] Mamontov A E and Prokudin D A 2013 *Meth. Appl. Anal.* **20** 179–95
[2] Nigmatulin R I 1990 *Dynamics of Multiphase Media* vol 1 (New York: Hemisphere)
[3] Rajagopal K L and Tao L 1995 *Mechanics of Mixtures* (Singapore: World Scientific Publishing)
[4] Mamontov A E and Prokudin D A 2017 Modeling viscous compressible barotropic multi-fluid flows *J. Phys.: Conf. Ser.*
[5] Kazhikhov A V and Petrov A N 1978 *Dinamika Sploshnoy Sredy* **35** 61–73
[6] Petrov A N 1982 *Dinamika Sploshnoy Sredy* **56** 105–21
[7] Zlotnik A A 1995 *Mathematical Notes* **58** 885–9
[8] Mamontov A E and Prokudin D A 2017 *Siberian Electr. Math. Reports* **14** 388–97
[9] Kucher N A, Mamontov A E and Prokudin D A 2012 *Siberian Math. J.* **53** 1075–88
[10] Mamontov A E and Prokudin D A 2014 *Izvestiya: Mathematics* **78** 554–79
[11] Mamontov A E and Prokudin D A 2016 *Siberian Electr. Math. Reports* **13** 541–83

[12] Mamontov A E and Prokudin D A 2016 *Siberian Electr. Math. Reports* **13** 664-93
[13] Mamontov A E and Prokudin D A 2016 *Siberian Math. J.* **57** 1044–54
[14] Mamontov A E and Prokudin D A 2017 *Siberian Math. J.* **58** 113–27
[15] Prokudin D A and Krayushkina M V 2016 *J. Appl. Indust. Math.* **10** 417–28
[16] Prokudin D 2017 *Siberian Electr. Math. Reports* **14** 568–85
[17] Prokudin D A 2017 Global solvability of the initial boundary value problem for a model system of one-dimensional equations of polytropic flows of viscous compressible fluid mixtures *J. Phys.: Conf. Ser.*

IOP Conf. Series: Journal of Physics: Conf. Series **894** (2017) 012060 doi:10.1088/1742-6596/894/1/012060

Group classification applications for analysis of discrete models of flow in porous media

Pavel Markov[1,2]

[1] Tyumen state university, Tyumen, Russia
[2] TyumD ITAM SB RAS, Tyumen, Russia

E-mail: markov.pv@mail.ru

Abstract. This work is about examples of applications of group classifications for continuous (differential equations) and discrete models of flow in porous media. Difference schemes and discrete dynamical systems are considered as discrete models. Group classifications has been carried out on the basis of continuous groups (Lie groups) of transformations in n-dimensional Euclidean spaces. The example of explicit difference scheme for Buckley-Leverett equation has been demonstrated where the correspondence to one of classes of the obtained earlier classifications of discrete dynamical system has been shown. The example of fractal capillary network on the basis of Sierpinski triangle has been demonstrated for the group classification of discrete dynamical systems. The approach of continuous symmetry absence or presence proof has been demonstrated using classifications of discrete dynamical systems. Difference schemes with continuous symmetry groups for partial differential equations of parabolic type, which correspond to the equation of gas flow in porous media and Rapoport-Leas equation (generalized Buckley-Leverett equation), are considered. Examples of difference schemes for these cases are given and related to classes of the obtained earlier classifications. Applications of continuous symmetry groups for numerical solutions generations are demonstrated for these cases.

1. Introduction

Analysis using continuous groups of symmetries proved to be a powerful instrument of analysis of different types of equations. But in many cases of interest it is difficult to find continuous symmetries and use them for further research. This work represents several approaches of how to use known group classifications for different discrete models of flow in porous media as examples.

This work presents three following applications approaches for group classifications of discrete models which are discrete dynamical systems [1] and difference schemes [2]:

- Applications of known group classifications of discrete dynamical systems for difference schemes.
- Proof of presence or absence of continuous symmetries for some particular discrete model of interest on the basis of known group classifications.
- Numerical solutions generations using continuous symmetries of discrete models.

The used in this work definition of continuous groups (Lie groups) of transformations in n-dimensional Euclidean spaces can be found in [3], [4], [5], [6]. The definitions and the complete

Content from this work may be used under the terms of the Creative Commons Attribution 3.0 licence. Any further distribution of this work must maintain attribution to the author(s) and the title of the work, journal citation and DOI.
Published under licence by IOP Publishing Ltd

theory of continuous symmetries for discrete dynamical systems and difference schemes used in this paper can be found in [1], [7] and [2], [8], respectively.

2. Difference schemes as discrete dynamical systems

The purpose of this section is to demonstrate an example of application for group classifications of discrete dynamical systems. Let us consider Buckley-Leverett equation [9] which describes process of two-phase flow in one-dimensional porous media without taking into account capillary forces and with constant total rate of liquid

$$\frac{\partial S}{\partial t}+\frac{V}{\varphi}\frac{df}{dS}\frac{\partial S}{\partial x}=0, \tag{1}$$

where $V=V_o+V_w=const$ - constant total rate of two-phase flow, φ - constant coefficient of porosity, $S(t,x)$ - water saturation, $\mu_w,\mu_o=const$ - viscosities of oil and water, respectively, $f(S)$ - fractional flow rate which is expressed as

$$f(S)=\frac{K_w(S)/\mu_w}{K_w(S)/\mu_w+K_o(S)/\mu_o}, \tag{2}$$

$K_w(S)$ - relative permeability for water and $K_o(S)$ - relative permeability for oil. Group analysis of this type of equations has been done in the paper [10].

For the following functions of relative permeability and fractional flow rate

$$K_w(S)=S^2,\ K_o(S)=(1-S^2),\ \mu_o=\mu_w=\mu,\ f(S)=S^2, \tag{3}$$

let us construct an example of an implicit difference scheme for six spatial cells with an uniform spatial step and with a constant time step. This difference scheme is essentially a six-dimensional discrete dynamical system and can be written as

$$\begin{aligned}
&S_1^{n+1}=S_1^n-S_1^nV\varphi^{-1}2\Delta t\Delta x^{-1}\left(S_1^n-S_L\right),\ S_2^{n+1}=S_2^n-S_2^nV\varphi^{-1}\Delta t(2\Delta x)^{-1}\left(S_3^n-S_1^n\right),\\
&S_3^{n+1}=S_3^n-S_3^nV\varphi^{-1}\Delta t\Delta x^{-1}\left(S_4^n-S_3^n\right),\ S_4^{n+1}=S_4^n-S_4^nV\varphi^{-1}\Delta t\Delta x^{-1}\left(S_4^n-S_3^n\right),\\
&S_5^{n+1}=S_5^n-S_5^nV\varphi^{-1}\Delta t(2\Delta x)^{-1}\left(S_6^n-S_4^n\right),\ S_6^{n+1}=S_6^n-S_6^nV\varphi^{-1}2\Delta t\Delta x^{-1}\left(S_R-S_6^n\right),
\end{aligned} \tag{4}$$

where the superscript n is responsible for the time layer, the subscript is responsible for cells numbers on the spatial mesh, S_L and S_R are constant boundary saturations on the left and right boundaries, respectively. This discrete dynamical system has the two-parameter group of continuous symmetry for which infinitesimal operators prolonged to discrete variables have the following form

$$X_1=S_2^n\frac{\partial}{\partial S_2^n}+S_2^{n+1}\frac{\partial}{\partial S_2^{n+1}},\ X_2=S_5^n\frac{\partial}{\partial S_5^n}+S_5^{n+1}\frac{\partial}{\partial S_5^{n+1}}. \tag{5}$$

The left and right finite-difference derivatives used for equations 2 and 5 in (4), respectively, allow one to have this symmetry group. The discrete dynamical system (4) and the two-dimensional symmetry group represented by the operators (5) refer to the case 1 of the classification of two-dimensional discrete dynamical systems [11] taking into account generalization of the classification to multidimensional cases.

3. Proof of presence or absence of continuous symmetries

This section is intended to demonstrate applications of the obtained classifications of discrete dynamical systems to search for new symmetries or, in particular, to prove their absence what is also a useful result.

It is necessary to make k iterations of resistance values on one of edges of the initial approximation when simulating flow in fractal capillary networks (as it was shown in [12]) in order to obtain a hydraulic resistance on one of edges of kth approximation of the fractal set. Iterations are performed with help of fractional-rational functions with the following conditions:

- the degree of the numerator polynomial is one greater than the degree of the denominator polynomial;
- the polynomials in the numerator and in the denominator are homogeneous functions, i.e. the total degrees of all monomials are equal to each other.

It has been shown that the mapping

$$x_{k+1} = \frac{x_k(7x_k + 4y_k)(x_k - y_k)}{2(2x_k - y_k)(3x_k + y_k)}, \qquad y_{k+1} = \frac{(x_k - y_k)(x_k^2 - 4x_k y_k - 2y_k^2)}{2(2x_k - y_k)(3x_k + y_k)}, \tag{6}$$

from the paper [1] for the mentioned above discreet model can be transformed to an one-dimensional discrete dynamical system. The similar result can be obtained for other fractal capillary networks [13] where the mapping for the considered model can written as

$$x_{k+1} = K(x_k, y_k)L(x_k, y_k)^{-1}, \qquad y_{k+1} = M(x_k, y_k)N(x_k, y_k)^{-1}, \tag{7}$$

where K, L, M, N are polynomials satisfying the conditions described above. The discrete dynamical system given by the mapping of the form (7) admits a group of homogeneous dilations

$$x' = xe^a, \; y' = ye^a, \; X_1 = x\frac{\partial}{\partial x} + y\frac{\partial}{\partial y} \tag{8}$$

The canonical variables for the given transformation group have the form

$$u = yx^{-1}, \; v = \ln y, \tag{9}$$

where the operator X_1 has the form of $\frac{\partial}{\partial y}$. The mapping (7) with using of the variables change (9) is transformed to the following form where $\tilde{K}(u) = K(u^{-1}, 1)$ and so on:

$$u_{k+1} = \tilde{L}(u_k)\tilde{M}(u_k)\left(\tilde{K}(u_k)\tilde{N}(u_k)\right)^{-1}, \; v_{k+1} = v_k + \ln\left(\tilde{M}(u_k)\tilde{N}(u_k)^{-1}\right). \tag{10}$$

Let us next consider some infinitesimal operator

$$X_2 = \xi(x, y)\frac{\partial}{\partial x} + \eta(x, y)\frac{\partial}{\partial y} \tag{11}$$

and assume that this operator relates to some continuous symmetry group together with the symmetry (8) for the system (7). Than let us use, for example, the class 4 from the classification of two-dimensional discrete dynamical systems which has been obtained in the paper [11]. Than one finds an operator X_2 that together with X_1 is related to the class 4. For this one writes and solves the following system of equations

$$X_2 u = \xi(x, y)\left(-yx^{-2}\right) + \eta(x, y) \cdot x^{-1} = 0, \qquad X_2 v = \xi(x, y) \cdot 0 + \eta(x, y) \cdot y^{-1} = y. \tag{12}$$

The solution of the system (12) is

$$\xi(x,y) = xy, \;\; \eta(x,y) = 0. \tag{13}$$

Than one uses the system of equations from [7], [11] to check whether these continuous groups of transformations corresponding to (8) and (12) are the symmetry group for (7) or not:

$$K \cdot M \cdot (L \cdot N)^{-1} - xy\frac{\partial}{\partial x}\left(K \cdot L^{-1}\right) - 0 = 0, \qquad 0 - xy\frac{\partial}{\partial x}\left(M \cdot N^{-1}\right) - 0 = 0. \tag{14}$$

It arises from the second equation that $M/N = const$ what contradicts the conditions for the functions of the discrete dynamical system (7). Hence, the assumption of the presence of yet another one-parameter symmetry group is not true. Therefore, the system (7) cannot be transformed to systems of the class 4.

4. Numerical solutions generations using continuous symmetries

Method of numerical solutions generations using transformations from continuous groups of symmetries has been presented in [14]. Knowledge of a continuous symmetry from group classifications allows in some cases to calculate one numerical solution for certain initial and boundary conditions using some classical methods. Than numerical solutions for the remaining set of conditions can be obtained by transformations using the symmetry group providing these conditions are connected by the symmetry group. Advantages of this approach are a significant increase in the calculation speed by several orders and the possibility of accuracy control for subsequent calculations. The main disadvantage is that one must find continuous symmetries first for an equation together with boundary and initial conditions of interest but the wide spectrum of known group classifications can help to overcome these difficulties.

4.1. Example of equation of gas flow in porous media

The first example is a particular case for parabolic type of partial differential equations - the equation of gas flow in one-dimensional porous media [15] which is written as

$$\frac{\partial\left(\rho(P)\varphi(P)\right)}{\partial P}\frac{\partial P}{\partial t} - \frac{\partial}{\partial x}\left(\frac{K(P)\rho(P)}{\mu(P)}\frac{\partial P}{\partial x}\right) = 0, \tag{15}$$

where $\rho(P)$ - gas density, $\varphi(P)$ - porosity, $K(P)$ - absolute permeability, $\mu(P)$ - gas viscosity, $P(t,x)$ - pressure. This equation is similar to the corresponding heat conduction equations. Different types of heat conduction equations are well studied from the point of view of group analysis using continuous groups of point transformations [16]. The group classification for the differential case was carried out for the family of equations (15) and can be found in [17].

Let us consider a concrete example of equations (15) with the equation of state of ideal gas $\rho = \chi P$ at a constant temperature and the Klinkenberg relation for permeability [15], [18] which is given by the formula (16):

$$K(P) = K_1\left(1 + K_2 P^{-1}\right), \tag{16}$$

where - $\chi = M/RT = const$, K_1 - effective liquid permeability, K_2 - the slope ratio in coordinates $K(P)$ and P^{-1}, also let $\varphi = const$ and $\mu = const$. The equation (15) can be rewritten for these coefficients as

$$\frac{\partial P}{\partial t} - \gamma\frac{\partial}{\partial x}\left((P + K_2)\frac{\partial P}{\partial x}\right) = 0, \;\; \gamma = \frac{K_1}{\mu\varphi}. \tag{17}$$

The equation (17) is related to the case of class 2 on the page 148 from the known group classification [16] with respect to the variable change $\overline{P} = P + K_2$. This equation has the four-parameter continuous group of symmetry [16], [17] with infinitesimal operators

$$X_1 = \frac{\partial}{\partial t},\ X_2 = \frac{\partial}{\partial x},\ X_3 = 2t\frac{\partial}{\partial t} + x\frac{\partial}{\partial x},\ X_4 = -t\frac{\partial}{\partial t} + (P + K_2)\frac{\partial}{\partial x}. \tag{18}$$

The following explicit difference scheme is used for numerical calculations

$$\begin{aligned} &(P_k^{n+1} - P_k^n)/\tau^n - \gamma h^{-2}\left((P_{k+1}^n - P_k^n)(P_k^n + K_2) - (P_k^n - P_{k-1}^n)(P_{k-1}^n + K_2)\right) = 0, \\ &h = x_{k+1} - x_k = const,\ \tau^n = t^{n+1} - t^n = A \cdot h^2(\overline{P}^n + K_2)^{-1}, \end{aligned} \tag{19}$$

where A - coefficient depending on maximum time value, $\overline{P}^n$ - average pressure for nth step.

The transformations from groups of symmetries with operators X_1 and X_2 from (18) mean translations for time and spatial variables for solutions what in turn means simply moving the mesh without any changes to the difference equations. More interesting is to consider of the operators X_3 and X_4 from (18) for which the transformation groups have the form for X_3 and X_4, respectively:

$$\overline{t} = e^{2a}t,\ \overline{x} = e^a x,\ \overline{P} = P, \qquad \overline{t} = e^{-a}t,\ \overline{x} = x,\ \overline{P} = (P + K_2)e^a - K_2 \tag{20}$$

Figure 1 shows the action of the symmetry group with X_4 from (20) on the numerical solution which has been obtained by solving of the corresponding system of linear equations. The specified initial pressures of the left boundary determine the parameters for generations. Verification of accuracy by the exact solution showed a coincidence of solutions and that is why they are not shown in the figure. The exact solution is presented in [19], the used initial and boundary conditions for the difference scheme are chosen with respect to the exact solution for accuracy verification.

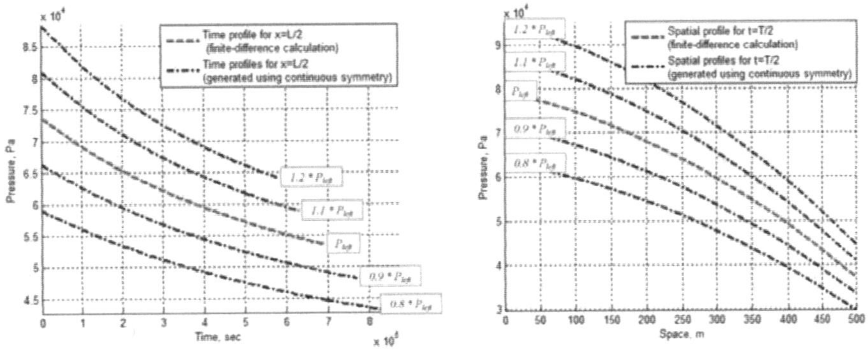

Figure 1. Profiles of pressure in time (left) and space (right) for generated solutions of (19)

4.2. *Example of Rapoport-Leas equation*

Rapoport-Leas partial differential equation describes two-phase flow in porous media with constant total liquid flow and with taking into account of capillary forces. It is also known as generalized Buckley-Leverett equation. The general form for one-dimensional porous media [20] is

$$\frac{\partial S}{\partial t} + \frac{K}{\varphi \mu_o}\frac{\partial}{\partial x}\left(K_o f \frac{dP_c}{dS}\frac{\partial S}{\partial x}\right) + \frac{V}{\varphi}\frac{df}{dS}\frac{\partial S}{\partial x} = 0, \tag{21}$$

$K = const$ - absolute permeability, $P_c = P_o - P_w$ - capillary pressure, other parameters are defined in the section 2.

The group analysis of the equation (21) has been presented in [16], [17], [21]. The equation (21) has the fairly obvious symmetry group of time and space transitions for arbitrary functions of relative permeability and capillary pressure. When these functions are specified in the form [14]

$$K_w(S) = S^N, \ K_o(S) = (1 - S^N), \ P_c(S) = -P_1 N^{-1} \ln\left|S^N (S^N - 1)^{-1}\right| + P_2, \ f(S) = S^N, \quad (22)$$

where P_1 and P_2 - scaling parameters, viscosities - $\mu_o = \mu_w = \mu = const$, N - dimensionless parameter with conditions $N \in \mathrm{R}$ and $N \neq 1$. The equation (21) has the following form with respect to the functions (22)

$$\frac{\partial S}{\partial t} - \alpha \frac{\partial}{\partial x}\left(S^{N-1}\frac{\partial S}{\partial x}\right) - \beta S^{N-1}\frac{\partial S}{\partial x} = 0, \ \alpha = \frac{KP_1}{\mu\phi}, \ \beta = -\frac{VN}{\phi}. \quad (23)$$

The equation (23) relates to the case of class 1 from table 3 according to the classifications from [21]. This equation has the three-parameter symmetry group which has the following infinitesimal operators and the form of the transformations, respectively:

$$X_1 = \frac{\partial}{\partial t}, \ X_2 = \frac{\partial}{\partial x}, \ X_3 = -t\frac{\partial}{\partial t} + \frac{S}{N-1}\frac{\partial}{\partial S},$$
$$\bar{t} = e^{-a_1}t + a_2, \ \bar{x} = x + a_3, \ \bar{S} = e^{\frac{a_1}{N-1}}S, \ a_1, a_2, a_3 \in \mathrm{R}. \quad (24)$$

The following implicit difference scheme is used to represent the method of numerical solutions generations

$$S_k^{n+1} = S_k^n + \alpha\Delta t\Delta x^{-2}\left(\left(S_k^n\right)^{N-1}\left(S_{k+1}^{n+1} - S_k^{n+1}\right) - \left(S_{k-1}^n\right)^{N-1}\left(S_k^{n+1} - S_{k-1}^{n+1}\right)\right) +$$
$$+\beta\Delta t\Delta x^{-1}\left(S_k^n\right)^{N-1}\left(S_k^{n+1} - S_{k-1}^{n+1}\right), \ \Delta t = t^{n+1} - t^n = const, \ \Delta x = x_{k+1} - x_k = const. \quad (25)$$

Figure 2 shows the results for numerical solutions generations using X_3. The parameters used for the calculations are presented in [14]. The particular exact solutions for the equation (23) are from [19]. It is used below to verify the numerical calculations obtained for the chosen difference scheme. The verification of accuracy showed good convergence of numerical solutions.

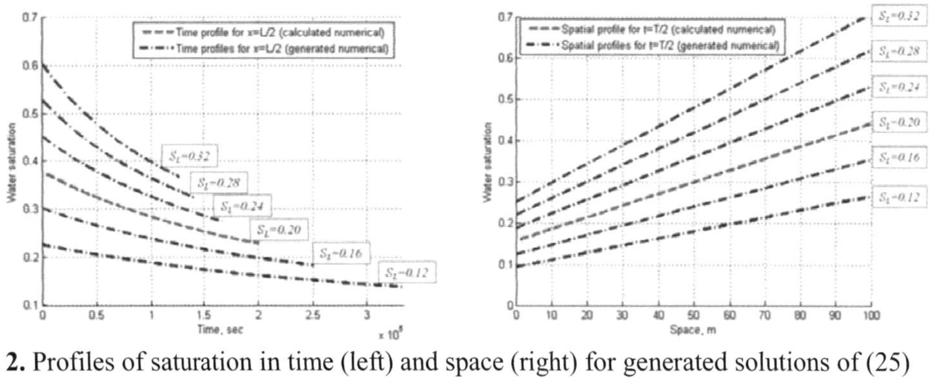

Figure. 2. Profiles of saturation in time (left) and space (right) for generated solutions of (25)

The calculation time using the generation algorithm with help of continuous symmetries is three orders lower than the calculation time using classical methods for difference schemes. It has been obtained during the numerical calculations using two examples above. This advantage allows to find solutions to many problems faster and more efficiently which require a large number of multivariate calculations for similar models: uncertainty analysis, history matching, upscaling and others.

5. Conclusions

This work is intended to demonstrate different approaches how to apply group classifications to analysis of discrete models for flow in porous media. Presented approaches of analysis of difference schemes as discrete dynamical systems, proof of absence or presence of continuous symmetries and numerical solutions generations are applied to the following models correspondently: the difference scheme for Buckley-Leverett equation, the fractal capillary network model on the basis of the Sierpinski triangle and two difference schemes for the gas flow equation and Rapoport-Leas equation.

These presented approaches can help to find applications for other group classifications of different types of equations and for different processes. First two approaches can help to find a class for particular discrete model and its equation of interest. The last one is intended to find numerical solutions by orders faster than other methods. Considered discrete models can be, for example, used for macroscale models of oil and gas reservoirs and microscale models of digital core samples.

Acknowledgments
The reported study was funded by RFBR according to the research project № 16-29-15119.

References

[1] Moritz B *et al* 1998 *J. Phys. A: Math. Gen* **31-36** 7379-7402
[2] Dorodnitsyn V 2011 *Applications of Lie Groups to Difference Equations* (Chapman and Hall, CRC) p 344
[3] Goloving S V and Chesnokov A A 2008 *Group Analysis of Differential Equations* (Novosibirsk: Novosibirsk State University) p 113
[4] Ibragimov N H 1993 *Handbook of Lie Group Analysis of Differential Equations, vol 1* (Taylor and Francis Group, CRC Press) p 448
[5] Olver P J 1986 *Applications of Lie Groups to Differential Equations* (Berlin-Heidelberg-New York-Tokyo: Springer-Verlag) p 500
[6] Ovsyannikov L V 2013 *Lectures on the Theory of Group Properties of Differential Equations* (World Scientific Publishing Company) p 156
[7] Maeda S 1987 *IMA J. Appl. Math.* **38**(2) 129-34
[8] Levi D *et al* 2001 *Czechoslovak Journal of Physics* **51**(4) 349-56
[9] Buckley S E and Leverett M C 1941 Mechanism of Fluid Displacement in Sands *Transactions AIME* (New York) p 107-116
[10] Nadjafikhah M 2010 *Advances in Applied Clifford Algebras* **20**(1) 71-7
[11] Markov P V 2013 *Rus. J. Nonlin. Dyn.* **9**(4) 641-9
[12] Adler P M 1992 *Porous Media: Geometry and Transports* (Boston: Butterworth-Heinemann) p 544
[13] Markov P V 2013 Using of continuous transformation groups in study of discrete dynamical systems *Mathematical and Informational Modeling vol* 13 (Tyumen: Tyumen State University) 172-85
[14] Markov P V and Rodionov S P 2015 *Automation, Telemechanization and Communication in Oil Industry* **12** 23-30
[15] Al-Hussainy R *et al* 1966 *Journal of Petroleum Technology* **18**(5) 624-36
[16] Lagno *et al* 2004 *Symmetry Analysis of Evolutional Type Equations* (Moscow-Izhevsk: Institute of computer research) p 392
[17] Baikov V A *et al* 1997 *Nonlinear Dynamics* **13** 395-409
[18] Klinkenberg L J 1941 *Drill. and Prod. Prac.* 200-13.
[19] Polyanin A D and Zaitsev V F 2004 *Handbook of nonlinear partial differential equations* (Boca Raton: Chapman & Hall/CRC) p 814
[20] Rapoport L A and Leas W J 1953 *Journal of Petroleum Technology* **5**(5) 139-48
[21] Oron A and Rosenau P 1986 *Physics Letters A* **118**(4) 172-6

IOP Conf. Series: Journal of Physics: Conf. Series **894** (2017) 012061 doi:10.1088/1742-6596/894/1/012061

The stability of homogeneous pulsating vertical seepage of mixture through porous media with solute immobilization

B S Maryshev[1,2]

[1]Institute of Continuous Media Mechanics UB RAS, Perm, Russia
[2]Perm State University, Perm, Russia

E-mail: bmaryshev@mail.ru

Abstract. The paper is devoted to the investigation of the linear stability of homogeneous pulsating vertical seepage of mixture through porous media with solute immobilization. The flow through porous media is modelled within the standard Darcy law and immobilization is described by the linear mobile/immobile media (MIM) model. It is known that the immobilization leads to oscilatory mode of convective instability due to transition of solute from the mobile phase to immobile. It means that immobilization effect to the flow pulsations should be non-trivial. The equations for the convection with immobilization is derived, the regime of homogeneous filtration is obtained and its stabilization is investigated. The neutral curves into the space of parameters of investigated system is ploted. The effect of homogeneous filtration stabilization under intensification of external seepage and dynamics interphase exchange is studied.

1. Introduction

The spatial structure of real porous media is rather complicated. The transport of solute particles in such a medium is not always governed by the Gaussian distribution law since the particles can stick to a solid matrix (immobilized), which slows down the solute transport through the porous medium.

In the present work, the immobilization effect is described within the MIM (mobile-immobile medium) model with linear sorption kinetics [1, 2].

The MIM approach is based on the assumption that solute can be divided to two phases: immobile (adsorbed or trapped solutes) and mobile (drifting with the flow). The evolution of concentration of solute in mobile phase is described by the classical diffusion equation including the term, which describes the transfer of solute into the immobile phase. The dependence of the rate of solute transfer into the immobile phase on the solute concentrations in two phases is linear and describes the kinetics of the "phase transition" [1, 2]. This type of MIM model is used for description of the effects at the small solute concentrations. Some problems where some additional effects are important is described in [3, 4, 5, 6, 7].

In our work, we study the stability of vertical seepage of a liquid mixture through a closed porous domain. The average difference of solute concentration is prescribed between the upper and lower boundaries. Thus difference initiates a convective motion in the gravity field. The development of convection in a closed box of porous medium is studied firstly in [8] where the

Content from this work may be used under the terms of the Creative Commons Attribution 3.0 licence. Any further distribution of this work must maintain attribution to the author(s) and the title of the work, journal citation and DOI.
Published under licence by IOP Publishing Ltd

stability of mechanical equilibrium is investigated. It is found that the lower level of instability is doubly degenerated. The dynamic properties of the convection in a porous medium in case of weak vertical homogeneous seepage is studied in [9, 10]. It was shown that vertical seepage leads to a monotonic instability mode realization. Into the work [11] the immobilization of solutes takes into account within the same problem. It is found that the transfer of the solute between phases provokes the oscillatory instability.

The objective of the present paper is to investigate the influence of modulation of vertical seepage to the stability of homogeneous filtration regime under the effect of solute immobilization. The paper consists of introduction, four Sections and conclusion. The introduction contains the motivation and short review of relevant scientific literature. In the first Section, the governing equations for the solutal convection within the linear MIM model is derived and the mathematical statement of problem is formulated. The second Section is devoted to the obtaing of base homogeneous solution and the discussion about its properties. In the third Section the equations for the small perturbations of the regime of homogeneous filtration is obtained and discussed. The fourth Section contains the numerical results of stability analysis. The conclusion summarizes the main findings of the present study.

2. Mathematical statement of problem

The two-dimensional flow of a mixture through square porous domain (of size L) is considered. Solute concentrations at the upper and lower boundaries of the domain oscillates near average values C_+ (upper) and C_- (lower) such that $C_- < C_+$ (see Figure 1). The lateral walls of the domain are assumed to be impermeable. The filtration velocity on the horizontal boundaries is prescribed $\mathbf{V} = (0, V\cos(\omega t))$.

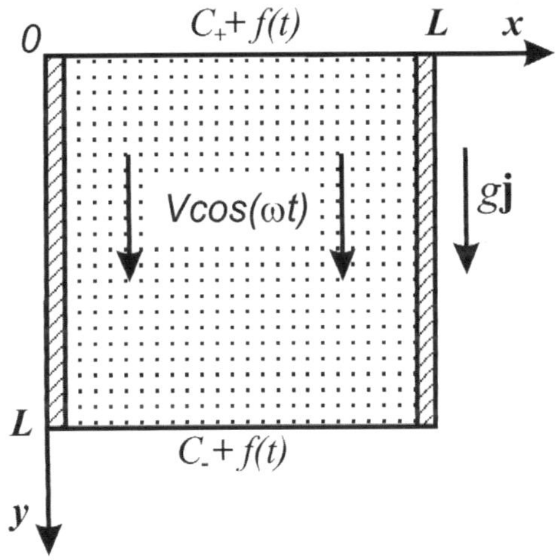

Figure 1. The problem configuration.

The described problem can be written within the Darcy-Boussinesq model with account of linear MIM model for solute immobilization as [2, 12]

$$
\begin{aligned}
\phi\frac{\partial}{\partial t}(c+q) &= -\mathbf{V}\cdot\nabla c + \phi D\nabla^2 c,\\
\frac{\partial q}{\partial t} &= \alpha(c-K_d q),\\
-\nabla p &= \frac{\eta}{\kappa}\mathbf{V} + \rho_l\beta_c c g\mathbf{j},\\
\nabla\cdot\mathbf{V} = 0, &\quad p = p' + \rho_l g z,\\
\mathbf{V}_\perp|_{y=0,L} = V\cos(\omega t), &\quad \mathbf{V}_\perp|_{x=0,L} = 0,\\
\left.\frac{\partial}{\partial x}c\right|_{x=0,L} = 0,\ c|_{y=0} = C_+, &\quad c|_{y=L} = C_-.
\end{aligned}
\tag{1}
$$

Here c is the volumic solute concentration in the mobile phase, q is the volumic solute concentration in the immobile phase, D is the effective diffusivity, α is the mass transport coefficient, K_d is solute distribution coefficient, $\mathbf{V}$ is the vector of filtration velocity, ω is the frequency of flow pulsations κ is the permeability of the porous medium, η is the dynamic viscosity, ρ is the fluid density, β_c is the coefficient of concentration expansion, p is the deviation of pressure from the hydrostatic one, g is the gravity acceleration, $\mathbf{j}$ is the unit vertical vector and ϕ is the porosity of the medium.

For nondimensionalization of the problem (1), we use the following scales for length, time, velocity, pressure and concentration: L, L^2/D, $\phi D/L$, $\phi D\eta/\kappa$, $C_0 = C_+ - C_-$.

In the dimensionless form the problem (1) reads

$$
\begin{aligned}
\frac{\partial}{\partial t}(c+q) &= -\mathbf{V}\cdot\nabla c + \nabla^2 c,\\
\frac{\partial q}{\partial t} &= ac - bq,\\
\nabla p + \mathbf{V} + Rp\, c\mathbf{j} &= 0,\\
\nabla\cdot\mathbf{V} &= 0,\\
\mathbf{V}_\perp|_{y=0,1} = Pe\cos(\Omega t), &\quad \mathbf{V}_\perp|_{x=0,1} = 0,\\
\left.\frac{\partial}{\partial x}c\right|_{x=0,1} = 0,\ c|_{y=0} = C_0, &\quad c|_{y=1} = 0.
\end{aligned}
\tag{2}
$$

Problem (2) includes the following dimensionless parameters: $Rp = C_0 g L\kappa\rho_l\beta_c/(D\eta\phi)$ is the solutal Rayleigh–Darcy number, $Pe = VL/D$ is the Péclet number, $\Omega = \omega L^2/D$, $a = \alpha K_d C_0 L^2/D$ and $b = \alpha L^2/D$ are dimensionless frequency of flow pulsations, adsorption and desorption rates respectively.

The fields of concentrations c and q and velocity $\mathbf{V}$ as the solution of problem in general case depends on x and y coordinates. But in the case of special choice of function $f(t)$ the solution of (2) is independent on x coordinate. This solution is named the regime of homogeneous filtration. The next Section is devoted to obtain the solution for homogeneous filtration.

3. Homogeneous filtration

The problem (2) allows the solution in the form $c = c_h = 1-y+f(t)$, $q = q_h = a/b\,(1-y)+g(t)$, $(V) = (V)_0 = (0, Pe\cos(\Omega t))$. The equations describing this regime of filtration take the following form

$$\begin{gathered}\partial_t (f+g) = Pe\cos(\Omega t),\\ \partial_t g = af - bg,\\ P = -Pe\cos(\Omega t)y + Rp\left(fy + y - y^2/2\right),\end{gathered} \tag{3}$$

where symbol ∂_t denotes the derivative with respect to time t, P_h is the variable describing pressure distribution.

The equations (3) are the system of ordinary differential equations. The solution of (3) can be written in the following form

$$\begin{gathered}f = \frac{aPe}{(a+b)^2+\omega^2}\left[\cos(\Omega t) - \frac{a+b}{\Omega}\right] + \frac{Pe}{\Omega}\sin(\Omega t),\\ g = \frac{aPe}{(a+b)^2+\omega^2}\left[\frac{a+b}{\Omega} - \cos(\Omega t)\right],\\ P_h = -Pe\cos(\Omega t)y + Rp\left(fy + y - y^2/2\right).\end{gathered} \tag{4}$$

The fields of concentration $c_h = 1-y+f(t)$, $q_h = a/b\,(1-y)+g(t)$ for the obtained functions f and g (4) is described the regime of homogeneous (in horizontal direction) filtration. The property of this regime is homogeneous pulsations in time of all fields with frequency Ω. The amplitude of pulsations depends on sorption parameters and it is proportional to the amplitude of external flow pulsations. The linear stability of this regime with respect to small perturbations is investigated in the next Section.

4. Linear stability problem

Let us proceed to the stability analysis of the homogeneous filtration. To this end, we represent the fields of velocity, pressure and solute concentrations in mobile and immobile phases as the sums of basic state fields and small perturbations: $q = q_h + Q$, $c = c_h + C$, $\mathbf{V} = \mathbf{V}_p + \mathbf{V}_0 = \mathbf{V}_p$, $p = P_h + P$.

We restrict ourselves to two-dimensional perturbations $\mathbf{V}_p = (u, w, 0)$. In this case, the system of linearized equations for small perturbations of the homogeneous filtration regime can be written as

$$\begin{gathered}\partial_t (C+Q) = -Pe\cos(\Omega t)\,\partial_y C + \partial_x^2 C + \partial_y^2 C,\\ \partial_t Q = aC - bQ,\\ u = -\partial_x P, \quad w + Rp\,C = -\partial_y P,\\ \partial_x u + \partial_y w = 0.\end{gathered}$$

The boundary conditions for perturbations are

$$\begin{gathered}u|_{x=0,1} = 0, \quad w|_{y=0,1} = 0,\\ \partial_x C|_{x=0,1} = 0, \quad C|_{y=0,1} = 0.\end{gathered}$$

Introducing the stream function ψ related to the velocity components as $u = -\partial_y\psi$ $w = \partial_x\psi$ and excluding the pressure we obtain

$$\begin{gathered}\partial_t (C+Q) = \partial_x^2 C + \partial_y^2 C - Pe\cos(\Omega t)\,\partial_y C\\ \partial_t Q = aC - bQ,\\ \partial_x^2\psi + \partial_y^2\psi - Rp\,\partial_x C = 0,\\ \psi\,|_{x=0,1} = 0, \quad \psi\,|_{y=0,1} = 0,\\ \partial_x C\,|_{x=0,1} = 0, \quad C\,|_{y=0,1} = 0.\end{gathered} \tag{5}$$

IOP Conf. Series: Journal of Physics: Conf. Series **894** (2017) 012061 doi:10.1088/1742-6596/894/1/012061

A solution to problem (5) can be found in the form of normal perturbations, which are periodic in the coordinate x: $Q, C \sim \cos(n\pi x)$, $\psi \sim \sin(n\pi x)$, where n is the integer number. We restrict to the case of $n = 1$, since it is known [8, 10, 13] that such perturbations correspond to the smallest possible critical values of the solutal Rayleigh–Darcy number. In this case, system (5) is represented as

$$\begin{aligned} \partial_t (C+Q) &= \partial_y^2 C - \pi^2 C - Pe\cos(\Omega t)\,\partial_y C, \\ \partial_t Q &= aC - bQ, \\ \partial_y^2 \psi &= \pi^2 \psi + Rp\,\partial_x C. \end{aligned} \tag{6}$$

The problem (6) is one dimensional system of partial differentional equations. This problem is solved numerically by implicit finite differences method of second order accuracy in space in the first order in time [14]. The stability problem is solved by the Floquet method [15]. The results of stability problem solution are presented in the next Section.

5. Results

The stability of homogeneous filtration regime with respect to two dimensional weak perturbations is investigated numerically by the Floquet method. We obtain the solution of (6) as the fields $C(y,t)$, $Q(y,t)$ and $\psi(y,t)$ for some time moment t. After that we calculate the same fields after one period of external flow modulation i.e. for time moment $t+T$ where $T = 2\pi/\Omega$. The fields $C(y,t)$ and $C(y,t)$ is compared for obtaining the multiplicator of perturbations. We use two different ways of multiplicator calculation: first is $\mu^* = \frac{\int_0^1 C(y,t+T)dy}{\int_0^1 C(y,t)dy}$ and $\mu = \frac{C(0.5,t+T)}{C(0.5,t)}$. If the multiplicator value $|\mu| > 1$ or $|\mu^*| > 1$ then the perturbations growth and they decay in opposite case [15]. As a result we plot the domains in space of system parameters where perturbations growth. The same domains for both ways of multiplicator calculation are identical, but the second way is simpler for realization because of that all presented plots are obtained for μ. All plots are demonstrated the neutral curves (the curve where $|\mu^*| = 1$) as the dependences of Rayleigh–Darcy number on another parameter. Everywhere we obtain the synchronous

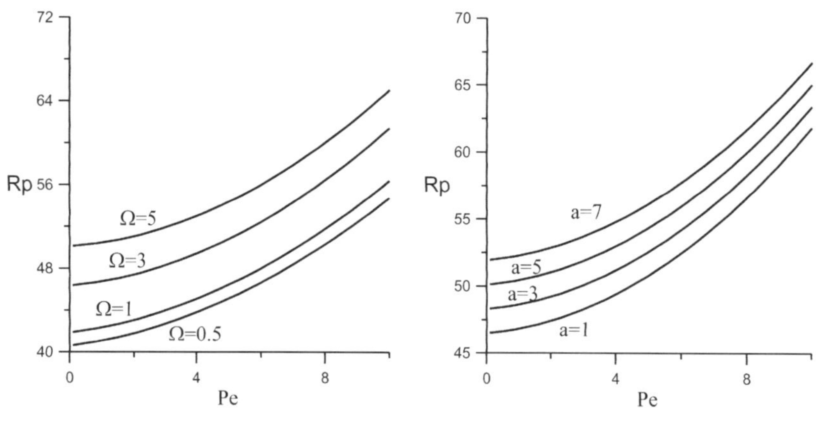

Figure 2. The neutral curves into the parameter plane Pe, Rp. The calculations are performed for the following values of parameters: $b = 5$ for both panels, $a = 5$ right panel, $\Omega = 5$ left panel. The instability of homogeneous filtration regime corresponds to the domain upper than curve.

perturbations (the frequency of pulsation for the field $C(y,t)$ equals to Ω). Due to that fact all

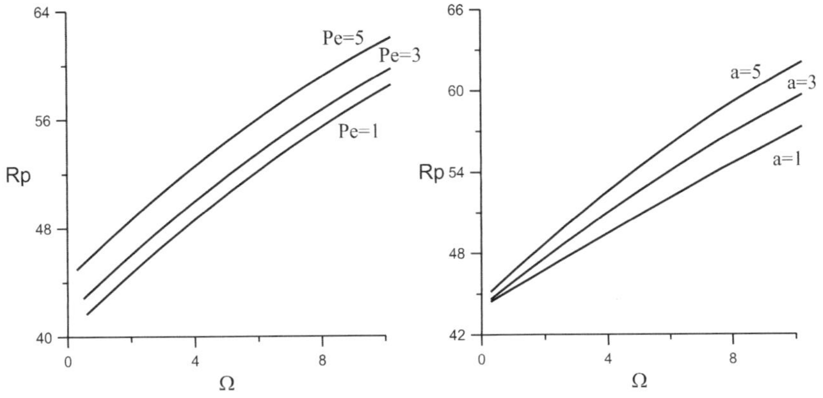

Figure 3. The neutral curves into the parameter plane Pe, Rp. The calculations are performed for the following values of parameters: $b = 5$ for both panels, $\Omega = 5$ left panel, $a = 5$ right panel. The instability of homogeneous filtration regime corresponds to the domain upper than curve.

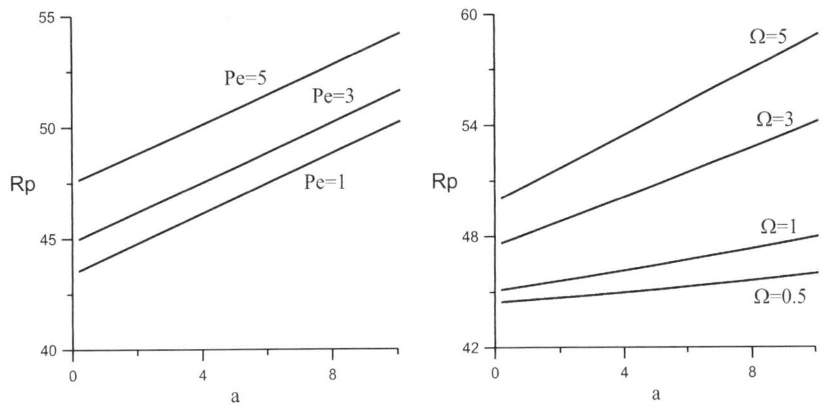

Figure 4. The neutral curves into the parameter plane a, Rp. The calculations are performed for the following values of parameters: $b = 5$ for both panels, $Pe = 5$ left panel, $\Omega = 3$ right panel. The instability of homogeneous filtration regime corresponds to the domain upper than curve.

presented neutral curves corresponds to the multiplicator value $\mu = 1$, the instability is observed into the domain upper neutral curve.

In the Figure 2 the neutral curves into the plane of parameters Pe, Rp is plotted. It is seen from Figure 2 that the increasing of pulsations amplitude (Pe) leads to the stabilization. This effect can be explained by the removal of perturbations from the porous domain by the external flux. It is well known effect (see for example [16]). The increasing of frequency (see Figure 3) also is cause of the stabilization effect for homogeneous filtration. It is linked to general growth to external flow intensity and additional the transition of solute between phases is induced by the pulsations. The enlargement of frequency speeds up the dynamics of phase transition, the solute accumulates in the immobile phase the field of mobile concentration becomes more homogeneous and as a result the system becomes more stable. The same reasoning can be apply for the explanation of stabilization under increasing of adsorption rate because of the

IOP Conf. Series: Journal of Physics: Conf. Series **894** (2017) 012061 doi:10.1088/1742-6596/894/1/012061

concentration in immobile phase is proportional to a (see for example equations (4)). The neutral curves in the plane of parameters a, Rp is presented in Figure 4.

6. Conclusion

The problem of exciting of convective motion into the close domain of porous media under pulsating seepage of mixture through the media with taking into account immobilization is investigated. The equations for the solutal convection with immobilization are derived and discussed. The solution for concentration fields is obtained for the regime of homogeneous filtration. The equations for small two-dimensional perturbation of homogeneous regime are derived and linear stability problem is formulated. The neutral curves into the space of system parameters are obtained and analyzed. It is found that both the intensification of external seepage and dynamics of interphase exchange lead to the stabilization of homogeneous filtration regime.

Acknowledgments

The work was supported by the Russian Science Foundation (grant 14-21-0090).

References

[1] Deans H A 1963 *Soc. Petrol. Eng. J.* **3** 49
[2] Van Genuchten M Th and Wierenga P J 1976 *Soil Sci Soc Am J* **40** 473
[3] Benson D A and Meerschaert M M 2009 *Adv. Water Resour.* **32** 532
[4] Maryshev B S, Joelson M, Lyubimov D V, Lyubimova T P and Néel M C 2009 *J. Phys A-Math. Theor.* **42** 115001
[5] Maryshev B S, Cartalade A, Latrille C, Joelson M and Néel M C 2013 *Comput. Math. Appl.* **66** 630
[6] Maryshev B S, Lyubimov D V, Lyubimova T P and Néel M C 2011 *Fluid Dyn.* **46** 148
[7] Maryshev B S, Lyubimov D V and Lyubimova T P 2016 *Transport Porous Med.* **111** 239
[8] Lyubimov D 1975 *J. Appl. Mech. Tech. Phys.* **16** 257
[9] Lyubimov D 1993 *Ch. Properties of Thermal Convection in Porous Medium. Instabilities in Multiphase Flows* (UK London: Plenum Publishing Co) p 289
[10] Bratsun D A, Lyubimov D V and Roux B 1995 *Physica D* **82** 398
[11] Maryshev B S, Lyubimova T P and Lyubimov D V 2016 *Int. J. Heat Mass Tran.* **102** 113
[12] Nield D A and Bejan A 2012 *Convection in Porous Media* (New York USA: Springer-Verlag)
[13] Lapwood E R 1948 *Proc. Camb. Philos. Soc.* **44** 508
[14] Fort T 1948 *Finite Differences and Difference Equations in the Real Domain* (Gloucestershire UK: Clarendon Press)
[15] Nayfeh A H 2008 *Perturbation Methods* (Weinheim Germany: Wiley-VCH)
[16] Sutton F M 1970 *Phys. Fluids* **13** 1931

IOP Conf. Series: Journal of Physics: Conf. Series **894** (2017) 012062 doi:10.1088/1742-6596/894/1/012062

Hydrodynamic dispersion in porous media with macroscopic disorder of parameters

D S Goldobin[1,2] **and B S Maryshev**[1,2]

[1]Institute of Continuous Media Mechanics UB RAS, Perm, Russia
[2]Perm State University, Perm, Russia

E-mail: Denis.Goldobin@gmail.com, bmaryshev@mail.ru

Abstract. We present an analytical derivation of the macroscopic hydrodynamic dispersion for flows in porous media with frozen disorder of macroscopic parameters: porosity and permeability. The parameter inhomogeneities generate inhomogeneities of filtration flow which perform fluid mixing and, on the large spacial scale, act as an additional effective diffusion (eddy diffusivity *or* hydrodynamic dispersion). The derivation is performed for the general case, where the only restrictions are (i) the spatial autocorrelation functions of parameter inhomogeneities decay with the distance r not slower than $1/r^n$ with $n > 1$, and (ii) the amplitudes of inhomogeneities are small compared to the mean value of parameters. Our analytical findings are confirmed with the results of direct numerical simulation for the transport of a passive scalar in inhomogeneous filtration flow.

1. Introduction

Real porous media are characterized by two scales of porosity: microscopic, the scale of small pore channels or granules in granular materials, and macroscopic, the scale of inhomogeneities in the porous medium structure. Macroscopic inhomogeneities are associated with various processes (washing out, crack formation, etc.) and with stones and other inclusions. These inhomogeneities are always present (see, for example, [1, 2]), even in the ideal case, where the porous medium is formed by identical spheres.

During the imposed displacement of the fluid through porous medium, these inhomogeneities lead to spatially irregular currents and are responsible for the dispersion of the fluid particles, which causes additional (to molecular diffusion) hydrodynamic diffusion of passive impurities (see, e.g., [1, 2, 3, 4, 5, 6, 7]). In contrast to the molecular diffusion, the convective dispersion is anisotropic and non-constant, proportional to the filtration flowrate, say, U. The isotropic medium is characterized by the convective dispersion with two main quantities: the longitudinal dispersivity, $D_{\|} = Ud_1$, and the lateral one, $D_{\perp} = Ud_2$. For typical displacement flowrates, convective dispersion is not less important than molecular diffusion (cf. [1, 2, 7]). In addition to anisotropy and dependence on the flowrate of displacement, an important feature of macroscopic convective dispersion is that it provides initial perturbations on the interface of immiscible displaced fluids when the displacement of this surface is unstable [8, 9]. Thus, any information on the fundamental characteristics of convective dispersion is important for research on various processes in nature (groundwater flow, chemical and biological impurities in soils, soil and rock erosion, formation of natural gas hydrate deposits, etc. [10, 11, 12, 13]) and industry (filtering devices, cooling of reactors, carbon capture and storage, etc. [14, 15]).

Content from this work may be used under the terms of the Creative Commons Attribution 3.0 licence. Any further distribution of this work must maintain attribution to the author(s) and the title of the work, journal citation and DOI.
Published under licence by IOP Publishing Ltd

The development of the theory of microscopic dispersion was the subject of series of works ([4, 5], a detailed review can be found in [2, 7]); later, the researchers turned to theoretical study of macroscopic dispersion, which is the result of irregular currents caused by large-scale inhomogeneities [16, 17, 18, 19, 20, 21]. The diffusivity of the random process is proportional to the square of the amplitude of the fluctuations and the length of the correlations. As result, the macroscopic diffusion can be more important than the microscopic diffusion, even with very weak macroscopic inhomogeneities, if the correlation length of the dispersing flows is sufficiently long. The paper [18] (which provoked the comment [22] and the answer [23]) became one of the fundamental works on the theory of macroscopic dispersion. This work had a crucial impact on all the subsequent development of this theory [19, 20, 21, 24], although it is not free of some limitations, beyond which we advance in our paper, and flaws in interpretation of analytical results. The assumption of ergodicity was adopted in the work unconditionally; however, it is incompatible with strong topological restrictions on the mutual diffusion of fluid particles for cases of one- and two-dimensional inhomogeneities (many of these flaws were realized and corrected, for example, in [19, 20, 21]).

In this paper, we consider the problem of hydrodynamic dispersion on macroscopic inhomogeneities in the imposed displacement of a fluid that saturates a porous medium. The heterogeneity of both permeability and porosity is considered (the latter was ignored in many previous studies), the most general variants of the statistical properties of inhomogeneities are allowed. The only restrictions are (i) the spatial autocorrelation functions of parameter inhomogeneities decay with the distance r not slower than $1/r^n$ with $n > 1$, and (ii) the amplitudes of inhomogeneities are small compared to the mean value of parameters. These assumptions are reasonable for real geophysical and technological systems. We do not impose any additional assumptions on the statistical properties of inhomogeneities (such as Gaussian distribution, etc.). In Sec. 1 we provide the mathematical formulation of the problem and derive the equation for fluid flow through an inhomogeneous porous medium. In Sec. 2, the expressions for the dispersivity tensor for isotropic case are obtained and discussed. For the verification of analytical results, the direct numerical simulation (DNS) is performed (Sec. 4). In Conclusion (Sec. 5), the main results are discussed.

2. Flow through porous medium with macroscopic inhomogeneity of parameters

It is known that particles transported through a system of channels (pores) are subject to additional dispersion due to the inhomogeneity of the velocity field across the channel. This effect can lead, among other things, to non-Fickian dispersion (anomalous diffusion, e.g., see [25]) due to immobilization of solute particles. However, for solutes with significant molecular diffusion, the concentration distribution tends to be uniform across the pores (as, e.g., in appendix of [26]), and the solute transport occurs with a velocity equal to (at microscopic level) the average fluid velocity in the pores $\mathbf{v}$, which is related to the filtration rate $\mathbf{u} = \varphi\mathbf{v}$, where φ is the porosity.

Flows in a porous medium where the fluid is displaced by an imposed pressure gradient are determined by the Darcy law and by the law of conservation of matter;

$$\mathbf{u} = -\frac{1}{\eta}K(\mathbf{r})\nabla p\,, \qquad \nabla\cdot\mathbf{u} = 0\,, \tag{1}$$

where $K(\mathbf{r})$ is the permeability of medium, η is the fluid viscosity and $\mathbf{r}$ is the spatial coordinate vector. The medium parameters are inhomogeneous in space; permeability

$$K(\mathbf{r}) = K_0(1 + \varepsilon\varkappa(\mathbf{r}))$$

and porosity

$$\varphi(\mathbf{r}) = \frac{\varphi_0}{1 + \varepsilon\mu(\mathbf{r})}\,,$$

IOP Conf. Series: Journal of Physics: Conf. Series **894** (2017) 012062 doi:10.1088/1742-6596/894/1/012062

where K_0 and φ_0 are the reference values of permeability and porosity, respectively; $\varkappa(\mathbf{r})$ and $\mu(\mathbf{r})$ are random functions of coordinates with zero mean values, $\langle\varkappa(\mathbf{r})\rangle = \langle\mu(\mathbf{r})\rangle = 0$; and $\varepsilon \ll 1$ is a small dimensionless parameter which quantifies the characteristic relative variation of medium parameters.

We assume that the externally imposed pressure drop creates the mean gradient $\nabla P_0 = -\eta K_0^{-1}\mathbf{U}$, where $\mathbf{U}$ is the mean enforced filtration flowrate. Further, we decompose pressure p into the mean-trend part and the fluctuating part P;

$$\nabla p = -\eta K_0^{-1}\mathbf{U} + \nabla P\,. \tag{2}$$

Substitution of (2) into the law of matter conservation yields the equation for the pressure fluctuation; to the leading order in ε,

$$\Delta P = \varepsilon\eta K_0^{-1}\mathbf{U}\cdot\nabla\varkappa(\mathbf{r})\,. \tag{3}$$

Equation (3) is a standard Poisson-type equation and its solution can be found by the Green's function method

$$P(\mathbf{r}) = \varepsilon\eta K_0^{-1}\mathbf{U}\cdot\int\nabla\varkappa(\mathbf{s})\,P_g(\mathbf{r}-\mathbf{s})\,\mathrm{d}^n\mathbf{s} = -\varepsilon\eta K_0^{-1}\mathbf{U}\cdot\int\varkappa(\mathbf{s})\,\nabla_{\mathbf{s}}P_g(\mathbf{r}-\mathbf{s})\,\mathrm{d}^n\mathbf{s},$$

where n is the dimensionality of the problem (2- or 3-d) and the subscript of the nabla-operator indicates the coordinates for differentiation. Function $P_g(\mathbf{r})$ is the Green's function for pressure fluctuations; it is $P_g(\mathbf{r}) = (2\pi)^{-1}\ln r$ for two-dimensional flow and $P_g(\mathbf{r}) = -1/(4\pi r)$ for three-dimensional flow (here $r = |\mathbf{r}|$). The expression for filtration flowrate is

$$\begin{aligned}\mathbf{u}(\mathbf{r}) = -\eta^{-1}K(\mathbf{r})\nabla p &\simeq \big(1+\varepsilon\varkappa(\mathbf{r})\big)\mathbf{U} + \varepsilon\nabla\left(\mathbf{U}\cdot\int\varkappa(\mathbf{s})\,\nabla_{\mathbf{s}}P_g(\mathbf{r}-\mathbf{s})\,\mathrm{d}^n\mathbf{s}\right)\\ &= \big(1+\varepsilon\varkappa(\mathbf{r})\big)\mathbf{U} - \varepsilon\mathbf{U}\cdot\int\varkappa(\mathbf{s}+\mathbf{r})\,\overline{\overline{\mathbf{G}}}(\mathbf{s})\,\mathrm{d}^n\mathbf{s}\,,\end{aligned}$$

where tensor $G_{ij}(\mathbf{r}) = \nabla_i\nabla_j P_g(\mathbf{r})$. Finally, the equation for fluid particle motion reads

$$\dot{\mathbf{r}} = \mathbf{v}(\mathbf{r}) = \frac{\mathbf{u}(\mathbf{r})}{\varphi(\mathbf{r})} \simeq \big(1+\varepsilon\{\mu(\mathbf{r})+\varkappa(\mathbf{r})\}\big)\mathbf{V} - \varepsilon\mathbf{V}\cdot\int\varkappa(\mathbf{s}+\mathbf{r})\,\overline{\overline{\mathbf{G}}}(\mathbf{s})\,\mathrm{d}^n\mathbf{s}\,, \tag{4}$$

where $\mathbf{V} \equiv \varphi_0^{-1}\mathbf{U}$ is the average velocity of fluid particles. In the next section we calculate the diffusivity tensor from the equation of particle motion.

3. Diffusivity tensor

The diffusivity tensor for random process $\mathbf{r}(t)$ is (cf. [27])

$$D_{ij} = \lim_{t\to\infty}\frac{1}{2t}\left\langle(r_i(t)-r_i(0)-\langle\dot{r}_i\rangle_t\,t)(r_j(t)-r_j(0)-\langle\dot{r}_j\rangle_t\,t)\right\rangle_t \tag{5}$$

where $\langle\cdot\rangle_t$ stands for the averaging in time and the subscripts i and j indicate spatial coordinates. Let the particle coordinate deviation from the average displacement be $q_i(t) \equiv r_i(t) - r_i(0) - V_i t$. The diffusivity tensor becomes [27]

$$D_{ij} = \lim_{t\to\infty}\frac{1}{2t}\left\langle q_i(t)\,q_j(t)\right\rangle_t = \frac{1}{2}\int_{-\infty}^{\infty}\left\langle\dot{q}_i(t)\,\dot{q}_j(t+\tau)\right\rangle_t\mathrm{d}\tau\,. \tag{6}$$

Substituting (4) into (6) with $r_i(t) - r_i(0) = V_i t + q_i(t)$, one finds

$$\begin{aligned}
2D_{ij} &= \varepsilon^2 V_i V_j \int_{-\infty}^{\infty} \mathrm{d}\tau \left\langle [\mu+\varkappa]\big(\mathbf{V}t + \mathbf{q}(t)\big)\, [\mu+\varkappa]\big(\mathbf{V}(t+\tau) + \mathbf{q}(t+\tau)\big)\right\rangle_t \\
&-\varepsilon^2 V_i V_k \int_{-\infty}^{\infty} \mathrm{d}\tau \left\langle [\mu+\varkappa]\big(\mathbf{V}t + \mathbf{q}(t)\big) \int \mathrm{d}^n\mathbf{s}\, \varkappa\big(\mathbf{V}(t+\tau) + \mathbf{q}(t+\tau) + \mathbf{s}\big)\, G_{kj}(\mathbf{s})\right\rangle_t \\
&-\varepsilon^2 V_k V_j \int_{-\infty}^{\infty} \mathrm{d}\tau \left\langle [\mu+\varkappa]\big(\mathbf{V}(t+\tau) + \mathbf{q}(t+\tau)\big) \int \mathrm{d}^n\mathbf{s}\, \varkappa\big(\mathbf{V}t + \mathbf{q}(t) + \mathbf{s}\big)\, G_{ik}(\mathbf{s})\right\rangle_t + \varepsilon^2 V_k V_m \\
&\times \int_{-\infty}^{\infty} \mathrm{d}\tau \left\langle \int \mathrm{d}^n\mathbf{s}_1\, \varkappa\big(\mathbf{V}t + \mathbf{q}(t) + \mathbf{s}_1\big)\, G_{ik}(\mathbf{s}_1) \int \mathrm{d}^n\mathbf{s}_2\, \varkappa\big(\mathbf{V}(t+\tau) + \mathbf{q}(t+\tau) + \mathbf{s}_2\big)\, G_{mj}(\mathbf{s}_2)\right\rangle_t, \quad (7)
\end{aligned}$$

where $[\mu + \varkappa](\mathbf{r}) \equiv \mu(\mathbf{r}) + \varkappa(\mathbf{r})$. One can introduce the correlation functions

$$C_{f,g}(\mathbf{s}_2 - \mathbf{s}_1) = \langle f(\mathbf{s}_1)\, g(\mathbf{s}_2)\rangle\,, \qquad C_f(\mathbf{s}_2 - \mathbf{s}_1) = \langle f(\mathbf{s}_1)\, f(\mathbf{s}_2)\rangle\,,$$

(the angle brackets without subscript indicate averaging over noise realizations) and rewrite expression (7) in a shorter form

$$\begin{aligned}
2D_{ij} &= \varepsilon^2 V_i V_j \int_{-\infty}^{\infty} \mathrm{d}\tau \left\langle C_{\mu+\varkappa}\big(\mathbf{V}\tau + \mathbf{q}(t+\tau) - \mathbf{q}(t)\big)\right\rangle_t \\
&-\varepsilon^2 V_i V_k \int_{-\infty}^{\infty} \mathrm{d}\tau \int \mathrm{d}^n\mathbf{s}\, G_{kj}(\mathbf{s}) \left\langle C_{\mu+\varkappa,\varkappa}\big(\mathbf{V}\tau + \mathbf{q}(t+\tau) - \mathbf{q}(t) + \mathbf{s}\big)\right\rangle_t \\
&-\varepsilon^2 V_k V_j \int_{-\infty}^{\infty} \mathrm{d}\tau \int \mathrm{d}^n\mathbf{s}\, G_{ik}(\mathbf{s}) \left\langle C_{\mu+\varkappa,\varkappa}\big(-\mathbf{V}\tau - \mathbf{q}(t+\tau) + \mathbf{q}(t) + \mathbf{s}\big)\right\rangle_t \\
&+\varepsilon^2 V_k V_m \int_{-\infty}^{\infty} \mathrm{d}\tau \int \mathrm{d}^n\mathbf{s}_1\, G_{ik}(\mathbf{s}_1) \int \mathrm{d}^n\mathbf{s}_2\, G_{mj}(\mathbf{s}_2) \left\langle C_{\varkappa}\big(\mathbf{V}\tau + \mathbf{q}(t+\tau) - \mathbf{q}(t) + \mathbf{s}_2 - \mathbf{s}_1\big)\right\rangle_t. \quad (8)
\end{aligned}$$

Further, we employ the assumption of smallness of macroscopic inhomogeneities of porosity and permeability, which allows one neglecting $|\mathbf{q}(t+\tau) - \mathbf{q}(t)|$ in equation (8). Indeed, $|\mathbf{q}(t+\tau) - \mathbf{q}(t)|$ is related to distortion of the homogeneous macroscopic flow and $\propto \varepsilon$; therefore, for vanishing ε the perturbation of the arguments of correlation functions in (8) should be either small for nonlarge τ or small compared to $V\tau$ for large τ. Hence, equation (8) reads

$$\begin{aligned}
D_{ij} = \frac{\varepsilon^2}{2V} V_i V_j \int_{-\infty}^{\infty} \mathrm{d}l\, C_{\mu+\varkappa}(l\mathbf{e}_U) &- \frac{\varepsilon^2}{2V} V_i V_k \int \mathrm{d}^n\mathbf{s}\, G_{kj}(\mathbf{s}) \int_{-\infty}^{\infty} \mathrm{d}l\, C_{\mu+\varkappa,\varkappa}(l\mathbf{e}_U + \mathbf{s}) \\
&- \frac{\varepsilon^2}{2V} V_k V_j \int \mathrm{d}^n\mathbf{s}\, G_{ik}(\mathbf{s}) \int_{-\infty}^{\infty} \mathrm{d}l\, C_{\mu+\varkappa,\varkappa}(l\mathbf{e}_U + \mathbf{s}) \\
+\frac{\varepsilon^2}{2V} V_k V_m \int \mathrm{d}^n\mathbf{s}_1\, G_{ik}(\mathbf{s}_1) \int \mathrm{d}^n\mathbf{s}_2\, G_{mj}(\mathbf{s}_2) &\int_{-\infty}^{\infty} \mathrm{d}l\, C_{\varkappa}(l\mathbf{e}_U + \mathbf{s}_2 - \mathbf{s}_1) + \mathcal{O}(\varepsilon^4)\,, \quad (9)
\end{aligned}$$

where $l = V\tau$ and $\mathbf{e}_U$ is the unit vector along the direction of the average imposed displacement flux, $\mathbf{e}_U = \mathbf{U}/U = \mathbf{V}/V$.

With expression (9), one can calculate the diffusivity for any small random inhomogeneity of permeability and porosity. However, the employed definition (5) is the definition for the diffusion in time. The spacial dispersivity of fluid particles equals to the temporal diffusivity only for an ergodic process. In the next section, the case of the inhomogeneities with isotropic correlation functions is considered and expression is validated by comparison to the results of direct numerical simulation. The issue of the system ergodicity is also discussed.

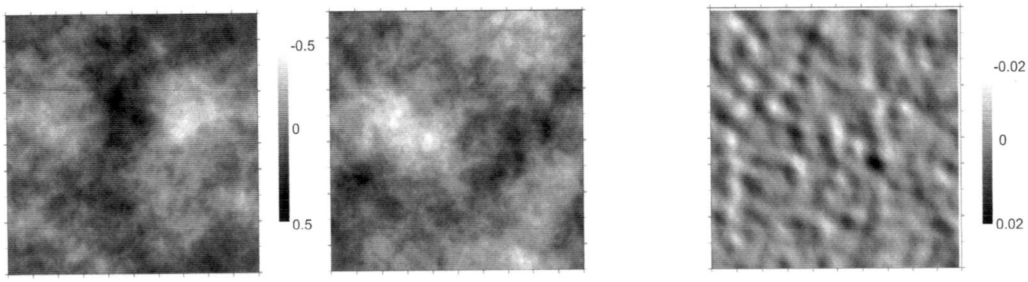

Figure 1. Sample fields of inhomogeneities of permeability $\varkappa(\mathbf{r})$ (left panel) and porosity $\mu(\mathbf{r})$ (middle panel) for the 2-d case and corresponding fluctuating part P of the pressure field (right panel) for average interstitial flow velocity $\mathbf{V} = (0.86, 0.52)$ (the direction corresponds to the golden section, which minimizes influence of periodicity-related resonances in the flow).

4. Isotropic statistical properties of inhomogeneities

For the case of isotropic correlation functions, $C_{f,g}(\mathbf{s}) = C_{f,g}(|\mathbf{s}|)$ and equation (9) yields

$$D_{ij} = \frac{\varepsilon^2}{2V} V_i V_j \int_{-\infty}^{\infty} \mathrm{d}l\, C_{\mu+\varkappa}(l) + \mathcal{O}(\varepsilon^4)\,.$$

The tensor D_{ij} is symmetric and can be rewritten into the diagonal form with components $D'_{11} = D_{\parallel}$, $D'_{22} = D_{\perp}$, and $D'_{12} = D'_{21} = 0$. In these terms,

$$D_{\parallel} = \frac{\varepsilon^2 V}{2} \int_{-\infty}^{\infty} C_{\mu+\varkappa}(l)\, \mathrm{d}l\,, \qquad D_{\perp} = \mathcal{O}(\varepsilon^4)\,. \tag{10}$$

In order to validate expressions (10), we performed direct numerical simulations for dimensionless equations (in which $\eta = 1$, $K_0 = 1$). Firstly, we generated numerically random functions $-0.5 < \varkappa(\mathbf{r}) < 0.5$ and $-0.5 < \mu(\mathbf{r}) < 0.5$ with isotropic statistical properties (see Figure 1). Secondly, equations (1) for pressure were solved by means of a standard finite difference method of second order accuracy for periodic boundary conditions (a sample field of pressure fluctuations is plotted in Figure 1). Thirdly, the velocity of fluid particles was calculated as

$$\mathbf{v} = \left(1 + \varepsilon\mu\right)\left(1 + \varepsilon\varkappa\right)\left(\mathbf{V} - \varphi_0^{-1}\nabla P\right)\,.$$

Finally, the advection-diffusion equation for passive scalar (C) was solved with prescribed molecular diffusivity D_m and calculated filtration velocity field $\mathbf{u}$:

$$\frac{\partial C}{\partial t} + \mathbf{v}\cdot\nabla C = \frac{D_\mathrm{m}}{\varphi(\mathbf{r})}\nabla\cdot\left(\varphi(\mathbf{r})\nabla C\right)\,. \tag{11}$$

The solution for equation (11) was obtained by the direct numerical simulation with the explicit finite difference scheme of second order accuracy in space and first order accuracy in time. This solution was further compared with the fundamental solution of diffusion equation with symmetric diffusivity tensor D_{ij}, which reads

$$C_*(x,y,t) = \frac{1}{4\pi t\sqrt{D_{11}D_{22} - D_{12}^2}} \exp\left[-\frac{D_{22}x^2 - 2D_{12}xy + D_{11}y^2}{4t\left(D_{11}D_{22} - D_{12}^2\right)}\right], \tag{12}$$

for the estimation of diffusivity tensor components.

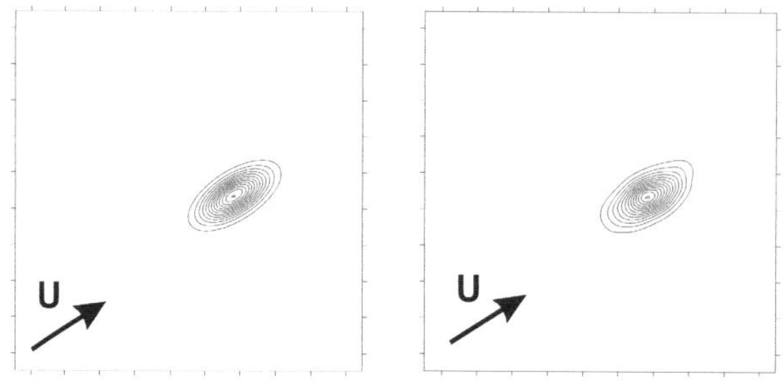

Figure 2. The isolines of the concentration field. The field C (numerical solution of (11)) is plotted in the right panel, the Gaussian approximation C_* (12) is plotted in the left panel. The direction of average filtration flow is indicated with the arrow, $\varphi_0^{-1}\mathbf{U} = \mathbf{V} = (0.86, 0.52)$. The calculation of $C(x, y, t)$ was performed with the pressure field presented in Figure 1 and molecular diffusivity $D_\mathrm{m} = 10^{-3}$. The components of diffusivity tensor D_{ij} were estimated by BFGS method ($D_{22} = 6.22 \cdot 10^{-3}$, $D_{11} = 1.43 \cdot 10^{-2}$, $D_{12} = 8.08 \cdot 10^{-3}$).

The parameter estimation was performed by the least squares method with residuals' sum

$$E = \sum_{x,y,t} \big(C(x, y, t) - C_*(x, y, t)\big)^2,$$

where summation was performed for all nodes of space mesh and for 30 time steps (from number 500 to number 530, after the transient process). The procedure of the error minimization with respect of components of D_{ij} was based on conventional Broyden-Fletcher-Goldfarb-Shanno (BFGS) algorithm [28]. The example of calculated fields C and C_* is presented in Figure 2.

In Table 1 the values of components of the diffusivity tensor estimated by BFGS method and compared to the diffusivity tensor calculated with expression (10), read as $D_\parallel = \varepsilon^2 V I$ with the integral $I = (1/2) \int_{-\infty}^{\infty} C_{\mu+\varkappa}(l)\, \mathrm{d}l$. Equation (11) is solved for the case of a small but non-zero molecular diffusivity. The latter is included into consideration for two reasons: firstly, the molecular diffusion is present for transport processes in the real porous media, and, secondly

Table 1. The analytical calculations with equation (10) are compared to the results of DNS for isotropic disorder in permeability and porosity for two-dimensional system with non-negligible molecular diffusivity D_m. Results are present for the average interstitial fluid velocity $\mathbf{V} = (0.86, 0.52)$; the integral $I = \frac{1}{2} \int_{-\infty}^{\infty} C_{\mu+\varkappa}(l)\, \mathrm{d}l$.

parameters		analytic			DNS estimation	
ε	D_m	I	$D_\parallel$	$D_\parallel + D_\mathrm{m}$	$D_{\parallel,\mathrm{net}}$	$D_{\perp,\mathrm{net}}$
10^{-1}	10^{-3}	1.82	$1.82 \cdot 10^{-2}$	$1.92 \cdot 10^{-2}$	$1.93 \cdot 10^{-2}$	$1.23 \cdot 10^{-3}$
10^{-1}	$5 \cdot 10^{-4}$	1.63	$1.63 \cdot 10^{-2}$	$1.68 \cdot 10^{-2}$	$1.67 \cdot 10^{-2}$	$6.12 \cdot 10^{-4}$
10^{-1}	10^{-4}	1.71	$1.71 \cdot 10^{-2}$	$1.72 \cdot 10^{-2}$	$1.72 \cdot 10^{-2}$	$2.17 \cdot 10^{-4}$
$5 \cdot 10^{-2}$	10^{-3}	1.58	$3.95 \cdot 10^{-3}$	$4.95 \cdot 10^{-3}$	$4.98 \cdot 10^{-3}$	$1.08 \cdot 10^{-3}$
$5 \cdot 10^{-2}$	$5 \cdot 10^{-4}$	1.72	$4.31 \cdot 10^{-3}$	$4.81 \cdot 10^{-3}$	$4.85 \cdot 10^{-3}$	$5.52 \cdot 10^{-4}$
$5 \cdot 10^{-2}$	10^{-4}	1.48	$3.72 \cdot 10^{-3}$	$3.82 \cdot 10^{-3}$	$3.81 \cdot 10^{-3}$	$1.68 \cdot 10^{-4}$
10^{-2}	10^{-3}	1.92	$1.92 \cdot 10^{-4}$	$1.19 \cdot 10^{-3}$	$1.19 \cdot 10^{-3}$	$1.00 \cdot 10^{-3}$
10^{-2}	$5 \cdot 10^{-4}$	1.73	$1.73 \cdot 10^{-4}$	$6.73 \cdot 10^{-4}$	$6.73 \cdot 10^{-4}$	$5.00 \cdot 10^{-4}$
10^{-2}	10^{-4}	1.67	$1.67 \cdot 10^{-4}$	$2.67 \cdot 10^{-4}$	$2.67 \cdot 10^{-4}$	$1.01 \cdot 10^{-4}$

(even more importantly mathematically), it brings the ergodicity into the system. The diffusivity components are presented in the coordinate system associated to the filtration velocity $\mathbf{U}$. $D_{\|,\mathrm{net}}$ and $D_{\perp,\mathrm{net}}$ are the net longitudinal and lateral diffusion coefficients contributed by both the molecular diffusion and the hydrodynamic dispersion. Provided the transport in the system is ergodic, the Brownian fluctuations of the particle position and the fluctuations owned by the macroscopic flow inhomogeneities are mutually independent; in this case, $D_{\|/\perp,\mathrm{net}} = D_{\|/\perp} + D_\mathrm{m}$. The results of the analytical theory and direct numerical simulations were found to be in good agreement up to as large values of ε as 0.1 (see Table 1).

5. Conclusion

The expression for macroscopic hydrodynamic dispersivity for inhomogeneous flows in porous medium has been derived analytically on the basis of the fact that the dispersion is produced by macroscopic inhomogeneities of porosity and permeability of the medium. The derivation has been performed for the most general case of statistical properties of inhomogeneities. The only restrictions are (i) relative smallness of inhomogeneities and (ii) finiteness of the autocorrelation length. The analytical results have been verified by comparison to the results of direct numerical simulation of the transport of a passive scalar in two-dimensional inhomogeneous filtration flow with a non-negligible molecular diffusivity.

Acknowledgments

The work was partially supported by the Grant of the President of Russian Federation (Grant No. MK-1447.2017.5).

References

[1] Collins R E 1961 *Flow of Fluids through Porous Materials* (New York: Reinhold)
[2] Sahimi M 1993 *Rev. Mod. Phys.* **65** 1393
[3] von Rosenberg D U 1956 *AIChE J.* **2** 55
[4] Saffman P G 1959 *J. Fluid Mech.* **6** 321
[5] Saffman P G 1960 *J. Fluid Mech.* **7** 194
[6] Dagan G 1986 *Water Resour. Res.* **22** 120S
[7] Dagan G 1987 *Annu. Rev. Fluid Mech.* **19** 183
[8] Saffman P G and Taylor G 1958 *Proc. Roy. Soc. Lond.* A **245** 312
[9] Lyubimov D V, Shklyaev S, Lyubimova T P and Zikanov O 2009 *Phys. Fluids* **21** 014105
[10] Davie M K and Buffett B A 2001 *J. Geophys. Res.* **106** 497
[11] Goldobin D S and Brilliantov N V 2011 *Phys. Rev.* E **84** 056328
[12] Hunter S J, Goldobin D S, Haywood A M, Ridgwell A and Rees J G 2013 *Earth Planet. Sci. Lett.* **367** 105
[13] Goldobin D S and Krauzin P V 2015 *Phys. Rev.* E **92** 063032
[14] Holloway S 1997 *Energy Conversion and Management* **38** S193
[15] Research Consortium for Methane Hydrate Resources in Japan 2008 *Japan's Methane Hydrate R&D Program, Phase 1 Comprehensive Report of Research Results*
[16] Buyevich Y A, Leonov A I and Safrai V M 1969 *J. Fluid Mech.* **37** 371
[17] Gelhar L W, Gutjahr A L and Naff R J 1979 *Water Resour. Res.* **15** 1387
[18] Gelhar L W and Axness C L 1983 *Water Resour. Res.* **19** 161
[19] Neuman S P, Winter C L and Newman C M 1987 *Water Resour. Res.* **23** 453
[20] Neuman S P and Zhang Y 1990 *Water Resour. Res.* **23** 887
[21] Dagan G 1991 *J. Fluid Mech.* **233** 197
[22] Cushman J H 1983 *Water Resour. Res.* **19**, 1641
[23] Gelhar L W and Axness C L 1983 *Water Resour. Res.* **19** 1643
[24] Dagan G 1988 *Water Resour. Res.* **24** 1491
[25] Maryshev B, Joelson M, Lyubimov D, Lyubimova T, Néel M-C N 2009 *J. Phys. A: Math. Theor.* **42** 115001
[26] Goldobin D S and Shklyaeva E V 2009 *J. Stat. Mech.: Theor. Exp.* P01024
[27] Gardiner C W 1997 *Handbook of Stochastic Methods* (Berlin Germany: Springer-Verlag Heidelberg)
[28] Nocedal J and Wright S J 2006 *Numerical Optimization* (New York USA: Springer-Verlag)

Invariant solutions in explicit form of the Boltzmann equation with a source term

S V Meleshko[1], Yu N Grigoriev[2], A Karnbanjong[1] and A Suriyawichitseranee[1]

[1] Suranaree University of Technology, Nakhon Ratchasima, Thailand
[2] Institute of Computational Technology, Novosibirsk, Russia

E-mail: sergey@math.sut.aac.th

Abstract. This paper is devoted to applications of the group analysis method to the Boltzmann equation with a source function. Exact solutions of the nonlinear kinetic Boltzmann equation with a source function in the case of an isotropic distribution function and Maxwell model of isotropic scattering were constructed. An equivalence Lie group is used for the construction. One of the transformations of the equivalence Lie group uniquely singles out a class of source functions which allows us to find invariant solutions of the Bobylev-Krook-Wu type in an explicit form. In particular, some of these solutions have a meaningful physical interpretation.

1. Introduction

The classical Boltzmann equation of the gas kinetic theory from mathematical point of view is a homogeneous integro-differential equation [1]. However, in order to model some physical situations one needs to include additional source terms into the classical equation, in particular, for describing presence of an external source (drain) under initiation of high threshold process by injection of 'hot' particles or 'running out' high energetic molecules from a force field, removal events [2, 3, 4] such as nonelastic interactions with a background host medium, and others. In the present paper the Boltzmann equation with a source term depending on the independent and dependent variables is consider.

The first attempt of construction of invariant solutions [5] of the Boltzmann equation with a source by the group analysis method was in [6], where similar to [7] the Boltzmann equation was reduced to a nonlinear differential equation for a moment generating function. Complete group classification of this equation with respect to a source function was presented in [8]. However, a reduction of the obtained invariant solutions into solutions of the original Boltzmann equation is an extremely complicated problem.

Therefore, it is more perspective to consider a Fourier image of the Boltzmann equation for an isotropic distribution function for which the well-known invariant Bobylev-Krook-Wu (BKW) solution [9] was found. For this goal a variant of the preliminary group classification method was developed in [10, 11]. An idea of a preliminary group classification method was given in [12], and its further developing is in [13, 14]. An equivalence Lie group and its nonequivalent subgroups, in particular, elements of an optimal system of subgroups, are applied for a preliminary group classification. It is obvious that for the classification with respect to a source function, one can

Content from this work may be used under the terms of the Creative Commons Attribution 3.0 licence. Any further distribution of this work must maintain attribution to the author(s) and the title of the work, journal citation and DOI.
Published under licence by IOP Publishing Ltd

use a Lie group and its optimal system of subgroups admitted by the homogeneous equation as an equivalence Lie group of the nonhomogeneous equation.

In [10], a group classification with respect to a source function which does not depend on a solution of the kinetic equation was studied. Directly solving the determining equation it was shown that the admitted infinitesimal generators of the nonhomogeneous equation are linear combinations of the generators of a complete admitted Lie algebra L_4 of the homogeneous equation which is obtained earlier in [15]. This result was expected, but its obtaining by a direct solving of the determining equation allowed stating that the obtained group classification is complete.

A more general and interesting from physics point of view case of source functions depending not only on independent but also a dependent variable was considered in [11]. Instead of solving a complicated determining equation, an admitted Lie group of a homogeneous equation was applied for a preliminary group classification of the nonhomogeneous equation. In this case the problem of completeness of the group classification is still an open problem. In the present paper it is shown that for equations with a source function depending on a solution of the kinetic equation it is not sufficient to exploit the generators used in [11]. An extension of the equivalence Lie group corresponding to L_4 allows obtaining invariant solutions of the Boltzmann equation with a source term depending on a solution in an explicit form.

2. The Boltzmann equation in isotropic case

It is considered the space homogeneous Boltzmann equation with a source term and Maxwell isotropic model of scattering for a distribution function $f(v,t)$ which is isotropic in the molecular velocity space. Its Fourier image in the velocity space $R^3(\mathbf{v})$ has the form [9]:

$$\frac{\partial\varphi(x,t)}{\partial t} + \varphi(x,t)\varphi(0,t) = \int_0^1 \varphi(xs,t)\varphi(x(1-s),t)\,ds + G(\varphi(x,t),t,x). \qquad (1)$$

Here the isotropic distribution function and its Fourier image are defined by the transformations:

$$f(v,t) = (2\pi)^{-3}\int \mathbf{dk}\exp(i\mathbf{kv})\varphi(k,t), \quad \varphi(x,t) \equiv \varphi(k^2/2,t) = \int \mathbf{dv}\exp(-i\mathbf{kv})f(v,t).$$

Respectively,

$$G(\varphi(x,t),t,x) = \int \mathbf{dv}\exp(-i\mathbf{kv})\tilde{G}(f(v,t),t,v)$$

is the Fourier image of the source function in the velocity space:

$$\mathbf{v} \in R^3(\mathbf{v})), \quad \mathbf{k} \in R^3(\mathbf{k})), \quad v = |\mathbf{v}|,\, k = |\mathbf{k}|.$$

A Cauchy problem (relaxation) is formulated for equation (1):

$$\varphi(x,0) = \varphi_0(x). \qquad (2)$$

The Cauchy problem (1), (2) determines evolutionary laws of particle number density

$$n(t) \equiv \varphi(0,t) = \int \mathbf{dv}f(v,t)$$

and energy

$$n(t)e(t) = \frac{\partial\varphi}{\partial x}(0,t),$$

which are written in the form

$$\frac{\partial\varphi(0,t)}{\partial t} = G(\varphi(0,t),t,0), \qquad (3)$$

IOP Conf. Series: Journal of Physics: Conf. Series **894** (2017) 012063 doi:10.1088/1742-6596/894/1/012063

$$\frac{\partial \varphi}{\partial t}(0,t)e(t) + \frac{\partial e(t)}{\partial t}\varphi(0,t) = \frac{\partial G}{\partial x}(\varphi(0,t),t,0) + \frac{\partial G}{\partial \varphi}(\varphi(0,t),t,0). \quad (4)$$

In the homogeneous case with $G \equiv 0$ the problem (1), (2) has the following conservation laws of the particle density and energy [9]:

$$\varphi(0,t) = 1, \quad \frac{\partial \varphi}{\partial x}(0,t) = -1. \quad (5)$$

In [15], it was shown that the complete Lie algebra L_4, admitted by the homogeneous equation (1), is defined by the generators

$$X_0 = x\partial_x, \quad X_1 = x\varphi\partial_\varphi, \quad X_2 = \varphi\partial_\varphi - t\partial_t, \quad X_3 = \partial_t.$$

The invariant BKW-solution [9] has the form

$$\varphi_B(x,t) = (1-y)\exp(y-x),\ y = x\exp(\lambda t),\ \lambda = -1/6,\ 0 \le \theta \le 2/5. \quad (6)$$

This solution is an invariant solution with respect to the generator $X_0 + \lambda^{-1}X_3$ [15].

3. Invariant solutions of the BKW-type

For the preliminary group classification of equation (1) in [11], admitted generators were sought in the form

$$X = c_0X_0 + c_1X_1 + c_2X_2 + c_3X_3.$$

One can note that in all presented their cases an invariant solution corresponding to the BKW-type depends on the self-similar variable $y = x\exp(-\lambda t)$ which is defined by the generator $X_0 + \lambda^{-1}X_3$:

$$\varphi(x,t) = \Psi(y).$$

This dependence contradicts to the presence in equation (1) a nonzero source function. However, the author of [2], assuming a representation of a solution similar to [16], found a generalized BKW-solution in an explicit form. In this case a source function in Fourier image had the form

$$G = -C_R\varphi(0,t)\varphi(t,x). \quad (7)$$

The principal difference of this source from the studies $G(\varphi(x,t),t,x)$ in [17] is in the dependence of the functional:

$$\varphi(0,t) = \int \mathbf{dv} f(v,t).$$

It brings an additional nonlocalness in a classifying equation. This nonlocalness could not be taken into account by only applying the generators of L_4 used as an equivalence Lie group in [17].

In [18], a transformation of time and a distribution function was presented. This transformation allowed mapping a nonhomogeneous equation (1) into homogeneous for which one can derive a BKW-solution in new variables, and thus, to avoid cumbersome computations [2]. This transformation has the form

$$\tilde{t} = \tau(t), \ \ \tilde{x} = x, \ \ \tilde{\varphi} = p(t)\varphi. \quad (8)$$

One can consider a more general form of the source term in equation (1)

$$G = G(\varphi(0,t), \varphi(x,t), t, x).$$

In [17], several ways of obtaining conditions for the transformation (8) to be an equivalence transformation are given. These conditions are

$$p(t)\tau'(t) = 1,$$

and

$$G(\varphi(0,t), \varphi(x,t), t, x) = \alpha(\varphi(0,t), t)\varphi(x,t). \tag{9}$$

In terms of particle number density $n(t) = \varphi(0,t)$ the conditions for the equivalence transformations (8) have the form

$$p'(t) + p(t)\alpha\,(t, n(t)) = 0,\;\; \tau' = p^{-1}(t),\;\; n'(t) = \alpha(t, n(t))n(t).$$

From the evolution law (3) one derives that $\alpha(t, n(t)) = n'(t)/n(t)$. Finally, one obtains

$$p = \frac{n_0}{n},\;\; \tau' = \frac{n}{n_0},\;\; \tau(t) = \int_0^t \frac{n(s)}{n_0} ds, \tag{10}$$

where n_0 is an arbitrary constant.

Moreover, from the energy evolution law (4) one finds for a linear source (9) that

$$\frac{\partial\varphi(0,t)}{\partial t} e(t) + \frac{\partial e(t)}{\partial t}\varphi(0,t) = \alpha(\varphi(0,t), t)e(t)\varphi(0,t).$$

Hence, from (3) one gets that the average energy (temperature) of particles is conserved in this case

$$e(t) = e_0 = const.$$

Thus, the transformation (8) with (10) is an equivalence transformation of equation (1) with a source (9) which maps it into an equation of the same form with the zero right hand side:

$$\tilde{\varphi}_\tau(x, \tau(t)) + n_0\tilde{\varphi}(x, \tau(t)) - \int_0^1 \tilde{\varphi}(xs, \tau(t))\tilde{\varphi}(x(1-s), \tau(t))\, ds = 0. \tag{11}$$

Choosing the constants according to the conservation laws (5):

$$n_0 = 1, \quad e_0 = -1,$$

one is able to rewrite the BKW-solution (6) of equation (11) in the transformed variables, and then return to the original variables. By this way one can construct generalized BKW-solutions for a class of sources (9).

According to the results of [15], the BKW-solution is determined by the generator

$$x\partial_x + \lambda^{-1}(p\partial_t - p'\varphi\partial_\varphi),$$

and a corresponding representation of a generalized invariant BKW-solution has the form

$$\varphi(x,t) = p^{-1}(t)\tau'(t)F(xe^{\lambda\tau(t)}).$$

Substitution of this representation into equation (1) with a source function from the class (9) leads to the reduced equation for the function F with the zero right hand side which gives a sought solution. Several examples of generalized BKW-solutions in an explicit form for some source functions of the class (9):

$$G_1(v,t) = -C_R n(t)f(v,t),\;\; G_2(v,t) = -C_r n^2(t)f(v,t),\;\; G_3(v,t) = -C_e f(v,t),$$

which model some physical phenomena are presented in [17]. One can study a source function of a more general form, for example,

$$G_k(v,t) = -(C_1 + C_2 n(t))^k f(v,t).$$

In particular, this form allows obtaining a combination of the presented above source functions.

4. Conclusion

Using the Lie algebra L_4 admitted by the Boltzmann equation in the case of an isotropic distribution function and a Maxwellian isotropic interaction potential an equivalence Lie group of equation with a source term is constructed. A class of source functions which vanish by the equivalence transformations is uniquely singled out. This allows finding generalized BKW-solutions in an explicit form.

At the same time there is a problem of further extension of equivalence Lie group. In particular, by transformations which can allow obtaining generalized BKW-solutions for other classes of source functions. A proof of absence of such transformations can be its alternative.

Acknowledgments

SVM and YuNG acknowledge RFFI (code of the project 17-01-00209a) for a partial financial support. AK thanks Walailak University, and Royal Thai Government Scholarship through the Ministry of Science and Technology of Thailand for financial support.

References

[1] Chapman S and Cowling T G 1952 *The Mathematical Theory of Non-uniform Gases* (Cambridge: The Cambridge University Press)
[2] Spiga G 1984 *Phys. Fluids* **27** 2599–600
[3] Boffi V C and Spiga G 1982 *J. Math. Phys.* **23** 2299–303
[4] Boffi V C and Spiga G 1982 *Phys. Fluids* **25** 1987–92
[5] Ovsiannikov L V 1978 *Group Analysis of Differential Equations* (Moscow: Nauka)
[6] Nonenmacher T F 1984 *J. Appl. Math. Phys. (ZAMP)* **35** 680–91
[7] Krook M and Wu T T 1976 *Phys. Rev. Lett.* **36** 1107–9
[8] Grigoriev Y N, Meleshko S V and Suriyawichitseranee A 2014 *Int. J. Non-Linear Mech.* **61** 15–8
[9] Bobylev A V 1975 *Dokl. Akad. Nauk SSSR* **225** 1041–4
[10] Suriyawichitseranee A, Grigoriev Y and Meleshko S 2015 *Commun. Nonlinear Sci. Numer. Simul.* **20** 719–30
[11] Long F S, Karnbanjong A, Suriyawichitseranee A, Grigoriev Y N and Meleshko S V 2017 *Commun. Nonlinear Sci. Numer. Simulat.* **48** 350–60
[12] Akhatov I S, Gazizov R K and Ibragimov N H 1991 *J. Math. Sci.* **55** 1401–50
[13] Ibragimov N H, Torrisi M and Valenti A 1991 *J. Math. Phys.* **32** 2988–95
[14] Dos Santos Cardoso-Bihlo E, Bihlo A and Popovych R O 2011 *Commun. Nonlinear Sci. Numer. Simul.* **16** 3622–38
[15] Grigoriev Y N and Meleshko S V 1987 *Dokl. AS USSR* **297** 323–7
[16] Bobylev A V 1975 *Dokl. AS USSR* **225** 1296–9
[17] Grigoriev Y N, Meleshko S V and Suriyawichitseranee A 2017 *J. Appl. Mech. Tech. Phys.* In press
[18] Santos A and Brey J J 1985 *Phys. Fluids* **29** 1750

IOP Conf. Series: Journal of Physics: Conf. Series **894** (2017) 012064 doi:10.1088/1742-6596/894/1/012064

Effect of porous material heating on the drag force of a cylinder with gas-permeable porous inserts in a supersonic flow

S G Mironov[1,2], T V Poplavskaya[1,2] and S V Kirilovskiy[1,2]

[1]Khristianovich Institute of Theoretical and Applied Mechanics SB RAS, 630090, Novosibirsk, Russia
[2]Novosibirsk National Research State University, 630090, Novosibirsk, Russia

E-mail: mironov@itam.nsc.ru

Abstract. The paper presents the results of an experimental investigation of supersonic flow around a solid cylinder with a gas-permeable porous insert on its front end and of supersonic flow around a hollow cylinder with internal porous inserts in the presence of heating of the porous material. The experiments were performed in a supersonic wind tunnel with Mach number 4.85 and 7 with porous inserts of cellular-porous nickel. The results of measurements on the filtration stand of the air filtration rate through the cellular-porous nickel when it is heated are also shown. For a number of experiments, numerical modeling based on the skeletal model of a cellular-porous material was carried out.

1. Introduction

In recent years work is underway to study the control of aerodynamic forces using gas-permeable porous inserts in the airframe of a supersonic aircraft [1-3]. The traditional aerodynamic control methods for the flow around supersonic aircraft do not always give the expected effect due to the appearance of the detachment phenomenon. The use of porous inserts makes it possible to eliminate undesirable effects and diversify the methods of flow control including using the thermal effect on flow in porous materials [4]. This approach allows to exclude mechanical movements of airframe elements or blowing of jets when creating a control action.

The paper presents the results of an experimental study of the effect of heating gas-permeable porous inserts from cellular-porous nickel on the rate of air filtration through them and on the drag force of cylindrical bodies with porous inserts in a supersonic flow. Separate experiments were supplemented by numerical modeling of flow on the basis of the skeleton model of a cellular-porous material.

2. Experimental setup and methods of diagnostics

Experiments on the effect of heating a gas-permeable cellular-porous nickel on the air filtration rate on the filtration stand described in [5] were previously performed. It was studied the air flow in a tubular channel with a densely installed sample of a porous material with heating from the outside of a metal tubular channel and when the sample was heated in a thermally insulated dielectric tubular channel by passing an electric current through it. The temperature of the porous sample was not measured, but the temperature of the air passing through the sample was measured. This scheme was chosen from the considerations that it is the heating of air that causes the change in the filtration rate. The design of the stand allowed independently to vary the pressure drop across the sample of the porous material and the

Content from this work may be used under the terms of the Creative Commons Attribution 3.0 licence. Any further distribution of this work must maintain attribution to the author(s) and the title of the work, journal citation and DOI.
Published under licence by IOP Publishing Ltd

Reynolds number calculated from the pore diameter and air filtration rate. Cell-porous nickel samples with a pore diameter d = 1, 2, 3 and 4 mm, a diameter of 15 mm and a length of 42 mm were used. The following data were determined in the experiments:

- dependences of the filtration rate on the pressure drop on a porous sample at a variation in the temperature of the air passing through the material for different pore diameters,
- dependences of the mass flow of air through a porous sample at a variation in the temperature of the air passing through the material for different pore diameters,
- dependences of the rate change of the filtration on the degree of heating of air from the Reynolds number Re_d^0 calculated from the pore diameter and the air filtration rate in the material without heating in the range of 30-600.

On the basis of these data, geometric characteristics of the skeleton model of porous-porous nickel were developed [5, 6], which was used in numerical simulations of supersonic flow around a cylinder with a front porous insert without heating [6] and with heating of the porous inserts in this work.

Experiments to study the effect of heating on the drag force of cylindrical models with porous inserts were performed in the supersonic wind tunnel T-327B of ITAM SB RAS at Mach numbers $M_\infty = 4.85$ and 7 and the corresponding unit Reynolds numbers $Re_{1\infty} = 2.7 \cdot 10^6$ and $1.5 \cdot 10^6 m^{-1}$. The stagnation temperature of the flow was 294 K. The wind tunnel T-327B with the Eiffel chamber has a shock-running system and the possibility of the air outflowing into a large vacuum volume pumped out by high-performance vacuum pumps. The pressure in the working section of the wind tunnel before starting is 10^{-2}torr. The force of aerodynamic drag was measured by tenso-balance on which the models were mounted. Registration of electric signals of devices and sensors was carried out by a multichannel ADC National Instruments PX.

In the experiments at $M_\infty = 4.85$ the model [1] of a solid cylinder with a front porous insert equal in length to two cylinder diameters with pores diameter of 1 mm was used.

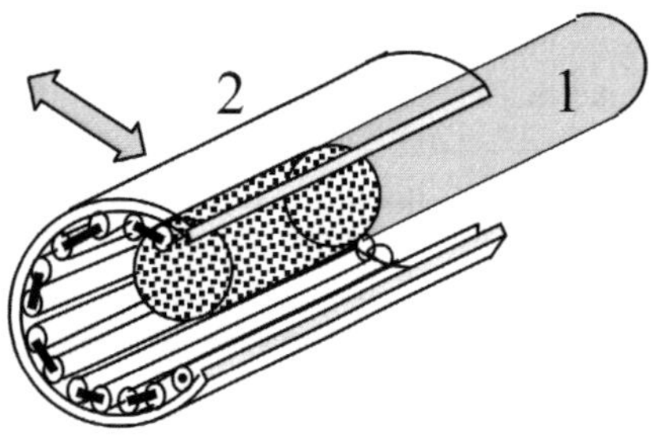

Figure 1. Scheme of the model heating: 1 – cylinder with front porous insert, 2 – retractable ohmic heater

On this model the influence of heating the porous material of the front insert on the aerodynamic drag of the model was studied. The diameter of the model was 15 mm, the total length of the model was 110 mm. The heating of the porous insert was carried out by the thermal radiation of the ohmic heater partially enclosing the front porous insert as shown in Fig. 1. The temperature of the porous insert was measured by a copper-constantan thermocouple at the insertion axis at a point half the length of the insert. When the maximum heating temperature of the insert is reached before the start of the wind tunnel the heater is taken out of the flow boundary. In the process of supersonic flow around the model, the temperature at the measurement point varied in the range from the maximum heating temperature before the wind tunnel was started up to the stagnation temperature.

In experiments at $M_\infty = 7$ a hollow thin-walled cylinder model with gas-permeable porous inserts was used (Fig. 2). The model consisted of two discs of cellular porous nickel (pore size 1 mm, 95% porosity) 24 mm in diameter and 10 mm thick inserted in a thin 46-mm-long fluoroplastic cylinder at a distance of 26 mm from each other. The cylinder was installed along the flow axis with the help of a holder with a small cluttering of the flow. To create a glowing electric discharge between porous disks the disks were connected through the ballast resistance with a 600-volt power supply source.

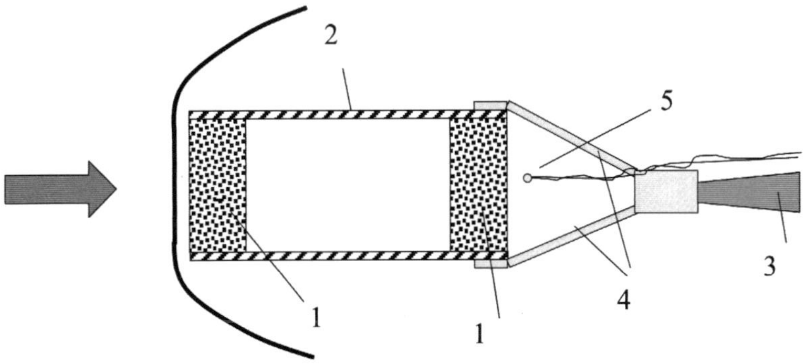

Figure 2. Scheme of the model: 1 – gas-permeable disks, 2 – cylinder, 3 – balance, 4 – model holder, 5 – thermocouple

The front disk served as the cathode, the rear disk as the anode. The discharge power was regulated by the ballast resistance in the range from 40 to 190 W. In the wake of the cylinder, on its axis and near the rear porous disk, there was a copper-constantan thermocouple used to measure the temperature of air passing through the porous disks. The temperature of the passing air served as an indicator of the heating of the porous material and the efficiency of heat exchange between of the porous material and the incoming air flow. During the operation of the wind tunnel the temperature of the air passing through the model increased with time and the growth rate of the temperature was depended on the power of the glow discharge. Measures have been taken to minimize the effect of the thermocouple and the electrical wires on the balance readings. During operation of the wind tunnel, the current in the discharge circuit, electric signals of the thermocouple and the balance were recorded. The output data of the measurements was the dependence of the drag force of the model on temperature of the air passing through the model. In addition the dependences of the air temperature behind the model on the discharge burning time and its power were determined. This model is ideologically close to the model of work [2], where the reduction in drag is based on the flow of part of the incident air flow into the rear region through the conical gas permeable front porous insert. In the model on Fig. 2 the flow of a portion of the incident airflow into the rear region behind the cylinder is controlled by heating the porous disks.

Preliminary experiments on the selection of modes of electric glow discharge, the distance between gas permeable porous disks and the pore diameter were carried out within the framework of studies of the excitation of gas micro-jets that outflow into low-pressure space.

3. Results and discussion

The main results of measurements at the filtration stand are set forth in [5]. In particular a significant increase in hydraulic resistance and a drop in the filtration rate in cellular-porous nickel was shown with an increase in the temperature of the air passing through the porous material for all pore diameters. With an equal heating temperature of the porous materials, the influence of heating on the hydraulic resistance is the greater, the smaller the pore diameter and the smaller the Reynolds number of Re_d^0, calculated from the pore diameter and the air filtration rate without heating the material. The processing of the measurement results gave an empirical dependence of the coefficient of reduction of the filtration rate on the degree of air heating γ[m/(s·grad)] from Re_d^0 and d:

$$\gamma = -\frac{1}{A - B/\sqrt{\mathrm{Re}_d^0}}, \quad A = 19.12 + 0.303 \cdot d^2, \quad B = 80.53 - 0.41 \cdot d .$$

IOP Conf. Series: Journal of Physics: Conf. Series **894** (2017) 012064 doi:10.1088/1742-6596/894/1/012064

Based on these data, experiments in the wind tunnel were performed on models with porous inserts with a pore diameter of 1mm where the greatest effect of heating on the drag force was expected.

In Fig. 3 shows the results of measurements and calculation of the dependence of the normalized coefficient of drag force of the cylinder with the front porous insert on the temperature of the porous insert at the measuring point T_{ins} for $M_\infty = 4.85$ (Fig. 1). The normalization was made on the drag coefficient at the insertion temperature of 294 K. Measurements were made only in the temperature

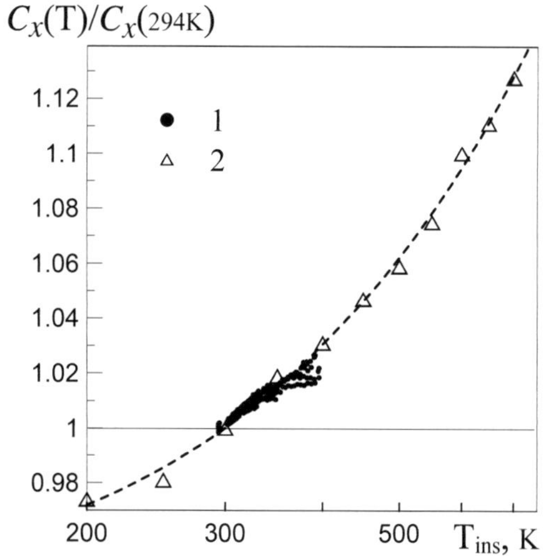

Figure 3. Dependence of the normalized coefficient of aerodynamic drag on the temperature of porous insert: 1 – experiment, 2 – numerical simulation

range of 300-400 K since the metal model intensively reflected the thermal radiation of the heater and the radiation was came out along the ends and into the cutting of the heater (see, Fig. 1). Numerical modeling within the framework of the skeleton model of a porous insert [5, 6] made it possible to obtain the dependence of the coefficient of drag force in over a wider range of temperatures. It can be seen that when the temperature of the porous insert is increased by 400 degrees, then an increase in the coefficient of aerodynamic resistance by ≅12% can be achieved. Calculations have shown that the increase in drag force is associated with the growth of viscous friction in the heated front layers of the porous insert. This leads to an increase in the angle of inclination of the head shock wave to the direction of the flow and to an increase in the momentum loss of the flow on the shock wave.

The obtained data allow to change the control scheme of the drag force of such model for the case of real flight in the atmosphere. In a real flight due to the high stagnation temperature the front porous insert will be heated to a temperature of more than 1000 K. In this case, when the insert is cooled with the liquid (for example, the fuel for aircraft engine), the drag force coefficient will decrease. In addition it is possible to additionally obtain a useful effect of the fuel pyrolysis into the gaseous state and increase the efficiency of the engine.

In Fig. 4 shows the results of measurements of the coefficient of drag force with variation of the temperature of the air transmitted through the model T_g, obtained for the model on Fig. 2. The results of measurements are presented in the form of increase in the coefficient of drag force in percents. The value of the coefficient of drag force is normalized to the coefficient of drag force at a temperature of 294 K. The graph shows the data obtained in different experiments, with different glow discharge

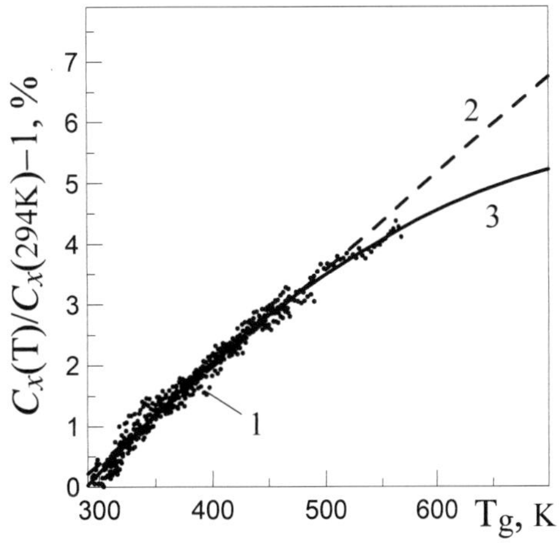

Figure 4. Dependence of the relative increase in aerodynamic drag on the temperature of air passing through the model: 1 – experiment, 2 – linear approximation, 3 – quadratic approximation

power and at different temperatures of the air passing through the model. The graph also shows the linear and quadratic approximations of the experimental data. It can be seen that when the air in porous inserts is heated to 700 K, the drag force of the model can be increased by 5% and a further increase in temperature T_g will only slightly increase this value. Most likely, this is due to the weak influence of pressure in the trailing edge of the model on drag force.

4. Conclusions

Experiments have been carried out to investigate the effect of heating of cellular-porous nickel on the hydrodynamic characteristics of a porous material with a one-dimensional flow of air through it. In experiments in a supersonic wind tunnel the effect of heating a porous material on the drag force of the cylinder with a front porous insert and on the drag force of the hollow cylinder with two porous inserts was studied.

It is shown that at an equal temperature of heating the porous material, its hydraulic resistance is the greater, the smaller the pore diameter and the smaller the Reynolds number calculated from the pore diameter and the air filtration rate without heating the material. An empirical relationship for the coefficient of proportionality between the filtration rate and the temperature of the air passing through the material as function of the Reynolds number is obtained.

For the solid cylinder with a front porous insert the experimental and calculated dependences of the normalized coefficient of drag force on the insert temperature are obtained.

For the hollow cylinder with porous inserts an experimental dependence of the increase in the normalized coefficient of drag force on the temperature of the air passing through the inserts was obtained. It is shown that the effect of heating on the drag force for this model is lower than for a cylinder with a front porous insert.

Acknowledgements

The work was supported by the Russian Foundation for Basic Research (Grant № 15-08-3867) and by the Russian Science Foundation (Grant № 17-19-01157).

References

[1] Fomin V M, Mironov G S and Serdyuk K M 2009 *Tech. Phys. Lett.* **35** 117-9
[2] Fomin V M, Zapryagaev V I, Lokotko A V, et al. 2010 *Appl. Mech. Tech. Phys.* **51** 65-73
[3] Fomin V M, Postnikov B V and Lomanovich K A 2015 *Tech. Phys. Lett.* **49** 686-8
[4] Mironov S G, Maslov A A and Tsyryulnikov I S 2014 *Tech. Phys. Lett.* **40** 868-70
[5] Mironov S G, Poplavskaya T V and Kirilovskiy S V 2017 *Siberian Phys. J.* **12** 66-74
[6] Poplavskaya T V, Kirilovskiy S V and Mironov S G 2016 *AIP Conf. Proc.* **1770** 030067

Burning stability of the coal-dust methane-air mixture in a recuperative burner

K M Moiseeva[1], D A Krainov[2] and A A Dement'ev[3]

[1]Tomsk State University, Tomsk, 634050 Russia
[2]Tomsk Polytechnic University, Tomsk, 634050 Russia
[3] Research and production center "Polus", Tomsk

Abstract. The physical mathematical model of combustion of the coal-dust methane-air mixture in a recuperative burner has been developed. The influence of the particles (inert and reacting) on the burning stability of coal-dust methane-air mixture with low concentration of methane has been determined. The regimes of stable combustion for lean methane-air mixture with coal suspension depending on the fuel flow rate at the inlet of the burner and the fuel content have been defined.

1. Introduction

In this paper we investigate the problem of combustion of the coal-dust methane-air mixture in a recuperative burner. In early papers [1-2] the aim was to investigate combustion of gas mixtures in a U-shaped recuperative burner. Combustion of a lean methane–air mixture in a slot burner with an inert inner insert was investigated in [3]. It was demonstrated that the combustion of a methane-air mixture (volume fraction of methane is 2.3% and up) might be initiated and supported by heat recovery. Under the normal conditions, the volume fraction of methane must be above 5.2 % to support the burning of the mixture [4]. A methane-air mixture is a by-product generated during coal mining. The mixture contains impurities and dust particles. The burning characteristics of such mixture differ from characteristics of a pure methane-air mixture. Coal dust can react in the air by itself, but under certain conditions. However a coal-dust methane-air mixture is able to ignite under the conditions at which coal dust and methane do not ignite separately. In the research [5] the authors showed that combustion of coal-dust in the air under the normal conditions is impossible without adding the small amount of methane or without preheat of the burner walls. The addition of methane to coal-dust-air mixture stabilizes the flame front. The presence of inert particles in a coal suspension can also change the nature of the burning process. Thus combustion of a coal-dust methane-air mixture in a recuperative burner is a complex process of heat transfer and chemical reactions.

This paper provides a numerical simulation on combustion of methane-air mixture with suspended coal dust in a fantail burner with an inert inner insert. The aim of the research is to determine the stable combustion conditions of the methane-air mixture depending on the content of the mixture (particle size and percentage composition of the inert particles) and the fuel flow rate at the inlet of the burner.

2. Mathematical model

A cold methane–air mixture with the methane mass content $a_{CH4,v}$, the gas phase temperature $T_{g,v}$ and the overall mass concentration $m_{dust} = \sum_{i=1..N} m_{dust,i}$ is fed to a preheated fantail burner at the rate u_v at the

Content from this work may be used under the terms of the Creative Commons Attribution 3.0 licence. Any further distribution of this work must maintain attribution to the author(s) and the title of the work, journal citation and DOI.
Published under licence by IOP Publishing Ltd

inlet (x = 0). It is supposed that the dust suspension has inert and reacting particles. Total number of particle fraction is N. Other assumptions of the mathematical model correspond to [6]. The mathematical formulation of the problem has the following form:

The energy equation for gas phase:

$$c_g\rho_g\frac{\partial T_g}{\partial t}+c_g\rho_g u\frac{\partial T_g}{\partial x}=\frac{\partial}{\partial x}\left(\lambda(T_g)\frac{\partial T_g}{\partial x}\right)+\sum_{i=1..N}\left[Gc_kT_{k,i}+\alpha_{k,i}n_{k,i}S_{k,i}\left(T_{k,i}-T_g\right)\right]+$$
$$+\frac{\alpha_S}{h}\left(T_{1,S}-T_g\right)+Q_1\rho_g^2a_{CH4}a_{O2}k_{01}\exp\left(-\frac{E_1}{R_uT_g}\right),\ T_{1,S}(x,t)=\begin{cases}T_S(x,t),\ x<L,\\ T_S(2L-x,t),\ x\geq L.\end{cases} \quad (1)$$

The energy equation for inner insert:

$$c_S\rho_S\frac{\partial T_S}{\partial t}=\lambda_S\frac{\partial^2 T_S}{\partial x^2}-\frac{\alpha_S}{h_S}\left(T_S-T_g(x,t)\right)-\frac{\alpha_S}{h_S}\left(T_S-T_g(2L-x,t)\right),\ 0\leq x\leq L. \quad (2)$$

The energy equation for the *i*-th fraction:

$$c_k\rho_{k,i}\frac{\partial T_{k,i}}{\partial t}+uc_k\rho_{k,i}\frac{\partial T_{k,i}}{\partial x}=\alpha_{k,i}S_{k,i}n_{k,i}\left(T_g-T_{k,i}\right)+Q_2G_i-G_ic_kT_{k,i},\ i=1..N. \quad (3)$$

The mass balance equation of methane in the mixture:

$$\frac{\partial a_{CH4}}{\partial t}+u\frac{\partial a_{CH4}}{\partial x}=\frac{\partial}{\partial x}\left(D(T_g)\frac{\partial a_{CH4}}{\partial x}\right)-k_{01}\rho_g a_{CH4}a_{O2}\exp\left(-\frac{E_1}{R_uT_g}\right). \quad (4)$$

The mass balance equation of oxidizer in the mixture:

$$\frac{\partial a_{O2}}{\partial t}+u\frac{\partial a_{O2}}{\partial x}=\frac{\partial}{\partial x}\left(D(T_g)\frac{\partial a_{O2}}{\partial x}\right)-\frac{\mu_{O2}\nu_{O2}}{\mu_{CH4}\nu_{CH4}}k_{01}\rho_g a_{CH4}a_{O2}\exp\left(-\frac{E_1}{R_uT_g}\right)-\frac{\mu_{O2}\nu_{O2}}{\mu_C\nu_C}\sum_{i=1..N}\frac{G_i}{\rho_g}. \quad (5)$$

The perfect-gas law:

$$p=\rho_gR_gT_g=const. \quad (6)$$

The gas continuity equation:

$$\frac{\partial\rho_g}{\partial t}+\frac{\partial(\rho_g u)}{\partial x}=\sum_{i=1..N}G_i. \quad (7)$$

The mass balance equation for particles of the *i*-th fraction:

$$\frac{\partial\rho_{k,i}}{\partial t}+\frac{\partial(\rho_{k,i}u)}{\partial x}=-G_i,\ i=1..N. \quad (8)$$

The equation of number concentration of the *i*-th fraction:

$$\frac{\partial n_{k,i}}{\partial t}+\frac{\partial n_{k,i}u}{\partial r}=0,\ i=1..N. \quad (9)$$

The equation of radius for *i*-th fraction:

$$r_{k,i}=\sqrt[3]{\frac{3\rho_{k,i}}{4\pi\rho_{k,i}^0n_{k,i}}},\ i=1..N. \quad (10)$$

The initial conditions:

$$T_g(x,0)=T_{g,b}(x),\ T_S(x,0)=T_{S,b}(x),\ T_{k,i}(x,0)=T_{k,b}(x),\ \rho_{k,i}(x,0)=\rho_{k,b}(x),$$
$$a_{CH4}(x,0)=a_{CH4,b}(x),\ a_{O2}(x,0)=a_{O2,b}(x),\ u(x,0)=u_{g,b}(x),\ n_{k,i}(x,0)=n_{k,b}(x). \quad (11)$$

The boundary conditions:

$$T_g(0,t)=T_{g,v},\ T_{k,i}(0,t)=T_{k,v},\ a_{CH4}(0,t)=a_{CH4,v},\ a_{O2}(0,t)=a_{O2,v},$$
$$u(0,t)=u_v,\ \rho_{k,i}(0,t)=\rho_{k,v},\ n_{k,i}(0,t)=\frac{\rho_{k,v}}{V_{k,i}\rho_k^0},\ \frac{\partial T_S(0,t)}{\partial x}=0. \quad (12)$$

$$\frac{\partial T_S(L,t)}{\partial x}=0, \qquad (13)$$

$$\frac{\partial T_g(2L,t)}{\partial x}=\frac{\partial a_{CH4}(2L,t)}{\partial x}=\frac{\partial a_{O2}(2L,t)}{\partial x}=0\ . \qquad (14)$$

Where: T is the temperature; c is the heat capacity; n is the number of particles per unit volume; R_u – the molar gas constant; R_g is the gas constant; r is radius; S is area; V is volume; $Q_{1,2}$, $E_{1,2}$, $k_{01,02}$ are the reaction heat; the energy of activation; and the constant of the chemical reaction rate for the gas phase and for particle surface; ρ is the density; ρ_k^0 is the density of a coal particle; a_{O2}, a_{CH4} are the mass concentration of oxidizer and methane in the mixture; μ_{O2}, μ_{CH4}, μ_C are the molar mass of oxygen, methane and carbon respectively; ν_{O2}, ν_{CH4}, ν_C is the mole number of oxygen, methane and carbon in the reaction; u is the velocity; h is width of burner canal; h_1 is width of the inner insert. Indexes: g is the gas parameters; k – the particle parameters; S is the inner insert parameters; b is the initial parameters, v is the parameters at the burner inlet; st is the values at temperature $T_g = 300\ K$, $i=1..N$ is the number of fraction; $\lambda=\lambda_{st}\left(T/T_v\right)^{2/3}$ is the thermal conductivity coefficient, $D\sim\lambda(T)$ is the diffusion coefficient, $\alpha_S=\lambda\, Nu_S/h$ is the gas-inner insert heat exchange coefficient, $\alpha_{k,i}=\lambda\, Nu_g/r_k$ is the gas-particles heat exchange coefficient. The heat exchange coefficient of gas and inner insert is determined by Nusselt number, Nu_S, [3]:

$$Nu_S=\begin{cases}0.979\left(\dfrac{h\,Re\,Pr}{x}\right)^{0.33}, & \dfrac{h\,Re\,Pr}{x}>1000,\\[2ex] 3.78+\left(Nu_*-3.78\right)\dfrac{h\,Re\,Pr\big/x-100}{900}, & 100\le\dfrac{h\,Re\,Pr}{x}\le1000,\\[2ex] 3.78, & \dfrac{h\,Re\,Pr}{x}<100.\end{cases}$$

where $\Pr=c\,\eta/\lambda$ is the Prandtl number, $\mathrm{Re}=\dfrac{\rho u\, h}{\eta}$ is the Reynolds number, Nu_* is the value of Nusselt number, when $\dfrac{h\,\mathrm{Re}\,\mathrm{Pr}}{x}=1000$, Nu_D is the diffusion Nusselt number.

The rate of particle mass changing is: $G=n_k S_k j_1 \rho_{O2}$, where $j_1=\beta_m k_{02}\exp\left(-E_2/R_u T_k\right)\big/\left[\beta_m+k_{02}\exp\left(-E_2/R_u T_k\right)\right]$ is the heterogeneous reaction rate, где $\beta_m=\lambda_g(T)Nu_D\big/\left(c_g\rho_g r_k\right)$ is the particle mass-transfer coefficient [7]. The rate of mass changing for inert particles $G_i=0$, the radius of an inert particle do not change with increasing temperature.

The problem (1) – (14) was solved numerically. The solution method corresponds to [6]. The energy equations (1) – (2), the mass balance equations of methane (4) and oxygen were solved by the second order finite-difference approximation method [8]. The gas continuity equation (7), the energy equation (3), the equation of number concentration (9) and the mass balance equation for particles (8) was approximated by an explicit difference scheme with upwind differences. The spatial step was set as in [5] and was equal to 10^{-5} m. The time step was determined by Courant stability criterion.

3. Results and discussion

The initial conditions (11) have been set according to the stable burning parameters of 6% methane-air mixture with the feed rate of 0.23 *m/s*. During the calculation the initial conditions are superseded by the parameters of the dust-laden gas with the lower concentration of methane. The calculations were held under the parameters taken from [6]. The size of the particles ranges from 0.1 to 10 *μm*, $r_k = 10^{-}$

$^7 \div 10^{-5}$ *m*. The mixture feed rate (u_v) and the volumetric content of methane ($a_{CH4,vol}$) at the inlet of the burner ranges from 0.1 to 0.5 *м/с*. and from 1 to 5% respectively. The dust mass concentration is $m_{dust} = 0.2$ *kg/m*^{3}. We numerically investigate combustion of the methane-air mixture with monodisperse suspension without inert particles and the methane-air mixture with bi-disperse suspension with 10% of inert particles. The aim of the calculations was to determine the boundary of stable combustion depending on the fuel flow rate at the inlet of the burner and the content of the mixture.

The results of two different calculations are shown in Figures 1-2. Figure 1a shows the temperatures of gas (curve 1), inert (curve 2) and reacting particles (curve 3) for the feed rate $u_v = 0.2$ *m/s* and particle radius $r_k = 10^{-6}$ *m*. Pic 1b shows temperatures for the feed rate $u_v = 0.16$ *m/s* and particle radius $r_k = 4 \cdot 10^{-6}$ *m*. The temperature of inert particles is close to the temperature of the gas. The reacting particles have the higher temperature due to the combustion. Moreover the largest particles have the higher temperature.

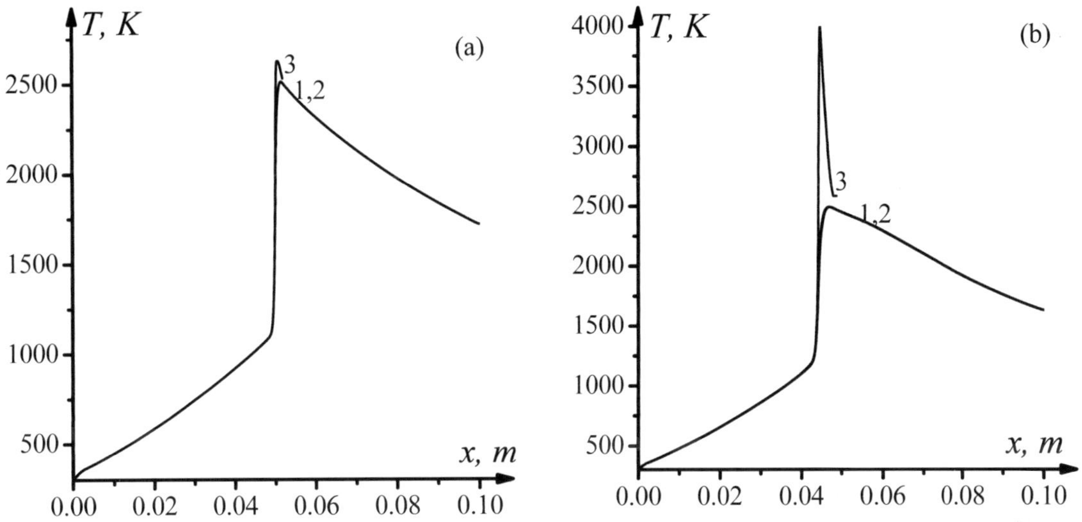

Figure 1. The temperatures of gas, inert and reacting particles.

Figure 2 provides data about combustion of the monodisperse and the bi-disperse coal suspension. The bi-disperse suspension has 10% of inert particles. According to the obtained result the presence of inert particles changes the boundary of stable combustion. The inert particles increase the mass of the mixture and pull heat from the area of combustion. The heat transfer leads to the decrease of the feed rate at which the mixture can ignite and burn steady.

The numerical simulation for 1% methane-air mixture with monodisperse reacting coal-dust suspension showed the possibility of stable combustion when the coal mass concentration is not less $m_{dust} = 0.068$ *kg/m*^{3} [6]. In this paper with the help of the model (1) – (14) we numerically investigate the stable combustion of a polydisperse coal-dust suspension when the coal mass concentration m_{dust} ranges from 0.02 to 0.05 *kg/m*^{3}. It is supposed that the suspension consists of 4 fractions. The sizes of the fractions are $r_{k,1} = 0.65 \cdot 10^{-6}$ *m*, $r_{k,2} = 1.3 \cdot 10^{-6}$ *m*, $r_{k,3} = 2 \cdot 10^{-6}$ *m*, $r_{k,4} = 3 \cdot 10^{-6}$ *m*. The masses of each fraction are $m_{dust,1} = 0.4\, m_{dust}$, $m_{dust,2} = 0.3\, m_{dust}$, $m_{dust,3} = 0.2\, m_{dust}$, $m_{dust,4} = 0.1\, m_{dust}$. It should be noted that the most particles in the suspension are micron in size. The result of the calculation is shown in Figure 3. According to the graph the mixture is able to burn at $a_{vol,CH4} = 1\%$ and $m_{dust} = 0.01$ *kg/m*^{3}. Moreover the lean gas mixture ($a_{vol,CH4} = 1\%$ and $m_{dust} = 0.01$ *kg/m*^{3}) burns steady at low feed rate at the inlet. The temperature of the burning front ranges from 1100 to 1300 *K*. Oxygen does not completely burn out at this composition of the mixture. Thus the stable regime of the low-temperature combustion is implemented, which can be used to burn-out the by-products generated during coal mining.

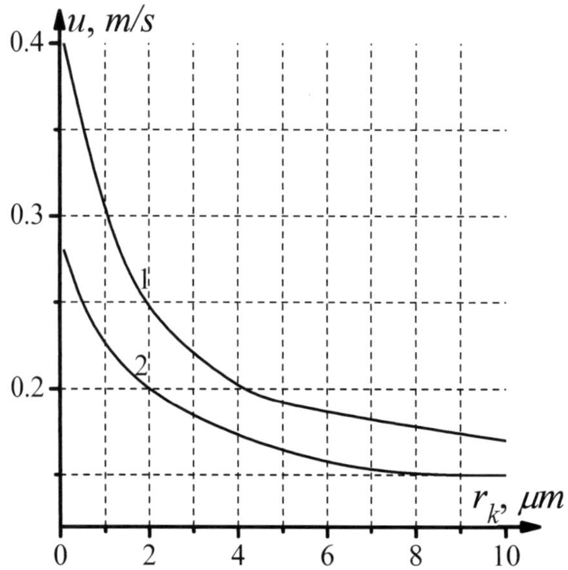

Figure 2. The boundary of stable combustion for coal-dust methane-air mixture. Curves: 1 – monodisperse suspension of reacting particles, 2 – bi-disperse suspension of reacting and inert particles (10% of total mass).

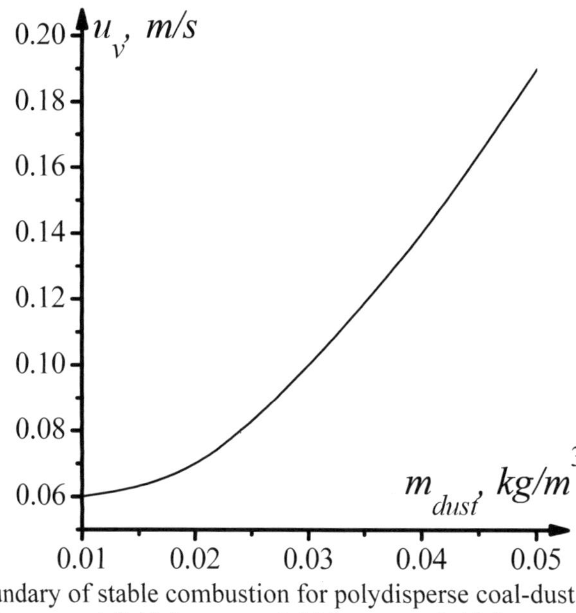

Figure 3. The boundary of stable combustion for polydisperse coal-dust methane-air mixture $r_{k,1} = 0.65 \cdot 10^{-6}$ m, $r_{k,2} = 1.3 \cdot 10^{-6}$ m, $r_{k,3} = 2 \cdot 10^{-6}$ m, $r_{k,4} = 3 \cdot 10^{-6}$ m; $a_{vol,\ CH4} = 1\%$.

4.Conclusions

We numerically investigated burning stability of the coal-dust methane-air mixture in a recuperative burner. We theoretically showed the possibility of the low-temperature burning for the lean gas mixture with the coal-dust mass concentration of 10 kg/m^3 and 1% methane volume content. The obtained results provide data about the influence of inert particles on burning stability of the coal-dust methane-air mixture. The presence of the inert particles decreases the range of the feed rate at the inlet at which the steady burning regime is implemented.

Acknowledgments
The research was supported by the Russian Foundation for Basic Research (project No. 16-38-00188).

References:

[1] Fursenko R V and Minaev S S 2005 *Combust., Expl.,Shock Waves* **41**(2) 133-139
[2] Ronney P D 2003 *Combust. Flame* **135**(4) 421-439
[3] Krainov A Yu and Moiseeva K M 2016 *Combust., Expl.,Shock Waves* **50**(5) 501-509
[4] Lewis B and Von Elbe G 1987 Combustion, Flames and Explosions of Gases (Moscow: Academic Press)
[5] Bradley D, Lawes M, Scott M J and Usta N 2001 *Combust. Flame* **124**(1-2) 82-105
[6] Krainov A Yu and Moiseeva K M 2016 Tomsk State University Journal of Mathematics and Mechanics **3**(41) 65-73
[7] Frank-Kamenetskiy D A 1987 *Diffusion and Heat Transfer in Chemical Kinetics* (Moscow: Nauka)
[8] Samarskii A A 1971 *Introduction to the Theory of FiniteDifference Schemes* (Moscow: Nauka)

IOP Conf. Series: Journal of Physics: Conf. Series **894** (2017) 012066 doi:10.1088/1742-6596/894/1/012066

Computational and experimental research of explosive meteorial devices with combined cumulative liners of the semi-sphere-cylinder shape

S V Fedorov, S V Ladov and Ya M Nikolskaya

Special Machinery Department, Bauman University, 5 2nd Baumanskaya Street, Moscow, 105005, Russia

Abstract. On the basis of numerical modeling within the two-dimensional axisymmetric problem of continuum mechanics and experimental studies, the features of the formation of high-speed compact elements using cumulative charges with liners of the combined form have been analysed.Such liners may have a jet-forming part in the form of a hemisphere, a slightly stretched semi-ellipsoid or a truncated sphere and an ellipsoid, as well as a cutting part in the form of a cylinder.The constructive solutions promoting increase in mass and high-speed parametrs of the compact element formed by explosion are proposed.The variants of steel combined cumulative liners as a part of an explosive device provided the formation of non-gradient elements with a mass from15 to a fraction of a gram moving at speeds from 7.5 to 10 km / s respectively are defined.Such fairly simple explosive devices can be used to simulate, in terrestrial conditions, single and group effects of micrometeorites and fragments of space debris on rocket and space equipment.

1. Introduction

In previous papers [1-3], the results of numerical simulation of the formation and motion of a non-gradient high-speed compact element which is formed during the operation of explosive projectile devices with combined shape charge liners were presented.

Such liners consist of a jet-forming part that forms the jet flow of material during the collapse on the axis and subsequent movement along the axis, and the cutting off part that separates its high-velocity part from the general flow of material.

2. Numerical calculations and experimental researches

Three main forms of the jet-forming part of such cumulative liners were studied: constant and degreasing (from top to base) thickness: hemispherical (HC-facing version), semi-ellipsoidal (SEC-facing version) and in the form of a truncated sphere (SC-facing version) .To cut off the low-velocity part of the jet, a cylindrical part of this liner adjacent to the jet-forming part was used.

As the base liner designed for the subsequent comparison, a PC-liner with a hemispherical jet-forming part of constant thickness was chosen for which experimental data were previously obtained [4]. Figure 1 shows different designs of such degressive thickness liners made of steel.

Content from this work may be used under the terms of the Creative Commons Attribution 3.0 licence. Any further distribution of this work must maintain attribution to the author(s) and the title of the work, journal citation and DOI.
Published under licence by IOP Publishing Ltd

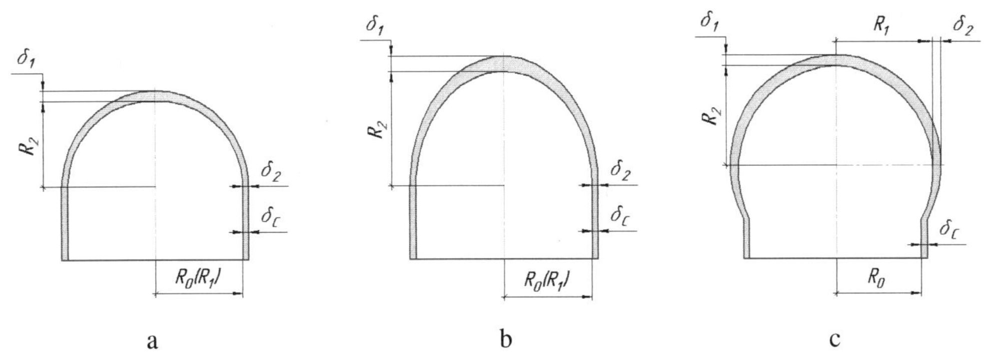

Figure 1. Constructional schemes for liners: a – HC-liners, b – SEC-liners; c – SC-liners

Numerical simulation was performed by using two methods. The first was developed in BMSTU and realize a computing algorithm of free Lagrangian points (it's name " ERUDIT " – "Heuristic Calculation of the Ordered Movement of Individual Points"). This method does not take into account the destruction of the material during the formation of a high-speed compact element. The second method is based on the ANSYS Autodyn program complex, in which it is possible to take into account the destruction of the material.

Figure 2 shows an example of the results of calculations the base variant of the PC-facing (Figure 1, a) by the method of the BMSTU (without taking into account the destruction of the material).

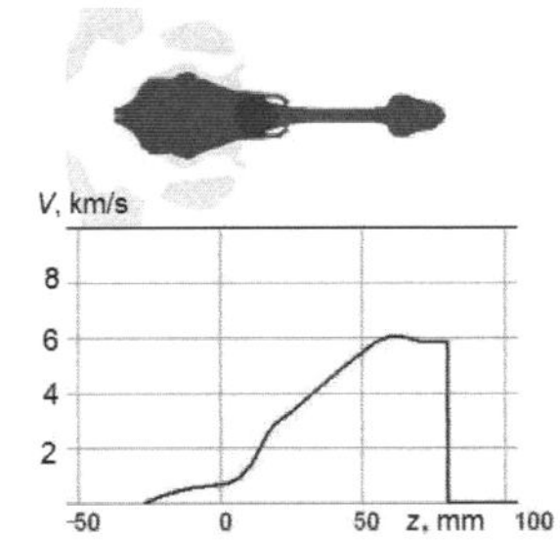

Figure 2. The process of forming an element in the program "ERUDIT" (base liner) for the moment of time 38 μs (without taking into account the criterion of destruction)

It can be seen that the stretched cumulative jet follows after the allocated gradientless high-speed compact element, that is a consequence of the absence of criterion of destruction of material in the calculations.As appears from the known results of the experiments [4] and numerical calculations taking into account model of destruction of material which are carried out according to the ANSYS Autodyn program instead of such "integral" jet the plume of the small particles scattering in the radial direction. This jet is gradually lagging behind from the moving compact element that moves faster than it.

Figure 3 shows a comparative pictures of the process of formation and movement of high-speed compact elements obtained from the results of calculations performed by the ANSYS Autodyn program without taking into account (a) and taking into account material destruction (b) [5].

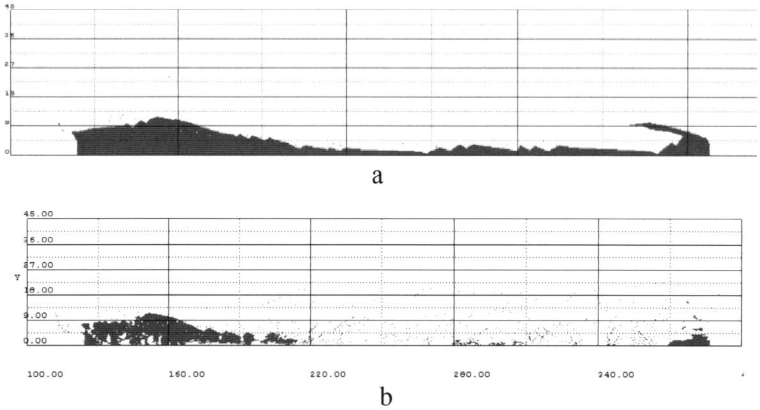

Figure 3. The process of formation and movement of a high-speed compact element in the ANSYS Autodyn program: a - without material destruction; b - taking into account the criterion of destruction

Table. 1 shows the comparative data of the calculations, given by both methods [5]. It can be seen that the discrepancy in velocities is less than 6%, which indicates the reliability of the results obtained.

Table 1. Comparative results of calculations for the "ERUDITE" and ANSYS Autodyn methods

Embodiment	**Geometric parameters liners, mm**						**Estimated speed of the element *V*, km/s**	
	R_0	δ_C	R_1	δ_1	R_2	δ_2	**«ERUDIT»**	**ANSYS Autodyn**
Base liner	24,0	2,5	24,0	2,5	24,0	2,5	6,0	6,0
HC-1	25,0	1,5	25,0	2,5	25,0	1,5	8,1	8,5
HC-2	25,3	1,2	25,3	2,5	25,3	1,2	10,1	10,5
SEC-1	-	1,2	-	2,4	30,6	1,2	7,85	7,4
SEC-2	-	1,0	-	2,4	26,6	1,0	10,4	9,8
SC-1	14,0	4,0	25,0	2,5	26,5	1,5	7,6	7,5
SC-2	15,0	3,0	25,0	3,0	26,0	1,5	8,1	8,0
SC-3	16,8	1,2	25,3	3,5	25,5	1,2	10,0	10,3

As it was pointed out in [1-3], the aim of the study was to identify the design of combined cumulative liners, which make it possible to increase the speed of throwing a compact steel element of gram mass to a level of 8 ... 10 km / s.From these positions influence of design characteristics of the liners on mass and high-speed parametrs of a high-speed compact element was investigated (Figure 1).

At the same time, the method of mass-velocity distributions (MVD), which was described in detail in [2], was used as the main tool for the results comparison. Figure 4 shows an example of this approach wich is given in the analysis of SC-lining of various geometric parameters.

The curves on Figure 4 represent following ordinates of each point correspond to the mass of material of the facing moving with an axial speed, wich is bigger than the value set by an abscissa of this point. Taking into account the provided rule of creation of MVD the curves corresponding to them always monotonically decrease with growth of speed and have a break on abscissa axis at a certain value which fixes the maximum speed of a jet flow– there are no material particles with a speed above this value in a jet flow. On the steepness of MVD in the vicinity of her point break it is possible to estimate the sizes of the head section of the flow– the more abruptly the curve goes from its extreme right point on the abscissa axis (towards lower velocities), the more massive this section is. With the use of the MVD method it is possible to trace clearly the fact of formation of an ungradient head section of the current is clearly traced, which, if preserved in its continuity, can later become a high-velocity compact element.

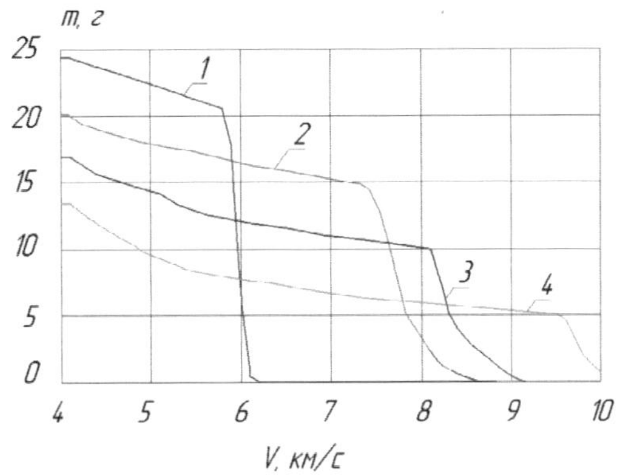

Figure 4. Mass-velocity distributions in the formation of a high-speed compact element from the SC-facing (Figure 1, c, Table 1):
1 – base liner (δ_1/δ_2 = 2,5/2,5 mm; δ_C = 3,2 mm);
2 –SC-1 (δ_1/δ_2 = 2,5/1,2 mm; δ_C = 4,0 mm);
3 –SC-2 (δ_1/δ_2 = 3,0/1,5 mm; δ_C = 3,0 mm);
4 –SC-3 (δ_1/δ_2 = 3,5/1,2 mm; δ_C = 1,2 mm);

In this case, a segment close to the vertical line on the MVD curve emerging from its extreme right-hand point on the abscissa axis (the presence of a vertical section means that some finite mass of the jet flow has the same axial velocity) is appeared. The ordinate of the upper point of this section, where the MVD curve undergoes a break at the transition to monotonous increase with decreasing velocity, corresponds to the mass of the formed compact element.

In the present study it was revealed that the main reason for increasing the speed of compact elements is the transition to the degressive thickness of the jet-forming (top) part of the cumulative liner. Simultaneously with the increase of liner thickness difference the gradient of the axial velocity in the head part of the jet also increases, which leads to a rapid decrease in the thickness of the head part due to its stretching and, as a consequence, the mass of the head part of the jet flow decreases [2].

Velocity increase can be explained by the spherical cumulation process [6, 7]. When the top part with degressive thickness collapses, conditions wich are closer to the spherically symmetric one are created, i.e. the effect of spherical cumulation (the concentration of energy in the inner layers of the collapsing spherical shell) increases. In the case of explosive compression of a hemispherical liner of constant thickness, these conditions are violated due to the advancing motion of its vertex part, which is associated with an earlier arrival of a detonation wave to a liner and the beginning of its loading. Thus, reducing the thickness of the peripheral part of the hemispherical liners leads to an increase of its velocity, and the effect of a liner "reversing" appears to a lesser extent, thereby providing better conditions for the realization of spherical cumulation.

To increase the mass of the head part of the jet flow, it was determined that the shaping of the slightly elongated half-ellipsoid shape (Figure 1, b) leads to a decrease in the velocity gradient at the head section. Reduction of the velocity gradient of the head part in turn leads to an increase in the transverse dimensions of the element being formed while maintaining its speed. From Table. 1, which summarizes the results of the calculations, it can be seen that the best speed results were obtained for the HC-2, SEC-2 and SC-3 variants, while in the SEC-2 version the mass of the elements is minimal and amounted to fractions of grams, whereas in the variants of SC-3, it has already increased up to 5 g. However, from the technological point of view, the most suitable variant for realization is a variant of SEC-liner with a hemispherical shape of the jet-forming part (Figure 1, a), for which a variant of the technological process was proposed in [8], namely, stampings by plastic metal.

To validate the numerical calculation results the experimental researches were carried out [9]. The model of the explosive throwing device for tests is presented on Figure 5. Cumulative liner of HC-1

IOP Conf. Series: Journal of Physics: Conf. Series **894** (2017) 012066 doi:10.1088/1742-6596/894/1/012066

option was investigated (Table 1, Figure 1, a). As the basic method of experimental research, an x-ray method with luminescent memory plates was used [10].

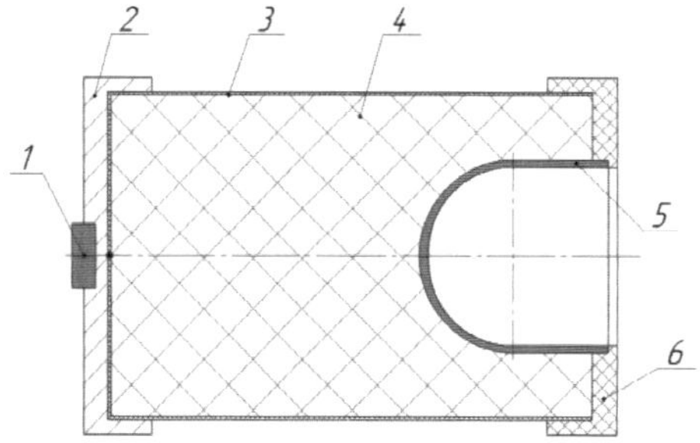

Figure 5. Model of an explosive throwing device for experimental researches: 1-place of initiation; 2-back cover; 3-case; 4-charge of explosives; 5-combined SEC-liner; 6-front cover

As X-ray studies have shown, explosive compression of SEC-liners of a degressive thickness produces, a flow of the high-velocity particles gradually dissipating in the radial direction with a leading particle moving along the explosive charge axis. The X-ray picture pattern shown in Figure 6 corresponds to the time when the leading particle was located at a distance of about 0.7 m from the base of the shape charge with the cumulative concavity. In this experiment, its speed was 7.8 km / s and it had the following dimensions – a length of about 8 mm, a diameter in the middle section of about 3 mm [9, 10].

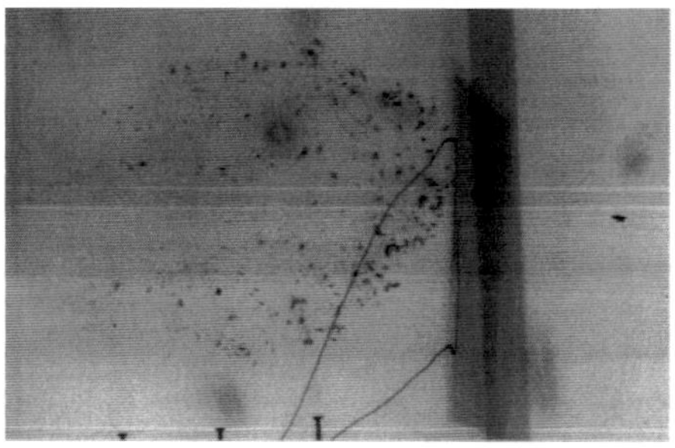

Figure 6. X-ray picture of the motion of high-velocity element formed during explosive reduction of a steel SEC-liner of a degressive thickness

It can be seen that at a certain distance from the leading element the scattered cloud of particles moves (coming in the recorded timepoint to the electrocontact sensor which is in the field of shooting). Obviously, the picture in Figure 6 is completely consistent with the principle of "work" SEC-liner. The leading particle is a cut off head portion of the jet flow of the material formed during the reduction of the hemispherical part of the SEC-liner, and the cloud of particles moving behind it is the result of material destruction during the collapse of the cutting off cylindrical part.

Based on the results of numerical calculations (see Table 1), a similar scheme with SEC-liner formed a compact element at a speed of 8.1 km/s. Thus, the discrepancy with experimental data on the speed of the formed compact element was 4%, which confirms the reliability of the numerical calculation method used by the authors.

3. Conclusion

As follows from the results of the experimental research, the transition to a SSC-liners of a degressive thickness made it possible to significantly increase the velocity of the cut off head part of the jet flow formed in the reduction of the hemispherical part in relation to the base SSC-liner of constant thickness (increment of the velocity from 35%). The significant decrease in the dimensions of the leading element, as compared to the dimensions of the compact element formed by the base SSC-liner, was also predicted in the results of numerical simulation [1-3]. Moreover, it was shown in these researches that an increase in the "massiveness" of the head part of the jet flow with an increased velosity provided by the degeneracy of the thickness of the jet-forming part of the combined liner, can be achieved by making this liner's part the shape of the truncated sphere (Figure 1, c) or by the elongated half-ellipsoid (Fig. 1, b). Experimental verification of these improvements is the subject of a further research.

Acknowledgements
The work was financially supported by the Ministry of Education and Science of the Russian Federation within the basic part of the state task under the section "Initiative Scientific Projects" (code 9.5330.2017 (BCh)).

References

[1] Fedorov S V, Bayanova Ya M and Ladov S V 2011 *Proc. 26-th Int. Symp. On Ballistics. Miami Florida, USA* **1** 253–265
[2] Fedorov S V Bayanova Ya M and Ladov S V 2015 *Combustion, Explosion, and Shock Waves* **51**(1) 130–142
[3] Selivanov V V, Fedorov S V, Nikolskaya Ya M and Ladov S V 2017 *Acta Astronautica* **135** 34–43
[4] Zhdanov I V, Knyazev A S and Malyarov D V 2010 *Work of Tomsk State University – Series of Physical-Mathematical* **276** 193–195.
[5] Bayanova Ya M, Fedorov S V and Ladov S V 2015 *Int.l conf.XVII Khariton's Top. Scient. Read. The collection of reports, Sarov. FSUE «VNIIEF»* 529–533.
[6] Fedorov S V 2016 *Combustion, Explosion, and Shock Waves* **52**(5) 600–612.
[7] Fedorov S V 2017 *Vestn. Mosc. Gos. Tekh. Univ. im. N.E. Baumana, Estestv. Nauki* **3** 71–92.
[8] Baburin M A, Baskakov V D, Zarubina O V, Ladov S V, Fedorov S V and Nikolskaya Ya M 2016 *Technology of metals* **11** 2–8
[9] Fedorov S V, Ladov S V, Nikolskaya Ya M, Baskakov V D, Baburin M A, Kurepin A E, Gorbunkov A A and Pyrozersky A S 2017 *Physics of Combustion and Explosion* **53**(4)
[10] Fedorov S V, Ladov S V, Nikolskaya Ya M, Baskakov V D, Baburin M A, Kurepin A E, Volkov A P, Gorbunkov A A and Pyrozersky A S *2017 Int. conf. XIX Khariton's Top. Scient. Read. The collection of reports, Sarov. FSUE «VNIIEF»* 213–214.

Numerical simulation of shock-free strong compression of 1D gas layer's problem subject to conditions on characteristic

S P Bautin[1] and N S Novakovskiy[1]

[1] Ural State University of Railway Transport (USURT), Yekaterinburg, Russia

E-mail: `SBautin@usurt.ru`, `n.s.novakovskiy@yandex.ru`

Abstract. The article describes results of one-dimensional gas layers strong compression problem's math modeling. The article provided numerical algorithm for solving the one-dimensional ideal gas shock-free strong compression problem in R. Mises configuration is proposed too. The method combines finite-difference method ROMB and tracking feature method. The method allows to calculate gas-dynamic characteristic (velocity, density, etc.) of ideal gas layer while time increase and provides better accuracy in comparison with other finite-difference method. The accuracy of the proposed method was demonstrated in calculations of test plane-symmetry problem. Exact solution and numerical one agree quite well. Numerical results of solving one-dimensional problems with different symmetry and gas characteristic is also shown. The main results of numerical simulations are shown in graphs and tables.

1. Introduction

The mathematical description of shock-free compression of ideal gas to any predefined density including infinite is of interest for controlled thermonuclear fusion [1, 2]. The idea of shock-free compression is very attractive as a way of attaining arbitrary high densities at minimal energy inputs due to the absence of shocks in the flow, i.e., unchanged initial entropy [3]. Attractiveness of the idea is seen from lots of publications on the topic (a comprehensive bibliography is provided in [4]). The mathematical theory of shock-free strong gas compression is developed in [4, 5]. In particular, for cylindrical ($\nu = 1$) and spherically ($\nu = 2$) symmetric layers of a polytropic gas with $\gamma > 1$, it is shown that the continuous joining of two flows gives a solution to shock-free strong compression of gas with a nonzero mass to any predefined density. The authors of [6] propose an algorithm which in reverse time simulates shock-free strong compression of 1D layers of initially resting homogeneous gas with initial density $\rho_0 = 1$ to any predefined finite and constant density $\rho_* > 1$ and then reconstructs, again in reverse time, the path of the gas compressing piston. Reference [7] presents calculations with the algorithm for a number of 1D problems with different symmetries, where the gas layers are compressed externally, i.e., the piston radius decreases. It should be mentioned particularly that the calculations, including the reconstruction of the compressing piston path, are done in reverse time. In [6, 7], the question of whether the reconstructed path implements when time increases is not discussed at all. The algorithm relies essentially upon the local theorem of existence of a solution in the vicinity of a singular point $(t = t_*, r = r_*)$ [4, 5]. For reconstructing the piston path in case of large masses of gas being compressed, it is necessary to depart rather far from $(t = t_*, r = r_*)$ and the question

Content from this work may be used under the terms of the Creative Commons Attribution 3.0 licence. Any further distribution of this work must maintain attribution to the author(s) and the title of the work, journal citation and DOI.
Published under licence by IOP Publishing Ltd

of whether the solution is true remains open. Reference [8] describes a reverse-time compression algorithm which is based on that one proposed in [6, 7] but modified for a non-decreasing piston radius. It presents direct-time calculations for the problem of interest by the ROMB finite difference scheme [9], using the piston path reconstructed in reverse-time calculation. That is, the calculations explicitly simulate the action of compression piston on resting gas that at last time gives the desired gas state - the required density in the entire mass. The direct-time ROMB calculations showed up the flow regions where the algorithm distorts solutions including the known exact solution for plane symmetry. Near the weak discontinuity the solution smears, i.e., the compression wave runs away from the exact solution. This often happens when finite difference methods are used to simulate compression and rarefaction waves. The calculation region can be reduced if take the right boundary not constant ($r = r_*$) but movable. In [10] another algorithm is proposed, which uses the known advantages of the ROMB scheme and the known data on the path of the weak discontinuity between gas at rest and gas in motion. The algorithm helps improve numerical accuracy near the characteristic which separates the compression wave and the background flow. However the solution of the problem under study is known to have a singularity at point ($t = t_*, r = r_*$) which in calculation manifests itself in the following way. While the overall gradient in the solution grows and reaches about 30, the gradient of the sought function near the boundary sound characteristic becomes as large as to prohibit further calculation. The study completes with derivation of analytic formulas for the Generalized Centered Wave (GCW) characterize and their use for calculation in the region between the compression piston and the boundary sound characteristic C^+ [11]. With the resulted analytic formulas the calculation region at times close to $t = t_*$ can be reduced to a portion where singularities in the solution are absent, i.e., to the region between the compression piston and the boundary GCW characteristic where the desired values of density are reached. This helped us accurately calculate compression of 1D gas layers to $t = t_*$.

2. Problem statement

Consider 1D symmetric isentropic ($s = const$) flows of a polytropic gas. Such flows are solutions of equations [4]:

$$\begin{cases} c_t + uc_r + \dfrac{(\gamma-1)}{2}c\left(u_r + \nu\frac{u}{r}\right) = 0, \\ u_t + \dfrac{2}{(\gamma-1)}cc_r + uu_r = 0. \end{cases} \tag{1}$$

Here $c = \rho^{(\gamma-1)/2}$ is sound velocity in gas; ρ is density; $\gamma > 1$ is adiabatic exponent in the

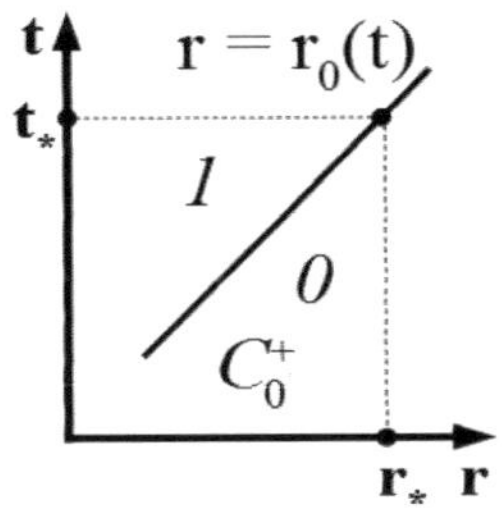

Figure 1. Problem statement.

equation of state $p = \rho^\gamma/\gamma$ where p is pressure; c and u are sought functions. The values $\nu = 0, 1, 2$ correspond to plane, cylindrical, spherical symmetry. At $\nu = 0$ $r = x_1$ and $u = v_1-$ is a projection of vector $\mathbf{V}$ onto the axis Ox_1. At $\nu = 1, 2$ $r = \sqrt{\sum\limits_{i=1}^{\nu+1} x_i^2} \geq 0$ and the gas velocity

u is a projection of vector $\mathbf{V}$ onto the radius vector in the plane x_1Ox_2 at $\nu = 1$ or the radius vector in the space of variables x_1, x_2, x_3 at $\nu = 2$.

Let a background flow $c(t,r) = 1$, $u(t,r) = 0$ be defined in a vicinity of point $(t = t_*,\ r = r_*)$ for $r_* > 0$. In Figure 1 it is shown in region 0, and in region 1 we are to find a flow with a predefined distribution of one of the hydrodynamic parameters which will be mated to the background flow via the weak discontinuity.

The weak discontinuity is the sound C^+ characteristic whose path with the above background values is defined by the equation: $r(t) = r_* + t - t_*$.

The region of the sought flow is limited by the straight line $t = t_*$ and the sound characteristic C_0^+ of the background flow. It consists of two parts: a lower triangle (denoted 1 in Figure 2) where the Riemann generalized centered wave is defined, and an upper triangle (denoted 2 in Figure 2) where the flow with a specified constant sound velocity at $t = t_*$ is defined. We need to find the law describing the path of a piston which with no shock compresses a resting homogeneous gas of density $\rho = 1$ to density $\rho = \rho(t_*, r) = \rho_* = \text{const}$, which remains constant within a compressed layer of width d_*, (Figure 2). The path is defined as a solution of the following Cauchy problem:

$$\begin{cases} \dfrac{dr}{dt} = u(t,r), \\ r|_{t=t_*} = r_* - d_*. \end{cases}$$

Here the function $u(t,r)$ is found from the solution of system (1).

The theorems proved in [4] assert that there is a nonzero mass m_0 between the piston path $r = r_p(t)$ and the line $r = r_*$, which can be compressed shock-free to any predefined density ρ_*. But the theorems do not determine the maximal value of m_0 and hence the maximal width d_* (Figure 2).

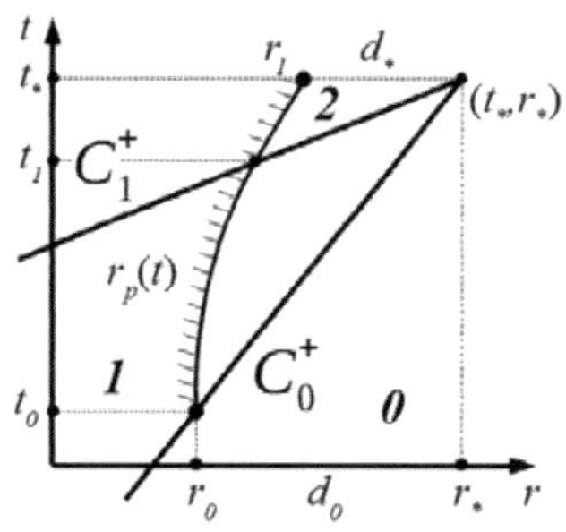

Figure 2. Three solution regions.

3. The reverse-time calculation method

The problem is numerically solved with an algorithm which exploits characteristic's method and relies upon results obtained in [4, 6, 12]. The main difference from the classical method of characteristics is that the solution is sought in reverse time and the singular point (r_*, t_*) is treated by creating not a single but a bundle of sound characteristics C^+ from it. Figure 3 a schematically shows how the characteristic grid in region 1 is calculated. A detailed description of the algorithm can be found in [6]. The path is reconstructed in reverse time from the point $(t = t_*, r = r_* - d_*)$ where d_* is the width of the compressed layer which defines its mass. Calculation with the algorithm from [6] runs until the piston path either crosses the characteristic C_0^+, or escapes from the region where the solution exists.

4. The direct-time calculation method

Consider the following 1D system of hydrodynamic equations in Lagrangian mass coordinates:

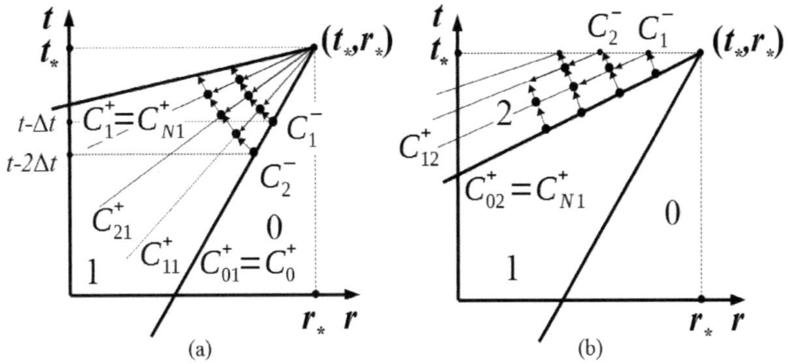

Figure 3. Solutions in region 1(a) and region 2 (b).

$$
\begin{aligned}
&\frac{\partial v}{\partial t} = \frac{\partial u}{\partial m},\\
&\frac{\partial u}{\partial t} = -\frac{\partial p}{\partial m},\\
&\frac{\partial E}{\partial t} + v\frac{\partial (pu)}{\partial r} = 0,
\end{aligned} \tag{2}
$$

here $t, r, m, \rho, v = \frac{1}{\rho}, \varepsilon, p$ are, respectively, time, mass, density, specific volume, specific internal energy, pressure, and $E = \varepsilon + u^2/2$ is total specific energy.

The system is closed with the equation of state for ideal gas $p = (\gamma - 1) \cdot \rho C_v T$, $\varepsilon = C_v T$, where C_v is specific heat at constant volume and T is gas temperature. System (2) is completed with the boundary and initial conditions

$$
\begin{aligned}
&u|_{r=r_L}(t) = u_p(t),\\
&u|_{r=r_R}(t) = 0,\\
&u(x,0) = 0,
\end{aligned} \tag{3}
$$

where r_L, r_R are the left and right boundaries of the section where system (2) is being solved. At $t = 0$: $r_L = r_p(0)$,$r_R = r_*$,. Here $u_p(t)$ is the velocity of the compression piston from the reconstruction of its path in reverse time.

System (2), (3) is solved with the ROMB scheme which is described in rather detail in [9]: pressure and Newton specific internal energy are linearized and iterations start. New variables (pressures P and velocities U at mesh nodes) are introduced through relating equations. After appropriate manipulation we come to a system of linear algebraic equations for P and U with a four-diagonal matrix which is solved with a method similar to stream sweeping [13].

5. Analytic representation of GCW characteristics

The analytic formulas for the sound GCW characteristics are derived under certain assumptions. It was obtained in [11] with help of analitical method from [14]. Below are the ordinary differential equations (ODE) that are solved to obtain the functional dependencies and the final formulas.

$$
\left[2 + \left(\frac{2\nu}{r_*}\ln c + \frac{\nu}{r_*} - 2\frac{\nu}{r_*}c\right)\tau\right]\frac{dc}{dt} = \frac{\nu}{r_*}c - \frac{\nu}{r_*}c\ln c, \gamma = 3, \tag{4}
$$

$$\left[4+\left(\frac{12\nu}{r_*}c+\frac{6\nu}{r_*}c^2+\frac{20\nu}{r_*}c\ln c\right)\tau\right]\frac{dc}{dt}=\frac{2\nu}{r_*}c^2-\frac{6\nu}{r_*}c-\frac{4\nu}{r_*}c^2\ln c, \gamma=\frac{5}{3} \tag{5}$$

$$\left[2\alpha+\left(\frac{4\nu\alpha(2+A)}{Br_*}c-\frac{8\nu\alpha}{Ar_*}\right)\tau\right]\frac{dc}{dt}=\left[\frac{2\nu\alpha}{r_*}\left(\frac{2c}{A}-\frac{Ac^2}{B}\right)\right], \gamma\neq 3,\frac{5}{3}, \tag{6}$$

where $A=3-\gamma$, $B=5-3\gamma$.

As $\tau\to 0$ and c is upper bound, the solution of equations (4)–(6) can be written explicitly in vicinity of point (t_*, r_*) :

$$c(\tau)=\exp\left[\left(\ln c_*-\tfrac{1}{3}\right)\exp\left(-\tfrac{3\nu}{2r_*}\tau\right)+\tfrac{1}{3}\right], \gamma=3, \tag{7}$$

$$c(\tau)=\frac{2B}{A^2\left[1-\left(1-\frac{2B}{A^2c_*}\right)e^{-\frac{2\nu}{Ar_*}\tau}\right]}, \gamma\neq 3,\frac{5}{3}. \tag{8}$$

6. Calculations

The reverse-time calculation algorithm and results of calculations according to the algorithm can be found in [8]. In this paper we only use the trajectories of compression piston was calculated with help of reverse-time method.

Now look at direct-time calculations. Figure 4 show results of calculation by the ROMB scheme without the explicit tracking of the weak discontinuity. It is seen that the ROMB scheme

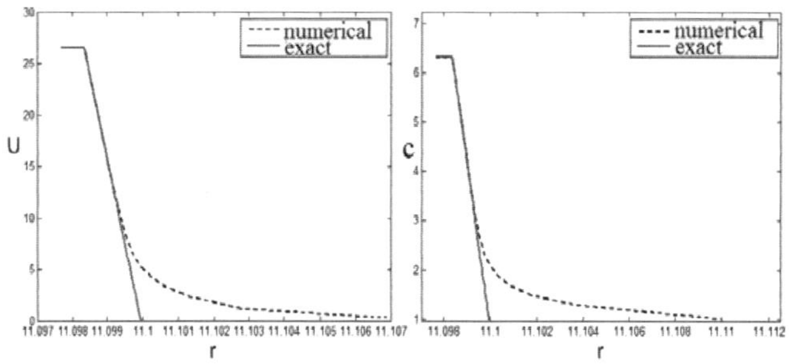

Figure 4. Velocity and sound speed at $t = 0.999995t_*$.

adequately reproduces the compression region but significantly smears the weak discontinuity. Of special note is a flat section seen on the left in the profiles of hydrodynamic parameters. It proves that the piston path reconstructed in reverse-time calculation does compress gas to the required density.

A test calculation was done to compare calculations by ROMB without singularities and by a hybrid method which calculates only the region between the compression piston and the C_0 characteristic. It is seen from Figure 5 that the latter describes the solution in the compression wave and weak discontinuity regions very much better than ROMB.

Calculations were done for cylindrical and spherical symmetries with different adiabatic exponents and initial masses. ROMB conservatism ensures energy conservation at the end of calculation at level close to computer zero. Energy conservation at each time step tn is checked by calculating energy disbalance with the formula

$$\Delta E^n=\frac{\sum\limits_{k=1,N} m_k\cdot(E_k^n-E_k^0)+W}{\sum\limits_{k=1,N} m_k\cdot E_k^n}, \tag{9}$$

IOP Conf. Series: Journal of Physics: Conf. Series **894** (2017) 012067 doi:10.1088/1742-6596/894/1/012067

where m is mass, E_k^n and E_k^0 are specific total energies at current and initial times, and W is the work done by the compression piston.

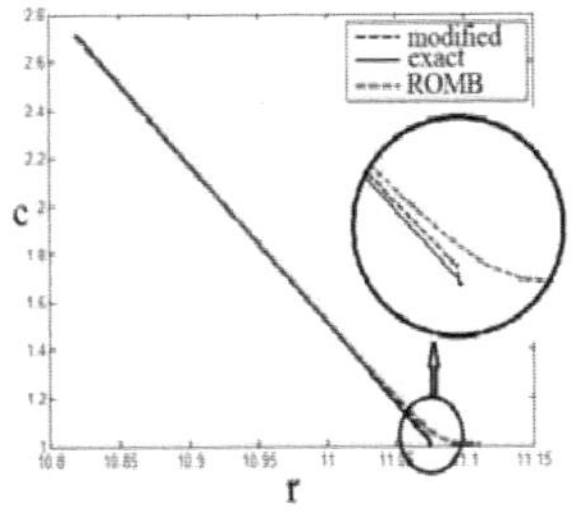

Figure 5. Comparison of two methods.

Data on energy disbalance, calculated according to (9), at different times from one of the calculations are provided in Table 1. Total energy disbalance is a result of inaccuracy in the determination of the current piston velocity through linear interpolation. The data of Table 1 show it to remain insignificant and almost unchanged through the entire calculation.

Table 1. Total energy disbalance at different time moments.

$0.95t^*$	$0.975t^*$	$0.99t^*$	$0.995t^*$	$0.998t^*$
$-5.0861 \cdot 10^{-5}$	$-8.7531 \cdot 10^{-5}$	$-2.9584 \cdot 10^{-4}$	$-3.3866 \cdot 10^{-4}$	$-4.6709 \cdot 10^{-4}$

Below are figures that demonstrate good accuracy of the modified method, and in the region where the compression wave and the resting gas join, too. Figure 6 shows calculated results for a cylindrical symmetric problem with $\gamma = 1.4, \nu = 1, m_* = 10$. The solution obtained in reverse-time calculation by the method of characteristics with recalculation is used as a reference. It should be noted that the region where the compression wave joints the resting gas is reproduced rather accurately.

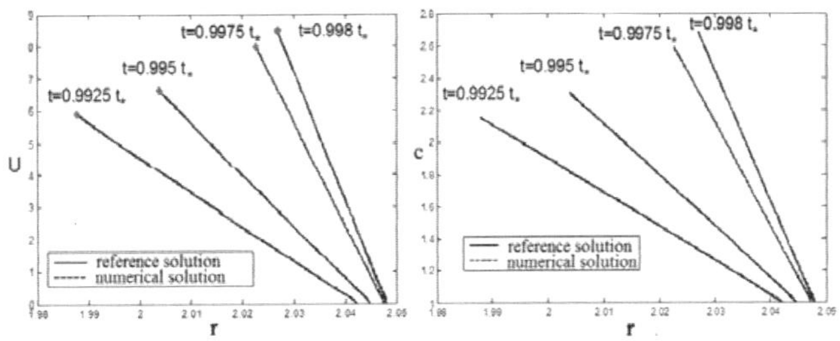

Figure 6. Results at different times for $\gamma = 1.4, \nu = 1, m_* = 10$.

The modified method and analytic formulas (7), (8) helped obtain in direct-time calculations the hydrodynamic parameters of a gas layer up to and inclusive $t = t_*$, in the region between the compression piston and the boundary GCW characteristic. Figure 7 shows their profiles at $t = t_*$, for $\gamma = 1.4, \nu = 2, \rho_* = 10^5, m_* = 10$. The solution is seen to agree well with the reference.

7. Conclusion

The results presented above allow the following inferences.

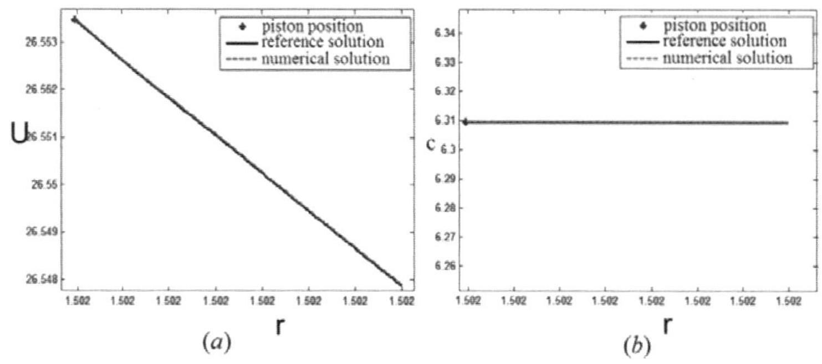

Figure 7. Sound speed (a) and gas velocity (b) at $t = t_*$.

1. The law describing the motion of a piston with a non-vanishing radius, which compresses a 1D layer of gas of a specified mass to a required density, was obtained in the reverse-time calculation. The algorithm proposed for solving such problems was tested. Reverse-time calculations allow the piston path to be reconstructed within 1-2 minutes on standard PC.

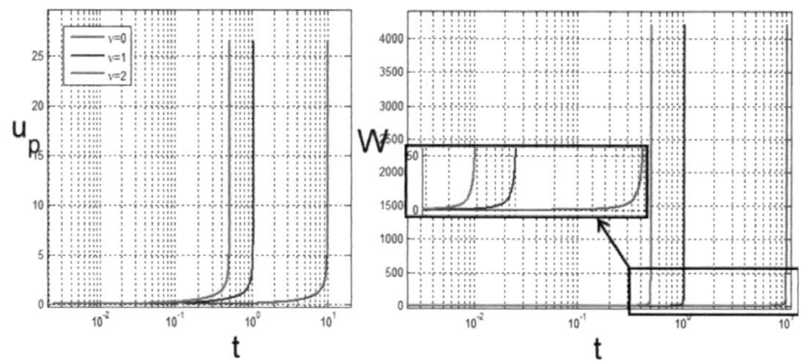

Figure 8. Compressing piston velocity and it's work as recommendations for physical experiment.

2. With the known $r = r_p(t)$, direct-time calculations were done for a number of problems by the ROMB method. Their results show that the numerical algorithm reconstructs the solution with good accuracy up to time $0.999t_*$ regardless of problem symmetry. Further calculations show that calculation accuracy degrades near the singular point where derivatives on the boundary sound characteristic start rapidly grow. The effect can be reduced by moving off the right boundary of the calculation region.

3. The basic error of calculation is confined within a rather narrow region about the weak discontinuity that separates the compression wave and the resting gas.

4. The implemented ROMB modification allows attaining higher accuracy of calculation up to $t = t*$.

5. The approach proposed not only helps reconstruct the path of compression piston for an arbitrary 1D gas layer, but also proves numerically that the piston does compress a specified mass of gas (including $m_* = 10$) to a desired density (including $\rho_* = 10^5$).

Figure 8 shows laws of compressing piston motion in time and it's work. These results therefore provide specific recommendations for physical experiment.

References

[1] Zababakhin E and Zababakhin I 1988 *Unlimited Cumulation Phenomena* (Moscow: Nauka)
[2] Nuckolls J 1984 *J. Advances in Physics* **143** 467
[3] Dolgolyova G and Zabrodin A 2004 *Energy Cumulation in Layer Systems and Shock-Free Compression Realization* (Moscow: Fizmatlit)
[4] Bautin S 2007 *Mathematical Modeling of Strong Gas Compression* (Novosibirsk: Nauka)
[5] Bautin S 1997 *Mathematical Theory of Shock-Free Strong Compression of Ideal Gas* (Novosibirsk: Nauka)
[6] Bautin S and Nikolayev Y 2000 *J. Computational Technologies* **5** 3
[7] Nikolayev Y 2001 *J. Computational Technologies* **6** 104
[8] Novakovskiy N 2016 *J. Mathematical Structures and Modeling* **3**(39) 93
[9] Gadzhiev A and Pisarev V 1979 *J. Computational Mathematics and Mathematical Physics* **19** 1288
[10] Novakovskiy N 2017 *J. Mathematical Structures and Modeling* **1**(41) 92
[11] Novakovskiy N 2017 *J. Mathematical Structures and Modeling* **2**(42) 94
[12] Rozhdestvensky B and Yanenko N 1968 *Systems of Quasi-Linear Equations and Their Hydrodynamic Applications* (Moscow: Nauka)
[13] Degtyaryov L and Favorsky A 1968 *J. Computational Mathematics and Mathematical Physics* **8** 679
[14] Bautin S and Deriabin S 2005 *Mathematical Modeling of Ideal Gas Flow into Vacuum* (Novosibirsk: Nauka)

IOP Conf. Series: Journal of Physics: Conf. Series **894** (2017) 012068 doi:10.1088/1742-6596/894/1/012068

Multiphase flow models for hydraulic fracturing technology

Andrei A. Osiptsov

Skolkovo Institute of Science and Technology (Skoltech), 3 Nobel Street, 143026, Moscow, Russian Federation

E-mail: a.osiptsov@skoltech.ru

Abstract. The technology of hydraulic fracturing of a hydrocarbon-bearing formation is based on pumping a fluid with particles into a well to create fractures in porous medium. After the end of pumping, the fractures filled with closely packed proppant particles create highly conductive channels for hydrocarbon flow from far-field reservoir to the well to surface. The design of the hydraulic fracturing treatment is carried out with a simulator. Those simulators are based on mathematical models, which need to be accurate and close to physical reality. The entire process of fracture placement and flowback/cleanup can be conventionally split into the following four stages: (i) quasi-steady state effectively single-phase suspension flow down the wellbore, (ii) particle transport in an open vertical fracture, (iii) displacement of fracturing fluid by hydrocarbons from the closed fracture filled with a random close pack of proppant particles, and, finally, (iv) highly transient gas-liquid flow in a well during cleanup. The stage (i) is relatively well described by the existing hydralics models, while the models for the other three stages of the process need revisiting and considerable improvement, which was the focus of the author's research presented in this review paper. For stage (ii), we consider the derivation of a multi-fluid model for suspension flow in a narrow vertical hydraulic fracture at moderate Re on the scale of fracture height and length and also the migration of particles across the flow on the scale of fracture width. At the stage of fracture cleanaup (iii), a novel multi-continua model for suspension filtration is developed. To provide closure relationships for permeability of proppant packings to be used in this model, a 3D direct numerical simulation of single phase flow is carried out using the lattice-Boltzmann method. For wellbore cleanup (iv), we present a combined 1D model for highly-transient gas-liquid flow based on the combination of multi-fluid and drift-flux approaches. The derivation of the drift-flux model from conservation olaws is criticall revisited in order to define the list of underlying assumptions and to mark the applicability margins of the model. All these fundamental problems share the same technological application (hydraulic fracturing) and the same method of research, namely, the multi-fluid approach to multiphase flow modeling and the consistent use of asymptotic methods. Multi-fluid models are then discussed in comparison with semi-empirical (often postulated) models widely used in the industry.

1. Introduction

The technology of hydraulic fracturing of a hydrocarbon bearing underground formation is based on injecting a fluid laden with rigid particles under a high pressure (up to several hundred bar) into the well to create fractures in the porous medium, which are filled with particles. After the end of pumping, fractures closed on packed granular material provide high-conductivity channels to transport hydrocarbons from reservoir to the well and all the way up to the surface.

Content from this work may be used under the terms of the Creative Commons Attribution 3.0 licence. Any further distribution of this work must maintain attribution to the author(s) and the title of the work, journal citation and DOI.
Published under licence by IOP Publishing Ltd

The well may be vertical (when a single bi-wing fracture is formed) or near-horizontal with several perforation clusters providing reservoir contact (the so-called multi-stage fracturing in low-permeability formations). The latter case gives rise to several transversal fractures.

With respect to different stages of the hydraulic fracturing technology, we consider four classes of multiphase flows that can be modelled within the multi-continua (or multi-fluid) approach [2]. In a more detail, we distinguish the following classes: (i) the flow of suspension of fluid with particles in a circular pipe at high Reynolds numbers during pumping, (ii) the flow of suspension in a narrow vertical hydraulic fracture at moderate Re during pumping [3, 4], (iii) suspension filtration through a packed of proppant particles in a closed fracture during cleanup [5], and (iv) multiphase gas-liquid flow with admixture of rigid particles in a circular pipe during well start-up, cleanup and testing in a wide range of the Reynolds numbers [6]. We discuss the advantages and limitations of the multi-fluid approach based on the simulation examples from each of the four classes of multiphase flows, in comparison with simplified semi-empirical approaches, e.g. the drift-flux model for well flows, the effective-fluid model for suspension transport in fractures, and the deep-bed filtration model. The talk ends up with recommendations for future research on the topic. A detailed review of the state of the art in numerical modeling of hydraulic fracturing can be found in [7], and the most recent review with a specific focus on fluid mechanics of hydraulic fracturing is [1].

2. Two-fluid model for suspension transport in a fracture

In this section, we present in a lumped form the two-fluid model for suspension flow and sedimentation in a hydraulic fracture. This model was derived in the multi-continua approach [2] using asymptotic methods in the lubrication approximation. The key assumption is that the width-to-length ratio of the fracture is a small parameter: $\varepsilon = w/L \ll 1$. The cross-flow particle concentration profile is assumed uniform. In the width-averaged variables, the equations are as follows [3]:

$$\frac{\partial wC}{\partial t} + \nabla(wC\mathbf{V}_p) = 0 \tag{1}$$

$$\frac{\partial w}{\partial t} = \nabla\left[\frac{w^3}{12\mu(C)}\left(\nabla P + \mathrm{Bu}\left[1 + C(\eta - 1)\right]\mathbf{e}_2\right) - wC\mathbf{V}_s\right] - 2v_l \tag{2}$$

$$\mathbf{V}_f = -\frac{w^2}{12\mu(C)}\left(\nabla P + \mathrm{Bu}\left[1 + C(\eta - 1)\right]\mathbf{e}_2\right), \quad \mathbf{V}_p = \mathbf{V}_f + \mathbf{V}_s \tag{3}$$

$$\mathbf{V}_s = -\frac{\mathrm{St}}{\mathrm{Fr}^2}\left(\frac{\eta - 1}{\eta}\right) f(C)\mathbf{e}_2, \; f(C) = \left(1 - \frac{C}{C_{max}}\right)^5, \; \mu(C) = \left(1 - \frac{C}{C_{max}}\right)^{-1.89}.$$

$$\mathrm{Bu} = \frac{\rho_f^0 g L^2}{\mu_0 U}, \quad \rho_p^0/\rho_f^0.$$

Here, Cartesian coordinate system Oxy is introduced in the cell plane, so that y-axis (with the basis vector $\mathbf{e}_2$) is vertical and origin O is located in the bottom left corner of the computational domain; C is the particle volume fraction; $w(x, y, t)$ is the width of the fracture (in hydraulic fracturing simulators, the width is available from solving the geomechanics problem of fracture growth [7]); $\mathbf{V}_f$, $\mathbf{V}_p$, $\mathbf{V}_s$ are the width-averaged velocities of the fluid and the particles, and the settling velocity of particles relative to the fluid; v_l is the velocity of fluid leak-off through the porous walls; differential operator '∇' acts in the (x, y) plane as we applied the averaging procedure along the cell width. The flow scales are as follows: L is the cell length, U is the scale of the injection velocity, d is the cell width scale, ρ_0^f is the fracturing fluid density, μ_0 and τ_0 are the fracturing fluid plastic viscosity and yield stress, respectively; g is the gravity acceleration; Bu is the Buoyancy number.

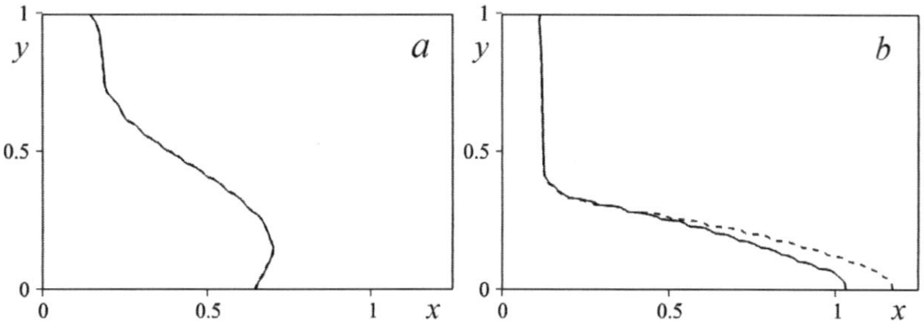

Figure 1. Particle concentration fronts in the fracture of elliptic cross-section at $t = 0.5$ for $Bu = 326$ (*a*) and $Bu = 3260$ (*b*). Result are obtained by the effective-fluid model (dashed line) and the two-fluid model (solid line).

Boundary and initial conditions for the hyperbolic equation for concentration (1) are given by:

$$t = 0 : C = 0, (x, y) \in [0, L/H] \times [0, 1]; \quad x = 0 : \; C = C_0, \; y \in [y_1, y_2]$$

Boundary conditions for the pressure equation (2) are as follows:

$$x = 0 : \; \frac{\partial p}{\partial x} = -\frac{12\mu(C)}{w^2}, \; y \in [y_1, y_2]; \; \frac{\partial p}{\partial x} = 0, y \in [0, y_1], [y_2, 1]$$

$$x = L/H : \; \frac{\partial p}{\partial y} = -Bu; \quad y = 0, \; 1 : \; \frac{\partial p}{\partial y} = -Bu(1 + C|_{y=0,1}(\eta - 1))$$

The particle settling velocity is given by an empirical formula, which is the generalization of the well-known Richardson-Zaki expression, taking into account the fact that particles slow down and stop completely when reaching the packed bed on the bottom. The existing effective-fluid models of suspension flows [8] contain an assumption that the *volume-averaged* suspension velocity is governed by the Poiseuille law, while in the present two-fluid model it is shown based on the derivation from the conservation laws that the Poiseuille law governs the *mass-averaged* velocity of the carrier fluid (3). Also, earlier models contained an assumption in the algebraic expression for the particle velocity (3) that the particles settle relative to the volume-averaged velocity of the suspension [8], and not relative to the fluid, as is the case on the two-fluid approach [3]. As a result, in contrast to the existing models the two-fluid model includes an additional term $-\nabla(wC\mathbf{V}_s)$ in the right-hand side of the pressure equation (2), which takes into account the two-speed effects.

3. Cross-flow inertial migration of particles in a fracture

Particle migration in fractures is essentially a multi-scale problem, spanning from the lift force on a single particle settling in a slot [9], through the inertial igration in a suspension flow in the horizotnal section of a fracture [10, 11] to the effects of cross-flow migration on the global transport and sedimentation in the entire fracture [12].

In thi seciton, we will consider in detail the inertial migration of particles in a dilute suspension flow through the entry region of a plane channel, which is important in application to modeling of proppant migration in the near-wellbore zone of the fracture. Within the two-fluid approach, an asymptotic one-way coupling model of the dilute suspension flow in the entry region of a channel is constructed in [10]. The carrier phase is a viscous incompressible Newtonian fluid, and the dispersed phase consists of identical non-colloidal rigid spheres. In the inter-phase momentum exchange, we take into account the drag force, the virtual mass force, the Archimedes force, and the inertial lift force with a correction factor due to the wall effect and an arbitrary particle slip velocity. The channel Reynolds number is high and the particle-to-fluid density ratio is of order unity or significantly larger unity. The solution is constructed using the matched asymptotic expansions method. The problem of finding the far-downstream cross-channel profile of particle number concentration is reduced to solving the equations of the two-phase boundary layer developing on the channel walls. The full Lagrangian approach is used to study the evolution of the cross-flow particle concentration profile.

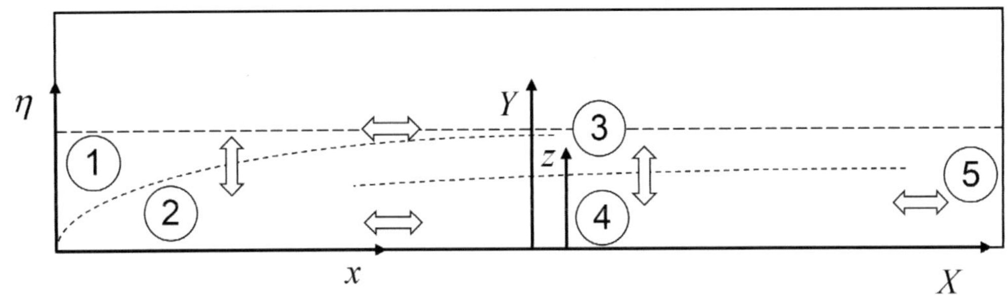

Figure 2. Asymptotic domains in the suspension flow in the horizontal section of an entry region of a fracture: *1* – the entry region, *2* – the boundary layer, *3* – the region of overlapping of the boundary layers, *4* – the lower sublayer, *5* – the far downstream region of the fully developed Poiseuille flow. Symbol ⇔ defines asymptotic matching of solutions in adjacent domains.

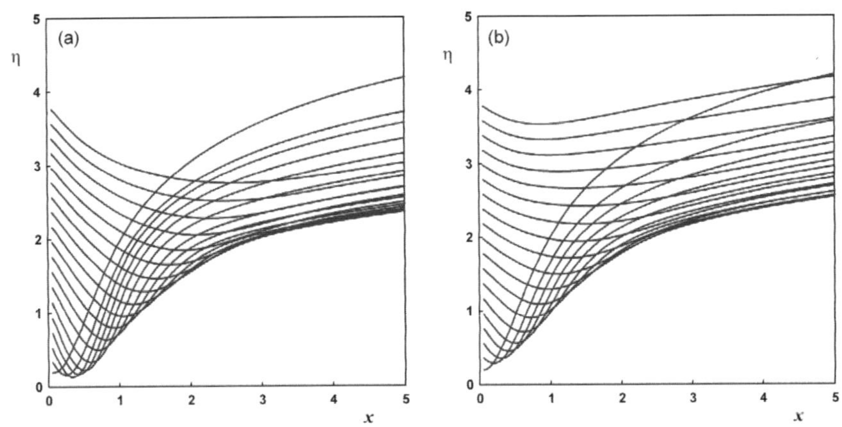

Figure 3. Trajectories of the particles in the boundary layer for the nondimensional intensity of the lift force $\kappa_0 = 20$, the particle-to-fluid density ratio $\xi = 3$ (*a*) and $\xi = 5$ (*b*).

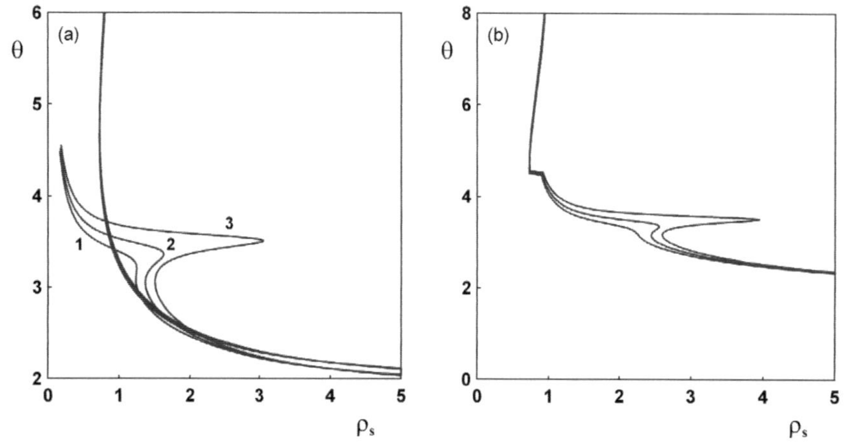

Figure 4. Far-downstream asymptotics of the cross-flow concentration profile ρ_s depending of the stream function θ in the boundary layer, $\kappa_0 = 20$ and $\xi = 5,\ 4.5,\ 4$ – curves *1-3*. Profiles in each layer of the fold (*a*) and the total concentration profile (*b*).

The system of equations of the two-phase boundary layer are as follows [10]:

$$\frac{\partial u}{\partial x} + \frac{\partial v}{\partial \eta} = 0, \quad \frac{\partial \rho_s u_s}{\partial x} + \frac{\partial \rho_s v_s}{\partial \eta} = 0, \tag{4}$$

$$u\frac{\partial u}{\partial x} + v\frac{\partial u}{\partial \eta} = \frac{\partial^2 u}{\partial \eta^2}, \quad \frac{\partial p}{\partial \eta} = 0,$$

$$u_s\frac{\partial u_s}{\partial x} + v_s\frac{\partial u_s}{\partial \eta} = F_{sx} = \frac{2\xi}{2\xi+1}D_0(u-u_s) + \frac{3}{2\xi+1}\left(u\frac{\partial u}{\partial x} + v\frac{\partial u}{\partial \eta}\right),$$

$$u_s\frac{\partial v_s}{\partial x} + v_s\frac{\partial v_s}{\partial \eta} = F_{s\eta},$$

$$F_{s\eta} = \frac{2\xi}{2\xi+1}\left[D_0(v-v_s) + \kappa_0 c_l\sqrt{\frac{\partial u}{\partial \eta}}(u-u_s)\right] + \frac{3}{2\xi+1}\left(u\frac{\partial v}{\partial x} + v\frac{\partial v}{\partial \eta}\right),$$

$$c_l(\chi, x, \eta) = c_l^{\infty}\left[1 - n\exp\left(-m_1(\chi)\frac{\eta}{x^{1/4}}\right)\right],$$

$$m_1 = m\frac{\sqrt{\varphi''(0)}}{(\varepsilon\lambda)^{1/4}}, \quad \chi = \frac{|u-u_s|}{(\varepsilon\lambda)^{1/4}}, \quad \xi = \rho_p^0/\rho_f^0$$

$$\kappa_0 = \frac{6.46}{12\sqrt[4]{18}\pi}Re_{s0}^{3/2}\left(\frac{\rho_s^0}{\rho}\right)^{1/4}, \quad D_0 = 1 + \frac{1}{6}Re_{s0}^{3/2}(u-u_s)^{3/2}, \quad Re_{s0} = \frac{2aU\rho}{\mu}.$$

Here the latter terms in the right-hand side of the momentum conservation equations for the particulate phase correspond to the virtual mass force and the Archimedes force. Additionally, φ is the Blasius function from the well known solution of the boundary layer problem [13]. External flow is uniform, hence the longitudinal pressure gradient is zero. Boundary conditions take the form:

$$x = 0: \quad u_s = \rho_s = 1, \quad v_s = 0; \tag{5}$$

$$\eta = 0: u = v = 0; \quad \eta \to \infty: \quad u \to 1.$$

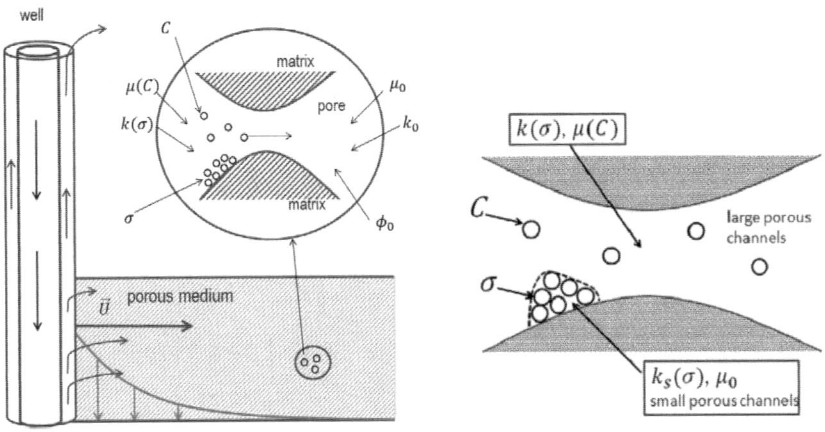

Figure 5. The sketch of suspension flow in a porous medium (a); a scheme of the porous space (b): large pore channels (1) and small pore channels (2).

The inertial migration in the entry region of a plane channel (a circular pipe) results in particle accumulation on two symmetric planes (an annulus) distanced from the walls, with a non-uniform concentration profile between the planes (inside the annulus) and particle-free layers near the walls. When the particle-to-fluid density ratio is of order unity, an additional local maximum of the particle concentration on inner planes (an inner annulus) is revealed. The inclusion of the corrected lift force makes it possible to resolve the non-integrable singularity in the concentration profile on the wall, which persisted in all previously published solutions for the dilute-suspension flow in a boundary layer [14]. The numerical results are compared with the tubular pinch effect observed in experiments, and a qualitative analogy is found.

In this case the trajectories shown in Fig. 3 are substantially different from the case of a dusty gas. The reason is that in the case of a suspension the particle-to-fluid density ratio is of order unity $\xi \sim 1$, and the terms due to the Archimedes force and the virtual mass force should be retained in the momentum conservation laws. These terms result in the formation of an additional local maximum in the cross-flow concentration profile (see Fig. 4,(*a*)).

4. Suspension filtration in proppant packings in a closed fracture

Suspension flow in the propped fracture is described within the three-continua approach: suspended partices (solid particles carried by the flow of a fluid within the porous space), trapped particles (solid particles, which are sedimented in pores), and carrier fluid (viscous incompressible Newtonian fluid). Suspended particles are characterised by the phase density ρ_p^{mob} and the mass-averaged velocity U_p^{mob}; the phase of trapped particles is characterized by the density ρ_p^{sed}; the carrier phase is characterised by the mass-averaged velocity U_f and density ρ_f. Particles have constant substance density ρ_p^0, and the fluid – a constant substance density ρ_f^0. Multi-continua modeling of suspension flows is applicable if the following hierarchy of scales exists: particle diamter is significantly smaller than the diameter of pore channels, while being significantly larger than the mean free path of the fluid molecules [2]. Large pore channels are the pore space excluding trapped particles and the void space between those. Small pore channels are pore space between trapped particles (Fig. 5, *b*).

Density of the particle phases are related to the porosity and the particle concentration by the following relations [15]:

$$\rho_p^{mob} = \rho_p^0 C \phi_c, \qquad \rho_p^{sed} = \sigma \rho_p^0, \qquad \rho_f = \rho_f^0(\phi_t - C\phi_c),$$

$$\phi_t = \phi_0 - \sigma, \qquad \phi_c = \phi_0 - \sigma/C_{max}, \tag{6}$$

where C is the particle volume fraction in the pore space available for the fluid to flow (in large pore channels), σ – the volume concentration of trapped particles in the entire volume of the porous medium, C_{max} is the maximum packing concentration (random close packing), ϕ_t – the porosity of the medium formed by the trapped particles and the matrix (large and small channels in total), ϕ_c – the porosity of the medium formed by the large pore channels; ϕ_0 – initial porosity of the medium.

The mass conservation equation for particles and fluid is written as [19]:

$$\frac{\partial \rho_p^{mob}}{\partial t} + \frac{1}{r^j}\frac{\partial(\rho_p^{mob} U_p^{mob} r^j)}{\partial r} = -q_s, \qquad \frac{\partial \rho_p^{sed}}{\partial t} + \frac{1}{r^j}\frac{\partial(\rho_p^{sed} U_p^{sed} r^j)}{\partial r} = q_s, \tag{7}$$

$$\frac{\partial \rho_f}{\partial t} + \frac{1}{r^j}\frac{\partial(\rho_f U_f r^j)}{\partial r} = 0.$$

Here $j = 0$ in a plane flow and $j = 1$ in a radial flow, q_s is the rate of trapping and mobilization of particles in the matrix, determined by the formulas [16]:

$$q_s = \rho_p^{mob} U_p^{mob} \lambda - \rho_p^0 \alpha \sigma \delta(U_s - U_{crit}), \tag{8}$$

$$\delta(U_s - U_{crit}) = \begin{cases} U_s - U_{crit}, & U_s > U_{crit} \\ 0, & U_s < U_{crit} \end{cases}.$$

Here λ is the colmatation coefficient (intensity of particle trapping) [20], α – the coefficient of mobilization of trapped particle, U_s is the filtration velocity in large and small pore channels, U_{crit} is the critical suspension velocity, above which trapped particles are mobilized.

In large pore channels with the permeability $k(\sigma)$, the suspension is flowing with the viscosity $\mu(C)$. In small pore channels with the permeability $k_s(\sigma)$, only the clean fluid is flowing with the dynamic viscosity μ_0. The momentum conservation equations (Darcy laws) for the carrier phase, suspended particles and the suspension as a whole are obtained by volume-averaging in large and small pore channels:

$$U = -\frac{k(\sigma)}{\mu(C)}\frac{\partial p}{\partial r}, \qquad U_s = -\left[\frac{k(\sigma)}{\mu(C)} + \frac{k_s(\sigma)}{\mu_0}\right]\frac{\partial p}{\partial r}, \tag{9}$$

$$U_f^{filtr} = -\left[(1-C)\frac{k(\sigma)}{\mu(C)} + \frac{k_s(\sigma)}{\mu_0}\right]\frac{\partial p}{\partial r}, \qquad U_p = -C\frac{k(\sigma)}{\mu(C)}\frac{\partial p}{\partial r},$$

$$U_f^{filtr} = U_f(\phi_t - C\phi_c).$$

Here U is the suspension filtration velocity in large and small pores, U_f^{filtr} is the fluid filtration velocity in large and small pore channels, $U_p = U_p^{mob} C\phi_c$ is the volume-averaged velocity of suspended particles in large pore channels, p – pressure.

The work [17] suggests the following relation between the permeability in large pore channels and the volume fraction of trapped particles:

$$k(\sigma) = k_0\left(1 - \frac{\sigma}{\phi_0 C_{max}}\right)^3. \tag{10}$$

Permeability in small pore channels:

$$k_s = k_{s0}\left(\frac{\sigma}{\phi_0 C_{max}}\right), \qquad k_{s0} = \frac{(1-C_{max})^3 d^2}{180 C_{max}^3}. \tag{11}$$

Here k_{s0} is the permeability of small pore channels during full packing of porous space ($\sigma = \phi_0 C_{max}$), determined from the Kozeny-Carman formula [21]; d is the diameter of particles in suspension.

Using the definitions introduced above, the conservation equations can be reformulated as:

$$\frac{\partial(C\phi_c)}{\partial t} + \frac{1}{r^j}\frac{\partial(CUr^j)}{\partial r} = -UC\lambda + \alpha\sigma\delta(U_s - U_{crit}), \tag{12}$$
$$\frac{\partial\sigma}{\partial t} = UC\lambda - \alpha\sigma\delta(U_s - U_{crit}),$$
$$\frac{\partial(U_s r^j)}{\partial r} = 0.$$

Classical deep-bed filtration models of suspension flow do not take into account the fact that the particles trapped in pores form a secondary porous medium (a random close packing of sedimented particles) with the permeability smaller than the permeability of the matrix. Filtration through the close packing of trapped particles is not considered in the classical models.

Below for comparison we present the classical model of deep-bed filtration in large pore channels without considering the flow in small pore channels in the packed bed of trapped particles ($\phi_c = \phi_t = \phi_0 - \sigma, U = U_s, k_s = 0$) [17, 22, 15, 24, 18]:

$$\frac{\partial(C\phi_c)}{\partial t} + \frac{1}{r^j}\frac{\partial(CUr^j)}{\partial r} = -CU\lambda(\sigma) + \alpha\sigma\delta(U - U_{crit}),$$
$$\frac{\partial\sigma}{\partial t} = CU\lambda(\sigma) - \alpha\sigma\delta(U - U_{crit}),$$
$$\frac{\partial(Ur^j)}{\partial r} = 0, \qquad U = -\frac{k(\sigma)}{\mu(C)}\frac{\partial p}{\partial r}. \tag{13}$$

Experiments on colmatation of core sample [23] have shown that the model (13) can be improved by introducing the following expression:$\lambda = \lambda_0(1 + \beta\sigma)$. Here, λ_0 is the original colmatation coefficient. Thus, the classical model (13) contains two tuning parameters λ_0 β.

Two versions of boundary conditions are considered: - (i) pressure is specified at the entry and at the outlet of the core sample (a): $r = r_0 : p = p_0, C = C_0$; $r = L : p = 0$. - (ii) the filtration velocity is specified at the inlet, and pressure – at the outlet (b): $r = r_0 : U_s = U_0, C = C_0$; $r = L : p = 0$.

Initial conditions: $t = t_0 : \phi = \phi_0, \sigma = 0$.

Permeability of mixed proppant pakcings from LBM simulations. Closure relations for permeability of the random close packing of proppant in a hydraulic fracture can be found from direct numerical simulation supported by proper conductivity experiments (see, for example, LBM simulations in mixed packing of proppant particles of various shape in [25]). Granular materials are important in many applications in petroleum and water supply industries. For example, during hydraulic fracturing, proppant particles are injected into a fracture to avoid its complete closure. In gravel packing, particulate material is injected into a well and placed in the borehole between a screen and the formation to prevent sand production from the reservoir. Granular mixtures of particles are also used in various other branches of industry, particularly in chemical and civil engineering. Because of their industrial importance, the properties of such materials have been the subject of extensive studies.

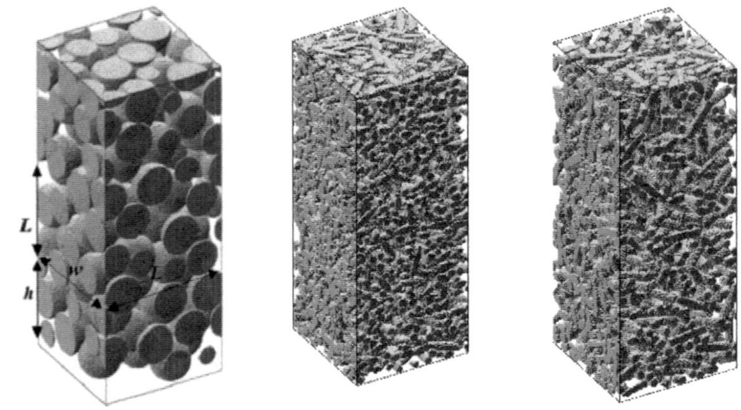

Figure 6. Spatially periodical packs of spheres *(a)*, ellipsoids *(b)*, and cylinders *(c)*. Aspect ratio of elongated particles is $l_{max}/l_{min} = 5$. Porosity is 0.42, 0.60, and 0.65, respectively.

The proppant packs are generated using the sequential deposition algorithm (Coelho et al., 1997). The pack is created by deposing particles into a box with a rectangular cross-section wL, where w is the fracture width (Fig. 6) and $H \gg L$. The length of the pack L is chosen equal to $6.4l_{max}$, where l_{max} is the maximum semi-axis of the particle. The binary samples are periodical in the longitudinal direction, while in the transverse direction they are either periodical or have flat walls. For the pack located between impermeable walls, the ratio of the fracture width to the proppant size w/l_{max} is in the range from 1.6 to 6.4. Using the lattice Boltzmann method developed in (Keehm, 2003) for slow single-phase flows in porous media, we conducted a series of numerical simulations to obtain the permeability tensors of different proppant packs. In our simulations, we consider the three-dimensional steady-state flow of a viscous incompressible Newtonian fluid through a granular packing. A series of numerical simulations is then carried out to obtain the permeability tensors for various random packs. As an example, permeability tensor $\mathbf{K}/100D$ for the pack of ellipsoids ($\phi = 0.65$):

41.09	-0.54	-0.07
-0.53	42.34	0.41
-0.05	0.41	43.52

We considered separately the case of a spatially periodical pack and the case of a pack confined between two impermeable walls. The results of numerical simulations demonstrate that the pack of cylinders provides the largest porosity and the highest permeability, while the pack of spheres has the lowest porosity and the smallest permeability. The porosity and permeability of the pack of ellipsoids are intermediate. All permeability tensors obtained in simulations are fairly isotropic. All numerical results are well approximated by the non-dimensional permeability-porosity relation, where permeability is scaled by the equivalent radius squared:

$$K/R_v^2 = 0.204\phi^{4.58}$$

A comparison with lab data is shown in Fig. 7. More details on the topic can be found in the paper [25].

5. Gas-liquid flows in a well during cleanup

In this section we will discuss the first-principles derivation of simplified drift-flux equations for gas-liquid flows in a well [26]. We consider a transient isothermal flow of a gas-liquid mixture in a long circular pipe with a variable inclination angle to the horizon. The flow is assumed to be axisymmetric and non-swirling. The liquid is a continuous carrier phase. The gas is a dispersed

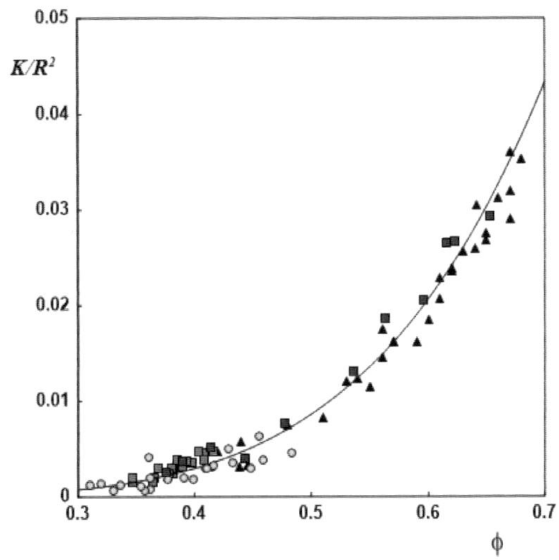

Figure 7. Non-dimensional permeability versus porosity. Solid line: power-law approximation (1), black triangles: simulations using lattice-Boltzmann method for Stokes equations, light-blue squares: results obtained using the finite-volume method for Navier-Stokes equations, green circles: lab tests, red squares: laboratory data on real proppants.

phase present in the form of monodisperse spherical bubbles suspended in the carrier fluid. The gas is compressible, and the liquid is incompressible. The cross-flow migration of the bubbles and their merging are not considered; however, we take into account the resulting nonuniform cross-flow profile of the gas volume fraction, which is formed as a result of the migration of the bubbles. The pressure difference in the bubbles and in the surrounding liquid due to surface tension is neglected. The bubble size is assumed to be much smaller than the spatial scales of variation of the fluid velocity, and the Reynolds numbers of the flow around individual bubbles are small.

The two-phase flow is considered within the approach of interpenetrating and interacting continua [2]. The problem is described by conservation laws in differential form, written for gas and liquid phases. The mass transfer between the phases is absent.

The balance laws of mass and momentum in the differential form for gas and liquid take the form [2]:

$$\frac{\partial}{\partial t}(\alpha_i \rho_i) + \nabla(\alpha_i \rho_i \mathbf{v}_i) = 0 \tag{14}$$

$$\alpha_i \rho_i \frac{d_i \mathbf{v}_i}{dt} = -\nabla p_i + \nabla \cdot \tau_i + \alpha_i \rho_i \mathbf{g} + n_b \mathbf{F}_{ij} \tag{15}$$

$$\frac{d_i}{dt} = \frac{\partial}{\partial t} + (\mathbf{v}_i \cdot \nabla).$$

Here, the indexes $i, j = g, l$ $(i \neq j)$ denote gas and liquid, respectively, α_i, ρ_i, and $\mathbf{v}_i$ are the volume fractions, densities and velocities of the phases, p_i and τ_i are the pressures and viscous stress tensors in each phase, and $\mathbf{g}$ is the gravity force acceleration. The momentum exchange between the phases is described by the term $\pm n_b \mathbf{F}_{ij}$, where $\mathbf{F}_{gl} = \mathbf{F}$ is the force exerted on a single bubble by the fluid, $\mathbf{F}_{lg} = -\mathbf{F}_{lg}$, and n_b is the number concentration of the bubbles.

For simplicity, the following calculations are performed for a vertical pipe, although the results may be generalized to the case of an inclined pipe, excluding flows in near-horizontal pipes.

It is assumed that the chaotic motion of the bubbles can be neglected, and the deviation of the bubble velocity from the mass-averaged velocity of the dispersed phase $\mathbf{v}_g$ is small, hence the stress tensor in the dispersed phase can be neglected [3]. The presence of the dispersed phase affects the stress tensor of the carrier phase. On the other hand, the bubbles of a compressible gas travel with a velocity different from that of the fluid, and the bubble volume fraction varies. Accordingly, the condition $\nabla \mathbf{v}_l = 0$ is no longer true. In this sense, the averaged liquid phase is compressible, in contrast to the liquid as a material. Therefore, the stress tensor of the carrier phase is written as for a viscous compressible fluid, with the coefficients of shear viscosity μ and bulk viscosity ζ dependent on the gas volume fraction:

$$\tau_l = 2\mu(\alpha_g)\left(\mathbf{e}_l - \frac{1}{3}\nabla \mathbf{v}_l \mathbf{I}\right) + \zeta(\alpha_g)\nabla \mathbf{v}_l \mathbf{I}.$$

Here, $\mathbf{e}_l$ is the strain rate tensor, and I is the unit tensor. To determine the dependences $\mu(\alpha_g)$ and $\zeta(\alpha_g)$ is a separate problem, which is usually solved for neutrally-buoyant particles, neglecting the phase slip. In what follows, we assume that $\mu(0) = \mu_0$, where μ_0 is the viscosity of the pure fluid, and $\zeta(\alpha_g)\nabla \mathbf{v}_l \to 0$ as $\alpha_g \to 0$.

The bubble radius is bounded by a limiting value R_c above which the bubble loses the spherical shape, becomes unstable, and is fragmented into smaller bubbles:

$$R \leq R_c \simeq \frac{1}{3}\left(\frac{\sigma}{\rho_l g}\right)^{\frac{1}{2}} \simeq 10^{-3}\mathrm{m}.$$

Here, σ is the gas-liquid surface tension. Under the above assumptions, the total force exerted on a bubble by the fluid is a superposition of the Stokes $\mathbf{F}_s$, the Archimedes $\mathbf{F}_A$, the added mass $\mathbf{F}_{am}$, and the Basset-Boussinesq $\mathbf{F}_{BB}$ forces. Accordingly, the forces on a bubble can be written in the form:

$$\mathbf{F} = \mathbf{F}_{St} + \mathbf{F}_A + \mathbf{F}_{am} + \mathbf{F}_{BB}$$

$$\mathbf{F}_{St} = 6\pi\mu R(\mathbf{v}_l - \mathbf{v}_g), \quad \mathbf{F}_A = \frac{4}{3}\pi R^3 \rho_l \left(\frac{d_l \mathbf{v}_l}{dt}\ \mathbf{g}\right), \quad \mathbf{F}_{am} = \frac{2}{3}\pi R^3 \rho_l \frac{d_g}{dt}(\mathbf{v}_l - \mathbf{v}_g)$$

For a transient flow in a well, the equations of the drift-flux model, known in the literature and implemented in the commercial reservoir simulator ECLIPSE (Schlumberger), take the form [27, 28]:

$$\frac{\partial}{\partial t}(A\alpha_i \rho_i) + \frac{\partial}{\partial z}(A\alpha_i \rho_i v_i) = 0 \tag{16}$$

$$\rho_m\left(\frac{\partial v_m}{\partial t} + v_m \frac{\partial v_m}{\partial z}\right) = -\frac{\partial p}{\partial z} + \rho_m g \cos\theta + \frac{2f\rho_m v_m |v_m|}{d} \tag{17}$$

$$v_g = C_0 v_m + v_d. \tag{18}$$

Here, A is the pipe cross-section, $v_m = \alpha_g v_g + \alpha_l v_l$ is the volume-averaged velocity of the mixture, $\rho_m = \alpha_g \rho_g + \alpha_l \rho_l$ is the mixture density, $f = f(\alpha_g, v_m, p)$ is the friction coefficient, d is the pipe diameter, and θ is the angle between the pipe axis and the vertical.

In the applications, a quasi–steady-state variant of the drift-flux model is widely used, in which the time-derivative in the momentum equation for the mixture is neglected and the total pressure difference is equal to the sum of the terms responsible for the gravity force, friction, and acceleration [27]. In the literature, this model is referred to as the no-pressure wave model, since it does not take into account the disturbance propagation with the transport velocity. In contrast to this quasi–steady-state formulation, we retain the time derivative of velocity in the momentum equation to take into account highly transient effects.

In the literature, formula (18) [2] (where $C_0 = C_0(\alpha_g, v_m, p)$ is the profile parameter which takes into account a non-uniform cross-flow profile of the bubble volume fraction and the carrier-phase velocity, and $v_d = v_d(\alpha_g, v_m, p)$ is the drift velocity) is called the drift-flux model relation.

There is also another known formulation of the drift-flux model [29], with the mixture momentum equation written as

$$\frac{\partial}{\partial t}(\alpha_g \rho_g v_g + \alpha_l \rho_l v_l) + \frac{\partial}{\partial z}(\alpha_g \rho_g v_g^2 + \alpha_l \rho_l v_l^2 + p) = Q_l + Q_g, \tag{19}$$

where Q_i are the source terms for each phase, and the drift-flux relation written as

$$v_g - v_l = \Phi(\alpha_g, v_g, p) \tag{20}$$

The asymptotic equations are derived in the long-channel approximation $\varepsilon \ll 1$, similar to the boundary-layer approximation, the narrow-channel approximation for fracturing flows [3], and the lubrication approximation for thin-film flows [30].

To come up with an algebraic relation between the phase velocities, additionally it is assumed that

$$\varepsilon \mathrm{St}\eta \ll 1, \quad \eta \gg 1, \quad \frac{\mathrm{St}}{\mathrm{Fr}^2}\eta \sim 1, \quad \varepsilon \mathrm{Re} \sim 1, \quad \varepsilon\xi_0 \ll 1.$$

Thus, the drift-flux model in the form (16)-(18) can be derived from conservation laws only in the following three cases: (i) small volume fraction of the dispersed phase $\alpha \ll 1$; (ii) no phase velocity slip $|C_0 - 1| \ll 1$, $\eta \mathrm{St}/\mathrm{Fr}^2 \ll 1$; and (iii) inertialess flows $\varepsilon \mathrm{Re} \ll 1$.

Our analysis demonstrates that the drift-flux model [27, 28] in the form (16)-(18) for the present flow configuration strictly follows from the conservation laws in the limited number of cases (i)-(iii) and represents essentially the effective-fluid model. The closure relations published in open literature are obtained from a calibration against a large body of experimental data [27, 28] for the governing parameters satisfying at least one of conditions (i)-(iii). At the same time, our analysis indicates that the drift-flux model [29] in the form (16), (19), and (20) is more general because it follows from the balance laws without any additional assumptions, besides the requirement of noninertial slip $\varepsilon \mathrm{St} \ll 1$.

6. Conclusions

We present a thorough review of a family of closely-related multi-scale models of multiphase flow constructed to describe all stages of the technology of hydraulic fracturing. These models were derived by the author from conservation laws within the multi-fluid approach using perturbation methods. The models cover suspension injection into the well to create primary fractures, suspension transport and sedimentation in a fracture, cross-flow inertial migration of particles in a horizontal section of a hydraulic fracture, filtration of suspensions of non-colloidal particles in a random close packing of proppant particles in the closed fracture, and gas-liquid flows in a well during cleanup and startup. In particular, our review covers:

1. A novel two-fluid model of suspension transport in a hydraulic fracture. It is shown that the model is different from the existing effective-fluid model of suspension flow by additional terms due to two-speed effects. These terms are important at high buoyancy numbers Bu, which corresponds to the case of high-rate slick-water fracturing. In the case of conventional fracturing with high-viscosity cross-linked gels, the models match. This model is then generalized to the case of a Bingham fluid flow and validated against four different sets of experiments on slumping, Saffman-Taylor fingering in Newtonian fluids, fingering and channeling in yield-stress fluids, and

suspension transport and sedimentation in a slot with formation of a packed bed of particles at the bottom.

2. A multi-scale model of particle migration in a dilute-suspension flow through the fracture, including the expression for the lift force on a particle settling in a horizontal flow through a vertical slot, the model of particle migration in the entry region of a plane channel with account for the Saffman lift force on a particle with the correction due to the wall effect, the model of migration of settling particles in a fully developed Poiseuille flow in a channel. Using the full Lagrangian approach, the particle concentration profile with an integrable singularity is obtained as a solution to the problem of the two-phase boundary layer. This achievement marks the first time that a self-consistent solution is obtained for the two-phase boundary layer problem within the dilute-suspension approximation, as all earlier solutions contained a non-integrable singularity thus making the model of dilute suspension inapplicable. New 2D width-averaged equations of suspension transport in a fracture are derived with account for the non-uniform cross-flow particle concentraiton profile formed as a result of the migration.

3. A three-continua model is constructed for the suspension flow in a porous medium with account for the effects of particle trapping and mobilization. The model takes into account the effect of trapped particles forming a secondary porous medium of random closed packing with smaller permeability. Fluid fluxe through packed trapped particles is taken into account. In order to close the model, for the first time in a wide range of porosity we derived a permeability-porosity correlation for the random close packing of non-spherical particles based on the 3D simulation of viscous flow using the lattice-Boltzmann method, validated against lab conductivity tests.

4. The derivation of asymptotic equations of the drift-flux model for a dilute gas-liquid disperse two-phase flow in a circular pipe is presented in the long-channel approximation. This asymptotic model is obtained as a limit of the full equations based on the balance laws, written for each phase in the multi-fluid approximation. The key assumptions are determined, which make it possible to derive the drift-flux model from the balance laws. This model contains an algebraic relation for the phase velocities and a single equation for the mixture momentum, written for the volume-averaged velocity of the mixture. To derive the drift-flux relation for the phase velocities, when one phase is continuous and the second dispersed, it is required to assume only that the characteristic length scale of the problem is significantly greater than the velocity relaxation length. To derive the single equation for mixture momentum, it is necessary to assume additionally that one of the following conditions is satisfied: (i) the dispersed-phase volume fraction is small, (ii) the phase velocity slip can be neglected, or (iii) the flow is inertialess, i.e. the mixture acceleration can be neglected. At the same time it is shown that the drift-flux model, in which a single mixture momentum conservation equation is obtained as the sum of two momentum conservation equations for the phases, follows from the conservation laws under one assumption that the characteristic length scale of the problem is significantly larger than the phase velocity relaxation length; hence, this model is more general.

Acknowledgements

Startup funds of Skolkovo Institute of Science and Technology are gratefully acknowledged.

References

[1] Osiptsov A 2017 Fluid mechanics of hydraulic fracturing: a review *J. Petroleum Sci. & Eng.* **156** 513–35.
[2] Nigmatulin R 1990 *Dynamics of multiphase media* (CRC Press).
[3] Boronin S and Osiptsov A 2010 Two-continua model of suspension flow in a hydraulic fracture *Doklady Physics* **55**(4) 199202.
[4] Boronin S, Osiptsov A and Desroches J 2015 Displacement of yield-stress fluids in a fracture *Int. J. Multiphase Flow* **76** 47–63.
[5] Boronin S, Osiptsov A and Tolmacheva K 2015 Multi-fluid model of suspension filtration in a porous medium *Fluid Dynamics* **50**(6) 759–68.

[6] Krasnopolsky B, Starostin A and Osiptsov A 2016 Unified graph-based multi-fluid model for gasliquid pipeline flows *Computers & Mathematics with Applications* **72**(5) 1244–62.
[7] Adachi J, Siebrits E, Peirce A and Desroches J 2007 Computer simulation of hydraulic fractures *Int. J. of Rock Mech. and Mining Sci.* **44**(1) 739–57.
[8] Pearson JRA 1994 On suspension transport in a fracture: framework for a global model *J. Non-Newtonian Fluid Mech.* **54** 503–13.
[9] Asmolov E and Osiptsov A 2009 The inertial lift on a spherical particle settling in a horizontal viscous flow through a vertical slot *Phys. Fluids* **21**(6) 063301.
[10] Osiptsov A and Asmolov E 2008 Asymptotic model of the inertial migration of particles in a dilute suspension flow through the entry region of a channel *Phys. Fluids* **20**(12) 123301.
[11] Asmolov E, Lebedeva N and Osiptsov A 2009 Inertial migration of sedimenting particles in a suspension flow through a Hele-Shaw cell *Fluid Dynamics*, **44**(3), pp.405-418.
[12] Boronin S and Osiptsov A 2014 Effects of particle migration on suspension flow in a hydraulic fracture *Fluid Dynamics* **49**(2) 208–21.
[13] Schlichting H 1960 *Boundary-layer theory* (New York: McGraw-hill. Vancouver).
[14] Foster M, Duck P and Hewitt R 2006 Boundary layers in a dilute particle suspension *Proc. R. Soc.* A **462** 136–68.
[15] Gruesbeck C and Collins R 1982 Entrainment and deposition of fine particles in porous media *SPE J.* **8430** 847–56.
[16] Bailey L, Boek E, Jaques S et al. 2000 Particulate Invasion From Drilling Fluids *SPE J.* **5**(4) 412–19.
[17] Boek E, Hall C and Tardy P 2012 Deep bed filtration modelling of formation damage due to particulate invasion from drilling fluids *Transport in Porous Media* **91**(2) 479–508.
[18] Guedes R, Al-Abduwani F, Bedrikovetsky P, Currie P 2009 Deep-bed filtration under multiple particle-capture mechanisms *SPE J.* **14**(3) 477–87.
[19] Sedov L 1972 *A Course in Continuum Mechanics* (Wolters-Noordhoff).
[20] Mikhailov N 1987 *Variation of physical properties of rocks in the near wellbore zone* [in Russian] (Moscow: Nedra).
[21] Van Der Hoef M, Beetstra R and Kuipers J 2005 Lattice-Boltzmann simulations of low-Reynolds-number flow past mono- and bidisperse arrays of spheres: results for the permeability and drag force *J. Fluid Mech.* **528** 233–54.
[22] Herzig J, Leclerc D and Le Goff P 1970 Flow of Suspensions through Porous Media *Ind. & Eng. Chem.* **62**(5) 9–34.
[23] Heertjes P and Lerk C 1967 The functioning of deep bed filters *Trans. Inst. Chem. Eng.* **45** 124–45.
[24] Civan F 2007 *Reservoir Formation Damage* (Gulf Professional Publishing: Elsevier Inc.)
[25] Osiptsov A 2017 Hydraulic fracture conductivity: effects of non-spherical proppant from lattice-Boltzmann simulations and lab tests *Advances in Water Resources* **194** 293-303.
[26] Osiptsov A, Sinkov K and Spesivtsev P 2014 Justification of the drift-flux model for two-phase flow in a circular pipe *Fluid Dynamics* **49**(5) 614–26.
[27] Hasan A and Kabir C 2002 *Fluid Flow and Heat Transfer in Wellbores* (Society of Petroleum Engineers: Richardson, Texas).
[28] Shi H, Holmes J, Durlofsky L et al. 2005 Drift-flux modeling of two-phase flow in wellbores *SPE J.* **10** 24–33.
[29] Evje S and Flatten T 2007 On the wave structure of two-phase models *SIAM J. Appl. Math.* **67** 487–511.
[30] Osiptsov A 2005 Non-isothermal lava flows over a conical surface *Fluid Dynamics* **40**(2) 221–32.
[31] Zhibaedov V, Lebedeva N, Osiptsov A and Sinkov K 2016 On the hyperbolicity of one-dimensional models for transient two-phase flow in a pipeline *Fluid Dynamics* **51**(1) 56-69.

Pulse X-ray device for stereo imaging and few-projection tomography of explosive and fast processes

E I Palchikov[1,2], A V Dolgikh[1,2], V V Klypin[1,2], I Y Krasnikov[1] and A M Ryabchun[1,2]

[1] Lavrentyev Institute of Hydrodynamics, Siberian Branch of the Russian Academy of Sciences, Russia
[2] Novosibirsk State University, Russia

e-mail: palchikov@hydro.nsc.ru

Abstract. This paper describes the operation principles and design features of the device for single pulse X-raying of explosive and high-speed processes, developed on the basis of a Tesla transformer with lumped secondary capacitor bank. The circuit with the lumped capacitor bank allows transferring a greater amount of energy to the discharge circuit as compared with the Marks-surge generator for more effective operation with remote X-ray tubes connected by coaxial cables. The device equipped with multiple X-ray tubes provides simultaneous X-raying of extended or spaced objects, stereo imaging, or few-projection tomography.

1. Introduction

X-ray imaging takes a prominent place among various methods for studying the movement and distribution of materials. In a number of cases, X-ray imaging happens to be virtually the only method that makes it possible to obtain reliable data on the internal structure of the object. In the case of single and fast or explosive processes, one usually applies pulse X-raying. The scientists of the Lavrentyev Institute of Hydrodynamics of the Siberian Branch of the Russian Academy of Sciences have been developing their own pulse X-ray devices of the PPX series (Portable Pulse X-ray) since the inception of the Institute. The PIR-4, PIR-600/1200, and PIR-100/240 devices have been developed for the operating voltages of 100 kV up to 1200 kV [1]. The most recent modern device of the series is PIR-200M (Figure 1).

Figure 1. General view of the PIR-200M pulsed X-ray device (left); internal layout (right).

Content from this work may be used under the terms of the Creative Commons Attribution 3.0 licence. Any further distribution of this work must maintain attribution to the author(s) and the title of the work, journal citation and DOI.
Published under licence by IOP Publishing Ltd

IOP Conf. Series: Journal of Physics: Conf. Series **894** (2017) 012069 doi:10.1088/1742-6596/894/1/012069

2. Principals of operation and characteristics

The PIR-200M pulse X-ray device consists of the following functional main components: a high-voltage unit, a remote control panel with a 15-meter long connecting cable, and external radiation head units with up 10-meter long high-voltage cables. Radiation heads allow for the use of pulse X-ray tubes of the types of IMA6-D, IMA5-320D, IRTP2-240 and IRTP3-D.

The operational principle of the X-ray device is as follows. The PIR-200M pulse X-ray device is based on a non-ferrous oil insulated Tesla transformer (Figure 2), an inhomogeneous forming line with ceramic capacitors, and a high-voltage cables. A nanosecond-long X-ray radiation is generated in a sealed vacuum tube with explosive emission.

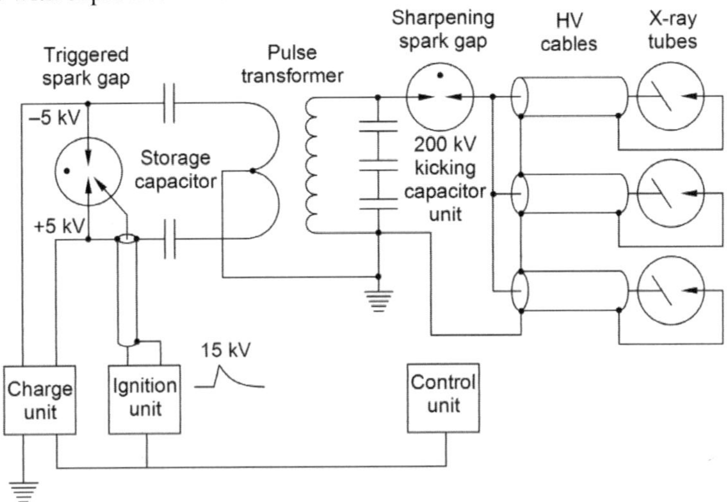

Figure 2. Circuit diagram of PIR-200M based on Tesla transformer.

The X-ray device is controlled via a remote control panel (Figure 3). The device can be started in three ways: manually, by pressing a button on the control panel, by an electric pulse signal of the amplitude from 5 V to 300 V, and by a contact sensor (by contacts short-circuit).

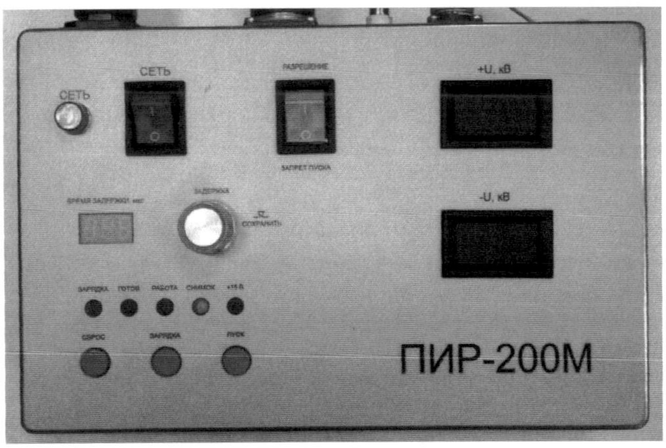

Figure 3. Control panel for the PIR-200M device.

The the X-ray device generates single bremsstrahlung pulses on the start command.

The main power supply of the PIR-200M is ~220 V 50 Hz line. The power consumption rate is not more than 50 W. The device contains a built-in battery of 12 V and 3.2 Ah, which allows for autonomous operation (up to 100 X-ray flashes and 4 hours in a standby mode).

The X-ray device has the following parameters:

- the duration of the X-ray pulse at half-height is 40 ns in the case of an IMA6-D tube and 80 ns in the case of an IMA5-320D tube;
- the delay since the start command until the X-ray radiation is 2 μs with dispersion of 50 ns;
- the output voltage is 200 kV;
- output dose per pulse for an IMA5-320D tube with a kovar window measured at 21 cm from focus is 75 mR;
- the device is powered by the alternate current line of 220 V, 50 Hz or by built-in battery 12 V;
- it is possible to start the device both by a TTL electric pulse of 5 V and a contact sensor;
- delay time from control panel built-in time generator – from 1 to 999 μs;
- the dimensions of the PIR-200M pulse X-ray device are 570×280×680 mm.

The device is designed to be in parallel connection with one to three external X-ray tubes (Figure 4). This technical solution is not available in any other domestic or foreign model. This allows for stereo imaging, few-projection tomography, and simultaneous X-raying of extended or spaced objects.

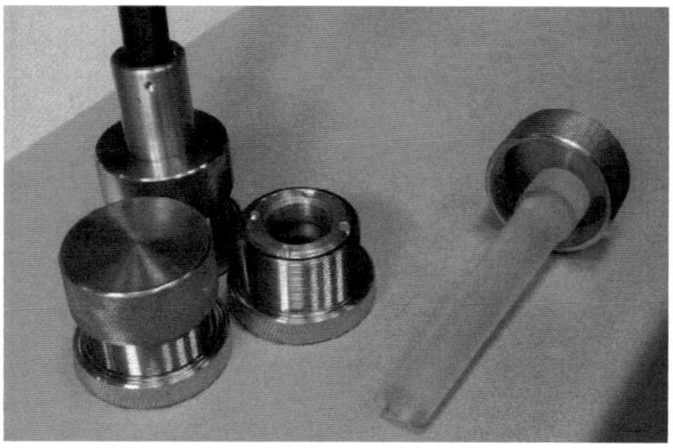

Figure 4. High-voltage connector unit of the PIR-200M consist of 3 sockets.

The table compares the characteristics of the PIR-200M device and those of the closest counterparts [2-4] intended for the single-flash radiography of fast processes. The table shows that the PIR-200M device has characteristics comparable to the best other counterparts, while having a number of advantages.

Table 1. Comparison with analogs.

	Pulserad 43733A	ScFl-300	Argument 150-200	PIR-200M
1. Operating voltage, kV	300	100-300	150-230	100-240
2. Current, kA	5	10	-	4
3. Pulse dose, mR[a]	8	9	0.15 ÷ 1.2 [b]	1,8 ÷ 8[b]
4. Flash duration, ns	30	20	4 ÷ 7	20 ÷ 80[c]
5. Focal spot size, mm	5	1	1.5 ÷ 2.3	2 ÷ 3[c]
6. Weight, kg	146	200	6,5 ÷ 17	55

[a] a distance of 1 m from the anode.
[b] for various voltages and different tubes.
[c] for different tubes.

3. Examples of application

The doses given in the tables can be used to obtain good-quality images per flash on distances of up to 2 meters on standard X-ray film with luminescent intensifying screens, on ImagePlate screens, or solid matrices (Figures. 5 and 6). The objects under control can move with velocities of up to 10 kilometers per second.

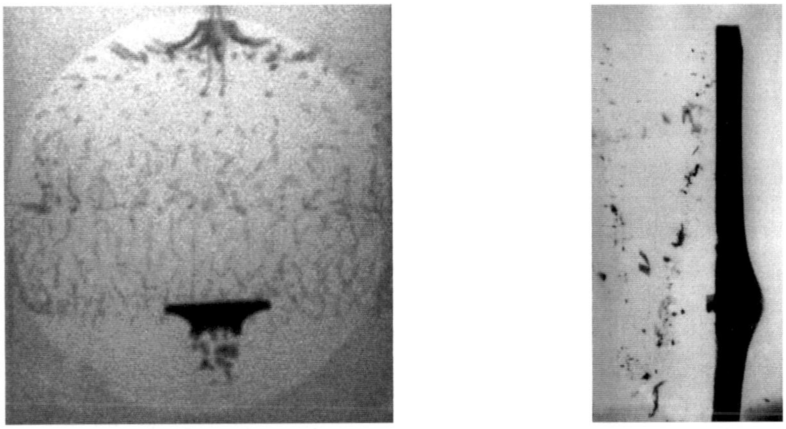

a b

Figure. 5. X-ray image of the exploding detonator; below is the coin perforated by the fragment of the detonator (a); interaction of the 7.62-mm bullet and the 12-mm steel target; the steel core and the shell expansion are seen (b).

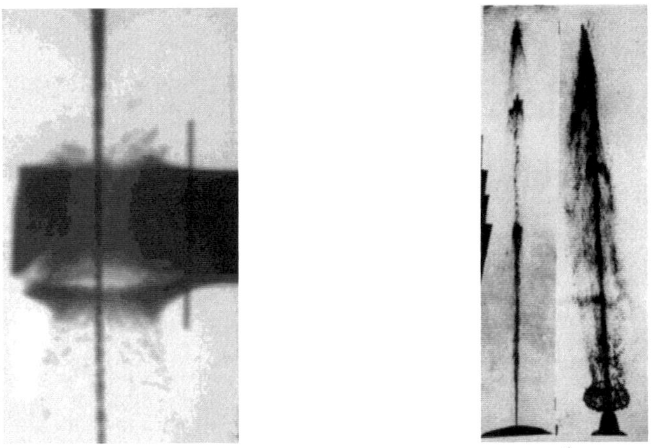

a b

Figure 6. (a) interaction of the copper shaped-charge jet with the aluminum target; (b) diesel fuel injection and spraying inside the combustion chamber.

The PIR-200M device is suitable for the X-raying of low-dense and medium-dense objects. It performed well in the X-raying of shaped-charge jets, the objects of inner, outer, and terminal ballistics, dynamic processes in brightly glowing media, foam, aerosols, and fragmentation fields.

4. Conclusions

This paper describes the development of a small device for the X-raying of single high-speed and explosive processes according to the characteristics not inferior to the best counterparts in the world.

When the device is equipped with several tubes, it is possible to perform the simultaneous X-raying of extended or spaced objects, stereo imaging, and few-projection tomography.

The PIR-200M device is capable for X-raying in a wide range of densities and thicknesses (from 0.1 mm of aluminum to 24 mm of steel).

The objects of X-raying can move with velocities of up to 10 km/s, and the image blurring does not exceed the space resolution of the detectors.

IOP Conf. Series: Journal of Physics: Conf. Series **894** (2017) 012069 doi:10.1088/1742-6596/894/1/012069

The fields of application of the PIR-200M device are terminal ballistics, wound ballistics, the study of the projectiles and armor of small arms, high-speed impact, the dynamic of shaped-charge jets, and the internal fragmentation dynamics in mine firing.

Acknowledgments
This work was conducted under support of SBRAS grant II.2. "Integration and development".

References

[1] Bichenkov E I and Palchikov E I 1997 *Combustion, Explosion and Shock Waves* **33** (3) 384–392
[2] Model 43733 A. Pulserad flash x-ray systems. URL: http://www2.l-3com.com/ati/pdfs/300kV.pdf
[3] Scandiflash for High Speed Rapidography. Hysbyborg, S-75590, Upsala, Sweden: Scanditronics, 1982. P. 1-6. URL: http://www.scandiflash.com/tabell.htm
[4] Avilov E A, Baltakov F N, Zavyalov N V, Kanunov I M, Polienko G A, Punin V T, Khoroshailo E S and Yuryev A L. 2003 *Proc. of Rus. Fed. Nucl. Center All-Rus. Res. Inst. Of Exp. Phys., Scient. Engin. Ed.*190–195

IOP Conf. Series: Journal of Physics: Conf. Series **894** (2017) 012070 doi:10.1088/1742-6596/894/1/012070

Correctness of the initial-boundary problem of the compressible fluid filtration in a viscous porous medium

A A Papin and M A Tokareva
Altai State University, Barnaul, Russia
E-mail: tma25@mail.ru

Abstract. The local solvability of initial-boundary value problem for the system of the equations of non stationary fluid motion in a viscous deformable medium in the field of gravity is proved.

1. Introduction

The process of filtration compressible fluid in a deformable viscous porous medium is described by a system of equations which includes the laws of conservation of mass for each phase, Darcy's law for fluid phase, taking into account the motion of a solid skeleton, the rheological law and the equation of conservation of momentum for system [1]–[3]. The solvability of this problem in some particular case is proved in [4]. The localization of solution for the filtration problem in an elastic medium is established in [5]. Similar problems with variable porosity were considered in [6]–[12].

2. Statement of the problem and formulation of the main result

A quasi-linear system of equations of composite type is considered:

$$\frac{\partial(1-\phi)\rho_s}{\partial t}+\frac{\partial}{\partial x}((1-\phi)\rho_s v_s)=0,\quad \frac{\partial(\rho_f\phi)}{\partial t}+\frac{\partial}{\partial x}(\rho_f\phi v_f)=0, \tag{1}$$

$$\phi(v_f-v_s)=-k(\phi)(\frac{\partial p_f}{\partial x}-\rho_f g), \tag{2}$$

$$\frac{\partial v_s}{\partial x}=-\frac{1}{\xi(\phi)}p_e,\quad p_e=p_{tot}-p_f, \tag{3}$$

$$\frac{\partial p_{tot}}{\partial x}=-\rho_{tot}g,\quad p_{tot}=\phi p_f+(1-\phi)p_s,\quad \rho_{tot}=\phi\rho_f+(1-\phi)\rho_s. \tag{4}$$

We seek a solution of this system in the domain $(x,t)\in Q_T=\Omega\times(0,T)$, $\Omega=(0,1)$, under the boundary and initial conditions

$$v_s\mid_{x=0,x=1}=v_f\mid_{x=0,x=1}=0,\qquad \phi\mid_{t=0}=\phi^0(x),\quad \rho_f\mid_{t=0}=\rho^0(x),\quad p_{tot}|_{x=0}=p^0(t). \tag{5}$$

Content from this work may be used under the terms of the Creative Commons Attribution 3.0 licence. Any further distribution of this work must maintain attribution to the author(s) and the title of the work, journal citation and DOI.
Published under licence by IOP Publishing Ltd

This initial-boundary problem describes 1D non-stationary isothermal motion of a compressible fluid in a viscous porous medium. Here ρ_f, ρ_s, v_f, v_s are, respectively, real density and velocity of solid and fluid phases, ϕ is the porosity, p_f, p_s are, respectively, pressures of the fluid and solid phases; p_e is the effective pressure, p_{tot} is the total pressure, ρ_{tot} is the density of the two-phase medium, g is the density of the mass forces; $k(\phi)$ is the coefficient of filtration, $\xi(\phi)$ is the coefficient of rock shear viscosity (specified function). The problem is written in the Eulerian coordinates x, t. The real density of the solid particles ρ_s is assumed constant. The unknown quantities are ϕ, ρ_f, v_f, v_s p_f, p_s. The system of equations (1)–(4) is closed by using the equation of state of the fluid phase $p_f = p_f(\rho_f)$.

We use the notation of functional spaces and the relevant norms adopted in [13]. In this paper by a solution of problem (1)–(5) we mean the set of functions $v_s \in C^{3+\alpha,1+\alpha/2}(Q_T)$ (ϕ, ρ_f, p_f, p_s) $\in C^{2+\alpha,1+\alpha/2}(Q_T)$, $v_f \in C^{1+\alpha,1+\alpha/2}(Q_T)$ such that $0 < \phi < 1$, $\rho_f > 0$, $p_f > 0$. These functions satisfy the equations (1)–(4), the initial and boundary conditions (5) and regarded as continuous functions in $\overline{Q}_T$.

Let us state the main results of the paper.

Theorem. *Suppose that the data of problem (1)–(5) satisfies the following conditions:*
1) the functions $k(\phi), \xi(\phi)$ and their derivatives up to the second order are continuous for $\phi \in (0,1)$ and satisfy the conditions

$$k_0^{-1}\phi^{q_1}(1-\phi)^{q_2} \le k(\phi) \le k_0\phi^{q_3}(1-\phi)^{q_4}, 1/\xi(\phi) = a_0(\phi)\phi^{\alpha_1}(1-\phi)^{\alpha_2-1}, \quad 0 < R_1 \le a_0(\phi) \le R_2,$$

where $k_0, \alpha_i, R_i, i = 1,2$ are positive constants, $q_1, ..., q_4$ are fixed real parameters, $p_f = R\rho_f$, $R = const > 0$,
2) the functions g, the initial and boundary functions $\phi^0, \rho^0, p^0(t)$ satisfy the following smoothness conditions: $g \in C^{1+\alpha,1+\alpha/2}(\bar{Q}_T), \phi^0 \in C^{2+\alpha}(\overline{\Omega}), \rho^0 \in C^{2+\alpha}(\overline{\Omega}), p^0(t) \in C^{1+\alpha/2}(0,T)$ and the matching conditions

$$((1-\phi^0)\frac{dp_f(\rho^0)}{dx} - \rho^0 g(x,0)) \mid_{x=0,x=1} = 0,$$

as well as satisfy the inequalities

$$0 < m_0 \le \phi^0(x) \le M_0 < 1, 0 < m_1 \le \rho^0(x) \le M_1 < \infty, 0 < g(x,t) \le g_0 < \infty, \quad x \in \overline{\Omega},$$

where m_0, M_0, m_1, M_1, g_0 are given positive constants.
Then problem (1)–(5) has a local solution, i.e. there exists a value of $t_0 \in (0,T)$ such that

$$v_s(x,t) \in C^{3+\alpha,1+\alpha/2}(\overline{Q}_{t_0}), (\phi(x,t), p_s(x,t), p_f(x,t), \rho_f(x,t)) \in C^{2+\alpha,1+\alpha/2}(\overline{Q}_{t_0}),$$

$$v_f(x,t) \in C^{1+\alpha,1+\alpha/2}(\overline{Q}_{t_0}).$$

Moreover, $0 < \phi(x,t) < 1$, $\rho_f(x,t) > 0$ in $\overline{Q}_{t_0}$.

3. Local solvability

Following [5], [14], we rewrite the system (1)–(5) in the mass Lagrangian variables. We obtain the following problem for unknown functions ρ_f, ϕ:

$$\frac{\partial}{\partial t}(a(\phi)\rho_f) - \frac{\partial}{\partial x}(K(\phi)b(\rho_f)\frac{\partial \rho_f}{\partial x} - \frac{K(\phi)}{1-\phi}\rho_f^2 g) = 0, \tag{6}$$

$$\frac{\partial G}{\partial t} = p_f - p_{tot}, \quad (1-\phi)\frac{\partial p_{tot}}{\partial x} = -\rho_{tot} g, \tag{7}$$

$$((1-\phi)\frac{\partial p_f(\rho_f)}{\partial x} - \rho_f g) \mid_{x=0,x=1} = 0, \quad \rho_f \mid_{t=0} = \rho^0(x), \quad \phi \mid_{t=0} = \phi^0(x), \tag{8}$$

IOP Conf. Series: Journal of Physics: Conf. Series **894** (2017) 012070 doi:10.1088/1742-6596/894/1/012070

where

$$a(\phi) = \frac{\phi}{1-\phi}, \quad K(\phi) = k(\phi)(1-\phi), \quad b(\rho_f) = \rho_f \frac{\partial p_f(\rho_f)}{\partial \rho_f}.$$

Lemma 1. *Let the data of problem (6)–(8) satisfy the conditions of the theorem. Then problem (6)–(8) has a unique local solution, i.e. there exists a value of t_0 such that*

$$(\phi(x,t), \rho_f(x,t), p_{tot}(x,t)) \in C^{2+\alpha,1+\alpha/2}(\overline{Q}_{t_0}).$$

Furthermore, $0 < \phi(x,t) < 1$, $\rho_f(x,t) > 0$ $\overline{Q}_{t_0}$.

The solvability of problem (6)–(8) is established by using the Tikhonov- Schauder fixed-point theorem [15].

Since the function $\psi = G(\phi)$ is strictly monotone, at $\phi \in (0,1)$, than the inverse function exists: $\phi = G^{-1}(\psi)$. Assuming that $\rho(x,t) = \rho_f(x,t) - \rho^0(x)$, $\omega(x,t) = G(\phi) - G(\phi^0)$. We represent the equations (6),(7) in the form

$$\frac{\partial}{\partial t}\Big(a(\omega)(\rho+\rho^0)\Big) = \frac{\partial}{\partial x}\left(K(\omega)b(\rho+\rho^0)\frac{\partial(\rho+\rho^0)}{\partial x} - \frac{K(\omega)}{1-\phi(\omega)}(\rho+\rho^0)^2 g\right), \tag{9}$$

$$\frac{\partial \omega}{\partial t} = p_f(\rho+\rho^0) - p_{tot}, (1-\phi(\omega))\frac{\partial p_{tot}}{\partial x} = -\rho_{tot} g. \tag{10}$$

Here $a(\omega) = \frac{\phi(\omega)}{1-\phi(\omega)}$, $K(\omega) = k(\phi(\omega))(1-\phi(\omega))$, $\phi(\omega) = G^{-1}(\omega + G(\phi^0))$. Moreover,

$$\rho\mid_{t=0} = \omega\mid_{t=0} = ((1-\phi(\omega))\frac{\partial(\rho+\rho^0)}{\partial x} - (\rho+\rho^0)g)\mid_{x=0,x=1} = 0, p_{tot}|_{x=0} = p^0(t).$$

For the Banach space, we choose the space $C^{2+\beta,1+\beta/2}(\overline{Q}_{t_0})$, where β is any number from the interval $(0,\alpha)$, $\alpha \in [0,1)$. Let

$$V = \{\bar\rho(x,t), \bar\omega(x,t) \in C^{2+\alpha,1+\alpha/2}(\overline{Q}_{t_0})| \quad \bar\rho\mid_{t=0} = \bar\omega\mid_{t=0} = ((1-\phi(\bar\omega))\frac{\partial(\bar\rho+\rho^0)}{\partial x} - (\bar\rho+\rho^0)g)\mid_{x=0,x=1} = 0,$$

$$\hat m_1 - \rho^0(x) \le \bar\rho(x,t) \le \hat M_1 - \rho^0(x) < \infty, \quad \hat m_1 = m_1(2+\frac{4g_0k_0}{R(1-M_0)})^{-1}, \quad \hat M_1 = m_1(2+\frac{4g_0k_0}{R(1-M_0)}),$$

$$G(\frac{m_0}{2}) - G(\phi^0) \le \bar\omega(x,t) \le G(\frac{M_0+1}{2}) - G(\phi^0) < \infty, \quad (x,t) \in Q_{t_0},$$

$$(|\bar\omega|_{1+\alpha,(1+\alpha)/2,Q_{t_0}}, |\bar\rho|_{1+\alpha,(1+\alpha)/2,Q_{t_0}}) \le K_1, \quad (|\bar\omega|_{2+\alpha,(2+\alpha)/2,Q_{t_0}}, |\bar\rho|_{2+\alpha,(2+\alpha)/2,Q_{t_0}}) \le K_1 + K_2\},$$

where K_1 is an arbitrary positive constant, while the positive constant K_2 will be given later. We note that on the set V following inequalities hold: $0 < \frac{m_0}{2} \le \phi(\bar\omega) \le \frac{M_0+1}{2} < 1$, $a(\bar\omega) > 0$, $K(\bar\omega) > 0$.

Let us construct an operator Λ mapping V in V. Suppose that $\bar\omega, \bar\rho \in V$. Using (10), we define the functions p_{tot}, ω by the equalities

$$p_{tot} = p^0(t) - \int_0^x g(\rho_s + (\bar\rho + \rho^0(\xi))\frac{\phi(\bar\omega)}{1-\phi(\bar\omega)})d\xi,$$

$$\omega = \int_0^t \left(R(\bar\rho(x,\tau)+\rho^0(x)) - p^0(\tau) + \int_0^x g(\rho_s + (\bar\rho+\rho^0(\xi))\frac{\phi(\bar\omega)}{1-\phi(\bar\omega)})d\xi\right)d\tau. \tag{11}$$

From the representation (11) it follows that smoothness ω and p_{tot} is determined by the smoothness of functions $\bar{\rho}, \bar{\omega}, \rho^0, p^0$ and g, and there exists a value $t_1 = t_1(m_0, M_0, m_1, M_1)$, such that for all $t_0 \le t_1$ the following inequality holds

$$0 < \frac{m_0}{2} \le \phi(x,t) \le \frac{M_0+1}{2}, \quad (x,t) \in Q_{t_0}. \tag{12}$$

In particular, we have an estimate

$$|\omega|_{2+\alpha,1+\alpha/2,Q_{t_0}} = C_1(m_0, M_0, m_1, M_1, K_1, T, |g|_{1+\alpha,\Omega}, |\rho^0|_{2+\alpha,\Omega}, |p^0|_{\alpha/2,[0,T]})(1 + t_0|\bar{\rho}_{xx}|_{\alpha,\alpha/2,\Omega}).$$

Taking into account (12) we also have the estimate for function $\omega(x,t)$: $G(\frac{m_0}{2}) \le \omega(x,t) + G(\phi^0) \le G(\frac{M_0+1}{2})$.

Using (9), $\bar{\rho}$ and $\bar{\omega}(x,t)$ we find the function $\rho(x,t)$ as a solution of the problem (here and elsewhere, we assume that the initial and boundary conditions are matched):

$$\frac{\partial}{\partial t}(a(\bar{\omega})(\rho+\rho^0)) = \frac{\partial}{\partial x}\left(K(\bar{\omega})b(\bar{\rho})\frac{\partial(\rho+\rho^0)}{\partial x} - \frac{K(\bar{\omega})}{1-\phi(\bar{\omega})}(\bar{\rho}+\rho^0)(\rho+\rho^0)g\right), \tag{13}$$

$$\rho\,|_{t=0} = 0, \quad ((1-\phi(\bar{\omega}))\frac{\partial(\rho+\rho^0)}{\partial x} - (\rho+\rho^0)g)\,|_{x=0,x=1} = 0.$$

The equation for $\rho(x,t)$ is uniformly parabolic. In view of the properties of $\bar{\omega}(x,t)$ and $\rho^0(x)$ the problem (13) has a classical solution [16]. In addition, we have the following estimate:

$$|\frac{1}{a(\bar{\omega})}\frac{\partial a(\bar{\omega})}{\partial t}| \le C_0(m_0, M_0, m_1, M_1, \max_{0\le t\le T}|p^0(t)|).$$

Imposing the additional smallness condition on the quantity t_0, we can state the following lemma.

Lemma 2. *There exists a value t_2, such that, when $t_0 \le \min(t_1, t_2)$, the classical solution of problem (13) satisfies the following inequality in Q_{t_0}*

$$0 < \hat{m}_1 \le \rho(x,t) + \rho^0(x) \le \hat{M}_1 < \infty.$$

Proof. Further, setting $U(x,t) = \rho(x,t) + \rho^0(x)$, we can express problem (13) in the form

$$\frac{\partial}{\partial t}(a(\bar{\omega})U) = \frac{\partial}{\partial x}\left(K(\bar{\omega})b(\bar{\rho})\frac{\partial U}{\partial x} - \frac{K(\bar{\omega})}{1-\phi(\bar{\omega})}(\bar{\rho}+\rho^0)Ug\right), \quad (\frac{\partial U}{\partial x} - \tilde{d}U)\,|_{x=0,x=1} = 0, \quad U\,|_{t=0} = \rho^0, \tag{14}$$

where $\tilde{d} = \frac{g}{(1-\phi(\bar{\omega}))R}$. First, we show that $U(x,t) \ge 0$, $(x,t) \in Q_{t_0}$. In equation (14), let us make the change $U(x,t) = -z(x,t)$. Then

$$z\frac{\partial a}{\partial t} + a\frac{\partial z}{\partial t} = \frac{\partial}{\partial x}(Kb\frac{\partial z}{\partial x} - \frac{K}{1-\phi(\bar{\omega})}(\bar{\rho}+\rho^0)gz).$$

Let

$$z^{(0)}(x,t) = \max\{z, 0\}, \quad z^{(0)}(x,t)\,|_{t=0} = \max\{-\rho^0, 0\} = 0,$$
$$\sigma_\varepsilon(x,t) = z^{(0)}(x,t)(|z^{(0)}(x,t)|^2 + \varepsilon)^{-1/2}, \varepsilon > 0.$$

Let us multiply the equation for the function z by σ_ε and then integrate over Ω. Following [4], we obtain the estimate

$$\int_0^1 az^{(0)}dx \le \varepsilon^{1/2}\int_0^t\int_0^1 |\frac{\partial a}{\partial \tau}|dxd\tau + \varepsilon^{1/2}\int_0^1 a\,|_{t=0}\,dx.$$

Passing to the limit as $\varepsilon \to 0$, we find that $z^{(0)} = 0$, i.e. $U \geq 0$.

The problem (14) can be represented as:

$$U_t - \tilde{a}_{11}U_{xx} + \tilde{a}_1 U_x + \tilde{a}U = 0, \quad (U_x - \tilde{d}U)|_{x=0,1} = 0, \tag{15}$$

where

$$\tilde{a}_{11} = \frac{Kb}{a}, \quad \tilde{a}_1 = \frac{d-(Kb)_x}{a}, \quad \tilde{a} = \frac{a_t + d_x}{a}, \quad \tilde{d} = \frac{g}{(1-\bar{\phi})R}.$$

Following [16], we pass from the function $U(x,t)$ to the new function $v(x,t)$ associated with the last one by the equality $U(x,t) = v(x,t)e^{\lambda t}$, while the constant λ will be given later. The function v satisfies the equation

$$v_t - \tilde{a}_{11}v_{xx} + \tilde{a}_1 v_x + (\tilde{a} + \lambda)v = 0.$$

Suppose that $\lambda > \max_{Q_{t_0}}[-\tilde{a}]$. Since $U \geq 0, g > 0$, it follows from the boundary conditions (15), that $U_x|_{x=0,1} = (\tilde{d}U)|_{x=0,1} \geq 0$, i.e. at $x = 0$ the maximum of the function can not be achieved. Suppose that $\omega(x,t) = v(x,t)\varphi(x)$, where

$$\varphi = -mx^2 + mx + 1 > 0, \quad m \equiv 2\max|\tilde{d}| = \frac{4g_0 k_0}{(1-M_0)R}.$$

The function ω is a solution of the problem

$$\omega_t - \tilde{a}_{11}\omega_{xx} + (\tilde{a}_1 + \frac{2a_{11}\varphi_x}{\varphi})\omega_x + (-2\tilde{a}_{11}\frac{\varphi_x^2}{\varphi^2} + \tilde{a}_{11}\frac{\varphi_{xx}}{\varphi} - \tilde{a}_1\frac{\varphi_x}{\varphi} + \tilde{a} + \lambda)\omega = 0,$$

$$\omega_x|_{x=0,1} = ((\frac{\varphi_x}{\varphi} + \tilde{d})\omega)|_{x=0,1}, \quad \omega_x|_{x=0} > 0, \quad \omega_x|_{x=1} < 0.$$

We choose

$$\lambda > \max\{\max_{Q_{t_0}}[2\tilde{a}_{11}\frac{\varphi_x^2}{\varphi^2} - \tilde{a}_{11}\frac{\varphi_{xx}}{\varphi} + \tilde{a}_1\frac{\varphi_x}{\varphi} - \tilde{a}], \max_{Q_{t_0}}[-\tilde{a}]\},$$

then the function ω reaches a positive maximum for $t = 0$. Therefore, we have an upper bound for U :

$$U \leq e^{\lambda t}M_1(1 + \frac{g_0 k_0}{(1-M_0)R}).$$

Then there exists a value $t_2 = \ln 2^{1/\lambda}$, such that for all $t \leq t_2$ we have the estimate from above for ρ from Lemma 2.

To obtain a lower estimate we represent equation (14) in the form $(z(x,t) = 1/U(x,t))$

$$z_t - \tilde{a}_{11}z_{xx} + \frac{2\tilde{a}_{11}}{z}(z_x^2) + \tilde{a}_1 z_x - \tilde{a}z = 0.$$

Applying a similar approach, we obtain the required estimate from below for ρ for any $t \leq t_2$.

Lemma 2 is proved.

In view of Lemma 2 and the properties of $\bar{\omega}$ we have the following estimates [16]:

$$|\rho|_{\alpha,\alpha/2,Q_{t_0}} \leq C_2,$$

$$|\rho|_{2+\alpha,1+\alpha/2,Q_{t_0}} \leq C_3\left(1 + |\rho^0|_{2+\alpha,\Omega} + |\bar{\rho}_x|_{\alpha,\alpha/2,Q_{t_0}} + |\bar{\omega}_t|_{\alpha,\alpha/2,Q_{t_0}} + |\bar{\omega}_x|_{\alpha,\alpha/2,Q_{t_0}}\right),$$

in which the constants C_2, C_3 depends on K_1, m_0, m_1, M_0, M_1. Therefore

$$|\rho|_{2+\alpha,1+\alpha/2,Q_{t_0}} \leq C_4(K_1, m_0, m_1, M_0, M_1).$$

Let $C_5 = \max\{C_1, C_4\}$. Choose K_2 so that $C_5 \leq \frac{K_1+K_2}{2}$. Then, for $t_0 < \min(t_1, t_2, (K_1+K_2)^{-1})$ we obtain

$$|\rho|_{2+\alpha,1+\alpha/2,Q_{t_0}} \leq K_1 + K_2, \quad |\omega|_{2+\alpha,1+\alpha/2,Q_{t_0}} \leq K_1 + K_2.$$

It remains to verify to conditions

$$|\rho|_{1+\alpha,(1+\alpha)/2,Q_{t_0}} \leq K_1, \quad |\omega|_{1+\alpha,(1+\alpha)/2,Q_{t_0}} \leq K_1.$$

Integrating equation (13) with respect to time, we obtain $|\rho|_{0,Q_{t_0}} \leq C_6 t_0$. From the equation (11) we obtain $|\omega|_{0,Q_{t_0}} \leq C_7 t_0$. Further, using for ρ, ω an inequality of the form [13]

$$|u|_{1+\alpha,(1+\alpha)/2,Q_{t_0}} \leq C|u|^c_{2+\alpha,1+\alpha/2,Q_{t_0}} |u|^{1-c}_{0,Q_{t_0}}, \quad c = (1+\alpha)(2+\alpha)^{-1},$$

we find that there exists a sufficiently small value of t_0, depending on K_1 and K_2, such that the required estimates hold: $|\rho|_{1+\alpha,(1+\alpha)/2,Q_{t_0}} \leq K_1$, $|\omega|_{1+\alpha,(1+\alpha)/2,Q_{t_0}} \leq K_1$.

Thus, the operator Λ maps the set V into itself for sufficiently small values of t_0. Using the estimates obtained above, we can easily show the continuity of the operator Λ in the norm of the space $C^{2+\beta,1+\beta/2}(\overline{Q}_{t_0})$. By the Tikhonov-Schauder theorem, there exists a fixed point $(\rho, \omega) \in V$ of the operator Λ. Uniqueness is established in the standard way [13].

Lemma 1 is proved.

Since $(\phi, \rho_f) \in C^{2+\alpha,1+\frac{\alpha}{2}}(Q_{t_0})$, we have:

$$v_s \in C^{3+\alpha,1+\frac{\alpha}{2}}(Q_{t_0}), \quad (p_f, p_s) \in C^{2+\alpha,1+\frac{\alpha}{2}}(Q_{t_0}), \quad v_f \in C^{1+\alpha,1+\frac{\alpha}{2}}(Q_{t_0}).$$

Theorem is proved.

4. Conclusion

We have studied the model of filtration of a compressible fluid in a viscous porous medium in the field of gravity. We establish the local solvability of the problem for the case in which the real densities of the solid phase are constant.

References

[1] Fowler A C and Yang X 1999 *J. Geophys. Res.* **104** 989-97
[2] Connolly J A D and Podladchikov Y Y 1998 *Geodin. Acta* **11** 55–84
[3] Morency C, Huismans R S, Beaumont C and Fullsack P 2007 *J. Geophys. Res.* **112** B10407
[4] Tokareva M A 2016 *J. Phys.: Conf. Ser.* **722** 012037
[5] Tokareva M A 2015 *J. Sib. Fed. Univ. Math. Phys.* **8** 467–77
[6] Prokudin D A and Krayushkina M V 2016 *J. Appl. Indust. Math.* **10** 417–28
[7] Yang X 2000 *Nonlin. Processes Geophys.* **7** 1–7
[8] Kucher N A, Mamontov A E and Prokudin D A 2012 *Sib. Math. J.* **53** 1075–88
[9] Papin A A 2008 *J. Appl. Mech. Tech. Phys.* **47** 527–36
[10] Simpson M, Spiegelman M and Weinstein M I 2007 *Nonlinearity* **20** 21–49
[11] Geng Y and Zhang L 2010 *Appl. Math. Comput.* **217** 1741–8
[12] Rudyak V Ya, Bocharov O B and Seryakov A V 2013 *Proceedings of the XLI Summer School-Conference Advanced Problems in Mechanics (APM-2013)* (St-Petersburg) 183-90
[13] Antontsev S N, Kazhikhov A V and Monakhov V N 1990 *Boundary-Value Problems of the Mechanics of Inhomogeneous Fluids* (Amsterdam: Noth-Holland)
[14] Akhmerova I G and Papin A A 2014 *Mathematical Notes* **96** 166–79
[15] Edwards R 1965 *Functional Analysis: Theory and Applications* (New York: Holt, Rinehart and Winston)
[16] Ladyzhenskaya O A, Solonnikov V A and Uraltseva N N 1968 *Linear and Quasilinear Eequations of Parabolic Type* (Providence, RI: American Mathematical Society)

IOP Conf. Series: Journal of Physics: Conf. Series **894** (2017) 012071 doi:10.1088/1742-6596/894/1/012071

On the mechanics of cerebral aneurysms: experimental research and numerical simulation

D V Parshin[1,2], I O Kuianova[2], A S Yunoshev[1,2], K S Ovsyannikov[3] and A V Dubovoy[3]

[1] Lavrentyev Institute of Hydrodynamics, Novosibirsk, Russia
[2] Novosibirsk State University, Novosibirsk, Russia
[3] Federal Neurosurgical Center, Novosibirsk, Russia

E-mail: danilo.skiman@gmail.com

Abstract. This research extends existing experimental data for CA tissues [1, 2] and presents the preliminary results of numerical calculations. Experiments were performed to measure aneurysm wall stiffness and the data obtained was analyzed. To reconstruct the geometry of the CAs, DICOM images of real patients with aneurysms and ITK Snap [3] were used. In addition, numerical calculations were performed in ANSYS (commercial software, License of Lavrentyev Institute of Hydrodynamics). The results of these numerical calculations show a high level of agreement with experimental data from previous literature.

1. Introduction

A cerebral aneursm (CA) is a vascular disease of the human brain, with an incidence of approximately 1 in 50 people [4]. Despite their relatively high occurrence, CAs are rarely diagnosed. It is of interest to neurosurgeons to know whether a given aneurysm is likely to rupture or not. The statistics inform us that approximately only 1-2% of unruptured aneurysms will rupture at some future point in time. To determine which aneurysms are likely to rupture, it is necessary to investigate: local stiffness of the vessel wall, the wall structure of the CA, the geometry of the CA, biochemical processes in the vessel wall, the hydrodynamic parameters of the flow through the CA, etc. It is a tremendous problem to consider all of these simultaneously, but doing so is nevertheless expedient: the accuracy of the prediction for aneurysm rupture increases with knowledge of the aneurysm. The mechanical properties of unruptured CAs were first considered in [2] and the most recent contributions to this field of research can be found in [5]. The hydrodynamical properties of unruptured CAs have been investigated in many studies, including, but not limited to, those in [6] and [7]. In all of the studies published, the results from different experimental approaches were coupled. This is because the mechanism of aneurysm formation is not currently well understood. The aim of recent experiments has been to obtain some biophysical markers to help predict future CA rupture, with this particular paper investigating different techniques for measuring the relevant data. In particular we desire to obtain values for measurement error and vessel elasticity, for the cases of three different patients presenting an CA.

Many researchers consider numerical calculations to be the best method for predicting the rupture of an aneurysm. Generally we agree with this, but only when the relevant calculations

Content from this work may be used under the terms of the Creative Commons Attribution 3.0 licence. Any further distribution of this work must maintain attribution to the author(s) and the title of the work, journal citation and DOI.
Published under licence by IOP Publishing Ltd

are performed in an adequate way. Firstly, the validity of the geometric model needs to be determined. In addition, sometimes boundary conditions may lead to interesting results [8]. However, it is of paramount importance to determine whether these boundary conditions accurately model reality, or whether they are unrepresentative of reality due to some shortcomings of the experiment.

2. Materials and methods

2.1. Population

This paper considers three cases, with each case being a tissue exhibiting an CA from distinct patients. Ethical considerations made by the Novosibirsk Centre of Neurosurgery dictate that cases will be denoted using the following convention: case X, where X represents the initial letter of the patient's surname. All patients in this study exhibited a saccular CA.

Table 1. Detailed patient information.

ID	Age	Gender	Smoking	A history of diabetes	Aneurysm location
CaseR	63	Female	No	No	MCA
CaseK	63	Female	No	Yes	MCA
CaseV	63	Female	No	No	ACA

Table 2. Cross-sectional area parameters for the samples.

ID	Mean thickness (mm)	Mean width (mm)	Mean cross-sectional area (mm2)
CaseR	0.2	3	0.6
CaseK	1.8	5	9
CaseV	0.85	3.75	3.19

3D models for these aneurysms were produced using ITK-Snap software [3], and the shape (look at Figure 1) of each sample was optimized for using in ANSYS software (licence of Lavrentyev Institute of Hydrodynamics).

2.2. Mechanical technique

Following neurosurgery treatments, tissues exhibiting an CA were harvested and delivered to the laboratory, being preserved using sodium 0.9%, +2°C-+5°C. As was shown in [9], results for refrigerated aneurysms vary slightly with respect to the temperature of storage. At the laboratory, the rupture of the tissues was investigated using a rupture machine (Zwick/Roell Z010, Germany). The tissue under consideration perishes very quickly, and so this affords only one series of experiments for each case of CA tissue. The speed of pulling was the same for all experiments and equal to 1mm per minute.

For each of the three cases, we conducted experiments using several different values of the pulling force. We now proceed to describe the method for these experiments. We were unable

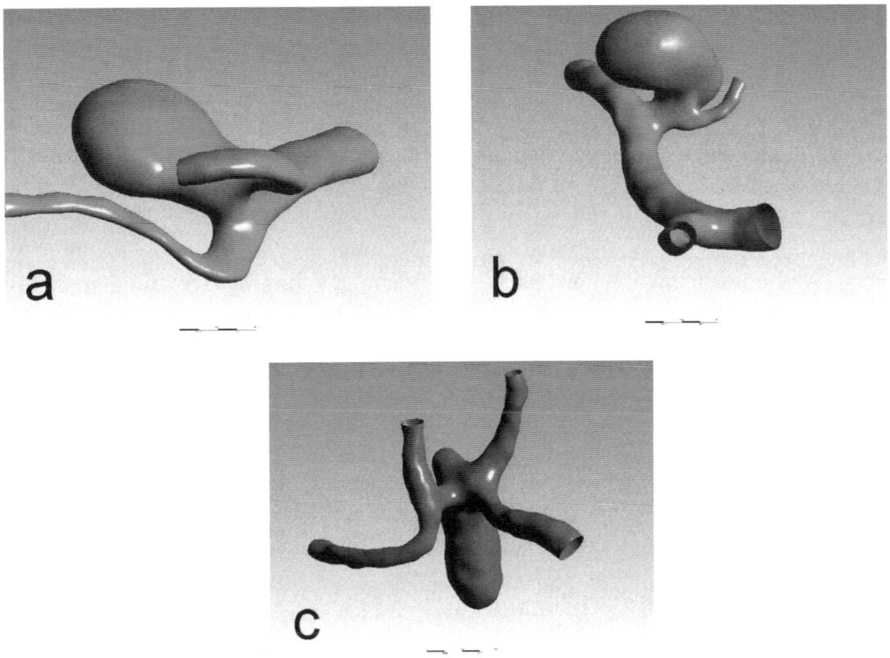

Figure 1. 3D Constructed model of samples. Case R (a), Case K (b), Case V (c).

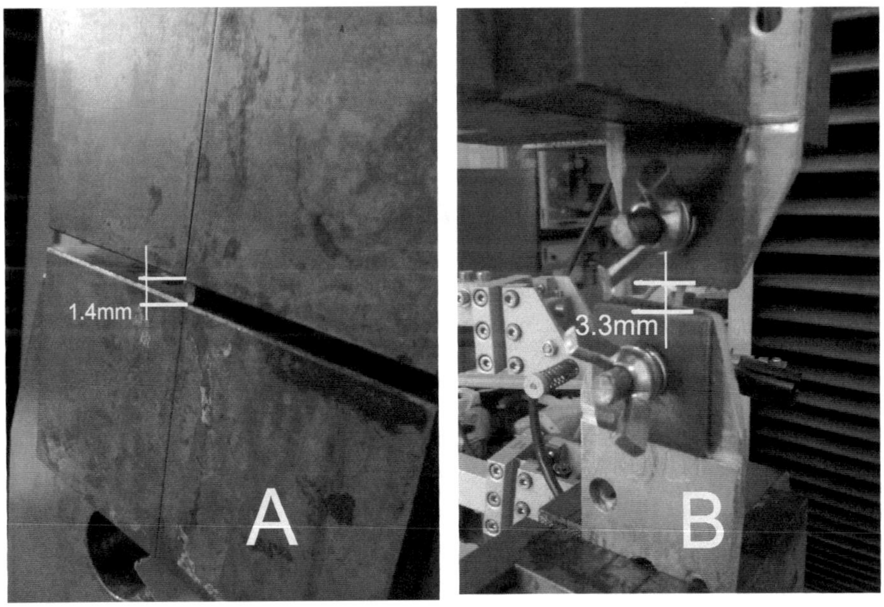

Figure 2. Method of mechanical experiment. A – tissue was fixed directly to the machine. B – tissue was fixed using anchoring device.

to find any prior research data regarding measurement error, but we are without doubt that such errors were present. We therefore calculated the error for each of the three cases of CA tissue. The first step in calculating this error was to perform measurements with the tissue present in the machine and afterwards to perform measurements, for the same value of force applied, without any material fixed in the machine. After each stage of the experiment, i.e. after subjecting a sample to a given force, the sample was hydrated with sodium 0.9% . We then calculated the mean average of the data for both sets of measurements (tissue present and tissue absent).

For the Cases R and K the aneurysm tissue was in the machine exactly as shown in Figure 2. The small dimensions of Case R aneurysm tissue only permitted us to perform one measurement for this sample. The purpose of this particular experiment was to determine mechanical properties of the aneurysm tissues, such as their Young's modulus, and compare this result with those previously determined in [1].

2.3. Numerical calculations

Numerical calculations were performed in ANSYS (commercial software, License of Lavrentyev Institute of Hydrodynamics), with the rigid walls model^{TM} being used for this stage of our research. The Navier-Stokes equations were solved with a steady inlet flow rate, and for all cases the blood flow rate was taken to be as in [10]. No slip condition was used for the walls. Blood was assumed to be a viscous incompressible fluid with rheology as described in [11]. The calculation was performed using a homogenous mesh.

3. Results

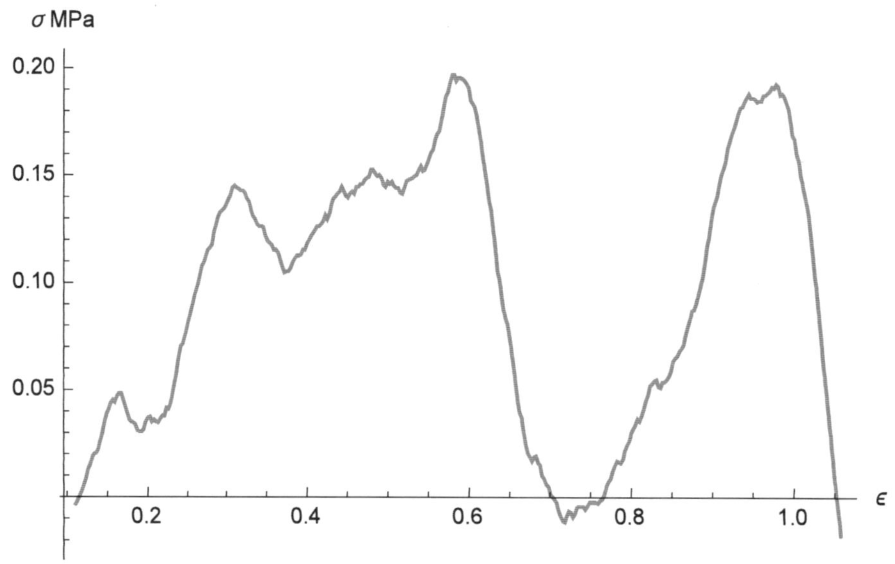

Figure 3. Stretch-strain graph for Case R Aneurysm.

The cross-sectional area of each sample was measured. This data is shown in Table 2. For the first experiment (Case R), the force was applied until the tissue ruptured. As can be seen in Figure 3 there are two local maximums, corresponding to the fact that the tissue ruptured in two distinct stages, due to the tissue consisting of two distinct layers. Firstly the tissue ruptured approximately midway between the fixing clips. This event corresponds to the point on the curve where a decrease occurs following the first local maximum. Following this, the limit of elasticity of the first layer was reached, and then the second layer also ruptured (corresponding to the second local maximum in Figure 3). For the second experiment nonelastic deformation is shown in Figure 4 was performed. The third sample was investigated with another technique

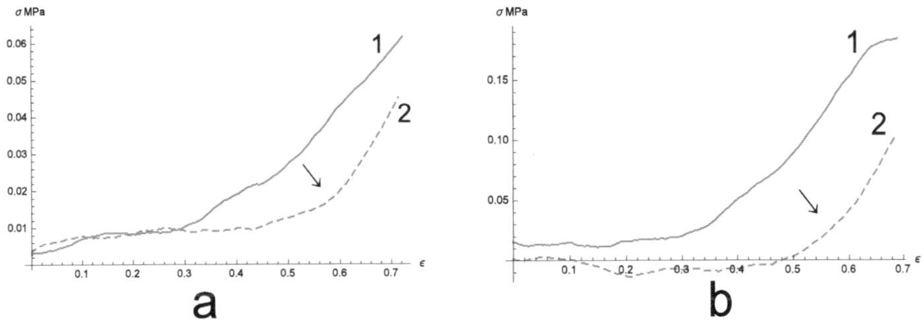

Figure 4. Stretch-strain graph for Case K (a) and Case V (b) Aneurysm. After the deformation 1 deformation 2 was performed

of anchoring. In Figure 4 nonelastic deformation is also shown. The ultimate stress for each sample was reached. The values of ultimate stress are $0.2, 0.19, 0.15 (MPa)$ for the cases Case R, Case K and Case V respectivly.

The idea of numerical calculations at this stage of the research is to show that the reconstructed

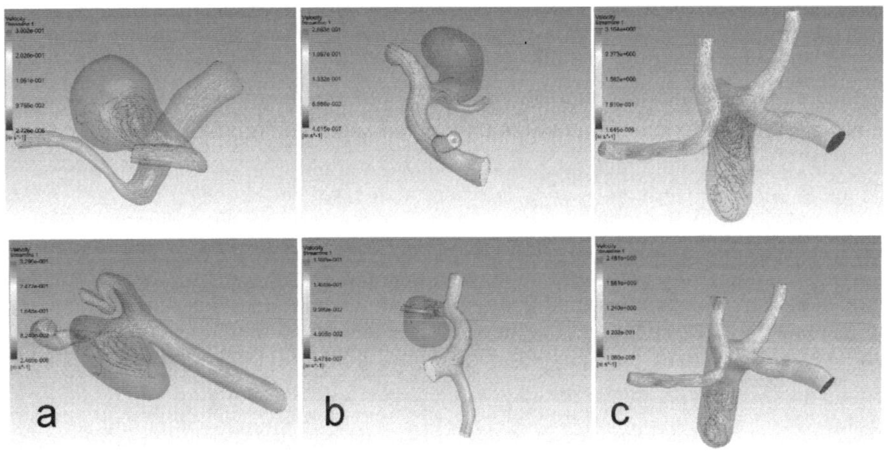

Figure 5. Velocity streamlines for cases Case R (a), Case K (b) and Case V (c). Maximum (top) and minimum (bottom) inlet blood flow rates are used for the calculations (correspondint to [10]).

3D models are plausible and able being used for numerical calculations. The values of hydrodynamical quantities (velocity, pressure) calculated through the vessel model show (look at Figure 5) have a good agreement to experiment data [10]. This fact is not obvious due to only good reconstruction gives plausible values of the pressure and velocity.

4. Discussion

In this study we discovered different behaviors of brain aneurysm tissue under the application of a pulling force. Despite the fact that the samples in Cases R and V have very similar character-

istic parameters (see Table 2), the ultimate stress curves are completely different. For Case K the value of the ultimate force, i.e. the force corresponding to ultimate stress, was as expected: the value of the ultimate force for Case K was higher than for Cases R and V. It would be of interest to conduct experiments with similar tissue, using the technique of Scott [2], in which tissues were investigated using the technique of sodium pressure.
The structural parameters of the wall wasn't considered. At the date data shows that the amount of collagen different type and fibber orientation give a food for thought. Implementation of coupled information may be very fruitfully to understand the direction of aneurysm development: whether it will stabilize or rupture.
The experiments presented in this paper did not consider the structural parameters of the vessel walls. Recently obtained data indicates that the amount of different types of collagen, and the orientation of fibres, may influence the results. Analysis of coupled information will be necessary to understand the future direction of aneurysm development i.e. whether an aneurysm with stabilise or rupture.
Our surgical practice shows that even very thin diverticula on the surface of aneurysms can be highly elastic. Therefore, considering only the parameters of an aneurysm's size and its elasticity is insufficient. The mechanical experiment for case R demonstrates this insufficiency: two distinct stages of rupturing were necessary for the complete rupture of a very thin aneurysm. For the case of a living patient, this result translates to the likely scenario of thrombosis of the aneurysm occurring following the first stage of rupture.
Our ongoing research will extend the experimental data obtained. We are of the opinion that rigid simulations performed in many previous research papers [1, 6] do not produce accurate models. Only FSI simulations tend to produce successful models of the biomechanical processes occurring in an aneurysm. However, it is debatable which the optimum type of FSI simulation to choose is. It needs to be determined how many factors, e.g. non-linear elasticity, non-homogeneous mesh grid, have to be considered to reach a sufficient level of accuracy for the model. Also time and expense constraints for the calculations need to be considered. We intend to obtain data for FSI for the samples under consideration, and to consider the effect that variations in the wall structure of aneurysms will have on the results.

5. Conclusions

This paper forms a preliminary part of our ongoing research. During this stage we have been building connections with neurosurgeons and have decided upon the logistical scheme for delivery and storage of the aneurysm tissue. Statistical data processing was applied to the data obtained in our experiments. The stretch-strain data, obtained by our experiments, shows a high level of agreement with those found in previous literature. The different data obtained by experiments in this paper will be used in our ongoing research. This research includes determining material parameters for use in ANSYS software, and determining the ultimate force required to rupture an aneurysm in FSI problems. numerical simulations, performed using the rigid walls model^{TM}, allow for the accurate construction of 3D models of CAs. Moreover this study represents our first step in producing transient patient specific numerical calculations for the FSI problem.

Acknowledgments

This work was supported by grant No 14.W03.31.0002 from the Government of Russian Federation.

References

[1] Cebral J R, Duan X, Chung B J, Putman C, Aziz K M and Robertson A M 2015 *AJNR Am. J.Neuroradiol.* **36** 1695–703
[2] Scott S, Ferguson G G and Roach M R 1972 *Can. J. Physiol. Pharmacol.* **50** 328–32
[3] Yushkevich P A, Piven J, Hazlett H C, Smith R G, Ho S, Gee J C and Gerig G 2006 *Neuroimage* **31** 1116–28
[4] Morita A, Kimura T, Shogima M, Sameshima T and Nishihara T 2010 *Neurol. Med. Chir.* **50** 777–87
[5] Robertson A M, Duan X, Aziz K M, Hill M R, Watkins S C and Cebral J R 2015 *Ann. Biomed. Eng.* **43** 1502–15
[6] Cebral J R, Mut F, Weir J and Putman C 2011 *AJNR Am. J. Neuroradiol.* **32** 145–51
[7] Khe A K, Chupakhin A P, Cherevko A A, Eliava S S and Pilipenko Yu V 2015 *Russ. J. Numer. Anal. Math. Modelling* **30** 277–87
[8] Larrabide I, Aguilar M L, Morales H G, Geers A J, Kulcsar Z, Rufenacht D and Frangi A F 2013 *AJNR Am. J.Neuroradiol.* **34** 816–22
[9] Stemper B D, Yoganandan N, Stineman M R, Gennarelli T A, Baisden J L and Pintar F A 2007 *J. Surg. Res.* **139** 236–42
[10] Zarrinkoob L, Ambarki Kh, Wahlin A, Birgander R, Eklund A and Malm J 2015 *J Cereb. Blood Flow Metab.* **35** 648–54
[11] Baskurt O K, Hardeman M R, Rampling M W and Meiselman H J 2007 *Biomedical and Health Research* **69**

Effect of wave action on near-well zone cleaning

V I Pen'kovskii and N K Korsakova

Lavrentyev Institute of Hydrodynamics of SB RAS, Russia

E-mail: penkov@hydro.nsc.ru, kors@hydro.nsc.ru

Abstract. Drilling filtrate invasion into the producing formation and native water accumulating of the near-well zone in well operation reduce the well productivity. As a result of that, depending on characteristic capillary pressure scale and differential pressure drawdown, oil production rate may become lower than expected one. In this paper, it is considered the hysteresis effects of capillary pressure after reversion of displacement. As applied to laboratory experiment conditions, the solution of problem of oil flow in formation model with a pressure drop on the model sides harmonically varied with time is presented. It was estimated a range of fluid vibration effective action on the near-well zone cleaning from capillary locking water. The plant simulating extraction of oil from formation using widely practised sucker-rod pump has been created. Formation model is presented as a slot filled with broken glass between two plates. In the process, natural oil and sodium chloride solution were used as working fluids. The experiments qualitatively confirm a positive effect of jack pumps on the near-well zone cleaning.

1. Introduction. Flow rate of producing well

Wells drill with use of a water clay solution, with consequent water filtrate invasion into the producing formation. Thereby, an oil or gas saturation of near-well zone reduces. When influx is made with a well bore pressure $p = p_c$ being less than initial strata pressure $p = p_0$, the penetrated water filtrate is not fully displaced. Part of water remains in porous medium in an immiscible capillary-locking state. In some cases, a stratal flow characteristic change near borehole is said to be skin effect [1].

Let us consider axisymmetric steady oil influx to well in a stratum with initial oil saturation $s = s_0$. Suppose that the stratum is hydrophilic, so some water is capillary locked in the near-well zone. We denote $r = R$ as coordinate of external boundary where $s = s_0$ and $p = p_0$. Let such boundary exist. So the Muscat-Leverett system of equations for steady flow takes the form:

$$v = -\frac{k}{\mu} f(s)\frac{\partial p}{\partial r}, \quad v_1 = -\frac{k}{\mu_1} f_1(s)\frac{\partial p_1}{\partial r}, \quad \frac{1}{r}\frac{\partial}{\partial r}(v \cdot r) = 0, \quad \frac{1}{r}\frac{\partial}{\partial r}(v_1 \cdot r) = 0,$$

$$p = p_1 + p_k^0 \phi(s).$$

Here p_k^0 is the capillary pressure, the functions $\phi(s), f$, and f_1 are specified on interval $(0,1)$ and should be defined experimentally. But it should be pointed out that until now the problem of one-valued definition of these functions at the ends of mentioned above interval has not been solved. On the borehole, where $r = r_c$, the phase pressures $p = p_1$ are the same and are equal to p_c, therefore

Content from this work may be used under the terms of the Creative Commons Attribution 3.0 licence. Any further distribution of this work must maintain attribution to the author(s) and the title of the work, journal citation and DOI.
Published under licence by IOP Publishing Ltd

$s(r_c)=0$. At steady regime of capillary locking the relation $p_0 - p_c = p_k^0\phi(s_0)$ is satisfied. In this case, $v_1 \equiv 0$, $p_1 \equiv p_c$, and for oil flow velocity we have the formula

$$v = -\frac{k}{\mu} f(s) p_k^0 \phi'(s) \frac{\partial s}{\partial r}.$$

Substituting this into the continuity equation for oil phase, we obtain the well production expression

$$q = r\frac{k}{\mu} f(s) p_k^0 \phi'(s) \frac{\partial s}{\partial r} = const.$$

By integration over r between the limits $[r_c, R]$ and over s from 0 to s_0 this formula rearranges in the form

$$q = I(s_0)\frac{k}{\mu} p_k^0 / \ln(R/r_c).$$

Here, $I(s_0)=\int_0^{s_0} f(s)d\phi(s)$ and, if we take the approximation $f(s)\approx s^{3.5}$, $\phi(s)=\left[s/(1-s)\right]^{1/2}$, this integral can be expressed in terms of elementary functions:

$$I\ (s) = \int_0^s f(s)\phi'(s)ds = \frac{1}{2}\int_0^s s^3(1-s)^{-3/2}ds = \frac{2}{5}\left(\frac{16-8s-2s^2-s^3}{\sqrt{1-s}} - 16\right).$$

If the capillary forces are negligible small (in this case water is also movable), so flow of fluids is homogeneous, $s \equiv s_0$, and for the oil influx q_0 we obtain the expression

$$q_0 = \frac{k}{\mu} f(s_0)\frac{p_0 - p_c}{\ln(R/r_c)} = \frac{k}{\mu} f(s_0) p_k^0 \phi(s_0)/\ln(R/r_c).$$

As can be seen, due to partial plugging of the near-well zone by water, the well production decreases in η times, where

$$\eta(s_0) = \frac{q}{q_0} = \frac{I(s_0)}{f(s_0)\phi(s_0)}. \quad (1)$$

This formula is valid for all reasonable initial values of oil strata saturation s_0 except vicinities of nil and unit where, as noted above, an approximation of experimental data of function ϕ is not precise. Determined by the formula (1) the plot of function is presented on Figure 1.

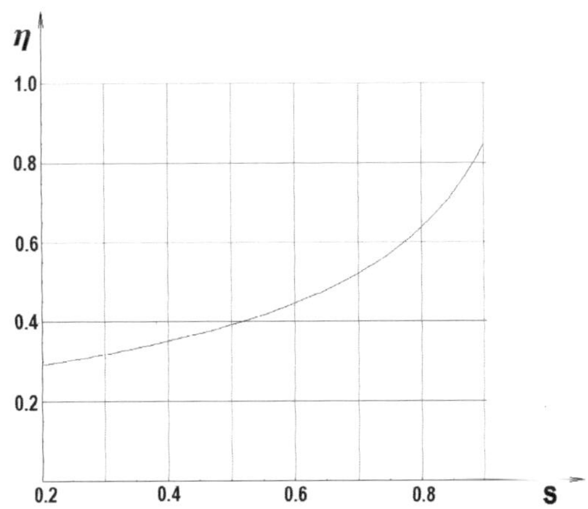

Figure 1. Relation of well productions.

Figure 2. . Oil meniscuses after a reversion of displacement.

2. Mixed wettable media. Reversion of displacement

Appearing in the formula for capillary pressure jump on phase boundary, parameter p_k^0 defines a degree of the rock wettability with water. Positive values of the parameter are typical for hydrophilic medium and negative ones are associated with hydrophobic rock. Most of oil-bearing rocks are mixed wettable. In such strata, pressures in different phases become equal in some inner points $s_* \in (0;1)$. A sign of parameter p_k^0 can change while immiscible displacement changes direction. As an illustration, in Figure 2 is presented the photo of oil inclusions in the water-filled quartz capillary (radius $r_k = 0,3*10^{-3}$ m) [2]. Experiments show that a convexity of meniscuses changes after a change of flow velocity direction. Capillary forces impede the displacement process because the energy of particle separation from capillary surface is in excess of the energy released in the course of its adhesion to the same surface.

Under change of displacement regime or its direction in the formations saturated with oil and water, a capillary pressure is subjected to the hysteresis phenomenon. A way of plotting of hysteresis loops of capillary pressure is described in paper [3].

3. Pressure wave propagation against the background of steady-state flow

Applying to laboratory experiment conditions, we consider one-dimensional oil flow in finite-size pattern with specified pressure drop $\Delta p = p_0 - p_c$ on the sides and harmonically varied with time pressure at the input. The process is described by problem:

$$\frac{\partial p}{\partial t} = a^2 \frac{\partial^2 p}{\partial x^2}; \quad p(0,t) = p_c + A\cos\omega t; \quad p(l,t) = p_0. \qquad (2)$$

Here $a^2 = k/(\beta\mu)$ is pressure conductivity factor, k, μ, β, ω, and l are permeability, viscosity, storage coefficient, vibration circular frequency of vibrations, pattern length respectively. Solution of problem (2) could be presented as sum of two functions $p(x,t) = p_1(x) + p_2(x,t)$, where $p_1 = p_c + \Delta px/l$ is a pressure distribution in steady flux, and $p_2 = \mathrm{Re}\{X(x)\exp(-i\omega t)\}$ is a periodic solution of pressure conductivity problem without initial data, satisfying boundary conditions $p_2(0,t) = A\cos\omega t; \; p_2(l,t) = 0$.

Solution of latest problem is sought in the form $p_2 = X(x)\exp(-i\omega t)$. Function $X(x)$ satisfies to the conditions:

$X''+\gamma^2 X=0;\ (\gamma=\gamma_0(1+i);\quad \gamma_0=\sqrt{\frac{\omega}{2a^2}});\ X(0)=A;\ X(l)=0$ and looks like

$X(x)=A(X_1+iX_2)=A\frac{\sin(\gamma(l-x))}{\sin\gamma l}$. Separating the real parts, we find

$p_2(x,t)=A(X_1(x)\cos\omega t+X_2(x)\sin\omega t)$, where

$$X_1=-2\frac{\cos\gamma_0 l\, sh\gamma_0 l\cos\gamma_0(l-x)sh\gamma_0(l-x)+ch\gamma_0 l\sin\gamma_0 l\sin\gamma_0(l-x)ch\gamma_0(l-x)}{\cos 2\gamma_0 l-ch2\gamma_0 l},$$

$$X_2=-2\frac{\sin\gamma_0 l ch\gamma_0 l\cos\gamma_0(l-x)sh\gamma_0(l-x)-sh\gamma_0 l\cos\gamma_0 l\sin\gamma_0(l-x)ch\gamma_0(l-x)}{\cos 2\gamma_0 l-ch2\gamma_0 l},$$

In the case of semibounded formation, one can analogously obtain the next solution:

$$p=p_1+p_2;\quad p_1=p_c+\Delta p\cdot erf\frac{x}{2a\sqrt{t}};\ p_2=A\exp(-\gamma_0 x)\cos(\omega t-\gamma_0 x).$$

$$x=0:\ p=p_c+A\cos\omega t;\ x\to\infty:\ p\to p_0.$$

Derivative of function $p(x,t)$ with respect to x is proportional to flow velocity

$$v/k_f=\frac{\partial p}{\partial x}(x,t)=\left[\Delta p\exp(-\frac{x^2}{4a^2t})\right]\Big/\left(a\sqrt{\pi t}\right)-A\exp(-\gamma_0 x)\gamma_0\cos(\omega t-\gamma_0 x)+ \qquad (3)$$
$$+A\exp(-\gamma_0 x)\gamma_0\sin(\omega t-\gamma_0 x).$$

Figure 3 gives the derivatives of fluid pressure as a functions of time at the distances 0, 0.25, 0.5, 0.75, 1.0, and 3 m from output of finite stratum (stratum length l = 100 m). Other parameters are γ_0 =1.67 1/m, $\omega=20$ 1/min, A=3 m, $\Delta p=5$ m.

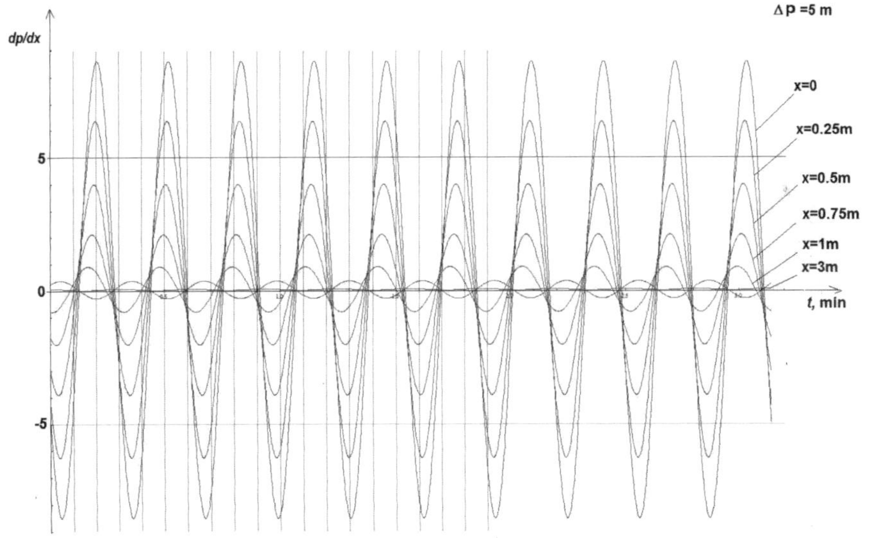

Figure 3. Subsidence of vibrational amplitude of fluid pressure with distance increasing from output.

Low curve on Figure 4 presents a theoretical distribution of oil saturation near borehole (end effect of capillary locking of water phase). Upper curve shows schematically a redistribution of both saturations at the vibration half-period after change of a flow direction. Meanwhile a part of pores is released from capillary-locking water and so the valuable product obtains access to borehole. "Swinging" helps to transform the near-well zone into neutrally-wettable one, impeding a development of capillary-locking effect.

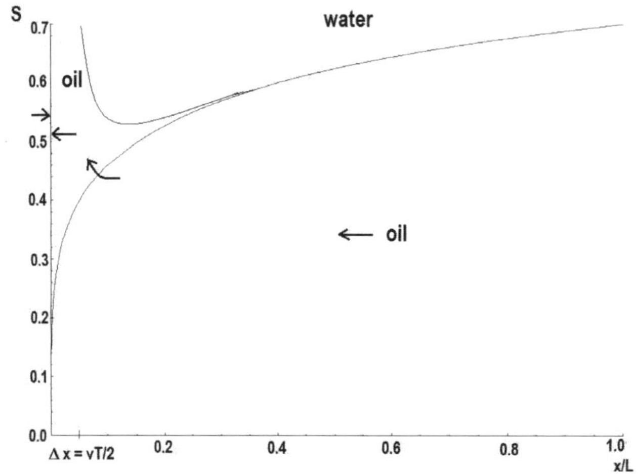

Figure 4. Mechanism of near-well zone cleaning from capillary-locking water.

4. Experiments

Laboratory experiments were performed on the flat model of stratum. Broken glass was taken as working porous material and filled the slot between two glass rectangle plates. Dimension of model is 0.4×0.2 m. Size of glass particles was in the range of $(2-2.5)\cdot10^{-4}$ m, slot depth was $h=2\cdot10^{-3}$ m. Porous volume of workspace occupied 43 % of total physical volume. Permeability of the pattern was 54 darcy. The model was saturated with oil through displacement of water up to obtaining the steady-state flow regime with the residual capillary-locking water formation in the neighborhood of outflow cross-section.

On output, the harmonic vibration of pressure were performed at specified parameters $\omega=2.5$ 1/sec, $A=0.075$ m, $\Delta p=0.1$ m, and vibration action time was 1 hour. Referring to Figures 5a,b, we have a state of zone with width 0.1 m near output cross-section of pattern before (shot a) and after (shot b) the vibration action on flow.

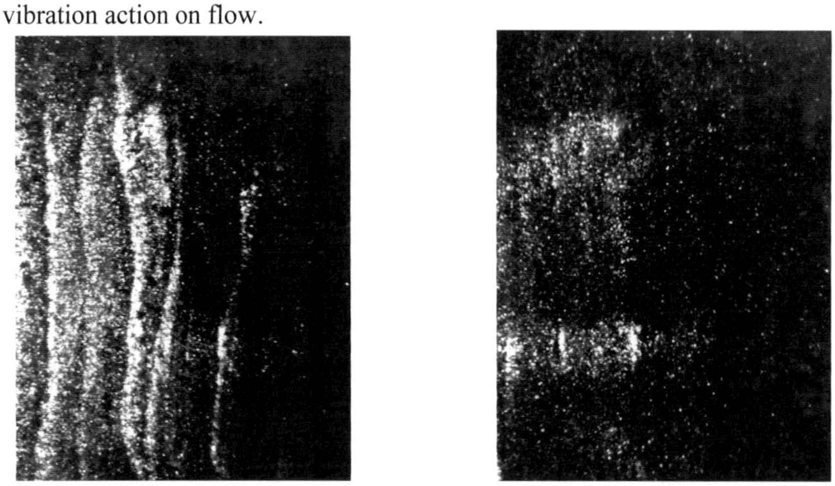

a b

Figure 5. State of zone near output before (a) and after (b) the vibration action.

IOP Conf. Series: Journal of Physics: Conf. Series **894** (2017) 012072 doi:10.1088/1742-6596/894/1/012072

5. Conclusions

As a rule, a real well production is less than expected one as defined, for example, by well-known formulas of Dupuit. Drop in well production depends on physicochemical properties of stratum and fluids contained in its porous space. Change of direction of immiscible fluid displacement can vary degree and behavior of wettability of rock porous surface. As the experiments demonstrated, the superposition of harmonic action and well liquid pressure on the output helps to clean the near-well zone from immiscible capillary-locking water. This fact proves an advantage of application of jack pumps in comparison with pumps that give the constant pressure drawdown.

References

[1] Earlougher R C Jr 1977 *SPE* New York
[2] Antontsev S N, Domanskii A V and Pen'kovskii V I 1989 *Lavrentyev Intitute of Hydrodynamics* Novosibirsk
[3] Pen'kovskii V I 1996 *Appl. Mech. Tech. Phys.* **37** 875

Suppression of a laminar separation by a spark discharge at a supersonic Mach number

P A Polivanov, A A Sidorenko and A A Maslov

Khristianovich Institute of Theoretical and Applied Mechanics SB RAS, Novosibirsk, Russia

E-mail: polivanov@itam.nsc.ru

Abstract. The paper deals with control the shock wave / laminar boundary layer interaction (SWBLI) by a spark discharge. Incident shock wave generated by a wedge induced the separation of the boundary layer developed on the flat plate at M=1.43. The inflow boundary layer state was laminar. The quantitative measurements were performed by PIV. It was found that a turbulent spot generated by a spark discharge suppresses the separation zone. But the thermal spot increases the loss of total pressure in the boundary layer. The effect of the discharge power on the zone of SWBLI has been studied.

1. Introduction

The modern trends in transonic aviation are connected with reducing of viscous drag. Therefore commercial airplanes of the next generations may be equipped with a laminar wing [1, 2]. Generic feature of the transonic flow near the airfoil is presence of local supersonic zone ended by a shock wave. Laminar boundary layer has weak resistance to adverse pressure gradients especially ones produced by shock waves, in contrast to the turbulent boundary layer [3]. This leads to significant flow separation that can eliminate all advantages of laminar flow. Flow separation on airfoil not only reduces the lifting force and increases drag, but also leads to additional adverse events, like unsteadiness [6, 7] including buffet [3]. It is known that separation of laminar flow usually has a greater length as compared to turbulent [4]. It was obtained [5] that for small supersonic Mach number losses of the total pressure and lifting force in SWBLI with laminar upstream boundary layer may be not worse than for turbulent conditions and this effect is more evident for high unit Reynolds numbers. The reason is fast turbulization of the shear layer in SWBLI for laminar case and significant reduction of the separation length.

Therefore it was decided that the use of electric discharge to excite perturbations should allow to reduce the size of a laminar separation bubble and suppress it. In the present paper the spark plasma actuator located upstream of the zone of interaction and controlling the laminar-turbulent transition has been studied.

2. Experimental setup

The experiments were performed in wind tunnel T-325 (ITAM SB RAS) for Mach number $M_\infty = 1.43$, total temperature 291 K and total pressure $P_0 = 0.55 \cdot 10^5$ Pa ($Re_1 = 8.5 \cdot 10^6$ m^{-1}). Test chamber of the wind tunnel has a rectangular cross section of 160x200 mm. The configuration of the experimental model is presented in Fig. 1. Experimental model consisted of a plate with sharp leading edge occupying full span of the test section and a wedge generating a shock wave. In this series of

Content from this work may be used under the terms of the Creative Commons Attribution 3.0 licence. Any further distribution of this work must maintain attribution to the author(s) and the title of the work, journal citation and DOI.
Published under licence by IOP Publishing Ltd

experiments, the wedge angle was 3-4°. This corresponds to regular reflection of a shock wave from a wall for the inviscid case. The blockage ratio of the test section was relatively high therefore the flow start was provided by extended cavities above the wedge and below the plate. To provide the same conditions near the leading edge the extension plates were installed downstream of the lower nozzle wall. The extension plates of length L = 100 mm were used. The experiments were done for laminar boundary layer. The figure shows that the distance from the flat plate leading edge to the center of the optical window is connected with length of the extension plate and is approximately equal to (380-L) mm. Position of the wedge was chosen to provide intersection of the shock wave with the model approximately in the center of the optical window.

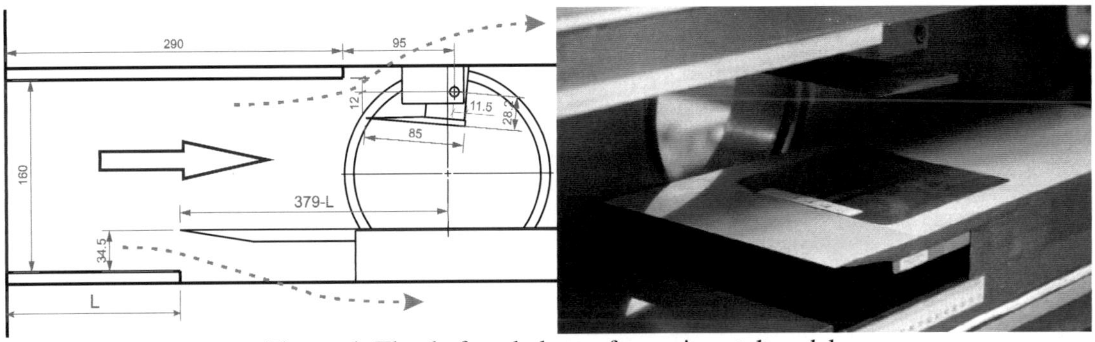

Figure 1. The draft and photo of experimental model

The main measurement techniques were PIV and Schlieren visualization. Seeding of DEHS microparticles with an average size of 1 micron were used for PIV measurement.

The plasma control devices were studied in the experiments. The spark discharge (SD) was chosen to achieve high concentrations of the energy in the plasma region. The model with installed SD actuator is shown in Fig. 2. It has ceramic insert made from MACOR holding a line of flush mounted electrodes. The electrodes are placed at distance 93 mm from the leading edge. There are three pairs of the electrodes with discharge gap of 4.5 mm. Distance between the neighbor electrode pairs is 14.5 mm. All tree discharge gaps are connected in-series. The capacitors connecting the interim electrodes to the ground were used to assist the breakdown (Fig. 3).

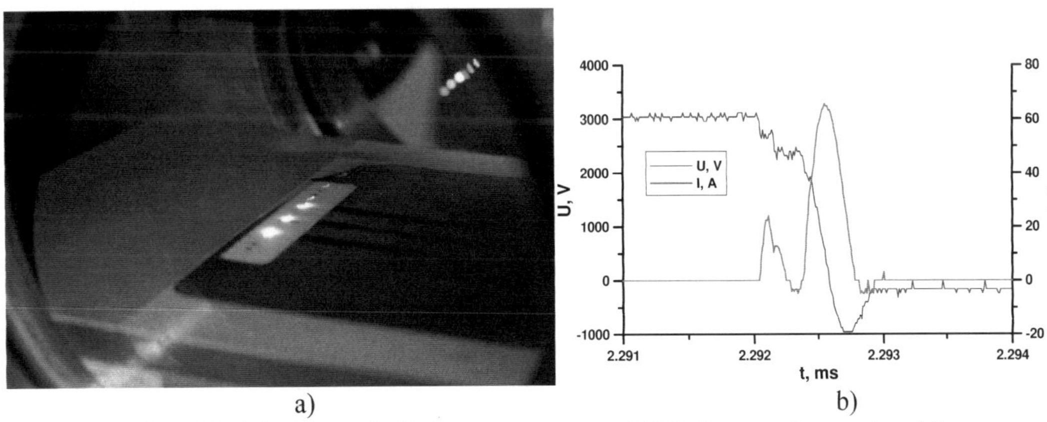

a) b)

Figure 2. a) Model with spark discharge actuator and b) Voltage and current oscillogram

Spark discharge actuator was fed by high voltage source using two transformers DAEHAN 15000V/30mA. The self-adapting scheme was used using a battery of capacitors C1 connected in parallel to the actuator (Fig. 3). This capacity is charged up to discharge level and consequently discharges. This process is periodic and the period depends on the environmental conditions and total capacity. The discharge repetition process is self-regulated therefore the frequency is not perfectly stable. In the following discussion this frequency is deduced as $f=1/T$ where T is averaged period of the discharges.

PIV measurements were synchronized with plasma discharge using time delay unit. The current pulse duration in the spark was less than 1 μs (Fig. 2b) and average power for one discharge gap was estimated as P_{dis} = 11 Watt. Figure 4 shows pulse energy and frequency for several values of capacity C1.

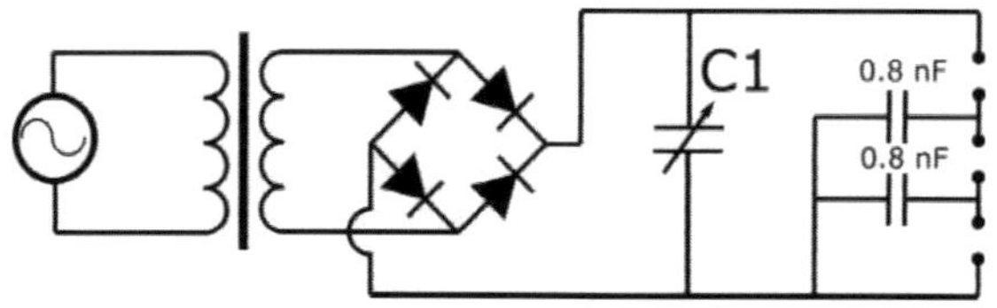

Figure 3. Stark discharge high voltage generator

3. Experimental results

Since some benefit was obtained for the transitional case [5] in comparison with the laminar one it was decided to excite in the upstream boundary layer disturbances similar to obtained in the laminar-turbulent transition. Artificial turbulization by the roughness did not result in any benefit therefore the spark discharge was applied.

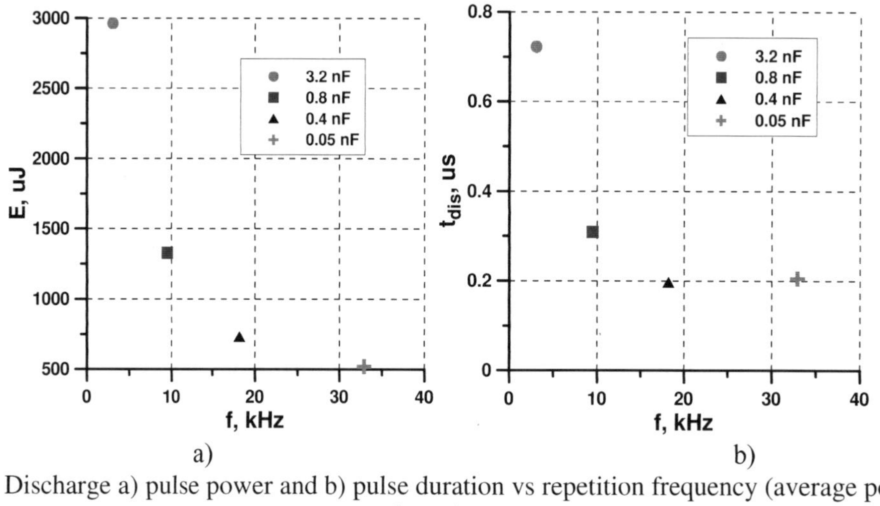

Figure 4. Discharge a) pulse power and b) pulse duration vs repetition frequency (average power was approximately 11 W)

The PIV data obtained for pulse energy of E_{dis} = 0.7 mJ (f = 18.2 kHz), β = 4° are shown in Fig. 5. High average frequency of the discharge means that the flow disturbances produced by the sparks travel downstream with small distance between them (≈15-20 mm). Since the high voltage system is self-regulated the breakdowns are not perfectly periodical. In the experiments PIV system was triggered by a discharge and the traces of the preceding discharges present in each single velocity distribution. However in the averaged data shown in Fig. 5 we can see only trace of the triggering breakdown since the delay between the sparks is not perfectly constant.

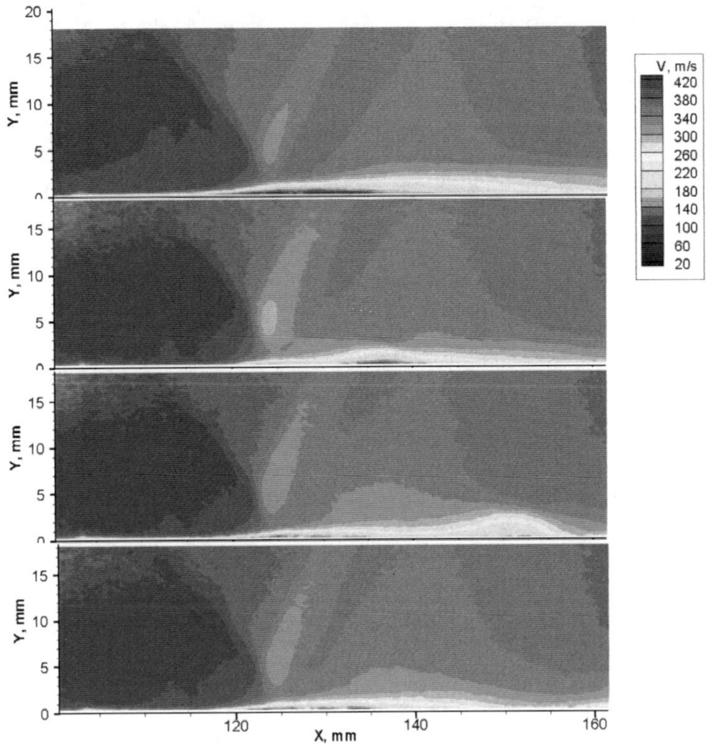

Figure 5. Mean velocity fields at different time lag (E_{dis} = 0.7 mJ, β = 4°). Top to bottom: reference, discharge 100-220 µs with Δt = 60 µs.

At the moment Δt = 100 µs the disturbance generated by the discharge passes the interaction. It can be seen that the compression waves upstream of the interaction are concentrated close to the interaction and they are more intense in comparison with the reference laminar case. This is evidence of diminishing or disappearance of the laminar separation zone upstream of the interaction. This is similar to the turbulent test case but the reflected shock wave is weaker. Fullness of the velocity profiles increases in the interaction and downstream.

The disturbance is represented by the area of low velocity in the boundary layer due to the hot spot with low density and high temperature. Decrease of velocity in this spot is amplified when it goes through the shock wave because of changes of the shocks configuration induced by the spot.

RMS of velocity pulsations for this test case is presented in Fig. 6. We can see high level of pulsations in the hot spot region. The hot spot travels downstream and the level of velocity pulsations decreases. It is necessary to note that level of the pulsations in the wake is generally decreased in comparison with the reference cease everywhere except the spot itself.

Figure 7 presents the comparison of laminar and turbulent (artificial turbulization, zig-zag trigger in position of X=100 mm) test cases and cases with discharge excitation with various power. It can be seen that discharge actuator allows to achieve less intensive shock wave and slow growth of the wake. The best result was obtained for the lower value of spark energy.

Variation of the momentum thickness for the case of E_{dis} = 0.7 mJ, β = 4° is presented in Fig. 8. It can be seen that the spark discharge is able to reduce the average momentum thickness in the wake by 30% in comparison with laminar case.

IOP Conf. Series: Journal of Physics: Conf. Series **894** (2017) 012073 doi:10.1088/1742-6596/894/1/012073

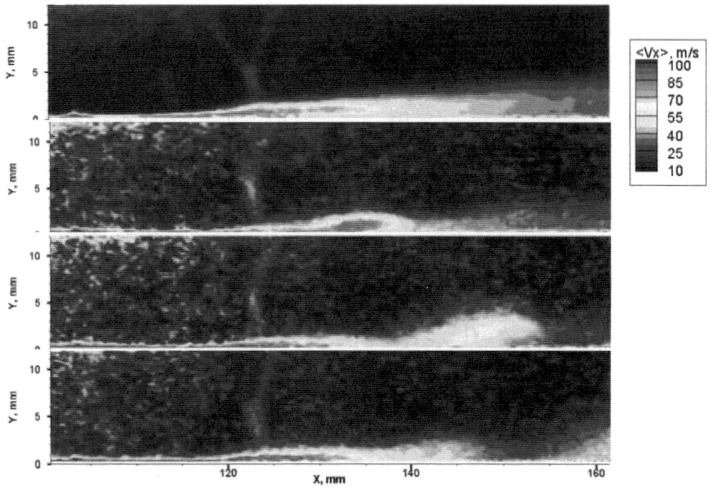

Figure 6. RMS of longitudinal velocity at different time lag (E_{dis} = 0.7 mJ, β = 4°). Top to bottom: reference, discharge 100-220 µs with Δt = 60 µs.

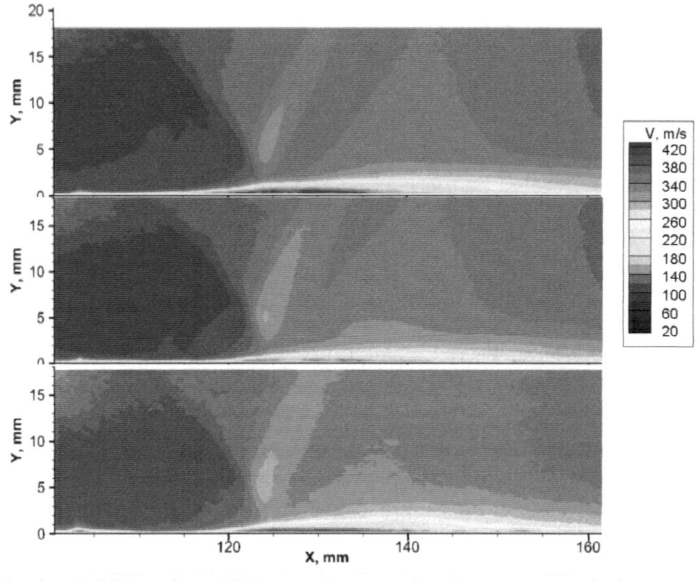

Figure 7. Mean velocity fields at $\beta = 4°$. Top to bottom: laminar case, turbulent case, average discharge case E_{dis} = 0.7 mJ, E_{dis} = 1.3 mJ

Comparison of the averaged θ distribution with instantaneous ones presented in the same figure shows that in the region of the hot spot there is increase of losses. Therefore, the positive effect of the disturbances provided by the discharge may be eliminated by the hot spot.

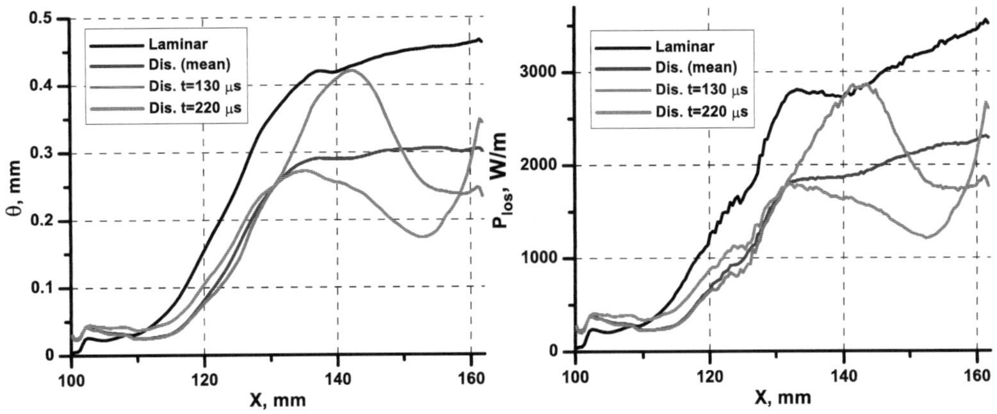

a) b)

Figure 8. Distribution of the a) momentum thickness and b) energy losses along SWBLI (E_{dis} = 0.7 mJ, $\beta = 4°$).

The flow control efficiency may be estimated basing on θ value at the end of measurement region as $\eta_{dis} = 0.5\rho \cdot U^3(\theta_{lam}-\theta_{dis})/P_{dis}$. For β = 4° the maximum value of efficiency η_{dis} = =225% was obtained for minimum spark energy E_{dis} = 0.7 mJ. If the spark energy was increased up to E_{dis} = 1.3 mJ the efficiency dropped to 167 %. Increase of spark energy up to 3 mJ resulted in negative efficiency -55%. This means that the disturbance generated by the spark is sufficient but there in negative effect provided by the hot spot increasing with the power. This conclusion agrees with results of computational study [8] where some optimum of pulse energy was found for the flow turbulization by a discharge. In the case of exceed of energy the effect diminishes due to heat spot formation. Since the flow parameters of study [8] and the presented experiments are close it can be assumed that decrease of energy by 10 will allow to keep positive effect and increase the control efficiency.

4. Conclusion

Spark discharge actuators were tested and found to be effective to excite powerful periodic disturbances and control the interaction region. Basing on the quantitative analysis it may be concluded that spark discharge actuator improves the average flow in the interaction region. It was found that all investigated values of pulse energy are sufficient for the generation of a turbulent spot and the suppression of the separation zone. But an increase the discharge power leads to an increase the effect of the heat spot which increases the losses in the boundary layer. The analysis shows that active flow control may be more effective in comparison with passive control by the roughness.

Acknowledgments

The work was supported by EU in framework of FP7 TFAST project and grant of Russian Foundation for Basic Research 16-08-01134A.

References

[1] Allison E, Kroo I, Sturdza P, Suzuki Y and Martins-Rivas H 2010 Aircraft Conceptual Design with Natural Laminar Flow *27th International Congress of the Aeronautical Sciences* (France, Nice, 2010)
[2] Green J E 2008 *AIAA Paper* 2008-3738
[3] Lee B H K 2001 Progress in Aerospace Sciences **37** 147-96
[4] Swoboda M and Nitschef W 1996 *Journal of Aircraft* **33** 100-8
[5] Polivanov P A, Sidorenko A A and Maslov A A 2015 *AIAA Paper* 2015-1974

[6] Dussage J P, Dupont P and Debieve J F 2006 *Aerospace Science and Technology* **10** 85-91
[7] Polivanov P A, Sidorenko A A and Maslov A A 2011 *Shock Waves* **21** 193-203
[8] Polivanov P A, Sidorenko A A and Maslov A A 2014 Artificial turbulization of the supersonic boundary layer by dielectric barrier discharge *29th International Congress of the Aeronautical Sciences* (Russia, Saint-Petersburg, 2014)

IOP Conf. Series: Journal of Physics: Conf. Series **894** (2017) 012074 doi:10.1088/1742-6596/894/1/012074

Simulation of a supersonic flow around a body with a frontal gas-permeable insert by using a skeleton model of a highly porous cellular material

T V Poplavskaya[1,2], **S V Kirilovskiy**[1,2] **and S G Mironov**[1,2]

[1]Khristianovich Institute of Theoretical and Applied Mechanics SB RAS, 630090, Novosibirsk, Russia
[2]Novosibirsk State University, 630090, Novosibirsk, Russia

Abstract. Numerical simulation of supersonic flow past a cylinder with a frontal gas-permeable insert is performed using the skeleton model of a highly porous cellular material. Numerical simulation was carried out within the framework of two-dimensional RANS equations written in an axisymmetric form. The skeleton model is a system of coaxial rings of different diameters, arranged in staggered order. The calculations were carried out in a wide range of determining parameters: Mach numbers M_∞ =3, 4.85 and 7, unit Reynolds numbers $Re_{1\infty}$ =$13.8\times10^5 \div 13.8\times10^6$ m^{-1}, the cylinder diameter 6÷40mm, the length of the porous insert 3÷45mm, the cell diameter of 1 and 3 mm. The results of the calculations are consistent with the available experimental data. The applicability of the skeleton model for the description of supersonic flow around axisymmetric bodies with front inserts from cellular-porous materials is shown.

1. Introduction

Gas-permeable porous materials in recent years have found application in promising methods of controlling supersonic aircraft [1-3]. Experimental studies in wind tunnels demonstrated the possibility of controlling supersonic flow and, in particular, controlling the aerodynamic drag of bodies by help of gas permeable highly porous cellular materials (HPCM) [1,4].HPCM (Figure 1a) is formed when the foamed liquid metals solidify. The cellulars of the frozen foam form a spatial carcass from the partitions between the contacting cellulars. The porosity (the ratio of the total volume of the porous sample to the volume of the skeleton) of such a material lies in the range from 76% to 98%. The material ceases to be gas permeable with less porosity, and with greater porosity, a skeleton of the partitions is not formed. There are some recent applications of gas-permeable porous materials and in supersonic aerodynamics to suppress acoustic waves. In particular, it was possible to suppress acoustic disturbances in the shock layer on a plate in nitrogen stream at M = 21 with the help of inserts of foamed nickel in experiments [5]. These phenomena are based on dissipation of the energy of acoustic disturbances, which are a dominating mode of instability of a hypersonic boundary layer, owing to friction in pores of the coating. It was shown in experiments [6] that the intensity of perturbations is attenuated by a sound-absorbing coating from the HPCM by 20% at the plate in the flow of vibrationally excited mixtures of CO_2 with air.

To increase the efficiency of the action of gas-permeable porous materials on a supersonic flow around various bodies and develop engineering methods for design of supersonic flying vehicles with porous control elements it is required to apply modern methods of numerical simulation of flow around bodies with porous inserts. At present, the main problem of numerical modeling is the optimal choice of the model of a gas permeable porous medium. Continuum models of a porous medium with a prescribed pressure gradient in the porous region as a function of the filtration rate in accordance with a certain law (e.g., the Darcy-Forchheimer law) are widely used for numerical simulations of the flow in a porous medium. Investigations [4,7] have demonstrated the results and the main difficulties in the use of continuum models of a porous medium during calculations. The fundamental problem of continuum models is that the scales of the structural elements of real porous materials (cell sizes) are comparable with the scale of porous inserts (the diameter of the model), but the continuum models are obtained for conditions of continuous media that do not have any structural elements inside themselves. Moreover, at supersonic velocities of the oncoming flow, there are drastic changes in the pressure gradient and filtration rate along the insert, which required the use of different filtration dependences along the porous insert.

Content from this work may be used under the terms of the Creative Commons Attribution 3.0 licence. Any further distribution of this work must maintain attribution to the author(s) and the title of the work, journal citation and DOI.
Published under licence by IOP Publishing Ltd

IOP Conf. Series: Journal of Physics: Conf. Series **894** (2017) 012074 doi:10.1088/1742-6596/894/1/012074

The way out of the situation is a direct simulation of the skeleton of porous materials and the flow of gas in them. The results of such studies are described, for example in [8, 9], where a low-velocity gas flow was modeled through a porous-porous material. They chose a skeleton model in the form of intersecting spheres (Fig.1b), which most closely reflects the spatial structure of cellular-porous materials. However, this approach requires the use of a huge number of cells of the calculated grid (of the order of several tens of millions), which is time-consuming for parametric engineering calculations. The exact form of the spatial structure of the material does not play an important role in supersonic flow around a skeleton of a porous material. Here the leading role is played by the porosity coefficient and the scale (diameter) of the cells of the cellular-porous material. This creates a certain opportunity for constructing simpler model skeletons of cellular-porous materials.

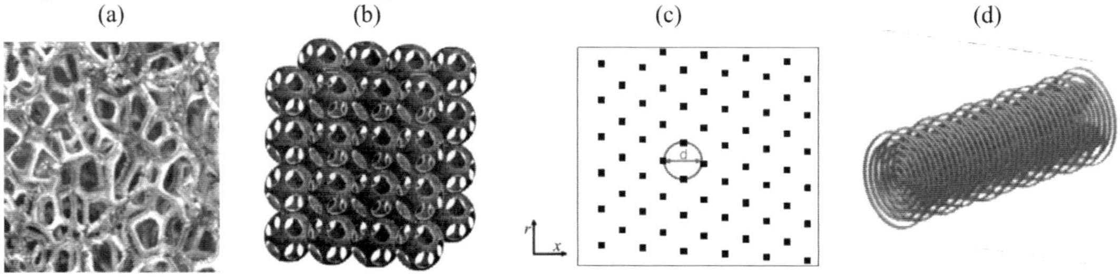

Figure 1. Enlarged image of a fragment of foamed nickel (a), computational skeleton model of the gas-permeable material of foamed nickel type (b), 3D image of the computational model of the gas-permeable insert (c)

In this paper, we solve the problem of the axisymmetric flow past a cylinder with a gas-permeable front insert made of LCPM by using a skeleton model of a highly porous cellular material.

2. Skeleton model of a gas-permeable porous medium

The challenge of the present activities was to perform numerical simulations of a supersonic flow around a streamwise aligned cylinder with a frontal porous insert without empirical data used in the computations. Mironov et al. [4] found that the main effect on the air flow in insert material pores is produced by the inertial term of the quadratic dependence responsible for flow interaction with partitions of the porous skeleton, i.e., for aerodynamics of the flow past pore partitions. This fact gives grounds to consider the flow around a cylinder with a porous insert as the flow around an integrated model consisting of the porous insert skeleton and the cylinder. In this approach, the porous medium is presented as a three-dimensional system of discrete elements modeling the skeleton of a real porous material. Further we term it as the skeleton model of a porous medium to distinguish it from traditional continuum models.

In this case, it is important to choose an appropriate geometry of the porous material skeleton, which would be consistent with conditions of a high-velocity air flow in the porous insert. In addition to the cell size and porosity value, one more criterion of a correct choice of the skeleton geometry is the correspondence of the predicted structure of the external flow field and the magnitude of wave drag reduction to experimental data.

A fairly simple geometry of the HPCM skeleton consisting of staggered elements with a square cross section (Figure 1c) was successfully applied for modeling the absorption of acoustic disturbances in a hypersonic flow past a flat plate with a porous insert [5,10]. A similar model of the skeleton of HPCM in the form of square or rectangular elements, twisted about the longitudinal axis, is proposed in this paper for solving of the task of axisymmetric flow around a cylinder with a gas-permeable front insert. Thus, the proposed skeleton model is a system of coaxial rings of different diameters arranged in staggered order (Figure 1d). One can ignore the azimuthal motion of the gas for axisymmetric inserts in flows at zero angle of attack and construct a model skeleton from ring elements without radial baffles, taking into account the absence of radial baffles in the characteristics of the ring elements and their arrangement. Some correction of such a model structure can be done by comparison with the data of the tube experiment. The shape and size of the elements of the HPCM skeleton were chosen from the condition of correspondence to the material porosity coefficient of 0.95, i.e., the total volume of the skeleton elements was only 5% of the porous insert volume. The distance between the skeleton elements in the axial directions was equal to the pore diameter d. In the radial direction, the distance between the elements was smaller than d to compensate for the absence of radial partitions of the model skeleton. The value of this difference was determined in a series of computations with variations of the distance between the elements and the shape of the elements themselves. Additional characteristics of the HPCM are the random arrangement of the skeleton elements and the length of material non-transparency. Therefore, the vertical row of elements was gradually shifted in the computational model to ensure the correspondence to the length of real HPCM non-transparency. This fact is illustrated in the image of the model porous insert in Fig. 1c as a periodically repeated step on the external boundary of the insert.

3. Experimental data

The investigations of the flow around a cylinder with a frontal porous insert made of cellular-porous nickel were performed in a T-327B supersonic blowdown wind tunnel based at ITAM SB RAS [4]. The model of the cylinder with the frontal gas-permeable porous insert made of the HPCM has the diameter D= 14.5 mm, the porous insert length was varied from zero to 45 mm, and the total length of the model was L=100mm. The length of the porous insert Δx normalized to the cylinder diameter D ($\Delta x/D$) was varied within 0.2÷3.1 with the use of a moving thin-walled tube closely adjacent to the cylinder. The model was mounted on a three-component strain-gauge balance, which measured the drag force generated by the model.

4. Numerical simulation

The problem of the axisymmetric flow past a cylinder with a gas-permeable front insert was solved within the framework of the two-dimensional RANS equations written in an axisymmetric form. The calculations were carried out using the k-ω SST turbulence model. The space derivatives were approximated by means of implicit schemes of the second order of accuracy. Convective fluxes were approximated by the Roe-FDS method of splitting with a TVD limiter for retaining solution monotonicity near discontinuities and local extremes of the flow. Integration with respect to time was performed by the third-order TVD Runge-Kutta scheme. The flow around the cylinder with model inserts was numerically simulated by the commercial package ANSYS Fluent.

The projection of the computational domain onto a plane is a rectangle, and its lower boundary coincides with the axis of symmetry of the cylinder. The computational domain consisted of several subdomains (region ahead of the body, region of the porous insert, and region behind the porous insert) in order to provide grid refinement near the bow SW and the cylinder surface and to minimize the number of cells in the computational domain. The porous subdomain was covered by a uniform rectangular computational grid. Up to ten processors of the Information-Computational Center of the Novosibirsk State University and the Siberian Supercomputer Center were used in the computations.

The calculations were carried out in a wide range of determining parameters: Mach numbers M_∞ =3, 4.85 and 7, unit Reynolds numbers$Re_{1\infty}$ =$13.8\times10^5 \div 13.8\times10^6$ m^{-1},the cylinder diameter D= 6÷40mm, the length of the porous insert Δx=3-45mm, the cell diameter of HPCM d=1 and 3mm.

Solving the problem, we obtained all gas-dynamic variables of the flow both outside the model and inside the porous material. The drag coefficient of the model was calculated as $C_x = F/\left(0.5 \cdot \rho_\infty u_\infty^2 S_m\right)$, where$\rho_\infty$, u_∞ are the free-stream density and velocity, S_m is the area of the model mid-section, and F is the aerodynamic force including the pressure force, the friction force, and the wave drag force. In this work, the force F was calculated from the difference in the total momentum of the flow ahead of the bow SW and in the plane of the rear end face of the cylinder (integral characteristics). Thus, we calculated the drag coefficients C_x of the cylinder with the frontal porous insert and C_{x0} of the solid cylinder.

5. Results

The computed pressure fields of flow past a solid cylinder and a cylinder with a porous insert located in front of it are shown in Figure 2a,b.The transition from the arc-shaped bow shock wave on the solid cylinder (Figure 2a) to the oblique shock (Figure 2b) on the cylinder with a gas permeable porous insert is clearly visible. The bow SW becomes “seated” onto the frontal plane of the cellular porous insert practically. The weaker oblique jumps caused by the outflow of air jets from the pores are arisen behind the bow shock. The fields of the streamlines for the flow around a solid cylinder and a cylinder with the frontal porous insert are in Figure 2 c,d. It is seen that the streamlines on cylinder with the frontal porous insert have a flux reversal angle less than on a solid cylinder. This leads to a weakening of bow shock wave and the appearance behind it of a sequence of weak shock waves on the trickle of flowing current. This gas flow redistribution in the cellular porous body leads to formation of a flow similar to the flow around a pointed body, resulting drag reduction of model in the supersonic flow.

The experimental and computed normalized drag coefficients of the model C_x/C_{x0} are plotted in Figure 3a as functions of the normalized length of the porous insert $\Delta x/D$. Figure 3 shows the experimental data [1] for two pore diameters d = 1 and 3 mm. The results of computations based on the skeleton model of a porous medium are shown in Figure 3a by the dashed curves. The solid curve illustrates the results for the flow computed by the two-zone continuum model for the pore diameter d = 3 mm [9]. It is seen significant (within 35%) drag reduction for large lengths of the gas-permeable porous insert ($\Delta x/D \geq 1.5$).

(a) (b)

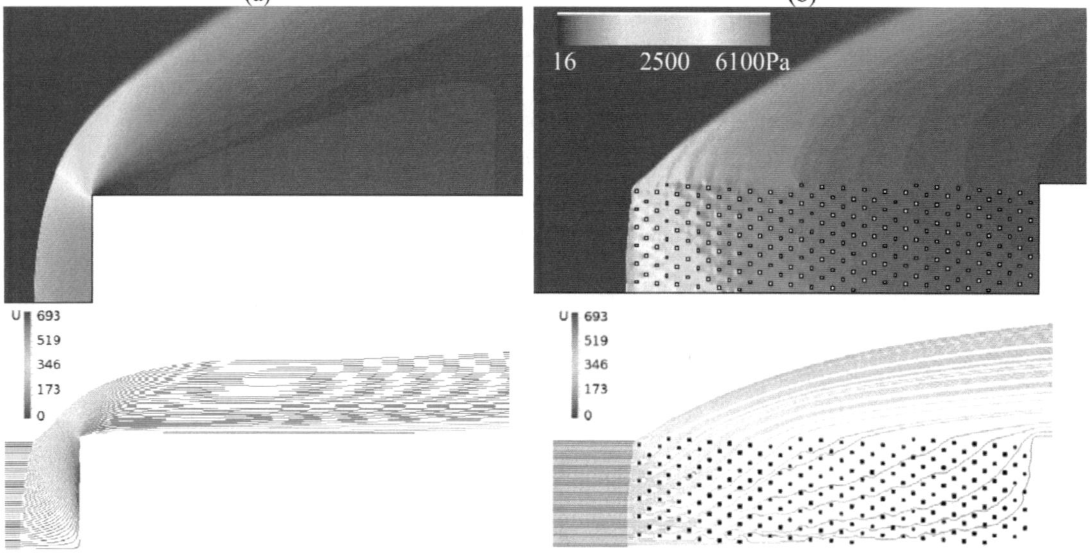

(c) (d)

Figure2. The computed pressure fields and the fields of the streamlines $M_\infty = 4.85$, p_∞=186Pa, T_∞=51K, $Re_{1\infty} = 2.7 \times 10^6 m^{-1}$: (a,c) – solid cylinder; (b,d) –cylinder with a frontal porous insert $\Delta x/D = 2$, D=14.5mm, d=1mm, L=100mm.

The experimental and computed normalized drag coefficients of the model C_x/C_{x0} as functions of the Reynolds number Re_D based on the free-stream parameters and the cylinder diameter D are compared in Figure 3b. The change of Reynolds number in the calculations was made by changing the static pressure of the oncoming flow calculated from the values of the stagnation pressure in the experiment. The results show a significant (within 11%) change in the aerodynamic drag at $Re_D < 0.4 \times 10^5$, where viscosity effects are likely to manifest themselves. The change of the drag coefficients of the model is insignificant in the rest of the Reynolds number range ($Re_D \geq 0.4 \times 10^5$).This indicates a weak dependence of the task on viscosity and the predominance of the process of inviscid flow inside a porous material in the creation of drag of the entire model as a whole.

(a) (b)

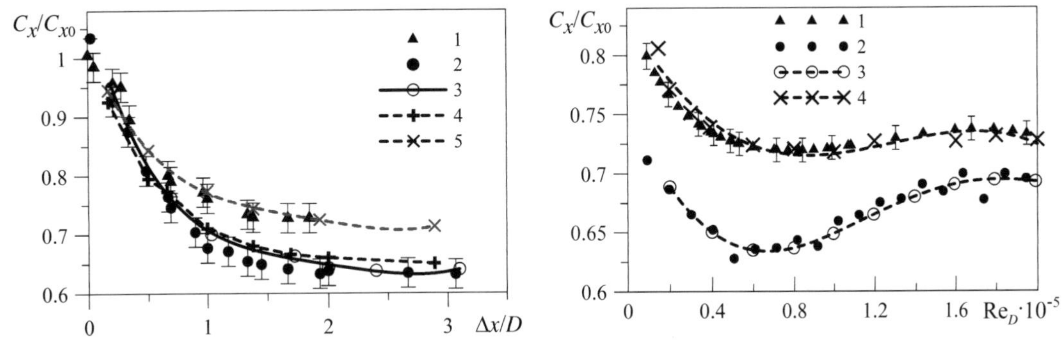

Figure 3. Experimental (1,2) and computed (3,4,5) normalized drag coefficient of the model versus the normalized length of the porous insert for $Re_{1\infty} = 2.7 \times 10^6$ m^{-1}(a) and versus the Reynolds number Re_D for $\Delta x/D = 2.0$ (b): 1, 5 – d=1mm; 2,3,4 – d=3mm; 3 – continuum model [9]; 4,5 – skeleton model. M_∞ = 4.85, p_∞=186Pa, T_∞=51K.

The calculations were made of the flow past a cylinder with a porous gas-permeable insert of a large length ($\Delta x/D$ = 2) with a pore diameter of 1 mm diameter model D = 14.5 mm that to determine the influence of Mach number on the decreasing of the drag coefficient. The flow parameters are as follows: stagnation pressure P_0=1atm, stagnation temperature T_0=290K, Reynolds number Re_D= 0.4×10^5. The results of these investigations in comparison with the experimental data are given in Table 1. It is seen that normalized coefficient of resistance of the model varies slightly (within 6%) with an increase in the Mach number from 3 to 7. Such a weak dependence is probably associated with a small thickness of the boundary layer growing on the elements of a porous material at such Mach numbers. This means that the flow in the pores is not blocked and the flow freely expires from the porous zone.

Table1. Computed and experimental data of normalized drag coefficient of the cylinder with a frontal porous insert $\Delta x/D$ = 2 for different Mach numbers.

C_x/C_{x0}	M_∞=3 $Re_{1\infty} = 2.7 \times 10^6$m^{-1}	M_∞=4.85 $Re_{1\infty} = 2.7 \times 10^6$m^{-1}	M_∞=7 $Re_{1\infty} = 2.7 \times 10^6$m^{-1}	M_∞=7 $Re_{1\infty} = 1.5 \times 10^6$m^{-1}
Experiment	–	0.739±0.025	-	0.83±0.025
CFD	0.733	0.737	0.778	0.818

(a)
(b)
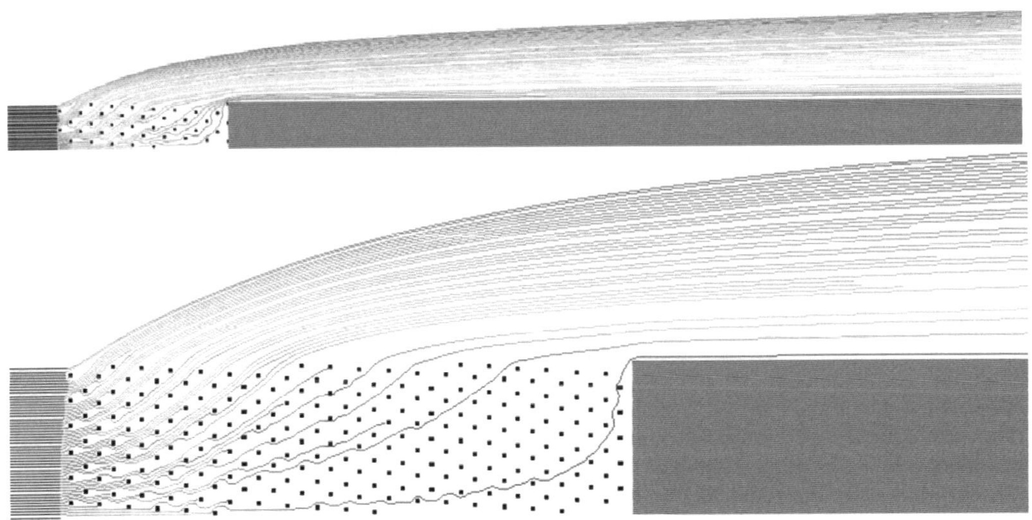

Figure 4. Thefieldsofthestreamlineson flow around cylinder with a frontal porous insert $\Delta x/D$ = 2, d=1mm, M_∞ = 7, p_∞=42.4Pa, T_∞=26.85K, $Re_{1\infty}$ = 2.7 $\times 10^6 m^{-1}$: (a) - D=6mm, (b) - D=14.5mm, (c) - D=20mm.

Figure 4 shows the fields of the streamlines for flow around cylinders of different diameters D = 6, 14.5 and 20 mm with a porous front insert $\Delta x/D$= 2, d=1mm at M_∞ = 7. It is seen that if the diameter of the model grows, the angle of deflection of the current lines from the model surface increases, which is associated with a more intensive outflow of gas from a porous zone of larger diameter. This leads to increasing of the angle of inclination of the bow shock to the direction of flow and, as a result, to increasing of the momentum loss of the oncoming stream. Therefore, the efficiency of the porous insert decreases with increasing of the model diameter and the aerodynamic drag increases.

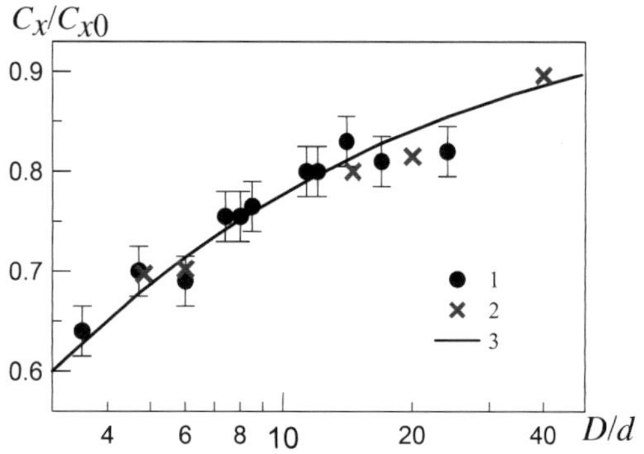

Figure5. Computed and experimental data of normalized drag coefficient of the cylinder with a frontal porous insert $\Delta x/D$ = 2 versus the ratio of the diameter of the model to the pore diameter:
1 – experimental data
2 – numerical data
3 – generalizing dependence (1) for M_∞=7

Figure 5 shows the calculated and experimental data of the normalized drag coefficient of a cylinder with a porous insert $\Delta x/D$= 2 for Mach number 7 versus the ratio of the model diameter to the pores diameter of the insert D/d. It is seen that the values of aerodynamic drag coefficient are approximated by a unified dependence

$$C_x / C_{x0} = 1 - 0.6879 / \sqrt{D/d} \quad (1)$$

and therefore, the parameter D/d can be considered a geometric similarity parameter for the aerodynamic drag coefficient of bodies with frontal porous inserts.

6. Conclusions

Numerical simulation of supersonic flow past a cylinder with a front insert from the HPCM using skeleton model of a porous medium over a wide range of Reynolds numbers, Mach numbers, lengths of the porous insert, pore diameters and the ratio of the model diameter to the pores diameter of the porous material is carried out.

It is shown that the skeleton model in the form of a system of coaxial rings of different diameters, located in a staggered order, makes it possible to obtain results consistent with the experimental data on the weight measurements of the aerodynamic drag.

In the parametric calculations, a weak dependence of aerodynamic drag coefficient of the cylinder with the front insert from the HPCM on the Mach numbers in the range 3-7 and on the unit Reynolds numbers in the range $13.8\times10^5 \div 13.8\times10^6$ m^{-1}. The influence of the ratio of the model diameter to the pores diameter (the geometric similarity criterion) on the aerodynamic drag coefficient is shown.

Acknowledgments

This work was supported by the Russian Foundation for Basic Research (project No. 015-08-03867).

References

[1] Fomin V M, Mironov S G and Serdyuk K M 2009 *Tech. Phys. Lett.* **35** 117-9
[2] Hunter C A, Viken S A, Wood R M and Bauer S X S *AIAA Paper* 2001-0249
[3] Fomin V M, Postnikov B V and Lomanovich K A 2015 *Tech. Phys. Lett.* **49** 686-8
[4] Mironov S G, Maslov A A, Poplavskaya T V and Kirilovskiy S V 2015 *J. Appl. Mech. Tech. Phys.* **56** 549-57
[5] Maslov A A, Mironov S G, Poplavskaya N V, Tsyryulnikov I S and Kirilovskiy S V 2012 *J. Appl. Mech. Tech. Phys.* **53** 162-72
[6] Kirilovskiy S V, Maslov A A, Poplavskaya T V and Tsyryulnikov I S 2015 *Technical Physics* **60** 645-55
[7] Bedarev I A, Mironov S G, Serdyuk K M, Fedorov A V and Fomin V M 2011 *J. Appl. Mech. Tech. Phys.* **52** 9-17
[8] Mitrichev I I, Koltsova E M and Zhensa A V 2012 *Fundam. Res., Techn. Sci.* **11** 440-6
[9] Kulakov S V 2000 *Perspektivnie Materiali* **3** 22-6
[10] Tsyryulnikov I S, Maslov A A, Mironov S G, Poplavskaya T V and Kirilovskiy S V 2015 *Technical Physics Letters* **41** 184-6

IOP Conf. Series: Journal of Physics: Conf. Series **894** (2017) 012075 doi:10.1088/1742-6596/894/1/012075

Boundary value problems for parabolic equations of high order with a changing time direction

S V Popov and V G Markov
North-Eastern Federal University, Yakutsk, Russia
E-mail: guspopov@mail.ru

Abstract. We analyze the solvability of boundary value problems for differential equations of high order with a changing direction of time. It is shown that a generalized solution exists in Sobolev spaces. Correctness of boundary value problems with a complete matrix of conjugation conditions is investigated. For such problems, it is shown that the Hölder classes of their solutions essentially depend on both the type of gluing conditions and on the non-integer index Hölder space.

1. Introduction

The paper concerns boundary-value problems for differential equations of high order of the form

$$\frac{\partial^{2m+1} u}{\partial t^{2m+1}} + \sum_{k=0}^{2m} \frac{\partial^k u}{\partial t^k} a_k(t) + Lu = f, \quad (x,t) \in Q = (a,b) \times (0,T), \tag{1}$$

where the operator L has the form $Lu \equiv \frac{1}{g(x)}(L_0 u + \lambda_0 u + Mu)$; the differential operator L_0 with respect to a variable x is of the form $L_0 u = \sum\limits_{i=0}^{2s} \alpha_i(x) \frac{\partial^i u}{\partial x^i}$ and M is its perturbation of the form $Mu = \sum\limits_{i=0}^{2s-1} b_i(x,t) \frac{\partial^i u}{\partial x^i}$.

The equation (1) is considered with the boundary conditions

$$U_j(u) \equiv \sum_{k=0}^{2s-1} \alpha_{jk} \frac{\partial^k}{\partial x^k} u(a,t) + \sum_{k=0}^{2s-1} \beta_{jk} \frac{\partial^k}{\partial x^k} u(b,t) = 0, \ j = 1, 2, ..., 2s, \tag{2}$$

$$u^{(i)}(x,0) = u^{(i)}(x,T), \quad i = 0, 1, \dots, 2m-1, \tag{3}$$

where α_{jk}, β_{jk} are some complex constant.

The function $g(x)$ can be equal to zero and change the sign on the interval (a,b). In particular, for $m = 0$, equations with varying direction of time are sufficiently well analyzed.

In this paper we consider the solvability of the problem (1)–(3). Note that similar equations arise in many areas of physics, mechanics and some other applications. Higher-order equations were considered in works [1–3].

Content from this work may be used under the terms of the Creative Commons Attribution 3.0 licence. Any further distribution of this work must maintain attribution to the author(s) and the title of the work, journal citation and DOI.
Published under licence by IOP Publishing Ltd

In §1 we formulate some notation, auxiliary statements and main results. In §2 we study parabolic equations with a changing time direction by using the theory of singular integral equations [4–7] as well as the systems of these equations [8].

Note that a large number of papers are devoted to the study of linear second-order equations. The general theory of boundary value problems for equations of mixed type with arbitrary coefficients and variety of type changes have been the subject of many studies [9, 10].

2. Notation and main result

The notation for the functional spaces that we shall use are standard: $W_p^s(Q)$ is the Sobolev space and $H^s(Q)$, $H_{x\ t}^{p,p/2s}(Q)$ are Hölder spaces. The symbol $(X,Y)_{\theta,p}$ denotes the space constructed by the method of real interpolation for Banach spaces X and Y [11].

We assume that the domain $D(L_0)$ of the operator L_0 consists of the functions $u(x) \in W_2^{2s}(a,b)$ satisfying (2) and inequality

$$Re(L_0u, u) \geq \delta_0 \|u\|^2_{W_2^s(a,b)}, \quad u \in D(L_0), \quad \delta_0 > 0, \tag{4}$$

where $(\cdot,\cdot)$ is the scalar product in $L_2(a,b)$. We denote by H_1 the closure of $D(L_0)$ in the norm $\|u\|_{H_1} = \|u\|_{W_2^s(a,b)}$. Assume that there exists a constant $c > 0$:

$$|(L_0u, v)| \leq c\|u\|_{H_1}\|v\|_{H_1}, \quad u, v \in D(L_0). \tag{5}$$

This implies the estimate

$$\|L_0u\|_{H_1'} \leq c\|u\|_{H_1}, \tag{6}$$

where H_1' is a negative space constructed from a pair $H_1, L_2(a,b)$.

We introduce the operator $\widetilde{L_0}u = \frac{1}{g(x)}(L_0 + \lambda_0)u$. Assume that $g(x) \in L_1(a,b)$. We set $F_1 = L_2(0,T;H_1)$. By a definition, the space $L_{2,g}(a,b)$ consists of measurable functions such that $\|u\|^2_{L_{2,g}(a,b)} = \int\limits_a^b |g(x)||u|^2\,dx < \infty$. We define auxiliary space $F_0 = L_2(0,T;L_{2,g}(a,b))$. The space F_0 becomes the Kerin space if we define an indefinite metric scalar product by the formula

$$[u,v]_0 = \int\limits_Q g(x)u(x,t)\overline{v(x,t)}dxdt,$$

$$(u,v)_0 = \int\limits_Q |g(x)|u(x,t)\overline{v(x,t)}dxdt.$$

We have

$$Re[\widetilde{L_0}u, u]_0 = Re\int\limits_Q (L_0 + \lambda_0)u\bar{u}dQ \geq \delta_0\|u\|^2_{F_1} + Re\lambda_0\|u\||^2_{L_2(Q)}, \quad u \in D(L_0). \tag{7}$$

By (6) and inequality (4), the operator $L_0 + \lambda_0$, $Re\lambda_0 \geq 0$ can be extended to isomorphism of F_1 and $F_1' = L_2(0,T;H_1')$. This assertion is a consequence of the Lax-Milgram theorem and inequalities (5) and (7).

We construct the space H_{-1} as the completion of $L_{2,g}(a,b)$ with the norm

$$\|u\|_{H_{-1}} = \sup_{v\neq 0}\frac{|[u,v]|}{\|v\|_{H_1}} = \|g(x)u\|_{H_1'},$$

IOP Conf. Series: Journal of Physics: Conf. Series **894** (2017) 012075 doi:10.1088/1742-6596/894/1/012075

where $[u,v]=\int\limits_a^b g(x)u(x)\overline{v(x)}dx$. We set $F_{-1}=L_2(0,T;H_{-1})$. Note that F_1 is densely embedded in F_0 by the condition on the function $g(x)$. Consider the operator acting from F_{-1} to F_1 with the domain of definition $L_2(0,T;H_1)$. Since $L_0+\lambda_0$ for $Re\lambda_0\geq 0$ is an isomorphism of $L_2(0,T;H_1)$ and $L_2(0,T;H_1')$, then $\widetilde{L_0}$ is an isomorphism of F_1 and F_{-1}.

The following assertion is valid [2, lemma 4.1, chapter 1]

Lemma 1. *Let $i\mathbb{R}\subset\rho(\widetilde{L_0})$. The following estimate of the resolvent $\|(\widetilde{L_0}+i\lambda)^{-1}f\|_{F_{-1}}\leq\frac{c}{1+|\lambda|}\|f\|_{F_{-1}}$ holds.*

Consider the operator $A=i\frac{\partial^{2m+1}}{\partial t^{2m+1}}$. Let H be a Hilbert space with the scalar product $(\cdot,\cdot)_H$. Put $D(A)=\{u\in W_2^{2m+1}(0,T,H):\ u^{(i)}(0)=u^{(i)}(T),\quad i=0,1,...,2m\}$.

Lemma 2. *The operator $A: L_2(0,T;H)\to L_2(0,T;H)$ is selfadjoint.*

We formulate a theorem on solvability of the problem (1)–(3). One has $L_{2,g}(a,b)\subset H_{-1}$. There is a self-adjoint operator A_0 in H_{-1} such that $D(A_0)=L_{2,g}(a,b)$, $\|A_0u\|_{H_{-1}}=\|u\|_{L_{2,g}(a,b)}$. Moreover A_0 is an isomorphism of $L_{2,g}(a,b)$ and H_{-1}. Then $\|v\|_{H_{-1}}=\|A_0^{-1}v\|_{L_{2,g}(a,b)}$. Equation (1) can be rewritten in the form

$$u^{(2m+1)}+\widetilde{L_0}u+M_0u=f, \tag{8}$$

where $M_0u=\sum\limits_{i=0}^{2m}u^{(i)}a_i(t)+\frac{1}{g(x)}\sum\limits_{j=0}^{2s-1}b_j(x,t)\frac{\partial^j u}{\partial x^j}$. In the following theorem, we consider the case when $M_0(u)\equiv 0$, that is the equation (8) has the form

$$u^{(2m+1)}+\widetilde{L_0}u=f. \tag{9}$$

Theorem 1. *Let $f\in L_2(0,T;F_{-1})$, $Re\lambda_0\geq 0$ and $M_0(u)\equiv 0$. Then there exists a unique solution of problem (9), (2), (3) from the class $u\in F_1$, $u^{(i)}\in(F_1,F_{-1})_{1-\theta_i,2}$ $(\theta_i=i/(2m+1), i=1,2,\dots,2m)$ and $u^{(2m+1)}\in F_{-1}$. The solution satisfies the estimate*

$$\|u\|_{F_1}+\sum_{i=1}^{2m}\|u^{(i)}\|_{(F_1,F_{-1})_{1-\theta_i,2}}+\|u^{(2m+1)}\|_{F_{-1}}\leq c\|f\|_{F_{-1}},$$

where c is some constant that does not depend on f.

We consider the problem of the solvability to (1)–(3). Assume that $\forall\varepsilon>0\ \exists c(\varepsilon)$, for which an inequality of the form

1) $\|\sum\limits_{j=0}^{2s-1}b_j(x,t)\frac{\partial^j u}{\partial x^j}\|_{F_1'}\leq\varepsilon\|u\|_{F_1}+c(\varepsilon)\|u\|_{L_2(Q)},\quad u\in D(L_0)$ takes place;

2) $a_j(t)\in L_{p_0}(0,T)$, $j=0,1,...,m$, $p_0>2$, $g(x)\in L_2(a,b)$.

Theorem 2. *Let the conditions of theorem 1 be satisfied and conditions 1), 2) hold. We fix $\varepsilon\in(0,\pi/2)$. Then there is $\lambda_1\geq 0$ such that for all $Re\lambda_0\geq\lambda_1$, $|\arg\lambda_0|\leq\pi/2-\varepsilon$, the problem (1)–(3) has a unique solution with properties mentioned in theorem 1.*

Remark 1. *The simplest conditions guaranteeing the fulfillment of condition 1) are as follows $b_j\equiv 0$ at $j\geq s$, $b_j(x,t)\in L_\infty((0,T)\times(a,b))$ at $j\leq m-1$.*

3. Parabolic equation

Consider the case $m = 0$, $s = 2$, corresponding to the parabolic equation fourth order with a changing direction of time associated with the application of the theory of singular integral equations [5–7] as well as systems of these equations [8]. Here, the results are refined and amplified, obtained in [12].

In the domain $Q^+ = \mathbb{R}^+ \times (0, T)$ we consider the system of equations

$$u_t^1 = Lu^1, \quad -u_t^2 = Lu^2 \quad \left(L \equiv -\frac{\partial^4}{\partial x^4}\right). \tag{10}$$

The solution of the system (10) is sought from the space Hölder $H_{x\,t}^{p,p/4}(Q^+)$, $p = 4l + \gamma$, $0 < \gamma < 1$, satisfying the initial conditions

$$u^1(x, 0) = \varphi_1(x), \quad u^2(x, T) = \varphi_2(x), \ x > 0, \tag{11}$$

and bonding conditions

$$T_1\,\vec{u}^{\,1}(0, t) = T_2\,\vec{u}^{\,2}(0, t), \tag{12}$$

where $\vec{u}^{\,k} = (u^k, u_x^k, u_{xx}^k, u_{xxx}^k)$, T_1, T_2 are nondegenerate matrices with constant real coefficients.

We assume that $\varphi_i(x) \in H^p(\mathbb{R})$ $(i = 1, 2)$. Then the functions

$$\omega_1(x, t) = \frac{1}{\pi} \int\limits_{\mathbb{R}} U_0(x, t; \xi, 0)\varphi_1(\xi)\,d\xi, \quad \omega_2(x, t) = \frac{1}{\pi} \int\limits_{\mathbb{R}} U_0(\xi, T; x, t)\varphi_2(\xi)\,d\xi$$

are solutions of the equations (10) satisfying conditions (11). We will use the integral representation of the solution for the system (10):

$$u^1(x, t) = \int\limits_0^t U_0(x, t; 0, \tau)\alpha_0(\tau)\,d\tau + \int\limits_0^t U_1(x, t; 0, \tau)\alpha_1(\tau)\,d\tau + \omega_1(x, t),$$

$$u^2(x, t) = \int\limits_t^T U_0(0, \tau; x, t)\beta_0(\tau)\,d\tau + \int\limits_t^T U_2(0, \tau; x, t)\beta_1(\tau)\,d\tau + \omega_2(x, t),$$

where U_i $(i = 0, 1, 2)$ are the fundamental and elementary solutions Bruno Pini.

The gluing conditions (12) can be rewritten in the form

$$\vec{u}^{\,1}(0, t) = (T_1^{-1} \cdot T_2)\,\vec{u}^{\,2}(0, t) = A\,\vec{u}^{\,2}(0, t), \tag{13}$$

where A is a nondegenerate matrix equal to $T_1^{-1}T_2$. Let the elements of the matrix C consist of real numbers a_{ij}. Consider the case of the matrix

$$A = \begin{pmatrix} a_{11} & 0 & 0 & a_{14} \\ 0 & a_{22} & 0 & a_{24} \\ 0 & 0 & a_{33} & a_{34} \\ 0 & 0 & 0 & a_{44} \end{pmatrix}. \tag{14}$$

We also note that in the case of symmetry of the matrix A, the conditions of work [12] are fulfilled. We assume that the coefficients a_{ij} of the matrix A satisfy the uniqueness condition for the boundary value problem (10), (11), (14).

IOP Conf. Series: Journal of Physics: Conf. Series **894** (2017) 012075 doi:10.1088/1742-6596/894/1/012075

Theorem 3. *Let the elements of the matrix A of the form* (14) *are different from zero, and* $a_{33} \neq \sqrt{2}a_{22}$, $\varphi_1, \varphi_2 \in H^p$ $(p = 4l + \gamma)$. *Then, if $4l$ conditions hold*

$$L_s(\varphi_1, \varphi_2) = 0, \quad s = 1, \ldots, 4l, \tag{15}$$

there exists a unique solution of the equation (10) *from space*

1) $H_{x\ t}^{p,p/4}$, *if* $0 < \gamma < 1 - 4\theta$;

2) $H_{x\ t}^{q,q/4}$, $q = 4l + 1 - 4\theta$, *if* $1 - 4\theta < \gamma < 1$;

3) $H_{x\ \ t}^{q-\varepsilon,(q-\varepsilon)/4}$, *if* $\gamma = 1 - 4\theta$, *where ε is an arbitrarily small positive constant, satisfying the conditions* (11), (13), $\theta = \frac{1}{\pi}\arctan|\frac{a}{b}| \in (0, \frac{1}{4})$, $a = a_{33}$, $b = a_{33} - \sqrt{2}a_{22}$.

Remark 2. *The proof of theorem 1 remains valid if in the general matrix of gluing* $A = \{a_{ij}\}$ *at least one of three elements* a_{13}, a_{14}, a_{24} *will be different from zero.*

If instead of the matrix A consider a matrix of the form

$$A = \begin{pmatrix} a_{11} & a_{12} & 0 & 0 \\ a_{21} & a_{22} & a_{23} & 0 \\ a_{31} & a_{32} & a_{33} & a_{34} \\ a_{41} & a_{42} & a_{43} & a_{44} \end{pmatrix},$$

then the theorem on unconditional solvability holds and there exists a unique solution of equation from space $H_{x\ t}^{p,p/4}(Q^+)$ (10) *satisfying the conditions* (11), (13), (15).

Acknowledgments

The work was supported by the Ministry of Education and Science of Russia (project 1.6069.2017/8.9).

References

[1] Kislov N V 1983 *Diff. Equat.* **19** 1427–36
[2] Egorov I E, Pyatkov S G and Popov S V 2000 *Nonclassical Operator–Differential Equations* (Novosibirsk: Nauka)
[3] Pyatkov S G 2002 *Operator Theory. Nonclassical Problems* (Utrecht, Boston, Köln, Tokyo: VSP)
[4] Tersenov S A 1985 *Parabolic Equations with Changing Time Direction* (Novosibirsk: Nauka)
[5] Gahov F D 1977 *Boundary Value Problems* (Moscow: Nauka)
[6] Mushelishvili N I 1968 *Singular Integral Equations* (Moscow: Nauka)
[7] Monakhov V N 1977 *Boundary-Value Problems with Free Boundaries for Elliptic Systems of Equations* (Novosibirsk: Nauka)
[8] Vekua N P 1977 *Systems of Singular Integral Equations* (Moscow: Nauka)
[9] Monakhov V N and Popov S V 2000 *Dynam. Cont. Med.* **116** 58–73
[10] Kislov N V and Pulkin I S 2000 *Moscow IPE* **6** 51–59
[11] Triebel H 1977 *Interpolation Theory, Function Spaces, Differential Operators* (Berlin: VEB)
[12] Popov S V 2005 *Reports Acad. Scien.* **400** 29–31

Global solvability of the initial boundary value problem for a model system of one-dimensional equations of polytropic flows of viscous compressible fluid mixtures

Dmitriy Prokudin[1,2]

[1] Lavrentyev Institute of Hydrodynamics, 630090 Novosibirsk, Russia
[2] Voronezh State University, 394018 Voronezh, Russia

E-mail: prokudin@hydro.nsc.ru

Abstract. We consider the initial boundary value problem for a model system of one-dimensional equations which describe unsteady polytropic motions of a mixture of viscous compressible fluids. We prove the global existence and uniqueness theorem for the strong solution without restrictions on the structure of the viscosity matrix except standard properties of symmetry and positiveness.

1. Statement of the problem, formulation of the result, the Lagrangian coordinates

We consider the problem of one-dimensional polytropic flow of a mixture which consists of $N \geq 2$ components. We work in the closure $\overline{Q}_T$ of the domain $Q_T = (0,1) \times (0,T)$, where $T > 0$ is an arbitrary positive number, and our aim is to find the density $\rho > 0$ of the mixture and the velocity u_i for each component of the mixture numbered by $i = 1, \ldots, N$, which satisfy the following system of equations, initial and boundary conditions [1]:

$$\partial_t \rho + \partial_x(\rho v) = 0, \quad v = \frac{1}{N}\sum_{i=1}^{N} u_i, \tag{1}$$

$$\rho\left(\partial_t u_i + v\partial_x u_i\right) + K\partial_x \rho^\gamma = \sum_{j=1}^{N} \mu_{ij}\partial_{xx}u_j + \sum_{j=1}^{N} a_{ij}(u_j - u_i), \quad i = 1, \ldots, N, \tag{2}$$

$$\rho|_{t=0} = \rho_0, \quad u_i|_{t=0} = u_{0i}, \quad i = 1, \ldots, N, \tag{3}$$

$$u_i|_{x=0} = u_i|_{x=1} = 0, \quad i = 1, \ldots, N. \tag{4}$$

Here v is the average velocity of the mixture; the values $K > 0$, $\gamma > 1$, $\mu_{ij} = \mu_{ji}$ and $a_{ij} = a_{ji} > 0$, $i,j = 1, \ldots, N$ are known constants, and the viscosity coefficients $\{\mu_{ij}\}_{i,j=1}^{N}$ compose the matrix $\mathbf{M} > 0$; the initial distributions ρ_0 and u_{0i}, $i = 1, \ldots, N$, are prescribed.

The aim of the paper is to prove the existence and uniqueness of the strong solution to the problem (1)–(4).

Content from this work may be used under the terms of the Creative Commons Attribution 3.0 licence. Any further distribution of this work must maintain attribution to the author(s) and the title of the work, journal citation and DOI.
Published under licence by IOP Publishing Ltd

Definition 1. *Strong solution to the problem (1)–(4) is called the collection of* $N+1$ *functions* $(\rho, u_1, \dots, u_N)$ *such that the equations (1), (2) are satisfied almost everywhere in* Q_T*, the initial data (3) are accepted for a.a.* $x \in (0,1)$*, the boundary conditions (4) are valid for a.a.* $t \in (0,T)$*, and the following inequality and inclusions hold*

$$\begin{gathered}\rho > 0, \quad \rho \in L_\infty(0,T;W_2^1(0,1)), \quad \partial_t\rho \in L_\infty(0,T;L_2(0,1)),\\ u_i \in L_\infty(0,T;W_2^1(0,1))\bigcap L_2(0,T;W_2^2(0,1)), \quad \partial_t u_i \in L_2(Q_T), \quad i=1,\dots,N.\end{gathered} \tag{5}$$

The main result of the paper is formulated as the following theorem.

Theorem 2. *Let the initial data in (3) satisfy the conditions*

$$\rho_0 \in W_2^1(0,1),\ \rho_0 > 0,\ u_{0i} \in W_2^1(0,1),\ u_{0i}|_{x=0} = u_{0i}|_{x=1} = 0,\ i=1,\dots,N\ (N \geq 2), \tag{6}$$

symmetric viscosity matrix $\boldsymbol{M}$ *is positive, the polytropic exponent* $\gamma > 1$*, all other numeric parameters* K*,* T *and* $a_{ij} = a_{ji}$*,* $i,j = 1,\dots,N$*, are positive.*

Then there exists the unique strong solution to the problem (1)–(4) in the sense of Definition 1.

Sketch of the proof of Theorem 2. The existence of unique strong solution to the problem (1)–(4) in a small time interval $[0,t_0]$ is proved in [2]. In order to extend this solution from the interval $[0,t_0]$ to the target interval $[0,T]$, we need to prove a priori estimates, in which the constants depend only on the input data of the problem and on the value T, but not on the small parameter t_0 (see, e.g., [3]). That is why we concentrate on the global estimates.

Lagrangian coordinates. While the problem (1)–(4) is studied, it is sometimes more convenient to use the Lagrangian coordinates. Let us consider $y(x,t) = \int\limits_0^x \rho(s,t)\,ds$ and t as new independent variables. Then the system (1), (2) turns into the form

$$\partial_t\rho + \rho^2\partial_y v = 0, \tag{7}$$

$$\partial_t u_i + K\partial_y\rho^\gamma = \sum_{j=1}^N \mu_{ij}\partial_y(\rho\partial_y u_j) + \frac{1}{\rho}\sum_{j=1}^N a_{ij}(u_j - u_i), \quad i=1,\dots,N. \tag{8}$$

The domain Q_T is mapped into the rectangular $\Pi_T = (0,d)\times(0,T)$, where $d = \int\limits_0^1 \rho_0(x)\,dx > 0$, and the initial and boundary conditions accept the form

$$\rho|_{t=0} = \widetilde{\rho}_0, \quad u_i|_{t=0} = \widetilde{u}_{0i}, \quad i=1,\dots,N, \tag{9}$$

$$u_i|_{y=0} = u_i|_{y=d} = 0, \quad i=1,\dots,N. \tag{10}$$

2. Global a priori estimates

Let us multiply the equations (2) by u_i, integrate the result over $(0,1)$ and sum over $i=1,\dots,N$. Due to (1), (4) and the condition $\mathbf{M} > 0$, we following relations hold

$$\sum_{i=1}^N \int\limits_0^1 \Big(\rho\partial_t u_i + \rho v\partial_x u_i\Big)u_i\,dx = \frac{1}{2}\frac{d}{dt}\left(\sum_{i=1}^N \int\limits_0^1 \rho u_i^2\,dx\right), \tag{11}$$

$$K\sum_{i=1}^N \int\limits_0^1 u_i\left(\partial_x\rho^\gamma\right)dx = -KN\int\limits_0^1 \rho^\gamma\left(\partial_x v\right)dx = \frac{KN}{\gamma-1}\frac{d}{dt}\left(\int\limits_0^1 \rho^\gamma\,dx\right), \tag{12}$$

$$\sum_{i,j=1}^{N}\mu_{ij}\int_0^1(\partial_{xx}u_j)u_i\,dx=-\sum_{i,j=1}^{N}\mu_{ij}\int_0^1(\partial_x u_i)(\partial_x u_j)\,dx\le -C_0(\mathbf{M})\sum_{i=1}^{N}\int_0^1|\partial_x u_i|^2\,dx, \tag{13}$$

$$\sum_{i,j=1}^{N}a_{ij}\int_0^1(u_j-u_i)u_i\,dx=-\frac{1}{2}\sum_{i,j=1}^{N}a_{ij}\int_0^1(u_i-u_j)^2\,dx, \tag{14}$$

and we come to the inequality

$$\frac{d}{dt}\sum_{i=1}^{N}\int_0^1\left(\frac{1}{2}\rho u_i^2+\frac{K}{\gamma-1}\rho^\gamma\right)dx+C_0\sum_{i=1}^{N}\int_0^1|\partial_x u_i|^2\,dx+\frac{1}{2}\sum_{i,j=1}^{N}a_{ij}\int_0^1(u_i-u_j)^2\,dx\le 0. \tag{15}$$

Let us agree that $C_k(\cdot)$, $k=0,1,\ldots$, stand for the quantities which take finite positive values and depend on the objects listed in the parentheses. When we integrate the inequality (15) over $(0,t)$ using (3), we obtain the estimate

$$\begin{gathered}\sum_{i=1}^{N}\int_0^1\left(\frac{1}{2}\rho u_i^2+\frac{K}{\gamma-1}\rho^\gamma\right)dx+C_0\sum_{i=1}^{N}\int_0^t\int_0^1|\partial_x u_i|^2\,dxd\tau+\\ +\frac{1}{2}\sum_{i,j=1}^{N}a_{ij}\int_0^t\int_0^1(u_i-u_j)^2\,dxd\tau\le\sum_{i=1}^{N}\int_0^1\left(\frac{1}{2}\rho_0u_{0i}^2+\frac{K}{\gamma-1}\rho_0^\gamma\right)dx,\end{gathered} \tag{16}$$

which leads to the fact that

$$\begin{gathered}\sum_{i=1}^{N}\left(\|\sqrt{\rho}u_i\|_{L_\infty(0,T;L_2(0,1))}+\|\partial_x u_i\|_{L_2(Q_T)}+\sum_{j=1}^{N}\|u_i-u_j\|_{L_2(Q_T)}\right)+\\ +\|\rho\|_{L_\infty(0,T;L_\gamma(0,1))}\le C_1\left(\left\{\|\sqrt{\rho_0}u_{0i}\|_{L_2(0,1)}\right\},\|\rho_0\|_{L_\gamma(0,1)},\{a_{ij}\},K,\mathbf{M},N,\gamma\right).\end{gathered} \tag{17}$$

The estimates (17), due to (10), lead to the inequality

$$\sum_{i=1}^{N}\|u_i\|_{L_2(0,T;L_\infty(0,1))}\le C_1. \tag{18}$$

The estimate (17) in the Lagrangian coordinates takes the form

$$\begin{gathered}\sum_{i=1}^{N}\left(\|u_i\|_{L_\infty(0,T;L_2(0,d))}+\|\sqrt{\rho}\partial_y u_i\|_{L_2(\Pi_T)}+\right.\\ \left.+\sum_{j=1}^{N}\|(u_i-u_j)/\sqrt{\rho}\|_{L_2(\Pi_T)}\right)+\|\rho\|_{L_\infty(0,T;L_{\gamma-1}(0,d))}\le C_2(C_1,\gamma).\end{gathered} \tag{19}$$

The next step is to prove the positiveness and boundedness of the density ρ. Here we use the equations (7), (8). Let us rewrite the equations (8) in the form

$$\sum_{j=1}^{N}\widetilde{\mu}_{ij}\partial_t u_j+K\left(\sum_{j=1}^{N}\widetilde{\mu}_{ij}\right)\partial_y\rho^\gamma=\partial_y(\rho\partial_y u_i)+\frac{1}{\rho}\sum_{j=1}^{N}\widetilde{\mu}_{ij}\left(\sum_{k=1}^{N}a_{jk}(u_k-u_j)\right),\quad i=1,\ldots,N, \tag{20}$$

IOP Conf. Series: Journal of Physics: Conf. Series **894** (2017) 012076 doi:10.1088/1742-6596/894/1/012076

where $\widetilde{\mu}_{ij}$ are the entries of the matrix $\widetilde{\mathbf{M}} = \mathbf{M}^{-1} > 0$, then sum (20) over $i = 1, \ldots, N$, and divide by N. Then we come to the equality

$$\partial_t V + \widetilde{K}\partial_y \rho^\gamma = \partial_y(\rho\partial_y v) + \frac{1}{N\rho}\sum_{i,j=1}^{N}\widetilde{\mu}_{ij}\left(\sum_{k=1}^{N} a_{jk}(u_k - u_j)\right), \tag{21}$$

where $V = \frac{1}{N}\sum_{i,j=1}^{N}\widetilde{\mu}_{ij}u_j$ and $\widetilde{K} = \frac{K}{N}\sum_{i,j=1}^{N}\widetilde{\mu}_{ij} > 0$. Let us use (7) in order to express

$$\rho\partial_y v = -\partial_t \ln\rho, \tag{22}$$

and substitute this relation into (21):

$$\partial_{ty}\ln\rho + \widetilde{K}\partial_y\rho^\gamma = -\partial_t V + \frac{1}{N\rho}\sum_{i,j=1}^{N}\widetilde{\mu}_{ij}\left(\sum_{k=1}^{N} a_{jk}(u_k - u_j)\right). \tag{23}$$

Let us multiply this equality by $\partial_y \ln\rho =: w$ and integrate over $y \in (0, d)$, then we obtain the equality

$$\frac{1}{2}\frac{d}{dt}\left(\int_0^d w^2\,dy\right) + \widetilde{K}\gamma\int_0^d \rho^\gamma w^2\,dy = -\int_0^d (\partial_t V)\,w\,dy + \frac{1}{N}\sum_{i,j,k=1}^{N}\widetilde{\mu}_{ij}a_{jk}\int_0^d \frac{(u_k - u_j)w}{\rho}\,dy. \tag{24}$$

We transform the first summand in the right-hand side of (24) via integration by parts and using (22):

$$-\int_0^d (\partial_t V)\,w\,dy = -\frac{d}{dt}\left(\int_0^d V w\,dy\right) + \int_0^d \rho(\partial_y v)(\partial_y V)\,dy, \tag{25}$$

and we estimate the second summand from above:

$$\frac{1}{N}\sum_{i,j,k=1}^{N}\widetilde{\mu}_{ij}a_{jk}\int_0^d \frac{(u_k - u_j)w}{\rho}\,dy \le C_3(\{a_{jk}\}, \widetilde{\mathbf{M}})\|1/\sqrt{\rho}\|_{L_\infty(0,d)}\|w\|_{L_2(0,d)}\times \times\sum_{j,k=1}^{N}\|(u_k - u_j)/\sqrt{\rho}\|_{L_2(0,d)}. \tag{26}$$

It is obvious, due to the equation (7) and the conditions (9), (10), that for every $t \in [0, T]$

$$\rho(z(t), t) = d \tag{27}$$

at least in one point $z(t) \in [0, d]$. Hence, we can use the representation

$$\frac{1}{\sqrt{\rho(y,t)}} = \frac{1}{\sqrt{\rho(z(t),t)}} + \int_{z(t)}^{y}\partial_s\rho^{-\frac{1}{2}}(s,t)\,ds = d^{-\frac{1}{2}} - \frac{1}{2}\int_{z(t)}^{y}\rho^{-\frac{1}{2}}(s,t)\partial_s\ln\rho(s,t)\,ds, \tag{28}$$

from which, using Hölder's inequality and (27), we obtain

$$\|1/\sqrt{\rho}\|_{L_\infty(0,d)} \le d^{-\frac{1}{2}} + \frac{1}{2}\|w\|_{L_2(0,d)}. \tag{29}$$

Hence, after the integration of (24) over $(0,t)$, using (25), (26) and (29), we come to the inequality

$$\|w\|^2_{L_2(0,d)} + 2\widetilde{K}\gamma \int_0^t \int_0^d \rho^\gamma w^2 \, dy d\tau \leq \|w_0\|^2_{L_2(0,d)} - 2\int_0^d Vw \, dy +$$

$$+2\int_0^d V_0 w_0 \, dy + 2\int_0^t \|\rho(\partial_y V)(\partial_y v)\|_{L_1(0,d)} \, d\tau + \tag{30}$$

$$+C_3 \sum_{j,k=1}^N \int_0^t \|(u_k - u_j)/\sqrt{\rho}\|_{L_2(0,d)} \|w\|_{L_2(0,d)} (2d^{-\frac{1}{2}} + \|w\|_{L_2(0,d)}) \, d\tau,$$

where $w_0 = w(0,t)$ and $V_0 = V(0,t)$. Using Cauchy's inequality and the estimate (19), we derive from the last formula that

$$\|w\|^2_{L_2(0,d)} \leq C_4 + C_5 \sum_{j,k=1}^N \int_0^t \|(u_k - u_j)/\sqrt{\rho}\|_{L_2(0,d)} \|w\|^2_{L_2(0,d)} \, d\tau, \tag{31}$$

where $C_4 = C_4\Big(C_2, C_3, \{\|\widetilde{u}_{0j}\|_{L_2(0,d)}\}, \|w_0\|_{L_2(0,d)}, \widetilde{\mathbf{M}}, N, T, d\Big)$ and $C_5 = C_5(C_3)$. Since (19) leads to the estimate

$$\sum_{j,k=1}^N \int_0^t \|(u_k - u_j)/\sqrt{\rho}\|_{L_2(0,d)} \, d\tau \leq C_6(C_2, T) \quad \forall t \in [0,T], \tag{32}$$

then the Gronwall lemma provides

$$\|w(t)\|_{L_2(0,d)} \leq C_6(C_4, C_5, C_6) \quad \forall t \in [0,T], \tag{33}$$

i. e. the norm of the derivative $\partial_y \ln \rho$ in $L_2(0,d)$ is bounded uniformly in $t \in [0,T]$. Hence, due to the representation (see the proof of (28))

$$\ln \rho(y,t) = \ln \rho(z(t),t) + \int_{z(t)}^y \partial_s \ln \rho(s,t) \, ds,$$

we have

$$|\ln \rho(y,t)| \leq |\ln d| + \sqrt{d}\|w\|_{L_2(0,d)} \leq C_7(C_6, d), \tag{34}$$

and consequently

$$0 < C_8^{-1}(C_7) \leq \rho(y,t) \leq C_8(C_7). \tag{35}$$

Now we possess the boundedness and positiveness of the density ρ, and the remaining a priori estimates can be obtained in the original (Eulerian) coordinates (x,t). Thus, from (33) and (35) we deduce

$$\|\partial_x \rho(t)\|_{L_2(0,1)} \leq C_9(C_6, C_8) \quad \forall t \in [0,T]. \tag{36}$$

Then, let us square the momentum equations (2) and sum over $i = 1, \ldots, N$, as a result we obtain

$$\sum_{i=1}^N \rho(\partial_t u_i)^2 + \frac{1}{\rho}\sum_{i=1}^N \left(\sum_{j=1}^N \mu_{ij}\partial_{xx}u_j\right)^2 - 2\sum_{i=1}^N (\partial_t u_i)\left(\sum_{j=1}^N \mu_{ij}\partial_{xx}u_j\right) =$$

$$= \sum_{i=1}^N \frac{1}{\rho}\left(\sum_{j=1}^N a_{ij}(u_j - u_i) - K\partial_x\rho^\gamma - \rho v \partial_x u_i\right)^2. \tag{37}$$

Let us introduce the function $\alpha(t)$ as

$$\alpha(t) = \sum_{i,j=1}^{N} \mu_{ij} \int_0^1 (\partial_x u_i)(\partial_x u_j)\, dx + \sum_{i=1}^{N} \int_0^t \int_0^1 \left(\rho(\partial_t u_i)^2 + \frac{1}{\rho} \left(\sum_{j=1}^{N} \mu_{ij} \partial_{xx} u_j \right)^2 \right) dx d\tau.$$

Then (37) and the inequalities (17), (35), (36) lead to the estimate

$$\begin{aligned} \alpha'(t) \le C_{10} + C_{11} \left(\sum_{j=1}^{N} \|u_j\|^2_{L_\infty(0,1)} \right) \left(\sum_{i,j=1}^{N} \mu_{ij} \int_0^1 (\partial_x u_i)(\partial_x u_j)\, dx \right) \le \\ \le C_{10} + C_{11} \left(\sum_{j=1}^{N} \|u_j\|^2_{L_\infty(0,1)} \right) \alpha(t), \end{aligned} \tag{38}$$

where $C_{10} = C_{10}(C_1, C_8, C_9, \{a_{ij}\}, K, N, \gamma)$ and $C_{11} = C_{11}(C_8, \mathbf{M})$, from which via the Gronwall lemma (see also (18)) it follows that

$$\alpha(t) \le C_{12} \left(C_1, C_{10}, C_{11}, \{\|u'_{0i}\|_{L_2(0,1)}\}, \mathbf{M}, N, T \right).$$

Using this and (35), we come to the inequality

$$\sum_{i=1}^{N} \left(\|\partial_x u_i\|_{L_\infty(0,T;L_2(0,1))} + \|\partial_{xx} u_i\|_{L_2(Q_T)} + \|\partial_t u_i\|_{L_2(Q_T)} \right) \le C_{13}, \tag{39}$$

where $C_{13} = C_{13}\,(C_8, C_{12}, \mathbf{M}, N)$. Finally, the continuity equation (1) and the inequalities (35), (36) and (39) provide

$$\|\partial_t \rho\|_{L_\infty(0,T;L_2(0,1))} \le C_{14}(C_8, C_9, C_{13}). \tag{40}$$

Theorem 2 is proved.

Acknowledgments

This research was supported by the Ministry of Education and Science of the Russian Federation (grant 14.Z50.31.0037).

References

[1] Mamontov A and Prokudin D 2017 *Sib. Elect. Math. Reports* **14** 388
[2] Prokudin D 2017 *Sib. Elect. Math. Reports* **14** 568
[3] Antontsev S N, Kazhikhov A V and Monakhov V N 1990 *Boundary Value Problems in Mechanics of Nonhomogeneous Fluids* (Amsterdam: North–Holland Publishing Co.)

IOP Conf. Series: Journal of Physics: Conf. Series **894** (2017) 012077 doi:10.1088/1742-6596/894/1/012077

Detonation velocity in poorly mixed gas mixtures

E S Prokhorov

Lavrentyev Institute of Hydrodynamics SB RAS, 15 prospect Lavrentyeva, Novosibirsk, 630090, Russia

E-mail: prokh@hydro.nsc.ru

Abstract. The technique for computation of the average velocity of plane detonation wave front in poorly mixed mixture of gaseous hydrocarbon fuel and oxygen is proposed. Here it is assumed that along the direction of detonation propagation the chemical composition of the mixture has periodic fluctuations caused, for example, by layered stratification of gas charge. The technique is based on the analysis of functional dependence of ideal (Chapman-Jouget) detonation velocity on mole fraction (with respect to molar concentration) of the fuel. It is shown that the average velocity of detonation can be significantly (by more than 10%) less than the velocity of ideal detonation. The dependence that permits to estimate the degree of mixing of gas mixture basing on the measurements of average detonation velocity is established.

1. Introduction

The velocity of detonation wave front is the most important and significant characteristic of explosive substance [1]. For explosive gases there are well-proven techniques (for example, described in [2]) for computation of parameters of steady-state ideal (without energy losses due to friction and heat exchange) detonation frequently called the Chapman-Jouget detonation [3]. The above-mentioned techniques are based on the assumption concerning chemical equilibrium of the composition of detonation combustion products (detonation products). The equilibrium computations are in good agreement with experimental data. Particularly, this concerns the velocity of detonation front D accurately measured in experiments.

The velocity of detonation D mainly depends on chemical composition of explosive gas constituting the mixture of gaseous fuel and oxidizer, which is usually oxygen or air. Taking into account only the value D, it is easy to estimate the values of other parameters of detonation products such as pressure, density, mass velocity and sound velocity [4].

It is evident that the constancy of value D in practice is available only in case of ideal mixing of gas mixture. When detonation wave propagates in a poorly mixed mixture with fluctuations of chemical composition, it is properly to speak of the average velocity of detonation $\overline{D}$ because the velocity of detonation front will also undergo fluctuations. The question to what extent $\overline{D}$ may differ from the velocity of detonation D in ideally mixed mixture remains open.

Such a problem is of interest because it concerns explosion safety problems, development and modernization of different technical devices with detonation products as working material [5, 6]. In such devices with continuous or pulse (pulsating) operation, the explosive gas mixture is formed

Content from this work may be used under the terms of the Creative Commons Attribution 3.0 licence. Any further distribution of this work must maintain attribution to the author(s) and the title of the work, journal citation and DOI.
Published under licence by IOP Publishing Ltd

under conditions of flow supply of fuel and oxidizer through two independent channels. It is evident that this does not provide complete mixing of all mixture components.

To estimate the degree of mixing influence on detonation parameters, a simplified scheme of computations is proposed in [7]. The computations imply that the mixture with inhomogeneous chemical composition is the system of elementary volumes uniformly distributed in space. Each volume consists of three parts: ideally mixed mixture, pure fuel and pure oxidizer. Here, pure components do not take part in chemical reaction since they are inert gas. By the degree of mixing we mean the fraction of ideally mixed mixture β in elementary volume. By [7], the change of detonation velocity D_β of inhomogeneous-composition mixture from β can be expressed in the form of approximation formula: $D_\beta / D = \sqrt{\beta}$, where D is the detonation velocity in an ideally mixed mixture.

In [8], by using the same assumptions and more accurate equilibrium computations of detonation in mixtures based on hydrocarbon fuels for values $\beta = 0.6 \div 1$, this dependence was refined: $D_\beta / D = \beta^{0.295\pm0.005}$. However, as the author of this work points out, the model of poorly mixed mixture used in computation is somewhat artificial since the division of elementary volume of the mixture into three parts, which are enormously different in properties, is sufficiently rough. It is more correct to consider the model when the first part of elementary volume contains the mixture of given composition, the second part contains a somewhat fuel-enriched explosive mixture and the third one contains a somewhat lean explosive mixture.

2. Statement of the problem

Let a plane detonation wave propagate along the coordinate axis x over(in) the explosive mixture consisting of gaseous hydrocarbon fuel (denote it by C_yH_z) and oxygen O_2. The chemical composition of the mixture can be described as

$$nC_yH_z + (1-n)O_2, \tag{1}$$

where n is a mole fraction (a relative molar concentration) of fuel in the mixture. In the case when this mixture is stoichiometric, n will equal to the value $n_{st} = 4/(4y+z+4)$.

For mixtures of type (1), the technique detailed in [2] makes it possible to compute the velocity of ideal detonation D while $n < n_*$ and the products of chemical reaction do not contain pure carbon C, i.e., when oxygen is still sufficient in the mixture to oxygenate C to carbon monoxide CO. It follows from (1) that the limiting value of the mole fraction of fuel is $n_* = 2/(y+2)$. Thus, in what follows we shall assume that within the interval $0 < n < n_*$ the functional dependence $D = D(n)$ is known. Note that the detonation velocity D coincides with sound velocity in a pure oxidizer (in this case, oxygen) as $n \to 0$.

Suppose that the gas mixture is poorly mixed and there are periodic fluctuations of the mole fraction of fuel n along the axis x. To describe the propagation of detonation wave in such a system, we divide the domain occupied by a gas mixture into elementary volumes located in sequence.

Let a longitudinal (along the axis x) dimension of these volumes be equal to L and coincide with spatial period of pulsations n. Then to describe the propagation of the mole fraction of fuel $n = n(x)$, it suffices to give this propagation inside one elementary volume. According to recommendations in [8], such a distribution can be represented as piecewise constant function

$$n(x) = \begin{cases} (1+\lambda)n, & \text{at} \quad 0 \le x < L/3 \\ n, & \text{at} \quad L/3 \le x < 2L/3 \\ (1-\lambda)n, & \text{at} \quad 2L/3 \le x \le L, \end{cases} \tag{2}$$

where dimensionless parameter λ ($\lambda \ge 0$) characterizes the amplitude of fluctuations of gas- mixture chemical composition. Thus, at $\lambda = 0$, the mixture will be assumed to be ideally mixed. It follows from (1) and (2) that fluctuations n must result in conformal changes of relative mole oxygen fraction within the range from $(1-n-\lambda n)$ to $(1-n+\lambda n)$ inside the elementary volume. Note that (2) satisfies the condition

$$\overline{n}=\frac{1}{L}\int_0^L n(x)dx=n\ , \tag{3}$$

where $\overline{n}$ is the average value of molar concentration of fuel inside the elementary volume and here deviations of $n(x)$ with respect to(from) $\overline{n}$ do not exceed the amplitude value λn for the fluctuations of chemical composition.

Combining dependences (2) and $D=D(n)$, for poorly mixed gas mixture ($\lambda>0$), it is possible to estimate the average detonation velocity $\overline{D}$ at length L by using an evident relation:

$$\frac{1}{\overline{D}}=\frac{1}{L}\int_0^L\frac{dx}{D}. \tag{4}$$

When varying the mole fraction of fuel n, such an approach permits to construct functional dependence $\overline{D}=\overline{D}(n)$ within the interval $0<n<\overline{n}_*$, where $\overline{n}_*=n_*/(1+\lambda)$. The latter relation is conditioned by the constraints of the used model since the value n must not exceed n_* at different degrees of mixing of gas mixture. However, these constraints are not so principal since it is always possible to use known results of detonation computations at $n>n_*$ (see, for example, [9]) performed by using the computer program "Safety" [10].

3. Results of computations

The confirmation of estimate of $\overline{D}$ is performed for detonation in fuel-oxygen mixture (FOM) based on methane $nCH_4+(1-n)O_2$ with spatial fluctuations of its chemical composition according to (2) under standard initial conditions: pressure is $p_0=1$ atm and temperature is $T_0=298.15$ K. For this mixture $n_{st}=1/3$ and the limiting value for the mole fraction of fuel is $n_*=2/3$.

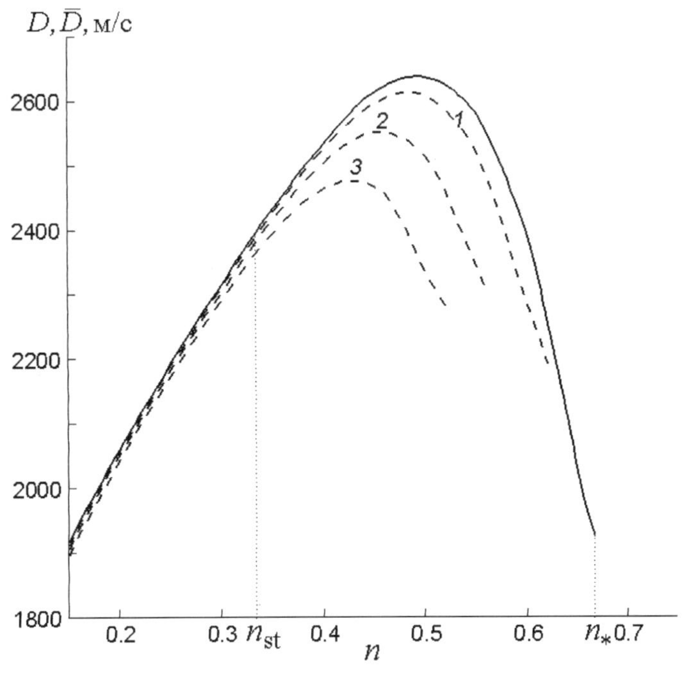

Figure 1. The velocity of ideal detonation D (solid line) and the average velocity of detonation $\overline{D}$ (dashed line) of poorly mixed mixture $nCH_4+(1-n)O_2$ versus the mole fraction of methane n at different values of dimensionless parameter λ: Curve 1 — $\lambda=0.1$; 2 — $\lambda=0.2$; 3 — $\lambda=0.3$

Figure 1 presents the computational dependences of ideal detonation velocity $D=D(n)$ and average detonation velocity $\overline{D}=\overline{D}(n)$ obtained by using (4) for different degrees of mixing of fuel and oxidizer. It is evident that at any value of the mole fraction of fuel n the average velocity of detonation (Curves 1-3) is less than the velocity of ideal detonation (shown by a solid line in the figure). At $n<0.3$, the deviations of $\overline{D}$ from D are slight (less than 1%) even in the case of relatively large fluctuations of chemical composition of mixture $\lambda=0.1\div0.3$. However, in the domain $n_{st}<n<n_*$ the difference between $\overline{D}$ and D becomes rather essential. This is due to the fact that within a specified range of molar concentrations of methane, function $D=D(n)$ behaves nonmonotonically and has a characteristic maximum approximately at $n=0.5$ when the velocity of ideal detonation D reaches the value of 2635 m/s. Note that in the vicinity of the extremum of function $D=D(n)$ relation $\overline{D}/D$ will be minimal. The table presents the values of $\overline{D}/D$ obtained for the extreme point ($n=0.5$) at different values of parameter λ. The worse the gas mixture is mixed (λ rises), the less its $\overline{D}$ is. Thus, at $\lambda=0.3$, the average velocity of detonation is less than detonation velocity D for ideally mixed mixture ($\lambda=0$) by more than 11 %.

An additional numerical research showed that dependences $D=D(n)$ and $\overline{D}=\overline{D}(n)$ obtained for FOM on the base of other hydrocarbon fuels, e.g., acetylene C_2H_2, propane C_3H_8 and butane C_4H_{10}, have analogous behavior. Computations for mixture $nC_2H_2+(1-n)O_2$ point to the presence of maximum $D\approx2934$ m/s in the vicinity of equimolar relation ($n=0.5$) and for mixture $nC_3H_8+(1-n)O_2$ the velocity of ideal detonation is maximal ($D\approx2609$ m/s) at mole fraction of fuel $n=0.29$. For mixture $nC_4H_{10}+(1-n)O_2$, at the extreme point we have $D\approx2601$ m/s at $n=0.23$. The corresponding results of computations of $\overline{D}/D$ with varying parameter λ for these FOMs are shown in the table 1.

Table 1. Detonation velocity in poorly mixed gas mixtures

Mixture	n	D, m/s	$\overline{D}/D$					
			$\lambda=0.05$	$\lambda=0.1$	$\lambda=0.15$	$\lambda=0.2$	$\lambda=0.25$	$\lambda=0.3$
$nCH_4+(1-n)O_2$	0.5	2635	0.997	0.989	0.974	0.954	0.923	0.886
$nC_2H_2+(1-n)O_2$	0.5	2934	0.985	0.967	0.949	0.930	0.913	0.890
$nC_3H_8+(1-n)O_2$	0.29	2609	0.999	0.995	0.989	0.979	0.967	0.953
$nC_4H_{10}+(1-n)O_2$	0.23	2601	0.999	0.996	0.991	0.984	0.975	0.964

In practical use of FOM detonation, it is domain $n_{st}<n<n_*$, where the nonmonotonical behavior of dependence $D=D(n)$ is most pronounced, that is the most demanded. Therefore, the deficiency of detonation velocity $\Delta D/D=1-\overline{D}/D$ observed in experiments may serve as a ground for the fact that the gas mixture is poorly mixed.

Figure 2 presents the computational results $\Delta D/D$ depending on the amplitude of fluctuations of mole fuel fraction λn for values n shown in the table. It is seen that computational data are well grouped near the curve (dashed line), which can be described by the approximation formula

$$\Delta D/D=(1.5\lambda n)^{1.5}. \tag{5}$$

Dependence (5) makes it possible to solve an inverse problem, namely: basing on the measurements of detonation velocity, to estimate the degree of mixing of mixture $\overline{\beta}$ that can be defined as $\overline{\beta}=1-\lambda n$.

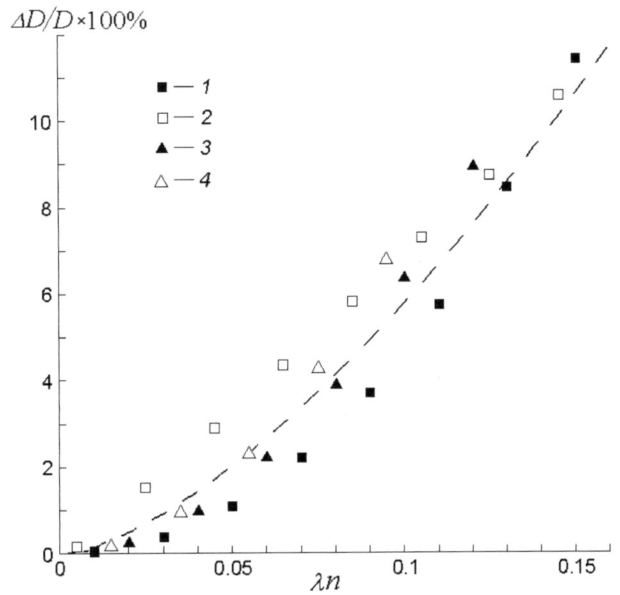

Figure 2. The estimate of deficiency of detonation velocity $\Delta D/D$ versus the amplitude of fluctuations of mole fuel fraction λn for different FOMs. Points: 1 – mixture $nCH_4+(1-n)O_2$; 2 — mixture $nC_2H_2+(1-n)O_2$; 3 — mixture $nC_3H_8+(1-n)O_2$; 4 — mixture $nC_4H_{10}+(1-n)O_2$. Dependence (5) is shown by a dashed lined.

4. Conclusion

The influence of periodic chemical-composition fluctuations of different intensity degree on the average detonation velocity in poorly mixed oxygen explosive mixtures based on hydrocarbon fuels is numerically investigated. There exists a range of relative molar fuel concentrations, which admit considerable deviations (to the smaller side) of the value of such a velocity from the detonation velocity of ideally mixed mixture. This is due to the fact that the computational dependence of ideal detonation velocity on the mole fuel fraction behaves nonmonotonically and has a characteristic maximum. For poorly mixed gas mixtures, the approximation formula describing the correlation between the average detonation velocity and the amplitude of fluctuations of chemical composition is obtained.

References

[1] Baum F M, Orlenko L P, Stanyukovich K P, Chelyshev V P and Shehter B I 1975 *Physics of Explosions* (Moscow: Nauka)

[2] Nikolaev Yu A., Topchiyan M E 1977 *Combust. Explos. Shock Waves* **13** 327

[3] Landau L D and Livshits E M *Course of Theoretical Physics, vol. 6: Fluids Mechanics* 1987 (Oxford: Pergamon)

[4] Prokhorov E S 2015 *J.Eng.Phys.Termophys.* **88** 730

[5] Bazhenova T V and Golub V V 2003 *Combust. Explos. Shock Waves* **39** 361

[6] Nikolaev Yu A, Vasil'ev A A and Ul'yanitskii V Yu 2003 *Combust. Explos. Shock Waves* **39** 382

[7] Bykovskii F A 1983 *Continuum dynamics* **62** 23

[8] Mikhalkin V N 1996 *Combust. Explos. Shock Waves* **32** 57

[9] Vasiliev A A, Drozdov M S and Khidirov S G 2006 *Combust. Explos. Shock Waves* **42** 746

[10] Vasiliev A A, Valishev A I, Vasil'ev V A and Panfilova L V 2000 *Combust. Explos. Shock Waves* **36** 358

IOP Conf. Series: Journal of Physics: Conf. Series **894** (2017) 012078 doi:10.1088/1742-6596/894/1/012078

Determination of optimum parameters of the technological process for plates forming from V95 and V-1461 alloys in creep applied in aircrafts constructed by "Sukhoi design bureau"

G A Raevskaya[1], **K Zakharchenko** [2,3], **A Larichkin** [2,4]

[1]Scientific Technical Center "NTC-Polzuchest", Novosibirsk, Russia
[2]Institute of Hydrodynamics named after M. A. Lavrentyev, SB RAS Novosibirsk, Russia
[3]Novosibirsk State Technical University, Novosibirsk, Russia
[4]Novosibirsk State University, Novosibirsk, Russia

E-mail: polsutchest@mail.ru

Abstract. The research is devoted to the scientific justification of metal processing by pressure with the help of thick monolithic plates forming (thickness 40 mm) from the V95 (analog 7475) (Al-Zn-Mg-Cu) and V-1461 (analog 2099) (Al-Cu- Li-Zn) alloys in creep and close-to-superplasticity. Optimum parameters of the technological process of plate forming are described. The effect of temperature on the magnitude of mechanical stresses (relaxation) during the tests of materials on pure bending is experimentally determined. Forming of thick plates (40 mm) on the UFP-1M unit, and the control of the obtained surface, in comparison with the given electronic model, made it possible to experimentally determine the time and number of forming stages. Mechanical properties of the material after the technological process and heat treatment are preliminary evaluated. The efficiency of using the obtained parameters of the technological process and treatment of metals by pressure in such methods in general is shown.

1. Introduction

Since the middle of the 20th century the high-strength V95och aluminum alloy (analog of 7475) has traditionally become the main material to manufacture power elements of the airframe. However, in recent years, the prospect of using a high-strength aluminum-lithium alloy of the V-1461 grade (analog of 2099) instead of V95och in the design of new aircrafts has been widely studied [1]. The mechanical properties of V-1461 exceed those of V95och: the density is 25% lower; modulus of elasticity – 9% higher; specific strength - 11% higher; yield strength - 14% higher; heat and corrosion resistance - 1.5-2 times higher [2].

One of the main objectives of the aviation industry development is to reduce the weight of an aircraft. It is possible to solve this problem in two ways: 1 – apply parts made from aluminum alloys of reduced density, doped with lithium, each per cent of which reduces the density by 3% and increases the modulus of elasticity by 6% [1, 3, 4]; 2) apply creep and close-to-superplasticity processes in the metals treatment by pressure, which ensures the resource conservation. This makes it possible to reduce the construction weight [5].

Content from this work may be used under the terms of the Creative Commons Attribution 3.0 licence. Any further distribution of this work must maintain attribution to the author(s) and the title of the work, journal citation and DOI.
Published under licence by IOP Publishing Ltd

Today Russian aviation plants are experienced in obtaining a number of structural elements due to irreversible creep deformations developing in time at elevated temperatures [6]. Metal treatment under pressure in creep is a new technology of deformation of monolithic panels with large thickness (more than 20 mm) in the panel of double Gaussian curvature. This forming method was developed by the Novosibirsk branch of NIAT (National Institute of Aviation Technologies) with the participation of Institute of Hydrodynamics named after M. A. Lavrentyev SB RAS and Novosibirsk Aircraft Plant named after Chkalov - a branch of PJSC Sukhoi Company [7-9].

The technological process of forming in creep and in close-to-superplasticity is implemented at the UFP-1M unit, which provides the specified accuracy of the product (Figure 1).

The purpose of the research is to scientifically justify the forming modes in creep and verify the conformity of the mechanical properties of the plate material after forming.

2. Samples, equipment and method

2.1. Determination of material parameters

In order to determine the optimum temperatures and efforts for plates forming, samples in the form of beams of rectangular section (8x19x300 mm) were used, cut from the plate, milled. The samples were deformed by pure bending to the values of the deflection of 23.5-26 mm on the base of 290 mm by applying a constant bending moment over the length of the sample. The maximum bending moment is 30000 kg·mm. The design of the equipment completely eliminates the loss of the active bending moment, the scheme is given in [11]. The test temperature of the samples is constant in time; the test temperature was in the range from 360 to 470 °C.

The pure bend ensures uniform heating of the samples to a specified temperature, a different loading level, an automatic heating rate, and control of the loading, deflection, and temperature.

The plate forming in creep was carried out at the UFP-1M unit.

2.2. The technological process of forming a plate of complex geometry

The technological process of plate forming (1800x800x40 mm) in creep at the UFP-1M unit is a method of deformation without deformation hardening. Forming is carried out by moving a multitude of punches (rods), the ends of which form a discrete surface of the matrix. The movements of each rod are set individually by a computer program using servo drivers. Such a scheme of equipment eliminates the need for having an individual matrix for each part, and makes the unit universal due to the possibility of adjusting the system to a given curvature.

Slow mode with a given kinematics of deformation of the workpiece makes it possible to reduce the forming efforts due to irreversible creep deformation of the material and to operatively control the technological process. The whole process of deformation lasted no more than 30 minutes, then the stress was relaxed for 30 minutes at a given temperature, after which the heating was switched off, and the cooling was slow.

Heat treatment of formed plates included hardening and ageing; the modes are given in [9].

The conformance of the surface of the formed plates to the specified theoretical model was controlled by means of a contactless coordinate measuring system based on a laser radar MV 224. The measurement error for distances up to 5 meters does not exceed 22.5 μm.

Figure 1. UFP-1M unit used for metal forming

The tensile-test after these technological operations was carried out on the Zwick / Roell Z100 unit at the Institute of Geology and Mineralogy SB RAS. The diameter of the samples is 10 mm, the base for deformation measurements is 70 mm.

3. Main results

3.1. Obtaining the parameters of the creep model

Figure 2 shows the relaxation curves of the maximum stress in rectangular beams (a) made from the V-1461T alloy and (b) from the V95 alloy for different temperatures. It follows that with increasing temperature the value of the maximum stress drops.

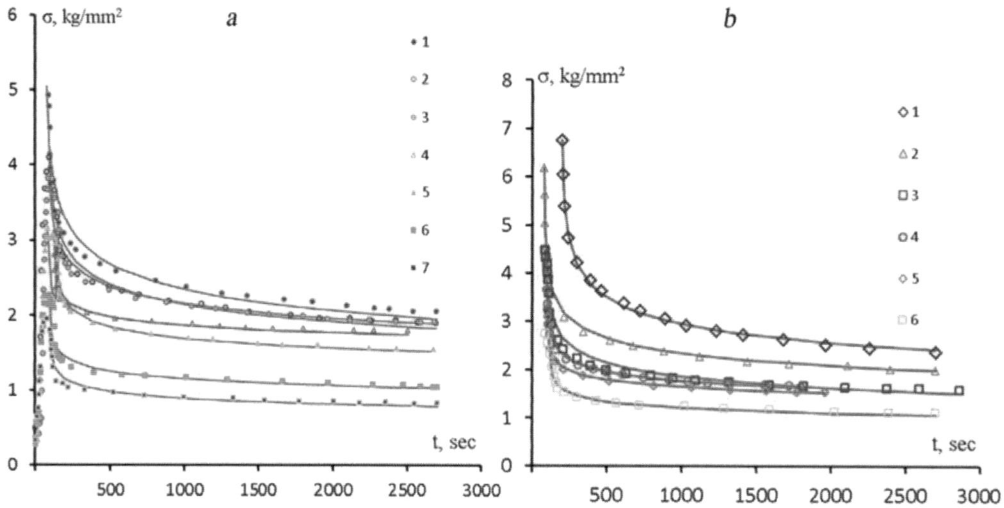

Figure 2. Dependence of stress on time in a rectangular beam (a) made from the V-1461 alloy and (b) made from the V95och2 alloy in the relaxation tests for different temperatures: (a) 1 - 420 °C, 2-430 °C, 3-440 °C, 4 - 450 °C, 5 – 460 °C, 6 - 470 °C, 7 - 480 °C; (b) 1 - 360 °C; 2 - 380 °C; 3, 4 - 410 °C; 5 - 420 °C; 6 - 430 °C.

It is possible to determine the parameters of the steady-state creep model from the stress relaxation diagrams. We take the analogue of the Maxwell model to describe the material - the material model is a serially connected spring and a damper: the elastic behavior of the material is modeled by a spring, so the relationship between stress and deformation is written $\sigma = E\varepsilon_e$, here ε_e is the elastic deformation, E is the elastic modulus, σ - stress. The inelastic behavior of the material is modeled by a damper as a power function of relationship of deformation rate and stress: $\dot{\varepsilon}_i = B_A\sigma^{n-1}$, here $\dot{\varepsilon}_i$ -

creep rate, B_A and n - parameters of the steady-state creep model. A refined model taking into account microstresses and damage was proposed in [12] for the transient creep under changing loads.

Since the total deformation for the Maxwell model is the sum of the elastic and inelastic deformation $\varepsilon = \varepsilon_e + \varepsilon_i$, then the relaxation process can be written as $\dot{\varepsilon}_e + \dot{\varepsilon}_i = 0$. Thus, we obtain the equation for stress: $\frac{d\sigma(t)}{dt} + E(T)B_A(T)\sigma(t)^{n(T)-1} = 0$. The initial condition $\sigma|_{t=t_0(T)} = \sigma_0(T)$. Here T – temperature. The solution is

$$\sigma(t) = \left(\sigma_0(T)^{2-n(T)} - (2-n(T))E(T)B_A(T)(t-t_0(T))\right)^{\frac{1}{2-n(T)}}. \tag{1}$$

Then using the least squares method the values of n and B_A were determined for each temperature. Figure 2 shows the curves (1) for each temperature.

Table 1. Model parameters for V-1461T1 at different temperatures

T, °C	B_A, ((kgf mm^{-2})$^{-n}$ h^{-1})	n	E, (kgf mm^{-2})	σ_{max}, (kgf mm^{-2})
420	$7.77 \cdot 10^{-7}$	7.67	6981.3	5.05
430	$8.36 \cdot 10^{-7}$	8.60	4719.0	4.09
440	$2.98 \cdot 10^{-7}$	9.76	4408.9	3.81
450	$7.15 \cdot 10^{-9}$	16.76	4128.4	3.11
460	$5.15 \cdot 10^{-7}$	12.18	3931.7	3.15
470	$5.80 \cdot 10^{-5}$	11.55	1902.1	2.26
480	$6.87 \cdot 10^{-4}$	10.54	1840.2	1.79

Table 2. Model parameters for V95ochT2 at different temperatures

T, °C	B_A, ((kgf mm^{-2})$^{-n}$ h^{-1})	n	E, (kgf mm^{-2})	σ_{max}, (kgf mm^{-2})
360	$3.75 \cdot 10^{-7}$	7.71	4250.3	6.83
380	$4.02 \cdot 10^{-7}$	8.42	6273.3	6.17
410	$3.80 \cdot 10^{-6}$	7.54	5952.0	4.50
410	$9.15 \cdot 10^{-7}$	10.01	5560.0	3.83
420	$6.31 \cdot 10^{-7}$	12.02	4279.6	3.39
430	$2.00 \cdot 10^{-5}$	10.50	4575.7	2.83

Tables 1 and 2 show the values of the parameters B_A and n depending on temperature, obtained on the basis of equation (1) for V-1461T2 and V95ochT2, respectively. The values of the modulus of elasticity and maximum effort are also given depending on temperatures, determined under the experimental data. Thus, the data on the relaxation of rectangular beams give the values of the creep model. The optimum temperature for forming a thick panel made from V-1461T1 was chosen to be $T = 470$ ° C, and for V95ochT2 - $T = 420$ ° C.

3.2. Comparison of the molded and target surfaces

The laser radar MV 224 allows determining the spatial coordinates of the characteristic points on the surface of the molded plate, which made it possible to calculate the deviations of the coordinates of these characteristic points from the target shape of the object. The analysis of the obtained data after plate forming made from V95 showed compliance; the maximum deviations, in accordance with the control points, are in the range from -9.27 to + 5.37 mm (Figure 3). Less than 2.5% of the plate is deformed with a deviation from the theory of less than 9.27 mm. More than 80% of the plate is deformed with a deviation of less than 1 mm.

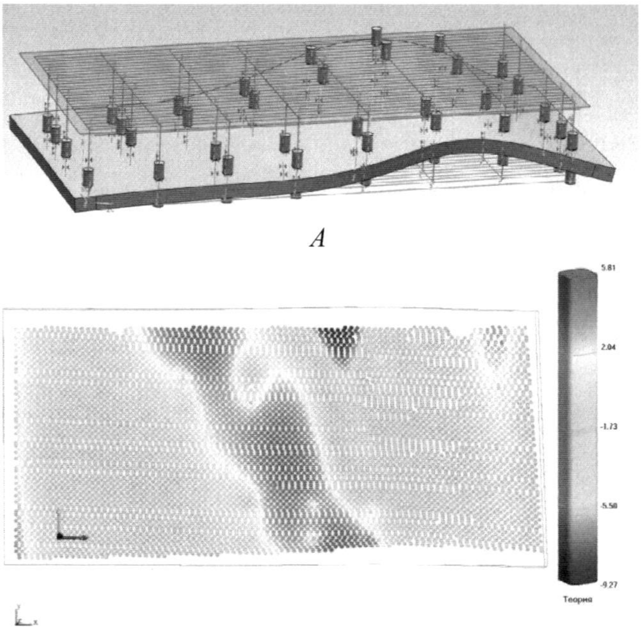

A

B

Figure 3. (a) The model of the required panel (theory). (b) deviations of the obtained surface from the surface specified in the theoretical model

3.3. Mechanical properties of samples from plates after forming

Table 3 shows the mechanical properties of samples from plates made from V95ochT2 after forming and heat treatment.

Samples cut both in the direction of rolling and across have practically the same values of the tensile strength σ_B and the yield strength $\sigma_{0,2}$ with minimum scatter of the experimental data.

Table 3. Mechanical properties of samples made from V95ochT2 under the tensile test in accordance with GOST 1497-84

Alloy, conditions	σ_B (kgf mm^{-2})	$\sigma_{0,2}$ (kgf mm^{-2})	δ, %	E, (kgf mm^{-2})
V95ochT2	49,95	42,81	9,0	7553,5

4. Discussion

The optimum temperature for V95 is the annealing temperature (T = 420 ° C), and for V-1461 - the plate rolling temperature (T = 470 ° C) [10]. A natural question may arise: will high temperatures of 420 ° C and 470 ° C lead to an increase in grain in the material structure? This temperature and the holding time (30-40 min) make it possible to avoid the process of collective recrystallization - the formation of a coarse crystal structure in the abovementioned alloys.

It should be noted that the described technical process of forming thick panels includes the subsequent heat treatment of the product - quenching and ageing, which means adding one more technical operation, compared with molding at the ageing temperature [13] and affects the mechanical properties of the materials positively (no more than two quenching are allowed PI-1.2.699-2007).

The choice of optimal modes of forming is dictated not only by the properties of materials, thickness of the samples (40 mm), but also by the power capabilities of the unit.

5. Conclusion

Optimum temperatures for forming thick plates made from V-1461 and V95 are 470 ° C and 420 ° C, respectively. It has been established that within 30-40 minutes of the process the following characteristics are ensured: steady creep, no springing, and almost complete relaxation of stresses. The experimental relaxation diagrams for various temperatures give the parameters of the steady-state creep model.

A program for the step-by-step forming plates has been developed, which excludes the possibility of a coarse-crystal structure development in the material, as well as wedging of the rods of the unit.

It is shown that the results of tensile tests for samples made from V95och2 after forming correspond to those indicated in [14].

Thus, the conducted research showed that after forming the plate at the UFP-1M unit, heat treatment and subsequent milling, it is possible to obtain a product of high quality with low material consumption.

Acknowledgments

This work was partially supported by the Russian Foundation for Basic Research (project codes 16-08-00483 A, 15-01-07631 A).

References

[1] Ogolodkov M 2013 Regularities of alteration structure and properties of rolled semi-finished products from alloy V-1461, depending on technological parameters of production and heat treatment *Abstract. VIAM. Moscow* p 27

[2] Khokhlatova L, Kolobnev N, Oglodkov M, Filatov A and Popova Yu 2014 *All materials. Encyclopedic reference book* **2** 16 - 22

[3] Khokhlatova L, Kolobnev N, Oglodkov M, Lukina E and Sbitneva S 2012 *Metallurgy and heat treatment of metals* **6** 20 - 4

[4] Khokhlatova L, Oglodkov M and Ponomarev E 2012 *Metallurgy of machine building* **3** 22 - 6

[5] Gorev B, Klopotov I, Raevskaya G and Sosnin O 1980 *j. Appl. mech. Tech. phys.* **5** pp 185 - 91

[6] Holman M 1989 *Journal of Mechanical Working Technology* **20** 477-88

[7] Miodushevsky P, Raevskaya G and Sosnin O 1992 The method of forming parts and device for its implementation *Patent for invention* RUS 2056197 6 p

[8] Klopotov I, Lyubashevskaya I, Raevskaya G, Rublevsky L and Sosnin 2002 O Molding device *Patent for the invention* RUS 2251464 22 07

[9] Larichkin A, Gorev B, Zakharchenko K, Kapustin V 2017 *Tekhnologiya Mashinostroeniya* **3** 5 - 9

[10] Yerisov Y, Grechnikov F and Oglodkov M 2015 *News of higher educational institutions Non-ferrous metallurgy* **6** 36 - 42

[11] Kolodezev V, Gorev B, Larichkin A and Shevtsova L 2017 *Tekhnologiya Mashinostroeniya* **2** 11 - 16

[12] Shutov A V, Larichkin A and Shutov V A 2017 *Angew Math. Mech.* DOI 10.1002/zamm.201600286 pp 1–17

[13] Troyanov V, Uksusnikov A, Senatorova O and Pushin V 2011 About the possibility of obtaining thermostable high-strength alloys of the Al-Zn-Mg-Cu system with nano-phase separation 1 121. *In the book: Second Moscow readings on problems of strength of materials, dedicated to the 80th anniversary of the birth of academician of the Russian Academy of Sciences Yu.A. Osipyana theses of the reports* p 152

[14] Fridlyander I, Senatorova O, Tkachenko E, Molostova I 2008 *All materials Encyclopedic reference book* **8** 17-21

IOP Conf. Series: Journal of Physics: Conf. Series **894** (2017) 012079 doi:10.1088/1742-6596/894/1/012079

Accounting for anisotropy of properties of substructural elements in study of strength of macroinhomogeneous media in the case of plane stressed state

B S Reznikov and K A Kuzmin

Novosibirsk State Technical University, Novosibirsk 630073, Russia

E-mail: kuzmin.2011@stud.nstu.ru

Abstract. In this paper on the mathematical model of a multiphase medium (with longitudinally transverse arrangement of phases) and the algorithm for calculating strength under combined thermomechanical loading the method is proposed for analyzing the limiting state of a composite, which allows one to take into account the internal structure of the material and the anisotropy of the elastic and thermal properties of substructural elements. For each element of the composition the strength condition of P. P. Balandin is used. It takes into account various limits of tensile strength and compression of phase materials. A numerical study of condition of initial destruction of composites and its various sections in space of parameters of the external action under isothermal loading in the case of plane stressed state is made.

1. Introduction

In modern composites, where essentially anisotropic crystalline inclusions, "whiskers", carbon and boron fibers, etc., are used as reinforcing elements, analysis of effect of anisotropy of properties of phase materials on initial destruction of structurally inhomogeneous medium is urgent task.

To study limiting state of a composite with longitudinally transverse arrangement of "M+N" phases (M is the number of longitudinal phases, N is the number of transverse phases, Fig. 1), we will use the defining relations from [1], obtained on the correctly formulated conjugation conditions for deformations, stresses and temperature at the interface boundaries [1-3].

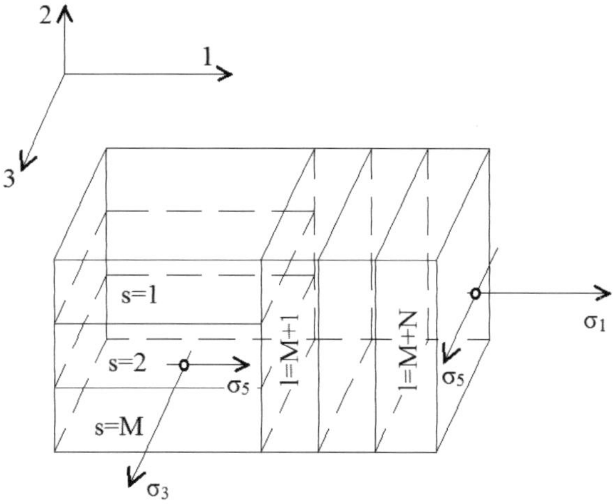

Figure 1. Characteristic structural element of composite with longitudinally transverse arrangement of "M + N" phases in the case of plane stressed state.

Content from this work may be used under the terms of the Creative Commons Attribution 3.0 licence. Any further distribution of this work must maintain attribution to the author(s) and the title of the work, journal citation and DOI.
Published under licence by IOP Publishing Ltd

2. The basic relations and algorithm for studying initial destruction of media with internal structure

For composite consisting of "M+N" phases (Figure 1), in case of action of homogeneous temperature field in case of plane stressed state (i.e., by loading in the plane $n_1 0 n_3$ by forces σ_1, σ_3, σ_5), taking into account the defining relations from [1-3], we have the following expressions for stresses in elements of composition:

in longitudinal phases

$$\sigma_j^{(s)} = \sum_{i=1,3} \psi_{ji}^{(s)} \cdot \sigma_i + \theta \cdot \psi_{jt}^{(s)}, \quad (1)$$

$$\sigma_4^{(s)} = \sigma_6^{(s)} = 0,\ \sigma_5^{(s)} = \psi_5^{(s)} \cdot \sigma_5,$$

$$j = 1,2,3;\ s = 1,2,\ldots, M;$$

in transverse phases

$$\sigma_1^{(l)} = \sigma_1, \quad (2)$$

$$\sigma_j^{(l)} = \sum_{i=1,3} \psi_{ji}^{(l)} \cdot \sigma_i + \theta \cdot \psi_{jt}^{(l)},$$

$$\sigma_4^{(l)} = \sigma_6^{(l)} = 0,\ \sigma_5^{(l)} = \sigma_5,$$

$$j = 2,3;\ l = M+1, M+2, \ldots, M+N.$$

The values $\psi_{ji}^{(s)}, \psi_{jt}^{(s)}, \psi_5^{(s)}, \psi_{ji}^{(l)}, \psi_{jt}^{(l)}$ are determined by effective compliance and linear thermal expansion coefficients of structurally inhomogeneous material, which were obtained in [1], $\theta = T-T_0$ is temperature increment, T_0 is temperature in initial state; we will use the matrix designations for stresses from [1-4] and notation from [1-3].

Studying initial destruction of multiphase composites, we will use the structural approach [5] and the strength condition P.P. Balandin [6] for each phase, which under the considered form of loading, taking into account (1), (2), has the form:

$$\sum_{j=1}^{3} \left(\sigma_j^{(m)}\right)^2 + 3\left(\sigma_5^{(m)}\right)^2 - \sigma_1^{(m)} \cdot \sigma_2^{(m)} - \sigma_1^{(m)} \cdot \sigma_3^{(m)} - \quad (3)$$

$$- \sigma_2^{(m)} \cdot \sigma_3^{(m)} + \left(\sigma_{(-)}^{(m)} - \sigma_{(+)}^{(m)}\right) \sum_{j=1}^{3} \sigma_j^{(m)} = \sigma_{(+)}^{(m)} \sigma_{(-)}^{(m)},$$

where $\sigma_{(+)}^{(m)}$ and $\sigma_{(-)}^{(m)}$ are the tensile and compressive strength limits of the material of the m-th phase, m = 1,2, ..., M, M + 1, ..., M + N.

To construct in space of external actions $\sigma_1 0 \sigma_3 \sigma_5 \theta$ hypersurface of initial destruction of macroinhomogeneous material and its various sections, we will use the algorithm from [7], that is, to take

$$\sigma_i = L_i \cdot p\ (\mathrm{i} = 1,3,5);\ \theta = L_t \cdot p,\ \sum_{i=1,3,5} |L_i| + |L_t| = 1, \quad (4)$$

where p>0 is load parameter, L_1, L_3, L_5, L_t - intensity of corresponding type of external influence. Using the expressions for stresses in substructural elements (1), (2), the strength condition (3), for fixed L_i (i = 1,3,5), L_t, we determine the value of $P_в^{(m)}$ parameter, corresponding to beginning of destruction of the m phase:

$$P_в^{(m)} = \frac{-B^{(m)} + \sqrt{\left(B^{(m)}\right)^2 + 4\sigma_{(+)}^{(m)} \sigma_{(-)}^{(m)} A^{(m)}}}{2A^{(m)}}, \quad (5)$$

where m = s, s = 1,2,…,M; m = l, l = M+1,M+2,…,M+N;

for quantity with index of «s» (s = 1,2,…,M) we have

$$A^{(s)} = \sum_{j=1}^{3}\left(\varphi_i^{(s)}\right)^2 + 3\left(L_5\psi_5^{(s)}\right)^2 - \varphi_1^{(s)}\varphi_2^{(s)} - \varphi_1^{(s)}\varphi_3^{(s)} - \varphi_2^{(s)}\varphi_3^{(s)}, \tag{6}$$
$$B^{(s)} = \left(\sigma_{(-)}^{(m)} - \sigma_{(+)}^{(m)}\right)\sum_{j=1}^{3}\varphi_j^{(s)},$$
$$\varphi_j^{(s)} = L_1\psi_{j1}^{(s)} + L_3\psi_{j3}^{(s)} + L_t\psi_{jt}^{(s)}\ (\mathrm{j} = 1,2,3);$$

for quantity with index of «l» (l = M+1,M+2,…,M+N) –

$$A^{(l)} = L_1^2 + \left(\varphi_2^{(l)}\right)^2 + \left(\varphi_3^{(l)}\right)^2 + 3L_5^2 - L_1\varphi_2^{(l)} - L_1\varphi_3^{(l)} - \varphi_2^{(l)}\varphi_3^{(l)}, \tag{7}$$
$$B^{(l)} = \left(\sigma_{(-)}^{(l)} - \sigma_{(+)}^{(l)}\right)\left(L_1 + \varphi_2^{(l)} + \varphi_3^{(l)}\right);$$
$$\varphi_j^{(l)} = L_1\psi_{j1}^{(l)} + L_3\psi_{j3}^{(l)} + L_t\psi_{jt}^{(l)}\ (\mathrm{j} = 2,3).$$

The magnitude of loading parameter $P_в$, corresponding to beginning of destruction of multiphase composite, is determined from solution of minimization problem:

$$P_в = \min\{P_в^{(m)}\},\ \mathrm{m} = 1, 2,\ldots, \mathrm{M}, \mathrm{M}+1, \mathrm{M}+2,\ldots,\mathrm{M}+\mathrm{N}. \tag{8}$$

Then, taking into account the first two relations from (4), we find external forces and the magnitude of temperature effect at which structurally inhomogeneous material begins to break down:

$$\sigma_{iв} = L_i p_в\ (\mathrm{i} = 1,3,5),\ \theta_в = L_i \cdot p_в. \tag{9}$$

Considering all possible values of L_i (1,3,5) and L_t from (4), we obtain hypersurface of strength of composite in space of external forces $\sigma_{1в}0\sigma_{3в}\sigma_{5в}\theta_в$.

It should be noted that the proposed approach makes it possible to determine the type of initial destruction (i.e., due to destruction of which substructural element destruction of composite material begins) for each type of external action.

3. Analysis of influence of anisotropy of properties of substructure elements on strength of composite

As the first example to consider structurally inhomogeneous material consisting of two longitudinal phases: M = 2, N = 0, s = 1,2. The first phase is transversally isotropic with the following parameters:

$$\overline{E}_1^{(1)} = 5,\ \overline{E}_3^{(1)} = 50,\ \nu_1^{(1)} = \nu_3^{(1)} = 0.2; \tag{10}$$
$$\overline{G}_1^{(1)} = \frac{\overline{E}_1^{(1)}}{2\left(1+\nu_1^{(1)}\right)},\ \overline{G}_3^{(1)} = 25;\ \overline{\alpha}_1^{(1)} = 0.3,\ \overline{\alpha}_3^{(1)} = 0.4;$$
$$\overline{\sigma}_{(+)}^{(1)} = 20,\ \overline{\sigma}_{(-)}^{(1)} = 2,$$

and the second phase is isotropic with parameters:

$$\overline{E}^{(2)} = 1;\ \overline{\sigma}_{(+)}^{(2)} = 1,\ \overline{\sigma}_{(-)}^{(2)} = 0.5;\ \nu^{(2)} = 0.2;\ \overline{\alpha}^{(2)} = 1. \tag{11}$$

Here and below the values with the bar are dimensionless:

$$\overline{E}_j^{(1)} = \frac{E_j^{(1)}}{E_0} (j = 1,3), \ \overline{E}^{(2)} = \frac{E^{(2)}}{E_0}; \tag{12}$$

$$\overline{\alpha}_{jj}^{(1)} = \frac{\alpha_{jj}^{(1)}}{\alpha_0} (j = 1,3); \ \overline{\sigma}_{(\pm)}^{(s)} = \frac{\sigma_{(\pm)}^{(s)}}{\sigma_0} (s = 1,2),$$

$$\overline{\sigma}_{j\text{в}} = \frac{\sigma_{j\text{в}}}{\sigma_0}, \ \overline{\theta}_{\text{в}} = \frac{\theta E_0 \alpha_0}{\sigma_0},$$

where $E_j^{(1)}, E^{(2)}, \nu_j^{(1)}, \nu^{(2)}, \alpha_{jj}^{(1)}, \alpha^{(2)}$ - Young's moduli, Poisson's, linear thermal expansion coefficients of materials of the 1st and 2nd phases; σ_0, E_0 – constant dimensions of stress, α_0 – constant [1/deg].

Figures 2 - 4 show different sections of hypersurface of initial destruction of composite with the specific volume content of first and second phases $\omega_1 = 0.2$ and $\omega_2 = 0.8$. For comparison, the dashed lines indicate the sections of hypersurface of initial destruction when both phases are different isotropic materials: for the second phase we have the parameters (11), and for the first phase the parameters from (10) with the "1" subscript. Numerical calculations have shown that the greatest qualitative and quantitative influence on strength of composite is anisotropy of Young's modulus of the first phase.

On the Fig. 2 - 4 parts of the curves marked with digits "1" and "2" correspond to destruction of the first and second phases, at the corner points we have simultaneous destruction of both phases.

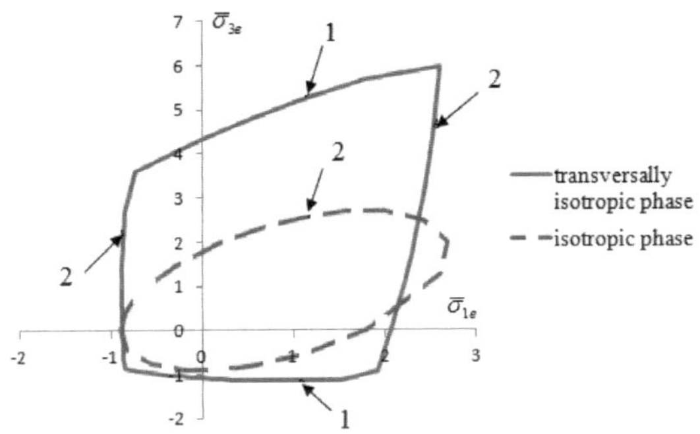

Figure 2. Section $\overline{\sigma}_{1\text{в}}0\overline{\sigma}_{3\text{в}} : \omega_1 = 0.2, \omega_2 = 0.8$.

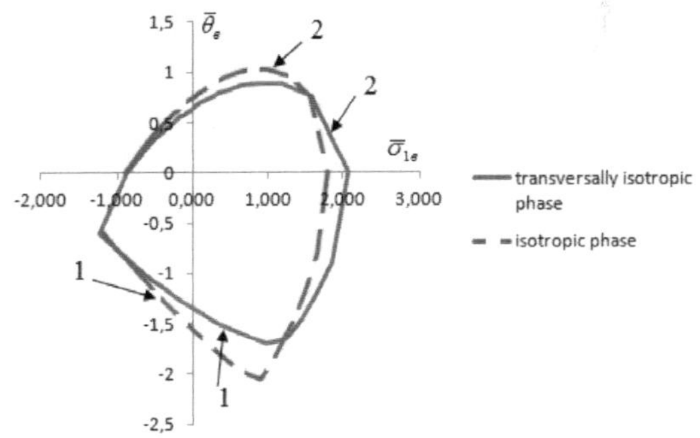

Figure 3. Section $\overline{\sigma}_{1\text{в}}0\overline{\theta}_{\text{в}} : \omega_1 = 0.2, \omega_2 = 0.8$.

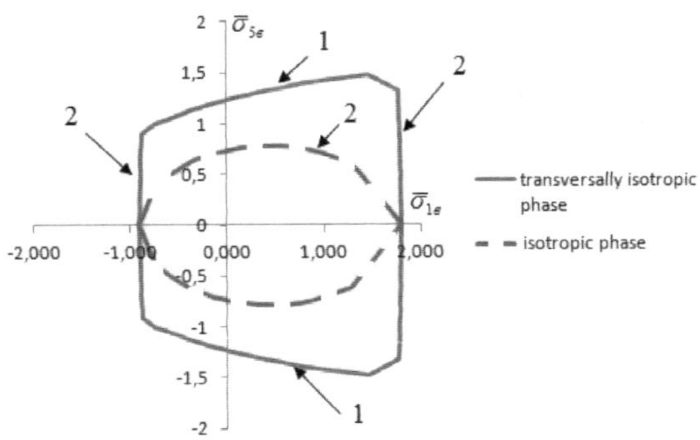

Figure 4. Section $\overline{\sigma}_{1\text{в}}0\overline{\sigma}_{5\text{в}} : \omega_1 = 0.2, \omega_2 = 0.8$.

To analyze the effect on the initial destruction condition of composite of the spatial arrangement of anisotropic phase and its specific volume content, calculations were carried out for three-phase composite materials with longitudinal arrangement of substructure elements: M = 2 (s = 1,2) and N = 1 (l = 3). In this case, the following parameters were adopted for transversely isotropic phase:

$$\overline{E}_1^{(m)} = 5,\ \overline{E}_3^{(m)} = 100;\ \nu_1^{(m)} = 0.2,\ \nu_3^{(m)} = 0.4; \tag{13}$$

$$\overline{G}_1^{(m)} = \frac{\overline{E}_1^{(m)}}{2\left(1+\nu_1^{(m)}\right)},\ \overline{G}_3^{(m)} = 50;\ \overline{\alpha}_{11}^{(m)} = 0.3,\ \overline{\alpha}_{33}^{(m)} = 0.7;$$

$$\overline{\sigma}_{(+)}^{(m)} = 20,\ \overline{\sigma}_{(-)}^{(m)} = 50.$$

For the isotropic phase for s = 2 we have the parameters (11), and for the other isotropic phase we have the following parameters:

$$\overline{E}^{(n)} = 5;\ \overline{\sigma}_{(+)}^{(n)} = 20,\ \overline{\sigma}_{(-)}^{(n)} = 5;\ \nu^{(n)} = 0.2;\ \overline{\alpha}_t^{(n)} = 0.3. \tag{14}$$

Values of the indices "m" and "n" in (13) and (14) depend on location of these phases in space, which will be indicated below.

To consider the two cases location of anisotropic phase in space:

1) the longitudinal arrangement of transversally isotropic phase
   m = s = 1 and n = l = 3;
2) the transverse location of transversally isotropic phase
   m = l = 3 and n = s = 1.

Figures 5 - 7 show different sections of hypersurface of initial destruction of three-phase composites.

The solid lines correspond to the case (15) for the following values of specific volume content of the phases: $\widetilde{\omega}_1 = 0.3;\ \widetilde{\omega}_2 = \widetilde{\omega}_3 = 0.35;$ and the dashed curves correspond to the case (16) with $\widetilde{\omega}_1 = \widetilde{\omega}_2 = 0.35;\ \widetilde{\omega}_3 = 0.3.$ Digits 1, 2, 3, as before, indicate the type of initial destruction.

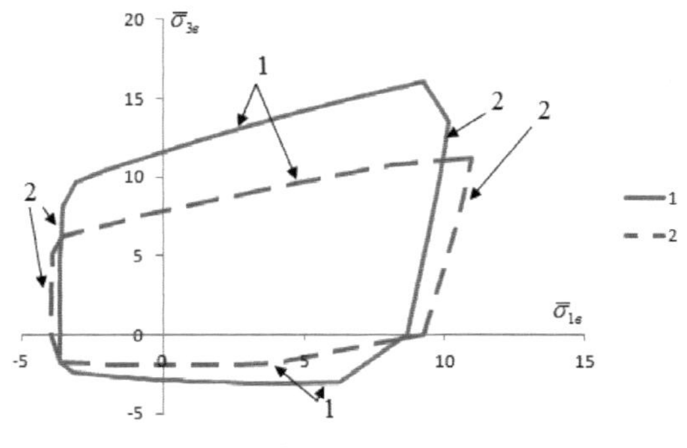

Figure 5. Section $\overline{\sigma}_{1e}0\overline{\sigma}_{3e}$.

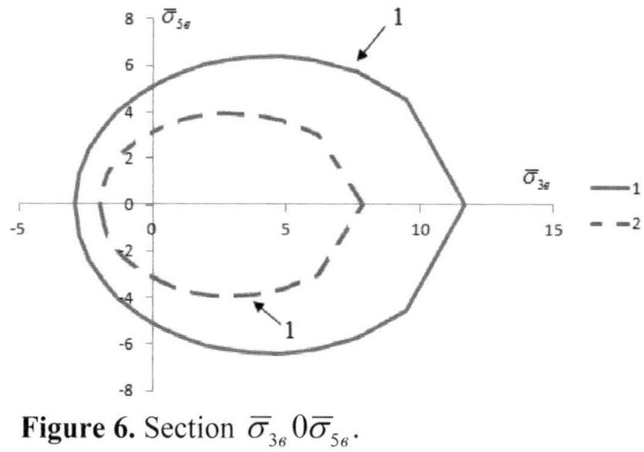

Figure 6. Section $\overline{\sigma}_{3e}0\overline{\sigma}_{5e}$.

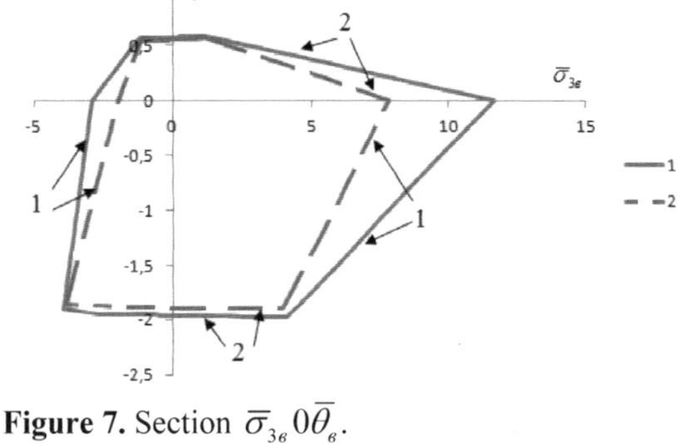

Figure 7. Section $\overline{\sigma}_{3e}0\overline{\theta}_{e}$.

IOP Conf. Series: Journal of Physics: Conf. Series **894** (2017) 012079 doi:10.1088/1742-6596/894/1/012079

The results of calculations of different sections of the strength condition of three-phase composites as function of the specific volume content of anisotropic phase are shown on Fig. 8 - 13. For the case (15), we have Fig. 8 - 11, the solid lines correspond to $\widetilde{\omega}_1 = 0.3;\ \widetilde{\omega}_2 = \widetilde{\omega}_3 = 0.35;$ dashed - $\widetilde{\omega}_1 = 0.5;\ \widetilde{\omega}_2 = \widetilde{\omega}_3 = 0.25;$ dot-dash - $\widetilde{\omega}_1 = 0.7;\ \widetilde{\omega}_2 = \widetilde{\omega}_3 = 0.15;$

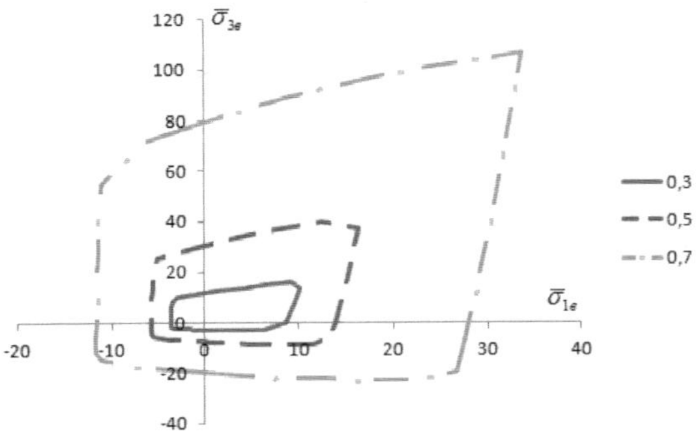

Figure 8. Section $\overline{\sigma}_{1e} 0 \overline{\sigma}_{3e}$.

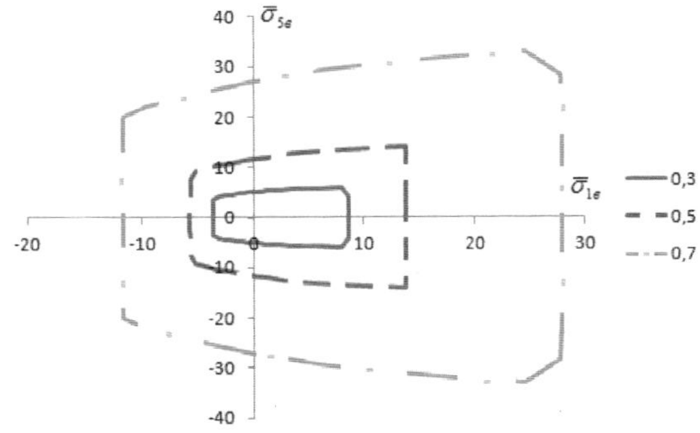

Figure 9. Section $\overline{\sigma}_{1e} 0 \overline{\sigma}_{5e}$.

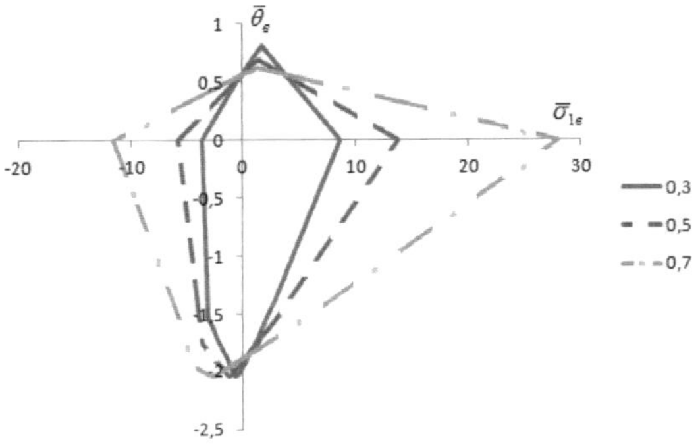

Figure 10. Section $\overline{\sigma}_{3e} 0 \overline{\sigma}_{5e}$.

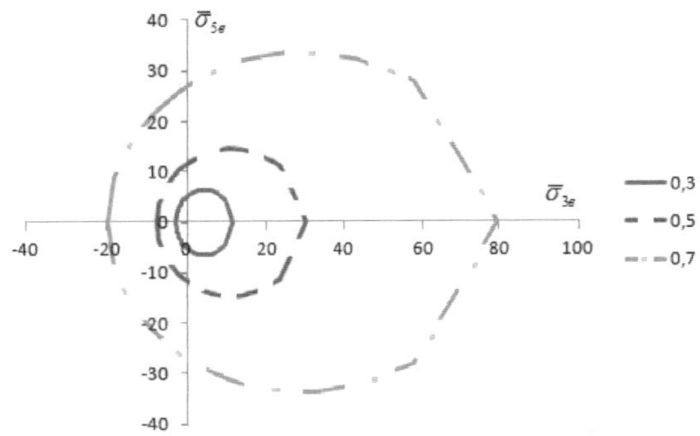

Figure 11. Section $\overline{\sigma}_{1e} 0 \overline{\theta}_{e}$.

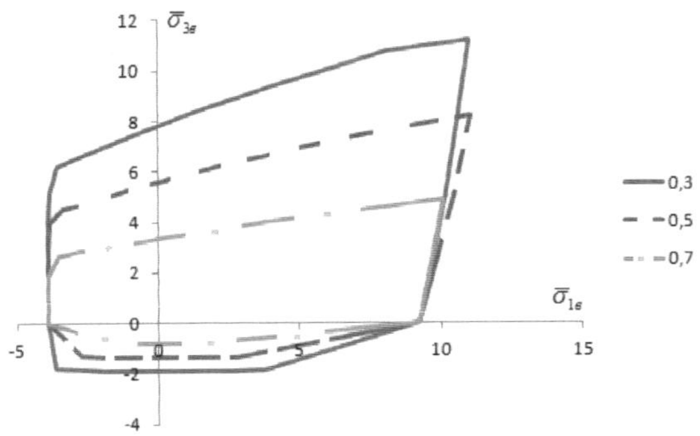

Figure 12. Section $\overline{\sigma}_{1e} 0 \overline{\sigma}_{3e}$.

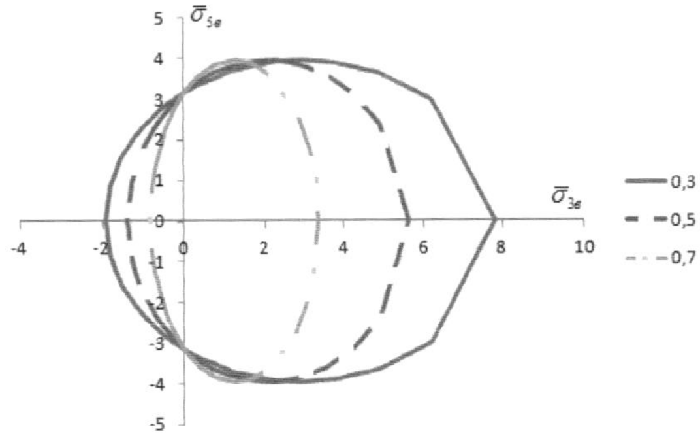

Figure 13. Section $\overline{\sigma}_{3e} 0 \overline{\sigma}_{5e}$.

Under the conditions (16), we have Fig. 12, 13, where solid lines correspond $\widetilde{\omega}_3 = 0.3;\ \widetilde{\omega}_1 = \widetilde{\omega}_2 = 0.35;$ dashed - $\widetilde{\omega}_3 = 0.5;\ \widetilde{\omega}_1 = \widetilde{\omega}_2 = 0.25;$ dot-dash - $\widetilde{\omega}_3 = 0.7;\ \widetilde{\omega}_1 = \widetilde{\omega}_2 = 0.15.$

The results on Fig. 5-13 show that strength of macroinhomogeneous media essentially depends both on location in space of anisotropic phase and on its specific volume content.

IOP Conf. Series: Journal of Physics: Conf. Series **894** (2017) 012079 doi:10.1088/1742-6596/894/1/012079

4. Conclusion

In this way, the proposed approach allows us to predict limiting state of composites, depending on the nature of anisotropy of substructure elements, their specific volume content and mutual arrangement. It allows us to purposefully design multiphase materials with the required strength characteristics in terms of operating conditions for constructions.

References

[1] Reznikov B S and Gobysh A V 2014 Prediction structure of multiphase size-stable composites under temperature influence *Reports of the 3rd All-Russian Conference Problems of Optimal Design of Constructions Novosibirsk State University of architecture and Civil Engineering (Sibstrin)* (Novosibirsk: Sibstrin) pp 345-52 (in Russian)

[2] Reznikov B S, Nikitenko A F and Kucherenko I V 2008 Prediction macroscopic properties of structurally inhomogeneous media *Report 1 Izvestia vuzov, Stroitelstvo* (Novosibirsk: Sibstrin) **2** pp 10-17 (in Russian)

[3] Reznikov B S and Gobysh A V 2013 Calculation effective coefficients of thermal expansion of macroinhomogeneous composites *Reports of Academy of Sciences of Higher School* **2(21)**, pp 139-49 (in Russian)

[4] Shermergor T D 1977 *The Theory of Elasticity of Microinhomogeneous Media* (Moscow: Nauka) (in Russian)

[5] Nemirovskiy Y V and Reznikov B S 1986 Strength of Structural Elements of Composite Materials (Novosibirsk: Nauka) (in Russian)

[6] Balandin P P 1937 On question of strength hypotheses *Vestnik inzhenerov i tekhnikov* **1** 19-24 (in Russian)

[7] Reznikov B S 1989 Predicting destruction of annular plates taking into account actual structure and stochastic nature of reinforced material *Krayevyye zadachi i ikh prilozheniya* (Cheboksary: Chuvash state university) pp 89–99 (in Russian)

IOP Conf. Series: Journal of Physics: Conf. Series **894** (2017) 012080 doi:10.1088/1742-6596/894/1/012080

Technology for formation of axisymmetric free jets with long laminar region

A Reshmin[1], A Sudarikova[1,2], S Teplovodskii[1] and J Zayko[1,2]

[1]Institute of Mechanics, Lomonosov Moscow State University, Moscow, Russia

[2]Lomonosov Moscow State University, Moscow, Russia

E-mail: alexreshmin@rambler.ru

Abstract. A new technology for formation of axisymmetric free jets with a long laminar region is presented in the work. Such jets are created by of a compact device and experimentally studied in the Reynolds number range from 2000 to 13000. The device is capable of creation a jet of 0.12 m diameter. Numerical simulations are conducted with the purpose to analyse the flow inside the device and its impact on the jet velocity profile. Also calculations are conducted in order to find a way for a jet laminar region prolongation. Based on the calculations, the device parameters are corrected and the jet laminar region length is magnified from the size of 5.5 to 6.5 jet diameters. Free jets with long initial laminar regions and the diameter more than 0.1 m are advantageous for detailed research of perturbation growth and transition to turbulence in round jets, and can also be used to organize air curtains in order to protect objects in medicine and high–accurate industries.

1. Introduction

Free jets and other shear flows often occur in nature and various technologies. The topic of turbulent submerged jets is thoroughly studied and it is the focus of various applications such as mixing, combustion, noise generation, and others [1–3]. Laminar jets are much less studied due to their immediate breakdown at normal conditions. Submerged laminar jets are unstable. The critical Reynolds number Re_{cr} of transition to turbulence for the flat submerged jets is equal to 4 [4], for axisymmetric – 37.9 [5]. Many works are devoted to the research on transition to turbulence in laminar free jets at the Reynolds number considerably superior to critical [5–7] and also to the control of free jets and the turbulence suppression in free shear flows [8–10]. Flow characteristics, such as velocity profile and intensity of turbulence, have a dominant role in these problems. For example, flow with low turbulence level occurs in the case of the Poiseuille velocity profile formation at air flow in a smooth pipe [11, 12]. In this way the velocity profile is defined only by Reynolds number and the channel length. Experimental results show that the free jet maintains its laminar character at distances of 10 jet diameters in the Reynolds number range from 600 to 7000 due to formation in the initial section of the submerged jet with a low velocity pulsation level and a velocity profile closed to parabolic. It is worth to notice that a velocity profile, which differs from the Poiseuille profile by less than 1%, forms in the pipe of the diameter d at its length $l = 0.065 \cdot d \cdot Re$ [13]. Therefore diameters of the studied jets do not exceed several centimeters.

Laminar initial regions of free jets of sufficient length could be used not only for fundamental research on transition to turbulence. Such regiones could accept to organize air curtains, which

Content from this work may be used under the terms of the Creative Commons Attribution 3.0 licence. Any further distribution of this work must maintain attribution to the author(s) and the title of the work, journal citation and DOI.
Published under licence by IOP Publishing Ltd

 doi:10.1088/1742-6596/894/1/012080

provide zones of clean air not mixing with the ambient medium. Local clean zones can be used in medicine, medical industry, microelectronics, and other technological processes.

In this study, we present a new method for the formation of free gas jets with the diameter of 0.12 m, in which the transition to turbulence occurs at the distance of 6.5 jet diameters (for optimal velocity regime) from the orifice. The forming device length is only of $\sim$ 1.5 jet diameter. The device is designed on the basis of the works [14, 15], where a similar design was used to study turbulent flows.

2. Experimental set–up

Figure 1 shows the experimental setup. The device is installed on a support (1). The equipment consists of two parts: a laminarization unit and a forming device. The flow enters the apparatus through the air line (2). A perforated plate (3) and a bushing with metal grids (4) are fixed inside the channel and serve for the flow laminarization. The perforated plate (3) has holes with diameter of 0.6 mm and it porosity is equal to 0.8. Wire diameter of the metal grids (4) is 30 μ m and the grid cell is 40 μ m. The intensity of velocity pulsations becomes less than 1% as the flow passing through them. Further air comes to the forming device. At this stage it is necessary to create a desired velocity profile at once, otherwise turbulence begins to develop directly in the forming device. A short round diffuser (5) is used for formation of a flow with the set velocity profile and low intensity of turbulence. The short diffuser is represented by an axisymmertric track that expands three times in the diameter while the track length is about the inlet diameter. A metal grids package (6) is located at the diffuser outlet. It slows the flow down and directs air to the diffuser wall. The diffuser shape and the metal grids permeation coefficient are chosen in a special way in ordrer to preven flow separation inside the diffuser. The experimental setup provides the velocity profile with almost constant velocity at the central jet part about 0.05 m in diameter and low turbulent intensity.

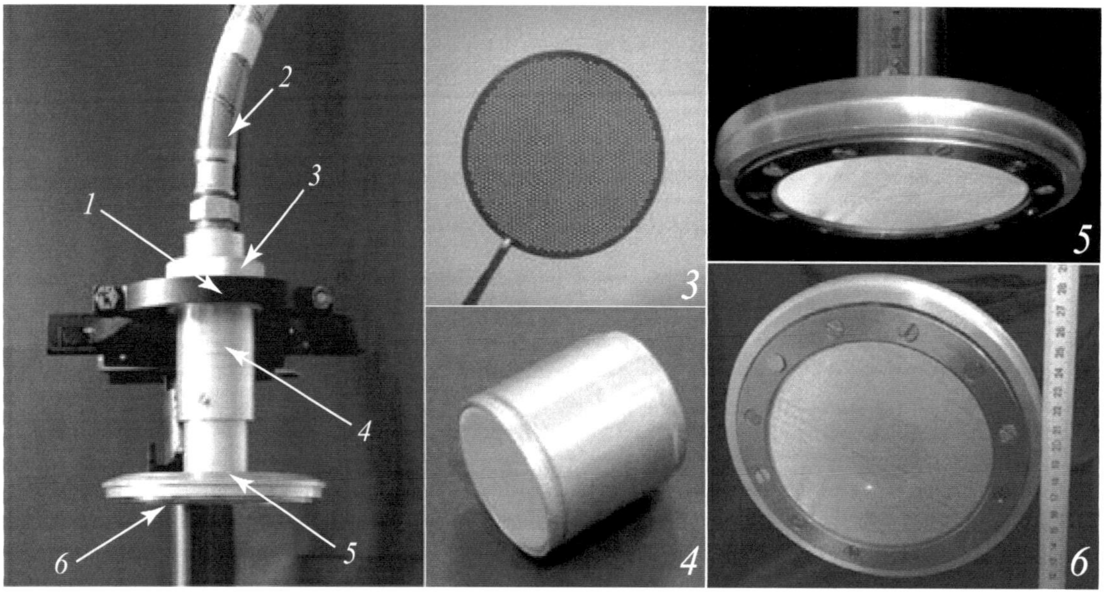

Figure 1. The forming device and its components: support 1, air line 2, perforated plate 3, bushing with metal grids 4, short round diffuser 5 with metal grids package at the outlet 6.

The dimensions of the experimental setup are summarized in Table 1. The distance between the perforated plate and the bushing is 0.03 m and the interval between the bushing and the diffuser inlet is 0.06 m. The jet flows to the atmosphere passing through the laminarization unit and the forming device.

Velocity profiles and turbulent fluctuations are measured by thermoanemometer DISA 56C01 CTA with small-size wire sensor Dantec Dynamics 55P11. The sensor of the thermoanemometer is located on the transfer equipment that allowed to measure flow parameters along the radial coordinate. The jet visualization is performed with an glycerin mist and laser KLM-532.

Table 1. Dimensions of the experimental set–up shown in Figure 1.

Part	Inlet diameter d, mm	Outlet diameter D, mm	Height h, mm
cylindrical channel	40	40	160
perforated plate	40	40	2
bushing	40	40	50
diffuser	40	120	40

3. Results

Three series of experiments were conducted. Velocity profiles measurements prove that the flow is axisymmetric with respect to the device axis and the jet parameters depend only on the radial coordinate r.

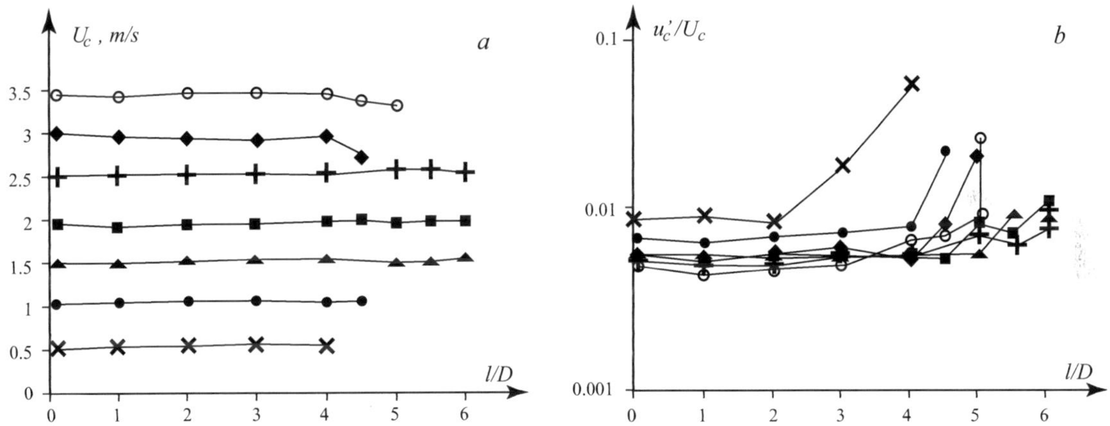

Figure 2. Experimental study of the jet: velocity U_c (a) and turbulent velocity fluctuations u_c'/U_c (b) at the jet axis.

The first series shows that velocity U_c and initial intensity of turbulence u_c'/U_c remain constant at the jet axis in a distance l of several jet diameters D downstream the flow at a fixed velocity regime. These characteristics start to change at the distance L increase: velocity decreases (Figure 2,a) and turbulent velocity fluctuations grow up (Figure 2,b). The flow characteristics are also noticed to remain unchanged in the jet core along diameter when they preserve at the jet axis.

The second experimental series is devoted to measurements of the jet laminar region length and detailed study of the flow characteristic behavior at different velocity regimes. The maximum distance from the device outlet at which velocity remains constant and turbulent velocity fluctuations remain less 1% is claimed to be the jet laminar region length l_{max}.

Numerical simulations are carried out with the purpose to analyse the flow inside the device and its impact on the jet velocity profile. The flow is assumed to be steady and laminar, hence, Navier–Stokes equations are solved. Calculated and measured velocity profiles at the distance of 5 mm downstream of the diffuser outlet are compared in order to check the numerical model. Figure 3 shows a good agreement of these profiles. Numerical simulations show that the flow inside the diffuser depends on the inlet velocity. The flow is attached to the wall at low velocities and small local separations appear when the velocity increases. This leads to the change of the jet velocity profile and jet unstable perturbations rising. As a result the diffuser geometry is modified for prevention of separations and the third series of experiments with the modified device takes place.

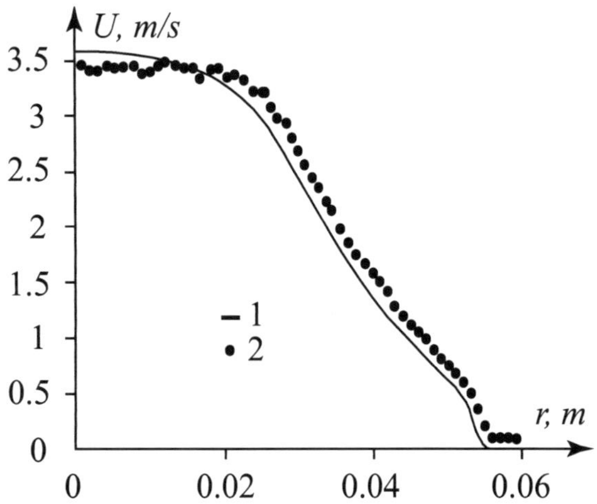

Figure 3. Comparison of the calculated (1) and experimental (2) velocity profiles at the distance of 5 mm from the diffuser outlet.

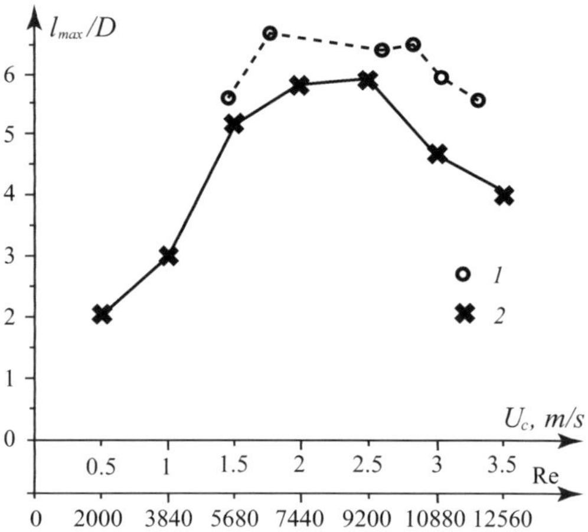

Figure 4. Free jet laminar region dimensionless length L_{max}/D versus velocity U_c and the Reynolds number for updated (1) and original (2) diffuser.

The third experimental series is analogous to the second one. Figure 4 shows the obtained jet laminar region length for the original diffuser and the length prolongation with the help of the updated unit versus the velocity U_c at the jet axis and the Reynolds number based on the jet diameter and average velocity. Relying on the experimental and numerical results the jet laminar region length is concluded to have the greatest value for the velocity regime $U_c \approx 2\ m/s$, and it reaches 5.5 jet diameters in case of the initial diffuser and 6.5 jet diameters in case of the modified one. Figure 5 shows the jet visualization at the velocity $U_c = 2.5\ m/s$.

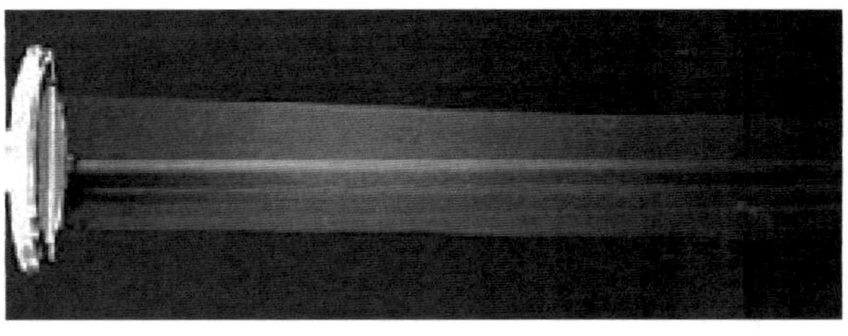

Figure 5. Free jets with laminar region length $l_{max} = 5.5D$ (rotated).

4. Conclusion

The main achievement of the work is a compact laminar jet formation device and the short round diffuser is its major part. The device size is of order of the jet diameter. Laminar jet of 0.12 m diameter at the Reynolds number *Re* 10000 based on the jet diameter and average velocity is obtained. Experimental results confirm laminar character of a flow at distance of 6.5 jet diameters from the forming device at the optimal velocity regimes for the device. The calculations allowing to optimize geometry of the diffuser channel for prolongtation of the jet laminar zone are conducted. Considerable length and diameter of the laminar site allows to conduct a detailed research of transition to turbulence in the submerged jets. Also results of the work are capable of assumption to form laminar free jets of a larger diameter with the aid of the proposed technology. Laminar sites of such jets can be used in medicine and high-precision industry for creation of local zones with the set characteristics.

5. Acknowledgments

The present study was funded by RFBR according to the research project No. 17-08-00115.

References

[1] Crow S and Champagne F 1971 *J. Fluid Mech.* **246(23)** 547–91
[2] List E 1982 *Ann. Rev. Fluid Mech.* **14** 189–212
[3] Zaitsev M, Kopiev V and Chernyshev S 2009 *Fluid Dyn.* **44** 587– 95
[4] Silcock G 1975. On the stability of parallel stratified shear flows *Ph.D. dissertation* (University of Bristol)
[5] Lessen M and Singh P 1973 *J. Fluid Mech.* **60** 433–57
[6] Andrade 1939 Proc. Phys. Soc. **51** 784
[7] Sato H 1959 *J. Fluid Mech.* **7** 53–80
[8] Zaman K and Hussain A 1981 *J. Fluid Mech.* **103** 133–59
[9] Hussain A and Hasan A 1985 *J. Fluid Mech.* **150** 159–68
[10] Sahu R, Chatterjee S, Biswas N, Mukhopadhyay A and Sen S 2014 *Int. J. Heat Fluid Fl.* **50** 160–68
[11] Lemanov V, Terekhov V, Sharov K and Shumeiko A 2013 *Tech. Phys. Lett.* **39** 421–3
[12] Kozlov G, Grek G, Sorokin A and Litvinenko Yu 2008 *Thermophysics and Aeromechanics* **15** 55–68
[13] Prandtl L and Tietjens O 1934 *Applied hydro- and aeromechanics* **2** (McGraw-Hill) p 22
[14] Navoznov O, Paveliev A and Yatsenko A 1972 *Fluid Dyn.* **7** 672–8
[15] Reshmin A, Teplovodskii S and Trifonov V 2012 *Fluid Dyn.* **47** 583 –9

IOP Conf. Series: Journal of Physics: Conf. Series **894** (2017) 012081 doi:10.1088/1742-6596/894/1/012081

Computer modelling of the mechanical behaviour of the cervical spine segment and intervertebral disk prosthesis

T V Kolmakova[1,2] and Y A Rikun[2]

[1] Institute of Strength Physics and Materials Science SB RAS, Tomsk, Russia
[2] Tomsk State University, Tomsk, Russia

E-mail: kolmakova@ftf.tsu.ru, yulirik.93@mail.ru

Abstract. The article describes the model of intervertebral disc prosthesis and the results of study of mechanical behavior of the cervical spine segment. The results show that degenerative changes in the intervertebral disc lead to a change in the location of the regions of maximum Mises stresses in cervical spine segment. The presented geometric model of the intervertebral disc prosthesis is based on the experimental sample of a ceramic endoprosthesis developed at the Institute of Strength Physics and Materials Science SB RAS.

1. Introduction

One of the most important constructions of the human body is the spine. Its structure allows to perform the functions of support and movement. The spine is divided into cervical, thoracic, lumbar and sacral regions [1]. The cervical spine moves most of all. An intervertebral disc (IVD) separates each vertebra, except in the upper cervical spine (C1 & C2), and in the sacrum and coccyx, where the vertebrae are fused together [2]. The discs resists spinal compression, permits limited bending, twisting, and sliding between vertebral bodies [1].

Degenerative changes of the intervertebral discs are the major cause of pain in the spine and neck in those of the middle and elder age [3]. A radical method of treatment in such cases is the replacement of the intervertebral disc with an implant. Adequate development and individual selection of implants play a decisive role in the treatment of a person. Improper selection of materials and design of implants can lead to deterioration of the bone tissue and the functioning of the spine as a whole. The problems of mechanics arising in the creation and selection of implants of biological tissues are solved on the basis of studying the structure, mechanical behavior and properties of the biological tissues themselves and their interaction with implants. The use of computer modeling methods makes it possible to more deeply understand the patterns of functioning of the spine in norm, at pathologies, at interaction with the implant, which facilitates the selection of a suitable prosthesis for a particular individual.

In this paper, the mechanical behavior of the model cervical spine segment at flexion forward is analyzed and a model of the intervertebral disc prosthesis is presented.

2. Model of the cervical spine segment

A geometric vertebra model was built based on the literature data about experimentally obtained dimensions [4].

Content from this work may be used under the terms of the Creative Commons Attribution 3.0 licence. Any further distribution of this work must maintain attribution to the author(s) and the title of the work, journal citation and DOI.
Published under licence by IOP Publishing Ltd

The simulation algorithm for the geometric vertebra model was developed and implemented in the ANSYS system with the use of the APDL language. This algorithm allows an automatic rebuilding of the model when input parameters are changed.

Figure 1 shows the geometric model of the cervical spine segment. The geometrical model includes the vertebrae C3 (1) and C4 (2), IVD (3), facet joints (4), interspinous ligament (5), vertebral arches (6), spinous processes (7), transverse processes (8), and articular processes of vertebrae (9).

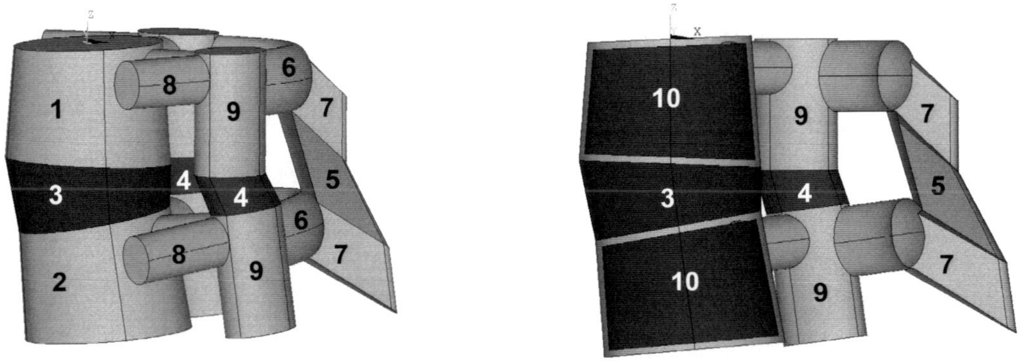

(a) (b)

Figure 1. Geometrical model of cervical spine segment: (a) lateral view, (b) sectional lateral view.

The presence of the cortical and cancellous tissue in the vertebrae was taken into account. In Fig. 1b, the cancellous tissue is marked by 10, a thin layer of the cortical tissue [2] covers vertebral bodies. It is considered that vertebral arches and processes of the vertebrae fully consist of compact bone tissue. The Z axis of the coordinate system is located along the segment axis. The X axis is directed in the anteroposterior direction of the spinal segment.

Materials of the cortical and cancellous tissues of the vertebral bodies, materials of the intervertebral disc, facet joints, interspinous ligament, vertebral arches, and processes of vertebrae are considered as isotropic linear elastic materials. Degenerative changes of the IVD were simulated through a reduction of the disc height (h) from 6 to 4.5 mm and an increase of Young's modulus (E) from 2.5 to 98 MPa according to the previous research data [5,6], which in reality results from the water content reduction in the disc [3].

The task was resolved within linear theory of elasticity. The stress-strain state of cervical spine segment was calculated with the help of ANSYS software using the finite-element method. The lower surface of the vertebral body of C4 was rigidly fixed. The upper surface of the C3 vertebra was loaded by the force of 1000 N. A bending moment of 7.5 N·mm [7] was applied to the central point of the upper surface of the C3 vertebral body in the negative direction of the X axis in flexion of the spine segment. Specified load corresponds to physiological flexion of the cervical spine segment.

3. Study of mechanical behaviour of the cervical spine segment

Figure 2 shows the Mises stress fields in the segments of the spine with different heights and the Young's modulus of the intervertebral disc.

It can be seen from the figure that the maximum stresses are localized in the areas of joints of the vertebra legs with articular processes of the vertebra C4 at Young's modulus of 2.5 MPa (Figure 2 (a, b)). The increase in Young's modulus of the disc leads to the localization of the maximum stresses in the areas of attachment of the vertebra legs to the bodies of the vertebrae C3, C4 and in the central part of the upper surface of the vertebra C3 (Figure 2 (c, d)). The change in the regions of localization of maximum Mises stresses in the segment is associated with a change in the character of the deformation of the segment at increase in the modulus of the intervertebral disc, as evidenced by the distribution of displacements along the axis of the segment, shown in Figure 3.

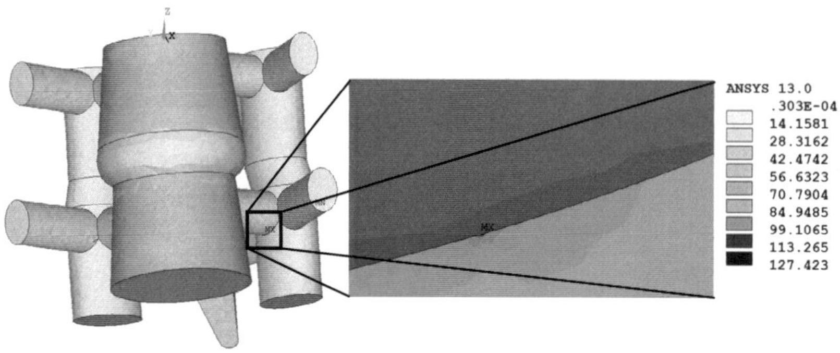

(a)

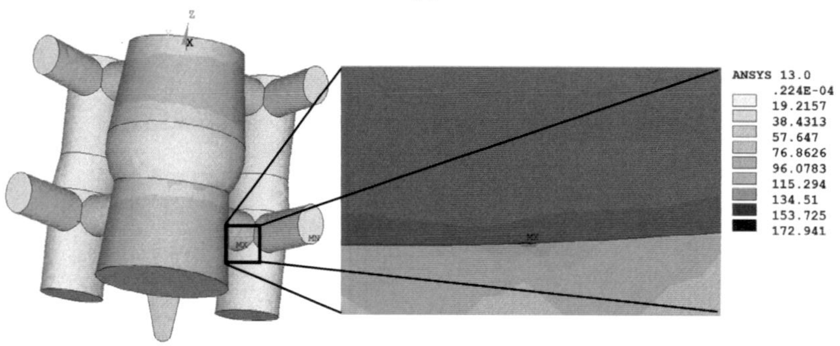

(b)

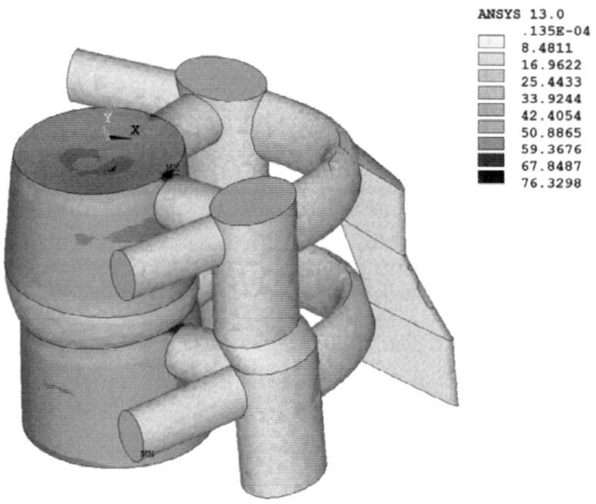

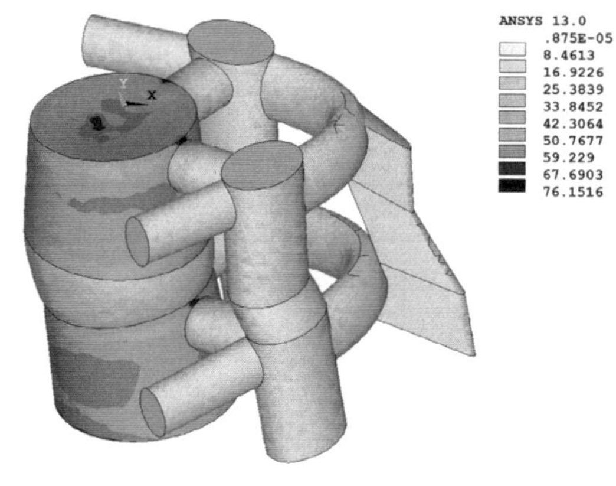

(c) (d)

Figure 2. Mises stress distribution (MPa) in the segment C3-C4 of the cervical spine: (a)- h=4.3 mm, E=2.5 MPa; (b)- h=6 mm, E=2.5 MPa; (c)- h=4.3 mm, E=98.0 MPa; (d)- h=6 mm, E=98 MPa.

It can be seen from Figure 3 that under Young's modulus of the disk 2.5 MPa, the greatest compression in the direction of the Z axis is observed in the anterior part of the vertebral body C3, affecting the upper part of the intervertebral disc, and stretching in the spinous process of the vertebra C3. As Young's modulus of the MTD increases, the greatest compression region shifts toward the center of the upper surface of the vertebra C3. Decreasing the MTD height, as well as increasing its Young's modulus, leads to a decrease in the degree of compression and stretching of indicated regions in the direction of the Z axis.

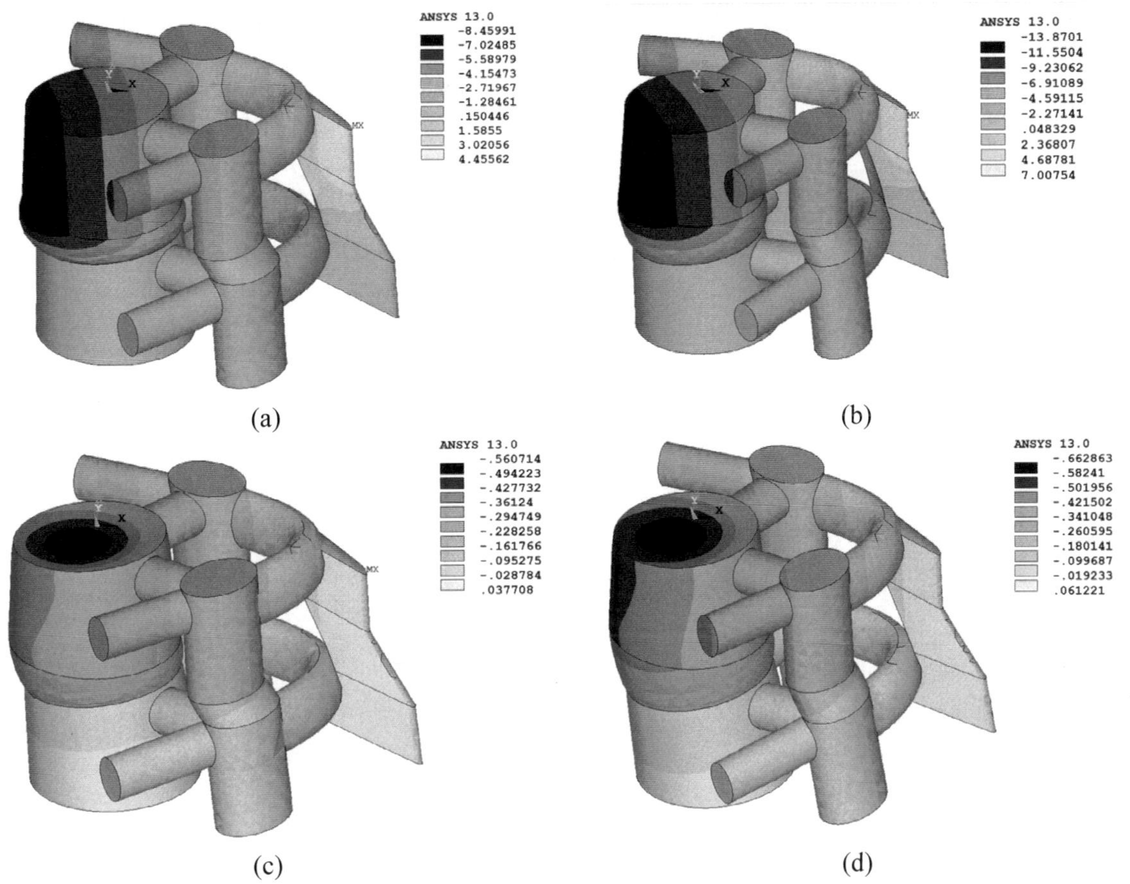

Figure 3. The displacement fields UZ (mm) of segments: (a)- h=4.3 mm, E=2.5 MPa; (b)- h=6 mm, E=2.5 MPa; (c)- h=4.3 mm, E=98.0 MPa; (d)- h=6 mm, E=98 MPa.

4. Model of the intervertebral disc prosthesis

The algorithm for constructing the geometric model of the intervertebral disc prosthesis is developed in the ANSYS programming language APDL. The geometric model of the intervertebral disc prosthesis is based on the experimental sample of a ceramic endoprosthesis developed at the Institute of Strength Physics and Materials Science SB RAS [8].

Figure 4 shows the geometrical model of the intervertebral disc prosthesis built in ANSYS. The Z axis is directed along the axis of the prosthesis (spine axis), the Y axis in the anteroposterior direction, the X axis in the lateral direction. Geometrical model includes two opposing plates: upper plate and lower plate. The upper and lower plates contain elements forming a pair of conjugation (Figure 5). It is assumed that the upper plate of the prosthesis will be attached to the lower surface of the upper vertebra, and the lower one to the upper surface of the lower vertebra respectively.

5. Conclusion

The following results and conclusions are obtained.

The increase in the Young's modulus of the intervertebral disc entails a change in the regions of Mises stress localization, which is associated with a change in the axial character of the deformation of the spine segment. The increase in the modulus of elasticity of the disc leads to the localization of the maximum stresses in the areas of attachment of the vertebrae legs to the vertebrae bodies C3, C4 and in the central part of the upper surface of the vertebra C3. An algorithm for constructing the geometric model of intervertebral disk prosthesis in the ANSYS programming language APDL is developed. The geometric model of the intervertebral disc prosthesis is based on the experimental sample of a ceramic endoprosthesis developed at the Institute of Strength Physics and Materials Science SB RAS. The constructed computer model is designed to study the behavior of the prosthesis in the segment of the spine under physiological loads and to study its effect on the stress-strain state of the spinal segment.

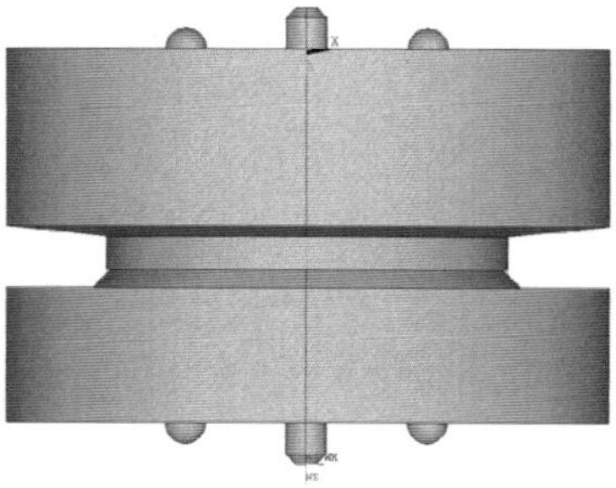

Figure 4. Geometrical model of the intervertebral disc prosthesis (front view).

(a) (b)

Figure 5. Geometric models of the structural components of the model prosthesis of the intervertebral disc: (a) the upper plate of the prosthesis, (b) the lower plate of the prosthesis.

Acknowledgments
The authors express their gratitude to Professor S. P. Buyakova and Professor S. N. Kulkov for providing the data for building a computer model of the intervertebral disc prosthesis.

References

[1] Latif M J A 2011 *Characterisation and Modelling of Spinal Facet Joints* (Leeds: University of Leeds) 168
[2] Panzer M B 2006 *Numerical modeling of the human cervical spine in frontal impact* (Waterloo: University of Waterloo) 248
[3] Wang Y, Chen H, Zhang L, Liu J and Wang Z 2012 *Chinese Journal of Traumatology* **15** 67
[4] Tan S H, Teo E C and Chua H C 2004 *European Spine Journal* **13** 137
[5] Ng H W, Teo E C and Lee V S 2001 *Journal of Biomechanics* **23** 155
[6] Yang H, Jekir M G, Davis M W and Keaveny T M 2016 *Journal of Biomechanics* **49** 1134
[7] Hueston S, Makola M, Mabe I and Goswami T 2012 Cervical spine anthropometric and finite element biomechanical analysis *Human Musculoskeletal Biomechanics* ed T Goswami (InTech, Rijeka, Croatia) pp. 107-158
[8] Buyakova S P, Sablina T Yu, Kulkov S N and Grigoriev M V 2016 Endoprosthesis of the intervertebral disc *Ru. Patent application* No. 2016147506

IOP Conf. Series: Journal of Physics: Conf. Series **894** (2017) 012082 doi:10.1088/1742-6596/894/1/012082

Leading coefficient's recovering problem for nonlinear convection–diffusion–reaction equation

Zh Yu Saritskaya[1] **and R V Brizitskii**[1,2]

[1] Far Eastern Federal University, Vladivostok, Russia

[2] Institute of Applied Mathematics FEB RAS, Vladivostok, Russia

E-mail: zhsar@icloud.com, mlnwizard@mail.ru

Abstract. The problem of diffusion coefficient identification for a nonlinear convection–diffusion–reaction equation, in which reaction coefficient has a rather common dependence on substance's concentration and on spacial variable is considered.

1. Introduction. The boundary value problem

Inverse problems' research for linear and nonlinear heat-and-mass transfer models has been an urgent issue during a long period of time. One of the main problems is an identification problem for unknown densities of boundary and distributed sources or of model's differential equations' coefficients or boundary conditions with the help of additional information about system's conditions, which is described by the model. The studying of such inverse problems can be reduced to the studying of corresponding extremal problems using the optimization method [1]. A number of papers is dedicated to the description of this method for heat-and-mass transfer and other hydrodynamic models (e.g. see [2–11]).

In this paper the diffusion coefficient's identification problem is studied for the nonlinear convection-diffusion-reaction equation in the case, when the reaction coefficient has a rather common dependence on substance's concentration and on spacial variable, is considered. The inverse coefficient problems for nonlinear heat-and-mass transfer models were considered in [5, 6] and [9, 10], also for semilinear convection-diffusion-reaction equation these problems were studied in [12, 13].

See [12–17] on studying of boundary value and extremum problems for the semilinear convection-diffusion-reaction model.

In a bounded domain $\Omega \subset \mathbf{R}^3$ with the boundary Γ the following boundary value problem is considered

$$-\mathrm{div}(\lambda(\mathbf{x})\nabla\varphi) + \mathbf{u}\cdot\nabla\varphi + k(\varphi,\mathbf{x})\varphi = f \text{ in } \Omega, \quad \varphi = \psi \text{ on } \Gamma. \tag{1}$$

Here function φ means polluting substance's concentration, $\mathbf{u}$ is a given vector of velocity, f is a volume density of external sources of substance, $\lambda(\mathbf{x})$ – a diffusion coefficient, function $k(\varphi,\mathbf{x})$ is a reaction coefficient, ψ is a given boundary function. This problem (1) will be called Problem 1 below.

In this paper, firstly, a Problem 1's global solvability is proved in a general case. An identification problem of the diffusion coefficient $\lambda(\mathbf{x})$ is reduced to the multiplicative control problem. It's solvability is proved for the common type reaction coefficient. For concrete

Content from this work may be used under the terms of the Creative Commons Attribution 3.0 licence. Any further distribution of this work must maintain attribution to the author(s) and the title of the work, journal citation and DOI.
Published under licence by IOP Publishing Ltd

reaction coefficients, as a rule, with a power dependence on the concentration, such as $k(\varphi, \mathbf{x}) = \beta_1(\mathbf{x})\varphi^2|\varphi| + \beta_2(\mathbf{x})\varphi^2 + \beta_3(\mathbf{x})$, the optimality systems for extremum problem are obtained.

While studying Problem 1 and optimal control problems, Sobolev spaces will be used: $H^s(D)$, $s \in \mathbf{R}$ $L^r(D)$, $1 \le r \le \infty$, where D is either a domain Ω or its boundary Γ. Inner products in $L^2(\Omega)$ and $H^1(\Omega)$ are denoted by $(\cdot,\cdot)$ $(\cdot,\cdot)_1$, inner products in $L^2(\Gamma)$ – by $(\cdot,\cdot)_\Gamma$, norm in $L^2(\Omega)$ – by $\|\cdot\|_\Omega$, norm and scalar product in $H^s(\Omega)$, $s > 0$ – by $\|\cdot\|_{s,\Omega}$ and $(\cdot,\cdot)_s$, semi-norm in $H^1(\Omega)$ – by $|\cdot|_{1,\Omega}$. Also let $\mathbf{Z} = \{\mathbf{v} \in \mathbf{L}^4(\Omega) : \operatorname{div}\mathbf{v} = 0 \text{ in } \Omega\}$.

Let $\mathcal{D}(\Omega)$ be the space of infinitely differentiable, finite functions in Ω, $H_0^1(\Omega)$ be the closure of $\mathcal{D}(\Omega)$ in $H^1(\Omega)$, $H^{-1}(\Omega) = H_0^1(\Omega)^*$. The Poincare-Friedrichs inequality holds: $|\varphi|_{1,\Omega}^2 \ge \delta\|\varphi\|_{1,\Omega}^2$ $\forall\varphi \in H_0^1(\Omega)$, where $\delta = \text{const} > 0$, $L_+^p(\Omega) = \{k \in L^p(\Omega) : k \ge 0\}$, $p \ge 3/2$, $H_{\lambda_0}^s(\Omega) = \{h \in H^s(\Omega) : h \ge \lambda_0 > 0\}$, $s \ge 0$.

Let us also remind that on the strength of Sobolev embedding theorem the space $H^1(\Omega)$ embeds in the space $L^s(\Omega)$ continuously at $s \le 6$ and compactly at $s < 6$, and the estimate takes place with some constant C_s, depending on s and Ω

$$\|\varphi\|_{L^s(\Omega)} \le C_s\|\varphi\|_{1,\Omega} \quad \forall\varphi \in H^1(\Omega). \tag{2}$$

It will be assumed that the following conditions hold:

(i) Ω is a bounded domain in $\mathbf{R}^3$ with boundary $\Gamma \in C^{0,1}$;

(ii) $\mathbf{u} \in \mathbf{Z}$, $f \in L^2(\Omega)$;

(iii) $\lambda \in H_{\lambda_0}^2(\Omega)$, $\psi \in H^{1/2}(\Gamma)$.

In the present paper the reaction coefficient $k(\varphi, \mathbf{x})$ is not an ordinary function of φ and $\mathbf{x}$, but k is an operator, acting from $H^1(\Omega) \to L_+^p(\Omega)$, where $p \ge 3/2$ and satisfying the following conditions:

(1) for any $w_1, w_2 \in B_r = \{w \in H^1(\Omega) : \|w\|_{1,\Omega} \le r\}$ the estimate takes place:

$$\|k(w_1,\cdot) - k(w_2,\cdot)\|_{L^p(\Omega)} \le L\|w_1 - w_2\|_{L^4(\Omega)},$$

where L is a constant, which depends on r, but doesn't depend on w_1, w_2;

(2) with a certain $s \ge 0$ the estimate holds

$$\|k(w,\cdot)\|_{L^p(\Omega)} \le C_{1p}\|w\|_{1,\Omega}^s \quad \forall w \in H^1(\Omega),$$

(3) $k(\varphi,\cdot)\,\varphi$ satisfies the monotonicity condition

$$(k(\varphi_1)\varphi_1 - k(\varphi_2)\varphi_2, \varphi_1 - \varphi_2) \ge 0 \quad \varphi_1, \varphi_2 \in H^1(\Omega).$$

Let us mention that power functions $k(w) = w^2$ and $k(w) = w^2|w|$ from [19, 20] also satisfy the conditions (1)–(3) (see [16]).

Further let us make an example of the operator $k(\varphi, \mathbf{x})$, satisfying (1)–(3) and not being an ordinary function of $(\varphi, \mathbf{x})$: $k(\varphi) = \varphi^2$ in a subdomain $Q \subset \Omega$ and $k(\varphi) = k_0$ in $\Omega \setminus \overline{Q}$, where $k_0 \in L_+^{3/2}(\Omega)$.

The following Green's theorem will be used (see [11, p. 128] for more details about it):

$$(\Delta\varphi, \eta) = -(\nabla\varphi, \nabla\eta) + \langle\partial\varphi/\partial n, \eta\rangle_\Gamma \quad \forall\varphi \in H^1(\Delta, \Omega),\ \eta \in H^1(\Omega), \tag{3}$$

$$(\mathbf{u}\cdot\nabla\varphi, \eta) = -(\mathbf{u}\cdot\nabla\eta, \varphi) \quad \forall\mathbf{u} \in \mathbf{Z},\ \varphi, \eta \in H_0^1(\Omega). \tag{4}$$

Here and below $\langle\cdot,\cdot\rangle_\Gamma$ means the relation of duality between $H^{1/2}(\Gamma)$ and $H^{-1/2}(\Gamma)$.

Let us denote bilinear forms $a_{\mathbf{u}}$ and a_0: $H_0^1(\Omega) \times H_0^1(\Omega) \to \mathbf{R}$ by formulae

$$a_{\mathbf{u}}(\varphi, \eta) = (\mathbf{u}\cdot\nabla\varphi, \eta) = \int_\Omega (\mathbf{u}\cdot\nabla\varphi)\eta d\mathbf{x},\ a_0(\varphi, \eta) = \lambda(\nabla\varphi, \nabla\eta) + a_{\mathbf{u}}(\varphi, \eta) + (k_0\varphi, \eta). \tag{5}$$

Using (2), (4) and Holder's inequality, it's not difficult to check that when conditions (i)–(iii) and $\lambda \in H^2_{\lambda_0}(\Omega)$, $k_0 \in L^p_+(\Omega), p \geq 3/2$ hold, then the forms, introduced in (5), are continuous and

$$|(\lambda(\mathbf{x})\nabla\varphi, \nabla\eta)| \leq \|\lambda(\mathbf{x})\|_{2,\Omega}\|\varphi\|_{1,\Omega}\|\eta\|_{1,\Omega},$$

$$|a_{\mathbf{u}}(\varphi, \eta)| \leq \gamma_1\|\mathbf{u}\|_{\mathbf{L}^4(\Omega)}\|\varphi\|_{1,\Omega}\|\eta\|_{1,\Omega} \ \forall(\varphi, \eta) \in H^1(\Omega)^2, \quad (6)$$

$$|(k_0\varphi, \eta)| \leq \gamma_p\|k_0\|_{L^p(\Omega}\|\varphi\|_{1,\Omega}\|\eta\|_{1,\Omega} \ \ \forall(\varphi, \eta) \in H^1(\Omega)^2, \quad (7)$$

$$a_{\mathbf{u}}(\varphi, \varphi) = 0, \ a_0(\varphi, \varphi) = \lambda(\nabla\varphi, \nabla\varphi) + (k_0\varphi, \varphi) \geq \lambda_*\|\varphi\|^2_{1,\Omega} \ \ \forall\varphi \in H^1_0(\Omega), \ \lambda_* \equiv \delta\lambda_0. \quad (8)$$

Here γ_1 or γ_p are some constants, which depend on Ω or on Ω and p correspondingly.

The proof of the solvability of the Problem 1 is based on the following theorem [21].

Theorem 1. *Let V be a reflexive separable Banach space. And let the operator $A : V \to V^*$ have the following properties:*
1) the operator A is bounded and semicontinuous, i.e. the form $(A(u+\lambda v), w)$ is continuous in $\lambda \in \mathbf{R}$ for all $u, v, w \in V$;
2) the operator A is monotonous, i.e. $(A(u) - A(v), u - v) \geq 0$ for all $u, v \in V$;
3) $(A(v), v)/\|v\|_V \to +\infty$ at $\|v\|_V \to \infty$.

Then the mapping $A : V \to V^*$ is surjective, i.e. for any $l \in V^*$ there is such $u \in V$ that $A(u) = l$.

Let us multiply the equation in (1) by $h \in H^1(\Omega)$ and integrate over Ω, applying the formula (3). Then this will be obtained:

$$(\lambda\nabla\varphi, \nabla h) + (k(\varphi, \mathbf{x})\varphi, h) + (\mathbf{u}\cdot\nabla\varphi, h) = (f, h) \ \ \forall h \in H^1_0(\Omega). \quad (9)$$

The function $\varphi \in H^1(\Omega)$, which satisfies (9), will be called a weak solution of problem 1. To prove the solvability of (9), represent the function φ in the form

$$\varphi = \tilde{\varphi} + \varphi_0, \quad (10)$$

where $\tilde{\varphi} \in H^1_0(\Omega)$ is an unknown function and the function $\varphi_0 \in H^1(\Omega)$ satisfies the conditions

$$\varphi_0 = \psi \text{ on } \Gamma, \ \ \|\varphi_0\|_{1,\Omega} \leq C_\Gamma\|\psi\|_{1/2,\Gamma}.$$

Substituting (10) into (9) we obtain

$$(\lambda\nabla\tilde{\varphi}, \nabla h) + (k(\tilde{\varphi} + \varphi_0, \mathbf{x})(\tilde{\varphi} + \varphi_0), h) + (\mathbf{u}\cdot\nabla(\tilde{\varphi} + \varphi_0), h) =$$

$$= (f, h) - (\lambda\nabla\varphi_0, \nabla h) - (\mathbf{u}\cdot\nabla\varphi_0, h) \ \ \forall h \in H^1_0(\Omega). \quad (11)$$

Adding to both sides of (11) the term $-(k(\varphi_0, \mathbf{x})\varphi_0, h)$, we obtain

$$(\lambda\nabla\tilde{\varphi}, \nabla h) + (k(\tilde{\varphi} + \varphi_0, \mathbf{x})(\tilde{\varphi} + \varphi_0) - k(\varphi_0, \mathbf{x})\varphi_0, h) + (\mathbf{u}\cdot\nabla(\tilde{\varphi} + \varphi_0), h) =$$

$$= (f, h) - (k(\varphi_0, \mathbf{x})\varphi_0, h) - (\lambda\nabla\varphi_0, \nabla h) - (\mathbf{u}\cdot\nabla\varphi_0, h) \ \ \forall h \in H^1_0(\Omega).$$

For the proof of problem (9)'s solvability introduce a nonlinear operator $A : H^1_0(\Omega) \to H^{-1}(\Omega)$ by formula

$$(A(\tilde{\varphi}), h) \equiv (\lambda(\mathbf{x})\nabla\tilde{\varphi}, \nabla h) + (k(\tilde{\varphi} + \varphi_0, \mathbf{x})(\varphi_0 + \tilde{\varphi}) - k(\varphi_0)\varphi_0, h) + (\mathbf{u}\cdot\nabla\tilde{\varphi}, h) \ \forall h \in H^1_0(\Omega).$$

Due to (6)–(8) and property (2) we show that the operator A is continuous:

$$|(A(\tilde{\varphi}_1) - A(\tilde{\varphi}_2), h)| \leq (\|\lambda\|_{2,\Omega} + \gamma_p LC_4\|\tilde{\varphi}_1\|_{1,\Omega} + \gamma_p C_{1p}C_s(\|\tilde{\varphi}_1\|^s_{1,\Omega} + C^s_\Gamma\|\psi\|^s_{1/2,\Gamma})+$$

$$+\gamma_1\|\mathbf{u}\|_{\mathbf{L}^4(\Omega)})\|\tilde{\varphi}_1-\tilde{\varphi}_2\|_{1,\Omega}\|h\|_{1,\Omega}. \quad (12)$$

By definition, $A(0) = 0$, then from (12) at $\tilde{\varphi}_2 = 0$ follows the boundedness of operator A.

The monotonicity of the operator A follows from property (2) and (8). From monotonicity of $k(\varphi,\cdot)\varphi$ and (8) follows that 3) is true.

Arguing further as in [16, 17], we are coming to the next theorem

Theorem 2. *Let conditions (i)–(iii) and (1)–(3) hold. Then there is a unique weak solution $\varphi \in H^1(\Omega)$ of the problem 1 and the following holds*

$$\|\varphi\|_{1,\Omega} \le C_*(\|f\|_\Omega + C_\Gamma(\|\lambda\|_{L^\infty(\Omega)} + \gamma\|\mathbf{u}\|_{\mathbf{L}^4(\Omega)}))\|\psi\|_{1/2,\Gamma} + \gamma_p C_{1p} C_\Gamma^s \|\psi\|^{s+1}_{1/2,\Gamma}) + C_\Gamma\|\psi\|_{1/2,\Gamma}.$$

2. Optimal control problem

There are a lot of situations in practice, when some parameters of Problem 1 are unknown and it's required to determine them together with the solution φ according to some information about the solution. On the capacity of the mentioned information about solution, the values $\varphi^d(\mathbf{x})$ of the concentration φ are often taken, which can be measured in points of some set $Q \subset \Omega$.

Below we study the case, when the diffusion coefficient $\lambda(\mathbf{x})$ is unknown and should be searched together with the solution φ.

To solve this inverse problem the optimization method will be used, according to which the considered problem is reduced to the solving of the control problem.

In compliance with this method let's divide the whole set of initial data of problem 1 into two groups: the group of fixed data, where functions $\mathbf{u}$, f, $k(\varphi,\mathbf{x})$ and ψ will be included, and the group of controls, which consists of function $\lambda(\mathbf{x})$, assuming that it can be changed in some set K.

Let us introduce the operator of the direct problem $F = (F_1, F_2) : H^1(\Omega) \times K \to Y = H^{-1}(\Omega) \times H^{1/2}(\Gamma)$ by the formulae

$$\langle F_1(\varphi,\lambda), h\rangle = (\lambda(\mathbf{x})\nabla\varphi, \nabla h) + (k(\varphi,\mathbf{x})\varphi, h) + (\mathbf{u}\cdot\nabla\varphi, h) - (f, h),$$

$$F_2(\varphi) = \varphi|_\Gamma - \psi.$$

Then (9) can be rewritten in the following form:

$$F(\varphi,\lambda) = 0. \quad (13)$$

Treating (13) as a conditional restriction on the state $\varphi \in H^1(\Omega)$ and on the control $\lambda \in K$, the problem of conditional minimization can be formulated as follows:

$$J(\varphi,\lambda) \equiv \frac{\mu_0}{2}I(\varphi) + \frac{\mu_1}{2}\|\lambda\|^2_{2,\Omega} \to \inf,\ F(\varphi,\lambda) = 0,\ (\varphi,\lambda) \in H^1(\Omega)\times K. \quad (14)$$

Here $I : H^1(\Omega) \to \mathbf{R}$ is a weakly semicontinuous below functional.

The set of possible pairs for problem (14) is denoted by $Z_{ad} = \{(\varphi,\lambda) \in H^1(\Omega)\times K : F(\varphi,\lambda) = 0, J(\varphi,\lambda) < \infty\}$ and let's suppose that these conditions hold

(j) $K \subset H^2_{\lambda_0}(\Omega)$ is a nonempty convex closed set;

(jj) $\mu_0 > 0$, $\mu_1 \ge 0$ and K is a bounded set or $\mu_0 > 0$, $\mu_1 > 0$ and functional I is bounded below.

The following cost functionals can be used in the capacity of the possible ones [11]:

$$I_1(\varphi) = \|\varphi - \varphi^d\|^2_Q = \int_\Omega |\varphi - \varphi^d|^2 d\mathbf{x},\ \ I_2(\varphi) = \|\varphi - \varphi^d\|^2_{1,Q}.$$

Here $\varphi^d \in L^2(Q)$ (or $\varphi^d \in H^1(Q)$) is a given function in some subdomain $Q \subset \Omega$.

Theorem 3. *Let conditions (i)–(iii), (1)–(3) and (j), (jj) hold, $I : H^1(\Omega) \to \mathbf{R}$ is a weakly semicontinuous below functional, and Z_{ad} is nonempty set. Then there is at least one solution $(\varphi, \lambda) \in H^1(\Omega) \times K$ of optimal control problem (14).*

Proof. Let $(\varphi_m, \lambda_m) \in Z_{ad}$ be a minimizing sequence for the functional J, for which the following is true:

$$\lim_{m\to\infty} J(\varphi_m, \lambda_m) = \inf_{(\varphi,\lambda)\in Z_{ad}} J(\varphi, \lambda) \equiv J^*.$$

That and the conditions of theorem for the functional J from (14) imply the estimate $\|\lambda_m\|_{2,\Omega} \leq c_1$. From Theorem 2 follows directly that $\|\varphi_m\|_{1,\Omega} \leq c_2$, where constants c_1, c_2 don't depend on m.

Then the weak limits $\varphi^* \in H^1(\Omega)$ and $\lambda^* \in K$ of some subsequences of the sequences $\{\varphi_m\}$ and $\{\lambda_m\}$ exist. Corresponding sequences will be also denoted by $\{\varphi_m\}$ and $\{\lambda_m\}$. With this in mind it can be considered that

$$\varphi_m \to \varphi^* \in H^1(\Omega) \text{ weakly in } H^1(\Omega) \text{ and strongly in } L^s(\Omega),\ s < 6, \tag{15}$$

$$\lambda_m \to \lambda^* \in K \text{ weakly in } H^2(\Omega) \text{ and strongly in } L^\infty(\Omega). \tag{16}$$

Let's show that $\mathcal{F}(\varphi^*, \lambda^*) = 0$, i.e.

$$(\lambda_* \nabla\varphi^*, \nabla h) + (k(\varphi^*, \mathbf{x})\varphi^*, h) + (\mathbf{u} \cdot \nabla\varphi^*, h) = (f, h)\ \ \forall h \in H_0^1(\Omega). \tag{17}$$

And it should be taken into account that the pair (φ_m, λ_m) satisfies the relations

$$(\lambda_m \nabla\varphi_m, \nabla h) + (k(\varphi_m, \mathbf{x})\varphi_m, h) + (\mathbf{u} \cdot \nabla\varphi_m, h) = (f, h)\ \ \forall h \in H_0^1(\Omega),\ m = 1, 2, \ldots. \tag{18}$$

Let's pass to the limit in (18) at $m \to \infty$. All linear summands in (18) turn into corresponding ones in (17).

From condition (1) follows that $k(\varphi_m, \cdot) \to k(\varphi, \cdot)$ strongly in $L^{3/2}(\Omega)$. Taking into account that $\varphi_m \to \varphi$ strongly in $L^3(\Omega)$ by (16), we obtain that

$$k(\varphi_m, \cdot)\varphi_m \to k(\varphi^*, \cdot)\varphi^* \text{ strongly in } L^1(\Omega)$$

and

$$(k(\varphi_m, \mathbf{x})\varphi_m, h) \to (k(\varphi^*, \mathbf{x})\varphi^*, h) \text{ at } m \to \infty\ \forall h \in H_0^1(\Omega).$$

From identity

$$(\lambda_m \nabla\varphi_m, \nabla h) - (\lambda^* \nabla\varphi^*, \nabla h) = ((\lambda_m - \lambda^*)\nabla\varphi_m, \nabla h) + (\lambda^* \nabla(\varphi_m - \varphi^*), \nabla h),$$

relations (16), (15) and Theorem 2 we obtain that $(\lambda_m \nabla\varphi_m, \nabla h) \to (\lambda^* \nabla\varphi^*, \nabla h)$ at $m \to \infty$ for all $h \in H_0^1(\Omega)$.

As the functional J is weakly semicontinuous below on $H^1(\Omega) \times H^2(\Omega)$, then from aforesaid follows that

$$J^* = \lim_{m\to\infty} J(\varphi_m, \lambda_m) = \underline{\lim}_{m\to\infty} J(\varphi_m, \lambda_m) \geq J(\varphi^*, \lambda^*) \geq J^*. \ \blacksquare$$

Let us $Y^* = H_0^1(\Omega) \times H^{-1/2}(\Gamma)$ is the dual space for $Y = H^{-1}(\Omega) \times H^{1/2}(\Gamma)$, where $H^{-1/2}(\Gamma) = H^{1/2}(\Gamma)^*$. We introduce the Lagrange multiplier $\mathbf{y}^* = (\theta, \zeta) \in Y^*$ and Lagrangian $\mathcal{L}$ by formula

$$\mathcal{L}(\varphi, \lambda, \mathbf{y}^*) = J(\varphi, \lambda) + \langle \mathbf{y}^*, F(\varphi, \lambda) \rangle =$$

$$= J(\varphi, \lambda) + \langle F_1(\varphi, \lambda), \theta \rangle_{H^{-1}(\Omega) \times H_0^1(\Omega)} + \langle \zeta, F_2(\varphi, \lambda) \rangle_\Gamma.$$

Arguing as in [16, 17] and using results of [18], we justify the Lagrange principle for the concrete reaction coefficient.

Let $k(\varphi, \mathbf{x}) = \beta_1(\mathbf{x})\varphi^2|\varphi| + \beta_2(\mathbf{x})$, where $\beta_1(\mathbf{x}) \in H_+^2(\Omega)$, $\beta_2(\mathbf{x}) \in L_+^2(\Omega)$.

Theorem 4. *Suppose that under conditions (i)–(iii) and (j), (jj), a pair* $(\hat{\varphi}, \hat{\lambda}) \in H^1(\Omega) \times K$ *is an element afforing a local minimum in problem* (14) *and let the functional* I *be continuously differentiable with respect to* φ *at the point* $\hat{\varphi}$. *Then, there exists a unique Lagrange multiplier* $\mathbf{y}^* = (\theta, \zeta) \in H_0^1(\Omega) \times H^{-1/2}(\Gamma)$ *for wich Euler–Lagrange equation*

$$\langle J'_\varphi(\hat{\varphi}, \hat{\lambda}), \tau \rangle + \langle F'_\varphi(\hat{\varphi}, \hat{\lambda})\tau, \theta \rangle = 0 \;\; \forall \tau \in H^1(\Omega)$$

is satisfied and equavivalent the relation

$$(\lambda \nabla \tau, \nabla \theta) + 4((\beta_1 \hat{\varphi}^2 |\hat{\varphi}| + \beta_2)\tau, \theta) + (\mathbf{u} \cdot \nabla \tau, \theta) + \langle \zeta, \tau \rangle_\Gamma = -(\mu_0/2)\langle I'_\varphi(\hat{\varphi}), \tau \rangle \;\; \forall \tau \in H^1(\Omega) \quad (19)$$

and minimum principle $\langle \mathcal{L}'_\lambda(\hat{\varphi}, \hat{\lambda}, \theta), \lambda - \hat{\lambda} \rangle \geq 0$ *holds for all* $\lambda \in K$, *which equivalent to the inequality*

$$\mu_1(\hat{\lambda}, \lambda - \hat{\lambda})_2 - ((\lambda - \hat{\lambda})\nabla \hat{\varphi}, \nabla \theta) \geq 0 \;\; \forall \lambda \in K. \quad (20)$$

The Problem 1's weak formulation (9), Euler–Lagrange equation (19) and minimum principle (20) are the optimality system for the extremum problem (14). This system describes the first-order necessary optimality conditions for problem (14) and will be used for studying of the uniqueness and stability of solutions to problem (14) in future papers.

Acknowledgments

The first author was supported by the Russian Science Foundation (project no. 14-11-00079), the second author was supported by Russian Foundation for Basic Research (project no. 16-01-00103-a).

References

[1] Tikhonov A N and Arsenin V Ya 1986 *Methods for Solving Ill-Posed Problems* (Moscow: Nauka)
[2] Ito K and Kunish K 1997 *Estimation of the convection coefficient in elliptic equations Inverse Problems* **14** 995–1013
[3] Agoshkov V I, Minuk F P, Rusakov A S and Zalesny V B 2005 *Study and solution of identification problems for nonstationary 2D- and 3D-convection–diffusion–reaction Russ. J. Numer. Anal. Math. Modelling* **20** 19–43
[4] Alekseev G V and Tereshko D A 1998 *On solvability of inverse extremal problems for stationary equations of viscous heat conducting fluid J. Inverse Ill-Posed Probl.* **6** 521–62
[5] Alekseev G V 2002 *Inverse extremal problems for stationary equations in mass transfer theory Comp. Math. Math. Phys.* **42** 363–76
[6] Alekseev G V 2001 *Solvability of inverse extremal problems for stationary heat and mass transfer equations Sib. Math. J.* **42** 811–27
[7] Nguyen P A and Raymond J P 2001 *Control problems for convection–diffusion equations with control localized on manifolds ESAIM: Control, Optimisation and Calculus of Variations* **6** 467–88
[8] Nguyen P A and Raymond J P 2011 *Pointwise control of the Boussinesq system Systems Control Lett.* **60** 249–55
[9] Alekseev G V, Soboleva O V and Tereshko D A 2008 *Identification problems for a steady-state model of mass transfer J. Appl. Mech. Tech. Phys.* **49** 537–47
[10] Alekseev G V, Vakhitov I S and Soboleva O V 2012 *Stability estimates in identification problems for the convection-diffusion-reaction equation Comp. Math. Math. Phys.* **52** 1635–49
[11] Alekseev G V 2010 *Optimization in Stationary Problems of Heat and Mass Transfer and Magnetohydrodynamics* (Moscow: Nauchnyi Mir)

[12] Brizitskii R V, Saritskaya Zh Yu and Byrganov A I 2016 *Multiplicative control problems for nonlinear convection-diffusion-reaction equation Siberian Electronic Mathematical Reports* **13** 352– 60

[13] Brizitskii R V Saritskaya Zh Yu 2016 *On solvability of inverse coefficient problems for nonlinear convection-diffusion-reaction equation J. Phys.: Conf. Series* **722** 012007

[14] Brizitskii R V and Saritskaya Zh Yu 2015 *Boundary value and optimal control problems for nonlinear convection-diffusion-reaction equation Siberian Electronic Mathematical Reports* **12** 447–56

[15] Alekseev G V, Brizitskii R V and Saritskaya Zh Yu 2016 *Stability estimates of extremum problem's solutions for nonlinear convection–diffusion–reaction equation Sib. J. Industrial Math.* **10** 155–67

[16] Brizitskii R V and Saritskaya Zh Yu 2016 *Stability of solutions to extremum problems for the nonlinear convectiondiffusionreaction equation with the Dirichlet condition Comp. Math. Math. Phys.* **56** 2011–22

[17] Brizitskii R V and Saritskaya Zh Yu 2017 *Stability of solutions of control problems for the convectiondiffusionreaction equation with a strong nonlinearity Diff. Eq.* **53** 485–96

[18] Fursikov A V 2000 *Optimal Control of Distributed Systems. Theory and Aplications* (Providence: American Mathematical Society)

[19] Kovtanyuk A E, Chebotarev A Yu, Botkin N D and Hoffman K-H 2015 *Unique solvability of a steady-state heat tranfer model Commun Nonlinear Sci. Number Simulat.* **20** 776–84

[20] Kovtanyuk A E and Chebotarev A Yu 2014 *Stationary free convection problem with radiative heat exchange Diff. Eq.* **50** 1592–99

[21] Lions J-L 1972 *Some Methods of Solutions of Nonlinear Boundary Value Problem* (Moscow: Nauka)

IOP Conf. Series: Journal of Physics: Conf. Series **894** (2017) 012083 doi:10.1088/1742-6596/894/1/012083

Spontaneous emission in the linear problem of shock wave disturbance in a non-classical case

E V Semenko[1,2] **and T I Semenko**[1,2]

[1] Novosibirsk State Technical University, Novosibirsk, Russia

[2] Novosibirsk State Pedagogical University, Novosibirsk, Russia

E-mail: semenko54@gmail.com

Abstract. The isentropic solutions of the linear problem of the shock wave disturbance are considered for a special (non-classical) case, where the both pre-shock and post-shock flows are subsonic. As it is known, only neutral stability is possible here, hence the spontaneous emission appears. The spontaneously emitted waves are studied and their general form is determined both for damped and oscillating initial values. For oscillating initial values both non-resonant and resonant cases are considered. Some interesting distinctions from the classical case are found. First, the spontaneously emitted acoustic waves are evanescent with respect to the variable normal to the shock but oscillate with respect to time. Second, only initial pre-shock vorticity wave may cause the resonance, while the resonance is impossible for initial acoustic both pre-shock and post-shock waves.

1. Introduction

A classical problem of the shock wave, i.e., of a discontinuous solution of the Euler equations, is considered, see for example the classical works [1, 2, 3, 4, 5, 6]. If the equation of state satisfies the condition $(\partial^2 V/\partial p^2)_s > 0$, where $V = 1/\rho$ is the specific volume, p is the pressure and s is entropy, then the entropy, pressure, and density increase across the shock (i.e., only a compression wave is possible), the pre-shock flow is supersonic, and the post-shock flow is subsonic, which in turn ensures the well-posedness of the Cauchy problem [1]. This case is called the classical one.

The case where the above-mentioned condition is not satisfied is considered as a non-classical problem. These non-classical cases (known as "anomalous" or "exotic" shock wave behavior), and also the existence (admissibility) of rarefaction shock waves, were studied by many researches, e.g. [6, 7, 8, 9, 10, 11, 12, 13], the list is far from being complete. Generally they investigate the existence of the piecewise-constant discontinuous solutions. The problem of disturbance of these solutions, and particulary their linear stability, usually isn't studied. The problem of shock wave disturbance for isentropic flows was studied in papers [11, 12]. As it was clarified, only neutral stability is possible there. At last, in author's paper [13] the isentropic solution of linear problem of shock wave disturbance was constructed and its properties were investigated.

The so-called spontaneous emission, i.e. the existence of refracted or reflected (emitted) plane waves at the absence of the incident plane waves, occurs in the case of the neutral stability. In the present paper, we announce some results of the spontaneous emission in the mentioned non-classical case. All investigations are based on the solution constructed in [13]. On the other

Content from this work may be used under the terms of the Creative Commons Attribution 3.0 licence. Any further distribution of this work must maintain attribution to the author(s) and the title of the work, journal citation and DOI.
Published under licence by IOP Publishing Ltd

hand, all considerations are absolutely similar to the papers [14, 15, 16] devoted to the classical case, and we use the same terms and notations as there.

2. Problem formulation

Since we consider the isentropic flow, the sought-for functions are density ρ, velocity $U = (U_x, U_y)$, pressure p, specific (per unit volume) internal energy ϵ, enthalpy w, and specific energy e. All sought-for values are considered as functions of (x, y, t). These functions satisfies the relations $p = p(\rho)$ (equation of state), $\epsilon = \epsilon(\rho)$, $w = \epsilon + p/\rho$, $e = \rho\epsilon + \rho|U|^2/2$ and Euler equations. We are looking for the discontinuous solutions, so the standard Rankine–Hugoniot conditions take place on the surface of discontinuity $x = f(y, t)$, which is the sought-for function too. According to the common terminology, the region $x < f(y, t)$ is called the pre-shock zone, and the region $x > f(y, t)$ is the post-shock zone.

In our (isentropic) case, outside the shock the energy conservation law is equivalent to the entropy conservation law $ds/dt = \partial s/\partial t + (U, \nabla s) = 0$ [1]; thus, it is automatically satisfied. Therefore, only three Euler equations remain outside the shock:

$$\frac{\partial\rho}{\partial t} + \operatorname{div}(\rho U) = 0, \quad \rho\frac{dU_x}{dt} + \frac{\partial p}{\partial x} = 0, \quad \rho\frac{dU_y}{dt} + \frac{\partial p}{\partial y} = 0, \quad x \neq f(y, t). \tag{1}$$

In turn, for discontinues functions, i.e., those on the shock, the energy conservation law is not equivalent to the entropy conservation law. That means we have four Rankine–Hugoniot conditions in the form

$$[J] = 0, \quad J[U_x] + [p] = 0, \quad [U_y + f'_y U_x] = 0, \quad \left[\frac{|U|^2}{2} + w\right] = f'_t[U_x], \tag{2}$$

where

$$J = J(y, t) = \rho(U_x - f'_t - f'_y U_y)\Big|_{x=f(y,t)}$$

is the flux through the shock and, as usual, the square brackets denote the jump of a value across the discontinuity.

3. Basic solution, linear problem

We are looking for a basic solution as a piecewise-constant solution in the form

$$f_0(y, t) \equiv 0\,; \; \rho_0 \equiv \rho_0^{\pm} = \text{const}\,, \; U_0 \equiv U_0^{\pm} = (u_x^{\pm}, u_y) = \text{const}\,, \; \pm x < 0$$

(hereinafter we use the plus and minus superscripts for the pre-shock and post-shock zones respectively). As it is shown in [13] for van der Waals gas, if the flux $J_0 = \rho_0^{\pm} u_x^{\pm}$ is small enough then such solutions exist and the pre-shock and post-shock flows are subsonic: $u_x^{\pm} < c^{\pm}$, where $c^{\pm} = \sqrt{p'(\rho^{\pm})}$ is the sound velocity.

For the problem, linearized at the basic solution, we use the following notations: the equation of the shock perturbation is $x = f(y, t)$ and the sought-for vector of disturbed quantities is

$$G(x, y, t) = \begin{pmatrix} \delta\rho \\ \rho_0\,\delta U_x/c \\ \rho_0\,\delta U_y/c \end{pmatrix} = \begin{pmatrix} \delta\rho^{\pm} \\ \rho_0^{\pm}\,\delta U_x^{\pm}/c^{\pm} \\ \rho_0^{\pm}\,\delta U_y^{\pm}/c^{\pm} \end{pmatrix} = G^{\pm}\,,$$

IOP Conf. Series: Journal of Physics: Conf. Series **894** (2017) 012083 doi:10.1088/1742-6596/894/1/012083

$\pm x \leq 0$. Note that we reduce the space of sought-for values to one dimension (density). Then the linearized system (1) takes the form

$$B^{\pm}G^{\pm} = 0\,, \quad B = \begin{pmatrix} \dfrac{d}{dt} & c\dfrac{\partial}{\partial x} & c\dfrac{\partial}{\partial y} \\ c\dfrac{\partial}{\partial x} & \dfrac{d}{dt} & 0 \\ c\dfrac{\partial}{\partial y} & 0 & \dfrac{d}{dt} \end{pmatrix},$$

$$\frac{d}{dt} = \frac{\partial}{\partial t} + (u, \nabla), \quad u = (u_x^{\pm}, u_y), \quad x \neq 0,$$

and the linearized system of the Rankine–Hugoniot conditions (2) is

$$A^{+}A_0^{+}G_{\Gamma}^{+} = A^{-}A_0^{-}G_{\Gamma}^{-} + F_0 f\,,$$

where $G_{\Gamma}^{\pm} = G^{\pm}(0, y, t)$,

$$A_0^{\pm} = \begin{pmatrix} u_x^{\pm} & c^{\pm} & 0 \\ c^{\pm} & u_x^{\pm} & 0 \\ 0 & 0 & u_x^{\pm} \end{pmatrix}, \quad A^{\pm} = \begin{pmatrix} 1 & 0 & 0 \\ u_x^{\pm} & c^{\pm} & 0 \\ 0 & 0 & c^{\pm} \\ 0 & u_x^{\pm}c^{\pm} & 0 \end{pmatrix},$$

$$F_0 = \begin{pmatrix} -[\rho_0](\partial/\partial t + u_y\partial/\partial y) \\ 0 \\ J_0[u_x]\partial/\partial y \\ -J_0[u_x](\partial/\partial t + u_y\partial/\partial y) \end{pmatrix}.$$

The initial conditions are $G(x, y, 0) = G_0(x, y) = G_0^{\pm}(x, y)$, $\pm x < 0$.

4. General form of the solution

To solve the problem, we use the so-called one-sided Fourier transform [13, 14, 15, 16]. We denote the dual (Fourier) variables for (x, y, t) as (ξ, η, ω) respectively. With respect to these Fourier variables the problem became algebraic, and all sought and given quantities become analytic (in generalized sense) with respect to ω in the upper half-plane $\operatorname{Im}\omega > 0$ and with respect to ξ in the upper half-plane $\operatorname{Im}\xi > 0$ for the pre-shock zone and in the lower half-plane $\operatorname{Im}\xi < 0$ for the post-shock zone. The full solution of that problem is described in [14] for classical case and in [13] for introduced non-classical one. We briefly describe its qualitative properties.

The whole solution is a sum of two terms. The first term is straightforwardly determined by the initial perturbations. The second term and the shock disturbance f are determined by initial perturbations by means of the solution of the boundary (i.e. on the shock) system of linear equations with respect to the Fourier variables (η, ω). Further we call it as the basic system. So this second term together with shock disturbance are the result of the interaction between the initial (incident upon the shock) waves and the shock itself. These waves induced by the interaction with the shock are refracted and reflected waves.

The matrix of basic system depends on the basic solution, and right hand side of the system depends on the basic solution and initial perturbations. The determinant of that system for any η has two real zeros $\omega_{\pm} = \eta(u_y \pm \alpha_0)$, where

$$\alpha_0 = \sqrt{\frac{-A(B^{+} + B^{-}) + \sqrt{A^2(B^{+} + B^{-})^2 + 4A(1 - A)AB^{+}B^{-}}}{2(1 - A)}},$$

$$A = \left(\frac{u_x^+ u_x^-}{c^+ c^-}\right)^2, \quad B^\pm = (c^\pm)^2 - (u_x^\pm)^2,$$

which corresponds to the case of neutral stability (of D'yakov–Kontorovich instability), and thus the solution of the basic system contains additional singular terms, e.g. in a form $I_1(\omega, \omega_\pm)$ (see Appendix). Just these singular terms produce the spontaneously emitted waves similar to the classical case [16].

Beside that, each solution naturally decomposes into the acoustic wave G_a (it is the solution of wave equation $d^2G_a/dt^2 - \Delta G_a = 0$ and it represents the density and potential component of the velocity field) and vorticity wave G_v (it is the solution of transfer equation $dG_v/dt = 0$ and represents the solenoidal component of the velocity field) [13, 14, 15, 16]. That decomposition takes place both for initial and for induced (refracted and reflected) waves. Since the induced wave in the pre-shock zone is acoustic only [13], we denote it as G_a^+. The induced wave in the post-shock zone contains both terms G_a^- and G_v^-.

We describe the general form of spontaneously emitted waves. To formulate the results, we use the zeros of symbols of the transfer operator $-id/dt$ (the symbol $Q^\pm = \xi u_x^\pm + \eta u_y - \omega$) and the wave operator $-d^2/dt^2 + \Delta$ (the symbol $P^\pm = (Q^\pm)^2 - (c^\pm)^2(\xi^2 + \eta^2)$).

5. Allocation of zeros; some terms and notations

In this sections, we drop the superscripts $\pm$ for brevity. We introduce the roots of P and Q directly for the complex ω, $\operatorname{Im}\omega \geq 0$.

We express the roots of P as

$$\xi_1(\eta, \omega) = \frac{u_x(\omega - \eta u_y) + c\sqrt{(\omega - \eta u_y)^2 - \eta^2(c^2 - u_x^2)}}{u_x^2 - c^2},$$

$$\xi_2(\eta, \omega) = \frac{u_x(\omega - \eta u_y) - c\sqrt{(\omega - \eta u_y)^2 - \eta^2(c^2 - u_x^2)}}{u_x^2 - c^2},$$

where the branch of the square root is chosen under the following condition: for $\omega = \eta u_y + i\alpha$ and $\alpha > 0$

$$\sqrt{(\omega - \eta u_y)^2 - \eta^2(c^2 - u_x^2)} = i\sqrt{\alpha^2 + \eta^2(c^2 - u_x^2)}.$$

The roots $\xi_{1,2}$ are analytic with respect to ω at $\operatorname{Im}\omega > 0$, at that $\operatorname{Im}\xi_1(\eta, \omega) < 0$ and $\operatorname{Im}\xi_2(\eta, \omega) > 0$.

The root of Q is marked by the subscript 3: $\xi_3(\eta, \omega) = (\omega - \eta u_y)/u_x$. Obviously, $\operatorname{Im}\xi_3 > 0$ at $\operatorname{Im}\omega > 0$.

Now we are ready to consider (formulate the results) the spontaneous emission for different kinds of initial perturbations. For simplicity, we consider the solution at constant η, i.e. we take the initial perturbation in a form $G_0^\pm(x, y) = G_0^\pm(x)e^{i\eta y}$, $\eta = \text{const}$, $\pm x < 0$. Further we denote the linear functionals as

$$\varphi(G_0) = \int_{-\infty}^{\infty} \varphi(x) G_0(x)\, dx,$$

where $\varphi(x)$ is some vector or scalar function determined by basic solution.

6. Damped initial value

Let the initial values be damped with respect to x, i.e. $G_0^\pm(x) \to 0$, $x \to \infty$. Then the right hand side of the basic system is regular with respect to ω and (see Appendix) the solution of this system contains only two singular terms in a form $A_+(\eta)I_1(\omega - \omega_+) + A_-(\eta)I_1(\omega - \omega_-)$

(remember that $\eta = \text{const}$), where $A_\pm = \varphi_\pm(G_0)$. So the straightforward computations show that (we display the singular terms only)

$$\hat{f}(\eta,\omega) = \Phi_0(G_0)I_1(\omega-\omega_+) + \Psi_0(G_0)I_1(\omega-\omega_-),$$

$$G_a^+ = \Phi_1(G_0)I_1(\xi-\xi_1^+(\eta,\omega_+))I_1(\omega-\omega_+) + \Psi_1(G_0)I_1(\xi-\xi_1^+(\eta,\omega_-))I_1(\omega-\omega_-),$$
$$G_a^- = \Phi_2(G_0)I_1(\xi-\xi_2^-(\eta,\omega_+))I_1(\omega-\omega_+) + \Psi_2(G_0)I_1(\xi-\xi_2^-(\eta,\omega_-))I_1(\omega-\omega_-),$$
$$G_v^- = \Phi_3(G_0)I_1(\xi-\xi_3^-(\eta,\omega_+))I_1(\omega-\omega_+) + \Psi_3(G_0)I_1(\xi-\xi_3^-(\eta,\omega_-))I_1(\omega-\omega_-).$$

After inverse Fourier transform, we obtain spontaneously emitted waves in a form

$$f(y,t) = \Phi_0(G_0)e^{i\eta(y-tu_y-t\alpha_0)} + \Psi_0(G_0)e^{i\eta(y-tu_y+t\alpha_0)};$$

$$G_a^+ = \Phi_1(G_0)e^{i\eta(y-tu_y-t\alpha_0)}e^{i\xi_1^+(\eta,\omega_+)x} + \Psi_1(G_0)e^{i\eta(y-tu_y+t\alpha_0)}e^{i\xi_1^+(\eta,\omega_-)x},$$

but

$$\xi_1^+(\eta,\omega_\pm) = \mp\alpha_1\eta + \beta_1|\eta|, \quad \alpha_1 = \frac{u_x^+\alpha_0}{(c^+)^2-(u_x^+)^2}, \quad \beta_1 = \frac{c^+\sqrt{(c^+)^2-(u_x^+)^2-\alpha_0^2}}{(c^+)^2-(u_x^+)^2} > 0,$$

since $\alpha_0^2 < (c^\pm)^2 - (u_x^\pm)^2$ [13], and then

$$G_a^+ = e^{\beta_1|\eta|x}\left(\Phi_1(G_0)e^{i\eta(-\alpha_1x+y-tu_y-t\alpha_0)} + \Psi_1(G_0)e^{i\eta(\alpha_1x+y-tu_y+t\alpha_0)}\right).$$

Absolutely similar

$$G_a^- = e^{-\beta_2|\eta|x}\left(\Phi_2(G_0)e^{i\eta(-\alpha_2x+y-tu_y-t\alpha_0)} + \Psi_2(G_0)e^{i\eta(\alpha_2x+y-tu_y+t\alpha_0)}\right),$$

$$\alpha_2 = \frac{u_x^-\alpha_0}{(c^-)^2-(u_x^-)^2}, \quad \beta_2 = \frac{c^-\sqrt{(c^-)^2-(u_x^-)^2-\alpha_0^2}}{(c^-)^2-(u_x^-)^2} > 0;$$
$$G_v^- = \Phi_3(G_0)e^{i\eta(\alpha_3x+y-tu_y-t\alpha_0)} + \Psi_3(G_0)e^{i\eta(-\alpha_3x+y-tu_y+t\alpha_0)},$$

$\alpha_3 = u_y\alpha_0/u_x^-$.

Thus, as we see, the spontaneously emitted acoustic pre- and post-shock waves $G_a^\pm$ are evanescent with respect to x but oscillate with respect to t. The spontaneously emitted post-shock vorticity wave G_v^- is usual plane wave.

7. Oscillating initial value, resonance

Let now $G_0(x) = e^{i\xi_0x}$, so its Fourier transform is $I_1(\xi,\xi_0)/2\pi i$ for the pre-shock zone and $\overline{I_1}(\xi,\xi_0)/2\pi i$ for the post-shock zone. Then the right hand side of the basic system has a singular term $CI_1(\omega,\omega_0)$, where C is a constant vector and ω_0 depends on η and ξ_0 (differently for acoustic and vorticity waves and also for pre- and post-shock zones). If $\omega_0 \neq \omega_\pm$ (non resonant case), then the solution of basic system has three singular terms (Appendix) $C_1I_1(\omega,\omega_0) + C_2I_1(\omega,\omega_+) + C_3I_1(\omega,\omega_-)$. The first term $C_1I_1(\omega,\omega_0)$ leads to the terms of the form $Ce^{i(\xi_0x+\eta y-\omega_0t)}$ in the induced waves. They are the forced oscillations (caused by the initial ones). The last terms $C_2I_1(\omega,\omega_+) + C_3I_1(\omega,\omega_-)$ produce spontaneously emitted waves of the same form as in previous section. Note that the coefficients before $I_1(\omega,\omega_0)$, $I_1(\omega,\omega_\pm)$ has a factor $\omega_0-\omega_\pm$ in denominator (Appendix) and so, generally speaking, the amplitudes of the forced oscillations and spontaneously emitted waves approaches infinity when $\omega_0 \to \omega_+$ or $\omega_0 \to \omega_-$.

When $\omega_0 = \omega_+$ or $\omega_0 = \omega_-$, the resonance occurs, i.e. the solution of basic system contains the term like $I_2(\omega, \omega_0)$ (Appendix). But as we have seen, $\operatorname{Im} \xi_j(\eta, \omega_\pm) \neq 0$, $j = \overline{1,2}$ for both pre- and post-shock zones, i.e. the resonance is impossible for initial acoustic perturbations (and $\omega_0(\xi_0, \eta)$ never approaches $\omega_\pm$ at any ξ_0). The resonance is possible for the initial vorticity perturbations, when $\xi_0 = \pm\eta\alpha_0 u_y/u_x$. But initial post-shock vorticity wave is not incident upon the shock, i.e. does not generate shock wave disturbance and induced waves [13], therefore the resonance takes place only for the initial pre-shock vorticity waves. So the waves caused the resonance have a form

$$G_v^+(x,y,t) = \begin{pmatrix} 0 \\ u_x^+ \\ \mp\alpha_0 u_y \end{pmatrix} e^{i\eta(y - tu_y \pm \alpha_0(t + xu_y/u_x^+))}, \quad x < 0,$$

and corresponding induced (forced and spontaneously emitted) waves have a form (we take $\omega_0 = \omega_+$ to be more specific)

$$f(y,t) = (\widetilde{\Phi}_{01}t + \widetilde{\Phi}_{02})e^{i\eta(y-tu_y-t\alpha_0)} + \widetilde{\Psi}_0 e^{i\eta(y-tu_y+t\alpha_0)},$$

$$G_a^+ = e^{\beta_1|\eta|x}\left((\widetilde{\Phi}_{11}x + \widetilde{\Phi}_{12}t + \widetilde{\Phi}_{13})e^{i\eta(-\alpha_1 x+y-tu_y-t\alpha_0)} + \widetilde{\Psi}_1 e^{i\eta(\alpha_1 x+y-tu_y+t\alpha_0)}\right),$$

$$G_a^- = e^{-\beta_2|\eta|x}\left((\widetilde{\Phi}_{21}x + \widetilde{\Phi}_{22}t + \widetilde{\Phi}_{23})e^{i\eta(-\alpha_2 x+y-tu_y-t\alpha_0)} + \widetilde{\Psi}_2 e^{i\eta(\alpha_2 x+y-tu_y+t\alpha_0)}\right),$$

$$G_v^- = (\widetilde{\Phi}_{31}x + \widetilde{\Phi}_{32}t + \widetilde{\Phi}_{33})e^{i\eta(\alpha_3 x+y-tu_y-t\alpha_0)} + \widetilde{\Psi}_3 e^{i\eta(-\alpha_3 x+y-tu_y+t\alpha_0)},$$

where $\widetilde{\Phi}$, $\widetilde{\Psi}$ are specific constant vectors depending on the basic solution and η.

8. Summary

We introduce some results of spontaneous emission for isentropic solutions of problem of shock wave disturbance in non-classical case. Namely, we indicate the general form of spontaneously emitted waves both for damped and oscillating initial values. For oscillating initial values we consider both non-resonant and resonant cases.

Some interesting distinctions from the classical case are found. First, in the non-classical case the spontaneously emitted acoustic waves are evanescent with respect to the variable x but oscillate with respect to t. Second, only initial pre-shock vorticity wave may cause the resonance. The resonance is impossible for initial acoustic both pre- and post-shock waves.

9. Appendix

Here we give some necessary details for the singular solutions of linear equations in the class of analytic (in generalized sense) functions.

The one-sided Fourier transform for functions $f_1(t) = 2\pi i e^{-i\omega_0 t}$ and $f_2(t) = 2\pi t e^{-i\omega_0 t}$ (at $t > 0$) is

$$I_1(\omega, \omega_0) = \begin{cases} \dfrac{1}{\omega - \omega_0}, & \operatorname{Im}\omega > 0, \\ \dfrac{1}{\omega - \omega_0} - \pi i \delta(\omega - \omega_0), & \operatorname{Im}\omega = 0, \end{cases}$$

$$I_2(\omega, \omega_0) = \frac{\partial I_1(\omega, \omega_0)}{\partial \omega_0} =$$

$$= \begin{cases} \dfrac{1}{(\omega - \omega_0)^2}, & \operatorname{Im}\omega > 0, \\ \dfrac{1}{(\omega - \omega_0)^2} + \pi i \delta'(\omega - \omega_0), & \operatorname{Im}\omega = 0, \end{cases}$$

respectively. These functions, though they are singular, are analytic (in generalized sense) in the upper half-plane [13, 14, 16].

Let $Z(\omega)$ is given regular (smooth) and analytic in upper half-plane function, $Y(\omega)$ is sought-for analytic in upper half-plane function. The solutions of linear equations have a form (everywhere $\varphi(\omega)$ is regular):

$$(\omega-\omega_0)Y(\omega)=Z(\omega)\Longleftrightarrow Y(\omega)=Z(\omega_0)I_1(\omega,\omega_0)+\varphi;$$

$$(\omega-\omega_0)Y(\omega)=Z(\omega)I_1(\omega,\omega_1)\Longleftrightarrow Y(\omega)=\frac{Z(\omega_1)}{\omega_1-\omega_0}I_1(\omega,\omega_1)+\frac{Z(\omega_0)}{\omega_0-\omega_1}I_1(\omega,\omega_0)+\varphi,$$

$\omega_1\neq\omega_0$ (non resonant case);

$$(\omega-\omega_0)Y(\omega)=Z(\omega)I_1(\omega,\omega_0)\Longleftrightarrow Y(\omega)=Z(\omega_0)I_2(\omega,\omega_0)+Z'(\omega_0)I_1(\omega,\omega_0)+\varphi,$$

(resonance).

References

[1] Landau L D and Lifchitz E M 1987 *Fluid Mechanics* (Course of Theoretical Physics, Vol. 6) 2nd ed (Oxford: Pergamon Press)
[2] D'yakov S P 1954 On the stability of shock waves *Sov. Phys. JETP* **27** 288–95
[3] D'yakov S P 1957 The interaction of shock waves with small perturbations *Sov. Phys. JETP* **33** 948–74
[4] Kontorovitch V M 1957 Reflection and refraction of sound by shock waves *Sov. Phys. JETP* **33** 1527–8
[5] McKenzie J F and Westphal K O 1968 Interaction of linear waves with oblique shock waves *Phys.Fluids* **11** 2350–62
[6] Zel'dovich Ya B and Raizer Yu P 2002 *Physics of Shock Waves and High-Temperature Hydrodynamic Phenomena* (Mineola, NY: Dover Publications)
[7] Kutateladze S S, Nakoryakov V E and Borisov A A 1987 Rarefaction waves in liquid and gas-liquid media *Annu. Rev. Fluid Mech.* **19** 577–600
[8] Bates J W and Montgomery D C 1999 Some numerical studies of exotic shock wave behavior *Phys.Fluids* **11** 462–75
[9] Zamfirescu C, Guardone A and Colonna P 2008 Admissibility region for rarefaction shock waves in dense gases *J. Fluid Mech.* **599** 363–81
[10] Zhao N, Mentrelli A, Ruggeri T and Sugiyama M 2011 Admissible shock waves and shock-induced phase transitions in a van der Waals fluid *Phys.Fluids* **23** 086101
[11] Benzoni-Gavage S 1998 Stability of multi-dimensional phase transitions in a van der Waals fluid *Nonlinear Analysis, Theory, Methods & Applications* **31** 243–63
[12] Benzoni-Gavage S 2001 Linear stability of propagating phase boundaries in capillary fluids *Physica D* **155** 235–73
[13] Semenko E V 2017 Linear problem of the shock wave disturbance in a non-classical case *Phys.Fluids* **29** 066101
[14] Semenko E V 2017 Linear problem of shock wave disturbance analysis. Part 1: general solution, incidence, refraction and reflection in general case *Siberian Electr. Math. Reports* **14** 451-72
[15] Semenko E V and Semenko T I 2017 Linear problem of shock wave disturbance analysis. Part 2: refraction and reflection of plane waves *Siberian Electr. Math. Reports* **14** 473-92
[16] Semenko E V and Semenko T I 2017 Linear problem of shock wave disturbance analysis. Part 3: refraction and reflection in the neutral stability case *Siberian Electr. Math. Reports* **14** 493-510

IOP Conf. Series: Journal of Physics: Conf. Series **894** (2017) 012084 doi:10.1088/1742-6596/894/1/012084

Shape derivative of the energy functional for the bending of elastic plates with thin defects

V V Shcherbakov[1,2]

[1] Lavrentyev Institute of Hydrodynamics, 630090 Novosibirsk, Russia
[2] Novosibirsk State University, 630090 Novosibirsk, Russia

E-mail: victor@hydro.nsc.ru

Abstract. The paper deals with an equilibrium problem for a homogeneous isotropic elastic plate with a thin rigid inclusion and interfacial crack. We provide an explicit formula for the first shape derivative of the energy functional in the direction of a given vector field by means of a volume integral. For specific examples of the vector field, we derive some representations of the formula in terms of path-independent contour integrals.

1. Setting of the problem

Let $\Omega \subset \mathbb{R}^2$ be a bounded domain with boundary $\partial\Omega$ of the class $C^{1,1}$ and let γ be a smooth curve without self-intersections. We assume that γ can be extended to a closed curve Σ of the class $C^{1,1}$ without self-intersections so that Ω is divided into two subdomains Ω_1 and Ω_2. The boundary of Ω_1 is Σ, and the boundary of Ω_2 is $\Sigma \cup \partial\Omega$. The outward pointing unit normals to Σ and $\partial\Omega$ are denoted by $\nu = (\nu_1, \nu_2)$ and $n = (n_1, n_2)$, respectively. Also, we put $\Omega_\gamma = \Omega \setminus \overline{\gamma}$ and assume that the domain Ω can be divided into two subdomains Ω_3 and Ω_4 with Lipschitz boundaries such that $\gamma \subset \partial\Omega_3 \cap \partial\Omega_4$ and $\mathcal{H}^1(\partial\Omega_i \cap \partial\Omega) > 0$, $i = 3, 4$, where $\mathcal{H}^1$ denotes the one-dimensional Hausdorff measure. The last condition guarantees that Friedrichs–Poincaré's inequality is satisfied in the non-Lipschitz domain Ω_γ. The geometrical setup is shown in Figure 1.

The domain Ω_γ is occupied by the middle plane of a homogeneous isotropic elastic plate, and the curve γ corresponds to a thin rigid inclusion. To describe the vertical deflection of the thin rigid inclusion, we introduce the space of rigid displacements

$$L(\gamma) = \{\, l(x) = a_0 + a_1 x_1 + a_2 x_2 \mid a_i \in \mathbb{R},\ i = 0, 1, 2,\ x \in \gamma \,\}.$$

In conformity with positive and negative directions of the normal ν, there are positive crack face γ^+ and negative crack face γ^-, and $v^\pm$ are the traces of a function v on $\gamma^\pm$. The inclusion γ is delaminated on γ^+ so that there is a crack on the interface between two media.

We are now in a position to formulate the equilibrium problem (see [1, 2]). For a given

Content from this work may be used under the terms of the Creative Commons Attribution 3.0 licence. Any further distribution of this work must maintain attribution to the author(s) and the title of the work, journal citation and DOI.
Published under licence by IOP Publishing Ltd

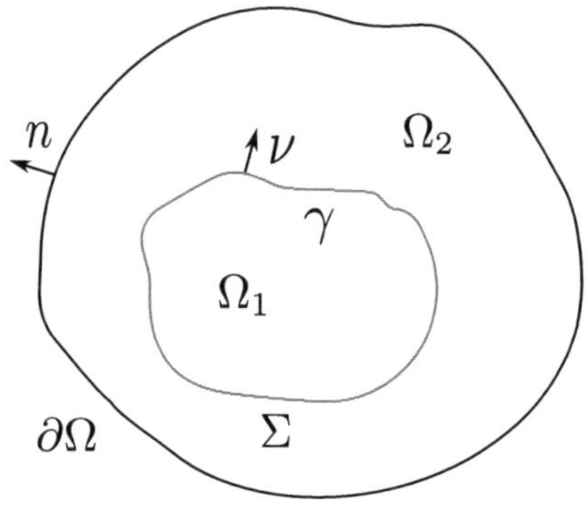

Figure 1. Geometrical setup of the problem. The middle plane of the plate.

external force $f \in C^1(\overline{\Omega})$, we want to find two functions w and $l_0 \in L(\gamma)$ satisfying

$$D\Delta^2 w = f \quad \text{in} \quad \Omega_\gamma, \tag{1}$$

$$w = \frac{\partial w}{\partial n} = 0 \quad \text{on} \quad \partial\Omega, \tag{2}$$

$$m_\nu^- = 0, \quad w^- = l_0 \quad \text{on} \quad \gamma, \tag{3}$$

$$m_\nu^+ = t_\nu^+ = 0 \quad \text{on} \quad \gamma, \tag{4}$$

$$\int\limits_\gamma t_\nu^- \bar{w}\, \mathrm{d}s_x = 0 \quad \forall \bar{w} \in K^0. \tag{5}$$

Here the set K^0 of admissible displacements is

$$K^0 = \left\{ w \in H^2_{\partial\Omega}(\Omega_\gamma) \mid w|_{\gamma^-} \in L(\gamma) \right\},$$

where the Sobolev space $H^2_{\partial\Omega}(\Omega_\gamma)$ is defined as

$$H^2_{\partial\Omega}(\Omega_\gamma) = \{ w \in H^2(\Omega_\gamma) \mid w = \partial w/\partial n = 0 \text{ on } \partial\Omega \}.$$

The bending moment m_ν and transverse force t_ν are given by the relations

$$m_\nu = D\left(\varkappa \Delta w + (1-\varkappa)\frac{\partial^2 w}{\partial \nu^2} \right),$$

$$t_\nu = D\frac{\partial}{\partial \nu}\left(\Delta w + (1-\varkappa)\frac{\partial^2 w}{\partial \tau^2} \right), \quad \tau = (\tau_1, \tau_2) = (-\nu_2, \nu_1),$$

where D is the flexural rigidity and $\varkappa$ is the Poisson ratio, $0 < \varkappa < 1/2$.

Equation (1) is a equilibrium equation, the boundary conditions (2) correspond to the clamped outer edge of the plate. The nonlocal boundary conditions (5) ensure the vanishing of the resultant force and moment along γ^-.

The boundary-value problem (1)–(5) admits a variational statement. Indeed, let us define the bilinear form

$$B(u,v) = D\left(u_{,11}v_{,11} + u_{,22}v_{,22} + \varkappa u_{,11}v_{,22} + \varkappa u_{,22}v_{,11} + 2(1-\varkappa)u_{,12}v_{,12}\right).$$

IOP Conf. Series: Journal of Physics: Conf. Series **894** (2017) 012084 doi:10.1088/1742-6596/894/1/012084

Subscripts following a comma represent differentiation, so that, for example, $u_{,ij} = \partial^2 u/\partial x_i \partial x_j$. Then problem (1)–(5) corresponds to minimization of the energy functional

$$\Pi(\Omega_\gamma; w) = \frac{1}{2} \int\limits_{\Omega_\gamma} B(w, w)\, \mathrm{d}x - \int\limits_{\Omega_\gamma} f w\, \mathrm{d}x$$

over the subspace K^0 and can be written in the form of the weak Euler–Lagrange equation

$$w \in K^0, \quad \int\limits_{\Omega_\gamma} B(w, \bar{w})\, \mathrm{d}x - \int\limits_{\Omega_\gamma} f \bar{w}\, \mathrm{d}x = 0 \quad \forall \bar{w} \in K^0. \tag{6}$$

Since the functional Π is coercive and weakly lower semicontinuous over the set K^0, equation (6) has a (unique) solution.

2. Shape derivative of the energy functional

We next turn our attention to the the first shape derivative of the energy functional in the direction of a given vector field. We apply the technique of shape sensitivity analysis in order to demonstrate that this derivative is well defined and derive an explicit representation to it.

For a small non-negative parameter δ, let us consider a one-parameter family of perturbations of the initial domain Ω_γ which is governed by a transformation T_δ. To be specific, we make use of the transformation in the form $T_\delta(x) = x + \delta V(x)$ with a vector field $V(x) = (V_1(x), V_2(x))$ of class $W^{2,\infty}(\mathbb{R}^2)^2$, or, in coordinate form,

$$y_1 = x_1 + \delta V_1(x), \quad y_2 = x_2 + \delta V_2(x), \quad x = (x_1, x_2) \in \Omega_\gamma. \tag{7}$$

Transformation (7) defines the perturbed domain $T_\delta(\Omega)$ and the perturbed thin rigid inclusion $\gamma_\delta = T_\delta(\gamma)$ for each fixed δ. For simplicity, we assume that the vector field V has the compact support in Ω. In this case, the outward pointing unit normal n of the domain Ω transforms to the outward unit normal n^δ of the domain $T_\delta(\Omega)$. We define the perturbed domain with the thin rigid inclusion as $\Omega_{\gamma_\delta} = T_\delta(\Omega) \setminus \overline{\gamma}_\delta$. We assume that for each δ the domain Ω_{γ_δ} satisfies the last condition from the first paragraph of Section 1.

The set of admissible displacements associated with the perturbed problem is

$$K^\delta = \left\{ w \in H^2_{\partial\Omega}(\Omega_{\gamma_\delta}) \mid w|_{\gamma_\delta^-} \in L(\gamma_\delta) \right\},$$

As before, there exists a unique solution of the weak Euler–Lagrange equation

$$w^\delta \in K^\delta, \quad \int\limits_{\Omega_{\gamma_\delta}} B(w^\delta, \bar{w})\, \mathrm{d}x - \int\limits_{\Omega_{\gamma_\delta}} f \bar{w}\, \mathrm{d}x = 0 \quad \forall \bar{w} \in K^\delta. \tag{8}$$

For the solution w^δ of equation (8), we can find the potential deformation energy

$$\Pi(\Omega_{\gamma_\delta}; w^\delta) = \frac{1}{2} \int\limits_{\Omega_{\gamma_\delta}} B(w^\delta, w^\delta)\, \mathrm{d}y - \int\limits_{\Omega_{\gamma_\delta}} f w^\delta\, \mathrm{d}y.$$

With these preliminaries, we introduce the first shape derivative of the energy functional in the direction of the vector field V as

$$\mathcal{D}_V \Pi(\Omega_\gamma; w) = \lim_{\delta \downarrow 0} \frac{\Pi(\Omega_{\gamma_\delta}; w^\delta) - \Pi(\Omega_\gamma; w)}{\delta}. \tag{9}$$

The question now is whether (9) is well defined and whether there exist simple formulae to calculate it. The first main result of our study is the following theorem.

Theorem 1. The first shape derivative of the energy functional in the direction of the vector field V exists and equals

$$\mathcal{D}_V\Pi(\Omega_\gamma;w)=\int\limits_{\Omega_\gamma}\Big(\frac{1}{2}B(w,w)\mathrm{div}V+B(w,\widetilde{V})-2D\left(V_{1,1}w_{,11}^2+\varkappa w_{,11}w_{,22}\mathrm{div}V\right.$$
$$+\left.(1-\varkappa)w_{,12}^2\mathrm{div}V+V_{2,2}w_{22}^2+(V_{1,2}+V_{2,1})w_{,12}(w_{,11}+w_{,22})\right)$$
$$-D((w_{,11}+\varkappa w_{,22})(V_{1,11}w_{,11}+V_{2,11}w_{,2})$$
$$+(\varkappa w_{,11}+w_{,22})(V_{1,22}w_{,1}+V_{2,22}w_2)+2(1-\varkappa)(V_{1,12}w_{,1}+V_{2,12}w_{,2})w_{,12})\Big)\,\mathrm{d}x$$
$$-\int\limits_{\Omega_\gamma}\left(\mathrm{div}(fV)+f\widetilde{V}\right)\,\mathrm{d}x,\quad(10)$$

where the vector field $\widetilde{V}$ is such that $\widetilde{V}(x)=a_1^0V_1(x)+a_2^0V_2(x)$, and the constants a_1^0 and a_2^0 correspond to the function l_0.

Remark. The first shape derivative of the energy functional in the direction of the vector field V is independent of the particular choice of the coordinate transformation T_δ in the following sense. If two vector fields V_1 and V_2 are such that the corresponding coordinate transformations T_δ^1 and T_δ^2 map the initial domain Ω_γ into the same perturbed domain Ω_{γ_δ} for each δ, then $\mathcal{D}_{V_1}\Pi(\Omega_\gamma;w)=\mathcal{D}_{V_1}\Pi(\Omega_\gamma;w)$ This easily follows from the definition of the shape derivative and the uniqueness of the weak solution $w^\delta\in K^\delta$.

Let us mention the main ingredients in the proof of Theorem 1. Following ideas of [3], we calculate limit (9) by transforming the integrals expressions defined in the perturbed domain Ω_{γ_δ} back to the initial domain Ω_γ. The transformation $T_\delta:\Omega_\gamma\to\Omega_{\gamma_\delta}$ induces an isomorphism between the spaces $H^2_{\partial\Omega}(\Omega_{\gamma_\delta})$ and $H^2_{\partial\Omega}(\Omega_\gamma)$ in the following way:

$$\mathbf{T}_\delta\,:\,H^2_{\partial\Omega}(\Omega_{\gamma_\delta})\to H^2_{\partial\Omega}(\Omega_\gamma)\,:\,w\mapsto w\circ T_\delta.$$

The lack of the one-to-one correspondence between the sets of admissible displacements K^δ and K^0 under the mapping $\mathbf{T}_\delta$ is the main obstacle in the proof of the theorem. To overcome this difficultly, we denote by K_δ the image of the admissible set K^δ under the mapping $\mathbf{T}_\delta$. Straightforward calculations yield that

$$K_\delta=\left\{w\in H^2_{\partial\Omega}(\Omega_\gamma)\mid w_{\gamma^-}\in L_\delta(\gamma)\right\},$$

where

$$L_\delta(\gamma)=\{\,l(x)=a_0+a_1x_1+a_2x_2+a_1\delta V_1(x)+a_2\delta V_2(x)\mid\ a_i\in\mathbb{R},\,i=0,1,2,\ x\in\gamma\,\}.$$

Hence, $\mathbf{T}_\delta$ is also isomorphism between the sets K^δ and K_δ. Denote by $w_\delta(x)$ the image of $w^\delta(y)$ under the mapping $\mathbf{T}_\delta$. The next lemmata play a key role in the derivation of the explicit formula for (9).

Lemma 1. Let $w\in K^0$ be the solution of problem (6). Then there exist functions w_δ^1 and w_δ^2 satisfying the inclusions

$$w+\delta w_\delta^1\in K_\delta,\quad w_\delta-\delta w_\delta^2\in K^0$$

and uniform in δ estimates

$$\|w_\delta^i\|_{H^2_{\partial\Omega}(\Omega_\gamma)}\le c,\quad i=1,2.\qquad(11)$$

We mention that the functions w_δ^1 and w_δ^2 can be defined explicitly by the relations

$$w_\delta^1 = a_1^0 V_1 + a_2^0 V_2, \quad w_\delta^2 = a_1^\delta V_1 + a_2^\delta V_2.$$

Using Lemma 1, we establish that w_δ is Hölder continuous with exponent 1/2.

Lemma 2. There exists a positive constant c such that

$$\|w_\delta - w\|_{H^2_{\partial\Omega}(\Omega_\gamma)} \le c\sqrt{\delta}.$$

Corollary. $w_\delta^2 \to \widetilde{V}$ strongly in $H^2_{\partial\Omega}(\Omega_\gamma)$ as $\delta \downarrow 0$.

We next apply transformation (7) to the integrals involved in $\Pi(\Omega_{\gamma_\delta}; \bar{w}^\delta)$ to obtain a functional $\Pi_\delta(\Omega_\gamma; \bar{w}_\delta)$ that defines over the space $H^2_{\partial\Omega}(\Omega_\gamma)$ and admits the following asymptotic expansion as $\delta \downarrow 0$:

$$\Pi_\delta(\Omega_\gamma; \bar{w}_\delta) = \Pi(\Omega_\gamma; \bar{w}_\delta) + \delta \mathcal{D}_V \Pi(\Omega_\gamma; \bar{w}_\delta) + o(\delta),$$

with the form $\mathcal{D}_V \Pi(\Omega_\gamma; \bar{w}_\delta)$ from (10).

Since the mapping $\mathbf{T}_\delta$ is an isomorphism between the sets K^δ and K_δ, it follows that

$$\frac{\Pi(\Omega_{\gamma_\delta}; w^\delta) - \Pi(\Omega_\gamma; w)}{\delta} = \frac{\Pi_\delta(\Omega_\gamma; w_\delta) - \Pi(\Omega_\gamma; w)}{\delta}.$$

Taking into account that w is a minimizer of $\Pi(\Omega_\delta, \cdot)$ over the set K^δ and w_δ is a minimizer of $\Pi(\Omega_\delta, \cdot)$ over the set K^δ, we invoke Lemma 1, for each $\delta > 0$, to deduce the following chain of inequalities:

$$\frac{\Pi_\delta(\Omega_\gamma; w_\delta) - \Pi(\Omega_\gamma; w_\delta - \delta w_\delta^2)}{\delta} \le \frac{\Pi_\delta(\Omega_\gamma, w_\delta) - \Pi(\Omega_\gamma, w)}{\delta} \le \frac{\Pi_\delta(\Omega_\gamma; w + \delta w_\delta^1) - \Pi(\Omega_\gamma; w)}{\delta}. \tag{12}$$

Applying Lemma 2 and its Corollary, we have no difficulty in calculating the limits as $\delta \downarrow 0$ of the right-hand and left-hand sides in (12) and checking that they are finite, coincide with each other, and equal to $\mathcal{D}_V \Pi(\Omega_\gamma; w_\delta)$ from (10). This completes the sketch of the proof of Theorem 1.

We conclude this section by mentioning the recent papers [4, 5, 6, 7], where a similar scheme for deriving of explicit formulae to the first shape derivative of the energy functional in the direction of a given vector field can be found for elastic models (2D elasticity and Timoshenko plates) with embedded rigid and semirigid inclusions and interfacial cracks.

3. Path-independent integrals

Our intention next is to consider specific examples of the vector field V that lead us to path-independent energy integrals via transformations of formula (10). To do this, it is necessary to have more information on the regularity of the transverse deflection w in comparison with the variational one. We next assume that the thin rigid inclusion γ is rectilinear, i.e., it lies in the line $x\nu = a$, $a =$ const, with the tips C_1 and C_2. In this case, the standard regularity theory for uniformly elliptic differential operators shows that the solution w of problem (6) possess H^4-regularity up to the faces γ^+ and γ^- except the thin inclusion tips C_1 and C_2.

In all cases below, we choose neighborhoods S and S_1 with Lipschitz boundaries ∂S and ∂S_1. We also assume that the boundaries of the domains $(S_1 \setminus S) \cap \Omega_\gamma$ also satisfy the Lipschitz condition and $f \equiv 0$ in $S \cap \Omega_\gamma$. Finally, denote by $q = (q_1, q_2)$ the outward normal vector to ∂S.

We first investigate the case when an integration path surrounds the whole rigid inclusion γ. Let the support of a smooth function η lies in a small neighborhood S_1 of γ and $\eta = 1$

in a neighborhood S of γ, $S \subset S_1$. Given a vector $p = (p_1, p_2)$, we consider the coordinate transformation (7) in the form

$$y_1 = x_1 + \delta p_1 \eta(x_1, x_2), \quad y_2 = x_2 + \delta p_2 \eta(x_1, x_2), \tag{13}$$

where $(x_1, x_2) \in \Omega_\gamma$ and $(y_1, y_2) \in \Omega_{\gamma_\delta}$. Transformation (13) corresponds to a translation of the thin rigid inclusion along the vector p, and the vector field V is determined by the formula $V = p\eta$. We next substitute the vector field V into (10) and integrate by parts there. Taking into account (1)–(5), we obtain the contour integral

$$\mathcal{I}_{p\eta} = D \int\limits_{\partial S} \Big(\frac{1}{2}(p_2 q_2 - p_1 q_1)(w_{,11}^2 - w_{22}^2) - (1-\varkappa) w_{,12}^2 pq - w_{,12}\left(p_2 q_1 M_1(w) + p_1 q_2 M_2(w)\right)$$
$$- \left(M_{1,1}(w) q_1 + M_{2,2}(w) q_2\right) \left(\frac{\partial w}{\partial p} + a_1^0 p_1 + a_2^0 p_2 \right)$$
$$+ 2(1-\varkappa) \left(w_{,112} \left(p_1 w_{,1} - \frac{a_1^0 p_1 + a_2^0 p_2}{2} \right) q_2 + w_{,122} \left(p_2 w_{,2} - \frac{a_1^0 p_1 + a_2^0 p_2}{2} \right) q_1 \right) \Big) \, \mathrm{d}x,$$

where

$$M_1(w) = w_{,11} + \varkappa w_{,22}, \quad M_2(w) = \varkappa w_{,11} + w_{,22}.$$

It is immediate from the construction that $\mathcal{I}_{p\eta}$ is a path-independent integral, i.e., it has the same value for all closed smooth curve surrounding the thin rigid inclusion γ.

The next step is to discuss a situation when the integration path surrounds only the thin inclusion tip. Let the support of a smooth function θ lies in a small neighborhood S_1 of thin inclusion tip C_1 and $\theta = 1$ in a neighborhood S of C_1, $S \subset S_1$. We consider the coordinate transformation (7) in the form

$$y_1 = x_1 + \delta \tau_1 \theta(x_1, x_2), \quad y_2 = x_2 + \delta \tau_2 \theta(x_1, x_2), \tag{14}$$

where $(x_1, x_2) \in \Omega_\gamma$ and $(y_1, y_2) \in \Omega_{\gamma_\delta}$. Transformation (14) corresponds to a local translation of the thin rigid inclusion along the vector τ, which is tangential to γ, and the vector field V is determined by the formula $V = \tau\theta$. Substituting the vector field V into (10) and integrating by parts there, we obtain the path-independent integral

$$\mathcal{I}_{\tau\theta} = D \int\limits_{\partial S} \Big(\frac{1}{2}(\tau_2 q_2 - \tau_1 q_1)(w_{,11}^2 - w_{22}^2) - (1-\varkappa) w_{,12}^2 \tau q - w_{,12}\left(\tau_2 q_1 M_1(w) + \tau_1 q_2 M_2(w)\right)$$
$$- \left(M_{1,1}(w) q_1 + M_{2,2}(w) q_2\right) \left(\frac{\partial w}{\partial \tau} + a_1^0 \tau_1 + a_2^0 \tau_2 \right)$$
$$+ 2(1-\varkappa) \left(w_{,112} \left(\tau_1 w_{,1} - \frac{a_1^0 \tau_1 + a_2^0 \tau_2}{2} \right) q_2 + w_{,122} \left(\tau_2 w_{,2} - \frac{a_1^0 \tau_1 + a_2^0 \tau_2}{2} \right) q_1 \right) \Big) \, \mathrm{d}x. \tag{15}$$

The path-independent integral (15) is an analogue of the well-known Eshelby–Cherepanov–Rice J-integral from fracture mechanics.

Acknowledgments

The work was supported by the Council on Grants of the President of the Russian Federation (grant No. MK-5173.2016.1) and by the Russian Foundation for Basic Research (grant No. 16-01-00610).

IOP Conf. Series: Journal of Physics: Conf. Series **894** (2017) 012084 doi:10.1088/1742-6596/894/1/012084

References

[1] Khludnev A M 2010 J. Appl. Industr. Math. **5** 582–94
[2] Khludnev A M 2012 *Europ. J. Mech. A-Solid* **32** 69–75
[3] Rudoy E M 2004 *Sib. Math. J.* **45** 388–97
[4] Itou H, Khludnev A M, Rudoy E M and Tani A 2012 *Z. Angew. Math. Mech.* **92** 716–30
[5] Lazarev N P 2015 *Z. Angew. Math. Phys.* **66** 2025–40
[6] Rudoy E M 2015 *Z. Angew. Math. Phys.* **66** 1923–37
[7] Shcherbakov V 2017 *Z. Angew. Math. Phys.* **68** 26

IOP Conf. Series: Journal of Physics: Conf. Series **894** (2017) 012085 doi:10.1088/1742-6596/894/1/012085

Numerical study of a mathematical model of internal erosion of soil

A Sibin[1]

[1] Altai State University, Barnaul, Russia

E-mail: sibin_anton@mail.ru

Abstract. The process of internal erosion in a three-phase saturated soil is studied. A mathematical model describing the process consists of the equations of mass conservation, Darcy's law and equation for capillary pressure. The original system of equations is reduced to a system of three equations for porosity, pressure and water saturation. Obtained equation for the water saturation is degenerate. The degenerate problem in an one-dimensional domain is solved numerically using the finite-difference method.

1. Introduction

The problems associated with soil erosion have been extensively studied throughout the last century. This process is of great importance in applied problems in agriculture: irrigation and drainage of agricultural fields [1] and canal irrigation of soil [2]. The erosion process must be taken into account in studies related to the forecast of the spread of contamination, filtration near reservoirs and other hydraulic structures [3]. Similar problems related to the process of soil erosion occur in other areas, including oil and gas production [4].

Several different approaches are existed for modeling the process of internal erosion and motion of fluidized soil particles. In the works of R. J. Garde [5], the process of motion of a mixture of gas and mobile solid particles is described without taking into account the process of internal suffusion. Is is noted that for the considering the erosion process the M.A. Biot model for water-saturated soils [6] can be used.

Fundamentally different approaches in erosion modeling are used in the series of works by S. Bonelli [7]. This approach involves the determination of an unknown interface between a solid skeleton and an one-velocity mixture of the water and fluidized solid particles with standard rheology. To determine the unknown boundary the common method is to introduce analogies of the kinematic and dynamic conditions in the problems with free boundaries.

The most complete models of the motion of water and moving solid particles in a porous medium are proposed in a series of works by I. Vardoulakis and his followers [8]. Soil in these works are considered as a multiphase medium on the one hand, on the other it is assumed that velocities of water and moving soil particles are proportional. This assumption is a consequence of the hypotheses that the pressures of moving phases are equal.

Last but not the least important thing of the modeling is that there is no single approach describing phase transitions. In this paper to guarantee the satisfaction of the physical maximum

Content from this work may be used under the terms of the Creative Commons Attribution 3.0 licence. Any further distribution of this work must maintain attribution to the author(s) and the title of the work, journal citation and DOI.
Published under licence by IOP Publishing Ltd

principles for porosity and phase concentration correct description of the phase transitions is considered.

2. Formulation of the problem

A mathematical model of internal erosion of soil in three-dimensional domain is formulated under the theory of multiphase filtration in a porous medium. Saturated soil consists of water ($i = 1$, first phase), fluidized solid particles ($i = 2$, second phase) and solid skeleton ($i = 3$, third phase) [8]. Constitutive equations of the model in the general case are the equations of mass conservation for each phase with phase transitions, generalized Darcy's law and equation of capillary pressure for water and fluidized solid particles [10]:

$$\frac{\partial \rho_i}{\partial t} + \nabla \cdot (\rho_i \vec{u}_i) = \sum_{j=1}^{3} I_{ij}, \quad i = 1,2,3, \quad I_{ji} = -I_{ij}, \quad \sum_{i,j=1}^{3} I_{ij} = 0; \tag{1}$$

$$\vec{v}_i = s_i \phi \vec{u}_i = -K_0(\phi)\frac{\overline{k}_{0i}}{\mu_i}(\nabla p_i + \rho_i^0 \vec{g}),\ i = 1,2; \quad p_2 - p_1 = p_c(s_1) \geq 0, \quad s_1 + s_2 = 1. \tag{2}$$

Here $\vec{u}_i$ is a velocity of the ith phase ($\vec{u}_3 = 0$, the third phase is considered motionless medium), ρ_i is a reduced density related to a true density ρ_i^0 and a volumetric concentration α_i by $\rho_i = \alpha_i \rho_i^0$ (identity $\sum_{i=1}^{3} \alpha_i = 1$ is a consequence of the definition of ρ_i), I_{ji} is a rate of mass transfer from the jth to the ith phase per unit volume in unit of time, $\vec{v}_i = \phi s_i \vec{u}_i$ is a filtration velocity of the ith phase (i=1,2), ϕ is porosity, s_1 and s_2 are water and fluidized solid particles saturations ($\alpha_1 = \phi s_1$, $\alpha_2 = \phi s_2$, $\alpha_3 = 1 - \phi$), K_0 is the filtration tensor; k_{0i} is a relative phase permeability ($\overline{k}_{0i} = \overline{k}_{0i}(s_i) \geq 0$, $\overline{k}_{0i}\big|_{s_i=0} = 0$), μ_i is a dynamic viscosity, p_i is a phase pressure (i=1,2); p_c is the capillary pressure, $\vec{g}$ is the acceleration vector due to gravity.

The system (1) – (2) is considered under the following assumptions for the densities: $\rho_i^0 = const$, $\rho_2^0 = \rho_3^0$, $I_{ij} = -I_{ji}$. For the phase transitions it is postulated that $I_{12} = I_{13} = 0$, $I_{23} = \rho_3^0 I$, and I is a given function of s_1, ϕ, v_1, v_2.

The most studied case of the described system is consindering porosity as a given function, then it can be shown that the system (1) – (2) reduces to the well-known Masket–Leverett equations of two–phase filtration for immiscible fluids. This study is concerned on the investigation of the problem with unknown and to be determined function of porosity.

In the one-dimensional domain and under the conditions set forth above the system (1) – (2) reads [11]

$$\phi\frac{\partial s}{\partial t} = \frac{\partial}{\partial x}(a(s,\phi)\frac{\partial s}{\partial x} + b(s)v + F(s,\phi)) - s\frac{\partial \phi}{\partial t}, \tag{3}$$

$$\frac{\partial v}{\partial x} = 0, \quad v = K(s,\phi)\frac{\partial p}{\partial x} + f(s,\phi), \tag{4}$$

$$\frac{\partial \phi}{\partial t} = I(s,\phi,p). \tag{5}$$

Here $s \equiv s_1$, $v = v_1 + v_2$ is a mixture filtration velocity (prescribed function of time) [5]. The rate of mass transfer is given as $I = \lambda\delta(s)R(\phi)\max\{|v(t)| - v_k, 0\}$ [2, 10], where $\delta(s) = 0$ if $s \geq 1$, $\delta(s) = 1 - s$ if $0 < s < 1$, $\delta(s) = 1$ if $s \leq 0$; $R(\phi) = 0$ if $\phi \geq 1$, $R(\phi) = \phi(1-\phi)$ if $0 < \phi < 1$, $R(\phi) = 0$ if $\phi \leq 0$, $\lambda > 0$ is a dimension constant $[1/m]$, v_k is the critical velocity. The critical velocity indicates whether the erosion process has started or hasn't. If the velocity

v reaches value of the critical speed then the theory suggests the erosion process will begin (the critical speed can be evaluated using experimental methods, see, e.g. [11]).

The coefficients of the system (3) – (4) have the form

$$a(s,\phi) = -K_0(\phi)\frac{k_{01}k_{02}}{k}\frac{\partial p_c}{\partial s} > 0, \quad k_{0i}(s) = \frac{\overline{k}_{0i}}{\mu_i} \geq 0, \quad k(s) = k_{01} + k_{02} > 0,$$

$$F(s,\phi) = K_0(\phi)\frac{k_{01}k_{02}}{k}(\rho_1^0 - \rho_2^0)g, \quad b(s) = \frac{k_{01}}{k} \geq 0.$$

The functions $s(x,t)$ and $\phi(x,t)$ satisfy the initial and boundary conditions

$$s(x,0) = s^0(x), \quad \phi(x,0) = \phi^0(x),$$

$$s(0,t) = s_0(t), \quad s(l,t) = s_l(t), \quad \frac{\partial p}{\partial x}(0,t) = p_0(t), \quad p(l,t) = p_l. \tag{6}$$

3. Method of the numerical solution

The system of the equations (3)–(5) is solved numerically for the initial and boundary conditions in the form (6) in a finite one-dimensional domain. The numerical study aims to investigate the behavior of the functions of porosity and water saturation as functions of space variable and time depending on given parameters of the problem.

The numerical solution of the problem (3) – (5) is obtained with the help of the finite-difference method. It is convenient to introduce a finite-difference grid with constant steps $x_i = ih$, $i = 0...N$ and $t_n = n\tau$, $n = 0...T$, where h is a mesh size, N is number of steps in space, τ is time step, T is number of steps in time. The convective term in the equation (3) is approximated by the directed difference. The equation (4) is approximated by an implicit scheme of the second order of accuracy. The result is the system of finite-difference equations

$$\phi_i^n\frac{s_i^{n+1} - s_i^n}{\tau} = a_{i+1/2}^n\frac{s_{i+1}^{n+1} - s_i^{n+1}}{h^2} - a_{i-1/2}^n\frac{s_i^{n+1} - s_{i-1}^{n+1}}{h^2} +$$

$$+\frac{(|G_i^n| + G_i^n)s_{i+1}^{n+1} - 2|G_i^n|s_i^{n+1} + (|G_i^n| - G_i^n)s_{i-1}^{n+1}}{2h} + F_{\phi i}^n\frac{\phi_{i+1}^n - \phi_{i-1}^n}{2h} - s_i^n\frac{\phi_i^{n+1} - \phi_i^n}{\tau}, \tag{7}$$

$$K_{i+1/2}^n\frac{p_{i+1}^n - p_i^n}{h^2} - K_{i-1/2}^n\frac{p_i^n - p_{i-1}^n}{h^2} + f_{si}^n\frac{s_{i+1}^n - s_{i-1}^n}{2h} + f_{\phi i}^n\frac{\phi_{i+1}^n - \phi_{i-1}^n}{2h} = 0. \tag{8}$$

The equation (5) is approximated by the Runge-Kutta scheme of the second order of accuracy. The value found in the first step in the form

$$\widetilde{\phi}_i^{n+1} = \phi_i^n + \tau I_i^n \tag{9}$$

is refined as

$$\phi_i^{n+1} = \phi_i^n + \tau\frac{I(\phi_i^n, s_i^n) + I(\widetilde{\phi}_i^n, s_i^n)}{2}. \tag{10}$$

Here $i = 1,...,N-1$, $\tau = 1,...,T-1$,

$$a_{i-1/2}^n = \frac{2K_0(\phi_{i-1}^n)a(s_{i-1}^n)K_0(\phi_i^n)a(s_i^n)}{K_0(\phi_{i-1}^n)a(s_{i-1}^n) + K_0(\phi_i^n)a(s_i^n)}, \quad a_{i+1/2}^n = \frac{2K_0(\phi_{i+1}^n)a(s_{i+1}^n)K_0(\phi_i^n)a(s_i^n)}{K_0(\phi_{i+1}^n)a(s_{i+1}^n) + K_0(\phi_i^n)a(s_i^n)},$$

$$F_{\phi i}^n = \frac{\partial F}{\partial \phi}(s_i^n, \phi_i^n), \quad f_{si}^n = \frac{\partial f}{\partial s}(s_i^n, \phi_i^n), \quad f_{\phi i}^n = \frac{\partial f}{\partial \phi}(s_i^n, \phi_i^n),$$

$$G_i^n = \frac{\partial F}{\partial s}(s_i^n, \phi_i^n) + v(t_n)\frac{\partial b}{\partial s}(s_i^n),$$

$$K_{i-1/2}^n = \frac{2K(\phi_{i-1}^n, s_{i-1}^n)K(\phi_i^n, s_i^n)}{K(\phi_{i-1}^n, s_{i-1}^n) + K(\phi_i^n, s_i^n)}, \quad K_{i+1/2}^n = \frac{2K(\phi_{i+1}^n, s_{i+1}^n)K(\phi_i^n, s_i^n)}{K(\phi_{i+1}^n, s_{i+1}^n) + K(\phi_i^n, s_i^n)},$$

$$I_i^n = \begin{cases} \lambda(1-\phi_i^n)(1-s_i^n)\phi_i^n(|v_{1i}^n| - |v_k|), |v_{1i}^n| \geq |v_k|; \\ 0, |v_{1i}^n| < |v_k|. \end{cases}$$

The numerical algorithm consists of several steps. First, the initial conditions for the porosity ϕ^0 and water saturation s_i^0 are substituted to the coefficients of the finite-difference scheme (8). Then, using sweep method in the scheme (8), values of the pressure p_i^0 are found. After that, using the values of the pressure, values of the water filtration rate v_{1i}^0 and moving soil particles velocity v_{2i}^0 are determined. Second, values of the porosity are calculated $\widetilde{\phi}_i^1$ in the each grid point (9). Then found values of the pressure and porosity are substituted to the coefficients of the finite-difference scheme (7). Then, using sweep method in the scheme (7) values of the water saturation s_i^1 are found. After that, using sweep method in the scheme (8) values of the pressure p_i^1 are determined in the next time step. Using found values of functions $\widetilde{\phi}_i^1$, s_i^1, p_i^0, p_i^1, the porosity value is refined in the first time step using formula (10). Finally, repeating the algorithm, values of porosity, pressure and concentration are found in next time step and so on.

4. Numerical results

The porosity and the water saturation are calculated for the following set of the parameters of the problem: $g = 0$ m/s^2, $v_k = 0$, $K_0(\phi) = \phi^3/(1-\phi^2)$, $k_{0i} = s_i^2$ if $0 \leq s \leq 1$, $k_{0i} = 0$ if $s_i \leq 0$, $k_{0i} = 1$ if $s_i \geq 1$. The initial and boundary conditions are taken in the form $\frac{\partial p_1}{\partial x}(0,t) = H$, $\frac{\partial p}{\partial x}(0,t) = H + \frac{k_{02}(s)}{k(s)}\frac{\partial p}{\partial s}\frac{\partial s}{\partial x}$, where H is water head and equals 2.4 in numerical calculations. At the points $x = 0$ there are no fluidized solid particles the boundary condition is $s(0,t) = s_0 = 1$. At the points $x = l$ the boundary conditions are $p(l,t) = 101325$ Pa and $s(l,t) = s_h(t)$, where s_h is function obtained by interpolating the experimental concentration value [9]. The initial condition for porosity is $\phi(x,0) = 3.5$, the one for s is assumed to be a linear function of x, $s(x,0) = x(s_h(0) - s_0)/l + s_0$. In presented results the numerical parameters are $h = 0.00123$, $\tau = 0.0001$, $N = 1000$, $T = 1000$.

Following values of the physical parameters are used in in numerical calculations [9]:

$$\rho_1^0 = 1000 \text{ kg/m}^3, \quad \rho_2^0 = 2600 \text{ kg/m}^3, \quad l = 123\,mm,$$

$$\mu_1 = 0.001787 \text{ kg/ms}^2, \quad \mu_2 = 0.003574 \text{ kg/ms}^2.$$

In numerical calculations mathematical model taken into account the capillary forces is studied. The Leverett function has the form [12] $p_c(s) = \gamma(1/s - 1)$, where γ – dimension parameter (in the calculations $\gamma = 0.001$). Obtained results of the calculations were compared with the experimental data presented in the paper [9]. A. Chetti and co-authors investigated a mathematical model of internal suffusion based on the advection-diffusion equation. The main assumption of the model studied by A. Chetti and co-authors was equality of the water and moving soil particles velocities. Comparison of results obtained in this paper with the experimental data showed that their model gives highly overestimated values of the eroded soil mass from the filtration region for soils with low susceptibility to suffusion ($\lambda = 4$ and $\lambda = 1$). I. Vardoulakis in his work proposed the relation $v_1 = \beta v_2$ $(0 < \beta < 1)$ for determining the velocities of mobile soil particles [8]. In his work it is postulated that the velocity of soil

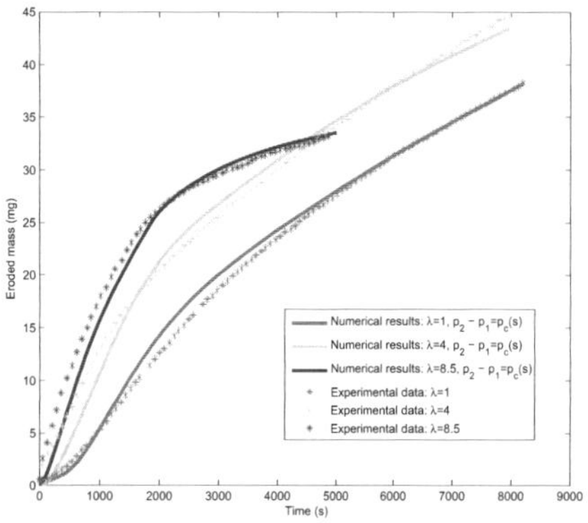

Figure 1. Effect λ on suffusion process.

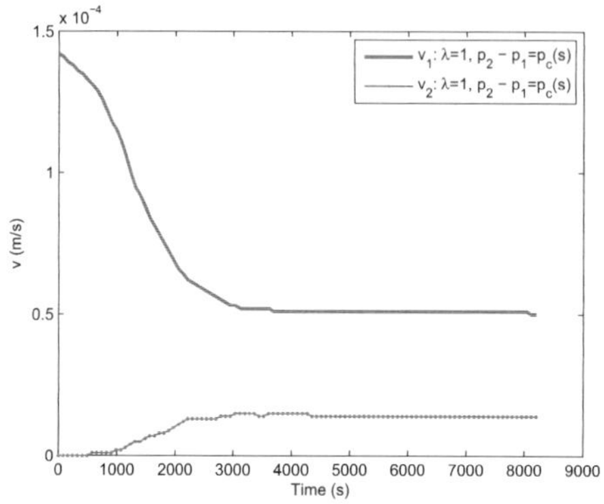

Figure 2. Velocity of water and fluidized solid particles.

particles is less than the rate of water filtration, but method how the value of the coefficient β could be chosen is not explained. In this paper the velocity of moving soil particles is determined during the solution of the problem using Darcy's law (2) [5]. The velocity modulus of the water filtration rate v_1 and the velocity modulus of the moving soil particles are shown in Figure 2. It is shown that the velocity of the moving soil particles is much less than the velocity of water. The comparison of the calculated values of the outward mass from the filtration region with the experimental data presented in [9] is shown in Figure 1. The comparison of the calculated values of the filtration rates of water and moving soil particles shows that the water velocity far exceeds the velocity of the mobile soil particles. Effect of the unsteady velocity of the moving soil particles is significant for soils that are slightly susceptible to suffusion ($\lambda = 4$ and $\lambda = 1$).

5. Conclusion

The degenerate one dimensional problem are solved numerically by the finite-difference method. Calculations results are compared with the experimental data. Using the Leverett J-function the moving phases velocities are found. The water and moving soil particles filtration rates are compared.

Acknowledgments

In conclusion the author express his profound gratitude to K. Shishmarev and A. Papin for their helpful comments and constructive suggestions.

References

[1] Vieira D A N and Dabney S M 2011 *Soil Tillage Research* **111**(2) 197-207
[2] Wilson G 2011 *Hydrol. Process* **25** 2354-64
[3] Einstein H A 1937 *Mitt. d. Versuchsanstaltf Wasserbau* (Zurich: Eidg. T. H.)
[4] Wang J, Walters D A, Settari A and Wan R G 2006 *Simulation of cold heavy oil production using an integrated modular approach with emphasis on foamy oil flow and sand production effects. World Heavy Oil Conference* (Calgary: AB)
[5] Garde R J and Ranga Raju K G 2000 *Mechanics of Sediment Transportation and Alluvial Stream Problems* (New Delhi: New Age International)
[6] Biot M A 1956 *Reprinted from the journal of the acoustical society of America* **28** 168-78
[7] Bonelli S 2012 *Erosion of Geomaterials* (London: John Wiley & Sons, Inc.)
[8] Vardoulakis I, Stavropoulou M and Papanastasiou R 2001 *J. Transp. Porous Med.* **45** 267-81
[9] Chetti A, Benamar A and Hazzab A 2016 *Transport in Porous Media* **113** 591-606

IOP Conf. Series: Journal of Physics: Conf. Series **894** (2017) 012085 doi:10.1088/1742-6596/894/1/012085

[10] Nigmatulin R I 1987 *Dynamics of multiphase media, Parts 1 and 2* (Moscow: Nauka)
[11] Papin A A and Sibin A N 2016 *J. Phys.: Conf. Ser.* **722** 1-8
[12] Bocharov O B, Rudyak V Ya and Seryakov A V 2014 *J. Min. Sci.* **50** 235-48

IOP Conf. Series: Journal of Physics: Conf. Series **894** (2017) 012086 doi:10.1088/1742-6596/894/1/012086

Generation of elastoplastic waves in shaped-charged (cumulative) jets and perforated barriers

V G Smelikov

JSC SPA "Bazalt", Moscow, Russia

Abstract. The article focus on making use of electron-ion tools, based on the energy-spend processes of thermobaric dissociation, atomization and ionization of the facing materials in shaped-charge liners that kick in during explosive jet and slug generation, shaping of penetrators, fragments, and as they go through obstacles. These provide sufficient grounds for qualifying the jet structure in terms of its ionization not as fourth state matter, i.e. plasma, but rather as 'third-and-a-half state', meaning the matter is still solid but approaching the pre-melting condition. The different ability of the matter to transfer mass in the jet and slug (vortex and antivortex) based on its state has been stressed with the speed variation and phase transition being an indicator. In addition, the research provides estimates of the specific energy required for steel armor atomization that equals ~ 9kJ/g (~70kJ/cm³) to the extent sufficient to destroy to breach an obstacle.

1. Introduction

Ya. I. Frenkel's theory of metals already known for more than 90 years prevents researchers of the present day from disregarding the contribution made by an electronic subsystem of pressure-bearing materials into changes of their thermodynamic characteristics of P, V and T. This contribution stresses the importance of proper physical estimation of negative pressure for cases of scaled non-linearity of connections between pressure and volume affected by thermobaric field created by shaped charge explosives, when a delta of dP pressure lags or surpasses a delta of dV volume. As a consequence, "... an electronic structure reinforces, surfaces and clears up a road for a massive flow of atomic and ionic flow" of a shaped-charge jet's solid-state plasma [2]. Resulting from rotary drawing or flow turning process and annealing, various textures of antitank grenades shaped-charge liner can be achieved, leading to directional anisotropy of pressure-bearing metals in a deformation area, out of which a cone shaped-charged jet runs and spins (Figure 1) and which is therefore a result of a concurrence of a dominant torsional and tensile strain of metals forming a jet alongside negative pressure (when a pressure by volume derivative is $dP/dV<0$), and less intense compressional and shear strain in a slug alongside positive pressure ($dP/dV>0$) and variation of dP/dT.

Texture and therefore shaped-charge jet anisotropy of properties testify a solid state of materials in shaped-charge jets. Restoring losses of shaped charge piercing performance or spin compensation [2] is also a direct evidence of such solid state. A synergy of explosive barothermic effect applied, materials in different part of a shaped-charged jet modify their states (during its "lifespan" of just 1 ms length) from pre-melting to trans-melting and hyper-melting ones, the latter being a superheated solid state due to a concurrence of rapid (up to ~300 ns) shaped-charge jet thermobaric ionization and even more rapid (up to ~10 ns) inertial atomization of a slug. A process of copper dissociation, rapidity of which being measured in a picosecond range of 10^{-12} s, completes within a thickness of a shock front. A pressure by volume derivative reversing its sign (at a point of $dP/dV=0$), pressure-loaded material passes through a metastable transient state creating oscillating loops visible on slugs' X-rayograms and moving shaped-charged jets, a hammer pattern (Figure 2) inside a barrier or compression elements, also visible on X-rayograms, if spraders of fixed conduction paths occur.

Content from this work may be used under the terms of the Creative Commons Attribution 3.0 licence. Any further distribution of this work must maintain attribution to the author(s) and the title of the work, journal citation and DOI.
Published under licence by IOP Publishing Ltd

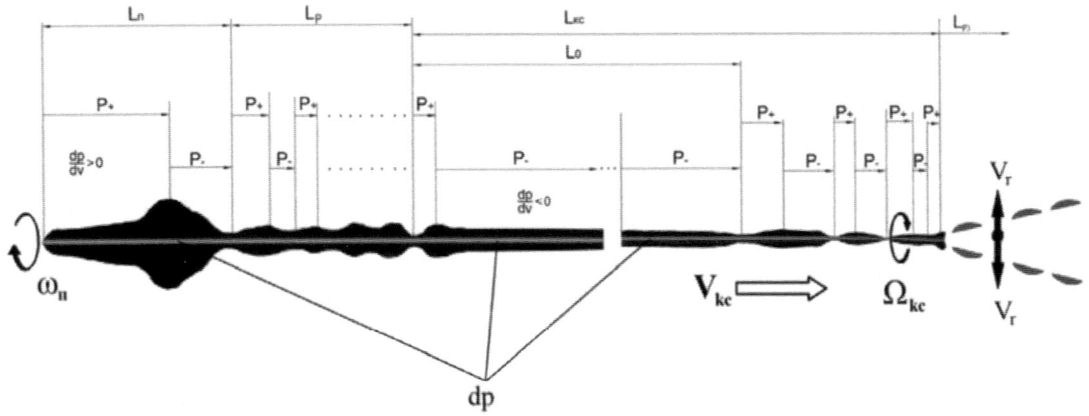

Figure 1. Elastoplastic waves in a slug and shaped-charged jets (an aggregated result of analyzing a large number of X-rayograms of antitank grenades shaped-charge jets), where Lп is a length of a slug, Lp is an acceleration area of shaped-charge jets, Lкс is a length of a shaped-charge jet, L_0 is a length of a shaped-charge jet before dividing into separate elements, P– is negative pressure and P+ is positive pressure, dP/dV - is a pressure by volume derivative, where dP is a delta of pressure, dV is a delta of volume, dρ - material discontinuity in a slug, shaped-charged jet and its elements, Lpэ is an area of elements' tranche division due to a domination of inertial forces on strength forces of solid shaped-charge jet elements.

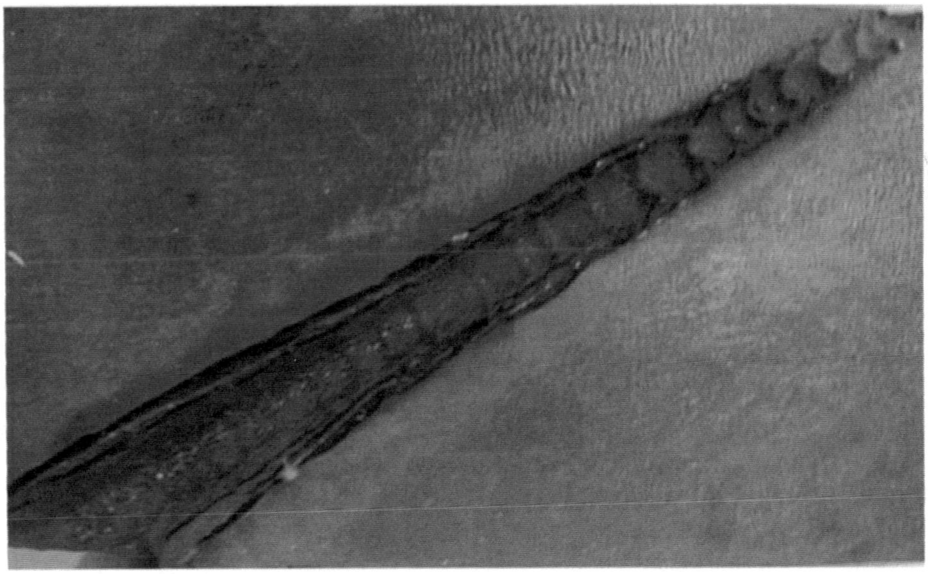

Figure 2. Elastoplastic waves on a barrier's surface. Fluctuation of a deformation area generated by a hammer pattern of a sharped-charged jet penetrating into a barrier.

Figures 1 and 2 illustrate two processes, first being a very beginning of a neck formation processes in a shaped-charged jet, or so-called hammer effect of a shaped-charged jet that "beats" inside a barrier with plasma frequency with corresponding wave length of ~ 5 to 25 mm; and second being an aligning effect in a broken barrier (refer to page 80 in [2]).

2. Harden jet evolution elongation - elastic slug compression deformation

Even "thinking in the context of hydrodynamic approximation" while analyzing rayograms and protonograms of shaped-charged jets, an obvious inference is that a shaped-charged jet, speaking metaphorically, "is not still water, but a rapid flow" that creates vortexes and antivortexes in solid plasma with viscosity of 10^5 P. Restrictions of incompressible flow models do not admit any uncontroversial explanation of achieving to throw out of bucket with viscous liquid (the Reynolds number) in microseconds, i.e. transferring at least a half of a shaped-charged liner mass into a shaped-charged jet of an antitank grenades within the same time interval. The work of S.K. Godunov and A.A. Deribas, published in the Combustion, Explosion, and Shock Waves scientific journal, which is a true handbook for ammunition engineers, in the 1970s, shows that the Reynolds number cannot exceed 4 during wave formation caused by explosive welding. This fact should have been taken into account when calculating the Navier-Stocks equation in the context of hydrodynamic approximation. For liquids the Reynolds number is Re=20…300, and vortexes easily appear. If water features viscosity of 10^{-2} Poise, then metal inside a shaped-charged jet features viscosity of 10^5 Poise, which was determined by A.D. Sakharov and V.N. Meneev in their works in the 1960s. A metal instability pattern appearing in a slug while passing through a spinodal (dP/dV=0) bears evidence of vortex formation being close to a wave top (refer to Figure 1), when metal being affected by deformation either loses or gains resistance, following fluctuations of electron density in electron wind outbursts inside an accelerating or deceleration shaped-charged jet.

Being in different aggregative and high-stress states due to effects of axial, radial or circumferential stress, materials composing shaped-charged jet cone boundaries show unequal orientation and inertial, incl. hammer and aligning effects of plastic deformation in different parts of a shaped-charged jet and slug under the impact caused by a barrier or in flight deceleration. Just in nanoseconds both a shaped-charged jet and slug acquire corresponding gradients of axial, radial and circumferential velocities, Laplacian and other operators of a partial differential field, including differential coefficients of P, V and T, incl. the first- and the second-order pressure by volume derivatives dP/dV, d^2P/d^2V influencing a crystal lattice reaction on powerful undulation of a solid object's volume and a Grüneisen parameter $\Gamma_v = -\frac{2}{3} - 0.5V(\frac{\ddot{P}}{\dot{P}})$, where $\dot{P}$ and $\ddot{P}$ are the first- and the second-order pressure by volume derivatives correspondingly. Coefficients of kinematic momentum also change rapidly, as their value abruptly decreases from front to tail parts of a jet. Shaped-charged jet radial stress also decreases notably from the periphery to the core and depend on the material density and jet velocity.

3. A synergy of barotermic effect or power engineering

Effectiveness of a shape charge is defined by collapsing energy of shaped-charge liner, capacity of applied explosive agents, axial symmetry of charged elements and other factors and also depends on density and pace of detonation of explosive agents. Almost the whole list of trotyl-, hexogen- and octogene containing explosive agents widely used in manufacture industry meet a range from 1.4 to 1.9 g/cm^3 in density, from 6.9 to 9.1 km/s in detonation velocity and from 4.2 to 5.86 kJ/g in chemical energy output.

An energy source for atomization and ionization processes in loaded materials of liners is an shock wave actual profile formed by a relatively coarse and thick (for trotyl-containing 0.7±0.1mm, for cyclonite-containing 0.64±0.04mm, and for octogene-containing 0.37±0.07mm) detonating area spreading in different explosive agents with different detonating pace and with evolution of specific amount of chemical energy per every unit of an explosive agent, and energy consumption goes far beyond mentioned "consumers" [2].

If we look at a shaped-charged liner as at a transformer of explosive agent energy expended for plastic volumetric form change of liner material into the state of a jet and slug, then, counting specific characteristics in kJ/g while estimating a total energy balance between explosive energy and energy expenditure, it is necessary to take into account energy expenditure on atomization, for instance, of copper (4.72 kJ/g) and its light ionization (11.72 kJ/g) along with table values of explosive agents energy content. For example, for octogene it is 5.86 kJ/g, for hexogen - 5.8 kJ/g, for trotyl - 4.23 kJ/g, for hexogen-TNT-based 50/50 explosives - 4.6 kJ/g, for octol 75/25 - 5.19 kJ/g, etc.). It is also necessary to estimate a mechanical momentum imposed to liner, which, according to the most optimistic estimations made by different sources, does not exceed 30% of a product of explosive agents mass by detonation velocity, i.e. $i=0{,}3\ m_{BB}D$.

That being said, a shaped-charged jet is not a flow of interconnected ions (or atoms of copper without an s-orbital) in a form of perfect fluid. Instead, it is a multicomponent structure of ions, atoms and molecules with metallic bindings formed by overlapping external electron sheaths; those parts being bound form clusters with correspondingly different mass, material ionization, atomization and dissociation ratio, which allows us to consider such a structure to the fourth state of matter or, more precisely, to the "third– and– a– half" state of a solid plasma as mentioned before.

In relation to such a representation of a metal state inside a jet, it is highly possible to increase an accuracy of calculation concerning energy balance, considering that mentioned above processes of atomization and ionization are accompanied by dissociation of molecules of shaped-charged stress-loaded material.

The biatomic copper energy of dissociation commensurable with its enthalpy of melting, as well as dissociation of impurities are due to calculation in the right part of the energy balance equation.Than the equation looks like the following:

$$A\sum_{1}^{k} m_{BB}\, Q_{BB} = \sum_{1}^{l} m_n\, E_a + \sum_{1}^{m} m_c\, E_u + \sum_{1}^{n} m_d\, E_d + \cdots,$$

A, the power selection ratio of a shaped-charge;

m_{BB}, m_n, m_c, m_d , correspondingly masses of an explosive agent, slug, jet and dissociated metals and impurities;

Q_{BB}, an energy content of explosive agents composing an explosive compound of a weapon;

E_a, E_u, E_d, correspondingly energy of atomization, ionization and dissociation of metals and impurities of shaped-charged materials.

On the grounds of the energy balance equation [2], an amount of energy required for atomization of armored steel was calculated and estimated as ~9 kJ/g (~70 kJ/cm^3), exceeding which breaking or penetration into a barrier is highly possible.

Jet and slug – vortex and antivortex

Incredible as it may seem, but physics and mechanics of discrete environment (as it can't be unbroken, if all the used materials are of industrial purity degree), as well as atomic-ionic deformation kinetics with its dislocational mechanisms of scaled plastic deformations of materials, allowed to increase mass characteristics of penetrating quality >1 mm of armor up to 1 g of explosive charge of antitank grenade. Herewith, industrial purity of M1 monomorphous copper has never exceeded 99.9% Cu. The necessity of compliance with scientific honesty, as well as of metal physical culture improvement for shaped-charged jets consisting of only three parts development, were subtly and accurately highlighted by academician V.M. Titov. Further research and improvements in this field could allow the scientific community exploring shaped-charged jets to go beyond speculations, anticipations and obvious fiction cultivated by some scientists in terminology of academician Kruglyakov E.P. [5]. Such speculations of the last two decades include the following:" a new theory of cumulation with relief shaped-charged liner", energy emergence from nowhere, as well as "metal explosion with bond energy exposure". Considering scientific honesty, mentioned above [3], there is no need to further increase the number of coefficients in the Lavrentiev equation and collect so-called numerical and engineer methods of a shaped-charged jet penetration capacity estimation that during the last several decades remain bivariate and do not include rapid vortex and antivortex bi- and trifurcated processes appearing in three-dimensional deformation areas in slugs, out of which solid shaped-charged jets

IOP Conf. Series: Journal of Physics: Conf. Series **894** (2017) 012086 doi:10.1088/1742-6596/894/1/012086

"come unscrewed" rapidly, in microseconds. Solutions of equations calculated with the help of modern computers remain approximate and apply only to two-dimensional spaces, where a vortex-antivortex twain of a shape-charge jet+slug type may appear [2], and they would have been needed to make into digital form the equation in full. Such an approximate nature of two-dimensional models and absence of equations applicable for three-dimensional spaces lead to the fact that further development of Frenkel's and Brigman's theories, as well as increase in productivity of modern computers could rapidly change the distance between the Schrödinger equation, Navier-Stokes equation and "forging workshop" like three-dimensional center of deformations in a slug, out of which a shaped-charged jet "comes unscrewed" in a "three-and-a-half" overheated, but still solid state of plasma bearing gradients of velocity, leaving energy vortexes and antivortex jet+slug, Laplacian and other operators and finally "hammer and aligning or conductor's effects" [2].

Reference

[1] Frenkel Ya I 1950 *Introduction to the Theory of Metals* (Moscow: Governmental Publishing for Technical and Theoretical Literature)
[2] Smelikov V G 2016 *Sustainability of limited cumulative effect* (Moscow: Ethnika Studio) p 125
[3] Titov V M 2000 *M.A. Lavrentiev's school* (Moscow: Combustion, Explosion and Shock Waves) **36**(6) 3-13
[4] Bridgeman P 1955 *Studies in large plastic flow and fracture: with special emphasis on the effects of hydrostatic pressure (*Moscow: Foreign literature) p 444
[5] Kruglyakov E P 2002 *"Uchenye s Bolshoi dorogi" (Scientists on the big road (*Moscow: Science Publishing) p 320

IOP Conf. Series: Journal of Physics: Conf. Series **894** (2017) 012087 doi:10.1088/1742-6596/894/1/012087

Multiscale Simulation of Porous Ceramics Based on Movable Cellular Automaton Method

A Smolin[1,2], I Smolin[1,2], G Eremina[1] and I Smolina[3]

[1] Institute of Strength Physics and Materials Science SB RAS, Tomsk, Russia
[2] National Research Tomsk State University, Tomsk, Russia
[3] Tomsk State University of Architecture and Building, Tomsk, Russia

E-mail: asmolin@ispms.tsc.ru

Abstract. The paper presents a model for simulating mechanical behaviour of multiscale porous ceramics based on movable cellular automaton method, which is a novel particle method in computational mechanics of solid. The initial scale of the proposed approach corresponds to the characteristic size of the smallest pores in the ceramics. At this scale, we model uniaxial compression of several representative samples with an explicit account of pores of the same size but with the random unique position in space. As a result, we get the average values of Young's modulus and strength, as well as the parameters of the Weibull distribution of these properties at the current scale level. These data allow us to describe the material behaviour at the next scale level were only the larger pores are considered explicitly, while the influence of small pores is included via the effective properties determined at the previous scale level. If the pore size distribution function of the material has N maxima we need to perform computations for $N-1$ levels in order to get the properties from the lowest scale up to the macroscale step by step. The proposed approach was applied to modelling zirconia ceramics with bimodal pore size distribution. The obtained results show correct behaviour of the model sample at the macroscale.

1. Introduction

Nowadays, ceramic materials are being widely used in medicine, electrical, electronics industries, body armor, etc. Advanced technologies of ceramics production are capable of creating materials with a very complex structure both of the porous space and the matrix itself, which in fact provides the material with high functional properties. For example, in the medical application, the porous structure of ceramics has to be interconnected in order to allow the bone tissue to fill the porous space [1]. Mainly, this leads to bimodal pore size distribution in the material. As a result, mechanical behaviour of the material can be nonlinear and even unpredictable.

The problem of predicting the physical and mechanical properties of porous materials has been being solved by many authors in various statements. The complexity of this problem consists, first of all, in the fact that the properties of advanced materials are mainly determined by their complex multiscale structure [2]. For the analytical solution of this problem, the most successful approaches are (i) the micromechanics of composites, which is based on the method of self-consistent field (definition of the property contribution tensor) as shown, for example, in [3]; and (ii) the method of random functions [4]. However, these approaches allow predicting only elastic, thermal, and electromagnetic properties of the material. Regarding the strength, the

Content from this work may be used under the terms of the Creative Commons Attribution 3.0 licence. Any further distribution of this work must maintain attribution to the author(s) and the title of the work, journal citation and DOI.
Published under licence by IOP Publishing Ltd

capability of these approaches is limited essentially to the periodic structure materials. At the same time, novel computational techniques allow correct simulation of the material behaviour from atomic scale up to the macroscale. Taking into account all the aspects mentioned above, we may conclude that for predicting the mechanical properties of the advanced ceramics it is promising to use multiscale computer simulation [5, 6].

2. Description of the model

At present, numerical methods of continuum mechanics are mainly used for simulating the mechanical behaviour of materials at meso and macroscale [7, 8]. However, recently methods based on the discrete representation of materials have been successfully developed and widely used [6, 9, 10]. One of them is the method of movable cellular automata (MCA), which assumes that the material consists of a set of elementary objects (automata), interacting with eahc other by the forces determined in accordance with the rules of many-particle approach. MCA allows one to simulate mechanical behaviour of a solid at different scales, including deformation, initiation, and development of damages, fracture and further interaction of fragments after failure [6, 11, 12, 13].

An automaton motion is governed by the Newton-Euler equations:

$$\begin{cases} m_i \dfrac{d^2 \vec{R}_i}{dt^2} = \sum\limits_{j=1}^{N_i} \vec{F}_{ij}^{pair} + \vec{F}_i^{\Omega} \\ \hat{J}_i \dfrac{d\vec{\omega}_i}{dt} = \sum\limits_{j=1}^{N_i} \vec{M}_{ij} \end{cases}, \tag{1}$$

where $\vec{R}_i$, $\vec{\omega}_i$, m_i and $\hat{J}_i$ are the location vector, rotation velocity vector, mass and moment of inertia of ith automaton, respectively, $\vec{F}_{ij}^{pair}$ is the interaction force of the pair of ith and jth automata, $\vec{F}_i^{\Omega}$ is the volume-dependent force acting on ith automaton and depending on the interaction of its neighbours with the remaining automata. In the latter equation, $\vec{M}_{ij} = q_{ij}(\vec{n}_{ij} \times \vec{F}_{ij}^{pair}) + \vec{K}_{ij}^{rot}$, here q_{ij} is the distance from the center of ith automaton to the point of its interaction with jth automaton, $\vec{n}_{ij} = (\vec{R}_j - \vec{R}_i)/r_{ij}$ is the unit vector directed from the centre of ith automaton to the jth one and r_{ij} is the distance between automata centres, $\vec{K}_{ij}^{rot}$ is the torque caused by relative rotation of automata in the pair.

The forces acting on automata are calculated using deformation parameters, i.e. relative overlap, tangential displacement and rotation, and conventional elastic properties of the material, i.e. shear and bulk moduli. A distinguishing feature of the method is calculating of forces acting on the automata within the framework of multi-particle interaction [13, 14] using volume-dependent force $\vec{F}_i^{\Omega}$, which among other advantages provides for an isotropic behaviour of the simulated medium.

A pair of the elements may be considered as a virtual bistable automaton having two stable states (bonded and unbonded), which permits simulation of fracture and coupling of fragments (or crack healing) by MCA. These capabilities are realized by means of the corresponding change of the state of the pair of automata. A fracture criterion used in simulation essentially depends on the physical mechanisms of material deformation and failure. An important advantage of the MCA formalism is that it makes possible direct application of conventional fracture criteria (Huber-Mises, Drucker-Prager, Mohr-Coulomb, etc.), which are written in the tensor form [14]. In this work a fracture criterion based on critical value of the equivalent stress was used. Switching of a pair of automata to an unbound state would result in a changeover in the forces acting on the elements; in particular, they would not resist moving away from one another.

Removing automata from initial dense packing allows the explicit taking into consideration voids or pores in the material.

Herein we present a multiscale approach based on MCA method for 3D modelling zirconia ceramics with bimodal pore size distribution [15]. In general, the approach is similar to 2D one described in [16], the main difference and advantage consists in employing Weibull distribution of the elastic and strength properties of a heterogeneous material at higher scales. The initial lowest scale of the proposed approach corresponds to the characteristic size of the small pores. At this scale, we model uniaxial compression of several representative samples with an explicit account of pores of the same size but with the unique position in space. As a result, we get the average values of Young's modulus and strength, as well as the parameters of the Weibull distribution of these properties at the current scale level. These data allow us to describe the material behaviour at the next scale level were only the large pores are considered explicitly. Simulations of the samples compression at this level provide us with the material properties used for final simulation at the macroscale.

3. Simulation results and discussion

Let us consider the simulation results of the samples compression at the lowest scale of the model corresponding to small pores of the ceramics which size is equal to 1 μm and comparable to the grain size of the material [15]. Here it is natural to use the size of automata to be equal to 1 μm too. The response function of automata was taken to be corresponding to the loading diagram for nanocrystalline $ZrO_2(Y_2O_3)$ with total porosity of 2% and an average pore size being equal to about the grain size [15]. Inter-automaton bond rupture criterion used in calculations was formulated as a threshold value for equivalent shear stress. A typical model sample with an explicit account of pores is shown in Figure 1a. The pores were obtained by removing the randomly selected automata from the original close fcc packing.

There were considered five representative samples with pores of the same size but with the unique position in space. Typical loading curves for these samples are shown in Figure 1b. Using these curves the elastic modulus in compression (the slope of the linear part of the curve) and the compression strength (its maximum value) were determined for each sample. These values are random variables due to random pore arrangement in the samples. As shown, for example, in [17] the time to occurrence of the "weakest link" of many competing failure processes is governed

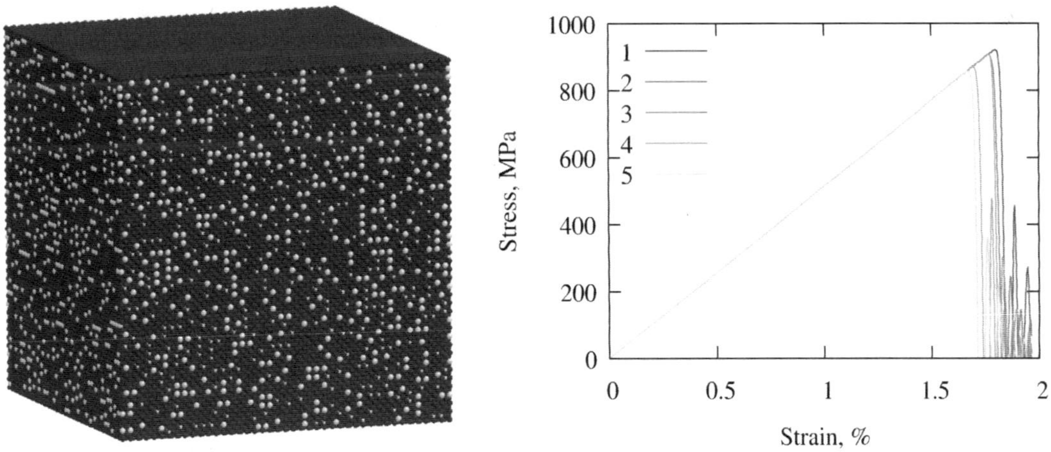

(*a*) (*b*)

Figure 1. (*a*) a typical model sample with 15 % of porosity and (*b*) the loading curves for five samples of the lowest scale level with different spatial distribution of pores.

by Weibull distribution model, which assumes the following cumulative distribution function

$$F(t) = 1 - \exp(-(t/\eta)^{\beta}) \tag{2}$$

where η is the scale parameter (also called as the characteristic life), and β is the shape parameter.

There are many commercial software products performing reliability or survival analysis based on the Weibull model, such as Weibull++, Visual-XSel, Statgraphics, Statistica, and others. In order to determine the parameters η and β these programs use several methods, the most important of which is the method of maximum likelihood estimation. But if the random sample size is small (five in our case), it is recommended to use the median rank regression, which is reduced to the transformation of Eq. 2 to a linear equation, and to the linear approximation of this equation by means of simple least-square regression. There is also free software for the analysis of large data based on the **R** statistical programming language, available at `https://www.r-project.org/` . In our work, we used a special package designed for the **R**, providing basic functionality needed to perform Weibull analysis available at `http://r-forge.r-project.org/projects/abernethy/` .

Figure 2a shows the plot of the normalized probability density function (solid line) for the strength of the samples with porosity of 15% obtained from the Weibull analysis of the presented modelling results. The same analysis was performed for the elastic modulus in compression of the model samples.

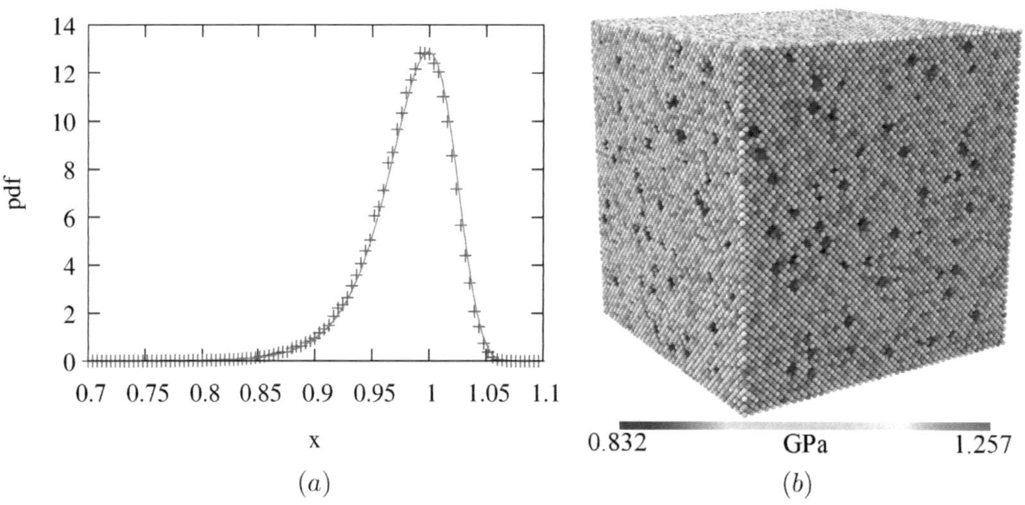

Figure 2. Distribution of the material strength at mesolevel: (a) normalized probability density function, (b) the spatial distribution in a model sample.

To verify the proposed approach against the behaviour of the real ceramics with hierarchical porous structure, let us consider the final results of simulation at the macroscale depicted in Figure 3. Figure 3a shows the loading curves for compression of two samples with the same Weibull distributions for elastic modulus and strength, but different scatter of these properties in space. One can see that this difference reveals itself only at the stage of macrocrack propagation (descending part of the curves). The main reason for such a difference is that the relative scatter of the elastic modulus is rather small compared to the strength scatter. That is why the elastic behaviour of different samples is very similar. Nevertheless, the strains when the macrocracks are generated in different samples differ, as well as the macrocrak paths. The failure pattern of

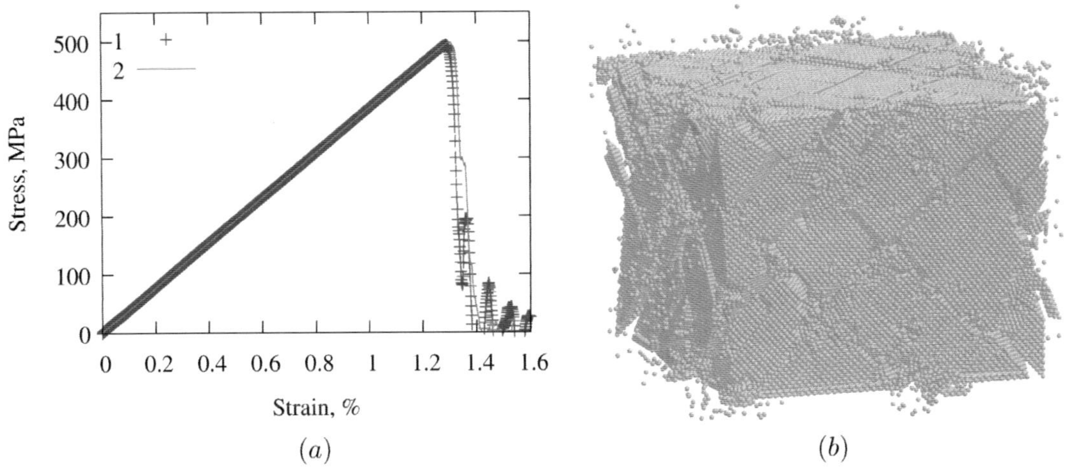

Figure 3. (*a*) loading curves and (*b*) a typical failure pattern of the model macrosamples.

the macrosamples shown in Figure 3b is typical for heterogeneous material and consists of a large number of small fragments. This is a result of a large number of local stress concentrators in the macrosample due to scatter of strength properties. In contrast to the samples with an explicit account of the pores, where crack paths are curved [16], in our macrosamples cracks propagate along a plane direction and just deviate at new stress concentrators. However, the obtained results allow us to conclude that the proposed approach is capable for correct simulating the ceramics with hierarchical porosity.

The obtained data provide us the ability to pass from one scale level to the next scale level (herein and after referred to as mesolevel) of the model where these data describe the effective properties of the material of large pore walls. This is a heterogeneous material with stochastically distributed elastic and strength properties. To model mechanical behaviour of samples at this scale, we scatter elastic and strength properties for each automaton according to the Weibull distribution obtained from the previous scale level. The automaton size at mesolevel corresponds to the representative sample of the previous scale and is equal to 60 μm. A typical model sample for this level is shown in Figure 2b, where the automaton colour corresponds to the strength of the automaton. The crosses in Figure 2a correspond to the distribution of automata strength over the sample in Figure 2b, so one can see that these random values are really described by the specified distribution law.

4. Conclusions

A 3D multiscale model for porous ceramics based on movable cellular automaton method is proposed in the paper. The important issue of the model is a determination of the Weibull distribution parameters for elastic and strength properties at the current scale from modelling uniaxial compression of the model samples with an explicit account of pores. These parameters allow correct transfer of the effective ceramics properties to the upper scale level of the model. The proposed approach was applied to modelling zirconia ceramics with bimodal pore size distribution. The obtained results show the correct behaviour of the model sample at the macroscale.

Acknowledgments
The investigation has been carried out at financial support of the Project No. III.23.2.3 of the Basic Research Program of State Academies of Sciences for 2017–2020.

References
[1] Buyakov A, Litvinova L, Shupletsova V, Kulbakin D and Kulkov S 2016 *AIP Conf. Proc.* **1760** 020008-1–4
[2] Froustey C, Naimark O B, Panteleev I A *et al* 2017 *Phys. Mesomech.* **20(1)** 31–42
[3] Sevostianov I 2013 *Effective properties of heterogeneous materials Solid Mechanics and Its Applications* vol 193 ed Kachanov M and Sevostianov I (London: Springer) pp 1–96
[4] Vildeman V E, Sokolkin Yu V and Tashkinov A A 1997 *Mechanics of Inelastic Deformation and Fracture of Composite Materials* (Moscow: Fizmatlit) [in Russian]
[5] Li G, Liang Z Z and Tang C A 2015 *Rock Mech. Rock Eng.* **48** 2235–62
[6] Dmitriev A I, Smolin A Yu, Popov V L and Psakhie S G 2009 *Phys. Mesomech.* **12(1-2)** 11–9
[7] Eremin M O 2016 *Phys. Mesomech.* **19(4)** 452–458.
[8] Smolin I Yu, Makarov P V, Eremin M O and Matyko K S 2016 *Proc. Struct. Integrity* **2** 3353–60
[9] Psakh'e S G and Zol'nikov K P 1998 *Combust. Expl. Shock Waves* **34(3)** 366–8
[10] Korchuganov A V, Zolnikov K P, Kryzhevich D S *et al* 2016 *Phys. Atomic Nuclei* **79(7)** 1193–8
[11] Smolin A Yu, Roman N V, Dobrynin S A and Psakhie S G 2009 *Phys. Mesomech.* **12(3-4)** 124–9
[12] Smolin A Yu, Roman N V, Konovalenko I S *et al* 2014 *Eng. Fract. Mech.* **130** 53–64
[13] Smolin A Yu, Shilko E V, Astafurov S V *et al* 2015 *Defence Technology* **11(1)** 18–34
[14] Shilko E V, Psakhie S G, Schmauder S *et al* 2015 *Comput. Mater. Sci.* **102** 267–85
[15] Kulkov S N, Buyakova S P and Maslovskii V I 2003 *Vestnik TGU* **13** 34–57 [in Russian]
[16] Konovalenko Ig S, Smolin A Yu and Psakhie S G 2013 *Frattura ed Integrità Strutturale* **24** 75–80
[17] Rinne H 2009 *The Weibull Distribution. A Handbook* (London: CRC Press)

IOP Conf. Series: Journal of Physics: Conf. Series **894** (2017) 012088 doi:10.1088/1742-6596/894/1/012088

Modeling the dynamics of polymer chains in water solution. Application to sensor design

V N Starovoitov[1,2] **and B N Starovoitova**[1]

[1]Lavrentyev Institute of Hydrodynamics, Novosibirsk, Russian Federation
[2]Novosibirsk State University

E-mail: starovoitov@hydro.nsc.ru, botagoz@hydro.nsc.ru

Abstract. This paper is devoted to a mathematical model of a chaotic dynamics of a polymer chain in water. The model consists of a parabolic equation that is derived according to the self-consistent field approach. This model is employed for the numerical simulation of a biological sensor that detects the presence of a specific protein in the fluid. The sensor is absolutely simple and seems to be new. Besides that, the suggested equation is interesting from the mathematical point of view. It includes a non-local operator of integration not only over the past time interval as in the problems with memory but also over the future time interval. It is unusual for parabolic problems.

1. Introduction

This paper is devoted to modeling and numerical simulations of a biosensor based on the chaotic motion of a polymer molecule in water. The biosensor detect the presence of a specific protein. The polymer chain is equipped with a special atom group (ligand) that can bind this protein. The behavior of the polymer chain changes if it capture the protein which is the spatial restriction for the random motion of the chain. In order to detect this change, we use the average end-to-end distance of the polymer chain. The measurement of this quantity is based on the following principle. Suppose that we have two fluorescent molecules. One of these molecules called donor is excited at its specific fluorescence excitation wavelength. This excited state is then nonradiatively transferred to a second molecule, the acceptor. The transfer occurs if the distance between the molecules does not exceed the so called Förster radius. Thus, the color of the light coming from the acceptor depends on the distance to the donor. If we place the donor and the acceptor at the ends of a polymer chain, we can detect whether the chain is stretched or not. To our knowledge, this principle was suggested in [1]. A possible sensor based on the this principle was considered in [2].

Besides the biological application, the development of the sensor mentioned above is interesting from the mathematical point of view. When modeling the chaotic motion of a polymer chain, we have obtained a parabolic equation whose coefficient depends on the integral of the solution over all the time interval where the problem is considered. In fact, in order to solve the problem we have to know "the future". Thus, this is a problem with a non-local operator that includes integration not only over the past time interval as in the problems with memory but also over the future time interval.

Content from this work may be used under the terms of the Creative Commons Attribution 3.0 licence. Any further distribution of this work must maintain attribution to the author(s) and the title of the work, journal citation and DOI.
Published under licence by IOP Publishing Ltd

2. Behavior of polymer chains in water

Here, we describe our approach to modeling of the behavior of polymer chains in water. General facts about polymers chains can be found in the books [3] and [4].

2.1. Ideal chains

The theory of dilute polymer solutions begins with a single linear flexible polymer chain in an infinite solvent medium. Within this simplified system, all intermolecular interactions between solute polymers and between the bonds of one chain are neglected. The only restriction is that the bond n is linked to the bonds $n-1$ and $n+1$. Besides that, all the bonds have the same length.

The instantaneous configuration of the entire chain can be specified by the coordinates of the elements composing the chain. If each bond is denoted by a vector $\boldsymbol{a}_j$, then the end-to-end vector of such a chain is $\boldsymbol{R} = \sum_{j=1}^{N} \boldsymbol{a}_j$, where N is the number of the segments (see Fig. 1). It is clear that, for a freely fluctuating chain, the average or mean end-to-end vector $\langle \boldsymbol{R} \rangle$ will be $\boldsymbol{0}$ as the chain can traverse in all directions with equal probability. Thus, $\langle \boldsymbol{R} \rangle$ does not serve as a good measure of the chain size since it will be zero irrespective of the number of the monomers. Another measure is the mean of the square of the end-to-end distance, or simply the "mean square end-to-end distance", denoted by $R = \langle \boldsymbol{R}^2 \rangle^{1/2}$. It is not difficult to deduce that $R^2 = Na^2$, where a is the length of the segments.

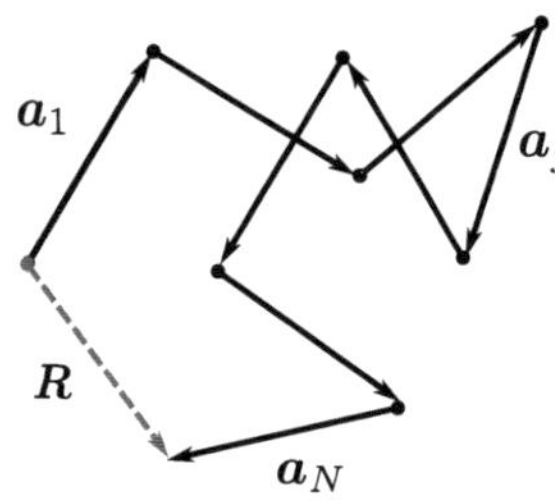

Figure 1. General flow pattern in the physical plane.

The most useful characteristic of the chain is the probability distribution of the square end-to-end distance. For ideal chains, this distribution is Gaussian and its density looks as follows

$$P_N(\boldsymbol{R}) = \Big(\frac{2\pi N a^2}{3}\Big)^{-3/2} exp\Big(-\frac{3\boldsymbol{R}^2}{2Na^2}\Big).$$

It is not difficult to see that $R^2 = \int \boldsymbol{R}^2 P_N(\boldsymbol{R})\, d\boldsymbol{R}$.

Let $p(\boldsymbol{x}, t)$ be the solution of the Cauchy problem:

$$\partial_t p - \frac{a^2}{6}\Delta p = 0, \quad p(\boldsymbol{x}, 0) = \delta(\boldsymbol{x}),$$

where $\delta(\boldsymbol{x})$ is the Dirac delta-function concentrated at the origin. Then $P_N(\boldsymbol{R}) = p(\boldsymbol{R}, N)$. Thus, we can find the density of probability as the solution of the problem above. In a more general case, if the chain is not absolutely free and is influenced by an external field with the potential φ, then we should solve the following problem:

$$\partial_t q - \frac{a^2}{6}\Delta q + \frac{\varphi}{kT} q = 0, \quad q(\boldsymbol{x}, 0) = \delta(\boldsymbol{x}),$$

where k is the Boltzmann constant and T is the temperature. In this case, q is not the density of probability but the statistical weight. In order to obtain the density of probability p, we have to normalize q:

$$p = q \left(\int q \, d\boldsymbol{x} \right)^{-1}.$$

For instance, if the motion of the chain is restricted by a body S, we take φ to be the indicator function of the set S: $\varphi(\boldsymbol{x}) = 0$ for $\boldsymbol{x} \notin S$ and $\varphi(\boldsymbol{x}) = +\infty$ for $\boldsymbol{x} \in S$. Thus, the body is considered as an energy barrier.

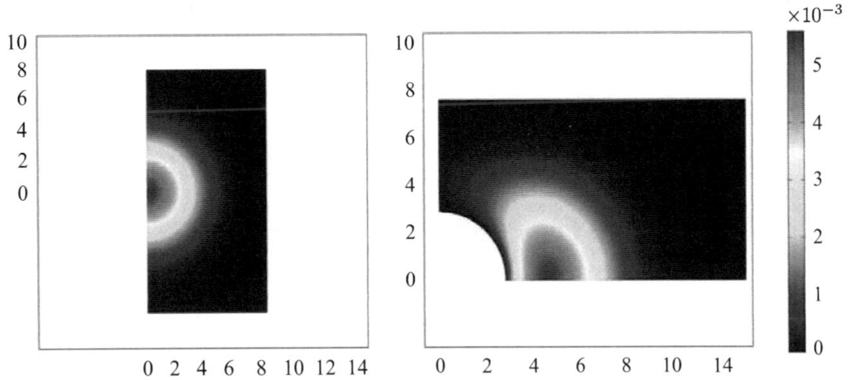

Figure 2. The distribution of the density of probability $P_N(\boldsymbol{x})$ for a free ideal chain (in the left picture) and for a restricted ideal chain (in the right picture).

In Fig. 2 (left), one can see the distribution of the density of probability $p(\boldsymbol{x})$ for a free ideal chain tethered to the point $\boldsymbol{x} = \boldsymbol{0}$. We should rotate the picture around the vertical axis in order to obtain the three-dimensional one. The calculations were done for $N = 100$ and $a = 0.4$. Suppose that a spherical body of radius 3 restricts the motion of the chain. For this case, the distribution of the probability is presented on the right picture. The chain is tethered to the point $\boldsymbol{x} = (0, 0, 3)$. The three-dimensional picture can be obtained by rotating around the horizontal axis. Pay attention to the fact that in the first case the maximum of the function p is at the point of tethering and in the second case it is moved approximately two units to the right.

2.2. Real chains

The behavior of real chains differs from that of ideal ones. The bonds composing real chains have non-zero volume and, besides that, interact through the surrounding fluid. Sometimes, the chains whose bonds have non-zero volume are called the chains with excluded volume since the domains occupied by various bonds cannot intersect. We will consider a single chain which does not interact with other chains.

Let us denote by a the length of the Kuhn segment of the chain (see [3, 4]) and by N the number of the Kuhn segments composing the chain. Further, we call them simply segments. One should distinguish the segments and the bonds of the chain. A segment can consist of several bonds. Notice also that in some papers the length of the Kuhn segment is considered as the persistent length of a chain.

In order to describe the behavior of real chains, we apply the self-consistent field approach (see [3, 4]). It is supposed that every segment of the chain produces an energy field in its neighborhood and, in such a way, acts on the other segments. The total energy field Φ is the sum of the segment fields. In order to calculate it, we have to know the density of the probability distribution ϱ of all segments in the space which is defined by the relation:

$$\varrho(\boldsymbol{x}) = \sum_{k=1}^{N} P_k(\boldsymbol{x}),$$

where $P_k(\boldsymbol{x})$ is the density of probability that $\boldsymbol{x}$ belongs to the k-th segment. Due to the excluded volume effect, ϱ cannot exceed ν^{-1} , where ν is the volume of one segment which can be found experimentally. We define Φ by the Flory – Huggins formula:

$$\Phi = -kT\big(\log(1 - \varrho\nu) + 2\chi\varrho\nu\big),$$

where χ is a constant called the polymer – solvent interaction parameter.

Thus, applying the differential equation approach described in the previous subsection, we obtain the following Cauchy problem:

$$\partial_t q - \frac{a^2}{6}\Delta q + \frac{\Phi + \varphi}{kT} q = 0, \quad q(\boldsymbol{x}, 0) = \delta(\boldsymbol{x}),$$

$$\varrho(\boldsymbol{x}) = \int_0^N p(\boldsymbol{x}, t)\, dt, \quad p = q\Big(\int q\, d\boldsymbol{x}\Big)^{-1}.$$

This problem has to be solved for $t \in (0, N)$ and $\boldsymbol{x} \in \mathbb{R}^3$. Here, φ is the energy potential of possible spatial restrictions. We employ this model for numerical simulations of a biological sensor.

It should be noted that the problem above is also interesting from mathematical point of view. The potential Φ depends on the values of the solution on the whole interval $(0, N)$. Thus, in order to solve the problem we have to know the "future". It is unusual for parabolic problems. The word "future" is enclosed in quotation marks since t is not the time variable but the arc length parameter on the chain.

3. The sensor design

3.1. General scheme

Let we have a polymer chain with two fluorescent molecules at the ends (see Fig. 3). One of these molecules called donor is excited at its specific fluorescence excitation wavelength. This excited state is then nonradiatively transferred to a second molecule, the acceptor. The transfer occurs if the distance between the molecules does not exceed the so called Förster radius. The color of the light coming from the acceptor depends on the distance to the donor. If we place the donor and the acceptor at the ends of a polymer chain, we can detect whether the chain is stretched or not. At one end, the chain has also a special atom group (ligand) that can bind a specific protein.

Suppose that some amount of such polymer chains is dissolved in water. The solution must be dilute in order to avoid the interaction of the chains, so we can consider the behavior of a single chain. Clearly, the behavior of the chain with captured protein differs from that of the free chain. The difference comes from the fact that the protein represents a spatial restriction for the chaotic motion of the chain and reduces the number of its possible configurations. Therefore, the protein changes the mean distance between the donor and the acceptor. This event can be detected since the color of the light coming from the acceptor depends on the distance to the donor. Thus, it is possible to detect the presence of the protein in the water.

To maximize the effect, we have to choose optimal values of such parameters as the length of the chain and its persistent length, the temperature, the Förster radius of the fluorescent molecules, and some other. Assuming medical and biological applications of the sensor being developed, the temperature must be in the range 10°C – 35°C. Besides that, the polymer has to be non-toxic.

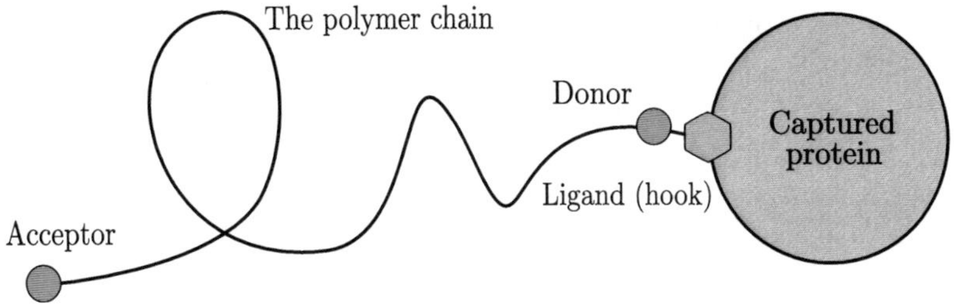

Figure 3. The general design of the sensor.

3.2. Numerical simulations

The simulations of the system were done for the polymer chain of polyethylene glycol, whose formula is

$$-(O - CH_2 - CH_2)_n-$$

This polymer is widely applied in biology and in medicine (see [5]). The necessary parameters of aqueous solutions of this polymer were taken from the works [5] – [10]. The following values of parameters were chosen: diameter of the protein 6 nm, the length of the Kuhn segment 0.6 nm, $\nu = 0.05\ nm^3$, the temperature is in the range 25°C – 28°C. It is assumed that the protein has the spherical form.

Suppose that the chain end with the ligand is at the point $\boldsymbol{x} = \boldsymbol{0}$. We would like to know the distribution of the density of probability for the other end (for the acceptor). In Fig. 4, the values of this quantity on the ray that starts out at the ligand and goes through the point of its maximum. The calculations were made for the chains of the length 100 (the left picture) and 200 (the right picture) segments. If the chain is free, the distribution of probability is spherically symmetric and depends only on the radial variable. The green dotted line shows this dependence. The red solid line corresponds to the case of the chain with the captured protein and represents the values of the density of probability on the straight line passing through the point orthogonal to the protein surface. The unit of the horizontal axis is nanometer. Notice that the density of probability is equal to zero at the origin due to the excluded volume effect. The first and the last segments of the chain cannot be in the same point simultaneously.

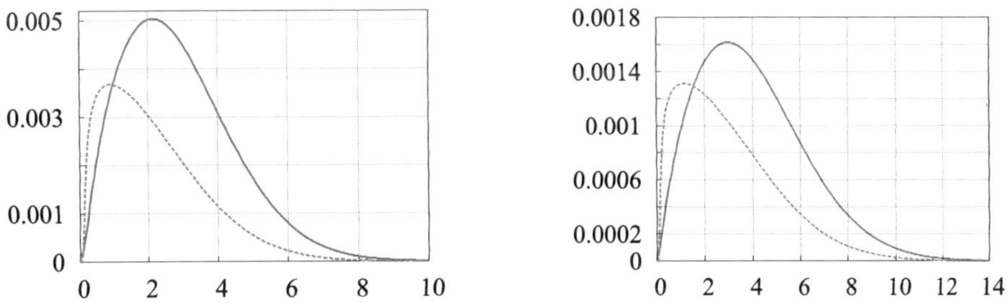

Figure 4. The density of probability $P_N(\boldsymbol{x})$ for $N = 100$ (left) and $N = 200$ (right). The green dotted line and the red solid line correspond to the free chain and to the chain with captured protein respectively.

As it was expected, for the chain with the captured protein, the maximum of the probability distribution has moved a few nanometers from the point of tethering. This effect can be detected. However, in order to choose the optimal length of the chain, we have to know another characteristics. It is the proportion of the time that the end of the chain (acceptor) is located outside of the sphere of the radius r with the center at $\boldsymbol{x} = \mathbf{0}$. Due to the ergodic theorem, it is exactly the probability of this event. In Fig. 5, the dependence of this probability on r (the horizontal axis) is shown.

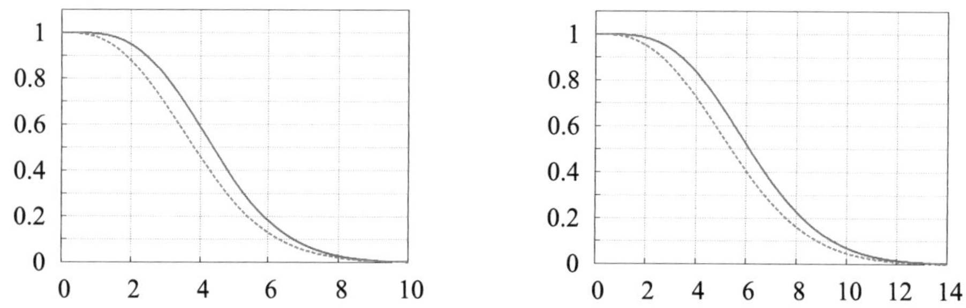

Figure 5. The probability of finding the acceptor outside of the sphere of the radius r (the horizontal axis). The graphs correspond to that in Fig. 4.

On the basis of these data, one can evaluate the intensity of the light coming from the acceptor. Assume that the acceptor is blue if the end-to-end distance of the chain does not exceed the Förster radius and is yellow in the other case. If the Förster radius (FR) is equal to 6 nm, we have the following diagrams for the chains consisting of 100 and 200 bonds (Fig 6). We see that for the chain of polyethylene glycol composed of 200 segments and for fluorescent

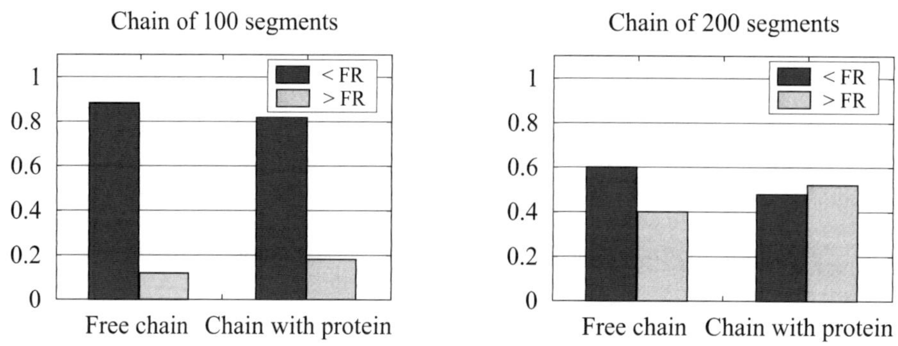

Figure 6. The probability of finding the acceptor outside of the sphere of the radius 6 nm (the yellow bar). The graphs correspond to that in Fig. 5.

molecules with the Förster radius 6 nm, the intensity of the yellow light coming from the solution increases on 12% if all chains have captured the protein molecules. This effect can be detected. Of course, the sensitivity of this sensor can be improved by a better choose of parameters.

References

[1] Hoffmann D and Moske M 2003 Verfahren zum Nachweis eines Analyts mit einem elasto-optischen Biosensor (European Patent EP1363124 24.12.2003)

[2] Botkin N D, Hoffmann K-H, Marx D, Starovoitov V N and Turova V L 2013 Modeling in the development of an elasto-optical biosensor based on nanostructures *International Journal of Biomathematics and Biostatistics* **2**(2) 191-202

[3] Grosberg A Yu and Khokhlov A R 1994 *Statistical physics of macromolecules* (New York : American Institute of Physics)

IOP Conf. Series: Journal of Physics: Conf. Series **894** (2017) 012088 doi:10.1088/1742-6596/894/1/012088

[4] Yamakawa H 1971 *Modern theory of polymer solutions* (New York: Harper and Row)
[5] Harris J M 1992 *Poly (ethylene glycol)* (New York)
[6] Kienberger F, Pastushenko V, Kada G, Gruber H, Riener C, Schindler H and Hinterdorfer P 2000 Static and dynamical properties of single poly (ethylene glycol) molecules investigated by force spectroscopy *Single Mol.* **1**(2) 123-128
[7] Zhu P, Yang H, Peng C and Zhang X 2001 The influence of solvent and temperature upon the aggregation of poly (ethylene glycol) *Macromol. Chem. Phys.* **202** 1380-1383
[8] Kawaguchi S, Imai G, Suzuki J, Miyahara A, Kitano T and Ito K 1997 Aqueous solution properties of oligo- and poly (ethylene oxide) by static light scattering and intrinsic viscosity *Polymer* **38**(12) 2885-2891
[9] Kinugasa S, Nakahara H, Fudagawa N and Koga Y 1994 Aggregative behavior of ploy (ethylene oxide) in water and methanol *Macromolecules* **27** 6889-6892
[10] Oesterhelt F, Rief M and Gaub H E 1999 Single molecule force spectroscopy by AFM indicates helical structure of poly (ethylene glycol) in water *New Journal of Physics* **1** 6.1-6.11

Acoustic properties of a layered porous elastic structure saturated with a Maxwell fluid

B N Starovoitova[1] **and V N Starovoitov**[1,2]

[1] Lavrentyev Institute of Hydrodynamics, Novosibirsk, Russia
[2] Novosibirsk State University, Novosibirsk, Russia

E-mail: botagoz@hydro.nsc.ru, starovoitov@hydro.nsc.ru

Abstract. We consider a one-dimensional layered structure that consists of an elastic material and a Maxwell fluid. It is assumed that the width of the layers is small compared with the size of the domain, where the problem is being solved. By employing the Nguetseng two-scale convergence technique, we have homogenized the equations that describe the mechanical system. For the homogenized equation, the limit cases of the Maxwell fluid with the small viscosity and with the small relaxation time are considered. The question of existence of the travelling wave solution for the homogenized equation is investigated.

1. Statement of the problem

In this paper, we consider a periodic layered structure that consists of alternating elastic and fluid layers. Our mail goal is to derive a homogenized macroscopic equations when the thickness of the layers tends to zero. A similar problem was investigated in [1]. Unlike that paper, we consider a Maxwell fluid instead of the usual viscous one. Besides that, we homogenize the non-stationary problem and then investigate the existence of the harmonic waves.

Suppose that the mechanical system is periodic in the direction of the variable x and does not depend on other variables. Denote by h the period of the structure and assume that the strips $0 < x < x_h$ and $x_h < x < h$ are occupied by the solid and by the fluid, respectively. Due to the periodicity, the same is true for the strips $kh < x < x_h + kh$ and $x_h + kh < x < (k+1)h$, where k is an arbitrary integer number. Let us write down the governing equations. Since we have supposed that the system depends only on the spatial variable x and on the time t, the corresponding equations will contain only these two variables. Besides that, we assume that all vector functions, such as the velocity and the displacement, will have only one non-zero component directed along the variable x.

The displacement u_S in the elastic part of the continuum satisfies the following equation:

$$\varrho_S \partial_t^2 u_S = (\lambda_S + 2\mu_S)\partial_x^2 u_S, \tag{1}$$

where $\varrho_S = const$ is the density, λ_S and μ_S are the Lame coefficients. The external balk forces are neglected, which is suitable for possible acoustic applications.

The Maxwell fluid is a continuum with the stress tensor of the form $-p_F + \sigma_F$, where p_F is the pressure. The tensor σ_F in our case has only one non-zero component denoted again by σ_F and defined as follows:

$$(1 + \tau_* \partial_t)\sigma_F = 2\mu_F \partial_x v_F,$$

Content from this work may be used under the terms of the Creative Commons Attribution 3.0 licence. Any further distribution of this work must maintain attribution to the author(s) and the title of the work, journal citation and DOI.
Published under licence by IOP Publishing Ltd

where v_F is the velocity of the fluid, $\mu_F = const$ its viscosity, and τ_* is a constant that has the dimension of time and is usually called the relaxation time. Originally, in the classical derivation of the Maxwell constitutive law, $\tau_* = 2\mu_F/E_F$, where E_F is the elastic modulus.

Since we intend to describe acoustic processes, we assume that the fluid is slightly compressible and the governing equations are linear:

$$\varrho_F \partial_t v_F = -\partial_x p_F + \partial_x \sigma_F,$$
$$\partial_t p_F = -\gamma \partial_x v_F,$$

where $\varrho_F = const$ is the density and γ is a positive constant that characterizes the compressibility of the fluid.

On the interface that separates the solid and fluid phases, we impose the usual continuity conditions for the velocity and for the internal forces:

$$v_F(t,x) = \partial_t u_S(t,x),$$
$$-p_F(t,x) + \sigma_F(t,x) = \mu_S \partial_x u_S(t,x)$$

for $x = kh$ and $x = x_h + kh$ with an arbitrary integer k and for all t.

It will be more convenient for us to rewrite the equations for the fluid in terms of the displacement u_F. Since in the framework of the linear approach $v_F = \partial_t u_F$, we after some calculation obtain that

$$\varrho_F \partial_t^2 u_F = \gamma\, \partial_x^2 u_F + \frac{2\mu_F}{\tau_*}\, \partial_x^2 (u_F - \mathcal{J}_t u_F) - \partial_x F_F, \tag{2}$$

where

$$\mathcal{J}_t u = \frac{1}{\tau_*} \int_0^t e^{(\tau - t)/\tau_*} u(\tau)\, d\tau$$

and

$$F_F(t,x) = e^{-t/\tau_*} G_F^0(x) + P_F^0(x), \quad G_F^0 = \frac{2\mu_F}{\tau_*} \partial_x u_F^0 - \sigma_F^0, \quad P_F^0 = p_F^0 + \gamma \partial_x u_F^0.$$

The upper index 0 means the initial value of the corresponding quantity at $t = 0$.

Equations (1) and (2) can be written as one equation. To do this, we introduce the function χ_h that is equal to 1 in the fluid part of the domain and to 0 in the solid. The subscript h shows that this function corresponds to the problem with the layers width h. The function

$$u_h(t,x) = u_F(t,x)\chi_h(x) + u_S(t,x)(1 - \chi_h(x))$$

satisfies the following equation:

$$\varrho_h\, \partial_t^2 u_h = \partial_x\big((\lambda_h + 2\mu_h)\partial_x u_h\big) - \partial_x\big(2\mu_h \chi_h \partial_x \mathcal{J}_t u_h\big) - \partial_x F_h, \tag{3}$$

where

$$\varrho_h = \varrho_F \chi_h + \varrho_S (1 - \chi_h), \quad \lambda_h = \gamma \chi_h + \lambda_S (1 - \chi_h), \quad \mu_h = \frac{\mu_F}{\tau_*} \chi_h + \mu_S (1 - \chi_h), \quad F_h = \chi_h F_F.$$

Equation (3) is understood in the distributional sense. This equation should be supplemented by the initial data $u_h(0,x) = u_h^0(x)$, $\partial_t u_h(0,x) = u_h^1(x)$ and by boundary conditions. Up to now, we have not specified the domain, where the problem is considered. It can be any finite interval of the real axis. We denote this interval by Ω and, without loss of generality, assume that $\Omega = (0, L)$, $L > 0$. We suppose that u_h satisfies the homogeneous boundary conditions:

$u_h(t, 0) = u_h(t, L) = 0$ for all t. In other case, we can make them homogeneous by changing the unknown function u_h and the right-hand side F_h. This problem will be referred as *Problem* A_h.

It is not difficult to prove that, under natural conditions on the initial data, Problem A_h has a unique weak solution $u_h \in L^2(0, T; H_0^1(\Omega))$ such that $\partial_t u_h \in L^\infty(0, T; L^2(\Omega))$. Here, T is an arbitrary positive number, L^p and H_0^1 are usual Lebesgue and Sobolev function spaces, respectively. This problem, however, is not appropriate for applications since its numerical realisation demands a lot of computer resources. The reason for this is that (3) is an equation with rapidly oscillating coefficients. Therefore, it would be better to homogenize the equation, i.e., to replace it by a new one whose coefficients do not oscillate and whose solution is close in some sense to the original one.

2. Homogenization of the problem

Suppose that $h/L = o(1)$, i.e., the domain Ω contains a large number of the layers. Without loss of generality, we can assume that L is finite and h is small. Generally speaking, we would have to rewrite the problem in the dimensionless form in which only the ratio of h and L as well as that of the other physical parameters makes sense. However, we will not do this since the dimensionless equation looks exactly the same. One can simply assume that all the quantities in (3) are dimensionless.

Thus, h is a small parameter in Problem A_h. In order to describe the homogenized continuum, one have to find equations for the principal term u of the expansion $u_h = u + O(h)$ as $h \to 0$. To do this, we employ the notion of two-scale convergence introduced in [2] and further developed in a number of papers (see, e.g., [3, 4]). We denote by Σ the interval $[0, 1]$ and say that a function is Σ-periodic if it has the period 1. A sequence of functions $\{v_h\}$ in $L^2([0, T] \times \Omega)$ is said to be two-scale converging to a function $v \in L^2([0, T] \times \Omega \times \Sigma)$ if

$$\lim_{h\to 0} \int_0^T \int_\Omega v_h(t, x)\phi(t, x, x/h)\, dx dt = \int_0^T \int_\Omega \int_\Sigma v(t, x, \xi)\phi(t, x, \xi)\, d\xi dx dt$$

for every smooth function $\phi(t, x, \xi)$ that is Σ-periodic in ξ. We denote this fact by $v_h \twoheadrightarrow v$.

Let us define the Σ-periodic function $\chi(\xi)$ that is equal to 0 in $(0, x_h/h)$ and to 1 in $(x_h/h, 1)$. It is not difficult to see that $\chi_h(x) = \chi(x/h)$ and $\chi(\xi) = \chi_h(h\xi)$. Similarly, we define the Σ-periodic functions: $\varrho(\xi) = \varrho_h(h\xi)$ and $\lambda(\xi) = \lambda_h(h\xi)$. These functions can be also defined exactly as in equation (3) but with χ_h replaced by χ. It is not difficult to see that $\chi_h \twoheadrightarrow \chi(\xi)$, $\varrho_h \twoheadrightarrow \varrho(\xi)$, and $\lambda_h \twoheadrightarrow \lambda(\xi)$. As for the limit of μ_h, we will consider two cases. In the first one, the viscosity μ_F of the Maxwell fluid is a finite positive constant. The second case deals with a weakly viscous Maxwell whose viscosity is small. This means that $\mu_F = \mu_* + o(1)$ as h is close to 0, μ_* is a positive constant in the first case and $\mu_* = 0$ in the second one. Thus, $\mu_h \twoheadrightarrow \mu$, where

$$\mu(\xi) = \frac{\mu_*}{\tau_*}\chi(\xi) + \mu_S(1 - \chi(\xi))$$

is a Σ-periodic function.

Let us turn to the initial data. Suppose that the sequences $\{u_h^1\}$, $\{\partial_x u_h^0\}$, $\{p_F^0\}$, and $\{\sigma_F^0\}$ are bounded in $L^2(\Omega)$ uniformly in h. The functions p_F^0 and σ_F^0 are assumed to be extended by zero to the solid part of the domain Ω. Then there exist functions $u^1 = u^1(x, \xi)$, $\sigma^0 = \sigma^0(x, \xi)$, $p^0 = p^0(x, \xi)$, $u^0 = u^0(x)$, and $\bar{u}^0 = \bar{u}^0(x, \xi)$ that are the two-scale limits of the corresponding sequences. Notice that $\partial_x u_h^0 \twoheadrightarrow \partial_x u^0 + \partial_\xi \bar{u}^0$. This implies that $F_h \twoheadrightarrow F$, where

$$F(t, x, \xi) = \chi(\xi)\big(e^{-t/\tau_*}G^0(x, \xi) + P^0(x, \xi)\big),$$

$$G^0 = \frac{2\mu_*}{\tau_*}(\partial_x u^0 + \partial_\xi \bar{u}^0) - \sigma^0, \quad P^0 = p^0 + \gamma\partial_x u^0 + \gamma\partial_\xi \bar{u}^0.$$

IOP Conf. Series: Journal of Physics: Conf. Series **894** (2017) 012089 doi:10.1088/1742-6596/894/1/012089

By using the two-scale convergence technique, it is possible to prove that there exist functions $u = u(t,x)$ and $\bar{u} = \bar{u}(t,x,\xi)$ such that, up to a subsequence, $u_h \twoheadrightarrow u$, $\partial_t u_h \twoheadrightarrow \partial_t u$, and $\partial_x u_h \twoheadrightarrow \partial_x u + \partial_\xi \bar{u}$. Notice that u does not depend on ξ and $\bar{u}$ is Σ-periodic in ξ. The functions u and $\bar{u}$ satisfy the following integral identities:

$$\int_0^T \int_\Omega \int_\Sigma \big(\varrho\, \partial_t u\, \partial_t \phi - (\lambda + 2\mu)(\partial_x u + \partial_\xi \bar{u})\, \partial_x \phi + 2\mu\chi(\partial_x \mathcal{J}_t u + \partial_\xi \mathcal{J}_t \bar{u})\, \partial_x \phi\big)\, d\xi dx dt$$
$$= -\int_\Omega \varrho\, u^1\, \phi^0\, d\xi dx dt - \int_0^T \int_\Omega \int_\Sigma F\, \partial_x \phi\, d\xi dx dt, \quad (4)$$

$$\int_\Sigma \big((\lambda + 2\mu)(\partial_x u + \partial_\xi \bar{u}) - 2\mu\chi(\partial_x \mathcal{J}_t u + \partial_\xi \mathcal{J}_t \bar{u}) - F\big)\, \partial_\xi \bar{\phi}\, d\xi = 0 \quad (5)$$

for arbitrary smooth functions $\phi = \phi(t,x)$ and $\bar{\phi} = \bar{\phi}(t,x,\xi)$ that are equal to zero at $t = T$ and for $x \in \partial\Omega$. In equation (4), $\phi^0 = \phi|_{t=0}$. We note also that $\mu\chi = \chi\mu_*/\tau_*$. Equation (5) is called *cell equation* and satisfied for almost all $(t,x) \in [0,T] \times \Omega$.

Although it is possible to deal with equations (4)–(5), this system is too complicated and says nothing about the mechanical properties of the homogenized continuum. Our next goal is to exclude the function $\bar{u}$ and the variable ξ from equation (4). To do this, we find an explicit expression for the function $\bar{u}$ from the cell equation and substitute it into (4). This procedure is based on the technique developed in [5].

Let us rewrite (5) as an equation in the Hilbert space $H = H^1_\#(\Sigma)/\mathbb{R}$ with the inner product $(u,v) = \int_\Sigma \partial_\xi u\, \partial_\xi v\, d\xi$. The subscript # means that it is a space of Σ-periodic functions. Let us define bounded self-adjoint operators $\mathcal{A} : H \to H$ and $\mathcal{B} : H \to H$ such that

$$(\mathcal{A}u, v) = \int_\Sigma (\lambda + 2\mu)\, \partial_\xi u\, \partial_\xi v\, d\xi, \quad (\mathcal{B}u, v) = \int_\Sigma 2\mu\chi\, \partial_\xi u\, \partial_\xi v\, d\xi$$

for all $u, v \in H$. We define also functions $a, b, f \in H$ such that

$$(a, v) = \int_\Sigma (\lambda + 2\mu)\, \partial_\xi v\, d\xi, \quad (b, v) = \int_\Sigma 2\mu\chi\, \partial_\xi v\, d\xi, \quad (f, v) = \int_\Sigma F\, \partial_\xi v\, d\xi$$

for all $v \in H$. Equation (5) is equivalent to the following equation in the space H:

$$\mathcal{A}\bar{u} - \mathcal{J}_t \mathcal{B}\bar{u} = -(a - b\mathcal{J}_t)\partial_x u + f. \quad (6)$$

Notice that the operator $\mathcal{B}$ is degenerate and its inverse exists only in the image of this operator $R(\mathcal{B})$ which is a closed subspace of H. Since $b \in R(\mathcal{B})$, there exists a unique $w \in R(\mathcal{B})$ such that $\mathcal{B}w = b$. Using this function we can write down the solution of equation (6) explicitly:

$$\bar{u}(t) = -\partial_x u\, \mathcal{A}^{-1} a + \int_0^t \partial_x u\, \mathcal{M}_{t-s}(w - \mathcal{A}^{-1} a)\, ds + \mathcal{A}^{-1} f + \int_0^t \mathcal{M}_{t-s} \mathcal{A}^{-1} f\, ds,$$

where

$$\mathcal{M}_t = \frac{1}{\tau_*} \mathcal{A}^{-1} \mathcal{B}\, e^{-t/\tau_*(\mathcal{I} - \mathcal{A}^{-1}\mathcal{B})}$$

and $\mathcal{I}$ is the identity operator in H. As a consequence, we find that

$$\mathcal{J}_t \bar{u} = -w\, \partial_x \mathcal{J}_t u + \int_0^t \partial_x u\, \mathcal{N}_{t-s}(w - \mathcal{A}^{-1} a)\, ds + \int_0^t \mathcal{N}_{t-s} \mathcal{A}^{-1} f\, ds,$$

where

$$\mathcal{N}_t = \frac{1}{\tau_*}\, e^{-t/\tau_*(\mathcal{I}-\mathcal{A}^{-1}\mathcal{B})}.$$

Substitution of these expressions into (4) gives the following integral identity:

$$\int_0^T \int_\Omega \big(\varrho_\theta\, \partial_t u\, \partial_t \phi - \alpha_\theta\, \partial_x u\, \partial_x \phi + \beta_\theta\, \partial_x(\mathcal{J}_t u)\, \partial_x \phi\big)\, dx dt$$
$$+ \int_0^T \int_\Omega \int_0^t \eta(t-s)\, \partial_x u(s)\, ds\, \partial_x \phi\, dx dt = -\int_\Omega \varrho\, u_\theta^1\, \phi^0\, dx dt + \int_0^T \int_\Omega \zeta\, \partial_x \phi\, dx dt, \quad (7)$$

where $\theta = \int_\Sigma \chi(\xi)\, d\xi$ is the porosity and

$$\varrho_\theta = \int_\Sigma \varrho(\xi)\, d\xi = \varrho_F \theta + \varrho_S(1-\theta), \quad \alpha_\theta = \lambda_\theta + 2\mu_\theta - (a, \mathcal{A}^{-1} a), \quad \beta_\theta = \frac{2\theta\mu_*}{\tau_*} - (b, w),$$

$$\lambda_\theta = \int_\Sigma \lambda(\xi)\, d\xi = \gamma\, \theta + \lambda_S(1-\theta), \quad \mu_\theta = \int_\Sigma \mu(\xi)\, d\xi = \frac{\mu_*}{\tau_*}\, \theta + \mu_S(1-\theta),$$

$$\eta(t) = -\big(a, \mathcal{M}_t(w - \mathcal{A}^{-1}a)\big) + \big(b, \mathcal{N}_t(w - \mathcal{A}^{-1}a)\big),$$

$$\zeta(t) = \big(a, \mathcal{A}^{-1} f + \int_0^t \mathcal{M}_{t-s} \mathcal{A}^{-1} f\, ds\big) - \big(b, \int_0^t \mathcal{N}_{t-s} \mathcal{A}^{-1} f\, ds\big) - F_\theta,$$

$$u_\theta^1(x) = \int_\Sigma \varrho(\xi)\, u^1(x,\xi)\, d\xi, \quad F_\theta(t,x) = \int_\Sigma F(t,x,\xi)\, d\xi.$$

Notice that ϱ_θ, α_θ, and β_θ are constants while $\eta = \eta(t)$ and $\zeta = \zeta(t,x)$.

Equation (7) can be also rewritten in the following form

$$\varrho_\theta\, \partial_t^2 u - \alpha_\theta \partial_x^2 u + \beta_\theta \partial_x^2 \mathcal{J}_t u = -\partial_x^2 \int_0^t \eta(t-s)\, u(s)\, ds - \partial_x \zeta \quad (8)$$

that should be understood in the distributional sense. This equation describes the behavior of the homogenized continuum. The terms with the coefficients α_θ and β_θ in this equation are related to the stresses. The integral term in the right-hand side represents a memory effect. This term arises due to the homogenization technique and is standard for non-stationary problems (see, e.g., [5, 6]). The kernel η, however, decreases rapidly, so, this memory is very short. The function ζ comes from the initial data and, generally speaking, from the boundary conditions.

3. Parameters of the homogenized continuum

In this section, we calculate the coefficients and the functions that are present in equation (8).

Calculation of a, b, and f.

Let us define a function $q \in H$ as the unique solution of the following equation:

$$\int_\Sigma \partial_\xi q\, \partial_\xi v\, d\xi = \int_\Sigma \chi\, \partial_\xi v\, d\xi \quad \text{for all} \quad v \in H.$$

It is not difficult to see that $a = a_* q$ and $b = 2q\mu_*/\tau_*$, where $a_* = \gamma - \lambda_S - 2\mu_S + 2\mu_*/\tau_*$. The function f also can be expressed in terms of the function q but we omit this representation. The function q can be easily find:

$$q(\xi) = \begin{cases} -(1-\theta)^2/2 + (1-\theta)/2 - \xi\theta, & \xi \in [0, x_h/h], \\ -(1-\theta)^2/2 - (1-\theta)/2 + \xi(1-\theta), & \xi \in [x_h/h, 1]. \end{cases}$$

Functions in H are defined up to an additive constant, for this reason, we have assumed that $\int_\Sigma q\,d\xi = 0$. This condition will be fulfilled further for all functions from H. Besides that, we have taken into account that $x_h/h = 1-\theta$.

Calculation of w.
The function w is in $R(\mathcal{B})$. As it follows from Proposition 4.10 in [5], this function is affine on the interval $[0, x_h/h]$. As a consequence of the definition of this function, it is affine also on $[x_h/h, 1]$. Thus, it is not difficult to find that

$$w(\xi) = \begin{cases} \theta/2 - \xi\theta/(1-\theta), & \xi \in [0, x_h/h], \\ \theta/2 - 1 + \xi, & \xi \in [x_h/h, 1]. \end{cases}$$

Calculation of α_θ and β_θ.
In order to calculate α_θ, we have to find $(a, \mathcal{A}^{-1}a)$. Since $a = a_* q$, it is enough to find $(q, \mathcal{A}^{-1}q)$. If $r = \mathcal{A}^{-1}q$, then r is the unique solution of the equation $\mathcal{A}r = q$. It is not difficult to find this solution: $r = q/r_*$, where $r_* = \theta(\lambda_S + 2\mu_S) + (1-\theta)(\gamma + 2\mu_*/\tau_*)$. As a consequence, we have that $(q, \mathcal{A}^{-1}q) = (q, r) = \theta(1-\theta)/r_*$ and $(a, \mathcal{A}^{-1}a) = a_*^2\theta(1-\theta)/r_*$. Thus,

$$\alpha_\theta = \Big(\frac{\theta}{\gamma + 2\mu_*/\tau_*} + \frac{1-\theta}{\lambda_S + 2\mu_S}\Big)^{-1}.$$

As for β_θ, it is not difficult to calculate that $\beta_\theta = 0$. It means that the term in equation (8) which corresponds to the Maxwell fluid vanishes, i.e., the homogenized continuum is purely elastic. Of course, the elastic modulus α_θ of this continuum depends on parameters of the original Maxwell fluid such as γ, τ_*, and μ_*. Notice also that, in the multi-dimensional case, β_θ is a non-zero fourth-rank tensor.

4. Asymptotic properties of the homogenized equation

If we neglect the influence of the initial data, then equation (8) with the calculated in the previous section coefficients takes the form:

$$\varrho_\theta\,\partial_t^2 u - \alpha_\theta\,\partial_x^2 u = -\partial_x^2 \int_0^t \eta(t-s)\,u(s)\,ds. \tag{9}$$

The left-hand side of this equation is the wave operator related to the elastic behavior of the homogenized continuum, however, (9) has no solutions of the travelling wave type. This is due to the memory term in the right-hand side of this equation. There are no travelling waves even for the weakly viscous fluid, i.e., for $\mu_* = 0$. The memory term is usual for homogenized non-stationary equations, but the one-dimensional case has special features. It is worth noting that the function η is positive and rapidly tends to zero. For small τ_*, it looks like the δ-function concentrated at 0. Really, it is possible to prove that

$$\int_0^t \eta(t-s)\,u(s)\,ds - \eta_\theta\,u(t) = o(1) \quad \text{in} \quad H \quad \text{as} \quad \tau_* \to 0,$$

where $\eta_\theta = -\big(a, \mathcal{A}^{-1}\mathcal{B}(\mathcal{A}-\mathcal{B})^{-1}(\mathcal{A}w - a)\big) + \big(b, (\mathcal{A}-\mathcal{B})^{-1}(\mathcal{A}w - a)\big)$. The constant η_θ can be calculated:

$$\eta_\theta = \theta\,\frac{m_S^2}{n_\theta^2}\Big(\frac{\tau_*}{2\mu_*} + \frac{1-\theta}{n_\theta}\Big)^{-1},$$

where $m_S = \lambda_S + 2\mu_S$, $n_\theta = (1-\theta)\gamma + \theta m_S$. We see that η_θ depends on the fraction $2\mu_*/\tau_*$. The same is true for α_θ. We consider three cases of possible values of the ratio $2\mu_*/\tau_*$ where τ_* tends to zero. In all these cases, equation (9) takes the form:

$$\varrho_\theta\, \partial_t^2 u - (\alpha_\theta - \eta_\theta)\, \partial_x^2 u = 0 \tag{10}$$

but the values of α_θ and η_θ will be different.

1. $\tau_* \to 0$, $\mu_* \to 0$, and $2\mu_*/\tau_* \to 0$. This is the case of usual weakly viscous fluid (not the Maxwell one). In this case

$$\alpha_\theta = \Big(\frac{\theta}{\gamma} + \frac{1-\theta}{\lambda_S + 2\mu_S}\Big)^{-1} \quad \text{and} \quad \eta_\theta = 0.$$

2. $\tau_* \to 0$, $\mu_* \to 0$, and $2\mu_*/\tau_* \to k_* \in (0, +\infty)$. This is another weakly viscous fluid. In this case

$$\alpha_\theta = \Big(\frac{\theta}{\gamma + k_*} + \frac{1-\theta}{\lambda_S + 2\mu_S}\Big)^{-1} \quad \text{and} \quad \eta_\theta = \theta\, \frac{m_S^2}{n_\theta^2}\Big(\frac{1}{k_*} + \frac{1-\theta}{n_\theta}\Big)^{-1},$$

3. If $\tau_* \to 0$ and $2\mu_*/\tau_* \to +\infty$, then

$$\alpha_\theta = \frac{\lambda_S + 2\mu_S}{1-\theta} = \frac{m_S}{1-\theta} \quad \text{and} \quad \eta_\theta = \frac{\theta}{1-\theta}\, \frac{m_S^2}{n_\theta}.$$

It is not difficult to see that $\alpha_\theta > \eta_\theta$ in all three cases. This means that equation (10) has solutions of the travelling wave type in all three cases. The velocity of the waves is equal to $\sqrt{(\alpha_\theta - \eta_\theta)/\varrho_\theta}$. It is the sound velocity. Notice that the travelling waves exists also in the third case where the viscosity of the fluid is not equal to zero.

Acknowledgments
This work was partially supported by Russian Foundation for Basic Research (Grants 15-01-01091 A).

References
[1] Shelukhin V V and Isakov A E 2012 Elastic waves in layered media: two-scale hogenization approach *Euro. Jnl of Applied Mathematics* **23** 691-707
[2] Nguetseng G 1989 Asymptotic analysis for a functional related to the theory of homogenization *SIAM Journal on Mathematical Analysis* **20**(3) 608-23
[3] Allaire G 1992 Homogenization and two-scale convergence *SIAM Journal on Mathematical Analysis* **23** 1482-18
[4] Zhikov V V 2004 On two-scale convergence *Journal of Mathematical Sciences* **120**(3) 1328-52
[5] Hoffmann K-H, Botkin N D and Starovoitov V N 2005 Homogenization of interfaces between rapidly oscillating fine elastic structures and fluids *SIAM Journal on Applied Mathematics* **65**(3) 983-1005
[6] Gilbert R P, Panchenko A and Vasilic A 2011 Homogenazing the acoustic of cancellous bone with an interstitial non-Newtonian fluid *Nonlinear Analysis: Theory, Methods and Applications* **74**(4) 1005-18

IOP Conf. Series: Journal of Physics: Conf. Series **894** (2017) 012090 doi:10.1088/1742-6596/894/1/012090

On some group properties of heat and mass transfer equations

I V Stepanova

Institute of Computational Modeling SB RAS, Krasnoyarsk, Russia

E-mail: stepiv@icm.krasn.ru

Abstract. Heat and mass transfer equations with variable transport coefficients are under study. The forms of unknown thermal conductivity, diffusion and Dufour coefficients are found by means of Lie group theory. It is shown that arbitrary elements have the power-law, logarithmic and exponential dependencies on temperature and concentration.

1. Introduction

The object of the study is the evolution equations describing heat and mass transfer in binary mixtures of liquids and gases. These equations in their classical form were known many years ago [1] and they were widely used in mathematical modelling. They are treated by means of numerical, qualitative and asymptotic analysis up to now. The Cauchy and boundary value problems for such equations are imposed and solved in a wide variety of statements. Many of these problems are included in the classical textbooks on mathematical physics.

In recent 20 years mathematical models of heat and mass transfer became more complex [2]. It is caused by high requirements to accuracy for describing physical processes and by the rapid development of computers as well. The classical heat and mass transfer models take into account two main processes. These are thermal conductivity and diffusion. Moreover, equation for temperature may not be connected with equation for mass distribution. In fact, the heat and mass transfer is more complex phenomena. Convection and cross effects influence essentially. Furthermore, for some cases it is impossible to avoid dependence of transport coefficients on temperature and concentration. Modern development of experimental equipment and computers allow taking into account collateral effects mentioned above more accurately.

We pay attention to study of heat and mass transfer equations taking into consideration Soret and Dufour effects. The Soret effect (or thermal diffusion) is the molecular transport of mass caused by the thermal gradient in a fluid mixture while the Dufour effect (or diffusive thermal conductivity) is the heat flux caused by mass concentration gradient in a fluid mixture. These effect are often called reciprocal ones. From the mathematical point of view temperature and concentration equations with these effects became essentially nonlinear and they can not be considered separately.

For the investigation of qualitative properties of these equations we need a powerful mathematical method. We use Lie symmetry analysis as one of the universal tool for the study of differential equations of arbitrary form. The basic principals of this theory and many examples of its application are described in books [3] and [4]. The number of papers devoted to analysis of

Content from this work may be used under the terms of the Creative Commons Attribution 3.0 licence. Any further distribution of this work must maintain attribution to the author(s) and the title of the work, journal citation and DOI.
Published under licence by IOP Publishing Ltd

simple and complex temperature and diffusion equations by means of Lie symmetry technique is growing permanently. Starting with the example of nonlinear thermal conductivity equation in book [3] and results of the preprint [5] many successive papers deal with the mathematical models with variable transport coefficients. Various classes of diffusion–convection–reaction equations with different types of coefficient dependence on unknown function are investigated in [6]–[8] and in the works cited therein. Papers [9] and [10] should be mention as the results of analysis of the group properties and exact solutions of nonlinear heat transfer with delay.

It is necessary to note that there are not so many papers devoted to the study of systems of heat and mass transfer equations. Nevertheless, the group properties of nonlinear reaction–diffusion equations with constant coefficients are studied in [11]. Group classification problem for constant and variable transport coefficients is solved for general convection equations with five arbitrary parameters in [12] and [13] respectively. The results of symmetry analysis of the heat and mass transfer equations taking into account the Soret effect without convection terms are presented in [14].

In this paper we study symmetry properties of heat and mass transfer equations which can not be considered separately. They are connected with each other by means of cross-effects. We find the forms of arbitrary transport coefficients and calculate admissible transformations of dependent and independent variables. These transformations allow to construct new exact solutions of the governing equations. The paper is organized as follows. In Section 2 we describe the mathematical model of mentioned above process. In Section 3 Lie symmetry analysis of the governing equations is carried out. The determining and classifying equations are derived. They are analyzed in detail in Section 4. The results of group classification are gathered into a table for the reader's convenience.

2. Governing equations

Due to taking into account Soret and Dufour effect the thermal and diffusive fluxes have the sophisticated forms

$$\mathbf{J}_c = -\rho(D\nabla C + D^\theta \nabla T) \quad \text{and} \quad \mathbf{J}_q = -(\kappa \nabla T + D^F \nabla C) \quad \text{respectively [1].}$$

Here thermal diffusion (D^θ) and Dufour (D^F) coefficients are presented in the most generalized forms as the functions of temperature T and concentration of the lighter component C, ρ is the density of the mixture. According to heat and mass transfer theory the mathematical model of this process can be written in the form

$$\frac{\partial T}{\partial t} = \operatorname{div}\Big(\kappa(T,C)\nabla T + D^F(T,C)\nabla C\Big), \tag{1}$$

$$\frac{\partial C}{\partial t} = \operatorname{div}\Big(D(T,C)\nabla C + D^\theta(T,C)\nabla T\Big). \tag{2}$$

Here t is the time, $\nabla = (\partial_{x^1}, \partial_{x^2}, \partial_{x^3})$, (x^1, x^2, x^3) is the coordinate vector. The thermal conductivity $\kappa(T,C)$ and diffusion coefficient $D(T,C)$ are positive and non-zero. The Dufour coefficient D^F is positive and the thermal diffusion coefficient D^θ has an arbitrary sign.

Dependence of transport coefficients on temperature and concentration and the cross effects make the governing equations new and interesting for analysis. On the basis of information about symmetry properties of equations (1) and (2) the heat and mass transfer processes can be studied more thoroughly.

3. Derivation of determining and classifying equations

The classical method for group analysis is properly described in well-known monographs of L.Ovsyannikov [3] and P.Olver [4]. Here we apply this technique step by step omitting cumbersome formulas given in the above mentioned books. We search for an admissible generator in the form

$$X = \xi^0 \frac{\partial}{\partial t} + \xi^1 \frac{\partial}{\partial x^1} + \xi^2 \frac{\partial}{\partial x^2} + \xi^3 \frac{\partial}{\partial x^3} + \eta^1 \frac{\partial}{\partial T} + \eta^2 \frac{\partial}{\partial C},$$

where coefficients ξ^i, $i = 0, \ldots, 3$, and η^j, $j = 1, 2$, depend on dependent T, C and independent t, x^i, $i = 1, 2, 3$, variables. We construct the second extension of the generator X with respect to formulas given in [3] and [4]. We apply the extended generator to system (1), (2) and require that the resulting equations are identically zero. There are variables T_t, C_t, T_{tx^i}, C_{tx^i}, $i = 1, 2, 3$, in the resulting equations. We eliminate them from equations (1), (2) and their differential consequences. Then we split the resulting equations with respect to the mixed partial derivatives of the third order and obtain that

$$\xi^0 = \xi^0(t), \quad \xi^i = \xi^i(t, x^i), \qquad i = 1, 2, 3.$$

Furthermore the coefficients at the mixed partial derivatives of the second order satisfy the Killing equations

$$\xi^2_{x^1} + \xi^1_{x^2} = 0, \qquad \xi^3_{x^1} + \xi^1_{x^3} = 0, \qquad \xi^3_{x^2} + \xi^2_{x^3} = 0. \tag{3}$$

Splitting with respect to derivatives $T_{x^ix^i}$ and $C_{x^ix^i}$, $i = 1, 2, 3$, gives the following equations

$$\xi^1_{x^1} - \xi^2_{x^2} = 0, \qquad \xi^1_{x^1} - \xi^3_{x^3} = 0, \qquad \xi^2_{x^2} - \xi^3_{x^3} = 0. \tag{4}$$

We also can obtain that

$$\eta^1_{TT} = \eta^1_{TC} = \eta^1_{CC} = 0, \qquad \eta^2_{TT} = \eta^2_{TC} = \eta^2_{CC} = 0 \tag{5}$$

if the following inequalities are fulfilled

$$D(D + \kappa) - D^\theta D^F \neq 0, \quad D^F \neq 0.$$

Equations (3)–(5) are determining and give preliminary information about the coefficients of the generator X. The functions ξ^i, $i = 0, \ldots, 3$, and η^j, $j = 1, 2$, are connected with the parameters of the governing equations through the following equalities

$$-\xi^i_t = D\Big(2\eta^2_{Cx^i} - \sum_{j=1}^{3} \xi^i_{x^jx^j}\Big) + 2D^\theta \eta^1_{Cx^i} + \left(\frac{\partial D}{\partial T} + \frac{\partial D^\theta}{\partial C}\right)\eta^1_{x^i} + 2\frac{\partial D}{\partial C}\eta^2_{x^i}, \tag{6}$$

$$-\xi^i_t = \kappa\Big(2\eta^1_{Tx^i} - \sum_{j=1}^{3} \xi^i_{x^jx^j}\Big) + 2D^F \eta^2_{Tx^i} + \left(\frac{\partial \kappa}{\partial C} + \frac{\partial D^F}{\partial T}\right)\eta^2_{x^i} + 2\frac{\partial \kappa}{\partial T}\eta^1_{x^i}, \tag{7}$$

$$2D\eta^2_{Tx^i} + 2D^\theta \eta^1_{Tx^i} - D^\theta \sum_{j=1}^{3} \xi^i_{x^jx^j} + \left(\frac{\partial D}{\partial T} + \frac{\partial D^\theta}{\partial C}\right)\eta^2_{x^i} + 2\frac{\partial D^\theta}{\partial T}\eta^1_{x^i} = 0, \tag{8}$$

$$2\kappa\eta^1_{Cx^i} + 2D^F \eta^2_{Cx^i} - D^F \sum_{j=1}^{3} \xi^i_{x^jx^j} + \left(\frac{\partial \kappa}{\partial C} + \frac{\partial D^F}{\partial T}\right)\eta^1_{x^i} + 2\frac{\partial D^F}{\partial C}\eta^2_{x^i} = 0, \tag{9}$$

$$\kappa(2\xi^1_{x^1} - \xi^0_t) + D^\theta \eta^1_C - D^F \eta^2_T = \eta^1 \frac{\partial \kappa}{\partial T} + \eta^2 \frac{\partial \kappa}{\partial C}, \tag{10}$$

IOP Conf. Series: Journal of Physics: Conf. Series **894** (2017) 012090 doi:10.1088/1742-6596/894/1/012090

$$D(2\xi^1_{x^1} - \xi^0_t) - D^\theta \eta^1_C + D^F \eta^2_T = \eta^1 \frac{\partial D}{\partial T} + \eta^2 \frac{\partial D}{\partial C}, \tag{11}$$

$$D^\theta(\eta^2_C - \eta^1_T - \xi^0_t + 2\xi^1_{x^1}) + (\kappa - D)\eta^2_T = \eta^1 \frac{\partial D^\theta}{\partial T} + \eta^2 \frac{\partial D^\theta}{\partial C}, \tag{12}$$

$$D^F(\eta^1_T - \eta^2_C - \xi^0_t + 2\xi^1_{x^1}) - (\kappa - D)\eta^1_C = \eta^1 \frac{\partial D^F}{\partial T} + \eta^2 \frac{\partial D^F}{\partial C}, \tag{13}$$

$$\eta^1_t = \kappa(\eta^1_{x^1x^1} + \eta^1_{x^2x^2} + \eta^1_{x^3x^3}) + D^F(\eta^2_{x^1x^1} + \eta^2_{x^2x^2} + \eta^2_{x^3x^3}), \tag{14}$$

$$\eta^2_t = D(\eta^2_{x^1x^1} + \eta^2_{x^2x^2} + \eta^2_{x^3x^3}) + D^\theta(\eta^1_{x^1x^1} + \eta^1_{x^2x^2} + \eta^1_{x^3x^3}). \tag{15}$$

There are 18 classifying equations in the above system because $i = 1, 2, 3$ in equations (6)–(9). In order to find the common part of Lie symmetries for equations (1) and (2) with arbitrary transfer coefficients we should vary the arbitrary elements and split with respect to them in the classifying equations (6) – (15). It results

$$\eta^1 = 0, \quad \eta^2 = 0$$

and

$$\xi^0 = c_0 t + c_1, \qquad \xi^1 = c_2 x^1 + c_3 x^2 + c_4 x^3 + c_5, \tag{16}$$

$$\xi^2 = -c_3 x^1 + c_2 x^2 + c_6 x^3 + c_7, \qquad \xi^3 = -c_4 x^1 - c_6 x^2 + c_1 x^3 + c_8, \tag{17}$$

where c_i, $i = 0, \ldots, 8$, are arbitrary constants. Consequently, assuming that one of constant c_i is equal to unit and the other constants are zero we obtain eight generators of the basic Lie algebra. They have the following form

$$L^0 = \langle \partial_t,\ \partial_{x^i},\ 2t\partial_t + \sum_{i=1}^{3} x^i \partial_{x^i},\ x^j \partial_{x^i} - x^i \partial_{x^j} \rangle,\ i, j = 1, 2, 3,\ i \neq j.$$

We have three types of transformations in the basic Lie group generated by the basic algebra L^0. These are transition with respect to time and spatial variables, dilatation of independent variables and rotations. Such transformations are typical for many equations in fluid mechanics. It is interested to note that there is no generator depending on arbitrary functions in the basic group.

Further we make some assumptions. Firstly, we suppose that the thermal diffusion coefficient D^θ vanishes for simplification of the problem. Such statement of the problem is justified at study of gas mixtures in which the Dufour effect is more essential then in liquid ones. Secondly, we consider the classified functions depending on two variables T and C. The cases of dependencies on T or C separately will be studied in further works. Thereby we solve the group classification problem with respect to the functions $D(T, C)$, $\kappa(T, C)$ and $D^F(T, C)$.

4. Analysis of the classifying equations

In this section we solve the group classification problem for equations (1) and (2) at $D^{\theta} \equiv 0$. We should find all possible forms of functions D, κ and D^F and extensions of the basic Lie algebra L^0.

It is necessary to note that equations (1) and (2) do not change their differential forms under following linear transformations

$$T' = AT + H, \ C' = AC + G, \tag{18}$$

where $A \neq 0$, H and G are constant. All of the classified functions are found with respect of transformations (18).

Under the assumption $D^{\theta} = 0$ equation (12) takes the form

$$(\kappa - D)\eta^2_T = 0.$$

If $\kappa = D$ we subtract equation (10) from equation (11) and obtain $\eta^2_T = 0$ also due to $D^F \neq 0$. That is why we may consider the case $\eta^2_T = 0$ only. Then three equations (8) transform into

$$\frac{\partial D}{\partial T}\eta^2_{x^i} = 0, \qquad i = 1, 2, 3.$$

Since we consider the case when D depends on T and C essentially then η^2 does not depend on x^i, $i = 1, 2, 3$, and t due to equation (15). We should find other coordinates of the generator X. We differentiate equation (11) with respect to x^i, $i = 1, 2, 3$, and subtract from equations (6) respectively. Further manipulations with obtained equalities lead to vanishing of all third order derivatives of ξ^1, ξ^2, ξ^3 with respect to x^1, x^2 and x^3. Using equations (3) and (4) we find out that the coordinates of the generator X satisfy equations (16) and (17). Corresponding to (5) the coefficients η^2 and η^1 are linear with respect to T and C. They are calculated from equation (6):

$$\eta^1 = c_9 T + c_{10} C + c_{11}, \qquad \eta^2 = c_{12} C + c_{13}.$$

System of classifying equations (6)–(15) reduces to three equations

$$\kappa(c_0 - 2c_2) + \Big(c_9 T + c_{10} C + c_{11}\Big)\frac{\partial \kappa}{\partial T} + \Big(c_{12} C + c_{13}\Big)\frac{\partial \kappa}{\partial C} = 0,$$

$$D(c_0 - 2c_2) + \Big(c_9 T + c_{10} C + c_{11}\Big)\frac{\partial D}{\partial T} + \Big(c_{12} C + c_{13}\Big)\frac{\partial D}{\partial C} = 0, \tag{19}$$

$$D^F(c_0 - 2c_2 + c_{12} - c_9) + \Big(c_9 T + c_{10} C + c_{11}\Big)\frac{\partial D^F}{\partial T} + \Big(c_{12} C + c_{13}\Big)\frac{\partial D^F}{\partial C} + (\kappa - D)c_{10} = 0.$$

For solution of the group classification problem we should consider equations (19) in an general form

$$A_1 F_i + \Big(A_2 T + A_3 C + A_4\Big)\frac{\partial F_i}{\partial T} + \Big(A_5 C + A_6\Big)\frac{\partial F_i}{\partial C} = 0, \qquad i = 1, 2,$$

$$F_3(A_1 + A_5 - A_2) + \Big(A_2 T + A_3 C + A_4\Big)\frac{\partial F_3}{\partial T} + \Big(A_5 C + A_6\Big)\frac{\partial F_3}{\partial C} + (F_1 - F_2)A_3 = 0,$$

where $F_1 = \kappa$, $F_2 = D$ and $F_3 = D^F$.

IOP Conf. Series: Journal of Physics: Conf. Series **894** (2017) 012090 doi:10.1088/1742-6596/894/1/012090

The solution of the equations considered depends on the coefficients A_i, $i = 1, \ldots, 6$. These coefficients are not equal to zero simultaneously. The basic role belongs to A_2 and A_5. If they are equal to each other or equal to zero we have different forms for functions F_i, $i = 1, 2, 3$.

In this way we should study the following cases.

1. Let $A_2 \neq A_5$.

1.1 If $A_2 \neq 0$ and $A_5 \neq 0$ then we set $A_4 = A_6 = 0$ using transformation (18).

1.2 If $A_2 = 0$ and $A_5 \neq 0$ then we set $A_6 = 0$ using transformation (18).

1.3 If $A_2 \neq 0$, $A_5 = 0$ then we set $A_4 = 0$ using transformation (18).

1.3.1 $A_6 \neq 0$.

1.3.2 $A_6 = 0$.

Cases 1.3.1 and 1.3.2 give different forms for classified functions.

2. Another case is $A_2 = A_5$ then we have three different variants.

2.1 If $A_2 \neq 0$ then we can set $A_4 = A_6 = 0$ using transformation (18).

2.2 If $A_2 = 0$ and $A_6 \neq 0$ we can not use transformation (18).

2.3 If $A_2 = 0$ and $A_6 = 0$ then we can set $A_4 = 0$ if $A_3 \neq 0$ using transformation (18).

As an example one of this cases is described here in detail. We consider case 1.3.1. After integrating equations for F_i, $i = 1, 2, 3$, we have

$$F_i = f_i(w')e^{-A_1C/A_6}, \qquad i = 1, 2,$$

$$F_3 = \Big(\frac{A_3}{A_2}e^{-\frac{A_2C}{A_6}}(f_1(w') - f_2(w')) + f_3(w')\Big)e^{(-\frac{A_1}{A_6}+\frac{A_2}{A_6})C},$$

where $w' = (T + A_3C/A_2 + A_2A_4 + A_3A_6)e^{-A_2C/A_6}$ is the argument of arbitrary functions f_1, f_2 and f_3. We chose $A = 1$ and $H = A_2A_4 + A_3A_6$ in transformation (18) for variable T and obtain simpler form for argument w'

$$w = \left(T + \frac{A_3}{A_2}C\right)e^{-\frac{A_2C}{A_6}}.$$

Using notation

$$\frac{A_3}{A_2} = \alpha, \quad -\frac{A_2}{A_6} = \varepsilon, \quad -\frac{A_1}{A_6} = \gamma$$

we have the unknown functions in the form

$$\kappa = f_1(w)e^{\gamma C}, \qquad D = f_2(w)e^{\gamma C}, \qquad D^F = (\alpha e^{\varepsilon C}(f_1(w) - f_2(w)) + f_3(w))e^{(\gamma-\varepsilon)C}.$$

Further we should substitute these forms of classified functions into equations (19) and calculate constants c_i, $i = 0, 2, 9-13$. Due to functions f_i, $i = 1, 2, 3$, are arbitrary we can split resulting equations and obtain

$$c_0 = 2c_2 - \gamma c_{13}, \quad c_9 = -\varepsilon c_{13}, \quad c_{10} = -\alpha\varepsilon c_{13}, \quad c_{11} = -\alpha c_{13}, \quad c_{12} = 0.$$

Thereby all constants are eliminated by means of c_{13}. When $c_{13} = 1$ and other constants are zero in equations (16) and (17) we have the following additional generator

$$-\gamma t\partial_t + \partial_C - \alpha\varepsilon C\partial_T - \varepsilon T\partial_T - \alpha\partial_T$$

which extents the basic Lie group.

Repeating the same procedure for the other six cases we get the exhausted result of group classification. Classified functions and admissible generators are gathered in Table 1, where the constants α, β, γ, ε are arbitrary ones such that $\varepsilon \neq 0$, $\delta \neq 0, -1$. The functions $f_i = f_i(w)$, $i = 1, 2, 3$, are arbitrary smooth non-zero functions. There is their argument w in the last column of Table 1. The eight generators of the basic Lie algebra L^0 are included into the third column of Table 1 for all presented cases. The following generators are involved to extensions of the Lie basic algebra L^0:

$$Z = t\partial_t,\ T^1 = T\partial_T,\ T^3 = \partial_T,\ C^1 = C\partial_C,\ C^2 = T\partial_C,\ C^3 = \partial_C.$$

As one can see the generators from the third column of Table 1 are the linear combinations of generators of time, temperature and concentration dilatation extent L^0 as well as temperature and concentration transition. There is the generator T^2 in this set also. It occurs due to taking into account Dufour effect in the governing equations. This generator displays relations between diffusion and thermal conductivity processes.

Table 1. Results of group classification for essential dependence of classified functions on T and C.

κ, D	D^F	generators	w
$e^{\gamma T/C} f_i$	$e^{\gamma T/C}((T/C)(f_2 - f_1) + f_3)$	$\gamma Z - T^2$	C
$e^{\gamma C} f_i$	$e^{\gamma C}(\alpha C(f_2 - f_1) + f_3)$	$\gamma Z - C^3 - \alpha T^2 + \beta T^3$	$T + \alpha C^2/2 - \beta C$
$e^{\gamma C} f_i$	$(\alpha e^{\varepsilon C}(f_1 - f_2) + f_3)e^{(\gamma-\varepsilon)C}$	$\gamma Z - C^3 + \alpha\varepsilon T^2 + \varepsilon T^1 + \alpha T^3$	$(T + \alpha C)e^{\varepsilon C}$
$C^\gamma f_i$	$C^{\gamma-1}(\alpha C(f_1 - f_2) + f_3)$	$\gamma Z - C^1 + \alpha T^2 + \beta T^3$	$T + \alpha C + \beta \ln C$
$C^\gamma f_i$	$\alpha C^\gamma (f_1 - f_2) + C^{\gamma-\delta-1} f_3$	$\gamma Z - C^1 + \alpha(1+\delta)T^2 + \delta T^1$	$(T + \alpha C)C^\delta$
$C^\gamma f_i$	$C^\gamma(\beta(f_1 - f_2)\ln C + f_3)$	$\gamma Z - C^1 + \beta T^2 - T^1$	$T/C + \beta \ln C$
$(T + \alpha C)^\varepsilon f_i$	$(T + \alpha C)^\varepsilon(\alpha(f_1 - f_2) + f_3(T + \alpha C))$	$\varepsilon Z - \alpha T^2 - T^1$	C

5. Conclusion

The equations describing heat and mass transfer in a binary mixture with thermodiffusion and diffusive thermal conduction are considered. Group classification with respect to three physical parameters is performed. The transformations left unchanged the structure of the governing equations are used for simplification of the form of classified functions. It should be noted that on the basis of group classification the governing equations can be studied more thoroughly with the use of the symmetry properties incorporated in these equations. New exact solutions and generalizations of the known solutions can be obtained.

Acknowledgments

The work is supported by Russian President Grant (project MK-4519.2016.1).

References

[1] Landau L D and Lifshitz E M 1987 *Course of Theoretical Physics* vol 6 (Butterworth Heinemahh)
[2] Andreev V K, Gaponenko Yu A, Goncharova O N and Pukhnachev V V 2012 *Mathematical Models of Convection* (De Gryuter Publ.)
[3] Ovsyannikov L V 1982 *Group Analysis of Differential Equations* (Academic Press)
[4] Olver P 1986 *Application of the Lie Groups to Differential Equations* (Springer-Verlag)
[5] Dorodnitsyn V A 1979 *Group Properties and Invariant Solutions of a Nonlinear Heat Equation With a Sorce or a Sink* (Keldysh Inst Appl Math of Academy of Sciences USSR)
[6] Vaneeva O O, Johnpillai A G, Popovych R O and Sophocleous C 2007 *J. Math. Anal. Appl.* **330** 1363–86

[7] Popovych R O and Ivanova N M 2004 *J. Phys. A* **37** 7547–65
[8] Demetriou E, Ivanova N M and Sophocleous C 2008 *J. Math. Anal. Appl.* **348** 55–65
[9] Meleshko S V and Moyo S 2008 *J. Math. Anal. Appl.* **338** 448–66
[10] Polyanin A D and Zhurov A I 2014*Applied Mathematics Letters* **37** 43–8
[11] Nikitin A G 2007 *J. Math. Anal. Appl.* **332** 666–90
[12] Stepanova I V 2013 *Commun. Nonlinear Sci. Numer. Simulat.* **18** 1341–46
[13] Ryzhkov I I 2008 *J. Sib. Fed. Univ. Math. Phys.* **1** 410–31
[14] Stepanova I V 2015 *Commun. Nonlinear Sci. Numer. Simulat.* **20** 684–91

IOP Conf. Series: Journal of Physics: Conf. Series **894** (2017) 012091 doi:10.1088/1742-6596/894/1/012091

Metal explosion chambers: designing, manufacturing, application

O I Stoyanovskii, B S Zlobin, A A Shtertser and Y P Meshcheryakov

Design and Technology Branch, Lavrentyev Institute of Hydrodynamics, Novosibirsk, 630090, Russia

E-mail: asterzer@mail.ru

Abstract. Designing of explosion chambers is based on research investigations of the chamber body stress-strain state, which is determined by numerical computation and experimentally by the strain gage technique. Studies show that chamber bottoms are the most loaded elements, and maximal stresses arise in chamber poles. Increasing the shell thickness around poles by welding-in an insert is a simple and saving way to solve this problem. There are structural solutions, enabling reliable hermetic closure and preventing leakage of detonation products from the chamber. Explosion chambers are employed in scientific research and in different industrial applications: explosive welding and hardening, synthesis of new materials, disposal of expired ammunition, and etc.

1. Introduction

Explosion chambers are the special facilities for containment of explosion effects, such as shock waves, fragments, toxic detonation products, and seismic waves. Metal explosion chambers can localize all the listed effects and can be mounted in ordinary industrial or laboratory premises. The review of published works on the concerned subject can be found in [1]. The nominal explosive charge mass for which chamber is designed is usually specified in trotyl (TNT) equivalent. The main part of explosion chamber is its body in a form of metal shell, which mostly consist of main cylindrical part and two bottoms (generally spherical or elliptical) attached to the said cylinder by welding or via the bayonet lock. Dynamic stresses in such composed shell were considered in [1], but for the case of their uniform distribution. In reality stress distribution in the shell is not uniform, and shell poles (bottom centres) are the mostly loaded places. The present studies are focused on the methods permitting to reduce the stresses in shell poles and to obtain close to uniform stress distribution. Besides, examples of explosion chambers and their application are given.

2. Theoretical considerations and stress metering

Hermetic shell is a main structural element of explosion chamber. Among the shells of different types, non-spherical ones are most common; as a rule they consist of a cylindrical part and two spherical or non-spherical, generally elliptic, bottoms. These shells have a number of technological advantages as compared to spherical ones, but at the same time have an increased weight resulting from the not uniform stress distribution due to the "swinging" effect arising at shell vibrations, even in case of uniform loading. Theoretical and experimental studies show that bottoms are the most loaded elements, and the maximal stresses determining the chamber lifetime emerge in bottom poles [2-4]. To

Content from this work may be used under the terms of the Creative Commons Attribution 3.0 licence. Any further distribution of this work must maintain attribution to the author(s) and the title of the work, journal citation and DOI.
Published under licence by IOP Publishing Ltd

make bottoms of not even thickness is a way to reduce stresses in poles and to obtain the more uniform distribution of stresses in the chamber shell. For this, there is the simplest and most saving method – to weld-in inserts of increased thickness into bottom poles. In this connection, the important problem is to find the radius of insert with the known thickness which would provide the most uniform stress distribution in the shell elements. Numerical analysis of the problem relating to the determination of the insert optimal radius has been carried out on the base of equations [5] with the use finite-element method [6].

Figure 1 presents the dependence of the maximal values of equivalent stresses σ_{eff} in the shell pole with embedded strengthening insert (curve 1) and in the insert – bottom conjugation point (curve 2) on the insert radius r. Calculations were made for the case when insert thickness is 1.6 times greater than the shell thickness. This dependence evidently has a periodical character (curve 1) and the amplitude of equivalent stress deviation from the equilibrium position decreases as the insert radius rises. As r decreases, the maximal and minimal values of the amplitude of stress oscillations practically do not change. Periodical dependence of the maximal effective stresses on the insert radius is caused by the vibrations of the insert as additional mass. The minimal and maximal values of the equivalent stress amplitudes correspond to the cases when the vibration phases of the bottom and insert are the same.

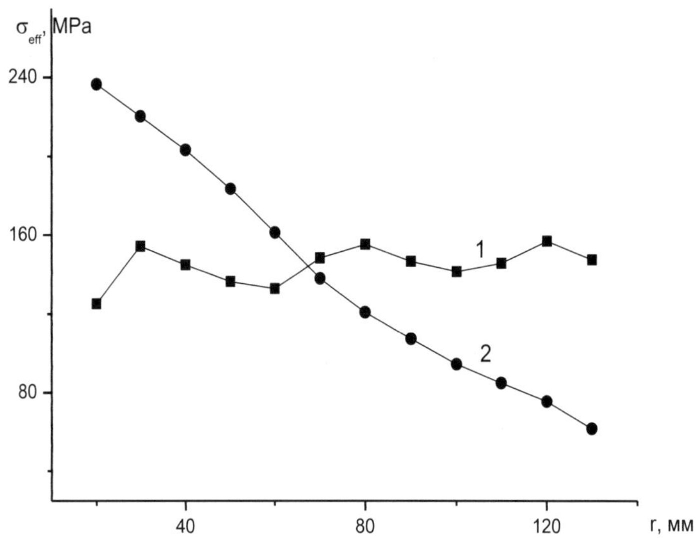

Figure 1. Maximal values of equivalent stresses versus the insert radius: in the insert center (1) and in the insert – bottom conjugation point (2).

The abscissa of curves 1 and 2 cross point in figure 1 corresponds to the optimal value of the insert radius which is about 30% of the bottom radius. As the insert radius is less than its optimal value, the effective stresses in the conjugation point are higher than those in the insert center (shell pole). And when the insert radius is greater than optimal one, effective stresses in the conjugation point are less than those in the insert center. Thus, the numerical simulation shows that there is the only one cross point of curves 1 and 2 (see Figure 1). It also follows from the calculations that insert of the optimal radius reduces maximal equivalent stresses proportionally to insert thickness/shell thickness ratio.

Measurements of dynamic stresses provided by the strain gauge technique confirm that there is really the stress concentration in shell poles. Maximal stresses in shell poles can be 1.5 – 2.0 times greater than those in the middle of cylindrical part. The method of insert embedding was verified experimentally. The strengthening insert in a form of disc was weld into one of the poles of cylindrical explosive chamber with two elliptical bottoms designed for 200 g TNT. Insert thickness was 1.7 times as much as the thickness of the rest part of shell. Measurements have shown that stress distribution becomes practically uniform, as maximal stresses in the strengthened pole (150 MPa) and in the middle of shell (137 MPa) differ only by 9.5%.

There is the other way to decrease stresses in poles, consisting in placing of cylindrical shield into the chamber body concentrically with the shell cylindrical part. The shield reduces stresses in the middle of the shell, in result stresses in poles go down because they depend on the amplitude of bending vibrations of shell cylindrical part. The shielding method was used in manufacturing of certain KV-2 and KV-5 chambers. Stress metering of KV-2 chamber equipped with the shield and without shield was performed. The shield thickness was 3 times less than that of the chamber body. Measurements have shown that without shield stresses in chamber pole reach the value of 233 MPa, whereas with installed shield they are equal to 181 MPa [7].

It should be noted that approaches, suggested in early investigations [8, 9] and in a sequel developed to the level of engineering practice [10] give rather good description of a chamber stress-strain state when the shell shape is close to spherical. But these approaches leave out of account bending waves travelling along the shell, and therefore omit stress concentration in a composed shell which is under study here. Thus, only numerical calculations can give reliable information on stress-strain state of composed shell under effect of explosion.

There is the useful formula for calculation the circumferetial stress σ in a spherical shell at explosion of high-explosive (HE) spherical charge [10, 12]

$$\sigma = \frac{2}{3\sqrt{1-\mu}} \cdot \frac{\rho_e r_e^3 \sqrt{Q_e}}{R^2 \delta} \cdot \sqrt{\frac{E}{\rho}} = \frac{2}{3\sqrt{1-\mu}} \cdot \frac{\rho_e r_e^3 c \sqrt{Q_e}}{R^2 \delta} \qquad (1)$$

where μ is the Poisson coefficient, E – the Young modulus, ρ – the density, c – the sound velocity of the shell material, δ – the shell thickness, R – the shell radius, ρ_e – the density of HE, r_e – the radius of HE charge, Qe – the specific heat of HE explosion. This formula can also be used in crude estimation of stresses in any composed non-spherical shell. Nevertheless, taking into account the stress concentrations, for chamber poles the value calculated by (1) must be multiplied by two.

3. Explosion chamber designs and applications

Explosion chambers of different modifications are used both in scientific research and in industry. For example, the chamber DVK-0.2 for 200 g TNT is designed for investigation of detonation products behind the detonation front with the use of synchrotron emission [1]. The chamber KV-0.2 (Figure 2) is made of stainless steel and is intended for explosive synthesis of new materials. Chamber shell has outward diameter of 860 mm and the length of 1100 mm. Stainless steel body enables production of pure materials without undesirable contaminants. This chamber can be used for explosive synthesis of nanodiamonds. The other chamber KVG-2 for 2 kg TNT is designed in 2016 for pulsed X-ray registration of explosion processes in vacuum (Figure 3). Its outward diameter and length are 1290 and 2550 mm, respectively. Special design of output valve and labyrinth sealing of chamber bayonet lock [11] permit to keep a vacuum and prevent leakage of detonation products from the chamber to laboratory or industrial premises. The output valve contains a destructible membrane with a thickness depending on an explosive charge value.

There are industrial chambers designed for explosive charge mass varying from 2 to 8 kg TNT and intended for explosive working of materials, including disposal of ammunition elements. Examples of such plants are given in [1], where explosion chambers KV-2, KV-5 and KVG-8 and their applications are described.

IOP Conf. Series: Journal of Physics: Conf. Series **894** (2017) 012091 doi:10.1088/1742-6596/894/1/012091

It should be noted that uniform distribution of stresses in chamber shell can be also achieved by the proper choice of chamber structure. The task is to prevent a propagation of bending waves from chamber cylindrical part to its bottoms, and hereby to avoid stress concentration in the chamber poles. For relatively small explosion chambers the design was developed, in which main body and chamber coverage are bound together by means of the special device. Such explosive chamber KIP-0.2 for 200 g TNT is shown on Figure 4. Overall dimensions of this chamber are 1030 x 850 x 1400 mm, weight 850 kg.

Figure 2. Explosion chamber KV-0.2.

Figure 3. Explosion chamber KVG-2.

Figure 4. Explosion chamber KIP-0.2.

The mentioned examples of explosion chambers does not present all modifications of equipment produced in Lavrentyev Institute of Hydrodynamics. Additional information can be found for example in [12]. There are the other companies, producing explosion chambers, e.g. DYNASAFE, Special Materials Corporation and a number of institutions in different countries. However, most of produced chambers are not designed for multiple explosions, but for one unauthorized blast. That's why they are lightweight and more cheap compared to metal chambers described above. Laboratory and industrial explosion chambers presented here are high-cycle, as they carry thousands and tens of thousands of explosions. About 100 explosion chambers of different modifications were made in Lavrentyev Institute of Hydrodynamics from 1976 to present time.

4. Conclusion

Numerical calculations and stress metering help to find construction solutions enabling to produce metal explosion chambers with uniform stress distribution in the chamber shell. Welded-in inserts, inner shields, and special joining devices between shell elements permit to reduce stresses in chamber poles. Metal explosion chambers enable the reliable containment of the effects caused by explosions, therefore they are successfully used in scientific research and in industry for explosive working of materials.

References

[1] Shtertser A, Stoyanovskii O, Zlobin B, Meshcheryakov and Yu Skornyakov Yu 2014 *Arch. Met. Mater.* **59** 1619-1624
[2] Adishchev V V and Kornev V M 1979 *Combust. Explos. Shock Waves* **15** 780-784
[3] Abakumov A I, Egunov V V, Ivanov A G, Ichaev A A, Tsypkin V I and Shitov A T 1984 *J. Appl. Mech. Tech. Phys.* **25** 455-458
[4] Sil'vestrov V V, Plastinin A V, Gorshkov N N and Stoyanovskii O I 1994 *Combust. Explos. Shock Waves* **30** 228-234
[5] Timoshenko S P 1972 *Kurs teorii uprugosti* ed E I Grigoluk (Kiev: Naukova Dumka) 501, in Russian
[6] Meshcheryakov Yu P 2006 *J. Appl. Mech. Tech. Phys.* **47** 419-425
[7] Pikarevskii A A and Stoyanovskii O I 2013 *J. Appl. Mech. Tech. Phys.* **54** 337-342
[8] Baker W E 1960 *J. Appl. Mech.* **27** 139-144
[9] Demchuk A F 1968 *J. Appl. Mech. Tech. Phys.* **9** 558-559
[10] Demchuk A F and Isakov V P 2006 *Metal explosion chambers: monograph* rev V V Pai, A I Lyamkin (Krasnoyarsk: Publishing Department of Krasnoyarsk State University) 297
[11] Meshcheryakov A A, Ermilov N P, Zlobin B S and Meshcheryakov Yu P 2016 *Application for a Russian Patent No 2016122745*
[12] Shtertser A A , Zlobin B S and Stoyanovskii O I 2015 *Appl. Mech. Mater.* **788** 306-312

IOP Conf. Series: Journal of Physics: Conf. Series **894** (2017) 012092 doi:10.1088/1742-6596/894/1/012092

Wave motion in a fluid under an inhomogeneous ice cover

I V Sturova and L A Tkacheva

Lavrentyev Institute of Hydrodynamics of SB RAS, Novosibirsk, Russia

E-mail: sturova@hydro.nsc.ru, tkacheva@hydro.nsc.ru

Abstract. This paper studies steady waves in fluid and in semi-infinite ice cover generated by a constant pressure distribution with a rectangular planform moving uniformly along the edge of ice cover at fixed distance. This load simulates the air-cushion vehicle (ACV). We consider two cases: (i) the surface of fluid is free outside of ice sheet, (ii) fluid is bounded by a solid vertical wall and the edge of ice cover can be either clamped or free. The fluid is assumed to be ideal incompressible and of finite depth. The ice sheet is modelled by elastic thin plate. The solution of linear hydroelastic problem is obtained by two methods: the Wiener-Hopf technique and matched eigenfunction expansions. The deflection of ice sheet and free surface elevation, as well as wave forces acting on ACV are investigated for different speeds of motion.

1. Introduction

Intensive development of the Arctic causes interest in solving new problems related to the interaction of dynamic perturbations in a fluid and in an ice cover. Previously considerable attention has been paid to three-dimensional flexural oscillations of an ice cover due to a moving pressure area [1], [2], and the ice cover was treated as infinitely extended isotropic, uniform elastic thin plate. In reality, the sea ice is strongly inhomogeneous. Cracks, polynyas and hummocks are the characteristic irregularities of dense ice [3]. The effect of such complex boundary conditions on the wave motion is in the initial stage of the study. The takeoff and landing of an airplane on the very large floating platform was investigated in [4]. The influence of the vertical wall at the clamped edge of the ice cover on the wave motion under moving load has been studied in [5]. The response of ice cover to an external load moving along a frozen channel is presented in [6].

In this paper, the solution of the steady problem for the fluid and semi-infinite ice cover under the action of a rectangle external load moving with uniform speed along the edge is given. Two cases are considered: (i) the surface of the fluid is free outside of the ice sheet, (ii) the fluid is bounded by a solid vertical wall and the edge of the ice sheet can be either clamped or free. The problem is formulated within linear hydroelastic theory. The fluid motion is potential. For the first case, the solutions obtained by two different methods (the Wiener-Hopf technique and matched eigenfunction expansions) are compared.

2. Mathematical formulation

Let us consider the statement of the problem in the case (i). The pressure distribution $P(x, y)$ moves with constant velocity U along the edge of the plate. The plate draft is ignored. We

Content from this work may be used under the terms of the Creative Commons Attribution 3.0 licence. Any further distribution of this work must maintain attribution to the author(s) and the title of the work, journal citation and DOI.
Published under licence by IOP Publishing Ltd

consider the moving Cartesian coordinate system (x, y, z) with x -axis directed perpendicular to the edge, the y -axis along the edge, and the z -axis vertically upwards. It is assumed that the plate is in contact with the water at all points.

The boundary-value problem for the velocity potential $\varphi(x, y, z)$ and the plate deflection or the free surface elevation $w(x, y)$ can be written as

$$\Delta_3\varphi = 0 \ \ (-\infty < x, y < \infty, \ -H \leq z \leq 0), \ \ \Delta_3 \equiv \Delta_2 + \partial^2/\partial z^2, \ \ \Delta_2 \equiv \partial^2/\partial x^2 + \partial^2/\partial y^2 \ , \tag{1}$$

$$gw - U\frac{\partial\varphi}{\partial y}\bigg|_{z=0} = 0 \ \ (x<0), \ \ D\Delta_2^2 w + \rho h U^2 \frac{\partial^2 w}{\partial y^2} + g\rho_0 w - \rho_0 U \frac{\partial\varphi}{\partial y}\bigg|_{z=0} = -P(x,y) \ \ (x>0), \tag{2}$$

$$\partial\varphi/\partial z\big|_{z=0} = -U\partial w/\partial y, \quad \partial\varphi/\partial z\big|_{z=-H} = 0, \tag{3}$$

$$\left(\frac{\partial^2}{\partial x^2} + \nu\frac{\partial^2}{\partial y^2}\right)w_+ = 0, \quad \frac{\partial}{\partial x}\left[\frac{\partial^2}{\partial x^2} + (2-\nu)\frac{\partial^2}{\partial y^2}\right]w_+ = 0, \ \ w_+ = w|_{x=0+}. \tag{4}$$

Here $D = Eh^3/[12(1-\nu^2)]$; E, ν, h, ρ are the Young's modulus, the Poisson's ratio, the thickness and the density of the ice sheet, respectively; ρ_0 is the fluid density; H is water depth; g is the acceleration due to gravity. In the far field a radiation condition should be imposed that requires the radiated waves to be outgoing.

For the case (ii), the fluid is restricted at the left by the rigid wall. Consequently, we have the boundary condition: $\partial\varphi/\partial x = 0$ at $x = 0$. The edge of the ice sheet can be free or frozen to the fixed vertical structure, then $w_+ = \partial w/\partial x|_{x=0+} = 0$.

We restrict our consideration to the constant pressure distribution in the rectangular planform: $P(x, y) = P_0$ in the domain $(|x - x_0| < a, \ |y| < b, \ x_0 > a)$ and zero otherwise. The forces F_x (side force) and F_y (wave resistance) acting on ACV and its non-dimensional values A_x, A_y are determined by formulas

$$(F_x, F_y) = -P_0 \int_{-b}^{b}\int_{x_0-a}^{x_0+a}(w_x, w_y)dxdy, \quad (A_x, A_y) = -\frac{g\rho_0}{2aP_0^2}(F_x, F_y).$$

3. Method of solution

The dimensionless variables and parameters are introduced

$$(x', y', z', a', b', x_0') = \frac{1}{H}(x, y, z, a, b, x_0), \ \beta = \frac{D}{\rho_0 g H^4}, \ \ F = \frac{U}{\sqrt{gH}}, \ \ \sigma = \frac{\rho h}{\rho_0 H}, \ \ P_0' = \frac{P_0}{\rho_0 g H}.$$

Below, the primes are omitted. We will seek the velocity potential and the displacement in the form $\varphi = UH\phi(x, y, z)$, $w = HW(x, y)$.

3.1. The Wiener-Hopf technique

We use the Fourier transform to the variables x and y in the form

$$\Phi_-(\alpha, s, z) = \int_{-\infty}^{\infty} \mathrm{e}^{-isy}dy \int_{-\infty}^{0} \phi(x, y, z)\mathrm{e}^{i\alpha x}dx, \ \ \Phi_+(\alpha, s, z) = \int_{-\infty}^{\infty} \mathrm{e}^{-isy}dy \int_{0}^{\infty} \phi(x, y, z)\mathrm{e}^{i\alpha x}dx.$$

From the Laplace equation (1) and the no-flux bottom condition (3), we have

$$\Phi(\alpha, s, z) = \Phi_- + \Phi_+ = C(\alpha, s)Z(\alpha, s, z), \ \ Z = \cosh[(z+1)\sqrt{\alpha^2+s^2}]/\cosh\sqrt{\alpha^2+s^2}, \tag{5}$$

where $C(\alpha, s)$ is unknown function. We introduce the functions $D_\pm(\alpha, s)$, $G_\pm(\alpha, s)$ in following manner:

$$D_- = \int_{-\infty}^{\infty} \mathrm{e}^{-isy}dy \int_{-\infty}^{0} \left(\phi_z + F^2\phi_{yy}\right)_{z=0} \mathrm{e}^{i\alpha x}dx, \ D_+ = \int_{-\infty}^{\infty} \mathrm{e}^{-isy}dy \int_{0}^{\infty} \left(\phi_z + F^2\phi_{yy}\right)_{z=0} \mathrm{e}^{i\alpha x}dx,$$

IOP Conf. Series: Journal of Physics: Conf. Series **894** (2017) 012092 doi:10.1088/1742-6596/894/1/012092

$$G_- = \int_{-\infty}^{\infty} \mathrm{e}^{-isy} dy \int_{-\infty}^{0} \Big[(\beta \Delta_2^2 + 1 + \sigma F^2 \partial^2/\partial y^2)\phi_z + F^2 \phi_{yy} \Big]_{z=0} \mathrm{e}^{i\alpha x} dx,$$

$$G_+ = \int_{-\infty}^{\infty} \mathrm{e}^{-isy} dy \int_{0}^{\infty} \Big[(\beta \Delta_2^2 + 1 + \sigma F^2 \partial^2/\partial y^2)\phi_z + F^2 \phi_{yy} \Big]_{z=0} \mathrm{e}^{i\alpha x} dx.$$

The functions with the indexes $+/-$ are analytical on α in the upper/lower half-plane, respectively. From the boundary conditions (2), we have

$$D_-(\alpha, s) = 0, \quad G_+(\alpha, s) = isQ(\alpha, s), \quad Q(\alpha, s) = 4P_0 \mathrm{e}^{i\alpha x_0} \sin(\alpha a) \sin(sb)/(\alpha s), \tag{6}$$

where $Q(\alpha, s)$ is the Fourier-transform of the function $P(x, y)$. Using (5), we can write

$$D(\alpha, s) = D_- + D_+ = C(\alpha, s) K_1(\alpha, s), \quad G(\alpha, s) = G_- + G_+ = C(\alpha, s) K_2(\alpha, s), \tag{7}$$

where $K_1(\alpha, s)$ and $K_2(\alpha, s)$ are the dispersion functions for the free surface waves and the flexural-gravity ones in a moving coordinate system, respectively,

$$K_1(\alpha, s) = \sqrt{\alpha^2 + s^2} \tanh \sqrt{\alpha^2 + s^2} - F^2 s^2,$$

$$K_2(\alpha, s) = [\beta(\alpha^2 + s^2)^2 + 1 - \sigma F^2 s^2] \sqrt{\alpha^2 + s^2} \tanh \sqrt{\alpha^2 + s^2} - F^2 s^2.$$

The dispersion relation for the free surface waves $\mathcal{K}_1(\gamma) \equiv \gamma \tanh \gamma - F^2 s^2 = 0$ has two real roots $\pm\gamma_0$ and the countable set of imaginary roots $\pm\gamma_m$ $(m = 1, 2, ...)$. The dispersion relation for the flexural-gravity waves $\mathcal{K}_2(\mu) \equiv (\beta\mu^4 + 1 - \sigma F^2 s^2)\mu \mathrm{th} \mu - F^2 s^2 = 0$ has two real roots $\pm\mu_0$, four complex roots $\pm\mu_{-1}$, $\pm\mu_{-2}$, $\mu_{-2} = -\bar{\mu}_{-1}$ (the bar denotes complex conjugation), and the countable set of imaginary roots $\pm\mu_m$ $(m = 1, 2, ...)$. Then the roots of the dispersion relations $K_n(\alpha, s) = 0$ $(n = 1, 2)$ are $\chi_m(s) = \sqrt{\gamma_m^2(s) - s^2}$ $(n = 1)$, $\alpha_m(s) = \sqrt{\mu_m^2(s) - s^2}$ $(n = 2)$. We take the values of these roots from the upper half-plane.

From the relations (6) and (7), we obtain

$$G_-(\alpha, s) + isQ(\alpha, s) = D_+(\alpha, s) K(\alpha, s), \quad K(\alpha, s) = K_2(\alpha, s)/K_1(\alpha, s). \tag{8}$$

In accordance with the Wiener-Hopf technique, we factorize the function $K(\alpha, s)$:

$$K(\alpha, s) = K_-(\alpha, s) K_+(\alpha, s), \quad K_\pm(\alpha, s) = \frac{(\alpha \pm \alpha_{-1})(\alpha \pm \alpha_{-2})}{\mu_{-1}\mu_{-2}} \prod_{j=0}^{\infty} \frac{(\alpha \pm \alpha_j)\gamma_j}{\mu_j(\alpha \pm \chi_j)},$$

where the functions $K_\pm$ are analytical in the upper/lower parts of the complex plane α, respectively. By dividing (8) into $K_-(\alpha, s)$, we have

$$\frac{G_-(\alpha, s)}{K_-(\alpha, s)} + 2P_0 \sin(sb) \frac{\psi(\alpha)}{K_-(\alpha, s)} = D_+(\alpha, s) K_+(\alpha, s), \quad \psi(\alpha) = \frac{\mathrm{e}^{i\alpha(x_0 + a)} - \mathrm{e}^{i\alpha(x_0 - a)}}{\alpha}.$$

Then we use the representation

$$\frac{\psi(\alpha)}{K_-(\alpha, s)} = L_-(\alpha, s) + L_+(\alpha, s), \quad L_\pm(\alpha, s) = \pm \frac{1}{2\pi i} \int_{-\infty \mp i\lambda}^{\infty \mp i\lambda} \frac{\psi(\zeta) d\zeta}{K_-(\zeta, s)(\zeta - \alpha)},$$

and as a result we obtain the equation

$$G_-(\alpha, s)/K_-(\alpha, s) + 2P_0 \sin(sb) L_-(\alpha, s) = D_+(\alpha, s) K_+(\alpha, s) - 2P_0 \sin(sb) L_+(\alpha, s).$$

The functions on the left and right sides of this equation are analytical in the lower and upper parts of the complex plane α, respectively. Then we have analytical function over the entire complex plane α. By the Liouville's theorem, this function is a polynomial. The degree of the polynomial is determined by the behavior of this function as $|\alpha| \to \infty$ and is equal to one. Consequently, we can write

$$D_+(\alpha, s)K_+(\alpha, s) - 2P_0 \sin(sb)L_+(\alpha, s) = 2P_0 \sin(sb)[a_1(s) + a_2(s)\alpha],$$

where $a_1(s)$ and $a_2(s)$ are unknown functions which are defined from the edge conditions. Then we have

$$\Phi(x, s, z) = \frac{P_0 \sin(sb)}{\pi} \int_{-\infty}^{\infty} \frac{\mathrm{e}^{-i\alpha x}[a_1 + a_2\alpha + L_+(\alpha, s)]Z(\alpha, s, z)d\alpha}{K_+(\alpha, s)K_1(\alpha, s)}.$$

Using the free-edge conditions (4), we obtain the system of two linear algebraic equations to define the coefficients $a_1(s)$ and $a_2(s)$. All integrals on α are evaluated by the residue method.

After solving this system, we find the free surface elevations and the ice deflections by performing inverse Fourier transform:
for $x < 0$

$$W(x, y) = -\frac{P_0 F^2}{\pi} \int_{-\infty}^{\infty} \mathrm{e}^{isy} s \sin(sb) \sum_{j=0}^{\infty} \frac{\mathrm{e}^{-i\chi_j x}\gamma_j[a_1 + a_2\chi_j + L_+(\chi_j, s)]}{\chi_j K_+(\chi_j, s)\mathcal{K}_1'(\gamma_j)} ds, \tag{9}$$

for $x > 0$

$$W(x, y) = -\frac{P_0}{2i\pi^2} \int_{-\infty}^{\infty} \frac{\mathrm{e}^{isy}\sin(sb)}{s} ds \int_{-\infty}^{\infty} \frac{\mathrm{e}^{-i\alpha x}\psi(\alpha)\sqrt{\alpha^2 + s^2}\tanh\sqrt{\alpha^2 + s^2}d\alpha}{K_2(\alpha, s)}$$

$$-\frac{P_0}{\pi} \int_{-\infty}^{\infty} \frac{\mathrm{e}^{isy}\sin(sb)}{s} \sum_{j=-2}^{\infty} \frac{\mathrm{e}^{i\alpha_j x}[a_1 - a_2\alpha_j - L_-(-\alpha_j, s)]K_+(\alpha_j, s)\mu_j \tanh\mu_j}{K_2'(\alpha_j, s)} ds, \tag{10}$$

where the prime denotes the derivative with respect to the first variable. Integrals on s are evaluated numerically.

3.2. Method of matched eigenfunction expansions

The solution of the boundary-value problem (1)-(4) at $x > 0$ is sought in the form

$$\phi(x, y, z) = \phi_0^+(x, y, z) + \phi_0^-(x, y, z) + \psi(x, y, z), \quad w(x, y) = w_0^+(x, y) + w_0^-(x, y) + v(x, y).$$

Here the functions $\phi_0^\pm$ and $w_0^\pm$ are the solution for the problem of the action of a rectangular pressure region centered at the point $(x = \pm x_0,\ y = 0)$ on an infinitely extended elastic plate. For the rectangular external load, the solutions for $\phi_0^\pm$ and $w_0^\pm$ have the form

$$\phi_0^\pm = -\frac{4P_0F}{\pi^2} \int_0^{\pi/2} \frac{d\theta}{\sin\theta} \int_0^{\infty} \frac{\cosh[k(z+1)]}{\cosh k Z(k, \theta)} \sin(ak\sin\theta)\sin(bk\cos\theta)$$
$$\times \cos[k(x \mp x_0)\sin\theta]\sin(ky\cos\theta)dk, \tag{11}$$

$$w_0^\pm = -\frac{4P_0}{\pi^2} \int_0^{\pi/2} \frac{d\theta}{\cos\theta\sin\theta} \int_0^{\infty} \frac{\tanh k}{Z(k, \theta)} \sin(ak\sin\theta)\sin(bk\cos\theta)$$
$$\times \cos[k(x \mp x_0)\sin\theta]\cos(ky\cos\theta)dk, \tag{12}$$

$$Z(k, \theta) = (1 + \sigma k \tanh k)[\omega^2(k) - F^2k^2\cos^2\theta], \quad \omega(k) = \sqrt{k\tanh k(\beta k^4 + 1)/(\sigma k \tanh k + 1)}.$$

IOP Conf. Series: Journal of Physics: Conf. Series **894** (2017) 012092 doi:10.1088/1742-6596/894/1/012092

The function $\omega(k)$ is a dispersion relation for flexural-gravity waves [1]. It is known that the phase velocity of flexural-gravity waves $c(k) = \omega(k)/k$ has the local minimum c_m. For a fluid of finite depth, we have always $c_m < 1$. The integrands in (9)-(12) have no singularities at the velocities of the load motion $F < c_m$ (subcritical case). This means that for an infinitely extended ice cover the wave motion is absent far from the load in this case. For $F > c_m$ (supercritical case), there are the rooted singularities in the integrands. By change of integration variable integrands become regular. In this case flexural-gravity waves are spreading in ice cover.

Using the Fourier transform to the variable y, the solutions for unknown functions $\phi(x, s, z)$, $w(x, s)$ at $x < 0$ and $\psi(x, s, z)$, $v(x, s)$ at $x > 0$ are sought in the form:

$$\phi(x,s,z) = -isF \sum_{m=0}^{\infty} \frac{E_m(s)}{\gamma_m \sinh \gamma_m} e^{-i\chi_m x} \cosh[\gamma_m(z+1)], \quad w(x,s) = \sum_{m=0}^{\infty} E_m(s) e^{-i\chi_m x},$$

$$\psi(x,s,z) = -isF \sum_{m=-2}^{\infty} \frac{V_m(s) e^{i\alpha_m x}}{\mu_m \sinh \mu_m} \cosh[\mu_m(z+1)], \quad v(x,s) = \sum_{m=-2}^{\infty} V_m(s) e^{i\alpha_m x}.$$

Unknown functions $E_m(s)$ and $V_m(s)$ are determined by the method of integral fitting from matching conditions of continuity of horizontal velocity and pressure on the plane $x = 0$:

$$\partial\phi/\partial x|_{x=0-} = \partial\phi/\partial x|_{x=0+}, \quad \partial\phi/\partial y|_{x=0-} = \partial\phi/\partial y|_{x=0+} \quad (|y| < \infty, \; -1 \le z \le 0).$$

4. Numerical results

The following input data are used for the ice sheet and the external load: $E = 5GPa$, $\nu = 1/3$, $\rho_0 = 10^3$ kg/m^3, $\rho = 900$ kg/m^3, $h = 2.5$ m, $P_0 = 10^3$ N/m^2, $a = 10$ m, $b = 20$ m, $x_0 = 50$ m, $H = 350$ m. The minimal phase velocity for these parameters is equal to $c_m \approx 21.86$ m/s. The critical velocity for the surface waves is equal to $U_c \equiv \sqrt{gH} \approx 58.60$ m/s. Calculations were performed for values of speed 10, 15, 20, 30, 50 and 60 m/s. Speeds of load 10, 15, 20 m/s are subcritical. At a speed of load 10 m/s waves aren't present neither in fluid, nor in a plate. Deformation of a free surface of fluid is small. At a speed of load 15 m/s, there are waves in the fluid which extend behind moving loading, but in a plate there are no waves. At a speed of load 20 m/s, wave amplitudes in fluid become greater and they excite the waves in a plate decaying with the distance from the edge.

Figure 1 (a)-(c) shows the deflection of the ice sheet $w(x, 0)$ for cases (i) and (ii) at the different speeds of load motion. The smallest and largest deflections take place for the vertical wall with the clamped and free edge, respectively. In the presence of a free surface, deflections of the ice cover take intermediate values between the values for the infinitely extended ice cover and the case of a vertical wall with a free edge. The isolines of $w(x, y)$ for the plate deflection and the free surface elevation at $U = 15$m/s are shown in Figure 1 (d). The deflections of the ice cover are practically symmetrical along the axis y. However in the region of pure water, a wave pattern is produced at $y < 0$. The dashed line shows the wedge within which, according to the linear theory of surface waves, there is a wave system at large distances downstream. At the given values of the load speed and the depth of the fluid, the half-angle of the aperture practically approaches 19.5°, that corresponds to the case of an infinitely deep fluid [8].

Figure 2 (a)-(d) shows the three-dimensional plots of wave surface at speeds $U = 20$, 30, 50, 60 m/s. At $U = 20$ m/s (subcritical case) waves in fluid generate waves in the ice cover near the edge which are damped away from it. The wave length is not changed. For supercritical case $U = 30$, 50, 60 m/s short elastic wave spread ahead and long gravity ones appear behind the loading region in the plate. As the speed of motion increases, the waves ahead of load become shorter and ones behind it become longer. At $U = 60$ m/s $> U_c$, the shadow zone is appeared behind the load region [7], that is the waves in this zone are absent. The wave wake in fluid is generated by flexural-gravity waves in ice cover.

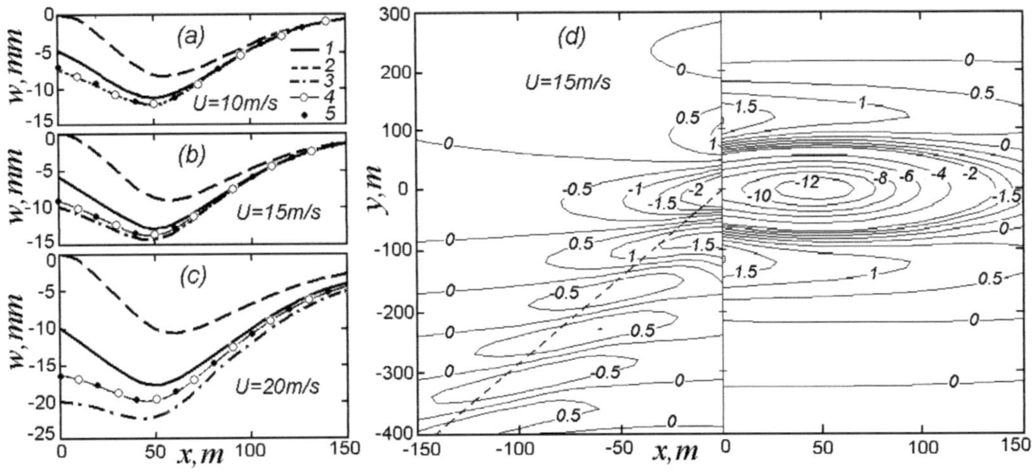

Figure 1. (a)-(c) Ice-sheet deflections along the line $y = 0$ at $U = 10, 15, 20$m/s for infinitely extended ice cover *(1)*, in the presence of a vertical wall with the clamped *(2)* and free *(3)* edge and in the presence of the free surface *(4, 5)* calculated by Wiener-Hopf technique and matched eigenfunction expansions, respectively. (d) Isolines of $w(x, y)$ in millimeters for the plate deflection ($x > 0$) and the free surface elevation ($x < 0$) at $U = 15$m/s.

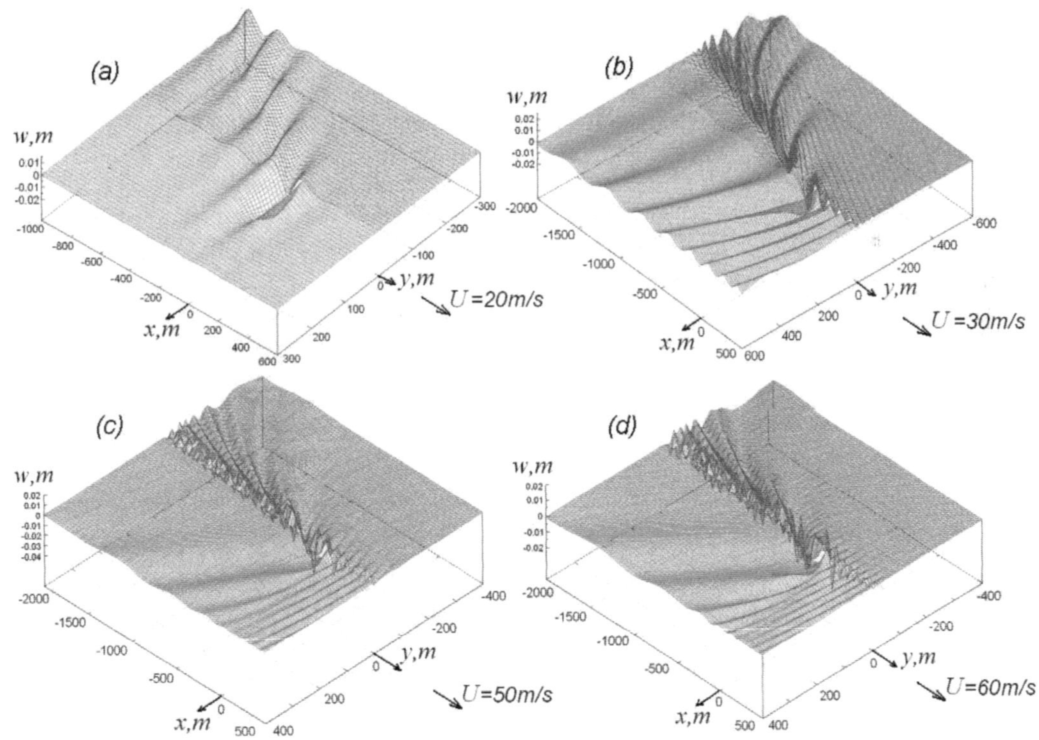

Figure 2. (a)-(d) Wave patterns of the free surface ($x < 0$) and the ice cover ($x > 0$) at speeds $U = 20,\ 30,\ 50,\ 60$m/s.

Figure 3 (a) shows vertical displacements of ice cover and the free surface at $x = 0$ (curve *1* and *2*), and the free surface at the distance 50 m from the edge (curve *3*) as functions of the coordinate y. It is seen that wave length in fluid is changed, short waves ahead of loading region become shorter and of less amplitude, but waves behind the loading zone become of great amplitude.

Non-dimensional values of wave forces acting on moving vehicle are presented on Figure 3 (b) as functions of the load speed. The greatest difference of the value A_y from its value for infinite ice cover is observed at near-critical values of speed $U \approx c_m$. The coefficient A_x changes the sign. This is due to the fact that minimal values of plate deflection are displaced back and the character of dependence of the plate deflection on the coordinate x is changed as the speed increased (Figure 3 (c)). Maximal values of the vertical displacements of ice cover are observed at near-critical values of speed $U \approx c_m$.

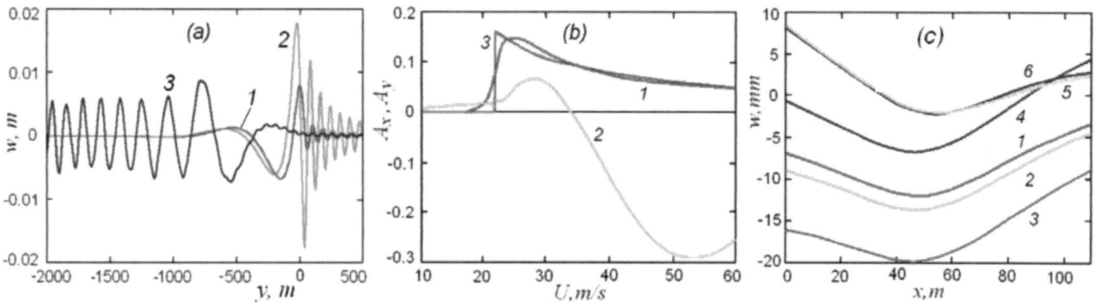

Figure 3. (a) The deflection of ice cover at the edge $w(0+, y)$ (*1*), free surface elevation at the edge w(0-,y) (*2*), and at the distance x_0 from the edge $w(-x_0, y)$ (*3*) as functions of the coordinate y. (b) The non-dimensional values of wave forces A_y (*1*) and A_x (*2*) for semi-infinite ice cover in the presence of free surface and corresponding values of wave resistance for infinite ice sheet (*3*) as functions of the load speed. (c) The ice cover deflection $w(x, 0)$ for different values of speed versus x: curves (*1-6*) correspond to $U = 10$, 15, 20, 30, 50, 60 m/s.

5. Conclusion

The 3-D linear hydroelastic problem on the waves induced by the uniformly moving load is solved for a semi-infinite ice sheet in two configurations: (i) the surface of the fluid is free outside of the sheet, (ii) an ice sheet is in contact with the fixed rigid vertical wall and the edge of the ice cover can be either clamped or free. The solution is obtained by two different methods: the Wiener-Hopf technique and matched eigenfunction expansions. Good agreement between these results is demonstrated. The wave pattern and magnitude of the ice-sheet deflection and free surface elevation are studied for different speeds of the load as well as the forces (wave resistance and side force) acting in horizontal directions on a moving vehicle.

References

[1] Squire V, Hosking R, Kerr A and Langhorne P 1996 *Moving Loads on Ice Plates* (Dordrecht: Kluwer Academic Publishers)
[2] Kozin V M, Pogorelova A V, Zemlyak V L, Vereschagin V Yu, Rogozhnikova E G, Kipin D Yu and Matiushina A A 2016 *Experimental and Theoretical Researches of the Dependence of Flexural-Gravity Waves in a Floating Plate on the Excitation Conditions* (Novosibirsk: Siberian Branch of Russian Academy of Sciences) [in Russian]
[3] Squire V A 2011 *Phil Trans R Soc Lond A* **369** 2813-31
[4] Kashiwagi M 2004 *J Mar Sci Technol* **9**(1) 14-23

[5] Brocklehurst P 2012 *Hydroelastic Waves and their Interaction with Fixed Structures* (PhD thesis, Univ East Anglia, UK)
[6] Shishmarev K, Khabakhpasheva T and Korobkin A 2016 *Appl Ocean Res* **59** 313-26
[7] Davys J W, Hosking R J and Sneyd A D 1985 *J Fluid Mech* **158** 269-87
[8] Wehausen J V and Laitone E V 1960 *Surface Waves* (Berlin: Springer) vol 9 pp 446-778

IOP Conf. Series: Journal of Physics: Conf. Series **894** (2017) 012093 doi:10.1088/1742-6596/894/1/012093

Wave propagation in channels and cracks with elastic walls

S V Sukhinin, V S Yurkovskiy, A P Konstantinov and A V Trilis

Lavrentyev Institute of Hydrodynamics, Novosibirsk, Russia

E-mail: sukhinin@hydro.nsc.ru

Abstract. In this paper wave propagation in the uniform and nonuniform elastic channels filled with fluid or gas is investigated. The weak discontinuities theory approach is proposed to determine the criteria of gradient catastrophe phenomena to occur for the cases of uniform and nonuniform channels. The results of the paper can be applied to studying of the nonlinear wave propagation phenomena in elastic pores, cracks and channels filled with fluid or gas.

1. Short review

The first fundamental research of hydroelastic wave propagation in pipeline systems was provided by N. E. Zhukovsky in the end of the nineteenth century. The results of his experimental and theoretical research are published in [1]. Historically, hydroelastic waves in channels with elastic walls were called hydraulic shock or hydro hummer. This terms come from above mentioned fundamental work "On the hydraulic shock in water pipes" [1]. The number of publications in this field is constantly growing. The theory of wave propagation in such pipeline systems can be found in [2]. A description of application of hydroelastic pipe waves to geophysical survey is provided in [3]. The main goal of the current research is to carry out a qualitative investigation of the nonlinear phenomena that occur in the hydroelastic wave propagation process in nonuniform channels and cracks. Nonlinear effects are usually connected to wave front steeping and result in strong discontinuity of the solution. In the present work the influence of the nonlinearity of the base equations and formation of the gradient catastrophe in the nonuniform elastic channels is studied using the theory of weak discontinuity (the discontinuity of the solution derivatives) propagation [4]. To the best of our knowledge such studies have not been conducted so far. The results of our investigation are protected by a Russian patent [5].

2. Problem definition and reference model

2.1. *Hypotheses and assumptions*

Let the uniform or nonuniform channel with elastic walls be filled with medium (fluid or gas). Then $\rho = \rho(p)$, $p = p(x,t)$ and $u = u(x,t)$ are the density, pressure and velocity of the medium, where x is the axis coordinate of the channel and t is time. Let the section area of the channel in the point x be equal to $A(p,x)$ (see Figure 1). We assume that density of the medium $\rho = \rho(p)$ and channel section area $A(p,x)$ are the known functions of pressure p and coordinate x. Let's also assume that $A(x,p) = a(x)b(p)$.

Content from this work may be used under the terms of the Creative Commons Attribution 3.0 licence. Any further distribution of this work must maintain attribution to the author(s) and the title of the work, journal citation and DOI.
Published under licence by IOP Publishing Ltd

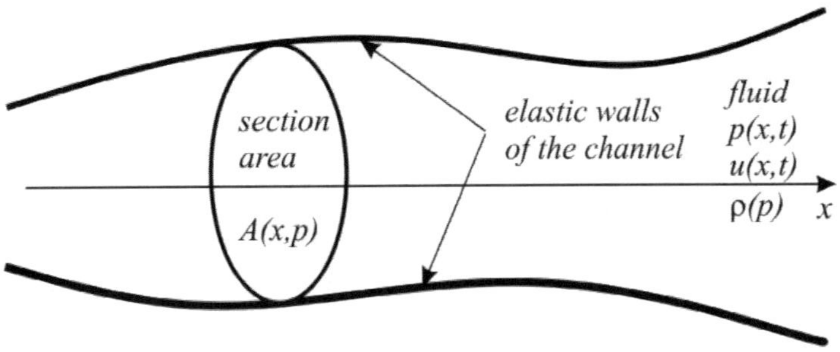

Figure 1. Elastic channel filled with fluid.

If the wavelength is considerably greater than the channel cross-section then the displacement transverse to the channel axis can be neglected. Waves satisfying these conditions are called longitudinal or hydraulic waves.

Mass and momentum conservation laws for the longitudinal waves are as follows

$$\begin{cases}(A\rho)_t + (A\rho u)_x = 0\\ u_t + uu_x + p_x/\rho = 0,\end{cases}$$

here and after $p = p(x,t)$ and $u = u(x,t)$ are unknown functions.

2.2. Mathematical model

The disturbance propagation in channels with elastic walls is described by the system of partial derivative equations (PDE) (1) which can be directly obtained from mass and momentum conservation laws

$$\begin{pmatrix} p \\ u \end{pmatrix}_t + \begin{pmatrix} u & \rho c^2 \\ 1/\rho & u \end{pmatrix}\begin{pmatrix} p \\ u \end{pmatrix}_x + \begin{pmatrix} u\rho c^2 A_x/A \\ 0 \end{pmatrix} = 0. \quad (1)$$

One can easily check that PDE system (1) is a quasi-linear hyperbolic PDE system, and the characteristic directions of the system are $u + c$ and $u - c$, where $c = \left(\frac{A}{(A\rho)_p}\right)^{1/2}$ is the local velocity of the hydroelastic wave in the channel. By direct evaluation one can obtain that the sound velocity in fluid c_f is greater than the velocity of the hydroelastic waves $c = c(p,x) < c_f = \left(\frac{\partial p}{\partial \rho}\right)^{1/2}$, assuming that the entropy is constant in the fluid and in the channel walls.

PDE system (1) describes the wave propagation in the pipes and channels with elastic walls including waves in the well fracturing channels which are widely used to increase the effective radius of the oil wells and to degas coal beds. Multiple applications of the hydraulic fracturing technology show the present paper is timely. Also, the PDE system (1) can be applied to model the wave propagation process in elastic media with open pores or cracks.

2.3. Characteristics.

The differential equation for the characteristics of system (1) is $\frac{dx}{dt} = u \pm c$. As $A = a(x)b(p)$ the velocity c of the hydoelastic waves depends only on the pressure in the fluid, $c = c(p)$, the characteristics of the state of rest are straight lines in the (x,t) space. The characteristics of the state of rest are satisfying the equation $x = c_0 t + \xi$, where $c_0 = c(p_0)$ is the velocity of the hydroelastic waves propagating in the state of rest, p_0 is the pressure in the fluid at rest, ξ is the initial position of the characteristic. It is convenient to apply the propagation theory of the weak discontinuities to predict the gradient catastrophe phenomena in the uniform and nonuniform elastic channels [3, 4].

2.4. Propagation of the weak discontinuity along state of rest characteristics

Let's assume that weak discontinuity of the solution is propagating along the state of rest characteristic $x = c_0 t + \xi$. Then the intensity of the weak discontinuity $\sigma = \sigma(t)$ is the solution of the Cauchy problem for the Riccati differential equation

$$\frac{\mathrm{d}}{\mathrm{d}t}\sigma + \sigma\frac{c_0}{2}\frac{A_x}{A} + \sigma^2\left(1 + \frac{1}{2}\left(\rho c^2\right)_p + \frac{1}{2}\rho_0^2 c_0^2\left(\frac{1}{\rho}\right)_p\right) = 0, \quad (2)$$

$$\sigma(0) = \sigma_0,$$

where $\begin{pmatrix}[p_x]\\ [u_x]\end{pmatrix} = \sigma\begin{pmatrix}\rho_0 c_0\\ 1\end{pmatrix}$; $[f]$ is the jump of function f on the characteristic $x = c_0 t + \xi$, and $\sigma(0) = \sigma_0$ is some initial condition for the intensity of the weak discontinuity. Generally, equation (2) can be analyzed numerically, but it is important to study some particular cases of the Cauchy problem (2).

Elastic properties of the system channel-fluid are defined by fluid compressibility and channel elasticity. Geometric properties of the channel can be defined by $\frac{A_x}{A}$ ratio: a uniform channel has $\frac{A_x}{A} = 0$, a V-shaped channel has $\frac{A_x}{A} = \frac{1}{x}$, a conical channel has $\frac{A_x}{A} = \frac{2}{x}$,an exponential channel $\frac{A_x}{A} = const$ (Fig. 2). In the present paper we study the basic cases of uniform, V-shaped and conical channels filled with an incompressible fluid.

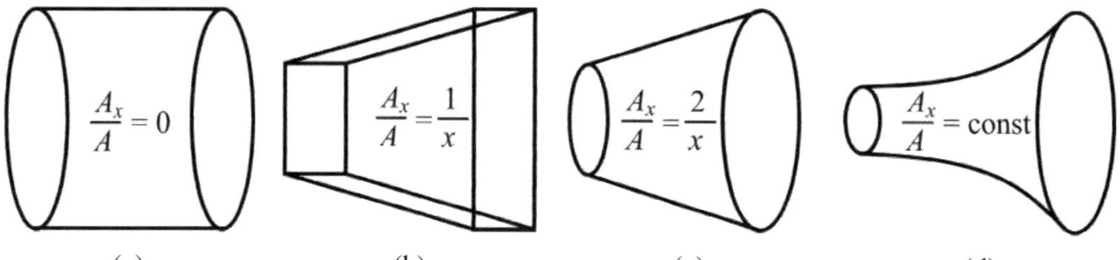

Figure 2. Different types of channels: (a) - uniform channel, (b) - V-shaped channel, (c) - conical channel, (d) - exponential channel.

2.4.1. Gradient catastrophe in the uniform channel filled with an incompressible fluid. In this case equation (2) transforms to

$$\frac{\mathrm{d}}{\mathrm{d}t}\sigma + \sigma^2 q = 0, \quad (3)$$

where $q = \left[1 + \frac{1}{2}\left(\rho c^2\right)_p\right]$. The solution of Cauchy problem (2) is given by the formula $\sigma = \frac{\sigma_0}{q t \sigma_0 + 1}$. According to the physical formulation of the problem $q > 0$, gradient catastrophe occurs if the initial intensity $\sigma_0 < 0$. The moment of gradient catastrophe is $t^* = -\frac{1_0}{q\sigma_0}$, while the coordinate is $x^* = c_0 t^* + \xi = -\frac{1_0}{q\sigma_0} c_0 + \xi$.

2.4.2. V-shaped channel filled with incompressible fluid. In this case equation (2) transforms to

$$\frac{\mathrm{d}}{\mathrm{d}t}\sigma + \sigma \frac{1}{2(t+\xi/c_0)} + \sigma^2 q = 0. \quad (4)$$

The solution of the Cauchy problem is given by formula (5)

$$\sigma(t) = \frac{\sigma_0 c_0 \sqrt{\xi}}{-2q\sigma_0 t c_0 \sqrt{\xi} - 2q\sigma_0 \xi^{\frac{3}{2}} - c_0\sqrt{c_0 t + \xi} + 2q\sigma_0 \xi \sqrt{c_0 t + \xi}}. \quad (5)$$

Gradient catastrophe occurs in $t^* = \frac{c_0 - 4q\xi\sigma_0}{4q^2\sigma_0^2\xi}$ at coordinate $x^* = c_0 \frac{c_0 - 4q\xi\sigma_0}{4q^2\sigma_0^2\xi} + \xi$ if the initial intensity $\sigma_0 < 0$.

2.4.3. Conical channel filled with an incompressible fluid.

In that case Riccati equation (2) changes to equation

$$\frac{\mathrm{d}}{\mathrm{d}t}\sigma + \sigma \frac{1}{(t+\xi/c_0)} + \sigma^2 q = 0 \quad (6)$$

and the solution of the Cauchy problem is given by formula

$$\sigma(t) = \frac{\sigma_0 \xi}{q\xi\sigma_0 c_0 t \ln(c_0 t + \xi) + q\xi^2\sigma_0 \ln(c_0 t + \xi) + c_0^2 t + c_0 \xi - \xi\sigma_0 q c_0 t \ln\xi - q\xi^2\sigma_0 \ln\xi}. \quad (7)$$

Gradient catastrophe occurs in $t^* = \frac{\xi}{c_0}\left(\exp\left(\frac{-c_0}{q\xi\sigma_0}\right) - 1\right)$ on the characteristic $x^* = c_0 t + \xi$ if the initial intensity $\sigma_0 < 0$.

2.5. Nonlinear effects and wave direction relative to V-shaped or conical channel expansion or reduction.

In the formulae (4-7) we considered the case that the channel cross-section expands. It means that gradient catastrophes, described by formulae (5, 7) occur for the waves propagating in the expanding channels. In case of channel cross-section reduces the sign of the coefficient related to σ changes to negative. From the mechanical point of view the influence of nonlinear effects increases for the waves that propagate in tapered channels, but we do not consider such effects in the present paper. In Cauchy problem (2), nonlinear effects of the system (1) are taken into account by the coefficient σ^2.

3. Conclusion

In the present paper criteria for the formation of gradient catastrophes in expanding channels filled with an incompressible fluid are determined. It should be noted that taking into account the compressibility of the fluid in the channel makes the Riccati equation more complex, but does not significantly change it. Methods for predicting gradient catastrophes in channels with elastic walls based on the theory of propagation of weak discontinuities have shown high efficiency.

In most of the cases the fluid flow rates in the pipes and channels are considerably smaller than the sound velocity in fluid and the velocity of the hydroelastic waves. The proposed approach can be used to describe nonlinear effects of wave propagation in channels and pipes with a fluid flow. Since the typical velocities of the fluid flow through pores and cracks are smaller than the velocity of hydroelastic wave propagation, all the above methods and approaches can be used to study nonlinear hydroelastic waves in the hydraulic fracturing channels.

Acknowledgments
This research is supported by the Russian Science Foundation (project no. 17-19-01389).

References

[1] Zhukovsky N E 1949 *On the hydraulic shock in water pipes* (Moscow-Leningrad: State publishing house of technical and theoretical literature)
[2] Lighthill J 2005 *Waves in fluids (*New York: Cambridge University Press)
[3] White J E 1983 *Underground Sound: Application of Seismic Waves (Methods in Geochemistry and Geophysics) (*New York: Elsevier)
[4] Rozdestvenskii B L and Yanenko N N *Systems of quasilinear equations and their applications to gas dynamics* (Providence-Rhode Island: American Mathematical Society. Translations of Mathematical Monographs)
[5] Sukhinin S V and Rymarenko K V 2012 *Method of hydraulic fracturing* RU Patent 2447278

IOP Conf. Series: Journal of Physics: Conf. Series **894** (2017) 012094 doi:10.1088/1742-6596/894/1/012094

Wave transmission and reflection at the boundary of phononic crystals

A P Konstantinov, S V Sukhinin and V S Yurkovskiy

Lavrentyev Institute of Hydrodynamics, Novosibirsk, Russia

E-mail: konstantinov@hydro.nsc.ru

Abstract. This paper deals with two problems concerning wave propagation in non-homogeneous media. The non-homogeneous media is modelled as one-dimensional phononic crystal. First problem is the determination of the low frequency passband. The second one is evaluation of the transmission ratio of the waves that propagate through the boundary of non-homogeneous periodic media. For the first problem low frequency passband boundaries are studied for the waves travelling along one-dimensional multicomponent periodic medium. And for the problem of wave transmission from the homogeneous medium into the one-dimensional multicomponent periodic medium dependencies of energy reflection on parameters of incident wave and medium properties are obtained and analysed.

1. Introduction

There are many studies showing that heterogeneous multicomponent composites can have promising vibration isolating and sound absorbing properties (see e.g. [1, 2]). This paper is a part of the study focusing on theoretical synthesis of the systems for broadband sound absorption and isolation starting from nano- and microstructure levels. Based on the results of works [3, 4], a vibration isolation and sound protection efficiency is illustrated through the study of the wave propagation in porous medium filled with the fluid and gas. Results of the study could be used to obtain new composite materials and finally to design the elements of interior and/or carrying structures of a transport vehicle. In the paper an approach is presented for mathematical modeling of such materials and numerical analysis of their behavior and protection properties under the broadband vibration and noise excitation.

There are some key points in design of the sound protection materials. It is well known that some material do not pass acoustic waves from certain frequency intervals [5]. In this paper we propose an approach to optimize the low frequency intervals of waves traveling through multicomponent periodic medium. Another approach is to operate the reflection properties of the sound protection materials. This feature of the multicomponent periodic medium is also discussed.

2. Problems definition and results

2.1. Waveguide problem for one-dimensional periodic medium

Let a heterogeneous one-dimensional periodic medium (Figure 1) consist of two components $M_1=\{\rho_1,c_1\}$ and $M_2=\{\rho_2,c_2\}$, where ρ and c are the density and the sound velocity of the corresponding medium. Let's accept that a chain of the inhomogeneities, consisting of the component M_1, is filled into the component M_2 (a matrix of the composite) and the linear concentration of coupled layers of

Content from this work may be used under the terms of the Creative Commons Attribution 3.0 licence. Any further distribution of this work must maintain attribution to the author(s) and the title of the work, journal citation and DOI.
Published under licence by IOP Publishing Ltd

components are equal k_1 and k_2 respectively ($k_1 + k_2 = 1$). This chain is spatially periodic; the smallest spatial period of the medium is equal to L.

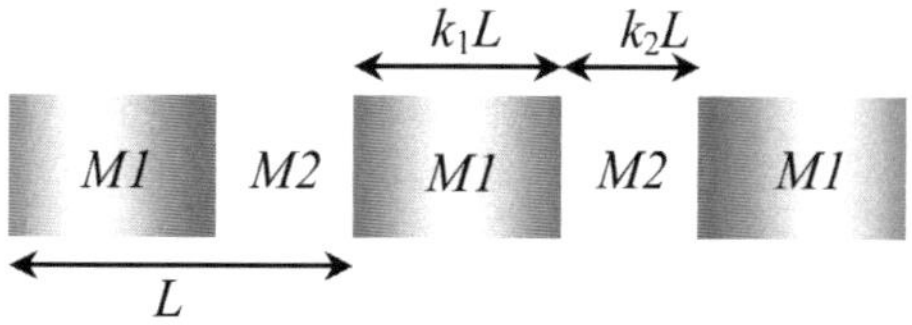

Figure 1. A monodisperse chain of inhomogeneities.

The steady-state acoustic oscillations of the pressure in the media are described by the following equations:

$$p_{xx}^{(1)} + \lambda_1^2 p^{(1)} = 0,$$
$$p_{xx}^{(2)} + \lambda_2^2 p^{(2)} = 0, \qquad (1)$$

where $p^{(1)}$ and $p^{(2)}$ are the acoustic pressures in these components, $\lambda_1 = \omega/c_1$, $\lambda_2 = \omega/c_2$, and $\omega = 2\pi f$ is the angular frequency. The conditions at the component boundary are:

$$p^{(1)} = p^{(2)},$$
$$\frac{p_x^{(1)}}{\rho_1} = \frac{p_x^{(2)}}{\rho_2}. \qquad (2)$$

Considering the symmetry properties of the medium the problem solution satisfy the Floquet theorem

$$p(x+L) = \exp(\mathrm{i}\xi)\, p(x). \qquad (3)$$

As shown in [3] non trivial solution of (1–3) for given ω exist if and only if the angular frequency ω belongs to the certain intervals. Boundaries of these intervals are defined by the dispersive properties of the periodic medium. The non-trivial solution of problem (1–3) will be called waveguide function. Frequency open intervals

$$I_n = \left(\inf_{0\le\xi\le\pi} (\omega_n(\xi)), \sup_{0\le\xi\le\pi} (\omega_n(\xi)) \right), \quad n \in N$$

are the passbands for the modes $\omega_n = \omega_n(\xi)$.

The complete equations system of given problem consist of equations (2) and the following equations derived from the Floquet theorem

$$p^{(1)}(-k_1 L)\exp(\mathrm{i}\xi) = p^{(2)}(k_2 L),$$
$$\frac{p_x^{(1)}(-k_1 L)}{\rho_1}\exp(\mathrm{i}\xi) = \frac{p_x^{(2)}(k_2 L)}{\rho_2}. \qquad (4)$$

Functions $p^{(i)}(x)$ satisfy (1) and have the form

$$p^{(i)}(x) = A^{(i)} \exp(\mathrm{i}\lambda_i x) + B^{(i)} \exp(-\mathrm{i}\lambda_i x), \quad i = 1, 2. \qquad (5)$$

The homogeneous system of linear equations (2), (4) has non zero solution if the determinant of this system equals to zero. Thus the dispersion relation for all waveguide modes is

$$4\cos(\xi) - (\theta+1)^2 \cos(\lambda_1 k_1 + \lambda_2 k_2) + (\theta-1)^2 \cos(\lambda_1 k_1 - \lambda_2 k_2) = 0, \qquad (6)$$

where $\theta = \dfrac{\rho_1 c_1}{\rho_2 c_2}$ and we assume that $\theta \le 1$.

The passband boundaries can be obtained from dispersion relation by substituting $\xi = 0$ or $\xi = \pi$ into (6).

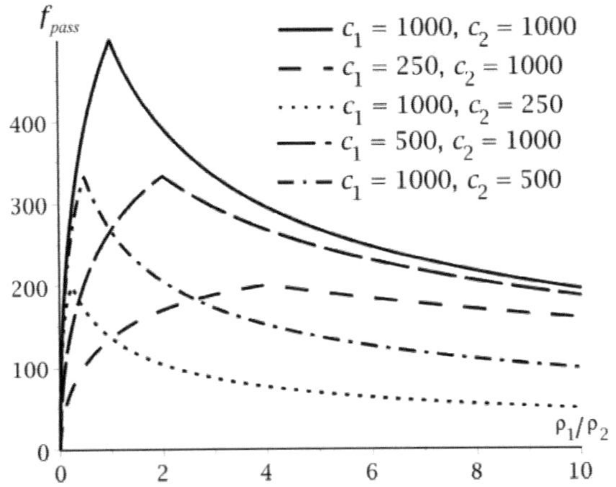

Figure 2. The first passband width dependence on densities ratio.

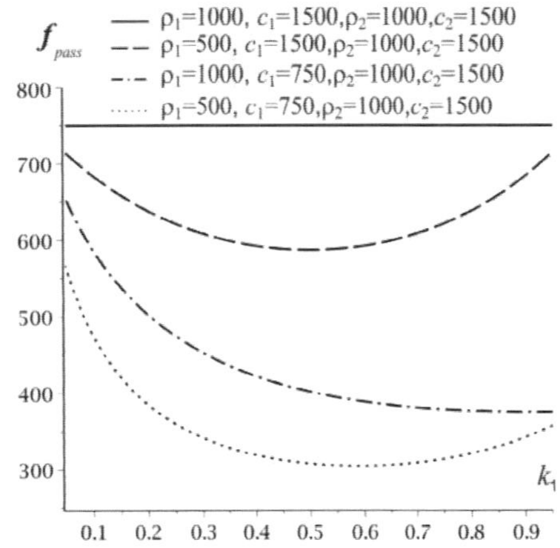

Figure 3. Concentration influence on the first passband width.

From the equation (6) we obtain

$$\operatorname{tg}\left(\frac{\lambda_1 k_1}{2}\right)\operatorname{tg}\left(\frac{\lambda_2 k_2}{2}\right)+\left(\operatorname{tg}\left(\frac{\lambda_1 k_1}{2}\right)\operatorname{tg}\left(\frac{\lambda_2 k_2}{2}\right)\right)^{-1}=\theta+\theta^{-1},$$
$$\operatorname{tg}^{-1}\left(\frac{\lambda_1 k_1}{2}\right)\operatorname{tg}\left(\frac{\lambda_2 k_2}{2}\right)+\operatorname{tg}\left(\frac{\lambda_1 k_1}{2}\right)\operatorname{tg}^{-1}\left(\frac{\lambda_2 k_2}{2}\right)=\theta+\theta^{-1}, \quad (7)$$

to determine the passbands boundaries. Let's consider the first low frequency passband properties. Since $\omega(0)=0$ we have the first passband width $\omega_{pass}=\omega(\pi)$ defined by

$$\operatorname{tg}\left(\frac{\omega L k_1}{2c_1}\right)\operatorname{tg}\left(\frac{\omega L k_2}{2c_2}\right)=\frac{\rho_1 c_1}{\rho_2 c_2}. \quad (8)$$

Solution of this equation satisfies inequality $\omega_{pass}\le\frac{2c_1}{L}\left(\frac{\rho_1}{\rho_2 k_1 k_2}\right)^{1/2}$ or $f_{pass}\le\frac{c_1}{\pi L}\left(\frac{\rho_1}{\rho_2 k_1 k_2}\right)^{1/2}$, which provides a good approximation in case of medium components are very contrast.

Analysis and computations of the equation (8) allow us to make the following conclusions:

- The width ω_{pass} of the first passband depends on the components density ratio and does not depend on the density magnitudes. Examples of this dependency are depicted on Figure 2.
- The concentration of the component with lower acoustic resistance increase from 0 to 0.5 implies decreasing of passband width. Further concentration increase may cause enlarging of the passband width as well as it reduction (see Figure 3).
- Sound velocity of the component with the lowest acoustic resistance has the most sufficient influence on passband width. This property is illustrated on Figure 4. One can notice a knee on the graphs at the argument value corresponding to $\rho_1 c_1/\rho_2 c_2=1$. The examples of passband width dependence on component sound velocity are given on Figures 4 and 5.
- The width of the first passband inversely proportional to the period length of inhomogeneous medium. For the computations presented on Figures 2 – 5 the period L was equal to unit.

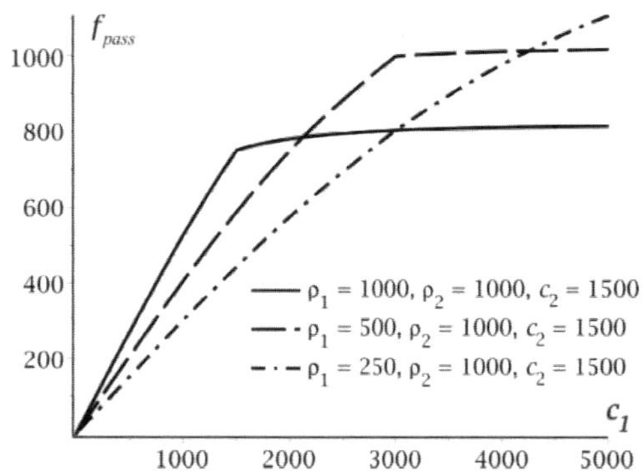

Figure 4. First passband dependence on first component sound velocity.

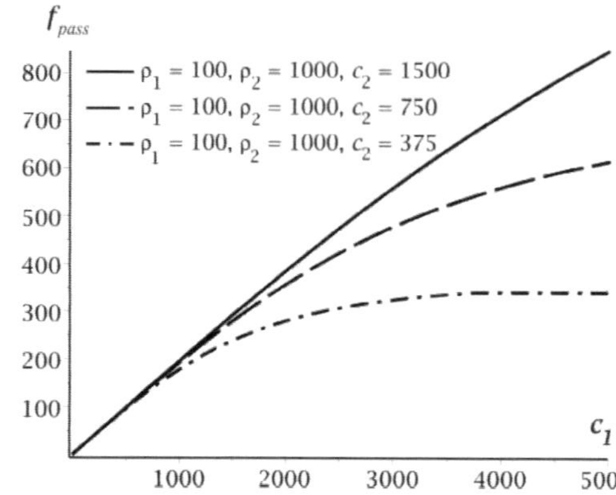

Figure 5. First passband dependence on first component sound velocity.

2.2. Wave transmission from the homogeneous medium into the composite.

The following problem is to evaluate the energy transmission of the wave traveling from homogeneous medium into the one-dimensional periodic composite (Figure 6), described in previous paragraph. As long as there is no energy dissipation, the sum of reflected and transmitted energy flow is equal to the energy flow of the incident wave:

$$W_R + W_T = W_I .$$

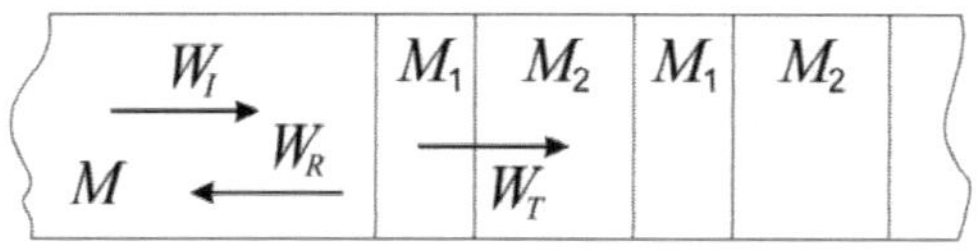

Figure 6. Transmission problem.

Let $p(x) = \exp(\mathrm{i}\lambda x) + R\exp(-\mathrm{i}\lambda x)$ be the acoustic pressure in the homogeneous medium $M=\{\rho,c\}$, $\lambda = \omega/c$. So the value R^2 signifies reflected energy flow, since energy flow of the incident wave is equal to unit. At the boundary of the homogeneous media and composite we have

$$p = p^{(1)}, \quad \frac{p_x}{\rho} = \frac{p_x^{(1)}}{\rho_1}. \tag{9}$$

Since $p^{(1)}(x)$ is the waveguide function (5) we have $B^{(1)} = \tau(\omega)A^{(1)}$. For given ω coefficient $\tau(\omega)$ is determined using dispersion relation (6) and equations (2) and (4). Solution of the system (9) provides

$$R = \frac{\tau(1+\theta)(1+\theta_0)+(1-\theta)(1-\theta_0)}{\tau(1+\theta)(1-\theta_0)+(1-\theta)(1+\theta_0)}, \tag{10}$$

where $\theta_0 = \rho c / \rho_1 c_1$. It should be noticed that R will not change if we use $p^{(2)}$ instead of $p^{(1)}$ under condition that parameters of periodic medium are the same. In other words wave propagation in composite does not depend on the composite orientation.

Results of the computations R^2 allow us to denote some properties of the wave propagation from homogeneous medium into composite. By default the following parameters of the medium were used for the computations: $L=0.01$, Air $\rho = 1.2, c = 340$, Water $\rho = 1000, c = 1400$, Aluminum $\rho = 2700, c = 5500$, Quartz $\rho = 2650, c = 5760$, Polypropylene $\rho = 950, c = 950$, Iron $\rho = 7800, c = 5900$ and k_1=0.5.

Graphs on Figures 7 and 8 shows that the reflection ratio depends on parameters of the composite as well as on parameters of the homogeneous media. The reflected energy flow is increasing while frequency of incident wave grows. When frequency reach passband boundary we obtain full reflection. The minimum of reflection ratio reaches at $\omega \to 0$.

Examples displaying reflection dependence on concentration of the composite components are presented on Figures 9 and 10. It shows that increase of component concentration with smaller acoustic resistance may do not change, decrease or even enlarge reflection ratio.

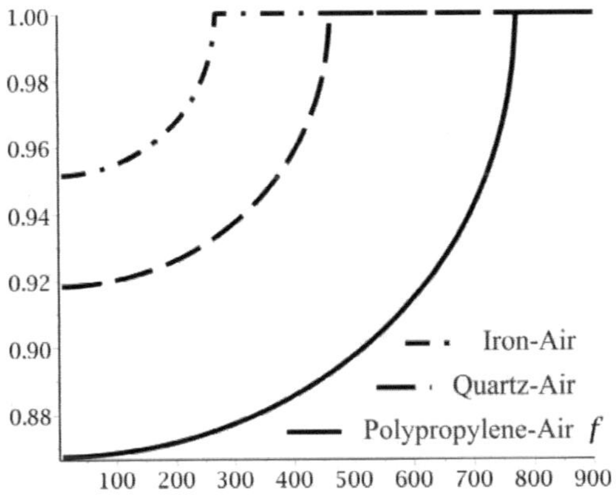

Figure 7. The reflected energy flow dependence on frequency for some composites. Homogeneous media is Air.

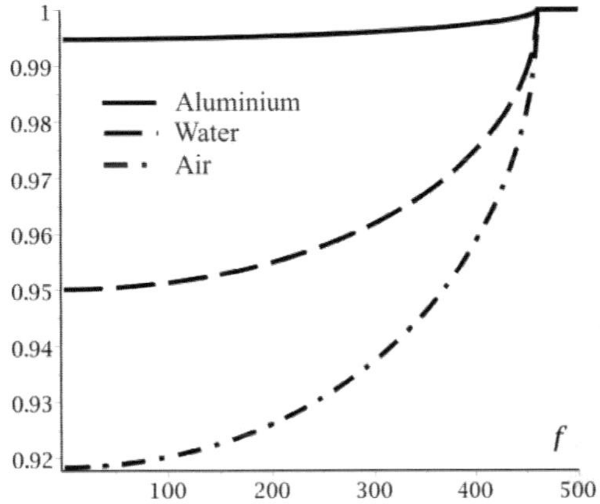

Figure 8. The reflected energy flow dependence on frequency for different homogeneous media. Composite is Air-Quarts, k_1=0.5.

Formula (10) shows that reflection ratio depends on the acoustic resistance of the homogeneous media. Graphs of the minimum of reflected energy ($\omega \to 0$) flow are shown on Figure 11. On this figure we can see that behavior of the reflection ratio is similar to the case of reflection at the boundary of two homogeneous media. In the case of two homogeneous media minimum of the reflection ratio is reached when the media have the same acoustic resistance. So using this analogy we can suggest the acoustic resistance of composite to be equal to the value of homogeneous medium acoustic resistance that corresponds to the minimal value of reflection ratio R

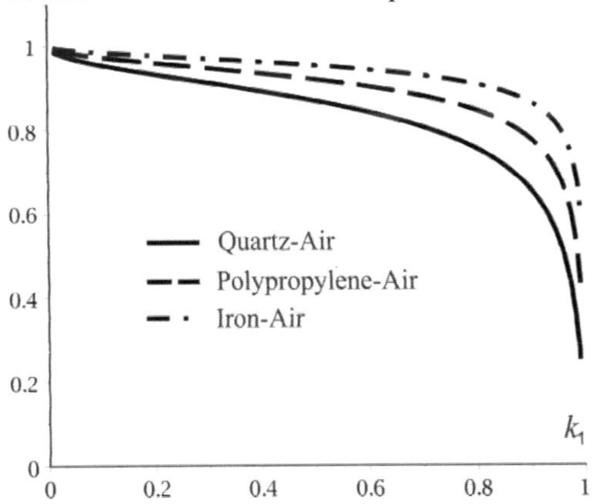

Figure 9. Examples of the concentration influence on reflected energy flow for some composites. Homogeneous media is Air.

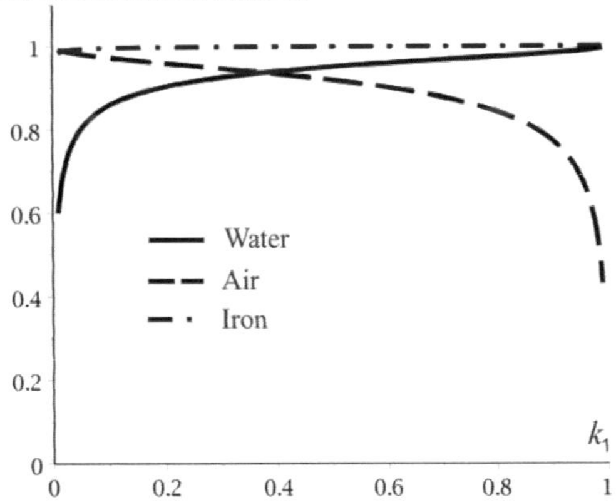

Figure 10. Examples of the concentration influence on reflected energy flow for different homogeneous media. Composite is Quartz-Air.

The minimal point of the reflection ratio R does not depend on frequency, but R at this point grows when frequency increases.

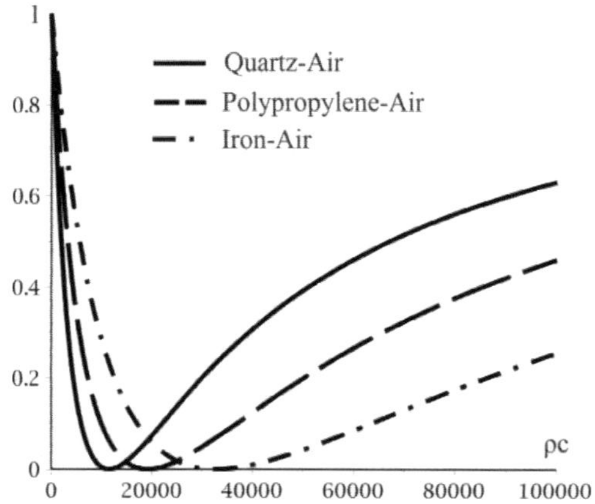

Figure 11. The reflected energy flow dependence on acoustic resistance of homogeneous media.

Acknowledgments
Research is supported by the Russian Science Foundation grant (project no. 17-19-01389).

References

[1] Deymier P A 2013 *Acoustic metamaterials and phononic crystals* (New-York: Springer)
[2] Madeo A and Gavrilyuk S 2010 *Propagation of acoustic waves in porous media and their reflection and transmission at a pure-fluid/porous-medium permeable interface Eur. J. Mech.* A *Solids* **29** p 897–910
[3] Sukhinin S V 2001 *Wave propagation and resonance phenomena in inhomogeneous media J. Appl. Mech. Tech. Phys.* **42** 411–9
[4] Sukhinin S V and Kondratenko D A 1998 *Wave propagation in strongly non homogeneous wave guide Proc. Int. Conf. on MMET (Kharkov)* p 2–5
[5] Brillouin L 1953 *Wave propagartion in periodic structures* (New-York: Dower)

IOP Conf. Series: Journal of Physics: Conf. Series **894** (2017) 012095 doi:10.1088/1742-6596/894/1/012095

Numerical solution of inverse problems in hydrodynamics

D A Tereshko

Institute of Applied Mathematics FEB RAS, Vladivostok, Russia

E-mail: ter@iam.dvo.ru

Abstract. The paper is devoted to inverse problems for the unsteady viscous incompressible fluid flow. In the problems under consideration we estimate the velocity vector on a part of the boundary using velocity measurements in a certain observation subdomain. Optimal control approach reduces this problem to the constrained minimization. For unsteady flow and the discrete measurement data we use discrete optimization methods to solve this problem. New numerical algorithm based on the finite dimensional minimization is proposed and numerical results are discussed.

1. Introduction

Inverse and optimization problems in hydrodynamics have a large number of applications in science and engineering. In inverse problems we must find distributed or boundary sources, coefficients or initial conditions using some information about the state. Usually these problems are formulated as minimization problems for suitable cost functionals and solved by applying a unified approach based on the constrained optimization theory [1]. Boundary control problems for the Navier-Stokes equations were considered in [2, 3, 4]. Optimization problems for heat convection equations were solved numerically in [5, 6, 7, 8, 9]. The same approach can be applied for inverse and parameter estimation problems (see for example [10, 11]).

In this paper we consider an unsteady channel flow of the viscous incompressible fluid over a forward facing step (see Figure 1). We assume that flow rate on the inlet and the main outlet are not equal. The bottom boundary of the channel contains unknown outlets. In our inverse problem we use velocity measurements in the downstream observation subdomain to find these outlets and estimate the boundary velocity. More precisely, the channel bottom wall has several holes. The positions of these holes are determined by the boundary velocity that was found numerically as the solution of the inverse problem under consideration.

To solve this problem we propose new numerical algorithm based on the finite dimensional minimization. The algorithm has several advantages over previously used methods. It does not use iterations and the first order necessary optimality conditions. The state and control are calculated simultaneously at the same time step. The vortex reduction problem for heat and electrically conducting fluid with temperature boundary control was solved by similar approach in [12]. In computational experiments we verify the efficiency of the algorithm and estimate the influence of the problem parameters (Reynolds number, size of the observation domain and a number of basis functions) on the accuracy of the inverse problem solution.

Content from this work may be used under the terms of the Creative Commons Attribution 3.0 licence. Any further distribution of this work must maintain attribution to the author(s) and the title of the work, journal citation and DOI.
Published under licence by IOP Publishing Ltd

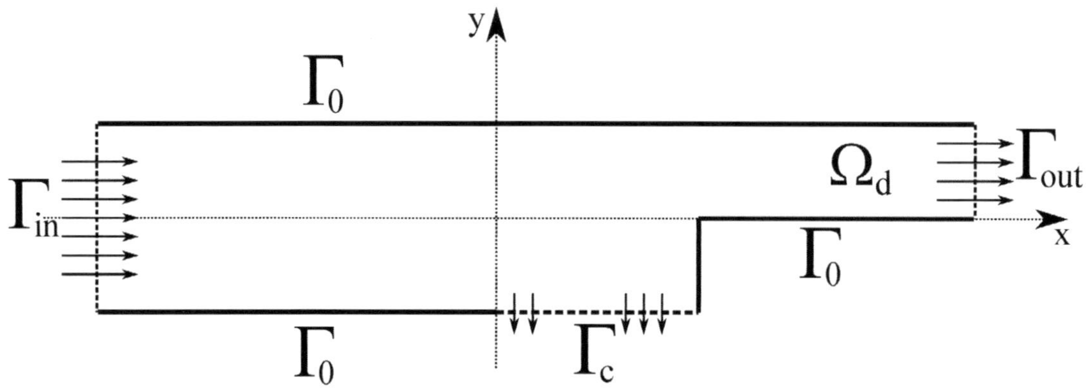

Figure 1. Geometry of the problem.

2. Problem statement

In the bounded two-dimensional domain Ω on the time interval $(0, t_{max})$ we consider the following dimensionless system of partial differential equations

$$\frac{\partial \mathbf{v}}{\partial t} + (\mathbf{v} \cdot \nabla)\mathbf{v} - \frac{1}{\mathrm{Re}}\Delta \mathbf{v} + \nabla p = \mathbf{0} \text{ in } \Omega, \tag{1}$$

$$\operatorname{div} \mathbf{v} = 0 \text{ in } \Omega, \quad \mathbf{v}|_{t=0} = \mathbf{v}_0 \text{ in } \Omega, \tag{2}$$

$$\mathbf{v} = \mathbf{v}_{\mathrm{in}} \text{ on } \Gamma_{\mathrm{in}}, \quad \mathbf{v} = \mathbf{0} \text{ on } \Gamma_0, \quad \mathbf{v} = \mathbf{g} \text{ on } \Gamma_c, \quad p\mathbf{n} - \frac{1}{\mathrm{Re}}\frac{\partial \mathbf{v}}{\partial \mathbf{n}} = \mathbf{0} \text{ on } \Gamma_{\mathrm{out}} \tag{3}$$

describing the unsteady viscous incompressible fluid flow in a channel. Here $\mathbf{v}$ and p are the dimensionless velocity and pressure, $\mathbf{n}$ is the unit outward normal vector. Reynolds number Re is dimensionless parameter of this problem. The boundary Γ consists of four parts: inlet section Γ_{in}, outlet section Γ_{out}, solid walls Γ_0 and control section Γ_c (see Figure 1). For the velocity vector $\mathbf{v}$ we prescribe the inflow parabolic profile $\mathbf{v}_{\mathrm{in}}$ on Γ_{in}, no-slip boundary condition on Γ_0 and "do nothing" boundary condition on the outlet Γ_{out}. The velocity boundary value $\mathbf{g}$ on the control section Γ_c will play the role of control.

3. Time discretization

In our inverse problem we want to obtain the velocity field $\mathbf{v}$ close to a given vector field $\mathbf{v}_d$ measured in the observation subdomain Ω_d using velocity boundary control $\mathbf{g}$ on Γ_c. We assume that values of the vector $\mathbf{v}_d$ are given only at fixed time points $t_1, t_2, \ldots, t_N = t_{max}$. So we have a discrete set of data and need to solve the discrete minimization problem.

At the beginning we split the time interval $(0, t_{max})$ into N parts (t_{n-1}, t_n) of length $\tau_n = t_n - t_{n-1}$, $n = 1, 2, \ldots, N$ and write the following semidiscrete approximation for the problem (1)–(3):

$$\frac{\mathbf{v}^n - \mathbf{v}^{n-1}}{\tau_n} + (\mathbf{v}^{n-1} \cdot \nabla)\mathbf{v}^n - \frac{1}{\mathrm{Re}}\Delta \mathbf{v}^n + \nabla p^n = \mathbf{0} \text{ in } \Omega, \tag{4}$$

$$\operatorname{div} \mathbf{v}^n = 0 \text{ in } \Omega, \quad \mathbf{v}^0 = \mathbf{v}_0 \text{ in } \Omega, \tag{5}$$

$$\mathbf{v}^n = \mathbf{v}_{\mathrm{in}}(t_n) \text{ on } \Gamma_{\mathrm{in}}, \quad \mathbf{v}^n = \mathbf{0} \text{ on } \Gamma_0, \quad \mathbf{v} = \mathbf{g}(t_n) \text{ on } \Gamma_c, \quad p^n\mathbf{n} - \frac{1}{\mathrm{Re}}\frac{\partial \mathbf{v}^n}{\partial \mathbf{n}} = \mathbf{0} \text{ on } \Gamma_{\mathrm{out}}. \tag{6}$$

Here $\mathbf{v}^n$, p^n $(n = 1, 2, \dots, N)$ are the new unknown functions that depend only on the space variables x, y. Let us note that we use an implicit difference scheme for time discretization. It ensures the stability of our numerical solutions even for large time intervals.

Multiplying equations in (4), (5) by the corresponding test functions, integrating over Ω and using Green's formula for certain terms we obtain a variational formulation of the problem (4)–(6). It consists in finding functions $\mathbf{v}^n$, p^n $(n = 1, 2, \dots, N)$ satisfying equations

$$\frac{(\mathbf{v}^n - \mathbf{v}^{n-1}, \mathbf{w})}{\tau_n} + ((\mathbf{v}^{n-1} \cdot \nabla)\mathbf{v}^n, \mathbf{w}) + \frac{1}{\mathrm{Re}}(\nabla \mathbf{v}^n, \nabla \mathbf{w}) - (p^n, \operatorname{div} \mathbf{w}) = 0 \ \ \forall \mathbf{w} \in \mathbf{W}, \tag{7}$$

$$(\chi, \operatorname{div} \mathbf{v}^n) = 0 \ \ \forall \chi \in X, \tag{8}$$

$$\mathbf{v}^n = \mathbf{v}_{\text{in}}^n \text{ on } \Gamma_{\text{in}}, \ \ \mathbf{v}^n = \mathbf{0} \text{ on } \Gamma_0, \ \ \mathbf{v}^n = \mathbf{g}^n \text{ on } \Gamma_c. \tag{9}$$

Here and below $\mathbf{W}$ and X are the test function spaces; the function $\mathbf{v}^0 = \mathbf{v}_0$ is determined by the initial condition in (2); $\mathbf{v}_{\text{in}}^n = \mathbf{v}_{\text{in}}(t_n)$, $\mathbf{g}^n = \mathbf{g}(t_n)$; $(\cdot, \cdot)$ and $(\cdot, \cdot)_{\Gamma_c}$ are the scalar products in $L^2(\Omega)$ and $L^2(\Gamma_c)$.

If we already know all values at time $t = t_{n-1}$ then this problem is linear for unknowns $(\mathbf{v}^n, p^n)$. We will use this linearity in the construction of our numerical algorithm. Applying the finite element method for space discretization we can solve the problem (7)–(9) numerically.

4. Discrete minimization

Let $\Omega_d \subset \Omega$ be the subset of Ω. It will play the role of the observation subdomain in our control problem for the system (7)–(9). At time step $t = t_n$ the problem of finding the boundary function $\mathbf{g}^n$ for the vector field $\mathbf{v}_d^n = \mathbf{v}_d(t_n)$ is reduced to the minimization of the quality functional

$$J(\mathbf{v}^n, \mathbf{g}^n) = \frac{1}{2} \int\limits_{\Omega_d} |\mathbf{v}^n - \mathbf{v}_d^n|^2 d\Omega + \frac{\mu}{2} \int\limits_{\Gamma_c} |\mathbf{g}^n|^2 d\Gamma, \tag{10}$$

that depends on the function $\mathbf{g}^n$ and corresponding velocity vector $\mathbf{v}^n$. Here $\mu = \text{const} \geq 0$ is the small regularization parameter.

The following method is based on the main idea of [2] where an optimal boundary control problem for the stationary Navier-Stokes equations was solved using the finite dimensional minimization approach. Here we develop an algorithm for the nonstationary problem.

We write the unknown control function $\mathbf{g}^n$ in the following form:

$$\mathbf{g}^n = \sum_{i=1}^{M} k_i \mathbf{g}_i \tag{11}$$

where $\{\mathbf{g}_i\}_{i=1}^M$ are the given basic functions on Γ_c while k_i, $i = 1, 2, \dots, M$ are the unknown coefficients that must be found at each time step. The corresponding velocity vector then can be written as

$$\mathbf{v}^n = \mathbf{w}^n + \sum_{i=1}^{M} k_i \mathbf{v}_i^n. \tag{12}$$

Here $\mathbf{w}^n$ is the solution of the problem (7)–(9) for $\mathbf{g}^n = \mathbf{0}$ in (9) while $\mathbf{v}_i^n$ $(i = 1, \dots, M)$ is the solution of the problem

$$\frac{(\mathbf{v}_i^n, \mathbf{w})}{\tau_n} + ((\mathbf{v}^{n-1} \cdot \nabla)\mathbf{v}_i^n, \mathbf{w}) + \frac{1}{\mathrm{Re}}(\nabla \mathbf{v}_i^n, \nabla \mathbf{w}) - (p_i^n, \operatorname{div} \mathbf{w}) = 0 \ \ \forall \mathbf{w} \in \mathbf{W},$$

$$(\chi, \operatorname{div} \mathbf{v}_i^n) = 0 \ \ \forall \chi \in X,$$

$$\mathbf{v}_i^n = \mathbf{0} \text{ on } \Gamma_{\text{in}}, \quad \mathbf{v}_i^n = \mathbf{0} \text{ on } \Gamma_0, \quad \mathbf{v}_i^n = \mathbf{g}_i \text{ on } \Gamma_c.$$

Then the functional (10) can be written as

$$J(\mathbf{v}^n, \mathbf{g}^n) = \frac{1}{2}\int_{\Omega_d} |\mathbf{v}^n - \mathbf{v}_d^n|^2 d\Omega + \frac{\mu}{2}\int_{\Gamma_c} |\mathbf{g}^n|^2 d\Gamma = \frac{1}{2}\int_{\Omega_d} \left| \mathbf{w}^n - \mathbf{v}_d^n + \sum_{i=1}^{M} k_i \mathbf{v}_i^n \right|^2 d\Omega$$

$$+\frac{\mu}{2}\int_{\Gamma_c} \left| \sum_{i=1}^{M} k_i \mathbf{g}_i \right|^2 d\Gamma = \frac{1}{2}\sum_{i=1}^{M}\sum_{j=1}^{M} a_{ij} k_i k_j - \sum_{j=1}^{M} b_j k_j + \frac{1}{2}c.$$

Here the coefficients

$$a_{ij} = (\mathbf{v}_i^n, \mathbf{v}_j^n)_{\Omega_d} + \mu(\mathbf{g}_i, \mathbf{g}_j)_{\Gamma_c}, \quad b_j = (\mathbf{v}_d^n - \mathbf{w}^n, \mathbf{v}_j^n)_{\Omega_d}, \quad c = \|\mathbf{v}_d^n - \mathbf{w}^n\|_{\Omega_d}^2 \tag{13}$$

can be easily calculated by known functions $\mathbf{g}_i$, $\mathbf{v}_i^n$, $\mathbf{v}_d^n$ and $\mathbf{w}^n$.

As a result, we have the finite dimensional minimization problem for variables k_i, $i = 1, 2, \ldots, M$. Solution of this problem can be found by solving the following system of linear algebraic equations:

$$\sum_{i=1}^{M} a_{ij} k_i = b_j, \quad j = 1, 2, \ldots, M. \tag{14}$$

Substituting a_{ij} and b_j in this system and using (12) we obtain the following equalities

$$(\mathbf{v}^n - \mathbf{v}_d^n, \mathbf{v}_j^n)_{\Omega_d} + \mu(\mathbf{g}^n, \mathbf{g}_j)_{\Gamma_c} = 0, \quad j = 1, 2, \ldots, M.$$

that have a clear meaning. If $\mu = 0$ then the difference $\mathbf{v}^n - \mathbf{v}_d^n$ must be orthogonal to all functions $\mathbf{v}_j^n$ in the observation subdomain Ω_d. Solving the system (14) and substituting k_i in (11) we find the boundary control $\mathbf{g}^n$ on the n-th time step.

5. Numerical algorithm

Proposed numerical algorithm can be written as follows.

Step 0. Choose M basic functions $\mathbf{g}_i$ on Γ_c. Set initial value $\mathbf{v}^0$. Set $n = 1$.

Step 1. Assuming that $\mathbf{v}^{n-1}$ is already known, solve $M+1$ linearized boundary value problems to find $\mathbf{v}_i^n$, $i = 1, \ldots, M$ and $\mathbf{w}^n$.

Step 2. Calculate a_{ij} and b_j by (13). Solve linear system (14) to find k_i, $i = 1, \ldots, M$.

Step 3. Calculate $\mathbf{g}^n$ by (11). Solve linear boundary value problem (7)–(9) to find $\mathbf{v}^n$, p^n.

Step 4. If $n < N$ then set $n := n + 1$ and go to step 1.

Let us note that this algorithm does not use the first order necessary optimality conditions (see [5, 6, 7]) and can be simply implemented. It can be efficiently parallelized because $M + 1$ boundary value problems to find $\mathbf{w}^n$ and $\mathbf{v}_i^n$, $i = 1, \ldots, M$ are solved independently. It must be noted that at each time step $t = t_n$ we find the control $\mathbf{g}^n$ and the state $(\mathbf{v}^n, p^p)$ simultaneously. A similar idea was used to optimize the viscous heat-conducting fluid flow by optimal heating or cooling of the boundary in [9].

6. Computational results

Now let us discuss some computational results for the inverse problem under consideration. In this problem we find the boundary velocity $\mathbf{g}$ on the control boundary section Γ_c to create the velocity $\mathbf{v}$ closed to the given vector field $\mathbf{v}_d$ in the subdomain Ω_d. The initial boundary value problem (1)–(3) is considered in the channel with a forward facing step (see Figure 1). All quantities are dimensionless. The size of the channel is determined by values $|x| < 4$, $|y| < 1$. The

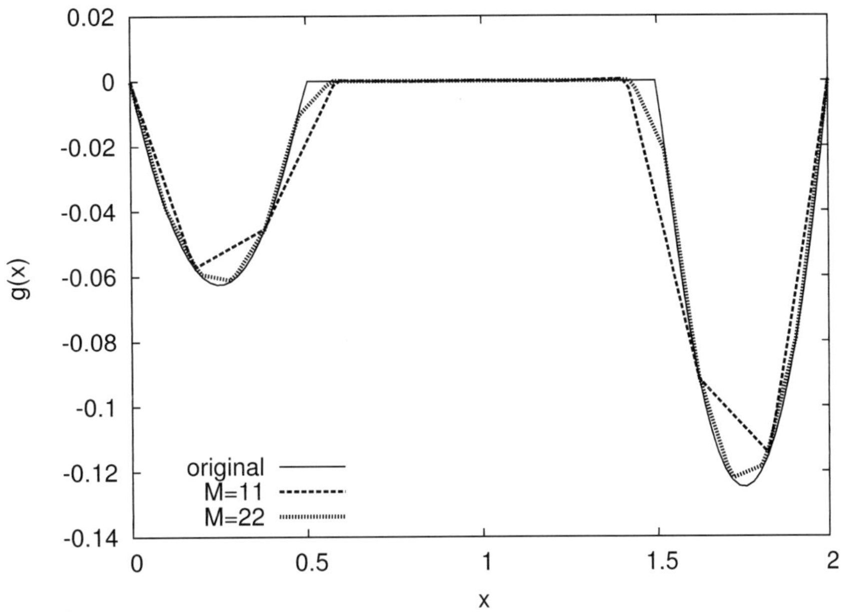

Figure 2. Original and estimated boundary velocity $g(x)$ on Γ_c.

step is placed at $x = 2$ and $\Gamma_c = \{(x, y) : 0 \leq x \leq 2, y = -1\}$. The dimensionless time interval $(0, t_{max})$ was chosen as $(0, 20)$ with $N = 100$. In these computations we used different Reynolds numbers Re from 50 to 500. The inflow velocity $\mathbf{v}_{\text{in}}$ is prescribed as $\mathbf{v}_{\text{in}} = (1 - y^2, 0)$. The observation subdomain Ω_d is defined by the inequality $x > x_d$ where x_d is a variable parameter of the inverse problem. In order to obtain "the measured velocity" $\mathbf{v}_d$ we solve numerically the initial boundary value problem (1)–(3) for a given boundary velocity $\mathbf{g}_d$ on Γ_c. Then we use this field $\mathbf{v}_d$ in our inverse problem and compare the obtained numerical solutions $\mathbf{g} = (0, g(x))$ and $\mathbf{v} = (v_x, v_y)$ with the original functions $\mathbf{g}_d = (0, g_d(x))$ and $\mathbf{v}_d = (v_{dx}, v_{dy})$. We assume that the boundary Γ_c contains two outlets and boundary velocity $g_d(x)$ is defined as

$$g_d(x) = \begin{cases} x(x - 0.5), & \text{if } 0 \leq x \leq 0.5 \\ 0, & \text{if } 0.5 < x < 1.5 \\ 2(x - 1.5)(x - 2), & \text{if } 1.5 \leq x \leq 2 \end{cases}$$

The open source software freeFEM++ (www.freefem.org) was used to solve the boundary value problems (7)–(9) by the finite element method. The main goal of the computational experiments was to determine the dependence of the solution accuracy on the choice of the problem parameters. The accuracy of the solution depends on the number of basis functions M. We use linear splines as basis functions $g_i(x)$. Original boundary velocity $g_d(x)$ and computed controls $g(x)$ for $M = 11$ and $M = 22$ at time point $t = 20$ with Re $= 100$, $x_d = 0$, $\mu = 10^{-6}$ are shown in Figure 2. It can be seen that this numerical solution has a good agreement with the original boundary velocity and allows us to find the location of the holes in the wall.

In order to estimate the accuracy of the inverse problem solution we use three errors

$$E_g = \|g - g_d\|/\|g_d\|, \quad E_{v_x} = \|v_x - v_{dx}\|/\|v_{dx}\|, \quad E_{v_y} = \|v_y - v_{dy}\|/\|v_{dy}\|.$$

The values of these errors for different numbers of basis functions M and Reynolds numbers Re are given in tables 1 and 2 respectively. It is clearly seen that in order to decrease the errors we must increase the number M and decrease the Reynolds number Re.

Table 1. Error values for different numbers of basis functions M (Re=100, $x_d = 0$).

	$M = 10$	$M = 15$	$M = 20$	$M = 25$	$M = 30$	$M = 35$
$E_g =$	4.1×10^{-3}	3.1×10^{-3}	1.2×10^{-3}	9.9×10^{-4}	8.7×10^{-4}	7.9×10^{-4}
$E_{v_x} =$	2.2×10^{-4}	1.7×10^{-4}	1.4×10^{-4}	1.1×10^{-4}	5.6×10^{-5}	4.1×10^{-5}
$E_{v_y} =$	1.2×10^{-4}	7.6×10^{-5}	5.2×10^{-5}	4.1×10^{-5}	2.6×10^{-5}	1.9×10^{-5}

Table 2. Error values for different Reynolds numbers Re ($M = 20$, $x_d = 0$).

	Re=50	Re=100	Re=200	Re=300	Re=400	Re=500
$E_g =$	1.2×10^{-3}	1.2×10^{-3}	1.3×10^{-3}	1.6×10^{-3}	1.9×10^{-3}	2.3×10^{-3}
$E_{v_x} =$	1.2×10^{-4}	1.4×10^{-4}	1.9×10^{-4}	4.1×10^{-4}	8.4×10^{-4}	1.3×10^{-3}
$E_{v_y} =$	4.6×10^{-5}	5.2×10^{-5}	8.8×10^{-5}	1.6×10^{-4}	2.6×10^{-4}	3.7×10^{-4}

Numerical experiments for different values of x_d show that the solution of inverse problem is more accurate if the observation subdomain Ω_d is located closer to the control boundary Γ_c. Based on the analysis of computational results we can choose optimal values of problem parameters and develop some recommendations for future applications.

Acknowledgments

This work was supported by the Russian Foundation for Basic Research (project no. 16-01-00365-a) and by the Far East Basic Research program (project no. 15-I-4-036).

References

[1] Gunzburger M D 2003 *Perspectives in Flow Control and Optimization* (Philadelphia: SIAM)
[2] Alekseev G V and Malikin V V 1994 *Numerical analysis of optimal boundary control problems for Navier-Stokes equations Comp. Fluid Dynamics J.* **3** 1–26
[3] De Los Reyes J C and Kunisch K 2005 *A semi-smooth Newton method for control constrained boundary optimal control of the Navier-Stokes equations Nonlinear Anal. Theor. Meth. Appl.* **62** 1289-316
[4] Ravindran S S 2017 *Finite element approximation of Dirichlet control using boundary penalty method for unsteady Navier-Stokes equations ESAIM: M2AN* **51** 825–49
[5] Lee H-C 2009 *Optimal control problems for the two dimensional Rayleigh-Benard type convection by a gradient method Japan J. Indust. Appl. Math.* **26** 93–121
[6] Alekseev G V and Tereshko D A 2010 *Extremum problems of boundary control for steady equations of thermal convection J. Appl. Mech. Tech. Phys.* **51** 510–20
[7] Alekseev G V and Tereshko D A 2011 *Two-parameter extremum problems of boundary control for stationary thermal convection equations Comp. Math. Math. Phys.* **51** 1539–57
[8] Alekseev G V and Tereshko D A 2013 *Control problems for heat-conducting viscous fluid flow in manufacturing processes Appl. Mech. Mater.* **372** 373–6
[9] Tereshko D 2016 *Discrete optimization of unsteady fluid flows CEUR Workshop Proc.* **1623** 293–302
[10] Wong J C-F and Protas B 2013 *Application of scaled nonlinear conjugate-gradient algorithms to the inverse natural convection problem Optim. Methods Software* **28** 159–85
[11] Tereshko D A 2015 *Numerical reconstruction of the boundary heat flux for the stationary equations of heat convection J. Appl. Indust. Math.* **9** 132–40
[12] Tereshko D 2016 *Control problem for unsteady magnetohydrodynamic flow of viscous heat conducting fluid Key Engineer. Mat.* **685** 23–6

Analogue of the Poiseuille flow for incompressible polymeric fluid with volume charge. Asymptotics of the linearized problem spectrum

A M Blokhin[1,2] **and D L Tkachev**[1,2]
[1]Sobolev Institute of Mathematics, Novosibirsk, Russia
[2]Novosibirsk State University, Novosibirsk, Russia

E-mail: `blokhin@math.nsc.ru`, `tkachev@math.nsc.ru`

Abstract. We deduce the formula of the asymptotics of the spectrum of the linear problem on the stability of stationary flows of an incompressible polymeric fluid with volume charge.

1. Introduction

In [1] stationary solutions of the electrohydrodynamic equations describing nonisothermic flows of an incompressible polymeric fluid with volume charge were studied. These stationary solutions are counterparts of the Poiseuille and Couette flows for the Navier-Stokes equations.

In this work we study equations derived by linearizing the system from [1] with respect to a basic solution describing stationary counterparts of Poiseuille flows. Our main achievement is the formula describing the asymptotics of the spectrum of this linear problem.

2. Preliminaries and main result

In [1] a mathematical model describing nonisothermic flow of an incompressible weakly conducting polymeric fluid with negative volume charge was proposed (for the description of polymeric media one used a new rheological model, see [2]). Formulation of the model in [1] was based on, for example, the results of [3], [4], [5] obtained for an inviscid incompressible fluid. In Cartesian coordinate system x, y and in a dimensionless form (the process of obtaining this form is described in detail in [1]) this mathematical model can be written as follows:

$$u_x + v_y = 0, \tag{1}$$

$$\frac{du}{dt} + P_x = \frac{1}{Re}\{(Za_{11})_x + (Za_{12})_y\} - \sigma q \Phi_x, \tag{2}$$

$$\frac{dv}{dt} + P_y = \frac{1}{Re}\{(Za_{12})_x + (Za_{22})_y\} + Ga\frac{Z-1}{\bar{\theta}} - \sigma q \Phi_y, \tag{3}$$

$$\frac{da_{11}}{dt} - 2A_1 u_x - 2a_{12}u_y + \frac{L_{11}}{\bar{\tau}_0(Z)} = 0, \tag{4}$$

$$\frac{da_{12}}{dt} - A_1 v_x - A_2 u_y + \tilde{K}_I\frac{a_{12}}{\bar{\tau}_0(Z)} = 0, \tag{5}$$

Content from this work may be used under the terms of the Creative Commons Attribution 3.0 licence. Any further distribution of this work must maintain attribution to the author(s) and the title of the work, journal citation and DOI.
Published under licence by IOP Publishing Ltd

$$\frac{da_{22}}{dt} - 2A_2 v_y - 2a_{12}v_x + \frac{L_{22}}{\bar{\tau}_0(Z)} = 0, \tag{6}$$

$$\frac{dZ}{dt} = \frac{1}{Pr}\Delta_{x,y} Z, \tag{7}$$

$$\frac{dq}{dt} = b(q^2 - (\nabla\Phi, \nabla q)), \tag{8}$$

$$\Delta_{x,y}\Phi = -q. \tag{9}$$

Here t is the time, u, v are the components of the velocity vector $\vec{\mathbf{u}}$, P is the pressure, a_{ij}, $i, j = 1, 2$ are the components of the symmetric anisotropy tensor of second kind, $Z = 1 + \bar{\theta}\nu$, $\bar{\theta} = \frac{\theta}{T_0}$, $\nu = \frac{T - T_0}{\theta}$ is the dimensionless deviation of the temperature T from the average temperature T_0, $\bar{\theta}$ is a constant having the same dimension as the temperature and described below, $L_{ii} = K_I a_{ii} + \beta(a_{ii}^2 + a_{12}^2)$, $i = 1, 2$; k, β $(0 < \beta < 1)$ are the phenomenological parameters of the rheological model (see [2]), $K_I = W^{-1} + \frac{\bar{k}}{3}I, \bar{k} = k - \beta, \tilde{K}_I = K_I + \beta I$, $I = a_{11} + a_{22}$ is the first invariant of anisotropy tensor, $A_i = a_{ii} + W^{-1}$, $i = 1, 2$, $\bar{\tau}_0(Z) = \frac{1}{Z\exp(\bar{E}_A\frac{Z-1}{Z})} = \frac{1}{\bar{\chi}_0(Z)}$, q is the density of the negative volume charge, Φ is potential of electric field, $\Delta_{x,y} = \frac{\partial^2}{\partial x^2} + \frac{\partial^2}{\partial y^2}$, $\frac{d}{dt} = \frac{\partial}{\partial t} + (\vec{\mathbf{u}}, \nabla)$, $Ga = \frac{Ra}{Pr}$ is the Grashof number, $\bar{E}_A = \frac{E_A}{T_0}$, the constants Re (the Reynolds number), W (the Weissenberg number), σ (the heat conductivity coefficient), b (the mobility coefficient), Ra (the Rayleigh number), Pr(the Prandtl number) and E_A (the activation energy) are described in detail in [1]– [5].

The independent and dependant variables t, x, y, u, v, P, a_{11}, a_{12}, a_{22}, q, Φ in system (1)–(9) are scaled to $\frac{l}{u_H}$, l, u_H, ρu_H^2, $\frac{W}{3}$, $\varepsilon_0\varepsilon\frac{V_0}{l^2}$, V_0, where l is a characteristic length, u_H is a characteristic velocity, $\rho(=\text{const})$ is the density, ε_0 is the electric constant, ε is the dielectric penetration, and the parameter V_0 is described below.

As in [1], our main problem is that for finding solutions of the mathematical model (1)–(9) describing a flow of a weakly conducting polymeric fluid in a plane horizontal condenser. The condenser (channel) of width l is limited by horizontal walls or, in other words, by electrodes (see Figure 1) which are maintained under a constant temperature difference θ and a constant potential difference V_0 applied to the electrodes (the cathode $y = 0$ and the anode $y = 1$ (l)). In this case system (1)–(9) should be supplemented with the following boundary conditions (see [1]– [5]):

$$\vec{\mathbf{u}} = 0,\ \nu = 1,\ \Phi = 0,\ q = -A\Phi_y,\ A > 0 \text{ on the cathode } y = 0, \tag{10}$$

$$\vec{\mathbf{u}} = 0,\ \Phi = 1,\ \nu = 0 \text{ on the anode } y = 1. \tag{11}$$

The injected charge on the cathode $(y = 0)$ is proportional to the magnitude of the electric field, and the injection is assumed to be weak (the positive coefficient $A << 1$).

We will below study the linear boundary value problem obtained resulting from the linearization of the original problem (1)–(11) with respect to the stationary solution u. This stationary solution is described in detail in [1] and for the case $k = \beta(\bar{k} = 0)$ has the following form:

$$u(t, x, y) = \hat{u}(y) = \frac{\hat{D}}{\bar{\theta}}\int_0^y F(s, \bar{C})(\hat{Z}(s) + \bar{C})ds,$$

$$v(t, x, y) = \hat{v}(y) = 0,$$

$$P(t, x, y) = \hat{P}(y) + P_0 - \hat{A}x, \quad \hat{P}(\frac{1}{2}) = 0,$$

$$\hat{P}(y) = \frac{1}{Re}(\hat{Z}(y)\hat{a}_{22}(y) - \hat{Z}(\frac{1}{2})\hat{a}_{22}(\frac{1}{2})) + Ga(y - \frac{y^2}{2} - \frac{3}{8}) + \sigma A\alpha^2(y - \frac{1}{2}),$$

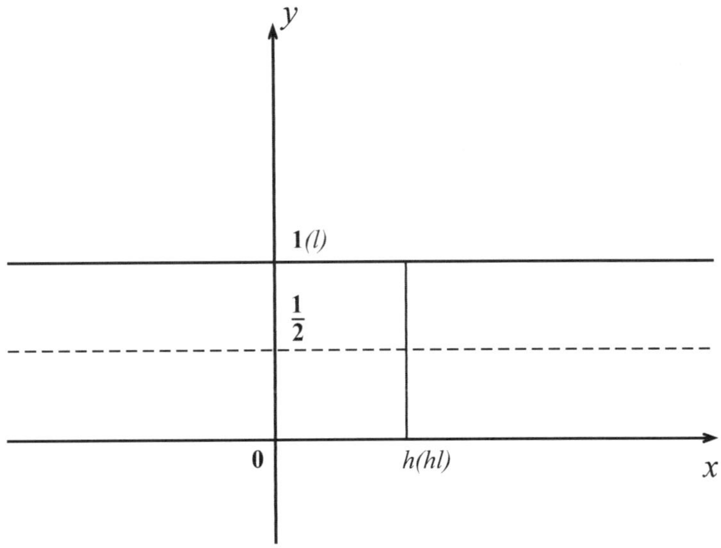

Figure 1. Plane condenser.

$$a_{11}(t,x,y) = \hat{a}_{11}(y) = \hat{a}_{22}(y) + \frac{2\hat{g}(y)}{\hat{A}_2(y)}, \quad \hat{g} = \hat{a}_{12}^2, \quad \hat{A}_2 = W^{-1} + \hat{a}_{22},$$

$$a_{22}(t,x,y) = \hat{a}_{22}(y) = \frac{-1 + \sqrt{1 - \hat{G}(y)}}{2\tilde{W}}, \quad \hat{G} = 4\tilde{W}^2\hat{q}, \quad \tilde{W} = W\beta,$$

$$a_{12}(t,x,y) = \hat{a}_{12}(y) = \frac{\hat{D}(\bar{C} + \hat{Z}(y))}{\bar{\theta}\hat{Z}(y)},$$

$$Z(t,x,y) = \hat{Z}(y) = 1 + \bar{\theta}(1-y),$$

$$q(t,x,y) = \hat{q}(y) = -\frac{3A^2}{(1+2Ay)^{\frac{1}{2}}((1+2A)^{\frac{3}{2}} - 1)},$$

$$\Phi(t,x,y) = \hat{\Phi}(y) = \frac{(1+2Ay)^{\frac{3}{2}} - 1}{(1+2A)^{\frac{3}{2}} - 1}, \text{ where} \tag{12}$$

$$F(y,\bar{C}) = \frac{\exp(\bar{E}_A \frac{\hat{Z}(y)-1}{\hat{Z}(y)})(1 + 2\tilde{W}(\hat{a}_{22}(y) + \frac{\hat{g}(y)}{\hat{A}_2(y)}))}{1 + W\hat{a}_{22}(y)},$$

$\hat{D} = Re\hat{A}$; $-\hat{A}$ ($\hat{A} = \frac{\Delta\hat{P}}{\rho u_H^2 h}$ is the positive constant and the dimensional value $\Delta\hat{P} > 0$) is the dimensionless drop of pressure on the segment h (see Figure 1), $\hat{P}_0$ ($=$ const > 0) is the pressure on the channel's axis $y = \frac{1}{2}$ for $x = 0$, $\alpha = \hat{\Phi}'(0) = \frac{3A}{(1+2A)^{\frac{3}{2}} - 1}$.

The constant $\bar{C}$ can be found from the nonlinear equation

$$\int_0^1 F(s,\bar{C})ds\bar{C} + \int_0^1 F(s,\bar{C})\hat{Z}(s)ds = 0.$$

Remark 2.1 Note that in (12) we can assign $\hat{D} = 1$ thanks to the choice of the characteristic parameter $u_H = \frac{l\Delta\hat{P}}{h\eta_0^*}$.

The influence of the parameters $\bar{\theta}$ ($\bar{\theta} > 0$ is the heating from below and $\bar{\theta} < 0$ is the heating from above), and $\bar{E}_A$ is shown in Figures 2 and 3. Moreover, in Figure 2 one has $\bar{k} = 0$, $\hat{D} = 1$, $W = 1$, $\beta = 0.5$, $\hat{E}_A = 0.01$, $\bar{\theta} = -0.4, 0, 0.4, 8$ whereas in Figure 3 $\bar{k} = 0$, $\hat{D} = 1$, $W = 1$, $\beta = 0.5$, $\bar{\theta} = 2$, $\bar{E}_A = 0.01, 1, 5$.

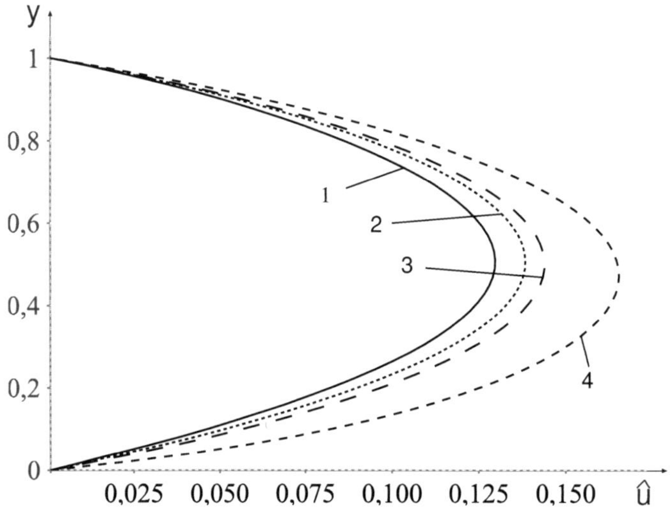

Figure 2. Curves: $1 \to \bar{\theta} = 8$, $2 \to \bar{\theta} = 0.4$, $3 \to \bar{\theta} = 0$, $4 \to \bar{\theta} = -0.4$.

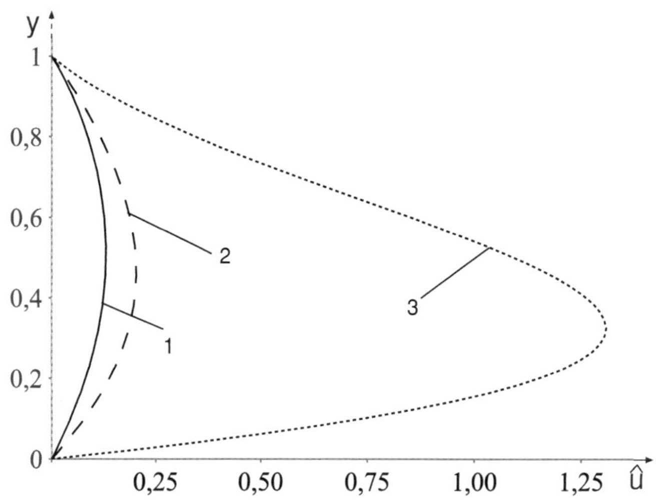

Figure 3. Curves: $1 \to \bar{E}_A = 0.01$, $2 \to \bar{E}_A = 1$, $3 \to \bar{E}_A = 5$.

Denoting small perturbations of the original values by capital letters and linearizing the original problem (1)–(11), we get

$$\vec{V}_t + \tilde{B}\vec{V}_x + \tilde{C}\vec{V}_y + R\vec{V} + \vec{\Gamma} = 0, \tag{13}$$

$$\Delta_{x,y}\Omega = \frac{1}{Re}\{\hat{Z}(a_{11} - a_{22})_{xx} + \frac{2\bar{\theta}\hat{q}}{\hat{A}_2}X_{xx} + 2(\hat{Z}a_{12} + \hat{a}_{12}\bar{\theta}X)_{xy}\} - 2\hat{u}'v_x + GaX_y + \sigma M_y, \tag{14}$$

$$X_t + \hat{u}X_x = \frac{1}{Pr}\Delta_{x,y}X + v, \tag{15}$$

$$Q_t + \hat{u}Q_x + b(\hat{\Phi}'Q_y + \hat{q}Q + \Phi_y) + v = 0, \tag{16}$$

$$\Delta_{x,y}\Phi = -\hat{q}'Q, \quad t > 0, \quad x \in R^1, \quad 0 < y < 1. \tag{17}$$

Here $\vec{V} = (u, v, a_{11}, a_{12}, a_{22})^T$, $\tilde{B} = \hat{u}I_5 + \tilde{B}_0$, I_5 is the unit matrix of fifth order, $\vec{\Gamma} = (\Gamma_1, \Gamma_2, \Gamma_3, \Gamma_4, 0)^T$, $X = \frac{Z}{\theta}$, $Q = \frac{q}{\hat{q}'}$, $\Omega = P - \frac{(\hat{Z}a_{22} + \hat{a}_{22}\bar{\theta}X)}{Re} + \sigma\hat{q}\Phi$, $M = \hat{q}'(\Phi - Q\hat{\Phi}')$,

$$\tilde{B}_0 = \left(\begin{array}{cc|ccc} \multicolumn{2}{c|}{O_2} & -\alpha_0^2 & 0 & \alpha_0^2 \\ & & 0 & -\alpha_0^2 & 0 \\ \hline -2\hat{A}_1 & 0 & & & \\ 0 & -\hat{A}_1 & & O_3 & \\ 0 & -2\hat{a}_{12} & & & \end{array}\right), \quad \tilde{C} = \left(\begin{array}{cc|ccc} \multicolumn{2}{c|}{O_2} & 0 & -\alpha_0^2 & 0 \\ & & 0 & 0 & 0 \\ \hline -2\hat{a}_{12} & 0 & & & \\ -\hat{A}_2 & 0 & & O_3 & \\ 0 & -2\hat{A}_1 & & & \end{array}\right),$$

$$R = \left(\begin{array}{cc|ccc} 0 & \hat{u} & 0 & \frac{\bar{\theta}}{Re} & 0 \\ 0 & 0 & 0 & 0 & 0 \\ \hline 0 & \hat{a}_{11}' & R_{33} & R_{34} & 0 \\ 0 & \hat{a}_{12}' & R_{43} & R_{44} & R_{45} \\ 0 & \hat{a}_{22}' & 0 & R_{54} & R_{55} \end{array}\right);$$

O_2 and O_3 are zero matrices of second and third order respectively,
$\alpha_0^2 = \frac{\hat{Z}(y)}{Re}$, $R_{33} = \bar{\chi}_0(\hat{Z})(W^{-1} + 2\beta\hat{a}_{11})$, $R_{44} = \bar{\chi}_0(\hat{Z})(W^{-1} + \beta\hat{I})$,
$R_{55} = \bar{\chi}_0(\hat{Z})(W^{-1} + 2\beta\hat{a}_{22})$, $R_{34} = 2(-\hat{u}' + \beta\bar{\chi}_0(\hat{Z})\hat{a}_{12})$,
$R_{43} = \beta\bar{\chi}_0(\hat{Z})\hat{a}_{12}$, $R_{45} = -\hat{u}' + \beta\bar{\chi}_0(\hat{Z})\hat{a}_{12}$, $R_{54} = 2\beta\bar{\chi}_0(\hat{Z})\hat{a}_{12}$,
$\hat{A}_1 = \hat{a}_{11} + W^{-1}$, $\hat{I} = \hat{a}_{11} + \hat{a}_{22}$,
$\Gamma_1 = \Omega_x - \frac{(\hat{a}_{12}\bar{\theta}X)_y}{Re} + \frac{(\hat{a}_{22} - \hat{a}_{11})\bar{\theta}}{Re}X_x$,
$\Gamma_2 = \Omega_y - \frac{\hat{a}_{12}\bar{\theta}}{Re}X_x - GaX - \sigma M$, $\Gamma_3 = r_1\bar{\theta}X$, $\Gamma_4 = r_2\bar{\theta}X$,
$r_1 = 2\bar{\chi}_0'(\hat{Z})\frac{\hat{a}_{12}(W^{-1} + \beta\hat{I})}{\hat{A}_2}$, $r_2 = \frac{\hat{A}_2 r_1}{2}$, $\bar{\chi}_0'(\hat{Z}) = \bar{\chi}_0(\hat{Z})\frac{\bar{E}_A + \hat{Z}}{\hat{Z}^2}$.

We now supplement system (13)–(17) with the boundary conditions

$$y = 0 \text{ (cathode)}: \; u = X = \Phi = 0, \; Q = -\frac{1}{A\alpha}\Phi_y, \Omega_y = \frac{1 + \bar{\theta}}{Re}(a_{12})_x - \sigma\alpha^2 A^2 Q, \tag{18}$$

$$y = 1 \text{ (anode)}: \; u = X = \Phi = 0, \quad \Omega_y = \frac{(a_{12})_x}{Re} - \sigma\frac{A^2\alpha^2}{1 + 2A}Q \tag{19}$$

and some initial data. Initial data must fulfill equations (14), (17), the boundary conditions (18), (19) as well as the incompressibility condition

$$div\mathbf{\vec{u}} = 0, \quad \mathbf{\vec{u}} = (u, v).$$

We will below search for solutions of the linear problem formulated above in the special form

$$\begin{cases} \vec{V} = \exp(\lambda t + i\omega x)\vec{V}(y), \\ (\Omega, X, Q, \Phi) = \exp(\lambda t + i\omega x)(\Omega^*(y), X^*(y), Q^*(y), \Phi^*(y)), \\ \lambda = \eta + i\xi; \quad \xi, \omega \in R^1 \end{cases} \tag{20}$$

(the hats are omitted below).

Let us assume that

$$\frac{\gamma_0}{\alpha_0} \neq \frac{1}{b\hat{\Phi}}, \quad y \in [0, 1]. \tag{21}$$

We are now in a position to formulate for the obtained spectral problem our main result on the asymptotic behavior of the spectrum points as $|\lambda| \to \infty$ for a fixed parameter ω.

Theorem 2.1 *If condition* (21) *holds, then the eigenvalues of the boundary value problem arising under the search of harmonic (with respect to x) solutions in form* (20) *for problem* (13)–(19) *(with some initial data) have the following asymptotic representation:*

$$\lambda_k = -\frac{1}{2\int_0^1 \frac{\gamma_0}{\alpha_0} d\xi}\left[\int_0^1 (d_{11} - d_{22})d\xi + 2ki\pi\right] + O(\frac{1}{|k|}),\ \gamma_0^2 = \frac{1}{\hat{A}_2},$$

where k is an integer, $|k| \to +\infty$, and the quantities d_{11}, d_{22} are defined as follows

$$d_{11} = \frac{\gamma_0}{\alpha_0}(\hat{u} + \hat{a}_{12}\alpha_0\gamma_0)i\omega - \frac{1}{2Re}\frac{1}{\alpha_0^2}\bar{\theta} + \frac{\gamma_0}{\alpha_0}\frac{R_{43}\hat{a}_{12}}{\hat{A}_2} + \frac{\gamma_0}{\alpha_0}\frac{R_{44}}{2} + \frac{\gamma_0}{\alpha_0}\frac{\hat{A}_2}{2}(\alpha_0\gamma_0)',$$

$$d_{22} = -\frac{\gamma_0}{\alpha_0}(\hat{u} - \hat{a}_{12}\alpha_0\gamma_0)i\omega - \frac{\bar{\theta}}{2Re\alpha_0^2} - \frac{\gamma_0 R_{43}\hat{a}_{12}}{\alpha_0\hat{A}_2} - \frac{R_{44}\gamma_0}{2\alpha_0} + \frac{\hat{A}_2\gamma_0}{\alpha_0}(\alpha_0\gamma_0)'.$$

The formulated result together with full proof will be published in [11].

Acknowledgments

This work is partially supported by RSBR grant 17-01-00791.

References

[1] Blokhin A M and Rudometova A S 2015 *Stationary solutions of equations describing nonisothermic electroconvention in weakly conducting incompressible polymeric fluid Journal of Applied and Industrial Mathematics* **9** 147–56

[2] Altukhov Yu A, Gusev A S and Pishnograi G V 2012 *Introduction Into Mesoscopic Theory of Flowing Polymeric Systems* (Barnaul: Alt. GPA)

[3] Mordvinov A N and Smorodin B L 2012 it Electroconvection for injection from cathode and heating from above Journal of Experimental and Theoretical Physics **114** 870–7

[4] Taraut A V and Smorodin B L 2012 *Electroconvection for autonomic unipolar injection and residua conductivity Journal of Experimental and Theoretical Physics* **115** 361–9

[5] Smorodin B L and Taraut A V 2014 *Dynamics of wave electroconvective flows in modulated field Journal of Experimental and Theoretical Physics* **118** 158–65

[6] Blokhin A M, Yegitov A V, Tkachev D L 2015 *Linear instability of solutions in a mathematical model describing polymer flows in an infinite channel Computational Mathematics and Mathematical Physics* **55** 848–73

[7] Blokhin A M and Tkachev D L 2014 *Linear asymptotic instability of a stationary flow of a polymeric medium in a plane channel in the case of periodic perturbations Journal of Applied and Industrial Mathematics* **8** 467–78

[8] Birkhoff G D 1950 *Collected Mathematical Papers* (New York: AMS)

[9] Brushlinski K V 1959 *On growth of mixed problem solution in case of incompleteness of eigenfunctions Izv. Akad. Nauk SSSR Ser. Mat.* **23** 893–912.

[10] Lomov S A 1981 *Introduction Into the General Theory of Singular Perturbations* (Moscow: Nauka) (in Russian)

[11] Blokhin A M, Tkachev D L, Yegitov A. V. 2017 *Asymptotics of the spectrum of the linear problem on the stability of stationary flows of an incompressible polymeric fluid with volume charge Journal of Computational and Applied Mechanics* (in press)

IOP Conf. Series: Journal of Physics: Conf. Series **894** (2017) 012097 doi:10.1088/1742-6596/894/1/012097

Numerical simulation of classical and rotating detonation waves in methane mixtures

A V Trotsyuk

Lavrentyev Institute of Hydrodynamics, Siberian Branch of the Russian Academy of Science, Novosibirsk 630090, Russia

E-mail: trotsyuk@hydro.nsc.ru

Abstract. A numerical simulations of a two-dimensional multi-front irregular structure of the detonation wave (DW) in methane-based mixtures at normal initial condition have been conducted. The computations have been performed in a wide range of channel height. From the analysis of the flow structure and the number of primary transverse waves in channel, the dominant size of the detonation cell for studied mixtures has been determined. We have simulated the cellular front structure in stoichiometric methane–air and methane–oxygen mixtures, and in a rich (equivalence ratio φ=1.5) methane-air mixture. Based on the fundamental studies of front structure of the classical propagating DW in methane mixtures, numerical simulation of continuous spin detonation of rich (φ=1.2) methane-oxygen mixture has been carried out in the cylindrical detonation chamber (DC) of the rocket-type engine. We studied the global flow structure in DC, and the detailed structure of the front of the continuous rotating DW. Integral characteristics of the detonation process - the distribution of average values of static and total pressure along the length of the DC, and the value of specific impulse have been obtained. The geometric limit of stable existence of rotating DW has been determined.

1. Introduction

One of the most important problems in the theory of dynamic systems is the formation and destruction of ordered gas-dynamic structures in a reactive medium. An example of such a self-organizing system is a multi-front (cellular) DW steadily propagating in a reacting gas mixture. The complex three-dimensional and time-dependent structure of the front of these waves propagating in constant cross-section channels shows some order, whose geometrical parameter is the size of the elementary cell of the DW – a_0. Based on this value, it is possible to determine such parameters of detonation as critical conditions of detonation combustion, the critical energy of initiation detonation propagation limits, etc., i.e. to estimate the detonation hazard of gaseous mixture.

Depending on the chemical composition of the gas, DW in some mixtures have very regular cellular structure, whereas in other mixtures DW shows very irregular chaotic cell structure. A very regular cellular DW structure in hydrogen–oxygen–argon mixtures was studied in our previous 2D numerical simulation [1]. A good agreement was obtained between the numerical results and experimental data on the size a_0 over a wide range of initial pressures and degrees of mixture dilution by argon. Then we have simulated the irregular detonation structure in stoichiometric methane–air [2] and methane–oxygen [3] mixtures.

Content from this work may be used under the terms of the Creative Commons Attribution 3.0 licence. Any further distribution of this work must maintain attribution to the author(s) and the title of the work, journal citation and DOI.
Published under licence by IOP Publishing Ltd

The objectives of this work are further development of the proposed approximate model of chemical kinetics to describe the detonation combustion of rich methane based gaseous mixtures, to perform a numerical simulation of the 2D irregular cellular structure of propagating DW in a rich methane–air mixture, and to study the flow structure in cylindrical DC with continuous rotating DW in a rich methane-oxygen mixture.

2. Governing equations and model of chemical kinetics

The dynamics of the compressible chemically reactive medium is described by the two-dimensional Euler equations.

The chemical reaction in the DW is described according to the two-stage model of the detonation kinetics (induction period and main heat release stage) first proposed in [4]. The duration of the induction period is determined by known empirical formula [5] for methane mixtures. The main heat release starts after the induction period.

Within the framework of this concept, a kinetics model of detonation combustion of methane-based mixtures $CH_4+a_1H_2+a_2O_2+a_3N_2+a_4Ar+a_5H_2O$ has been developed in our paper [6]. It is assumed that during the induction period all the methane molecule is completely decomposed to form CO molecules. We replace the real multistep chemical processes by single *brutto*–reaction of special form [6]. The molar mass of the gas μ in the main heat release phase is described by the kinetic equation [7,8]. The thermal effect of the chemical reaction and thermodynamic parameters of the gas are described by explicit algebraic formulas on density and temperature [8]. Then we adopt proposed model to describe a rich methane mixtures.

The calculations of CJ parameters and Von Neumann spike parameters for methane mixtures showed that this model is highly accurate. The model is consistent with the second law of thermodynamics. Constants of the model have a clear physical meaning and are calculated from the tabulated thermochemical parameters for the given mixture before the two-dimensional numerical calculation of the DW structure. The model is useful for multi-dimensional numerical simulations of detonation processes and is currently the only approximate model kinetics with high accuracy to calculate the irregular structure of the detonation wave.

The system of governing equations was closed by the well-known thermal equation of state for an ideal gas.

3. Numerical method

The resultant systems of equations were solved numerically using the code based on the Godynov-type finite-volume scheme [9] with the fourth-order MUSCL TVD reconstruction [10] and the advanced HLLC algorithm [11] for an approximate solution of the Riemann problem. In implementation of HLLC algorithm for the case of a chemically reacting mixture, the "energy relaxation method" [12] was used. This method eliminates the problem of numerical solution of the Riemann problem for a medium with a complicated nonlinear equation of state (including that with a variable ratio of specific heats). Integration in time was performed with second-order accuracy by using additive semi-implicit Runge-Kutta methods [13]. The time step was determined at each time layer of the solution from the stability condition [9]. In the present simulations, the values of the Courant number were CFL=0.3–0.4. In numerical simulations of propagating 2D detonation wave in straight channel, we use an adaptive moving grids technique in the x direction with local refinement in the vicinity of a leading shock front. The code is parallelized with MPI library using the domain decomposition technique.

4. Simulation of 2D structure of propagating DW in a rich methane–air mixture

The DW propagation in a 2D straight-line channel was studied. The slip conditions (impermeable solid wall) [9] were imposed on the upper, lower, and left boundaries of the channel, and the right boundary was subjected to the condition of an undisturbed initial state of the gas. The DW was initiated near the left closed end of the channel and propagated from left to right along the x axis. The size and energy of the initiation source [1-3] were chosen sufficiently large to ensure a supercritical regime of DW initiation in methane mixture.

The detonation cell size for the methane-based mixture was determined in a procedure similar to that used in the numerical study of stoichiometric hydrogen mixtures [1] with regular cellular structure. The initial height of the channel *H* was chosen rather arbitrarily; usually, it was assumed to be slightly higher than the presumed size of the cell a_0. Further variation of *H* in the calculations affected the number of primary transverse waves (TWs), remaining at the DW front after its initiation and the establishment of a self-sustained Chapman–Jouguet detonation regime, thus, the cell size was determined. The two transverse waves form the detonation cell.

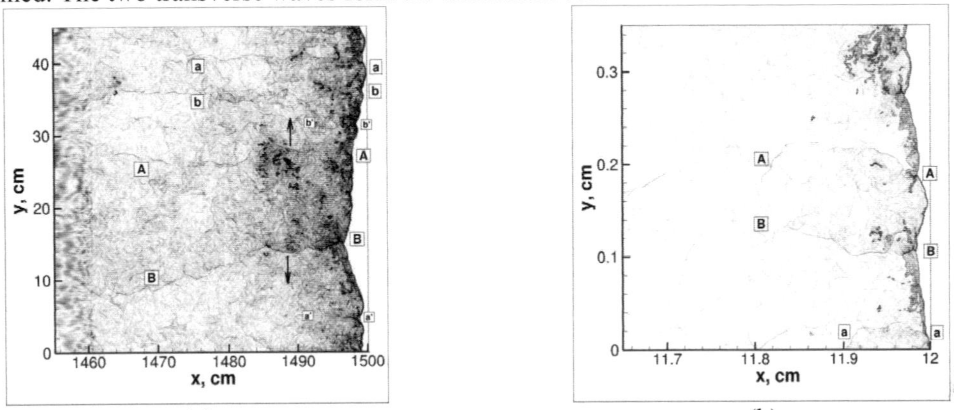

(a) (b)

Figure 1. DW multi-front structure: (a) rich methane-air mixture; (b) stoichiometric methane-oxygen mixture.

According to this method, the dominant size of the detonation cell at normal initial conditions for stoichiometric methane-air mixture determines to be 34±1 cm, and 0.3÷0.35 cm for methane-oxygen mixture [2,3]. These values are in good agreement with all available experimental data.

For a rich methane–air $1.5CH_4+2Air$ mixture at normal initial conditions numerical simulation has been performed for equivalence ratio $\varphi=1.5$. This value of φ corresponds to the mole fraction of methane in mixture $c= 0.136$, which is near the upper (rich) limit for detonability. We have modeled the DW structure in channels with a height H = 40 cm, 45 cm, 50 cm, 60 cm, 65 cm, 70 cm. The total number of numerical cells in the *x* and *y* directions were N_x=2000, N_y=2000 respectively. Calculations with varying *H* showed that the cellular structure of the DW with two main (primary) transverse waves exists in sufficiently wide range of *H*.

The results of the simulation of the rich methane–air mixture are presented in Figure 1a. Figure 1a shows the DW structure in a channel of height H=45 cm with the wave front located at x_{fr}=1500 cm. This is the numerical Schlieren visualization of the flow field. For comparison, the front structure for stoichiometric methane-oxygen mixture in a channel height $H=a_0$=0.35 cm at x_{fr}=12 cm is shown on Fig. 1b. For both value of *H* there is a system of the two main (primary) TWs, denoted by AA and BB, that forms the detonation cell. At showed time, TWs have approximately equal size and intensity, and move in the opposite direction to each other, in antiphase.

Moreover, the DW front has a system of secondary (smaller) TWs, denoted by 'aa', 'bb', etc. of different size, intensity, and direction of motion. There are also the third order smaller TWs 'a′a′', 'b′b′', etc. on leading shock. The movement all these perturbations of the DW leading shock front is rather random.

Figure 2a shows a front-propagation velocity (solid line) as a function of DW front position after the initiation in a channel H=45 cm. Denotation has reached aquasi-steady propagation regime at the last 9÷10 meters. The average DW velocity obtained on this base is D_{aver} =1876 m/s (dashed line). This value is slightly higher than the Chapman–Jouguet velocity (dash-dot line) D_{CJ}=1840 m/s derived from thermodynamic calculations. Approximately the same absolute difference in the velocity of the multifront DW in a self-sustained CJ regime with D_{CJ} was observed for hydrogen-oxygen mixtures with regular cell structure [1] and stoichiometric methane mixtures with irregular structure [2,3].

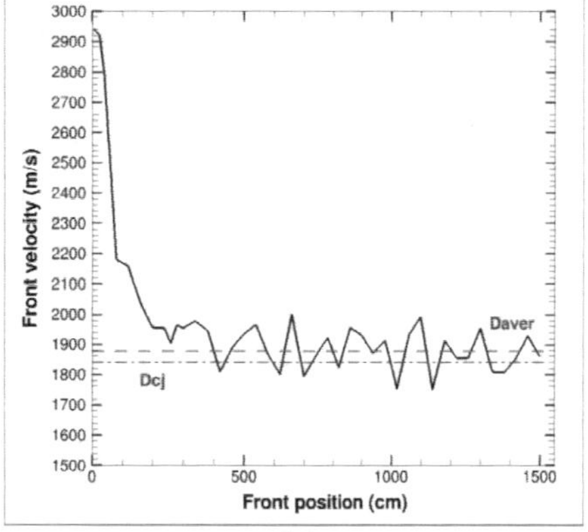

(a)

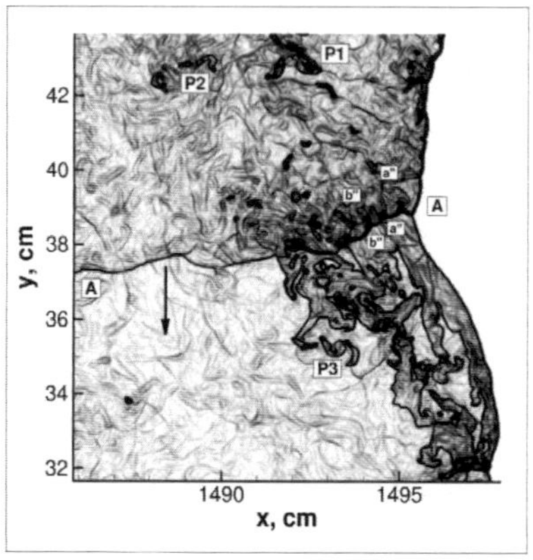

(b)

Figure 2. A rich methane-air mixture: (a) front-propagation velocity (solid line) as a function of front position in channel H=45 cm; (b) magnified fragment of the primary transverse wave AA in channel H=50 cm.

Figure 2b presents an enlarged fragment of the primary transverse wave AA t in the position y≈38 cm for DW channel H=50 cm. Figure 2b clearly shows the fine cellular structure of the front of the main transverse wave. The fine TWs on the front of primary wave AA are denoted as ‘a′′a′′′’, ‘b′′b′′′’. The multifront (cellular) structure of the transverse waves had been observed earlier experimentally on soot imprints [14]. The secondary transverse waves that form the fine structure of the leading DW front (see Fig.1), had been observed in the experiments [15]. Thus, this simulation reproduces a real characteristic feature of the detonation of hydrocarbon mixtures – instability of the front of the transverse waves itself and instability the leading shock front of the DW.

The fine (cellular) structure of the transverse waves has been found previously in our simulations of DW in stoichiometric methane–air [2] and methane– oxygen [3] mixtures. Furthermore, Figure 2b shows a significant amount of unburned pieces of gas at a considerable distance behind the DW leading shock front, denoted as P1 and P2. This phenomenon was not observed in simulations of regular mixtures based on hydrogen [1].

To study grid convergence of the numerical solutions the calculations have been performed with twice as many number of numerical cells N_x=4000, N_y=4000 for channel height H = 45 cm, 50 cm, and 65 cm. The simulations revealed a flowfield with two approximately equal primary transverse waves for H = 45 cm and 50 cm. Based on these calculation results, we conclude that grid convergence of the numerical solutions was obtained in our study. The transverse size of the dominant detonation cell for rich methane-air mixture determines to be a_0 = 45÷50 cm. Empirical data on the size of the cell a_0 for a given mixture composition are not available in literature, it is necessary to carry out large-scale experiments for this mixture.

5. Simulation of continuous rotating DW in a rich methane-oxygen mixture

Numerical investigation of continuous spin detonation (CSD) in a rich methane-oxygen 1.2 CH_4+2O_2 (φ = 1.2) in the annular cylindrical detonation chamber (DC) has been conducted according physical and mathematical 2D model, see [16,17]. The described above kinetics model of detonation combustion of methane mixtures has been used. In order to compare numerical simulation results with experiments in annular cylindrical DC (average diameter d_c=3,5 cm, the annular gap Δ = 0.5 cm, DC length L_c=9 cm) [17], two-dimensional time-dependent calculations has been done under the same geometric and input governing physical parameters, as in the experiments [17], Chapter 2, Table 2.1. Total mass flow rate of a fresh 1.2 CH_4+2 O_2 mixture from manifold (reciver) in the experiment was

G_0 = 85 g/s, which ensure the specific mass flow rate in the annular channel of the DC $g_{\Sigma 0} = G_0/S_c = \rho \cdot u_y =$ 154.6 kg/s·m^2 on the inlet boundary y=0. In the experiments the stagnation pressure and temperature in injection manifold (receiver) were p_m=6.81 atm, T_m=293.15 K. For given manifold data this value of $g_{\Sigma 0}$ at critical regime of inflow from the system of Laval micro-nozzles at the inlet wall of DC [17] is provided with the value of S_*/S_c=0.1026, where S_c – DC annular channel cross-sectional area, S_* – the total throat area of micro-nozzles, ρ– mixture density, u_y–velocity on y axe. In experiments [17] with this mixture a single rotating DW has been obtained in the given DC, respectively a two-dimensional computational domain width has been selected $W=\pi \cdot d_c$=10.9956 cm. A periodic boundary conditions have been set at the left and right boundaries x=0 and $x=W$ [16]. On the outlet boundary $y= L_c$ a value of counterpressure p_{out}=1 atm has been set.

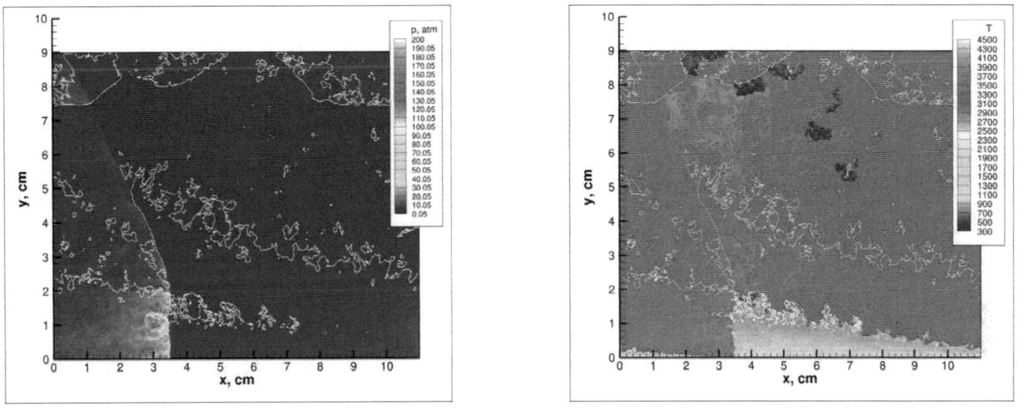

(a) (b)

Figure 3. Structure of rotating detonation wave in a detonation chamber of diameter d_c=3,5 cm: (a) pressure (atm) flow field; (b) temperature (K) flow field.

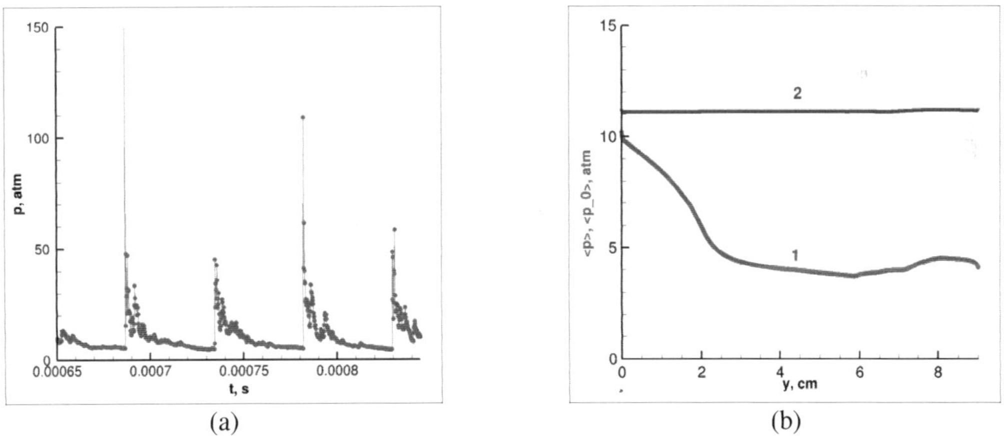

(a) (b)

Figure 4. Spin detonation parameters: (a) static pressure p (atm) history at the point x=0, y=0, see Fig.3; (b) distribution of average static pressure $\langle p \rangle$ (curve 1) and total pressure $\langle p_0 \rangle$ (curve 2) along an axial length of a detonation chamber.

Numerical simulations have shown that for given manifold parameters (p_m=6.81 atm, T_m=293.15 K) a stable CSD regime with a single rotating transverse DW (TDW) cannot exist in the given DC. After the numerous numerical calculations it has been found that self-sustained continuous rotating DW can exist at the manifold parameters p_m=40 atm, T_m=1000 K.

Figure 3 shows the flow structure with a single TDW in this DC after the establishing of steady spinning regime. The pressure and temperature flow fields are shown for time moment t=844,6 μs, see

Figure 4a. A white curve on these flow fields shows an instant position of Mach number $M=u_y/c=1.0$ isoline, where c is a speed of sound. Figure 3 clearly shows the cellular (multifront) structure of the front of this TDW, that is similar to a multifront structure of TWs on the front of propagating DW, see Figure 2b. Analogous on Figure 3b one can see a number of the unburned pockets of gas. These pockets lead to variation in time the value of specific flow rate $g_{\sum Out}$ on outlet boundary.

Figure 4a shows the dependence of static pressure (atm) on time (s) on the inlet boundary of the DC. This is a virtual pressure gauge data at a fixed point (x= 0, y=0). Analysis of the obtained pressure history shows, that within 194.6 μs of the detonation process the TDW is on a steady rotation regime with an average time period of rotation $\langle\Delta t\rangle=47.6 \pm 0.2$ μs.

Figure 4b presents the distribution of calculated values of the averaged static $\langle p(y)\rangle$ and total $\langle p_0(y)\rangle=\langle p+\rho\cdot u_y^2\rangle$ pressures along the length of DC channel. The data for the drawing of these curves has been taken from the simulation of detonation process shown in Figure 4a. It is seen that the static pressure (curve 1) decreases along the length of the chamber. Moreover, in the main heat release zone ($0 < y < 2.0$ cm) it is observed rapid decline of pressure from about 10 atm to 6 atm, and then it with minor fluctuations decreases slowly to a value of about 4 atm at the outlet of the channel ($y=L_c=9$ cm). The length of the main heat release zone, obtained from this graphics correlates very well with the average height of TDW $\langle h\rangle$. The curve 2 shows that the average total pressure is constant (with very minor fluctuations) along the length of the DC. This total pressure behavior is completely consistent with the laws of gas dynamics.

Numerical simulations have shown that a reduction of the stagnation temperature to T_m=900 K at p_m=40 atm leads to a complete breakdown of detonation combustion process and disappearance of TDW in DC. The reduction of stagnation pressure to p_m=35 atm at T_m=1000 K still gives stationary in general detonation regime with rotating TDW. The TDW shape is significantly changing over time and has greatly distorted form with 1 or 2 transverse waves on its front. This mode can be considered as marginal, insofar further decrease of stagnation pressure to p_m=30 atm at T_m=1000 K leads to the complete destruction of the CSD regime.

6. Concluding Remarks

The two-stage model of the kinetics of detonation combustion of methane-based mixtures with an oxygen and air has been developed. The use of this kinetic model for modelling of the two-dimensional cellular structure of detonation wave provides realistic results. The dominant size of the detonation cell for stoichiometric methane-air mixture determines to be 34±1 cm, and 0.3÷0.35 cm for methane-oxygen mixture [2,3]. These values are in good agreement with all available experimental data. A numerical simulation of 2D structure of the DW in rich (φ=1.5) methane-air mixture has been performed. The size of the detonation cell for this mixture determines to be a_0=45÷50 cm.

The use of the proposed model of chemical kinetics and the same two-dimensional numerical code to calculate CSD in rich methane-oxygen mixture in the annular cylindrical DC produced very interesting results. It has been found that it is impossible to obtain a stable CSD regime with the parameters in the manifold, at which in experiment was obtained stably rotating single TDW. In [17] it was obtained that the primary control parameter, which determines the existence of the CSD regime is the specific mass flow rate $\langle g_{\Sigma,In}\rangle$ of fresh mixture at DC inlet boundary. In experiments [17] this value was $\langle g_{\Sigma,In}\rangle$=154.6 kg/s·m^2 for the manifold parameters p_m=6.81 atm, T_m=293.15 K. In the numerical simulations the stable CSD regime with single TDW has been obtained at the substantially greater (in several times) value of $\langle g_{\Sigma,In}\rangle$=470 kg/s·m^2 ($p_m$=40 atm, T_m=1000 K) in the same DC, and the marginal regime has been realized at $\langle g_{\Sigma,In}\rangle$=411.4 kg/s·m^2 ($p_m$=35 atm, T_m=1000 K).

Numerical simulation also showed that in addition to the specific flow rate $\langle g_{\Sigma,In}\rangle$, the second governing parameter is the stagnation temperature T_m. Indeed, reducing the stagnation temperature from T_m=1000 K to T_m=900 K at p_m=40 atm leads to termination of the CSD.

A significant discrepancy has been obtained for the TDW velocity. In the numerical simulation the value of the average velocity is $\langle D\rangle/D_{CJ}$=0.931, and in experiment [17] this value is $\langle D\rangle/D_{CJ}$=0.75. In our opinion, such low experimental value of the TDW velocity raises the question whether this shock is in a reality the detonation wave in the classical consideration according to the ZND theory. But the TDW height is in very good agreement with the experimental data.

Thus, the developed two-stage kinetic model and 2D numerical code give excellent results in the modelling of irregular cellular structures in methane-based mixtures for classical DW in a straight 2D channel. However, in the case of simulations of the rotating DW in the annular cylindrical DC the significant discrepancies have been obtained between the calculated and experimental values of governing parameters under which the stable CSD regime is realized.

Acknowledgements

The authors gratefully acknowledge the financial contribution from the Russian Foundation for Basic Research (Grant No. 17-03-01351). The computations were performed on an MVS-100K supercomputer at the Joint SuperComputer Center (JSCC) of the Russian Academy of Sciences (Moscow).

References

[1] Trotsyuk A V 1999 Numerical simulation of the structure of two-dimensional gaseous detonation of an H_2–O_2–Ar mixture *Combust. Explo. Shock Waves* **35** 549–58

[2] Fomin P A, Trotsyuk A V and Vasil'ev A A 2015 Numerical study of cellular detonation structures of methane mixtures *J. Loss. Prevent. Proc.* **36** 394–403 DOI: 10.1016/j.jlp.2015.03.012

[3] Trotsyuk A V and Vasil'ev A A 2014 Numerical study of cellular structure of detonation of a methane–oxygen mixture *Modern Science. Researches, Results, Technologies (ISSN 2076-6866)* № 1 (14) 139–45

[4] Korobeinikov V P, Levin V A, Markov V V and Chernyi G G 1972 Propagation of blast waves in a combustible gas *Acta Astronaut.* **17** 529–37

[5] Soloukhin R I 1969 Measurement methods and basic results of shock-tube experiments *Proc. 7th Int. Symp. on Shock Tubes* (Novosibirsk: SB Academy of Sciense USSR Press)

[6] Fomin P A, Trotsyuk A V and Vasil'ev A A 2014 Approximate model of chemical reaction kinetics for detonation processes in mixture of CH_4 with air *Combust. Sci. Technol.* **186** 1716–35 DOI: 10.1080/00102202.2014.935643

[7] Nikolaev Yu A 1978 Model of chemical reactions at high temperatures. *Combust. Explo. Shock Waves* **14** 468–71

[8] Nikolaev Yu A and Zak D V 1988 Agreement of models of chemical reactions in gases with the second law of thermodynamics *Combust. Explo. Shock Waves* **24** 461–4

[9] Godynov S K 1976 *Numerical Solution of the Multi-Dimensional Problems of Gaseous Dynamics* (Moscow: Nauka Press) p 400

[10] Yamamoto S and Daiguji H 1993 Higher-order-accurate upwind schemes for solving the compressible Euler and Navier-Stokes equations *Comput. Fluids* **22** 259–70

[11] Batten P, Leschziner M A and Goldberg U C 1997 Average-state Jacobians and implicit methods for compressible viscous and turbulent flows. *J. Comput. Phys.* **137** 38–78

[12] Coquel F and Perthame B. 1998 Relaxation of energy and approximate Riemann solvers for general pressure laws in fluid dynamics *SIAM J. Numer. Anal.* **35** 2223–49

[13] Shen J W and Zhong X 1996 Semi-implicit Runge-Kutta schemes for non-autonomous differential equations in reactive flow computations *AIAA 27th Fluid Dynamics Conf., New Orleans, LA, June 17-20, 1996* AIAA Paper No.96–1969

[14] Manzhalei V I and Mitrofanov V V 1973 The stability of detonation shock waves with a spinning configuration *Combust. Explo. Shock Waves.* **9** 614–20

[15] Manzhalei V I 1977 Fine structure of the leading front of a gas detonation. *Combust. Explo. Shock Waves* **13** 402–4

[16] Zhdan S A Bykovskii F A and Vedernikov E F 2007 Mathematical modeling of a rotating

detonation wave in a hydrogen-oxygen mixture *Combust. Explo. Shock Waves* **43** 449–59
[17] Zhdan S A and Bykovskii F A 2013 *Continious Spin Detonation* (Novosibirsk: Lavrentyev Institute of Hydrodynamics Press) p 422

The time-space relationship between strain, microstrain temperature fields and microseismic emission parameters in geomodels with hole under external loading

O M Usol'tseva[1], **V I Vostrikov**[1], **P A Tsoi**[1,2], **V N Semenov**[1] **and O A Persidskaya**[1]

[1]Chinakal Institute of Mining, Novosibirsk, Russia
[2]Novosibirsk State Technical University, Novosibirsk, Russia

E-mail: usoltseva57@mail.ru

Abstract. A complex analysis of evolution of microseismic emission signals, temperature field and microdefofmation field under uniaxial loading before the destruction of prismatic samples from artificial geomaterial and rocks allowed to establish the relationship between the patterns of signal changes and the level of loading. The evolution of deformation process, development of microdamages and the formation of main rupture fracture lead to significant transformation of spectral composition of microseismic emission signals, microdeformation field, and the temperature field. In the region of future main discontinuity the temperature increases, localization of maximum microdeformations occurs and velocities of microdeformations components increase. Generation of powerful low-frequency harmonics at loads approaching the peak, can serve as a harbinger of a rupture on the surface and, consequently, destruction of the geomaterial.

1. Introduction

A literature review over the past three years has shown that there are a number of papers devoted to the study of the specific change in the parameters of the microdeformation field by the speckle method [1, 2, 5, 6] and acoustic emission signals (AE) [3, 4]. In the experiments on uniaxial compression of argillite [1] using a speckle method, heterogeneous local strain fields that correlate well with the microstructure of argillites are illustrated. Based on the monitoring of seismic signals and the use of the speckle method [2] for slip processes on gypsum samples with inhomogeneous contact surfaces (smooth and rough), it was noted that a distinct peak in the amplitude of the AE signal wave can be considered as a failure "precursor". According to the speckle – method, slips are identified as jumps in the displacement field along the crack. Two measurement methods with simultaneous measurement of stresses, deformations and AE signals [3] were compared under uniaxial compression of granite. It was concluded that the AE method is more preferable for assessing the level of damage. An extensive series of experimental studies of rock specimens of different genotypes during their mechanical loading and theoretical justification of the evolution laws concerning acoustic emission signals are considered in paper [4]. The use of speckle – photographic methods to study deformation and fracture of rocks was justified [5], an analysis of the deformation field in the process of loading prior to failure, and zones of maximum deformations localization was given. The deformation evolution of granitic samples with measurement of acoustic emission signals and method of digital image correlation were presented in paper [6]. A good agreement between the location of AE signals and the trajectory of a crack is shown.

Content from this work may be used under the terms of the Creative Commons Attribution 3.0 licence. Any further distribution of this work must maintain attribution to the author(s) and the title of the work, journal citation and DOI.
Published under licence by IOP Publishing Ltd

In this paper, a laboratory study of the deformation process in artificial layered geomaterials prior to destruction with simultaneous measurement of stresses, deformations and MSE signals is given. The analysis of synchronized experimental data made it possible to determine the specific change in the parameters of microseismic signals depending on the stage of the deformation process in rock samples, and also to reveal the evolution dynamics of microdefections and the main discontinuity domain.

2. Experimental procedure and meashurement equipment

In the experimental research it was used Instron-8802 servohydraulic testing machine for loading specimens by axial force. The test samples were made of argillite. The samples were rectangular prisms 50x50x20 mm with a hole diameter 15 mm in the center (Fig. 1). Tests of rock specimens were carried out under uniaxial strain-controlled compression prior to failure at a rate of 0.1÷6 mm/min. Microstrains were recorded using automated digital speckle photography analyzer ALMEC-tv at frequency of 27 frames per second and spatial resolution not less than 1 μm. The processing output is the coordinates and displacements of the specimen surface points and timing, which allows calculating strain tensor components [7]. Four microseismic KD 91 sensors were installed on four lateral faces of the cube to record the MSE signals. The measurement of the temperature field was carried out using by computer thermal imager TKVr-SVIT 101, the accuracy of the measurement is 0.03°K. Figure 2 illustrates general view of the experiment. Figure 3 presents the curve of the specimen stress and macrostrain and the specimen stress and time.

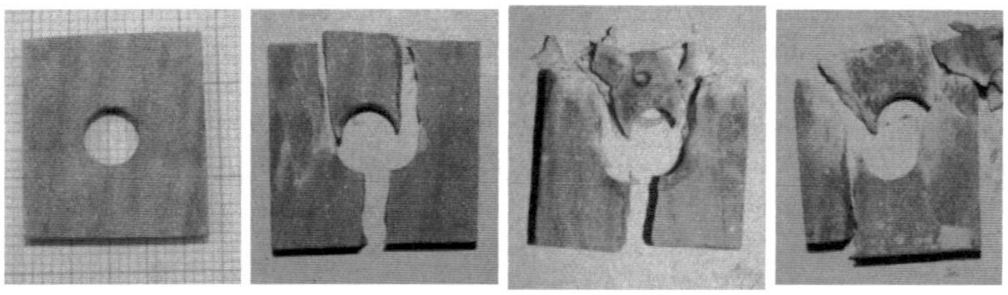

Figure 1. Rock specimens before and after fracture.

Figure 2. General view of the experiment: *1*—optical/tv measurement complex ALMEC-tv; *2*—rock specimen; *3*—compression grips of the Instron-8802.

3. Experimental data analysis

The uniaxial strain-controlled compression used prismatic specimens of argillite at the mobile grip displacement rate of 0.1÷6 mm/min. Figure 3 shows the curves of the specimen stress and macrostrain and the specimen stress and time. Ultimate strength of the specimen was 39.1 MPa, ultimate strain along x, measured by the machine cross beam unit displacement, was 0.076.

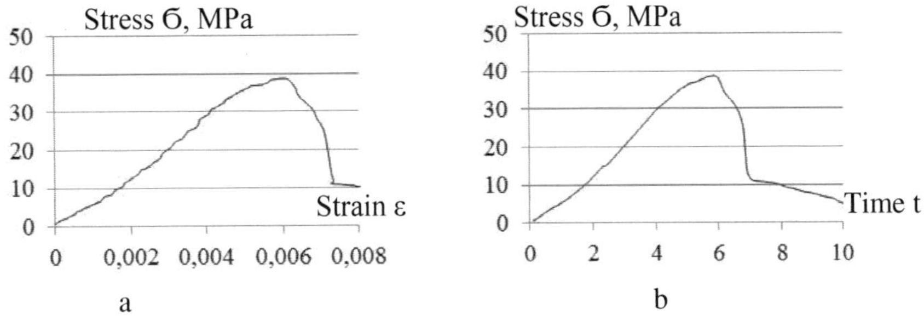

Figure 3. The curve of stresses and strain in the rock specimen in the direction of *x* (a) and the stress and time curve (b).

A large number of microseismic signals were recorded under uniaxial compression prior to destruction of prismatic samples. Figure 4 shows the diagram "$б/б^B$ –t/t_m" obtained from uniaxial compression data, where $б/б^b$ is the ratio of the current stress value to the ultimate strength value $б^b$, and t/t_m is the ratio of the current experimental time to the value of t at $б=б^B$. The typical signals recorded at points 1, 2, 3 are shown in Fig. 5.

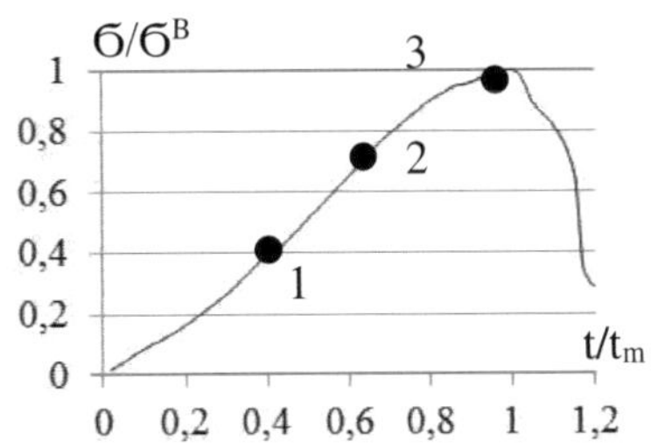

Figure 4. The diagram "$б/б^B$ –t/t_m" obtained from uniaxial compression of a prismatic sample.

Based on the analysis of a large data bank on the MSE signals, the following regularities are revealed:
– At stress levels $б/б^B$=0,3÷0,4 (point 1), the first MSE signals are recorded, the acceleration magnitude is 2÷4 m/sec², the broadband frequency is 8 ÷ 24 kHz (Fig. 5a, b).
– At stress levels $б/б^B$=0,5÷0,8 (point 2), the amplitude of the MSE signal increases, its magnitude reaches a value of 0.35 m/s². At the same time, the frequency spectrum is somewhat narrowed and shifted towards low frequencies of f=10÷14 kHz (Fig. 5c, d).
– At the last deformation stage at a stress close to $б/б^B$=0.8÷1 (point 3), the number of microdamages and MSE signals increases significantly, their energy increases, the frequency spectrum is further narrowed and shifted to low frequencies, up to f=8÷10 kHz.
A powerful low – frequency signal of 8÷9 kHz is generated at the time of the main discontinuity initiation (Fig. 5 d, e).

In previous studies [7] the detailed analysis into the micro-level stress–strain state distribution and propagation over acting faces of rock specimens subject to uniaxial loading until failure, using automated digital speckle photography analyzer ALMEC-tv, have shown that:

• under uniaxial stiff loading of prismatic sandstone, marble and sylvinite specimens on the Instron-8802 servohydraulic testing machine at the mobile grip displacement rate 0.02–0.2 mm/min, at a certain level of stressing, low-frequency micro-deformation processes originate in the specimens due to slow (quasi-static) force;

• the amplitude of that deformation-wave processes greatly depends on the micro-loading stage:

—at the elastic deformation stage, under the specimen stress lower than half ultimate strength of the specimen, there are no oscillations of microstrains;

—at the nonlinearly elastic deformation stage, under stress varied from 0.5 to 1 ultimate strength of the specimens, the amplitudes of microstrains grow, including the descending stage 3; the oscillation frequency $f = 0.5 \div 4$ Hz;

—at the residual strength stage, the amplitudes of the microstrains drop abruptly (3–5 times) as against stages 2 and 3;

• in the elements of the scanned specimen surface in the region with the incipient crack, the microstrain rate amplitudes ε'_y are a few times higher than in the undamged surface region of the same specimen. Deformation rate greatly grows with increase in the load.

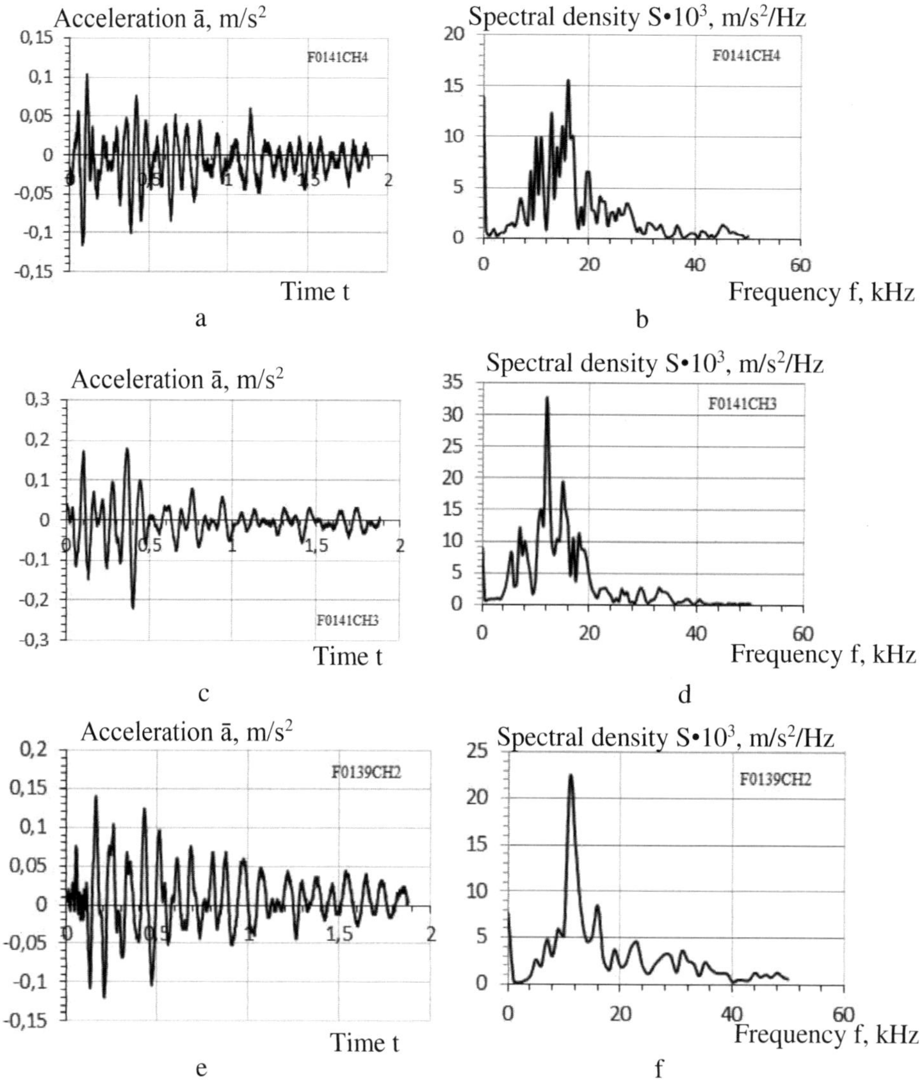

Figure 5. The MSE signals registered at points 1 (a, b), 2 (c, d), 3 (e, f) of the diagram $\sigma/\sigma^B - t/t_m$

Fig. 6 shows mapping of the scanned surface by the microstrain ε_y in longitudinal direction y of a specimen at points 1, 2, 3 in the stress–time curve in Fig. 4. In the surface of specimen deformation scan in longitudinal direction in Fig. 5, the negative values of deformation (shortening) correspond to blue color. The zero strains are shown by black color. The positive values of deformation (elongation) comply with red color. White color shows strains over 0.02 mm. In the cross direction deformation scans in Fig. 6, white color also marks the strain above 0.02.

At the first stage (point 1) of loading at voltages not exceeding 0.25 ÷ 0.5 of the peak load, the field of microdeformations is chaotically inhomogeneous, the vibrations of the components of microdeformations are practically absent (Fig. 6a). At the second stage of deformation (point 2), when the voltage takes values from 0.4 ÷ 0.5 to 0.7 ÷ 0.8, the field of microdeformations becomes more non-uniform, zones of maximum microdeformation occur, the values of which exceed the average values over the surface of the sample (Fig. 6 b). At the third stage of deformation (point 3) at the stresses from 0.8 of the ultimate strength to the peak load, the zones of maximum microdeformations are localized in a certain volume of the rock sample, which indicates the beginning of the formation of the main discontinuity, whose exit location on the sample surface can be determined at loads less than the peak (Fig. 6 c).

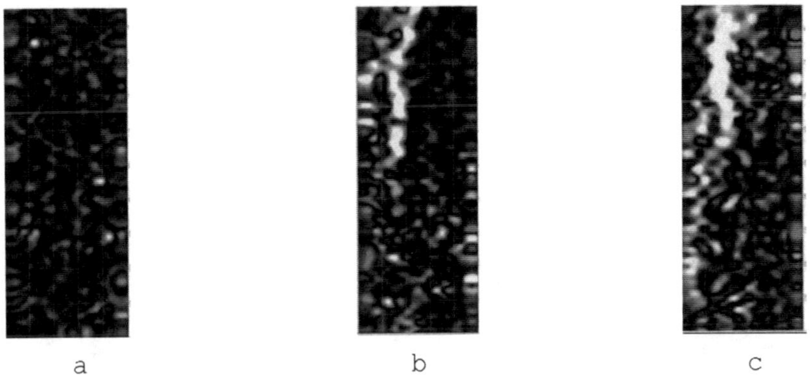

Figure 6. Mapping of the scanned surface by the microstrain ε_y in longitudinal direction x of a part of specimen at the compression moment at points 1 (a), 2 (b), 3 (c) of the diagram "$\sigma/\sigma^B - t/t_m$ "(Fig. 4).

The microdeformation rates at the second and third deformation stages were estimated for two different region types of the scanned sample surface: 1 – region with a subsequently formed crack, 2 – region of the nondestructed material. A typical diagram of the change in the rates of microdeformations is shown in Fig. 7. It can be seen that in the elements of the scanned sample surface in the regions with a future crack (region 1), the amplitude of the microdeformation rate ε_y exceeds several times the rate observed in surface region of the indestructible geomaterial (region 2). In a number of cases, there is a tendency for a significant increase in the deformation rate with increasing loading.

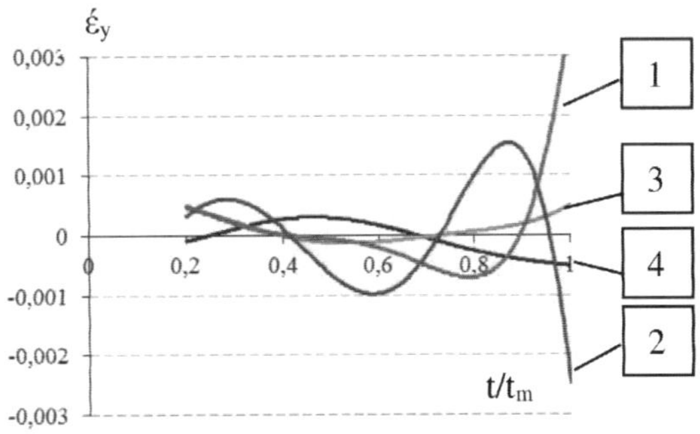

Figure 7. Typical dependences of the lateral microdeformation rates on the time for element 0.5 x 0.5 mm in size for the region with a future crack (1, 2) and region of the nondestructed material (3, 4).

Temperature measurement on the sample surface was carried out using the computer thermal imager TKVr-SVIT 101, the accuracy of the measurement is 0.03°K. Analysis of temperature data shows that at the beginning stage of loading, temperature field exhibits no obvious change. With the loading

IOP Conf. Series: Journal of Physics: Conf. Series **894** (2017) 012098 doi:10.1088/1742-6596/894/1/012098

development, strains focus around the hole gradually, over and under of the hole appear tensile strain along the direction of loading and grow gradually.

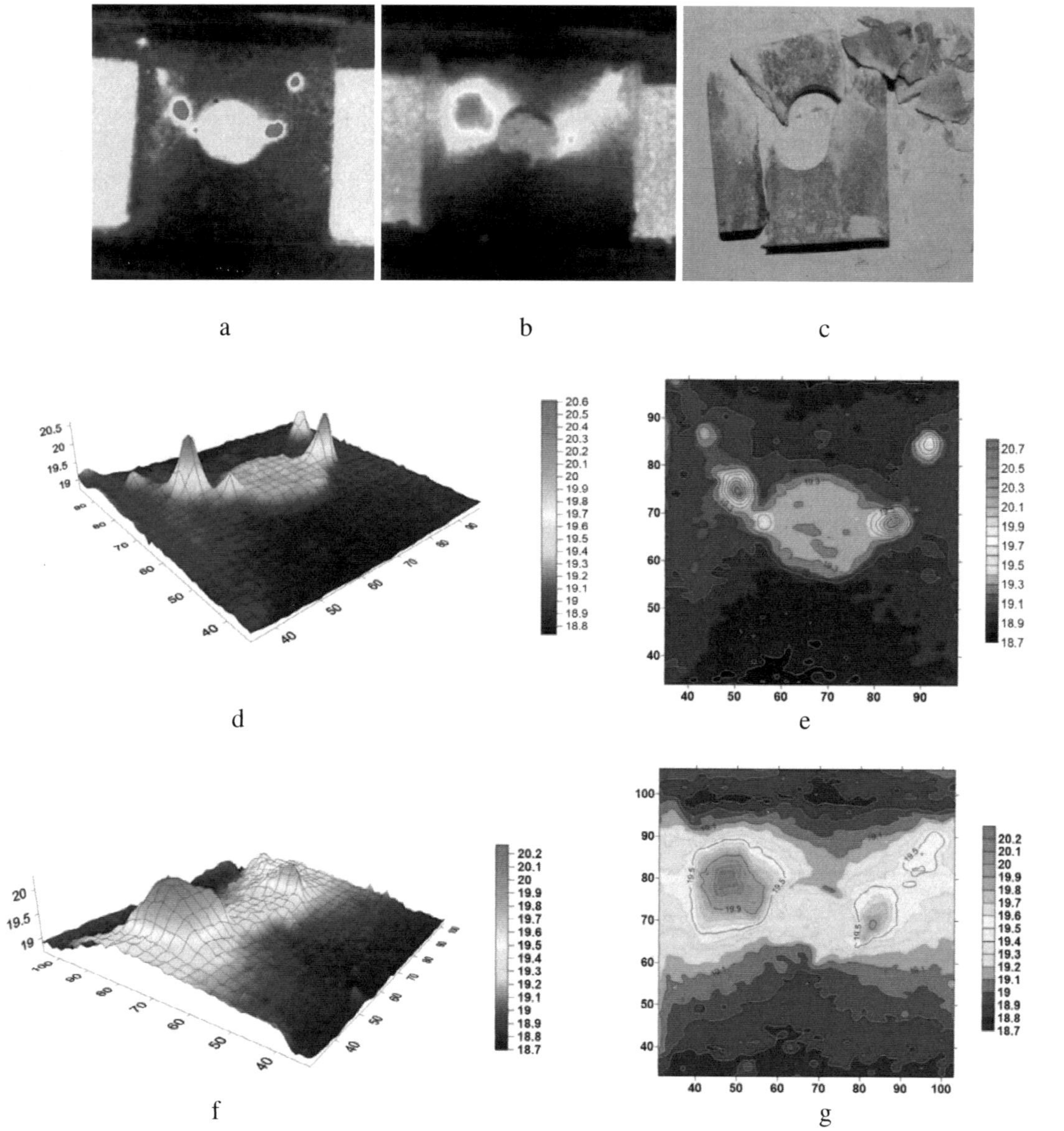

Figure 8. Temperature field on the sample surface at loading moments t=0.8 t_m (a, d, e) and t=0.9 t_m (b, f, g), photograph of the corresponding destroyed rock sample (c).

At the values t=0.8 t_m the temperature increases in the region with an X-shaped shape near the hole. When the loading time is t=0.9 t_m, the red area to the left of the hole increases in size. On the diagram of y-component of the microdeformation it is a correspondence - this is the region of higher values of tensile microdeformations. With further increase in the load, a main crack appears at the top left of the hole, then from the top right, at the last moment - a vertical crack under the slope and the sample is completely destroyed. The temperature rise in the zones of localization of the maximum microdeformation is 1.5°K. With an increase of loading rate from 0.1 mm per second to 6 mm per second, these patterns of temperature field variation manifest more contrastively.

4. Conclusion

Complex analysis of the evolution of microseismic emission signals, microdeformations, temperature field, strain and deformation showed that the evolution of crack formation in the rock samples is completely satisfactorily characterized by each of the methods.

There is precise relationship between the patterns of change of the parameters of microseismic emission signals, the components of microdeformations field, their velocities, the temperature field in correlation with the stress-strain diagram.

The character of the change of parameters of these physical fields makes it possible to determine the location of the exit of main discontinuity to the sample surface at loads less than the peak when the sample still preserves its integrity.

Acknowledgments

This study has been carried with partial financial support of the Russian Foundation for Basic Research, projects no. 15–05–07566, using equipment of the Shared Use Center for Geomechanical, Geophysical and Geodynamic Measurements, Siberian Branch, Russian Academy of Sciences.

Reference

[1] Wang L, Bornert M, Heripre E, Chanchole S, Pouya A and Halphen B 2015 *Rock Mechanics and Rock Engineering* **48(5)** 1913-26
[2] Hedayat A, Pyrak – Nolte L J and Bobet A. 2014 *Rock Mechanics and Rock Engineering* **47(5)** 1575-87
[3] Kim J S, Lee K S, Cho W J, Choi H J and Cho G C 2015 *Rock Mechanics and Rock Engineering* **48(2)** 495-508
[4] Shkuratnik V L, Novikov E A and Oshkin R O 2014 *Journal of Mining Science* **50(2)** 249 - 255
[5] Zuev L B, Barannikova S A, Nadezhkin M V and Gorbatenko V V 2014 *Journal of Mining Science* **50(1)** 43-49
[6] Shi Y, He Q, Liu S and Wu L 2010 *Progress In Electromagnetics Research Symposium Proceedings (Xi'an, China)* 114-18
[7] Oparin V N, Vostrikov V I, Usol'tseva O M, Tsoi P A and Semenov V N 2015 *Journal of Mining Science* **51(3)** 624-33

Mathematical description of ignition, combustion and propagation of detonation in reacting gas mixtures in the presence of micro- and nanoparticles

A V Fedorov and D A Tropin

Khristianovich Institute of Theoretical and Applied Mechanics SB RAS, 630090, Novosibirsk, Russia

d.a.tropin@itam.nsc.ru

Abstract. The physical and mathematical models for the description of the processes of ignition, combustion and propagation of detonation in mixtures of hydrogen-oxygen, methane-oxygen and silane-air in the presence of inert micro- and nanoparticles were proposed. On the basis of these models the dependencies of detonation velocity deficit vs the size and concentration of inert micro- and nanoparticles were found. Two modes of detonation flows in gas suspensions of reactive gases and inert nanoparticles were revealed: - propagation of weak detonation wave in the gas suspension, - destruction of the detonation process. It was determined that the mechanisms of detonation suppression by micro- and nanoparticles are closed and lies in the splitting of a detonation wave to frozen shock wave and ignition and combustion wave. Concentration limits of detonation were calculated. It turned out that in the transition from microparticles to nanoparticles the detonation suppression efficiency does not increase.

1. Introduction

The addition of inert particles in the reactive gas mixture is one of methods of modification and control of combustion and detonation in such mixtures. For example, the addition of chemically inert microparticles reduces the velocity of detonation [1-3], change the length of the chemical reaction zone of the detonation wave (DW) [4]. The authors of the papers [5-6] have been defined the concentration limits of detonation in a mixture of hydrogen-oxygen, methane-oxygen and methane-hydrogen-oxygen while suppressing DW by clouds of inert microparticles. Furthermore, in [5-6] it has been shown that the most effective in detonation suppressing the smaller diameter particles. In [7-8], it was shown that the addition of inert microparticles destroys the cell structure of the DW in some model mixture. The detonation cell size in a stoichiometric silane-air mixture with the microparticles are calculated in [9]. It is revealed that the increasing of the mass fraction of particles in the mixture leads to increasing the detonation cell size.

Thus, the processes of ignition, combustion, propagation and interaction of DW with clouds of inert microparticles was studied in details. In this paper the interaction of DW in mixture of hydrogen-oxygen, methane-oxygen and silane-air with clouds of inert micro- and nanoparticles are calculated.

2. Physical and mathematical formulation of the problem

Let's consider a shock tube filled with a gas mixture of hydrogen-oxygen or methane-oxygen or silane-air, as well as the cloud of inert micro- and nanoparticles of SiO_2 located in the low pressure chamber

Content from this work may be used under the terms of the Creative Commons Attribution 3.0 licence. Any further distribution of this work must maintain attribution to the author(s) and the title of the work, journal citation and DOI.
Published under licence by IOP Publishing Ltd

(LPC) at a certain distance from the diaphragm, separating the high and low pressure chambers. Conditions in LPC: pressure equal to 1 atm, temperature of gas mixture equal to 300K. After the rupture of the diaphragm in the LPC the DW is initiated, which prior to the interaction with the cloud enters the Chapman-Jouget regime. Mathematical model of the mechanics of the reacting gas mixtures and inert particles represents the system of equations of the dynamics of the gas mixture and solid particles and has the form shown in our previous studies [5-6].

For very small particles when the particle size becomes comparable with the mean free path of the gas molecules, the expression of the drag force should be complemented with the Cunningham correction factor: $C_c = 1 + \frac{2\lambda}{d}\left[1.257 + 0.4e^{-\frac{1.1d}{2\lambda}}\right]$, where d - the particle diameter, $\lambda = \frac{k_b T}{\sqrt{2}\pi d_g^2 p}$ - mean free path of gas molecules, p, T, - pressure and temperature of gas, k_B - Boltzmann constant, d_g - the diameter of the surrounding gas molecules. Drag force has following form:

$$f = \frac{3\mathrm{m}_2\rho_{11}}{4\mathrm{d}}\frac{C_D}{C_c}\left|u_1 - u_2\right|\left(u_1 - u_2\right),$$

where C_D - drag coefficient, m_2 - particles volume concentration, ρ_{11} - gas mixture true density, u_1, u_2 - gas mixture and particles velocity, respectively.

This model which takes into account the Cunningham correction factor, will be valid only within a specific range of particle sizes (the sizes change from 10 nm to 100 microns). In this range of particle diameters Cunningham correction factor in the structure of the detonation wave changes its value from 1 (for particles with the diameter ranging from 10 to 100 microns) to 1.2 (for 1 micron particles), in gas suspensions with nanoparticles (for particles with the diameter ranging from 10 to 100 nm) Cunningham correction factor monotonically increases to value $C_c \sim 13$. When the particle size becomes close to the size of the molecules of the surrounding gas, the model has to be unfair, because the phenomenological approach no longer works.

The characteristic times of thermal relaxation for the nanoparticles are determined by the change in the mode of the particles flow from the continuum to free-molecular [10]. In the continuum flow mode (at $\mathrm{Kn} < 0.01$) characteristic time of thermal relaxation describes by the following dependence $\tau_T^{cont} = d^2 \rho_{22} c_{v2} / 6\lambda_1 \mathrm{Nu}$. For free-molecular flow regime (at $\mathrm{Kn} > 10$) [11]: $\tau_T^{fm} = \frac{\rho_{22} c_{p2} d}{6\alpha p}\sqrt{\frac{8\pi\mu T}{R}}\left(\frac{\gamma - 1}{\gamma + 1}\right)$, where $Kn = \frac{\lambda}{d}$ - Knudsen number. Here ρ_{22}, c_{p2} - the true density and heat capacity of the particles, μ - molar mass of ambient gas, α - the accommodation coefficient. In the range of Knudsen numbers $0.01 \div 10$ there is a transitional flow regime [10]. Here we use an approximation formula of the form $\tau_T^{tr} = [(\lg \mathrm{Kn} + 2)\tau_T^{fm} + (1 - \lg \mathrm{Kn})\tau_T^{cont}]/3$.

It should also be noted that for microparticles with diameters ranging from 1 to 100 μm continuous flow regime is realized, for nanoparticles with diameters ranging from 10 to 100 nm transition regime is realized.

To describe the chemical reactions in the reacting gas mixture we will use the detailed kinetics models: for the hydrogen-oxygen mixture - model [6, 12, 13], taking into account 38 reactions for 8 component; for methane-oxygen mixture - model [5, 14], taking into account the 92 reaction for 15 components, for silane-air mixture - model [15, 16], taking into account 140 reactions for 25 components.

3. The calculation results

3.1. Hydrogen-oxygen mixture

First of all, let's consider the processes of DW interaction in a hydrogen-oxygen mixture with inert nanoparticles. Figure 1 shows the dependences of the detonation velocity deficit ($\eta = \frac{D}{D_{CJ}}$, where D - the detonation velocity in a mixture of gas and particles, D_{CJ} - Chapman-Jouget detonation velocity: $D_{CJ} = 2750$ m/s in hydrogen-oxygen mixture, $D_{CJ} = 2440$ m/s in methane-oxygen mixture and $D_{CJ} = 1820$ m/s in silane-air mixture) vs the volume concentration of inert particles for the particle size ranging from 10 nm to 100 microns. The data for micro particles are taken from [6]. From figure 1 it is seen that in the case of DW propagation in mixture of a gas and particles with the diameter change from 10 nm to 1 micron detonation velocity deficit decreases to values $\eta = 0.45$ while for the micro particles with the diameters ranging from 10 microns to 100 microns this parameter decreases only to values $\eta = 0.75$ after which the failure and suppression of DW are occurs. Thus, for all considered sizes of the particles, exists the two flow regimes: 1. stationary propagation of weakened DW at velocities D lower than D_{CJ} (in the micro particles with a DW velocity deficit ranging from 0.75 to 1, in the nanoparticles with a DW velocity deficit ranging from 0.45 to 1); 2. the suppression of DW. Possible types of detonation flows in gas suspensions of reactive gases and inert micro particles, as well as the scenario of DW suppression by micro particles described in detail in [1-2, 5-6], in which the suppression means splitting of DW to frozen shock wave (SW) and lagging front of ignition and combustion. In the case of DW suppression by nanoparticles it is also seen the decay of the DW to frozen SW and front of ignition and combustion. I.e. the mechanisms of detonation suppression by micro- and nanoparticles are the same.

Furthermore, it should be noted that the concentration limits of detonation, i.e. minimum volume concentration of inert particles that suppress the DW, in hydrogen-oxygen mixture are close for the particles with diameters of 10 nm, 100 nm, 1 micron, 10 micron. For 10 nm and 100 nm particles the critical (minimum) volume concentration of particles – $m_2^* = 6 \cdot 10^{-4}$, for 1 micron – $m_2^* = 4 \cdot 10^{-4}$, for 10 microns – $m_2^* = 5 \cdot 10^{-4}$. Thus it is seen that the efficiency of DW suppression in hydrogen-oxygen mixture in the transition from microparticles to nanoparticles does not increase, dependences goes at each other.

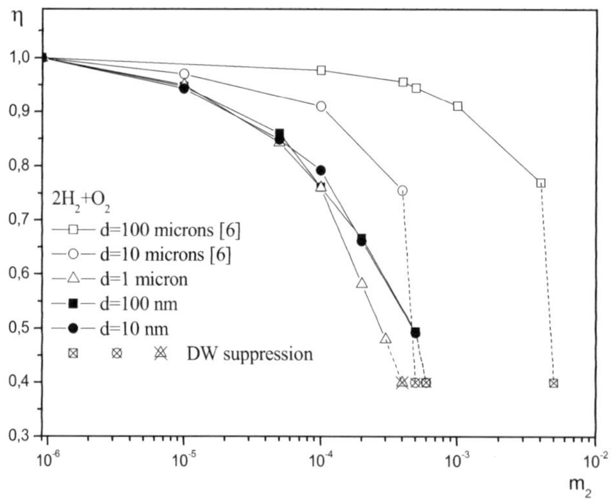

Figure 1. The dependencies of DW velocity deficit in the stoichiometric hydrogen-oxygen mixture vs the inert particles volume concentration. Comparison of the effectiveness of DW weakening and suppression by micro- and nanoparticles.

3.2. Methane-oxygen mixture

IOP Conf. Series: Journal of Physics: Conf. Series **894** (2017) 012099 doi:10.1088/1742-6596/894/1/012099

Figure 2 shows the dependence of the detonation velocity deficit in methane-oxygen mixture on volume concentration of inert particle with diameters ranging from 10 nm to 100 microns. The data for micro particles are taken from [5]. It is seen that all the above-described flow regimes in the hydrogen-oxygen mixture is valid for the methane-oxygen mixture. However, for a gas suspension with 1 micron particles, the decreasing of the DW velocity deficit is observed only to value $\eta = 0.8$, after which the failure and suppression of DW are occurs. In addition, the efficiency of DW suppression in methane-oxygen mixture in the transition from microparticles to nanoparticles does not increase too. Critical volume concentrations of inert particles are as follows: for 10 nm – $m_2^* = 1.5 \cdot 10^{-3}$, for 100 nm – $m_2^* = 1.3 \cdot 10^{-3}$, for 1 micron – $m_2^* = 2 \cdot 10^{-4}$, for 10 microns – $m_2^* = 2 \cdot 10^{-3}$.

If we compare the effectiveness of DW suppression in mixtures of hydrogen-oxygen and methane-oxygen, it is seen that the micro particles with the diameter ranging from 10 microns to 100 microns, and from 10 nm to 100 nm effectively suppress the detonation in hydrogen-oxygen mixture, whereas the 1 micron particle effectively suppress the detonation in methane-oxygen mixture.

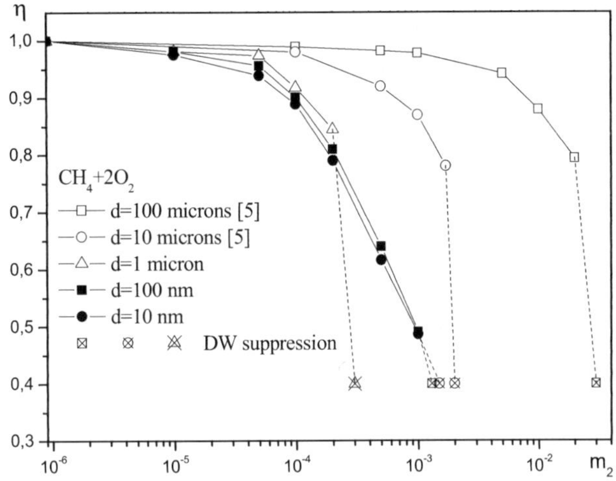

Figure 2. The dependencies of DW velocity deficit in the stoichiometric methane-oxygen mixture on the inert particles volume concentration. Comparison of the effectiveness of DW weakening and suppression by micro- and nanoparticles.

3.3. Silane-air mixture

It is seen in figure 3 that in the case of DW propagation in silane-air mixture and particles with the diameter ranging from 10 nm to 10 micron detonation velocity deficit decreases to values $\eta = 0.6$ after which the failure and suppression of DW are occurs. It should be noted that considered mathematical model is not valid for gas suspensions with particles volume concentration greater than 0.1. So in gas suspension with 100 micron particles in considered range of particles volume concentrations (till value $m_2 = 0.1$) detonation velocity deficit decreases only to value $\eta = 0.85$ and no suppression is observed.

Besides that, it is seen, that the efficiency of DW suppression in silane-air mixture in the transition region (from 10 nm to 100 nm) particles does not increase, dependences goes at each other.

Critical volume concentrations of inert particles suppressing the DW in silane-air mixture are follow: for 10 nm – $m_2^* = 5 \cdot 10^{-4}$, for 100 nm – $m_2^* = 5 \cdot 10^{-4}$, for 1 micron – $m_2^* = 1.5 \cdot 10^{-3}$, for 10 microns – $m_2^* = 4 \cdot 10^{-2}$.

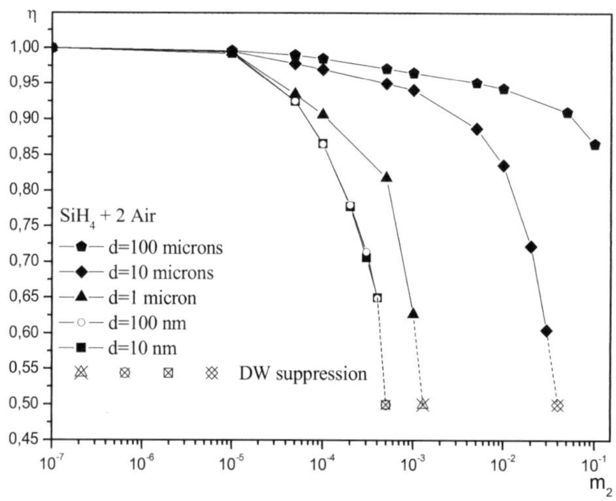

Figure 3. The dependencies of DW velocity deficit in the stoichiometric silane-air mixture on the inert particles volume concentration.

4. Conclusions

The physical and mathematical models describing the processes of ignition, combustion and detonation in mixtures of hydrogen-oxygen, methane-oxygen and silane-air in the presence of inert micro- and nanoparticles were proposed. Based on these models the dependencies of detonation velocity deficit vs the size and concentration of inert micro - and nanoparticles were found.

It was revealed that the two types of detonation flows exists in the mixture of a gas and nanoparticles, as well as in gas suspensions with micro particles: 1. stationary propagation of weakened DW at velocities less than Chapman-Jouget velocity; 2. the DW suppression. Furthermore, it is determined that the mechanisms of detonation suppression by micro- and nanoparticles are quite similar.

Concentration limits of detonation were calculated. It turned out that in the transition from microparticles to nanoparticles the detonation suppression efficiency does not increase.

Acknowledgments

The work was supported by Russian Science Foundation (project No. 16-19-00010) and Russian Foundation for Basic Research (grants No. 15-08-01947-a, No. 17-08-00634-a). Part of the work concerning calculations in the silane-air mixture were carried out on the basis of the RSF project, in the hydrogen-oxygen and methane-oxygen mixtures were carried out on the basis of the RFBR projects.

References

[1] Fedorov A V, Fomin P A, Fomin V M, Tropin D A and Chen J R 2012 *Mathematical Analysis of Detonation Suppression by Inert Particles* (Kao Tech Publishing, Kaohsiung, Taiwan) p 143 ISBN 978-986-88423-0-4

[2] Fedorov A V, Fomin P A, Fomin V M, Tropin D A and Chen J R 2011 *Physical - mathematical modeling of detonation suppression by clouds of fine particles* (Novosibirsk: Institute of Theoretical and Applied Mechanics SB RAS; NSUACE (Sibstrin)) p 156

[3] Fomin P A and Chen J R 2009 *Combustion, Explosion, and Shock Waves* **45**(3) 303–13

[4] Borisov A A, Gel'fand B E, Gubin S A and Kogarko S M 1975 *Combustion, Explosion and Shock Waves* **11**(6) 774–8

[5] Tropin D A and Fedorov A V 2014 *Combustion, Explosion and Shock Waves* **50**(5) 542–6

[6] Fedorov A V and Tropin D A 2013 *Combustion, Explosion and Shock Waves* **49**(2) 178–87

[7] Papalexandris M V 2004 *J. Fluid Mech.* **507** 95–142

[8] Shafiee H and Djavareshkian M H 2014 CFD *International Journal of Computer Theory and Engineering* **6**(6) 466–71
[9] Fedorov A V, Fomin P A, Tropin D A and Chen J R 2014 *Proceedings of the universities. Building* **9-10** 108–25
[10] Sundaram D S, Yang V and Zarko V E 2015 *Combustion, Explosion and Shock Waves* **51**(2) 173–96.
[11] Filippov A V and Rosner D E 2000 *International Journal of Heat and Mass Transfer* **43**(1) 127–38
[12] Bedarev I A and Fedorov A V 2006 *Combustion, Explosion and Shock Waves* **42**(1) 19–26
[13] Tien J H and Stalker R J 2002 *Combustion and Flame* **130** 329–48
[14] Westbrook C K and Urtiew P A 1983 *Combustion, Explosion and Shock Waves* **19(6)** 753–66
[15] Tropin D A and Fedorov A V 2015 *Combustion, Explosion and Shock Waves* **51**(4) 431–8.
[16] Tropin D A and Fedorov A V 2017 *Journal of Loss Prevention in the Process Industries* DOI: 10.1016/j.jlp.2017.02.029

Reduced detonation kinetics and detonation structure in one- and multi-fuel gaseous mixtures

P A Fomin[1], A V Trotsyuk and A A Vasil'ev

Lavrentyev Institute of Hydrodynamics SB RAS Novosibirsk 630090 Russia

E-mail: [1]pavel_fomin_new@mail.ru

Abstract. Two-step approximate models of chemical kinetics of detonation combustion of (i) one-fuel (CH_4/air) and (ii) multi-fuel gaseous mixtures (CH_4/H_2/air and CH_4/CO/air) are developed for the first time. The models for multi-fuel mixtures are proposed for the first time. Owing to the simplicity and high accuracy, the models can be used in multi-dimensional numerical calculations of detonation waves in corresponding gaseous mixtures. The models are in consistent with the second law of thermodynamics and Le Chatelier's principle. Constants of the models have a clear physical meaning. Advantages of the kinetic model for detonation combustion of methane has been demonstrated via numerical calculations of a two-dimensional structure of the detonation wave in a stoichiometric and fuel-rich methane-air mixtures and stoichiometric methane-oxygen mixture. The dominant size of the detonation cell, determines in calculations, is in good agreement with all known experimental data.

1. Introduction

Methane-air mixture is typical example of one-fuel gaseous mixture. The interest in modeling of detonation processes in gaseous mixtures containing methane has recently increased (explosion safety problems in mining industry). Numerical calculation of such processes needs the adequate models of chemical kinetics.

Typically the detailed systems of equations of chemical kinetics in hydrocarbons (including methane)-air mixtures are cumbersome. Therefore, approximate (reduced) kinetic models are frequently used for numerical calculations of multifront detonation waves (see, for example, calculation of detonation wave structure in methane-air mixture [1]). In contrast with the reality, in the frames of such models the heat release of chemical reaction does not depend on pressure and temperature and molar mass and adiabatic index of gaseous mixture assume to be constant. Therefore there is still an urgent need to develop simple but highly accurate approximate kinetic models for detonation combustion of methane, hydrogen peroxide and ozone.

One of the most suitable approximate models of chemical kinetics in one-fuel mixtures of hydrogen with oxygen, water vapor, and inert diluents is the model, presented in [2-5]. This model allows one to calculate molar mass evolution with the help of one differential equation, energy release and changes in thermodynamic parameters of the gas after the induction period with the help of several explicit algebraic formulas. The calculation of the induction period of chemical reaction is not included in the framework of this model. The model is also applicable for calculating thermodynamic parameters of the gas in the chemical equilibrium state. Explicit algebraic formulas are obtained for such calculations. The model is widely used to solve a wide range of research and application problems (for

Content from this work may be used under the terms of the Creative Commons Attribution 3.0 licence. Any further distribution of this work must maintain attribution to the author(s) and the title of the work, journal citation and DOI.
Published under licence by IOP Publishing Ltd

IOP Conf. Series: Journal of Physics: Conf. Series **894** (2017) 012100 doi:10.1088/1742-6596/894/1/012100

example, modeling of cellular structure of detonation waves [6], gas-droplet detonation [7], detonation application of coatings [8], pulse detonation engine [9], and explosion processes in bubble systems [10]).

In our articles [11-13] we have presented the reduced model of chemical kinetics in methane-air mixtures: $CH_4 + a_1O_2 + a_2N_2$. This model is the further modification of the model [2-5] for detonation combustion of hydrogen-air mixtures and has similar advantages: clear physical meaning of constants, simplicity, high accuracy, consistency with the second law of thermodynamics and Le Chatelier's principle. The model was successfully used for 2-D numerical calculations of detonation wave structure [12, 13]. Unfortunately, this model is valid only for $a_1 \geq 0.8$. In the present work a new modification of this model will be presented. This model will be valid for $a_1 \geq 0.5$.

There are not simple and physically grounded reduced models of chemical kinetics of detonation combustion of multi-fuel gaseous mixtures up to now. In the present work such models for CH_4/H_2/air and CH_4/CO/air will be presented for the first time.

2. Reduced model of chemical kinetics of detonation combustion of methane

We will consider the following mixture: $CH_4 + a_1O_2 + a_2N_2$, $a_1 \geq 0.5$. We will not consider the mixture with $a_1 > 0.5$, because fuel concentration for such mixture is well below the Lower Flammable Limit and out of detonation limit as well.

The model has two steps. The first step is induction period and the second one corresponds to main heat release. Note, that two-step model of chemical kinetic (including the induction step and the step of main heat release) for description of detonation processes in gaseous mixtures was used for the first time in [14].

As is commonly done, we calculate the value of the induction parameter Y during the induction period of the chemical reaction in the gas mixture with variable pressures and temperatures by the formula [15]:

$$Y = \int_0^{t_i} \frac{dt}{\tau_i}$$

where τ_i is the induction period at constant parameters. At the beginning of the induction period, $Y = 0$, and at the end of it (the moment in time $t = t_i$), $Y = 1$. Semi-empirical relations have to be used to calculate τ_i. The value of τ_i for the methane–air mixture will be calculated, for example, by the formula, recommended in [16] as the best for calculations of the cell size of the detonation wave.

We replace the real multistep process occurring during the induction period with some overall reaction chosen from the following general considerations. We take into account that by the end of the induction period, the temperature increase (and, hence, the total heat release due to chemical reactions) is low. Therefore, the overall reaction is chosen so that its thermal effect is much smaller than the maximum possible thermal effect of the complete recombination of the reaction products to form CO_2 and H_2O molecules. Note that under this condition, the thermal effect of the overall reaction will also be much smaller than the thermal effect at the Chapman–Jouguet point.

We also take into account that the induction period involves chemical reactions leading to the formation and increase in the number of active centers and decomposition of heavy molecules. In our case, the heavy molecules are methane molecules. It is therefore assumed that during the induction period, all methane molecules decompose to form CO molecules. Single C atoms in the products of the overall reaction are neglected since the mixture contains a more than sufficient amount of oxidizer.

Thus, we assume that during the induction period, each methane molecule of the mixture is involved in the chemical reactions:

$$CH_4 + a_1O_2 \rightarrow CO + (2a_1 - 1)OH + (2.5 - a_1 - \alpha)H_2 + 2\alpha H \text{ if } 0.5 \leq a_1 < 1,$$

$$CH_4 + O_2 \rightarrow CO + OH + 1.5H_2 \text{ if } 1 \leq a_1 .$$

IOP Conf. Series: Journal of Physics: Conf. Series **894** (2017) 012100 doi:10.1088/1742-6596/894/1/012100

The value α should be calculated from the condition: the thermal effect of the reaction at the end of the induction period Q_τ is significantly smaller than the maximum possible heat effect of the chemical reaction Q_{max} per one mole of CH_4. It is easy to show that $Q_\tau / Q_{max} <<1$ for $1 \le a_1$. If $0.8 \le a_1 \le 1$ then $Q_\tau / Q_{max} \le 0.1$ at $\alpha = 0$. If a_1 equals to 0.5, 0.6 and 0.7, then $Q_\tau / Q_{max} \le 0.1$ at α = 0.097, 0.067 and 0.036 respectively.

Let β be the fraction of un-decomposed methane. In the initial state, $\beta = 1$; during the induction period, β decreases monotonically; and at the end of the induction period, $\beta = 0$. According to the above, the chemical composition of the gas undergoes the following change during induction period.

For $1 \le a_1$,

$$CH_4 + a_1O_2 + a_2N_2 \rightarrow \beta CH_4 + (1\text{-}\beta)CO + (1\text{-}\beta)OH + (1.5\text{–}1.5\beta)H_2 + (a_1\text{-}1+\beta)O_2 + a_2N_2.$$

For $0.5 \le a_1 < 1$,

$$CH_4 + a_1O_2 + a_2N_2 \rightarrow \beta CH_4 + (1-\beta)CO + (1-\beta)(2a_1 - 1)OH + [(1-\beta)(2.5 - a_1 - \alpha)]H_2 +$$
$$+ (1-\beta)2\alpha H + a_1\beta O_2 + a_2N_2.$$

We note that the value of β affects only the wave parameter profiles in the induction zone. The wave velocity, flow parameters at the Chapman-Jouguet point, and zone of main heat release do not depend on β. Therefore, if the above conditions are satisfied, the particular form of the formula for calculating β affects the wave parameter profiles in the induction zone only quantitatively. As a rule, the chemical reaction rate increases at the end of the induction period. It is therefore reasonable to choose a formula for calculatingβ that it meets the above conditions, and the rate of decrease as the movement grew mixture in the induction zone. For example, the quantity β can be defined as

$$\beta = 1 - Y .$$

Within this kinetic model, the chemical composition of the gas during the induction period is calculated by explicit algebraic formulas. Therefore, the molar mass and the internal energy of the gas can be easily calculated by the well-known standard algorithms using the thermodynamic parameters of individual species.

After induction period (the second step of reaction) the model includes a kinetic equation for the molar mass of the gas [4]

$$\frac{d\mu}{dt} = 4K_+ \frac{\rho^2}{\mu}(1 - \frac{\mu}{\mu_{max}})^2 - AT^{3/4}(1 - \exp(-\Theta/T))^{3/2}\rho(\frac{\mu}{\mu_{min}} - 1)\exp(-E/\Re T)$$

and formulas for the specific internal energy of the mixture [2-5]:

$$U(T,\mu) = U_{therm}(T,\mu) + U_{chem}(\mu),$$

$$U_{therm}(T,\mu) = [\frac{3}{4}(\frac{\mu}{\mu_a} + 1) + \frac{3}{2}(\frac{\mu}{\mu_a} - 1)\frac{\Theta/T}{\exp(\Theta/T) - 1}]\frac{\Re T}{\mu},$$

$$U_{chem}(\mu) = E(\frac{1}{\mu} - \frac{1}{\mu_{min}}),$$

where ρ, T and μ are the density, temperature, and the mean molar mass of the gaseous mixture; $\Re$ is the universal gas constant; μ_a, μ_{min}, μ_{max} are the molar masses of gas in the atomic, completely dissociated and completely recombined states; A and K_+ are the rate constants of dissociation and recombination of the generalized reaction products; θ is the effective excitation temperature of the vibrational degrees of freedom of the molecules; E is the mean dissociation energy of the reaction products; U is the total specific internal energy of the gas, U_{therm} and U_{chem} are thermodynamic and chemical parts of U; t is the time. The algorithm for calculation the model constants are presented in [12, 13].

The first and second terms in the kinetic equation correspond to recombination and dissociation reactions.

The model takes into account that the energies of dissociation of diatomic molecules (H_2, O_2, OH) into atoms are approximately equal to the energy of dissociation of H_2O molecule into OH and H and CO_2 molecule into CO and O. Thus, the recombination reaction heats are assumed to be identical and noted as E. It is taken into account that under detonation temperatures molecules CO and N_2 practically not dissociate into atoms.

It is known that recombination reactions occur almost without the activation energy. Therefore, the activation energy of recombination reactions in the model is assumed to be zero. As a result, the first term in the kinetic equation has no exponent. The activation energy in the second term equals to E. The reason is that the difference between the activation energy of the dissociation reaction and that of the reverse recombination reaction equals the chemical reaction heat.

For a specified caloric equation of state, the rate constants of direct and reverse chemical reactions in the kinetic equation cannot be arbitrary but should be correlated with allowance for the second law of thermodynamics. Therefore, the form of the temperature-dependent pre-exponent of the dissociation rate constant was obtained from the condition that thermodynamic potentials of the gas have an extremum in chemical equilibrium ($d\mu/dt = 0$).

The recombination rate constants differ from each other insignificantly. In the kinetic model, therefore, they are assumed to be identical and equal to K_+. Formula for calculation U_{therm} was derived by analyzing the mean number of degrees of freedom of molecules versus temperature. Formula for calculation U_{chem} is based on the assumption that the dissociation energy of diatomic molecules and the energy of dissociation of H_2O molecule into OH and H and CO_2 into CO and O equal to E.

The presented model can also be used to calculate thermodynamic parameters of the gas in chemical equilibrium ($d\mu/dt = 0$ in).

3. Reduced models of chemical kinetics of detonation combustion of multi-fuel gaseous mixtures CH_4/H_2/air and CH_4/CO/air

In case of multi-fuel gaseous mixtures CH_4/H_2/air and CH_4/CO/air it is assumed that during induction period in detonation wave H_2 and CO are not involved in chemical transformation. Therefore, in these mixtures the same chemical process in the induction zone as in CH_4/Air mixture occurs.

To calculate induction period of multi-fuel gaseous mixtures under consideration, an algorithm, presented in [17], should be used. Constants of Arrhenius formulas for calculation induction period of chemical reaction in one-fuel mixtures of CH_4, H_2 or CO with O_2/Air assumed to be known.

After induction period the same formulas for calculation internal energy, molar mass and adiabatic index, as for CH_4/Air mixtures, should be used.

4. Results of calculations of detonation wave structure in $CH_4/Air/O_2$ mixtures

Advantages of the kinetic model for detonation combustion of methane has been demonstrated via numerical calculations of a two-dimensional structure of the detonation wave in a stoichiometric and fuel-rich ($1.5CH_4$+2Air) methane-air mixtures and stoichiometric methane-oxygen mixture. The numerical algorithm, presented in [4, 5] is used. According to calculations, qualitative structures of detonation waves in all these mixtures are identical. Typical results of calculations are presented in Figures 1 and 2; H is the thickness of a channel, a_0 is the detonation cell size, D_{aver} is the average velocity of detonation wave front; D_{CJ} is Chapman-Jouguet detonation wave velocity (calculation, based on detailed equations of chemical equilibrium). Figure 1: $T_0 = 298$ K, $P_0 = 1$ atm, $H = a_0 = 45$ cm, $x_{front} = 1500$ cm. $D_{aver} = 1876$ m/s, $D_{CJ} = 1840$ m/s. Figure 2: $T_0 = 298$ K, $P_0 = 1$ atm, $H = a_0 = 50$ cm.

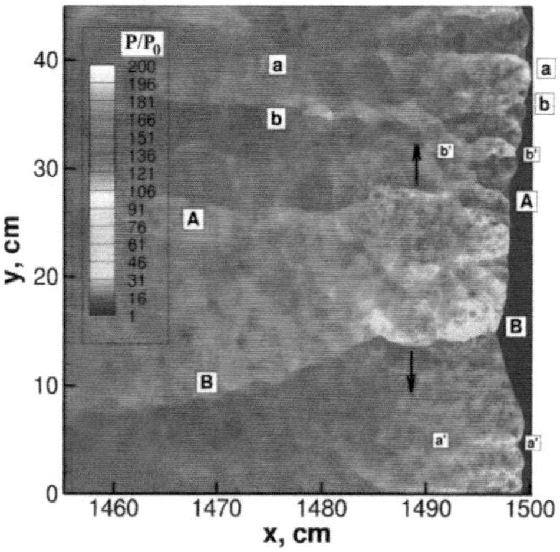

Figure 1. Detonation wave structure (normalized pressure P/P_0 field) in fuel-rich ($1.5CH_4 + 2Air$) mixture. AA and BB – primary transverse waves; aa and bb – secondary transverse waves; a'a' and b'b' – tertiary transverse waves.

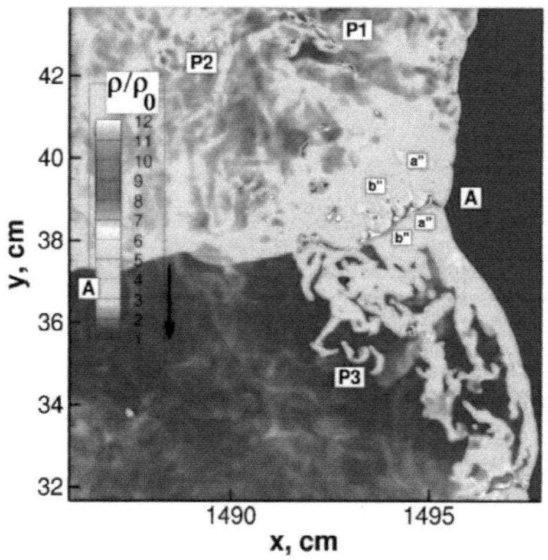

Figure 2. "Thin" structure of transverse wave (normalized density ρ/ρ_0 field (enlarged) in fuel-rich ($1.5CH_4 + 2Air$) mixture. AA – primary transverse wave; a"a" and b"b" – secondary transverse waves on the primary transverse wave AA. P1, P2 and P3 – pockets of unburnt mixture.

The calculations reproduce an irregular cell structure with all of its main features; random inconsistent movement of the main transverse waves; numerous secondary transverse waves, forming the hierarchy of perturbations of decreasing size at the leading shock front of detonation wave; regions of unburned mixture at a considerable distance behind detonation front; fine (cellular) structure of the transverse wave. These secondary waves are responsible for the small-cell network inside the main cell in soot imprints in experiments. The fine structure is manifested in the typical shape of transverse-wave trajectories recorded in soot imprints. Results of calculations of parameters and dominant cell size of detonation waves reveal good agreement with available experimental data.

5. Conclusions

Approximate models of chemical kinetics of detonation combustion of (i) one-fuel (CH_4/air) and (ii) multi-fuel (CH_4/H_2/air and CH_4/CO/air) gaseous mixtures are developed. Owing to the simplicity and high accuracy, the models can be used in multi-dimensional numerical calculations of detonation waves in corresponding gaseous mixtures.

The validity of the model for $CH_4/Air/O_2$ mixture are shown. For this purpose numerical calculations of detonation wave structure in stoichiometric CH_4/Air and CH_4/O_2 mixtures and fuel-rich CH_4/Air mixture are performed. Chemical reaction in these calculations is described by the presented model of chemical kinetics. The calculations reproduce an irregular cell structure with all of its main features; random inconsistent movement of the main transverse waves; numerous secondary transverse waves, forming the hierarchy of perturbations of decreasing size at the leading shock front of detonation wave; regions of unburned mixture at a considerable distance behind detonation front; fine (cellular) structure of the transverse wave. These secondary waves are responsible for the small-cell network inside the main cell in soot imprints in experiments. The fine structure is manifested in the typical shape of transverse-wave trajectories recorded in soot imprints. Results of calculations of parameters and dominant cell size of detonation waves reveal good agreement with available experimental data.

Further research is planned to apply the proposed model of the detonation kinetics for numerical 2-D calculations of detonation wave parameters and structure in multi-fuel mixtures CH_4/H_2/Air and CH_4/CO/Air.

Acknowledgments
The authors gratefully acknowledge the financial support from Russian Foundation for Basic Research (Grant № 17-03-01351).

References

[1] Oran E S, Weber J W, Stefaniw E I, Lefebvre M H and Anderson J D 1998 *Combust. Flame* **113** 147-163
[2] Nikolaev Yu A 1978 *Combust. Explos. Shock Waves* **3** 468–471
[3] Nikolaev Yu A and Fomin P A 1982 *Combust. Explos. Shock Waves* **1** 53–58
[4] Nikolaev Yu A and Zak D V 1988 *Combust. Explos. Shock Waves* **4** 461–464
[5] Fomin P A and Trotsyuk A V 1995 *Combust. Explos. Shock Waves* **4**, 455–457.
[6] Trotsyuk A V 1999 *Combust. Explos. Shock Waves* **5** 549–558
[7] Nikolaev Yu A and Fomin P A 1984 *Combust. Explos. Shock Waves* **4** 447–455
[8] Gavrilenko T P, Grigoriev V V, Zhdan S A, Nikolaev Yu A., Boiko V M and Papyrin A N 1986 *Combust. Flame* **66** 121–128
[9] Zhdan S A, Mitrofanov V V and Sychev A I 1994 *Combust. Explos. Shock Waves* **5** 657–663
[10] Fomin P A and Vasil'ev A A. 2001 *J. Loss. Prev. Proc. Ind.* **6** 521–526
[11] Fomn P A, Fedorov A.V. and Tropin D.A. 2014 *Combust. Explos. Shock Waves* **1** 87–96
[12] Fomin P A, Trotsyuk A V and Vasil'ev A A 2014 *Combust. Sci. Technol.* **10-11** 1716–1735
[13] Trotsyuk A V, Fomin P A and Vasil'ev A A 2015 *J. Loss. Prev. Proc. Ind.* **36** 394–403
[14] Korobeinikov V P, Levin V A, Markov V V and Chernyi G G 1972 *Acta Astronaut.* **17** 529–537
[15] Strickland-Constable R F 1949 *Proc. Third Symp. on Combustion and Flame, and Explosion Phenomena*, **33**(1) 229
[16] Soloukhin R I 1970 *Proc. 7th Int. Shock Tubes Symposium* (University of Toronto Press. Toronto and Buffalo) 662–706
[17] Vasil'ev A A 2007 *Combust. Explos. Shock Waves* **3** 282–285

Parameter identification problems for thin inclusions in elastic bodies

A M Khludnev[1,2] and T S Popova[3]

[1] Lavrentyev Institute of Hydrodynamics, Novosibirsk, Russia
[2] Novosibirsk State University, Novosibirsk, Russia
[3] North-Eastern Federal University, Yakutsk, Russia

E-mail: khlud@hydro.nsc.ru, ptsokt@mail.ru

Abstract. The paper concerns an identification of rigidity parameters for thin inclusions located inside elastic bodies. A delamination of the inclusions is assumed thus providing a crack between inclusions and the elastic matrix. Inequality type boundary conditions are imposed at the crack faces to exclude a mutual penetration. We consider elastic as well as rigid inclusions and solve an optimal control problem for finding a rigidity parameter minimizing a suitable cost functional. The cost functional characterizes a displacement of the inclusion, and a rigidity parameter serves as a control function. We prove a solution existence of the problems formulated.

1. Setting the problem

Let $\Omega \subset \mathbb{R}^2$ be a bounded domain with smooth boundary Γ, and $\gamma_0 = (-1,0) \times \{0\}$, $\gamma_0 \subset \Omega$, $\Omega_\gamma = \Omega \setminus \bar{\gamma}_0$, $\gamma_e = (0,1) \times \{0\}$, $\gamma = \gamma_0 \cup \gamma_e \cup \{(0,0)\}$. We assume that γ_e does not belong to Ω, see Figure 1. Denote by $\nu = (0,1)$ a unit normal vector to γ; $\tau = (1,0)$ is a tangential vector.

Let $B = \{b_{ijkl}\}, i,j,k,l = 1,2$, be a given elasticity tensor with the usual properties of symmetry and positive definiteness,

$$b_{ijkl} = b_{jikl} = b_{klij}, \quad i,j,k,l = 1,2, \; b_{ijkl} \in L^\infty(\Omega),$$

$$b_{ijkl}\xi_{ij}\xi_{kl} \ge c_0|\xi|^2 \quad \forall \; \xi_{ij}, \; c_0 = const > 0 \; .$$

In what follows, we identify functions defined on γ with functions of the variable $x_1 \in \mathbb{R}$; $x = (x_1, x_2) \in \mathbb{R}^2$. Summation convention over repeated indices is used; all functions with two lower indices are assumed to be symmetric in these indices.

Let $f = (f_1, f_2) \in L^2(\Omega)^2$, $g \in L^2(\gamma)$ be given functions corresponding to external forces acting on the elastic body and the inclusion, respectively. The domain Ω_γ is a region filled with an elastic material, and γ corresponds to an inclusion crossing the boundary Γ at the point $(0,0)$. We assume that a delamination of the inclusion takes place. This means that a crack is located between the inclusion and the elastic matrix. To fix a cracking we assume that the positive (with respect to ν) side of γ_0 is delaminated.

We first provide a formulation of an equilibrium problem for the elastic body with the elastic inclusion γ. We want to find a displacement field $u^\lambda = (u_1^\lambda, u_2^\lambda)$, a stress tensor $\sigma^\lambda = \{\sigma_{ij}^\lambda\}, i,j = 1,2$, defined in Ω_γ, and a thin inclusion displacement v^λ defined on γ such that

$$-\text{div } \sigma^\lambda = f; \;\; \sigma^\lambda - B\varepsilon(u^\lambda) = 0 \text{ in } \Omega_\gamma, \tag{1}$$

Content from this work may be used under the terms of the Creative Commons Attribution 3.0 licence. Any further distribution of this work must maintain attribution to the author(s) and the title of the work, journal citation and DOI.
Published under licence by IOP Publishing Ltd

$$\lambda v^{\lambda}_{,1111} = [\sigma^{\lambda}_{\nu}] + g \text{ on } \gamma_0; \ \lambda v^{\lambda}_{,1111} = g \text{ on } \gamma_e, \tag{2}$$

$$u^{\lambda} = 0 \text{ on } \Gamma; \ v^{\lambda}_{,11} = v^{\lambda}_{,111} = 0 \text{ for } x_1 = -1, 1, \tag{3}$$

$$v^{\lambda} = u^{\lambda -}_{\nu}, \ [u^{\lambda}_{\nu}] \ge 0, \ \sigma^{\lambda +}_{\nu}[u^{\lambda}_{\nu}] = 0, \ \sigma^{\lambda +}_{\nu} \le 0, \ \sigma^{\lambda \pm}_{\tau} = 0 \text{ on } \gamma_0, \tag{4}$$

$$v^{\lambda}(0) = [v^{\lambda}_{,1}(0)] = [v^{\lambda}_{,11}(0)] = 0. \tag{5}$$

Here, λ is a positive rigidity parameter of the elastic inclusion, $[p] = p^+ - p^-$ is a jump of a

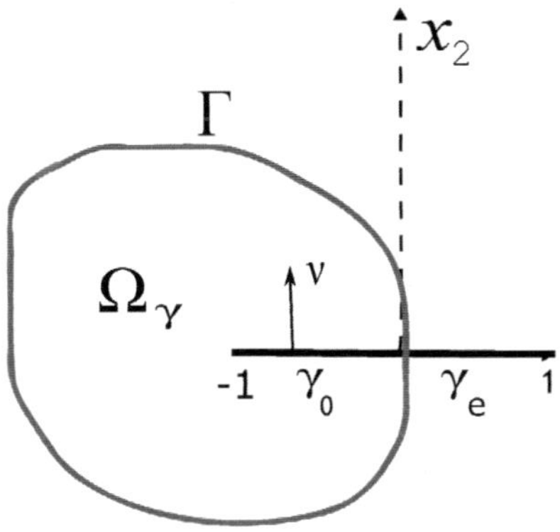

Figure 1. Elastic body with thin inclusion.

function p on γ, where $p^{\pm}$ are the traces of p on the faces $\gamma^{\pm}$. The signs $\pm$ correspond to positive and negative directions of ν; $u_\nu = u\nu$; $[w(0)] = w(0+) - w(0-)$; $\varepsilon(u) = \{\varepsilon_{ij}(u)\}$ is a strain tensor,

$$\sigma_\nu = \sigma_{ij}\nu_j\nu_i, \ \sigma_\tau = \sigma_{ij}\nu_j\tau_i, \ \varepsilon_{ij}(u) = \frac{1}{2}(u_{i,j} + u_{j,i}), \ i, j = 1, 2 \,.$$

We use notations $p_{,i} = \frac{dp}{dx_i}$, and write the equilibrium equations (2) for the elastic inclusion γ. The term $[\sigma^{\lambda}_{\nu}]$ in the first equation of (2) describes forces acting on γ_0 from the surrounding elastic body.

Relations (1) are equilibrium equations and Hooke's law, respectively. According to the first relation of (4), the vertical (along the axis x_2) displacements of the elastic body on γ_0^- coincide with displacements of the inclusion. The first inequality of (4) provides a mutual nonpenetration between the crack faces. All the rest relations of (4) are standard for contact problems with unknown set of a contact, see [1, 2, 3]. The second group of boundary conditions of (3) fits to zero moments and zero transverse forces at the tip points of the inclusion γ. Relations (5) are junction conditions at the point $(0, 0)$, see, for example, [4, 5].

The problem (1)-(5) can be formulated as a variational one. To this end, introduce Sobolev spaces

$$H^1_\gamma(\Omega_\gamma) = \{\phi \in H^1(\Omega_\gamma) \,:\, \phi = 0 \text{ on } \gamma\}, \ H^{2,0}(\gamma) = \{w \in H^2(\gamma) \,:\, w(0) = 0\}$$

and a set of admissible displacements

$$K = \{(u, v) \in H^1_\gamma(\Omega_\gamma)^2 \times H^{2,0}(\gamma) \ : \ [u_\nu] \ge 0 \ \text{ on } \ \gamma_0; \ u^-_\nu|_{\gamma_0} = v\}.$$

IOP Conf. Series: Journal of Physics: Conf. Series **894** (2017) 012101 doi:10.1088/1742-6596/894/1/012101

Consider the energy functional

$$\pi_\lambda(u,v) = \frac{1}{2}\int\limits_{\Omega_\gamma} \sigma(u)\varepsilon(u) - \int\limits_{\Omega_\gamma} fu + \frac{\lambda}{2}\int\limits_{\gamma} v_{,11}^2 - \int\limits_{\gamma} gv$$

and the minimization problem:

$$\text{find } (u^\lambda, v^\lambda) \in K \text{ such that } \pi_\lambda(u^\lambda, v^\lambda) = \inf_K \pi_\lambda. \tag{6}$$

Here and below we write $\sigma(u)\varepsilon(u)$ instead of $\sigma_{ij}(u)\varepsilon_{ij}(u)$. A solution of the problem (6) satisfies the variational inequality

$$(u^\lambda, v^\lambda) \in K, \tag{7}$$

$$\int\limits_{\Omega_\gamma} \sigma(u^\lambda)\varepsilon(\bar{u} - u^\lambda) - \int\limits_{\Omega_\gamma} f(\bar{u} - u^\lambda) + \tag{8}$$

$$+\lambda \int\limits_{\gamma} v_{,11}^\lambda(\bar{v}_{,11} - v_{,11}^\lambda) - \int\limits_{\gamma} g(\bar{v} - v^\lambda) \geq 0 \ \forall (\bar{u}, \bar{v}) \in K.$$

The problem (7)-(8) has a unique solution. Moreover, it can be checked that problem formulations (1)-(5) and (7)-(8) are equivalent provided that the solutions are smooth, see [3].

2. Rigid Inclusion

In this section, we formulate an equilibrium problem in the case when the inclusion γ is a rigid one, see Figure 1. This means that a displacement field on γ has a special structure. Introduce a space of infinitesimal rigid displacements

$$L(\gamma) = \{l \ : \ l(x_1) = cx_1, \ x_1 \in (-1,1), \ c \in \mathbb{R}\}.$$

An equilibrium problem for the elastic body occupying the domain Ω_γ with the rigid inclusion γ can be formulated as follows [6]. Find functions $u = (u_1, u_2)$, $\sigma = \{\sigma_{ij}\}, i, j = 1, 2$, $l_0 \in L(\gamma)$, such that

$$-\text{div } \sigma = f \text{ in } \Omega_\gamma, \tag{9}$$

$$\sigma - B\varepsilon(u) = 0 \text{ in } \Omega_\gamma, \tag{10}$$

$$u = 0 \text{ on } \Gamma; \ l_0 = u_\nu^- \text{ on } \gamma_0, \tag{11}$$

$$[u_\nu] \geq 0, \ \sigma_\nu^+ \leq 0, \ \sigma_\tau^\pm = 0, \ \sigma_\nu^+[u_\nu] = 0 \text{ on } \gamma_0, \tag{12}$$

$$\int\limits_{\gamma_0} [\sigma_\nu] l + \int\limits_{\gamma} gl = 0 \ \forall l \in L(\gamma). \tag{13}$$

According to (11), a displacement l_0 of the inclusion γ coincides with the vertical displacement of the elastic body on γ_0^-. Non-local boundary condition (13) guarantees that a principal moment of forces acting on γ is equal to zero.

The problem (9)-(13) admits a variational formulation. To this end, we introduce a set of admissible displacements

$$K_\infty = \{(u, l) \in H_\gamma^1(\Omega_\gamma)^2 \times L(\gamma) \ : \ [u_\nu] \geq 0 \ \text{ on } \ \gamma_0; \ u_\nu^-|_{\gamma_0} = l\}.$$

Then it is possible to check that the problem (9)-(13) is equivalent to the following variational inequality (provided that the solutions are quite smooth)

$$(u, l_0) \in K_\infty, \tag{14}$$

$$\int_{\Omega_\gamma} \sigma(u)\varepsilon(\bar{u} - u) - \int_{\Omega_\gamma} f(\bar{u} - u) - \int_\gamma g(\bar{l} - l_0) \geq 0 \ \ \forall (\bar{u}, \bar{l}) \in K_\infty. \tag{15}$$

The problem (14)-(15) has a unique solution. We omit a checking of this fact; moreover, checking the equivalence of the problem formulations (14)-(15) and (9)-(13) is also omitted.

To conclude this section, we should underline that the problem (14)-(15) is a limit one for the family of problems (7)-(8) provided that $\lambda \to \infty$. Therefore, for any fixed $\lambda > 0$ the model (1)-(5) describes an equilibrium state of the elastic body with the elastic inclusion γ. This model reduces to the model (9)-(13) provided that $\lambda \to \infty$ what describes an equilibrium of the elastic body with the rigid inclusion γ.

3. Optimal Control Problem

This section concerns an investigation of an optimal control problem for an elastic body with thin delaminated inclusions, and a rigidity parameter λ being a control function. For any fixed $\lambda > 0$ we can find a solution of the problem

$$(u^\lambda, v^\lambda) \in K, \tag{16}$$

$$\int_{\Omega_\gamma} \sigma(u^\lambda)\varepsilon(\bar{u} - u^\lambda) - \int_{\Omega_\gamma} f(\bar{u} - u^\lambda) + \tag{17}$$

$$+\lambda \int_\gamma v^\lambda_{,11}(\bar{v}_{,11} - v^\lambda_{,11}) - \int_\gamma g(\bar{v} - v^\lambda) \geq 0 \ \ \forall (\bar{u}, \bar{v}) \in K.$$

For $\lambda = \infty$, we can find a solution of the problem (14)-(15). Hence, for a finite λ we have an elastic inclusion γ, and $\lambda = \infty$ corresponds to the rigid inclusion γ.

A cost functional will characterize a displacement of the inclusion part γ_e, and λ is a control parameter, $\lambda \in [\lambda_0, \infty]$. Let $v_* \in H^2(\gamma_e)$ be a given function. For each $\lambda \in [\lambda_0, \infty]$ we can find a solution (u^λ, v^λ) of the suitable boundary value problem. Define a cost functional for $\lambda \in [\lambda_0, \infty]$,

$$G(\lambda) = \|v^\lambda - v_*\|_{H^2(\gamma_e)}.$$

Optimal control problem is formulated as follows

$$\inf_{\lambda \in [\lambda_0, \infty]} G(\lambda). \tag{18}$$

We see that a solution of the optimal control problem (18) allows us to find a rigidity parameter that minimizes a difference on γ_e, in a suitable sense, between the solution v^λ and the given function v_*. It is very important that we compare solutions of different models corresponding to finite and infinite values of λ.

Theorem. There exists a solution of the optimal control problem (18).

Proof. Let $\lambda_n \in [\lambda_0, \infty]$ be a minimizing sequence. For any λ_n we can find a unique solution of the problem like (16)-(17) provided λ_n is finite, or of the problem (14)-(15) for $\lambda = \infty$. We can assume that the sequence is converging. There are two possible cases:

1. $\lambda_n \to \lambda_*$, $n \to \infty$, $\lambda_n \in [\lambda_0, \infty)$, $\lambda_* \in \mathbb{R}$;
2. $\lambda_n \to \infty$, $n \to \infty$, $\lambda_n \in [\lambda_0, \infty)$.

If $\lambda_n = +\infty$ for $n \geq n_0$, then a solution of the problem (18) clearly exists. We consider two cases separately.

Case 1. Assume that $\lambda_n \to \lambda_*$, $n \to \infty$, $\lambda_n \in [\lambda_0, \infty)$, $\lambda_* \in \mathbb{R}$. For every n we find a solution of the problem

$$(u^n, v^n) \in K, \tag{19}$$

$$\int\limits_{\Omega_\gamma} \sigma(u^n)\varepsilon(\bar{u} - u^n) - \int\limits_{\Omega_\gamma} f(\bar{u} - u^n) + \tag{20}$$

$$+\lambda_n \int\limits_\gamma v^n_{,11}(\bar{v}_{,11} - v^n_{,11}) - \int\limits_\gamma g(\bar{v} - v^n) \geq 0 \;\; \forall\, (\bar{u}, \bar{v}) \in K.$$

It is possible to pass to the limit in (19)-(20) as $n \to \infty$. The following estimates take place uniformly in $n \geq n_0$, see [7],

$$\|u^n\|^2_{H^1_\gamma(\Omega_\gamma)^2} + \|v^n\|^2_{H^{2,0}(\gamma)} \leq c. \tag{21}$$

By (21), we can assume that as $n \to \infty$

$$u^n \to u \text{ weakly in } H^1_\gamma(\Omega_\gamma)^2, \;\; v^n \to v \text{ weakly in } H^{2,0}(\gamma).$$

This convergence allows us to pass to the limit in (19)-(20) as $n \to \infty$ which gives

$$(u, v) \in K, \tag{22}$$

$$\int\limits_{\Omega_\gamma} \sigma(u)\varepsilon(\bar{u} - u) - \int\limits_{\Omega_\gamma} f(\bar{u} - u) + \tag{23}$$

$$+\lambda_* \int\limits_\gamma v_{,11}(\bar{v}_{,11} - v_{,11}) - \int\limits_\gamma g(\bar{v} - v) \geq 0 \;\; \forall\, (\bar{u}, \bar{v}) \in K.$$

Relations (22)-(23) mean that $(u, v) = (u^{\lambda_*}, v^{\lambda_*})$. Thus, we have

$$\inf_{\lambda \in [\lambda_0, \infty]} G(\lambda) = \lim G(\lambda_n) = \lim\inf \|v^n - v_*\|_{H^2(\gamma_e)} \geq$$

$$\geq \|v - v_*\|_{H^2(\gamma_e)} = G(\lambda_*) \geq \inf_{\lambda \in [\lambda_0, \infty]} G(\lambda),$$

and the existence proof is complete in the case 1.

Case 2. Assume that $\lambda_n \to \infty$, $n \to \infty$, $\lambda_n \in [\lambda_0, \infty)$. In this case, for any n, the solution (u^n, v^n) satisfies (19)-(20). From (19)-(20) we obtain the relation

$$\int\limits_{\Omega_\gamma} \sigma(u^n)\varepsilon(u^n) - \int\limits_{\Omega_\gamma} f u^n + \lambda_n \int\limits_\gamma (v^n_{,11})^2 - \int\limits_\gamma g v^n = 0. \tag{24}$$

From (24) we can derive for $n \geq n_0$

$$\|u^n\|^2_{H^1_\gamma(\Omega_\gamma)^2} \leq c, \;\; \|v^n\|^2_{H^{2,0}(\gamma)} \leq c, \tag{25}$$

and moreover, from (24) it follows for $n \geq n_0$

$$\int\limits_\gamma (v^n_{,11})^2 \leq \frac{c}{\lambda_n}. \tag{26}$$

By (25)-(26), we assume that

$$u^n \to u \text{ weakly in } H^1_\gamma(\Omega_\gamma)^2, \; v^n \to v \text{ weakly in } H^{2,0}(\gamma), \tag{27}$$

$$v(x_1) = a_1 x_1, \; x_1 \in (-1, 1). \tag{28}$$

Denote $l_0(x_1) = a_1 x_1$, $x_1 \in (-1, 1)$. We take any fixed element $(\bar{u}, \bar{l}) \in K_\infty$. It is clear that $(\bar{u}, \bar{l}) \in K$. Substitute $(\bar{u}, \bar{l})$ as a test function in (20) and pass to the limit as $n \to \infty$. It provides

$$(u, l_0) \in K_\infty,$$

$$\int_{\Omega_\gamma} \sigma(u)\varepsilon(\bar{u} - u) - \int_{\Omega_\gamma} f(\bar{u} - u) - \int_\gamma g(\bar{l} - l_0) \geq 0 \;\; \forall (\bar{u}, \bar{l}) \in K_\infty.$$

Thus, the limit functions u, v from (27)-(28) satisfy the condition $(u, l_0) = (u^\infty, v^\infty)$, compare with (14)-(15). By this, we obtain

$$\inf_{\lambda \in [\lambda_0, \infty]} G(\lambda) = \lim G(\lambda_n) = \liminf \|v^n - v_*\|_{H^2(\gamma_e)} \geq$$

$$\geq \|l_0 - v_*\|_{H^2(\gamma_e)} = G(\infty) \geq \inf_{\lambda \in [\lambda_0, \infty]} G(\lambda).$$

The proof of Theorem is complete.

References

[1] Khludnev A M and Negri M 2012 *Z. Angew. Math. Mech.* **92** 341–54
[2] Khludnev A M and Leugering G R 2014 *Math. Mech. Complex Systems* **2** 1–21
[3] Khludnev A M 2015 *Z. Angew. Math. Mech.* **95** 1256–67
[4] Khludnev A M and Popova T S 2016 *Arch. Appl. Mech.* **86** 1565–77
[5] Khludnev A M and Popova T S 2016 *Quart. Appl. Math.* **74** 705–18
[6] Khludnev A M 2015 *J. Appl. Math. Mech.* **79** 493–9
[7] Khludnev A M 2017 *J. Opt. Theory Appl.* **172** 281–97

IOP Conf. Series: Journal of Physics: Conf. Series **894** (2017) 012102 doi:10.1088/1742-6596/894/1/012102

Cellular detonations in nano-sized aluminum particle gas suspensions

TA Khmel

Khristianovich Institute of Theoretical and Applied Mechanics SB RAS, Institutskaya 4/1, 630090, Novosibirsk, Russia

E-mail: khmel@itam.nsc.ru

Abstract. Formation of cellular detonation structures in monodisperse nano-sized aluminum particle – oxygen suspensions is studied by methods of numerical simulations of two-dimensional detonation flows. The detonation combustion are described within the semi-empirical model developed earlier which takes into account transition of the regime of aluminum particle combustion from diffusion to kinetic for micro-sized and nano-sized particles. The free-molecular effects are considered in the processes of heat and velocity relaxation of the phases. The specific features of the cellular detonation of nanoparticle suspensions comparing with micron-sized suspensions are irregular cellular structures, much higher pick pressure values, and relatively larger detonation cells. This is due to high value of activation energy of reduced chemical reaction of aluminum particle combustion in kinetic regime.

1. Introduction

The combustion of nano-dispersed aluminum powders, including in the detonation mode, attracts great interest in view of the wide prospects for their use. The problems of detonations in aluminum particles gas suspensions have been extensively studied experimentally and theoretically for the micro-dimensional particle size range.

There are several theoretical models of detonation of aluminum dust in air, detonation products of gaseous mixtures, and oxygen [1 - 3]. In [4, 5], a model of heterogeneous detonation of monodisperse gaseous suspensions of aluminum particles in oxygen with a particle diameter of 1-10 μm was developed. The model has been verified by experimental data [6] on detonation velocity. Also agreements with known experimental data on reaction zone length, initiation energy, and detonation cell size was obtained in calculations within the framework of the model. The data on the propagation of cellular detonations in planar channels in monodisperse and polydisperse suspensions were obtained in [7, 8] and analogous problems for volumes of changing geometry were considered in [9, 10, 11].

In recent years, interest in research in the field of nanodispersed media has increased. A comprehensive review of new experimental data on the ignition, melting, heat transfer, and combustion of submicron and nanoscale aluminum particles is given in [12].

In [14], using the experimental data [12, 13], a semi-empirical physical and mathematical model of detonation of nanoscale aluminum particles in oxygen was constructed. The model takes into account the free-molecular effects in the description of the flow around and heat transfer of nanoscale particles with gas. It is expressed in the dependence of the characteristic times of the velocity and temperature relaxation processes on the Knudsen number. The simplest analysis of stationary detonation structures

Content from this work may be used under the terms of the Creative Commons Attribution 3.0 licence. Any further distribution of this work must maintain attribution to the author(s) and the title of the work, journal citation and DOI.
Published under licence by IOP Publishing Ltd

showed that in the nanoscale range of particle diameters the burning times are much longer (order of magnitude more for 100-nm particles) than the times of thermal and velocity relaxation. It means that the combustion proceeds essentially in equilibrium mixture with respect to the velocities and temperatures of the phases. In addition, the specific feature of nanoscale particles burning is the transition from the diffusion combustion regime inherent in micro-dimensional suspensions [2] to the kinetic regime [12], characterized by much higher values of the activation energy of the combustion reaction [15]. These properties bring nano-sized aluminum suspensions closer to gas mixtures. In this connection, it is of interest to what extent this will affect the characteristics of cellular detonation. The purpose of this paper is to analyze cellular detonation in gas suspensions of nanoscale aluminum particles on the basis of numerical simulation of two-dimensional flows within the framework of the model [14].

2. Physical and mathematical model

The Euler equations for two-dimensional flow in dilute gas particle suspensions follow from conservation laws for mass, momentum, and energy (the subscripts 1, 2 indicate gas and reactive particles, respectively):

$$\frac{\partial \boldsymbol{W}}{\partial t}+\frac{\partial \boldsymbol{F}}{\partial x}+\frac{\partial \boldsymbol{G}}{\partial y}=\boldsymbol{\Gamma},\quad \boldsymbol{W}=\begin{pmatrix}\boldsymbol{W}_1\\ \boldsymbol{W}_2\end{pmatrix},\ \boldsymbol{F}=\begin{pmatrix}\boldsymbol{F}_1\\ \boldsymbol{F}_2\end{pmatrix},\ \boldsymbol{G}=\begin{pmatrix}\boldsymbol{G}_1\\ \boldsymbol{G}_2\end{pmatrix},\ \boldsymbol{\Gamma}=\begin{pmatrix}-\boldsymbol{\Gamma}_2\\ \boldsymbol{\Gamma}_2\end{pmatrix},$$

$$\boldsymbol{W}_1=\begin{pmatrix}\rho_1\\ \rho_1u_1\\ \rho_1v_1\\ \rho_1E_1\end{pmatrix},\ \boldsymbol{F}_1=\begin{pmatrix}\rho_1u_1\\ p+\rho_1u_1^2\\ \rho_1u_1v_1\\ \rho_1u_1E_1+pu_1\end{pmatrix},\ \boldsymbol{G}_1=\begin{pmatrix}\rho_1v_1\\ \rho_1u_1v_1\\ p+\rho_1v_1^2\\ \rho_1v_1E_1+pv_1\end{pmatrix}. \tag{1}$$

$$\boldsymbol{W}_2=\begin{pmatrix}\rho_2\\ \rho_2u_2\\ \rho_2v_2\\ \rho_2E_2\end{pmatrix},\ \boldsymbol{F}_2=\begin{pmatrix}\rho_2u_2\\ \rho_2u_2^2\\ \rho_2u_2v_2\\ \rho_2u_2E_2\end{pmatrix},\ \boldsymbol{G}_2=\begin{pmatrix}\rho_2v_2\\ \rho_2u_2v_2\\ \rho_2v_2^2\\ \rho_2v_2E_2\end{pmatrix},\ \boldsymbol{\Gamma}_2=\begin{pmatrix}-J\\ f_{2x}\\ f_{2y}\\ q_2+f_{2x}u_2+f_{2y}v_2\end{pmatrix}.$$

The system is enclosed by the equations of state and relationships for mass, momentum, and heat exchange between the gas and the particles

$$\vec{f}=\frac{\rho_2}{\tau_u}(\vec{u}_1-\vec{u}_2),\ q=\frac{\rho_2c_{v,2}}{\tau_T}(T_1-T_2),\ J=\frac{\rho_2}{\tau_\xi}(\xi-\xi_k),$$

$$p=\rho_1RT_1,\ E_1=c_{v1}T_1+0.5u_1^2,\ E_2=c_{v2}T_2+Q+0.5u_2^2. \tag{2}$$

Here indexes 1 and 2 refer to gas and particle, respectively. We use the following notations: p is pressure; $\rho_i=m_i\rho_{ii},u_i,E_i,T_i,c_{v,i},m_i$, are mean densities, velocities, total energy per unit mass, temperatures, heat capacities, and volume concentrations of the components, $\xi=\rho_2/\rho$ is mass particle concentration, ξ_k is fraction of unburnt particles, ρ_{ii} are true densities, R is the reduced gas constant, Q is heat effect of particle combustion in the frame of reduced chemical kinetics. The value of Q has been determined to agree with experimental data by Strauss [6] on detonation velocity. Since weak dependence of detonation velocity on particle diameter for micro-sized and nano-sized aluminum particles [13] we use the same data as in [7-11].

The interphase interactions are presented by parameters J (the mass transfer), f (the interphase interaction force), and q_2 (the heat transfer). The processes of velocity relaxation and heat transfer in gas-laden nanoscale particles are characterized by a transition from a continual to a free-molecular regime of particle interaction with gas molecules. If the particle size is comparable with the mean free path of gas molecules, then the particle drag force in the gas flow is determined with regard for the Cunningham correction factor [12]:

$$\tau_u = 4d\rho_{22}C_c / 3c_D\rho_{11}|\vec{u}_1 - \vec{u}_2|,\ C_c = 1 + 2\mathrm{Kn}[1.257 + 0.4\exp(-1.1/2\mathrm{Kn})]. \quad (3)$$

Here $\mathrm{Kn} = RT_1 / (\sqrt{2}\pi d_g^2 N_A p d_2)$ is the Knudsen number that is the ratio of the mean free path of the molecular in gas to the particle diameter, d_g, m_g are the gas molecular size and mass, N_A is the Avogadro number. The drag coefficient c_D is determined similar to micron-sized particle detonations with regard for supersonic flow around particle in detonation structure using the formula [7-11]

$$c_D(\mathrm{Re}, M_{12}) = \left(1 + \exp(-\frac{0.43}{\mathrm{M}_{12}^{4.67}})\right)\left(0.38 + \frac{24}{\mathrm{Re}} + \frac{4}{\sqrt{\mathrm{Re}}}\right), \quad (4)$$

$$\mathrm{Re} = \frac{\rho_{11}d|\vec{u}_1 - \vec{u}_2|}{\mu_1},\ \mathrm{M}_{12} = \frac{|\vec{u}_1 - \vec{u}_2|\sqrt{\rho_{11}}}{\sqrt{\gamma_1 p}},$$

where μ_1 is the gas dynamic viscosity and γ_1 is the adiabatic exponent.

The heat exchange between the gas and particles for submicron and nano-sized particles is determined following to changing the flow regime of particles from the continuum at Knudsen numbers below 0.01 to the free-molecular one at $\mathrm{Kn} > 10$ [12]. Presentations for the characteristic times of heat relaxation in the regimes mentioned have a different form

$$\tau_T^{\ cont} = d^2\rho_{22}c_{v2} / 6\lambda_1\mathrm{Nu},\ \mathrm{Nu} = 2 + 0.6\mathrm{Re}^{\ 1/2}\mathrm{Pr}^{1/3}, \quad (5)$$

$$\tau_T^{\ fm} = \frac{d\rho_{22}c_{v2}}{6\alpha p}\sqrt{\frac{8\pi m_g T_1}{k_B}}\left(\frac{\gamma_1 - 1}{\gamma_1 + 1}\right). \quad (6)$$

Here Nu, Pr are the Nusselt and Prandtl numbers (usually, Pr=0.7), λ_1 is the gas heat conductivity, k_B is the Boltzman constant, α is the accommodation coefficient. In the particle size range 10 nm - 1 μm, the Knudsen number varies from about 0.01 to about 10, respectively, the flow regime is transitional. For the transition region we suggest a logarithmic interpolation formula [14].

$$\tau_T^{\ tr} = [(\log\mathrm{Kn} + 2)\tau_T^{\ fm} + (1 - \log\mathrm{Kn})\tau_T^{\ cont}] / 3 \quad (7)$$

The processes of ignition and combustion of nanoscale aluminum particles are described using the known empirical data given in [12, 15]. A temperature criterion for the ignition of micron-sized particles with an ignition temperature of about 900 K (which is close to the melting point of Al) was adopted in works by A.V. Fedorov et al [7-11]. According to numerous experimental data on aluminum nanoparticle ignition presented in [12], we take for particles of size 50 ÷ 250 nm the ignition temperature equal to the melting point $T_{ign} = T_{\mathrm{melt}}$=930 K.

For describing the detonation combustion of a stoichiometric suspension of aluminum with dispersion 50 nm ≤ d ≤ 250 nm we assume the following approximation for the characteristic time of combustion in oxygen [12, 15, 14]

$$\tau_{\xi} = \tau_0 (d / d_0)^{0.3} \exp(E_a / RT_1)(p / p_*)^{-m} \quad (8)$$

with constants: d_0=80 nm, p_*=8 atm; $E_a = 60$ kJ/mol, $m = 0.5$, τ_0 =0.2 µs. For micron-sized and submicron suspensions we use the detonation model [7] with activation energy of the Arrhenius type kinetics 32 kJ/mol (half the data for aluminum nanoparticles), thus reflecting the transitional combustion regime

$$\tau_{\xi} = \tau_0 (d / d_0)^{0.3} \exp(E_a / RT_2) \quad (9)$$

with d_0=3.5 µm, $E_a = 32$ kJ/mol, $\tau_0 = 0.294$ µs. In the particle size range 50 nm - 250 nm, the values E_a, m and the relaxation parameters differ significantly from the data for the micrometer particles. A specific feature of the stationary detonation structures in nano-sized particle gas suspension is a significant difference in the scales of the relaxation zones (velocity and thermal) and the combustion zone. Also in a nano-sized suspension there is a significant difference in the scales of combustion zones of strong (over-compressed) and attenuated detonation waves due to large value of the activation energy. This may affect the characteristics of cellular detonations.

The numerical technique is based on the conservative flux-splitting schemes: the TVD scheme by Harten for gas and the Gentry-Martin-Daly scheme for particles. The numerical method has been tested earlier and applied for 2-D numerical simulations of the shock wave and detonation flows in [7-11]. The step of the finite-difference grid was determined according to the scale of the relaxation zones, varied in the test calculations. The calculations were performed using IBM PC Intel Core2Quad.

3. Formation of cellular detonation structures

We consider the problem of detonation initiation under interaction of a planar shock wave with a cloud of particles dispersed in a plane channel. A small perturbation of particle density on the cloud corner is posed to stimulate the transverse wave generations [7].

Figures 1-5 present the maximal pressure history fields that reveal the trajectories of the triple points of cellular detonations. All the pictures in Figs. 1 - 5 are built in a single shadow scale.

All the figures show that the system of transverse waves does not fold immediately, but after a while the propagation of detonation in the channel. Initially, weak transverse waves are amplified, traces of their collisions form cells. The geometric characteristics of the developed cellular structure depend on the particle size, the width of the channel, and also the parameters determining the extent of the combustion zone (in particular, the activation energy).

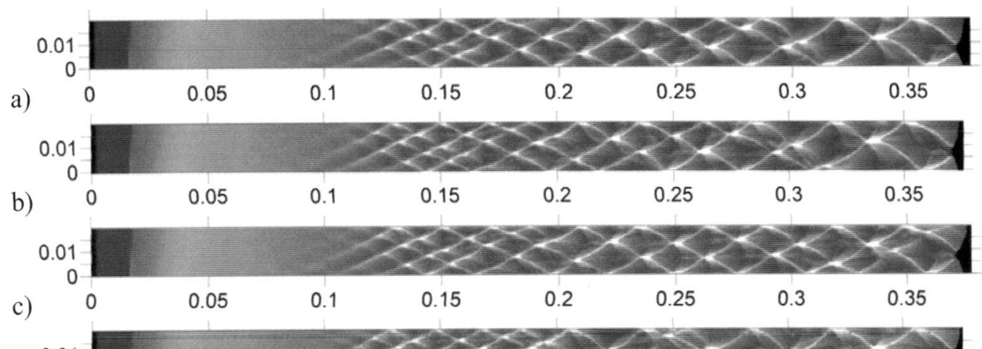

Figure 1. The results of test calculations.

Figures 1 show the results of test calculations for the suspension of particles with a diameter of 300 nm. In the fragments a) and b), the calculations were carried out without taking into account the Cunningham correct factor and dependence on the Knudsen number in heat exchange processes, the grid spacing was 0.0001 m (a) and 0.000075 m (b). The fragments c) and d) show the results, where these corrections were taken into account, the grid spacing is 0.000075 m (c) and 0.00005 m (d). As can be seen, for the suspensions of particles with diameters from 300 nm and higher, the free-molecular effects in the description of the flow around particle and heat transfer regime are insignificant. For calculations in the 300 - 600 nm range, the step was 0.000075 m, for particles of smaller diameters, the optimal step of the finite-difference grid was determined in similar way.

Figures 2 show examples of calculations for the suspension of 1-μm particles within the kinetics of micron-size particles (9) at $E_a = 32$ kJ/mol. As can be seen, the transverse waves formed from small perturbations are amplified as the front propagates, but the distances between them practically remain unchanged. It can also be noted that the trajectories of triple points are slightly different from straight lines and certainly smeared. The smearing character is due to the influence of relaxation zones comparable with the combustion zone of micron-size particles [5, 7]. The cell size correlates with the predictions of the acoustic theory [16], which is confirmed by calculations in [7]. The maximum pressures in the collision of triple points are of the order of 15 MPa, as was obtained in the calculations of [7].

Figures 3 show examples of calculations of cellular detonation formation in 1-cm channel for nano-sized gas suspensions of 100 nm (Fig. 3a) and 200 nm (Fig. 3b) particles obtained in the framework of the kinetic regime of aluminum combustion (8) at $E_a = 60\,\text{kJ / mol}$. Here one can trace the similarities with the flow patterns of cellular gaseous detonation: the trajectories of triple points are much more clear-cut and more curved than in micron-sized suspensions (Fig. 2). It can also be noted that the initially formed system of transverse waves in the process of front propagation is significantly rearranged, the number of transverse waves decreases, the distance between them increases.

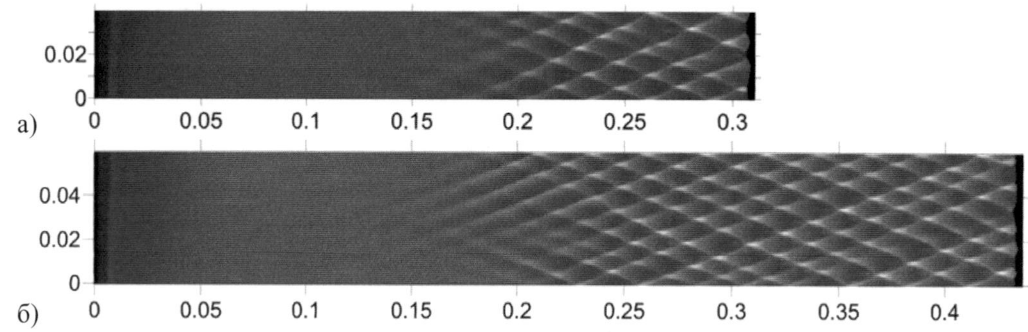

Figure 2. Cellular detonations in a micron-sized mixture.

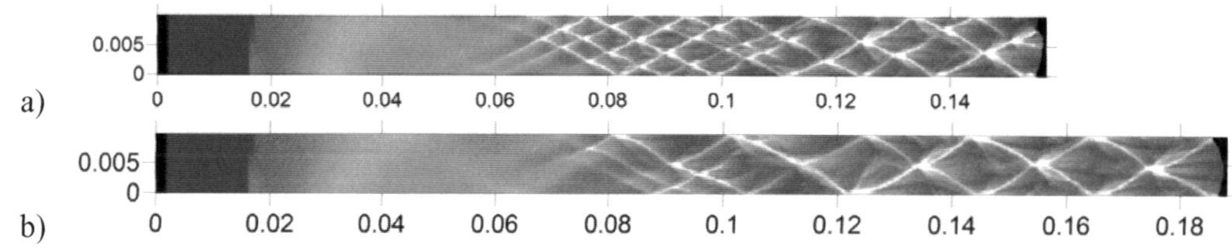

Figure 3. Cellular detonations in a nano-sized mixture.

The effect of the channel width on the number of cells formed is shown in Fig. 4 on the example of calculation for a suspension of particles with a diameter of 600 nm. Here it can be seen that the irregularity of cellular structures takes place not only at the stage of cell formation, but also in the process of further propagation of detonation. Although the forming cell is much larger than at the initially formed transverse waves, they may then divide with subsequent re-establishing. The result of calculations in a rather wide channel (4-cm channel, d=300 nm) is shown in Fig. 5, where the irregularity of the cellular structure formed in the channel is more obvious.

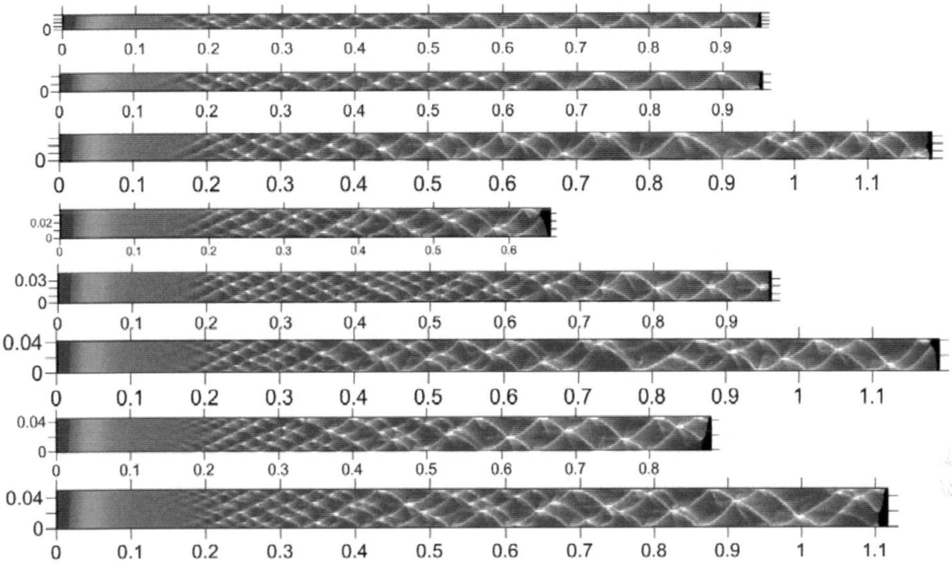

Figure 4. Channel width effect on cellular detonations formation and propagation.

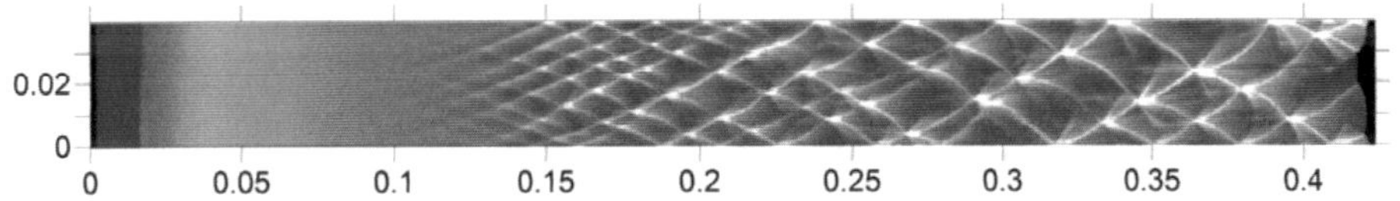

Figure 5. Evolution of the transverse waves.

Thus, in nano-dispersed aluminum-oxygen mixtures small-sized detonation cells are initially formed from small perturbations, and the distance between successive transverse waves agrees with the predictions of the acoustic theory [16, 7]. Further, as the front propagates and the transverse waves amplify, the cells are rearranged and enlarged. Trajectories of the triple points are highly curved, maximum pressures in the collision of transverse waves reach 40-50 MPa, which is three times higher than the values for micron-sized particles. Cellular structures are characterized by non-uniformity and irregularity. For micron-sized particles with a lower value of the activation energy of the reduced combustion kinetics (about 32 kJ/mol) [4], this behavior was not observed. From this it can be concluded that the properties listed above of cellular detonation of nano-dispersed mixtures are due to much higher values (60 kJ/mol) of the activation energy of the reduced kinetics of aluminum particle combustion.

4. Conclusions

Thus, nano-dispersed aluminum-oxygen mixtures small-sized detonation cells are initially formed from small perturbations, and the distance between successive transverse waves agrees with the predictions of the acoustic theory [16, 7]. Further, as the front propagates and the transverse waves amplify, the cells are rearranged and enlarged. Trajectories of the triple points are highly curved, maximum pressures in the collision of transverse waves reach 40-50 MPa, which is three times higher than the values for micron-sized particles.

Formation of cellular detonation structures in monodisperse nanosized aluminum particle – oxygen suspensions is studied by methods of numerical simulations of two-dimensional detonation flows.

The detonation combustion are described within the semiempirical model developed earlier which takes into account transition of the regime of aluminum particle combustion from diffusion to kinetic for micro-sized and nanosized particles. The free-molecular effects are considered in the processes of heat and velocity relaxation of the phases. It is established that for particles 300 nm and more coarse it is possible to neglect the free-molecular corrections in gas particle flow and heat transfer.

Common properties and specific features of the flows of suspensions of micro-sized or submicron particles and suspensions of nanoscale particles are established. The specific feature of the cellular detonation formation in nanoparticle suspensions is sufficient reconstruction of the primary weak transverse waves in the process of the detonation front propagation and their amplification.

The cellular detonation structures in developed detonation flow are irregular and characterized by much higher pick pressure values than in micro-disperse mixtures and relatively larger detonation cells. It seems that this is due to much higher value of activation energy of reduced chemical reaction of aluminum particle combustion in kinetic regime than in transition regime of combustion of micron-sized particles or diffusion regime of combustion of coarse aluminum particles.

Acknowledgements

The work was supported by the Russian Foundation for Basic Research (grant № 16-08-00778).

References

[1] Medvedev A E, Fedorov A V and Fomin V M 1984 *Comb., Explosion, and Shock Waves* **20** 127
[2] Veyssiere B and Khasainov B 1991 *Combustion and Flame* **85** 241
[3] Zhang F, Gerrard K and Ripley R C 2009 *Journal of Propulsion and Power* **25** 845
[4] Fedorov A V 1992 *Comb., Explosion, and Shock Waves* **28** 277
[5] Fedorov A V, Fomin VM and Khmel TA 1999 *Shock Waves* **9** 313
[6] Strauss W A 1968 *AIAA Journal* **6** 1753
[7] Fedorov A V and Khmel T A 2005 *Comb., Explosion, and Shock Waves* **41** 435
[8] Fedorov A V and Khmel T A 2008 *Comb., Explosion, and Shock Waves* **44** 163
[9] Kratova Yu V, Fedorov A V and Khmel T A 2009 *Comb., Explosion, and Shock Waves* **45** 591
[10] Kratova Yu V, Fedorov A V and Khmel T A 2011 *Comb., Explosion, and Shock Waves* **47** 572
[11] Kratova Yu V, Khmel T A and Fedorov A V 2016 *Comb., Explosion, and Shock Waves* **52** 74
[12] Sundaram D S, Yang V and Zarko V E 2015 *Comb., Explosion, and Shock Waves* **51** 173
[13] Zhang F, Murray S B and Gerrard R B 2006 *Shock Waves* **15** 313
[14] Fedorov A V, Khmel T A 2016 *Nonequilibrium Processes in Physics and Chemistry II*, ed A M Starik and S M Frolov (Moscow: Torus-Press) p 341
[15] Bazyn T, Krier H and Glumac N 2006 *Combustion and Flame* **145** 703
[16] Barthel H O 1974 *Physics of Fluids* **17** 1547

Interaction of long-wave perturbations with a shock wave on a wedge and method of mode decomposition of supersonic flow pulsations

I S Tsyryulnikov[1,2], **S V Kirilovskiy**[1,2] **and T V Poplavskaya**[1,2]

[1]Khristianovich Institute of Theoretical and Applied Mechanics SB RAS, Novosibirsk, Russia
[2]Novosibirsk State University, Novosibirsk, Russia

Abstract. The problem of the interaction of long-wave perturbations of an external flow with a shock wave formed in supersonic flow past a wedge was considered in this paper. The formulas for the amplitudes of long-wave disturbances are obtained from the general analytical solution for the perturbation amplitudes. This data are compared with the results of numerical simulation. It is shown that pulsations of different modes can be separated from the total pulsation field of the incoming flow due to the different amplitude of the transformation of the perturbations of various type behind the shock wave. A method for such a separation is proposed and its approbation is carried out in a numerical experiment.

1. Introduction

To create and operate high-speed aircraft, it is necessary to study the interaction of perturbations of the incoming flow with a shockwave arising at the leading edges of bodies in a supersonic flow. Each of the modes of incoming perturbations, namely the entropy, acoustic, and vortex modes, induce a perturbation-superposition of all these three species [1] behind the shock wave in the case of a shock wave isolated from the body.If an shock wave arises in a flow around a rigid body, the situation is much more complicated due to the attachment of the shock wave to the front edge of the body and the reflection and refraction of perturbations between the surface of the body and the shock wave [2]. In this case, each of the types of waves of the oncoming stream can generate pulsations of all perturbation modes behind the shock wave.

Today, the receptivity and evolution of perturbations in supersonic boundary layers are widely studied by means of wind tunnels. For such studies, the accurate knowledge is needed not only about the intensity of pulsations in the core of the supersonic flow of the wind tunnel, but also about the mode composition of these perturbations.This determines the urgency of developing methods and creating devices for isolating pulsations of certain perturbations modes from the total supersonic flow pulsations field. Parametric calculations of the pressure pulsation fields during the interaction of pulsations of an external flow with a shock wave formed in supersonic flow around a wedge of length L were carried out in [3] in a wide range of propagation angles of long-wave (λ>L) perturbations. Received numerical data on the coefficients of transformation of long-wavelength external perturbations on an shock wave have been represented in the form of analytical dependences on the shock wave inclination angle, the perturbation propagation angle, and the Mach number of the incident flow [3]. These dependences can be used to solve the task of mode decomposition of perturbations in supersonic flows. The range of admissible root-mean-square amplitudes of oscillations of vortex, entropy, and acoustic modes in the free stream generated in IT-302M was obtained by using the pressure fluctuations measured on the model surface and the calculated conversion factors [4]. However, the results of [4] are based on the assumptions about the zero angle of propagation of the

Content from this work may be used under the terms of the Creative Commons Attribution 3.0 licence. Any further distribution of this work must maintain attribution to the author(s) and the title of the work, journal citation and DOI.
Published under licence by IOP Publishing Ltd

modes of external flow disturbances and the absence of a phase connection between them. Moreover, only time-averaged pulsation amplitudes of different modes were obtained.

In this paper, we propose a method of the mode decomposition of perturbations in real time of measurements with an arbitrary distribution of the perturbations over the propagation angles. The magnitude of a certain disturbances mode of the incoming flow is determined from the pressure pulsations on the surfaces of the wedges and the coefficients of conversion of the long-wave length perturbations of various modes to the shock wave taken from analysis of the analytical solution [2]. In this paper, this method of the modal decomposition of perturbations of the oncoming stream is approbated in real time in a numerical experiment.

2. Analytic solution for the case of long-wave perturbations

The general analytical solution for the perturbation amplitudes in a region bounded by a shock wave and a wedge surface in a supersonic flow was obtained in [2] for perturbations of the external flow in the form of plane monochromatic waves. The solution is given in the form of an infinite series of terms containing Bessel functions of the first kind with a set of coefficients, the determination of whose values can be calculate from recurrent relations. Unfortunately, in the case of large values of the ratio of the longitudinal coordinate of the wedge surface to the perturbation wavelength ($x/\lambda \gg 1$), calculations become extremely sensitive to the number of taken terms of the series. This significantly complicates the analysis of this solution in far-downstream behavior. However, in the case of long-wave perturbations with $x/\lambda \ll 1$, it's possible to decompose the general solution obtained in this work for the small parameter x/λ. As a result of this analysis, the equation for pressure perturbations generation behind the shock wave on a wedge in a supersonic flow was obtained:

$$p' = -e^{i\omega t}\frac{\gamma \overline{p}\overline{u}}{2\overline{c}\overline{c}}a_1. \tag{1}$$

Here $\overline{p}$, $\overline{u}$, $\overline{c}$ are the pressure, velocity and sound speed behind shock wave, γ is the adiabatic index of the gas,

$$a_1 = 2\frac{\overline{c}}{\overline{u}\eta_2}\left[\left(B_1 - (A_2B_1 - A_1B_2)\frac{\overline{u}_{\infty\tau}}{\eta_1}\right)R_0^{(2)} - A_1R_0^{(3)} + A_1(A_2B_1 - A_1B_2)\frac{\overline{u}_{\infty\tau}}{\eta_1}\mathrm{tg}\,(\varphi - \alpha)R_0^{(1)}\right],$$

A_i, B_i are functions containing the parameters of the unperturbed flow from [2], φ is the shock wave angle of inclination, α is the angle of inclination of the wedge,

$$\eta_1 = A_1\mathrm{tg}\,(\varphi - \alpha)(\overline{u}_{\infty n} - \overline{u}_n) + A_2\overline{u}_{\infty\tau},$$

$$\eta_2 = \overline{u}_{\infty\tau}(A_2B_1 - A_1B_2)\frac{A_3}{\eta_1} - (A_3B_1 - A_1B_3).$$

Functions $\overline{u}_{\infty n}, \overline{u}_n, \overline{u}_{\infty\tau}$ are unperturbed normal to shock components of the flow velocity ahead and behind the shock wave, and the tangential velocity component, respectively. Values are normalized in the same way as in [2].

Functions

$$R_0^{(1)} = u'_{\infty\tau},$$

$$R_0^{(2)} = (A_1 + A_2)u'_{\infty n} + A_2\overline{u}_{\infty n}\rho'_\infty + \frac{p'_\infty}{2\gamma\overline{u}_{\infty n}M_\infty^2},$$

$$R_0^{(3)} = \left[\frac{1}{\overline{\rho}}(B_1 - 1) + \frac{\overline{u}_{\infty n}}{\overline{u}_{2n}}\right]u'_{\infty n} + \left[\frac{\overline{u}_{\infty n}}{\overline{\rho}}(B_1 - 1) - \frac{1}{(\gamma - 1)\overline{u}_n M_\infty^2}\right]\rho'_\infty + \frac{p'_\infty}{(\gamma - 1)\overline{u}_n M_\infty^2}$$

linearly depend on the amplitudes of normal and tangential to shock wave velocities $u'_{\infty n}$ and $u'_{\infty\tau}$, amplitudes of density ρ'_∞ and pressure p'_∞ perturbations. $\overline{\rho}$ is the density of the unperturbed flow. The analytical dependencies obtained from the general solution allow to determine the transformation coefficients of the acoustic g_a, entropy g_e, and vortex g_v mode in the pressure pulsation on the wedge surface in the following form:

$$g_a = A^* + B^* cos\left(\frac{\pi}{2} - \varphi^* - \theta\right); g_e = C^*; \ g_v = \gamma B^* sin\left(\frac{\pi}{2} - \varphi^* - \theta\right). \quad (2)$$

Here A^*, B^*, C^*, φ^* are functions of the unperturbed flow parameters. It should be noted that the transformation coefficients obtained from the results of numerical simulation [3] are consistent with the general solution obtained for the case of long-wave perturbations.

3. Numerical simulation of supersonic flow past a wedge

Numerical simulation of disturbed supersonic flow around a wedge was carried out using the ANSYS Fluent software package with the addition of built-in UDF modules for introducing perturbations of various modes. The system of two-dimensional Navier-Stokes equations for viscous and inviscid cases was solved. An explicit scheme of second-order accuracy with respect to space with Roe-FDS by the method of splitting convective flows and the explicit Runge-Kutta method with respect to time was used. The procedure of the numerical method is described in more detail in [5,6].

The length of wedge generatrix was L=0.1 m. The computational domain was a rectangle with the part of its lower side consist the wedge. The left (input) boundary was located some distance upstream from the leading edge of the wedge, the height of the computational domain was chosen from the condition that the bow shock wave did not interact with the upper boundary of the computational region. The right boundary is shifted from the rear end of the model, so that the flow in the output section is supersonic. On the left and upper bounds of the computational domain, the parameters of the supersonic incoming flow p_∞, M_∞, T_∞ were set, and the downwash condition was used at the right output boundary. Non-slip conditions and constant temperature of T_w= 300K was set to calculate the viscous flow on the surface of the wedge. The angle of the wedge solution was taken into account by the corresponding direction of the velocity vector of the incident flow u_∞ at an angle α to the wedge. The computational area was covered with a rectangular grid.

After calculation of the main unperturbed flow, the flow with pressure, velocity and temperature perturbations was calculated. For this, a superposition of the main flow (quantities with the indices "∞") and plane monochromatic waves of the corresponding mode was set at the input boundary of the computational domain: $p = \overline{p}_\infty + p'_\infty$; $T = \overline{T}_\infty + T'_\infty$; $u = \overline{u}_\infty + u'_\infty$ similar to [5]. The amplitude of the disturbances normalized to the pressure in the unperturbed flow, A_0=0.03. The angle of propagation of external disturbances had a starting point from the direction of the velocity vector of the running flow.

The flow around the wedge was considered under the following conditions:

I– $\overline{p}_\infty$=494Pa, $\overline{T}_\infty$=98K, M_∞=7.8, inviscid case

II– $\overline{p}_\infty$=494Pa, $\overline{T}_\infty$=98K, M_∞=7.8, $Re_{1\infty}=4\cdot10^6 m^{-1}$

III– $\overline{p}_\infty$=4237 Pa, $\overline{T}_\infty$=177K, M_∞=5.56, $Re_{1\infty}=10.38\cdot10^6 m^{-1}$

for the wedge angles α = 5 ÷15°and wide range of angles of propagation of external disturbances θ = -60 ÷ 60°.

Figures 1 and 2 show the results of numerical simulation of the action of a fast acoustic wave with a frequency f= 50 kHz (or $L/\lambda = 2.88$) on a wedge with α = 10° in comparison with analytical solution [2]. The parameters of the unperturbed flow for obtaining the analytical solution corresponded to the case I (M_∞=7.8, α = 10°, $L/\lambda = 2.88$). Figure 1 shows the fields of instantaneous pressure pulsations, obtained in numerical simulation of viscous flow (Figure 1a) and inviscid flow (Figure 1b) and according to theory [2] (Figure 1c) at M_∞=7.8. Figure 1b shows the field of instantaneous pressure pulsations in numerical simulation with parameters (I), and Figure 1c - with parameters (II). The agreement of the amplitudes of instantaneous pulsations and its phase velocities is evident for all cases.

Figure 2 shows a comparison of the rms pressure pulsations amplitudes on the wedge surface normalized to the amplitude of pressure pulsations in the incoming stream, depending on the parameter x/L. It is seen that the pulsation amplitudes are in good agreement with [2] and the calculated data for an inviscid case over the whole surface of the wedge. In the case of calculation taking into account the viscosity, the amplitude of the pressure pulsations for $x/L > 0.15$ is slightly higher. However, all amplitudes are practically the same up to x/L=0.15 which corresponds to the case of long-wave disturbances ($x/\lambda < 0.4$). This allows us to use equations (1) and (2) to calculate the perturbation transformation coefficients in viscous supersonic flows. Moreover, the approximation of long-wave perturbations can be used for the mode decomposition of perturbations in viscous supersonic flows.

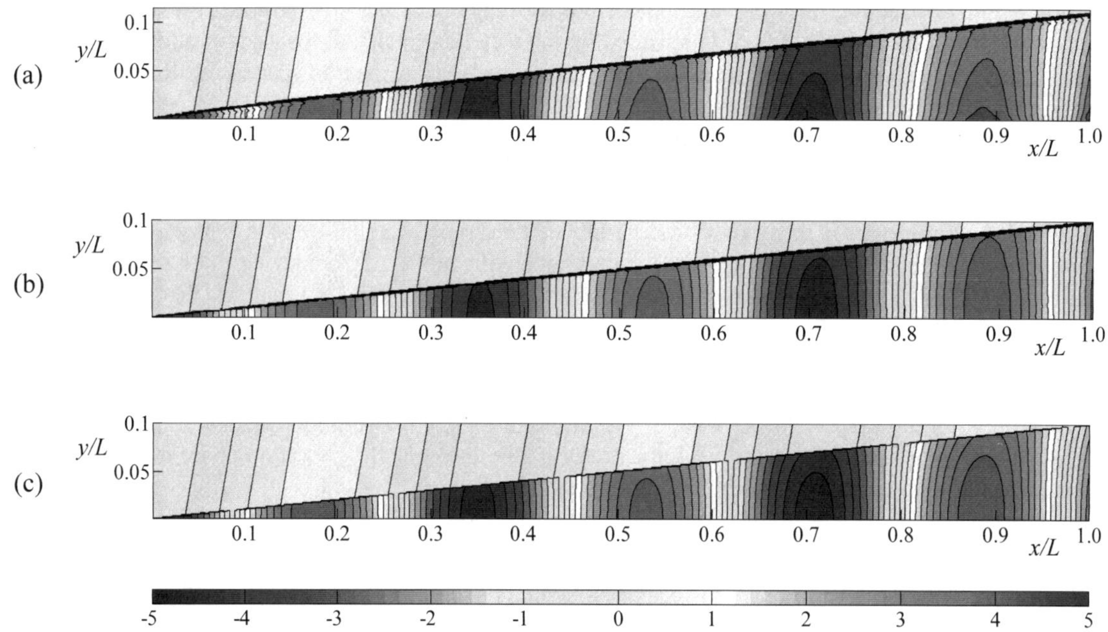

Figure 1. Fields of instantaneous pressure pulsations: (a) – viscous case II, (b) – inviscid case I, (c) – general solution [2].

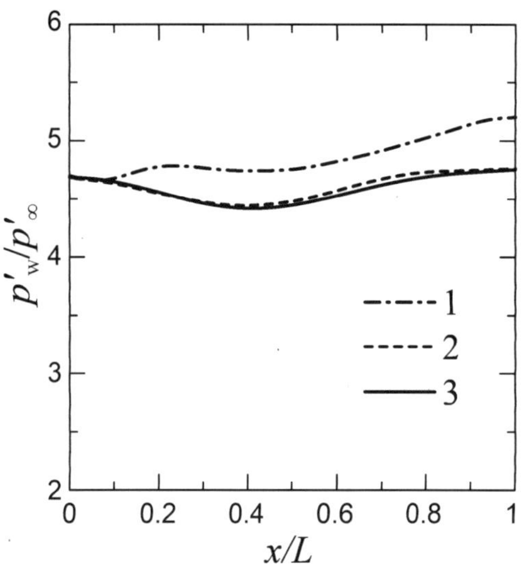

Figure 2. Normalized rms amplitudes of pressure pulsations on the wedge surface: 1 – viscous case II, 2 – inviscid case I, 3 – general solution [2].

4. Mode decomposition of disturbances

IOP Conf. Series: Journal of Physics: Conf. Series **894** (2017) 012103 doi:10.1088/1742-6596/894/1/012103

The idea of the method of mode decomposition consists in the possibility of separation of the pulsations of individual modes in the incoming flow by relations (1) and measured pressure pulsations on elements of the special device. This device contains several elements with a sharp leading edge, located at different angles of attack. On the surfaces of these elements pressure pulsations are measured. The difference in the angles of inclination of the surfaces of the device leads also a difference in the magnitude of the conversion coefficients of the different modes of perturbations of the external flow, obtained from (2). Then, according to the relations (1), for various angles of inclination of the device surfaces, a system of equations is computed for calculating the pulsations of the individual modes of the incoming flow.

This method is suitable for long-wave perturbations of the incoming flow under the condition $\lambda \gg H$, where H is the largest device size. Potentially this allows to separation pressure p'_∞, velocities $u'_{\infty x}$, $u'_{\infty y}$, $u'_{\infty z}$ and entropy s'_∞ fluctuations, which requires using of measuring sensors on 5 different surfaces of the device. However, the analysis of relations (2) shows that $B^* sin\varphi^{**}$ and C^* are linearly dependent quantities in the inviscid case, and the solution is determined only by four independent variables. In this case, it is sufficient to use two pairs of wedges or plates symmetric with respect to the direction of the incoming flow and located at the angle of attack for the mode decomposition of the perturbations: $\alpha_1^+, \alpha_1^-, \alpha_2^+, \alpha_2^-$ (here $\alpha_1 \neq \alpha_2$). Analysis of the solution for the three-dimensional case from [7] shows that the perturbation transformation coefficients do not depend on the velocity fluctuations in the transverse direction to the plane formed by the velocity vector and the normal to the shock wave. Therefore, one pair of wedges α_1^+ and α_1^- for determination of $u'_{\infty y}$ is located so that the normal to the surface was lay in the (*x*, *y*) plane. The second pair of wedges α_2^+ and α_2^- is located so that the normal to the surface was lay in the (*x*, *z*) plane. The instantaneous values of pulsations superposition of the oncoming stream are:

$$p'_\infty = \frac{\sigma_1 C_2^* - \sigma_2 C_1^*}{C_2^* A_1^* - C_1^* A_2^*}, \frac{B^* C_p}{C^* c_\infty} u'_{\infty x} + s'_\infty = \frac{C_p \; \sigma_1 A_2^* - \sigma_2 A_1^*}{p_\infty C_1^* A_2^* - C_2^* A_1^*}, u'_{\infty y} = \frac{c_\infty \quad \delta_1}{p_\infty \gamma B_1^* cos\varphi_2^*}, u'_{\infty z} = \frac{c_\infty \quad \delta_2}{p_\infty \gamma B_2^* cos\varphi_2^*}, \tag{3}$$

where $\sigma = \frac{p'_{w+} + p'_{w-}}{2}$, $\delta = \frac{p'_{w+} - p'_{w-}}{2}$ – half-sum and half-difference of pressure pulsations on symmetric surfaces. To separate pulsations $u'_{\infty x}$ and s'_∞ the model addition is required by another measuring device for determination of pulsations which are different from pressure pulsations. Such a device, for example, can be a mass flow pulsation sensor (hot-wire anemometer).

In this paper, the approbation of the method of mode decomposition of perturbations was carried out in numerical experiment. For this, calculations of pressure perturbation were made on the surface of the wedge generatrix at the point x_0= 0.0646m with the angle of attack α_1 = 5° and α_2 = 15° on time. A double calculation was made for each angle of attack in case of external flow disturbances having a propagation angle θ which is different from 0: using wave action with propagation angle of θ and -θ to simulate symmetrical wedges. The values of the pressure pulsations at the same time points of the calculation were processed using the relations (3) to obtain the instantaneous values of the disturbance amplitudes of the external flow. Figure 3 shows the data of the computational experiment of the mode decomposition under the conditions of the oncoming stream III (see item 3) and under action of fast acoustic wave at frequency 10 kHz ($x_0/\lambda =$ 0.38) and with the propagation angle θ = 0. Figure 3a shows the calculated pressure pulsations on the wedges surfaces (analog of the experimental values), Figure 3b – result of the restoration of pressure pulsations in the oncoming flow and Figure 3c – result of restoration of the combination of pulsations of the longitudinal component of velocity and entropy according to relations (3).

It is seen that the reconstructed amplitudes of perturbation are in good agreement with the initial ones: deviations in amplitude p'_{∞} are less than 2%. Recovered values $u'_{\infty y}$ are also consistent with the initial pulsations in the investigated range of propagation angles θ to 60°. The next stage of investigation will be the experimental implementation of this method on the basis of measurements of pressure pulsations by special device under the action of controlled perturbations of a supersonic flow of known amplitude and known mode composition.

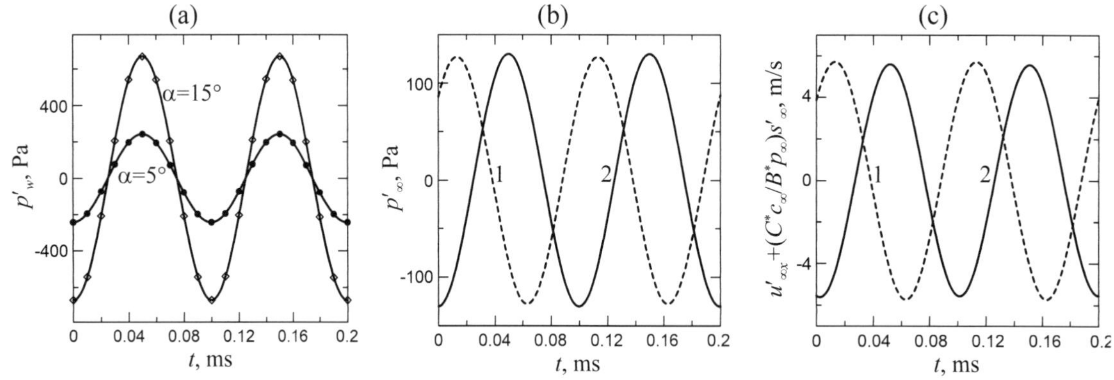

Figure 3. Calculated pressure pulsations on wedges surfaces (a), restored pressure pulsations in the incoming stream (b) and restored complex of pulsations of longitudinal component of velocity and entropy (c): 1 - calculated values under the conditions of the oncoming flow (initial), 2 - reconstructed values.

5. Conclusion

Dependences of the transformation coefficients of long-wave disturbances were obtained from the general solution of the inviscid problem of the interaction of perturbations of an incoming supersonic flow with a shock wave on the wedge.

Numerical simulation of perturbed supersonic flow past a wedge in a viscous and inviscid case was carried out. It is shown that the transformation coefficients for inviscid and viscous cases coincides under the conditions of the action of long-wave disturbances.

The method of real time decomposition pulsations of different modes of the incoming flow from the total pulsation field (mode decomposition) was proposed and its validation in a numerical experiment was performed.

Acknowledgments

The work was supported by the Russian Foundation for Basic Research (project No. 16-08-00674).

References

[1] McKenzie J F and Westphal K O 1968 *Phys. Fluids.* **11** 2350-2362
[2] Duck P W, Lasseigne D G and Hussaini M Y 1993 *NASA CR- 191528 ICASE Report No. 93-61* 1-44
[3] Tsyryulnikov I S, Kirilovsky S V and Poplavskaya T V 2016 *Technical Physics Letters.* **42**(11) 1094-98
[4] Tsyryulnikov I S, Kirilovskiy S V and Poplavskaya T V 2016 *AIP Conference Proceedings (Perm)* vol 1770

[5] Kirilovskiy S V, Maslov A A, Poplavskaya T V and Tsyrulnikov I S 2015 *Technical Physics.* **60**(5) 645-55
[6] Kirilovskiy S V, Poplavskaya T V, I S Tsyrulnikov I S and Maslov A A 2017 *Thermophysics and Aeromechanics.* **24**(3) 433-42
[7] Duck P W, Lasseigne D G and Hussaini M Y *NASA CR-201698 ICASE Report No. 97-26* 1-27

Study of deformation-strength properties of sandstone under multistage triaxial compression

P A Tsoi[1,2] **and I P Homenok**[1]

[1]Chinakal Institute of Mining, Novosibirsk, Russia
[2]Novosibirsk State Technical University, Novosibirsk, Russia

E-mail: paveltsoy@mail.ru

Abstract. A multistage test under conventional triaxial compression is an alternative way (instead of triaxial tests on a set of samples) to describe the mechanical strength (the envelope of the Coulomb-Mohr half-circles) of rocks. This paper is based on the experimental study of sandstones under multistage triaxial compression. Empirical dependencies of the modulus of elasticity, Poisson's ratio, the specific energy intensity on each of the stages are discussed. The Coulomb-Mohr envelopes are constructed. Tests of rocks under multistage triaxial compression can be suitable in case of a very limited number of specimens.

1. Introduction

The design of geo-engineering objects is accompanied by processes involving cyclic loads-unloads and leading to the mechanical destruction of rocks. Rock micro-cracks, induced by excavation, softening and accompanying changes in the natural stress state, can cause degradation of strength and a change in elastic properties [1]. For this reason, computational models based on the use of conventional mechanical properties may be the cause of erroneous estimated stress state forecasts.

Another problem is the difficulty in extracting the unbroken core from petroleum deposits. Therefore, to determine the strength properties of such cores, a small number of specimens must be tested. The tests with cyclic (multistage) loads make it possible to build a strength passport (Mohr-Coulomb envelope) involving only one specimen [2].

Therefore, the description of the behavior of the elastic and strength properties of rocks under static cyclic loads, as well as the determination of the strength characteristics of the core from the data of a single specimen test, is an actual problem when drilling wells in the case of a limited number of specimens.

The data of the mechanical experiments that compiled the presented work are based on cyclic (multistage) tests, in which the lateral pressure was varied depending on the stage of the loading program of the geomaterial being studied. Such or related tests are devoted to a number of the international research studies [3-9].

2. Experimental tests

2.1. Specimens

For the studies, selected (by properties) cores of sandstones were chosen from the special rock collection.

Content from this work may be used under the terms of the Creative Commons Attribution 3.0 licence. Any further distribution of this work must maintain attribution to the author(s) and the title of the work, journal citation and DOI.
Published under licence by IOP Publishing Ltd

IOP Conf. Series: Journal of Physics: Conf. Series **894** (2017) 012104 doi:10.1088/1742-6596/894/1/012104

Cylindrical specimens of the desired diameter (30mm) were made using the CPM 400 Coretest System. To ensure the parallelism of the loaded surfaces and required length (60 mm), drilled cylindrical specimens were cut on the DTS-430 machine.

Six specimens (Figure 1, Figure 2) were tested under the multistage triaxial compression.

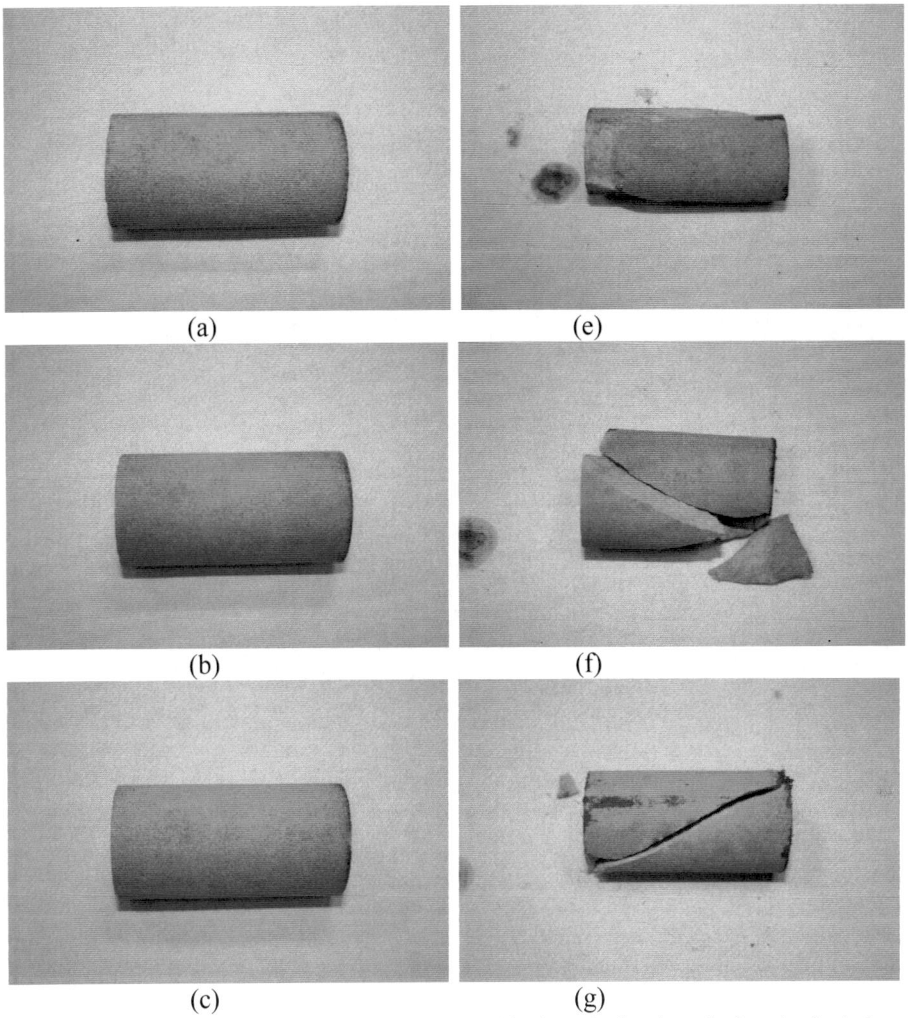

Figure 1. Sandstone specimens No1-2, 1-3, 1-4 before (a, b, c) and after (e, f, g) the multistage tests.

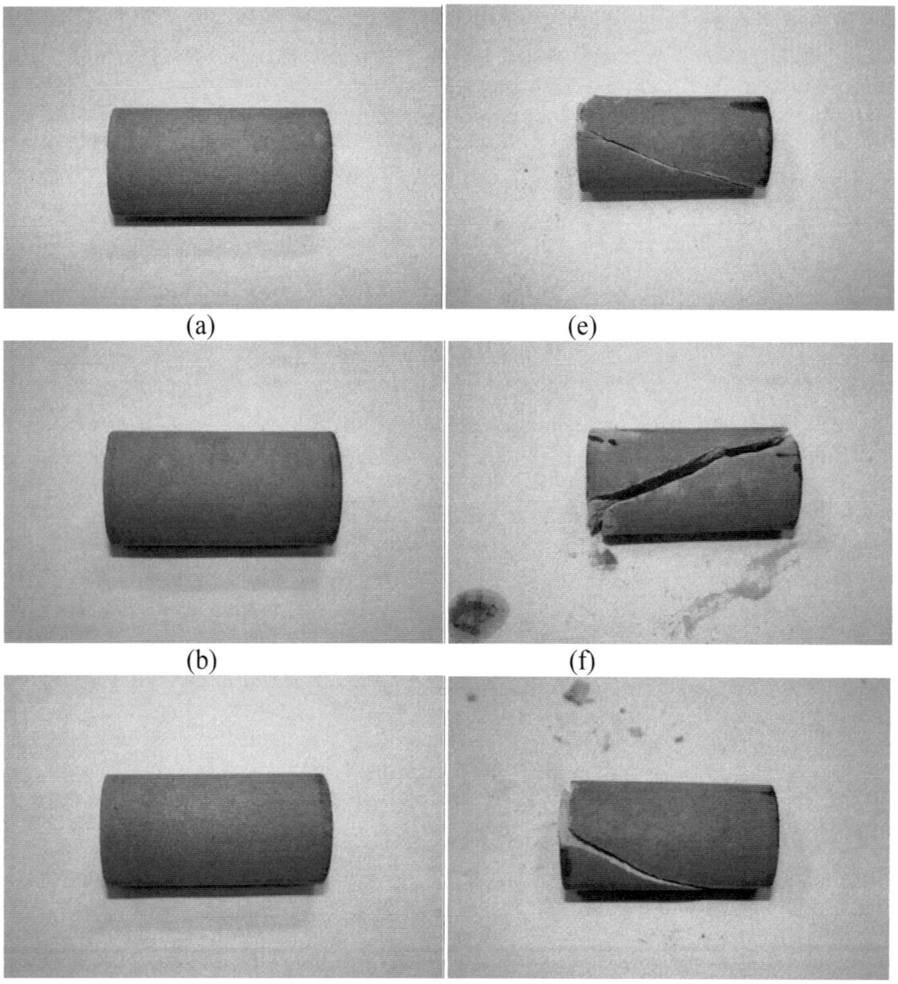

(a) (e) (b) (f) (c) (g)

Figure 2. Sandstone specimens No2-1, 2-2, 2-3 before (a, b, c) and after (e, f, g) the multistage tests.

2.2. Multistage triaxial test description

During the triaxial multistage loading of specimen the axial stress value reached the compressive elastic limit at each loading stage at certain lateral pressures. The values of last ones were 0.5, 1.5, 7, 9.5, 11 MPa, respectively, for 1st, 2nd, 3rd, 4th, 5th loading stages.

Each specimen was placed in the triaxial cell. The specimens were jacketed by rubber cup to prevent oil from seeping into the pores during tests. The oil serves as a distributor for uniform lateral pressure: due to the absence of shearing stresses in the liquid, the lateral pressure through the liquid is transmitted to the normal direction along each point of the sample wall. The axial load was applied to the specimen ends.

The procedure for testing rock specimens under the multistage triaxial compression consists of several steps [2]. First the lateral pressure is applied, and then the axial load gradually increases until the plastic strains begin. After this, gradual unloading of the specimen till the equality of lateral and axial pressures is carried out. Then the lateral pressure increases to the second value of lateral pressure, the axial load reaches the second value of plastic strain. This procedure is repeated. The specimen is brought to destruction at the test's end. Obtained "stress σ_z-strain ε_z or ε_φ" curve is shown in Figure 3.

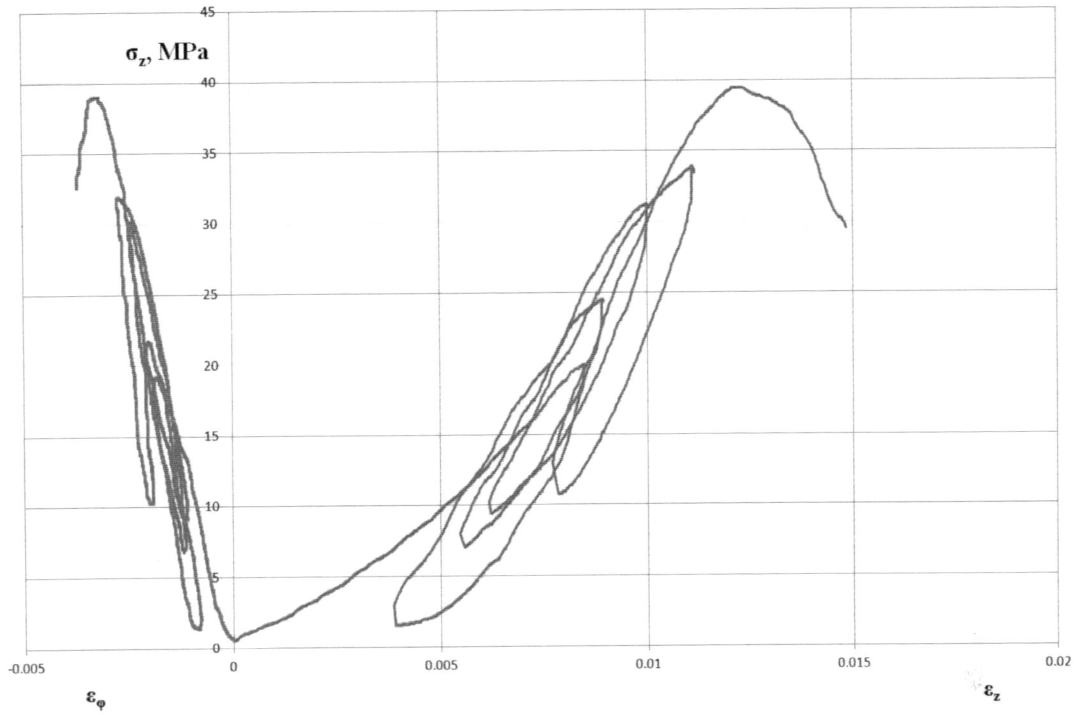

Figure 3. “Stress-strain” curve for sandstone specimen No 1-2 after the multistage test.

2.3. Experimental data-processing

To determine the modulus of elasticity (GPa), the standard ASTM D 7012-04 [10] was used. Approximately linear part in the "stress-strain" curve was used to calculate modulus of elasticity.

Poisson’s ratio ν was calculated by the formula below in the same linear part as modulus of elasticity:

$$\nu = -\frac{\varepsilon_\varphi}{\varepsilon_z}.$$

In addition to the elastic and strength characteristics, the specific energy intensity (kJ / m^2) was calculated for each loading stage. Specific energy intensity is the specific portion of energy accumulated in the sample at each stage of loading until the compressive elastic limit. This characteristic is described by the integral below:

$$w = \int \left(\sigma_z(u_z) - \sigma_\phi^i\right) \cdot du_z = \left[\begin{array}{c} \int \left(a_1 u_z^2 + a_2 u_z + a_3 - \sigma_\phi^i\right) \cdot du_z \\ \int \left(b_1 u_z^3 + b_2 u_z^2 + b_3 u_z + b_4 - \sigma_\phi^i\right) \cdot du_z \end{array} \right.,$$

where $\sigma_z(u_z)$ is the polynomial dependence of the axial stress σ_z on the axial displacements u_z; σ_φ^i is the value of the lateral pressure on the i-th (i = 1, ..., 5) loading stage.

Interpreting the compressive elasticity limits as the compressive strength limits for given lateral pressures, the envelopes of Mohr-Coulomb half-circles were obtained by using standard technique [11].

For each sample, the values of the elastic limit, the elasticity modulus, the Poisson's ratio, the specific energy intensity and envelopes of Mohr-Coulomb circles under multistage triaxial compression were obtained at each loading stage (see tables 1-5).

Based on the values of the coefficient of determination, we revealed the strong, moderate and weak dependencies of the considered mechanical characteristics on the stage number (No). Stage No varied from 1 to 5.

Table 1. Empirical dependence of compressive elastic limit σ_y on Stage No.

Specimen No	Dependence of compressive elastic limit σ_y (MPa) on Stage No	Coefficient of determination R^2
1-2	$\sigma_y = 4.15 \cdot StageNo + 14.85$	0.99
1-3	$\sigma_y = 4 \cdot StageNo + 15.2$	0.97
1-4	$\sigma_y = 2.7 \cdot StageNo + 21.6$	0.95
2-1	$\sigma_y = 4.15 \cdot StageNo + 15.45$	0.93
2-2	$\sigma_y = 2.8 \cdot StageNo + 15$	0.97
2-3	$\sigma_y = 1.3 \cdot StageNo + 19.6$	0.34

Table 2. Empirical dependence of elasticity modulus E on Stage No.

Specimen No	Dependence of elasticity modulus E (GPa) on Stage No	Coefficient of determination R^2
1-2	$E = 0.242 \cdot StageNo + 1.846$	0.88
1-3	$E = 0.158 \cdot StageNo + 2.648$	0.94
1-4	$E = 0.014 \cdot StageNo + 2.066$	0.02
2-1	$E = 0.215 \cdot StageNo + 2.101$	0.73
2-2	$E = 0.096 \cdot StageNo + 2.458$	0.67
2-3	$E = 0.103 \cdot StageNo + 1.791$	0.44

Table 3. Empirical dependence of Poisson ratio ν on Stage No.

Specimen No	Dependence of Poisson ratio ν on Stage No	Coefficient of determination R^2
1-2	$\nu = 0.006 \cdot StageNo + 0.212$	0.9
1-3	$\nu = 0.007 \cdot StageNo + 0.211$	0.72
1-4	$\nu = 0.008 \cdot StageNo + 0.202$	0.57
2-1	$\nu = 0.008 \cdot StageNo + 0.242$	0.89
2-2	$\nu = 0.008 \cdot StageNo + 0.248$	0.94
2-3	$\nu = 0.003 \cdot StageNo + 0.261$	0.45

Table 4. Empirical dependence of specific energy intensity W on Stage No.

Specimen No	Dependence of specific energy intensity W (kJ/m^2) on Stage No	Coefficient of determination R^2
1-2	$W = -0.426 \cdot StageNo + 4.266$	0.33
1-3	$W = 0.428 \cdot StageNo + 2.01$	0.43
1-4	$W = -0.907 \cdot StageNo + 7.761$	0.16
2-1	$W = -0.396 \cdot StageNo + 4.866$	0.69
2-2	$W = -0.316 \cdot StageNo + 4.22$	0.36
2-3	$W = 0.703 \cdot StageNo + 1.425$	0.42

Table 5. Envelopes of Mohr-Coulomb half-circles.

Specimen No	Dependence of shear stress τ (MPa) on normal stress σ (MPa)	Range of normal stress σ (MPa)
1-2	$\sigma = 0.234 \cdot \tau + 7.039$	$0 \le \sigma \le 34.942$
1-3	$\sigma = 0.159 \cdot \tau + 8.508$	$0 \le \sigma \le 34.767$
1-4	$\sigma = 0.138 \cdot \tau + 9.741$	$0 \le \sigma \le 37.469$
2-1	$\sigma = 0.321 \cdot \tau + 6.511$	$0 \le \sigma \le 39.094$
2-2	$\sigma = 0.107 \cdot \tau + 7.659$	$0 \le \sigma \le 32.594$
2-3	$\sigma = 0.139 \cdot \tau + 6.085$	$0 \le \sigma \le 33.182$

Strong dependencies on the stage number prevail for the compressive elastic limit. Moderate dependencies are observed for the elasticity modulus and the Poisson ratio. Weak dependencies are obtained for the specific energy intensity. Envelopes of Mohr-Coulomb half-circles are approximated by the linear relationships σ(τ) in certain range of σ.

3. Conclusion

In this study, specific tests were carried out for multistage triaxial compression of sandstone specimens, as a result of which a Mohr-Coulomb envelope for each of the sandstone samples was constructed, the modulus of elasticity, the Poisson ratio, the compression elastic limit and the specific energy intensity depending on the stage number were determined.

Based on the processed experimental data, the following conclusions can be drawn:

• With each subsequent loading stage, a monotonous increasing tendency (strong, moderate) of the elastic limit, the elasticity modulus, the Poisson ratio is traced, which indicates the existence of a variation of these parameters (under loading-unloading conditions), which is recommended for taking into account during the theoretical calculations of the stress-strain state.

• There is no common clear pattern of variation in the specific energy intensity (weak dependency on stage number).

• Envelopes of Mohr-Coulomb half-circles constructed on the basis of the conducted multistage tests with triaxial compression are suitable for describing the strength properties of sandstones in the case of core deficiency and are an alternative to the traditional technique for constructing a strength envelope from the experimental data of three tests (splitting tensile, uniaxial compression, conventional triaxial compression).

• Multistage triaxial compression makes it possible to describe the stepwise mechanical behavior of rock from an elastic state to a plastic state.

Acknowledgments
We appreciate the contribution of the Shared Use Center of Geomechanical, Geophysical, and Geodynamic Measurements, Siberian Branch, Russian Academy of Sciences to the present research work.

References

[1] Song H, Zhang H, Fu D and Zhang Q 2016 *Int. J. of Rock Mech. & Mining Sci.* **88** 157-164
[2] Myers M T and Sharf-Aldin M H 2015 *Proc. ARMA-2015-767, 49th U.S. Rock Mechanics* 11 p
[3] Gatelier N, Pellet F and Loret B 2002 *Int. J. of Rock Mech. & Mining Sci.* **39(3)** 335-354
[4] Yoshinaka R, Tran T V and Osada M 1997 *Int. J. of Rock Mech. & Mining Sci.* **34(3-4)** 354.e1-354.e14
[5] Yoshinaka R, Tran T V and Osada M 1998 *Int. J. of Rock Mech. & Mining Sci.* **35(7)** 941-955
[6] Zhang J C, Zhou S H, Xu S H and Fang L G 2013 *Int. J. of Rock Mech. & Mining Sci.* **58** 103-110
[7] Ma L.-j., Liu X.-y., Wang M.-y., et al. 2013 *Int. J. of Rock Mech. & Mining Sci.* **62** 34-41
[8] Youn H., Tonon F. 2010 *Int. J. of Rock Mech. & Mining Sci.* **47(4)** 678-684
[9] Bro A. 1997 *Int. J. of Rock Mech. & Mining Sci.* **34(1)** 143-145
[10] ASTM D 7012-04. Standard test method for compressive strength and elastic moduli of intact rock core specimens under varying states of stress and temperatures. 8 p.
[11] State Standard (GOST) 21153.8-88. The rocks. Method for determining the ultimate strength limit at conventional triaxial compression

IOP Conf. Series: Journal of Physics: Conf. Series **894** (2017) 012105 doi:10.1088/1742-6596/894/1/012105

Quantitative analysis of transverse non-uniformity of liquid film at the initial stage of annular-dispersed flow

S Isaenkov[1,2], A Cherdantsev[1,2], M Cherdantsev[1], D Markovich[1,2]

[1]Kutateladze Institute of Thermophysics, Novosibirsk, Russia
[2]Novosibirsk State University, Novosibirsk, Russia

Abstract. The process of formation of disturbance waves in annular gas-liquid flow was studied using three-dimensional configuration of brightness-based laser-induced fluorescence technique in two ducts of different geometry in the proximity of the inlet. The process was found to consist of three stages: formation of initial two-dimensional high-frequency waves; fragmentation into localised chaotic 3D-waves and formation of large-scale quasi-2D disturbance waves. A method of quantitative characterisation of the degree of transverse coherence of film surface was developed and used to estimate the coordinates of the borders of the three stages. All the stages occur closer to the inlet at higher superficial gas velocities and lower liquid flow rates. It was found that the shape of the duct does not exert significant influence of the distance of existence of two-dimensional initial waves, whilst the disturbance waves are formed closer to the inlet in the circular duct.

1. Introduction

Annular flow is a common regime of two-phase gas-liquid flow occurring at high gas flow rates. In this flow regime liquid phase travels as a film along the duct walls together with gas stream flowing through the core of the duct. At high gas and liquid flow rates part of liquid is torn from the film surface and travels in form of droplets inside the gas core. In such a case the interface is always covered by large-scale disturbance waves coexisting with small-scale ripple waves. The entrainment occurs mainly due to disruption of ripples on top of disturbance waves by the gas stream (see [1]).

The disturbance waves appear not far from the inlet. Frequency of disturbance waves decreases with downstream distance due to coalescence of disturbance waves in [2], whilst their velocity grows with distance and is stabilised far from the inlet in [3]. Average characteristics of disturbance waves far from the inlet were studied in a large number of experimental works; many of them are summarised in the review paper by Azzopardi [4].

The process of disturbance waves formation is less studied. It was found by Zhao et al. in [5] that prior to disturbance waves formation the film surface is covered by low-amplitude high-frequency initial waves. Alekseenko et al. [6] applied spatiotemporally resolved measurements of film thickness to observe that the disturbance waves are created due to coalescence of the initial waves. These measurements were performed in two-dimensional representation, i.e., along one longitudinal section of the duct.

Clark et al. in [7] found that presence of gas stream enhances formation of large quasi-two-dimensional waves on film surface. Nevertheless, the film surface in annular flow remains essentially three-dimensional. The disturbance waves, being the most two-dimensional waves in such a system, are of limited transverse size (see [8]) and they are non-uniform by amplitude in transverse direction [9]. Zhao et al. in [5] observed that the transverse coherence of disturbance waves increases with

Content from this work may be used under the terms of the Creative Commons Attribution 3.0 licence. Any further distribution of this work must maintain attribution to the author(s) and the title of the work, journal citation and DOI.
Published under licence by IOP Publishing Ltd

downstream distance. Thus, studying the process of disturbance waves formation only in frames of 2D approach might lead to loss of important information.

The first qualitative results on formation of disturbance waves in 3D space were obtained by Alekseenko et al. (see [10]). It was found that the initial waves are two-dimensional with flat fronts normal to flow direction and covering the whole width of the duct. At some distance from the inlet the initial 2D waves are broken into small-scale localised 3D waves, which are very different by shape and size. The 3D waves undergo multiple acts of coalescence, which leads to formation of large-scale quasi-two-dimensional disturbance waves.

In the present work, a new method of quantitative estimation of the degree of three-dimensionality of film surface is proposed. The method is applied to the obtained fields of film thickness to study the downstream evolution of two-dimensionality of film surface at the initial stage of annular flow.

2. Experimental setup and measurement technique

The experiments were conducted in two vertical ducts of different geometry: in a circular pipe with inner diameter of 11.7 mm and in a rectangular duct with cross-section of 50 mm by 5 mm (Fig. 1). The length of each of the ducts was about 70 cm. The liquid was supplied onto the duct walls through a tangential slot distributor with width of 0.5 mm. In the circular duct liquid film covered the whole inner surface of the pipe; in the rectangular duct only one of the wider walls was covered by liquid. The wetted perimeter was approximately the same in both cases. The advantage of the rectangular duct consists in absence of optical distortions caused by the duct's curvature and in possibility to observe the whole wetted area. The latter is not true for the circular duct, in which only 1/4 of wetted perimeter can be observed by a single camera. The main disadvantage of the rectangular duct is presence of the side walls which might slow down the disturbance waves.

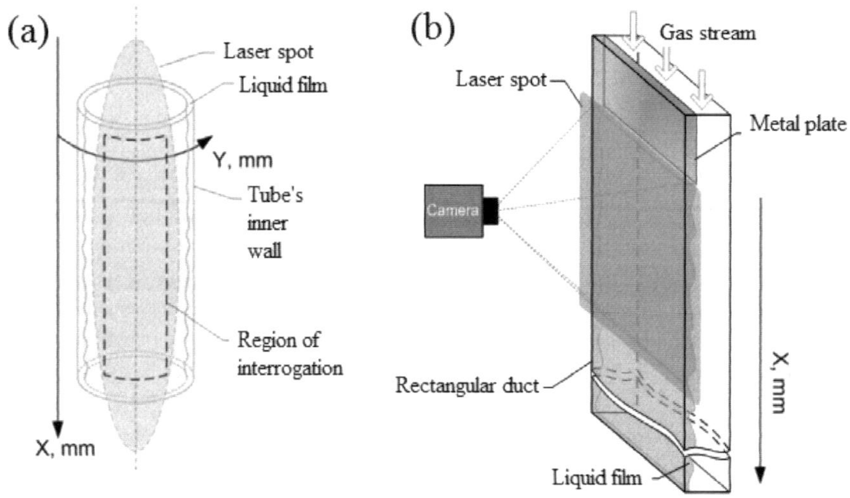

Figure. 1. 3D-BBLIF method in circular (a) and rectangular (b) ducts.

Adiabatic air-water flow at nearly atmospheric pressure was studied in both cases. Superficial gas velocity, V_g, and liquid Reynolds numbers, Re_L, were used as flow parameters. Re_L was defined as:

$$\mathrm{Re}_L = Q(\pi d\nu)^{-1}, \qquad \mathrm{Re}_L = Q(l\nu)^{-1}$$

for circular and rectangular duct, respectively. Here Q is volumetric liquid flow rate, ν - kinematic viscosity of liquid (water at working temperature of 16°C has viscosity of $1.15*10^{-6}$ m^2/s), d - inner pipe diameter, l - width of the wider wall of the rectangular duct.

IOP Conf. Series: Journal of Physics: Conf. Series **894** (2017) 012105 doi:10.1088/1742-6596/894/1/012105

	Circular duct	Rectangular duct
Longitudinal size of RoI (mm)	80	200
Transverse size of RoI (mm)	12	50
Spatial resolution, mm/pixel	0.1	0.2
Sampling frequency (kHz)	10	1
Liquid Reynolds numbers	140-400	140-400
Gas velocity (m/s)	22-57	14-43

Table 1.

To measure the local film thickness, 3D-configuration of brightness-based laser-induced fluorescence technique was applied. This method enables one to make measurements of film thickness resolved in both longitudinal and transverse coordinates and in time with high spatial and temporal resolution. Small amount (10 mg/l) of Rhodamine 6G dye was dissolved in working liquid. For excitation of fluorescence, liquid was illuminated by a continuous 2 W laser with wavelength of 532 nm. The laser beam was spread into a spot illuminating the whole region of interrogation (RoI). Brightness of re-emitted fluorescent light was measured by high-speed camera Photron Fastcam SA5 equipped with orange optical filter to cut-off the reflected laser light. The size of the RoI, spatial resolution, sampling frequency and the range of flow parameters for both ducts are given in Table 1.

The measured brightness of fluorescent light, $I(x,y,t)$, can be converted into the local film thickness, $h(x,y,t)$, where x is longitudinal coordinate, y - transverse coordinate and t is time. I and h are related as follows:

$$I(x,y,t) = C(x,y)\cdot[1-e^{-\alpha h(x,y,t)}]\cdot[1+K\cdot e^{-\alpha h(x,y,t)}]+D(x,y).$$

Here α is coefficient of absorption of the laser light by the fluorescent dye, which is measured separately; K is the coefficient of reflection of the laser light from water-air interface, which is equal 0.02 at moderate interface slopes; $D(x,y)$ is the dark level of camera and $C(x,y)$ is the compensation matrix, created to compensate the non-uniformity of laser illumination and to create a reference value of brightness corresponding to a reference value of film thickness.

3. Results and discussion

As mentioned above, the process of disturbance waves' formation consists of three stages, characterised by different degrees of two-dimensionality (see [10]):

Stage 1. Formation and development of the initial 2D waves;

Stage 2. Decay into 3D waves.

Stage 3. Formation of quasi-2D disturbance waves.

Figure 2 shows the waves typical for these three stages marked on instantaneous fields of film thickness in circular (a) and rectangular (b) ducts.

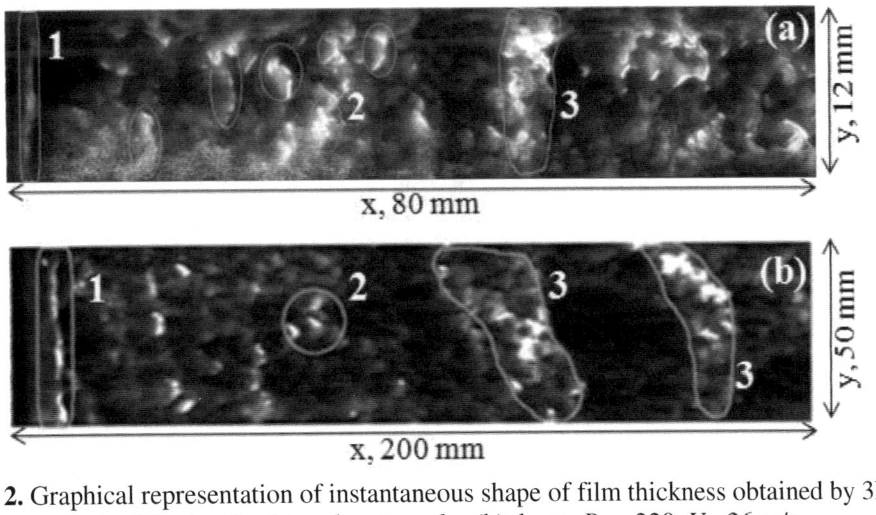

Figure 2. Graphical representation of instantaneous shape of film thickness obtained by 3D-BBLIF method in circular (a) and rectangular (b) ducts. Re_L=220, V_g=26 m/s.

Automatic identification of these three stages requires a method of quantitative estimation of two-dimensionality (or "transverse coherence") of film surface. This method was derived on the basis of cross-correlation analysis. Similar approach was employed by Zhao et al. in [5], who used cross-correlation between the signals of conductance probes placed around the circumference of the pipe.

Present experimental data are obtained in form of 3D matrices, $h(x, y, t)$. The film is considered to be fully two-dimensional when all the temporal records of film thickness, obtained at the same x coordinate but different y coordinates, are identical. Vice versa, absence of correlation between such records would correspond to complete three-dimensionality of film surface. The degree of two-dimensionality is then obtained in form of a 3D-matrix $b(x, y_i, y_j)$, which was defined as maximum value of normalised cross-correlation function between the two temporal records of film thickness, obtained at the same downstream coordinate, x, and different transverse coordinates, y_i and y_j:

$$b(x, y_i, y_j) = \frac{\max_\tau\left(\sum_{t=1}^{T} h(x, y_i, t)h(x, y_j, t-\tau)\right)}{A}.$$

Here T is duration of a whole record, τ - variable temporal delay between the two records and

$$A = \max\left(\sum_{t=1}^{T} h(x, y_i, t)^2, \sum_{t=1}^{T} h(x, y_j, t)^2\right).$$

Variation of τ was introduced in order to minimise possible influence of constant time shift between the two records, i.e., the case when the waves are two-dimensional but slightly inclined to the horizon. Such distortion of the wave fronts is expected to occur next to the side walls of the rectangular duct. Normalisation by A in this form was introduced to compensate for the difference in the amplitudes of the waves. Thus, the present approach enables us to minimise the influence of both curved/inclined wave fronts and persistent change in waves' amplitudes across the waves and to focus on the transverse size of the waves.

For a fixed value of $x=x_0$, $b(x_0,y_i,y_j)$ is a square matrix symmetrical relative to the main diagonal; each element of this diagonal is equal 1. The value of b decreases with the distance from this line; the rate of decrease depends on two-dimensionality of the film. Fig. 3(a) shows an example of b-matrix at constant y_i=24 mm together with 3 profiles $b(y_j)$ obtained at different values of x (Fig. 3b). The rate of decrease is rather low close to the inlet, where the initial waves are approximately two-dimensional (see Fig. 3b at x=6 mm). In the range of 3D-waves (x=50 mm in Fig. 3b) b reaches low values very quickly. After the disturbance waves are formed (see x=180 mm in Fig. 3b) the values of b at larger $dy=|y_j-y_i|$ grow again. For further analysis, a constant value of dy was selected so that b in the region of 3D-waves would be about 0.1 for all flow regimes. At the same time, dy must be small enough to allow analysis of maximum number of pairs of records, $\{y_i,y_j\}$. The optimum value of dy in such a case should be equal to maximum transverse size of the 3D-waves, L, which was found to be about 5 mm (see Fig. 3b for x=50 mm).

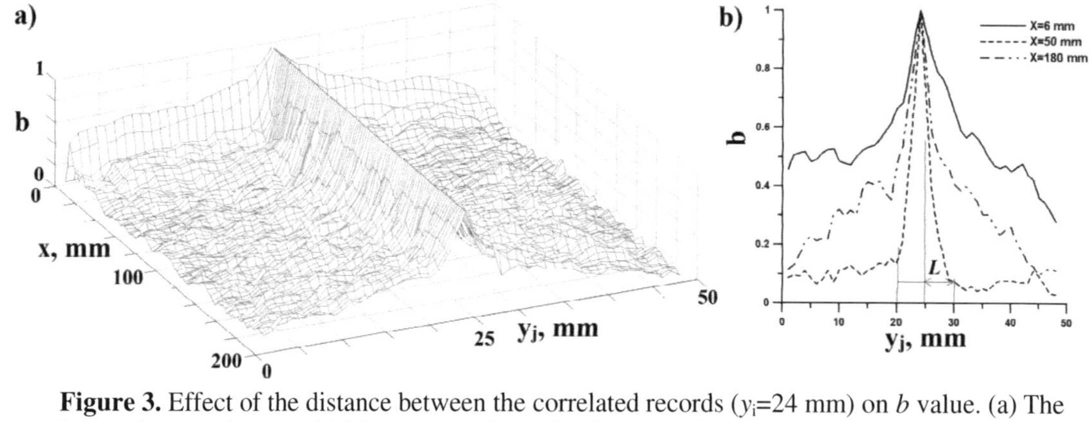

Figure 3. Effect of the distance between the correlated records (y_i=24 mm) on b value. (a) The whole set of x-coordinates; (b) Three selected profiles $b(y_j)$ at different stages. Re_L=220, V_g=26 m/s.

Figure 4 shows examples of downstream evolution of two-dimensionality coefficient at constant dy=5 mm (i.e., section of b-matrix defined as $b(x,y_i,y_i+5\text{mm})$) for two flow regimes. This matrix is quite uniform across y_i coordinate (see Fig. 4a-b), so the average of this matrix over y_i was used for further analysis. At moderate gas velocities (see Fig. 4a,c) a prominent peak of two-dimensionality is observed near the inlet, corresponding to the Stage 1; transition to Stage 2 (the fragmentation of 2D-waves into 3D-waves) is accompanied by decrease of two-dimensionality to the value of about 0.1. At the Stage 3 the transverse coherence is recovered gradually with downstream distance, denoting the formation of disturbance waves and their growth in transverse direction. The picture is similar in the case of high gas velocities (Fig. 4 b, d), except for two moments. First, the initial peak is nearly absent in this case, which means that the initial waves are broken into 3D-waves almost immediately. Second, the third stage (recovery of transverse coherence due to formation of disturbance waves) occurs much closer to the inlet as well.

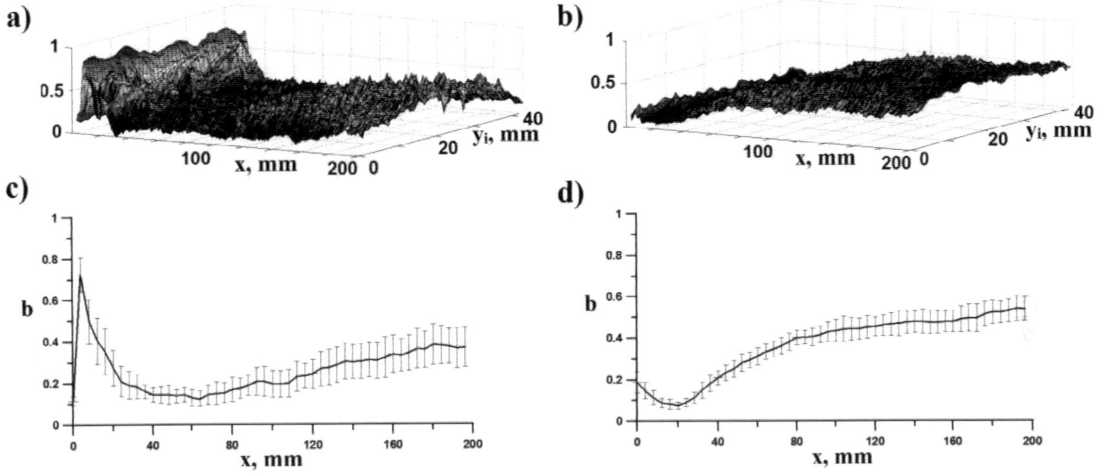

Figure 4. (a, b) The effect of y-coordinate at constant dy for the whole set of x-coordinate. (c,d) Dependence of cross-correlation averaged over y on x. (a,c) - Re_L=220, V_g=26 m/s, (b,d) - Re_L=400, V_g=43 m/s. Rectangular duct.

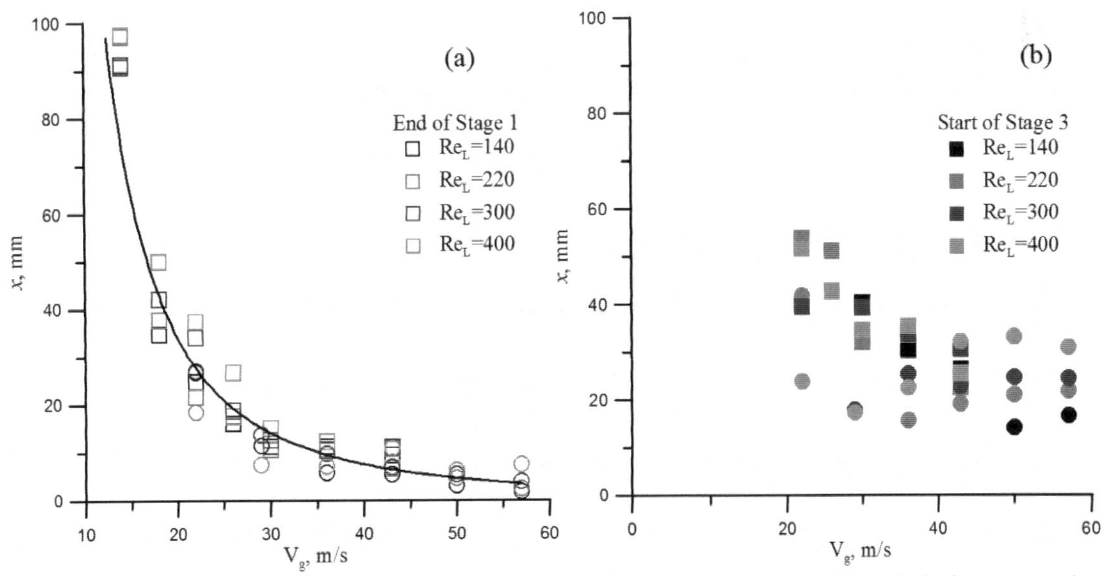

Figure 5. Measured x-coordinates of the end of Stage 1 (a) and beginning of Stage 3 (b) in rectangular (squares) and circular (circles) ducts.

Based on dependencies $b(x)$ it is possible to identify the borders between the Stages 1-2 and 2-3 for each combination of flow rates. The borders were defined as the points where linear extrapolations of particular areas of $b(x)$ dependencies cross the level b=0.1. The first selected area was the steep negative slope just behind the maximum of two-dimensionality of the initial waves. The second area was the steepest positive slope right after the shallow plateau around the value b~0.1.

The measurements results are shown in Fig. 5. The spatial length of Stage 1 decreases with increasing gas velocity; it reaches large values of ~100 mm for the lowest V_g (Fig. 5a). There is an effect of Re_L as well: the lower is Re_L, the faster 2D-waves are disrupted into 3D-waves, but this effect is weak in comparison to that of gas velocity. The most interesting thing is that the data for both ducts lie on the same line; the difference does not overcome the difference due to liquid flow rate. The whole set of data can be approximated by a single dependence $x \sim V_g^{-2}$ (see the solid line in Fig. 5a). Thus, the shape of the duct does not strongly affect the process of disruption of 2D waves.

The coordinate of start of Stage 3 also decreases with gas velocity and increases with liquid Reynolds number (Fig. 5b). In the rectangular duct, the effect of Re_L is less expressed, being comparable to the uncertainty of the method; at the same time, the effect of V_g is prominent in the used V_g range. The data obtained in circular duct at the highest gas velocities, seem to satisfy the general trend observed in the rectangular duct, though the effect of Re_L is much more evident in this region whilst the effect of V_g nearly disappears. The most striking difference is that the formation of disturbance waves starts much earlier at moderate V_g in the circular duct. This may be explained by the effect of the side walls of the rectangular duct, slowing down the disturbance waves.

4. Conclusions

The process of formation of disturbance waves in annular gas-liquid flow was studied using brightness-based laser-induced fluorescence technique in three-dimensional approach. The measurements were performed in two ducts of different shape in a wide range of air and water flow rates. The process of disturbance waves formation consists of three stages, which differ, in particular, by three-dimensionality of film surface. A method of quantitative characterisation of degree of transverse coherence of liquid film was developed to identify the stages automatically. It was found that the length of each stage of the process decreases with gas velocity but increases with liquid flow rate. The length of the area of existence of initial 2D-waves was found to be nearly independent on the duct's shape. At the same time, formation of disturbance waves may occur closer to the inlet in the circular duct in the range of moderate gas velocities.

Acknowledgments

The work was supported by Russian Science Foundation (project RSF 16-19-10449).

References

[1] Woodmansee D E and Hanratty T J 1969 *Chem. Eng. Sci.* **24** 299
[2] Hall Taylor N S and Nedderman R M 1968 *Chem. Eng. Sci.* **23** 551
[3] Wolf A, Jayanti S and Hewitt G F 2001 *Chem. Eng. Sci.* **56** 3221
[4] Azzopardi B J 1997 *Int. J. Mult. Flow* **23** 1
[5] Zhao Y, Markides C N, Matar O K and Hewitt G F 2013 *Int. J. Mult. Flow* **55** 111
[6] Alekseenko S V, Cherdantsev A V, Cherdantsev M V, Isaenkov S V and Markovich D M 2015 *Int. J. Mult. Flow* **77** 65
[7] Clark W W, Campbell G B, Hills J H and Azzopardi B J 2001 *4th Int. Conf. Mult.* (New Orleans, USA)
[8] Cherdantsev A V, Hann D B and Azzopardi B J 2014 *Int. J. Mult. Flow* **67** 52
[9] Belt R J, Van't Westende J M C, Prasser H M and Portela L M 2010 *Int. J. Mult. Flow* **36** 570
[10] Alekseenko S, Cherdantsev A, Cherdantsev M, Isaenkov S and Markovich D 2016 *MATEC Web Conf.* **84** 00001

IOP Conf. Series: Journal of Physics: Conf. Series **894** (2017) 012106 doi:10.1088/1742-6596/894/1/012106

The unidirectional motion of two heat-conducting liquids in a flat channel

V K Andreev[1,2] **and E N Cheremnykh**[1,2]

[1] Institute of Computational Modelling, Krasnoyarsk, Russia
[2] Siberian Federal University, Krasnoyarsk, Russia

E-mail: andr@icm.krasn.ru, elena_cher@icm.krasn.ru

Abstract. The unidirectional motion of two viscous incompressible liquids in a flat channel is studied. Liquids contact on a flat interface. External boundaries are fixed solid walls, on which the non-stationary temperature gradients are given. The motion is induced by a joint action of thermogravitational and thermocapillary forces and given total non - stationary fluid flow rate in layers. The corresponding initial boundary value problem is conjugate and inverse because the pressure gradients along axes channel have to be determined together with the velocity and temperature field. For this problem the exact stationary solution is found and a priori estimates of non - stationary solutions are obtained. In Laplace images the solution of the non - stationary problem is found in quadratures. It is proved, that the solution converges to a steady regime with time, if the temperature on the walls and the fluid flow rate are stabilized. The numerical calculations for specific liquid media good agree with the theoretical results.

1. Statement of the problem

We consider the system of two incompressible immiscible liquids with the interface $y = 0$. The parameters of liquids moving in strips $-h_1 < y < 0$ and $0 < y < h_2$, $-\infty < x < \infty$ are denoted by the indices "1" and "2", respectively; ρ_j, ν_j, χ_j, c_{p_j}, β_j, k_j are the densities, kinematic viscosities, coefficients of thermal diffusivity, specific heat of liquids, volumetric expansion and thermal conductivity, respectively. Further, it is assumed that these parameters are positive constants. The motion of liquids is described by the Oberbeck - Boussinesq equation system. Following the works [1, 2], we set

$$u_j = w_j(y,t), \quad \theta_j = -a_j(y,t)x + T_j(y,t), \quad p_j = -b_j(y,t)x + P_j(y,t), \quad j = 1,2. \tag{1}$$

Substitution (1) in the Oberbeck - Boussinesq equation system leads to equations

$$\begin{gathered} a_{jt} = \chi_j a_{jyy}, \quad b_{jy} = \rho_j g \beta_j a_j, \quad w_{jt} = \nu_j w_{jyy} + \frac{1}{\rho_j} b_j, \\ T_{jt} = \chi_j T_{jyy} + a_j w_j, \quad P_{jy} = \rho_j g \beta_j T_j, \end{gathered} \tag{2}$$

where u_j is the projection of the velocity vector on the axis x, p_j is the deviation of pressure from hydrostatic one, $g =$ const is the acceleration of gravity force, θ_j is the temperature. Further, a problem for determining the velocity field (i. e. for w_j and a_j) will be considered only. The functions T_j is the solution of the conjugate problem that is analogous to the problem for a_j.

Content from this work may be used under the terms of the Creative Commons Attribution 3.0 licence. Any further distribution of this work must maintain attribution to the author(s) and the title of the work, journal citation and DOI.
Published under licence by IOP Publishing Ltd

The functions b_j and P_j are found from the second and fourth equations (2) with known a_j and T_j.

The conjugate boundary value problem for the functions $a_j(y,t)$ has the form

$$a_{jt} = \chi_j a_{jyy}, \tag{3}$$

$$a_j(y,0) = a_{j0}(y), \tag{4}$$

$$a_1(-h_1,t) = A_1(t), \quad a_2(h_2,t) = A_2(t), \tag{5}$$

$$a_1(0,t) = a_2(0,t), \quad k_1 a_{1y}(0,t) = k_2 a_{2y}(0,t) \tag{6}$$

with given functions $a_{j0}(y)$, $A_j(t)$, $j = 1,2$, which for the smooth solutions must satisfy the matching conditions $a_{10}(-h_1) = A_1(0)$, $a_{20}(h_2) = A_2(0)$, $a_{10}(0) = a_{20}(0)$, $k_1 a_{10y}(0) = k_2 a_{20y}(0)$. Relations (6) result from the equality of temperatures and heat fluxes on the interface $y = 0$ [3].

Let us formulate the problem for $w_j(y,t)$. Due to (2)

$$b_j(y,t) = \rho_j g \beta_j \int_0^y a_j(z,t)\,dz + C_j(t), \tag{7}$$

with arbitrary functions $C_j(t)$, then

$$w_{jt} = \nu_j w_{jyy} + g\beta_j \int_0^y a_j(z,t)\,dz + \frac{1}{\rho_j} C_j(t), \tag{8}$$

$$w_j(y,0) = w_{j0}(y), \tag{9}$$

$$w_1(-h_1,t) = 0, \quad w_2(h_2,t) = 0. \tag{10}$$

The dynamic condition at the interface [3] is reduced into two conditions:

$$\rho_2\nu_2 w_{2y}(0,t) - \rho_1\nu_1 w_{1y}(0,t) = -a_1(0,t)æ, \quad p_1(0,t) = p_2(0,t). \tag{11}$$

From the latter condition and the representations (1), (7) we obtain the equalities $C_1(t) = C_2(t) \equiv C(t)$, $P_1(0,t) = P_2(0,t)$. Besides,

$$w_1(0,t) = w_2(0,t) \tag{12}$$

and the initial data for a smooth solution must be agreed: $w_{10}(-h_1) = 0$, $w_{20}(h_2) = 0$, $A_{10}(0) = A_{20}(0)$, $\rho_2\nu_2 w_{20y}(0) - \rho_1\nu_1 w_{10y}(0) = -A_{10}(0)æ$, $w_{10}(0) = w_{20}(0)$.

If the function $C(t)$ is given, then the equation (8) with the boundary conditions (10) – (12) are the problem for the velocities. The aim of this paper is to study the inverse problem. Therefore, it is necessary to impose additional condition. We assume that the total fluid flow rate in layers is given

$$\int_{-h_1}^0 w_1(y,t)\,dy + \int_0^{h_2} w_2(y,t)\,dy = Q(t). \tag{13}$$

Similar formulations of problems arise under studying some technological processes, namely: thermal stabilization of power plants or cooling of electronic devices, no - contact coating technologies or alloying of steel [4, 5, 6].

2. Stationary flow

As before, we assume that $A_1^s = \text{const} \neq 0$ on the bottom wall $y = -h_1$. Now, we give the form of the stationary solution. The following dimensionless variables and parameters are introduced: $\xi = y/h_1$, $-1 \leqslant \xi \leqslant 0$, $\eta = y/h_2$, $\leqslant \eta \leqslant 1$, $k = k_1/k_2$, $h = h_1/h_2$, $\chi = \chi_1/\chi_2$, $\rho = \rho_1/\rho_2$, $\nu = \nu_1/\nu_2$, $a = A_2^s/A_1^s = \text{const}$,

$$\text{Ra}_1 = \frac{g\beta_1 A_1^s h_1^4}{\nu_1\chi_1}, \quad \text{Ra}_2 = \frac{g\beta_2 A_1^s h_2^4}{\nu_2\chi_2}, \quad \bar{C}^s = \frac{h_1^3 C^s}{\rho_1\nu_1\chi_1}, \quad \text{Ma} = \frac{æ A_1^s h_2^2}{\rho_2\nu_2\chi_2},$$

then

$$a_1^s(\xi) = A_1^s\left[1 + \frac{h(a-1)(1+\xi)}{k+h}\right], \quad a_2^s(\eta) = A_1^s\left[1 + \frac{(a-1)(h+k\eta)}{k+h}\right], \tag{14}$$

$$w_1^s(\xi) = \frac{\chi_1}{h_1}\left\{-\frac{\xi^2}{2}\bar{C}^s - \frac{\text{Ra}_1}{6}\left[\xi^3 + \frac{h(a-1)}{k+h}\left(\xi^3 + \frac{\xi^4}{4}\right)\right]\right\} + \alpha_1\xi + \alpha_2,$$

$$w_2^s(\eta) = \frac{\chi_2}{h_2}\left\{-\frac{\rho\nu\chi}{2h^3}\bar{C}^s\eta^2 - \frac{\text{Ra}_2}{6}\left[\eta^3 + \frac{(a-1)}{k+h}\left(h\eta^3 + \frac{k\eta^4}{4}\right)\right]\right\} + \alpha_3\eta + \alpha_2, \tag{15}$$

$$\alpha_1 = \frac{\chi_1 h}{h_1(h+\rho\nu)}\left\{\frac{\bar{C}^s}{2}\left(\frac{\rho\nu}{h^2} - 1\right) + \frac{\text{Ra}_1}{6}\left[1 + \frac{3h}{4}\frac{(a-1)}{k+h}\right] + \right.$$
$$\left. + \frac{h}{6\chi}\text{Ra}_2\left[1 + \frac{(a-1)}{k+h}\left(h + \frac{k}{4}\right)\right] + \frac{h}{\chi}\text{Ma}\left[1 + \frac{h(a-1)}{k+h}\right]\right\},$$

$$\alpha_2 = \alpha_1 + \frac{\chi_1}{h_1}\left\{\frac{\bar{C}^s}{2} - \frac{\text{Ra}_1}{6}\left[1 + \frac{3h}{4}\frac{(a-1)}{k+h}\right]\right\},$$

$$\alpha_3 = \frac{\chi_2 h}{h_2(h+\rho\nu)}\left\{\frac{\rho\nu}{h}\left[\frac{\chi\bar{C}^s}{2h}\left(\frac{\rho\nu}{h^2} - 1\right) + \frac{\chi\text{Ra}_1}{6h}\left(1 + \frac{3h}{4}\frac{(a-1)}{k+h}\right) + \right.\right.$$
$$\left.\left. + \frac{\text{Ra}_2}{6}\left(1 + \frac{(a-1)}{k+h}\left(h + \frac{k}{4}\right)\right)\right] - \text{Ma}\left[1 + \frac{h(a-1)}{k+h}\right]\right\}.$$

In the considered inverse problem, with given total fluid flow rate $Q^s = \text{const}$ (see (13)), dimensionless constant $\bar{C}^s$ is defined as follows

$$\bar{C}^s = \frac{6h^3\chi^{-1}}{\alpha_5}\left\{\bar{Q}^s - \chi\text{Ra}_1\left[\frac{1}{24} + \frac{h(a-1)}{30(k+h)} - \left(\frac{1}{6} + \frac{h(a-1)}{8(h+k)}\right)\left(1 + \frac{1-\alpha_4}{h}\right)\right] + \right.$$
$$\left. + \text{Ra}_2\left[\frac{1}{24} + \frac{(a-1)(k+5h)}{120(k+h)} - \alpha_4\left(\frac{1}{6} + \frac{(a-1)(k+4h)}{24(h+k)}\right)\right] + \frac{\text{Ma}(1-2\alpha_4)}{2(k+h)}\right\}, \tag{16}$$

where $\bar{Q}^s = Q^s/\chi_2$ is dimensionless flow, $\alpha_4 = (\rho\nu + h(h+2))(\rho\nu + h)^{-1}/2$, $\alpha_5 = h^2(2h + 3 - 3\alpha_4) + \rho\nu(1 + 3\alpha_4)$. We have $a = 1$ at equal temperature gradients on the walls $y = -h_1$, $y = h_2$ and the all formulas are simplified.

Figure 1 present the profiles of dimensionless velocities $\bar{w}_j^s = w_j^s h_j \chi_j^{-1}$ dependending on the thickness of the first layer for the system water ($j = 1$) – liquid CO_2 ($j = 2$). The physical parameters of the system are the following: $\rho_1 = 0.99821 \cdot 10^3\,\text{kg/m}^3$, $\nu_1 = 1.0038 \cdot 10^{-6}\,\text{m}^2/\text{s}$, $k_1 = 0.5984\,\text{kg}\cdot\text{m/s}^3\cdot\text{K}$, $\chi_1 = 1.44152 \cdot 10^{-7}\,\text{m}^2/\text{s}$, $\beta_1 = 0.18 \cdot 10^{-3}\,\text{K}^{-1}$, $\rho_2 = 0.867191 \cdot 10^3\,\text{kg/m}^3$, $\nu_2 = 2.0076 \cdot 10^{-6}\,\text{m}^2/\text{s}$, $\chi_2 = 1.1221 \cdot 10^{-7}\,\text{m}^2/\text{s}$, $k_2 = 0.1101056\,\text{kg}\cdot\text{m/s}^3\cdot\text{K}$, $\beta_2 = 1.0188 \cdot 10^{-3}\,\text{K}^{-1}$, $æ = 0.0001989\,\text{N/m}\cdot\text{K}$, $\sigma = 72.86 \cdot 10^{-3}\,\text{N/m}$. If the thicknesses of the layers are the same $h_1 = h_2 = 0.001\,\text{m}$, a zone with the reverse flow arises in the second layer and velocity profile is a parabolic one. The flow is the Poiseuille one. With increasing thickness of the first layer the reverse flow is formed only in the first layer and the velocity profile becomes linear. Such flow is the Couette flow.

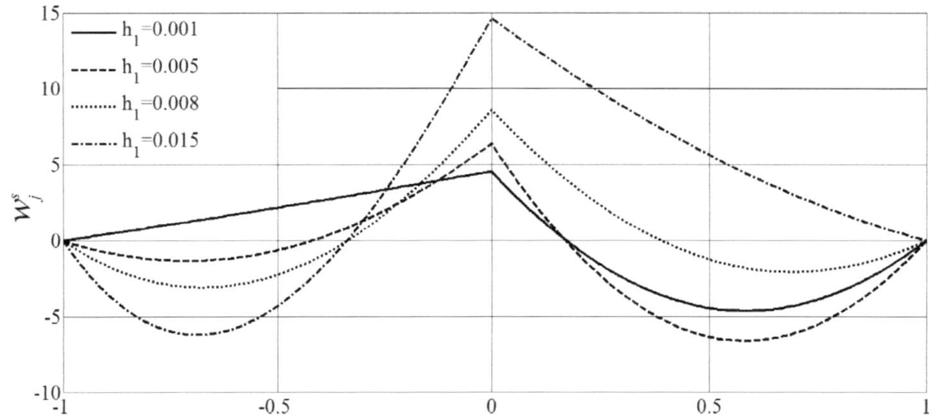

Figure 1. Stationary profile of dimensionless velocities $\bar{w}_j^s$; $A_1^s = 2^oC/\text{m}$; $Q^s = 0$; $h_2 = 0.001\text{m}$

3. A priori estimates

In solving the initial conjugate initial boundary value problems (3) – (6), (8) – (13) the method of variables separation is not applicable. Therefore, for the study the qualitative properties of the solutions, we use the method of a priori estimates [7]. Using this technique we can show that the solution of initial boundary value problem (3) – (6) tends to stationary one (14) if the integrals

$$\int_0^\infty e^{\delta t}|A_j(t) - A_j^s|\,dt, \quad \int_0^\infty e^{\delta t}|A_j'(t)|\,dt, \quad \delta = \frac{1}{M}\min_j\left(\frac{k_j}{\chi_j}\right), \quad j = 1,2, \tag{17}$$

converge and we have estimates for the convergence rate are valid

$$|a_j(y,t) - a_j^s(y)| \leqslant \left(\frac{8\chi_j}{k_j^2}\,F(t)K(t)\right)^{1/4} + |A_j(t) - A_j^s|. \tag{18}$$

The last estimates are uniform ones with respect to $y \in [-h_1, 0]$ for $j = 1$ and $y \in [0, h_2]$ at $j = 2$ and $t \in [0, \infty)$. In (17) M is the constant from the Friedrichs inequality [8]; the function $F(t)$ in (18) is limited and $K(t) \sim e^{-2\delta t}$.

For the prolem (8) – (13) we get similar result. The solution of initial boundary value problem (8) – (13) tends to stationary one (15), (16) if the integrals (17) and

$$\int_0^\infty e^{\delta t}|A_j''(t)|\,dt, \quad \int_0^\infty e^{\delta_1 t}|Q(t) - Q^s|\,dt, \quad \int_0^\infty e^{\delta_1 t}|Q'(t)|\,dt, \quad \int_0^\infty e^{\delta_1 t}|Q''(t)|\,dt, \quad \delta_1 = \frac{1}{M}\min_j\left(\frac{1}{\rho_j}\right) \tag{19}$$

converge and we have estimates for the convergence rate are valid

$$|w_j(y,t) - w_j^s(y)| \leqslant \left\{ \begin{array}{l} \dfrac{\sqrt{3}}{18}\dfrac{æh_1}{\rho_1\nu_1}\,|a_1(0,t) - a_1^s(0)| \\[2ex] \dfrac{16h_2}{9}\,|Q(t) - Q^s| \end{array} \right. + \left(\frac{8\nu_j}{\mu_j^2}\,F_2(t)E(t)\right)^{1/4}. \tag{20}$$

Estimates (20) are the uniform ones with respect to $y \in [-h_1, 0]$ at $j = 1$ and $y \in [0, h_2]$ at $j = 2$; $t \in [0, \infty)$. And for the function $C(t)$ we have

$$|C(t) - C^s| \leqslant c\Bigg[|A_1(t) - A_1^s| + \left(\frac{8\chi_1}{k_1^2}\,F(t)K(t)\right)^{1/4} + |A_1'(t)| + \left(\frac{8\chi_1}{k_1^2}\,F_1(t)K_1(t)\right)^{1/4} + \\ + \left(\frac{8\nu_1}{\mu_1^2}\,F_2(t)E(t)\right)^{1/4} + \left(\frac{8\nu_1}{\mu_1^2}\,F_3(t)E_1(t)\right)^{1/4}\Bigg] \quad (21)$$

with positive constant c depending on the input data. In (20), (21) the functions $F_1(t)$, $F_2(t)$, $F_3(t)$ are limited and $E(t), E_1(t) \sim e^{-2\delta_1 t}$.

4. The solution of conjugate problem by the Laplace transformation method. Analysis of the numerical results

For more accurate information on behavior $a(y,t)$ and $w(y,t)$ the Laplace transformation is applied. As a result, we get to the boundary value problem for images $\hat{a}_j(y,p)$ of functions $a_j(y,p)$

$$\hat{a}_{1\xi\xi}(\xi,p) - p\hat{a}_1(\xi,p) = -a_{10}(\xi), \\ \hat{a}_{2\eta\eta}(\eta,p) - p\chi h^{-2}\hat{a}_2(\eta,p) = -\chi h^{-2}a_{20}(\eta), \quad (22)$$

$$\hat{a}_1(-1,p) = \hat{A}_1(p), \quad \hat{a}_2(1,p) = \hat{A}_2(p), \\ \hat{a}_1(0,p) = \hat{a}_2(0,p), \quad kh^{-1}\hat{a}_{1\xi}(0,p) = \hat{a}_{2\eta}(0,p) \quad (23)$$

and images $\hat{w}_j(y,t)$ of functions $w_j(y,t)$

$$\hat{w}_{1\xi\xi}(\xi,p) - p\,\mathrm{Pr}_1^{-1}\hat{w}_1(\xi,p) = -\left(\mathrm{Pr}_1^{-1}w_{10}(\xi) + Ra_1\int_0^\xi \hat{a}_1(z,p)\,dz + \hat{C}(p)\right), \\ \hat{w}_{2\eta\eta}(\eta,p) - \frac{p\chi}{h^2\,\mathrm{Pr}_2}\hat{w}_2(\eta,p) = -\left(\frac{\chi}{h^2\mathrm{Pr}_2}w_{20}(\xi) + Ra_2\int_0^\eta \hat{a}_2(z,p)\,dz + \frac{\rho\nu\chi}{h^3}\hat{C}(p)\right), \quad (24)$$

$$\hat{w}_1(-1,p) = \hat{w}_2(1,p) = 0, \quad \chi h^{-1}\hat{w}_1(0,p) = \hat{w}_2(0,p), \\ \hat{w}_{2\eta}(0,p) - \rho\nu\chi h^{-2}\hat{w}_{1\xi}(0,p) = -\mathrm{Ma}\hat{a}_1(0,p), \quad \chi\int_{-1}^0 \hat{w}_2(\xi,p)\,d\xi + \int_0^1 \hat{w}_2(\eta,p)\,d\eta = \hat{Q}(p). \quad (25)$$

The problems (22), (23) and (24), (25) are written in a dimensionless form with the characteristic scales h_1^2/χ_1^{-1}, A_1^s, χ_j/h_j^{-1} for time and functions $a_j(y,t), w_j(y,t)$, respectively. Upon that $A_1^s = \max\limits_t |A_1(t)| > 0$, $\mathrm{Pr}_j = \nu_j/\chi_j$ are the Prandtl numbers. In deriving equations (22), (23) the initial data (4), (9) are used. In (23), (25) $\hat{A}_1(p)$, $\hat{A}_2(p)$ and $\hat{Q}(p)$ are images of given functions $A_1(t)$, $A_2(t)$ and $Q(t)$, respectively (see conditions (5) and (13)).

The general solution of equations (22) has the form

$$\hat{a}_1(\xi,p) = m_1\mathrm{sh}\sqrt{p}\xi + m_2\mathrm{ch}\sqrt{p}\xi - \frac{1}{\sqrt{p}}\int_{-1}^\xi a_{10}(z)\mathrm{sh}\sqrt{p}(\xi - z)\,dz, \\ \hat{a}_2(\eta,p) = m_3\mathrm{sh}\sqrt{p\chi}h^{-1}\eta + m_4\mathrm{ch}\sqrt{p\chi}h^{-1}\eta - \frac{1}{h\sqrt{p\chi^{-1}}}\int_0^\eta a_{20}(z)\mathrm{sh}\sqrt{p\chi}h^{-1}(\eta - z)\,dz. \quad (26)$$

Similarly, the solution of equations (24) has the form

$$\hat{w}_1(\xi,p) = C_1\mathrm{sh}\alpha_1\xi + C_2\mathrm{ch}\alpha_1\xi - \frac{1}{\alpha_1}\int\limits_0^{\xi} f_1(z,p)\mathrm{sh}\alpha_1(\xi - z)\,dz - \frac{\hat{C}(p)}{\alpha_1}(\mathrm{ch}\alpha_1\xi - 1),$$

$$\hat{w}_2(\eta,p) = C_3\mathrm{sh}\alpha_2\eta + C_4\mathrm{ch}\alpha_2\eta - \frac{1}{\alpha_2}\int\limits_0^{\eta} f_2(z,p)\mathrm{sh}\alpha_2(\eta - z)\,dz - \frac{\rho\nu\mathrm{Pr}_2\hat{C}(p)}{ph}(\mathrm{ch}\alpha_2\eta - 1),$$

$$f_1(\xi,p) = \mathrm{Pr}_1^{-1}w_{10}(\xi) + \mathrm{Ra}_1\int\limits_0^{\xi}\hat{a}_1(z,p)\,dz, \quad f_2(\eta,p) = \frac{\chi}{h^2\mathrm{Pr}_2}w_{20}(\xi) + \mathrm{Ra}_2\int\limits_0^{\eta}\hat{a}_2(z,p)\,dz, \tag{27}$$

$$\alpha_1 = \sqrt{p\mathrm{Pr}_1^{-1}}, \quad \alpha_2 = \sqrt{p\chi\mathrm{Pr}_2^{-1}h^{-1}}.$$

The values m_k, C_k, $k = 1,..,4$, including in (26), (27), and the function $\hat{C}$ are determined from boundary conditions (23), (25) have complex expressions and are not given here.

Suppose that $\lim\limits_{t\to\infty} A_j(t) = A_j^s$ and $\lim\limits_{t\to\infty} Q(t) = Q^s$. Using the obtained representations for $\hat{a}_j(y,p)$, $\hat{w}_j(y,p)$ and $\hat{C}(p)$ we can prove the limit equalities

$$\lim_{t\to\infty} a_j(y,t) = \lim_{p\to 0} p\hat{a}_j(y,p) = a_j^s(y), \quad \lim_{t\to\infty} w_j(y,t) = \lim_{p\to 0} p\hat{w}_j(y,p) = w_j^s(y),$$
$$\lim_{t\to\infty} C(t) = \lim_{p\to 0} p\hat{C}(p) = C^s, \tag{28}$$

where $a_j^s(y), w^s(y)$ and C^s are given by formulas (14) – (16). It is in agreement with the conclusions of the Section 3.

Let us apply the method of the numerical inversion of the Laplace transformation to obtain some results for the velocity functions. Let $Q(t) = Q^s = 0$ (the motion is caused by the action of the thermogravitational forces) and the longitudinal temperature gradient on the walls is distributed according to the law $A_j(t) = \gamma_j^1 + \gamma_j^2 e^{-\gamma_j^3 t}\sin(\gamma_j^4 t), j = 1,2$, where the coefficients γ_j^2, γ_j^4 correspond to the amplitude and frequency of the oscillations, respectively. If $\gamma_j^1 \neq 0, \gamma_j^3 > 0$ then takes place convergence of the solution to the stationary regime, according to equations (28). If $\gamma_j^3 \leqslant 0$, then limits of the function $A_j(t)$ at $t \to \infty$ do not exist then the solution does not tend to a stationary regime. Figure 2 and Figure 3 correspond to cases $A_1(t) = 1 - 5e^{-0.01t}\sin(0.1t)$ and $A_1(t) = 2\sin(0.1t)$, respectively. As expected, the solution does not tend to a stationary regime with increasing time in the latter example.

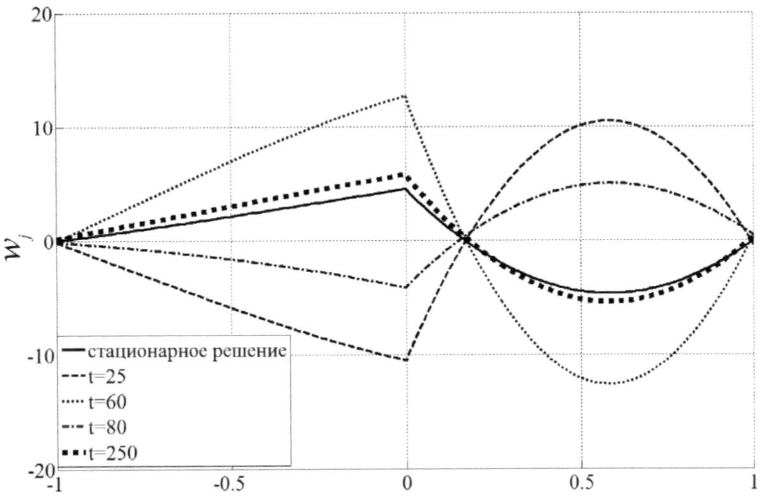

Figure 2. Dimensionless velocities profile $w_j(y,t)$ at $A_1(t) = 1 - 5e^{-0.01t}\sin(0.1t)$, $A_2(t) = 0$, $h_1 = h_2 = 0.001$m

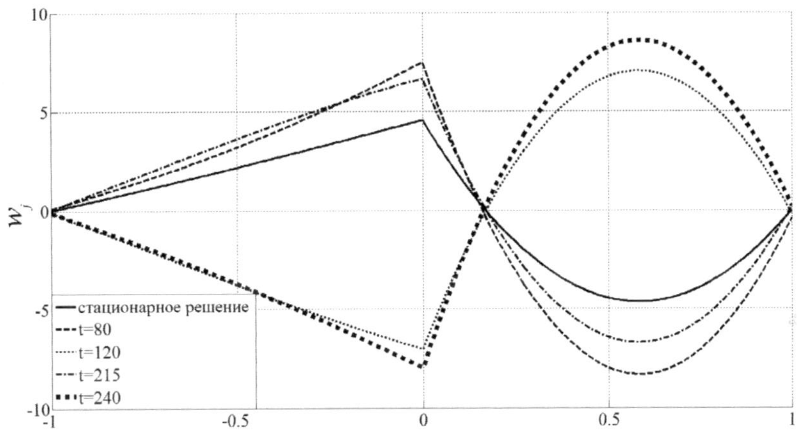

Figure 3. Dimensionless velocities profile $w_j(y,t)$ at $A_1(t) = 2\sin(0.1t)$, $A_2(t) = 0$, $h_1 = h_2 = 0.001$m

References

[1] Pukhnachev V V 2011 Non-stationary analogues of the Birikh Solution *Izvestiya AltGU* **1/2** 62-9
[2] Andreev V K, Gaponenko Yu A, Goncharova O N and Pukhnachev V V 2012 *Mathematical Models of Convection* (Berlin/Boston: Walter de Gruyter GmbH & CO KG)
[3] Andreev V K, Zakhvataev V E and Ryabitskii E A 2000 *Thermocapillary Instability* (Novosibirsk: Nauka)
[4] Andreev V K and Bekezhanova V B 2010 *Stability of Nonisothermal Fluids* (Krasnoyarsk: Sibirsk. Federal. Univ.)
[5] Nepomnyashchy A A, Velarde M G and Colinet P 2001 *Interfacial Phenomena and Convection* (Boca Raton: Chapman & Hall) p 365
[6] Sobhan C B and Peterson G P 2008 *Microscale and Nanoscale Heat Transfer: Fundamentals and Engineering Applications* (Boca Raton/London/New York: CRC Press Taylor and Francis Group) p 440
[7] Andreev V K and Cheremnykh E N 2016 The joint creeping motion of three viscid liquids in a plane layer: A priori estimates and convergence to steady flow *Journal of Applied and Industrial Mathematics* **XIX** 3-17
[8] Andreev V K 2009 On Inequalities of the Friedrichs type for Combined Domains *J. of Sib. Fed. Un. Math. and Phys.* **2(2)** 146-57

IOP Conf. Series: Journal of Physics: Conf. Series **894** (2017) 012107 doi:10.1088/1742-6596/894/1/012107

Kinematic-wave model of viscous fingers with mixing layer

A A Chesnokov[1,2], V Yu Liapidevskii[1,2] and I V Stepanova[1,3]

[1] Novosibirsk State University, Novosibirsk, Russia
[2] Lavrentyev Institute of Hydrodynamics SB RAS, Novosibirsk, Russia
[3] Institute of Computational Modelling SB RAS, Krasnoyarsk, Russia

E-mail: chesnokov@hydro.nsc.ru, liapid@hydro.nsc.ru, stepiv@icm.krasn.ru

Abstract. A new one-dimensional kinematic model of viscous fingers growth is proposed. This model is based on the assumption of an intermediate layer formation. The shear instability of the flow is developed in this layer due to intensive mixing of liquids of different viscosities. The thickness of the intermediate layer between outer ones is determined by the velocity difference in the layers. The results of the calculation of the growth rate of the fingers in the framework of this model are in a good agreement with the calculations carried out on the basis of more general two-dimensional equations as well as with the known Koval model.

1. Introduction

It is well known that a displacement process involving two fluids is often unstable if the displacing fluid has lower viscosity than the displaced one. The resulting instability developing at the interface between two fluids is often called the viscous fingering [1, 2]. A number of works devoted to understanding of various aspects of such form of instability is available in the literature [3, 4]. It should be noted that a numerical modelling of the displacement process requires the wasteful expenses of computer time at high Peclet numbers. Moreover, it is difficult to reproduce the detailed fingering pattern. As it was shown in [5, 6] it is simpler to describe the concentration of solute averaged across the fingers. In such cases, the mixing zone is an important feature for determining the extent of mixing. Despite the considerable work in this field was presented, the spreading and growth of the mixing zone is the important opened question. There are the several empirical models for the modelling of the evaluation of mixing zone in unstable, miscible displacements. An empirical model has been suggested by Koval [7] to give a basis for computation of miscible displacement. This model suffers from adoption of empiricism. The involved principal parameters have insignificant or indirect physical value. However, an experimental study of the mixing zone growth in miscible viscous fingering at high Peclet numbers [8] shows that the trend is consistent with Koval's model [7]. Further development of averaged models of fingers formation is represented in [9, 10]. All these models are based on the assumption that pressure is constant in the transverse direction of the main flow, as well as an empirical information about the displacing and displaced fluids distribution in the region of intensive viscous fingering.

In this paper we propose a 1D kinematic-wave model of viscous fingers taking into account the formation of an intermediate mixing layer. This model includes an empirical parameter

Content from this work may be used under the terms of the Creative Commons Attribution 3.0 licence. Any further distribution of this work must maintain attribution to the author(s) and the title of the work, journal citation and DOI.
Published under licence by IOP Publishing Ltd

depending on the viscosity ratio and related with the thickness of the mixing zone. The velocity of propagation and the thickness of the viscous finger in the framework of the kinematic-wave model coincide with the corresponding calculations on the basis of the 2D equations [10]. This makes it possible to predict the parameters of viscous fingers avoiding the time-consuming calculations. We also show that the proposed 1D model is in a good agreement with the known Koval model [7].

2. Equations of motion

A Newtonian weakly-compressible flow in a Hele-Shaw cell (the area between two parallel plates separated by a small gap of constant thickness b in the z direction) is described by the equations

$$(\rho\mathbf{v})_t + \mathrm{div}(\rho\mathbf{v}\otimes\mathbf{v}) + \nabla p = \mu\mathbf{v}_{zz}, \quad \rho_t + \mathrm{div}(\rho\mathbf{v}) = 0, \quad \mathbf{v}\big|_{z=\pm b/2} = 0. \tag{1}$$

Here p is the pressure, ρ is the density, and μ is the dynamic viscosity. The second derivatives of the velocity vector $\mathbf{v} = (u, v, w)$ with respect to the variables x and y are negligible compared to the derivatives in the vertical direction z. They are assumed to be zero in the suggested model. We put on that the velocity field can be represented as

$$u = \frac{3}{2}\Big(1 - \Big(\frac{2z}{b}\Big)^2\Big)u'(t,x,y), \quad v = \frac{3}{2}\Big(1 - \Big(\frac{2z}{b}\Big)^2\Big)v'(t,x,y), \quad w = 0.$$

The functions p and ρ are supposed to be not depend on z. Averaging equations (1) through the gap we obtain (primes are omitted)

$$\begin{aligned} &(\rho u)_t + (\beta\rho u^2 + p)_x + (\beta\rho uv)_y = -\mu u, \\ &(\rho v)_t + (\beta\rho uv)_x + (\beta\rho v^2 + p)_y = -\mu v, \\ &\rho_t + (u\rho)_x + (v\rho)_y = 0. \end{aligned} \tag{2}$$

Here and below μ denotes the viscosity of the fluid divided by the medium permeability $b^2/12$ (further in the text we will call it simply "viscosity"); coefficient β is equal to $6/5$.

We are interested to describe the process of displacement of a fluid with the viscosity μ_1 by less viscous fluid with viscosity μ_2. We utilize for this description equations (2) with variable viscosity depending on the concentration function c. This function is scaled such that it is equal to unity in the displaced fluid ($\mu = \mu_2$) and zero in the displacing one ($\mu = \mu_1$). We should add the transport equation for the concentration c to the governing equations. It has the divergent form (the diffusion terms are assumed to be neglected)

$$(\rho c)_t + (u\rho c)_x + (v\rho c)_y = 0. \tag{3}$$

Following [11] we use a monotonic relationship between the viscosity and the concentration in the form $\mu(c) = \mu_1^{1-c}\mu_2^c$. In order to close model (2), (3) we should specify the equation of state $p = p(\rho)$. A weak compressibility of the medium given by the equation of state $p = p(\rho)$ provides hyperbolicity of the model. If the condition $u^2 + v^2 \ll p'(\rho)$ keeps then the results depend on the choice of $p = p(\rho)$ insignificantly. Therefore, for the numerical simulation of the 2D flows we assume

$$p(\rho) = a^2\rho^2/2 \quad (a^2 = c_0^2/\rho_0), \tag{4}$$

where the constants ρ_0 and c_0 mean characteristic density and speed of sound in the medium.

System (2)–(4) was recently proposed in [10] for modelling of the viscous fingers taking into account the inertia of fluid. It may be important for the high finger velocities. It was also pointed out that for the process of unidirectional displacement the pressure variation in the transverse direction is small ($p_y = 0$).

2.1. Layered flow

We study an incompressible Hele-Shaw flow in a two-dimensional rectangular domain. The size of the domain is $L \times H$. We assume $\rho = 1$ without loss of generality. We replace the second of momentum equation in (2) by $p_y = 0$. It corresponds to transverse flow equilibrium. We consider the class of viscosity-stratified flows

$$u = u_i(t,x), \quad c = c_i = \text{const}, \quad \big(y \in (y_{i-1}, y_i), \quad i = 1,\dots,N\big),$$

wherein $0 = y_0 < y_1(t,x) < \ldots < y_N(t,x) = H$. In this case equations (2) and (3) take the form (see [10, 12] for the details)

$$\begin{aligned} &u_{it} + \beta u_i u_{ix} + p_x = -\mu_i u_i, \quad h_{it} + (u_i h_i)_x = 0, \quad (i = 1,...,N) \\ &\sum_{i=1}^{N} h_i = H, \quad \sum_{i=1}^{N} u_i h_i = Q. \end{aligned} \tag{5}$$

Here $h_i(t,x) = y_i(t,x) - y_{i-1}(t,x)$ is the depth of i–th liquid layer of viscosity μ_i and velocity $u_i(t,x)$, $Q =$ const is the total flow rate through the cell. In deriving equations (5) the kinematic condition at the layers interface is utilized.

For the average velocity $U = Q/H$ of fluid injection, one can use the following simplified version of governing equations (5). The simplification is based on the replacement of the momentum equations by the corresponding Darcy laws:

$$p_x = -\mu_i u_i \quad (i = 1,...,N). \tag{6}$$

The remaining equations of system (5) do not vary. In what follows, we assume that $H = 1$ and $Q = 1$.

2.2. Two-layer flow

It is interesting to note that in the case of two-layer flow system (5), where the momentum equations are replaced by Darcy laws (6), is reduced to the so-called naïve Koval model [9]

$$\frac{\partial h_1}{\partial t} + \frac{\partial}{\partial x}\left(\frac{M h_1}{M h_1 + 1 - h_1}\right) = 0. \tag{7}$$

Here $M = \mu_2/\mu_1$ is the viscosity ratio. Indeed, taking into account the relations

$$\mu_1 u_1 = \mu_2 u_2, \quad h_1 + h_2 = 1, \quad u_1 h_1 + u_2 h_2 = 1$$

we can represent equation $h_{1t} + (u_1 h_1)_x = 0$ as equality (7).

If $M > 1$ we construct the solution of (7) in the form of a simple wave

$$h_1(\xi) = \frac{1}{M-1}\left(\sqrt{\frac{M}{\xi}} - 1\right), \quad \xi = \frac{x - x_0}{t} \qquad \big(M^{-1} \le \xi \le M\big).$$

The previous formulae give the solution of equation (7) with discontinuous initial data

$$h_1\big|_{t=0} = \begin{cases} 1, & x < x_0 \\ 0, & x > x_0. \end{cases}$$

Let us note, that for $M < 1$ solution of the same Cauchy problem is given by the shock wave moving with the average flow velocity $D = U = 1$. In this case the Saffman–Taylor instability does not develop.

IOP Conf. Series: Journal of Physics: Conf. Series **894** (2017) 012107 doi:10.1088/1742-6596/894/1/012107

It is known that the growth rate of the viscous fingers calculating with the help of the naïve Koval model is significantly higher than it is observed experimentally. Koval [7] postulates empirically that equation (7) is valid if the viscosity ratio M is replaced by an effective viscosity ratio M_e, where

$$M_e = \left(M^{1/4}c_e + 1 - c_e\right)^4, \quad c_e = 0.22. \tag{8}$$

It should be noted that formula (8) is used in practice [8] but not justified theoretically. Below we propose a way to correct model (7) by using an intermediate mixing layer.

3. Three-layer flow with mixing

We use the notation (μ_1, u_1, h_1), $(\bar{\mu}, v, \eta)$ and (μ_2, u_2, h_2) for the fluid viscosities, velocities and depths in the displacing, intermediate and the displaced layers respectively ($\mu_1 < \bar{\mu} < \mu_2$). The cell height H and the flow rate Q are assumed to be unity as before. Then within the kinematic model ($p_x = -\mu_i u_i$) we have

$$\mu_1 u_1 = \bar{\mu} v = \mu_2 u_2, \quad h_1 + \eta + h_2 = 1, \quad u_1 h_1 + v\eta + u_2 h_2 = 1. \tag{9}$$

Following [13, 14] we consider the equilibrium between generation and dissipation of energy of the small-scale motions in a shear flow. In this case the thickness of the intermediate layer can be expressed in the form

$$\eta = \frac{\delta}{\bar{\mu} v}\left((u_1 - v)^2 + (u_2 - v)^2\right), \tag{10}$$

where δ is an empirical parameter. Here we do not present the derivation of relation (10). Therefore, this formula can be regarded as an additional hypothesis reducing the equations of fingers evolution in the development of the Saffman–Taylor instability to the Hopf equation for the averaged thickness of a finger.

Let us define the middle-line $z = h_1 + \eta/2$ and formulate the law of mass conservation for the liquid layer of depth z

$$z_t + (\psi(z))_x = 0, \quad \text{where } \psi = u_1 h_1 + v\eta/2. \tag{11}$$

Using the notation

$$a_1 = 1 - \frac{\bar{\mu}}{2\mu_1} - \frac{\bar{\mu}}{2\mu_2}, \qquad a_2(z) = \frac{\bar{\mu}}{\mu_2} + \left(\frac{\bar{\mu}}{\mu_1} - \frac{\bar{\mu}}{\mu_2}\right)z,$$

$$a_3 = \left(\left(\frac{\bar{\mu}}{\mu_1} - 1\right)^2 + \left(\frac{\bar{\mu}}{\mu_2} - 1\right)^2\right)\frac{\delta}{\bar{\mu}}$$

and formulas (9) we express the fluid velocity v in the mixing layer by means of the variables η and z:

$$v = (a_1\eta + a_2(z))^{-1}.$$

In this case formula (10) takes the form $\eta = a_3 v$ and allows one to obtain the dependence $\eta(z)$

$$\eta(z) = \frac{-a_2(z) + \sqrt{a_2^2(z) + 4a_1 a_3}}{2a_1}.$$

Substitution of $\eta(z)$ into the previous formula expresses velocities v and $u = \bar{\mu}v/\mu_1$ as the functions of the middle-depth z. As the result we derive the relation for the flux ψ in the layer of depth z:

$$\psi(z) = \left(\frac{\bar{\mu}}{\mu_1}z + \left(1 - \frac{\bar{\mu}}{\mu_1}\right)\frac{\eta(z)}{2}\right)\frac{1}{a_1\eta(z) + a_2(z)}. \tag{12}$$

IOP Conf. Series: Journal of Physics: Conf. Series **894** (2017) 012107 doi:10.1088/1742-6596/894/1/012107

Note that it is necessary to require the fulfilment of the inequalities

$$a_2^2(z) + 4a_1a_3 \geq 0, \quad z - \eta(z)/2 \geq 0, \quad z + \eta(z)/2 \leq 1. \tag{13}$$

The last two inequalities are the consequence of $h_1 \geq 0$ and $h_1 + \eta \leq 1$. To determine the depth $z = z(t, x)$ we use conservation law (11) with function $\psi(z)$ given by formula (12).

Let us construct a solution of equation (11) with piecewise constant initial data $z(0, x) = 1$ for $x < x_0$ and $z(0, x) = 0$ for $x > x_0$. This formulation means that the domain $x < x_0$ is filled by the liquid of viscosity μ_1 ($h_1 = 1$, $\eta = h_2 = 0$) whereas the liquid of viscosity μ_2 ($h_2 = 1$, $h_1 = \eta = 0$) fills the domain $x > x_0$. Since the fluid of viscosity μ_1 moves in the positive direction of the x-axis with the average velocity $U = 1$ then if $\mu_1 < \mu_2$ the Saffman–Taylor instability develops at the interface.

Due to constraints (13) the function $\psi(z)$ does not define on the whole interval $[0, 1]$. Typical behavior of the function $\psi(z)$ is given in Figure 1 (curve *1*) for the following parameters $\mu_1 = 2$, $\mu_2 = 8$, $\bar{\mu} = \sqrt{\mu_1\mu_2} = 4$ and $\delta = 1$. Therefore, in order to solve the problem we should define the function $\psi(z)$ on the entire interval $[0, 1]$. Taking into account that $\psi(z)$ is the flow rate in the layer of the depth $z = h_1 + \eta/2$ the conditions $\psi(0) = 0$ and $\psi(1) = 1$ should be satisfied.

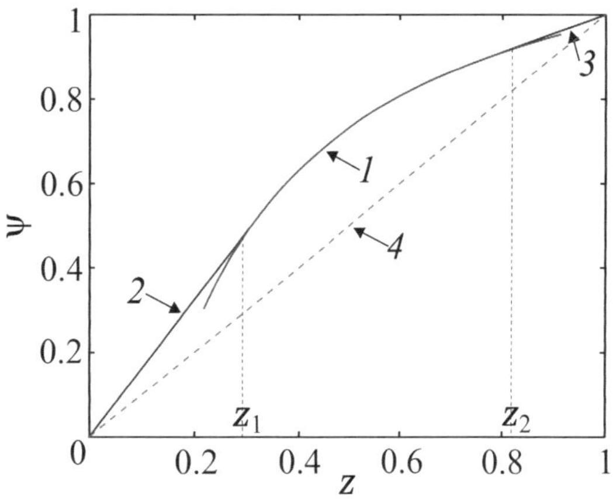

Figure 1. Flux $\psi(z)$ for model (11) (curve *1*) and its expansion.

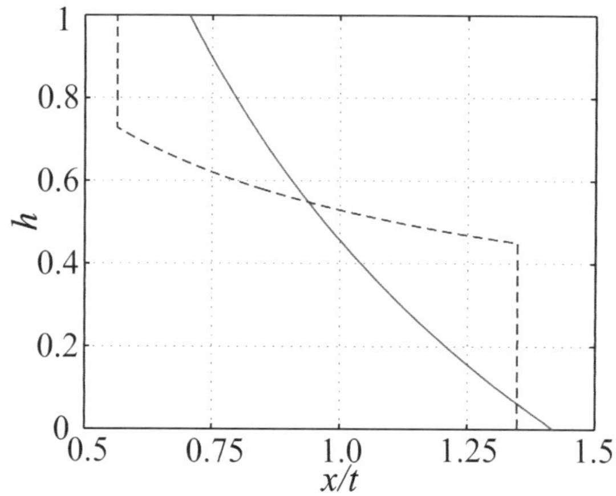

Figure 2. Comparison between the Koval model (solid curve) and model (11) with $\delta = 1.8$ (dashed) for $M = 4$.

We assume that a function $\tilde{\psi}(z)$ is the extension of $\psi(z)$ on the entire interval $[0, 1]$. Note that the function $\psi(z)$ can be extended in various ways. If the plot of constructed extension $\tilde{\psi}(z)$ lies between the convex hull of the plot $\psi(z)$ (Figure 1, line *1*) and the diagonal of the unit square connecting points $(0, 0)$ and $(1, 1)$ (Figure 1, line *4*) then the solution of the problem does not depend on the choice of extension [15]. Let us construct the tangents to the plot of $\psi(z)$ from the origin and from the point $(1, 1)$ (z_1 and z_2 denote the abscissae of the points of tangency). Thus, the function $\tilde{\psi}(z)$ is given by formula (12) on the interval $[z_1, z_2]$. On the intervals $[0, z_1)$ and $(z_2, 1]$ it is represented by the tangent lines. Using the function $\tilde{\psi}(z)$ we construct a self-similar solution of equation (11) with discontinuous initial data. This solution is represented by the simple wave

$$\xi = \tilde{\psi}'(z), \quad \tilde{\psi}'(z_2) < \xi < \tilde{\psi}'(z_1), \quad \xi = (x - x_0)/t.$$

This wave is joined to the strong discontinuities $\xi_1 = \tilde{\psi}'(z_1)$ and $\xi_2 = \tilde{\psi}'(z_2)$ (see [15] for details).

IOP Conf. Series: Journal of Physics: Conf. Series **894** (2017) 012107 doi:10.1088/1742-6596/894/1/012107

3.1. A comparison with the Koval model

Let us verify kinematic-wave model (11) by means of comparing with the well-approved Koval model (7). The effective viscosity ratio M_e given by formula (8) is used instead of M. We choose the following parameters: $\mu_1 = 2$, $\mu_2 = 8$. It means that the viscosity ratio $M = 4$ and $M_e = 1.417$ according to (8). Self-similar solution of the Koval model for these parameters is shown in Figure 2 (solid curve). Corresponding solution of the scalar conservation law (11) (with $\delta = 1.8$) is presented in Figure 2 by the dashed curve. As we can see in the framework of these models the growth rates of viscous fingers are similar. Although the parameter δ is the function of M, the values of $\delta(M)$ vary weakly for $M \in (1, 8)$. Moreover, proposed model (11) describes the behavior of the finger near its tip better than the Koval model.

It should be noted that another kinematic-wave model has been recently obtained in [10] by taking into account the friction between the fluid layers of different viscosities. This effect is important when the finger become thin at the vicinity of the tip.

3.2. A comparison with 2D model (2)–(4)

We compare the results of numerical simulations of the viscous fingers formation based on two-dimensional hyperbolic system of equations (2)–(4) with the results obtained by using kinematic-wave model (11). The calculations are performed in the Hele-Shaw cell with sizes 20×2 in the coordinate system moving with the average flow velocity $U = 1$ with respect to the x-axis. The viscosities of the fluids are equal to $\mu_1 = 2$ and $\mu_2 = 8$. At the initial moment of time the interface $x = x_0 = 10$ is perturbed with respect to the formula $x = x_0 + 4\Delta_1(\exp(-10(y-1)^2) - 1/2)$ where Δ_1 is the resolution of the uniform grid in x. For discretization with respect to x and y we use 400 and 50 nodes respectively. On the left and right boundaries of the computational domain the reflection conditions are imposed. We assume that in the framework of kinematic model (11) in the mixing layer the variable c is equal to $1/2$. So, the viscosity in this layer is given by formula $\bar{\mu} = \sqrt{\mu_1\mu_2}$ by virtue of the fact in two-dimensional model (2)–(4) the viscosity is $\mu(c) = \mu_1^{1-c}\mu_2^c$. The calculations are carried out at $\beta = 1$, $\rho_0 = 1$ and $c_0 = 75$. In this case the change in density is not more than 0.15%. At the same time the condition $p_y = 0$ is fulfilled with high accuracy. It means that the above proposed one-dimensional model is suitable for describing a such flow. The given perturbation of the interface leads to the formation of a single finger. It is symmetric with respect to the line $y = 1$. The results of the concentration c calculations using model (2)–(4) at time $t = 10$ and $t = 20$ are shown in Figure 3 .

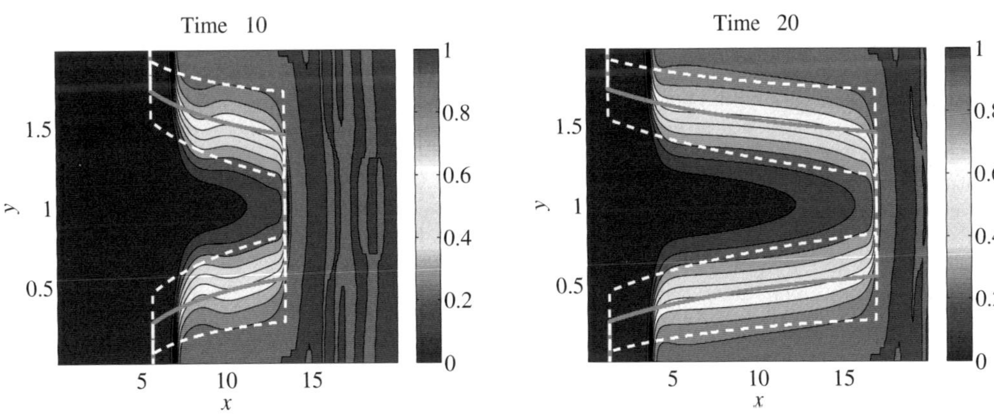

Figure 3. The growth of viscous finger: comparison of the results for two-dimensional equations (2)–(4) and one-dimensional kinematic-wave model (11) with $\delta = 1.8$.

Self-similar solutions of kinematic-wave model (11) with $\delta = 1.8$ is shown in Figure 3 at $t = 10$ and $t = 20$ by solid (the middle-line $1 \pm z$) and dashed (the outer $1 \pm h_1$ and inner $1 \pm (h_1 + \eta)$ boundaries of the finger) curves. The self-similar variable ξ is replaced by $\xi + U$, $U = 1$. It corresponds to a transition into the moving coordinate system. As we can see, the velocity of viscous finger propagation and its transverse diffusion are in a good agreement with calculations provided by equations (2)–(4).

4. Conclusion

The main result of this paper is the construction of a new kinematic-wave model of viscous finger growth (11). The model is based on the assumption about an intermediate mixing layer formation between two layers of liquid with different viscosities and velocities. Relation (10) gives the thickness of the mixing layer by means of the fluid velocity in the outer layers. This formula is not derived in the paper and can be considered as an additional hypothesis. It allows describing the evolution of the viscous fingers by using Hopf-like equation. The proposed model (11) is verified with the help of the comparison with the well-tested Koval model (7) and (8). The results are also confirmed by the numerical simulation on the basis of 2D equations (2)–(4). Figure 3 shows that the velocity of the viscous finer propagation are the same for the both models.

Thus, the suggested 1D model predicts the propagation velocity of the viscous fingers to a sufficient degree of accuracy and do not require large processing power. The model also provides an additional information concerning to the transverse diffusion of viscous finger.

Acknowledgments

The work is supported by Russian Science Foundation (project 15-11-20013).

References

[1] Saffman P G and Taylor G 1958 *Proc. Roy. Soc. London. A* **245** 312-29
[2] Homsy G M 1987 *Annu. Rev. Fluid Mech.* **19** 271-311
[3] Manickam O and Homsy G M 1991 *J. Fluid Mech.* **288** 75-102
[4] Tanveer S 2000 *J. Fluid Mech.* **409** 273-308
[5] Zimmerman W B and Homsy G M 1992 *Phys. Fluids A* **4** 2348-59
[6] Yang Z M, Yortsos Y C and Salin D 2002 *Adv. Water Res.* **25** 885-98
[7] Koval E J 1963 *Soc. Petrol. Eng. J.* **450** 145-54
[8] Malhotra S, Sharma M M and Lehman E R 2015 *Phys. Fluids* **27** 014105
[9] Booth R J S 2010 *J. Fluid Mech.* **655** 527-39
[10] Chesnokov A A and Liapidevskii V Yu 2017 *Int. J. Nonlin. Mech.* **89** 168-76
[11] Azaiez J and Singh B 2002 *Phys. Fluids* **14** 1559–71
[12] Chesnokov A A and Stepanova IV 2015 *Appl. Math. Comput.* **265** 320-8
[13] Liapidevskii V Yu and Teshukov V M 2000 *Mathematical Models of Propagation of Long Waves in a Non-Homogeneous Fluid* (Novosibirsk: Siberian Branch of Russian Academy of Sciences) [in Russian]
[14] Liapidevskii V Yu and Chesnokov A A 2014 *J. Appl. Mech. Tech. Phys.* **55** 299-310.
[15] Rozhdestvenskij V Yu and Yanenko N N 1983 *Systems of quasilinear equations and their applications to gas dynamics* (Am. Math Soc) [translated from Russian]
[16] Nessyahu H and Tadmor E 1990 *J. Comp. Phys.* **87** 408-63

IOP Publishing
doi:10.1088/1742-6596/894/1/012108

Motion of gas in highly rarefied space

Yu A Chirkunov

Novosibirsk State University of Architecture and Civil Engineering, 630008 Novosibirsk, Russia

E-mail: chr101@mail.ru

Abstract. A model describing a motion of gas in a highly rarefied space received an unlucky number 13 inthe list of the basic models of the motion of gas in the three–dimensional space obtained by L.V. Ovsyannikov. For a given initial pressure distribution, a special choice of mass Lagrangian variables leads to the system describing this motion for which the number of independent variables is less by one. Hence, there is a foliation of a highly rarefied gas with respect to pressure. In a strongly rarefied space for each given initial pressure distribution, all gas particles are localized on a two–dimensional surface that moves with time in this space We found some exact solutions of the obtained system that describe the processes taking place inside of the tornado. For this system we found all nontrivial conservation laws of the first order. In addition to the classical conservation laws the system has another conservation law, which generalizes the energy conservation law. With the additional condition we found another one generalized energy conservation law.

1. Introduction

The basic models of gas motion in three–dimensional space were obtained in [1]. A model describing the motion of a gas in a highly sparse space got an unlucky No 13 in the list of these models. The study of these model was begun in [2] and was continued in [3, 4]. In this report we will use materials of these articles.

The model of the thermal motion of a gas in a three–dimensional rarefied space is determined by the following system of differential equations [2–4]

$$\mathbf{u}_t + (\mathbf{u}\cdot\nabla)\,\mathbf{u} + \frac{1}{\rho}\nabla p = \mathbf{0},\ \rho_t + \mathbf{u}\cdot\nabla\rho + \rho\,\mathrm{div}\,\mathbf{u} = 0,\ p_t + \mathbf{u}\cdot\nabla p = 0 \tag{1}$$

where t is a time, $\mathbf{x} = \mathbf{x}(x, y, z) \in R^3$, $\mathbf{u} = \mathbf{u}(t, \mathbf{x}) \in R^3$ is a velocity vector, $\rho = \rho(t, \mathbf{x})$ is a density, $p = p(t, \mathbf{x})$ is a pressure.

Suppose the variables $\boldsymbol{\xi} = (\xi, \eta, \zeta)$ are initial values of variables $\mathbf{x} = \mathbf{x}(t, \boldsymbol{\xi}) = (x, y, z)$, i.e. $\mathbf{x}(0, \boldsymbol{\xi}) = \boldsymbol{\xi}$.

With the help of these variables, the system (1) is written in the form of a vector equation:

$$\left|\frac{\partial\mathbf{x}}{\partial\boldsymbol{\xi}}\right|^{-1}\left(\frac{\partial\mathbf{x}}{\partial\boldsymbol{\xi}}\right)^{\mathrm{T}}\mathbf{x}_{tt} + \frac{1}{\rho_0}p_{\boldsymbol{\xi}} = \mathbf{0},\ \rho = \rho_0\left|\frac{\partial\mathbf{x}}{\partial\boldsymbol{\xi}}\right|^{-1},\ p_t = 0 \tag{2}$$

where $\rho_0 = \rho_0(\boldsymbol{\xi})$ is an initial density distribution. A solution of the last equation of the system (2) is $p = p(\boldsymbol{\xi})$, where $p(\boldsymbol{\xi})$ is a given function that determines the initial pressure distribution. Since the pressure is stored in the particle, it is Lagrange variable.

Content from this work may be used under the terms of the Creative Commons Attribution 3.0 licence. Any further distribution of this work must maintain attribution to the author(s) and the title of the work, journal citation and DOI.
Published under licence by IOP Publishing Ltd

IOP Conf. Series: Journal of Physics: Conf. Series **894** (2017) 012108 doi:10.1088/1742-6596/894/1/012108

After the change of variables

$$\xi' = \xi'(\boldsymbol{\xi}),\ \eta' = \eta'(\boldsymbol{\xi}),\ \zeta' = p(\boldsymbol{\xi}) \tag{3}$$

such that

$$\frac{\partial(\xi', \eta', p)}{\partial(\xi, \eta, \zeta)} = \rho_0(\boldsymbol{\xi}) \tag{4}$$

the system (2) is transformed to the equivalent vector equation (the strokes are omitted)

$$\mathbf{x}_{tt} + \mathbf{x}_\xi \times \mathbf{x}_\eta = 0 \tag{5}$$

The velocity vector and density are determined by the formulae

$$\mathbf{u} = \mathbf{x}_t,\ \frac{1}{\rho} = \left|\frac{\partial \mathbf{x}}{\partial \boldsymbol{\xi}}\right|$$

The equation (5) does not contain the variable ζ' and the derivatives $\partial_{\zeta'}\mathbf{x}$. Thus, the special choice of the Lagrange mass variables, by the formulae (3) and (4), makes it possible to transform the system (1) to vector equation (5) containing only three independent variables. The variable $\zeta' = p$ is a parameter that, according to (3) and (4) defines the variables of equation (5). It means that there is a stratification of a highly rarefied gas with respect to pressure. Namely, in a strongly rarefied space for each given initial pressure distribution, at each instant of time all the gas particles are localized on the two–dimensional surface S_t, defined by equation $\mathbf{x} = \mathbf{x}(t, \xi, \eta)$. The surface S_t moves with time in this space. At each point of the surface S_t, the acceleration vector is collinear to the normal vector to this surface.

In the cylindrical variables $r = \sqrt{x^2 + y^2}$, $\varphi = \arctan \frac{y}{x}$, z the system (5) is written as

$$r_{tt} - r\varphi_t^2 = r\left(z_\xi \varphi_\eta - z_\eta \varphi_\xi\right),\ \left(r^2 \varphi_t\right)_t = r\left(r_\xi z_\eta - r_\eta z_\xi\right),\ z_{tt} = r\left(\varphi_\xi r_\eta - \varphi_\eta r_\xi\right) \tag{6}$$

In further we assume that the initial pressure distribution in a highly rarefied gas is given.

2. Exact solutions describing the processes taking place inside of the tornado

2.1.

The solution of the system (6) for which $z = \pm r$ is defined by the formulas

$$r = t\alpha(\varphi) + \beta(\varphi),\ \varphi = \varphi(\xi, \eta),\ z = \pm r \tag{7}$$

where $\varphi(\xi, \eta)$, $\alpha(\varphi)$ and $\beta(\varphi)$ are arbitrary functions.

The solution (7) describes the following dynamic process: each gas particle is arranged on a circular cone $z = \pm\sqrt{x^2 + y^2}$ and move along its generator with a constant velocity

$$u = \alpha(\varphi)\mathbf{e}_r + 0 \cdot \mathbf{e}_\varphi \pm \alpha(\varphi)\mathbf{e}_z$$

where $\mathbf{e}_r$, $\mathbf{e}_\varphi$, $\mathbf{e}_z$ are the orts of cylindrical coordinate system.

If $\alpha < 0$, then each particle of the gas moves toward the apex of the cone, and if $\alpha > 0$, then each gas particle moves in a direction from apex of the cone.

IOP Conf. Series: Journal of Physics: Conf. Series **894** (2017) 012108 doi:10.1088/1742-6596/894/1/012108

2.2.
Let k and m are arbitrary constants, $k > 0$. System (6) has only two solutions, for which $\varphi = kt + m$. The first solution is

$$r = f(\alpha)\exp(kt) + \frac{c}{f(\alpha)}\exp(-kt),\ \varphi = kt + m,\ z = \alpha t + g(\alpha) + \frac{2k^2 f(\alpha)}{f'(\alpha)}h(\xi,\eta) \tag{8}$$

and the second solution is

$$r = f(\alpha)\exp(-kt),\ \varphi = kt + m,\ z = \alpha t + g(\alpha) - \frac{2k^2 f(\alpha)}{f'(\alpha)}h(\xi,\eta) \tag{9}$$

where $f(\alpha) > 0$, $g(\alpha)$, $\alpha = \alpha(\xi,\eta)$ are arbitrary functions, c is an arbitrary positive constant,

$$g(\alpha) = \begin{cases} \int \frac{\mathrm{d}\xi}{(\alpha_\eta)_{\alpha=const}}, & \text{for } \alpha_\eta \neq 0, \\ \int \frac{\mathrm{d}\eta}{(\alpha_\xi)_{\alpha=const}}, & \text{for } \alpha_\xi \neq 0. \end{cases}$$

Vector of velocity $\mathbf{u} = u_r\mathbf{e}_r + u_\varphi\mathbf{e}_\varphi + u_z\mathbf{e}_z$ and vector of acceleration of each particle of gas for these solutions are defined by the formulas:
for the first solution

$$u_r = k\left(f(\alpha)\,\mathrm{e}^{kt} - \frac{c}{f(\alpha)}\,\mathrm{e}^{-kt}\right),\ u_\varphi = k\left(f(\alpha)\,\mathrm{e}^{kt} + \frac{c}{f(\alpha)}\,\mathrm{e}^{-kt}\right),\ u_z = \alpha(\xi,\eta),$$

$$a_r = 0,\ a_\varphi = 2k^2\left(f(\alpha)\,\mathrm{e}^{kt} - \frac{c}{f(\alpha)}\,\mathrm{e}^{-kt}\right),\ a_z = 0,$$

for the second solution

$$u_r = -kf(\alpha)\,\mathrm{e}^{-kt},\ u_\varphi = kf(\alpha)\,\mathrm{e}^{-kt},\ u_z = \alpha(\xi,\eta),$$

$$a_r = 0,\ a_\varphi = -2k^2 f(\alpha)\,\mathrm{e}^{-kt},\ a_z = 0.$$

The solution (8) for $c = 0$ describes the movement of the gas particles along helical lines obtained by stretching along an axis z unwinding logarithmic spirals. At that, the components u_r, u_φ of the velocity vector and component a_φ of the acceleration vector tend to infinity when $t \to \infty$.

The solution (9) describes the movement of the gas particles along helical lines obtained by stretching along an axis z twisting logarithmic spirals. At that, the components u_r, u_φ of the velocity vector and component a_φ of the acceleration vector tend to zero when $t \to \infty$.

The solutions (7), (8) and (9) can be used at the study of the processes taking place inside of the tornado.

2.3.
The main Lie algebra of the system (5) was obtained in [2]. The following operators

$$X_1 = t\partial_t + \xi\partial_\xi + \eta\partial_\eta,\ X_2 = t\partial_t - 2\mathbf{x}\cdot\partial_{\mathbf{x}},\ X_3 = \eta\partial_\xi - \xi\partial_\eta$$

are contained in this algebra. Let α and β are arbitrary constants.

The invariant $\langle X_1 + \alpha X_2,\ \beta X_2 + X_3\rangle$–solution of the system (5) describes the gas flow, in which the Euler coordinates of each gas particle change according to the law

$$x = \left(\xi^2 + \eta^2\right)^{-\alpha} \mathrm{e}^{2\beta \arctan\frac{\eta}{\xi}}\,\mathbf{w}(\lambda),\ \lambda = t\left(\xi^2 + \eta^2\right)^{-\frac{\alpha+1}{2}} \mathrm{e}^{\beta \arctan\frac{\eta}{\xi}}.$$

The function $\mathbf{w}$ is a solution of the vector equation

$$\mathbf{w}'' + 2\beta\lambda\mathbf{w}' \times \mathbf{w} = \mathbf{0}$$

The main mechanical characteristics of this gas flow are:

$$|\mathbf{x}| = \frac{1}{(\xi^2+\eta^2)^\alpha}\left(\frac{c_1^2t^2\,\mathrm{e}^{\beta\arctan\frac{\eta}{\xi}}}{(\xi^2+\eta^2)^{\alpha+1}} + \frac{c_2t}{(\xi^2+\eta^2)^{\frac{\alpha+1}{2}}} + c_3\right)^{\frac{1}{2}} \mathrm{e}^{\frac{5}{2}\beta\arctan\frac{\eta}{\xi}},$$

$$|\mathbf{u}| = \frac{c_1\,\mathrm{e}^{3\beta\arctan\frac{\eta}{\xi}}}{(\xi^2+\eta^2)^{\frac{3\alpha+1}{2}}},$$

$$|\mathbf{a}| = \frac{2c_1\,|\beta|\,t}{(\xi^2+\eta^2)^{\frac{5\alpha+3}{2}}}\left(\frac{c_1^2t^2\,\mathrm{e}^{\beta\arctan\frac{\eta}{\xi}}}{(\xi^2+\eta^2)^{\alpha+1}} + \frac{c_2t}{(\xi^2+\eta^2)^{\frac{\alpha+1}{2}}} + c_3\right)^{\frac{1}{2}} \mathrm{e}^{\frac{11}{2}\beta\arctan\frac{\eta}{\xi}}\sin\theta,$$

where c_1, c_2 and c_3 are arbitrary constants.

If $c_2^2 - 4c_1^2c_3 > 0$, then this solution exists only in two cases:

(i) either

$$t < -\frac{1}{2c_1^2}\left(c_2 + \left(c_2^2 - 4c_1^2c_3\right)^{\frac{1}{2}}\right)\left(\xi^2+\eta^2\right)^{\frac{\alpha+1}{2}}\mathrm{e}^{-\beta\arctan\frac{\eta}{\xi}},$$

(ii) either

$$t > -\frac{1}{2c_1^2}\left(c_2 - \left(c_2^2 - 4c_1^2c_3\right)^{\frac{1}{2}}\right)\left(\xi^2+\eta^2\right)^{\frac{\alpha+1}{2}}\mathrm{e}^{-\beta\arctan\frac{\eta}{\xi}}.$$

If $c_2^2 - 4c_1^2c_3 < 0$, then this solution exists for all t, ξ, η.

Each particle of gas with the passage of the time moves away from the origin by a nonlinear law, the absolute value of the velocity vector of each particle is constant, and the absolute value of the acceleration vector nonlinearly depends on the time. The acceleration vector is perpendicular to the velocity vector. In particular, such processes are characteristic for the motion of gas particles inside of the tornado [5–7].

3. Conservation laws

The conservation law of the first order for the system (5) is a vector [8–15]

$$\mathbf{A} = \mathbf{A}(t,\xi,\eta,\mathbf{x},\mathbf{x}_t,\mathbf{x}_\xi,\mathbf{x}_\eta) = (A_0, A_1, A_2)$$

such that

$$(\mathbf{D}\cdot\mathbf{A})_{\{5\}} = 0,$$

where $\{5\}$ is a manifold, which is defined by the equations of the system (5) in the extended space, $\mathbf{D} = (D_t, D_\xi, D_\eta)$ and D_t, D_ξ, D_η are operators of total differentiation by t, ξ, η.

In addition to the classical conservation laws (an energy conservation law, a momentum conservation law, a angular momentum conservation law, a conservation law, which determines the law of motion of centre of mass) the system (5) has an additional generalized energy conservation law

$$\begin{aligned} A_0 &= t\left(\frac{1}{2}|\mathbf{x}_t|^2 + \frac{1}{3}(\mathbf{x}_\xi\times\mathbf{x}_\eta)\cdot\mathbf{x}\right) + \left(\frac{1}{3}\mathbf{x} + \frac{5}{6}(\xi\mathbf{x}_\xi + \eta\mathbf{x}_\eta)\right)\cdot\mathbf{x}_t, \\ A_2 &= -\frac{1}{3}t\,(\mathbf{x}_\xi\times\mathbf{x}_t)\cdot\mathbf{x} - \frac{5}{12}\eta\,|\mathbf{x}_t|^2. \end{aligned} \tag{10}$$

Other conservation laws are possible under the additional conditions imposed to the nature of the motion of the gas. In particular, let, the motion of the gas is such that at each time in each point (ξ, η) the velocity vector is perpendicular to the vector $\xi \mathbf{x}_\xi + \eta \mathbf{x}_\eta$, lying in the plane tangent to the surface S_t. This means that a differential connection

$$\mathbf{x}_t \cdot (\xi \mathbf{x}_\xi + \eta \mathbf{x}_\eta) = 0$$

is added to the system (5). In this case, another one generalized energy conservation law is added to the classical conservation laws and conservation law (10)

$$A_0 = t^2 \left(\frac{1}{2} |\mathbf{x}_t|^2 + \frac{1}{3} (\mathbf{x}_\xi \times \mathbf{x}_\eta) \cdot \mathbf{x} \right) + \frac{1}{3} t \left(2\mathbf{x} + 5 \left(\xi \mathbf{x}_\xi + \eta \mathbf{x}_\eta\right)\right) \cdot \mathbf{x}_t - \frac{1}{3} |\mathbf{x}_t|^2 ,$$

$$A_1 = \frac{1}{3} t^2 (\mathbf{x}_\eta \times \mathbf{x}_t) \cdot \mathbf{x} - \frac{5}{6} t \xi |\mathbf{x}_t|^2 , \; A_2 = -\frac{1}{3} t^2 (\mathbf{x}_\xi \times \mathbf{x}_t) \cdot \mathbf{x} - \frac{5}{6} t \eta |\mathbf{x}_t|^2 .$$

4. Conclusion and discussion

The main models of the motion of the gas in three–dimensional space were obtained in [1]. The model describing motion of the gas in the rarefied space, got unlucky number 13 in the list of these models. This model can be used, in particular, in the study of the processes taking place inside of tornado.

For a given initial distribution of the pressure a specific selection of mass Lagrange variables leads to a reduction of the differential equations describing this motion to the system, for which the number of independent variables is less on the unit. Consequently, there is foliation of gas in the rarefied space with respect to the pressure. In the rarefied space for any given initial distribution of the pressure, all the gas particles are localized on the two–dimensional surface. This surface over time moves in this space. At each point of the surface, the vector of acceleration is collinear to the vector of normal to this surface.

We found some exact solutions of the obtained system that describe the processes taking place inside of the tornado. For this system we found all nontrivial conservation laws of the first order. In addition to the classical conservation laws the system has another conservation law, which generalizes the energy conservation law. With the additional condition we found another one generalized energy conservation law.

Acknowledgments

The reported study was funded by RFBR according to the research project No. 16–01–00446 a.

References

[1] Ovsyannikov L V 1994 The "PODMODELI" program. Gas dynamics *J. Appl. Math. and Mech.* **58(4)** 601–27
[2] Chirkunov Yu A 2009 The conservation laws and group properties of the equations of gas dynamics with zero velocity of sound *J. Appl. Math. and Mech.* **73(4)** 421–25 (*Preprint* doi:10.1016/j.jappmathmech.2009.08.009)
[3] Chirkunov Yu A 2016 Exact solutions of the system of the equations of thermal motion of gas in the rarefied space *Int. J. Non–Linear Mech.* **83** 9–14 (*Preprint* doi:10.1016/j.ijnonlinmec.2016.03.008)
[4] Chirkunov Yu A, Pikmullina E O 2017 Invariant submodels of the model of thermal motion of gas in a rarefied space *Int. J. Non–Linear Mech.* **93** (*Preprint* doi:10.1016/j.ijnonlinmec.2017.05.005)
[5] Liu S, Wang Z, Gong Z, Chen F, Peng Q 2006 Physically based modeling and animation of tornado *J. Zhejiang Univ. Science* A **7(7)** 1099–106
[6] Haan F L, Sarkar P P, Gallus W A 2008 Design, construction and performance of a large tornado simulator for wind engineering applications *Engineering Structures* **30** 1146–59
[7] Shtern V, Hussain F 1993 Hysteresis in a swirling jet as a model tornado *Phys. Fluids* **5(9)** 2183–95
[8] Olver P J 1986 *Applications of Lie Groups to Differential Equations* (New York: Springer–Verlag)

[9] Chirkunov Yu A 2007 *Group Analysis of Linear and Quasi–Linear Differential Equations* (Novosibirsk: NSUEM)[In Russian]

[10] Chirkunov Yu A, Khabirov S V 2012 *Elements of Symmetry Analysis of Differential Equations of Continuum Mechanics* (Novosibirsk: NSTU)[In Russian]

[11] Chirkunov Yu A 2009 Method of A–operators and conservation laws for the equations of gas dynamics *J. Appl. Mech. and Techn. Phys.* **50(2)** 213–9 (*Preprint* doi:10.1007/s10808-009-0029-7)

[12] Chirkunov Yu A 2009 On group properties and conservation laws for second-order quasi-linear differential equations *J. Appl. Mech. and Techn. Phys.* **50(3)** 413–8 (*Preprint* doi:10.1007/s10808-009-0055-5)

[13] Chirkunov Yu A 2010 Conservation laws and group properties of equations of isentropic gas motion *J. Appl. Mech. and Techn. Phys.* **51(1)** 1–3 (*Preprint* doi:10.1007/s10808-010-0001-6)

[14] Chirkunov Yu A and Medvedev S B 2013 Conservation laws for plane steady potential barotropic flow *Europ. J. of Appl. Math.* **24(06)** 789–801

[15] Chirkunov Yu A 2012 Nonlocal conservation laws for the equations of stationary irrotational isentropic plane movement of gas *J. Appl. Math. and Mech.* **76(2)** 199–204 (*Preprint* doi:10.1016/j.jappmathmech.2012.05.011)

IOP Conf. Series: Journal of Physics: Conf. Series **894** (2017) 012109 doi:10.1088/1742-6596/894/1/012109

Deflection of ice cover caused by an underwater body moving in channel

K A Shishmarev[1], T I Khabakhpasheva[2] and A A Korobkin[3]

[1] Altai State University, Barnaul 656049, Russia
[2] Lavrentyev Institute of Hydrodynamics of SB RAS, Novosibirsk 630090, Russia
[3] University of East Anglia, Norwich NR4 7TJ, UK

E-mail: shishmarev.k@mail.ru

Abstract. Deflections and strains in an ice cover of a frozen channel caused by an underwater body moving under the ice with a constant speed along the channel are studied. The channel is of rectangular cross section, the fluid in the channel is inviscid and incompressible. The ice cover is clamped to the channel walls. The ice cover is modeled by a thin viscoelastic plate. The underwater body is modeled by a three-dimensional dipole. The intensity of the dipole is related to the speed and size of the underwater body. The problem is considered within the linear theory of hydroelasticity. For small deflections of the ice cover the velocity potential of the dipole in the channel is obtained by the method of images in leading order without account for the deflection of the ice cover. The problem of moving dipole in the channel with rigid walls provides the hydrodynamic pressure on the upper boundary of the channel, which corresponds to the ice cover. This pressure distribution does not depend on the deflection of the ice cover in the leading approximation. The deflections of the ice and strains in the ice plate are independent of time in the coordinate system moving together with the dipole. The problem is solved numerically using the Fourier transform, method of the normal modes and the truncation method for infinite systems of algebraic equations.

1. Introduction

The problem of the ice cover response to applied external loads has been well studied for an unbounded ice cover [1] and for a semi-infinite ice cover clamped to a vertical wall [2]. The external load is usually modelled as a point pressure or smooth localized pressure distribution moving with constant speed along the ice cover. The problems were studied within the linear theory of hydroelasticity. The ice deflections and strain distributions in the ice cover were determined. A main practical goal of those studies was to answer the question whether ice can be broken or not by the applied moving pressure and where this would happen, close to the load or at a distance from it.

The ice deflection and strains in the ice plate covering a channel for a localized external load moving along the channel were investigated in [3, 4]. The effect of the channel parameters and characteristics of the external load on the ice deflections and strain distribution in the ice cover were studied. The hydroelastic waves propagating along a frozen channel were studied in [5, 6] within the model of elastic ice plate.

Small oscillations of a two-dimensional body in a liquid were studied by Sturova and Tkacheva [7, 8, 9] within the linear wave theory for finite floating ice plates. In their studies, the upper

Content from this work may be used under the terms of the Creative Commons Attribution 3.0 licence. Any further distribution of this work must maintain attribution to the author(s) and the title of the work, journal citation and DOI.
Published under licence by IOP Publishing Ltd

boundary of the liquid consisted of either two semi-infinite intervals of free surfaces and a floating elastic plate or a finite interval of the free surface between two semi-infinite elastic plates of different thickness. The oscillating body was placed either under the ice or under the free surface. The problem was reduced to an integral equation for an unknown distribution of mass sources on the equilibrium position of the body surface.

In the present paper, the problem of ice cover deflection caused by an underwater body moving along a frozen channel is considered. This problem is approximately identical to the problem of a stationary body placed in a uniform current under the ice cover in a channel. The moving body is modelled by a three-dimensional dipole. A dipole moving at a constant speed in an unbounded fluid generates the flow and the pressure corresponding to a rigid sphere moving at the same speed. The radius of the sphere is related to the speed of the dipole and its intensity. A dipole moving in a channel with rigid walls also represents approximately a rigid sphere if the dipole intensity is small and the sphere radius is much smaller than the distance from the dipole to the walls of the channel. For a dipole of large intensity, a corresponding shape of a rigid body moving along the channel can be identified. This shape depends on the distances of the dipole from the walls. The velocity potential of a dipole placed in a rectangular channel with rigid walls is determined below by the method of images. The potential of several dipoles can be determined by superposition of the potentials of individual dipoles. A system of dipoles can be designed in such a way that these dipoles accurately represent a body of a given shape. By using the known velocity potential and the unsteady Bernoulli equation, the hydrodynamic loads acting on the channel walls are determined. For a dipole of small intensity, the linearized Bernoulli equation is used. This case corresponds to the motion of a small sphere along the channel covered with ice. The hydrodynamic loads on the ice cover caused by a moving underwater body are determined in the leading order without account for the ice deflection which is assumed small. The ice deflection is determined thereafter as the ice response to the applied loads caused by the moving body. The deflections of the ice cover are assumed small. However the strains in the cover can be large enough to cause ice breaking. The Kelvin–Voigt model of viscoelastic ice plate is used in the present study to estimate the strains in the ice cover. In this model, hydroelastic waves decay with a distance from a load due to the damping effect. The viscoelastic ice model is closer to reality than the elastic ice plate model.

2. Formulation of the problem

The deflection of an ice sheet caused by the motion of a dipole under the ice in the positive x-direction along a channel is considered. The channel is of rectangular cross section with depth H and width $2L$. The channel is of infinite extent in the x-direction. The channel is occupied with the inviscid and incompressible liquid of density ρ_l. The flow caused by both the dipole and the deflection of the ice is potential. The ice sheet is of constant thickness h_i and rigidity $D = Eh_i^3/[12(1-\nu^2)]$, where E is Young's modulus of ice and ν is Poisson's ratio. The ice sheet is clamped to the walls of the channel at $y = \pm L$.

The deflection of the ice sheet, $w(x, y, t)$, is governed by the equation of thin viscoelastic plate [10]

$$Mw_{tt} + D\left(1 + \tau\frac{\partial}{\partial t}\right)\nabla^4 w = p(x, y, 0, t) \tag{1}$$

$$(-\infty < x < \infty,\ -L < y < L,\ z = 0),$$

where $\tau = \eta/E$ is the retardation time, η is the viscosity of the ice, $\nabla^4 = \frac{\partial^4}{\partial x^4} + 2\frac{\partial^4}{\partial x^2 \partial y^2} + \frac{\partial^4}{\partial x^4}$, $M = \rho_i h_i$ is the mass of the ice plate per unit area, ρ_i is the ice density, $p(x, y, 0, t)$ s the hydrodynamic pressure acting on the lower surface of the ice plate, and t is the time. The ice sheet is frozen to the walls, which is modelled by the clamped conditions,

$$w = 0, \quad w_y = 0 \qquad (-\infty < x < \infty, y = \pm L). \tag{2}$$

The hydrodynamic pressure on the ice-liquid interface is given by the linearized Bernoulli equation,

$$p(x, y, 0, t) = -\rho_l \varphi_t - \rho_l g w \quad (-\infty < x < \infty, \ -L < y < L), \tag{3}$$

where g is the gravitational acceleration and $\varphi(x, y, z, t) = \varphi^D(x, y, z, t) + \varphi^E(x, y, z, t)$ is the total potential of the flow, φ^D is the velocity potential of a 3D dipole moving in the channel with rigid walls and φ^E is a correction potential accounting for the deflection of the ice cover. The velocity potential $\varphi^E(x, y, z, t)$ satisfies Laplace's equation in the flow region, the linearized kinematic condition on the ice-liquid interface and the boundary conditions of impermeability on other boundaries of the channel:

$$\varphi_z^E = w_t(x, y, t) \quad (z = 0), \qquad \varphi_y^E = 0 \quad (y = \pm L), \qquad \varphi_z^E = 0 \quad (z = -H). \tag{4}$$

The potential $\varphi^D(x, y, z, t)$ represents the dipole of intensity I placed at the point with coordinates x_0, y_0, z_0, which may depend on time t, between rigid walls at $z = -H$, $z = 0$, $y = L$ and $y = -L$. This potential satisfies the boundary conditions

$$\varphi_z^D = 0 \quad (z = 0, z = -H), \qquad \varphi_y^D = 0 \quad (y = \pm L). \tag{5}$$

The potential of the dipole oriented in the negative x-direction in unbounded three-dimensional fluid is [11]

$$\phi(x, y, z, t) = \frac{I(x - x_0)}{4\pi r^3}, \quad r(x, x_0, y, y_0, z, z_0) = \sqrt{(x - x_0)^2 + (y - y_0)^2 + (z - z_0)^2}. \tag{6}$$

Parameters x_0, y_0, z_0 and I can be functions of time. The potential of the dipole in the channel with the boundary conditions (5) is derived using the method of images [12]. The potential which behaves as (6) when $r \to 0$ and satisfies the boundary conditions at $y \pm L$ is given by

$$\phi^y(x, x_0, y, y_0, z, z_0, t) = -\frac{Ix}{4\pi}\left(\frac{1}{r^3(x, x_0, y, y_0, z, z_0)} + \sum_{n=1}^{\infty}\left(\frac{1}{r^3(x, x_0, y, y_0 + 4nL, z, z_0)} + \right.\right.$$

$$+\frac{1}{r^3(x, x_0, y, 2L - y_0 + 4(n-1)L, z, z_0)} + \frac{1}{r^3(x, x_0, y, y_0 - 4nL, z, z_0)} +$$

$$\left.\left. + \frac{1}{r^3(x, x_0, y, -2L - y_0 - 4(n-1)L, z, z_0)}\right)\right),$$

where the method of images for two vertical walls was used. Then the potential $\varphi^D(x, y, z, t)$ is obtained as an infinite series of the potentials ϕ^y reflected from the walls $z = -H$ and $z = 0$:

$$\varphi^D(x, y, z, t) = \phi^y(x, x_0, y, y_0, \hat{z}, \hat{z}_0, t) + \sum_{m=1}^{\infty}\left(\phi^y(x, x_0, y, y_0, \hat{z}, \hat{z}_0 + 2mH, t) + \right.$$

$$\phi^y(x, x_0, y, y_0, \hat{z}, H - \hat{z}_0 + 2(m-1)H, t) + \phi^y(x, x_0, y, y_0, \hat{z}, \hat{z}_0 - 2mH, t) +$$

$$\left. + \phi^y(x, x_0, y, y_0, \hat{z}, -H - \hat{z}_0 + 2(m-1)H, t)\right), \tag{7}$$

where $\hat{z} = z + H/2$, $\hat{z}_0 = z_0 + H/2$. Below, y_0 and z_0 are constant and $x_0 = Ut$.

The plate equation (1) and the Bernoulli equation (3) give

$$Mw_{tt} + D\left(1 + \tau\frac{\partial}{\partial t}\right)\nabla^4 w = -\rho_l \varphi_t^D - \rho_l \varphi_t^E - \rho_l g w, \tag{8}$$

IOP Conf. Series: Journal of Physics: Conf. Series **894** (2017) 012109 doi:10.1088/1742-6596/894/1/012109

where $\varphi_t^D(x, y, 0, t)$ is obtained by differentiating (7) in time and φ_t^E is to be determined as a solution of the boundary problem for Laplace's equation with the boundary conditions (4). Therefore, equation (8) describes the ice deflection caused by the "external" load $-\rho\varphi_t^D(x, y, 0, t)$ moving along the channel. Equations (6) and (7) show that φ^D depends on $x - x_0$ but not on x and x_0 separately. Then

$$\varphi^D = \varphi_1^D(x - x_0, y, y_0, z, z_0)$$

and

$$\varphi_t^D = -U\varphi_{1x}^D(x - x_0, y, y_0, z, z_0)$$

for $x_0 = Ut$.

The term with $\tau\partial/\partial t$ in the equation of viscoelastic plate (1), (8) describes the damping of ice plate oscillations, so hydroelastic waves decay far away from the moving load, where $|(x - x_0)/L| \to \infty$.

The formulated problem is considered in the case of the dipole moving with constant speed along the channel in the x-direction, all other parameters of the dipole are constant. Let us introduce non-dimensional variables denoted by tilde. The half-width of the channel L is taken as the length scale, the ratio L/U as the time scale. The non-dimensional depth of the channel H/L is denoted by h. The moving system of nondimensional coordinates $(\widetilde{x}, \widetilde{y}, \widetilde{z})$ with the origin on the upper channel wall at the centre of the dipole is introduced by

$$\widetilde{y} = \frac{y}{L}, \quad \widetilde{x} = \frac{x - Ut}{L}, \quad \widetilde{z} = \frac{z}{L}, \quad \widetilde{t} = \frac{U}{L}t.$$

We are concerned with a steady-state solution in the moving coordinate system,

$$w(x, y, t) = w(\widetilde{x}L + Ut, L\widetilde{y}, t) = w_{sc}\,\widetilde{w}(\widetilde{x}, \widetilde{y}),$$

$$\varphi^E(x, y, z, t) = \varphi^E(\widetilde{x}L + Ut, L\widetilde{y}, L\widetilde{z}, t) = \varphi_{sc}^E\,\widetilde{\varphi}^E(\widetilde{x}, \widetilde{y}, \widetilde{z}),$$

$$\varphi^D(x, y, z, t) = \varphi^D(\widetilde{x}L + Ut, L\widetilde{y}, L\widetilde{z}, t) = \frac{I}{L^2}\,\widetilde{\varphi}^D(\widetilde{x}, \widetilde{y}, \widetilde{z}),$$

where w_{sc}, φ_{sc} are the scales of the ice deflection and the velocity potential φ^E correspondingly. The scales are chosen as $w_{sc} = (UI)/(L^3 g)$ and $\varphi_{sc}^E = (U^2 I)/(gL^3)$.

In the non-dimensional variables the problem reads (tildes are omitted below)

$$\alpha h \mathrm{Fr}^2\, w_{xx} + \beta\Big(1 - \varepsilon\frac{\partial}{\partial x}\Big)\nabla^4 w + w = h\mathrm{Fr}^2 \varphi_x^E + \varphi_x^D \quad (-\infty < x < \infty, \quad -1 < y < 1, \quad z = 0), \tag{9}$$

$$w = 0, \quad w_y = 0 \qquad (y = \pm 1), \tag{10}$$

$$\nabla^2 \varphi^E = 0 \qquad (-\infty < x < \infty, \ -1 < y < 1, \ -h < z < 0), \tag{11}$$

$$\varphi_z^E = -w_x \quad (z = 0), \qquad \varphi_y^E = 0 \quad (y = \pm 1), \qquad \varphi_z^E = 0 \quad (z = -h), \tag{12}$$

$$\varphi_z^D = 0 \quad (z = 0, z = -h), \qquad \varphi_y^D = 0 \quad (y = \pm 1), \tag{13}$$

$$w, \varphi \to 0 \qquad (|x| \to \infty). \tag{14}$$

Here $\beta = D/(\rho_l g L^4)$, $\varepsilon = (\tau U)/L$, $\alpha = M/(\rho_l L) = (\rho_i h_i)/(\rho_l L)$ and $\mathrm{Fr} = U/\sqrt{gH}$ is the Froude number. The solution of the problem (9) – (14) depends on five non-dimensional parameters h, α, β, ε, Fr, as well as on the position of the dipole (x_0, y_0, z_0), which describe the aspect ratio of channel, characteristics of ice and the dipole. We shall determine the deflection w and strain distribution in the ice sheet for some given values of these parameters. Note that the scales of the deflection and the velocity potential directly proportional to the intensity of the dipole I.

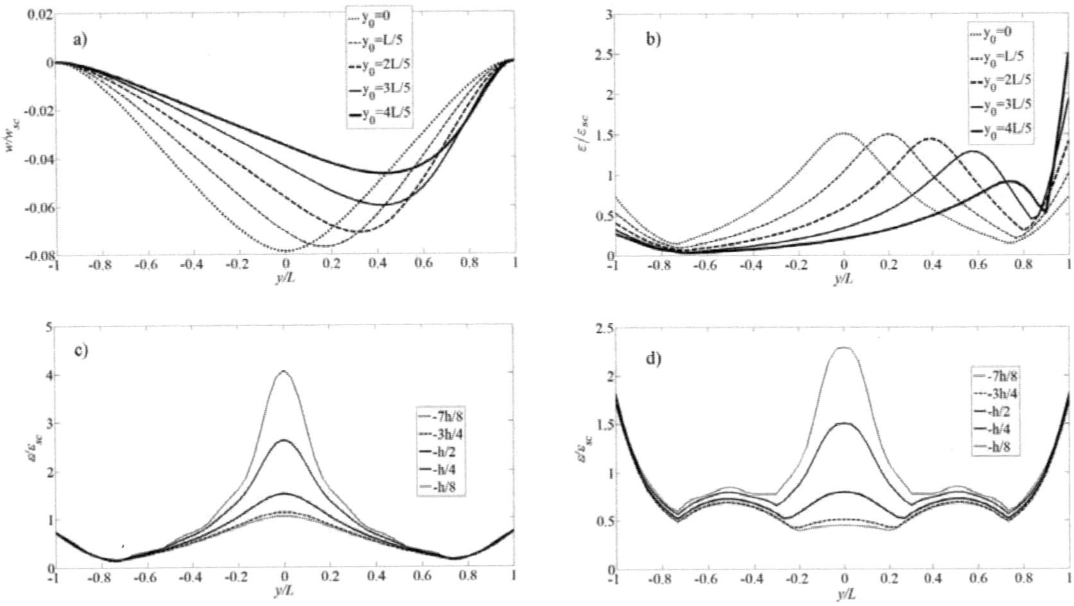

Figure 1. The nondimensional deflection *a)* and scaled strains *b)* in the ice cover across the channel at $x = 0$ for $z_0 = -h/2$ and different horizontal coordinate y_0 of the dipole, the speed of the dipole is $U = 3\text{m/s}$. Scaled strains in the ice cover across the channel at $x = 0$ for $y_0 = 0$ and different vertical positions of the dipole, the speed of the dipole is $U = 3\text{m/s}$ *c)* and $U = 6\text{m/s}$ *d)*.

3. Numerical results

The hydroelastic behaviour of the ice cover in a channel, which is caused by a dipole moving under the ice, is investigated numerically for a freshwater ice with density $\rho_i = 917$ kg/m^3, Young's modulus $E = 4.2 \cdot 10^9$ N/m^2, Poisson's ratio $\nu = 0.3$ and the retardation time $\tau = 0.1$ s. The thickness of the ice plate is 10 cm, the half width of the channel is 10 m and the water depth is 2 m. The coordinates of the dipole are (Ut, y_0, z_0). The speed U and the position of the dipole with respect to the walls of the channel vary in the calculations. In contrast to the problem of a dipole moving under ice sheets of infinite extent, there are infinitely many critical speeds of hydroelastic waves in a frozen channel. Correspondingly, there are many values of the speeds of a moving dipole at which the stresses in the ice cover are amplified. The ranges of the speeds U and the widths L are chosen according to the critical speeds of hydroelastic waves propagating in the frozen channel, which were computed in [5], and the characteristic length $(D/\rho_l g)^{1/4}$ of the ice cover, which is equal to 2.48 m in the case under consideration. The lowest critical speed is 5.38 m/s for $\tau = 0$ s and is equal approximately to 5.5 m/s for $\tau = 0.1$ s for the channel considered in this paper and in [3]. We shall investigate elastic behavior of the ice cover for speeds of the dipole smaller or greater than the first critical speed.

The nondimensional deflection and strains scaled with $\varepsilon_{sc} = (h_i UI)/(2gL^5)$ across the channel at $x = 0$ for the speed of the dipole $U = 3$ m/s are shown in Figures 1*a* and 1*b*. The dipole is located at $y_0 = 0, L/5, 2L/5, 3L/5, 4L/5$, $z_0 = -h/2$ in these figures. For the subcritical speed, $U = 3$ m/s, the ice response is localised near the dipole, because the hydroelastic waves propagating with this speed along a channel do not exist [5]. When the dipole approaches the wall, the maximum deflection decreases, see Figure 1*a*, the strains decrease near the dipole and at the farther wall, and increase at the nearest wall, see Figure 1*b*, for this subcritical speed. Scaled strains across the channel for different vertical coordinate of the dipole, $z_0 = -7h/8, -3h/4, -h/2, -h/4, -h/8$, and $y_0 = 0$ are shown in Figures 1*c* and 1*d*. The speed of the dipole is 3 m/s in Figure 1*c* and is 6 m/s in Figure 1*d*. This speed is greater than the first critical speed and is smaller than the second critical speed for the channel. Note that the strains are always maximum near the dipole for subcritical speed of the dipole. In contrast, for supercritical speeds of the dipole, the maximum strain can be achieved either near the dipole or on the wall depending on the position of the dipole in the channel.

4. Conclusion

The mathematical model of steady hydroelastic waves in the ice cover of a channel caused by a dipole moving in the liquid has been considered. The formulation of the problem was presented and discussed. The hydrodynamic pressure on the ice-liquid interface caused by the dipole moving in a rectangular channel with rigid walls has been derived. The ice deflection was neglected at leading order. This hydrodynamic pressure was treated as an external pressure on the upper surface of the ice cover. The problem of a moving underwater dipole was reduced to the problem of an external load moving on ice along the channel. The later problem was solved by the method described in [3]. With the help of the Fourier transform along the channel the original problem was reduced to the problem of the wave profile across the channel, which was solved by the method of normal modes.

The problem was solved numerically. It has been shown that displacement of the dipole across the channel towards a wall increases the strain on the wall nearest to the dipole and decreases the strain on the other wall. For subcritical speeds of the dipole strains are maximum near the dipole. For supercritical speeds the strains near the dipole decrease and can be smaller than strains on the walls when the dipole moves deeper from the ice cover.

Acknowledgments

This work was partially supported by the grant RFBR 16-08-00291 "Hydroelastic and thermodynamic effects with interaction of poroelastic ice and structures".

References

[1] Squire V, Hosking R, Kerr A and Langhorne P 1996 *Moving loads on ice* (Amsterdam: Kluwer Academic Publishers)
[2] Brocklehurst P 2012 *Hydroelastic waves and their interaction with fixed structures* (Norwich: University of East Anglia)
[3] Shishmarev K, Khabakhpasheva T and Korobkin A 2016 *Appl. Ocean Res.* **59** 313–26
[4] Shishmarev K 2015 *The news of Altai State University* **85**(1-2) 189–94 (in Russian)
[5] Korobkin A, Khabakhpasheva T and Papin A 2014 *Eur. J. Mech. B-Fluids* **47** 166-75
[6] Batyaev E A and Khabakhpasheva T I 2015 *Fluid Dynamics* **6** 84–101
[7] Sturova I V and Tkacheva L A 2016 *Proc. 3rd Int. Sc. Conf. Polar Mechanics* (Vladivostok: FEFU) pp 976–85
[8] Sturova I V 2015 *J. Fluid. Mech.* **784** 373–95
[9] Tkacheva 2015 *J. Appl. Mech. Tech. Phys.* **56**(6) 173–86 (in Russian)
[10] Shishmarev K, Khabakhpasheva T and Korobkin A 2015 *Proc. 5th Int. Conf. Hydroelastisity in Marine Technology* (Zagreb: VIDICI d.o.o.) pp 149–60
[11] Miloh T 1991 *Mathematical Approaches in Hydrodynamics* (Philadelphia: Society for Industrial and Applied Mathematics)
[12] Newman J N 1977 *Marine Hydrodynamics* (Cambridge: The MIT Press)

Finite strain transient creep of D16T alloy: identification and validation employing heterogeneous tests

A V Shutov[1,2] **and A Yu Larichkin**[1,2]
[1] Lavrentyev Institute of Hydrodynamics, 630090 Novosibirsk, Russia
[2] Novosibirsk State University, 630090 Novosibirsk, Russia

E-mail: alexey.v.shutov@gmail.com

Abstract. A cyclic creep damage model, previously proposed by the authors, is modified for a better description of the transient creep of D16T alloy observed in the finite strain range under rapidly changing stresses. The new model encompasses the concept of kinematic hardening, which allows us to account for the creep-induced anisotropy. The model kinematics is based on the nested multiplicative split of the deformation gradient, proposed by Lion. The damage evolution is accounted for by the classical Kachanov-Rabotnov approach. The material parameters are identified using experimental data on cyclic torsion of thick-walled samples with different holding times between load reversals. For the validation of the proposed material model, an additional experiment is analyzed. Although this additional test is not involved in the identification procedure, the proposed cyclic creep damage model describes it accurately.

1. Introduction

Numerous experimental observations carried out on metals subjected to elevated temperatures indicate that creep is a loading history dependent anisotropic phenomenon [1, 2, 16]. An accurate modelling of the creep behaviour is important for the analysis of various metal forming applications and for the safety analysis of high-temperature structural components operating under cyclic loading [6]. A creep-induced anisotropy has to be taken into account in the analysis of creep under rapidly changing loadings. One of the most popular approaches to the description of transient creep occurring shortly after load changes is based on backstresses (see, among others, [14, 11, 25, 26]). Backstresses are stress-like quantities which represent internal stress fields, acting on the microstructural level. In metals, backstresses may represent a resistance to dislocation motion caused by dislocation pile-ups [25] (atomistic scale), while other backstresses are caused by plastic strain incompatibilities between grains (mesoscopic scale) [4]. Within the phenomenological approach, the evolution of backstresses is typically assumed in a hardening/recovery format. The hardening is governed by the creep strain rate. The recovery can be strain-controlled, also known as dynamic recovery (cf. [15]). Alternatively, it can be temperature-controlled, which is also known as static recovery (cf. [14, 11, 25]). In the material model used here, both types of recovery are combined (see [23] for the justification of this combined approach).

There is only a small number of publications dealing with finite strain creep under rapidly changing loadings. Although different phenomenological modelling approaches to the kinematic

Content from this work may be used under the terms of the Creative Commons Attribution 3.0 licence. Any further distribution of this work must maintain attribution to the author(s) and the title of the work, journal citation and DOI.
Published under licence by IOP Publishing Ltd

hardening can been used, we employ here the multiplicative split of the deformation gradient; it has several advantages, as discussed in [20]. More specifically, we employ a nested multiplicative split proposed in [13]. The seminal idea of the nested split was successfully used for metal plasticity with nonlinear kinematic hardening [19, 3, 12] as well as its extensions to shape memory alloys [8], thermo-plasticity [18], ductile damage [21], evolution of microstructure [24] and other physically nonlinear phenomena. One of the goals of the current study is to promote this approach by presenting practical identification and validation procedures. Toward that end, experiments on deformation of D16T aluminium alloy (similar to AlCuMg2 and 24ST4)are considered. Here we focus on heterogeneous experiments, since they can be performed under more general conditions than homogeneous tests. In the particular case of torsion tests, experiments on thick-walled tubes allow one to induce larger strains than in experiments on thin-walled samples. However, an improved flexibility of the heterogeneous testing comes at the cost of increased computational efforts during the identification of parameters. In the current study, the torsion tests are simulated numerically using the nonlinear finite element method (direct problem). The material parameters are identified by the minimization of a least square error functional, reflecting the deviation of the FEM simulation results from the experimental data (inverse problem). After the parameters are identified, they are validated by simulating an additional torsion test. A good correspondence between experimental and theoretical results is observed both in identification and validation.

2. Material model of cyclic creep in the finite strain range

The material model used here is a slight modification of the model proposed in [23]. Here, for simplicity, we consider isothermal processes only.[1] Just as in [23], we employ the Rabotnov damage variable $\omega \in [0, 1]$. Here, $\omega = 0$ and $\omega = 1$ characterize respectively the virgin state and the fully destroyed state [9, 17]. A rheological motivation of the model is shown in Figure 1a. This motivation represents an idealized device which is build of two elastic springs, two viscous dashpots and one rate-independent dashpot. The model kinematics is based on the nested multiplicative split of the deformation gradient, originally proposed by Lion in [13] for metal plasticity. The deformation gradient $\mathbf{F}$ is a product of a creep-related part $\mathbf{F}_{\mathrm{cr}}$ and an elastic part $\mathbf{F}_{\mathrm{e}}$; the creep part itself is decomposed multiplicatively into a certain dissipative part $\mathbf{F}_{\mathrm{ii}}$ and a conservative part $\mathbf{F}_{\mathrm{ie}}$

$$\mathbf{F} = \hat{\mathbf{F}}_{\mathrm{e}} \mathbf{F}_{\mathrm{cr}}, \quad \mathbf{F}_{\mathrm{cr}} = \check{\mathbf{F}}_{\mathrm{ie}} \mathbf{F}_{\mathrm{ii}}. \tag{1}$$

The multiplicative decompositions are summarized in a commutative diagram shown in Figure 1b.

In order to formulate the constitutive equations on the reference configuration, we introduce the right Cauchy-Green tensor $\mathbf{C}$, the right Cauchy-Green tensor of creep $\mathbf{C}_{\mathrm{cr}}$, and the inelastic right Cauchy-Green tensor of substructure $\mathbf{C}_{\mathrm{ii}}$

$$\mathbf{C} := \mathbf{F}^{\mathrm{T}} \mathbf{F}, \quad \mathbf{C}_{\mathrm{cr}} := \mathbf{F}^{\mathrm{T}}_{\mathrm{cr}} \mathbf{F}_{\mathrm{cr}}, \quad \mathbf{C}_{\mathrm{ii}} := \mathbf{F}^{\mathrm{T}}_{\mathrm{ii}} \mathbf{F}_{\mathrm{ii}}. \tag{2}$$

Using these quantities, the second Piola-Kirchhoff stress $\tilde{\mathbf{T}}$ is computed through (cf. [23])

$$\tilde{\mathbf{T}} = (1-\omega)^{n_{\mathrm{e}}} \Big[\frac{k}{10} \left((\det \mathbf{C})^{5/2} - (\det \mathbf{C})^{-5/2} \right) \mathbf{C}^{-1} + \mu\, \mathbf{C}^{-1} (\overline{\mathbf{C}} \mathbf{C}^{-1}_{\mathrm{cr}})^{\mathrm{D}} \Big], \tag{3}$$

where $k > 0$ and $\mu > 0$ are the elasticity parameters of the virgin material, $n_{\mathrm{e}} > 0$ is a material parameter governing the deterioration of elastic properties,[2] the symbol $(\cdot)^{\mathrm{D}}$ stands for the

[1] Thus, we assume that the temperature is constant in space and time. The model can be generalized to the thermo-mechanical case using the procedure discussed in [18].
[2] In the classical Kachanov approach, $n_{\mathrm{e}} = 1$.

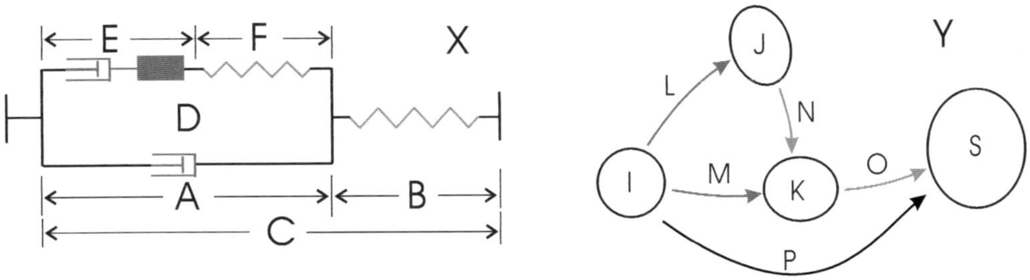

Figure 1. Rheological model of cyclic creep with partial deformations (a); commutative diagram of the nested multiplicative split of the deformation gradient (b).

deviatoric part of a tensor, by the overline $\overline{(\cdot)}$ we denote the unimodular part of a tensor. The used in (3) ansatz for the hydrostatic component of the stress complies with basic plausibility restrictions (cf. [7]); if necessary, it can be replaced by any alternative. Next, a backstress operating on the reference configuration is computed by

$$\tilde{\mathbf{X}} = (1-\omega)^{n_e}\frac{c}{2}\,\mathbf{C}_{\mathrm{cr}}^{-1}(\mathbf{C}_{\mathrm{cr}}\mathbf{C}_{\mathrm{ii}}^{-1})^{\mathrm{D}}, \tag{4}$$

where $c \geq 0$ is a constant pertaining to the intact material. Further, an effective stress tensor, which essentially governs the creep, is defined by the equation

$$\tilde{\mathbf{\Sigma}} := \mathbf{C}\tilde{\mathbf{T}} - \mathbf{C}_{\mathrm{cr}}\tilde{\mathbf{X}}. \tag{5}$$

Using it, we define the magnitude of the driving force:

$$\mathfrak{F} := \mathfrak{N}(\tilde{\mathbf{\Sigma}}^{\mathrm{D}}), \quad \text{where} \quad \mathfrak{N}(\mathbf{A}) := \sqrt{\mathrm{tr}(\mathbf{A}^2)}. \tag{6}$$

A Mandel-like backstress on the reference configuration is introduced through

$$\tilde{\mathbf{\Xi}} := \mathbf{C}_{\mathrm{cr}}\tilde{\mathbf{X}}. \tag{7}$$

The evolution of $\mathbf{C}_{\mathrm{cr}}$ is governed by the flow rule of J_2 type

$$\dot{\mathbf{C}}_{\mathrm{cr}} = 2\,\frac{\lambda}{\mathfrak{F}}\tilde{\mathbf{\Sigma}}^{\mathrm{D}}\,\mathbf{C}_{\mathrm{cr}}, \tag{8}$$

where $\lambda \geq 0$ is the creep strain rate. The material rate $\dot{\mathbf{C}}_{\mathrm{ii}}$ is given by (cf. [23])

$$\dot{\mathbf{C}}_{\mathrm{ii}} = \left(\varkappa_{\mathrm{dynam}}\,\mathfrak{N}(\mathbf{C}_{\mathrm{cr}}^{-1}\,\dot{\mathbf{C}}_{\mathrm{cr}}) + 2\varkappa_{\mathrm{stat}}\right)\tilde{\mathbf{\Xi}}^{\mathrm{D}}\,\mathbf{C}_{\mathrm{ii}}, \tag{9}$$

where $\varkappa_{\mathrm{dynam}}$ and $\varkappa_{\mathrm{stat}}$ are respectively the parameters of static and dynamic recovery. Note that these parameters do not depend on the damage ω. Although this assumption is very simple, it allows one to achieve a good accuracy in some applications (cf. the next section). Probably, one of the most simple assumptions governing the creep rate λ is provided by the Norton creep law combined with Kachanov's approach to creep damage

$$\lambda = A(1-\omega)^{-m}(\mathfrak{F}/\sigma_0)^n, \tag{10}$$

where $A \geq 0$, $m \geq 0$, and $n \geq 0$ are material parameters; $\sigma_0 := 1$ MPa is not a material constant. The evolution of the damage variable is described by the following modification of the classical Kachanov-Rabotnov relation (cf. [10, 2])

$$\dot{\omega} = B(1-\omega)^{-l}\Big(\mathfrak{N}((\mathbf{C}\tilde{\mathbf{T}})^{\mathrm{D}})/\sigma_0\Big)^k, \tag{11}$$

where $B \geq 0$, $l \geq 0$, and $k \geq 0$ are material constants. Here, in contrast to the damage evolution rule used in [23], the damage growth is governed by the norm of the deviatoric Kirchhoff stress, which is equal to $\mathfrak{N}\big((\mathbf{C}\tilde{\mathbf{T}})^{\mathrm{D}}\big)$. As will be seen from the next section, this assumption allows us to obtain reasonable results in a wide range of cyclic creep scenarios. In order to close the system of constitutive equations, the following initial conditions need to be specified:

$$\mathbf{C}_{\mathrm{cr}}|_{t=0} = \mathbf{C}^0_{\mathrm{cr}}, \quad \mathbf{C}_{\mathrm{ii}}|_{t=0} = \mathbf{C}^0_{\mathrm{ii}}, \quad \omega|_{t=0} = \omega^0. \tag{12}$$

Here, $\omega^0 \in [0, 1)$ is the initial damage. Note that an appropriate choice of the initial values $\mathbf{C}^0_{\mathrm{cr}}$ and $\mathbf{C}^0_{\mathrm{ii}}$ allows one to change the reference configuration and to introduce initial anisotropy.

For $n_{\mathrm{e}} = 1$, the introduced model is a special case of the creep damage model presented in [23] with exception of the damage evolution rule (11). Similar to the proof presented in [23], it is possible to show that this model is thermodynamically consistent and w-invariant. Obviously, this model does not account for the damage-induced dilatation of the material due to void nucleation and growth. This drawback can be overcome using the modelling steps described in [21].

The numerical implementation of this model is essentially based on the efficient time stepping scheme presented in [22]. The algorithm exactly preserves the incompressibility condition, and it is first order accurate. In the damage-free case, it is unconditionally stable. The model is implemented into MSC.MARC employing the user-defined material subroutine via the Hypela2 interface.

3. Parameter identification and validation

In this section we use a series of torsion tests carried out in [5] on the D16T aluminium alloy (similar to AlCuMg2 and 24ST4) at 250 ^{0}C. The experiments are conducted on thick-walled tubular samples. The advantage of this experimental set-up is that it allows one to achieve large strains without loss of stability and without the undesired influence of friction. The sample dimensions in the gage area are as follows: sample length $L = 70$ mm, inner radius $r_i = 5$ mm, outer radius $r_0 = 10$ mm. Since the stress distribution within the sample is heterogeneous, the analysis of the experiment requires a solution of a boundary value problem. The corresponding FEM mesh and boundary conditions are the same as discussed in [23].

The initial state of the material is assumed to be stress-free, undeformed, and isotropic. Therefore we set the following initial conditions

$$\mathbf{C}_{\mathrm{cr}}|_{t=0} = \mathbf{C}^0_{\mathrm{cr}} = \mathbf{1}, \quad \mathbf{C}_{\mathrm{ii}}|_{t=0} = \mathbf{C}^0_{\mathrm{ii}} = \mathbf{1}, \tag{13}$$

where $\mathbf{1}$ is the identity tensor. The material parameters are identified from two different torsion experiments. In both tests, a piecewise constant torque is applied with $|M| = 67.75$ N $\cdot$ m. In the first test, the applied torque is reversed every 24 hours, in the second - every 96 hours. The experimentally measured twist angle of the sample is plotted in Figure 2 for different holding times between load reversals. Some of the parameters are determined by general considerations (cf. [23]). The remaining parameters are identified by the minimization of the least square error functional, reflecting the deviation of the simulated twist angle from the corresponding experimental result. The identification results are summarized in Table 1. There, parameters which appeared in the minimization problem are marked by the asterisk *. As can be seen from Figure 2, the obtained set of material parameters provides a good accuracy.

Now we proceed to the validation of the parameters from Table 1. An additional cyclic creep torsion test with $|M| = 67.75$ N $\cdot$ m and holding times of 48 hours between load reversals is simulated using these parameters. The corresponding experimental data and simulation results are shown in Figure 3. *Although these data are ignored during parameter identification, a very good correspondence between simulation and experiment is observed.*

Table 1. Material parameters of the D16T alloy.

parameter	value	brief explanation	equation
k	73500 MPa	bulk modulus of intact material	(3)
μ	28200 MPa	shear modulus of intact material	(3)
c	20000 MPa	shear modulus of intact substructure	(4)
* n_e	35.24 [-]	impact of damage on the elastic properties	(3), (4)
* A	$3.574 \cdot 10^{-13}$ h^{-1}	parameter of Norton's law	(10)
n	5 [-]	parameter of Norton's law (exponent)	(10)
* m	68.49 [-]	impact of damage on the creep rate	(10)
* $\varkappa_{dyam}$	0.0586 MPa^{-1}	dynamic recovery coefficient	(9)
$\varkappa_{stat}$	0.0 $MPa^{-1}h^{-1}$	static recovery coefficient	(9)
* B	$3.428 \cdot 10^{-13}$ h^{-1}	damage evolution parameter	(11)
l	5 [-]	damage evolution parameter	(11)
k	5 [-]	damage evolution parameter	(11)
ω_0	0.01 [-]	initial damage	$(12)_3$

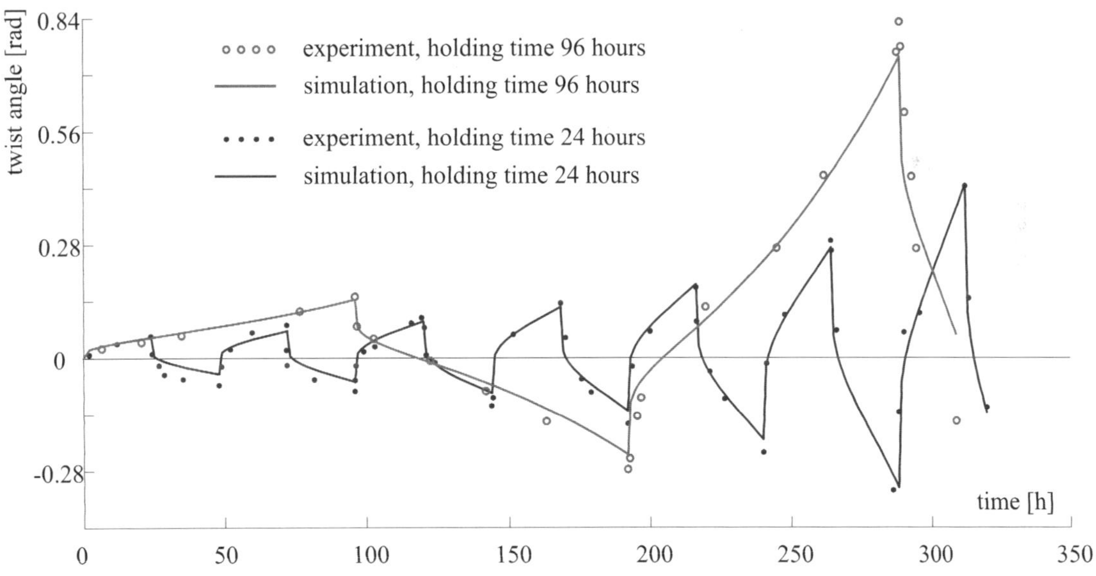

Figure 2. Cyclic torsion of thick-walled tubular specimens with different holding times: experimental data from [5] and corresponding simulation results visualizing the identified material parameters.

4. Discussion and conclusions

Intuitively, the torsion test with holding times of 48 hours is located "between" the tests with holding times of 24 and 96 hours. In other words, we have checked the ability of the

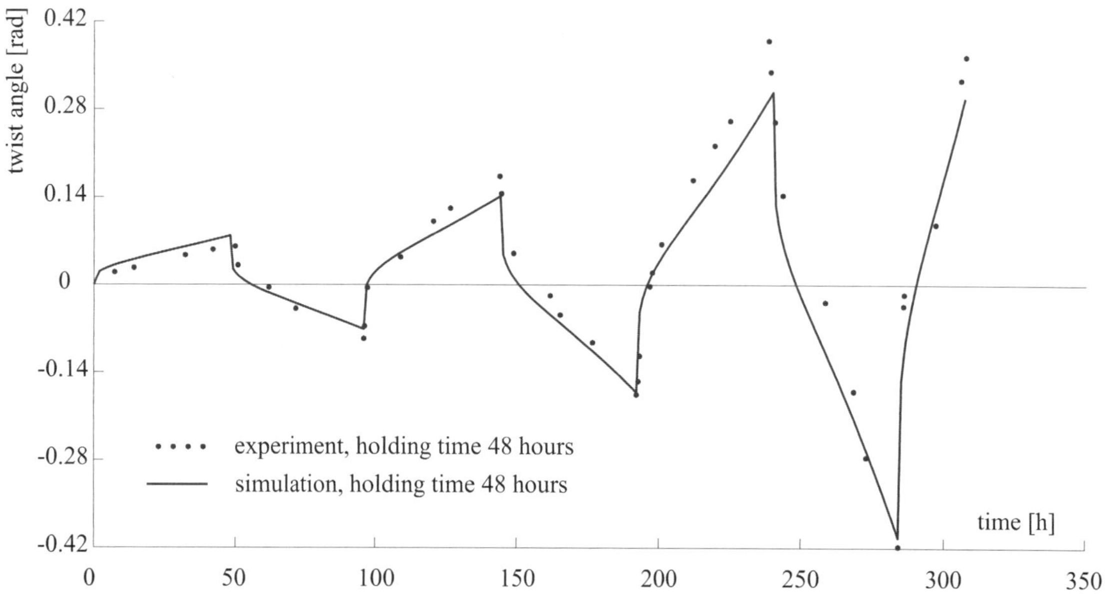

Figure 3. Validation of the material model using additional experiment on cyclic torsion of a thick-walled tubular specimen: experimental data from [5] and simulation results using the previously identified material parameters.

phenomenological material model to "interpolate between" known experimental results. In general, dealing with purely phenomenological models, the extrapolation procedure is expected to be less accurate than the interpolation. This aspect and the reliability of the obtained set of parameters will be analyzed in the follow-up study.

In the current study a practical example of parameter identification and validation is presented, basing on the evaluation of heterogeneous test data. It is shown that a rather complex finite strain cyclic creep damage model can be calibrated using relatively simple tests.

Acknowledgments

The financial support provided by the RFBR (grant number 17-08-01020) is acknowledged.

References

[1] Altenbach H, Altenbach J and Zolochevsky A 1995 *Erweiterte Deformationsmodelle und Versagenskriterien der Werkstoffmechanik* (Stuttgart: Deutscher Verlag für Grundstoffindustrie)

[2] Altenbach H and Naumenko K 2007 *Modeling of Creep for Structural Analysis (Foundations of Engineering Mechanics)* (Springer-Verlag Berlin Heidelberg)

[3] Bröker C and Matzenmiller A 2014 On the generalization of uniaxial thermoviscoplasticity with damage to finite deformations based on enhanced rheological models *Technische Mechanik* **34** 3-4 142 – 165

[4] Feaugas X 1999 On the origin of the tensile flow stress in the stainless steel AISI 316L at 300 K: back stress and effective stress *Acta Materialia* **47**(13) 3617–3632

[5] Gorev B V 1979 On the estimation of creep and long-term strength of structural elements using the method of characteristic parameters. Message 1. (in Russian) *Problems of Strength* **4** 30–36

[6] Gorash Y, Altenbach H and Lvov G 2012 Modelling of high-temperature inelastic behaviour of the austenitic steel AISI type 316 using a continuum damage mechanics approach *The Journal of Strain Analysis for Engineering Design* **47**(4) 229–243

[7] Hartmann S and Neff P 2003 Polyconvexity of generalized polynomial-type hyperelastic strain energy functions for near-incompressibility *International Journal of Solids and Structures* **40** 2767–2791

[8] Helm D 2001 *Formgedächtnislegierungen, experimentelle Untersuchung, phänomenologische Modellierung und numerische Simulation der thermomechanischen Materialeigenschaften* (Kassel: Universitätsbibliothek Kassel)

[9] Kachanov L M 1958 Time of the rupture process under creep conditions (in Russian) *Izv. AN SSSR. Otd. Tekh. Nauk* **8** 26–31

[10] Kachanov L M 1986 *Introduction to Continuum Damage Mechanics* (Dordrecht: Martinus Nijhoff Publishers)

[11] Kawai M 1995 Creep hardening rule under multiaxial repeated stress change *JSME Int. J.* **38**A 201–212

[12] Kießling R, Landgraf R, Scherzer R and Ihlemann J 2016 Introducing the concept of directly connected rheological elements by reviewing rheological models at large strains *International Journal of Solids and Structures* **97–98** 650–667

[13] Lion A 2000 Constitutive modelling in finite thermoviscoplasticity: a physical approach based on nonlinear rheological elements *International Journal of Plasticity* **16** 469–494

[14] Malinin N N and Khadjinsky G M 1972 Theory of creep with anisotropic hardening *Int. J. Mech. Sci.* **14** 235–246

[15] Naumenko K, Altenbach H and Kutschke A 2010 A combined model for hardening, softening, and damage processes in advanced heat resistant steels at elevated temperature *International Journal of Damage Mechanics* **20** 578–597

[16] Naumenko K, Kutschke A, Kostenko Y and Rudolf T 2011 Multi-axial thermo-mechanical analysis of power plant components from 912%Cr steels at high temperature *Eng. Fract. Mech.* **78** 1657–1668

[17] Rabotnov Yu N 1969 *Creep Problems in Structural Members* (Amsterdam: North-Holland)

[18] Shutov A V and Ihlemann J 2011 On the simulation of plastic forming under consideration of thermal effects *Materials Science & Engineering Technology* **42**(7) 632–638

[19] Shutov A V and Kreißig R 2008 Finite strain viscoplasticity with nonlinear kinematic hardening: Phenomenological modeling and time integration *Computer Methods in Applied Mechanics and Engineering* **197** 2015–2029

[20] Shutov A V and Ihlemann J 2014 Analysis of some basic approaches to finite strain elasto-plasticity in view of reference change *International Journal of Plasticity* **63** 183–197

[21] Shutov A V, Silbermann C B and Ihlemann J 2015 Ductile damage model for metal forming simulations including refined description of void nucleation *International Journal of Plasticity* **71** 195–217

[22] Shutov A V 2016 Efficient implicit integration for finite-strain viscoplasticity with a nested multiplicative split *Computer Methods in Applied Mechanics and Engineering* **306**(1) 151–174

[23] Shutov A V, Larichkin A Yu, Shutov V A 2017 Modelling of cyclic creep in the finite strain range using a nested split of the deformation gradient *ZAMM* DOI: 10.1002/zamm.201600286

[24] Silbermann C B, Shutov A V, Ihlemann J 2014 Modeling the evolution of dislocation populations under non-proportional loading *International Journal of Plasticity* **55** 58-79

[25] Zolochevsky A, Sklepus S, Kozmin Yu, Kozmin A, Zolochevsky D and Betten J 2004 Constitutive equations of creep under changing multiaxial stresses for materials with different behavior in tension and compression *Forschung im Ingenieurwesen* **68** 182–196

[26] Zolochevsky A and Voyiadjis G Z 2005 Theory of creep deformation with kinematic hardening for materials with different properties in tension and compression *International Journal of Plasticity* **21** 435–462

IOP Conf. Series: Journal of Physics: Conf. Series **894** (2017) 012111 doi:10.1088/1742-6596/894/1/012111

Influence of bed material entrainment and non-Newtonian rheology on turbulent geophysical flows dynamics. Numerical study

M E Eglit [1], A E Yakubenko [2] and T A Yakubenko [2]

[1] Lomonosov Moscow state university, Moscow, Russia
[2] Institute of Mechanics, Moscow, Russia

E-mail: m.eglit@mail.ru

Abstract. This paper deals with the mathematical and numerical modeling of the propagation stage of geophysical gravity-driven flows, such as snow avalanches, mudflows, and rapid landslides. New mathematical models are presented which are based on full, not-depth-averaged equations of mechanics of continuous media. The models account for three important issues: non-Newtonian rheology of the moving material, entrainment of the bed material by the flow, and turbulence. The main objective is to investigate the effect of these three factors on the flow dynamics and on the value of the entrainment rate. To exclude the influence of many other factors, e.g., the complicated slope topography, only the motion down a long uniform slope with a constant inclination angle is studied numerically. Moreover, the entire flow from the front to the rear area was not modeled, but only its middle part where the flow is approximately uniform in length. One of the qualitative results is that in motion along homogeneous slope the mass entrainment increases the flow velocity and depth while the entrainment rate at large time tends to become constant which depends on the physical properties of the flow and the underlying material but not on the current values of the flow velocity and depth.

1. Introduction

This paper deals with mathematical and numerical modeling of dense geophysical gravity-driven mass flows, such as snow avalanches, mudflows, and rapid landslides. These flows can pose grave hazards to people and property. Knowledge of their dynamic parameters and run-out distance, i.e., the boundaries of the affected area, is needed for properly land planning in mountains and design of defense constructions. Numerical modeling may provide one of the tools to obtain this knowledge.

Mathematical models of dense geophysical flows can be divided into three groups according to the complexity level and the extent account for different flow characteristics. The simplest models treat the flow as a moving material point, i.e., the internal structure is not accounted for. In models of the second group the mass flow is a flow of a continuous medium but only the depth-averaged velocities are calculated besides the flow depth and width. Models of the third group describe the structure of the flow not only along its body, but also in a direction normal to the slope. They are based on full, not-depth-averaged equations of mechanics of continuous media. The development and use of such models have been launched recently [1-4]. This became possible due to the appearance of new measuring techniques and increase of the capabilities of computers. To construct these new models one should know 1) the equations which relate stresses and deformation characteristics inside the flow,

Content from this work may be used under the terms of the Creative Commons Attribution 3.0 licence. Any further distribution of this work must maintain attribution to the author(s) and the title of the work, journal citation and DOI.
Published under licence by IOP Publishing Ltd

i.e., the rheological properties of the moving medium, 2) the equation which determines the entrainment of mass from the slope by the flow since dense gravity-driven flows always entrain bed material while descending steep slopes [5, 6], and 3) the equations which describe the characteristics of turbulence since most of large gravity-driven flows are turbulent.

The paper presents new models of the third group. The models take into account all three factors mentioned above. The first point is the formulation of the constitutive rheological relations for the flow material. These relations are different for flows of different physical nature: dry and wet dense snow avalanches, powder avalanches, mudflows, debris flows, volcanic lava, landslides, water flows etc. The simplest model for dense snow avalanches is the model of incompressible isotropic linear-viscous (Newtonian) fluids with high value of the viscosity coefficient [2, 7-9]. The components τ_{ij} of the viscous stresses tensor in this model are linear functions of the components of the strain rate tensor,e_{ij}, and vanish at $e_{ij} = 0$: $\tau_{ij} = 2\mu e_{ij}$, μ being the viscosity coefficient. However, observations and measurements indicate the necessity to introduce more sophisticated models: measured velocity profiles differ from those for Newtonian fluids [9-14]. Besides, many geophysical flows, e.g., avalanches and landslides can stop at an incline, being at rest with non-zero tangential stress. The latter fact means that the flow material have a yield limit: it doesn't move or moves without deformations as a rigid body until the stress intensity is less than the yield stress. To describe the rheological properties of the flow material here we use the so-called Herschel–Bulkley model, which can correspond to both linear and nonlinear viscous (power-law) fluids, as well as media with a yield stress, in particular, the Bingham fluid, by the appropriate choice of the coefficients. The Bingham and Herschel–Bulkley models have been proposed as possible rheological models for snow avalanches, mudflows, lava flows, and landslides, for example in [1-4, 11, 13]. The second point is including bed material entrainment. Field evidence shows that a flow's volume can increase many fold as a consequence of entrainment [5, 6]. It is known that the size, speed, and destructive potential of geophysical mass flows are strongly influenced by entrainment of bed material. Here we model the effect of the basal entrainment assuming the following hypothesis [6, 9]: the entrainment occurs, when the shear stress on the flow bottom reaches the value of the bed material shear strength. The last point in constructing the mathematical model is the account for turbulence of the flow. We deal with the Reynolds's averaged equations and flow parameters. To obtain the closed system of equations we employed the three-parameters turbulence model developed by Luschik, Paveliev and Yakubenko [15, 16] which was successfully applied to simulate turbulent flows in pipes and channels with account for heat transfer, mass exchange through porous walls and nonlinear temperature dependence of the fluid viscosity coefficient.

Simulations of flows with different rheological properties have been done by the code developed by the authors. The main objective was to investigate the effect of flow rheology, bed material entrainment and turbulence on the flow dynamics and on the value of the entrainment rate. To exclude the influence of many other factors, e.g., the complicated slope topography, only the motion down a long uniform slope with a constant inclination angle was studied. Moreover, the entire flow from the leading edge to the tail was not modeled, but only its middle part where the flow is approximately uniform in length. The statement of the problem and examples of simulations results are given in the next sections.

2. Statement of the problem

We consider the flow on an infinitely long uniform slope with a constant angle θ . The fluid is incompressible, with a bulk density $\rho = \rho_0 = const$. The origin of the Cartesian coordinate system is taken on the free surface of the flow, the x axis is parallel to the slope and the z axis is normal to it and downward directed (Figure 1). The flow bottom AB moves down if the entrainment of the bed material occurs.

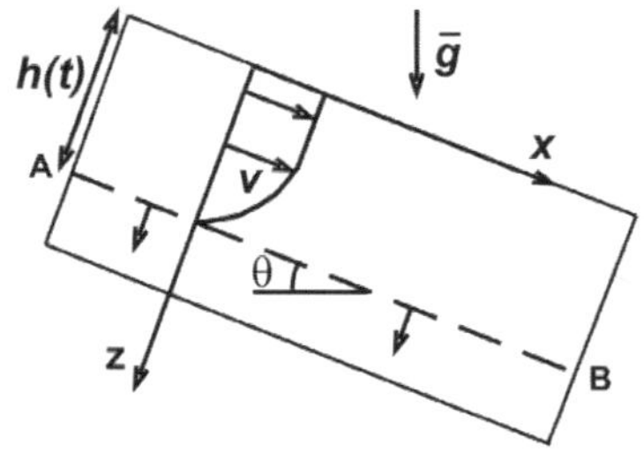

Figure 1. The scheme of the flow.

The Reynolds-averaged flow parameters including the streamwise velocity depend only on z and time t , while the flow depth h varies with time if the flow entrains the bed material. The function $h(t)$ is to be determined. The Reynolds equation reads

$$\frac{\partial\langle \mathrm{v}_x\rangle}{\partial t}=g\sin\theta+\frac{1}{\rho}\frac{\partial \mathrm{T}_{xz}}{\partial z}\ \text{at}\ 0\le z\le h(t). \tag{1}$$

Here, $\mathrm{T}_{xz}=\langle\tau_{xz}\rangle+\rho T$ is the total shear stress, i.e., the sum of the averaged molecular stress $\langle\tau_{xz}\rangle$ and the turbulent stress ρT, $T=-\langle \mathrm{v}'_x\mathrm{v}'_z\rangle$. The primes denote the fluctuated velocity components, while the angular brackets stand for Reynolds averaging. Below we omit the angular brackets in denoting the average values. Boundary condition at the free surface $z=0$ is $T_{xz}=0$; at the bottom $z=h(t)$ the no-slip condition is assumed: $\mathrm{v}_x=0$ at $z=h(t)$. Besides, if the flow entrains the bed material, then, in accordance with the assumed hypothesis on the entrainment mechanism, an additional boundary condition at the bottom is set: $|T_{xz}|=\tau_c$ at $z=h(t)$, where τ_c is the shear strength of the bed material. This additional condition serves to calculate the increasing flow depth.
The full system of equations consists of the Reynolds equation (1), rheological relation which determines the molecular stress τ_{xz} and its average, and equations for turbulent stress and other turbulence characteristics. As it was mentioned in the Introduction, we assumed the Herschel – Bulkley models to approximate rheological relations for the moving media. For a laminar simple shear flow along the x axis ($\mathrm{v}_x=\mathrm{v}_x(z,t),\ \mathrm{v}_y=\mathrm{v}_z=0$) the Herschel – Bulkley equation is

$$\text{if}\ |\tau_{xz}|\le\tau_0\ \text{then}\ \frac{\partial \mathrm{v}_x}{\partial z}=0\ ;\ \text{if}\ |\tau_{xz}|>\tau_0\ \text{then}\ |\tau_{xz}|=\tau_0+K\left|\frac{\partial \mathrm{v}_x}{\partial z}\right|^n . \tag{2}$$

Here τ_0 is the yield limit, K, n are coefficients called the consistency and the power-law index respectively. This equation and its generalisation for complex flows have been proposed to describe the behaviour of different non-Newtonian flows met in nature and in engineering practice. In particular, it corresponds to a Newtonian fluid if $\tau_0=0,\ n=1$; to a power-law fluid if $\tau_0=0,\ n\ne1$, and to a Bingham fluid if $\tau_0\ne0,\ n=1$. Examples of the coefficients values are 1) $\rho=200\mathrm{kg/m^3}$, τ_0=0, n=1, $K/\rho=10^{-3}\mathrm{m^2/s}$ (dry snow avalanche as a Newtonian flow [7]; 2) $\rho=500\mathrm{kg\ /m^3}$, $\tau_0/\rho=2\mathrm{m^2/s^2}$, $n=2$ $K/\rho=89\cdot10^{-6}\mathrm{m^2}$ - dense avalanche [10]; 3) $\tau_0/\rho=1.5\mathrm{m^2/s^2}$, $n=0.33$, $K/\rho=0.1\mathrm{m^2/s^{2\text{-}n}}$ - mudflow, kaolin concentration >10% [11]; 4) $\rho=1130\mathrm{kg/m^3}$, $\tau_0=10\mathrm{Pa}$, $n=0.8$, $K=0.03\mathrm{Pa\cdot s}^n$ - a suspension of mineral particles, concentration 3-5% [13].

IOP Conf. Series: Journal of Physics: Conf. Series **894** (2017) 012111 doi:10.1088/1742-6596/894/1/012111

For turbulent flows the problem arises in averaging the nonlinear rheological relation. We further assume that in a turbulent shear flow the average value of the molecular stress is given by the relation of the same structure as (2). This assumption is the simplest possible hypothesis until detailed experimental data about fluctuations of the viscosity coefficient appear.

The three-parameter turbulence model developed by Luschik, Paveliev and Yakubenko which we referred to as LPY model includes the differential equations for three parameters, namely, T - the turbulent stress divided by the bulk density, $E = 0.5\langle v'_x v'_x + v'_y v'_y + v'_z v'_z \rangle$ - the turbulent energy density, and the parameter ω related to the turbulence scale L by the formula $\omega = E / L^2$. The equations of the LPY model for the flow in consideration are

$$\frac{\partial E}{\partial t} = -(c\sqrt{E}L + c_1\nu)\frac{E}{L^2} + \frac{\partial}{\partial z}\left(D_E \frac{\partial E}{\partial z}\right) + T\frac{\partial \mathrm{v}}{\partial z} ,$$

$$\frac{\partial T}{\partial t} = -(c_5\sqrt{E}L + c_6\nu)\frac{T}{L^2} + \frac{\partial}{\partial z}\left(D_\tau \frac{\partial T}{\partial z}\right) + c_7 E\frac{\partial \mathrm{v}}{\partial z} ,$$

$$\frac{\partial \omega}{\partial t} = -(2c\sqrt{E}L + 1.4c_1\nu f_\omega)\frac{\omega}{L^2} + \frac{\partial}{\partial z}\left(D_\omega \frac{\partial \omega}{\partial z}\right) + \left[\frac{T}{E} + 2c_4 \mathrm{sign}\left(\frac{\partial \mathrm{v}}{\partial z}\right)\right]\omega\frac{\partial \mathrm{v}}{\partial z} ,$$

$$D_\phi = a_\phi\sqrt{E}L + \alpha_\phi\nu, \quad (\phi = E, T, \omega), \quad f_\omega = 1 - \frac{1}{2c_1}\left(\frac{L}{E}\frac{\partial E}{\partial z}\right)^2 .$$

Here, $\mathrm{v} = \mathrm{v}_x$ and $\nu = \mu / \rho$ is the kinematic coefficient of effective molecular viscosity. The values of the dimensionless coefficients are as follows: $c = 0.3$, $c_1 = 5\pi/4$, $c_4 = 0.04$, $c_5 = 3c$, $c_6 = 9c_1$, $c_7 = 0.2$, $a_E = a_\omega = 0.06$, $a_\tau = a_E c_5 / c$, $\alpha_E = \alpha_\tau = 1$, $\alpha_\omega = 1.4$. These values of the coefficients were successfully used without any variations in simulations of different flows, including the cases in which the viscosity coefficient μ was not constant but was a function of the variable temperature [16]. However, the applicability of these values for non-Newtonian flows needs experimental justification. The boundary conditions for T, E, ω were at the bottom $z = h(t)$: $T = 0$, $E = 0$, $\frac{\partial E}{\partial z} = 0$; at the open surface $z = 0$: $T = 0$, $\frac{\partial E}{\partial z} = 0$, $\frac{\partial \omega}{\partial z} = 0$.

3. Results of simulations

The computational code was developed on the base of an implicit finite difference scheme with iterations. Simulations of flows with different rheological properties moving with and without entrainment down slopes with different inclination angles and different strength of the bed material have been done [4]. Below only the results concerning the comparison of Newtonian and Bingham flows behavior are presented. The features of all other investigated flows are similar. The rheological relations for both flows are

$$\text{if } |\tau_{xz}| < \tau_0, \text{ then } \frac{\partial \mathrm{v}_x}{\partial z} = 0 \text{ ; if } |\tau_{xz}| \geq \tau_0, \text{ then } |\tau_{xz}| = \tau_0 + \mu\left|\frac{\partial \mathrm{v}_x}{\partial z}\right| .$$

Here τ_0 is the yield stress (equal to zero for Newtonian fluid), $\mu = \mathrm{const}$ - the viscosity coefficient. One of the typical features of Bingham flows is the appearance of the quasi-solid parts in the flow, where the fluid moves without deformations. Since according to the boundary condition at the upper surface $\tau_{xz} = 0$, there exists a layer $0 \leq z \leq r(t)$ adjacent to the flow surface where $|\tau_{xz}| \leq \tau_0$. This layer moves without deformation as a solid (Figure 2). This layer is referred to a plug layer. The depth $r(t)$ of the plug layer is not known in advance. It can be calculated by the condition $|\tau_{xz}| = \tau_0$ at $z = r(t)$.

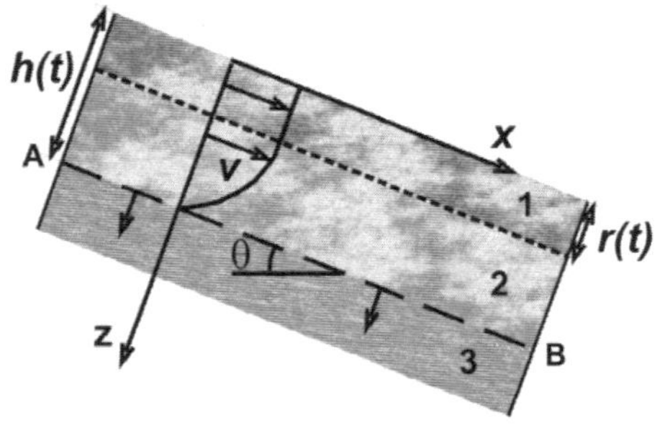

Figure 2. A scheme of a Bingham flow on an inclined plane: 1 – the plug layer, 2 – the shear layer; 3 – the bottom material; AB – the destruction front; $h(t)$ – the flow depth; θ – the slope angle; $r(t)$ – the plug layer depth.

The effect of entrainment of the bed material is demonstrated on Figures 3, 4 for turbulent flows of a linearly viscous fluid. The calculations were performed with the following values of the input parameters: $\tau_0 = 0$, $\tau_c / \rho = 1.5\text{м}^2 / \text{с}^2$, $\mu / \rho \equiv \nu = 0.001\text{м}^2/\text{с}$, $h_0 = 1\text{м}$ where h_0 is the initial flow depth.

Figure 3. Newtonian flow. (a) - velocity profiles 10 seconds after the start of entrainment, (b) - the average velocity vs. time. 1 - motion without mass entrainment, 2 - with entrainment.

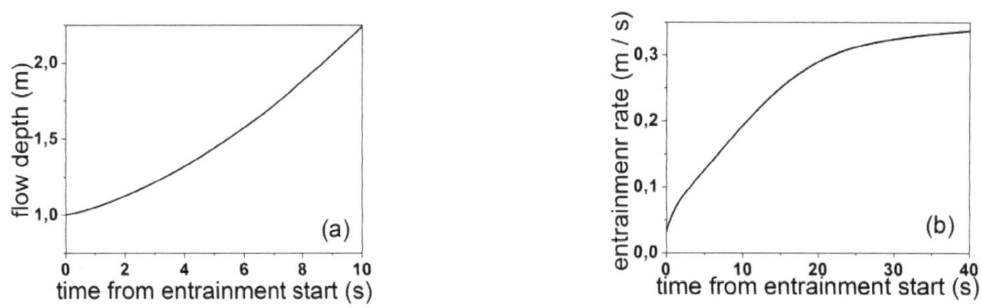

Figure 4. Newtonian flow with mass entrainment. Dependence of (a) - the flow thickness and (b) - the entrainment rate on time.

The graphs in Figures 3 and 4 show that in motion down a long homogeneous slope with mass entrainment, the velocity increases. At large time the dependence of the flow thickness and maximum and average velocities on time tends to be linear, and the entrainment rate tends to a constant (Figure 4(b)). Thus, for large time the entrainment rate does not depend on the velocity and thickness of the flow. Such an asymptotic behaviour is observed if the thickness of the layer accessible for the entrainment is sufficiently large. Otherwise, after the entire available layer is exhausted, the motion tends to become stationary, corresponding to the new value of the thickness.

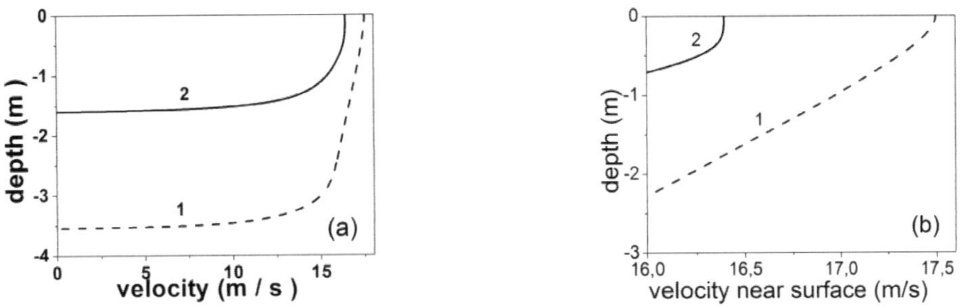

Figure 5. a) Velocity profiles in flows with mass entrainment. b) The shape of the velocity profiles near the upper surface. (1) - Newtonian, (2) – Bingam ($\tau_0 / \rho = 0.05 \text{m}^2 / \text{s}^2$) fluid.

Figure 5(b) demonstrates a plug layer near the upper surface of the Bingham flow.

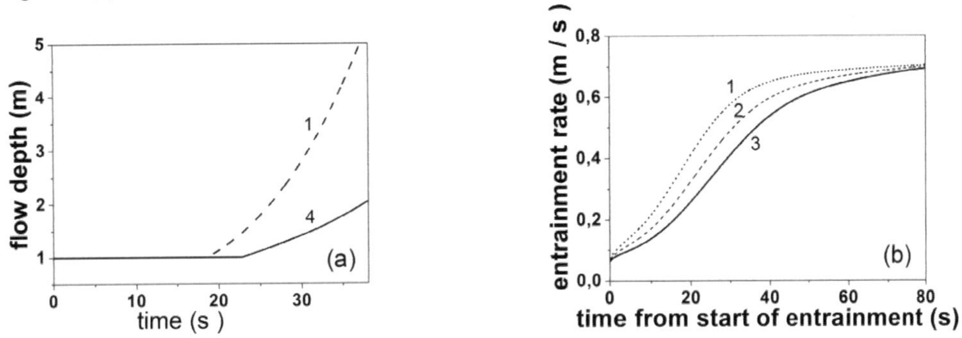

Figure 6. Influence of the yield limit value (a) - on the start of the bed material entrainment, (b) - on the magnitude of the entrainment rate. (1) - (4) - $\tau_0 / \rho = 0,\ 0.01,\ 0.02,\ 0.05 \text{m}^2/\text{s}^2$.

Figure 6 (a) shows that the yield stress reduces the flow velocity, mass entrainment rate, and the entrainment starts later comparing to the linearly viscous flow. However, if the entrainment lasts for a long time the effect of the yield stress on the entrainment rate weakens (Figure 6 (b)).

4. Conclusion

This paper presents models for slope flows based on equations of continuum mechanics that are not averaged over depth. Models take into account the nonlinear rheological properties of moving media, the entrainment of the underlying material, and the turbulent nature of the motion. Numerical investigation of the effect of these three factors is performed for flow at long homogeneous slope. The main conclusions resulting from analysis of the simulations are as follows:

- By entrainment the underlying material, the velocity and thickness of the flow increase.
- When moving along homogeneous slope with entrainment, at large times from the start of the entrainment, regardless of the rheological properties of the flow, the velocity on the surface of the flow, the depth-averaged velocity and the flow depth grow linearly with time until exhausted the entire available mass.
- For all the models studied, the entrainment rate of the bottom material tends with time to a constant, the magnitude of which depends only on the slope angle and the physical properties of the flow and slope materials, but not on the current velocity or depth.

- The presence of the yield point leads to a later start of entrainment, a lower flow velocity and an entrainment rate compared to flow of media that do not have a yield point.

Acknowledgments
This work was supported by the Russian Foundation for Basic Research (Projects 15-01-00361, 15-01-08023).

References

[1] Eglit M E and Yakubenko A E 2014 Numerical modeling of slope flows entraining bottom material *Cold Reg. Sci. Technol.* **108** 139–148
[2] Bovet E, Chiaia B and Preziosi L 2010 A new model for snow avalanche dynamics based on non-Newtonian fluids. *Meccanica* **45** 6 753-765
[3] Zaiko Yu S 2016 Numerical modeling of downslope flows of different rheology *Fluid Dyn.* **51**(4) 443 - 450
[4] Eglit M E and Yakubenko A E 2016 The effect of bed material entrainment and non-Newtonian rheology on dynamics of turbulent slope flows *Fluid Dyn.* **51** (3) 299–310
[5] Sovilla B, Burlando P and Bartelt P 2006 Field experiments and numerical modeling of mass entrainment in snow avalanches *J. Geophys. Res.* **111** F03007
[6] Hungr O, McDougall S and Bovis M 2005 Entrainment of material by debris flows. *Debris-Flow Hazards and Related Phenomena* ed Jakob M and Hungr O (Praxis-Springer, Berlin Heidelberg) 135-158.
[7] Dent J D and Lang T E 1980 Modeling of Snow Flow *J. Glaciology* **26 (94)** 131
[8] Nishimura K and Maeno N 1989 Contribution of viscous forces to avalanche dynamics *Annals of Glaciology* **13** 202
[9] Issler D and Pastor Peréz M 2011 Interplay of entrainment and rheology in snow avalanches; a numerical study *Annals of Glaciology* **52(58)** 143-147
[10] Kern M F, Tiefenbacher J and McElwaine J 2004 The rheology of snow in large chute flows *Cold Reg. Sci. and Techn.* **39** 181–192.
[11] Coussot P 1997 *Mudflow Reology and Dynamics* (Balklema Publ., Rotterdam)
[12] Issler D 2003 Experimental information on snow avalanches. *Dynamic Response of granular and porous materials under large and catastrophic deformations.* Ed Hutter K and Kirchner N (Springer Belin) 109–160
[13] Chhadra R P and Richardson J F 2008 *Non-Newtonian Flow and Applied Rheology: Engineering Applications* (Elsevier Oxford) 518
[14] Naaim M Naaim-Bouvet F Faug T and Bouchet A. 2004 Dense snow avalanche modeling: flow, erosion, deposition and obstacle effects *Cold Reg. Sci. Technol.* **39** 193–204.
[15] Lushchik V G, Paveliev A A and Yakubenko A E 1978 Three-parameter model of shear turbulence *Fluid Dynamics* **13** (3) 350-362
[16] Leontiev A I, Lushchik V G and Yakubenko A E 2009 A heat-insulated permeable wall with suction in a compressible gas flow *Int. J. Heat Mass Transfer* **52** (17–18) 4001

IOP Conf. Series: Journal of Physics: Conf. Series **894** (2017) 012112 doi:10.1088/1742-6596/894/1/012112

Lee wave breaking region: the map of instability development scenarios

S N Yakovenko

Khristianovich Institute of Theoretical and Applied Mechanics, Novosibirsk, Russia

E-mail: yakovenk@itam.nsc.ru

Abstract. Numerical study of a stably stratified flow above the two-dimensional cosine-shaped obstacle has been performed by DNS and LES. These methods were implemented to solve the three-dimensional Navier–Stokes equations in the Boussinesq approximation, together with by the scalar diffusion equation. The results of scanning in the wide ranges of physical parameters (Reynolds and Prandtl/Schmidt numbers relating to laboratory experiment cases and atmospheric or oceanic situations) are presented for instability and turbulence development scenarios in the overturning internal lee waves. The latter is generated by the obstacle in a flow with the constant inflow values of velocity and stable density gradient. Evolution of lee-wave breaking is explored by visualization of velocity and scalar (density) fields, and the analysis of spectra. Based on the numerical simulation results, the power-law dependence on Reynolds number is demonstrated for the wavelength of the most unstable perturbation.

1. Introduction

The numerical investigations of instability and turbulence development mechanisms in the overturning lee waves were performed in [1, 2] where the DNS data for Re = 4000 and Pr = 1 were presented, with the grid resolution sufficient to capture the fine-scale transition features [1] and the subsequent turbulence behavior [2]. In the preceding and present studies, Reynolds and Prandtl/Schmidt number are Re = UH/ν and Pr = ν/κ for a stably stratified flow with the constant values of inflow density gradient and velocity U where the 2D cosine-shaped obstacle of height H was inserted. Such a phenomenon has also been explored in laboratory experiments in water tanks with towed body [3-5]. The secondary density-field instability generated after internal lee wave overturning reveals a range of spanwise modes. The smallest mode has the wavelength $\lambda_y \sim 0.5H$ for Re = 4000 and Pr = 1 [1, 2], relating to perturbations of Rayleigh–Taylor instability (RTI) type, growing massively and resulting in convective mushroom-like structures, with associated Kelvin–Helmholtz instability rolls. At late times of transition to turbulence the smaller-scale vortices transform into the bigger vortex structures, and another dominant mode ($\lambda_y \sim 2.5H$) arises when the large-scale toroidal vortex structures can be visualized [1, 4, 5, 6]. This larger-scale mode can be associated with the most unstable perturbation of the initial laminar 2D vortex pair in the place of lee wave overturning [1, 2]. Visualization of large structures is complicated by the presence of a cascade of finer-scale eddies, leading to the formation of the inertial "–5/3" range and the dissipative range of spanwise spectra with steeper slopes. For Pr >> 1, the earlier and faster growth of RTI with finer-scale structures is seen [7].

The intention of the present paper is to numerically simulate the lee wave overturning region for various Reynolds, Prandtl/Schmidt numbers ($50 \le \mathrm{Re} \le 40\,000$, $1 \le \mathrm{Pr} \le 700$) and to examine the influence of these parameters on the details of instability development and transition to turbulence. For

Content from this work may be used under the terms of the Creative Commons Attribution 3.0 licence. Any further distribution of this work must maintain attribution to the author(s) and the title of the work, journal citation and DOI.
Published under licence by IOP Publishing Ltd

example, a decrease in Re allows us to improve the resolution due to smaller grid-size ratios to the Kolmogorov microscale at lower Re, which in turn gives more accurate statistics in the turbulent patch. Calculations for various Re can reveal a critical value (Re_1) for the onset of instability. On the other hand, for large values of Re > Re_2 and depending on the grid resolution, the DNS method (which transforms here into the implicit LES approach) will become inadequate, and it is necessary to introduce the models of SGS viscosity and scalar diffusion to parameterize small eddies of sub-grid scales (SGS). Variations of Prandtl number value will also help us to found critical levels, in particular, for determining the DNS applicability limit. It makes sense to consider separately the real situations of non-isothermal flows in the atmosphere (where Pr ~ 1) and water reservoirs (Pr ~ 7), as well as saline water in oceanic estuaries (Pr ~ 700 for diffusion of salted water in still water).

2. Numerical model

Numerical simulation of a stratified flow with the obstacle is based on the continuity, Navier–Stokes and scalar (density) equations with the Boussinesq approximation of buoyancy effects. All details of the governing equations non-dimensionalized by inflow velocity U and obstacle height H, and their numerical realization for DNS cases are given in [1, 2].

To overcome the problems of insufficient resolution, SGS models of Smagorinsky type for velocity and scalar equations are used, with the standard value of Smagorinsky constant $C_s = 0.1$, and the same boundary conditions, sponge layers, refined uniform grid as in [1, 2]. The SGS Prandtl/Schmidt number is $\mathrm{Pr}_{sgs} = 0.3$ which is enough to efficiently remove the excessive numerical noise distorting the scalar field plots and spectra both in ILES with $C_s = 0$ and in LES with $C_s = 0.1$ and $\mathrm{Pr}_{sgs} = 1.0$ [7].

3. Computation results

The DNS results at fixed Pr = 1 and lower Re values (Re < 4000) than in [1, 2] demonstrate that, for instance, for Re = 2000 (Figures 1–2) multiple peaks in spanwise spectra (corresponding to the wavelength $0.5H < \lambda_y \leq 5H$ of spanwise periodicity) occur at times $20 < t < 32$, during perturbation growth. The wavelength of the most unstable perturbation corresponding to the RTI mode during the exponential growth stage ($23 < t < 27$) is $\lambda_y \approx 0.63H$ according to Figures 1–2. The large-structure mode $\lambda_y \approx 3.3H$ arises at $t > 40$ during the stage the quasi-steady turbulent patch forming due to lee wave breaking. The larger RTI structures (in comparison with those at Re = 4000 in [1]) are evidently occur due to stronger molecular diffusion effects which also lead to instability development delay, as well as smaller density drop between heavier and lighter fluid layers (Figure 1).

The noted diffusion effects are seen for lower Re cases too. In particular, for Re = 1000 (Figure 3a) perturbation growth takes place later ($25 < t < 40$) and with $\lambda_y \approx 1.1H$ as the most unstable perturbation wavelength during the exponential growth stage ($30 < t < 35$). For Re = 500 the spectra peak with wavelength $\lambda_y \approx 2H$ (corresponding to the late RTI growth at $45 \leq t \leq 50$) dominates almost persistently. For all different Re values studied here (as for the Re = 4000 case tested in [1, 2]), the large-structure enlargement happens with time.

For Re = 200 the spanwise instability magnitude is low and does not lead to the turbulence, having no inertial spectra range (Figure 3b), i.e. the critical value is within $200 < \mathrm{Re}_1 < 500$ at Pr = 1.

In DNS for Pr = 1 and higher Re > 4000 than in [1, 2], the numerical noise becomes evident at large-wavenumber parts of spectra at late times corresponding to turbulent stages where the inertial range ($k^{-5/3}$) should be followed by larger slopes at higher k corresponding to dissipative ranges for both velocity and scalar spectra. In contrast, after the inertial range at $0.5 < k_y < 2.5$ we observe lower spectra slopes at higher wave numbers $2.5 < k_y < 10$ (Figure 3c) which indicates to excessive spurious numerical noise appearing because of poor resolution (or absence of SGS models to suppress such a noise). Therefore, the second critical value at Pr = 1 is within $4\,000 < \mathrm{Re}_2 < 10\,000$, i.e. finer grids or LES tools are needed to predict the small-scale flow features adequately as confirmed by Figure 3(d).

IOP Conf. Series: Journal of Physics: Conf. Series **894** (2017) 012112 doi:10.1088/1742-6596/894/1/012112

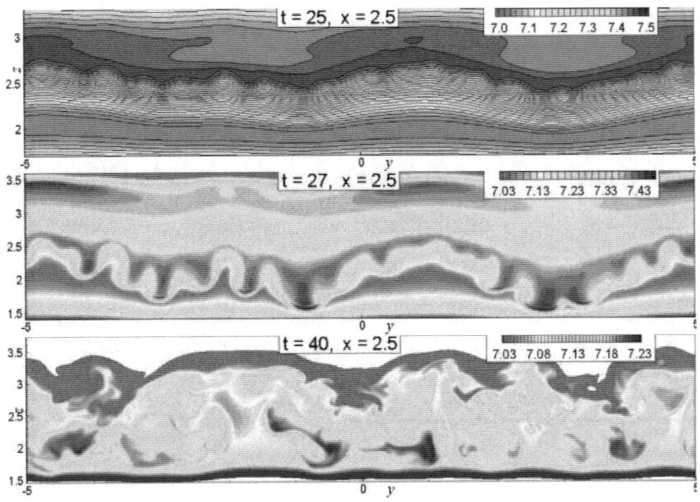

Figure 1. Instantaneous contours of scalar at Re = 2000 and Pr = 1 for $t = 25$, 27, 40 and $x = 2.5$.

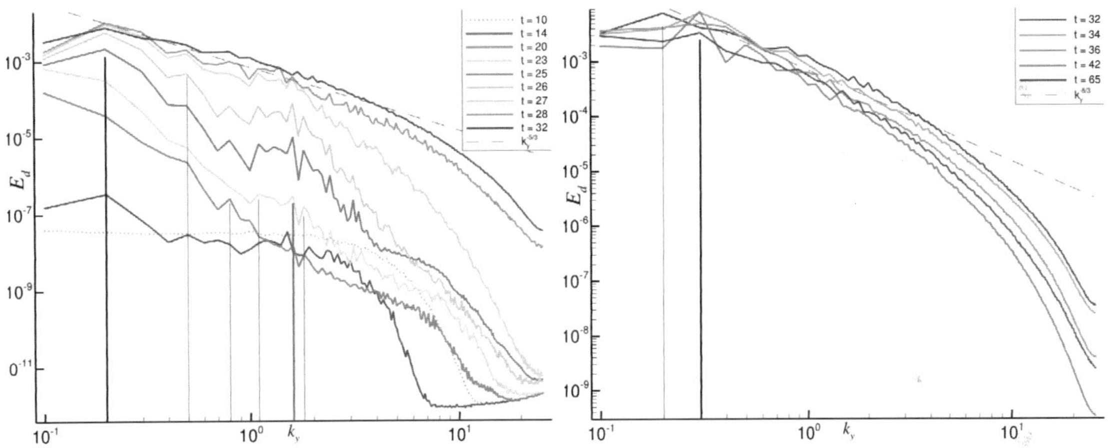

Figure 2. Spanwise spectra of scalar variance at Re = 2000 and Pr = 1 (averaged at $1.25 \le x \le 5.00$, $1.25 \le z \le 3.75$) for white-noise perturbation decay and RTI growth (left), turbulence stage (right).

In contrast to the case of Re = 200 and Pr = 1, runs with the higher Prandtl number Pr = 700 at both Re = 100 and 200 (Figures 4, 5) do show the RTI growth. For Re = 100 one can see existence of well-organised periodic structures (Figure 5) which may exist during long periods almost without change. This may be similar to measurement [4, 5] at Re = 150 where the spanwise periodicity was associated with toroidal vortices. The spanwise spectra (Figure 4a) do not have convincing inertial ranges, so the case of Pr = 700 and Re = 100 can be considered to have significant instability, but 'undeveloped' turbulence. In LES at Re = 200 (Figure 4b) the scalar spectra at the turbulent stage ($t > 60$) reveal both the inertial range ($k^{-5/3}$) and the dissipative one with higher slopes.

For intermediate value Pr = 7, the instability development at Re = 200 is still seen due to the RTI growth (which occurs quite late), in contrast to that at Pr = 1, with unsteady and intermittent in time of the resulting turbulent state (Figure 6). At Pr = 7 we found $100 < Re_1 < 200$, and $2000 < Re_2 < 4000$.

Note, the interesting features of spanwise periodicity at Pr = 7 and Re = 100 where scalar contours illustrate steady triangular cells surviving for long periods (Figure 7), and spanwise spectra do show weak instability without turbulence (Figure 6a). The similar cells appear for Re = 200, but during short periods at $t \sim 70$. Such a feature may demonstrate another scenario of instability growth (rather than RTI with mushroom-like structures), for instance the Rayleigh–Benard convection.

IOP Conf. Series: Journal of Physics: Conf. Series **894** (2017) 012112 doi:10.1088/1742-6596/894/1/012112

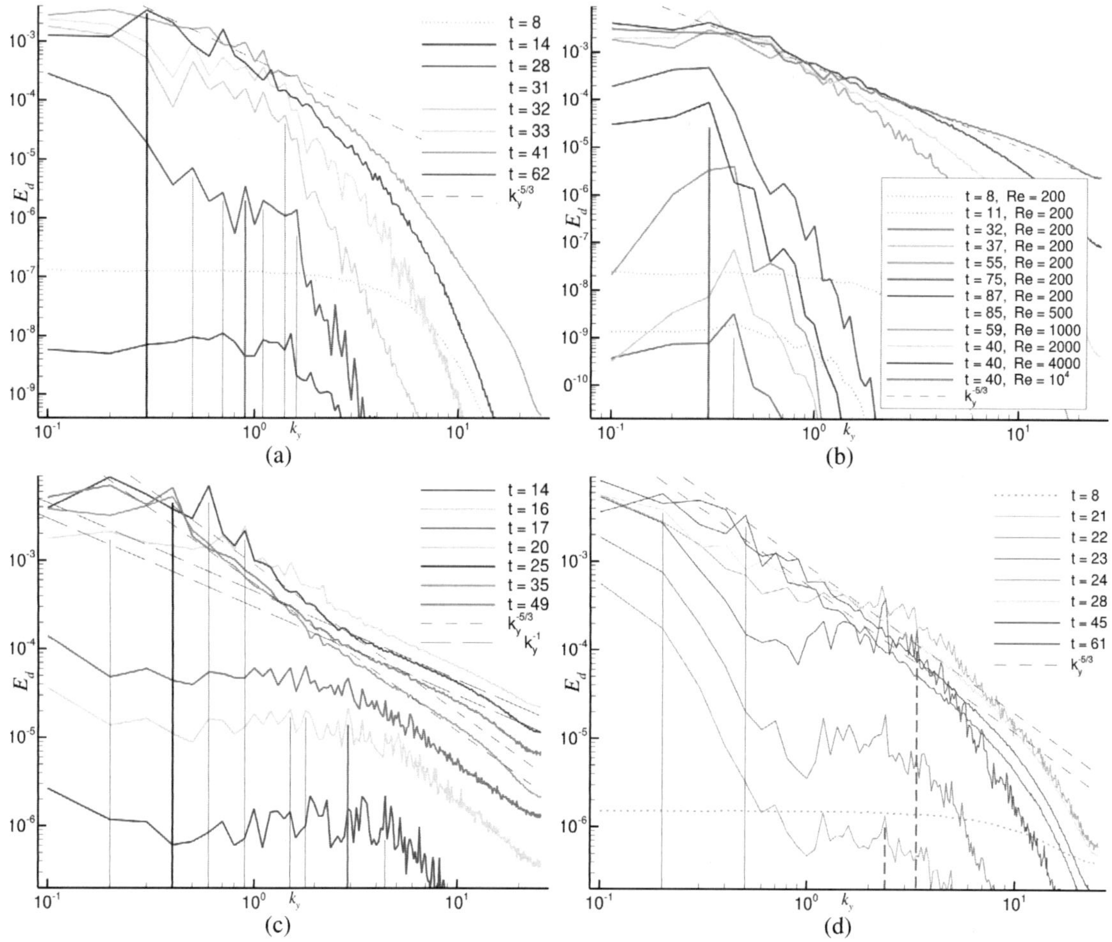

Figure 3. Spanwise spectra (as in Figure 2) for Pr = 1 in DNS at Re = 1000 (a), DNS at Re = 200 (compared to "turbulent stage" curves for other Re) (b), ILES at Re = 10^4 (c), LES at Re = 10^4 (d).

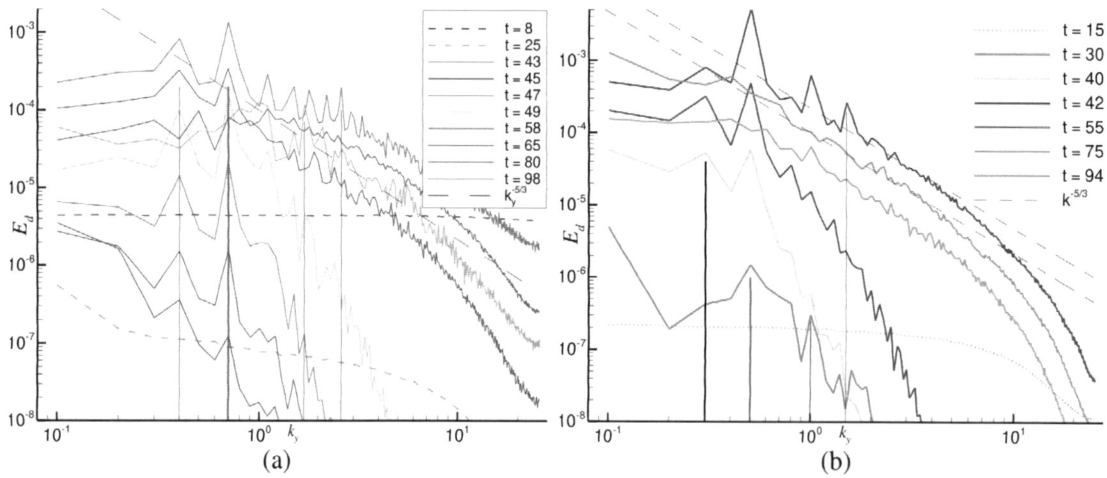

Figure 4. Spanwise spectra (as in Figures 2, 3) for Pr = 700: (a) ILES, Re = 100, (b) LES at Re = 200.

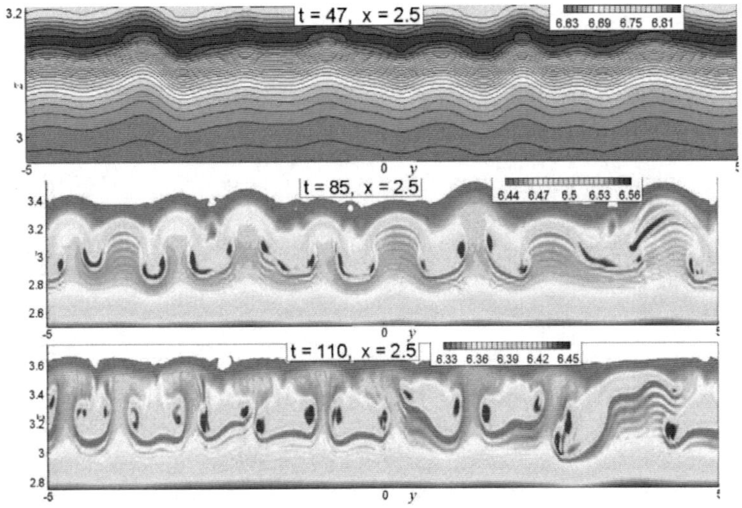

Figure 5. Instantaneous contours of scalar at Pr = 700 and Re = 100 for $t =$ 47, 85, 110 and $x = 2.5$.

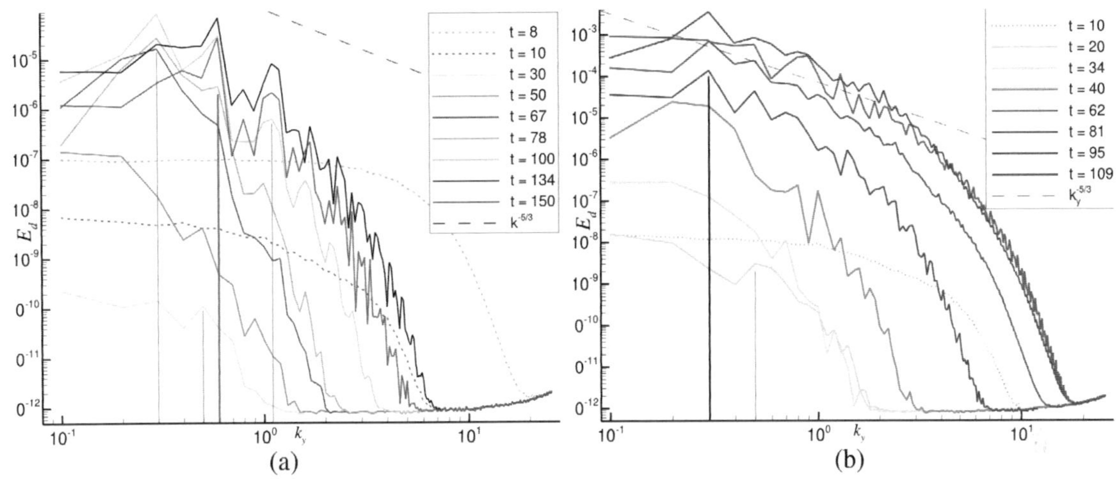

Figure 6. The same spectra as in Fig. 4, 5 for Pr = 7 in DNS at Re = 100 (a), Re = 200 (b).

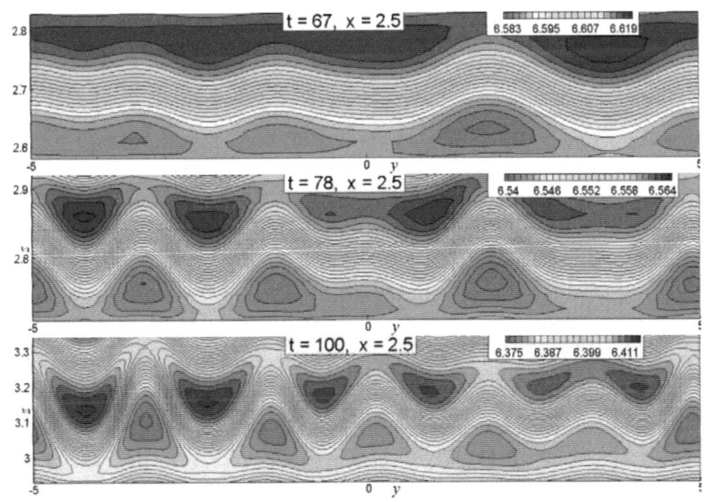

Figure 7. Scalar contours in DNS at Re = 100 and Pr = 7 for $t =$ 67, 78, 100 and $x = 2.5$.

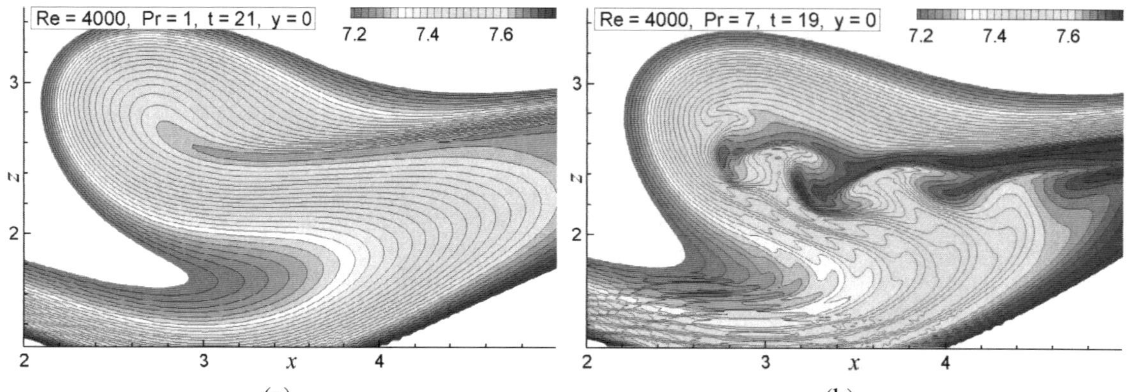

Figure 8. Scalar contours at $y = 0$ in DNS: (a) Re = 4000, Pr = 1, $t = 21$; (b) Re = 4000, Pr = 7, $t = 19$.

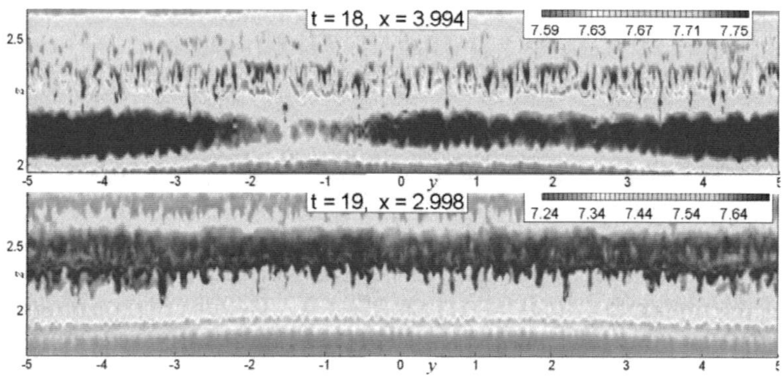

Figure 9. Contours of scalar in spanwise sections for different times in LES at Re = 40 000, Pr = 700.

It should be noted that at for Re $\geq$ 4000 and Pr $\geq$ 7 the RTI periodicity is seen not only along the span, but also in the streamwise direction as shown by visualization of scalar contours at $y = 0$ (Figure 8). The similar effects are observed for Re $\geq 10^4$ and Pr = 1. Although, in both cases we have the under-resolved density field resulting in excessive numerical noise, this evidence may indicate to another route of instability growth than that discussed in [1] with quasi-two-dimensional structures.

For high Re cases one can also see the multiple rows of the small-scale perturbations of RTI type (Figure 9) instead of the single row for lower Re cases (Figures 1, 5, 7). The similar picture is given in [8] where several unstable layers in overturning wave region have been observed, with the convective elements appearing and growing in the spanwise sections, creating a multilayered structure of vortices.

The LES results at high Re and Pr show that in all cases a SGS model (used in [7]) efficiently suppresses numerical noise in spectra and scalar contours, allowing us to view very fine convective structures along the span (Figure 9). Although, the SGS models lead to slight delay in the wave breaking development, working in fact together with molecular viscosity and diffusion.

From LES with higher Re, one can also see that the results become almost independent on Pr or Re. This means that the SGS diffusion coefficients in fact are much larger than the molecular diffusion ones. Note, at finer grids (when SGS viscosity and diffusion coefficients will be smaller), the Pr and Re dependence will be more visible. Nevertheless, at very high Re levels corresponding to geophysical flows, it is natural to omit the molecular viscosity/diffusion coefficients as done e.g. in [8].

Finally, the results for instability development scenarios can be summarized in Figure 10 where the most unstable perturbation wavelengths deduced from density snapshot visualizations and spectra peak analysis are given by symbols (each relates to one computation at different Re or/and Pr). One can see (red dashed line in Figure 10) that the analytical expression [9]

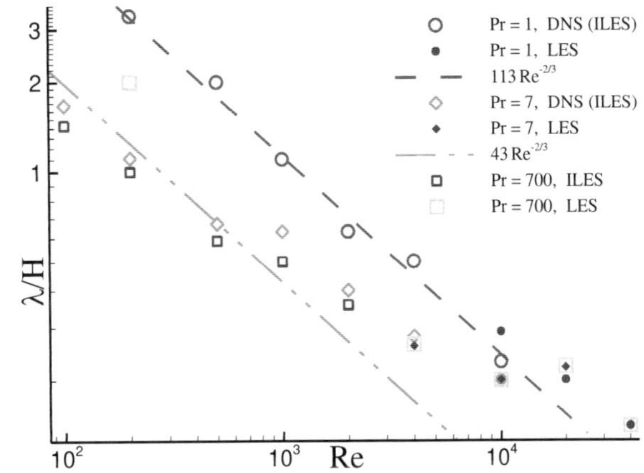

Figure 10. The map for the wavelength of the most unstable perturbation versus Re and Pr.

$$\lambda = \lambda' = 4\pi H[(2\Delta\rho_H/\Delta\rho)(\mathrm{F}_H)^2\mathrm{Re}^{-2}]^{1/3} \quad (1)$$

written in dimensionless form [1] at Pr = 1 with the typical density drop, $\Delta\rho \approx 10^{-3}\Delta\rho_H$, between the layers of heavier and lighter fluids related to earlier RTI stages is satisfied for dependence on Re.

Note, results at Pr = 700 are affected by the density field under-resolution, even at Re = 200. As the result, the numerical wiggles can be seen in growing mushroom-like structures (Figures 1, 5, 8), and the numerical dissipation effect is a possible reason of obvious over-prediction of the most unstable perturbation wavelength (Figure 10), as well as the SGS model effect in the case of LES computations.

4. Conclusions

The results for scenarios of instability development in overturning lee waves have been presented for $50 \leq \mathrm{Re} \leq 40\,000$ and $1 \leq \mathrm{Pr} \leq 700$. The range of Pr relates to density stratification in atmospheric and oceanic flows, whereas the range of Re corresponds to that in measurements [3-5] and concurrent simulations [6, 10]. The typical phenomena in nature have much higher levels of Re. However, the preceding and present studies replicate well the cases in natural conditions, except for the instability development stage which can reveal various routes with different structures at different Re and Pr. The scenarios viewed here allow us to extrapolate results to geophysical cases with much higher Re values.

Acknowledgments
The author thanks Prof I.P. Castro and Dr T.G. Thomas for help in studies of breaking internal waves. The work was performed with use of the Iridis computing resources at the University of Southampton. The research was partially supported by RFBR (project No. 17-01-0332a).

References

[1] Yakovenko S N, Thomas T G and Castro I P 2014 *J. Fluid Mech.* **760** 466.
[2] Yakovenko S N, Thomas T G and Castro I P 2011 *J. Fluid Mech.* **677** 103.
[3] Castro I P and Snyder W H 1993 *J. Fluid Mech.* **255** 195
[4] Eiff O F and Bonneton P 2000 *Phys. Fluids* **12** 1073
[5] Eiff O, Huteau F and Tolu J 2005 *Dyn. Atmos. Oceans* **40** 71
[6] Gheusi F, Stein J and Eiff O F 2000 *J. Fluid Mech.* **410** 67.
[7] Yakovenko S N, Thomas T G and Castro I P 2014 *Progress in Turbulence V: Proceedings of the iTi Conference in Turbulence 2012* ed A Talamelli et al. *Springer Proceedings in Physics* **149** (Springer) 233
[8] Afanasyev Ya D and Peltier W R 1998 *J. Atmos. Sci.* **55** 19
[9] Youngs D L 1984 *Physica* D **12** 32
[10] Paisley M F and Castro I P 1996 *Dyn. Atmos. Oceans* **23** 309

IOP Conf. Series: Journal of Physics: Conf. Series **894** (2017) 012113 doi:10.1088/1742-6596/894/1/012113

Numerical modeling of compressible fluid flow through elastic porous medium

A A Yanchenko[1,3], E Romenski[2,3] and A K Khe[1,3]
[1] Lavrentyev Institute of Hydrodynamics, Novosibirsk, Russia
[2] Sobolev Institute of Mathematics, Novosibirsk, Russia
[3] Novosibirsk State University, Novosibirsk, Russia

E-mail: arseny@protonmail.com

Abstract. A numerical technique for solving an initial-boundary value problem of the single compressible fluid flow through elastic porous medium is presented. The model of compressible fluid flow in elastic porous medium is derived within the theory of hyperbolic thermodynamically compatible systems of conservation laws that provides a formulation of mathematically and physically well-posed governing equations of physical processes. Numerical method is based on the WENO–Runge–Kutta algorithm. For the boundary flux evaluation the characteristic decomposition and linearized boundary conditions are used. A two-dimensional problem of the fluid inflow into the saturated porous rectangular domain from its left boundary is studied numerically.

1. Introduction

Mathematical and numerical modeling of a fluid flow in a porous medium is of interest in many scientific areas and industrial applications. Despite of many research efforts there is still no a conventional approach allowing one to formulate a continuum model of multiphase flow in deformable porous media. For slow flows in the filtration regime there is a well established Darcy approach which gives the resulting governing equations of parabolic type for many kinds of filtration flows. This approach is not applicable if one need to model processes with finite speed of the fluid flow or flow in porous deformable elasto-plastic medium. There are some studies in the development of such models which are more general than the Darcy models, see for example [1], but they are still far from being generally accepted. A present research continues the development of mathematical and computational model for a compressible fluid flow in a porous medium [2, 3, 4, 5]. The model is derived in the framework of the theory of hyperbolic thermodynamically compatible systems of conservation laws [6] that gives a way to formulate mathematically and physically well-posed governing equations of physical processes in different areas of continuum mechanics (see, for example [7] and references therein). In this paper we apply the WENO–Runge–Kutta numerical method for the solution of an initial-boundary value problem of the single compressible fluid flow through elastic porous medium and present some numerical results showing the applicability of the model to real problems.

2. Formulation of the model

The governing differential equations for modeling a fluid flow through elastic porous medium are derived within the theory of thermodynamically compatible systems of conservation laws [6].

Content from this work may be used under the terms of the Creative Commons Attribution 3.0 licence. Any further distribution of this work must maintain attribution to the author(s) and the title of the work, journal citation and DOI.
Published under licence by IOP Publishing Ltd

IOP Conf. Series: Journal of Physics: Conf. Series **894** (2017) 012113 doi:10.1088/1742-6596/894/1/012113

These equations can be found in [2]. Formulation of the governing system consists of several key steps:

(i) Introduce a set of physical variables characterizing the medium.

(ii) Define a set of thermodynamically compatible conservation laws written in terms of generating potential and variables, which are used to construct the system of governing equations.

(iii) Establish a relationship between the generating potential and variables with the physical variables and thermodynamic potential (internal energy).

(iv) Introduce source terms responsible for the dissipation and phase interaction in the system of governing equations.

In accordance with (i), we introduce a set of physical variables which characterize the elastic porous medium saturated by the compressible fluid. The fluid flow in the porous medium can be considered as the mixture of two phases, in which each phase (fluid and elastic porous) can be characterized by its own parameters of state. Note that we consider processes in which the temperature variations are negligibly small and consider the isentropic approximation of the model. We characterize the element of the mixture with the following state parameters:

α_1 — volume fraction of the elastic medium;
ρ — mass density of the mixture;
c_1 — mass fraction of the elastic medium;
u^i — velocity of the mixture;
w^i — relative velocity of motion of the elastic porous medium with respect to the fluid;
F_{ij} — elastic deformation gradient of the mixture.

Volume and mass fractions of the liquid can be calculated as $\alpha_2 = 1 - \alpha_1$, $c_2 = 1 - c_1$, respectively. The parameters characterizing the state of the mixture are related to the phases' state parameters by the relations:

$$\rho = \alpha_1\rho_1 + \alpha_2\rho_2, \quad c_1 = \alpha_1\rho_1/\rho, \quad c_2 = \alpha_2\rho_2/\rho, \quad u^i = c_1u_1^i + c_2u_2^i, \quad w^i = u_1^i - u_2^i,$$

where ρ_1, ρ_2 — mass densities of the elastic porous medium and liquid; u_1^i, u_2^i — their velocities.

The governing equations of the fluid flow in porous medium presented in [2] are formulated first in terms of the generating potential and variables in a conservative form and are hyperbolic if the generating potential is a convex function of generating variables. It turns out that for solution of specific problems it is more convenient to use individual phase parameters and the so-called generalized internal energy E, which is a function of α_1, ρ, c_1, w^k, F_{ij}. We call E the equation of state of the mixture and define it as a mass average of the phase internal energies $e_1(\rho, F), e_2(\rho)$, supplemented with the kinematic energy of the relative motion

$$E(\alpha_1, \rho, c_1, w^k, F) = c_1e_1(\rho_1, F) + c_2e_2(\rho_2) + \frac{1}{2}c_1c_2w^iw^i.$$

Thus, the thermodynamically compatible system of conservation laws for the fluid flow through elastic porous medium presented in [2] reads as

$$\begin{gathered}
\frac{\partial\rho\alpha_1}{\partial t} + \frac{\partial\rho\alpha_1u^k}{\partial x_k} = \lambda(p_1 - p_2), \quad \frac{\partial\rho}{\partial t} + \frac{\rho u^k}{\partial x_k} = 0, \quad \frac{\partial\rho_1\alpha_1}{\partial t} + \frac{\partial\rho_1\alpha_1u_1^k}{\partial x_k} = 0,\\
\frac{\partial(u_1^k - u_2^k)}{\partial t} + \frac{\partial(u_1^iu_1^i/2 - u_2^iu_2^i/2 + e_1 + p_1/\rho_1 - e_2 - p_2/\rho_2)}{\partial x_k} = e_{klj}u^l\omega_j - \chi E_{w^k},\\
\frac{\partial(\alpha_1\rho_1u_1^i + \alpha_2\rho_2u_2^i)}{\partial t} + \frac{\partial(\alpha_1\rho_1u_1^iu_1^k + \alpha_2\rho_2u_2^iu_2^k + (\alpha_1p_1 + \alpha_2p_2)\delta_{ik} - \alpha_1\tilde{\sigma}_{ik})}{\partial x_k} = 0,\\
\frac{\partial\rho F_{ij}}{\partial t} + \frac{\partial(\rho F_{ij}u^k - \rho F_{kj}u^i)}{\partial x_k} = 0,
\end{gathered} \tag{1}$$

where $p_1 = (\rho_1)^2 \partial e_1/\partial \rho_1$, $p_2 = (\rho_2)^2 \partial e_2/\partial \rho_2$ are pressures of the elastic porous medium and fluid, respectively, $\tilde{\sigma}_{ik}$ are shear stresses of the mixture; χ is an interfacial friction coefficient, λ is a coefficient characterizing the rate of phase pressures relaxation to the common value; e_{klj} is the Levi–Civita symbol, δ_{ik} is the Kronecker symbol; ω_j is the artificial variable introduced to write the equation for the relative velocity in a divergence form

$$\frac{\partial w^k}{\partial x_i} - \frac{\partial w^i}{\partial x_k} = -e_{ki\alpha}\omega_\alpha.$$

Now we choose a particular form of the phase internal energies. The internal energy for the fluid is defined by the stiffened gas equation of state (isentropic case)

$$e_2(\rho_2) = \frac{(C_2)^2}{\gamma_2(\gamma_2-1)}\left(\frac{\rho_2}{\rho_{20}}\right)^{\gamma_2-1} + \frac{(C_2)^2\rho_{20} - \gamma_2 p_{20}}{\gamma_2\rho_2},$$

where C_2 is the sound speed, ρ_{20} is the reference density, p_{20} is the reference pressure, and γ_2 is the polytropic coefficient.

The internal energy for the porous medium is defined in the following way:

$$e_1(\rho_1, F) = e_l(\rho_2) + e_s(F),$$

where

$$e_l(\rho_1) = \frac{(C_1)^2}{\gamma_1(\gamma_1-1)}\left(\frac{\rho_1}{\rho_{10}}\right)^{\gamma_1-1} + \frac{(C_1)^2\rho_{10} - \gamma_1 p_{10}}{\gamma_1\rho_1},$$

and C_1, ρ_{10}, p_{10}, γ_1 have the same meaning as for the internal energy e_2 of the fluid. $e_s(F)$ is defined according to [8]

$$e_s(F) = \frac{\mu}{8\rho_{10}}(J_2 - 1),$$

where μ is the Lame coefficient and

$$J_2 = \mathrm{tr}(g^2), \quad g = \frac{G}{(\det G)^{1/3}}, \quad G = B^{-1}, \quad B = FF^{\mathrm{T}}.$$

Thus, using the above phase internal energies we obtain the thermodynamically compatible system of conservation laws which describes compressible fluid flow through the elastic porous medium.

3. Numerical method

We apply the presented model to the solution of two-dimensional test problems for the purpose of verification of the model. Since the governing equations form a hyperbolic system of conservation laws, the standard WENO–Runge–Kutta numerical method [9] can be directly applied for solving two-dimensional version of equations. The 2D version of system (1) consists of 11 equations, which formally can be written as

$$\frac{\partial \mathbf{U}}{\partial t} + \frac{\partial \mathbf{F}(\mathbf{U})}{\partial x_1} + \frac{\partial \mathbf{G}(\mathbf{U})}{\partial x_2} = \mathbf{Q}(\mathbf{U}), \tag{2}$$

where $\mathbf{U}$ is a vector of conservative variables, $\mathbf{F}(\mathbf{U}), \mathbf{G}(\mathbf{U})$ are fluxes and $\mathbf{Q}(\mathbf{U})$ is a vector of source terms. Application of the WENO–Runge–Kutta method of the third order in space and fifth order in space for the solution of the initial value problem for the presented model can be found in [3]. Below we apply the same method for the initial-boundary value problem. For the flux evaluation at intercells we use the Lax–Wendroff flux whereas for the boundary flux evaluation we use characteristic decomposition and linearized boundary conditions. Note that here we use the instantaneous pressure relaxation assuming the characteristic scale of pore space is small. The numerical treatment of the instantaneous pressure relaxation consists in correction of the phase volume fraction α_1 and densities ρ_1, ρ_2 by solution of algebraic equation $p_1 = p_2$ in the end of each time step at each mesh cell.

3.1. Boundary value problem

Let us consider a two-dimensional problem of the fluid inflow into the saturated porous rectangular domain from its left boundary. To construct necessary left boundary condition we rewrite system of equations (2) in quasilinear form (in terms of physical variables)

$$A_0(\mathbf{V})\frac{\partial \mathbf{V}}{\partial t} + A_1(\mathbf{V})\frac{\partial \mathbf{V}}{\partial x_1} + A_2(\mathbf{V})\frac{\partial \mathbf{V}}{\partial x_2} = \mathbf{P}(\mathbf{V}), \qquad (3)$$

where $\mathbf{V}$ is a vector of physical variables. Then we suppose that solution on the previous step ($\mathbf{V}^n$) is known. For the characteristic decomposition applied to the boundary conditions treatment we linearize system (3) about the constant value of $\mathbf{V}^n$ (the solution is sought in the following form $\mathbf{V} = \mathbf{V}^n + \mathbf{v}$). We also neglect the source terms and arrive at the following system of linear equations at the left boundary:

$$\widetilde{A}_0(\mathbf{V}^n)\frac{\partial \mathbf{v}}{\partial t} + \widetilde{A}_1(\mathbf{V}^n)\frac{\partial \mathbf{v}}{\partial x_1} = 0.$$

The latter can be rewritten as

$$\frac{\partial \mathbf{v}}{\partial t} + \hat{A}_1(\mathbf{V}^n)\frac{\partial \mathbf{v}}{\partial x_1} = 0, \qquad (4)$$

where $\hat{A}_1(\mathbf{V}^n) = \widetilde{A}_0^{-1}(\mathbf{V}^n)\widetilde{A}_1(\mathbf{V}^n)$. Assuming that initially both phases are at rest and the medium moves from the left to the right, we have 3 incoming and 8 outgoing characteristics. Three incoming characteristics correspond to two longitudinal velocities (C_1, C_2) and one transverse velocity (C_s). Among the eight outgoing characteristics there are three of them corresponding to C_1, C_2, C_s (sound characteristics) and five contact characteristics (equal to the velocity of the mixture u_1 in x_1 direction). Thus there are three Riemann invariants that are transported from border cell to the left boundary: r^1, r^2, r^3. To set the left boundary condition correctly we need 8 more conditions. We set them in the following manner: define phases velocities and set shear stress in x_2-direction equal to zero (this condition should be linearized)

$$\begin{aligned} u_1^1 &= u_{10}^1, \\ u_2^1 &= u_{20}^1, \\ \Sigma_{12} &= \tilde{\sigma}_{12} = 0, \end{aligned} \qquad (5)$$

For setting the rest values the structure of R^{-1} must be considered, where R is the matrix of right eigenvectors of $\hat{A}_1(\mathbf{V}^n)$. We can only set the conditions that are lineary independent of previously set ones ($r^1(\mathbf{v}) = r^1(\mathbf{V}^n)$, $r^2(\mathbf{v}) = r^2(\mathbf{V}^n)$, $r^3(\mathbf{v}) = r^3(\mathbf{V}^n)$ and (5)). Those can be the following conditions

$$\alpha_1^{n+1} = \alpha_1^n, \quad F_{11}^{n+1} = F_{11}^n, \quad F_{12}^{n+1} = F_{12}^n, \quad F_{22}^{n+1} = F_{22}^n, \quad (u_1^2)^{n+1} = (u_1^2)^n.$$

We suppose that these values varies so slowly that they can be taken from the previous time step.

Thus we have non-degenerate system of linear algebraic equations for boundary values of the parameters of state. The solution of this system gives vector of physical variables $\mathbf{V}^{n+1}$ on the left boundary on the new time step.

On the right boundary non-reflecting conditions are set, on the top and bottom boundaries we set periodical boundary conditions.

4. Examples

In this section we consider a two-dimensional problem of the fluid inflow into the saturated porous rectangular 3 cm long domain from its left boundary. We take sandstone-like material as the elastic porous medium with material parameters

$$C_1 = 3800 \text{ m/s}, \quad \rho_{10} = 2650 \text{ kg/m}^3, \quad p_{20} = 10^5 \text{ Pa}, \quad \gamma_1 = 2.8, \quad \mu = 21.3 \text{ GPa}.$$

Fluid is a water with constants

$$C_2 = 1500 \text{ m/s}, \quad \rho_{20} = 1000 \text{ kg/m}^3, \quad p_{20} = 10^5 \text{ Pa}, \quad \gamma_2 = 2.4.$$

Constant χ is equal to 10^5 s^{-1}. Initially both phases are at rest:

$$\begin{gathered} p_1 = p_2 = 10^5 \text{ Pa}, \\ \rho_1 = 2650 \text{ kg/m}^3, \quad \rho_2 = 1000 \text{ kg/m}^3, \\ \alpha_1 = 0.5, \\ u_1^1 = u_2^1 = u_1^2 = u_2^2 = 0, \\ F_{11} = F_{22} = 1, \quad F_{12} = F_{21} = 0. \end{gathered}$$

We consider three different test problems with the following left boundary conditions:

- (Test 1) Liquid velocity is equal to elastic porous medium velocity: $u_1^1 = u_2^1 = 1$ cm/s.
- (Test 2) Liquid flows into fixed porous medium: $u_1^1 = 0$, $u_2^1 = 1$ cm/s.
- (Test 3) Liquid and elastic porous medium flow with different velocities: $u_1^1 = 1$ cm/s, $u_2^1 = 2$ cm/s.

Below all the data are displayed at four time moments: $t_1 = 0.1776315 \cdot 10^{-5}$ s, $t_2 = 0.4144735 \cdot 10^{-5}$ s, $t_3 = 0.6513155 \cdot 10^{-5}$ s, $t_4 = 0.8881575 \cdot 10^{-5}$ s. In the figures the pressure and stress are measured in bars (1 bar = 10^5 Pa) whereas the velocities are measured in centimetres per second.

One can see the correct propagating wave structure for all test problems. The difference in left boundary velocities results in the difference in wave fields which depends also on the interfacial friction coefficient. The results of computations look physically reasonable. The model and numerical method require further verification and comparison with available experiments

IOP Conf. Series: Journal of Physics: Conf. Series **894** (2017) 012113 doi:10.1088/1742-6596/894/1/012113

4.1. Test 1

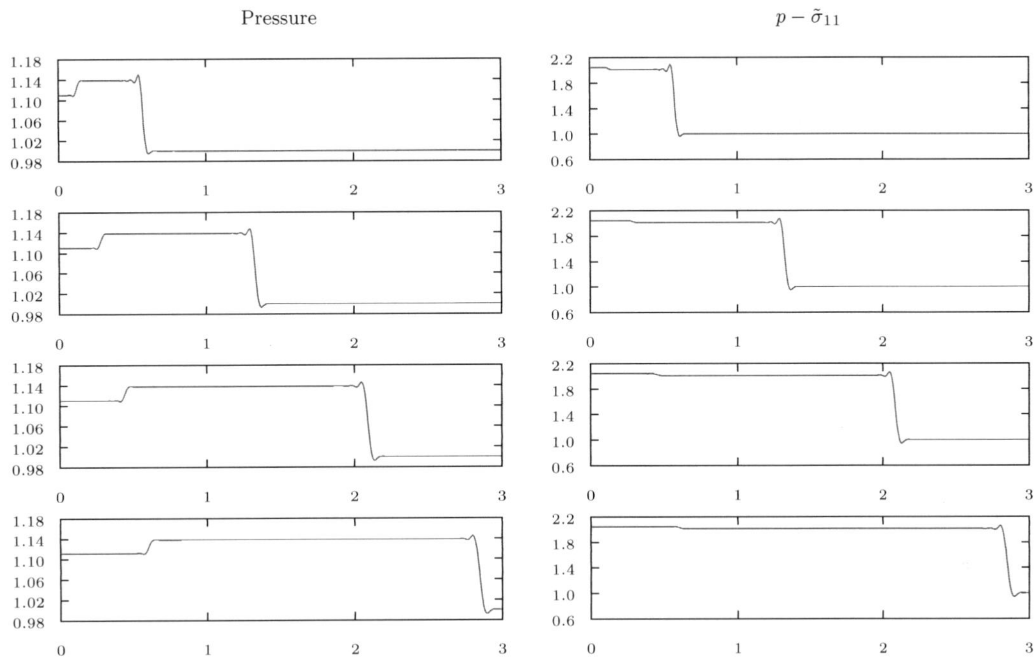

Figure 1. Test 1. Pressure and stress in the skeleton ($p - \tilde{\sigma}_{11}$).

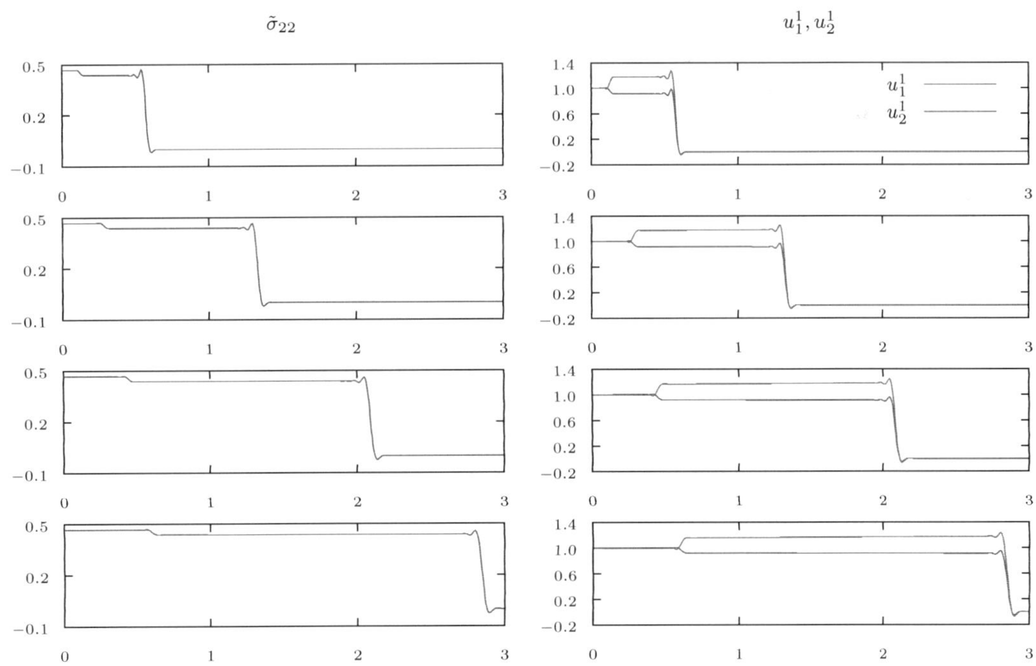

Figure 2. Test 1. Shear stress in x_2-direction and velocities in x_1-direction.

IOP Conf. Series: Journal of Physics: Conf. Series **894** (2017) 012113 doi:10.1088/1742-6596/894/1/012113

4.2. Test 2

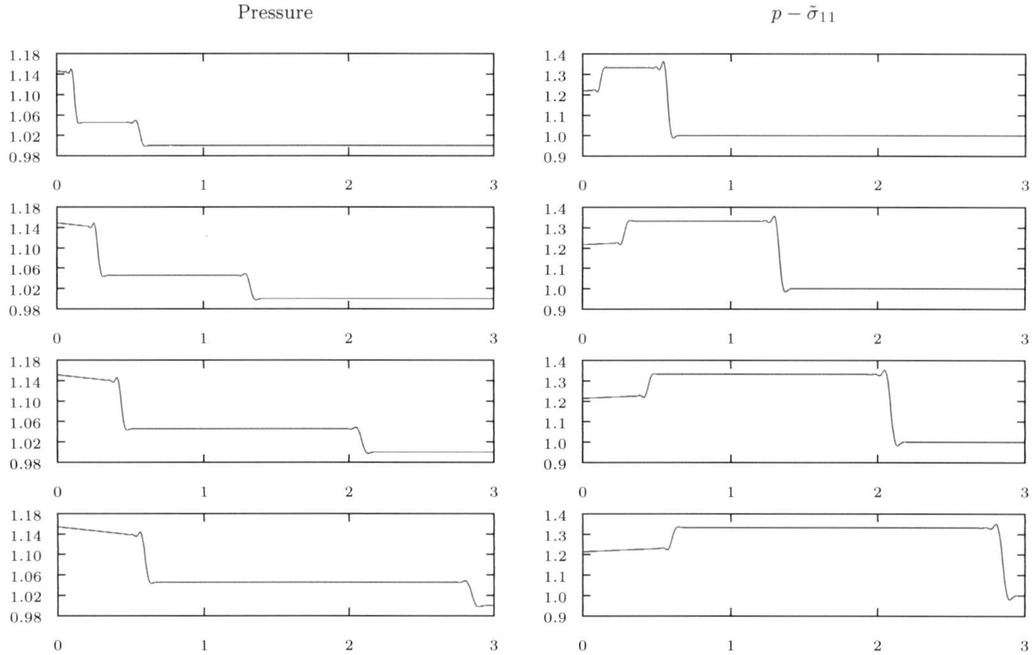

Figure 3. Test 2. Pressure and stress in the skeleton ($p - \tilde{\sigma}_{11}$).

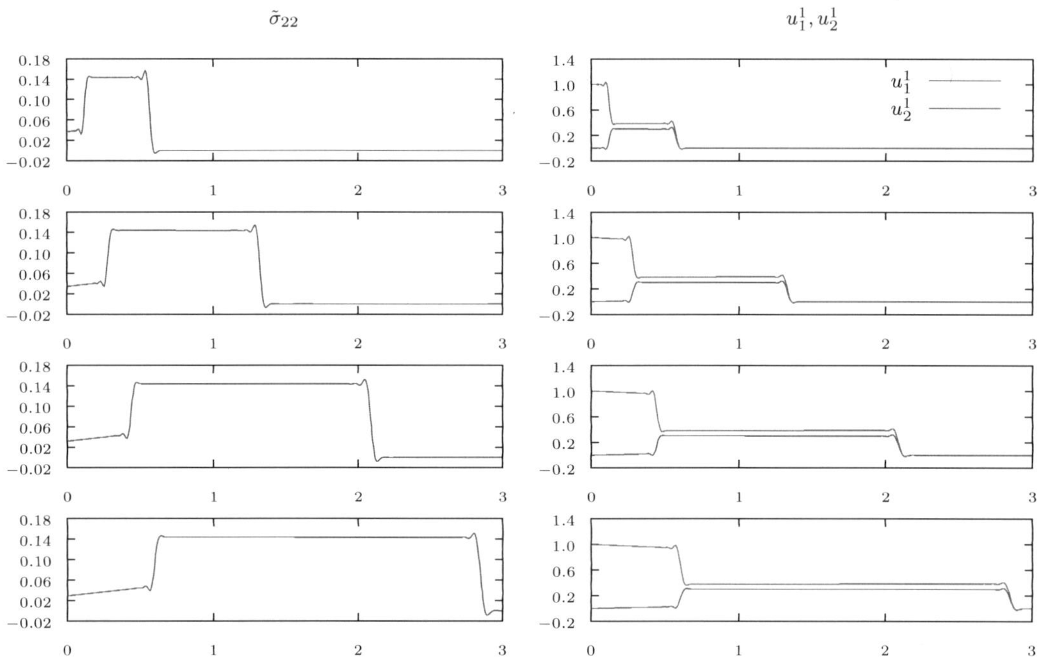

Figure 4. Test 2. Shear stress in x_2-direction and velocities in x_1-direction.

4.3. Test 3

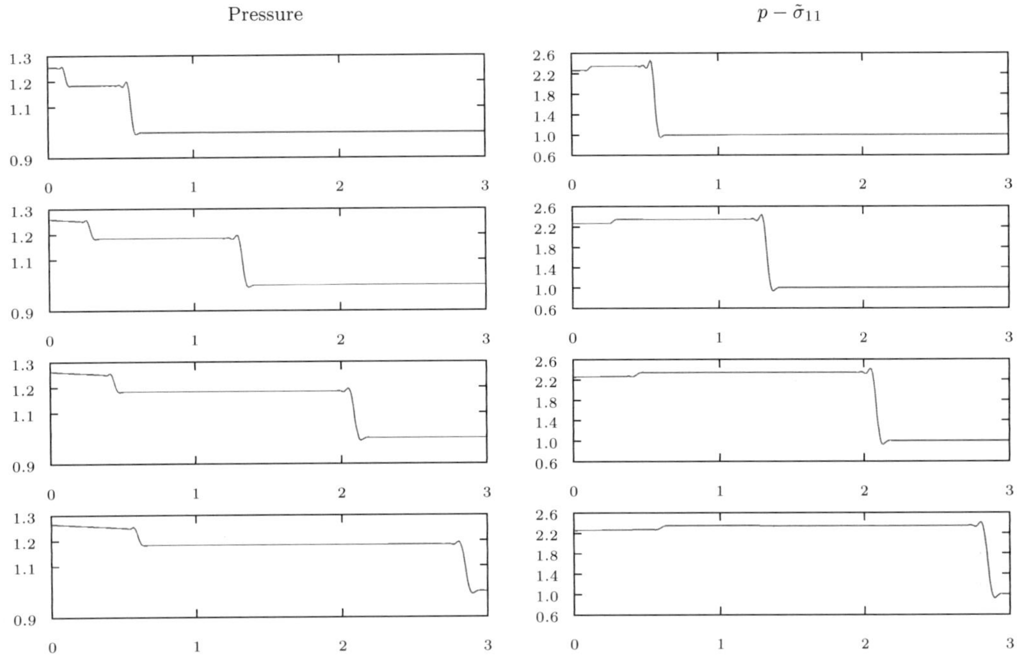

Figure 5. Test 3. Pressure and stress in the skeleton ($p - \tilde{\sigma}_{11}$).

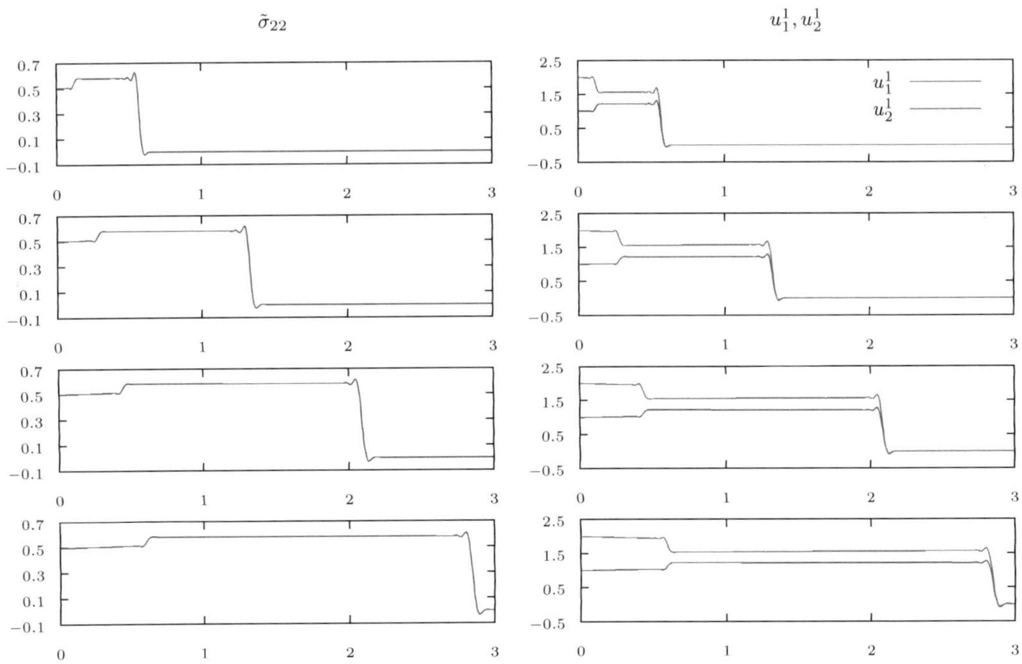

Figure 6. Test 3. Shear stress in x_2-direction and velocities in x_1-direction.

5. Conclusion

The numerical method for initial-boundary value problems for the governing equations of the model of compressible fluid flow through elastic porous medium is developed. The model is based on the theory of thermodynamically compatible systems of conservation laws and governing equations of the model form a hyperbolic system of divergence form equations. The WENO–Runge–Kutta method with characteristic decomposition at the boundary is applied for the numerical solution of the test problems of fluid inflow from the boundary into the saturated porous medium. The numerical computations give physically reasonable results and further research will include the improvement of the numerical method and verification of the model.

Acknowledgments

Work of ER was supported by the Russian Foundation for Basic Research (Grants No 16-29-15131, 15-05-01310). Work of AAY and AKK was supported by the Grant of Government of the Russian Federation No 14.W03.31.0002.

References

[1] Khoei A R and Mohammadnejad T 2011 *Comput. Geotech.* **38** 142–166
[2] Romenski E 2013 *Int. Conf. to Honour Professor E.F. Toro - Proc. of the Int. Conf. on Numerical Methods for Hyperbolic Equations: Theory and Appl.* 193–200
[3] Perepechko Y V, Romenski E I and Reshetova G V 2014 *11th World Congress on Computational Mechanics, WCCM 2014, 5th European Conference on Computational Mechanics, ECCM 2014 and 6th European Conference on Computational Fluid Dynamics, ECFD 2014* 4630–41
[4] Yanchenko A A, Romenski E, Khe A K and Chupakhin A P 2017 *International Conference "Computational and Applied Mathematics 2017"* 198–99
[5] Yanchenko A A, Chupakhin A P and Romenski E 2016 *All-Russian Conference "Nonlinear waves: theory and new applications" dedicated to the 70th birth anniversary of corresponding member of RAS V.M. Teshukov"* 121
[6] Godunov S K and Romenskii E I 2003 *Elements of continuum mechanics and conservation laws* (New York: Kluwer Academic/Plenum Publishers)
[7] Dumbser M, Peshkov I, Romenski E and Zanotti O 2016 *J. Comput. Phys.* **314** 824–62
[8] Gavrilyuk S, Ndanou S and Hank S 2016 *J. Elast.* **124** 133–41
[9] Shu C W 1997 *NASA/CR-97-206253* ICASE. report No. 97-65

IOP Conf. Series: Journal of Physics: Conf. Series **894** (2017) 012114 doi:10.1088/1742-6596/894/1/012114

Special vortex in relativistic hydrodynamics

A P Chupakhin[1,2] **and A A Yanchenko**[1,2]

[1] Lavrentyev Institute of Hydrodynamics, Novosibirsk, Russia

[2] Novosibirsk State University, Novosibirsk, Russia

E-mail: chupakhin@hydro.nsc.ru

Abstract. An exact solution of the Euler equations governing the flow of a compressible fluid in relativistic hydrodynamics is found and studied. It is a relativistic analogue of the Ovsyannikov vortex (special vortex) investigated earlier for classical gas dynamics. Solutions are partially invariant of Defect 1 and Rank 2 with respect to the rotation group. A theorem on the representation of the factor-system in the form of a union of a non-invariant subsystem for the function determining the deviation of the velocity vector from the meridian, and invariant subsystem for determination of thermodynamic parameters, the Lorentz factor and the radial velocity component is proved. Compatibility conditions for the overdetermined non-invariant subsystem are obtained.

A stationary solution of this type is studied in detail. It is proved that its invariant subsystem reduces to an implicit differential equation. For this equation, the manifold of branching of solutions is investigated, and a set of singular points is found.

1. Introduction

Equation of relativistic hydrodynamics governing the flow of a compressible fluid (gas dynamics) have the form [1]

$$\begin{aligned}&(\Gamma\rho)_t + \nabla\cdot(\Gamma\rho\vec{u}) = 0,\\ &(\Gamma^2\rho w\vec{u})_t + \nabla\cdot(\Gamma^2\rho w\vec{u}\times\vec{u}) + \nabla p = 0,\\ &(\Gamma^2\rho w - p)_t + \nabla\cdot(\Gamma^2\rho w\vec{u}) = 0.\end{aligned} \tag{1}$$

In (1) vector $\vec{u}$ — gas velocity in laboratory coordinate system, p, ρ — pressure and density, $w = 1 + \frac{\gamma}{\gamma-1}p/\rho$ — enthalpy, $\Gamma = (1-|\vec{u}|^2)^{-1/2}$ — Lorentz factor are function of independent variables — the time t and the spatial coordinates $\vec{x} = (x^1, x^2, x^3)$. A coordinate system is chosen such that the speed of light $c = 1$ and for the velocity modulus $|\vec{u}| = (\sum_{i=1}^{3} {u^i}^2)^{1/2}$ inequality $|\vec{u}| < 1$ is satisfied.

Equations (1) are written in the form of conservation laws. This is convenient for investigating the correctness of initial-boundary value problems and numerical modelling [2, 3, 4]. To construct and analyse the exact solution "special vortex", another form of equations (1) is more convenient.

Lemma 1 *The Euler equations of compressible fluid for relativistic hydrodynamics can be written in the form*

$$\begin{aligned}&b\,\mathrm{D}\vec{u} + p_t\,\vec{u} + \nabla p = 0,\\ &\mathrm{D}a + a\,\mathrm{div}\,\vec{u} = 0,\\ &a\,\mathrm{D}d = p_t.\end{aligned} \tag{2}$$

Content from this work may be used under the terms of the Creative Commons Attribution 3.0 licence. Any further distribution of this work must maintain attribution to the author(s) and the title of the work, journal citation and DOI.
Published under licence by IOP Publishing Ltd

IOP Conf. Series: Journal of Physics: Conf. Series **894** (2017) 012114 doi:10.1088/1742-6596/894/1/012114

In (2) $a = \Gamma\rho$, $b = \Gamma^2\rho w$, $d = b/a = \Gamma w$, $D = \partial_t + \vec{u}\cdot\nabla$ — total derivative.

A special vortex (Ovsyannikov vortex) is a solution of the differential equations, which is partially invariant with respect to the group of rotations $SO(3)$ in the space $\mathbb{R}^3(\vec{x}) \times \mathbb{R}^3(\vec{u})$. In this representation, the group $SO(3)$ has invariants $r = |\vec{x}|^{1/2}, |\vec{u}|$ and $\vec{x}\cdot\vec{u}$. Since there are only two invariants depending on the three velocity components (u^1, u^2, u^3), we can construct a partially invariant solution. Invariant independent variables are t and r. The choice of a non-invariant function is less obvious. L. V. Ovsyannikov in [5] proposed to take this function as an angle $\omega = \omega(t, r, \theta, \varphi)$ measuring the deviation of the velocity vector $\vec{u}$ from meridian of the sphere $r =$ const. We introduce spherical coordinates (r, θ, ϕ) and (U, V, W):

$$x = r\sin\theta\cos\varphi, \quad y = r\sin\theta\sin\varphi, \quad z = r\cos\theta,$$

such that $r \geqslant 0$, $0 \leqslant \theta \leqslant \pi$, $0 \leqslant \varphi \leqslant 2\pi$; U, V, W — radial, altitude and longitude components of the velocity. We introduce new coordinates for the velocity components, namely polar coordinates for tangent component $\vec{u}_\tau = (V, W)$ of the velocity vector:

$$V = H\cos\omega, \quad W = H\sin\omega, \tag{3}$$

such that $H = \sqrt{V^2 + W^2}$, $\omega = \arctan W/V$. Special vortex is a solution of equations (2) in which a special dependence of the functions on the independent variables is realized, namely:

$$\begin{aligned} &U = U(r,t), \quad \Gamma = \Gamma(r,t), \quad H = H(r,t), \\ &p = p(r,t), \quad \rho = \rho(r,t), \quad w = w(r,t), \\ &\omega = \omega(t, r, \theta, \varphi). \end{aligned} \tag{4}$$

From the general theory of group analysis of differential equations [6], after substitution of the representation (4) in (2), we obtain composition of a factor-system in the form of a union of an overdetermined system of differential equations for function ω and a system of equations for invariant functions U, Γ, H, p, ρ, w.

Relativistic hydrodynamics equations are written in various forms, adapted to solve specific problems [4]. In [2] they are reduced to a symmetrized form. Some particular solutions of these equations are investigated, for example, in [7, 8, 9].

2. Special vortex equations for relativistic gas dynamics

After substitution of (4) in (2), written in spherical coordinates (r, θ, ϕ) and (U, H, ω), we obtain following statement

Lemma 2 *Special vortex equations for relativistic gas dynamics are represented as a union of invariant subsystem*

$$\begin{aligned} &ad(\mathrm{D}_0 U - \frac{1}{r}H^2) + p_t U + p_r = 0, \\ &\mathrm{D}_0 H + \frac{U}{r}H + p_t H = 0, \\ &a\mathrm{D}_0 d = p_t, \end{aligned} \tag{5}$$

where $\mathrm{D}_0 = \partial_t + U\partial_r$ *— invariant part of the total derivative, and overdetermined subsystem for function* ω*:*

$$\begin{aligned} &k\sin\theta \mathrm{D}_0\omega + \sin\theta\cos\omega\omega_\theta + \sin\omega\omega_\varphi + \cos\theta\sin\omega = 0, \\ &\sin\theta\sin\omega\omega_\theta - \cos\omega\omega_\varphi = h\sin\theta + \cos\theta\cos\omega, \end{aligned} \tag{6}$$

where

$$k = r/H, \quad h = k(a^{-1}\mathrm{D}_0 a + r^{-2}(r^2 U)_r). \tag{7}$$

It is remarkable that (6) and (7) coincide exactly with the corresponding equations for classical gas dynamics [5, 10]. But function h is different, it is related to other physical quantities. This fact emphasizes the extremely successful choice of variables made by Ovsyannikov in [5]. A consequence of this fact is

Lemma 3 (Ovsyannikov compatibility condition) *Compatibility condition of overdetermined system (6) has the form*

$$k\mathrm{D}_0 h = h^2 + 1. \tag{8}$$

Equations (8) complement invariant subsystem (5). Thus, all mathematical results proved for the special vortex in classical gas dynamics [5, 10] are carried over to special vortex in relativistic gas dynamics.

3. Stationary special vortex for relativistic gas dynamics

Consider a stationary special vortex, i.e. solution which is partially invariant with respect to the group $\langle \partial_t, SO(3)\rangle$. In representation (4) and in equations of Lemma 2, it is necessary to remove the dependence on time, then $\mathrm{D}_0 = U\frac{d}{dr}$. Consider, in the following, monatomic gas with $\gamma = 5/3$, which is interesting for applications. Equations (5) can be partially integrated and the following Lemma takes place

Lemma 4 *For stationary special vortex for relativistic gas dynamics invariant subsystem is reduced to ordinary implicit differential equation*

$$F(R, h, p; m_0, s_0) \equiv q^{3/2} - R^2 p\Big(3m_0 + s_0\frac{p^2}{1+h^2}\Big)q + 3m_0^2 R^4 p^2 q^{1/2} - m_0^3 R^6 p^3 = 0, \tag{9}$$

where $p = dh/dR$*;* $s_0 > 0$*,* $0 < m_0 < 1$ *— constants characterizing physics of the problem,* $R > 1$ *— normalized distance,* $q = q(R, h, p)$ *has the form*

$$q(R, h, p) = R^2(R^2 - 1)p^2 - (1 + h^2)^2. \tag{10}$$

Invariant functions have following form in terms of function h *and its derivative* h_R*:*

$$U = \frac{1+h^2}{R^2 h_R}, \quad a = \frac{\text{æ}_0}{a_0}\frac{h_R}{\sqrt{1+h^2}},$$
$$\Gamma = \Big[1 - \frac{1}{R^2}\Big(1 + \Big(\frac{1+h^2}{R h_R}\Big)^2\Big)\Big]^{-1/2}, \tag{11}$$
$$d = d_0, \quad H = \frac{1}{R}, \quad w = d_0/\Gamma,$$

where a_0*,* æ_0*,* d_0 *are constant.*

Thus, the determination of the special vortex for relativistic gas dynamics reduces to solving equation (9) and overdetermined system (6). This system is integrated in finite form in [5], its solution ω is given by an implicit function. Geometric interpretation and domain investigation are given in [11].

Equation (9) belongs to a class of equations that are unresolved with respect to derivatives. Now such equations are usually called implicit differential equations. The present state of the theory is presented in [12], detailed historical review is given in [13, 14]. The specificity of equations of this type is the existence of manifold of branching solutions, the presence of trajectories bundle starting from singular points of different degrees of degeneracy [12]. Applications of this theory to gas dynamics can be found, for example, in [5, 15].

4. Investigation of singular points of equation (9)

Implicit equation (9) can be resolved with respect to derivative p at all points of $\mathbb{R}^3(R, h, p)$, except for the points of manifold

$$F(R, h, p) = 0, \quad F_p(R, h, p) = 0. \tag{12}$$

Curve (12) is called a criminant of equation (9), it is manifold of branching of integral curves. It consists of singular points of equation (9), called regular singular points. In general, equation (9) can have singular points of two types — folded and collected. Folded singular points can be found from the system of equations

$$F(R, h, p) = 0, \quad F_p(R, h, p) = 0, \quad F_R(R, h, p) + pF_h(R, h, p) = 0. \tag{13}$$

Collected singular points are determined from the system of equations

$$F(R, h, p) = 0, \quad F_p(R, h, p) = 0, \quad F_{pp}(R, h, p) = 0. \tag{14}$$

Projection of the criminant curve onto plane $\mathbb{R}^2(h, R)$ is called a discriminant curve.

The essence of the geometric approach to study of implicit differential equations, proposed by Poincaré, is to raise the equation to a vector field

$$R_\tau = F_p, \quad h_\tau = pF_p, \quad p_\tau = -(F_R + pF_h), \tag{15}$$

where τ — new parameter along integral curve. With this interpretation, the integral curves of equation (9) are located on different sheets of surface $F = 0$. These integral curves can be projected onto the plane $\mathbb{R}^2(h, R)$ with overlapping.

The investigation of the singular points of equations (9) is connected with large computational difficulties. The solution of the systems of equations (13) and (14), determined by the resultants of the corresponding polynomials, can be obtained by means of a system of symbolic and numerical computations. Even with the use of such computing systems, the calculation can be quite long.

5. Numerical techniques

We will deal with F^2 because it is more convenient from the computational point of view (everywhere below by F we mean F^2).

5.1. Discriminant curve

The discriminant curve is given by the following system of equations

$$F(R, h, p) = 0, \quad F_p(R, h, p) = 0. \tag{16}$$

To solve this and other similar systems, the following technique is used. Since the equations of the system (16) are polynomials in p, we can eliminate p by constructing the resultant of two these polynomials. Denote by $\Phi_1 = F$, $\Phi_2 = F_p$. Then

$$R_{12}(R, h) = \text{Resultant}(\Phi_1, \Phi_2). \tag{17}$$

Further, in the (R, h)-plane, we can numerically solve the algebraic equation $R_{12}(R, h) = 0$ for h. Thus, for given m_0, s_0 and R, we can calculate the discriminant curve. As a result of the numerical experiment, the following statement was obtained.

Theorem 1 (about discriminant curve) *The discriminant curve consists of one or two components.*

The discriminant curve always has its "main" component, but for small s_0 ($s_0 \approx 10^{-2}$ and less) an additional ("secondary") component of the discriminant curve appears (Figure 1).

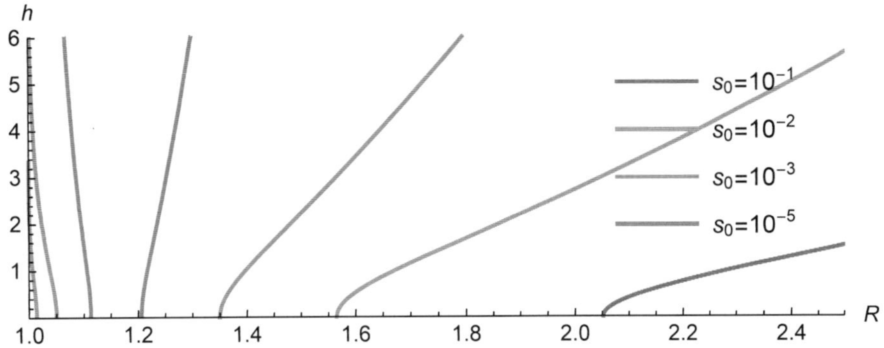

Figure 1. Discriminant curves for fixed $m_0 = 0.5$ and $s_0 = 10^{-1}$, 10^{-2}, 10^{-3}, 10^{-5}.

5.2. Folded singular points

Folded singular points are found from the system of equations (13). To solve this system, we apply the same technique as for system of equations (16). Denote by $\Phi_1 = F$, $\Phi_2 = F_p$, $\Phi_3 = F_R + pF_h$. In order to solve this system, we construct the following resultants

$$\begin{aligned} R_{12}(R,h) &= \text{Resultant}(\Phi_1, \Phi_2), \\ R_{13}(R,h) &= \text{Resultant}(\Phi_1, \Phi_3). \end{aligned} \tag{18}$$

Intersection of curves $R_{12} = 0$ and $R_{13} = 0$ gives us a set of folded singular points. To find $R_{12}(R,h)$, it is necessary to calculate determinant of order 19 ($11! = 39{,}916{,}800$ terms, without zeros). After simplification, the degree of the polynomial obtained is 60 and 108 for R and h, respectively. To find $R_{13}(R,h)$ we calculate determinant of order 20 ($15! = 1{,}307{,}674{,}368{,}000$ terms, without zeros). The degree of $R_{13}(R,h)$ is 64 for R and 170 for h.

As a result of the numerical experiment, the following statement was obtained.

Theorem 2 (existence and uniqueness of folded singular points) *For any values of parameters s_0 and m_0 there is a unique folded singular point.*

Figure 2 shows the standard arrangement of R_{12} and R_{13}.

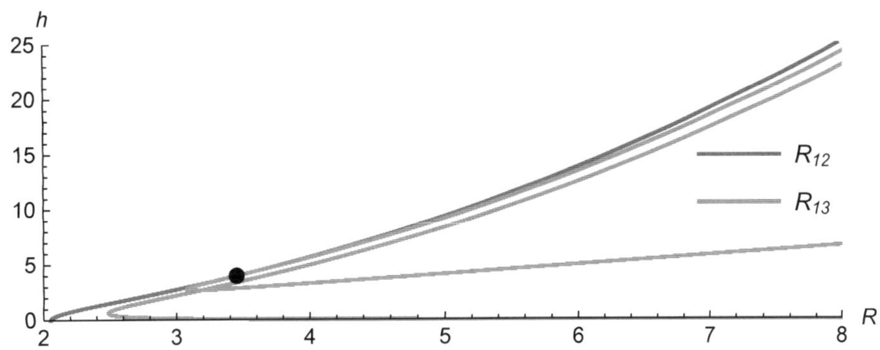

Figure 2. Folded singular point for $m_0 = 0.5$ and $s_0 = 0.1$.

IOP Conf. Series: Journal of Physics: Conf. Series **894** (2017) 012114 doi:10.1088/1742-6596/894/1/012114

5.3. Collected singular points

Collected singular points are found from the system of equations (14). Note that we can use previous result — $R_{12} = \text{Resultant}(\Phi_1, \Phi_2)$, see (18). Denoting $\Phi_4 = F_{pp}$, we obtain one more resultant for finding collected singular points

$$R_{14}(R, h) = \text{Resultant}(\Phi_1, \Phi_4). \quad (19)$$

Intersection of curves $R_{12} = 0$ and $R_{14} = 0$ gives us a set of collected singular points. To find $R_{14}(R, h)$ we calculate determinant of order 18 (11! terms). After simplification, the degree of $R_{14}(R, h)$ is 88 for R and 144 for h.

As a result of numerical simulation we obtained the following statement.

Theorem 3 (classification of collected singular points) *For all parameters $s_0 > 0$ and $0 < m_0 < 1$ only one of the following cases is realized:*

(i) there are no collected points,

(ii) there is one collected point (on the "main" component),

(iii) there are two collected points (one on the "main" component and one on the "secondary" component).

It is found that when s_0 moves from 0 to 1, the curve $R_{14} = 0$ moves to the direction of increasing R. The curve $R_{12} = 0$ has exactly the same behaviour when m_0 is moving from 0 to 1. For identical m_0 and s_0 they coincide. The third case is realized for small s_0 ($s_0 \approx 10^{-5}$ and less).

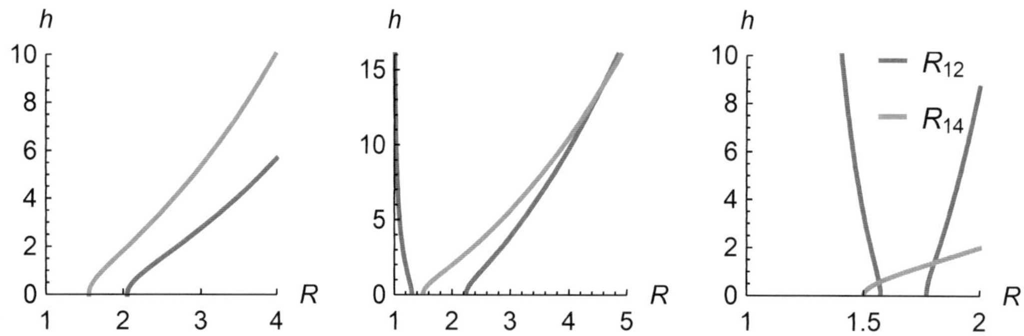

Figure 3. Zero, one and two collected singular points. The intersection points of curves R_{12} and R_{14} are collected singular points.

All calculations are performed in Wolfram Mathematica.

Acknowledgments

The research was supported by Grant of Government of Russian Federation No 14.W03.31.0002.

References

[1] Landau L D and Lifshitz E M 1959 *Fluid Mechanics, Course in theoretical physics Vol. 6* (Oxford: Pergamon press)

[2] LeFloch P G and Ukai S A 2008 *ArXiv e-prints* (*Preprint* 0812.4001)

[3] Godunov S K 2012 *J. Comput. Math. Math. Phys.* **52**(5) 916-29

[4] Gourgoulhon E 2006 An introduction to relativistic hydrodynamics, Stellar Fluid Dynamics and Numerical Simulations *EAS Publ. Ser.* **21** 43-79

[5] Ovsyannikov L V 1995 *J. Appl. Mech. Tech. Phys.* **36**(3) 45-52

[6] Ovsyannikov L V 1982 *Group analysis of Differential Equations* (New York: Academic Press)
[7] Borshch M S and Zhdanov V I 2007 *SIGMA* **3** 116
[8] Alexa C and Vrinceanu D 1996 *ArXiv e-prints* (*Preprint* 9609060)
[9] Perucho M, Marti J M and Hanasz M 2005 *ArXiv e-prints* (*Preprint* 0510426)
[10] Cherevko A A and Chupakhin A P 2005 *Lavrentyev Institute of Hydrodynamics SB RAS* Preprint N **1**
[11] Golovin S V 2005 *J. Phys. A Math. Gen* **38** 4501-16
[12] Arnold V I 2006 *Ordinary differential equations* (Berlin/Heidelberg: Springer-Verlag)
[13] Remizov A O 2006 *Contemp. Math. Fundam. Direct.* **19** 131-70
[14] Davydov A A 1985 *Funct. Anal. Appl.* **19**(2) 1-10
[15] Barlukova A M and Chupakhin A P 2012 *J. Appl. Mech. Tech. Phys.* **53**(6) 11-24

On the dispersion of liquid in coaxial supersonic gas jet

S V Poplavski[1], V M Boiko[1], V V Lotov[1] and A U Nesterov[1]

[1]Khristianovich Institute of Theoretical and Applied Mechanics SB RAS, Novosibirsk, Russia

Abstract. The aim of this work was to study the dispersion of liquids in gas jets in connection with the creation of high productivity nozzles. For effective combustion of fuel, systems with intensive air supply to the spray of a liquid are promising. In connection with this, a supersonic coaxial jet was experimentally studied with a central supply of liquid beyond the slit of the confuser nozzle at the modes Npr = 4 and Npr = 6. New data are obtained on the structure of the gas-liquid jet: the gas velocity field, the shadow visualization of the geometry and wave structure of the jet with and without liquid, the velocity profiles of the liquid phase, the dispersion of the droplets. The spatial distribution of the concentration of the spray was first determined. From these data, the parameters of the dispersion processes are obtained in terms the We numbers. A physical model of a supersonic coaxial gas-liquid jet with a central fluid supply is proposed.

1. Introduction

Dispersing of liquids is an important component of a wide range of technological processes in many industries. These are modern fire-fighting systems, chemical treatment of machine parts and application of polymer coatings, fuel injectors, irrigation systems in agrotechnics, pharmacology, medical sprays, etc., A whole complex of other physical processes takes place here in addition to the proper dispersion of the liquid: drop and gas velocity relaxation, intense interfacial heat transfer, phase transitions, and droplet coagulation [1]. Basic data on the aerodynamic destruction of single drops were obtained in shock waves and flows with jumps of gas dynamic parameters [1-4]. These metabolic processes are also inherent in large-scale atmospheric phenomena of transport and evolution of the water state in rain clouds and storm fronts [5]. Experimental data on the structure and dynamics of gas-liquid flows are critically important for verifying the numerical modeling of liquid dispersion in gas jets [6]. Thus, the relevance of the study of nonequilibrium gas-liquid systems is not limited to technical applications. The study of dispersion of liquids and the dynamics of sprays in gradient gas flows is a fundamental item within the framework of the problem of interphase heat-mass transfer in nonequilibrium heterogeneous systems of technogenic and natural origin.

Many of the above applications use so-called pneumatic nozzles, for example, in the oil industry for high flow burners (~ $1 m^3/h$). They are used to incinerate oil waste in unconfined fields to avoid spillage to the ground. For spraying liquids with high flow rates, the most promising ones are pneumatic nozzles, and in particular nozzles like Y-jet [7]. There the primary decay of the liquid occurs in the mixing chamber by means of lateral gas jets, followed by secondary breaking up when it flows out of the mixing chamber. The velocity lag of the drops from the gas is an important condition for their secondary destruction, which provides the necessary Weber number for the required dispersion. This determined the choice of supersonic gas flow regimes in this paper and the feed of liquid along the axis of the jet beyond the nozzle cut for a more detailed observation of the disintegration of the liquid jet. Such a configuration gives a coaxial jet with a supersonic gas shell and

Content from this work may be used under the terms of the Creative Commons Attribution 3.0 licence. Any further distribution of this work must maintain attribution to the author(s) and the title of the work, journal citation and DOI.
Published under licence by IOP Publishing Ltd

a gas-liquid core. This is a complex flow with a large number of parameters, a study of it has not been carried out yet. The structure of two-phase core of a gas-liquid jet is the subject of this study.

2. The destruction of a fluid in a coaxial gas flow and the characteristics of a gas-liquid jet

Hypothetically, the destruction of a jet of liquid in a coaxial gas flow occurs in two stages according to the following scheme. The interphase boundary instability develops with a sufficient difference in the velocity of the gas and the liquid jet, leading to a primary disintegration of the liquid into fragments of various shapes and sizes [8] (Figure 1). The subsequent secondary drop decay takes place according to one of the known aerodynamic mechanisms depending on the Weber number $We=\rho u^2 d/\sigma$ [1, 2, 4], ρ and u – are the density and relative velocity of the gas, d and σ are the size and surface tension of the drop.

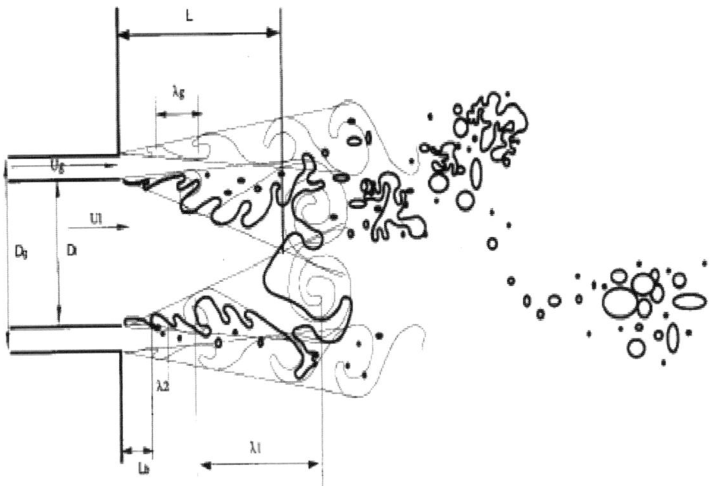

1. vibrational breakup: We ~ 12;
2. bag breakup: 12 < We < 50;
3. bag-and-stamen breakup: 50 < We < 100;
4. sheet stripping: 100 < We < 250;
5. wave crest stripping: We > 250;
6. catastrophic breakup: We > 250.

Figure 1. Disintegration of a liquid jet in a coaxial gas jet [8] and mechanisms of secondary drops breakup depending on the Weber number We [4].

With regard to the breakup mechanisms of droplet destruction (4 and 5 in Figure 1), in [2] it was shown by direct observation by the fast-acting shadow method that the change in the mechanism of 4 by 5 occurs at a number We ~ 1000. But in gas-liquid jets neither primary nor secondary destruction is not available for direct observation due to the high concentration of liquid. Therefore, there always remain questions about the conditions under the number of We of the drops, when they appear in the flow after the primary disintegration of the liquid jet, and what are the mechanisms for the secondary droplets breakup. Somewhat better is the case with the determination of the parameters of the spray at a distance from the nozzle, where the concentration decreases significantly. But here the high-speed lag of the phases is minimal, the dispersion of the drops stops, and the state of the spray is close to the final one. Thus, the final state of the spray in gas-liquid jets depends on the factors that should be attributed to the peculiarities of aerodynamic loading of droplets in practically important but unexplored cases.

In these experiments, the main breaking up of droplets should occur in gradient sections of the flow in supersonic regions of the jet with a characteristic alternation of gas acceleration and braking, where the number We is maximized [3]. One of the unexplored questions of the structure of supersonic gas-liquid jets is whether these structures (cells) are preserved in the presence of a liquid. This is the question of the qualitative picture of the jet, and what quantitative parameters of the gas-liquid jet are needed for its exhaustive description.

IOP Conf. Series: Journal of Physics: Conf. Series **894** (2017) 012115 doi:10.1088/1742-6596/894/1/012115

The question of the required characteristics of a gas-liquid jet can be put in two ways. In the applied statement, it is important what conditions are necessary for the generation of the spray with the required parameters. Such a formulation presupposes the availability of data on the flow in a wide range of parameters that are obtained in academic studies. But here the problem is posed differently: what is the final state of the spray under the given experimental conditions. In both statements of the question the same parameters of the flow are of interest. What is the set of physical parameters of a gas-liquid jet sufficient for its description?

The most substantiated list is provided by the energy approach. It can be shown that in the simplest case without taking into account heat transfer and phase transitions, the energy of the spray $E(x)$ in a unit of time in the cross section x is added from the work of the gas to increase the surface of the liquid S with the free energy $E_S=S\sigma$ and for the growth of the kinetic energy of the droplets:

$$E(x) = \frac{3\pi}{2} v_0 \frac{D_0^2}{d(x)} \sigma + \frac{\pi}{8} \rho_l V_l^3(x) D^2(x) \beta(x) \tag{1}$$

Here v_0 is the initial velocity of the liquid in the jet with the initial diameter D_0, σ and ρ_l are the surface tension and density of the liquid, $\beta(x)$ and $V_l(x)$ are the volume concentration and droplet velocity, $d(x)$ is droplets size at the distance x from the nozzle, $D(x)$ is the diameter of the two-phase jet core. Thus, to describe the state of the spray, the following data are needed: the geometry of the main jet structures and, in particular, $D(x)$, the liquid dispersion $d(x)$, the gas and liquid phase velocities u and $V_l(x)$, as well as the volume concentration of the spray $\beta(x)$ along the jet.

3. Experiments and methods of diagnosis

The experiments were performed on the "Gas-liquid stand" installation of ITAM SB RAS, which allows generating high-velocity gas jets with a high concentration of liquid sprays in a wide range of gas velocities. The air jet was formed at the outflow from the convergent nozzle at the modes with nozzle pressure ratio Npr = 4 and Npr = 6 (Figure 2). Nozzle cut diameter 14 mm.

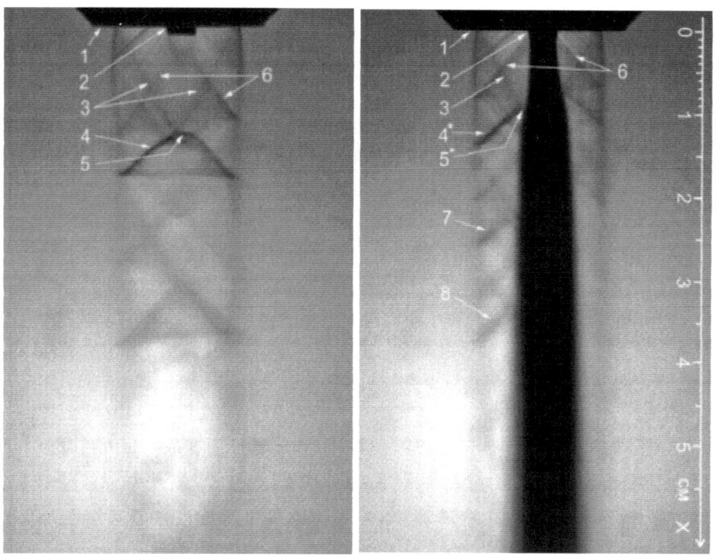

(a) (b)

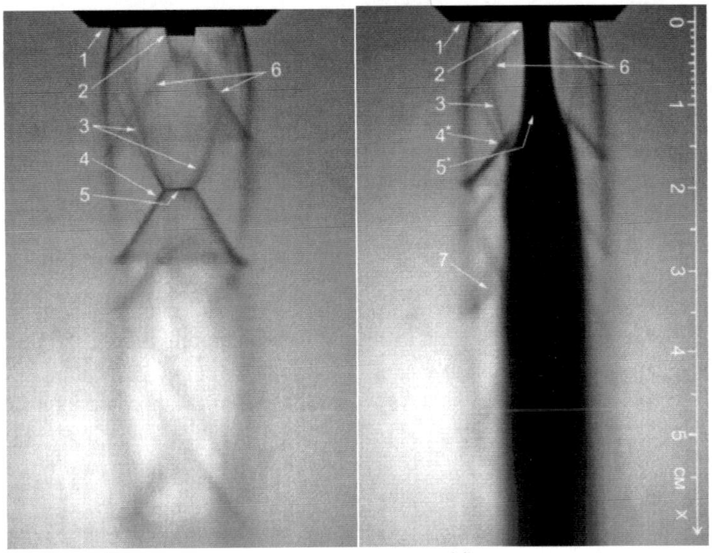

(c) (d)

Figure 2. The wave structure of the near region of a supersonic underexpanded jet at the modes Npr = 4 (a, b) and Npr = 6 (c, d) in the absence (a, c) and in the presence of liquid (b, d); 1 - nozzle cut, 2 - tube, 3 - incoming jump, 4 - reflected jump, 4* - jump near gas-liquid region broadening, 5 - Mach disk, 5* - liquid jet disintegration region, 6 - tail jump on aerodynamic track behind central body, 7, 8 - jumps similar to 4*.

The parameters of the gas-liquid jet were studied by four optical methods.

1. Shadow visualization for recording the geometry and wave structure of the jet.
2. PIV method for recording the gas velocity field.
3. Laser Doppler anemometer (LDA) with direct spectral analysis [9, 10] on the basis of the WS-7 spectrum analyzer [11, 12] for measuring the velocity of the spray particles.
4. The Malvern "Spray Tech" device for recording the dispersity of the spray.

Shadow visualization was performed by a system of two TAL-125R telescopes, a 125 mm lens diameter, a focal length of 1125 mm. The images were recorded with a digital camera with an exposure of 125 μs. Shadow images of the jet at a distance of 300 mm from the cut of the nozzle are obtained. Figure 2 shows a shadow visualization of the near field of a supersonic underexpanded jet from a convergent nozzle (1) with a central tube (2) in the absence and presence of a liquid. Here, the main wave structures of the gas jet are seen: a hanging (coming) jump (3), a reflected jump (4), and a Mach disk (5) with Npr = 4 (Figure 2a) and Npr = 6 (Figure 2c).

The complex used basically meets the diagnostic requirements shown in the previous section, except for measurements of the spray concentration β. As will be shown below, this parameter can also be indirectly determined from the master data.

4. Concentration of the spray

Estimation of the volume concentration of the spray β in the gas-droplet jet was carried out according to the following procedure. Within the framework of linear optics, it can be shown that for a known optical path L in a suspension of particles with a size d, the dispersion medium becomes opaque to the parallel beam if the concentration is $\beta \geq 2d/3L$. Consequently, on the transparency boundary, we have:

$$\beta = \frac{2d}{3L} \tag{2}$$

The dispersion composition of the spray in the jet was studied by the Malvern "Spray Tech" serial device, which operates on the principle of analyzing the angular scattering of light on droplets. The available dispersion data refers to $x > 10$ mm near the axis of the jet, and are $d \sim 12$ μm for the Npr4 regime and $d \sim 7$ μm for the Npr6 regime, and no significant variability is observed along the jet. But flow regimes with bimodal dispersion of the spray were detected.

Instrumental control of the thickness of the two-phase region L at the transparency boundary is based on the determination of the transmission pattern of light on shadow images in arbitrary sections of the jet. Passage profiles are the result of digital photometry on a given segment of a two-dimensional digital brightness field (shadow image). They can be obtained by various programs for processing graphic data; here we used the program "Image Pro".

The measured diameter $D(x)$ and the thickness of the translucent region $\delta(x)$ allow us to estimate the optical thickness $L(x)$ as a chord of the circular section of the jet of radius $R=D/2$, provided this chord is tangent to the coaxial circle of radius $r = R - \delta$ (radius of the opaque jet core):

$$L(x) = 2\sqrt{2R\delta - \delta^2} = 2\sqrt{D(x)\delta(x) - \delta^2(x)} \qquad (3)$$

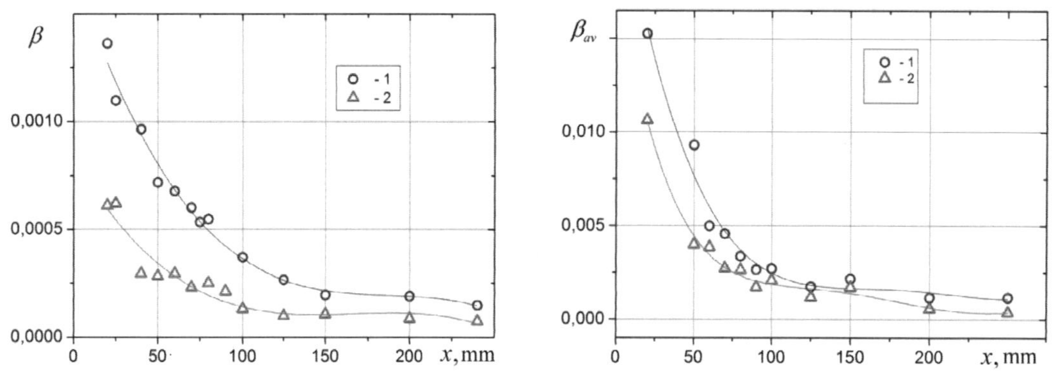

(a) (b)

Figure 3. a) The volume concentration of the spray β at the transparency boundary of the supersonic coaxial gas-liquid jet according to (2), (3); b) the average volume concentration of the spray β_{av} in the cross sections x of the coaxial supersonic gas-liquid jet according to (4); 1 – Npr=4, 2 – Npr=6.

Figure 3a shows the results of calculating the volume concentration β for the measured values of $D(x)$ and $\delta(x)$ at different sections x along the jet. The estimation of the average concentration of the spray is obtained from the considerations of conservation of liquid mass along the jet. Namely, for a known flow of liquid Q_0 into the flow (in the given experiments $Q_0 \approx 5*10^{-5}$, m³/s), the average concentration of the spray β_{ev} in an arbitrary section of diameter $D(x)$ is expressed as follows:

$$\beta_{ev}(x) = \frac{4 \quad Q_0}{\pi D^2(x) V(x)} \qquad (4)$$

Here $V(x)$ is the velocity of the spray particles in the cross section x. The velocity profile $V(x)$ is obtained by the LDA method. The average concentration profile along the jet $\beta_{ev}(x)$, calculated from (4), is shown in figure 3b.

5. Discussion of results

From the comparison of Figure 3a and Figure 3b, one can see, firstly, that two different approaches give a similar profile of the spray concentration in x, and this confirms the structure of the flow as a whole. Secondly, with the general trend of decay, the average concentration is more than an order of magnitude greater than the concentration at the transparency boundary in the entire range of x studied. This suggests that the bulk of the liquid is concentrated in the jet core and is the supplier of a finely dispersed spray at its periphery over the entire apparent extent of the jet. The sharp decrease in the mean concentration, as well as the local one, ends in the region $x \sim$ 100-120 mm, but the shadow visualization does not show any noticeable features in this section of the jet. There is only a feature of the velocity profile of the spray (figure 4a), namely, at $x \sim$ 120 mm, a droplet speed maximum of $\sim$ 250 m/s is observed with Npr4 and $\sim$ 320 m/s with Npr6.

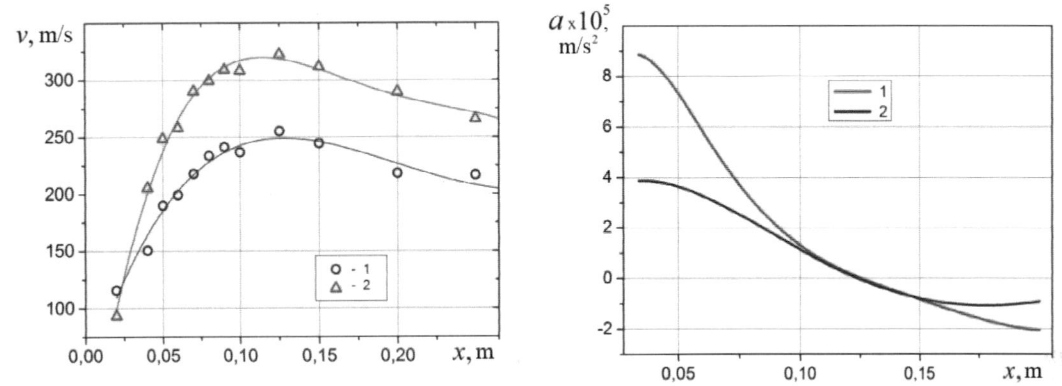

(a) (b)

Figure 4. a) Speed profiles of spray particles along the axis of the jet (LDA method); 1 - Npr4, 2 - Npr6, points - experiment, curves - approximation by a polynomial of the 5th degree; b) acceleration of spray particles along the axis of the jet; 1 - Npr6, 2 - Npr4, curves - differentiation of velocity approximation in Figure 4a.

The deceleration of the spray behind the point of maximum droplet velocity (x > 120 mm) is a consequence of the gas braking due to the loss of dispersion of the liquid phase and the acceleration of the crushing products. It is also seen from Figure 4a that the greatest acceleration of a drop is experienced at the initial portion of the jet, where the maximum phase velocity difference occurs. The magnitude of the droplet acceleration and the character of its variation with respect to x, obtained by differentiating the velocity approximation, are shown in Figure 4b.

The obtained data on the main parameters of the gas-liquid jet allow us to draw indirect conclusions about the mechanisms of secondary destruction of droplets in the flow. The known velocity of the gas and droplets enables to estimate the We number of the gas-liquid flow at different modes and at different points of the spray. Figure 5 shows the distribution of the number We along the axis of the jet in various modes. Taking into account the effective Weber numbers of $12<We<100$, we can assume that the secondary crushing mechanism is a "bag" (the 2-nd or 3-rd type in Figure 1). It is for this type of crushing of droplets that the bimodal structure of the dispersed composition of the crushing products takes place. The lower boundary of the numbers We for the destruction of liquid by aerodynamic mechanisms is shown in the graph of the horizontal line.

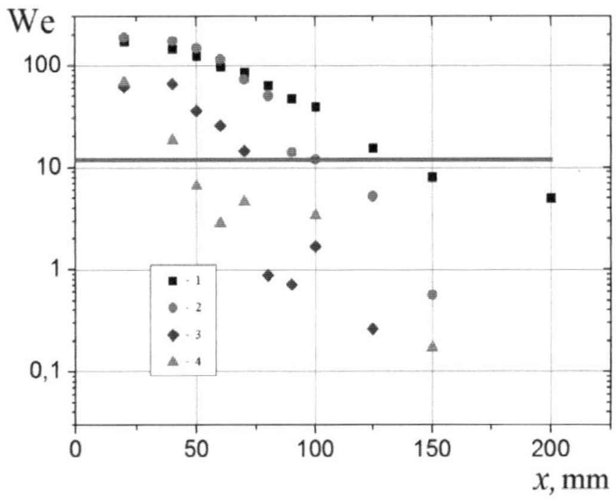

Figure 5. Weber's numbers in gas-liquid jet; the horizontal line represents the lower boundary of the We numbers for aerodynamic dispersion of droplets; 1 - Npr = 1.5, 2 - Npr = 2, 3 - Npr = 4, 4 - Npr = 6

6. Conclusion

The supersonic coaxial jet was experimentally studied with a central supply of liquid beyond the confuse nozzle at the modes with nozzle pressure ratio Npr = 4 and Npr = 6. At the first time the energy principle has been formulated to determination efficient set of physical parameters of gas-liquid jets. According to this approach, data on the jet structure, gas and liquid velocity profiles, the dispersion composition and concentration of the spray have to be obtained. One-moment local measurements of the velocity and size of droplets in saturated gas-liquid flows were performed for the first time. A method is proposed for estimating the liquid concentration in a jet by shadow imaging, supplemented with data on the dispersed composition and the speed of the spray. Flow regimes with bimodal dispersion of the spray were detected. The role of the "bag breakup" mechanism in the formation of the bimodal structure of droplet size distribution is revealed.

References

[1] Gelfand B E 1996 *Prog. Energy Combust.* Sci. 22 201–65
[2] Boiko V M and Poplavski S V 2012 *Combustion, Explosion, and Shock Waves.* **48**(4) 440-45
[3] Boiko V M, Zapryagaev V I, Pivovarov A A and Poplavski S V 2015 *Combustion, Explosion, and Shock Waves.* **51**(5) 587-96
[4] Hsiang L P and Faeth G M 1993 *Int. j. multiphase flow.* **19**(5) 721-35
[5] Arkhipov V A, Kozlov E A, Zharova I K, Titov S S, Usanina A S 2016 *Arabian Journal of Geo-sciences, Special Issue, Dust.* **9**(2) 9-114
[6] Arkhipov V A, Boiko V M, Goldin V D, Maslov E A, Orlov S E, Poplavskiy S V, Zharova I K and Usanina A S 2016 *IOP Conf. Series: Materials science and Engineering.* **124** 012076
[7] Couto H S, Carvalho J A and Bastos-Netto Jr. 2002 *IV Asian – Pacific Intern. Symp. of Combustion and Energy Utilization.* 310-15
[8] Lasheras J C and Hopfinger E J 2000 *Annual review of fluid mechanics.* **32** 275-308
[9] Boiko V M, Papyrin A N and Poplavski S V 1980 *Optika Spektr.* **48**(2) 356-62
[10] Poplavski S V, Boiko V M and Nesterov A U 2016 *AIP Proceedings of the XVIII International conference on the methods of aerophysical research, Russia, Perm,* doi: 10.1063/1.4963958
[11] "High Precision Wavelength Meters: WS_HighFinesse Angstrom Series", http://www.toptica.de/products/itemlayer/59/wavelengthmeter_2007.pdf.
[12] Kobtsev S, Kandrushin S and Potekhin A 2007 *Applied Optics.* **46**(23) 5840-43

IOP Conf. Series: Journal of Physics: Conf. Series **894** (2017) 012116 doi:10.1088/1742-6596/894/1/012116

Effect of additives on CuO/Al nanothermite properties

V V Gordeev, M V Kazutin and N V Kozyrev

Institute for Problems of Chemical and Energetic Technologies, Biysk, Russia

E-mail: gordeev.vladimir92@yandex.ru

Abstract: The effect of additives on CuO/Al nanothermite properties is reported. The incorporation of additives resulted in the decreased sensitivity. It had an adverse impact on gas dynamic characteristics of CuO/Al nanothermite, up to the loss of the unique capability of burning in narrow channels.

1. Introduction

A metal fuel mixed with a less reactive metal oxide as the oxidizer at the nanoscale is commonly called nanothermite. Such a system is characterized by a high reaction temperature and a small amount of exhaust products. In contrast to micron-sized thermite systems, nanothermites exhibit a high burning rate and are capable of burning in microchannels of 0.1 mm and less [1-4].

A lot of reports have shown the promising outlook of using nanothermites in pyrotechnics (incendiary and priming formulations, etc.), microelectromechanical systems (MEMS), dedicated fastening and connection of materials (nanoscale welding), etc. [1,5-6]. The practical application is hindered by extremely high sensitivities to friction and electrostatic discharge comparable to some priming explosives [6-9]. The incorporation of additives into nanothermite composites is considered to be one of the options to solve this challenge. For instance, introducing fluoropolymers contributes to decreasing the electric spark sensitivity [10], in contrast to additives such as soot or graphite [11]. The incorporation of carbon nanotubes or molybdenum disulfide can decrease the mechanical susceptibility [11-12]. Along with that, reports focused on reducing the sensitivity of nanothermites often do not provide information on how formulation variations influence explosion performance of systems studied.

The present work aimed at investigating the effect of additives on explosion performance and sensitivity of CuO/Al nanothermite composites.

2. Materials and Method

2.1. Materials

A CuO/Al nanothermite system and a formulation based thereon were examined herein. The characteristics of the formulation components are summarized in Table 1.

The following additives were studied: SKI-NL M isoprene rubber (Technical Specifications TU No. 38.03.1.031, hereinafter “SKI”), PMS-20 polymethylsiloxane (Russian State Standard GOST No.13032-77, hereinafter “PMS”), F-42L fluoropolymer (Russian State Standard GOST No. 25428-82, hereinafter “FL-42”), SKF-26 fluororubber (Russian State Standard GOST No. 18376-79, hereinafter “SKF-26”); TAUNIT-M multilayer carbon nanotubes (made by Nanotekhtsentr, 8-15 nm outer diameter, 4-8 nm inner diameter, 2 μm average length) and single-layer carbon nanotubes

Content from this work may be used under the terms of the Creative Commons Attribution 3.0 licence. Any further distribution of this work must maintain attribution to the author(s) and the title of the work, journal citation and DOI.
Published under licence by IOP Publishing Ltd

(OSUNT-75 powder with a specific surface of 440 m^2/g, containing 75% single-layer tubes 1.4-1.6 nm in diameter, up to 1 μm long, manufactured by Uglerod ChG company, hereinafter "OSUNT").

Table 1. Basic characteristics of nanosize components.

Component	Mean particle size, nm	Content, %	Manufacturer
ALEX Aluminum powder, Al	50-70	85.0	Peredovyye Poroshkovyye Tekhnologii, Tomsk
Copper oxide powder (II), CuO	30-110	99.5	Plazmoterm, Moscow

2.2. Fabrication of composites and test specimens

A general procedure for the fabrication of composites involved the following steps:

- Prearrangement of the components: nanopowders were preliminary deagglomerated; a low-viscosity solution or suspension of the additive in organic solvent was prepared;
- Ultrasonic mixing of the powders in the solution/suspension of the additive;
- The solvent was evaporation to a residual content of about 50% relative to the weight of the solid components;
- The pyrotechnic mixture was ground under the layer of the volatile liquid being not a binder solvent or plasticizer;
- The suspension was vacuum dried to give a composite that represents a powder of agglomerated particles of the target composition.

To determine the burning rate at bulk density, the powders were loaded into polymer tubes of 2.4 mm i.d. with a wall thickness of 0.2 mm.

To measure the burning rate in the thin layer, the powders examined at bulk density were placed between two steel plates with dimensions of 10x20 mm, the layer thickness being adjusted by special assemblies whose schematic diagrams for the burning rate measurement are illustrated in Figure 1.

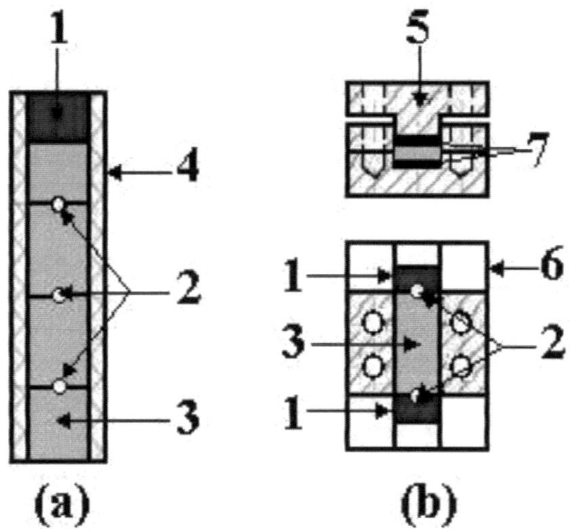

Figure 1. Schematic diagrams of assemblies for the burning rate measurement: a) in polymer tube at bulk density; b) in thin layer (1 – pyrotechnic igniter, 2 – ionization detectors, 3 – test specimen, 4 – polymeric shell, 5 – punch, 6 – steel support, 7 – steel plates).

2.3. *Test methods*

The calculated characteristics (heat of combustion Q, adiabatic flame temperature T_{ad}) of the studied compositions were obtained using the REAL software package [13]. On the basis of the calculations made, basic composites were formulated.

The friction sensitivity in shock shear was determined on a K-44-II friction tester pursuant to GOST No. 50835-95 (lower limit, $LL_{friction}$, kgf/cm^2) and impact sensitivity on a K-44-III impact tester pursuant to GOST No. 4545-88 (lower limit, LL_{impact}, mm).

The burning rate was recorded by ionization detectors according to the diagrams shown in Figure 1. The detectors were at a distance of 14–15 mm when the burning rate was measured in the tube, and were at a distance of 20 mm when the burning rate was measured in the thin layer. All the test specimens were thermally initiated from one of the end faces of the explosive charge.

To determine the minimum energy of electrostatic spark (W_{min}), the test specimen at bulk density was placed into a polymer tube about 4.5 mm in height. Afterwards, spark discharge of the specified energy was transmitted from the live capacitor through the specimen layer located between the electrodes. The response or its absence was recorded. The experimental results were recorded as follows: $W_1 < W_{min} < W_2$, where W_1 – is the spark energy at which no ignition occurs in any of the twenty experiments; W_2 – is the spark energy at which ignition takes place in each of the twenty tests. The lower measurement limit of this method was 0.022 mJ.

The TMD for each mixture as a weighted average of the pure solid densities of the all reactants. The bulk densities for these experiments ranged from 12% to 24% TMD.

3. Results and Discussions

The first phase of the research was to investigate the effect of different polymer additives on the CuO/Al nanothermite properties. The results are summarized in Table 2.

Table 2. Properties of CuO/Al nanothermite and composites based thereon with polymer additives.

No	Composition, wt.-%.	Calcd data [a]			Experimental data				
					U, m/s		Sensitivity		
		Q, kJ/g	T_a, K	TMD, %	in tube ∅ 2.4 mm	in layer 0.1 mm	friction, kgf/cm^2	impact, mm	W_{min}, mJ
1	CuO/Al 76/24	3.84	3930	21	570-650	30-70	400	>500	<0.022
2	CuO/Al/SKI 73/21/6	3.82	3190	23	0.5-0.7	0	1600	>500	<0.022
3	CuO/Al/PMS 76/21/3	3.82	3560	22	150-250	0	800	>500	<0.022
4	CuO/Al/F-42 76/21/3	4.10	3980	24	250-350	10-50	450	>500	<0.022
5	CuO/Al/SKF-26 76/21/3	4.11	4075	23	300-350	10-50	450	>500	<0.022

[a] Aluminum nanopowder was considered in the calculations as mixed 85/15% Al/Al_2O_3

The broad spread of the experimental values of the burning rate is due to the convective mechanism of the reaction front propagation and is, as practice shows, typical of nanothermite composites at low bulk factors [14-15].

The incorporation of the polymer additives resulted naturally in the decreased burning rate. However, this decrease is by no means significant: the rate remains quite high, except for specimen No.2 having higher polymer content. A decrement in the explosive charge thickness led naturally to the decreased burning rate for all of the composites. The fluoropolymer additives (specimens No. 4 and 5) inserted into the basic nanothermite diminished the burning rate in the thin layer negligibly, whereas the introduction of hydrocarbon (SKI) and silicon (PMS) rubbers (specimens No. 2 and 3), which are inert with respect to the nanothermite components, led to the burning interruption.

The friction sensitivity of the thermites was decreased to an acceptable processability level by introducing the inert SKI and PMS, whereas fluoropolymers did not almost decrease the sensitivity. Together with that, the minimum energy to ignite the nanothermite composites by electrostatic discharge was extremely high, and the incorporated polymer additives did not alter this parameter.

The conventional way of reducing the electrostatic stimulus sensitivity is to insert electrically conductive powders into high-energy composites. We examined a series of nanothermite systems with carbon nanotubes (CNT) added. The study results are given in Table 3.

Table 3. Characteristics of CuO/Al/CNT nanothermite composites.

No.	Composition, wt.-%.	Q, kJ/g	Experimental data				
			TMD, %	Burning rate U, m/s		Sensitivity	
				in tube ∅ 2.4 mm	in layer 0.1 mm	friction, kgf/cm^2	W_{min}, mJ
1	CuO/Al 76/24	3.84	21	570–650	30–70	400	<0.022
2	CuO/Al/MUNT 69/22/9	3.54	24	–	20–60	400	<0.022
3	CuO/Al/OSUNT 69/22/9	3.54	24	–	20–60	400	<0.022
4	CuO/Al/MUNT/SKF 68/19/10/3	3.71	12	50–100	0 [a]	2000	$0.032< W_{min}$ $W_{min} < 7.36$

[a] At a layer thickness of 0.5 mm

It follows from the data obtained that incorporating carbon nanotubes did not bring about the desired outcome: the electric spark sensitivity remained that of the basic nanothermite composite. The application of that type of additives did not help decrease the friction sensitivity as well. The burning rate in the thin layer diminished marginally compared to that of the basic composite even at a relatively high CNT content of up to 9 wt.-%.

The incorporation of the fluoropolymer (specimen No.4, Table 3) into the CuO/Al/CNT system appreciably lowered the sensitivity values, but such a composite was unable to burn even at a layer thickness of 0.5 mm.

It can be said on a whole that embedding the polymers and carbon nanotubes into the nanothermite system assists in decreasing the sensitivity parameters, but in this case, the resultant composites lose their unique detonation properties right up to the lost capability of burning in the narrow channels.

4. Conclusions

Incorporating the inert polymers (PMS, SKI) into the CuO/Al nanothermite system discernibly diminishes the friction sensitivity, but with that, the critical combustion thickness rises. The use of the fluoropolymer additives (F-42, SKF) has an influence on the latter parameter to the lowest extent, but the friction sensitivity cannot measurably be achieved as well. None of the additives examined enables the acceptable electrostatic discharge sensitivity to be attained. The appropriate sensitivities to friction and electric spark are achieved by simultaneously introducing carbon nanotubes and fluoropolymer (MUNT and SKF) into the nanothermite system; however the resultant composite drastically loses the burning rate and is unable to propagate the flame front even at a layer thickness of 0.5 mm.

IOP Conf. Series: Journal of Physics: Conf. Series **894** (2017) 012116 doi:10.1088/1742-6596/894/1/012116

References

[1] Rossi C, Zhang K, Esteve D, Alphonse P, Tailhades P and Vahlas C 2007 *J. Microlelectro. Sys.* **16** 919
[2] Puszynski J A, Bulian C J and Swiatkiewicz J J 2007 *J. Prop. and Pow.* **23** 698
[3] Bockmon B S, Pantoya M L, Son S F, Asay B W and Mang J T 2005 *J. Appl. Phys.* **98** 064903
[4] Son S F, Asay B W, Foley T J Yetter R A, Wu M H and Risha G A 2007 *J. Prop. and Pow.* **23** 715
[5] Wang H, Jian G, Egan G C, Zachariah M R 2014 *Comb. and Flame.* **161** 2203
[6] D G Piercey and T M Klapotke 2010 *Cent. Euro. J. Energ. Mat.* **7** 115
[7] Apperson S. J. 2010 *PhD dissertation* 91 (Missouri: University of Missouri)
[8] Stamatis D, Jiang X, Beloni E and Dreizin E L 2010 *Prop. Expl. Pyro.* **35** 260
[9] Matyas R and Pachman J 2013 *Primary Explosives* (Springer) pp. 29-32
[10] Foley T, Pacheco A, Malchi J, Yetter R and Higa K 2007 *Prop. Expl. Pyro.* **32** 431
[11] Bach A, Gibot P, Vidal L, Gadiou R and Spitzer D 2015 *J. Energ. Mat.***33** 260
[12] Kelly D G, Beland P, Brousseau P and Petre C F 2015 *J. Energ. Mat.* 1-15
[13] Belov G V 2002 *Termodinamicheskoye modelirovaniye: metody, algoritmy, programmy* [Thermodynamics modeling: methods, algorithms, programs] (Moscow: Nauchnyi Mir) p. 184
[14] Gordeev V V, Kozyrev N V and Kazutin M V 2016 *Polzunovsky Vestnik* **1** 30
[15] Gordeev V V, Kazutin M V, Kozyrev N V and Glaktionov V S 2016 *Polzunovsky Vestnik* **1** 37

IOP Conf. Series: Journal of Physics: Conf. Series **894** (2017) 012117 doi:10.1088/1742-6596/894/1/012117

Transonic streamline of symmetric wing under the influence unilateral oscillations characterized by the spectrum of two frequencies

V P Zamuraev[1,2] and A P Kalinina[1,2]

[1]Khristianovich Institute of Theoretical and Applied Mechanics, Novosibirsk, Russia
[2]Novosibirsk State University, Novosibirsk, Russia

E-mail: zamuraev@itam.nsc.ru, kalinina@itam.nsc.ru

Abstract. Forced high-frequency vibrations of the airfoil surface part with the amplitude almost equal to the sound velocity can change significantly the lift force of the symmetric profile streamlined at zero angle of attack. The oscillation consists of two harmonics. The ratio of harmonics frequencies values is equal to 2. The present work shows that the aerodynamic properties depend significantly on the specific energy contribution of each frequency.

1. Introduction

The influence of energy supply on the transonic flow around airfoils can be used to improve their aerodynamic characteristics [1, 2] and to reduce flutter [3, 4]. The appearance of spontaneous and forced oscillations connected with the phenomena of flutter and buffeting influences the flow around the airfoils. The possibility of development of these phenomena was investigated in [3-5], but the streamline of the airfoil was not considered. The study in [1] concerns controlling the transonic flow around the profiles with the help of pulse periodic supply of heat energy. The possibility of significant influence of such effects on shock-wave structure and significant changes in aerodynamic properties was shown. However, formation of near surface high-temperature zone resulting from energy supply can be an obstacle for using this method for control. High frequency forced oscillations of the surface element of the profile were investigated in [2]. The effectiveness of such a method for controlling transonic flow around wing was shown. The influence of forced oscillations on the transonic streamline requires more detailed parametric studies.

When studying flutter or buffeting [3-5], the development of these processes is usually simulated only in the very beginning. The result of the studies is the answer to the question whether these phenomena occur. The numerical simulation is carried out by using the unsteady Euler equations [3] or Navier – Stokes [4, 5] ones. Obtaining the periodic solution is not the goal of these studies. Even in the absence of oscillations, the transonic streamline of the profile is a complex task including the modeling of viscous and thermal phenomena. It should be noted that the physical models for this phenomena are constantly being improved [6]. The computational complexity increases when considering the oscillations. For example, using the direct numerical simulation based on unsteady Navier – Stokes equations could lead to the numerical oscillations [5]. Therefore search of modes with occurrence of buffeting is performed by means of turbulence models [4, 5] without analyzing the streamline of the airfoil.

Content from this work may be used under the terms of the Creative Commons Attribution 3.0 licence. Any further distribution of this work must maintain attribution to the author(s) and the title of the work, journal citation and DOI.
Published under licence by IOP Publishing Ltd

The parametric study of the influence of the forced oscillation on the transonic streamline of the profile based on the unsteady Euler equations was performed in [7]. The calculation was carried out until obtaining the periodic solution. It was assumed that the oscillations do not cause any flutter or buffeting. It is shown that in case of certain parameters of single-mode vibration in the form of a standing wave, it is possible to improve of the aerodynamic characteristics (see the Figure 1).

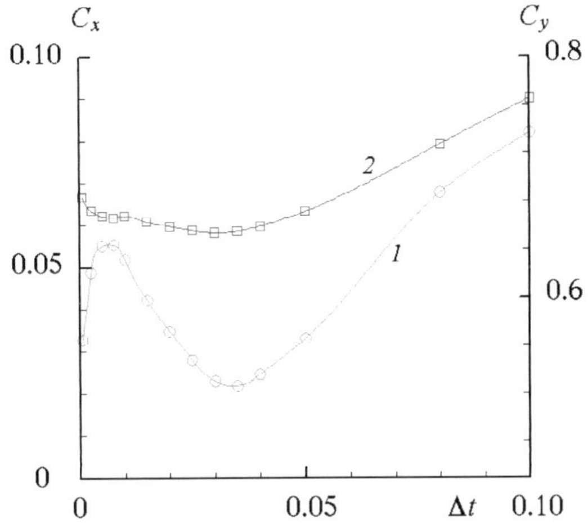

Figure 1. The coefficients of the lifting force C_y (*1*) and the wave drag C_x (*2*) depending on the time period Δt

The curves in Figure 1 are obtained at constant amplitude of velocity oscillations, which is close in magnitude to the flow velocity. The aerodynamic characteristics dependences on the period of oscillation are not monotonic.

However, the question about affect of the multimode oscillation on the profile streamline arises. This paper presents the results of parametric study of the influence of two-mode oscillation localized in a zone of the profile surface.

2. The problem formulation

The airfoil NACA 0012 streamlined by ideal gas with the adiabatic index $\gamma = 1.4$ and the Mach number of the incident flow $M_\infty = 0.85$ and zero angle of attack (the streamline without separation zones) is considered. A small section of the surface profile with one of its sides makes an oscillatory motion. The two-frequency oscillations are considered. Use was made of two-dimensional Euler equations in conservative form as a mathematical model to describe the unsteady planar flow of inviscid gas. This approach allows evaluating the max value of the effect of oscillation, because the dissipative phenomena aids in decreasing of this effect. The calculation region is a rectangle with an inner border corresponding to the contour under consideration. On the left, upper and lower boundaries of this region are the conditions of undisturbed flow; the extrapolation is used on the right boundary.

The change of the initial geometry $f_0(x)$ on the contour part with the boundaries x_1 and x_2 is described by the formula:

$$f(x,t) = f_0(x) + (A_1 \sin(\omega_1 t) + A_2 \sin(\omega_2 t)) \sin\left(\pi \frac{x - x_1}{x_2 - x_1} \right).$$

Here A_i and ω_i ($i = 1, 2$) are the amplitude and frequency of the i harmonic; all the dimensions are related to the chord length of the profile b, x coordinate is directed along the chord of the profile ($0 \leq x \leq 1$); the time t is related to b/a_∞ (a_∞ is the sound velocity in the incoming flow). The oscillation frequency ω_i is associated with a period of Δt_i by the ratio $\omega_i = 2\pi/\Delta t_i$, $i = 1, 2$. In the case of single-

IOP Conf. Series: Journal of Physics: Conf. Series **894** (2017) 012117 doi:10.1088/1742-6596/894/1/012117

mode oscillation the second amplitude A_2 was supposed equal to zero: $A_2 = 0$. The formulation of the problem for single mode oscillations is presented in [7].

The finite-volume scheme of the third order reducing the total variation of the flow parameters considering the surface motion is used.

3. Results

The simulation is performed for two-harmonic oscillation with frequencies ω_1 and ω_2. The total energy was assumed equal to the energy of single-mode oscillations with amplitude A and frequency ω_1 [7]. The following equality takes place

$$A_1^2\omega_1^2 + A_2^2\omega_2^2 = A^2\omega_1^2.$$

The frequency values differ in twofold:

$$\omega_2 = 2\omega_1.$$

The low frequency is equal to the frequency of the single-harmonic oscillation. The choice of the ratio between the frequencies is determined by the possibility of resonant excitation of multiple frequencies. The ratio of the contribution to the total energy of the individual harmonics is following:

$$\frac{A_2^2\omega_2^2}{A_1^2\omega_1^2} = k, \quad k = 1, 0.5, 0.1.$$

Figure 2 shows the dependence of the wave drag coefficient C_x and the lift force coefficient C_y from the the period $\Delta t = \Delta t_1$ for all values of k.

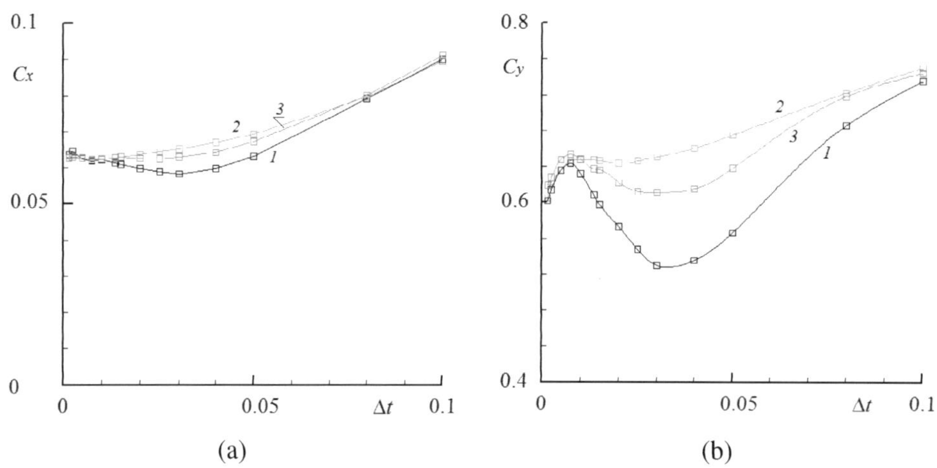

Figure 2. The coefficients of the wave drag C_x (a) and the lift force C_y (b) depending on the period Δt: curves *1*, *2* and *3* correspond to k = 1, 0.5, 0.1 respectively.

The case with the same specific contribution to the total energy of the individual harmonics ($k = 1$, curves 1) at low frequency gives almost the same values for the aerodynamic coefficients as the single-harmonic oscillations [7]. Both of coefficients change non-monotonically in dependence of the oscillation period for each value of relative contribution of the harmonics. The coefficient of wave drag varies much less than the lift force coefficient. The change of k from 1.0 to 0.5 for C_y gives an increase in minimum (at $\Delta t \approx 0.03$) for 24% and increase C_x for 12%. The value of the aerodynamic quality of the profile increases.

The distribution of pressure coefficient and the Mach number for one - and two-harmonic oscillations in the zero phase for relatively large oscillation period $\Delta t = 0.08$ are shown in Figure 3. The positions of the shock waves almost coincide. The differences in aerodynamic characteristics in this case are insignificant and do not exceed 5%.

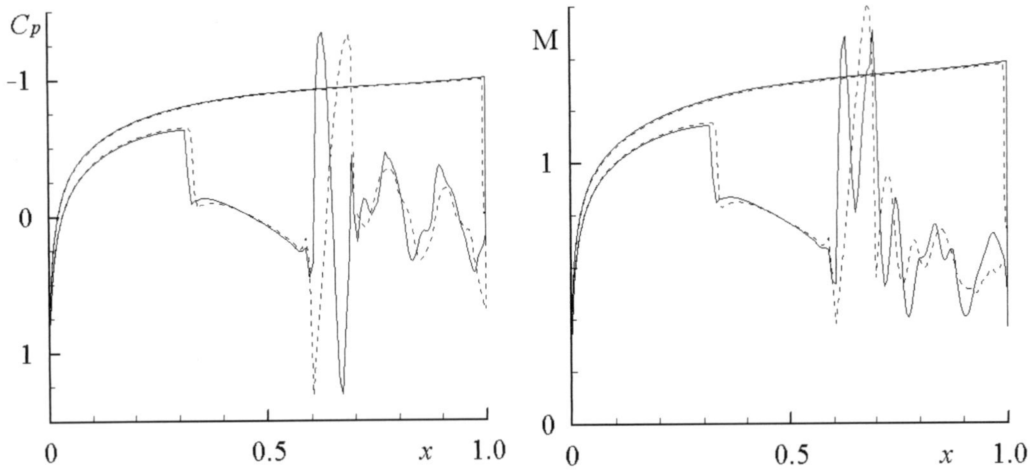

(a) (b)

Figure 3. Distribution of the pressure coefficient (a) and the Mach number (b): the dashed line is the oscillation with a single frequency, the solid line corresponds to two-harmonic oscillation.

The distribution of the Mach number for one- and two-harmonic oscillation in the zero phase for the period $\Delta t = 0.03$ and $\Delta t = 0.015$ are shown in Figure 4. These periods correspond to the local minimum (Figure 4a, $\Delta t_1 = 0.03$, $\Delta t_2 = 0.015$, $k = 1$) and the maximum of C_y (Figure 4b, $\Delta t_1 = 0.015$, $\Delta t_2 = 0.0075$; $k = 1$). In the first of these cases ($\Delta t_1 = 0.03$) the closing shock localized on the side profile without oscillations does not achieve the trailing edge. Differences in shock-wave flow structure between the single-harmonic and two-harmonic oscillations for sufficiently small period ($\Delta t_1 = 0.015$) is relatively small. The differences in aerodynamic characteristics are also small.

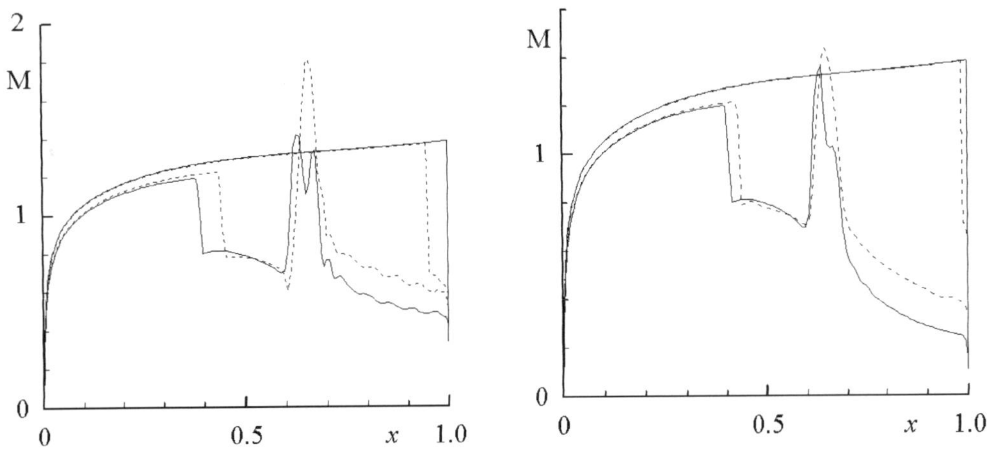

(a) (б)

Figure 4. The distribution of the Mach number along the contour of the profile: (a) – $\Delta t = 0.015$, (b) – $\Delta t = 0.015$; the dashed line for the oscillation with a single-harmonic oscillation, solid line for the two-harmonic oscillation.

However, there is no difference in the flow structure of fluctuations. This can be seen in Figure 3 and Figure 4, and Figure 5, that shows the distribution the Mach number for the variant in Figure 4b.

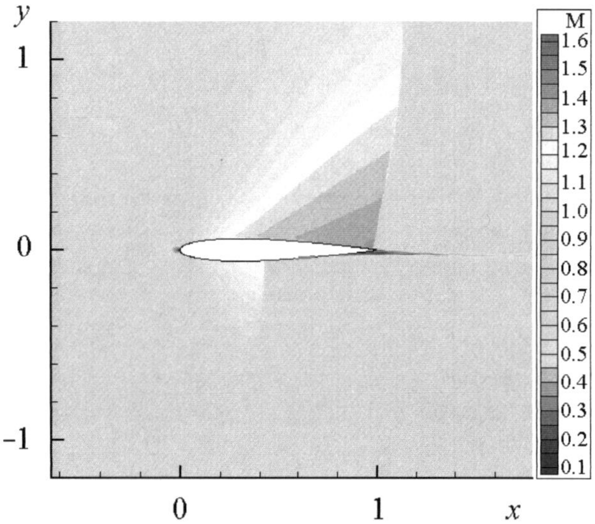

Figure 5. The part of the Mach number distribution at $\Delta t = 0.015$.

The closing shock wave on the upper part of the picture is localized on the back edge of the profile. On the lower side the shock wave is shifted upstream. This increases the lift force. Attention must be paid to the fact that shocks practically don't change their position within the period (see Figure 6 taken from [8]).

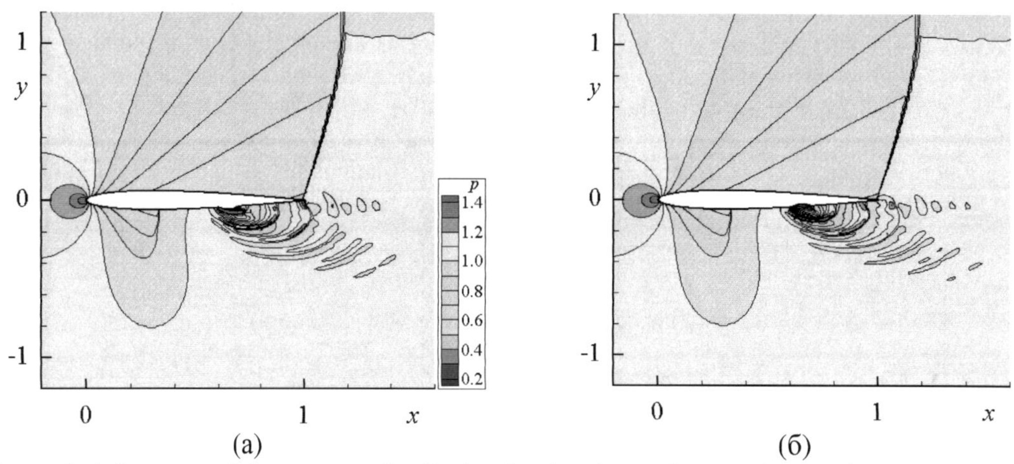

Figure 6. A fragment of the pressure distribution for the phase $\pi/2$ (*a*) and $3\pi/2$ (*б*) at $\Delta t = 0.08$.

It can be seen in Figure 6 that the positions of shock waves in opposite phase at relatively low frequencies of oscillations remain constant.

4. Conclusion

Thus, the studies show that the impact of two-harmonic oscillations on the transonic flow around a wing profile is determined by the total power, if both of frequencies correspond to the monotonic part of the dependence of aerodynamic characteristics on the period of oscillations. In the other case the lift force coefficient depends significantly from the relative energy contribution of the individual harmonics.

Acknowledgments
The work was performed under the project of SB RAS III.22.6.2.
The authors gratefully acknowledge Christopher Hearsey and Valeriya Hearsey for their assistance.

References
[1] Aulchenko S, Zamuraev V and Kalinina A 2011 *Technical Physics* **11** 13
[2] Aulchenko S, Zamuraev V and Kalinina A 2014 *Journal of Engineering Physics and Thermophysics* **87** 635–47
[3] Ishmuratov F, Kuzmina S and Mosunov V 1999 *TsAGI Science Journal* **30** 151
[4] Vedeneev V, Kolotnikov M, Makarov P and Firsanov V 2011 *Vestnik of Samara State Aerospace University Named after Academician S. P. Korolev (National Research University)* 2011 **3** 47
[5] Lipatov I and Vignes F 2014 *Scientific Bulletin of Moscow state technical University of civil aviation* 2014 **200** 133
[6] Bogdanov A, Diesperov V, Zhuk V and Korolev G 2015 *Proc. Russian Congress on fundamental problems of theoretical and applied mechanics* ed. Akhmetov D (Kazan: Kazan) p. 3778
[7] Zamuraev V and Kalinina A 2016 *Journal of Physics: Conference Series* 2016 **722** 012018
[8] Zamuraev V and Kalinina A 2017 *Journal of Engineering Physics and Thermophysics* **90** 599

Study of the geometry effect of the channel with variable cross section under forming transonic region in the supersonic flow with energy supply

V P Zamuraev[1,2] and A P Kalinina[1, 2]

[1]Khristianovich Institute of Theoretical and Applied Mechanics, Novosibirsk, Russia
[2]Novosibirsk State University, Novosibirsk, Russia

E-mail: zamuraev@itam.nsc.ru, kalinina@itam.nsc.ru

Abstract. This work concerns the process of deceleration of supersonic flow up to the transonic velocities in the channel. This process is connected with the problem of combustion organizing of fuels in the various ramjet engines. The influence of mode of energy supply and the type of channel geometry (axisymmetric or planar channel) is studied as well as the other factors. The similarity of gas dynamic structure formed under hydrogen burning and under pulse periodic energy supply is studied.

1. Introduction

One of the actual problems concerning organization of fuels combustion in various ramjet engines is decreasing of sizes of device. Especially it concerns the cases when it is necessary to organize combustion in the supersonic flow, because the size of a device increases simultaneously with the Mach number.

The simplified model for investigation of the main features of this problem is a channel with an increasing cross section. The energy is supplied into various zones of this channel. The formulation of the problem can be different. One of them is based on the Euler equations with simple function of heat release. The other variants of models analyze chemical kinetics and gas-dynamic basing on the Navier – Stocks equations. One of the combustion organization ways in such a system is described below. It is supposed that the channel includes two sufficiently long parts with the transversal section of constant area. The first of them is used to decrease the rate of flow up to the value almost equal to the sound velocity, but higher than it is. The amount of fuel which is supplied into this section is comparatively small. It is supposed that the main part of the fuel must be supplied into the expanding part of the channel located after the considered part of the channel with constant area. The scheme of such device is shown in Figure 1. It was described in [1]. The jet of compressed air is supplied transversely to the flow from the hole in the channel wall. The jet creates the effect of throttle. Moreover, it is creates a weak shock wave. By such a manner, the flow velocity decreases, and the ignition of fuel takes place in the front of the shock wave, and the process of burning begins to move upstream.

Content from this work may be used under the terms of the Creative Commons Attribution 3.0 licence. Any further distribution of this work must maintain attribution to the author(s) and the title of the work, journal citation and DOI.
Published under licence by IOP Publishing Ltd

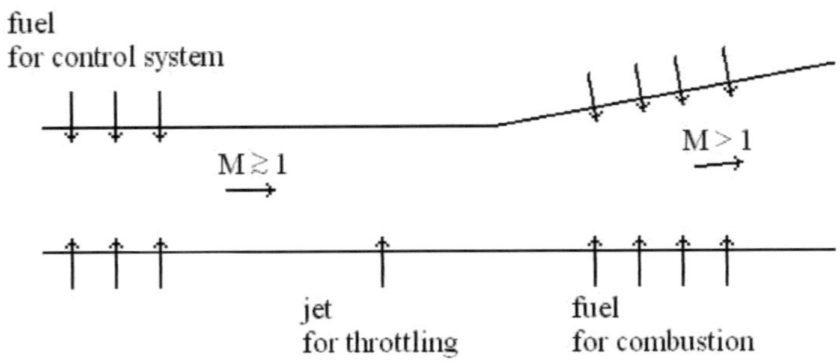

Figure 1. The common scheme of fuel combustion organizing.

There are many difficulties of organizing such a process. This work is devoted to the study of the process in the first part of channel with constant area. The main aim is controlling the flow to create the stable transonic region in the sufficiently large region.

The main questions to be answered to solve this problem are the following: where the main part of energy must be supplied; what is the value of the average power; what are the main properties of the mode of the energy supplying which are characterized by the maximal effectiveness, and how long the process of the energy supply can continued. It is very important to avoid locking the channel: the amount of the energy must depend on the pressure in the channel. When the pressure increases sufficiently to lock the channel, the supply of energy must be finished. It is realized by means of the transversal jet. The value of pressure in the gas generator must be lower than the pressure corresponding to the channel locking. The channel is characterized by planar or by axisymmetric geometry. The axisymmetric geometry has advantages in the durability. So, it is important to determine if the conclusions for the planar channel can be used for the axisymmetric channel.

We concern only a part of questions in this work. Using the parametrical study based on the unsteady Euler equations and impulse periodic energy supply we estimated the average power which is necessary to decelerate the flow. It is necessary to consider a question about quantity of gas-dynamic pulses for creating a zone of transonic velocities. Nevertheless, the solution of the Navier – Stokes equations closed by the SST k-ω turbulence model with a simple kinetic model of hydrogen burning is also used to represent the common features of this process. The purpose of our work is restricted by clearing up some important aspects of the problem. The other purpose is to offer some possible ways to solve the problem considered.

2. The results of the parametrical study based on the unsteady Euler equations

2.1. The formulation of the inviscid problem

This formulation is based on the unsteady two-dimensional Euler equations. This formulation of the problem is described in more detail in [2, 3]. The planar or axisymmetric channel is under consideration. The area of the transversal section grows simultaneously with coordinate x along the channel. As it was mentioned above, the channel contains two parts with the constant area of the transversal section. In the first part the Mach number is about M=2. This part contains the energy supply sources working in a pulse-periodic mode. These zones are located near the wall of channel. The jet of compressed air is also supplied into this part of channel. The system studied is schematically presented in the upper part of Figure 2a and Figure 2c.

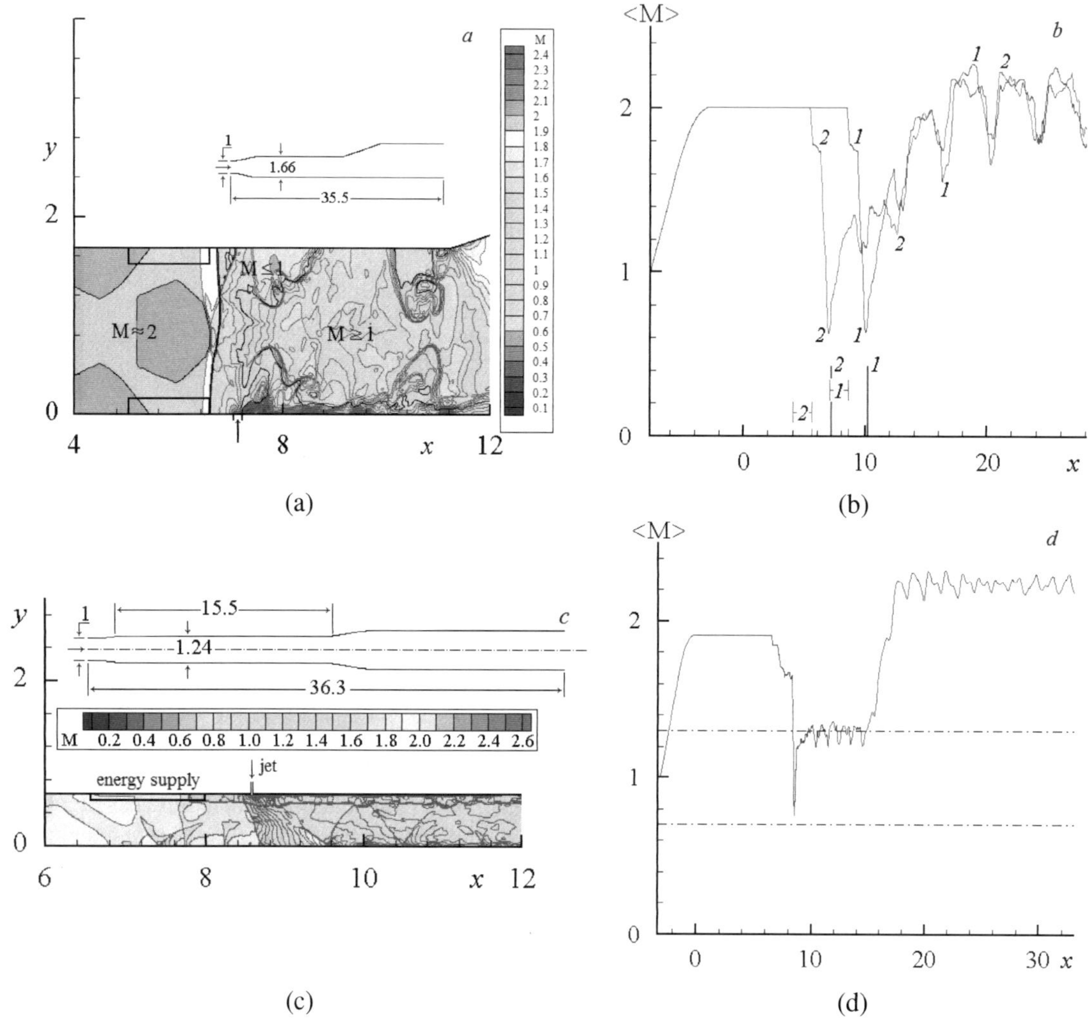

Figure. 2. The system studied and the Mach distribution: (a,c) – the Mach number distribution in the planar and axisymmetric channel respectively; (b,d) – the Mach Number which is averaged over the transversal section in dependence on the longitudinal coordinate x in the planar and axisymmetric channel respectively.

The maximum frequency of the energy supply is so high that every portion of gas receives energy when travelling through the energy zones. The minimum frequency of the energy supply is determined by the condition when the portion of gas which received the energy in the previous period did not leave the channel until the next energy supply.

This simple model can be considered for energy supply by the plasma actuators. However, this model could be used for any process, when energy release in such a mode takes place.

2.2. The transonic area in the channel for axisymmetric and planar case

The transonic area in our case corresponds to the Mach number into the range $0.7<M<1.4$. This mode can be created for the channel of planar geometry as well as for the axisymmetric channel. The Mach number averaged over the transversal section, as the function of the longitudinal coordinate x, is shown for the planar geometry in Figure 2b and for the axisymmetric geometry in Figure 2d. The fragments of corresponding distributions of the Mach number are presented in Figure 2a and Figure 2c for the planar or axisymmetric case respectively. The energy supplied is equal to the energy realized during burning of the poor hydrogen-air mixture with the excess-air coefficient equal approximately to 10.

In such a way, we can see that formation of the transonic region is possible for axisymmetric geometry as well as for planar geometry of the channel.

The difficulty of creating stable transonic region is the following: the transonic mode is not stable in common case. A little disturbance is enough to convert it into subsonic mode. The jet of compressed air is used for solving this problem. This jet flows from the gas generator at the pressure equal to the pressure arising when maximum energy is supplied into the channel. By the way, when the pressure in channel is equal to the pressure in gas generator, the injecting of compressed air is stopped. But with decreasing of pressure in the channel the compressed air begins to flow to the channel again. These correlations we can see in Figure 3a.

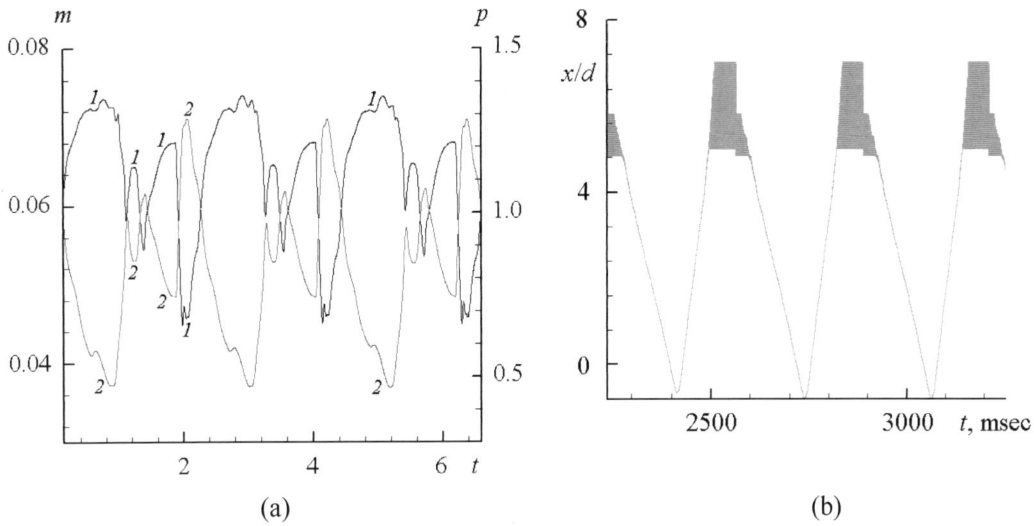

Figure 3. (a) – dynamics of the jet consumption m (curve *1*) and pressure p (curve *2*) at the point close to the jet aperture; (b) – the coordinate of disturbance with respect of time.

The pressure in the channel in the region of the jet and consumption of compressed air as the functions of time are presented in Figure 3a. Moreover, increasing of pressure must decelerate the flow and aids in ignition process as it is described in [1].

Moreover, the existence of the jet sufficiently decelerates the upstream motion of disturbance. Let us consider the following dependence for the system which is similar to the system in Figure 2a. For numerical simulation, we consider the coordinate of pressure change on the some sufficiently great value within one grid cell as a coordinate of disturbance. We can observe the coordinate of disturbance for a long time corresponding to more than one gas dynamic pulses of compressed air. The sizes of the system are mentioned below. Let us consider the planar channel with size of about 30 mm and the total length of 500 mm. The total period of gas dynamic impulse consists of 30 msec. The time for maximal pressure in the gas generator (2 atm) is about 15 ms, and the time for the atmospheric pressure is 15 ms. The power of energy supply was proportional to the pressure in the gas generator. The intensity of energy supply is equal to the heat release during the burning of the poor air-hydrogen mixture with the excess air ratio 2.8 at the temperature 1000 K. In Figure 3b we can observe this time dependence. As we can see from Figure 3b, the disturbance upstream motion is sufficiently slower than the downstream one.

2.3. The range of power for the transonic area

There are three cases of the shock wave structure formation during energy supply. The first of them corresponds to reconstruction of the steady solution between the pulses of energy. This reconstruction takes place upstream the energy supply. The second of them corresponds to the case of locking of the channel when the shock wave moves to the narrowing part of the channel. The other case is characterized by the following properties: the disturbance change its position but remains in the part of channel with cross section constant area. We named the range of the power between the mode of locking and the mode of reconstruction as corridor of transonic regime. This corridor as function of period of energy supply is presented in Figure 4.

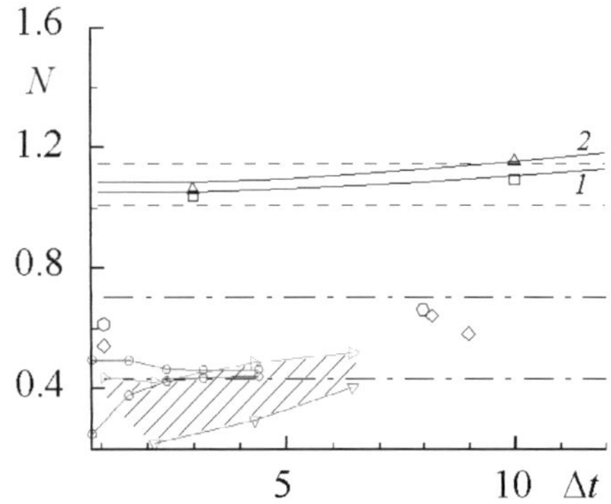

Figure 4. The chart is "normalized power N" – "period Δt of the energy supply".

The upper corridor (the lines *1*, *2*) corresponds to the case of transversal energy supply. The dashed corridor and point-dashed corridor correspond to the analytical estimation for the transversal and surface energy supply respectively. The yellow corridor corresponds to the axisymmetric channel with the annular jet. The shaded region corresponds to planar channel. The axisymmetric channel is equal to the planar channel according to the sections area and to their length. We can see that both corridors are in the same range of parameters but the corridor for the axisymmetric channel is narrower. Moreover, the power corresponding to the transonic corridor for transversal energy sources exceeds twice the power for the near-wall sources with jet.

3. The results of the calculation based on the unsteady Navier-Stokes equations closed by the turbulence model

3.1. The separation zone near the jet

The CFD simulation based on the two-dimensional unsteady Navier – Stokes equations closed by the SST turbulence model was performed for the planar channel with jet without heat release (the transversal size of the section of constant area is 30 mm). The flow after the channel travels to the receiver with low pressure. The flow in the channel was formed under the pressure of about 3 atm in the prechamber. The jet of compressed air is formed under the pressure of 2 atm. The flow in the inlet is characterized by the turbulence intensity of 0.025% and turbulence length scale of 0.00025 m. The numerical scheme of the third order with using high-order term relaxation is used. The distribution of density is presented in Figure 5 at the beginning of jet injection (the time from the injection beginning is about 700 μs).

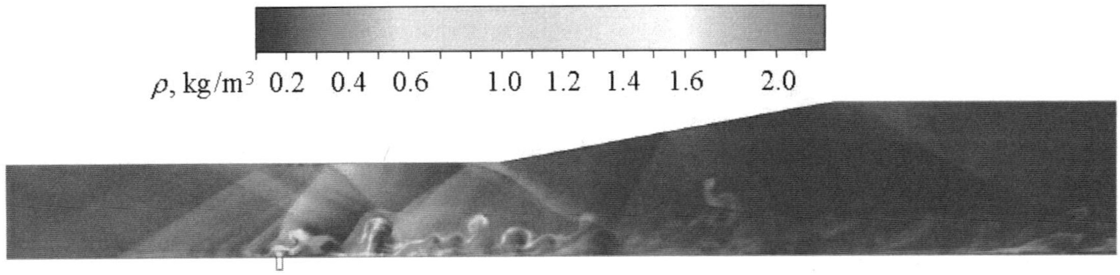

Figure 5. Density distribution for the channel from Figure 2a.

There is a sufficiently large separation zone near the jet. Figure 6 shows the experimental results of the authors [4]. In [4] the supersonic flow and flux from the pulse detonator (PD) with hole of compact shape are formed under the similar conditions: the Mach number in the flow is equal to 2 and the pressure in the PD is 2 atm (the time 1500 μs corresponds to the value of pressure into PD of 2 atm). The temperature of flow and jet is differs from the case of our numerical simulation. The transversal size in [4] is 136 mm, i.e. is fourfold larger than the transversal size in our case. The common view of Figure 5 and Figure 6 is qualitatively similar. In our case the separation zone is larger than that in [4], because we have a long gap in comparison with a small zone of compact shape. The jet diameter in [4] is 4.8 mm in comparison with transversal size of gap of 2 mm. It should be noted that the separation zones play the important role in the process of ignition and combustion.

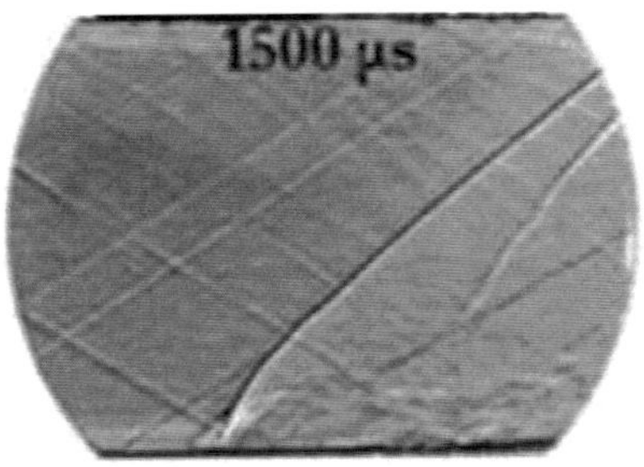

Figure 6. The experimental results of the authors [4] for the supersonic flow with the Mach number M=2 and the compact flux produced by the pulse detonator (PD).

3.2. Nuimerical simulation of the hydrogen combustion

The burning of homogeneous air-hydrogen mixture when injecting jet is simulated in [5], the other similar systems were considered in [6, 7]. The results of the CFD simulation based on the two-dimensional unsteady Navier – Stokes equations is presented. The SST k-ω turbulence model is used. The planar channel with the same parameters as in the case from Figure 5 is considered. The hydrogen flows at the beginning of the constant cross section. The parameters of turbulence were changed: the turbulence intensity was 0.1% and turbulence length scale was 0.001 m. The process of burning is very complex and sometimes may be followed by special effects [8, 9]. Here the burning is modeled by one reaction for the qualitative analysis. Hydrogen flows through the holes into the channel wall. These holes were located opposite to each other. The Mach number distribution is presented in Figure 7. This distribution is qualitatively similar to the results [7] and to the case with a pulse periodic energy supply (Figure 2). These results of burning are similar to the ones of impulse-periodic energy supply. For this reason some conclusions concerning the case with impulse- periodic energy supply can be applied for the case of burning.

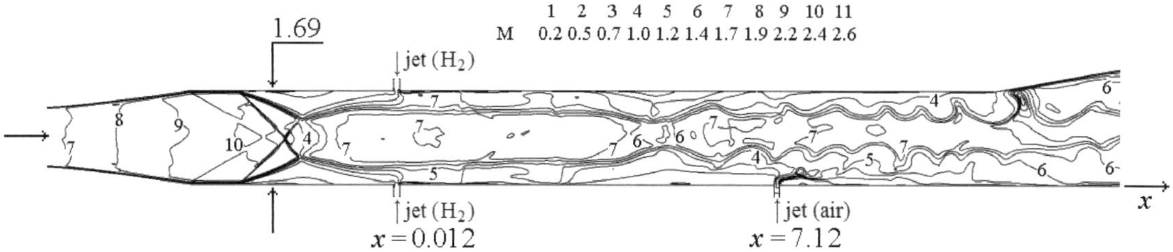

Figure 7. The Mach number distribution with marked isolines for burning of hydrogen.

4. Conclusion

The data analysis indicates the following dependences which can be useful to organize the process of fuel combustion in the supersonic flow at M=2 in the channel. The near-wall energy supply is more effective for the planar channel as well as for the axisymmetric one. Injecting compressed air at the appropriate pressure of the gas generator helps to increase the time of channel locking. There is a quality similarity in the gas dynamic structure between the near-wall burning and the pulse periodic near-wall energy supply when injecting compressed air. So, the majority of conclusions for pulse periodic energy supply can be applied in the case of burning at the following condition: there must be approximate equality between the value of transversal length of hydrogen flux and the similar size for the zone of energy supply.

Acknowledgments

The work was performed under the project of SB RAS III.22.6.2 and with the financial support of the RFBR (grant No. 17-08-00183). The authors gratefully acknowledge Christopher Hearsey and Valeriya Hearsey for their assistance.

References

[1] Tretyakov P, Zabaykin V and Prokhorov A 2015 *Proc. Russian Congress on fundamental problems of theoretical and applied mechanics* ed. Akhmetov D (Kazan: Kazan) p. 3778
[2] Zamuraev V and Kalinina A 2016 *Thermal processes in engineering* **7** 292
[3] Zamuraev V and Kalinina A 2016 *Journal of Engineering Physics and Thermophysics* **89** 688
[4] Timothy Ombrello and Campbell Carter 2015 *Journal of Propulsion and Power* **2** 654
[5] Surzhikov S 2014 *Physical-chemical kinetics in gas dynamics* **16** 3
[6] Egorian A and Kraiko A 2016 *Thermophysics and Aeromechanics* **2** 307
[7] Hudic G, Zhukov V, Manukovskiy K, Novikov N, Rykov Yu and Feodoritova O 2016 *Numerical simulation of high speed combustion using OpenFOAM package* 10 (Preprint IPM im. M. V. Keldysh) p 32
[8] Tupikin A, Tretyakov P, Denisova N, Zamashikov V and Kozulin V 2016 *Combustion, Explosion, and Shock Waves* **2** 49
[9] Tupikin A, Tretyakov P and Venediktov V 2017 *Combustion, Explosion, and Shock Waves* **1** 38

IOP Conf. Series: Journal of Physics: Conf. Series **894** (2017) 012119 doi:10.1088/1742-6596/894/1/012119

About control of convection intensity in chemical reacting equilibrium gas

I B Palymskiy[1], V I Palymskiy[2], P A Fomin[3] and A V Trifanov[4]

[1] Siberian State University of Telecommunications and Information Sciences, Novosibirsk, Russia
[2] Kutateladze Institute of Thermophysics, Novosibirsk, Russia
[3] Lavrentyev Institute of Hydrodynamics, Novosibirsk, Russia
[4] Rzhanov Institute of Semiconductor, Novosibirsk, Russia

E-mail: palymsky@hnet.ru

Abstract. The problem of the Rayleigh-Benard convection in a gas in a state of chemical equilibrium is considered in the frames of Boussinesq approximation. The method of linear analysis is used. Horizontal boundaries are assumed to be isothermal and free from shearing stresses. Thermodynamic parameters of a gas are described by the model of chemical equilibrium. Stoichiometric hydrogen-oxygen mixture is considered. It is shown that owing to recombination and dissociation processes the Rayleigh number in the gas under consideration equals to the corresponding number in chemically inert gas, multiplied with some additional factor. The critical Rayleigh number weakly depends on the pressure. Another and more effective way to control of the convection is to change temperature. The minimal value of the convection intensity is realized at very low and very high temperatures. At such temperatures the mixture under consideration can be considered as inert gas. An increase of temperature between its minimal and maximal values led to a shift of chemical equilibrium. As the result, the convection intensity pass through a local maximum.

1. Introduction

The Rayleigh–Benard convection is the classical field of fluid mechanics where a mathematical model based on the Boussinesq approximation and corresponding numerical methods were developed [1–4]. Traditionally, the major attention was paid to modeling convection in chemically inert gases and liquids. However, convective processes also occur in chemically reacting gases: combustion of gas mixtures in industrial reactors, engineering devices, and experimental facilities. Convection in a chemically equilibrium gas is fairly frequent case of convection in gases accompanied by chemical reactions, which occurs if the characteristic time of the chemical reaction is much smaller than the characteristic time of the convective process. It can be naturally expected that the qualitative and quantitative characteristics of convection are determined to a large extent by the dependence of the thermal expansion coefficient of the medium on temperature because it directly determines the buoyancy force responsible for convection.

The thermal expansion coefficient of a chemically inert gas is monotonic and inversely related to the temperature. In the case of a chemically equilibrium gas, the dependence of this coefficient on

Content from this work may be used under the terms of the Creative Commons Attribution 3.0 licence. Any further distribution of this work must maintain attribution to the author(s) and the title of the work, journal citation and DOI.
Published under licence by IOP Publishing Ltd

temperature changes considerably and becomes non-monotonic with a noticeable local maximum [5-7].

Authors in [8, 9] were the first to formulate and solve numerically a two-dimensional problem for Rayleigh-Benard convection in a chemically equilibrium gas, as well as considered linear and non-linear regimes. The molar mass and thermodynamic parameters of the gas were calculated according to a high accuracy model of chemical equilibrium [5-7]. Nevertheless, the physical and mathematical model provided in [8,9] is quite complicated and inconvenient. Some development of this physical and mathematical model may be found in [10]. However, results of the linear analysis [10] need an additional improvement.

In laboratory scales, the gas compressibility can be neglected and the convection media can be considered as incompressible. Moreover, if the domain has a certain geometry and small supercriticality, convection develops as two-dimensional rolls [4], which allows us to consider two-dimensional convective flows of a viscous incompressible fluid even at the first stage of our investigation.

Following [8-10], we consider a convective two-dimensional and non-stationary flow of a viscous incompressible fluid in the horizontal layer heated from below in the Boussinesq approximation. The boundaries of the layer are assumed isothermal and free from shearing stresses. Thermodynamic parameters of a chemical equilibrium gas are calculated according to the model, presented in [5-7].

In the frames of present work the linear analysis of convective instability will be investigated and the problem of control of Rayleigh-Benard convection by means of pressure and temperature will be considered.

2. Nomenclature

The following notations will be used:

Table 1. A list of notations used in this article.

Notation		Description
ρ, P, T and μ	-	the density, pressure, absolute temperature and molar mass of the gas;
R	-	the universal gas constant;
μ_{min}, μ_{max}	-	the molar masses of the gas in the completely dissociated and completely recombined states;
A, K_+	-	the rate constants of dissociation and recombination of the generalized reaction products;
θ	-	the effective excitation temperature of the vibrational degrees of freedom of the molecules;
E	-	the mean dissociation energy of the reaction products;
β	-	the thermal expansion coefficient of the reacting equilibrium gas;
$\beta_{in} = 1/T$	-	the thermal expansion coefficient of the inert ideal gas;
u, v	-	the velocities in x and y directions;
ν, χ	-	the kinematic and thermal diffusivity coefficients;
T_h, T_c	-	the temperatures of the lower and upper boundaries;
H	-	the thickness of the layer;
ψ	-	the stream function ($\psi_y = u$ and $\psi_x = -v$);
ω	-	the vorticity $\omega = v_x - u_y$;
$Ra = gH^3\delta T/(T\chi\nu)$	-	Rayleigh number defined for inert gas;
Ra_{cr}, α_m	-	the critical values of Rayleigh and wave numbers;
$Pr = \nu/\chi$	-	Prandtl number;

λ, α, πm	-	the eigenvalue and the wavenumbers of linear problem in x and y directions;
$Q = T\text{-}T_e$	-	the temperature deviation from linear equilibrium distribution.

3. Kinetic model

To describe chemical equilibrium in the gas mixture, we use a highly accurate kinetic model [5-7]. According to this approach, the density of a chemical equilibrium ideal gas can be considered as a function of the absolute temperature and pressure:

$$\rho = \frac{p\mu}{RT},\ \mu(T) = \mu_{\max} + 0.5\cdot(f - \sqrt{f^2 + 4f(\mu_{\max} - \mu_{\min})}),$$
$$f = \frac{AR\mu_{\max}^2 T^{1.75}}{4K_+ P\mu_{\min}}\cdot(1 - \exp(-\Theta/T))^{1.5}\exp(-E/RT)). \tag{1}$$

The molar mass of the gas can vary within the interval $\mu_{min} \le \mu \le \mu_{max}$. We have $\mu \to \mu_{max}$ with decreasing temperature and $\mu \to \mu_{min}$ with increasing temperature.

We assume that the pressure P in the gas mixture (external pressure) is constant. Then, the gas density ρ, as is seen from system (1), depends on the absolute temperature T only. Note that the molar mass and density have monotonic dependence from absolute temperature [8,9]. The more detail description of used kinetic model may be found in [10].

The present paper refers to a stoichiometric hydrogen-oxygen mixture. The numerical values of the constants were as follow: $E = 459.2\cdot kJ/mole$, $\Theta = 4000K$, $K_+ = 6\cdot 10^8 m^6/(kmole^2\cdot sec)$, $A = 5.1\cdot 10^{10} m^3/(kmole\cdot sec\cdot K^{3/4})$, $\mu_{min} = 6kg/kmole$, $\mu_{max} = 18kg/kmole$ [5-7].

Considering the expression for f at comparatively high temperatures T, we have the following

$$T \to \infty,\ f \sim T^{0.25}\ (E/R \ll T),\ \mu = \mu_{\min} + (\mu_{\max} - \mu_{\min})^2/f + O(f^{-2}).$$

The thermal expansion coefficient β can be found dependent on the density or the molar mass as

$$\beta = -\frac{1}{\rho}\cdot\frac{\partial\rho}{\partial T} = -\frac{d}{dT}\ln(\mu/T) = \frac{1}{T} - \frac{1}{\mu}\frac{d\mu}{dT}.$$

Notice that the asymptotic approximation of the molar mass μ to constants as the temperature is reduced or increased allows us to find the asymptotics for the thermal expansion coefficient as $\beta = 1/T$.

4. Mathematical model

Convection of incompressible fluids in the Boussinesq approximation is described by the following system of equations [1,2]:

$$\omega_t + \frac{1}{\Pr}(\psi_y\omega_x - \psi_x\omega_y) = \Delta\omega + C\cdot Ra\mathrm{Q}_x,\qquad \Delta\psi = -\omega,$$
$$\mathrm{Q}_t + \frac{1}{\Pr}(\psi_y\mathrm{Q}_x - \psi_x\mathrm{Q}_y) = \frac{1}{\Pr}\Delta\mathrm{Q} - \frac{1}{\Pr}\psi_x,\ C = \frac{\beta(T)}{\beta_{in}(T)} = T\cdot\beta(T). \tag{2}$$

We choose the following reference values for the layer: H for length, H^2/ν for time, χ/H for velocity, $\rho_0\nu\chi/H^2$ for pressure, $\rho_0 = \rho(T)$ for density, and $\delta T = T_h - T_c$ for temperature, Q is temperature deviation from equilibrium linear distribution and $\Delta f = f_{xx} + f_{yy}$ is the Laplace operator.

According to the buoyancy force in the first equation of system (2), we see that the convection rate is defined by $C\cdot Ra$. The Rayleigh criterion specifies the ratio between the buoyancy force and the viscosity and depends on the temperature gradient or the degree of heating, while the dimensionless factor C depends on the recombination and dissociation processes under the temperature given.

Figures 1 - 3 were plotted with $P = 1atm$.

The solid curves in Figures 1 and 2 show the thermal expansion coefficient $\beta = \beta(T)$ and factor $C = C(T)$. We see there asymptotics (dashed curve 1) with close values at high temperature. The temperature interval where the thermal expansion coefficient β and factor C demonstrate an anomalous (not-monotonic) behavior with noticeable local maximum are also seen. As it is shown below, it is this anomalous behavior that causes a possibility of convective instability amplification in comparison with inert gas.

Figure 2 shows that when the temperature falls or rises, the convection intensity decreases up to convection intensity level in an ideal inert gas. Note that the value of local maximum C has very weakly dependency from pressure and vanishes in a practically linear way with respect to $lg(P)$ as

$$C(P) = 3.9241 - 0.44393 \cdot lg(P/1atm).$$

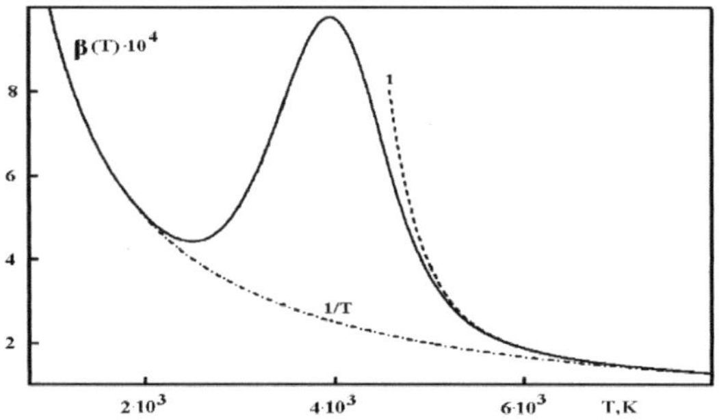

Figure1. The thermal expansion coefficient.

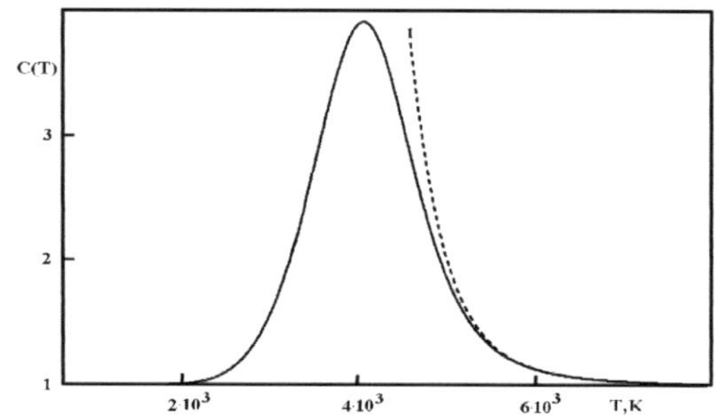

Figure 2. The factor $C = C(T)$.

5. Linear analysis

Consider the linear analog of system (2) and seeking harmonic solution in the time exponential form [2,10], we find:

$$\lambda_{1,2} = \frac{1+\Pr}{2\Pr} S \pm \sqrt{\left(\frac{1-\Pr}{2\Pr}\right)^2 S^2 + \frac{C \cdot Ra \cdot \alpha^2}{S \Pr}},$$

where $S = \alpha^2 + m^2\pi^2$. Here eigenvalue λ is matched by amplitude losses and $\lambda < 0$ by amplitude rises in harmonic solutions. Formulas for an inert medium correspond to written here expressions with $C = 1$.

Assuming eigenvalue λ in the last equation is equal to 0, one can obtain the Rayleigh number corresponding to the neutral curve:

$$Ra = \frac{(m^2\pi^2 + \alpha^2)^3}{C \cdot \alpha^2}.$$

The critical value of the Rayleigh number (minimum of Ra versus α with $m = 1$) can be found by the formula [1]:

$$Ra_{cr} = \frac{27}{4C}\pi^4 = \frac{657.511}{C}, \quad \alpha_{cr} = 2.221.$$

And if $C - 1$ is small (for $|T - \Theta| >> 1$), one can obtain that:

$$Ra_{cr} = \frac{657.511}{C} = 657.511 \cdot (2 - C).$$

The factor 657.511 is identified here as the critical value of the Rayleigh number for the convection in a chemically inert media [1,2], and the factor $1/C$ as the function of temperature and pressure is due to recombination and dissociation processes.

Figure 3 shows the critical Rayleigh number as a function of T (solid curve), its asymptotic (dashed curve 1) and $Ra_{cr} = 657.511\cdot(2\text{-}C)$ (dashed curve 2). It can be seen that in domain ($3000K \le T \le 5000K$, $T \approx \Theta = 4000K$) the chemical reactions in gas considerably decrease the critical Rayleigh number.

Taking into account at a high temperature the expression for f, we can obtain the far-field asymptotics for β, C, Ra_{cr} and increment λ as:

$$f = \frac{AR\mu_{\max}^2\Theta^{1.5}}{4K_+P\mu_{\min}}T^{0.25} \ (T \to \infty)$$

$$\beta = \frac{1}{T}(1+C_1), \quad C = 1+C_1, \quad Ra_{cr} = 657.511\cdot(1-C_1),$$

$$\lambda_r = \lambda_{in} \pm \frac{Ra\,\alpha^2 C_1}{S^{0.5}\sqrt{(1-\Pr)^2S^3+4Ra\,\alpha^2\Pr}}, \quad C_1 = \frac{(\mu_{\max}-\mu_{\min})^2K_+P}{AR\mu_{\max}^2\Theta^{1.5}T^{0.25}}.$$

Here λ_r and λ_{in} are increments for the reacting and inert gases correspondingly. The dependences given show that β and C approach asymptotics with the rise of the temperature quite slowly, which means that the corresponding boundary is fuzzy.

The convective processes intensiveness can be influenced. The critical Rayleigh number Ra_{cr} is weakly depending on the pressure, the intensiveness decreases as the pressure grows, changing from the maximum values at low pressure to minimal values at high pressure. Another more cardinal way to affect the intensiveness is to change the temperature. The intensiveness of convective processes is equal to its minimal value (inert media) at high and low temperatures and its maximal value in between. Of course, it is possible to change both the pressure and the temperature.

To study the possibility to control convective processes by changing the pressure and the temperature in more details, let us consider the instability domain on P-T plane in Figure 4.

A curve 1 in Figure 4 correspond to the temperatures at which the maximum convection intensiveness at given pressure is observed (maximum C with respect to T corresponds to curve Ra_{cr}). Curves 2 and curve 3 correspond to the top and low boundaries of domain where reacting-inert gas transitions are observed.

It is obvious that the highest increase in temperature occurs at the upper boundary of the reacting-inert gas transition (curve 2), whereas the other curves demonstrate a substantially smaller change of temperature.

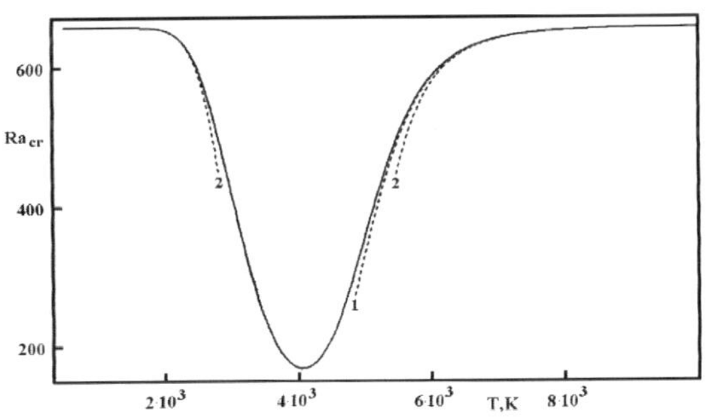

Figure 3. The neutral curve on the T-Ra plane.

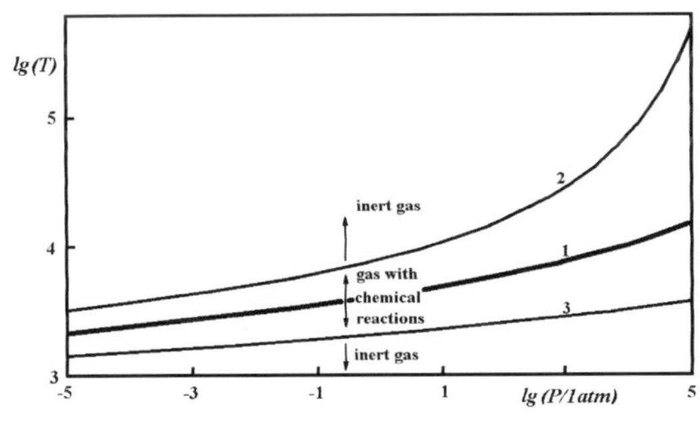

Figure 4. The instability domain on the P-T plane.

6. Conclusion

Given recombination and dissociation processes, the Rayleigh number in the gas under consideration equals to the corresponding number in chemically inert gas, multiplied with some additional factor $C = C(T)$. Hence, all the regimes and characteristics of convective flows of a chemical equilibrium gas can be derived from the corresponding system of equations of convection in a chemically inert gas by means of re-determination of Rayleigh number.

The neutral curve on the plane *T-Ra* shows the significant decreasing of the critical Rayleigh number in the reacting gas. Note that the critical Rayleigh number (minimal value with respect to *T*) weakly depends on the pressure. Another and more effective way to control of the convection is to change temperature. The minimal value of the convection intensity is realized at very low and very high temperatures. At such temperatures the mixture under consideration can be considered as inert gas. An increase of temperature between its minimal and maximal values led to a shift of chemical equilibrium. As the result of dissociation and recombination processes, the convection intensity pass through a local maximum.

7. Discussion

In the present work the far-field high-temperature asymptotes for factor C, critical Rayleigh number and the increment λ are derived. It would be of certain interest to investigate also the low-temperature asymptotic curves.

The performed linear analysis shows the existence of the anomalous (non-monotonic) behavior of the thermal expansion coefficient on temperature. It seems to be possible that such anomalous behavior must bring to the secondary instability. It would be interesting to investigate such kind of instability.

Acknowledgements

We acknowledge financial support of this work by Russian Foundation for Basic Research Gr. №15-08-05166 and Gr. №17-58-53100.

References

[1] Gershuni G Z and Zhukhovitskii E M 1976 Convective stability of incompressible fluids (Jerusalem: Israel Program for Scientific Translations)

[2] Palymskiy I B 2011 Turbulent Rayleigh-Benard convection. Numerical method and results of simulation (Turbulentnaya konvektziya Rehleya-Benara. Chislennyi metod i resultaty raschetov) (Germany: LAP) p 224

[3] Palymskiy I B 2014 Vorticity scale and integral values of Rayleigh-Benard convection Comp. Therm. Sc. **6**(2) 113-127

[4] Getling A V 1998 Rayleigh-Bénard convection: structures and dynamics (Singapore: World Scientific Publishing Co) p 249

[5] Nikolaev Yu A 1978 Model of the kinetics of chemical reactions at high temperatures Comb., Expl. Sh. Wav. **14**(4) 468-471

[6] Nikolaev Yu A and Fomin P A 1982 Analysis of equilibrium flows of chemically reacting gases Comb., Expl. Sh. Wav. **18**(1) 53-58

[7] Fedorov A V, Fomin P A, Fomin V M, Tropin D A and Chen J R 2012 Mathematical analysis of detonation suppression by inert particles (Kaohsiung: Kao Tech Publishing) p 124

[8] Palymskiy I B, Fomin P A and Hieronymus H 2007 The Rayleigh–Benard convection in gas with chemical reactions Sib. J. Num. Math. **10**(4) 371-383

[9] Palymskiy I B, Fomin P A and Hieronymus H 2008 Rayleigh-Benard convection in a chemical equilibrium gas (simulation of detonation wave initiation) App. Math. Mod. **32**(5) 660-676

[10] Palymskiy I B, Palymskiy V I and Fomin P A Rayleigh-Benard convection in a chemically active gas in the chemical equilibrium state Comb., Expl. Sh. Wav. **53**(2) 123-133

Calculation of heat sink around cracks formed under pulsed heat load

G G Lazareva[1,2,3], A S Arakcheev[2,4], I V Kandaurov[2], A A Kasatov[2], V V Kurkuchekov[2], A G Maksimova[3], V A Popov[3], A A Shoshin[2], A V Snytnikov[1], Yu A Trunev[2], A A Vasilyev [2] and L N Vyacheslavov[2]

[1] Institute of Computational Mathematics and Mathematical Geophysics SB RAS, Novosibirsk, Russia
[2] Institute of Nuclear Physics SB RAS, Novosibirsk, Russia
[3] Novosibirsk State University, Novosibirsk, Russia
[4] Novosibirsk State Technical University. Novosibirsk, Russia

E-mail: lazareva@ssd.sscc.ru, asarakcheev@gmail.com

Abstract. The experimental and numerical simulations of the conditions causing the intensive erosion and expected to be realized infusion reactor were carried out. The influence of relevant pulsed heat loads to tungsten was simulated using a powerful electron beam source in BINP. The mechanical destruction, melting and splashing of the material were observed. The laboratory experiments are accompanied by computational ones. Computational experiment allowed to quantitatively describe the overheating near the cracks, caused by parallel to surface cracks.

1. Introduction

The plasma-facing components (PFC) of the reactor tokamak prototype ITER will be subjected to intense heat load during the discharge [1]. The load will consist of continuous and pulsed parts. The continuous one is result of plasma flow along scrap off layer (up to 20 MW/m^2) and the pulsed loads are results of type I ELMs, breakdowns, etc. The energy release of the pulsed heat load can be about 80 MJ/m^2 during 1 μs. Currently experimental facilities cannot produce the conditions of the pulsed plasma loads. So the quasi-stationary plasma accelerators [2,3], electron beam sources [4,5] and lasers [6] are used for experimental simulations of the pulsed heat loads. The obtained experimental data on the erosion of materials become basis of the computational models for predictions of the tokamak operational regimes.

The proper operation of fusion reactor assumes the heat sink from the surface of the PFC. The formation of the parallel to surface cracks can make significant obstacle for the heat sink and excessively increase the surface temperature (see Figure 1a). The crack formation was observed by post-mortem analysis in many experimental works [7,8], although numerical modeling of the surface overheating above the cracks was not carried out. The overheating near cracks was measured at the BETA facility in BINP. The tungsten samples were heated by submillisecond powerful electron beam and were measured by in-situ optical diagnostics at the experiments [9].

Content from this work may be used under the terms of the Creative Commons Attribution 3.0 licence. Any further distribution of this work must maintain attribution to the author(s) and the title of the work, journal citation and DOI.
Published under licence by IOP Publishing Ltd

2. Statement of the problem

The cracks form in the surface layer of tungsten in a result of thermal expansion during pulsed heat load. It was found experimentally that not only perpendicular to the surface cracks appear after pulsed heating, but also the parallel to the surface cracks form. The latter may significantly affect the heat sink from the irradiated surface of material. The aim of this work is to calculate the heat sink at pulsed heat loads near parallel to the surface of the material crack. In particular, we attempt to reproduce the measured temperature distribution on the surface and to predict the typical dependencies of the suppression of the heat sink on the geometry of cracks.

The computational domain is the rectangular area of the sample cross-section with a subsurface crack (see Figure 1a). The photo demonstrates the parallel to surface crack at the depth of ~250 microns. The left part of the sample surface had a weak mechanical contact with a deeper part of material and was lost during the preliminary cutting. On the boundary the heat flux is specified. The perpendicular to heated surface crack is located on the boundary of the computational region $x = 0$ and transits to the parallel to surface crack (see Figure 1b).

a 

b 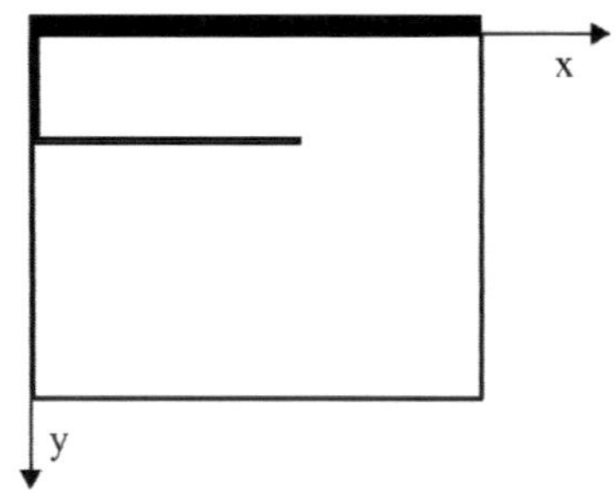

Figure 1. Cross-section of the tungsten sample made with the electron microscope (a). Scheme of the computational domain (b).

The thermal conductivity equation (1) was solved in the area using method [10].

$$c(T)\rho(T)\frac{\partial T}{\partial t} = div\lambda(T)gradT + Q, \quad (1)$$

where T is the temperature, $c(T)$ is the specific heat capacity, $\rho(T)$ is the density, $\lambda(T)$ is the thermal conductivity, Q is the heat source.

The initial and the boundary conditions are of the form:

$$T(0,x,y) = T_0, \quad (n,\nabla T)\big|_\gamma = 0, \quad \left.\frac{\partial T}{\partial y}\right|_{y=0} = \frac{W_\Gamma(t,x)}{\lambda(T)},$$

where $W_\Gamma(t,x)$ is the heat flow power (see Figure 2a), with a duration of 100-200 µs.

The subsurface crack set condition $(n,\nabla T)\big|_{cr} = 0$. The typical run time is of 1000 µs. The following characteristic values were used: $t = 1\mu s$, $x = 10^{-2}$ mm, $T = 10^3$ K, $\lambda = 10^{-2}\dfrac{W}{mm\cdot K}$, $\rho = 10^{-5}\dfrac{kg}{mm^3}$, $c = 10^7\dfrac{W\cdot\mu s}{kg\cdot K}$, $W_\Gamma = 10^3\dfrac{W}{mm^2}$.

Measurement of thermal characteristics of refractory metals is rather a difficult task. So generally reference books and articles present approximated or theoretically predicted dependencies with the accuracy about 10% or worse. Thermal conductivity and specific heat of solid tungsten are taken from [11]. The thermal conductivity of liquid tungsten is estimated from [12-14]. In the temperature range of density, specific heat capacity and thermal conductivity are specified as dependencies on the temperature (2-4).

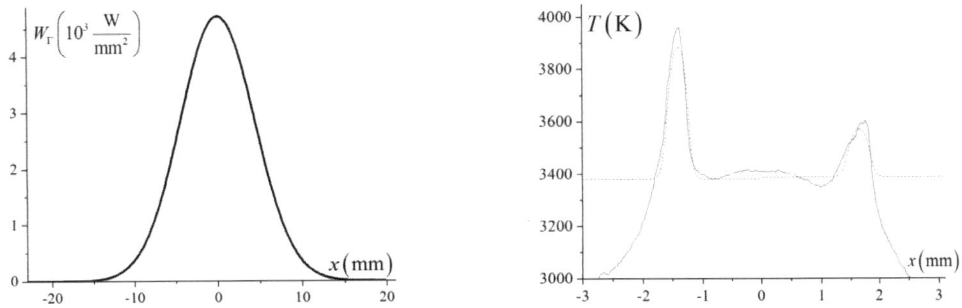

a b

Figure 2. Distribution of the surface power of heat flow (a) and temperature (b): experimental data (solid line) and calculated results (dotted line).

The density (see Figure 3a):

$$\rho(T)=\begin{cases}\begin{bmatrix}19.25 - 2.66207\cdot 10^{-4}\cdot(T - 273.15)\\ -3.0595\cdot 10^{-9}\cdot(T - 273.15)^2\\ -9.5185\cdot 10^{-12}\cdot(T - 273.15)^3\end{bmatrix}\cdot 10^{-6}\ \mathrm{kg/mm^3}, & 300\mathrm{K}\le T\le 3695\mathrm{K},\\ \begin{bmatrix}16.267 - 7.679\cdot 10^{-4}\cdot(T - 3695)\\ -8.091\cdot 10^{-8}\cdot(T - 3695)^2\end{bmatrix}\cdot 10^{-6}\ \mathrm{kg/mm^3}, & 3695\mathrm{K}\le T\le 6000\mathrm{K},\end{cases} \quad (2)$$

The thermal conductivity (see Figure 3b):

$$\lambda(T)=\begin{cases}\begin{bmatrix}149.441 - 45.466\cdot 10^{-3}\ T\ +\ 13.193\cdot 10^{-6}T^2 -\\ -1.484\cdot 10^{-9}T^3 + 3.866\cdot 10^{6}T^{-2}\end{bmatrix}\cdot 10^{-3}\dfrac{\mathrm{W}}{\mathrm{mm\cdot K}}, & 300\mathrm{K}\le T\le 3695\mathrm{K},\\ \begin{bmatrix}66.6212 + 0.02086(T\ -\ 3695) -\\ -3.7585\cdot 10^{-6}(T\ -\ 3695)^2\end{bmatrix}\cdot 10^{-3}\dfrac{\mathrm{W}}{\mathrm{mm\cdot K}}, & 3695\mathrm{K}\le T\le 6000\mathrm{K},\end{cases} \quad (3)$$

The specific heat capacity (see Figure 3c):

$$c(T)=\begin{cases}\begin{bmatrix}21.868372 + 8.068661\cdot 10^{-3}\cdot T -\\ -3.756196\cdot 10^{-6}\cdot T^2 + 1.075862\cdot 10^{-9}\cdot T^3 +\\ +1.406637\cdot 10^{4}\cdot T^{-2}\end{bmatrix}\cdot\dfrac{10^{-6}}{0.186}\dfrac{\mathrm{W\cdot\mu s}}{\mathrm{kg\cdot K}}, & 300\mathrm{K}\le T\le 3080\mathrm{K},\\ \left[2.022 + 1.315\cdot 10^{-2}\cdot \mathrm{T}\right]\cdot\dfrac{10^{-6}}{0.186}\dfrac{\mathrm{W\cdot\mu s}}{\mathrm{kg\cdot K}}, & 3080\mathrm{K}\le T\le 3695\mathrm{K},\\ 51.3\cdot\dfrac{10^{-6}}{0.186}\dfrac{\mathrm{W\cdot\mu s}}{\mathrm{kg\cdot K}}, & 3695\mathrm{K}\le T\le 6000\mathrm{K},\end{cases} \quad (4)$$

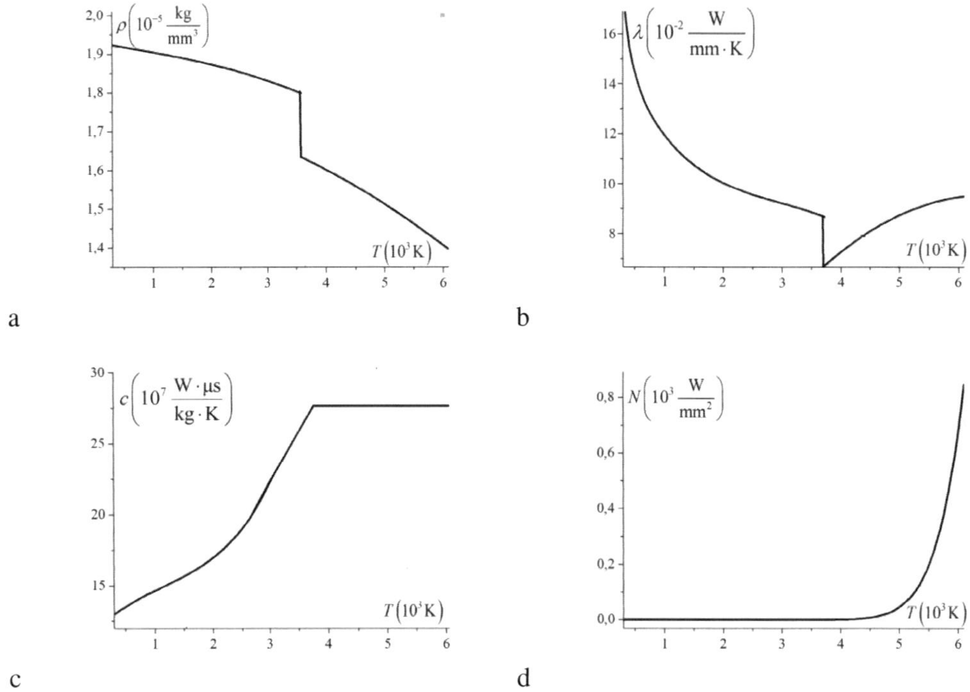

Figure 3. Plots of the temperature-dependent density (a), thermal conductivity (b), specific heat capacity (c) capacity and the energy losses caused by evaporation (d).

3. Melting and evaporation processes

The enthalpy of the phase transition L_m [15] is added in equation (1) to take into account the melting (see Figure 4) and the evaporation:

$$\left(c(T)\rho(T)+L_m\delta\left(T,\Delta\right)\right)\frac{\partial T}{\partial t}=div\lambda(T)gradT+Q,$$

$$\delta\left(T,\Delta\right)=\begin{cases}\frac{1}{2\Delta}, & \left|T-T_m\right|\leq\Delta,\\ 0, & \left|T-T_m\right|>\Delta,\end{cases}$$

$$x_m=\eta_m(t),\quad \left[\lambda_m\frac{\partial T}{\partial x}\right]=-L_m\frac{\partial\eta_m}{\partial t},$$

where the melting temperature $T_m=3695$ K, and the melting heat $L_m=51.1\cdot10^5\,\frac{\mathrm{W}\cdot\mu\mathrm{s}}{\mathrm{mm}^3}$.

The following expression was taken for the energy losses caused by evaporation: $N(T_\Gamma)=L_e\cdot\frac{1}{S}\frac{dm}{dt}$ (see Figure 3d). The complete heat flux is $W_*(t,x)=W_\Gamma(t,x)-N(T_\Gamma)$, where $P(T)$ - saturated vapor pressure, $\frac{1}{S}\frac{dm}{dt}$ - the rate of mass losses caused evaporation:

$$L_e = 4.482 \cdot 10^{12} \ \frac{\mathrm{W} \cdot \mu\mathrm{s}}{\mathrm{kg}},$$

$$\frac{1}{S}\frac{dm}{dt} = P(T)\sqrt{\frac{M}{2\pi RT\big|_{y=0}}} = \exp\left(26.19104 - \frac{83971.3}{T\big|_{y=0}}\right)\sqrt{\frac{0.184}{2\pi 8.314 T\big|_{y=0}}}10^{-12}\,\frac{\mathrm{kg}}{\mathrm{mm}^2\cdot\mu\mathrm{s}}.$$

To take into account the deep heating the possibility of the change of the heat flow through the boundary for the heat source inside a thin layer within the region is provided:

$$Q = \begin{cases} 0, & \left.\dfrac{\partial T}{\partial y}\right|_{y=0} = \dfrac{W_\Gamma(t,x,y)}{\lambda(T)}, \\ \begin{cases} \dfrac{W_\Gamma(t,x,y)}{y^W{}_0 c(T)\rho(T)}, & 0 \le y \le y^W{}_0, \\ 0, & y_0^W \le y. \end{cases} & \left.\dfrac{\partial T}{\partial y}\right|_{y=0} = 0. \end{cases}$$

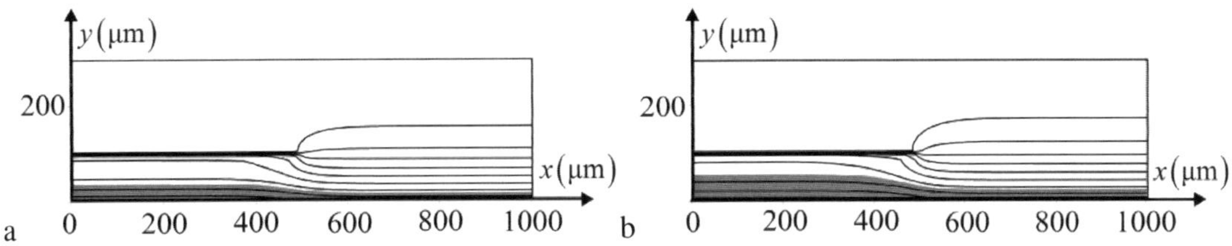

Figure 4. Temperature distribution on cross-section at time 200 μs (a) 230 μs (b). Area of the melt is grayed out. Step between temperature level lines is 500 K, $W_\Gamma = 2.5 \cdot 10^3 \ \frac{\mathrm{W}}{\mathrm{mm}^2}$.

4. Results of numerical simulation

The results of computational experiments match measurement data (see Figure 2b). Calculation parameters: the duration of exposure of the beam 186, ISS, time of measurement 200 μs from the beginning of beam exposure time - 10 μs, the left crack depth of about 0.12 mm, the left crack length of about 0.2 mm, the right crack depth of about 0.15 mm, the right crack length of about 0.145 mm. The heating was assumed to be uniform $W_\Gamma = 3 \cdot 10^3 \ \frac{\mathrm{W}}{\mathrm{mm}^2}$.

Given the isotherms on the cross section at the time of measurement 200 μs and 400 μs from the start of the beam (see Figure 4) for exposure of a beam of 400 μs. In the case of duration of exposure of the beam 186 of ISS at time 400 microseconds from the start of the beam material have time to cool down and the temperature reaches the melting point.

The calculations confirmed the hypothesis that nonuniformity of the surface temperature of tungsten associated with the presence of subsurface cracks.

The extension of the model involves the inclusion of thermodynamic coordinated conservation equations for solid, liquid and gas phases of a substance [16-18] in the axisymmetric statement of the problem with the purpose of studying the processes happening during melting and evaporating of the substance.

5. Conclusions

The experimental and numerical simulations of the conditions causing the intensive erosion and expected to be realized infusion reactor were carried out. The influence of relevant pulsed heat loads to tungsten was simulated using a powerful electron beam source in BINP. The mechanical destruction, melting and splashing of the material were observed. The laboratory experiments are accompanied by computational ones. Computational experiment allowed to quantitatively describe the overheating near the cracks, caused by parallel to surface cracks.

Acknowledgments
This work was supported by Program of RAS (project N 0315-2015-0010), by Russian President grant (project SP-4010.2016.1) and by Russian Science Foundation (project N 17-79-20203).

References

[1] Pitts R A, Carpentier S, Escourbiac F, Hirai T, Komarov V, Lisgo S et al 2013 A full tungsten divertor for ITER: Physics issues and design status *J. Nucl. Mater.* **438** S48-S56.
[2] Klimov N, Podkovyrov V, Zhitlukhin A, Kovalenko D, Bazylev B, Landman I, Pestchanyi S, Janeschitz G, Federici G, Merola M, Loarte A, Linke J, Hirai T and Compan J. 2009 Experimental study of PFCs erosion under ITER-like transient loads at plasma gun facility QSPA *J. Nucl. Mater.* **390-391** 721-726.
[3] Tereshin V I, Bandura A N, Byrka O V et al 2007 Application of powerful quasi-steady-state plasma accelerators for simulation of ITER transient heat loads on divertor surfaces *Plasma Phys. Control. Fusion* **49** A231-A239.
[4] Wirtz M et al 2013 Thermal shock response of deformed and recrystallised tungsten *Fus. Eng. Design* **88** 1768–1772.
[5] Arakcheev A S, Burdakov A V, Trunev Y A, Kandaurov I V, Kasatov A A, Kurkuchekov V V, Mekler K I, Popov V A, Shoshin A A, Skovorodin D I, Vasilyev A A and Vyacheslavov L N 2016 Heating of tungsten target by intense pulse electron beam *AIP Conf. Proc.* **1771** 060016
[6] Huber A, Arakcheev A, Sergienko G, Zlobinski M et al 2014 Investigation of the impact of transient heat loads applied by laser irradiation on ITER-grade tungsten *Phys. Scripta* **T159** 014005.
[7] Pintsuk G and Loewenhoff 2013 The Impact of microstructure on the plasma performance of industrial and high-end tungsten grades *Journal of Nuclear Materials* **438** S945-S948.
[8] Wirtz M, Cempura G, Linke J, Pintsuk G and Uytdenhouwen I 2013 Thermal shock response of deformed and recrystallised tungsten *Fusion Engineering and Design* **88:9** 1768-1772.
[9] Vasilyev A A et al 2016 In-situ imaging of tungsten surface modification under ITER-like transient heat loads *Nucl. Matter Energy*. DOI:10.1016/j.nme.2016.11.017
[10] Konovalov A N and Popov Yu P 2015 Explicitly solvable optimal discrete models with controlled disbalance of the total mechanical energy for dynamical problems of linear elasticity *Siberian Math. J.* **56**(5) 872–878.
[11] Davis JW and Smith PD 1996 ITER material properties handbook *Journal of Nuclear Materials* **233** 1593-1596.
[12] Ho C Y, Powell R W and Liley P E 1972 Thermal Conductivity of Elements *J. Phys. Chem. Ref. Data* **1** 279-421.
[13] Pottlacher G 1999 Thermal conductivity of pulse-heated liquid metals at melting and in the liquid phase *J. of Non-Crystalline Solids* **250** 177-181.
[14] Taluts S G 2001 Experimental study of the thermophysical properties of transition metals and alloys based on iron at high temperatures (abstract. dis. d-RA Fiz.-Mat. Sciences: 01.04.14, Yekaterinburg: Publ. house UGGA) p. 38.
[15] Samarskii A A and Vabishchevich P N 2003 *Computational heat transfer* (Moscow: editorial URSS) p. 784.
[16] Godunov S K and Kulikov I M 2014 Computation of discontinuous solutions of fluid dynamics equations with entropy nondecrease guarantee *Comp. Math. and Math. Phys.* **54:6** 1012–1024.
[17] Godunov S K, Kiselev S P, Kulikov I M and Mali V I 2013 Numerical and experimental simulation of wave formation during explosion welding *Proc. of the Steklov Institute of*

Mathematics **281**(1) 12–26.
[18] Godunov S, Kiselev S, Kulikov I and Mali V 2014 The numerical simulation of shock-wave processes in the elastic and plastic materials at different (atomic, meso and thermodynamic) structural levels (Izhevsk: Institute of computer researches) p. 289.

Static and dynamic overdetermined problems in elasticity, plasticity and post-limit deformation

A I Chanyshev[1,2], O E Belousova[1] and I M Abdulin[1]

[1]Chinakal Institute of Mining Siberian Branch of Russian Academy of Sciences, Novosibirsk, Russia
[2]Novosibirsk State University of Economics and Management, Novosibirsk, Russia

Abstract. Some overdetermined problems, formulated for the Laplace equation in a circle (for arbitrary class of functions, not necessarily analytical) and a half-plane, the heat equation, the one-dimensional wave equation, the elasticity equations for planar deformation, the plasticity and deformations theory problem for a plane with a circular hole, dynamic elasticity theory problems for a half-plane and a half-space with simultaneously known on one its boundary the Dirichlet condition, and the Neumann condition are investigated. Analytical and numerical solutions are constructed, stress-strain states, thermal and other states are restored, internal structure of the body, concentrated sources are determined.

1. Introduction

There are various statements of boundary-value problems: Dirichlet, Neumann, Robin [1-3]. The paper deals with boundary value problems in Cauchy's formulation, when both the Dirichlet and the Neumann conditions are described on the known contour. Some examples of analytical solutions are presented.

2. Laplace equation for the half-plane

Let the Laplace equation for the half-plane have the form:

$$\frac{\partial^2 u}{\partial x^2}+\frac{\partial^2 u}{\partial y^2}=0 \ \ (-\infty<x<\infty,\ y\le 0), \tag{1}$$

with boundary conditions:

$$u\big|_{y=0}=2g_1(x),\ \left.\frac{\partial u}{\partial y}\right|_{y=0}=2g_2'(x). \tag{2}$$

General solution $u=f(z)+\overline{f(z)}$ [4], where $f(z)=g_1(z)-ig_2(z)$, $z=x+iy$, g_1, g_2 – given boundary functions.

3. Elasticity theory plane problem for a half-plane

Boundary conditions at $y=0$ take the following form

$$\sigma_y=f_1(x),\ \tau_{xy}=f_2(x),\ u_x=f_3(x),\ u_y=f_4(x), \tag{3}$$

Content from this work may be used under the terms of the Creative Commons Attribution 3.0 licence. Any further distribution of this work must maintain attribution to the author(s) and the title of the work, journal citation and DOI.
Published under licence by IOP Publishing Ltd

IOP Conf. Series: Journal of Physics: Conf. Series **894** (2017) 012121 doi:10.1088/1742-6596/894/1/012121

where f_1, f_2, f_3, f_4 – arbitrary boundary functions. For this special problem, there is a solution in Kolosov-Muskhelishvili potentials:

$$\varphi(z)=\frac{\int\limits_{\Gamma}[f_1(z)-if_2(z)]dz}{1+\aleph}+\frac{2\mu}{1+\aleph}[f_3(z)+if_4(z)]+C_1,$$

$$\psi(z)=\frac{\aleph}{1+\aleph}\int\limits_{\Gamma}[f_1(z)+if_2(z)]dz-\frac{z}{1+\aleph}[f_1(z)-if_2(z)]- \tag{4}$$

$$-\frac{2\mu}{1+\aleph}[f_3(z)-if_4(z)+z(f_3'(z)+if_4'(z))]+C_2,$$

where $\aleph$ is the Muskhelishvili parameter. A similar solution has other specified problems.

4. One-dimensional case of the wave equation

Consider the classic formulation of the one-dimensioned wave equation. In the beginning, we consider a simplified situation. We have a homogeneous semi-infinite rod immersed into the soil. For it there is an equation describing the motion of its elements:

$$\frac{\partial^2 u}{\partial x^2}=\frac{1}{a^2}\cdot\frac{\partial^2 u}{\partial t^2}, \tag{5}$$

where $a=\sqrt{\rho/E}$ (ρ – density, E – Young's modulus), u – particle displacement.

There is a classic statement of the problem, when along the axis of the rod at some fixed time t_0 function of displacement and velocity function are assumed to be given (derivative of displacement function with reference to time t).

Let the initial conditions be given in the form:

$$u\Big|_{t=t_0}=\alpha(x),\ \frac{\partial u}{\partial t}\Big|_{t=t_0}=\alpha\cdot\beta'(x), \tag{6}$$

where $\beta'(x)$ – derivative of the function $\beta(x)$ with reference to x (function $\beta(x)$ is considered to be known).

Solution of (5) has the form

$$u=1/2[\alpha(x-at)-\beta(x-at)]-1/2[\alpha(x+at)+\beta(x+at)]. \tag{7}$$

The characteristic triangle ABC (Figure 1) appears in a view of two progressive waves: the one going up is the first group of characteristic lines (6), another wave going down is the second. It is clear that on the boundary of the rod CD the information only from the first characteristic item will come as a difference

$$1/2[\alpha(-at)-\beta(-at)] \text{ (at } x=0).$$

Another group of characteristic lines: $1/2[\alpha(x+at)+\beta(x+at)]$ will go down to infinity.

To solve the problem in the triangle CBD it is necessary to put on a CD boundary problem (Dirichlet; Neumann, Robin).

There is another formulation of the problem. We assume that on the boundary $x=0$ (Figure 1) the very function of the displacement and its derivative with reference to coordinate x (deformation) is set. In other words, the function and its normal derivative to the surface are defined on the boundary $x=0$ (Cauchy problem on coordinate!). It is assumed that these functions are independent from each other, and at the same time depend on the variable t.

The feature of this approach, unlike the previous one, is that these functions can be measured on the boundary $x=0$ at any time t. Knowing their distribution on the boundary $x=0$, our purpose is to find their distribution along the axis of the rod at any time t.

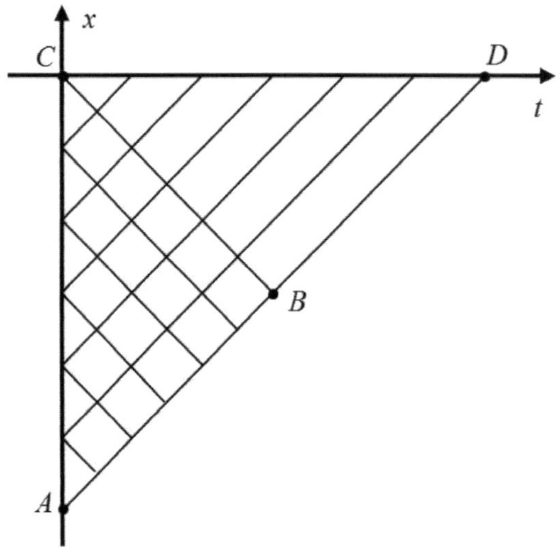

Figure 1. Characteristics of the first problem.

Let us consider the solution of the following problem. The following conditions are set on the boundary $x=0$:

$$u\big|_{x=0}=\alpha(t),\ \frac{\partial u}{\partial t}\bigg|_{x=0}=\frac{\beta'(t)}{\alpha}. \quad (8)$$

D'Alembert solution in our case is

$$u=\frac{1}{2}\left[\alpha\left(t-\frac{x}{a}\right)+\alpha\left(t+\frac{x}{a}\right)\right]+\frac{1}{2}\left[\beta\left(t+\frac{x}{a}\right)-\beta\left(t-\frac{x}{a}\right)\right]. \quad (9)$$

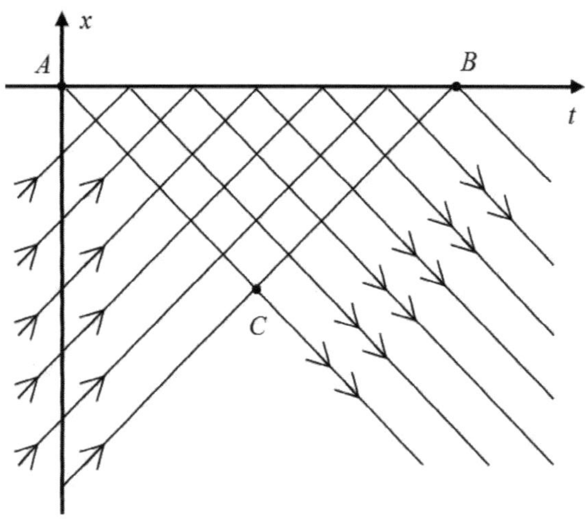

Figure 2. Decision in characteristics.

There are also two waves – the one that comes to the boundary $x=0$, and the other, that leaves the boundaries (Figure 2).

5. The problem of elasticity (plane strain case)

To simplify the situation, we consider not the spatial pattern but the plane strain case. In the coordinate system xOy we have a boundary with the equation $y=0$ where the normal stress σ_y, tangent stress τ_{xy}, normal displacement u_y, tangent u_x are given. All these quantities are considered to be the functions of the coordinate x and time t:

$$\sigma_y = \sigma_y(x,t),\ \tau_{xy} = \tau_{xy}(x,t),\ u_y = u_y(x,t),\ u_x = u_x(x,t). \tag{10}$$

The procedure of determination of tension, deformations and displacements on the boundary $y=0$ and below it is the following: we find the derivative $\partial u_x / \partial x$ thereby on boundary $y=0$, thus the deformation ε_x is found; then, according to Hooke's law in case of plane strain on this boundary the tension σ_x from the relation $\varepsilon_x = \frac{1-\nu^2}{E}\sigma_x + \frac{\nu(1+\nu)}{E}\cdot\sigma_y$ is found. And in this case $\sigma_x = \frac{\nu\sigma_y}{1-\nu} + \frac{E\varepsilon_x}{1-\nu^2}$.

According to the known stresses σ_y and σ_x from Hooke's law we find deformation ε_y: $\varepsilon_y = -\frac{\nu(1+\nu)}{E}\sigma_x + \frac{1-\nu^2}{E}\cdot\sigma_y$. To find a derivative $\partial u_x / \partial y$ we consider one more relation of Hooke's law $\frac{1}{2}\left(\frac{\partial u_x}{\partial y} + \frac{\partial u_y}{\partial x}\right) = \frac{\tau_{xy}}{2\mu} = \frac{\tau_{xy}(1+\nu)}{E}$ from which $\frac{\partial u_x}{\partial y} = \frac{2\tau_{xy}(1+\nu)}{E} - \frac{\partial u_y}{\partial x}$.

Next, vector components $\omega_z = \frac{1}{2}\left(\frac{\partial u_y}{\partial x} - \frac{\partial u_x}{\partial y}\right)$ are found. Further, consider the layer $y=-h$. For this purpose we use the balance equations. From which we find

$$\tau_{xy}(x,-h,t) = \tau_{xy}(x,0,t) - h\cdot\left(\frac{u_x^n - 2u_x^n + u_x^n}{\Delta t^2} - \frac{\sigma_x(x+\Delta x) - \sigma_x(x)}{\Delta x}\right),$$

$$\sigma_y(x,-h,t) = \sigma_y(x,0,t) - h\cdot\left(\frac{u_y^n - 2u_y^n + u_y^n}{\Delta t^2} - \frac{\tau_{xy}(x+\Delta x) - \tau_{xy}(x)}{\Delta x}\right).$$

Here, derivatives $\partial\sigma_x / \partial x$ and $\partial\tau_{xy} / \partial x$ are also calculated at $t=t_0$. To find u_x and u_y on a layer $y=-h$ we use formulas for calculation ε_{xy}, ε_y. We act in the same way for all other times. That is on the layer $y=-h$ all components of stresses and strains, components of a displacement vector are calculated for times $t=t_0+\tau$, $t=t_0+2\tau,\ldots$, $t=t_0+h\tau$.

Then we go down to the layer $y=-2h$. Here the equilibrium equations are used again. On the layer $y=-h$ we have displacements u_x at the time points $t=t_0+\tau, t=t_0+2\tau, t=t_0+3\tau$ etc. On this layer it is required to make the second derivative from the displacement u_x. This can be done at the time $t=t_0+\tau$: $\left.\frac{\partial^2 u_x}{\partial t^2}\right|_{t=t_0+\tau} = \frac{u_x^2 - 2u_x^1 + u_x^0}{\tau^2}$. Therefore, for a layer $y=-h$ all stresses and strains referred by the time $t=t_0+\tau w$ are considered. As the stress σ_x at $y=-h$ is known, the derivative $\partial\sigma_x / \partial x$ at

$t = t_0 + \tau w$ is known as well. From the balance equation the derivative $\partial\tau_{xy}/\partial x$ is found again and as τ_{xy} is known at $y=-h$, we find τ_{xy} on a layer $y=-2h$ for time $t=t_0+\tau$. In the same way all quantities $(\sigma_x, \sigma_y, \tau_{xy}, \varepsilon_x, \varepsilon_y, \varepsilon_{xy})$, including u_x, u_y, u_z on this layer, are found. All these calculations are repeated for all other times. Further going down to $y=-3h$ layer, the second derivatives from u_x, u_y with reference to time t in a time point $t=t_0+2\tau$ are also determined. On a layer $y=-2h$ the derivative $\left.\dfrac{\partial^2 u_x}{\partial t^2}\right|_{t=t_0+2\tau} = \dfrac{u_x^3 - 2u_x^2 + u_x^1}{\tau^2}$ is formed.

Process thus repeats. For verification of the proposed scheme the test is constructed.

6. Test problem

Suppose that at the origin of some coordinate system xOy we have a source of cylindrical disturbance. Environment deformation law around the source is described by equation:

$$\frac{\partial^2 u}{\partial r^2} + \frac{1}{r}\frac{\partial u}{\partial r} - \frac{u}{r^2} = \frac{1}{a^2}\frac{\partial^2 u}{\partial t^2}, \tag{11}$$

where $a^2 = \dfrac{(1-\nu)E}{\rho(1+\nu)(1-2\nu)}$, ρ – density, E – Young's modulus, ν – Poisson's ratio.

The solution of (11) is found by separation of variables method.

For the analysis we take a displacement function g as:

$$u = C_1 J_1(\lambda r)\sin(a\lambda t), \tag{12}$$

where $J_1(\lambda r)$ – Bessel function of the 1st kind. Then

$$\begin{aligned}
\varepsilon_\varphi &= \frac{u}{r} = \lambda C_1 \sin(a\lambda t)\frac{J_1(\lambda r)}{\lambda^2},\\
\varepsilon_r &= \frac{\partial u}{\partial r} = \lambda\left[J_0(\lambda r) - \frac{J_1(\lambda r)}{\lambda r}\right]C_1\sin(a\lambda t),\\
\sigma_r &= \frac{E}{(1+\nu)(1-2\nu)}\left[(1-\nu)\lambda J_0(\lambda r) - (1-2\nu)\lambda\frac{J_1(\lambda r)}{\lambda r}\right]C_1\sin(a\lambda t),\\
\sigma_\varphi &= \frac{E}{(1+\nu)(1-2\nu)}\left[\nu\lambda J_0(\lambda r) - (1-2\nu)\lambda\frac{J_1(\lambda r)}{\lambda r}\right]C_1\sin(a\lambda t).
\end{aligned} \tag{13}$$

Moving to Cartesian coordinates x, y we consider the trace of the solution (12), (13) on the boundary $y=H$. In dimensionless quantities it has the form:

$$\begin{cases}
\tilde{\sigma}_y = \dfrac{1}{2(1-\nu)}\left\{J_0(\sqrt{\tilde{x}^2+1}) - \left[(1-2\nu)J_0(\sqrt{\tilde{x}^2+1}) - 2(1-2\nu)\dfrac{J_1(\sqrt{\tilde{x}^2+1})}{\sqrt{\tilde{x}^2+1}}\right]\dfrac{\tilde{x}^2-1}{\tilde{x}^2+1}\right\}\tilde{C}_1\sin(\tilde{t}),\\
\tilde{\tau}_{xy} = \dfrac{(1-2\nu)}{2(1-\nu)}\left\{(1-2\nu)J_0(\sqrt{\tilde{x}^2+1}) - 2(1-2\nu)\dfrac{J_1(\sqrt{\tilde{x}^2+1})}{\sqrt{\tilde{x}^2+1}}\right\}\dfrac{2\tilde{x}}{\tilde{x}^2+1}\tilde{C}_1\sin(\tilde{t}).
\end{cases} \tag{14}$$

$$\begin{cases}
\tilde{u}_x = \tilde{C}_1 J_1(\sqrt{\tilde{x}^2+1})\dfrac{\tilde{x}}{\sqrt{\tilde{x}^2+1}}\sin(\tilde{t}),\\
\tilde{u}_y = \tilde{C}_1 J_1(\sqrt{\tilde{x}^2+1})\dfrac{1}{\sqrt{\tilde{x}^2+1}}\sin(\tilde{t}),
\end{cases} \tag{15}$$

Task: let on the boundary $y = H$ traces of the test solution be set. Figure 3 shows the dependencies of various quantities (σ_y, τ_{xy}, u_x, u_y) on x. The numerical and analytical solutions coincide.

7. Creating solutions and examples

It is visible that pictures are identical. The solid line in Figure 3 shows dependencies for the point with coordinates on $y = H$, the dashed line for $H-2$, the dotted one for $H-4$, the dotted-dashed line for $H-6$.

Analyzing the curves it is possible to make the conclusion: on the ordinate where there is a source of perturbation firstly the tangent stresses turns into zero, secondly, vertical displacements u_y are accepted as constants. In the reviewed example, u_y is equal to zero. As for abscissa of perturbation source, at approximation to the perturbation source along the straight line $y = y_0$ stress σ_y increases up to the maximum value with reduction of x to zero (the point zero can be taken for the perturbation source

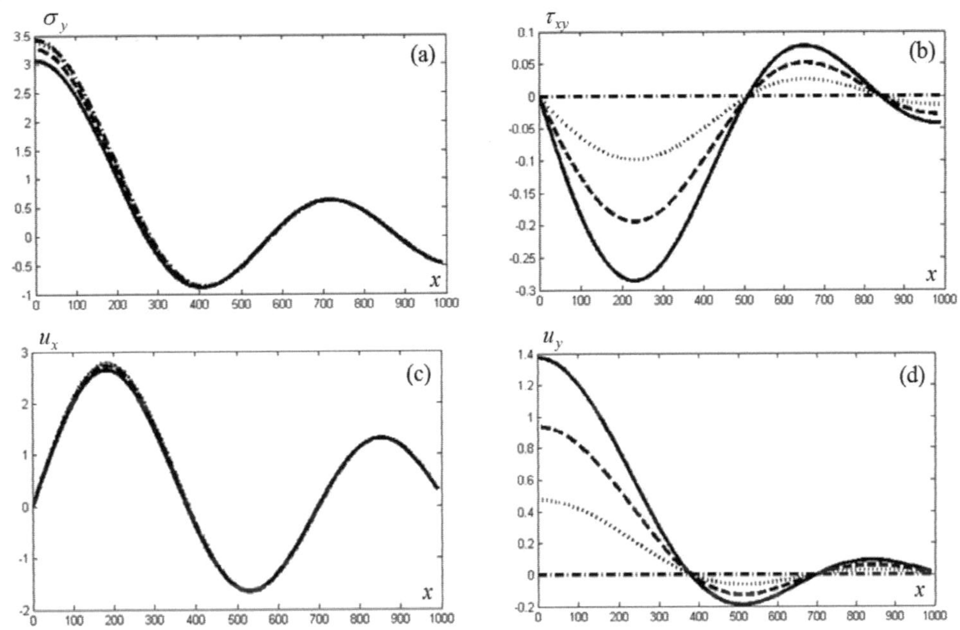

Figure 3. Analytical and numerical solution: a) $\sigma_y(x)$, b) $\tau_{xy}(x)$, c) $u_x(x)$, d) $u_y(x)$.

abscissa). These features can serve as a criterion for defining the location of perturbation source.

Let us consider another example. Let the half-plane be free ($\sigma_y = 0, \tau_{xy} = 0$) and, besides, on its boundary distribution of displacements is set as following:

$$u_x = \frac{Ax\sin(\omega t)}{x^2 + b^2}, \quad u_y = \frac{AH\sin(\omega t)}{x^2 + b^2},$$

where $t \geq 0$. Using the proposed scheme of the solution, we obtain required distributions; they are considered in the dimensionless variables ($A = 1, H = 1, b = 1, \omega = 1$). In Figure 4 curves of these values are shown for dimensionless time $t = \pi/2$.

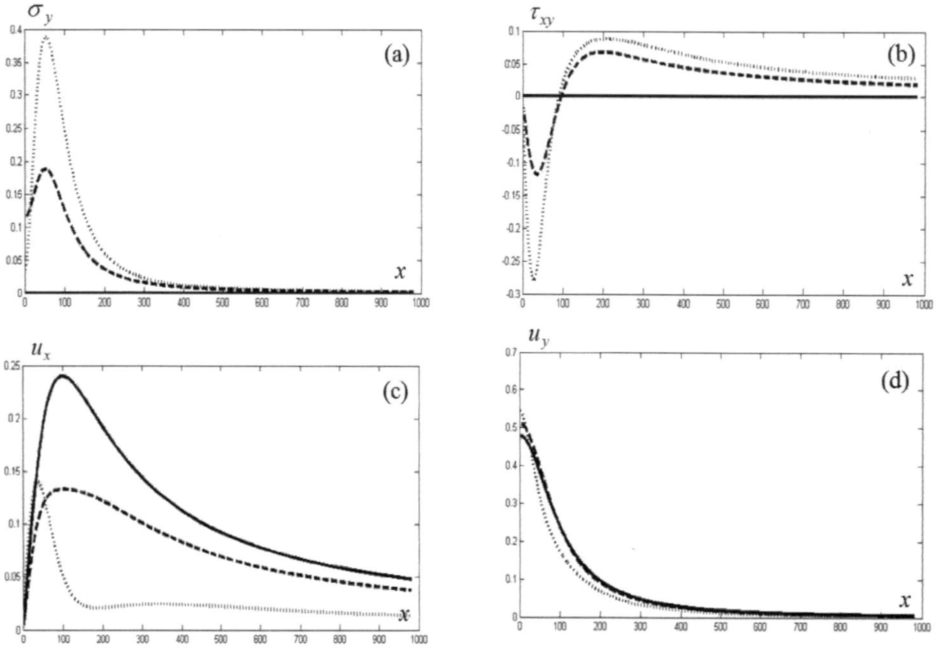

Figure 4. Numerical solution of the task of the offered scheme: a) $\sigma_y(x)$, b) $\tau_{xy}(x)$, c) $u_x(x)$, d) $u_y(x)$.

8. Conclusions

The solutions of some basic boundary value problems with initial conditions are developed.
Finite-difference algorithms for Cauchy problems numerical solutions are constructed. Its verification is carried out using the available analytical solutions. The problem of distinguishing features in numerical solutions with application of solutions for lumped sources in an infinite medium is considered. The problem of boundary functions smoothing (constructed from its discrete values on the boundary) is solved with correlation analysis usage to ensure correctness of assigned problems solutions (stable dependence on the input data).

Reference

[1] Alekseev A S, Avdeev A V, Fatianov A G and Tcheverda V A 1991 Closed cycle of mathematical modeling of wave processes in vertically inhomogeneous media (direct and inverse problems) *Matem. Mod.* **3**(10) 80–94 (In Russian)
[2] Hadamard J 1978 Cauchy problem for linear partial differential equations of hyperbolic type (Moscow: Nauka Publishing House) (In Russian)
[3] Mikhailenko B G and Reshetova G V 2006 Numerical - analytical method of solving the problem of seismic and acoustic-gravity waves for the inhomogeneous model Earth – Atmosphere *Siberian Journal of Numerical Mathematics* **9**(1) 37-46 (In Russian)
[4] Muskhelishvili N I 1966 Some Basic Problems of the Mathematical Theory of Elasticity (Moscow: Nauka Publishing House) (In Russian)

IOP Conf. Series: Journal of Physics: Conf. Series **894** (2017) 012122 doi:10.1088/1742-6596/894/1/012122

The study of stationary flow in the core of thin vortex ring

V F Kopiev[1], S A Chernishev[1], R V Akinshin[1,2] and M A Yudin[1,2]

[1]Central Aerohydrodynamics Institute, Moscow, Russia
[2]Moscow Institure of Physics and Technology, Dolgoprudny, Russia

E-mail: vkopiev@mktsagi.ru

Abstract. The theoretical investigation of stationary vortex ring in an ideal incompressible fluid is carried out. To obtain the stationary solution for vortex ring with uniform vorticity distribution, Fraenkel's procedure is used. Further, based on finding solution Fraenkel's procedure is modified to receive stationary vortex ring with isochronous vorticity distribution. The solutions of the uniform and isochronous problems are obtained in the form of an asymptotic expansion with respect to the small parameter characterised by the ratio of the core and ring diameters.

1. Introduction

The study of a vortex ring dynamics is of great interest due to the fact that it allows to investigate the complicated mechanisms of instabilities, which can play a key role in the problem of turbulence excitation.

Vortex ring can serve as a simple dynamic model of the large-scale structures observed in shear flows. Moreover, it is probably the easiest vortex element that can be created experimentally and studied theoretically. At the same time a vortex ring is 3-dimensional flow in which the effects connected with curvature of vortex lines become apparent. This fact permits one to investigate directly many problems of vortex dynamic and acoustic including processes of formation of perturbations, transfer to turbulence near the core, the connection of sound field with incompressible vortex dynamic.

As it is known a vortex ring is a vortex torus,moving with ellipsoidal region filled with fluid, called an atmosphere of the ring. In a case of a laminar vortex ring there is potential flow in the region of an atmosphere and in a case of a turbulent vortex ring the region of an atmosphere is filled with turbulent perturbations. The careful description of a model of vortex rings was made in [1], where, first of all, a self-induced movement of a vortex thread, rolling in a ring, was examined. The representation of a vortex as a thin circular thread in an ideal fluid is in accordance with a known hydrodynamic phenomena and it allows one to connect the middle parameters of the vortex ring – its shape and the vorticity with the propagation velocity. In a case of a laminar ring in an ideal fluid these parameters prove to be constant during a movement of the vortex that means the vortex will live unlimitedly long, not changing its form and the velocity.

Content from this work may be used under the terms of the Creative Commons Attribution 3.0 licence. Any further distribution of this work must maintain attribution to the author(s) and the title of the work, journal citation and DOI.
Published under licence by IOP Publishing Ltd

The existence and the uniqueness of solving for a stationary thin vortex ring were proved by Fraenkel in [2]. The proof relies on mapping of 2D vortex flow with circular current lines to 3D flow with some another distribution of a vorticity inside the core. This mapping is sought as a series in a parameter of thickness determined as a ratio of the core cross-section dimension a to the ring radius R (fig.1) $\mu=\frac{a}{R}<<1$, which is supposed to be small. The examples of solving this problem for

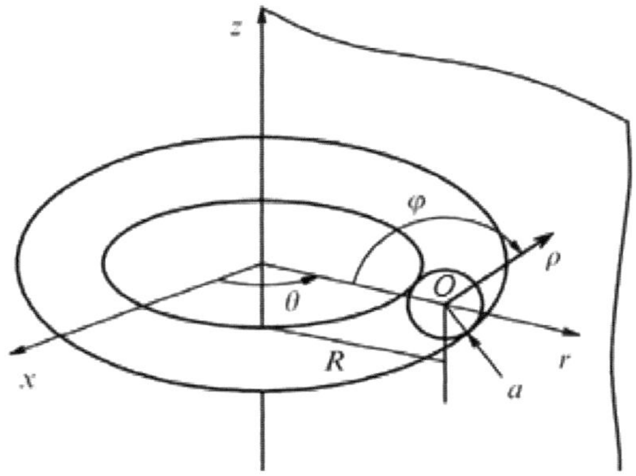

Figure 1. Vortex ring.

different distributions of a vorticity in the core of a vortex ring were considered in [3]. In particular, in a case of the uniform vortex ring (accordingly to [4] in uniform distribution the magnitude of the vorticity is proportional to the distance to the ring symmetry axis) the velocity field and the boundary shape were found in two approximations in the parameter of thickness [3]. However, a uniform vortex ring is not the isochronous one: more detail solution shows that the period of rotation of fluid particles inside the core of the vortex appears not to be the same. The main aim of this paper is to find the algorithm of solving the task for the isochronous vortex ring. Importance of this investigation is that all non-trivial eigen-oscillations of the isochronous vortex ring relates to a discrete spectrum while rings with other distributions of a vorticity contain continuous part. This fact means that the isochronous vortex ring is the simplest 3D vortex to study the problem of vortex ring stability.

Because of the fact that concrete 2D flow with circular streamlines which turns into the isochronous vortex ring is unknown in advance, Fraenkel's procedure needs to be essentially modified on every step connected with calculation a 2D profile of the vorticity. The flow with the isochronous distribution of the vorticity differs from the flow with the uniform one only in the second approximation in the parameter of thickness and it gives an opportunity to choose the uniform distribution as a start point to build the solution for the isochronous ring. In the first section of this paper the uniform solution is built in which the terms are found up to the third approximation in the small parameter. In the second part the flow with the isochronous distribution of the vorticity is found in the same approximation. Higher approximations in the small parameter than was obtained in [4] will be required to investigate eigen-oscillations near the accumulation point.

2. The statement of the problem

The coordinate system moving with the vortex ring is used. Consider the cylindrical coordinates r,θ,z which coincide with the axis of the vortex ring and the polar coordinates ρ,φ in the cross-section of the vortex with the centre in the stagnation point (fig.1). Taking into account the axis symmetry of the flow, components of the velocity field are expressed through the ψ called the stream function

$$\mathbf{v}(r,z)=\left(-\frac{1}{r}\frac{\partial\psi}{\partial z};0;\frac{1}{r}\frac{\partial\psi}{\partial r}\right) \tag{1.1}$$

There is a connection between the stream function and the vorticity in accordance with [1]

$$\psi(r,z)=-\frac{1}{2}\tilde{W}r^2+\frac{1}{4\pi}\iint_{\sigma}\omega rr'dr'dz'\int_0^{2\pi}\frac{\cos(\theta)}{\left((z-z')^2+r^2+r'^2-2rr'\cos(\theta)\right)^{1/2}}d\theta \quad (1.2)$$

Where ω is a contravariant component of the vorticity. The velocity $\tilde{W}$ of the ring and the region σ of the vortex core are defined during solving this problem.

For the stationary field of the vorticity $\mathbf{\Omega}=(0,\omega,0)$ the Helmholtz equation is $(\mathbf{v}\nabla)\mathbf{\Omega}=0$. Due to the axis symmetry this equation can be rewritten as $\mathbf{v}\cdot grad\left(\frac{\omega}{r}\right)=0$, from which the equation $\frac{\omega}{r}=F(\psi)$ follows, where ψ is determined in (1.2). In particular, for the uniform distribution of the vorticity one can obtain $F(\psi)=\text{const}$

In this investigation only thin vortex rings are considered. Determine the parameter of thickness ε which is supposed to be small, $\varepsilon \ll 1$. Analogously to [2] determine the coordinates S,φ, where $\varepsilon RS \equiv \rho$. These coordinates are connected with the cylindrical coordinates with the axis z coinciding with the axis of the ring by the following relations

$$r(\varepsilon,S,\varphi)=R(1-\varepsilon S\cos\varphi),\ z(\varepsilon,S,\varphi)=\varepsilon SR\sin\varphi \quad (1.3)$$

Determine streamlines using parameter s

$$S=s+q(\varepsilon,s,\ \varphi) \quad (1.4)$$

So that on the streamlines $s=const$, $0\le s\le 1$, where $s=1$ corresponds to the core boundary of the vortex ring. Note the streamlines in the first approximation in the case of the thin vortex ring are circular and $q(\varepsilon,s,\ \varphi)=O(\varepsilon)$

Following [2], introduce the contravariant component of the vorticity in the form $\frac{\omega}{r}=\frac{\Omega(s)}{R}$, $s\le 1$ and represent the stream function as $\psi=\varepsilon^2R^3\left(\Psi(s)+B(\varepsilon)\right)$, where

$$S=s+q(\varepsilon,s,\ \varphi), |\mathbf{s}-\hat{\mathbf{s}}|=\left(s^2+\hat{s}^2-2s\hat{s}\cos(\varphi-\hat{\varphi})\right)^{1/2} \quad (1.5)$$

Where $\tilde{W}=\varepsilon^2RW(\varepsilon)$. Fraenkel proved in [3] that for each two-dimensional distribution of vorticity $\Omega(s)$, the only stationary solutions turn out to exist. Transfer to the coordinates s, φ in (1.2)

$$\begin{gathered}\Psi(s)+B(\varepsilon)=-\frac{1}{2}W(\varepsilon)\left(1-\varepsilon(s+q)\cos\varphi\right)^2\\ +\frac{1}{2\pi}\int_0^{2\pi}\int_0^1 \mathrm{G}(\varepsilon,s+q,\varphi,\hat{s}+\hat{q},\hat{\varphi})\left(1+\frac{\hat{q}}{\hat{s}}\right)\left(1+\frac{\partial\hat{q}}{\partial\hat{s}}\right)\hat{\Omega}\hat{s}d\hat{s}d\hat{\varphi}\end{gathered} \quad (1.6)$$

$$\mathrm{G}(\varepsilon,\mathrm{S},\varphi,\hat{\mathrm{S}},\hat{\varphi})=\frac{1}{R^2}r^{1/2}(\varepsilon,S,\varphi)\hat{r}^{3/2}(\varepsilon,S,\varphi)\left(\left(\frac{2}{k}-k\right)E(k)-\frac{2}{k}K(k)\right)$$

$$k=\left(\frac{4r(\varepsilon,S,\varphi)r(\varepsilon,\hat{S},\hat{\varphi})}{\left(r(\varepsilon,\hat{S},\hat{\varphi})+r(\varepsilon,S,\varphi)\right)^2+\left(z(\varepsilon,\hat{S},\hat{\varphi})-z(\varepsilon,S,\varphi)\right)^2}\right)^{1/2}$$

IOP Conf. Series: Journal of Physics: Conf. Series **894** (2017) 012122 doi:10.1088/1742-6596/894/1/012122

Where $E(k)$ and $K(k)$ – are complete elliptic integrals of the first and second kinds. Introduce all initial magnitudes in series expansion in the parameter of thickness ε. We relate terms ε^n and $\varepsilon^n \ln\varepsilon$ to the same $O(\varepsilon^n)$.

$$B(\varepsilon)=\sum_{m=0}^{\infty} B_m \varepsilon^m$$

$$\frac{1}{2}W(\varepsilon)\left(1-\varepsilon(s+q)\cos\varphi\right)^2=\sum_{m=0}^{\infty} W_m \varepsilon^m$$

$$q(\varepsilon,s)=\sum_{m=1}^{\infty} q_m \varepsilon^m$$

3. The algorithm of finding the stationary flow

The condition of constancy of the parameter s on the streamlines and independence of the left part (1.6) from the coordinate φ allows us to get secular equations to find the initial approximations for the shape of the streamlines of the stationary flow $q_n(s,\varphi)$, which will be signed for the brevity as q_n and the constants $B_n(\varepsilon)$, $W_n(\varepsilon)$. Getting the secular equations consists in linearization of (1.6) in the small parameter ε and this way was described by Fraenkel in [3]. The problem of finding a concrete solution with certain 2 – dimensional distribution of a vorticity $\Omega(s)$ consists in obtaining the magnitudes q, B, W.

4. Calculating the period of rotation of fluid particles inside the core

It is convenient to use a new parameter μ, which is defined as a ratio of vortex core equivalent radius $a=\frac{\sqrt{S}}{\pi}$ (S is an area of vortex ring core)to the vortex ring radius R, so that $\varepsilon(\mu)=\mu-\frac{61}{128}\mu^3+O(\mu^5)$. Following [4] find the period of rotation of fluid particles on different streamlines. Further, we define the coordinates $\sigma(\rho,\varphi),\psi(\rho,\varphi)$ in the core cross-section of the ring, which slightly differ from the coordinates ρ,φ, respectively. We require that for the coordinates $\sigma(\rho,\varphi)$ and $\psi(\rho,\varphi)$ the following equations to be fulfilled

$$V_0^\sigma=0,\ V_0^\psi=V_0^\psi(\sigma),\ |g|^{1/2}=\sigma$$

Where V_0^σ, V_0^ψ are the contravariant components of the velocity field, g_{ij} is the metric tensor of the coordinate system $\sigma,\ \psi, p$, where $p=R\theta$ (Figure 1).

Making use of the equations $V_0^\sigma=\frac{\partial\sigma}{\partial\rho}V_0^\rho+\frac{\partial\sigma}{\partial\varphi}V_0^\varphi$, $V_0^\psi=\frac{\partial\psi}{\partial\rho}V_0^\rho+\frac{\partial\psi}{\partial\varphi}V_0^\varphi$, $|g(\rho,\varphi,p)|^{1/2}=\frac{\partial(\sigma,\psi)}{\partial(\rho,\varphi)}|g(\sigma,\psi,p)|^{1/2}$, where $p=R\theta$ (Figure 1) we can get explicit expressions for σ,ψ.

It is not difficult to obtain that on the different streamlines $\sigma,\ p=const$ the period of rotation of fluid particles proves to be

$$T = \frac{2\pi}{V_0^{\psi}(\sigma)} \tag{4.1}$$

The coordinates σ, ψ for the uniform vortex ring ($\Omega(s) = \Omega_0 = const$) can be get as series in μ

$$\sigma = \rho - \frac{5}{8}\frac{\mu\rho^2\cos\varphi}{a} + \mu^2\left(\frac{45\rho^3}{256a^2} + \left(\frac{3}{8}\rho\ln\frac{8}{\mu} - \frac{15}{64}\rho - \frac{\rho^3}{256a^2}\right)\cos 2\varphi\right) + \mu^3\sigma_3 + O(\mu^4)$$

$$\psi = \varphi + \frac{7}{8}\frac{\mu\rho\sin\varphi}{a} + \mu^2\left(\frac{15}{64} - \frac{3}{8}\ln\frac{8}{\mu} + \frac{11\rho^2}{128a^2}\right)\sin 2\varphi + \mu^3\psi_3 + O(\mu^4)$$

Where

$$\sigma_3 = \frac{\cos\varphi\left(a^2\rho^2\left(96\ln\frac{8}{\mu} - 60\right) - 1219\rho^4\right) + \cos 3\varphi\left(a^2\rho^2\left(192\ln\frac{8}{\mu} - 264\right) - 13\rho^4\right)}{4096a^3}$$

$$\psi_3 = \frac{3\left[\sin\varphi\left(a^2\rho\left(368\ln\frac{8}{\mu} - 230\right) - 356\rho^3\right) + \sin 3\varphi\left(a^2\rho\left(56\ln\frac{8}{\mu} - 47\right) - 4\rho^3\right)\right]}{1024a^3}$$

In the coordinates σ, ψ for the contravariant components of the velocity field and the metric tensor we obtain

$$V_0^{\sigma} = 0,\ V_0^{\psi} = \frac{\Omega_0}{2} - \Omega_0\frac{21\sigma^2}{64a^2}\mu^2 + O(\mu^4),\ |g|^{1/2} = \sigma \tag{4.2}$$

Using (4.1) and (4.2) we get

$$T = \frac{2\pi}{\Omega_0\left(\frac{1}{2} - \mu^2\frac{21\sigma^2}{64a^2}\right)} + O(\mu^4) \tag{4.3}$$

As it is seen for the uniform vortex ring the period of fluid particles orbits is not a constant over the whole vortex core cross-section and it is in full accordance with [5]. Thus, the flow with the uniform vorticity is not isochronous.

5. The isochronous vortex ring

Determine the stationary flow for the isochronous vortex ring. In the case of the uniform vortex ring the distribution of the vorticity $\Omega(s)$ is assigned beforehand in accordance with (1.5) but to find the stationary flow of the isochronous vortex ring we need to determine $\Omega(s)$ for this case from the condition of constancy of the contravariant component of the velocity field V_0^{ψ} on different streamlines $\sigma,\ p = \text{const}$.

Since the period of fluid particle rotation in the uniform vortex ring differs from the constant in the second approximation, an amendment to the vorticityfor the isochronous vortex ring has the order μ^2. Obviously, the coordinates σ, ψ also will be changed. Determine new coordinates σ^*, ψ^*, which corresponds to the conditions $V^{\sigma^*} = 0$, $V^{\psi^*} = const$ and $|g|^{1/2} = \sigma^*$. In the first approximation these coordinates for the isochronous vortex ring match with σ, ψ for the uniform vortex ring founded earlier. So we seek the vorticity as $\Omega_T^p = \Omega_0 + \mu^2\omega_2(\sigma^*)$.

Using $\sigma^* = \rho - \mu \dfrac{5\rho^2 \cos\varphi}{8a} + O(\mu^2)$ we obtain

$$\Omega_T^p = \Omega_0 + \mu^2 \omega_2(\rho) + O(\mu^3) \tag{5.1}$$

It appears to be convenient to find the contravariant components of the velocity field in the following form

$$\begin{aligned} V^\rho &= \mu\Omega_0 \frac{5\rho^2}{16a}\sin\phi + \mu^2 g_2 \sin 2\phi + O(\mu^3) \\ V^\phi &= \frac{\Omega_0}{2} + \mu\Omega_0 \frac{7\rho}{16a}\cos\phi + \mu^2 h_2 \cos 2\phi + \mu^2 h_2 + O(\mu^3) \end{aligned} \tag{5.2}$$

Taking into account (5.1), (5.2) and that $\Omega^s = \dfrac{a}{a\rho - \mu\rho^2\cos\phi}\left(\dfrac{\partial \rho^2 V^\phi}{\partial \rho} - \dfrac{\partial V^\rho}{\partial \phi}\right)$, we get

$$\omega_2(\rho) = \frac{1}{\rho}\frac{\partial\left(\rho^2 \hat{h}_2\right)}{\partial \rho} \tag{5.3}$$

Analogously to (5.2) we represent the coordinates σ^*, ψ^* as

$$\begin{aligned} \sigma^*(\rho,\varphi) &= \rho - \mu\frac{5\rho^2}{8a}\cos\varphi + \mu^2 a_2 \cos 2\varphi + \mu^2 \hat{a}_2 + O(\mu^3) \\ \psi^*(\rho,\varphi) &= \varphi + \mu\frac{7\rho}{8a}\sin\varphi + \mu^2 b_2 \sin 2\varphi + O(\mu^3) \end{aligned} \tag{5.4}$$

Using the rules of the transformation for the contravariant components of the velocity field, the metric tensor and ((5.1), (5.2), (5.3), (5.4)), we can obtain $a_2, \hat{a}_2, b_2$ and $\hat{h}_2$. As a result $\omega_2(\rho) = \dfrac{21\rho^2\Omega_0}{16a^2}$. Thus, we get

$$\Omega(s) = \Omega_0\left(1 + \varepsilon^2 \frac{21}{16}s^2\right) + O(\varepsilon^3) \tag{5.5}$$

As it is seen, the vorticity inside the core of the vortex ring can be represented as a series in ε: $\Omega(\varepsilon, s) = \Omega_0(s) + \varepsilon\Omega_1(s) + \ldots$. Than in the left part of (2.4) adding terms $\Psi_n(s)$ appear and they are determined by the expression (1.5) from adding terms of the vorticity $\Omega_0(s) + \varepsilon\Omega_1(s) + \ldots$. Knowing the distribution of the vorticity in the core we can obtain the boundary shape of the core and the velocity field in the core (the algorithm of finding these magnitudes was described above). Note that the expression for the vorticity (5.5) determines the flow in the isochronous vortex ring up to the third approximation $T = \dfrac{4\pi}{\Omega_0} + O(\mu^4)$.

Introduce also the mapping of the coordinates which corresponds to the isochronous distribution of the vorticity in the vortex ring

$$\sigma^* = \rho - \mu\frac{5\rho^2\cos\varphi}{8a} + \mu^2\left(\frac{45\rho^3}{256a^2} - \left(\frac{15}{64}\rho + \frac{\rho^3 + 60a^2\rho}{256a^2} - \frac{3}{8}\rho\ln\frac{8}{\mu}\right)\cos 2\varphi\right) + \mu^3\sigma_3^* + O(\mu^4)$$

$$\sigma_3^* = -\frac{\rho^2}{4096a^3}\left(\left(3\left(369\rho^2 + 20a^2 - 32a^2\ln\frac{8}{\mu}\right)\cos\varphi + \left(13\rho^2 + +264a^2 - 192a^2\ln\frac{8}{\mu}\right)\cos 3\varphi\right)\right)$$

$$\psi^* = \varphi + \mu \frac{7\rho \sin\varphi}{8a} + \mu^2 \frac{\left(11\rho^2 + 30a^2 - 48a^2 \ln\frac{8}{\mu}\right)\sin 2\varphi}{128a^2} + \mu^3 \psi_3^* + O\left(\mu^4\right)$$

$$\psi_3^* = \frac{\rho}{1024a^3}\left(2\left(464\rho^2 + 345a^2 - 552a^2 \ln\frac{8}{\mu}\right)\sin\varphi + 3\left(4\rho^2 + 47a^2 - 56a^2 \ln\frac{8}{\mu}\right)\sin 3\varphi\right)$$

6. Conclusion

The vortex ring with uniform and isochronous distributions of the vorticity was examined in this paper. Based on Fraenkel's procedure for both cases the shape of the streamlines and the velocity field inside the core of the vortex ring are calculated up to μ^3.

Fraenkel's procedure [2,3] which allows to find the stationary flow in a thin vortex ring was essentially modified to determine the flow in the isochronous vortex ring. The main aim of finding the following terms is to obtain more precise expressions for oscillations of the vortex ring to reveal eigen-frequencies nearer the stagnation point obtained in [4]. There is a reasonable assumption that in this region of frequencies a new instability exists [5]. Earlier it was impossible to confirm this assumption because of shortage of finding approximations.

Acknowledgements

This work was supported by the Russian Science Foundation (project №17-11-01271).

References

[1] Lamb H 1932 *Hydrodynamics* (London: Cambridge University Press) p 738
[2] Fraenkel L E 1970 *Proc. Roy. Soc. Lond. A (Cambridge)* **316** 29–62
[3] Fraenkel L E 1972 *J. Fluid Mech.* **2** 119–35
[4] Kopiev V F and Chernyshev S A 1997 *J. Fluid Mech.* **341** 19–47
[5] Kopiev V F and Chernyshev S A 2000 *Physics-Uspekhi* **170** 713–42

IOP Conf. Series: Journal of Physics: Conf. Series **894** (2017) 012123 doi:10.1088/1742-6596/894/1/012123

High temperature creep of steel 09G2S under non-stationary loading

S V Iyavoynen[1], I A Banshchikova[1], I V Lubashevskaya[1,2] and M A Legan[1,3]

[1] Lavrentyev Institute of Hydrodynamics SB RAS, Novosibirsk, Russia
[2] Novosibirsk State University, Novosibirsk, Russia
[3] Novosibirsk State Technical University, Novosibirsk, Russia

E-mail: svetaiyavoynen@gmail.com

Abstract. Experimental analysis of deformation of steel 09G2S under creep conditions has been carried out to determine the optimal modes of high temperature forming of steel shells. The efficient temperature range in terms of saving of the plasticity resource was defined. The material parameters including scalar damage parameters have been determined by taking into account the description of the deformation of a material under non-stationary loading conditions. A satisfactory description of the unsteady creep processes of the material 09G2S in the temperature range from 700 to 800°C has been received.

1. Introduction

The work of structural elements under non-stationary loading conditions under long-term temperature influence is one of the major issues appearing in the material rupture time analysis. This is a relevant question in the field of shipbuilding and aircraft building industries, engineering and other industries. Non-stationary processes have a significant impact on the deformation processes, strength and component life of structural elements. The study takes into consideration in these processes changes of remaining life time.

To describe the materials creep processes, we used the determined relations characterizing considered non-stationary state [1-9].

2. Theory

2.1. Experimental data on the deformation of a material in the creep mode

Tensile tests were carried out under the slow mode of deformation in order to determine the optimal modes for forming of steel 09G2S. The experiments were carried out on cylindrical specimens. The strain was determined by the logarithmic dependence $\varepsilon = \ln(L/L_0)$, where L is the current and L_0 is the initial length of the specimens. Three-four experiments were carried out for each fixed temperature level: 700, 730, 750, 770 and 800°C, both at constant stress and with an overload to other stress levels at a constant temperature. The external load of the axial tension P is corrected during experiment taking into account cross section area of the specimens so as the true stress $\sigma = P/S$ has remained constant. This correction is based on the assumption that the material is an incompressible medium, i.e. its volume is not changed ($V = S_0 L_0 = SL$, where S_0 and S are cross-sectional area at initial time

Content from this work may be used under the terms of the Creative Commons Attribution 3.0 licence. Any further distribution of this work must maintain attribution to the author(s) and the title of the work, journal citation and DOI.
Published under licence by IOP Publishing Ltd

and at current time, respectively). All experiments were brought to the destruction of the specimens in order to determine the limiting value of the strain ε^*.

An analysis of the experimental data showed that up to deformation value $\varepsilon \approx 10\%$, the strain rate is a function of only the stress and temperature $\dot{\varepsilon} = \varphi(\sigma, T)$, i.e. the material acts like a viscous medium and strain rate is not dependent on any hardening or softening parameters. The material softening is observed when $\varepsilon > 10\%$.

Diagram of the change in the limiting strains as a function of temperature are shown in Figure 1. From the graphic, it is concluded that the most efficient temperature range from the point of view of the magnitude of the "plastic reserve" should be seen as $T \approx 730^0C$.

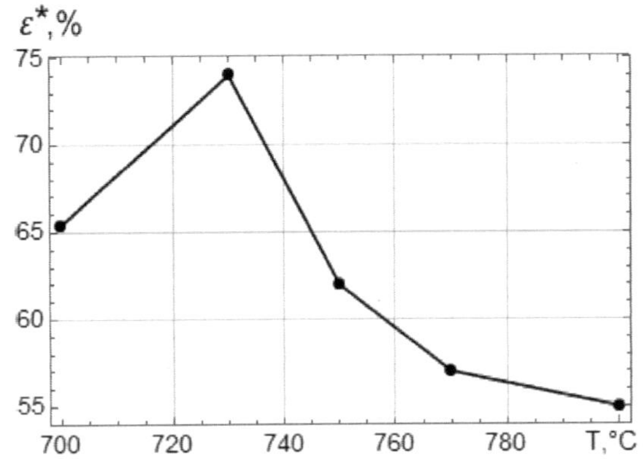

Figure 1. The dependence of limiting deformation at destruction point from temperature.

3. The determination of the material constants of creep taking into account damage material

The experimental data on the creep at constant stress are analyzed to determine the constants of the material. A mathematical description of the forming process is based on the kinetic equations with the parameter of damage by Rabotnov [1, 3–5, 11]. The constitutive equations describing of the creep processes and long-term strength are used with the same softening coefficient m in both equations [1, 4, 5, 8–11]:

$$\begin{cases} \dfrac{d\varepsilon}{dt} = \dfrac{\varphi(\sigma)}{(1-\omega)^m} \\ \dfrac{d\omega}{dt} = \dfrac{\psi(\sigma)}{(1-\omega)^m} \end{cases}. \tag{1}$$

For plotting the approximation curves were used functional dependences such as

$$\varphi(\sigma,T) = B_\varepsilon(T)\sigma^{k(T)} \text{ and } \psi(\sigma,T) = B_\omega(T)\sigma^{n(T)}. \tag{2}$$

The scalar parameter of damage ω is determined by the ratio of the strain during the creep process of the material to its corresponding value at the moment of destruction time in the experiments, i.e. $\omega = \varepsilon/\varepsilon_*$.

We integrate (1) under $\sigma = const$ to determine constants B_ε, n

$$\varepsilon = \frac{B_\varepsilon}{B_\omega}\sigma^{n-k}\left(\left(1-(1+m)B_\omega\sigma^k t\right)^{1/(m+1)} - 1\right), \tag{3}$$

and after taking the logarithm of the equation at the steady-state creep stage, we obtain the equation of a straight line in the $\ln(\sigma) - \ln(\dot{\varepsilon})$ plane:

$$\ln\dot{\varepsilon} = \ln B + n\ln\sigma. \tag{4}$$

Figure 2 presents the experimental data that are denoted by dots in the coordinates $\ln\sigma - \ln\dot{\varepsilon}$.

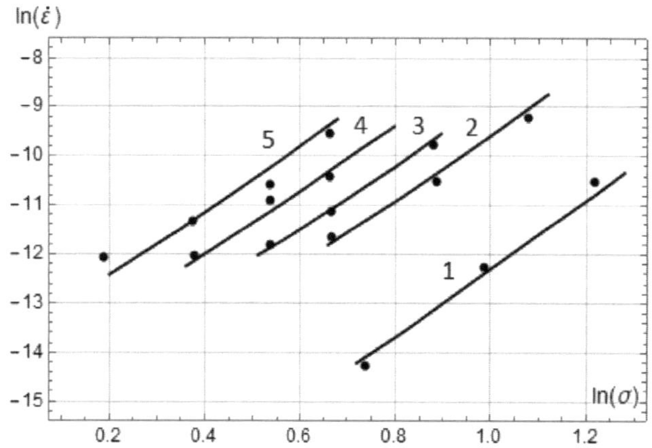

Figure 2. Experimental data 1-5 on the logarithmic plane correspond to fixed temperatures 700, 730, 750, 770 and 800°C respectively.

Having a groups of experimental points (1-5) corresponding to fixed temperatures (700, 730, 750, 770 and 800°C respectively in Figure 2), can approximate experimental data of direct line in the plane $\ln\sigma - \ln\dot{\varepsilon}$ which integrally reflects the strain-strength behavior of the material corresponding to the temperatures. We believe that all the lines are parallel, i.e. the exponent n reflecting the slope of the lines, is temperature independent. The constants n and B_ε can be determined.

The softening coefficient m is determined after the experimental curves are reduced to a single curve by scaling of the strain and time to ε_* and t_* [5–8]. Similarly to the second equation (2), the constants B_ω and k are determined. All the creep constants of the material obtained using the system of equations (1) are indicated in Table 1.

Table 1. Creep constants of St. 09G2S.

T, °C	B_ε, (MPa$^{-n}\cdot$s^{-1})	B_ω, (MPa$^{-k}\cdot$s^{-1})	n	k	m
700	$3.12\cdot10^{-15}$	$2.09\cdot10^{-11}$	5.5	3.5	1.105
730	$1.97\cdot10^{-14}$	$3.97\cdot10^{-14}$	5.5	5.4	0.65
750	$2.97\cdot10^{-14}$	$6.04\cdot10^{-13}$	5.5	4.82	1.054
770	$5.8\cdot10^{-14}$	$1.39\cdot10^{-13}$	5.5	5.4	0.89
800	$1.12\cdot10^{-13}$	$2.39\cdot10^{-11}$	5.5	4.2	0.44

Figure 3 shows diagrams of the constants of creep in the form of strain versus the time when specimens are stretched under 700, 730, 750, 770 and 800°C.The dots denote experimental data, and the curves are obtained using the system of equations (1). It shows that with the use of the system (1) a satisfactory agreement of numerical and experimental data was obtained.

4. Analysis of an unsteady creep process of steel 09G2S

The experiments carried out confirmed that with deformations not exceeding 10%, the softening is insignificant. Comparison of the experiments carried out under a constant stress and piecewise constant stress indicates that the overload has a slight effect on the strain rate towards the rate increase.

Figure 4 presents the experiments under the constants stress (solid line) and the experiments with one stage–overload at the same values of the stress and temperatures (dash line) for different strain values at the time of the overload. It can be seen that the rate after the overload in comparison with the

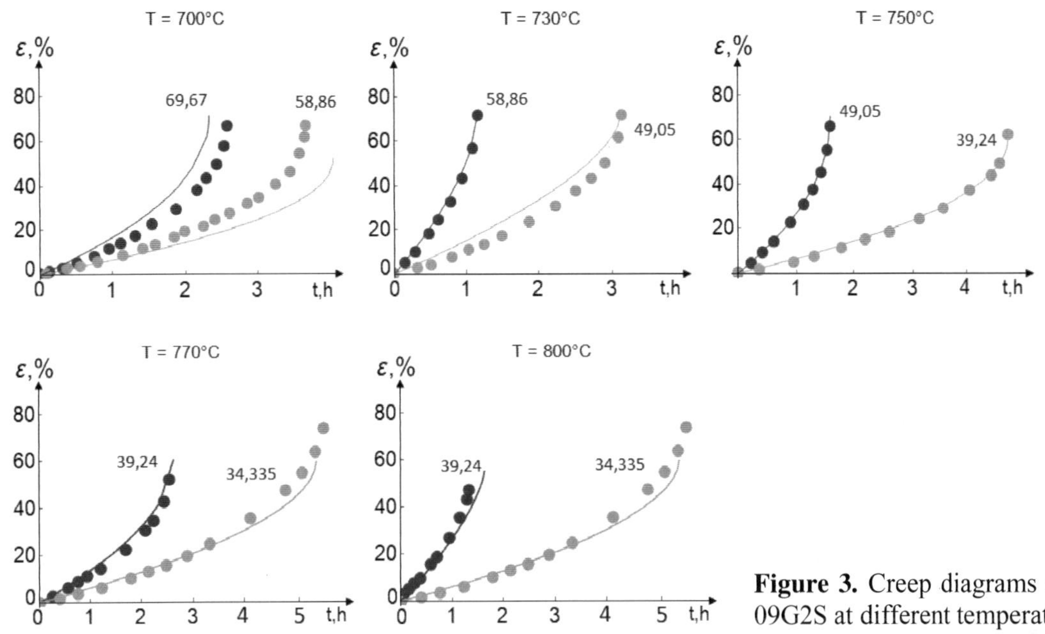

Figure 3. Creep diagrams for st. 09G2S at different temperatures.

stationary experiment increased insignificantly. This fact demonstrates that time is a weak softening factor.

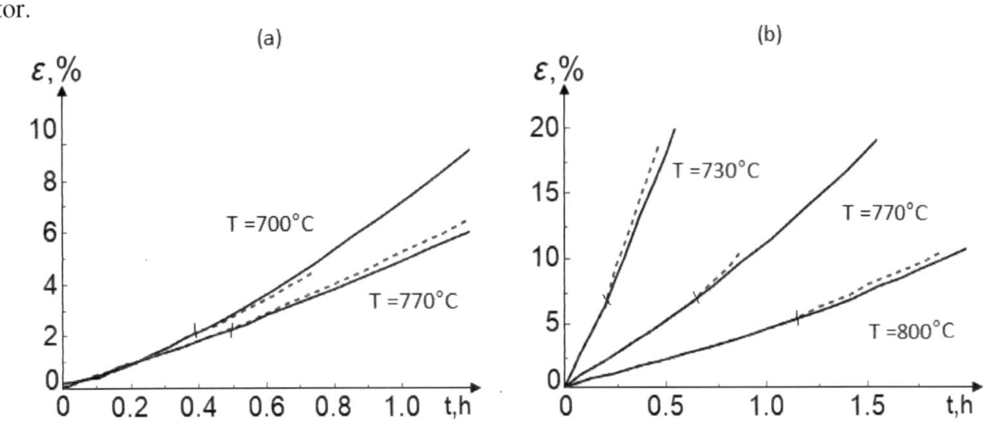

Figure 4. The influence of the overloading on strain rate: (a) under $\varepsilon \approx 2\%$; (b) under $\varepsilon \approx 5\%$.

Kinetic equations of creep and damage were applied to the description of the non-stationary process of the strain. To describe the non-stationary creep process [1, 2, 4, 6, 8, 12, 13] at the moments in time of overload with the stress σ_1 to the stress of σ_2, the parameter of damage $\omega_1 = 1-(1-\tau)^{1/(m+1)}$ is determined by the constants of the softening material m and the duration of experiment $\tau = t(m+1)\psi(\omega)$. The deformation $\varepsilon = \omega_1 B_\varepsilon(T)/B_\omega(T)\sigma^{n_1-k_1}$ is calculated. When the overloading to the next level of loading σ_2, the process starts with the accumulated damage level ω_1 and the corresponding time. The considered models are isotropic. The use of isotropic models to finite-strain transient creep was discussed in [14].

The results of 5 experiments and comparison of experimental (points) and numerical (solid curves) creep data with the overloads to other stress levels as temperatures ranges from 700 to 800°C are shown in Figure 5. Moments of overload are marked by vertical arrows.

IOP Conf. Series: Journal of Physics: Conf. Series **894** (2017) 012123 doi:10.1088/1742-6596/894/1/012123

The non-stationary experiments were carried out according to the following schemes:

a) starting with $\sigma_1 = 53.955$ MPa for $t < t_1$, $t_1 = 80$ min; at $t = t_1$ the overload up to $\sigma_2 = 58.86$ MPa with a holding time $t_2 = 30$ min; then after overload up to $\sigma_3 = 63.765$ MPa, the stresses are kept constant to destruction, total long-term experiment time $t_{*1} \approx 4$ h;

b) starting with $\sigma_1 = 39.24$ MPa for $t < t_1$, $t_1 = 112$ min; at $t = t_1$ the overload up to $\sigma_2 = 58.86$ MPa, the stresses are kept constant to destruction, total long-term experiment time $t_{*2} \approx 2.8$ h;

c) starting with $\sigma_1 = 34.335$ MPa for $t < t_1$, $t_1 = 120$ min; at $t = t_1$ the overload up to $\sigma_2 = 49.05$ MPa, the stresses are kept constant to destruction, total long-term experiment time $t_{*3} \approx 3.13$ h;

d) starting with $\sigma_1 = 29.43$ MPa for $t < t_1$, $t_1 = 64$ min; at $t = t_1$ the overload up to $\sigma_2 = 34.335$ MPa with a holding time $t_2 = 51$ min, after the overload up to $\sigma_3 = 39.24$ MPa and during $t_3 = 8$ min then after overload up to $\sigma_4 = 42.183$ MPa the creep occurred to destruction, total long-term experiment time $t_{*4} \approx 2.89$ h;

e) starting with $\sigma_1 = 24.525$ MPa for $t < t_1$, $t_1 = 115$ min; after the overload up to $\sigma_2 = 29.43$ MPa and during $t_2 = 46$ min, after the overload up to $\sigma_3 = 34.335$ MPa and during $t_3 = 30$ min, then after overload up to $\sigma_4 = 39.24$ MPa the creep occurred to destruction, total long-term experiment time $t_{*5} \approx 3.79$ h.

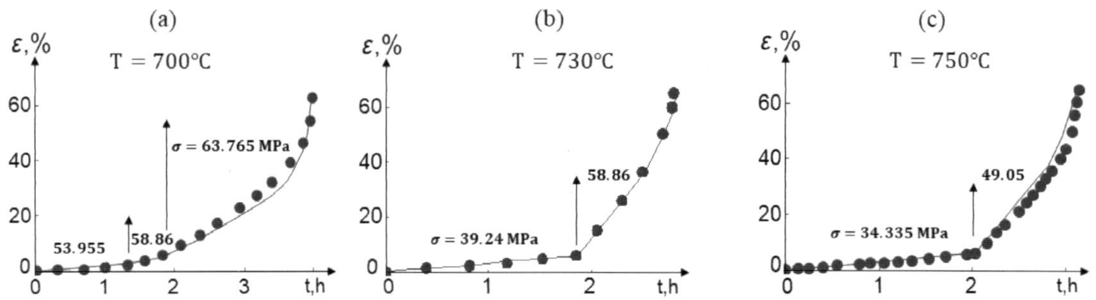

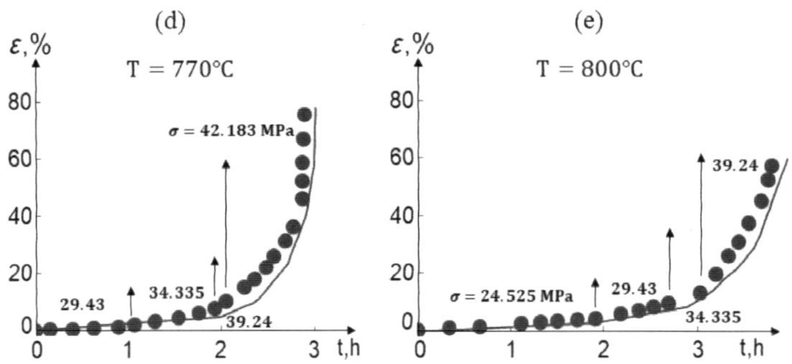

Figure 5. Experimental (points) and approximation (curves) diagrams under non-stationary stress: (a) T=700°C; (b) T=730°C; c) T=750°C; (d) T=770°C; e) T=800°C.

From the analysis of the primary creep curves obtained for all the tested specimens, it is observed that failure of material under steady-state conditions occurs when the creep strain ε_* range of 55 to 75% is reached (Figure 3). The same limiting values ε_* are observed under conditions of non-stationary regimes.

5. Conclusion

The material constants under steady-state loading conditions were identified to describe the unsteady creep process. Simulation of stationary and nonstationary creep processes with damage was carried out. A satisfactory agreement between numerical and experimental data was obtained.

Acknowledgments

This work was partially supported by the Russian Foundation for Basic Research (project codes 15-01-07631-a and 16-08-00713-a).

References

[1] Rabotnov Yu N 1969 *Creep Problems in Structural Members* (Amsterdam: North-Holland) p 836
[2] Kachanov L M 1960 *Teorija Polzuchesti [The Theory of Creep]* (Moscow, Fizmatgiz) p 455 (in Russian)
[3] Sosnin O V 1973 *Strenght of Materials* **5**(5) 132–44
[4] Nikitenko A F 1997 *Polzuchest' i Dlitel'naya Prochnost' Metallicheskikh Materialov [Creep and Long-term Strength of Metal Materials*] (Novosibirsk: Novosibirsk State University of Architecture and Civil Engineering) p 278 (In Russian)
[5] Sosnin O V , Gorev B V and Nikitenko A F 1986 *Energeticheskii Variant Teorii Polzuchesti [Energy Variant of the Creep Theory]*(Novosibirsk: Lavrentyev Institute of Hydrodynamics SO AN SSSR) p 95 (In Russian)
[6] Gorev B V , Klopotov I D and Lyubashevskaya I V 1998 *Theor. and Appl. Fract. Mech.* **29**(1) 1–10
[7] Sosnin O V, Nikitenko A F and Gorev B V 2010 *J. Appl. Mech. Tech. Phy.* **51**(4) 608–14
[8] Gorev B V, Lyubashevskaya I V, Panamarev V A and Iyavoynen S V 2014 *J. Appl. Mech. Tech. Phy.* **55**(6) 1020–30
[9] Banshchikova I A, Gorev B V and Legan M A 2016 *J. Phys.: Conf. Ser.* **754** 082001
[10] Gorev B V, Zakharova T E and Klopotov I D 2002 *Physical Mesomechanics* **5**(2) 17–22
[11] Lokoshenko A M 2016 *Polzuchest' i Dlitel'naya Prochnost' Metallov [Creep and Long-Term Strength of Metals]*(Moscow: Moscow State Industrial University) p 504 (In Russian)
[12] Lokoshchenko A M and Shesterikov S A 1982 *J. Appl. Mech. Tech. Phy.* (2) 160–63
[13] Oding I A and Fridman Z G 1959 *Zavod. Lab.* **25**(3) 329–32
[14] Shutov A V, Larichkin A Yu and Shutov V A 2017 *ZAMM* DOI:10.1002/zamm.201600286

Deformation and fracture in one-dimensional creep problems

A M Kovrizhnykh and S A Kovrizhnykh

Chinakal Institute of Mining, Novosibirsk, Russia

E-mail: amkovr@mail.ru

Abstract. It is assumed that irreversible deformation is a result of shearing in certain planes. In perpendicular directions to these planes, normal strain undergoes change that is proportional to the associated shear. This approach allows accounting for growth of fractures and pores in the background of increasing creep strains without using Kachanov–Rabotnov's kinetic equation for development of damage. Material begins failing when maximum shear creep strain reaches critical value, which initiates drop in shear strength. Using the model based on the maximum shear stress and the exponential law, the authors solve problems on deformation and failure of an elastic–creeping body at the stages of unstable and stable creep.

1. Introduction

A shear model of damage accumulation in deforming solids is proposed, where irreversible strain is considered to result from shearing over planes of the maximum shear stress [1-3]. A solid material starts fracturing when maximum shear reaches peak value, which provokes shear strength loss. Based on the maximum shear stress model and exponential law, the author illustrates solution of problems about deformation and fracture of a body under elastic creep at the stages of unsteady-state and steady-state creep in terms of cylindrical and spherical holes under internal pressure in an infinite body, as well as in cases of pure bending of a rectangular cross section bar and torsion of a round bar. Stresses, creep strains, fracture start time and complete failure time, as well as the fracture front's position and velocity any time are determined for the considered structural elements.

2. Creep deformation and creep failure of cylindrical and spherical holes

At first we consider plane strain of an infinite body's material around a cylindrical hole with radius $r = a$. Denote by σ_r, σ_θ, σ_z radial, hoop and axial stresses, respectively. The surface $r = a$ experiences constant pressure $\sigma_r = -p$, while the material is stress-free at infinity: $\sigma_r = \sigma_\theta = 0$ when $r \to \infty$. An elastic problem involves inequality $\sigma_\theta > \sigma_z > \sigma_r$, that is why irreversible creep strain will represent shear in the line of $\tau_{\max} = (\sigma_\theta - \sigma_r)/2$ [1-3]. This shear under creep, the shearing rate and the maximum shear stress will be considered to correlate according to the exponential law [4]:

$$\gamma_c = \Omega(t)\left(\frac{\sigma_\theta - \sigma_r}{2}\right)^n, \quad \dot{\gamma}_c = \frac{d\Omega(t)}{dt}\left(\frac{\sigma_\theta - \sigma_r}{2}\right)^n, \tag{1}$$

where γ_c – maximum shear creep strain. By projecting shear γ_c on principal axes of stresses and including elastic strains, we have:

Content from this work may be used under the terms of the Creative Commons Attribution 3.0 licence. Any further distribution of this work must maintain attribution to the author(s) and the title of the work, journal citation and DOI.
Published under licence by IOP Publishing Ltd

$$\varepsilon_\theta = \frac{\gamma_c}{2} + \frac{1-\nu}{2\mu}\sigma_\theta - \frac{\nu}{2\mu}\sigma_r,\ \varepsilon_r = -\frac{\gamma_c}{2} + \frac{1-\nu}{2\mu}\sigma_r - \frac{\nu}{2\mu}\sigma_\theta. \tag{2}$$

If we ignore the elastic strains in (2) and take (1), boundary conditions for σ_r, the equilibrium equation and strain compatibility conditions, then we find:

$$\sigma_\theta = -p\left(\frac{a}{r}\right)^{2/n}\left(1-\frac{2}{n}\right),\ \sigma_r = -p\left(\frac{a}{r}\right)^{2/n},\ \gamma_c = \Omega(t)\frac{p^n a^2}{n^n r^2}. \tag{3}$$

Now we change over to dimensionless values with the aim to accounting for elastic strains:

$$\gamma = \frac{\mu}{p}\gamma_c,\ \tau = \frac{\sigma_\theta - \sigma_r}{2p},\ \tilde{\sigma}_\theta = \frac{\sigma_\theta}{p},\ \tilde{\sigma}_r = \frac{\sigma_r}{p},\ \tilde{t} = \mu p^{n-1}\Omega(t),\ \tilde{r} = r/a. \tag{4}$$

Expressing $(\sigma_\theta - \sigma_r)/2$ in (1) in terms of γ_c, including the constitutive relations (2), equilibrium equation and strain compatibility condition will yield a differential equation:

$$\frac{\partial\gamma}{\partial\tilde{r}} + \frac{2(1-\nu)}{n\gamma}\left(\frac{\gamma}{\tilde{t}}\right)^{1/n}\frac{\partial\gamma}{\partial\tilde{r}} + \frac{2\gamma}{\tilde{r}} + \frac{4(1-\nu)}{\tilde{r}}\left(\frac{\gamma}{\tilde{t}}\right)^{1/n} = 0. \tag{5}$$

When $n \to \infty$ a subcase ensues from (5) in the form of the perfectly plastic body equation derived in [3]. Let us find solution of (5) for γ and $\tau = (\gamma/\tilde{t})^{1/n}$ which coincides with the elastic solution when $\tilde{t} = 0$ and with the solution of (3) when $\tilde{t} \to \infty$:

$$\gamma + 2(1-\nu)\left(\frac{\gamma}{\tilde{t}}\right)^{1/n} = \left[2(1-\nu) + \frac{\tilde{t}}{n^n}\right]\frac{1}{\tilde{r}^2}. \tag{6}$$

The exponential law for dimensionless variables is of the form of: $\gamma = \tilde{t}\,\tau^n$. With the pre-assigned values of the arguments $\tilde{r}$, $\tilde{t}$ as well as with ν and n, we determine γ and τ from (6). Then, taking the equilibrium equation and the boundary condition $\tilde{\sigma}_r = -1$ for $\tilde{r} = 1$, integration with respect to $\tilde{r}$ yields $\tilde{\sigma}_r$ and $\tilde{\sigma}_\theta$.

Let us determine starting time of the cylindrical hole fracture, position of the fracture front $r = c$ and its velocity without regard to elastic strains. The hole undergoes constant internal pressure, creep strains increase and at $t = t_0$ when $\gamma_c = \gamma_*$ the hole contour $r = a$ ruptures. The fracture starting time t_0 is found from (3):

$$\Omega(t_0) = \Omega_0 = \gamma_*\frac{n^n}{p^n},\quad t_0 = \Omega^{-1}(\Omega_0).$$

Use (3) and (2) to find radial displacement velocity of an arbitrary point in the creep area $r \geq c > a$ at $t > t_0$:

$$\dot{\gamma}_c = \frac{d\Omega(t)}{dt}\frac{p^n c^2}{n^n r^2} = 2\frac{\dot{u}}{r},\quad \dot{u} = \frac{d\Omega(t)}{dt}\frac{p^n c^2}{2n^n r}. \tag{7}$$

Replacing a by c in (3) will allow solution for stresses. When $t > t_0$ we have the creep area $r > c$ and the fracture area $a \leq r \leq c$. Taking into account (2) at the fracture front when $r = c$, we have that $\dot{u} = \gamma_*\dot{c}/2$; on the other hand, we have (7). We conclude that the radial velocity $\dot{u}$ is continuous at the fracture front any time and arrive at:

$$d\Omega(t) = \frac{\gamma_* n^n}{p^n}\frac{dc}{c}. \tag{8}$$

After integrating of (8) with $c = a$ at $t = t_0$, we obtain:

$$\Omega(t) = \Omega_0\left(1 + \ln\frac{c}{a}\right),\ \frac{c}{a} = \exp\left(\frac{\Omega(t)}{\Omega_0} - 1\right),\ \frac{dc}{dt} = \frac{a}{\Omega_0}\exp\left(\frac{\Omega(t)}{\Omega_0} - 1\right)\frac{d\Omega(t)}{dt}. \tag{9}$$

As for the deformation and failure of a spherical hole under constant internal pressure, it follows from analyzing the elastic solution that $\sigma_\theta = \sigma_\varphi > \sigma_r$, and the creep strain is a result of shears in the lines of the principal shear stresses $(\sigma_\theta - \sigma_r)/2$ and $(\sigma_\varphi - \sigma_r)/2$ [3]. This stress state is termed here as the total creep, and the constitutive relations in this case are:

$$\varepsilon_\theta = \frac{\gamma_c}{2} + \frac{1-\nu}{E}\sigma_\theta - \frac{\nu}{E}\sigma_r, \; \varepsilon_r = -\gamma_c - \frac{2\nu}{E}\sigma_\theta + \frac{1}{E}\sigma_\theta. \tag{10}$$

For the shear γ_c we have the exponential law (1). When elastic strains are negligible, differentiating (10) with respect to time produces $2\dot{\varepsilon}_\theta = -\dot{\varepsilon}_r = \dot{\gamma}_c$. We place the values of $\dot{\varepsilon}_\theta$ and $\dot{\varepsilon}_r$ in the strain rate compatibility condition and obtain an equation for $\dot{\gamma}_c$ and its general solution:

$$\frac{\partial \dot{\gamma}_c}{\partial r} + \frac{3\dot{\gamma}_c}{2r} = 0, \quad \dot{\gamma}_c = \frac{C(t)}{r^3}.$$

Substitution of the found solution in (1) gives $(\sigma_\theta - \sigma_r)/2$. After that, with the use of the equilibrium equation and boundary conditions for the stress σ_r, we find the other stresses and γ_c:

$$\sigma_\theta = -p\left(\frac{a}{r}\right)^{3/n}\left(1 - \frac{3}{2n}\right), \; \sigma_r = -p\left(\frac{a}{r}\right)^{3/n}, \; \gamma_c = \Omega(t)\left(\frac{3p}{4n}\right)^n \frac{a^3}{r^3}. \tag{11}$$

In order to solve this problem with elastic strains, we transfer to dimensionless values via replacing μ by E in (4). Expressing $(\sigma_\theta - \sigma_r)/2$ from (1) in terms of γ_c as well as including the constitutive relations (10), equilibrium equation and the strain compatibility condition results in the differential equation for γ:

$$\frac{\partial \gamma}{\partial \tilde{r}} + \frac{4(1-\nu)}{n\gamma}\left(\frac{\gamma}{\tilde{t}}\right)^{1/n}\frac{\partial \gamma}{\partial \tilde{r}} + \frac{3\gamma}{\tilde{r}} + \frac{12(1-\nu)}{\tilde{r}}\left(\frac{\gamma}{\tilde{t}}\right)^{1/n} = 0. \tag{12}$$

When $n \to \infty$ a subcase ensues from (12) in terms of the perfectly plastic body equation derived in [3]. By solving (12), we find integrals for γ and $\tau = (\gamma/\tilde{t})^{1/n}$ which coincide with the elastic solutions at $\tilde{t} = 0$ and with the solution (11) when $\tilde{t} \to \infty$:

$$\gamma + 4(1-\nu)\left(\frac{\gamma}{\tilde{t}}\right)^{1/n} = \left[3(1-\nu) + \tilde{t}\left(\frac{3}{4n}\right)^n\right]\frac{1}{\tilde{r}^3}. \tag{13}$$

Having the pre-assigned arguments $\tilde{r}$, $\tilde{t}$ and the values ν, n, it is possible to determine γ and τ from (13). After that, based on the equilibrium equation and boundary condition $\tilde{\sigma}_r = -1$ for $\tilde{r} = 1$, integration with respect to $\tilde{r}$ yields $\tilde{\sigma}_r$ and $\tilde{\sigma}_\theta$.

Let us determine the fracture starting time t_0, without regard to elastic strains, for the spherical hole contour $r = a$ at $t = t_0$ when $\gamma_c = \gamma_*$. As a result, we have from (11):

$$\Omega(t_0) = \Omega_0 = \gamma_*\left(\frac{4n}{3p}\right)^n, \quad t_0 = \Omega^{-1}(\Omega_0). \tag{14}$$

Based on the condition that the radial displacement velocity of points of the fracture front is continuous, we derive formulas from (9), where Ω_0 for the spherical hole is found from (14).

3. Pure bending of a rectangular cross section bar

Axial strain of a bar exposed to bending under condition of creep is written as follows [4]:

$$\varepsilon_z = \varepsilon_e + \varepsilon_c = y(\kappa_e + \kappa_c) = \frac{\sigma_z}{E} + \Omega(t)\sigma_z^n,$$

where ε_e and ε_c – elastic strain and creep strain, respectively; κ_e and κ_c – the bar axis bends due to elasticity and creep, respectively. We know that $\Omega(t) = 0$ at $t = 0$ and it is possible to neglect elastic strains when $\Omega(t) \to \infty$, thus, we find κ_e and κ_c:

$$\kappa_e = \frac{M}{EI_x},\ \kappa_c = \Omega(t)\left(\frac{M}{I_{nx}}\right)^n,\ I_x = \frac{4bh^3}{3},\ I_{nx} = \frac{4nbh^{2+1/n}}{2n+1},$$

where M – bending moment; E – elastic modulus; I_x, I_{nx} – moment and generalized moment of inertia of cross section about neutral axis x. Let us introduce the following dimensionless values:

$$\sigma = \frac{\sigma_z}{M/W_x},\ e = \frac{E\varepsilon_c}{M/W_x},\ \tilde{t} = E\Omega(t)\left(\frac{M}{W_x}\right)^{n-1},\ \tilde{y} = \frac{y}{h},$$

where $W_x = I_x / h$ – section modulus. Considering the aforesaid, we derive a non-dimensional axial stress equation:

$$\sigma + \tilde{t}\sigma^n = \tilde{y}\left[1 + \tilde{t}\left(\frac{2n+1}{3n}\right)^n\right],$$

wherefrom, given $\tilde{t}$, $\tilde{y}$ and n, we can find $\sigma(\tilde{y},\tilde{t})$ and, in the next place, $e = \tilde{t}\sigma^n$.

Let us discuss determining the fracture start time and the full rupture time for a pure bending bar under creep. Ignore elastic strains and, from the Bernoulli hypothesis and exponential law, obtain the following relations:

$$\varepsilon_c = y\kappa_c,\ \kappa_c = \Omega(t)\left(\frac{M}{I_{nx}}\right)^n,\ d\kappa_c = d\Omega(t)\left(\frac{M}{I_{nx}}\right)^n. \quad (15)$$

The pre-failure strain ε_c achieves maximum at $y = h_0$. The rupture starting time is determined based on (15) and $\varepsilon_c = \varepsilon_*$ at $t = t_0$:

$$\kappa_c = \varepsilon_* / h_0 = \Omega(t_0)\left(\frac{M}{I_{nx}}\right)^n,\ \Omega(t_0) = \Omega_0 = \frac{\varepsilon_*(4nb)^n h_0^{2n}}{(2n+1)^n M^n},\ t_0 = \Omega^{-1}(\Omega_0).$$

Now we assume that a material under creep resists in the same manner a tension fracture and a compression fracture. In the case in question, the neutral layer of the bar is motionless and the fracture fronts, that are at the same time free boundaries of the bar, begin moving from two sides $y = \pm h(t)$. At an arbitrary time, we have $d\kappa_c = -\varepsilon_* dh/h^2$ on one side fracture front and the third equality from (15) on the other side. As a result, we obtain the following differential equation for $\Omega(t)$:

$$d\Omega(t) = -\frac{\varepsilon_*(4nb)^n h^{2n-1} dh}{(2n+1)^n M^n}. \quad (16)$$

Integration of (16), considering that $h(t_0) = h_0$ at $t = t_0$, yields:

$$\Omega(t) = \Omega_0\left[1 + \frac{1}{2n}\left(1 - \frac{h^{2n}}{h_0^{2n}}\right)\right],\ \frac{h(t)}{h_0} = \left[1 + 2n\frac{\Omega_0 - \Omega(t)}{\Omega_0}\right]^{\frac{1}{2n}}. \quad (17)$$

Find the bar fracture start time from (17) for $h = 0$. Differentiation of the second equation in (17) with respect to time produces the fracture front velocity.

4. Torsion deformation and fracture of a round bar under creep

Then we consider torsion deformation and fracture of a round bar under creep. Denote axial and tangential lines by z and θ, respectively. Under a torsion torque M_z, we have that:

$$\gamma_z = \frac{\tau_z}{\mu} + \gamma_c = \frac{\tau_z}{\mu} + \Omega(t)\tau_z^n = r(\omega_e + \omega_c),$$

where $\tau_z = \tau_{z\theta}$; $\gamma_z = \gamma_{z\theta}$ – shear strain; $\gamma_c = \gamma_{z\theta}^c$ – shear strain caused by creep; ω_e and ω_c – torsion angle components generated by the bar elasticity and creep under torsion. Taking into account that $\Omega(t) = 0$, $\tau_z = \mu r\omega_e$ at $t = 0$ and ignoring elastic strains when $\Omega(t) \to \infty$, determine ω_e and ω_c:

$$\omega_e = \frac{M_z}{\mu I_r},\ \omega_c(t) = \Omega(t)\left(\frac{M_z}{I_{nr}}\right)^n,\ I_r = \frac{\pi a^4}{2},\ I_{nr} = \frac{2\pi a^{3+1/n}}{3+1/n},$$

where μ – elastic shear modulus; I_r and I_{nr} – polar moment and generalized polar moment of inertia. Now introduce dimensionless values:

$$\tau = \frac{\tau_z}{M_z/W_r},\ \gamma = \frac{\mu\gamma_c}{M_z/W_r},\ \tilde{t} = \mu\Omega(t)\left(\frac{M_z}{W_r}\right)^{n-1},\ \tilde{y} = \frac{y}{h},$$

where $W_r = I_r/a$ – polar section modulus. Based on the aforesaid, we obtain the dimensionless shear stress equation:

$$\tau + \tilde{t}\tau^n = \tilde{r}\left[1 + \tilde{t}\left(\frac{3n+1}{4n}\right)^n\right], \tag{18}$$

where from (18), with $\tilde{t}$, $\tilde{r}$ and n pre-assigned, find $\tau(\tilde{r},\tilde{t})$ and then $\gamma = \tilde{t}\tau^n$.

Now let us analyze the torsion and creep fracture of the spheric bar without regard to elastic strains. First, we determine the fracture start time and complete fracture time in the bar. Let a_0 and a be initial and current radii of the bar. During rupture, the current radius $r = a(t)$ coincides with the fracture front that is moving toward the bar center. For the torsion angle $\omega_c(t)$ per unit length of the bar is independent of r and depends on t only, we have that:

$$\gamma_c = r\omega_c(t),\quad \dot{\gamma}_c = r\dot{\omega}_c,\quad \omega_c(t) = \Omega(t)\left(\frac{M_z}{I_{nr}}\right)^n. \tag{19}$$

The creep strain γ_c achieves maximum when $r = a$, so, fracture sets on at the surface $r = a_0$ at $t = t_0$, where t_0 – the fracture starting time. Then, from (19) we obtain:

$$\frac{\gamma_*}{a_0} = \Omega(t_0)\left(\frac{M_z}{I_{nr}}\right)^n,\quad \Omega(t_0) = \Omega_0 = \frac{\gamma_* a_0^{3n}}{M^n}\left(\frac{2\pi}{3+1/n}\right)^n,\quad t_0 = \Omega^{-1}(\Omega_0).$$

When $t \geq t_0$ the bar surface $r = a(t)$ ruptures and the torsion angle grows with time. It follows from (19) that:

$$d\omega_c = d\Omega(t)\left(\frac{3+1/n}{2\pi}\right)^n \frac{M_z^n}{a^{3n+1}}. \tag{20}$$

On the other hand, based on the formula (19), we find the torsion angle alteration at the fracture front $d\omega_c = -\gamma_* da/a^2$. Equation (20) as well as the condition of continuous velocity of the torsion angle alteration at the fracture front and in the creep area yield a differential equation to determine $\Omega(t)$:

$$d\Omega(t) = -\frac{\gamma_*}{M_z^n}\left(\frac{2\pi}{3+1/n}\right)^n a^{3n-1}da.$$

Given that $a = a_0$ at $t = t_0$, the above equation integrates into:

$$\Omega(t) = \Omega_0\left[1 + \frac{1}{3n}\left(1 - \frac{a^{3n}}{a_0^{3n}}\right)\right],\ a = a_0\left[1 + 3n\frac{\Omega_0 - \Omega(t)}{\Omega_0}\right]^{\frac{1}{3n}}, \tag{21}$$

where from we find the complete fracture time in the bar under torsion for $a = 0$. Differentiating the second equality in (21) with respect to time gives the fracture front velocity.

5. Conclusion

New solutions have been obtained for determining creep strains, as well as position and velocity of fracture front in the problems on deformation and failure of cylindrical and spherical holes under internal pressure at the stages of unsteady-state and steady-state creep. The problems on pure bending of a rectangular cross section bar and on torsion of a round bar provide stresses, creep strains and the times of fracture start and complete destruction. In case of the steady-state creep, the complete fracture time to the fracture start time ratio for the bar exposed to torsion matches the result obtained [4, 5].

References

[1] Kovrizhnykh A M 1986 *Izvestia AN SSSR Mechanics of Solids* **4** 140-6
[2] Kovrizhnykh A M 1987 *Izvestia AN SSSR Mechanics of Solids* **6** 98-106
[3] Kovrizhnykh A M 2009 *Journal of Mining Science* **6** 29-39
[4] Kachanov L M 1960 *The Theory of Creep* (Moscow: Nauka) p 455
[5] Rabotnov Yu N 1969 *Creep Problems in Structural Members* (Amsterdam: North-Holland) p 752
[6] Kovrizhnykh A M, Baryshnikov V D, Manakov A V and Nikitenko A F 2007 *Journal of Applied Mechanics and Technical Physics* **48**(6) 871-7
[7] Kovrizhnykh A M 2009 *Mechanics of Solids* **44**(2) 269-76

IOP Conf. Series: Journal of Physics: Conf. Series **894** (2017) 012125 doi:10.1088/1742-6596/894/1/012125

Numerical simulation of stress-strain state of electrophoretic shell molds

A V Sviridov, V I Odinokov, E A Dmitriev, A I Evstigneev and O V Bashkov

Komsomolsk-na-Amure State Technical University, Komsomolsk-on-Amur, Russia

Abstract. In the foundry engineering, castings obtained in one-piece non-gas-generating high-refractory electrophoretic shell molds (ShM) by investment patterns (IP) have an increased rejects percentage associated with low deformation resistance and crack resistance of the SM at different stages of their formation and manufacturing. Crack resistance of the ShM based on IP depends mainly on their stress-strain state (SSS) at various stages of mold forming. SSS decrease significantly improves their crack resistance and decreases their rejects percentage of castings occurring due to clogging and surface defects. In addition, the known methods of decreasing the SSS are still poorly understood. Thus, current research trends are to determine SSS at each stage of ShM forming and develop the ways to decrease it. Theoretical predicting of crack formation in multiple-layer axisymmetric shell molds is given in the work [1], and SSS of multiple-layer axisymmetric shell molds is given in the work [2]. Monolayer electrophoretic ShM had a lack of concern in this field, thus it became an argument for the present workMathematical Model of ShM SSS

1. Engineering Formulation of the Problem

Mathematical model of ShM SSS is based on heat transfer equations that allow calculation of distribution of thermal fields within the mold and casting, and, based on equations of continuum mechanics, calculation of distribution of stresses and strains in the structure of the ShM during solidification of the casting. To construct a mathematical model, the numerical method developed by Professor V I Odinokov, has been used.

Ceramic shell mold, single-layer, one-piece. Curvilinear orthogonal coordinates around the perimeter of the area are shown on the Figure 1 (a), (b). Cross-section of the casting, axially symmetrically shaped in the vertical plane, has a view shown in Figure 1. Where I-I is the axis of rotation. The mold is exposed from inside to ferrostatic pressure of molten metal – P_1. During simulation, two options of filling the mold were considered. In option one, casting process and forming casting formation to a preheated mold were simulated – the case of casting to the ShMs obtained by investment patterns. In option two, casting process and forming casting formation to a cold shell mold were simulated – the case of casting to the shell molds obtained by permanent models.

During development of the mathematical process model, it was taken into consideration that the simulated system at the initial stage is a two-component: liquid metal – mold; as time goes, due to change in temperature fields, the system evolves into a three-component one – liquid metal I – (LM), solid metal II – (SM), mold III. It is modeling of interaction of these three components in temperature field changing over time at a pressure P_1 that is an aim of this paper.

Content from this work may be used under the terms of the Creative Commons Attribution 3.0 licence. Any further distribution of this work must maintain attribution to the author(s) and the title of the work, journal citation and DOI.
Published under licence by IOP Publishing Ltd

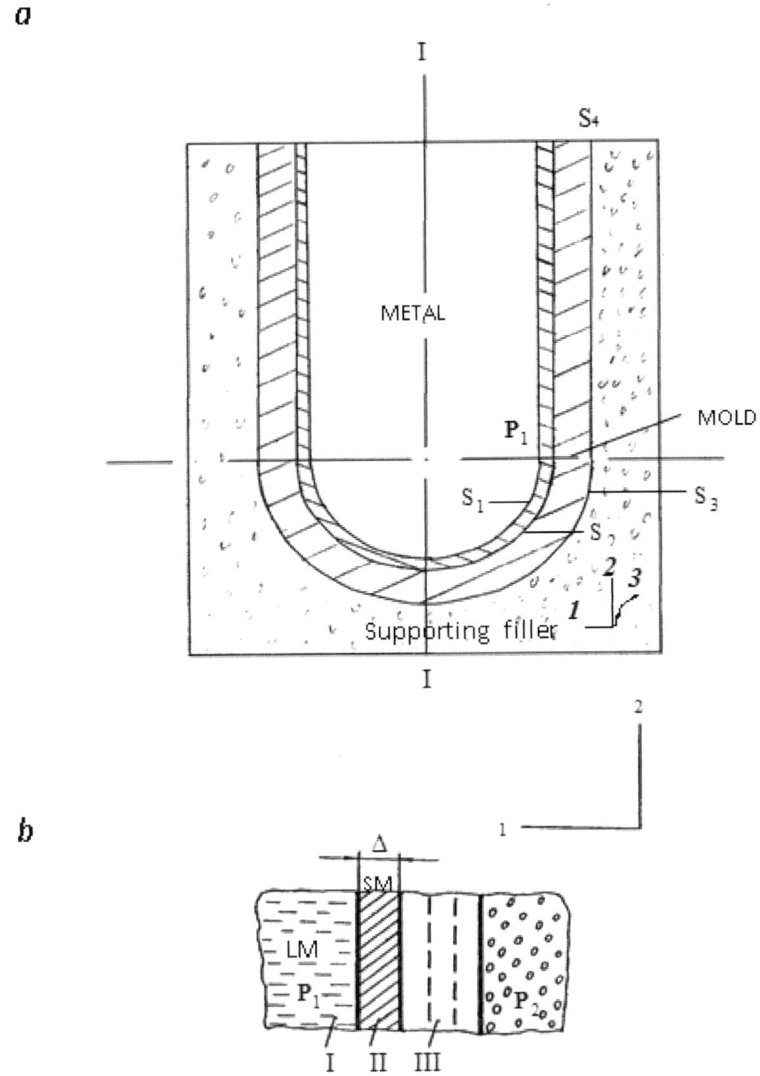

Figure 1. Modeling diagram of ShM SSS: a – casting and ShM cross-section in the horizontal plane; b – the system areas layout (I – liquid metal, II – solidified metal, III – mold).

2. Mathematical Formulation of the Problem

An axially symmetric body of revolution is considered. Movement is considered being slow.There is a three-component system. Deformable medium, solidified metal (area II), and mold (area III) are isotropic materials. The process is transient. Using the theory of small elastic-plastic deformations and Euler coordinate system, we can write a system of equations for each of the areas. Namely, we have

Area I:

$$\begin{aligned} &\sigma_{11} = \sigma_{22} = \sigma_{33} = \sigma = P_1, \\ &\dot{\theta} = a_1 \Delta \theta. \end{aligned} \tag{1}$$

Areas II, III:

$$\sigma_{ij,j} = 0,$$
$$\sigma_{ij} - \sigma\delta_{ij} = 2G_p\varepsilon^*_{ij}; \; \varepsilon^*_{ij} = \varepsilon_{ij} - \frac{1}{3}\varepsilon\delta_{ij}; \; \varepsilon = \varepsilon_{ii}, \tag{2}$$
$$\varepsilon_{ii} = 3k_p\sigma + 3\alpha_p\left(\theta_n - \theta^*_p\right); \; \varepsilon_{ij} = 0{,}5\left(U_{i,j} + U_{j,i}\right),$$
$$\dot{\theta} = a^*_p\Delta\theta.$$

where U_i are displacements, ε_{ij} are strains, σ is hydrostatic stress, $G_p\ (p = II, III)$ is shear modulus of the metal $(p = II)$ and mold $(p = III)$ $(p = III)$, δ_{ij} is Kronecker symbol, k_p are volume compressibility coefficients, α_p are linear expansion coefficients, a^*_p are thermal diffusivity coefficients, $(p = II, III)$, θ_n is current temperature, τ is time, θ^*_ρ are initial temperatures in the areas $(p = II, III)$.

During cooling the liquid metal, provided that the metal temperature is $\theta_M \leq \theta_K$ (θ_K is crystallization temperature), thickness of the solidified layer is determined based on solution of interphase transfer equation:

$$\frac{d\theta_1}{dn}\lambda_1 - \frac{\theta_2}{dn}\lambda_2 = \frac{d\Delta}{d\tau}L\rho \tag{3}$$

where θ_1 and θ_2 are temperatures, respectively, in solid and liquid phases, λ_1 and λ_2 are thermal conductivity coefficients in the respective phases, Δ is thickness of crust, L is latent heat of fusion, ρ is density, n is normal line to the boundary of the two phases.

Crystallization process time, τ_*, is divided into small steps $\Delta\tau_n$ (n is a time step number). At each time step $\Delta\tau_n$, thickness of the solid phase Δ_n ($\Delta = \sum_1^n \Delta_n$) is calculated. Thus, it is assumed that temperature in the solid phase along thickness Δ_n varies linearly and temperature gradient in the liquid phase is zero. In view of this, solution of equation (3) gives the following relation for determining the thickness of crystallized crust Δ_n at time step $\Delta\tau_n$

$$\Delta_n = C\sqrt{\tau}, \; C = \sqrt{\frac{2\Delta\theta_1\lambda_1}{\rho L}}, \tag{4}$$

where $\Delta\theta_1$ is differential temperature in the solid phase near solidification front.

2.1. Numerical Diagram for Solving the Problem

To solve system (2), the numerical method described in paper [3] was used. According to this method, the strain area is divided into a finite number of orthogonal curvilinear elements (Figure 2, a).

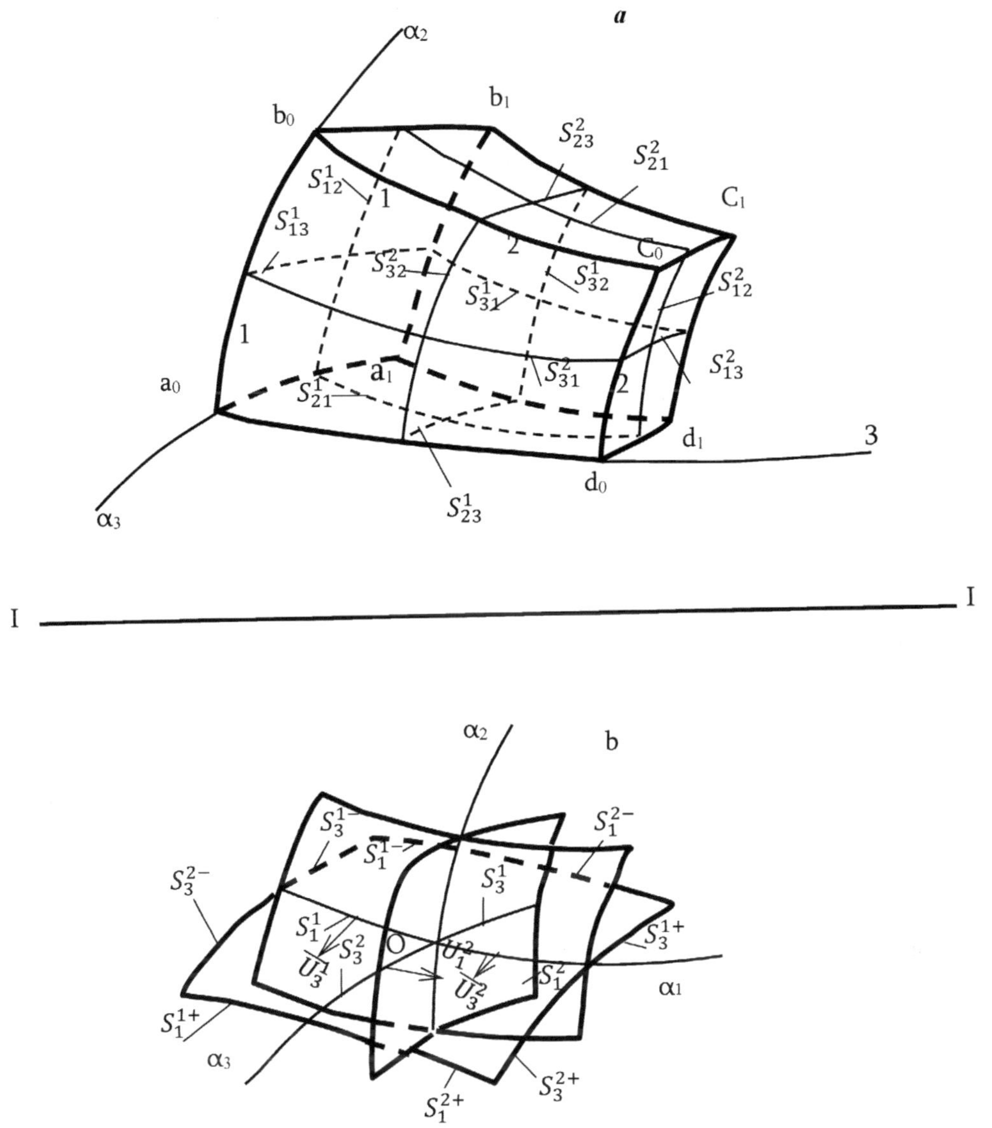

Figure 2. Strain area: a – area division into finite orthogonal elements; b – shear strains.

In case of axial symmetry (Figure 1, *a*), the following conditions take place:

$$\sigma_{31} = \sigma_{32} = 0;\ \varepsilon_{13} = \varepsilon_{23} = 0;\ \mathrm{U}_3 = 0 \tag{5}$$

In accordance with paper [3] and due to (5), equations (2) and values ε_{ij} can be written as

$$\begin{aligned} &S_{13}\Delta S_{12}(\sigma_{11} - \sigma_{22}) + S_{12}\Delta S_{13}(\sigma_{11} - \sigma_{33}) + 0.5\Delta\sigma_{11}S_{12}S_{13} + \\ &+ 0.5\Delta\sigma_{12}S_{21}S_{23} + (S_{21}\Delta S_{23} + 2S_{23}\Delta S_{21})\sigma_{12} = 0, \\ &S_{21}\Delta S_{23}(\sigma_{22} - \sigma_{33}) + S_{23}\Delta S_{21}(\sigma_{22} - \sigma_{11}) + 0.5\Delta\sigma_{22}S_{23}S_{21} + \\ &+ 0.5\Delta\sigma_{21}S_{12}S_{13} + (S_{12}\Delta S_{13} + 2S_{13}\Delta S_{12})\sigma_{21} = 0; \end{aligned} \tag{6}$$

$$\begin{aligned}&\sigma_{11}-\sigma_{22}=2G_p(\varepsilon_{11}-\varepsilon_{22}),\\&\sigma_{22}-\sigma_{33}=2G_p(\varepsilon_{22}-\varepsilon_{33}),\\&\sigma_{12}=G_p\cdot\varepsilon_{12};\end{aligned}\tag{7}$$

$$\varepsilon_{11}+\varepsilon_{22}+\varepsilon_{33}=3k_p\sigma+3\alpha_p(\theta_n-\theta_p^*),\tag{8}$$

where $\sigma_{ij}=0.5(\sigma_{ij}^1+\sigma_{ij}^2)$, $\Delta\sigma_{ij}=\sigma_{ij}^2-\sigma_{ij}^1$, $(i,j=1,2,3)$.

Reached equations (6) formulate by themselves equations of balance, equations (7) – equations of state (or the Hook's law), and (8) - mass-conservation equation.

$$\begin{aligned}&\varepsilon_{11}=\frac{2\Delta U_1}{S_{21}}+\frac{2U_2}{S_{21}}\cdot\frac{\Delta S_{21}}{S_{12}};\quad \varepsilon_{22}=\frac{2\Delta U_2}{S_{32}}+\frac{2U_1}{S_{12}}\cdot\frac{\Delta S_{12}}{S_{21}};\\&\varepsilon_{33}=\frac{2U_1}{S_{13}}\cdot\frac{\Delta S_{13}}{S_{31}}+\frac{2U_2}{S_{23}}\cdot\frac{\Delta S_{23}}{S_{32}},\end{aligned}\tag{9}$$

where $U_i=U_i^1+U_i^2$, $\Delta U_i=U_i^2-U_i^1$, $(i=1,2)$.

In this case, $S_{ik}^j,(i,k=1,2,3;\ j=1,2)$ is arc value of the midline located on the face j along the coordinate α_k and perpendicular to the coordinate α_i. Values of S_{ik}^j are calculated as averages of arc values of the face boundary. Designations of $S_{ij}=S_{ij}^1+S_{ij}^2$, $\Delta S_{ij}=S_{ij}^2-S_{ij}^1$ are introduced. Location of σ_{ij} and U_i on the element faces is shown in Figure 2.

Equations (6), (9) are written considering that

$$\frac{\partial U_1}{\partial x_3}=0,\ \frac{\partial\sigma_{3i}}{\partial x_3}=0;\ i=1,2,3;$$

for bodies of revolution, the following takes place

$$\Delta S_{31}=0;\ \Delta S_{32}=0;\ \frac{\Delta\overline{U}_1}{S_3}=0,\ \frac{\Delta\overline{U}_2}{S_3}=0;\ U_3=0;$$

on surface x_1x_3 : $S_1^+-S_1^-=0$; on surface x_2x_3 : $S_2^+-S_2^-=0$, shear values ε_{ij} $(i\neq j)$ can be written for node (0) in the following form

$$\varepsilon_{12}^0=\frac{2\Delta\overline{U}_2}{S_1}-0.5\overline{U}_2\frac{S_2^+-S_2^-}{S_1S_2}+\frac{2\Delta\overline{U}_1}{S_2}-0.5\overline{U}_1\frac{S_1^+-S_1^-}{S_1S_2},\tag{10}$$

where $S_i=S_i^1+S_i^2$; $\Delta\overline{U}_i=\overline{U}_i^2-\overline{U}_i^1$(Fig 2, b); $S_i^+=S_i^{1+}+S^{2+}_i$; $S_i^-=S_i^{1-}+S_i^{2-}$; values $\overline{U}_i^j$ are calculated as averages of values U_i on faces of the element. Paper [3] proves that the difference analogue of system (6)–(8) taking into account (9), in case of presence of initial and boundary conditions, is determinable. Dimensionality of system (6)–(8) is significantly reduced upon implementation of the following operations:

1. Differences $(\sigma_{ii}-\sigma_{jj})$ in equations (6) can be expressed via (7).
2. Mass conservation equation can be rewritten in the recurrent form taking into account expressions (9) in the form of:

$$U_1^2=U_1^1+[A],$$

where $[A]$ is operator not containing U_1^2; sense of area rotation by $x_1(\rightarrow)$, by $x_2(\uparrow)$.

3. Shear expressions ε_{ij} $(i \neq j)$ are determined by the inner grid nodes in accordance with formulas (10); $i=1,\ j=2$.

4. σ_{ij} $(i \neq j)$ are determined by the inner grid nodes of equations of state: $\sigma_{12}^{0} = G_{p}^{0}\varepsilon_{12}^{0}$.

5. σ_{ij} are determined by external grid nodes from the boundary conditions, and on the contact surfaces – from the friction law.

6. σ_{ij} are determined on faces of the elements as the averages of values σ_{ij} in nodes of the faces.

7. The first equation (6) can be rewritten in the form of a recurrence:

$\sigma_{11}^{1} = \sigma_{11}^{2} + [B]$,

where $[B]$ is operator not containing σ_{11}^{1}; sense of area rotation $x_1(\leftarrow), x_2(\downarrow)$.

8. From the system of equations – the second equation in system (6) and equation $\sigma_{22} - \sigma_{11} = 2\lambda(\varepsilon_{11} - \varepsilon_{22})$ – values σ_{22}^{1} and σ_{22}^{2} are defined for the element and the equations having the following are made:

$$F_3 = \left(\sigma_{22}^{2}\right)_{IJ} - \left(\sigma_{22}^{1}\right)_{I,J+1} = 0$$

for internal faces, where J is an index of the element by coordinate x_2.

Thus, if variables $X = \left\{U_2, U_1\big|_{x_1=0}, \sigma_{11}\big|_{x_1=x_1^*}\right\}$ are considered as independent, then along a sequence of (1÷7), the dependent variables can be determined via X (x_1^* is the final value of the coordinate x along the curved area).

Equivalent system of equations consists of:

$$F_1 = \left(U_1^2 - U_1^*\right)\Big|_{x_1=x_1^*} = 0,\ F_2 = \left(\sigma_{11}^{1} - \sigma_{11}^{*}\right)\Big|_{x_1=0} = 0,\ F_3 = 0,$$

where U_1^* are known from the boundary conditions displacements U_1 on the boundary of the area $(x_1 = x_1^*)$; σ_{11}^{*} are known from boundary conditions stresses σ_{11} on the boundary of the area $(x_1 = 0)$. Moreover, number of equations $F_1 = 0$ equals the number of unknown variables $\sigma_{11}\big|_{x_1=x_1^*}$, number of equations $F_2 = 0$ equals the number of unknown variables $U_1\big|_{x_1=0}$.

In order to solve the thermal conductivity equation, a numerical method developed in paper [3] is used. According to this method, for each internal k-th element (Figure 2, *a*), a system of thermal conductivity is written in a difference form, and iterative procedure is made that, given that the heat flux on x_3 is zero, is represented by iterative formula

$$\theta_k = \frac{\theta_k^* + t_{12}\theta_1^+ + t_{11}\theta_1^- + t_{22}\theta_2^+ + t_{21}\theta_2^-}{1 + t_{12} + t_{11} + t_{22} + t_{21}}, \tag{11}$$

where

$$t_{12} = \frac{2\left(\lambda_k + \lambda_1^+\right)}{S_{21} + S_{21}^+} F_1^2 \frac{\Delta\tau}{C_k\gamma_k V_k};\ t_{11} = \frac{2\left(\lambda_k + \lambda_1^-\right)}{S_{21} + S_{21}^-} F_1^1 \frac{\Delta\tau}{C_k\gamma_k V_k};$$

$$t_{22} = \frac{2\left(\lambda_k + \lambda_2^+\right)}{S_{12} + S_{12}^+} F_2^2 \frac{\Delta\tau}{C_k\gamma_k V_k};\ t_{21} = \frac{2\left(\lambda_k + \lambda_2^-\right)}{S_{12} + S_{12}^-} F_2^1 \frac{\Delta\tau}{C_k\gamma_k V_k};$$

$$F_i^j = S_{ik}^j \cdot S_{ip}^j;\ i \neq k \neq p;\ i,k,p = 1,2,3;\ j = 1,2;\ V_k = \frac{S_{13}S_{12}\left(S_{21} + S_{31}\right)}{16},$$

θ_k^* is the average temperature in the k-th element at the beginning of the time step; $\Delta\tau$, λ_k, θ_k, C_k, γ_k are averages of thermal conductivity, temperature, heat capacity and specific gravity in the k-th element at the end of the time step; $\Delta\tau$; $\lambda_i^-, \theta_i^- \ (i=1,2)$ are, respectively, thermal conductivity and temperature coefficients in the element following the k-th element for the coordinate x_i in the negative direction; λ_i^+, θ_i^+ are equivalent parameters in the positive direction x_i.

$$S_{21}^- = S_{21}^{1-} + S_{22}^{2-};\ S_{21}^+ = S_{21}^{1+} + S_{21}^{2+};$$

S_{ij}^{1+} $(i \neq j; i,j = 1,2)$ is arc length of the S_{ij}^1-th element following the k-th element in the positive direction for the coordinate x_j; S_{ij}^{1-} is the same, but in the negative direction on x_j; θ_i^+ is temperature of the element following the k-th element in the direction x_i; θ_i^- is the same, but in the negative direction x_i.

In paper [3] the convergence of the iterative procedure (11) was proved.

The initial conditions of the problem:

$\Delta|_{\tau=0} = 0$ is lack of a solid metal phase;

$\theta_I^*|_{\tau=0} = \theta_0$ is temperature of the cast liquid metal;

$\theta_{III}^*|_{\tau=0} = \theta^*$ is initial mold temperature; θ_{II} is sand temperature; θ_{II}^* is metal crystallization temperature. The boundary conditions of the problem: at the symmetry axis

$$U_1 = 0,\ \sigma_{21} = 0,\ q_{\Pi} = 0.$$

on surfaces S_1, S_3

$$\sigma_{11}|_{S_1} = -P_1;\ \sigma_{12}|_{S_1} = 0;\ U_1|_{S_3} = 0;\ \sigma_{21}|_{S_4} = 0;\ \sigma_{22}|_{S_4} = 0,$$
$$\sigma_{12}|_{S_3} = -\psi \frac{V_{c\kappa}}{V^*} \cos(n_1 x_1);\ q_n|_{S_3} = \alpha^* (\theta_n - \theta_{II}), \tag{12}$$

where $V_{c\kappa}$ is sliding velocity of the mold material relative to sand, V^* is normalizing speed, ψ is the parameter characterizing the friction conditions between the mold and sand, $q_n|_{S_3}$ is heat flux density normal to the surface S_3, α^* is heat transfer coefficient between the ceramic mold and sand.

2.2. Problem Solution Algorithm

1. Cooling time τ^* is divided into a finite number of steps: $\tau^* = \sum \Delta\tau_n$, n – a time step number.

2. The area of interest is divided into a finite number of orthogonal elements.

3. The initial and boundary conditions are given for the elements forming the area of interest, as well as constants of physical-mechanical properties of materials.

4. Arc lengths of elements $S_{ik}^j \ (i,k = 1,2; i \neq k; j = 1,2)$ are calculated.

5. Temperature field at time step $\Delta\tau_n$ is determined by numerical solution of the thermal conductivity equation using the iterative formula (11) in case of presence of initial and boundary conditions for a given time step.

6. If the temperature in the area (I) at the surface S_2 $\theta|_{S_2} \leq \theta_{II}^*$, then thickness of crystallized crust Δ_n is calculated by the formula (4). If $\theta|_{S_2} > \theta_{II}^*$, then operation from clause 7 is performed.

7. System of equations (2) is solved with the difference analogues (6), (7), (9) and developed technique described in paper [3]. Stress σ_{ij} and displacement $U_i\ (i, j = 1,2)$ fields are determined.

8. Time step is made. If $\sum \Delta\tau_n < \tau^*$, then operation from clause 4 is performed. If $\sum \Delta\tau_n = \tau^*$, calculating process is finished.

3. Conclusion

Formulas reached in the work [3] for the optional element limited by orthogonal area were used in this work. Specific feature of previous works in this field is their connectivity to multiple-layer shell molds but present is on simulation on monolayer electrophoretic ShM. Problem on process simulation was solved. The numerical scheme and problem algorithm were developed.

At the present time a range of works on process flow optimization in the fields of machinery, metallurgy and foundry engineering is elaborated, for example steel pouring process into different structure types of multiple-layer shell molds. The solutions of these problems are based on technology specified in this work. Proven numerical approach was used for the solution of formulated equation system.

Acknowledgements. The paper was financially supported by a grant from Russian Science Foundation (RSF) No. 16-19-10149 dated 26.05.16, "Monitoring Systems with Distributed Adaptive Fiber Optic Acoustic Emission Sensors"

References

[1] Evstigneev A I, Odinokov V I, Sviridov A V, Dmitriev E A and Petrov V V Mathematical modelling of stress-strain state of multilayer shell molds *Materials Science Forum* **857** 565-572

[2] Evstigneev A I, Odinokov V I, Sviridov A , Dmitriev E A and Petrov V V Theoretical prediction of crack formation in axisymmetric multilayer shell molds *Materials Science Forum* **857** 573-577

[3] Odinokov V I 1995, *Numerical Study of Deformation Process by Coordinate-Free Method* (Vladivostok: Dalnauka) p 168

IOP Conf. Series: Journal of Physics: Conf. Series **894** (2017) 012126 doi:10.1088/1742-6596/894/1/012126

Cylinder instability in bounded flow with constant vorticity

A G Petrov[1] and M A Yudin[2,3]

[1]Institute for Problems in Mechanics of the Russian Academy of Science, Moscow, Russia
[2]Central Aerohydrodynamics Institute, Moscow, Russia
[3]Moscow Institute of Physics and Technology, Dolgoprudny, Russia

E-mail: pertovipmech@gmail.com

Abstract. The stability of the system consisting of an inner moving cylinder, an outer fixed cylinder and circular flow with constant vorticity between them is considered. Solution of the problem is obtained in form of a Loran series with respect to the distance between axis of inner and outer cylinder. The solution can be obtained with any given accuracy by including more terms in Loran series. The Lagrangian of the system is obtained and the analysis of the system stability with respect to non-linear approximation is derived.

1. Introduction

This paper is an extension of a previous studies [1], in which the linear problem of cylinder stability in a bounded circular flow with arbitrary vorticity was solved. The stability of inner cylinder with radius a in non-viscous incompressible flow with constant vorticity, bounded by the outer cylinder with radius R, in non-linear approximation is investigated (Figure 1).

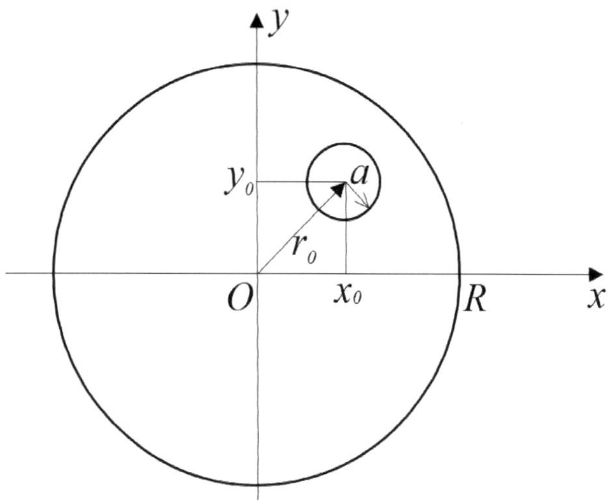

Figure 1. The flow under consideration.

Content from this work may be used under the terms of the Creative Commons Attribution 3.0 licence. Any further distribution of this work must maintain attribution to the author(s) and the title of the work, journal citation and DOI.
Published under licence by IOP Publishing Ltd

Let the coordinate origin coincide with the center of outer cylinder. Than the position of inner cylinder can be described by two coordinates of its center x_0, y_0. The inner cylinder is moving with the velocity $\dot{x}_0, \dot{y}_0$. This movement is caused by the hydrodynamic forces acting on the cylinder.

Generalized coordinates are used to obtain the velocity field and the Lagrangian of the system. A non-linear analysis of cylinder dynamics is derived using the obtained Lagrangian.

2. Principal equations

The objective of the study is to obtain Lagrangian $L(x_0, y_0, \dot{x}_0, \dot{y}_0)$ of the system. The dynamic of the inner cylinder has the form

$$\frac{d}{dt}\frac{\partial L}{\partial \dot{x}_0} - \frac{\partial L}{\partial x_0} = 0, \quad \frac{d}{dt}\frac{\partial L}{\partial \dot{y}_0} - \frac{\partial L}{\partial y_0} = 0$$

The Lagrangian of the circular flow with circulation Γ and constant vorticity Ω_0 can be calculated as follows [2, 3]

$$L = \rho_0 \int_S \left(\frac{V^2}{2} - \left(\mathbf{V} - \mathbf{V}^*\right)\mathbf{V} \right) dxdy \tag{1}$$

where ρ_0 is fluid density, $\mathbf{V}$ is velocity vector field, $V = |\mathbf{V}|$, $\mathbf{V}^*$ is velocity of imaginary incompressible continuum where particle positions are uniquely determined by the boundary ∂S of the flow region S.

Velocity field $\mathbf{V}^*$ can be obtained using the mapping between some fixed region $\mathbf{X}^* \in \tilde{S}$ (for example annulus) and the flow region $\mathbf{X} \in S$ – $\mathbf{X}\left(x_0, y_0, \mathbf{X}^*\right)$ with Jacobian $\det\left\| \partial \mathbf{X} / \partial \mathbf{X}^* \right\| = 1$. The coordinates of the inner cylinder center varies in time $x_0(t), y_0(t)$. Thus velocity $\mathbf{V}^*$ can be found as

$$\mathbf{V}^* = \frac{\partial \mathbf{X}}{\partial x_0}\dot{x}_0 + \frac{\partial \mathbf{X}}{\partial y_0}\dot{y}_0$$

The total Lagrangian of system is the sum of fluid Lagrangian (1) and the kinetic energy of inner cylinder $\Lambda = L + M\,\dfrac{\dot{x}_0^2 + \dot{y}_0^2}{2}$.

3. Velocity field

The velocity field can be obtained as sum of velocities of circular flow $\mathbf{V}_\Gamma$, flow with constant vorticity $\mathbf{V}_\Omega$ and potential flow $\mathbf{V}_\mathbf{P}$

$$\mathbf{V} = \mathbf{V}_\mathbf{P} + \mathbf{V}_\Gamma + \mathbf{V}_\Omega \tag{2}$$

Let's consider the rotated coordinate system (Figure 2)

$$x = x'\frac{x_0}{r_0} - y'\frac{y_0}{r_0}, \quad y = x'\frac{y_0}{r_0} + y'\frac{x_0}{r_0}$$

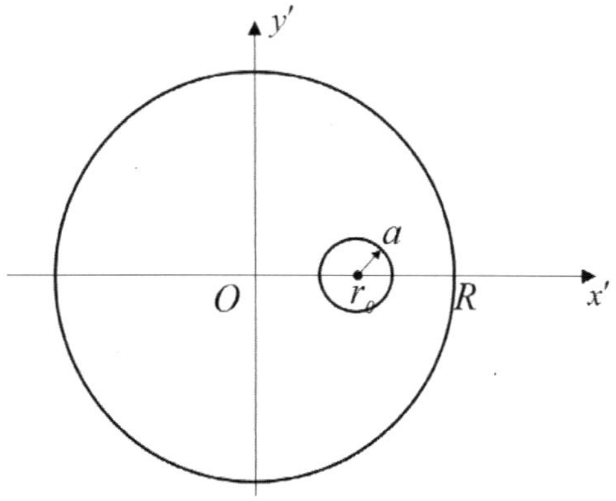

Figure 2. Flow in rotated coordinate system.

This coordinate system is used to simplify the following calculations.

The first term in eq. (2) is the potential velocity field that is caused by the movement of inner cylinder. A complex potential of the flow can be represented as sum of two complex potentials

$$W_P = W_x \dot{x}'_0 + W_y \dot{y}'_0 \,,$$

Where W_x and W_y are complex potential functions corresponding to the movement of the cylinder along Ox' and Oy' axes.

Each component can be expressed as Loran series

$$W_x(z') = A_1\left(z' + \frac{R^2}{z'}\right) + A_2\left(z'^2 + \frac{R^4}{z'^2}\right) + A_3\left(z'^3 + \frac{R^6}{z'^3}\right) + O\left(r_0^3\right)$$

$$W_y(z') = i\left(A_1\left(z' - \frac{R^2}{z'}\right) + A_2\left(z'^2 - \frac{R^4}{z'^2}\right) + A_3\left(z'^3 - \frac{R^6}{z'^3}\right)\right) + O\left(r_0^3\right)$$

The imaginary parts of complex potentials W_x, W_y on the outer cylinder ($|z'| = R$) are equal to zero. On the surface of the inner cylinder $z' = \rho_0 + a(\cos\alpha + i\sin\alpha), 0 \le \alpha \le 2\pi$ normal to the surface of the inner cylinder velocities of fluid and cylinder have to be equal. This boundary condition can be expressed as $\mathrm{Im}(W_x(z_\alpha) - z_\alpha) = 0$, $\mathrm{Im}(iW_x(z_\alpha) - iz_\alpha) = 0$. Using it, we obtain coefficients $A_1, A_2 \ldots$ by formulas

$$A_1 = \frac{a^2}{a^2 - R^2} + \frac{2a^4 R^2}{\left(a^2 - R^2\right)^3 \left(a^2 + R^2\right)} r_0^2$$

$$A_2 = -\frac{a^2 R^2}{\left(R^2 - a^2\right)^2 \left(a^2 + R^2\right)} r_0$$

$$A_3 = \frac{a^2 R^2 \left(a^4 + R^4\right)}{\left(a^2 - R^2\right)^3 \left(a^6 + 2a^4 R^2 + 2a^2 R^4 + R^6\right)} r_0^2$$

The velocity of potential flow can be obtained by taking derivative of complex potential

$$U_P - iV_P = \frac{dW_P}{dz'} \tag{3}$$

The second term in eq. (2) is the velocity field of flow with circulation. The conformal mapping $q(z') = \frac{z' - z_1}{z' - z_2}$, which maps the flow region S into an annulus $q_1 < |q| < q_2$, is used to obtain the complex potential.

Point z_2 is inverse to the point z_1 with respect to both inner cylinder and outer cylinder:

$$z_1 = \frac{-a^2 + R^2 + r_0^2 - \sqrt{-4R^2 r_0^2 + (-a^2 + R^2 + r_0^2)^2}}{2r_0}$$

$$z_2 = \frac{-a^2 + R^2 + r_0^2 + \sqrt{-4R^2 r_0^2 + (-a^2 + R^2 + r_0^2)^2}}{2r_0}$$

The radii q_1, q_2 can be also found

$$q_1 = \frac{r_0 - a - z_1}{r_0 - a - z_2}, \quad q_2 = \frac{R + z_1}{R + z_2}$$

The complex potential in q plane is well-known solution of circulation flow in a region between coaxial cylinders, which complex potential is expressed as $W_\Gamma(z') = -i\frac{\Gamma}{2\pi}\ln q$. Thus, it can be obtained

$$W_\Gamma(z') = -i\frac{\Gamma}{2\pi}\ln\frac{z' - z_1}{z' - z_2}$$

Complex velocity of circular flow is founded by taking derivative

$$U_\Gamma - iV_\Gamma = \frac{dW_\Gamma}{dz'} \tag{4}$$

The third term in eq. (2) corresponds to the flow with constant vorticity in the whole flow region. It is convenient to use rotating with angular velocity ω coordinates to find solutions for first and second terms in eq. (2). Thus in rotating coordinates we express the inner cylinder velocity as $\dot{x}_0'$, $\dot{y}_0' - \omega r_0$. Thus, in (3) the velocity of the inner cylinder has to be changed: $\dot{y}_0' \to \dot{y}_0' - \omega r_0$. The third term can be written as

$$U_\Omega - iV_\Omega = -\omega y' - i\omega x' \tag{5}$$

Using eqs. (2 - 5) we can find the velocity field

$$U' - iV' = \frac{dW_\Gamma}{dz'} + \dot{x}_0' \frac{dW_x}{dz'} + \left(\dot{y}_0' - \omega r_0\right)\frac{dW_y}{dz'} - \omega y' - i\omega x'$$

4. Velocity of continuum

The mapping $\mathbf{X}\left(x_0, y_0, \mathbf{X}^*\right)$ has to be found to determine the velocity $\mathbf{V}^*$. Let region $\tilde{S}$ to be annulus $a < r^* < R$. Than the condition $\det\left\| \partial\mathbf{X} / \partial\mathbf{X}^* \right\| = 1$ can be rewritten as $r^* dr^* d\varphi^* = r dr d\varphi$ or $d\left(r^{*2}\right)d\varphi^* = d\left(r^2\right)d\varphi$.

For the convenience, we introduce the coordinates $\rho = r^2 - a^2$, $\rho^* = r^{*2} - a^2$. Than the condition $d\left(r^{*2}\right)d\varphi^* = d\left(r^2\right)d\varphi$ can be represented as $d\rho^* d\varphi^* = d\rho d\varphi$. We introduce generating function $F\left(\varphi, \rho^*\right)$ $\left(\rho = \frac{\partial F}{d\varphi},\quad \varphi^* = \frac{\partial F}{\partial \rho^*}\right)$. We will seek the generating function in the form $F(\varphi, \tilde{\rho}) = (\varphi + \Phi(\varphi))\rho^*$, $\rho = (1 + \Phi'(\varphi))\rho^*$, $\varphi^* = \varphi + \Phi(\varphi)$.

On the outer boundary of the flow region we can find that $\rho = r^2 - a^2 = r_0^2 \cos 2\varphi + R^2 - a^2 - 2r_0 \cos\varphi \sqrt{R^2 - r_0^2 \sin^2\varphi}$ and $\rho^* = R^2 - a^2$. By using the condition $\rho = (1 + \Phi'(\varphi))\rho^*$, we can obtain

$$\Phi'(\varphi) = \frac{1}{R^2 - a^2}\left(r_0^2 \cos 2\varphi - 2r_0 \cos\varphi \sqrt{R^2 - r_0^2 \sin^2\varphi}\right)$$

The generating function is

$$F\left(\varphi, \rho^*\right) = \rho^* \left(\varphi + \frac{\left(-R^2 \arcsin\left(\frac{r_0 \sin\varphi}{R}\right) + \frac{r_0^2 \sin(2\varphi)}{2} - r_0\left(\sqrt{R^2 - r_0^2 \sin^2\varphi}\right)\sin\varphi \right)}{R^2 - a^2} \right)$$

The components of velocity $\mathbf{V}^*$ can be found by taking derivatives

$$V_{x'}^* = \left(\frac{\partial X}{\partial x_0}\dot{x}_0' + \frac{\partial X}{\partial y_0}\dot{y}_0' \right), \quad V_{y'}^* = \left(\frac{\partial Y}{\partial x_0}\dot{x}_0' + \frac{\partial Y}{\partial y_0}\dot{y}_0' \right)$$

$$X = x_0' + r(r^*, \varphi^*)\cos\left(\varphi(r^*, \varphi^*) + \varphi_0'\right),$$

$$Y = y_0' + r(r^*, \varphi^*)\sin\left(\varphi(r^*, \varphi^*) + \varphi_0'\right)$$

After calculations of derivatives we rewrite the velocity $\mathbf{V}^*$ in the rotated coordinates x', y' by setting to zero the y component of inner cylinder displacement, which can be clearly seen on Figure 2

5. Lagrangian calculation

If we represent velocity as a series in r_0, we will find out that it can be represented in the following way

$$V_{x'} = C_0 + C_{11}\sin\varphi + C_{12}\cos\varphi + C_{22}\sin 2\varphi + \ldots$$

$$V_{y'} = D_0 + D_{11}\sin\varphi + D_{12}\cos\varphi + D_{22}\sin 2\varphi + \ldots$$

All of the terms in integral (1) can be represented as trigonometric series. This representation significantly simplifies the calculation of integral. For example

$$\int_0^{2\pi} V_{x'}^2 d\varphi = 2\pi C_0^2 + \pi C_{11}^2 + \pi C_{12}^2 + \pi C_{22}^2 + \ldots$$

Finally, we obtain the Lagrangian of fluid

$$L = \rho_0 \frac{2\pi a^2 \left(a^2 + R^2\right)\left(\dot{x}_0^2 + \dot{y}_0^2\right)}{4\pi\left(R^2 - a^2\right)} + \rho_0 \frac{2\pi\left(\Gamma\left(a^2 + R^2\right) + 4\pi a^2 R^2 \omega\right)\left(\dot{x}_0 y_0 - \dot{y}_0 x_0\right)}{4\pi\left(R^2 - a^2\right)} +$$
$$+\rho_0 \frac{\left(\Gamma + 2\pi\omega a^2\right)\left(\Gamma + 2\pi\omega R^2\right)\left(x_0^2 + y_0^2\right)}{4\pi\left(R^2 - a^2\right)}$$

Total Lagrangian of the system can be expressed as a sum of Lagrangian of fluid and the kinetic energy of moving inner cylinder

$$\Lambda = \rho_0 \frac{2\pi a^2 \left(a^2 + R^2\right)\left(\dot{x}_0^2 + \dot{y}_0^2\right)}{4\pi\left(R^2 - a^2\right)} + \rho_0 \frac{2\pi\left(\Gamma\left(a^2 + R^2\right) + 4\pi a^2 R^2 \omega\right)\left(\dot{x}_0 y_0 - \dot{y}_0 x_0\right)}{4\pi\left(R^2 - a^2\right)} +$$
$$+\rho_0 \frac{\left(\Gamma + 2\pi\omega a^2\right)\left(\Gamma + 2\pi\omega R^2\right)\left(x_0^2 + y_0^2\right)}{4\pi\left(R^2 - a^2\right)} + M\frac{\dot{x}_0^2 + \dot{y}_0^2}{2}$$

6. Analysis

The term with $\left(\dot{x}_0 y_0 - \dot{y}_0 x_0\right)$ in Lagrangian can be considered as the gyroscopic forces acting on the cylinder. They can be eliminated by introducing rotating coordinates

$$x_0 = X_0 \cos(\Omega t) + Y_0 \sin(\Omega t)$$
$$y_0 = X_0 \sin(\Omega t) + Y_0 \cos(\Omega t)$$

where $\Omega = -\frac{\rho_0\left(\Gamma\left(a^2 + R^2\right) + 4\pi a^2 R^2 \omega\right)}{2\left(M\left(R^2 - a^2\right) + \pi\rho_0 a^2\left(a^2 + R^2\right)\right)}$ is determined from the condition of disappearing of gyroscopic terms in Lagrangian Λ of the system.

Thus, total Lagrangian of the system can be rewritten as the sum of two terms, one of which is proportional to the velocity square and can be treated as the total kinetic energy of the system T, and the other term Π that is independent on velocities and can be treated as total potential energy of the system.

$$\Lambda = T - \Pi, \quad T = \frac{M\left(R^2 - a^2\right) + \pi a^2\left(a^2 + R^2\right)}{2\left(R^2 - a^2\right)}\left(\dot{X}_0^2 + \dot{Y}_0^2\right),$$

$$\Pi = \frac{\Gamma^2\left(-2M + \pi\left(a^2 + R^2\right)\right) + 4\Gamma\pi\left(M - \pi a^2\right)\left(a^2 + R^2\right)\omega + 8\pi^2 a^2\left(-M + \pi a^2\right)R^2\omega^2}{\left(8\pi\left(M\left(R^2 - a^2\right) + \pi a^2\left(R^2 + a^2\right)\right)\right)}\left(X_0^2 + Y_0^2\right)$$

The system is stable if potential part is positive ($\Pi > 0$) and unstable otherwise. Potential part appears to be positive when inner cylinder mass is less than critical

$$M < M_*, \quad M_* = \frac{\pi\rho_0\left(\Gamma^2 a^2 + \Gamma^2 R^2 + 4\Gamma\pi a^4\omega + 4\Gamma\pi a^2 R^2\omega + 8\pi^2 a^4 R^2\omega^2\right)}{2\left(\Gamma + 2\pi a^2\omega\right)\left(\Gamma + 2\pi R^2\omega\right)}$$

When mass of the cylinder is greater than critical one, the system is unstable.

The middle flow between two coaxial cylinders with constant vorticity is a well-known Couette flow that can be expressed in terms of circulation and vorticity as well as in terms of inner and outer cylinder rotational speed ω_1 and ω_2. Thus we can rewrite the stability condition as a function of rotational velocity. Comparing two different representation of Couette flow, we derive

$$\omega = \frac{\omega_2 R^2 - \omega_1 a^2}{R^2 - a^2}, \quad \Gamma = 2\pi \frac{(\omega_1 - \omega_2) R^2 a^2}{(R^2 - a^2)}$$

So, the condition of stability can be expressed as

$$M < M_*, \quad M_* = \frac{\pi\rho_0 a^2 \left(-2a^2\omega_1\omega_2 + R^2\left(\omega_1^2 + \omega_2^2\right)\right)}{\left(R^2 - a^2\right)\omega_1\omega_2}$$

The system is unstable, if $M > M_*$.

The case $M = M_*$ can be studied by including more terms in series expansion of velocity field $\mathbf{V_P}, \mathbf{V_\Gamma}, \mathbf{V_\Omega}$ and velocity $\mathbf{V}^*$. Using the expansions, we obtain terms that can not be found in linear approximation. Similarly to the previous case we introduce the rotating coordinates to eliminate the gyroscopic terms in total Lagrangian of the system Λ .The addition terms to the Lagrangian Λ :

$$\Delta\Lambda = \Delta T - \Delta\Pi, \quad \Delta T = \frac{2\left(\pi\rho_0 a^4 R^4 \left(X_0^2 + Y_0^2\right)\left(\dot{X}_0^2 + \dot{Y}_0^2\right)\right)}{\left(R^2 - a^2\right)^3 \left(a^2 + R^2\right)},$$

$$\Delta\Pi = -\frac{\rho_0 \left(M - \pi\rho_0 a^2\right)^2 \left(\Gamma\left(a^2 + R^2\right) + 4\pi a^2 R^2 \omega\right)^2 \left(X_0^2 + Y_0^2\right)^2}{8\pi\left(R^4 - a^4\right)\left(M\left(-a^2 + R^2\right) + \pi\rho_0 a^2\left(a^2 + R^2\right)\right)^2}$$

Thus, in the case of the inner cylinder mass equals the critical mass, the potential energy term in total Lagrangian is negative (Π is equal to zero and the additional term $\Delta\Pi$ is always negative) and the system is unstable.

7. Conclusion

The Lagrangian of the system, consisting of inner free cylinder, outer fixed cylinder and circular flow with constant vorticity between them, was calculated. The Lagrangian was found in form of the series in displacement of the centre of inner cylinder r_0 . The stability dependence on mass of the inner cylinder was investigated. The stability condition matches with the previous papers [1] and is extended for the case $M = M_*$ by the non-linear analysis, that was derived by including higher order terms in series expansions into consideration.

Acknowledgments
This work was supported by the Russian Science Foundation (project № 14-19-01633).

References

[1] Kopiev V F, Chernyshev S A and Yudin M A 2017 The instability of a cylinder in the circulation flow of incomprecible ideal fluid *J. Appl. Math. and Mech.* **81** 216-229
[2] Petrov A G 1983 Hamilton's principle and certain problems of dynamics of perfect fluid *J. Appl. Math. and Mech.* **47** 30-36
[3] Petrov A G 2009 *Analiticheskaya gidrodinamika* Moskva: FIZMATLIT p 520

IOP Conf. Series: Journal of Physics: Conf. Series **894** (2017) 012127 doi:10.1088/1742-6596/894/1/012127

Optimization of oil-mixture "hot" pumping in main oil pipelines

T T Bekibayev, U K Zhapbasbayev and G I Ramazanova
Kazakh-British Technical University, Almaty, Kazakhstan
E-mail: timur_bekibaev@mail.ru

Abstract. The optimal conditions for high pour point and high-viscosity oil transportation through the main oil pipeline sections are determined. The optimization of "hot" pumping is investigated by determining the energy-saving operating conditions of pumps and heating furnaces. The objective function of optimality for the pipeline section with several stations is determined by a minimum costs of energy consumed by pumps and heating furnaces. The problem solving algorithm is constructed by new dynamic programming approach. The problem is divided into many overlapping subtasks with finding the optimal substructure. The object of each subtask is the cost function of pumps and heating furnaces at the stations.

1. Introduction

The method of "hot" pumping with associated heating remains a reliable way of pumping of high pour point oil and high-viscosity oil [1]. Objective functions and optimization problems of "hot" pumping are considered in scientific works [2-4]. Unlike the well-known works, the optimization of "hot" pumping is investigated by determining the optimal operating conditions for pumping units and heating furnaces.

2. Statement of the problem

The method of oil-mixture "hot" pumping at the section of main oil pipeline is made by the operation of pumping units and heating furnaces. The objective function of optimality is defined in the form of a minimum of the sum of costs of oil-mixture pumping and heating:

$$\sum_{i=1}^{n}\left(z_i^{el}\sum_{j=1}^{m_i^{pum}} c_{ij}^{pum} N_{ij}^{PU}(\mathbf{k}_{ij}) + z_i^{fl}\sum_{j=1}^{m_i^{fur}} g_{ij}^{fur} Q_{ij}^{fl}\right) \to min \tag{1}$$

where n – the number of stations, m_i^{pum}/m_i^{fur} – the quantity of pumps/furnaces at i-th station, z_i^{el}/z_i^{fl} – the electricity cost (tenge/(kW·h))/fuel (tenge/kg) at i-th station; $c_{ij}^{pum}/g_{ij}^{fur}$ – the integer variable that has value 1 if pump/furnace is in operation, and 0 otherwise; N_{ij}^{PU} – the power consumption of j-th pumping unit at i-th station (kW); $\mathbf{k}_{ij}$ – the ration of the rotor speed to nominal speed for this pump; Q_{ij}^{fl} – the speed of fuel delivery to j-th furnace of i-th station (kg/h).

The criterion (1) is considered together with the safety conditions of pumping:

1) Set point chart satisfaction - pressure limit at the inlet/outlet of the pump station (PS);

Content from this work may be used under the terms of the Creative Commons Attribution 3.0 licence. Any further distribution of this work must maintain attribution to the author(s) and the title of the work, journal citation and DOI.
Published under licence by IOP Publishing Ltd

2) Safety condition of heating furnaces operating;
3) Avoidance of oil temperature decrease below the critical (pour point) in the pipeline.
The safety condition of the heating furnaces requires the following restriction:

$$Q^{\text{fl}}_{min,ij} \leq Q^{\text{fl}}_{ij} \leq Q^{\text{fl}}_{max,ij} \tag{2}$$

The pressure drop for pumping through group of pumps ΔP^{gr} is determined by the formula:

$$\Delta P^{gr}(Q, \mathbf{k}) = \begin{cases} 0, & c_{oper} = 0 \\ \rho g H\left(\frac{Q}{c_{oper}}, \mathbf{k}\right), & c_{oper} > 0 \end{cases} \tag{3}$$

where c_{oper} – the number of operating pumps in the group, ρ – the density of oil being pumped, $H(Q/c_{oper}, \mathbf{k})$ – the dependence of pressure on the flow rate of any pump in the group.

With a fixed oil flow rate Q the pressure and power consumption functions generated by pumps or a group of pumps, in addition to the parameter $\mathbf{k}$, will have an additional argument on the temperature at the inlet to the pump (we will denote T_i^{pum}).

The pressure coefficients and efficiency characteristics of pumps depend on the viscosity of the oil being pumped, which in turn depends on its temperature. Therefore, for the pressure drop and power consumption, the dependencies are valid:

$$\Delta P^{gr}_{ij} = \Delta P^{gr}\left(\mathbf{k}_{ij}, T_i^{pum}\right), N^{\text{PU}}_{ij} = N^{\text{PU}}_{ij}\left(\mathbf{k}_{ij}, T_i^{pum}\right).$$

The pressure drop ΔP^{gr}_{ij} satisfies the pressure balance equation [5]:

$$P_{init} + \sum_{i=1}^{n}\sum_{j=1}^{m_i^{gr}} \Delta P^{gr}_{ij}(T_i^{pum}) = \sum_{i=1}^{n}\left(\Delta P_i^{\text{OHS}} + \Delta P_i^{\text{PR}} + \Delta P_i^{sec}\right) + \Delta P^{bp} + P_{fin} \tag{4}$$

where P_{init} – the initial pressure; m_i^{gr} – the quantity of pumping groups on PS; ΔP^{gr}_{ij} – the increase in pressure created by the j-th group of pumps at the i-th PS; ΔP_i^{PR} – the pressure loss after pressure regulator (PR); ΔP_i^{sec} – the pressure loss in pipeline with taking into account the difference in hydrostatic pressure at the i-th and $(i+1)$-th PS at the oil-mixture flow rate Q; ΔP^{bp} – the backpressure value at the input of terminal station; P_{fin} – the pressure required for oil-mixture pumping to final tank, ΔP_i^{OHS} – the pressure loss at oil heating stations (OHS).

In the case of "hot" pumping, the pressure loss varies depending on the oil temperature at the outlet of the station. Therefore, for the safety of pumping the pressure parameters $max\_\Delta P_i^{sec}$ are temperature functions:

$$\Delta P_i^{sec} = \Delta P_i^{sec}(T_i + \Delta T_i^{\text{OHS}}),\ max\_\Delta P_i^{sec} = max\_\Delta P_i^{sec}(T_i + \Delta T_i^{\text{OHS}})$$

where T_i – the oil temperature at the inlet to the i-th station, ΔT_i^{OHS} – the amount of heating at the oil heating station (OHS), $(T_i + \Delta T_i^{\text{OHS}})$ – the outlet temperature from the station.

The temperature T_{i+1} at the inlet of the next station can be calculated at a known flow rate Q, using the value $(T_i + \Delta T_i^{\text{OHS}})$. The value T_1 is equal to the oil temperature at the outlet of the tanks of the initial station. The pressure limits for safe pumping at the inlet P_k^{in} and at the outlet from the PS to Pressure Regulator (PR) P_k^{out1} for the k-th station satisfies the condition:

$$P_k^{in} = P_{init} + \sum_{i=1}^{k-1}\sum_{j=1}^{m_i^{gr}} \Delta P^{gr}_{ij} - \sum_{i=1}^{k-1}\left(\Delta P_i^{\text{OHS}} + \Delta P_i^{\text{PR}} + \Delta P_i^{sec}\right) \geq P_k^{in\,min} \tag{5}$$

$$P_k^{out1} = P_k^{in} + \sum_{j=1}^{m_k^{gr}} \Delta P^{gr}_{kj} - \Delta P_k^{\text{OHS}} \leq P_k^{out1\,max} \tag{6}$$

As is known, the heating furnaces are located in different ways relative to pumping unit. If the furnaces are located after the pump, then $T_k^{pum} = T_k$. If the furnaces are installed before the pump, then $T_k^{in\,pum} = T_k + \Delta T_k^{\mathrm{OHS}}$ and the conditions of non-cavitation operating for each l-th group of pumps of the k-th station are supplemented with a new term:

$$P_{kl}^{in\,\mathrm{gr}} = P_k^{in} - \Delta P_k^{\mathrm{OHS}} + \sum_{j=1}^{l-1} \Delta P_{kj}^{gr} \geq P_{kl}^{minin\,\mathrm{gr}} \tag{7}$$

For the flow rate Q_i^{fur} of the heated oil through each i-th furnace there is a dependence of type:

$$Q_i^{\mathrm{fur}} = Q_i^{\mathrm{fur}} \left(\Delta P^{\mathrm{OHS}}, \boldsymbol{g}^{\mathrm{fur}} \right) \tag{8}$$

where ΔP^{OHS} – the value of pressure drop in bypass pipe, $\boldsymbol{g}^{\mathrm{fur}} = (g_1^{\mathrm{fur}}, g_2^{\mathrm{fur}}, \ldots, g_m^{\mathrm{fur}})$ – the vector, signifying the operation of each furnace. The dependence $Q_i^{\mathrm{fur}} \left(\Delta P^{\mathrm{OHS}}, \boldsymbol{g}^{\mathrm{fur}} \right)$ can be determined taking into account the furnace construction at OHS.

For each heating furnace there is also a dependence of the value of its heating ΔT^{fur} on the oil flow through the furnace Q^{fur} and fuel supply rate Q^{fl} to this furnace:

$$\Delta T^{\mathrm{fur}} = C^{\mathrm{fur}} Q^{\mathrm{fur}} Q^{\mathrm{fur}} \tag{9}$$

where the constant C^{fur} depends on the furnace construction, its condition, composition of fuel, time of the year. The value of C^{fur} is proportional to the efficiency coefficient of heating furnace.

Naturally, there is a certain restriction on the temperature of the oil being pumped, to which it can be safely heated in a furnace, so that it does not decompose into separate fractions or evaporate. Let's call this value T_{oil}^{max}. In order to take this into account in the task model, the following restriction is introduced for each j-th furnace at the i-th station:

$$T_i + \Delta T_{ij}^{\mathrm{fur}} = T_i + C_{ij}^{\mathrm{fur}} Q_{ij}^{\mathrm{fur}} Q_{ij}^{\mathrm{fl}} \leq T_{\mathrm{oil}}^{max} \tag{10}$$

Since heated streams with the flow rates Q_i^{fur} from each furnace are then mixed with a non-heated bypass stream with a flow rate $\left(Q - \sum Q_{ij}^{\mathrm{fur}} \right)$, the total heating at the outlet from the OHS with a flow rate Q is determined by the known mixing formula for flows with a different weighting coefficients:

$$\Delta T^{\mathrm{OHS}} = \sum_{i=1}^{m^{\mathrm{fur}}} g_i^{\mathrm{fur}} \frac{Q_i^{\mathrm{fur}}}{Q} \Delta T_i^{\mathrm{fur}} \tag{11}$$

Using the equalities (8), (9), (11) we write the dependence of the heating value at OHS on the values ΔP^{OHS}, g_i^{fur} and Q_i^{fl}:

$$\Delta T^{\mathrm{OHS}} = \frac{1}{Q} \sum_{i=1}^{m^{\mathrm{fur}}} g_i^{\mathrm{fur}} C_i^{\mathrm{fur}} Q_i^{\mathrm{fl}} \left(Q_i^{\mathrm{fur}} \left(\Delta P^{\mathrm{OHS}}, \boldsymbol{g}^{\mathrm{fur}} \right) \right)^2$$

For each type of the oil being pumped the lower limit of the cooling temperature is set at the section (lets denote T_{oil}^{min}). Usually its value is related to the pour point temperature. Therefore, there is a need to enter a restriction:

$$T(x) > T_{\mathrm{oil}}^{min} \tag{12}$$

where $T(x)$ – the value of the oil temperature at the section. For assign of the optimum mode with heating, the objective function and pressure limitations (5) - (7) are formulated, as well as temperature restrictions (2), (10), (12). The main parameters of the task are the following variables:

1) c_{ij}^{pum} – optimal combination of pumps;
2) g_{ij}^{fur} – optimal combination of heating furnaces;
3) Q_{ij}^{fl} – necessary fuel supply to the furnace;
4) $\Delta P_i^{\mathrm{OHS}}$ – necessary pressure drop at OHS.

3. Algorithm of solution

Using the solution of the problem for n pumps, it is possible to effectively find solutions for n+1 pumps. A network of the operation states of pumping units is constructed. Each node of the network contains data on the number of pumps used and their parameters, pressure reduction in the PR. The nodes of the network are connected, based on the pressure characteristics of the pumps and the rotational speeds of their rotors. The transition of the subtask solution to the solution of the general problem is found and the correctness of the approach is proved.

The object of each subtask is the function of the dependence of the supplied power consumption $S(P,T)$ on the created differential pressure of the pump. Naturally, $P \geq 0$. The optimal pressures are associated with optimal pumping temperatures. In the search for a solution, instead of a continuous function $S(P,T)$, its discrete variant is used. The pressure value is represented discretely with a sufficiently small step $\varepsilon_P = 0.01$ bar, the temperature value is - $\varepsilon_T = 0.05°$C.

The solution of the problem is stored in a discrete array $Info(P,T)$, which contains a list of necessary pumps for each value of P.

The cost function $S(P,T)$ and the array of solutions $Info(P,T)$ for the pump have the form:

$$S(P) = \begin{cases} +\infty, & P \neq P_{pum} \\ zN^{PA}(Q), & P = P_{pum} \end{cases} ; \; Info(P) = \begin{cases} \emptyset, & P \neq P_{pum} \\ (\mathrm{pump's\,number}), & P = P_{pum} \end{cases} \quad (13)$$

where z – the electricity cost at the station (tenge/(kWh)), N^{PU} – the power consumption of the pumping unit (kW), Q – the flow rate, which passes through the pump.

The pressure drop P_{pum}, created by pump is determines as:

$$P_{pum} = [\rho g H(Q,T)]$$

where the operator [ ] means rounding to the nearest rational number with step ε_P.

The initial cost function is defined as follows:

$$S_{in}^{st,1}(P,T) = \begin{cases} 0, & P = P_{init} \text{ and } T = T_{init} \\ +\infty, & otherwise \end{cases} \quad (14)$$

We denote by the union of two functions S^A and S^B the function $S(P)$, which has a value for each P:

$$S(P,T) = S^A(P,T) \cup S^B(P,T) = min(S^A(P,T), S^B(P,T))$$

Similarly, we denote by the union of two arrays $Info^A$ and $Info^B$ the array $Info(P,T)$, which has a value for each P:

$$Info(P,T) = Info^A(P,T) \cup Info^B(P,T) = \begin{cases} Info^A(P,T), & S^A(P,T) \leq S^B(P,T) \\ Info^B(P,T), & S^A(P,T) > S^B(P,T) \end{cases}$$

We denote by the "overlaying" of function S^B to function S^A the function $S(P,T)$, which has a value for each P:

$$S(P,T) = S^A(P,T) \leftarrow (S^B) = min\left(S^A(P,T), S^A(P-P^*,T) + S^B(P^*,T)\right)$$

where the value of variable P^* for a specific value P determines as:

$$P^* = \text{argmin}_{P^* \in [0,P]}\left(S^A(P-P^*,T) + S^B(P^*,T)\right)$$

Similarly, we denote by "overlaying" of the array $Info^B$ to the array $Info^A$ the array $Info(P,T)$, which has a value for each P:

$$Info(P,T) = Info^A(P,T) \leftarrow Info^B(P,T) =$$
$$= \begin{cases} Info^A(P,T), S^A(P,T) \leq S^A(P-P^*,T) + S^B(P^*,T) \\ Info^A(P-P^*,T) + Info^B(P^*,T), \text{ otherwise} \end{cases}$$

In the heating mode the cost functions $S_v^{fur,i}$ and the solution arrays $Info_v^{fur,i}$ for each i-th furnace must be considered, the arguments of which are: the first argument P indicates the pressure drop at OHS, the second argument T is the oil temperature at the inlet to the furnace, an additional third argument is ΔT - the increase in the temperature of the total flow rate Q with help of this furnace. The cost function of the furnace has an additional index v - the number of selection of operating furnaces at the OHS from the set m^{fur}. Naturally, $v \leq 2^{m^{fur}} - 1$.

The cost function for the operating furnace and its array of solutions are calculated as following:

$$S_v^{\text{fur},i}(P,T,\Delta T) = \begin{cases} z^{\text{fur}} Q_i^{\text{fl}}, & Q_{i,v}^{\text{fl}}(P,\Delta T) \in \left[Q_{min,i}^{\text{fl}}, Q_{max,i}^{\text{fl}}\right] \text{ and } T + \Delta T^{\text{fur}} \leq T_{oil}^{max} \\ +\infty, & \text{otherwise} \end{cases}$$

$$Q_{i,v}^{\text{fl}}(P,\Delta T) = \frac{\Delta T^{\text{fur}}(P,\Delta T)}{C_i^{\text{fur}} Q_i^{\text{fur}}(P, \boldsymbol{g}_i^{fur}(v))}; \quad \Delta T^{\text{fur}}(P,\Delta T) = \frac{Q}{Q_i^{\text{fur}}(P, \boldsymbol{g}_i^{\text{fur}}(v))}\Delta T; \quad (15)$$

$$Info_v^{\text{fur},i}(P,T,\Delta T) = \begin{cases} (\text{furnace's number}) + Q_i^{\text{fl}}, & Q_{i,v}^{\text{fl}}(P,\Delta T) \in \left[Q_{min,i}^{\text{fl}}, Q_{max,i}^{\text{fl}}\right] \\ \text{"the furnace doesn't operate"}, & \text{othewise} \end{cases}$$

where $Q_{i,v}^{\text{fl}}$ – the necessary fuel consumption for the i-th furnace for the selection v, ΔT^{fur} – the flow heating temperature through this furnace.

The vector $\boldsymbol{g}_i^{\text{fur}}(v)$ signifies the operating combination of heating furnaces for the selection v. In the obtained solution, the restrictions on the fuel consumption of the furnace (2) and the maximum heating in the furnace (10) are automatically taken into account, since in the determination of $S_v^{\text{fur},i}(P,T,\Delta T)$ the operation "overlaying" is performed on the values of $Q_{i,v}^{\text{fl}}$ and ΔT^{fur}.

After calculating of all $S_v^{\text{fur},i}$ and $Info_v^{\text{fur},i}$, it is necessary to define the cost function S_v^{OHS} and an array of solutions $Info_v^{\text{OHS}}$ of the total furnace operating at OHS for each selection v. The values S_v^{OHS} and $Info_v^{\text{OHS}}$ are determined by the "overlaying" of the functions of all operating heating furnaces:

$$\begin{aligned} S_v^{\text{OHS}}(P,T,\Delta T) &= S_v^{\text{fur},1} \leftarrow S_v^{\text{fur},2} \leftarrow \ldots \leftarrow S_v^{\text{fur},g_{oper}(v)} \\ Info_v^{\text{OHS}}(P,T,\Delta T) &= Info_v^{\text{fur},1} \leftarrow Info_v^{\text{fur},2} \leftarrow \ldots \leftarrow Info_v^{\text{fur},g_{oper}(v)} \end{aligned} \quad (16)$$

where $g_{oper}(v)$ the quantity of operating furnaces for the selection v.

The operation overlaying for the cost functions of furnaces A and B will have the following form (similarly for the array of solutions):

$$S^{\text{fur},A}(P,T,\Delta T) \leftarrow (S^{\text{fur},B}) = \min_{\Delta T^*>0} \Big(S^{\text{fur},A}(P,T,\Delta T) + (S^{\text{fur},B}(P,T,\Delta T - \Delta T^*) \Big) \quad (17)$$

For any triple of the values $(P,T,\Delta T)$ the optimal operating of OHS means the optimal selection of operating furnaces from m^{fur} furnaces. Therefore, the optimal cost function for OHS is the "union" of all possible cost functions from the $(2^{m^{fur}}-1)$ selections (similarly for the array of solutions):

$$S^{\text{OHS}}(P,T,\Delta T) = \bigcup_{v=1}^{2^{m^{\text{fur}}}-1} S_v^{\text{OHS}}(P,T,\Delta T) \quad (18)$$

In the case of non-operating OHS (i.e. when none of the OHS is operating) the calculated cost function and the array of solutions need to be adjusted:

$$S^{\text{OHS}}(0,T,0) = 0,\ Info^{\text{OHS}}(0,T,0) = \text{"the furnaces don't operate"}$$

Since, unlike a group of pumps, when the oil passes through the OHS, the flow temperature may increase and the pressure may drop, the "overlaying" of the heating cost function S^{OHS} to the cost function of the entire station looks as follows (similar to the array of solutions):

$$S(P,T) = S^{\text{sec}}(P,T) \leftarrow (S^{\text{OHS}}) =$$
$$= \min_{P^*,\Delta T^* \geq 0} \left(S^{\text{sec}}(P+P^*, T-\Delta T^*) + S^{\text{OHS}}(P^*, T-\Delta T^*, \Delta T^*) \right)$$

Naturally, if the heating furnaces are located after the pumps, then the cost function at the inlet to the station and its array of solutions the cost function S^{OHS} first must be overlaid, and only then the cost functions of the pump groups. Otherwise, vice versa. If there are no heating furnaces at any station, there is no need to calculate the function S^{OHS} for it and carry out an "overlay" on the cost function of the station.

The pressure drop and the temperature drop at the section depend on the initial temperature. To properly "cut" the cost functions and arrays of solutions at the outlet from the station and determine the cost functions and an array of solutions at the inlet of the next station, calculate the discrete functions $\Delta P_i^{sec}(T)$, $\Delta T_i^{sec}(T)$ and $max\_\Delta P_i^{\text{sec}}(T)$, $max\_\Delta T_i^{\text{sec}}(T)$ for each i-th section with step ε_T.

Moreover, the function $max\_\Delta T_i^{\text{sec}}(T)$ determines the maximum value of the temperature drop between the start point and a point on the i-th section.

Therefore, functions $P_{out}^{min,i}(T)$, $P_{out}^{max,i}(T)$ can be determined for the minimum and maximum permissible pressure at the outlet from station:

$$P_i^{out2} \geq P_{out}^{min,i} = max(P_{i+1}^{in\,min} + \Delta P_i, max\_\Delta P_i)$$

$$P_i^{out2} \leq P_{out}^{max,i} = min(P_i^{out2\,max}, P_i^{pipe\,max})$$

Wherein, the condition (12) of the minimum necessary temperature at the outlet from the station takes the form:

$$T_{out}^{min,i}(T) = T_{oil}^{min} + max\_\Delta T_i^{sec}(T)$$

IOP Conf. Series: Journal of Physics: Conf. Series **894** (2017) 012127 doi:10.1088/1742-6596/894/1/012127

Cutting of cost function at the outlet from station $S^{st}(P,T)$ is determined by using the functions $P_{out}^{min}(T)$, $P_{out}^{max}(T)$, $T_{out}^{min}(T)$ and is denoted by the function $S^{cut.st}(P,T)$, which has a value for each P and T:

$$S^{\mathrm{cut.st}}(P,T) = CUT(S^{\mathrm{st}}(P,T), P_{out}^{min}, P_{out}^{max}, T_{out}^{min}) = \begin{cases} +\infty, & P \notin [P_{out}^{min}(P), P_{out}^{max}(P)] \\ +\infty, & T < T_{out}^{min}(T) \\ S^{\mathrm{st}}(P,T), & \text{otherwise} \end{cases}$$

The array of solutions is "cut" in a similar way. In the cost function and the array of solutions in the outlet from the OHS will be supplemented with pressure drop on the furnace and in case of its "cutting", the changed pressure conditions (5) and (6) are automatically taken into account.

The cost function at the inlet of the next station $S_{in}^{next\ st}(P,T)$ is determined by the "shift" operation using the cost function at the outlet from the station $S_{out}^{next\ st}(P,T)$ through $\Delta P^{sec}(T)$ and $\Delta T^{sec}(T)$. This function has the meaning for each P and T:

$$S_{in}^{next\ st}(P,T) = SHIFT(S_{out}^{\mathrm{st}}(P,T), \Delta P^{\mathrm{sec}}, \Delta P^{\mathrm{sec}}) =$$
$$= \min_{T^*}\left(S_{out}^{\mathrm{st}}\left(P + \Delta P^{\mathrm{sec}}(T^*), T + \Delta T^{\mathrm{sec}}(T^*)\right)\right)$$

The "shift" operation for the array of solutions is carried out in a similar way.

Thus, all operations for finding the optimal mode of the "hot" pumping method are defined above. These operations are carried out for all stations except the last in the order of their location on the pipeline section:

$k = 1$

1)For $i = 1$ till $m^{gr,k}$ calculate $S_i^{gr,\,k}$;

2)If $m^{fur,k} > 0$, then calculate S_k^{OHS} using the $(15) - (18)$

3) $S_{out}^{st,k}(P,T) = S_{in}^{st,k}(P,T) \leftarrow (S_k^{OHS}) \leftarrow (S_1^{gr,k}) \leftarrow \ldots \leftarrow (S_{m^k}^{gr,k})$

or $S_{out}^{st,k}(P,T) = S_{in}^{st,k}(P,T) \leftarrow (S_1^{gr,k}) \leftarrow \ldots \leftarrow (S_{m^k}^{gr,k}) \leftarrow (S_k^{OHS})$; (19)

4) Calculation of the functions $\Delta P_k^{sec}(T), \Delta T_k^{sec}(T), max\_\Delta P_k^{sec}(T), max\_\Delta T_k^{sec}(T)$;

5) $S_{out}^{st,k}(P,T) = CUT(S_{out}^{st,k}(P,T), P_{out}^{min,k}, P_{out}^{max,k}, T_{out}^{min,k})$;

6) $S_{in}^{st,k+1}(P,T) = SHIFT(S_{out}^{st,k}(P,T), \Delta P_k^{sec}, \Delta T_k^{sec})$;

7) $k = k+1$. If $k \neq n+1$, then turn to point 1, otherwise loop termination.

After the conducted loop (19) with the initial condition (14) the optimal final pressure and temperature are calculated as:

$$(P^{ans}, T^{ans}) = argmin_{T \geq T^{fin}, P \geq P^{fin}}\left(S_{in}^{st,\,n+1}(P,T)\right)$$

The minimum sum of costs for productivity Q is the value of the function $S_{in}^{st,n+1}(P^{ans}, T^{ans})$. The optimal combination of operating pumps and heating furnaces, as well as their operating modes and the necessary pressure drops ate the OHS will be stored in the cell of the array $Info_{in}^{st,\,n+1}(P^{ans}, T^{ans})$.

IOP Conf. Series: Journal of Physics: Conf. Series **894** (2017) 012127 doi:10.1088/1742-6596/894/1/012127

References

[1] Abramzon L S 1979 Optimum parameters of hot pipelines operation *Oil industry* **2** 53-54
[2] Agapkin V M, Krivoshein B L and Yufin V A 1981 *Heat and hydrodynamic calculations of pipelines for oil and oil products* (Moscow: Nedra) p 256
[3] Yablonsky V S, Novoselov V F, Galeev V B and Zakirov G Z 1965 *Design, operation and repair of oil product pipelines* (Moscow: Nedra) p 410
[4] Zhapbasbayev U K, Makhmotov E S, Ramazanova G I, Bekibaev T T and Rziev S A 2015 Calculation of the optimum pumping temperature for oil transportation *Science & Technologies: Oil and Oil Products Pipeline Transportation* **20** 61-66
[5] Tugunov P I, Novoselov V F, Korshak A A and Shammazov A M 2002 *Typical calculations for the design and operation of gas and oil pipelines* (Ufa: DesignPoligraphService) p 658

IOP Conf. Series: Journal of Physics: Conf. Series **894** (2017) 012128 doi:10.1088/1742-6596/894/1/012128

On the experimental analysis of dissipative processes under cyclic loading of metals

V I Kapustin[1] **and K V Zakharchenko**[1,2]

[1]Novosibirsk State Technical University, Novosibirsk, Russia
[2]Lavrentyev Institute of Hydrodynamics, Novosibirsk, Russia

E-mail address: macler06@mail.ru

Abstract. The article is devoted to the experimental study of dissipative material properties tested under homogeneous stress-strain conditions, which realized on the gage part of a smooth sample subjected to cyclic loading with increasing amplitude. A new method to estimate inelastic material properties under strain lower than the nominal strength is proposed. At the initial stages of loading, the method makes it possible to find out irreversible changes in sample characteristics and define material elastic limit under cyclic loading. Strain, longitudinal and lateral stress, as well as temperature of the gage part of a sample are the sample characteristics measured simultaneously. They make it possible to distinguish the components connected with reversible and irreversible strain processes associated with fatigue damage of material.

1. Introduction

Data on inelastic stress of metals obtained at stresses in the range between the proportional elastic limit and yield strength of the material are often applied in the preliminary estimation of fatigue damage resistance characteristics. In particular, Coffin's formula [1] is widely used to describe the fatigue stress accumulation process. The formula includes inelastic stress characteristics which is a width of hysteresis loop. There are several methods in which damage accumulation is connected with the change of various integral parameters of metal energy dissipation, based on measurement of the absorption coefficient, logarithmic decrement of oscillation and temperature inside damage nucleus [2 - 4]. The phenomenon of elasto-plastic loosening of the material and the material model describing the formation of micropores are used in some approaches [5, 6].

There is a necessity in experimentally derived dependencies from general inelasticity concepts, for example, to identify equations describing the real inelastic strain process, which follows degradation of the material under cyclic loadings.

Modern concepts of reversible processes deal with the term "hysteresis loop", which is widely used to describe inelastic strain cycle.

Indeed, the sigma-epsilon diagrams in Figure 1 are not closed loops (and the stress process is not cyclic) because the "loops" are broken, and the presented trajectories are closed (coincide) only at one point. From the point of view of thermodynamics such a cycle is irreversible.

The hysteresis loop for plastic materials is known to be open at stresses below the endurance limit [8]. Then a question arises: "Why does the irreversible stress not lead to destruction?" The research

Content from this work may be used under the terms of the Creative Commons Attribution 3.0 licence. Any further distribution of this work must maintain attribution to the author(s) and the title of the work, journal citation and DOI.
Published under licence by IOP Publishing Ltd

which takes into account the kinetics of phase transformations in stress of metastable austenitic steels [9] shows the correlation between the density of microcracks and the amount of martensite.

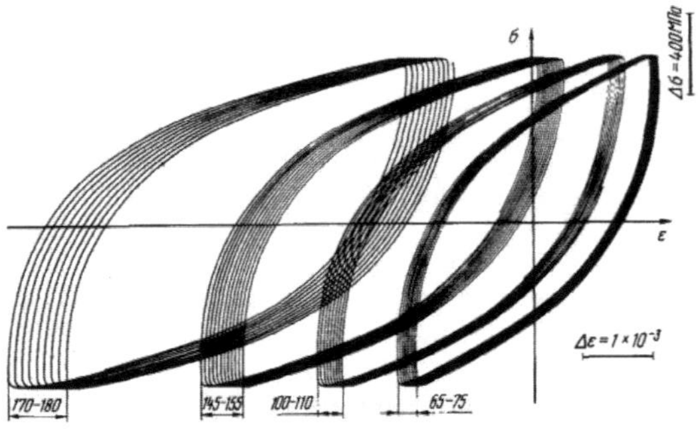

Figure 1. Stress-strain diagrams during cyclic loading.

The research presents methods to define the characteristics of materials describing dissipative processes. The approach is based on the concepts of reversible and irreversible processes [10]. The main goal of the research is to show the possibility of experimental determination of the material's characteristics describing its fatigue damage from the point of physics of strength.

In the research it is required to obtain the following effects under periodic loading: nonlinear stress when loaded beyond the limit of proportionality; changes in volume or shape, deterministic change of material microstructure; dissipative heating of the material.

For the research, the metastable steel 08X18H10T of austenitic class with the following chemical composition was chosen [11]: C not more than 0.08%, Si not more than 0.8%, Mn not more than 2%, Cr 17-19%, Ni 9-11%, Ti not more than 0.7%, S not more than 0.02%, P not more than 0.035%, Cu not more than 0.3%, the basic element is - Fe. The mechanical characteristics were obtained under quasistatic soft loading at a speed of 150 N/s. Modulus of elasticity is 200700 MPa, proportional elastic limit is 280 MPa, conventional yield strength is 302 MPa, temporary stress resistance is 586 MPa. Choosing austenitic steel we relied on the ability of austenite to transform isothermally into strain-induced martensite under mechanical stresses.

2. Samples, equipment and methods

For the research, samples of rectangular cross section of 6 x 10 mm^2 with a gage length of 50 mm were used which allowed fixing two extensometers for measuring axial and transverse strains.

The sample surface, intended to measure the temperature with the help of a thermal imagery camera, was preliminary blackened covering with a thin layer of soot, thus ensuring a blackness coefficient of the surface close to 1. To eliminate the influence of external thermal fields on the measurements radiant temperature screens were used.

The samples were deformed by zero-to-compression stress cycle with a stress amplitude increasing in proportion to time. The selected zero-to-compression stress cycle made it possible to obtain the stressed state characteristics and the sample surface temperature in the part of maximum stresses of a cycle and after unloading at the end of each loading cycle. Thus, the beginning of a spontaneous irreversible dissipation process was identified by the change in characteristics and was characterized by the maximum limit stress in the cycle.

The loading frequency is permanent and equal to 0.5, 1, 2, 4 Hz. The test duration is 680 seconds. During testing, the maximum stress in the cycle is 276 MPa.

During testing, the loading, axial and transverse stress, radiation temperature of the gage part of a sample are measured synchronously. The samples are loaded and the stress is recorded using the INSTRON equipment, and temperature is measured by thermal camera TKVr-IFP SVIT. The experiments are made at a room temperature.

IOP Conf. Series: Journal of Physics: Conf. Series **894** (2017) 012128 doi:10.1088/1742-6596/894/1/012128

3. The main results

According to the data of [12-14], the volume fraction of the magnetic α' phase in the process of cyclic loading of steel 08X18H10T increases in the course of material degradation that leads to a change of the elastic characteristics in fatigue. If the sample is deformed, for example, by a zero-to-compression stress cycle with a frequency of 4 Hz, recording the characteristics at the moment of unloading, then it is possible to define not only some limited critical stress, but also the property of a material when loaded above this stress.

Figure 1 shows the patterns of a smooth sample stress. Let us remark here, that the represented diagrams are not smooth curves, which are usually obtained when constructing sigma-epsilon diagrams, but a collection of points placed nearby.

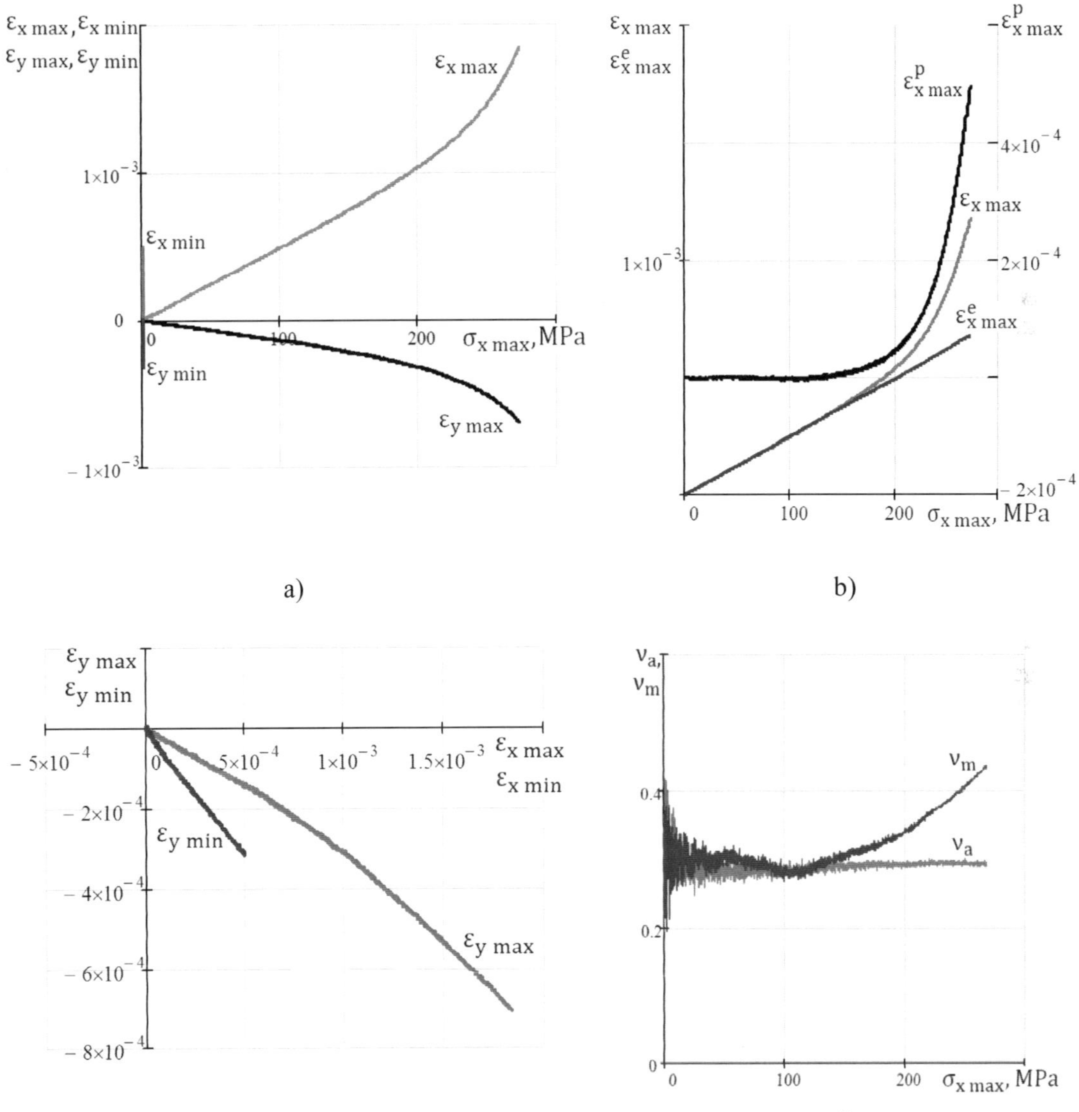

Figure 2. Dependencies of the sample stress with a zero-to-compression stress cycle.

IOP Conf. Series: Journal of Physics: Conf. Series **894** (2017) 012128 doi:10.1088/1742-6596/894/1/012128

In Figure 2 a) – the interdependence of extreme axial $\varepsilon_{x\max}$, $\varepsilon_{x\min}$ and transverse $\varepsilon_{y\max}$, $\varepsilon_{y\min}$ stresses on extreme stresses $\sigma_{x\max}$, $\sigma_{x\min}$ MPa in the cycle is shown. It can be seen from the figure that periodic unloading make it possible to identify inelastic phenomena at stresses less than the conventional yield strength and proportional elastc limit. In Figure 2 b) the linear $\varepsilon^{e}_{x\max}$ and nonlinear $\varepsilon^{p}_{x\max}$ components of total stress are singled out. They make it possible to identify the critical stress level at which the dissipative processes are activated.

Figure 2 c) represents the parametric stress diagrams– in coordinates - maximum and minimum axial and transverse stresses where the parameters of the diagrams are the maximum and minimum loading, respectively. A section of elastic and inelastic stress can be distinguished in the diagram, and the unloading branch will show the change in the volume or shape of the sample after stress. For the extreme point in the diagram corresponding to unloading after the maximum loading, the coefficient of transverse stress is 0.622.

Figure 2 d) clearly shows change in the coefficient of transverse stress for the components of the total stress cycle: amplitude - $v_a = \left|\dfrac{\varepsilon_{y\max} - \varepsilon_{y\min}}{\varepsilon_{x\max} - \varepsilon_{x\min}}\right|$ and average stress in the cycle - $v_m = \left|\dfrac{\varepsilon_{y\max} + \varepsilon_{y\min}}{\varepsilon_{x\max} + \varepsilon_{x\min}}\right|$. The ratio of stress amplitudes over the entire stress range does not change and is equal to 0.285. So, the material reacts to the loading as a perfectly elastic material in the entire range of studied stresses in the cycle. The coefficient of transverse stresses for the average stress in the cycle increases then loading for a maximum stress in the cycle exceeds 100 MPa.

Dependence of the temperature change in the cycle is represented in Figure 3. The amplitude of the temperature T_a grows linearly as well as the stress and the stress amplitude in the cycle (Figure. 3 a) and does not depend on the loading frequency.

The nonlinear dependence of the average component of the temperature cycle T_m on stress, (Figure 3 b), is an indicator of irreversible energy dissipation in the sample and is observed in the same stress area as inelastic stress.

Division of the characteristics describing material deformation process into amplitude (linear) and average (nonlinear dependence) components allows presenting the material thermomechanical properties in the form of a surface (Figure 4).

Received reactions can be considered as a slow change in the sample stressed state under conditions of the asymmetric stress cycle and resembles the creep of a material under the periodic loading.

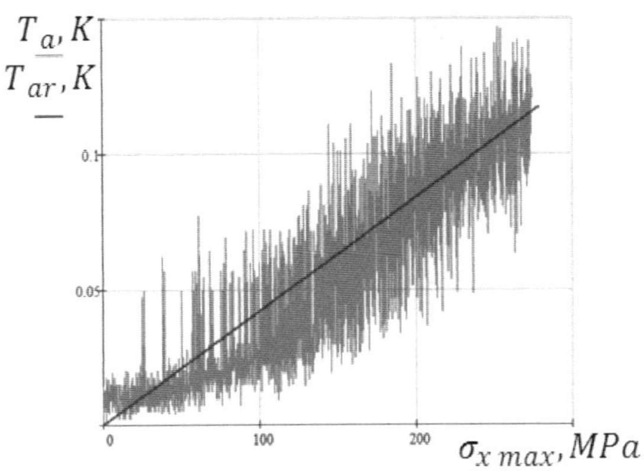

a)

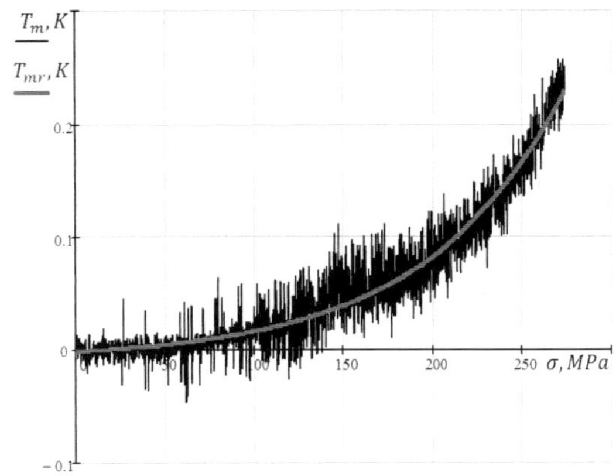

b)

Figure 3. Patterns of sample temperature variations with a zero-to-compression stress cycle.

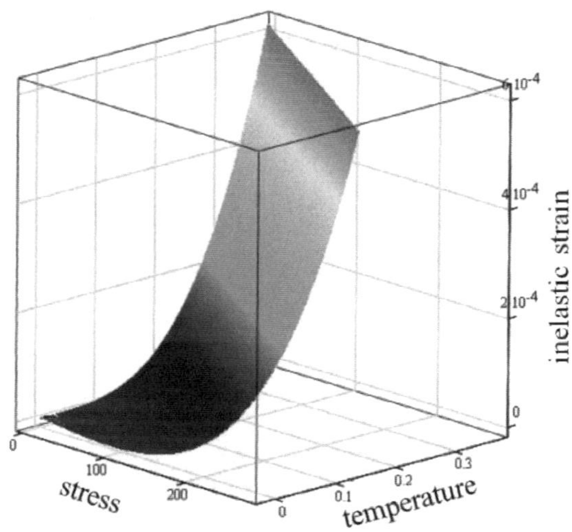

Figure 4. Diagram of material thermomechanical reactions to cyclic loading.

4. Discussion

The results are obtained under homogeneous conditions; they characterize the behaviour of the material at a point and do not contradict the known practice of tests for the material strength and structural elements.

The small inelastic components of the total stress (microplastic stress - irreversible motion of separate dislocations) during periodic stress are different from the residual stresses observed in creep where the corporate movement of the elements of the material's structure is observed (plastic stress in the whole volume of the material - an increase in the dislocation density by more than two orders and interaction among them) [15]. In contrast to the creep observed under quasistatic loading, small inelastic stresses during periodic loading are followed by local structural changes [16], which in the literature are referred to as martensitic transformations of materials [17]. Despite the local nature of the transformations, they are unidirectional and, for a considerable number of stress cycles, can change the initial properties of the material throughout the entire gage part of the sample's surface.

Fatigue damage - formation of a microcrack - occurs in the place where the coherence of the lattice of the material self-organized structure is disturbed. A crack can develop if under the action of mechanical stresses the next evolution of the material takes place on its top and conditions for martensitic transformations and destruction of the material on its top will be created. The martensitic structure is brittle and breaks delicately, which is observed on the surface of fatigue fractures. These arguments are confirmed by the established facts of martensitic transformations in the sample surface and data from factographic studies of the samples in which the traces of transformations phases of the material known as fish-eye and bird's-eye are observed at the site of failure [18], as well as data on the change in the transverse strain coefficient of the samples [16].

In the authors' opinion, the results of the presented material tests can be claimed to characterize the material not only for identification of material model parameters, in which the material properties must vary from cycle to load cycle, but also for practical applications where it is necessary to measure the material's ability to deform without irreversible processes.

5. Conclusion

It has been shown experimentally that the phenomenon of dissipation under cyclic soft loading is caused by a process of monotonous accumulation of the average component of the strain cycle, ratchetting, and leads to heating of the material at stresses much lower than the yield strength of the material. Despite this, the linear dependence of the stress cycle amplitude from the stress amplitude remains in the researched area.

Acknowledgments

This work was partially supported by the Russian Foundation for Basic Research (project code 16-08-00483 A).

References

[1] Coffin L F 1962 Low-cycle fatigue: a review *Appl. Mater.Res.* **1** 129-141

[2] Kapustin V I, Gileta V P and Tereshin E A 2010 On determination of elastic limits by dissipative heating of materials *Journal of Applied Mechanics and Technical Physics* **51** 393-397

[3] Diaz F A, Patterson E A, Tomlinson R A and Yates J R 2004 Measuring stress intensity during fatigue cracking using thermoelasticity *Fatigue and fracture of engineering materials and structures* **27** 571-583

[4] Luong M P 1995 Infrared Thermographics scanning of fatigue in metals *Nuclear Engineering and Design* **158** 363-376

[5] Novozhilov V V, Kadashevich Yu I and Rybakina O G 1988 Disintegration and prospects for the construction of the strength criterion under complex loading taking creep into account *From Bulletin. The Academy of Sciences of the USSR. MTT* **5** 108-114

[6] Shutov A V, Silbermann C B and Ihlemann J 2015 Ductile damage model for metal forming simulations including refined description of void nucleation *International Journal of Plasticity* **71** 195-217

[7] Kotsanda S 1990 *Fatigue fracture of metals* (Moscow: Metallurgiya) p 623

[8] Terent'ev V F 2003 Microscopic and macroscopic deformation of metallic materials below the fatigue limit *Russian metallurgy (Metally)* **5** 445-451

[9] Sosnin O V 2003 Change of grain structure and phase composition of austenitic steel during fatigue loading *Materialovedenie* **1** p 27

[10] Prigozhin I and Condépudi D 2002 *Modern thermodynamics. From heat engines to dissipative structures* (Moscow: Mir) p 461

[11] Sorokina V G 1989 *List of steels and alloys* (Moscow: Mechanical Engineering) p 640

[12] Mishakin V V, Klyushnikov V A and Gonchar A V 2015 Relation between the deformation energy and the Poisson ratio during cyclic loading of austenitic steel *Technical Physics. The*

Russian Journal of Applied Physics **60** 665-668

[13] Terent'ev V F, Dobatkin S V, Prosvirnin D V, Bannikh I O, Rybalchenko O V and Raab G I 2008 Fatigue strength of austenitic steel X18H10T after equal-channel angular pressing *Stress and destruction of materials* **10** 30-38

[14] Gorkunov E S, Zadvorkin S M, Kokovikhin E A, Tueva E A, Subachev Yu V, Goruleva L S and Podkopytova A V 2001 The effects of deformations by rolling and uniaxial tension on the structure and the magnetic and mechanical properties of Armco iron, steel 12X18H10T, and a Steel 12X18H10T-Armco Iron-Steel 12X18H10T composite material *Russian Journal of Nondestructive Testing* **47** 369-380

[15] Wu C H, Hsu J and Chen C H 1998 The effect of surface stress on the stability of surfaces of stressed solids *Acta mater* **46** 3755-3760

[16] Kapustin V I, Gileta V P, Zakharchenko K V and Popelyukh A I 2012 Study of the Regularities of Periodic Deformation of Metallic Materials *Industrial Laboratory. Materials Diagnostics* **78** 50-55

[17] Terent'ev V F and Petukhov A N *Fatigue of high-strength metal materials* (Moscow: IMET RAS CIAM) P 515

[18] Shaniavski A A 2007 *Modeling of fatigue cracking of metals. Synergetics for aviation* (Ufa: LLC Monography) P 500

Modelling the motion of a non-free system of rigid bodies using the Lagrange equations of the first kind

A R Zagidulin[1,2], **E G Podruzhin**[1] **and G I Rastorguev**[1]

[1] Novosibirsk State Technical University, Novosibirsk, Russia
[2] Siberian Aeronautical Research Institute named after S. A. Chaplygin, Novosibirsk, Russia

E-mail: zagidulin@corp.nstu.ru

Abstract. A technique for modelling the motion of a non-free system of rigid bodies is proposed (the mutual position of bodies is limited by superimposed holonomic constraints) using Lagrange equations of the first kind. In the paper, the proposed technique was used to simulate the compression of the amortization strut of the aircraft landing gear.

1. Introduction

Traditionally, Lagrange equations of the second kind in generalized coordinates were used to solve the motion of a non-free system of rigid bodies, which made it possible to reduce the dimensionality of the system of equations being solved. The main drawback of this approach was the fact that, depending on the choice of generalized coordinates, the type of the system being solved changed, that is, the equations were determined by an individual approach to the problem being solved. In the case of a change (complication) in the system of rigid bodies, it was necessary to obtain again resolving equations. Modeling with the proposed approach allows representing the model of the system as a set of objects: solids, power factors and mechanical connections, which ensure the modularity and extensibility of the model. This in turn makes it possible to automate the process of constructing the model [1–5]. Calculations were carried out on the work capacity of the amortization struts during the drop testing of the landing gear of specific aircraft in Siberian Aeronautical Research Institute named after S. A. Chaplygin (Tu-204SM, MS-21, Ka-62). When compared with the results of measurements of the main parameters carried out during the tests, the high accuracy of the analytical calculations obtained using the model was confirmed.

2. Mathematical foundations

The motion of an arbitrary system of rigid bodies with holonomic constraints relative to the fixed coordinate system *Oxyz* is considered. For each body, a mobile coordinate system $C\xi\eta\zeta$ is defined, rigidly connected with the body, starting at the center of mass of the body and with axes coinciding with the principal axes of inertia of the body.

The inertial characteristics of the body are given by the mass m_i and the inertia tensor:

$$\mathbf{I}_{0i} = \begin{pmatrix} I_i^{(\xi)} & 0 & 0 \\ 0 & I_i^{(\eta)} & 0 \\ 0 & 0 & I_i^{(\zeta)} \end{pmatrix}.$$

Content from this work may be used under the terms of the Creative Commons Attribution 3.0 licence. Any further distribution of this work must maintain attribution to the author(s) and the title of the work, journal citation and DOI.
Published under licence by IOP Publishing Ltd

The center of mass of the body is determined by its position vector:

$$\mathbf{r}_i = \begin{pmatrix} x_i & y_i & z_i \end{pmatrix}^T$$

and the rotation quaternion:

$$q_i = \left(\cos(\alpha_i/2),\ \ \mathbf{u}_i \sin(\alpha_i/2)\right),$$

where α_i is the angle of rotation of the body, $\mathbf{u}_i$ is the axis of rotation (vector of unit length).

The angular position of the body can also be represented as a rotation matrix:

$$\mathbf{R}_i = \begin{pmatrix} 1-2b_i^2-2c_i^2 & 2a_ib_i-2w_ic_i & 2a_ic_i+2w_ib_i \\ 2a_ib_i+2w_ic_i & 1-2a_i^2-2c_i^2 & 2b_ic_i-2w_ia_i \\ 2a_ic_i-2w_ib_i & 2b_ic_i+2w_ia_i & 1-2a_i^2-2b_i^2 \end{pmatrix},$$

where w_i is the scalar (real) part of the rotation quaternion; a_i, b_i and c_i are components of the vector (imaginary) part of the rotation quaternion.

The position of the body can be represented as a transition matrix in homogeneous coordinates:

$$\mathbf{T}_i = \begin{pmatrix} \mathbf{R}_i & \mathbf{r}_i \\ \mathbf{0} & 1 \end{pmatrix}.$$

Equations of motion of the center of mass of the body:

$$\mathrm{d}v_i/\mathrm{d}t = m_i^{-1}\left(\mathbf{f}_i + \mathbf{f}_i^c\right),$$

where v_i is the velocity of the center of mass; $\mathbf{f}_i$ is the total vector of active forces acting on the body; $\mathbf{f}_i^c$ is the total vector of the constraint reactions acting on the body.

Equations of rotational motion (dynamic Euler equations) in a fixed coordinate system:

$$\mathrm{d}\boldsymbol{\omega}_i/\mathrm{d}t = \mathbf{I}_i^{-1}\left(\boldsymbol{\tau}_i - \boldsymbol{\omega}_i \times (\mathbf{I}_i\boldsymbol{\omega}_i) + \boldsymbol{\tau}_i^c\right),$$

$$\mathbf{I}_i = \mathbf{R}_i \mathbf{I}_{0i} \mathbf{R}_i^T,$$

where $\boldsymbol{\omega}_i$ is the angular velocity of the body; $\boldsymbol{\tau}_i$ is the total vector of the moments of the active forces acting on the body; $\boldsymbol{\tau}_i^c$ is the total vector of constraint reaction moments acting on the body.

Equations of motion of the system of rigid bodies can be written in the following form:

$$\mathrm{d}\mathbf{V}/\mathrm{d}t = \mathbf{M}^{-1}\left(\mathbf{F} + \mathbf{F}^c\right),$$

$$\mathbf{V} = \begin{pmatrix} \mathbf{v}_1 \\ \boldsymbol{\omega}_1 \\ \vdots \\ \mathbf{v}_n \\ \boldsymbol{\omega}_n \end{pmatrix}, \mathbf{M} = \begin{pmatrix} m_1\mathbf{E}_3 & \mathbf{0} & \cdots & \mathbf{0} & \mathbf{0} \\ \mathbf{0} & \mathbf{I}_1 & \cdots & \mathbf{0} & \mathbf{0} \\ \vdots & \vdots & \ddots & \vdots & \vdots \\ \mathbf{0} & \mathbf{0} & \cdots & m_n\mathbf{E}_3 & \mathbf{0} \\ \mathbf{0} & \mathbf{0} & \cdots & \mathbf{0} & \mathbf{I}_n \end{pmatrix}, \mathbf{F} = \begin{pmatrix} \mathbf{f}_1 \\ \boldsymbol{\tau}_1 - \boldsymbol{\omega}_1 \times (\mathbf{I}_1\boldsymbol{\omega}_1) \\ \vdots \\ \mathbf{f}_n \\ \boldsymbol{\tau}_n - \boldsymbol{\omega}_n \times (\mathbf{I}_n\boldsymbol{\omega}_n) \end{pmatrix}, \mathbf{F}^c = \begin{pmatrix} \mathbf{f}_1^c \\ \boldsymbol{\tau}_1^c \\ \vdots \\ \mathbf{f}_n^c \\ \boldsymbol{\tau}_n^c \end{pmatrix},$$

where n is the number of rigid bodies in the system; $\mathbf{E}_3$ is the unit matrix of dimensions 3×3.

Ideal holonomic constraints are superposed on the positions of the bodies. One such constraint between *i*th and *j*th bodies can be described by the scalar function C_k ($\mathbf{r}_i$, q_i, $\mathbf{r}_j$, q_j). Constraint functions for the system of rigid bodies are combined into a column vector $\mathbf{C}$ with a dimension d equal to the number of constraints in the system.

Unilateral constraint, restricting movement in only one direction, is given as [5]:

$$C_k\left(\mathbf{r}_i, q_i, \mathbf{r}_j, q_j\right) \geq 0.$$

Bilateral constraint is given as:

$$C_k\left(\mathbf{r}_i, q_i, \mathbf{r}_j, q_j\right) = 0.$$

Consider the *k*th constraint function that restricts the displacement of *i*th and *j*th bodies by a given distance L between two points of these bodies:

$$C_k = \left(\left(\mathbf{p}_j - \mathbf{p}_i\right)^2 - L^2\right)/2,$$

where $\mathbf{p}_i$ and $\mathbf{p}_j$ are position vectors of the points in the fixed coordinate system.

Time derivative of the constraint function:

$$\mathrm{d}C_k/\mathrm{d}t = \left(\mathbf{p}_j - \mathbf{p}_i\right)\left(\mathbf{v}_j + \boldsymbol{\omega}_j \times \left(\mathbf{p}_j - \mathbf{r}_j\right) - \mathbf{v}_i - \boldsymbol{\omega}_i \times \left(\mathbf{p}_i - \mathbf{r}_i\right)\right)$$

or

$$\mathrm{d}C_k/\mathrm{d}t = \begin{pmatrix} -\mathbf{d}^T & -\left(\left(\mathbf{p}_i - \mathbf{r}_i\right)\times\mathbf{d}\right)^T & \mathbf{d}^T & \left(\left(\mathbf{p}_j - \mathbf{r}_j\right)\times\mathbf{d}\right)^T \end{pmatrix} \begin{pmatrix} \mathbf{v}_i & \boldsymbol{\omega}_i & \mathbf{v}_j & \boldsymbol{\omega}_j \end{pmatrix}^T,$$

$$\mathbf{d} = \left(\mathbf{p}_j - \mathbf{p}_i\right).$$

Denoting $\mathbf{J}_{ki} = \begin{pmatrix} -\mathbf{d}^T & -\left(\left(\mathbf{p}_i - \mathbf{r}_i\right)\times\mathbf{d}\right)^T \end{pmatrix}$, $\mathbf{J}_{kj} = \begin{pmatrix} \mathbf{d}^T & \left(\left(\mathbf{p}_j - \mathbf{r}_j\right)\times\mathbf{d}\right)^T \end{pmatrix}$, we obtain:

$$\mathrm{d}C_k/\mathrm{d}t = \begin{pmatrix} \mathbf{J}_{ki} & \mathbf{J}_{kj} \end{pmatrix}\begin{pmatrix} \mathbf{V}_i & \mathbf{V}_j \end{pmatrix}^T.$$

Generalizing for any constraint function:

$$\mathrm{d}\mathbf{C}/\mathrm{d}t = \mathbf{J}\mathbf{V},\ \mathbf{J} = \begin{pmatrix} \mathbf{J}_{11} & \cdots & \mathbf{J}_{1n} \\ \vdots & \ddots & \vdots \\ \mathbf{J}_{d1} & \cdots & \mathbf{J}_{dn} \end{pmatrix},$$

where $\mathbf{J}$ is the Jacobi matrix for the vector $\mathbf{C}$ of constraint functions.

The vector of constraint reactions:

$$\mathbf{F}^c = \mathbf{J}^T\boldsymbol{\lambda},$$

where $\boldsymbol{\lambda} = \begin{pmatrix} \lambda_1 & \cdots & \lambda_d \end{pmatrix}^T$ is the vector of indeterminate Lagrange multipliers.

Consider the kth constraint function that bounds the position of the point $\mathbf{p}_i$ of the ith body by the plane of the jth body passing through the point $\mathbf{p}_j$:

$$C_k = \mathbf{n}_j\left(\mathbf{p}_j - \mathbf{p}_i\right),$$

where $\mathbf{n}_j$ is the normal to the given plane of the jth body.

Time derivative of the constraint function:

$$\mathrm{d}C_k/\mathrm{d}t = \mathbf{n}_j\left(\mathbf{v}_j + \boldsymbol{\omega}_j \times \left(\mathbf{p}_j - \mathbf{r}_j\right) - \mathbf{v}_i - \boldsymbol{\omega}_i \times \left(\mathbf{p}_i - \mathbf{r}_i\right)\right) + \left(\boldsymbol{\omega}_j \times \mathbf{n}_j\right)\left(\mathbf{p}_j - \mathbf{p}_i\right),$$

$$\mathrm{d}C_k/\mathrm{d}t = \mathbf{n}_j\mathbf{v}_j + \left(\mathbf{p}_j - \mathbf{r}_j\right)\times\mathbf{n}_j\boldsymbol{\omega}_j - \mathbf{n}_j\mathbf{v}_i - \left(\mathbf{p}_i - \mathbf{r}_i\right)\times\mathbf{n}_j\boldsymbol{\omega}_i + \mathbf{n}_j \times \left(\mathbf{p}_j - \mathbf{p}_i\right)\boldsymbol{\omega}_j,$$

$$\mathrm{d}C_k/\mathrm{d}t = \begin{pmatrix} -\mathbf{n}_j^T & -\left(\left(\mathbf{p}_i - \mathbf{r}_i\right)\times\mathbf{n}_j\right)^T & \mathbf{n}_j^T & \left(\left(\mathbf{p}_i - \mathbf{r}_j\right)\times\mathbf{n}_j\right)^T \end{pmatrix}\begin{pmatrix} \mathbf{v}_i & \boldsymbol{\omega}_i & \mathbf{v}_j & \boldsymbol{\omega}_j \end{pmatrix}^T,$$

$$\mathbf{J}_{ki} = \begin{pmatrix} -\mathbf{n}_j^T & -\left(\left(\mathbf{p}_i - \mathbf{r}_i\right)\times\mathbf{n}_j\right)^T \end{pmatrix},\ \mathbf{J}_{kj} = \begin{pmatrix} \mathbf{n}_j^T & \left(\left(\mathbf{p}_i - \mathbf{r}_j\right)\times\mathbf{n}_j\right)^T \end{pmatrix}.$$

Consider the kth vector constraint function that limits the position of the point of the ith body to the point of the jth body (ball joint):

$$\mathbf{C}_k = \mathbf{p}_j - \mathbf{p}_i,$$

where $\mathbf{p}_i$ and $\mathbf{p}_j$ are position vectors of the points in the fixed coordinate system.

Time derivative of the constraint function:

$$\mathrm{d}\mathbf{C}_k/\mathrm{d}t = \mathbf{v}_j + \boldsymbol{\omega}_j \times \left(\mathbf{p}_j - \mathbf{r}_j\right) - \mathbf{v}_i - \boldsymbol{\omega}_i \times \left(\mathbf{p}_i - \mathbf{r}_i\right),$$

$$\mathrm{d}C_k/\mathrm{d}t = \begin{pmatrix} -\mathbf{E}_3 & -\left[\mathbf{p}_i - \mathbf{r}_i\right]_\times^T & \mathbf{E}_3 & \left[\mathbf{p}_j - \mathbf{r}_j\right]_\times^T \end{pmatrix}\begin{pmatrix} \mathbf{v}_i & \boldsymbol{\omega}_i & \mathbf{v}_j & \boldsymbol{\omega}_j \end{pmatrix}^T,$$

$$\mathbf{J}_{ki} = \begin{pmatrix} -\mathbf{E}_3 & -\left[\mathbf{p}_i - \mathbf{r}_i\right]_\times^T \end{pmatrix},\ \mathbf{J}_{kj} = \begin{pmatrix} \mathbf{E}_3 & \left[\mathbf{p}_j - \mathbf{r}_j\right]_\times^T \end{pmatrix},$$

where $[\mathbf{a}]_\times$ s is a skew-symmetric matrix.

The numerical integration of the equations of motion of a system of solids will be considered using the fourth-order Runge-Kutta method.

Equations of motion:

$$d\mathbf{V}/dt = \mathbf{M}^{-1}\mathbf{F} + \mathbf{M}^{-1}\mathbf{J}^{T}\boldsymbol{\lambda},$$

$$d\mathbf{r}_i/dt = \mathbf{v}_i,\ dq_i/dt = \boldsymbol{\omega}_i q_i/2,\ \forall i \in [1,n].$$

The velocity vector at the next step of integration is calculated as:

$$\mathbf{V}(t_c + \Delta t) = \mathbf{V}(t_c) + \Delta t(\mathbf{k}_1 + 2\mathbf{k}_2 + 2\mathbf{k}_3 + \mathbf{k}_4)/6,$$

where $\mathbf{k}_1$, $\mathbf{k}_2$, $\mathbf{k}_3$ and $\mathbf{k}_4$ are the right-hand sides of the differential equation calculated in four steps.

Introduce notations:

$$\mathbf{r} = (\mathbf{r}_1 \ \cdots \ \mathbf{r}_n)^T,\ \mathbf{r} = (q_1 \ \cdots \ q_n)^T,\ \mathbf{r}^{(1)} = \mathbf{r}(t_c),\ \mathbf{q}^{(1)} = \mathbf{q}(t_c),\ \mathbf{V}^{(1)} = \mathbf{V}(t_c),$$

$$\mathbf{F}^{(1)} = \mathbf{F}\left(t_c, \mathbf{r}^{(1)}, \mathbf{q}^{(1)}, \mathbf{V}^{(1)}\right),\ \mathbf{J}^{(1)} = \mathbf{J}\left(\mathbf{r}^{(1)}, \mathbf{q}^{(1)}\right),$$

then

$$\mathbf{k}_1 = \mathbf{M}^{-1}\mathbf{F}^{(1)} + \mathbf{M}^{-1}\left(\mathbf{J}^{(1)}\right)^T \boldsymbol{\lambda}^{(1)}.$$

To determine $\mathbf{k}_1$, it is necessary to calculate the values of the indeterminate Lagrange multipliers $\boldsymbol{\lambda}^{(1)}$. Let's write the velocity expression for the second stage:

$$\mathbf{V}^{(2)} = \mathbf{V}^{(1)} + \Delta t(\mathbf{k}_1)/2.$$

Using the expression for the derivative of constraint functions vector, we get:

$$\mathbf{J}^{(1)}\mathbf{M}^{-1}\left(\mathbf{J}^{(1)}\right)^T \boldsymbol{\lambda}^{(1)} = d\mathbf{C}/dt - \mathbf{J}^{(1)}\left(2\mathbf{V}^{(1)}/\Delta t + \mathbf{M}^{-1}\mathbf{F}^{(1)}\right).$$

This system of algebraic equations can be solved by the Gauss-Seidel iterative method, using additional conditions:

$$dC_k/dt = 0 \text{ for bilateral constraint,}$$

$$dC_k/dt \geq 0 \text{ for unilateral constraint.}$$

Then $\mathbf{k}_1$, $\mathbf{V}^{(2)}$, $\mathbf{r}^{(2)}$, $\mathbf{q}^{(2)}$, $\mathbf{F}^{(2)}$, $\mathbf{J}^{(2)}$ is calculated:

$$\mathbf{r}_i^{(2)} = \mathbf{r}_i^{(1)} + \Delta t \mathbf{v}_i^{(1)}/2,\ q_i^{(2)} = q_i^{(1)} + \Delta t \boldsymbol{\omega}_i^{(1)} q_i^{(1)}/4,$$

$$\mathbf{F}^{(2)} = \mathbf{F}\left(t_c + \Delta t/2, \mathbf{r}^{(2)}, \mathbf{q}^{(2)}, \mathbf{V}^{(2)}\right),\ \mathbf{J}^{(2)} = \mathbf{J}\left(\mathbf{r}^{(2)}, \mathbf{q}^{(2)}\right).$$

The following steps are calculated in a similar way:

$$\mathbf{k}_2 = \mathbf{M}^{-1}\mathbf{F}^{(2)} + \mathbf{M}^{-1}\left(\mathbf{J}^{(2)}\right)^T \boldsymbol{\lambda}^{(2)},$$

$$\mathbf{r}_i^{(3)} = \mathbf{r}_i^{(1)} + \Delta t \mathbf{v}_i^{(2)}/2,\ q_i^{(3)} = q_i^{(1)} + \Delta t \boldsymbol{\omega}_i^{(2)} q_i^{(1)}/4,\ \mathbf{V}^{(3)} = \mathbf{V}^{(1)} + \Delta t(\mathbf{k}_2)/2,$$

$$\mathbf{F}^{(3)} = \mathbf{F}\left(t_c + \Delta t/2, \mathbf{r}^{(3)}, \mathbf{q}^{(3)}, \mathbf{V}^{(3)}\right),\ \mathbf{J}^{(3)} = \mathbf{J}\left(\mathbf{r}^{(3)}, \mathbf{q}^{(3)}\right),$$

$$\mathbf{k}_3 = \mathbf{M}^{-1}\mathbf{F}^{(3)} + \mathbf{M}^{-1}\left(\mathbf{J}^{(3)}\right)^T \boldsymbol{\lambda}^{(3)},$$

$$\mathbf{r}_i^{(4)} = \mathbf{r}_i^{(1)} + \Delta t \mathbf{v}_i^{(3)},\ q_i^{(4)} = q_i^{(1)} + \Delta t \boldsymbol{\omega}_i^{(3)} q_i^{(1)}/2,\ \mathbf{V}^{(4)} = \mathbf{V}^{(1)} + \Delta t(\mathbf{k}_3),$$

$$\mathbf{F}^{(4)} = \mathbf{F}\left(t_c + \Delta t, \mathbf{r}^{(4)}, \mathbf{q}^{(4)}, \mathbf{V}^{(4)}\right),\ \mathbf{J}^{(4)} = \mathbf{J}\left(\mathbf{r}^{(4)}, \mathbf{q}^{(4)}\right),$$

$$\mathbf{k}_4 = \mathbf{M}^{-1}\mathbf{F}^{(4)} + \mathbf{M}^{-1}\left(\mathbf{J}^{(4)}\right)^T \boldsymbol{\lambda}^{(4)},$$

$$\mathbf{r}_i(t_c + \Delta t) = \mathbf{r}_i^{(1)} + \Delta t\left(\mathbf{v}_i^{(1)} + 2\mathbf{v}_i^{(2)} + 2\mathbf{v}_i^{(3)} + \mathbf{v}_i^{(4)}\right)/6,$$

$$q_i(t_c + \Delta t) = q_i^{(1)} + \Delta t\left(\boldsymbol{\omega}_i^{(1)} q_i^{(1)} + 2\boldsymbol{\omega}_i^{(2)} q_i^{(1)} + 2\boldsymbol{\omega}_i^{(3)} q_i^{(1)} + \boldsymbol{\omega}_i^{(4)} q_i^{(1)}\right)/12,$$

$$\mathbf{V}(t_c + \Delta t) = \mathbf{V}^{(1)} + \Delta t(\mathbf{k}_1 + 2\mathbf{k}_2 + 2\mathbf{k}_3 + \mathbf{k}_4)/6.$$

3. Model of the main landing gear leg of the Tu-204SM aircraft

Figure 1 shows the model of the main landing gear leg of the Tu-204SM aircraft [6]. The model consists of six rigid bodies: two pairs of wheels 1 and 2, the bogie 3, the shock absorber rod 4, the piston of the second shock absorber chamber 6 and the shock absorber cylinder together with the weight of the airframe per one leg 5.

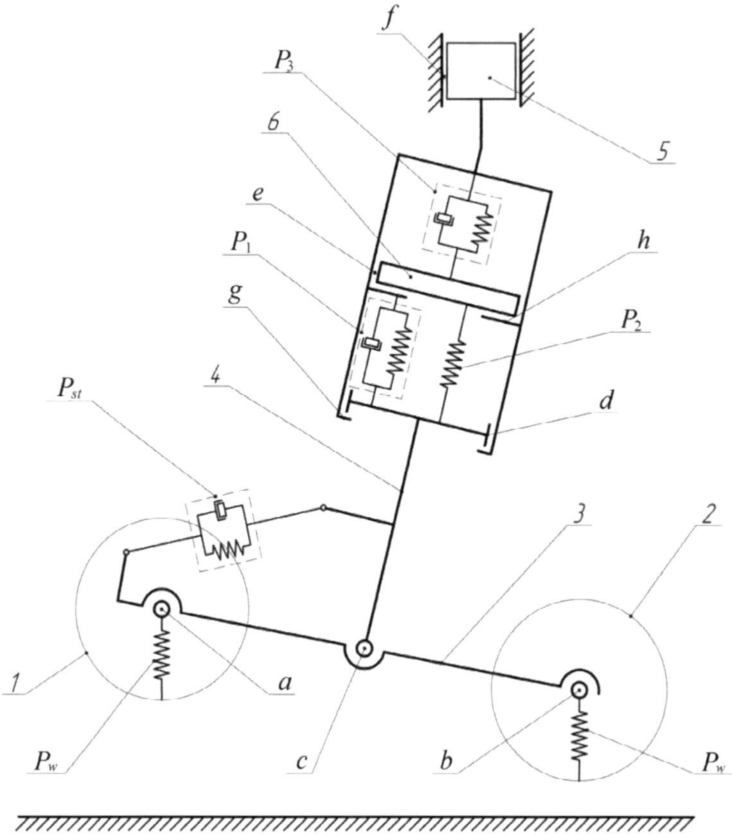

Figure 1. Model of the main landing gear leg.

Joints "*a*" and "*b*", connecting pairs of wheels with the bogie, as well as the joint "*c*" connecting the bogie with the shock absorber rod, are hinges. Each hinge connection restricts the two degrees of freedom of the system and is determined by two bilateral constraints. Sliding joint "*d*" connect the shock absorber rod and joint "*e*" connect the piston of the second chamber with the cylinder. A shock absorber cylinder is fixed in the sliding fit "*f*". The sliding joint also restricts the two degrees of freedom of the system and is determined by two bilateral constraints. Stops "*g*" and "*h*" restrict one degree of freedom each and are modeled by the unilateral constraints. Thus, the model includes 14 constraints.

The following six active forces are introduced in the model: two forces of compression P_w, force P_{st} by a stabilizing damper and three forces in the shock absorber – P_1, P_2 and P_3. The force P_1 includes the frictional force in the axle-box and the force of the hydrodynamic resistance when the liquid flows from chamber 3 to chamber 1 (Figure 2). The force P_2 is the force of compression of the gas spring of the chamber 1. The force P_3 includes the force of compression of the gas spring of the chamber 2 and the frictional force of the piston.

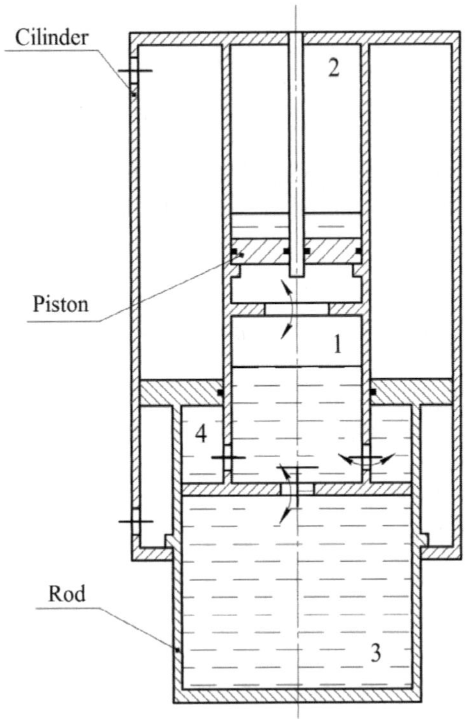

Figure 2. Scheme of the shock absorber.

Figure 3 shows theoretical and experimental load curves for damping systems of the strut. The parameters of the landing impact of the aircraft landing gear calculated by the proposed method agree with the test data within the experimental error, which confirms the predictive capabilities of the mathematical model.

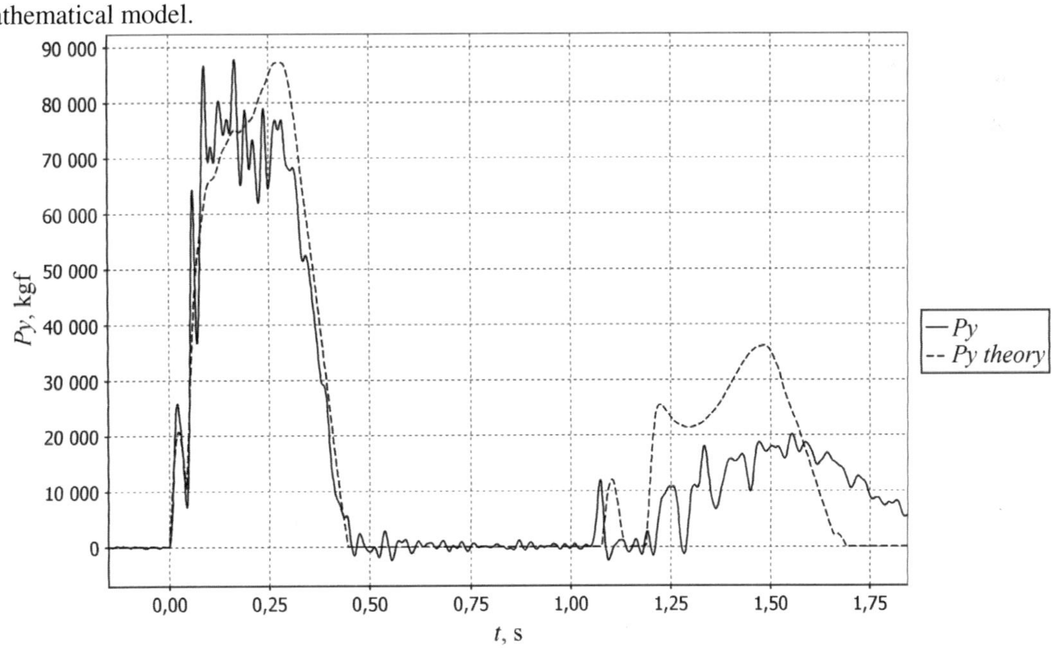

Figure 3. Vertical load diagram.

IOP Conf. Series: Journal of Physics: Conf. Series **894** (2017) 012129 doi:10.1088/1742-6596/894/1/012129

4. Conclusion

The method proposed in this paper differs from the existing methods in that it is, first of all, universal. If another system of rigid bodies is considered, one does not need to rewrite the equations of motion in the generalized coordinates and determine these coordinates by excluding the "redundant" coordinates using the constraint equations. The dimensionality of the system changes, but the form of equations remains the same. This universal approach is more algorithmic and simple to implement numerically. As an advantage of the method, the method can be used to solve a wide range of problems of rigid body system dynamics.

References

[1] Gantmakher F R 1966 *Lectures in analytical mechanics* (Moscow: Nauka) p 300
[2] Lotstedt P 1982 Mechanical systems of rigid bodies subject to unilateral constraints *SIAM Journal on Applied Mathematics* **42** 281–296
[3] Cline M B 2002 *Rigid Body Simulation with Contact and Constraints* (The University of British Columbia) p 102
[4] Baraff D 1996 Linear-Time Dynamics using Lagrange Multipliers *COMPUTER GRAPHICS Proceedings, Annual Conference Series (New Orleans)* ed J Fujii (New York: ACM Press) pp 137–146
[5] Anitescu M and Potra F A 2002 A time-stepping method for stiff multibody dynamics with contact and friction *Int. J. Numer. Meth. Engng* **55** 753–784
[6] Podruzhin E G and Zagidulin A R 2013 Model of the compressed landing gear damper strut of a trunk-route plane *Scientific Bulletin of Novosibirsk State Techn. Univ* **51** 144–154

IOP Conf. Series: Journal of Physics: Conf. Series **894** (2017) 012130
doi:10.1088/1742-6596/894/1/012130

Study of gas dynamics of the intersecting jets in a semi-open channel

V M Abashev[1], I V Eremkin[1], N P Zhivotov[1], V P Zamuraev[2,3], A P Kalinina[2,3], P K Tretyakov[2] and A V Tupikin[2]

[1]Moscow Aviation Institute (National Research University), Moscow, Russia
[2]Khristianovich Institute of Theoretical and Applied Mechanics, Novosibirsk, Russia
[3]Novosibirsk State University, Novosibirsk, Russia

E-mail: zamuraev@itam.nsc.ru, kalinina@itam.nsc.ru

Abstract. The subject of the study is axisymmetric channel with a hemispherical bottom. The compressed air flows into the channel through the holes located on the surface of the hemispherical bottom. The ratio of all holes area to the outlet area of the channel is equal to 0.25. The experimental and numerical study of jets intersecting at a single point is performed. The countermoving jets and the focused jets are investigated. The distribution of pressure over the surface of the channel and the total pressure at the output of the channel are obtained. The main experimental result is as follows: the total pressure recovery coefficient for the focused jets is twice more than that for the countermoving jets.

1. Introduction

Slot nozzles are of practical interest as an alternative use of the Laval nozzle [1]. The gas dynamics of separate off-design supersonic jet is studied extensively [2-8], but interaction of multiple jets is a more complex problem. The gas dynamics structure of the flow has a significant impact on the processes of fuels ignition and combustion. Therefore, the gas dynamic structure formed by the multiple jets deserves additional study. Some of the results of the effusion of the jets from the holes of a few millimeters are presented in [1]. The area of all the holes of the channels from [1] is tenfold smaller than the output area of the channel. This work is dedicated to the effusion of intersecting jets from the holes, which total area is fourfold smaller than the area of the channel. The paper presents the results of the experimental and numerical study of jets gas dynamics. The semi-open channels are characterized by sufficiently thick walls to create directional jets. The compressed air flows from the prechamber through the holes. The jets are focused at a single point on the axis of the channel. It is shown that the direction of the jets has a significant effect on the total pressure recovery coefficient.

2. Results and Discussion

The shadow scheme with a circular knife was applied for registering the flow structure. There were fluctuations of the gas dynamics parameters at low pressures in the prechamber p_f. The supersonic mode is established when the pressure in the prechamber is greater than a certain critical value.

Numerical modeling was carried out basing on unsteady three-dimensional Euler equations. Experimental data and calculation results are given in Table 1 and Table 2 for the countermoving jets

Content from this work may be used under the terms of the Creative Commons Attribution 3.0 licence. Any further distribution of this work must maintain attribution to the author(s) and the title of the work, journal citation and DOI.
Published under licence by IOP Publishing Ltd

and focused jets respectively. The pressure at the test points, the total pressure p_0 and static pressure p_S on the axis in the outlet section of channel, and the Mach number are presented in these tables. The total pressure for a direct shock wave and static pressure were measured in the experiment to obtain the Mach number and the total pressure on the axis at the outlet section of the channel. The total pressure for a shock wave, and all other values were obtained in the calculation process for the case of numerical modeling.

Let us consider the case of the countermoving jets. The compressed air flows to the channel through four holes located opposite each other in pairs in two mutually perpendicular planes. The test points for measuring the pressure and the pressures values in these points are shown in Figure 1a and Figure 1b respectively.

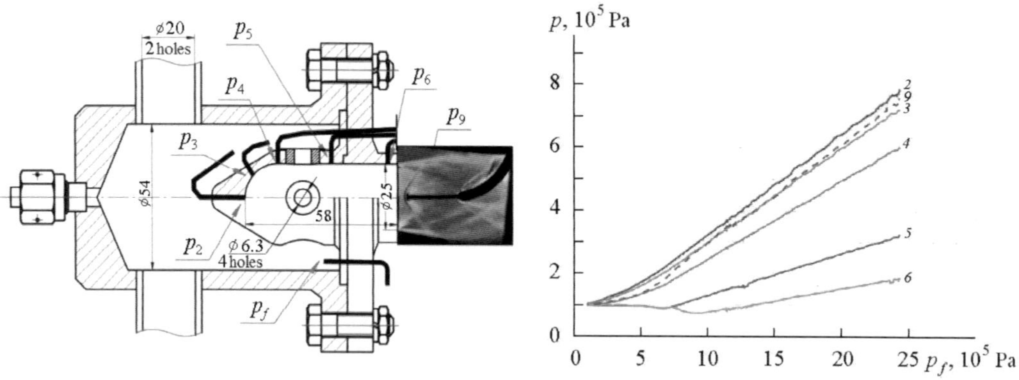

(a) (b)

Figure 1. The case of the countermoving jets: (a) – the model with countermoving jets; (b) – the experimental dependences of the pressure for the test points from the prechamber pressure.

The plane with the test points is located between the planes containing the opposite holes at the same distance from each other. Table 1 contains the data for two values of the pressure in the prechamber. The graphs in Figure 1b indicate the linear character of the static pressure dependences on the walls under the following conditions: the value of the pressure in the prechamber must be more than some certain value. We named this type of solution an automodel solution. The main features of this automodel solution are following: the constant Mach number, the linear character of the dependences of static pressure on the walls as function of the pressure in the prechamber. The unsteady subsonic flow is formed at low pressures in the prechamber. While the pressure of the prechamber achieves some critical value, the supersonic flow is formed. The data in Table 1 are averaged with respect to time. There are fluctuations of flow parameters in the experiment as well as in the numerical simulation. The maximum fluctuation during calculation does not exceed 20%. It should be noted a good agreement of the experimental data and calculation results.

The difference between the results of measurements at the point p_2, p_3, p_4 and the data of three-dimensional calculations is shown in Table 2. This difference is within the range of the fluctuations. The Mach number is also practically independent on the pressure in the prechamber and coincides with the value obtained in the experiment.

Table 1. Pressure (10^5 Pa) at the test points and the Mach number for the countermoving jets.

	p_f	p_2	p_3	p_4	p_5	p_6	p_9	p_0	p_S	M
Experiment	15	4.8	4.38	3.6	2	1.07	4.38	5.61	-	1.65

3D simulation	15	4.89	3.82	3.31	1.36	1.77	4.7	5.72	0.95	1.69
Experiment	24	7.69	7.15	6	3.15	1.84	7.38	8.38	-	1.65
3D simulation	24	8.21	5.56	5.61	2.54	2.36	6.84	8.21	1.67	1.69

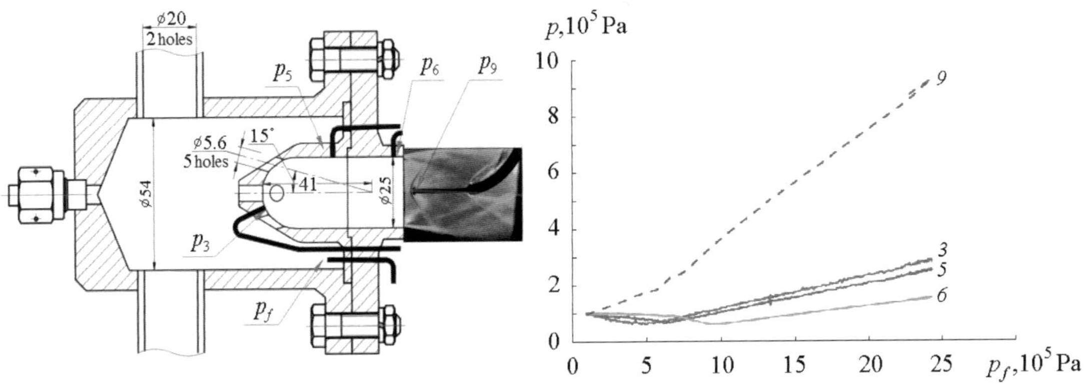

(a) (b)

Figure 2. The case of the focused jets: (a) – the model with focused jets; (b) – the experimental dependences of the pressure for the test points from the prechamber pressure.

The plane with the test points is located between the planes containing the opposite holes on the same distance from each other. The graphs in Figure 2 indicate the linear character of the static pressure dependences on the walls under the following conditions: the value of the pressure in the prechamber must be more than a certain value. The unsteady subsonic flow is formed at low pressures in the prechamber. While the pressure of the prechamber achieves some critical value, the supersonic flow is formed. We can point out that the experimental data as well as the calculation ones have the same properties as in the case of countermoving jets. We can also name automodel the solution for the focused jets. The data in Table 2 are averaged with respect to time. The fluctuation of the parameters is less than for the case of countermoving jets: their value consists of 10% for the value of the pressure p_f=15 atm and the fluctuations of parameters are small to negligible for the pressure p_f=24 atm. As we can see from Figure 1b and Figure 2b, the supersonic mode formation takes place at the pressure values in the prechamber, which are less than the ones for the countermoving jets.

Table 2. Pressure (10^6 Pa) at the test points and the Mach number for the focused jets.

	p_f	p_3	p_5	p_6	p_9	p_0	p_S	M
Experiment	1.5	1.77	1.62	1.0	5.54	7.52	-	2.05
3D simulation	1.5	1.41	2.03	1.3	4.4	5.8	0.84	1.92
Experiment	2.4	2.77	2.46	1.46	9.15	13.6	-	2.05
3D simulation	2.4	2.61	1.88	2.06	7.31	10.4	1.24	2.07

The difference between the experimental and the calculation results consists of 25% with exeption of the Mach number. The coinciding of experimental and calculation value of the Mach number is very good.

As we can see from Figure 3, the experimental values of the loss of total pressure in the case of the focused jets are almost twice less than for the countermoving jets. We should point out that the calculation values of the loss of total pressure differs from the experimental values in 25%: loss of total pressure in the case of the focused jets is almost 1.5 fold less than for the countermoving jets. This difference must be investigated in detail in the future, but we can suppose some reasons for this phenomenon. The main of them is the heterogeneous distribution of the flow parameters values.

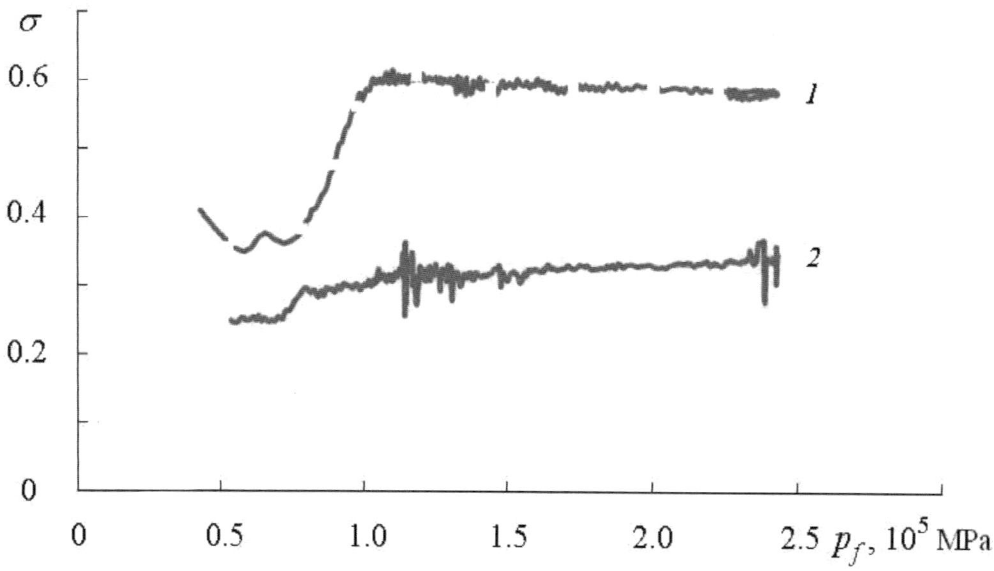

Figure 3. The dependences of the total pressure recovery coefficient $\sigma = p_0 / p_f$ from the prechamber pressure for the focused holes (blue line) and the opposite holes (red line)

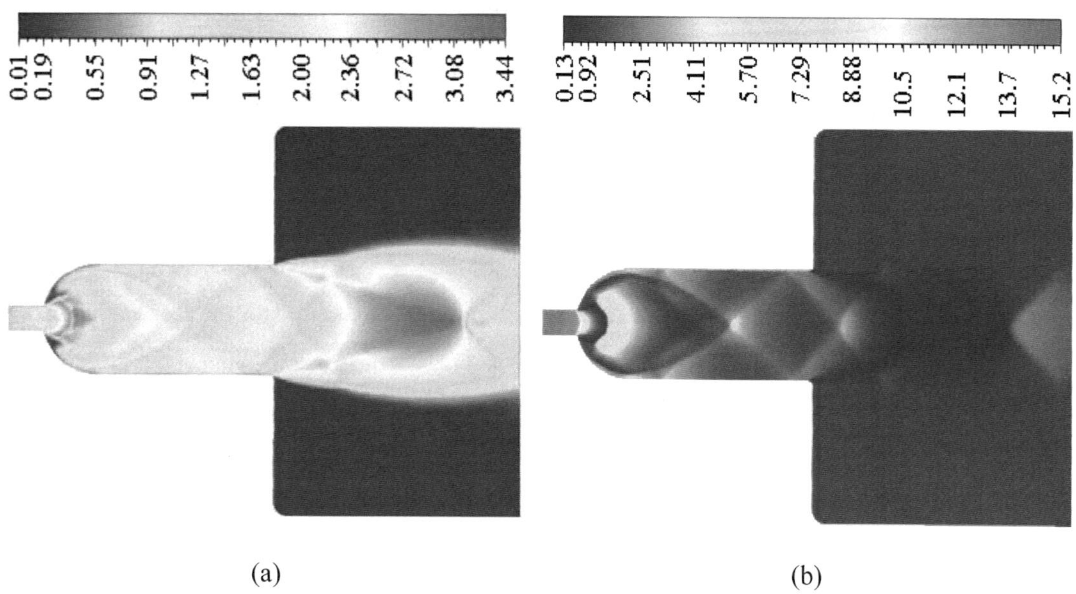

Figure 4. The distributions for the focused jets at the prechamber pressure p_f=2.4 MPa in the plane of measurements: (a) – the Mach number; (b) – the pressure

The calculations results shown in Figure 4 are distribution of the Mach number (Figure 4a) and the pressure in the plane of measurements (Figure 4b) at pressure in the prechamber $p_f = 2.4$ MPa. We can see that the flow is sufficiently heterogeneous.

The distribution of the Mach number, static pressure and total pressure along the diameter of the output section of the channel is presented in Figure 5 for the case considered. We can see from Figure 5 that the flow at the nozzle is fully supersonic, but the maximum difference between parameters consists of 15%.

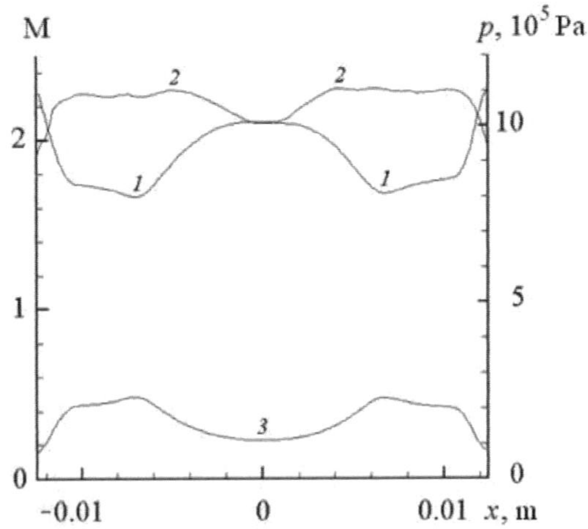

Figure 5. The distribution of the Mach number (*1*), total pressure (*2*) and static pressure (*3*) along the diameter of the output section of the channel.

3. Conclusion

Data analysis indicates that semi-open channel can be used as a supersonic nozzle. The loss of total pressure depends considerably on the process of jets mixing. This process is determined by the ratio of area of all the holes to the output area of the channel as well as location and direction of the jets.

The total pressure recovery coefficient for the focused jets is twice more than the same parameter for the countermoving jets according to the experimental estimation, but for the case of numerical simulation the difference is about 1.5 fold.

Acknowledgments
The authors gratefully acknowledge Christopher Hearsey and Valeriya Hearsey for their assistance.

References

[1] Abashev V, Eremkin I, Zhivotov N, Zamuraev V, Kalinina A, Tretyakov P and Tupikin A 2016 *AIP Conference Proceding* 1770 030078
[2] Glotov G and Faiman M 1971 *TsAGI Science Journal* **2** 69–75
[3] Ivanov M and Kraiko A 1975 *Numerical methods of continuum mechanics* **6** 2
[4] Avduevskii V and Ivanov A 1970 *Fluid Dynamics* **3** 125-131
[5] Antsupov A and Blagosklonov V 1976 *TsAGI Science Journal* **1781** 1–25
[6] Isaev S and Lipnitskii Yu 2012 *Journal of Engineering Physics and Thermophysics* **85** 1357–71
[7] Kosarev V and Klinkov S 2016 *Thermophysics and Aeromechanics* **23** 321
[8] Kosarev V and Klinkov S 2017 *Thermophysics and Aeromechanics* **24** 221

Rationing and shock testing on-board equipment of spacecrafts

S A Orlov[1]**, K A Matveev**[2] **and G I Rastorguev**[2]

[1] JSC "Information Satellite Systems" Reshetnev Company, Zheleznogorsk, Russia
[2] Novosibirsk State Technical University, Novosibirsk, Russia

E-mail: office@iss-reshetnev.ru

Abstract. The article addresses the issues of valuation and testing of on-board equipment for high-intensity shocks, discusses the problems of various shock testing facilities use and compares approaches to the problem in our country and abroad. The decision to use the specific mechanical test facilities for the on-board equipment testing is taken by its developer based on the assessment of on-board equipment resistance to the impact (resulting from own experience and numerical modeling).

1. Introduction

Currently, both Russian and foreign spacecrafts are produced in accordance with unpressurized scheme. The power base of such spacecraft is cellular panels and composite materials structures. As a result, there are certain changes in the rigidity and damping properties of a spacecraft, the way in which the on-board equipment is attached and, consequently, the load on the spacecraft itself and the on-board equipment. At the same time, the existing normative documentation, which governs the loads on the on-board equipment and the adjustment methods, has remained virtually unchanged for more than 40 years [1].

2. Research question

To date, the problems arising from the adjustment of on-board equipment for quasi-static and vibratory (both harmonic and random vibrations) loads have largely been solved [1–6]. One of the most complex types of mechanical loads on a spacecraft and on-board equipment where methodological problems remain are the increased intensity shock inputs. The source of such loads is the shock inputs (primarily pyroshocks) on the part of the carrier rocket (stages separation, cone discharge etc.), as well as shocks from its own pyrotechnics (spacecraft separation, solar array release and antenna deployment, etc.) [2–5]. It should be noted that the approaches to the rate making and adjustment of the on-boar equipment for shock inputs in Russia and abroad have significant differences. The USA and European shock requirements are set in the form of shock response spectra, regardless of the industry, while in Russia there is still widespread shock rate making in the form of single pulses (more often in the form of a half-sine pulse). There are no real inputs that would appear as single pulses on the spacecraft and on-board equipment. In the Russian aerospace industry the concept of shock response spectra has been used for more than 30 years, gradually supplanting rate making in the form of single impulses. The equipment developers are constantly faced with the requirements for the application of components whose shock input is specified as single pulses, while the hardware requirements are already set in the form of shock response spectra.

Content from this work may be used under the terms of the Creative Commons Attribution 3.0 licence. Any further distribution of this work must maintain attribution to the author(s) and the title of the work, journal citation and DOI.
Published under licence by IOP Publishing Ltd

3. Shock rating

The shock response spectrum (SRS) is "the graph of the maximum acceleration response to the specified input effect of the oscillator ensemble with the specified values of the acceleration constant of the oscillators own frequencies". In calculations, if not otherwise specified, the single-degree-of-freedom systems with viscosity damping are usually used (if not specified, the value of acceleration constant $Q = 10$ is assumed) [7]. Deduced from the definition, there are positive and negative SRS. That is, spectra that show the maximum positive/negative acceleration. There is also residual SRS and primary SRS, i.e. spectra which do not account for the oscillator response during the shock input or, on the contrary, assess the response only during the shock input. Various algorithms for the solution of the equation (e.g. [8]) are used in the calculation of the SRS. There are also some approximate solutions for the residual SRS on the basis of the Duhamel's integral [9, 10, 11]. Additionally, Smallwood algorithm for SRS computing has become a regular procedure recently. This requires the calculation of the SRS to be carried out at intervals of no more than 1/12 octave.

Next, the concept of the maximax SRS (primary and residual SRS) is used to rate and compute the required SRS, the amplitude value of which being assumed modulo.

Both in Russian and foreign literature there are three stages in the rating of shock input on the on-board equipment:

1. Modelling of the applied shock propagation in spacecraft frame using finite-element models and/or statistical energy analysis, with subsequent rating of loads at on-board equipment.
2. Computational and experimental approach when the measurement results for the autonomous tests of subsystems (antennas, separation systems, solar arrays, etc.) and numerical modeling provide a refined forecast of the on-board equipment load and the shock rating.
3. Confirmation of the specified (rated or normalized) inputs based on the results of the load measurements at the regular pyrotechnics application.

The first stage is that of the preliminary design. Previous projects "heritage" in the form of databases on damping, shock input values of pyrotechnics, etc. is implemented for the construction of finite-elements and statistical energy analysis models.

The second stage is that of the operational design when the results of the first phase are refined if necessary.

The third stage is the verification phase of the entire project, confirming the correctness of the decisions taken in phases 1 and 2. The results of the measurements supplement the existing database.

There are currently experimental data on ~ 20 spacecrafts built on three different platforms, which have different mass, dimensions, number and type of pyrotechnics.

Standard shock input modes to on-board equipment of spacecrafts in our country are represented in Table 1 [11]. The applied shock on all equipment is divided into 6 zones (S0, S1, S2, S3, S4, S5) depending on the distance to the source of shock (Table 1).

Table 1. Model regimes impact.

Zone	Distance from the shock input source (m)
S0	To 0.3
S1	0.3–0.7
S2	0.7–1.5
S3	1.5–3
S4	3–8
S5	8–15

The shock loads are set for each of the three mutually perpendicular axes in the form of shock response spectra (with the acceleration constant of the resonators $Q = 10$).

IOP Conf. Series: Journal of Physics: Conf. Series **894** (2017) 012131 doi:10.1088/1742-6596/894/1/012131

Table 2. The values of the shock spectrum of accelerations.

Zone	Frequency range (Hz)	
	35–1000	1000–10000
	Shock response spectrum value (g)	
S0	~ 10000 and more	
S1	25–5000	5000
S2	13–2500	2500
S3	5–1000	1000
S4	3–500	500
S5	3–100	100

As seen from Table 2, the maximum shock response spectra for the equipment being adjusted is accepted at 5000g and its minimum value is 100g. In addition to that, S5 zone is usually overlapped with a reaction to the vibration input and the need to carry out shock tests on S5 zone modes is very rare. S0 zone is the one of ~ 0.3 m within the pyrotechnics location area where only shock insensitive equipment can be installed.

It is noteworthy that the test modes presented in Table 2 include the data for some pressurized spacecrafts, which reduces the "transition frequency" to 1 kHz. The more appropriate test modes are the ones shown in Table 3 with the transition frequency of to 2.5 kHz. In the frequency range preceding the transition frequency, the load in Tables 2 and 3 is changed by linear law at logarithmic scale by frequency and amplitude of the shock response spectra. The loads for S3–S5 zones are to be taken in accordance with Table 2. The safety factor used to generate the load modes of on-board equipment is 2.

Table 3. Shock spectra of accelerations.

Zone	Frequency range (Hz)	
	35–2500	2500–10000
	Shock response spectrum value (g)	
S0	From 100 to 10000 and more	10000 and more
S1	50–5000	5000
S2	25–2500	2500

In general, the transition rate of the normalized shock response spectra generated by the implementation of pyrotechnics in the Western specification is ~ 2–3 kHz (e.g. [15]).

Both the ECSS and NASA standards differentiate only 3 zones in the shock rating. The applied shocks on the spacecraft board equipment are divided into three zones: near-field (up to 15 sm), middle (15–60 sm) and far-field (over 60 sm). These zones differ in both the magnitude (amplitude) of the applied shocks and the rated frequency range. The near-field zone includes shocks with the amplitudes of more than 5000 g and up to100 kHz by frequency; medium zone corresponds to those from 1000 g to 5000 g by amplitude and from 10 to 100 kHz by frequency and far-field zone corresponds to those of less than 1000 g by amplitude and 10 kHz by frequency [13,16].

Undoubtedly, the development of a zone rating methodology is an exceptional task. A large number of zones leads to difficulty in devices borrowing from one project to another due to a greater probability of a mismatch between requirements and specifications; and a small number of zones can lead to an unreasonable increase in the mass of the devices.

4. Shock testing

There are a number of features in carrying out the shock testing in the form of the shock response spectra shown in Tables 2 and 3, so we will look further at the use of a variety of testing facilities to apply shock input by the SRS method while testing the spacecraft on-board equipment.

To date, a sufficiently wide range of shock testing facilities is used in conducting the high-intensity shock tests (the exposure levels reach a few thousand "g"). These can be roughly divided into three groups:

- Mechanical shock testing facilities with a free falling Table and an accelerated Table drop.
- Mechanical pendulum (piling) shock testing facilities.
- Pyrotechnic testing facilities.

Mechanical shock test systems with a free falling Table and with an accelerated Table drop are focused mainly on setting the acceleration of the shock pulse.

Most often in Russia, there are shock testing facilities with a free falling Table produced by Lansmont (United States) and with an accelerated Table drop produced by Tira (Germany). In the RF Tira products under the brand VSTS gained wider distribution. The effectiveness of such facilities is determined primarily by the used programming tools as well as by the seismic mass and the type of seismic mass shock absorber. A significant drawback of the falling Table facilities is their dimensions (primarily the height of the stand) with relatively small desktop sizes for the subject equipment and their high cost. Merits include a wide range of reproduced acceleration pulses (both amplitude and duration) and relatively small "residual" noise (less than 20 per cent, which meets the requirements of domestic standards, e.g. [17]). Like the falling Table facilities, the Table with the accelerated drop of the Table is designed to create an acceleration pulse, but its dimensions are much smaller and, consequently, the working mass and instruments dimension ranges are not wide.

Pile shock stands were the first equipment to be used for shock testing. They have not changed in essence until today. The stand consists of a hammer, which creates the necessary shock input, a frame with a hammer suspension system fixing a braked anvil, a rotating traverse to fix the hammer, the counterweight, the anvil suspension system, the anvil for the installation of the subject equipment, the pneumodampers of the anvil and the hammer.

The levels of impact of the test object are determined by the impact velocity of the hammer against the anvil, its mass and braking mechanics, and by the mass of the test object mounted on the anvil. The shape of the impulse is determined by the size and shape of the striker, as well as the crusher installed between the hammer and the anvil. Such stands allow for the reproduction and input of nonstationary vibration. This is, for example, a well-known domestic stand K200 [18]. Pendulum shock test facilities are being further developed into stands with resonance panels and resonance beams. The merits of such facilities include the ability to test sufficiently massive and positional equipment in a wide range of impacts. Often, the stands are geared towards creating the necessary shock response spectra resulting from the nonstationary vibration caused by the impact. Shortcomings include the difficulty of adjusting and the size of the stands. Such stands are usually custom-designed and have high costs. It should be noted that the regulatory documentation does not recommend the use of such stands for testing the shock sensitive on-board equipment [19].

5. Conclusion

The decision to use the specific mechanical test facilities for the on-board equipment testing is taken by its developer based on the assessment of on-board equipment resistance to the impact (resulting from own experience and numerical modeling).

Most of the sources, both in our country and abroad, recommend the use of equipment shock test by pyrotechnic device when creating high intensity shock inputs which corresponds to the actual physics of the on-board equipment load and does not result in unreasonable loading of the equipment structure and components from the test methodology used [1, 2, 6, 9, 13, 16]. Methodologies and equipment for shock testing have now been developed with the use of a special pyrotechnic device [20–22]. The device has several regulatory parameters that create the desired shock input. These are the thickness and material of the pin; mass, material and shape of the striker; the power of generator; the initial volume of the cavity; the piston stroke prior to the impact. With a large number of adjusTable parameters it is possible to create shock inputs in all the required amplitude and frequency range given in Tables 2 and 3.

References

[1] Orlov S A and Kopytov V I 2013 Rate Making and Development of Spacecraft Onboard Equipment to Mechanical Environment *Vestnik SibGAU* **47** 125–129
[2] Gladkij V F 1975 *Stability, Vibration and Safety of Aircraft Construction* (Moscow: Nauka) p 456
[3] MIL-STD-1540D 1999 *Product verification requirements for launch, upper-stage and space vehicles* p 308
[4] ECSS-E10-03A 2002 *Space engineering. Testing*
[5] Calvi A 2011 *Spacecraft Loads Analysis. An Overview* (ESA / ESTEC) p 126
[6] ECSS-E-HB-32-26A 2013 *Space engineering. Spacecraft mechanical loads analysis handbook* p 636
[7] GOST R53190 2009 *Shock Testing with Reproduced Shock Response Spectra* (Moscow: Standartinform) p 24
[8] Doyar O P 1982 Shock Response Spectra Computation Algorithm *sb. Dinamika sistem. Chislennye metody issledovanija dinamicheskih sistem* (Chisinau: Nistru) pp 126–128
[9] Harris C M and Piersol A G 2002 *Shock and Vibration Handbook. Fifth Edition* (McGraw-Hill) pp 735-749
[10] Harris S and Kridi Ch 1973 *Shock Load Handbook* (Leningrad: Sudostroenie) p 360
[11] Orlov S A 2011 Development of Methodologies for the Rating and Testing of Spacecraft On-Board Equipment for the Mechanical Loads of the Zero-Lift Trajectory *Nauchnyj vestnik NGTU* **44** 137–148
[12] Smallwood D O *An Improved Recursive Formula for Calculating Shock Response Spectra* (Albuquerque, New Mexico: Sandia National Laboratories)
[13] ECSS-E-HB-32-25A 2015 *Space engineering. Mechanical shock design and verification handbook* p 541
[14] Orlov S A, Kopytov V I and Matveev K A 2013 Formation of High Intensity Shock Inputs for Spatial Structures *Izv. Vyssh. uchebnyh zavedenij. Fizika. Vol. 56* **7/3** 197-200
[15] Arian space Inc. 2004 *Ariane5 User Manual Issue 4 Revision0* p 303
[16] NASA-STD-7003 1999 *Pyroshock test criteria* p 26
[17] GOST R51371-99 2000 *Test Methods for Resistance to Mechanical External Influences of Machinery, Appliances and Other Technical Products. Shock Testing* (Moscow: IPK. Izdatel'stvo standartov) p 24
[18] Kruglov Ju A and Tumanov Ju A 1986 *Shock and Vibratory Protection of Equipment* (Leningrad: Mashinostroenie) p 222
[19] OST 92-5100-2002 2002 *Space-Complex Equipment. General Technical Specifications* (Federal'noe kosmicheskoe agentstvo) p 174
[20] Orlov S A and Orlov A S 2010 Patent 2394217 RF, MPK G01M 7/08. *Pyrotechnic Device for the Creation of Shock Inputs* p 8
[21] Orlov S A and Orlov A S 2010 Patent 2386939 RF, MPK G01M 7/08. *The Method of Shock Input Testing* p 11
[22] Orlov S A, Matveev K A and Rastorguev G I 2017 Patent 2616353 RF, MPK G01M 7/08. *Shock Input Tester for Appliances and Equipment* p 10

The numerical modelling of MHD astrophysical flows with chemistry

I Kulikov[1,2], I Chernykh[1,3] and V Protasov[2]

[1] Institute of Computational Mathematics and Mathematical Geophysics SB RAS, Novosibirsk, Russia
[2] Novosibirsk State Technical University, Novosibirsk, Russia
[3] Novosibirsk State University, Novosibirsk, Russia

E-mail: kulikov@ssd.sscc.ru

Abstract. The new code for numerical simulation of magnetic hydrodynamical astrophysical flows with consideration of chemical reactions is given in the paper. At the heart of the code – the new original low-dissipation numerical method based on a combination of operator splitting approach and piecewise-parabolic method on the local stencil. The chemodynamics of the hydrogen while the turbulent formation of molecular clouds is modeled.

1. Introduction

Magnetic field plays a key role in formation and dynamics of astrophysical objects. Thus, on cosmological scales the influence of weak magnetic field, with order μG, on the dynamics of hydrodynamic instabilities and ram-pressure mechanism in galactic clusters [1] was studied; primarily radial orientation of the magnetic field in Virgo cluster outside the central area was defined [2], and comparison of magnetic field with radio observation was carried out [3]. The structure of magnetic field in spiral arms of the M51 galaxy was investigated in[4], and the evolution of the disk galaxy with consideration of the influence of magnetic field was described in [5]. The consideration of the influence of magnetic field plays an important role in the evolution of interstellar turbulent flows where the magnetic field is sufficiently strong [6, 7]. For the problems of evolution of MHD instabilities, the power spectrum [8], subalfven flows [9], and starburst rate [10] were studied, and the comparison of different codes on the problem of supersonic turbulence was made in [11]. For the problems connected with the stellar wind, the MHD simulation is also necessary. Thus, the turbulence in stellar wind was described in [12], one-dimensional MHD model of interaction of stellar wind with 67P/Churyumov-Gerasimenko comet in [13], Halley comet [14], and also gas planet [15, 16] was built. In addition, the similar problems of interaction between stellar wind and stars [17] are worth noting.

The new low-dissipation numerical scheme for solving equations of magnetic gas dynamics with consideration of chemical processes and its software implementation are presented in the paper. The numerical method is based on a hybrid method developed earlier with the combination of operator splitting approach and Godunov method in its basis [18-29]. The piecewise-parabolic method on local stencil is used in all stages of the scheme to get low dissipation of the solution. The special algorithm for building the local parabola had allowed us to fully eliminate the using of TVD reconstructions of the numerical solution in the region

Content from this work may be used under the terms of the Creative Commons Attribution 3.0 licence. Any further distribution of this work must maintain attribution to the author(s) and the title of the work, journal citation and DOI.
Published under licence by IOP Publishing Ltd

of discontinuous solutions. We specifically do not claim the new method as a method of high order accuracy, because this term is not fully formulated in a case of discontinuous solutions [30]. In the first section, the numerical method is defined, and its software implementation is briefly described. In the second section, the one- and two-dimensional tests are shown. The third section is devoted to simulation of 3D MHD flows.

2. Numerical method description

In this paper, we will only consider the MHD flows, and the model problems will be considered just in MHD approximation. Thus, the system of equations of gravitational multicomponent magnetic gas dynamics in 3D cartesian coordinates taking into account the function of heating and cooling is used:

$$\frac{\partial \rho}{\partial t} + \nabla \cdot (\rho \mathbf{v}) = 0$$

$$\frac{\partial \rho_i}{\partial t} + \nabla \cdot (\rho_i \mathbf{v}) = \mathcal{S}_i$$

$$\frac{\partial \rho \mathbf{v}}{\partial t} + \nabla \cdot (\rho \mathbf{v}\mathbf{v} - \mathbf{B}\mathbf{B}) = -\nabla p^* - \rho \nabla \Phi$$

$$\frac{\partial \rho E}{\partial t} + \nabla \cdot ((\rho E + p^*)\,\mathbf{v} - \mathbf{B}\,(\mathbf{B} \cdot \mathbf{v})) = -(\rho \mathbf{v} \cdot \nabla \Phi) + \Gamma - \Lambda$$

$$\frac{\partial \rho \varepsilon}{\partial t} + \nabla \cdot (\rho \varepsilon \mathbf{v}) = -(\gamma - 1)\,\rho \varepsilon \nabla \cdot (\mathbf{v}) + \Gamma - \Lambda$$

$$\frac{\partial \mathbf{B}}{\partial t} = \nabla \times (\mathbf{v} \times \mathbf{B})$$

$$\triangle \Phi = 4\pi G \rho$$

the condition of divergency free of magneic field is

$$\nabla \cdot (\mathbf{B}) = 0$$

where $\rho = \sum_i \rho_i$ is density, ρ_i is density of each component of the gaseous mixture, $\mathcal{S}_i$ is formation rate of i-th component of the mixture, $\mathbf{v}$ is velocity, $\mathbf{B}$ is magnetic field, $p = \rho\varepsilon(\gamma - 1)$ is pressure, $\rho\varepsilon$ is internal energy, $p^* = p + \mathbf{B}^2/2$ is full pressure, γ is adiabatic index, $\rho E = \rho\varepsilon + \rho\mathbf{v}^2/2 + \mathbf{B}^2/2$ is full mechanical energy, Φ is gravitational potential, Γ is heating function, Λ is cooling function, G is gravitational constant.

The numerical method for solving the system of gravitational multicomponent magnetic hydrodynamics is based on a combination of operator splitting approach and Godunov method with using of the piecewise-parabolic method on a local stencil. This method consists of the following stages:

- (i) eulerian stage, where the equations for density, impulse, full and internal energy are solved without consideration of advective terms and functions of heating and cooling, but with consideration of work of gravitational force;
- (ii) recomputation of magnetic field with conservation of condition $\nabla \cdot (\mathbf{B}) = 0$ with using of Flux-CT scheme;
- (iii) lagrangian stage, where the advection of density, momentum, full, and internal energy happens;
- (iv) solving of homogeneous differential equations in each cell of computational domain to determine the concentration of the gas mixture;
- (v) consideration of subcell cooling/heating processes;

IOP Conf. Series: Journal of Physics: Conf. Series **894** (2017) 012132 doi:10.1088/1742-6596/894/1/012132

(vi) regularization of numerical solution;

(vii) solving the Poisson equation to evaluate the gravitational potential.

Before proceeding to the detailed description of each stage let us describe two procedures, on which the Eulerian and Lagrangian stages are based – the procedure of building the local parabolas, which will be used in solution of the Riemann problem at each stage; and the procedure of using the fourth-order Runge-Kutta method that used on the Eulerian and Lagrangian stages separately.

2.1. Chemistry

Chemical reactions for the i-th component of the mixture are considered in the following form:

$$\frac{dn_i}{dt} = C_i\left(T, n_j\right) - D_i\left(T, n_j\right) n_i$$

where C_i – speed of construction of i-th component, D_i – speed of destruction of i-th component. To solve such differential equations the scheme of inverse differentiation is used:

$$n_i^{t+\tau} = n_i^t + \tau \frac{C_i - D_i n_i^t}{1 + \tau D_i}$$

We understand that it is, possibly, not the best way and, for example, using the code KROME [31] is more efficient but a similar approach was successfully used in works [32, 33].

2.2. Subgrid physics

To consider the subgrid physics the following equations are solved:

$$\frac{\partial \rho E}{\partial t} = \Gamma - \Lambda$$

$$\frac{\partial \rho \varepsilon}{\partial t} = \Gamma - \Lambda$$

in each cell with using of Euler method for solving the ODE. There is no matter to use complex way of approximation of Runge-Kutta type because the values of heating and cooling functions are constant while the time step in each cell of the computational domain.

3. Chemodynamics of evolution of MHD turbulence of the interstellar medium

The problem of chemodynamics of evolution of MHD turbulence of the interstellar medium was examined in full formulation with consideration of self-gravitation. To do this, the region $[256pc]^3$ with vertical component of magnetic field, uniform initial concentration of atoms $n = 5$ cm^{-3}, initial random perturbation with speed $v_{rms} = 10$ km/s, initial value of plasma parameter $\beta_{th} = 8\pi p_0/B_0^2 = 25$, initial value of turbulent plasma parameter $\beta_{turb} = 8\pi\rho v_{rms}^2/B_0^2 = 25$, alfvenic Mach number $\mathcal{M} = 3.52$, was considered. The following eight reactions from [32], was examined:

(i) Molecular hydrogen formation [34];

(ii) Molecular hydrogen first dissociation [35];

(iii) Molecular hydrogen second dissociation [36];

(iv) Molecular hydrogen photodissociation [32];

(v) Cosmic Ray ionization [32];

(vi) Collision ionization [37];

IOP Conf. Series: Journal of Physics: Conf. Series **894** (2017) 012132 doi:10.1088/1742-6596/894/1/012132

(vii) Radiative recombination [38];

(viii) EI recombination on grains [39].

Effective adiabatic index was used in the following form:

$$\gamma = \frac{5n_H + 5n_e + 7n_{H_2}}{3n_H + 3n_e + 5n_{H_2}}$$

where n_H is the concentration (in cm^{-3}) of hydrogen, n_e is the concentration of electrons, n_{H_2} is the concentration of molecular hydrogen. Behaviour of concentration of different forms of hydrogen, which mostly was ionized, and molecular hydrogen was a several thousandth of a percent (see fig. 1), was modeled with using of code ChemPAK [40] for specific values of temperature $T = 1000$ K and $T = 5000$ K, and also for specific concentration of atomic neutral hydrogen. In numerical experiment concentrations behaved in a similar way.

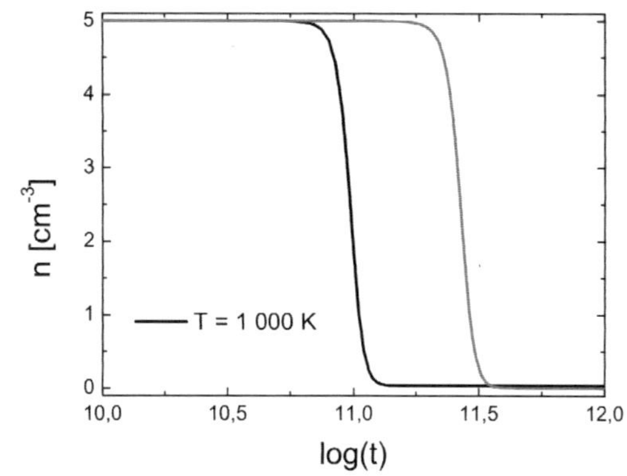

Figure 1. Behaviour of concentrations with temperature of region $T = 1000$ K and $T = 5000$ K. Rapid process of ionization within the time period $10^{11} < t < 10^{11.5}$ s for both temperatures is taking a place.

The results of the simulation are shown in figure 2. The formation of some small waves of density in the moments $t = 10$ and $t = 14$ Myr are visible in the figure. However, then the clusterization process is accelerated, that leads to the formation of clouds. Of course, we couldn't say is it possible to simulate the most known part of Carina Nebula, but, in our opinion, some kind of finger-like formations were obtained during the simulation.

Also dependence of the Alfven speed on the gas density (see fig. 3a), and dependence of cosine of an angle of collinearity between velocity and vector of magnetic field on the gas density (see fig. 3b). It is clear from the figures, that for Alfven Mach number the correlation $\mathcal{M} \sim n^2$ is traced (it is shown with white line), and the most part of the cloud $n > 10cm^{-3}$ is in the over-Alfven region (see fig. 3a). The reason of emergence of such mode is in the magnetic turbulent interstellar medium in trans Alfven mode $\mathcal{M} \sim 1$ with $n \sim 1$. With such densities (see fig. 3b) contours of the cosine of an angle of collinearity between velocity and vector of magnetic field forms saddle-like structure, which means that the compression is along force lines of magnetic field. Then, further increase of mass and density of the cloud happening due to the influence of self-gravitation. In its turn, in dense clouds, turbulence is just over Alfven with Mach number $\mathcal{M} > 100$.

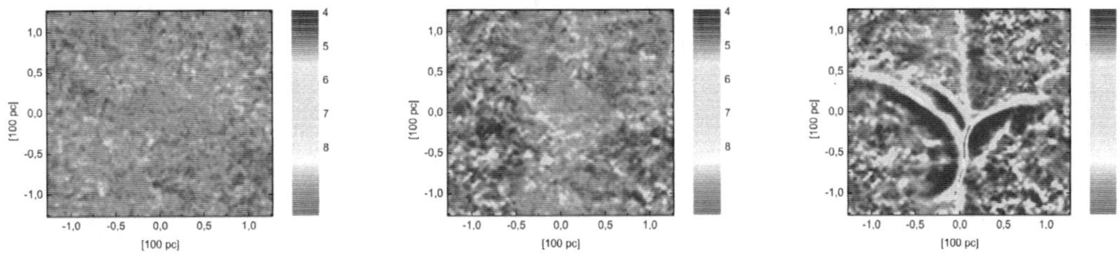

Figure 2. Problem of chemodynamics of evolution of MHD turbulence of interstellar medium. Concentration of the gas in cm^{-3} in moments of time $t = 10$ Myr (left), $t = 14$ Myr (middle), $t = 15$ Myr (right) are given in figure. After the process of hydrogen ionization, the cloud structures are formed in the numerical experiment (512^3 grid size was used).

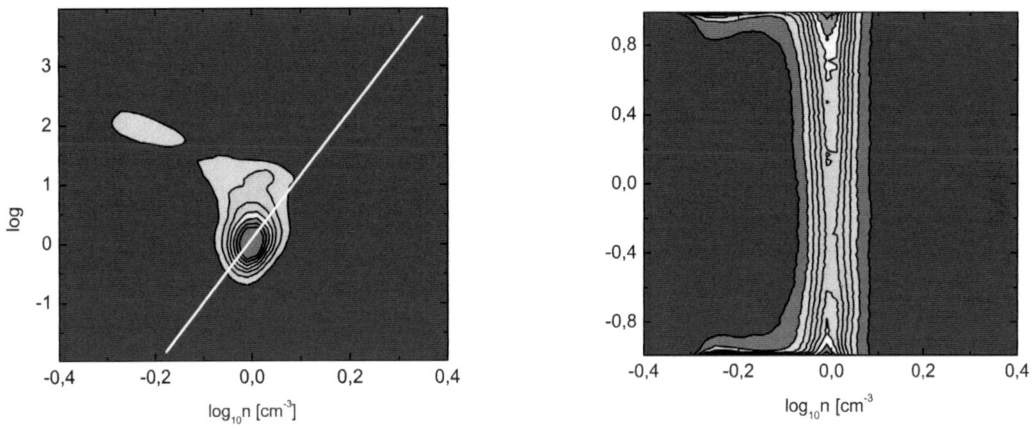

(a) (b)

Figure 3. The problem of chemodynamics of evolution of MHD turbulence of the interstellar medium. Dependence of Alfven speed on gas density (a), and dependence of cosine of an angle of collinearity between velocity and vector of magnetic field on gas density (b) are shown in figures.

4. Conclusion

In the paper, the new code for numerical simulation of magnetic hydrodynamical astrophysical flows with consideration of chemical reactions is given. New original low-dissipation numerical method, based on a combination of operator-splitting approach and piecewise-parabolic method on a local stencil, for solving equations of magnetic-hydrodynamics is described. Chemodynamics of hydrogen during the process of turbulent formation of molecular clouds were modeled.

This work is a part of the common joint "Hydrodynamical Numerical Modelling of Astrophysical Flow at the Peta- and Exascale", developed by our team at the Siberian Supercomputer Center ICMMG SB RAS.

Acknowledgments

The research work was supported by the Grant of the President of Russian Federation for the support of young scientists number MK – 1445.2017.9, RFBR grant 15-01-00508, and 16-07-00434.

Refences

[1] Bruggen M 2013 *Astronomische Nachrichten* **334**(6) 543-547
[2] Pfrommer C and Dursi J 2010 *Nature Physics* **6** 520-526
[3] Xu H et al 2012 *The Astrophysical Journal* **759** 40
[4] Fletcher A, Beck R, Shukurov A, Berkhuijsen E and Horellou C 2011 *Monthly Notices of the Royal Astronomical Society* **412**(4) 2396-2416
[5] Pakmor R and Springel V 2013 *Monthly Notices of the Royal Astronomical Society* **432** (1) 176-193
[6] Perez and Boldyrev S 2010 *The Astrophysical Journal Letters* **710** L63-L66
[7] Mason J, Perez J, Cattaneo F and Boldyrev S 2011 *The Astrophysical Journal Letters* **735** L26
[8] Beresnyak A 2014 *The Astrophysical Journal Letters* **784** L20
[9] McKee C, Li P and Klein R 2010 *The Astrophysical Journal* **720** 1612-1634
[10] Federrath C and Klessen R 2012 *The Astrophysical Journal* **761** 156
[11] Kritsuk A et al 2011 *The Astrophysical Journal* **737** 13
[12] Galtier S and Buchlin E 2007 *The Astrophysical Journal* **656** 560-566
[13] Mendis D and Horanyi M 2014 *The Astrophysical Journal* **794** 14
[14] Ogino T, Walker R and Ashour-Abdalla M 1988 *Journal of Geophysical Research* **93** (A9) 9568-9576
[15] Johnstone C et al 2015 *Astronomy & Astrophysics* **577** A122
[16] Shematovich V, Bisikalo D, Barabash S and Stenberg G 2014 *Solar System Research* **48** (5) 317-323
[17] Villaver E, Manchado A and Garcia-Segura G 2012 *The Astrophysical Journal* **748** 94
[18] Tutukov A, Lazareva G and Kulikov I 2011 *Astronomy Reports* **55** (9) 770-783
[19] Kulikov I, Lazareva G, Snytnikov A and Vshivkov V 2009 *Lecture Notes in Computer Science* **5698** 414-422
[20] Vshivkov V, Lazareva G, Snytnikov A, Kulikov I and Tutukov A 2011 *The Astrophysical Journal Supplement Series* **194** 47
[21] Vshivkov V, Lazareva G, Snytnikov A, Kulikov I and Tutukov A 2011 *Journal of Inverse and Ill-posed Problems* **19** 151-166
[22] Godunov S and Kulikov I 2014 *Computational Mathematics and Mathematical Physics* **54** (6) 1012-1024
[23] Kulikov I 2014 *The Astrophysical Journal Supplements Series* **214** 12
[24] Kulikov I, Chernykh I, Snytnikov A, Glinskiy B and Tutukov A 2015 *Computer Physics Communications* **186** 71-80
[25] Kulikov I, Chernykh I, Glinskiy B, Weins D and Shmelev A 2015 *Proceedings - 2015 IEEE/ACM 15th International Symposium on Cluster, Cloud, and Grid Computing, CCGrid 2015* 1131-1134
[26] Kulikov I, Chernykh I, Snytnikov A, Protasov V, Tutukov A and Glinsky B 2014 *In Parallel Programming: Practical Aspects, Models and Current Limitations (ed. M. Tarkov)* 71-116
[27] Kulikov I and Vorobyov E 2016 *Journal of Computational Physics* **317** 318-346
[28] Protasov V and Kulikov I 2015 *Bulletin of the Tomsk Polytechnic University, Geo Assets Engineering* **326** 61-70
[29] Protasov V, Serenko A, Nenashev V, Kulikov I and Chernykh I 2016 *Journal of Physics: Conference Series* **681** 012022
[30] Godunov S, Manuzina Yu and Nazareva M 2011 *Computational Mathematics and Mathematical Physics* **51** 88-95
[31] Grassi T et al 2014 *Monthly Notices of the Royal Astronomical Society* **439** (3) 2386-2419
[32] Glover S and Mac Low M 2007 *The Astrophysical Journal Supplement Series* **169** (2) 239-268
[33] Anninos P, Zhang Y, Abel T and Norman M 1997 *New Astronomy* **2** (3) 209-224
[34] Hollenbach D and McKee C 1979 *The Astrophysical Journal Supplement Series* **41** 555-592
[35] Lepp S and Shull J 1983 *The Astrophysical Journal* **270** 578-582
[36] Martin P, Keogh W and Mandy M 1998 *The Astrophysical Journal* **499** 793-798
[37] Abel T, Anninos P, Zhang Y and Norman M 1997 *New Astronomy* **2** (3) 181-207
[38] Ferland G, Peterson B, Horne K, Welsh W and Nahar S 1992 *The Astrophysical Journal* **387** 95-108
[39] Weingartner J and Draine B 2001 *The Astrophysical Journal* **563** 842-852
[40] Chernykh I, Stoyanovskaya O and Zasypkina O 2009 *Chemical Product and Process Modeling* **4** (4) 3
[41] Draine B 1978 *The Astrophysical Journal Supplement Series* **36** 595-619
[42] Draine B and Bertoldi F 1996 *The Astrophysical Journal* **468** 269-289

Optimal error functional for parameter identification in anisotropic finite strain elasto-plasticity

A V Shutov[1,2], A A Kaygorodtseva[1,2] and N S Dranishnikov[1,2]
[1] Lavrentyev Institute of Hydrodynamics, Novosibirsk, Russia
[2] Novosibirsk State University, Novosibirsk, Russia

E-mail: alexey.v.shutov@gmail.com

Abstract. A problem of parameter identification for a model of finite strain elasto-plasticity is discussed. The utilized phenomenological material model accounts for nonlinear isotropic and kinematic hardening; the model kinematics is described by a nested multiplicative split of the deformation gradient. A hierarchy of optimization problems is considered. First, following the standard procedure, the material parameters are identified through minimization of a certain least square error functional. Next, the focus is placed on finding optimal weighting coefficients which enter the error functional. Toward that end, a stochastic noise with systematic and non-systematic components is introduced to the available measurement results; a superordinate optimization problem seeks to minimize the sensitivity of the resulting material parameters to the introduced noise. The advantage of this approach is that no additional experiments are required; it also provides an insight into the robustness of the identification procedure. As an example, experimental data for the steel 42CrMo4 are considered and a set of weighting coefficients is found, which is optimal in a certain class.

1. Introduction

In the current study, a viscoplastic material model of Shutov and Kreißig is considered [6]. It is suitable for large strain metal forming applications.[1] Within the model, the material properties are described by a set of material parameters. In general, the inverse problem of parameter identification is ill posed. For that reason, some regularized strategies of material parameter identification were presented in [10] for the considered material model. In this work we seek for an optimal parameter identification strategy which provides material parameters least dependent on the measurement errors.

2. Material model of finite strain viscoplasticity

Let us recall the constitutive equations of the model proposed by Shutov and Kreißig (cf. [6]), which accounts for nonlinear isotropic and kinematic hardening. We start with the classical multiplicative decomposition of the deformation gradient $\mathbf{F}$ into the inelastic and elastic parts

$$\mathbf{F} = \hat{\mathbf{F}}_{\mathrm{e}}\mathbf{F}_{\mathrm{i}}. \tag{1}$$

Next, following [3], two nested multiplicative decompositions are introduced

$$\mathbf{F}_{\mathrm{i}} = \mathbf{F}_{1\mathrm{e}}\mathbf{F}_{1\mathrm{i}}, \quad \mathbf{F}_{\mathrm{i}} = \mathbf{F}_{2\mathrm{e}}\mathbf{F}_{2\mathrm{i}}, \tag{2}$$

[1] Further extensions of this model are reported in [11] and references cited therein.

Content from this work may be used under the terms of the Creative Commons Attribution 3.0 licence. Any further distribution of this work must maintain attribution to the author(s) and the title of the work, journal citation and DOI.
Published under licence by IOP Publishing Ltd

where $\mathbf{F}_{1\mathrm{i}}$ and $\mathbf{F}_{2\mathrm{i}}$ are dissipative parts of the inelastic deformation gradient $\mathbf{F}_{\mathrm{i}}$; $\mathbf{F}_{1\mathrm{e}}$ and $\mathbf{F}_{2\mathrm{e}}$ are conservative parts of $\mathbf{F}_{\mathrm{i}}$. The local deformation history is described by the right Cauchy-Green tensor $\mathbf{C} := \mathbf{F}^{\mathrm{T}}\mathbf{F}$. The current state of the material is described by internal variables

$$\mathbf{C}_{\mathrm{i}} := \mathbf{F}_{\mathrm{i}}^{\mathrm{T}}\mathbf{F}_{\mathrm{i}}, \quad \mathbf{C}_{1\mathrm{i}} := \mathbf{F}_{1\mathrm{i}}^{\mathrm{T}}\mathbf{F}_{1\mathrm{i}}, \quad \mathbf{C}_{2\mathrm{i}} := \mathbf{F}_{2\mathrm{i}}^{\mathrm{T}}\mathbf{F}_{2\mathrm{i}}. \tag{3}$$

By definition, these tensors are symmetric and positive definite; due to the incompressibility condition, they are unimodular. Let ψ be the Helmholz free energy per unit mass. Assume

$$\psi = \psi_{\mathrm{el}}(\mathbf{C}\mathbf{C}_{\mathrm{i}}^{-1}) + \psi_{\mathrm{kin1}}(\mathbf{C}_{\mathrm{i}}\mathbf{C}_{i1}^{-1}) + \psi_{\mathrm{kin2}}(\mathbf{C}_{\mathrm{i}}\mathbf{C}_{i2}^{-1}) + \psi_{\mathrm{iso}}(s - s_{\mathrm{d}}), \tag{4}$$

where functions $\psi_{\mathrm{el}}(\mathbf{C}\mathbf{C}_{\mathrm{i}}^{-1})$, $\psi_{\mathrm{kin1}}(\mathbf{C}_{\mathrm{i}}\mathbf{C}_{i1}^{-1})$, $\psi_{\mathrm{kin2}}(\mathbf{C}_{\mathrm{i}}\mathbf{C}_{i2}^{-1})$ and $\psi_{\mathrm{iso}}(s - s_{\mathrm{d}})$ describe the energy storage due to macroscopic elastic strain as well as kinematic and isotropic hardening. The functions ψ_{el}, ψ_{kin1}, and ψ_{kin2} are isotropic. To be definite, we employ the following ansatz:

$$\rho_{\mathrm{R}}\psi_{\mathrm{el}}(\mathbf{A}) = \frac{k}{2}(\ln\sqrt{\det\mathbf{A}})^2 + \frac{\mu}{2}(\mathrm{tr}\overline{\mathbf{A}} - 3), \tag{5}$$

$$\rho_{\mathrm{R}}\psi_{\mathrm{kin1}}(\mathbf{A}) = \frac{c_1}{4}(\mathrm{tr}\overline{\mathbf{A}} - 3), \quad \rho_{\mathrm{R}}\psi_{\mathrm{kin2}}(\mathbf{A}) = \frac{c_2}{4}(\mathrm{tr}\overline{\mathbf{A}} - 3), \tag{6}$$

$$\rho_{\mathrm{R}}\psi_{\mathrm{iso}}(s_{\mathrm{e}}) = \frac{\gamma}{2}(s_{\mathrm{e}})^2, \quad \overline{\mathbf{A}} := (\det\mathbf{A})^{-1/3}\mathbf{A}, \tag{7}$$

where $\mathbf{A}$ is an arbitrary second rank tensor; s_{e} is an arbitrary scalar; k, μ, c_1, c_2, γ are material parameters; ρ_{R} is the mass density in the reference configuration. Using the Coleman-Noll procedure, the second Piola-Kirchhoff stress is computed through

$$\widetilde{\mathbf{T}} = 2\rho_{\mathrm{R}}\frac{\partial\psi_{\mathrm{el}}(\mathbf{C}\mathbf{C}_{\mathrm{i}}^{-1})}{\partial\mathbf{C}}|_{\mathbf{C}_{\mathrm{i}}=const}. \tag{8}$$

Partial backstresses $\widetilde{\mathbf{X}}_1$, $\widetilde{\mathbf{X}}_2$ and the overall backstress $\widetilde{\mathbf{X}}$ are evaluated by

$$\widetilde{\mathbf{X}}_1 = 2\rho_{\mathrm{R}}\frac{\partial\psi_{\mathrm{kin1}}(\mathbf{C}_{\mathrm{i}}\mathbf{C}_{1\mathrm{i}}^{-1})}{\partial\mathbf{C}_{\mathrm{i}}}|_{\mathbf{C}_{1\mathrm{i}}=const}, \quad \widetilde{\mathbf{X}}_2 = 2\rho_{\mathrm{R}}\frac{\partial\psi_{\mathrm{kin2}}(\mathbf{C}_{\mathrm{i}}\mathbf{C}_{2\mathrm{i}}^{-1})}{\partial\mathbf{C}_{\mathrm{i}}}|_{\mathbf{C}_{2\mathrm{i}}=const}, \quad \widetilde{\mathbf{X}} = \widetilde{\mathbf{X}}_1 + \widetilde{\mathbf{X}}_2. \tag{9}$$

The isotropic hardening R is defined by

$$R = \rho_{\mathrm{R}}\frac{\partial\psi_{\mathrm{iso}}(s - s_{\mathrm{d}})}{\partial s}|_{s_{\mathrm{d}}=const}. \tag{10}$$

A viscous overstress f, which depends on the loading rate, is given by

$$f := \mathfrak{F} - \sqrt{\frac{2}{3}}(K + R), \quad \mathfrak{F} := \sqrt{\mathrm{tr}[(\mathbf{C}\widetilde{\mathbf{T}} - \mathbf{C}_{\mathrm{i}}\widetilde{\mathbf{X}})^{\mathrm{D}}]^2}, \tag{11}$$

where K stands for the initial yield stress, $(\cdot)^{\mathrm{D}}$ denotes a deviatoric part, $\mathfrak{F}$ is a driving force of the inelastic flow. The inelastic multiplier λ_{i} equals the rate of the inelastic flow; it is computed using the Perzyna law

$$\lambda_{\mathrm{i}} = \frac{1}{\eta}\Big\langle\frac{f}{k_0}\Big\rangle^m, \quad \langle x\rangle := \max(x, 0), \tag{12}$$

where η and m are viscous parameters; $k_0 = 1$ MPa is used to obtain a non-dimensional quantity in the bracket. The inelastic flow is governed by the following evolution equations

$$\dot{\mathbf{C}}_{\mathrm{i}} = 2\frac{\lambda_{\mathrm{i}}}{\mathfrak{F}}(\mathbf{C}\widetilde{\mathbf{T}} - \mathbf{C}_{\mathrm{i}}\widetilde{\mathbf{X}})^{\mathrm{D}}\mathbf{C}_{\mathrm{i}}, \tag{13}$$

$$\dot{\mathbf{C}}_{1\mathrm{i}} = 2\lambda_{\mathrm{i}}\varkappa_1(\mathbf{C}_{\mathrm{i}}\widetilde{\mathbf{X}}_1)^{\mathrm{D}}\mathbf{C}_{1\mathrm{i}}, \quad \dot{\mathbf{C}}_{2\mathrm{i}} = 2\lambda_{\mathrm{i}}\varkappa_2(\mathbf{C}_{\mathrm{i}}\widetilde{\mathbf{X}}_2)^{\mathrm{D}}\mathbf{C}_{2\mathrm{i}}, \tag{14}$$

$$\dot{s} = \sqrt{\frac{2}{3}}\lambda_{\mathrm{i}}, \quad \dot{s}_{\mathrm{d}} = \frac{\beta}{\gamma}\dot{s}R. \tag{15}$$

Here, $\varkappa_1$, $\varkappa_2$, and β are material parameters; $\dot{(\cdot)}$ is a material time rate. If the initial state is isotropic, undeformed, and stress free, then we may assume the following trivial initial conditions:

$$\mathbf{C}_{\mathrm{i}}|_{t=0} = \mathbf{C}_{1\mathrm{i}}|_{t=0} = \mathbf{C}_{2\mathrm{i}}|_{t=0} = \mathbf{1}, \quad s|_{t=0} = s_{\mathrm{d}}|_{t=0} = 0. \tag{16}$$

This model is thermodynamically consistent, objective and w-invariant.[2] It is free from nonphysical shear oscillations. Robust and efficient numerical algorithms are available [12].

3. Experimental data for the steel 42CrMo4

In the current study we use experimental data reported in [9] for the high-strength steel 42CrMo4. Torsion tests carried out on thin-walled tubular specimens are considered. In Figure 1, the shear stresses are plotted versus the prescribed shear strain for two different loading programs (see [9] for additional details). In the as-received state the material is isotropic. Figure 1 indicates that the material exhibits a distinct Bauschinger effect. Thus, the material model should employ the nonlinear kinematic hardening.

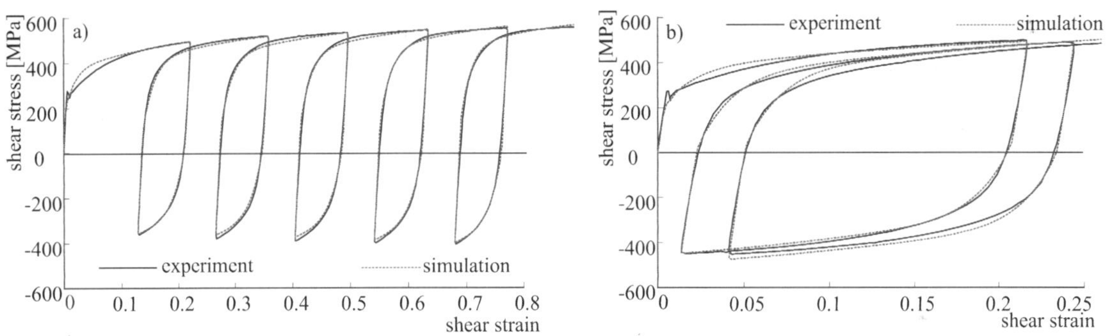

Figure 1. Experimental data and simulation results for torsion of thin-walled tubes made from 42CrMo4 steel: shear stresses versus shear strains for two different loading programs.

4. Classical strategy of material parameter identification

Some of the parameters can be identified by general considerations (cf. [9]). These pre-identified parameters are summarized in Table 1.

Table 1. Pre-identified parameters for the 42CrMo4 steel.

k	μ	η	m	K
135600 MPa	52000 MPa	500000 s	2.26	335 MPa

[2] A general definition of the w-invariance is given in [7]. A proof of the w-invariance for the current model is provided in [8].

Now we need to identify six hardening parameters: γ, β from the Voce isotropic hardening law and c_1, c_2, $\varkappa_1$, $\varkappa_2$ of the Armstrong-Frederick kinematic hardening rule. Since the mentioned hardening mechanisms are coupled, all the hardening parameters need to be identified simultaneously. Following the classical approach, we build an error functional

$$\Phi(\overline{p}) = \Phi(\gamma, \beta, c_1, \varkappa_1, c_2, \varkappa_2) = \sum_{i=1}^{N_{exp}} \omega_i [Exp_i - Mod_i(\overline{p})]^2, \tag{17}$$

where N_{exp} is the number of points on the experimental stress-strain curves, Exp_i are measured shear stresses, Mod_i are the corresponding simulation results (model response), $\omega_i \geq 0$ are the weighting coefficients.[3] For this error functional, an optimal set of parameters is identified:

$$\overline{p}_*(\omega_1, ...\omega_{N_{exp}}) = (\gamma, \beta, c_1, \varkappa_1, c_2, \varkappa_2) = \arg\min \Phi. \tag{18}$$

The minimum of Φ is found using the gradient-based Levenberg-Marquard algorithm (cf. [4]).

Within the classical identification procedure, all data points are equally weighted ($\omega_i \equiv 1$). The identified hardening parameters are presented in Table 2 and the corresponding simulation results are shown in Figure 1. One drawback of this approach is that it says nothing about the influence of the measurement errors on the resulting parameters.

5. Strategy with optimal choice of weighting coefficients

Let $p^*(\omega_i)$ be the solution of the optimization problem (17)—(18) with weighting coefficients ω_i, $i = 1, 2, ..., N_{exp}$. In this section we formulate a problem of finding the weighting coefficients ω_i which are optimal in a certain sense.

5.1. Introduction of noise to experimental data

Let us replace the experimental data Exp_i by the noisy data $Exp_i + Noise_i$, where

$$Noise_i = Noise_i^{nonsyst} + \varepsilon_1 Mode1_i + \varepsilon_2 Mode2_i. \tag{19}$$

Here, $Noise_i^{nonsyst} \in \Phi(0, \sigma)$ is a non-systematic error; ε_1, $\varepsilon_2 \in \Phi(0, \sigma)$ and vectors $Mode1_i$ and $Mode2_i$ are known a priory and correspond to systematic measurement errors (see Figure 2).[4]

5.2. Analytical expression for material parameters corresponding to noisy data

First, consider the error functional (17) built from the noise-free data. The minimizing set of parameters and the Jacobian are

$$\overline{p}_*(\omega_1, ...\omega_{N_{exp}}) = \arg\min \Phi, \quad J = \frac{\partial Mod(\overline{p})}{\partial \overline{p}}|_{\overline{p}_*}. \tag{20}$$

Next, the model prediction is linearized near p_*:

$$Mod(\overline{p}) \approx Mod^{lin}(\overline{p}) = Mod(\overline{p}_*) + J(\overline{p} - \overline{p}_*). \tag{21}$$

The error functional, which corresponds to the j-th realisation of (19), takes the form

$$\Phi^{(j)}(\overline{p}) = \sum_{i=1}^{N_{exp}} \omega_i [Exp_i + Noise_i^{(j)} - Mod_i^{lin}(\overline{p}_*)]^2. \tag{22}$$

[3] Obviously, the error functional (17) is related to the L_2 norm of the residual. The use of other norms for the identification of material parameters was discussed in [5].

[4] The idea of introducing artificial errors in order to estimate the level of uncertainty was considered, among others, in [2]. Other engineering applications are given in [1].

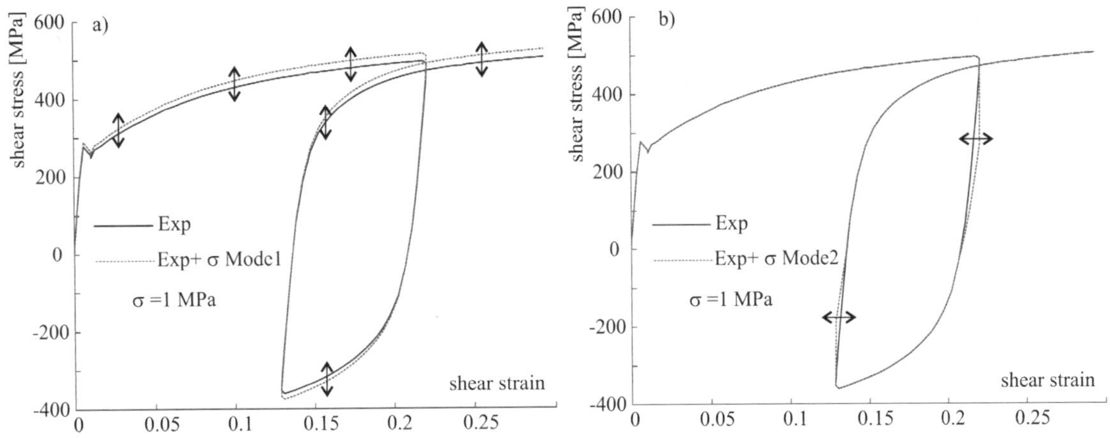

Figure 2. Various modes of measurement errors: error due to false measurement of stresses (a); error due to a play in the testing assembly, which appears after load reversal (b).

Denote by $\overline{p}^{(j)}$ the solution corresponding to the j-th realisation of the stochastic model:

$$\overline{p}^{(j)} = \arg\min \Phi^{(j)}. \tag{23}$$

The set of vectors $\overline{p}^{(j)}$ for a sufficiently large number of realisations will be referred to as "cloud of parameters". The size of the "cloud" is a measure for the parameter dependence on the measurement errors. In the following we seek for the weighting factors w_i, which minimize the size of the "cloud". The advantages of this problems statement are as follows: (i) There is no need for additional experiments. (ii) The form of the "cloud" provides insight into the correctness of the parameter identification problem.

A closed form expression for $\overline{p}$ pertaining to the linearized model response Mod^{lin} is available:

$$\overline{p} = -\frac{1}{2}(J^{\mathrm{T}}W^2J)^{-1}(J^{\mathrm{T}}W^2A + AW^2J), \tag{24}$$

where $A := Mod(\overline{p}_*) - J\overline{p}_* - Exp - Noise, \quad W = \mathrm{diag}(\omega_1, \omega_2, ..., \omega_{N_{exp}})$.

Let us show that for the noise-free data ($Noise = 0$) the analytical solution (24) coincides with $\overline{p}_*$. Toward that end we consider the original problem (17)—(18) (for simplicity assume that all the weights are equal to 1):

$$\Phi(\overline{p}) = ||Mod(\overline{p}) - Exp||^2, \quad \overline{p}_* = \arg\min \Phi \quad \Rightarrow \quad \frac{\partial \Phi}{\partial \overline{p}}|_{\overline{p}_*} = 0. \tag{25}$$

$$\frac{\partial \Phi}{\partial \overline{p}} = 2(Mod(\overline{p}) - Exp)^{\mathrm{T}} \frac{\partial Mod(\overline{p})}{\partial \overline{p}} = 2(Mod(\overline{p}) - Exp)^{\mathrm{T}} J, \quad Mod_* = Mod(\overline{p}_*), \tag{26}$$

$$2(Mod_* - Exp)^{\mathrm{T}} J = 0. \tag{27}$$

On the other hand, for the linearized problem (22) with $Noise = 0$ we have

$$\Phi^{lin}(\overline{p}) = ||Mod_* + J(\overline{p} - \overline{p}_*) - Exp||^2, \quad \Phi^{lin}(\overline{p}_*) = \Phi(\overline{p}_*). \tag{28}$$

Recalling (27), we arrive at

$$\frac{\partial \Phi^{lin}}{\partial \overline{p}}|_{\overline{p}_*} = 2(Mod_* - Exp)^{\mathrm{T}} J = 0. \tag{29}$$

Thus, in the noise-free case ($Noise = 0$), the linearized problem (29) has the same solution $\overline{p}_*$ as the original problem (27). This justifies the linearization of Mod.

Let N_{noise} be the number of realizations of (19). The size of the "cloud" is defined as

$$Size(\omega_1, ..., \omega_{N_{exp}}) := \frac{1}{N_{noise}} \sum_{j=1}^{N_{noise}} \sqrt{\sum_{k=1}^{6} \left(\frac{p_k^{(j)}}{p_k^*} - 1 \right)^2}, \quad \overline{p}_* = \overline{p}_*(\omega_1, ... \omega_{N_{exp}}). \tag{30}$$

5.3. Strategies with non-trivial weighting coefficients

The material under consideration exhibits a clear Bauschinger effect. Thus, on the stress-strain curve there is a zone of a smooth re-plastification, which we call "plastic knee", Figure 3. In other words, the "knee" corresponds to the transient hardening, occurring shortly after load reversals. Experimental data near the "knee" are very important for identification of parameters of kinematic hardening. This motivates an identification strategy where the experimental data near the "knee" are given a larger weight than the remaining experimental data.

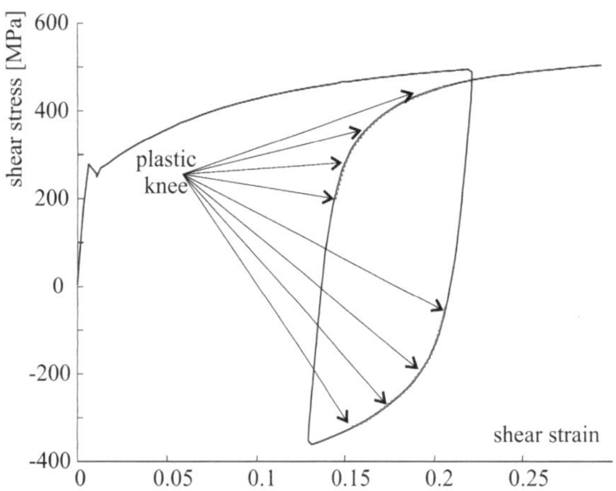

Figure 3. A domain called "plastic knee" on the stress-strain curve.

By "knee 9/4" and "knee 4" denote identification strategies, where the weighting coefficients in the "knee" region equal $\omega_i = 9/4$ and $\omega_i = 4$, respectively. The identified material parameters for different strategies are summarized in Table 2. Computations of the size of the "cloud" (cf. Table 3) using $N_{noise} = 1000$ realisations reveal that the strategy "knee 4" yields results, which are least sensitive to the introduced noise.

Table 2. Identified hardening parameters using different strategies.

parameter	classical ($\omega_i \equiv 1$)	"knee 4"	"knee 9/4"
γ	141.18 MPa	165.29 MPa	157.81 MPa
β	0.0375 [-]	0.2569 [-]	0.1832 [-]
c_1	1692.6 MPa	1985.8 MPa	1875.7 MPa
c_2	23255.1 MPa	23386.9 MPa	23481.1 MPa
$\varkappa_1$	0.0038419 1/MPa	0.0038660 1/MPa	0.0038567 1/MPa
$\varkappa_2$	0.0045175 1/MPa	0.0047926 1/MPa	0.0046769 1/MPa

Table 3. Size of the "parameter cloud" for $\sigma = 1$ MPa.

	classical	"knee 4"	"knee 9/4"
Size	0.3106	0.0739	0.1042

6. Conclusion

A new method is suggested, which allows one to choose optimal strategies of material parameter identification. The method is based on the two-scale approach: on the lower scale, the material parameters are identified by minimizing the error functional Φ; on the upper scale, optimal weighting coefficients are found by minimizing the size of the "parameter cloud". For the material under consideration, an optimal identification strategy is found, among the three considered possibilities. As an alternative to (19), other stochastic models of noise can be used depending on the specific application. In the follow-up research, a broader class of identification strategies will be analyzed. The method is especially promising when experiments of different types are to be efficiently combined within a single error functional.

Acknowledgments

The financial support provided by the RFBR (grant number 17-08-01020) is acknowledged.

References

[1] Doltsinis I 1999 *Stochastic Analysis of Multivariate Systems in Computational Mechanics and Engineering* (Barcelona: CIMNE)

[2] Harth T, Schwan S, Lehn J and Kollmann F G 2004 Identification of material parameters for inelastic constitutive models: statistical analysis and design of experiments *International Journal of Plasticity* **20** 1403–1440

[3] Lion A 2000 Constitutive modelling in finite thermoviscoplasticity: a physical approach based on nonlinear rheological elements *International Journal of Plasticity* **16** 469–494

[4] Lourakis M I 2005 A brief description of the Levenberg-Marquardt algorithm implemented by levmar *Foundation for Research and Technology* **4** 1–6.

[5] Seibert T, Lehn J, Schwan S, Kollmann F G 2000 Identification of material parameters for inelastic constitutive models: Stochastic simulations for the analysis of deviations *Continuum Mechanics and Thermodynamics* **12** 95–120

[6] Shutov A V and Kreißig R 2008 Finite strain viscoplasticity with nonlinear kinematic hardening: Phenomenological modeling and time integration *Computer Methods in Applied Mechanics and Engineering* **197** 2015–2029

[7] Shutov A V and Ihlemann J 2014 Analysis of some basic approaches to finite strain elasto-plasticity in view of reference change *International Journal of Plasticity* **63** 183–197

[8] Shutov A V, Pfeiffer S and Ihlemann J 2012 On the simulation of multi-stage forming processes: invariance under change of the reference configuration *Materialwissenschaft und Werkstofftechnik* **43**(7) 617–625

[9] Shutov A V, Kuprin C, Ihlemann J, Wagner M F-X and Silbermann C 2010 Experimentelle Untersuchung und numerische Simulation des inkrementellen Umformverhaltens von Stahl 42CrMo4 *Materialwissenschaft und Werkstofftechnik* **41** 765–775

[10] Shutov A V and Kreißig R 2010 Regularized strategies for material parameter identification in the context of finite strain plasticity *Technische Mechanik* **30**(1-3) 280–295

[11] Shutov A V, Silbermann C B and Ihlemann J 2015 Ductile damage model for metal forming simulations including refined description of void nucleation *International Journal of Plasticity* **71** 195–217

[12] Shutov A V 2016 Efficient implicit integration for finite-strain viscoplasticity with a nested multiplicative split *Computer Methods in Applied Mechanics and Engineering* **306**(1) 151–174

IOP Conf. Series: Journal of Physics: Conf. Series **894** (2017) 012134 doi:10.1088/1742-6596/894/1/012134

Thermodynamics of micropolar fluids with variable concentration of polar particles

V V Neverov and V V Shelukhin
Lavrentyev Institute of Hydrodynamics, Novosibirsk, Russia
Novosibirsk State University, Novosibirsk, Russia

E-mail: NeverovVladim@gmail.com, shelukhin@hydro.nsc.ru

Abstract. In this paper we derive equations of micropolar fluid with variable concentration of polar particles. Such fluids have non-symmetrical stress tensor and exhibit micro-rotations. To illustrate the derived equations, we consider a pressure driven flow between two parallel planes.

1. Introduction

There are fluids which exhibit microrotational effects and microrotational inertia. Examples are animal blood, cements, cutting transport by drilling mud, and granular flows. Recent examples concern the proppant flow within the hydrofracture [1]. One can tackle such fluids by applying the notion of the Cosserat continua. The theory of micro-polar fluids goes back to [2, 3, 4] where gyration tensor, inertial spin, conservation of micro-inertia and objectivity of micro-deformation rate tensors are derived and discussed. For an overview of developed theories we refer the reader to [1, 5, 6, 7, 8, 9, 10] Within this theory, each material point is treated like a rigid body. In the theory of classical fluids, the local stress state is characterized completely by the Cauchy stress tensor T. To characterize stresses in the micro-polar fluid, one should take also into account the couple stress tensor N. Local deformations are characterized also by two tensors which allow to calculate the velocity gradient and the spin gradient.

In the present paper, we derive a mathematical model allowing for variable concentration of polar particles. To this end, we apply the basic thermodynamic principles and formulate conservation laws and constitutive equations in such a way that they agree with the non-negative sign of entropy production. Earlier, thermodynamics of the micropolar fluid was studied without effects of convection and inertia [11]. We emphasize that the concentration flux obeys a generalized Fick law, i. e. it depends not only on gradients of the functions like concentration, temperature, and pressure, but on the particle gyration as well.

To test our model, we address the so-called "tubular pinch effect" of Ségre and Silberberg [12] showing that for the laminar flow of neutrally buoyant dilute suspension of rigid spheres through a circular tube, particles tend to migrate towards a concentric annular region with the mean radius $r_m = 0.6\,R$, where R is the tube radius. To this end, we apply the equations derived to study one-dimensional steady flows in a channel driven by a pressure gradient for different constitutive parameters and discover that the Ségre-Silberberg effect is genuinely nonlinear one since it disappears if the gyration coefficient in the generalized Fick law vanishes.

The paper is organized as follows. First, we recall thermodynamics of classical compressible fluids without dissipation. Next, we consider thermodynamics of compressible micropolar fluids

Content from this work may be used under the terms of the Creative Commons Attribution 3.0 licence. Any further distribution of this work must maintain attribution to the author(s) and the title of the work, journal citation and DOI.
Published under licence by IOP Publishing Ltd

IOP Conf. Series: Journal of Physics: Conf. Series **894** (2017) 012134 doi:10.1088/1742-6596/894/1/012134

without and with dissipation. The equations derived are applied to one-dimensional steady flows between two parallel planes.articularly, we analyze the Ségre-Silberberg effect.

2. Thermodynamics of micropolar fluid

First, we remind classical conservation laws for fluids without dissipation:

$$\rho_t + \operatorname{div}\mathbf{j} = 0, \quad \mathbf{j}_t + \operatorname{div}\Pi = 0, \quad \eta_t + \operatorname{div}(\eta\mathbf{v}) = 0, \tag{1}$$

where

$$\mathbf{j} = \rho\mathbf{v}, \quad \Pi = p\mathrm{I} + \mathbf{j}\otimes\mathbf{v}. \tag{2}$$

Here, $\rho$is the density, p is the pressure, $\mathbf{v}$ is the velocity, η is the specific entropy, I is the identical matrix. The symbol $\otimes$ means tensor product of two vectors, $(\mathbf{a}\otimes\mathbf{b})_{ij} = a_i b_j$. Given a matrix A, the vector div$A$ is defined by the formula $(\operatorname{div}A)_i = \partial A_{ij}/\partial x_j$.

Let $E_0 = E_0(\rho,\eta)$ be the internal energy, then we can write the Gibbs law as follows:

$$dE_0 = \theta d\eta + \mu d\rho\,. \tag{3}$$

One can define the absolute temperature θ and the chemical potential μ as derivatives:

$$\theta = \frac{\partial E_0}{\partial\eta}, \quad \mu = \frac{\partial E_0}{\partial\rho}.$$

The thermodynamic pressure is given by the expression [13, 14]

$$p = -E_0 + \theta\eta + \mu\rho. \tag{4}$$

It follows from (1) - (4) that total energy $E = E_0 + \rho\mathbf{v}^2/2$ satisfies the conservation law

$$E_t + \operatorname{div}\mathbf{Q} = 0, \tag{5}$$

where the energy flux is equal to

$$\mathbf{Q} = \mathbf{j}\frac{|\mathbf{v}|^2}{2} + \eta\theta\mathbf{v} + \rho\mu\mathbf{v}. \tag{6}$$

Our goal is to generalize (1)-(6) for micropolar fluids with dissipation.

We remind that in Cosserat's continuum each material point with Lagrange coordinates $(t,\boldsymbol{\xi})$ is characterized by the position vector $\mathbf{x}(t,\boldsymbol{\xi})$ and three orthogonal director - vectors $\mathbf{d}_i(t,\boldsymbol{\xi})$, $i = 1,2,3$ whose orientation is controlled by the orthogonal tensor $\mathrm{Q}(t,\boldsymbol{\xi})$:

$$\mathbf{d}_i(\boldsymbol{\xi},t) = \mathrm{Q}(\boldsymbol{\xi},t)\langle\mathbf{d}_i(\boldsymbol{\xi},0)\rangle, \quad \mathrm{Q}\cdot\mathrm{Q}^* = \mathrm{Q}^*\cdot\mathrm{Q} = I.$$

With the use of the material derivative $\dot{\mathrm{Q}}(x,t)$, one can define the skew-symmertic gyration tensor

$$\Omega(\mathbf{x},t) = \dot{\mathrm{Q}}(\mathbf{x},t)\cdot\mathrm{Q}^*(\mathbf{x},t) \equiv \mathrm{Q}_t(t,\boldsymbol{\xi})\cdot\mathrm{Q}^*(t,\boldsymbol{\xi}),$$

which enjoys the formula

$$\Omega\langle\mathbf{h}\rangle = \mathbf{w}\times\mathbf{h}, \quad \forall\,\mathbf{h}\in\mathbb{R}^3, \quad (\Omega\langle\mathbf{h}\rangle)_i \equiv \Omega_{ij}h_j,$$

where $\mathbf{w}(\mathbf{x},t)$ - gyration vector,

$$2\mathbf{w} = \mathbf{e}_i\times\Omega\langle\mathbf{e}_i\rangle = \boldsymbol{\varepsilon} : \Omega^*.$$

Here $\{\mathbf{e}_i\}_1^3$ is any orthogonal basis in $\mathbb{R}^3$ and ε is third order Levi Chivita tensor

$$\begin{aligned}\varepsilon\langle\mathbf{a},\mathbf{b},\mathbf{c}\rangle &= \mathbf{a}\cdot(\mathbf{b}\times\mathbf{c}), \quad \mathbf{e}_i\times\mathbf{e}_j=\varepsilon_{sij}\mathbf{e}_s,\\ \varepsilon_{sij} &\equiv \varepsilon\langle\mathbf{e}_s,\mathbf{e}_i,\mathbf{e}_j\rangle, \quad (\varepsilon:\Omega)_i\equiv\varepsilon_{ijk}\Omega_{jk}.\end{aligned}$$

Scalar product $A:B$ for two 3×3 matrices A and B is defined by $A:B=A_{ij}B_{ij}$.

We assume that all particles are balls. We denote by Θ the microinertia tensor which satisfies the equation [5]

$$\frac{d}{dt}\Theta=\Omega\Theta-\Theta\Omega.$$

For balls, $\Theta=J\cdot I$, where J is the itertia constant.

Naturally, for Cosserat continuum without dissipation, microrotation of two different particles do not interact. Hence,

$$\frac{d\mathbf{w}}{dt}\equiv\mathbf{w}_t+\nabla\mathbf{w}\langle\mathbf{v}\rangle=0, \quad (\rho c)_t+\operatorname{div}(\rho c\mathbf{v})=0, \quad (\nabla\mathbf{w})_{ij}=\frac{\partial w_i}{\partial x_j}; \tag{7}$$

here c is the mass concentration of polar particles.

We use equations (1) to write system (7) as the conservation laws

$$J(\rho c\mathbf{w})_t+\operatorname{div}M=0, \quad (\rho c)_t+\operatorname{div}(\rho c\mathbf{v})=0, \tag{8}$$

where $M=\rho Jc\mathbf{w}\otimes\mathbf{v}$.

Equations (1)-(6), (8) together constitute a model of micropolar fluid without dissipation:

$$\rho_t+\operatorname{div}\mathbf{j}=0, \quad \mathbf{j}_t+\operatorname{div}\Pi=0, \quad \eta_t+\operatorname{div}(\eta\mathbf{v})=0, \tag{9}$$

$$J(\rho c\mathbf{w})_t+\operatorname{div}M=0, \quad (\rho c)_t+\operatorname{div}(c\mathbf{j})=0. \tag{10}$$

Let $E_0(\rho,\eta,c)$ be the internal energy. Starting from the Gibbs law

$$dE_0=\theta d\eta+\mu d\rho+zd(\rho c)\,. \tag{11}$$

one can define chemical potentials μ,z and temperature θ as follows

$$\theta=\frac{\partial E_0}{\partial\eta}, \quad \mu=\frac{\partial E_0}{\partial\rho}, \quad z=\frac{\partial E_0}{\partial(\rho c)}.$$

Pressure is given by the formula [14]

$$p=-E_0+\theta\eta+\mu\rho+z\rho c. \tag{12}$$

It follows from (9)-(12) that the total energy

$$E=E_0+\frac{\rho|\mathbf{v}|^2}{2}+\frac{Jc\rho|\mathbf{w}|^2}{2} \tag{13}$$

satisfies the energy equation

$$E_t+\operatorname{div}\mathbf{Q}=0, \tag{14}$$

here

$$\mathbf{Q}=\rho\mathbf{v}\frac{|\mathbf{v}|^2}{2}+\rho\mathbf{v}Jc\frac{|\mathbf{w}|^2}{2}+\eta\theta\mathbf{v}+\rho\mu\mathbf{v}+\rho zc\mathbf{v}, \tag{15}$$

IOP Conf. Series: Journal of Physics: Conf. Series **894** (2017) 012134 doi:10.1088/1742-6596/894/1/012134

To equip the model (9) - (15) with dissipation,we write the conservation laws

$$\rho_t + \operatorname{div}\mathbf{j} = 0, \quad \mathbf{j}_t + \operatorname{div}(\Pi + \Pi_1) = 0, \quad \eta_t + \operatorname{div}\left(\eta\mathbf{v} + \frac{\mathbf{q}}{\theta}\right) = \frac{R}{\theta}, \tag{16}$$

$$J(\rho c\mathbf{w})_t + \operatorname{div}(M + M_1) = \mathbf{f}_w, \quad (\rho c)_t + \operatorname{div}(c\mathbf{j} + \mathbf{l}) = 0, \tag{17}$$

$$E_t + \operatorname{div}(\mathbf{Q} + \mathbf{Q}_1) = 0, \tag{18}$$

with unknown functions $\xi = \{\Pi_1, M_1, \mathbf{l}, \mathbf{q}, \mathbf{f}_w, \mathbf{Q}_1, R\}$. Here R is entropy production and $\mathbf{q}$ is the heat flux. Conservation laws (16)-(18) should agree with the thermodynamic relations (11)-(13). We assume as in [14, 15], that the energy equation (18) results from the system (16)-(18), (11)-(13) In this way we can determine the unknowns ξ.

We calculate E_t using conservation laws (16)-(17) for elimination the time derivatives ρ_t, η_t and etc. We have

$$E_t = \rho_t\frac{|\mathbf{v}|^2}{2} + \mathbf{j}\cdot\mathbf{v}_t + J(\rho c)_t\frac{|\mathbf{w}|^2}{2} + J\rho c\mathbf{w}\cdot\mathbf{w}_t + \theta\eta_t + \mu\rho_t + z(\rho c)_t =$$

$$-\frac{|\mathbf{v}|^2}{2}\operatorname{div}(\rho\mathbf{v}) + \mathbf{v}\cdot[\mathbf{v}\operatorname{div}\mathbf{j} - \operatorname{div}(\Pi + \Pi_1)] - J\frac{|\mathbf{w}|^2}{2}\operatorname{div}(c\mathbf{j} + \mathbf{l}) +$$

$$\mathbf{w}\cdot[-\operatorname{div}(M + M_1) + J\mathbf{w}\operatorname{div}(c\mathbf{j} + \mathbf{l}) + \mathbf{f}_w] + \theta\left(\frac{R}{\theta} - \operatorname{div}(\eta\mathbf{v}) - \operatorname{div}\frac{\mathbf{q}}{\theta}\right) -$$

$$-\mu\operatorname{div}\mathbf{j} - z\operatorname{div}(c\mathbf{j} + \mathbf{l}).$$

With the help of the equality

$$\mathbf{u}\cdot\operatorname{div}A = \operatorname{div}(A^*\langle\mathbf{u}\rangle) - A : \nabla\mathbf{u}, \quad (\nabla\mathbf{u})_{ij} = \frac{\partial u_i}{\partial x_j},$$

we calculate that

$$E_t + \operatorname{div}\left[\mathbf{Q} + \Pi_1^*\langle\mathbf{v}\rangle + M_1^*\langle\mathbf{w}\rangle + \mathbf{q} + \mathbf{l}\left(z - \frac{J|\mathbf{w}|^2}{2}\right)\right] =$$

$$\Pi_1 : \nabla\mathbf{v} + M_1 : \nabla\mathbf{w} + \mathbf{l}\cdot\nabla\left(z - \frac{J|\mathbf{w}|^2}{2}\right) + \frac{\mathbf{q}\cdot\nabla\theta}{\theta} + \mathbf{w}\cdot\mathbf{f}_w + R.$$

Thus, we define R and $\mathbf{Q_1}$ as follows

$$R = -\Pi_1 : \nabla\mathbf{v} - M_1 : \nabla\mathbf{w} - \mathbf{l}\cdot\nabla\left(z - \frac{J|\mathbf{w}|^2}{2}\right) - \frac{\mathbf{q}\cdot\nabla\theta}{\theta} - \mathbf{w}\cdot\mathbf{f}_w,$$

$$\mathbf{Q}_1 = \Pi_1^*\langle\mathbf{v}\rangle + M_1^*\langle\mathbf{w}\rangle + \mathbf{q} + \mathbf{l}\left(z - \frac{J|\mathbf{w}|^2}{2}\right).$$

We choose Π_1, M_1, $\mathbf{q}$, $\mathbf{l}$ and $\mathbf{f}_w$ in such a way that the inequality $R \geq 0$ holds.

Let B_s and B_a be symmetric and skew-symmetric parts of the matrix B:

$$B_s = \frac{B + B^*}{2}, \quad B_a = \frac{B - B^*}{2}, \quad B^d = B - \frac{\operatorname{tr}B}{3}I.$$

It is clear that

$$B = B_s + B_a = (B_s)^d + \frac{\operatorname{tr}B}{3}I + B_a.$$

IOP Conf. Series: Journal of Physics: Conf. Series **894** (2017) 012134 doi:10.1088/1742-6596/894/1/012134

For any skew-symmetric matrix A, one can find unique vector $\mathbf{a}$ such that $A\langle\mathbf{h}\rangle = \mathbf{a}\times\mathbf{h} \quad \forall\mathbf{h}$. One can verify easily that $A_1 : A_2 = 2\mathbf{a_1}\cdot\mathbf{a_2}$ if the linear maps A_i, $i = 1, 2$, are defined by the formulas $A_i\langle\mathbf{h}\rangle = \mathbf{a_i}\times\mathbf{h}$. Particularly,

$$\left(\nabla\mathbf{v} - (\nabla\mathbf{v})^*\right)\langle\mathbf{h}\rangle = \operatorname{rot}\mathbf{v}\times\mathbf{h} \quad \forall\mathbf{h}\in\mathbb{R}^3, \quad \operatorname{rot}\mathbf{v} = -\varepsilon : \nabla\mathbf{v}.$$

We introduce two strain rate tensors $A = \nabla\mathbf{w}$ and $B = \nabla\mathbf{v} - \Omega$. It is clear that one can write skew-symmetric matrix B_a as follows

$$B_a\langle\mathbf{h}\rangle = \left(\frac{\operatorname{rot}\mathbf{v}}{2} - \mathbf{w}\right)\times\mathbf{h}.$$

We chose Π_1, M_1, $\mathbf{q}$ and $\mathbf{f}_w$ in the form

$$\Pi_1 = -2\eta_1 B_s^d - 2\eta_2 B_a - 2\eta_3 \operatorname{tr} B\cdot I, \quad \mathbf{q} = -\varkappa\nabla\theta,$$

$$M_1 = -2\beta_1 A_s^d - 2\beta_2 \operatorname{tr} A\cdot I - 2\beta_3 A_a + M_2, \quad \mathbf{f}_w = 2\eta_2\left(\operatorname{rot}\mathbf{v} - 2\mathbf{w}\right).$$

In general, the viscosities η_i and β_i depend on mass concentration c. Hence,

$$\Pi_1 : \nabla\mathbf{v} + M_1 : \nabla\mathbf{w} + \mathbf{w}\cdot\mathbf{f}_w + \frac{\mathbf{q}\cdot\nabla\theta}{\theta} = -r^2 + M_2 : \nabla\mathbf{w},$$

where

$$r^2 = 2\eta_1\left|B_s^d\right|^2 + 2\eta_2\left|B_a\right|^2 + 2\eta_3\left|\operatorname{tr} B\right|^2 + 2\beta_1\left|A_s^d\right|^2 + 2\beta_3\left|A_a\right|^2 \tag{19}$$

$$+2\beta_2\left|\operatorname{tr} A\right|^2 + \varkappa|\nabla\theta|^2/\theta.$$

We notice that

$$\mathbf{l}\cdot\nabla\mathbf{w}^2 = 2(\mathbf{w}\otimes\mathbf{l}) : \nabla\mathbf{w},$$

hence, the entropy production becomes

$$R = r^2 + \nabla\mathbf{w} : (J\mathbf{w}\otimes\mathbf{l} - M_2) - \mathbf{l}\cdot\nabla z.$$

We denote

$$\mathbf{b} = \mathbf{w} - \frac{\operatorname{rot}\mathbf{v}}{2} \equiv \frac{\varepsilon : B}{2},$$

and choose

$$\nabla z = -\mathbf{l}/\alpha_1 - \alpha_2\mathbf{l}\times\mathbf{b} + \alpha_3\operatorname{rot}\mathbf{w} + \alpha_4\operatorname{rot}\mathbf{w}\times\mathbf{b}, \tag{20}$$

$$M_2 = J\mathbf{w}\otimes\mathbf{l} + \alpha_3\epsilon : \mathbf{l} + \alpha_4\epsilon : (\mathbf{l}\times\mathbf{b}), \tag{21}$$

here $\alpha_i = \alpha_i(c)$ are diffusion coefficients, so

$$R = r^2 + \mathbf{l}^2/\alpha_1 - \alpha_3[\nabla\mathbf{w} : (\epsilon : \mathbf{l}) + \mathbf{l}\cdot\operatorname{rot}\mathbf{w}] \tag{22}$$

$$-\alpha_4\{\nabla\mathbf{w} : [\epsilon : (\mathbf{l}\times\mathbf{b})] + \mathbf{l}\cdot(\operatorname{rot}\mathbf{w}\times\mathbf{b})\} = r^2 + \mathbf{l}^2/\alpha_1.$$

Hence, the conditions

$$\varkappa \geq 0, \quad \eta_i \geq 0, \quad \beta_i \geq 0, \quad \alpha_1 > 0$$

insure the inequality $R \geq 0$.

We introduce the matrix

$$S_b = \alpha_1^{-1} I - \alpha_2\epsilon : \mathbf{b},$$

and write equation (20) in the form

$$\mathbf{l} = S_b^{-1}\langle -\nabla z + \alpha_3 \mathrm{rot}\,\mathbf{w} + \alpha_4 \mathrm{rot}\,\mathbf{w}\times\mathbf{b}\rangle \tag{23}$$

It is a generalization of the classical Fick law [13] in which $\alpha_2 = \alpha_3 = \alpha_4 = 0$.

It results from (11) and (12) that

$$\nabla p = \eta\nabla\theta + \rho\nabla\mu + \rho c\nabla z.$$

We assume that chemical potential μ depends on p, θ and c. Hence,

$$\nabla\mu = \mu_p\nabla p + \mu_\theta\nabla\theta + \mu_c\nabla c.$$

And we can write instead of (23) the following equality

$$\mathbf{l} = -S_b^{-1}\langle \gamma_3\nabla c + \gamma_1\nabla p + \gamma_2\nabla\theta - \alpha_3 \mathrm{rot}\,\mathbf{w} - \alpha_4 \mathrm{rot}\,\mathbf{w}\times\mathbf{b}\rangle, \tag{24}$$

where

$$\nabla z = \gamma_1\nabla p + \gamma_2\nabla\theta + \gamma_3\nabla c,$$

$$\gamma_1 = \frac{1-\rho\mu_p}{\rho c}, \quad \gamma_2 = -\frac{\eta+\rho\mu_\theta}{\rho c}, \quad \gamma_3 = -\frac{\mu_c}{c}.$$

For classical fluids, when $\alpha_2 = \alpha_3 = \alpha_4 = 0$, equation (24) is usually written in the form

$$\mathbf{l} = -\rho D(\nabla c + k_\theta\nabla\theta + k_p\nabla p),$$

here D is the diffusion coefficient [15]. This is why we denote

$$\gamma_1 = \frac{\rho D k_p}{\alpha_1}, \quad \gamma_2 = \frac{\rho D k_\theta}{\alpha_1}, \quad \gamma_3 = \frac{\rho D}{\alpha_1}.$$

Let us introduce a stress tensor T and a couple stress tensor N:

$$T = -pI + S, \quad S = -\Pi_1, \quad N = -M_1 + J\mathbf{w}\otimes\mathbf{l},$$

where S is the viscous part of stress tensor. One can verify that rate of entropy production is given by the formula

$$R = S : B + N : A + \frac{\mathbf{l}^2}{\alpha_1} + \alpha_3\mathbf{l}\cdot(\epsilon : A) + \frac{\alpha_4\mathbf{l}}{2}(\varepsilon : A)\times(\varepsilon : B). \tag{25}$$

We arrive at the following conservation laws:

$$\rho_t + \mathrm{div}\,\rho\mathbf{v} = 0, \tag{26}$$

$$(\rho\mathbf{v})_t + \mathrm{div}(\rho\mathbf{v}\otimes\mathbf{v}) = -\nabla p + \mathrm{div}\,S, \tag{27}$$

$$J\{(\rho c\mathbf{w})_t + \mathrm{div}[\mathbf{w}\otimes(\rho c\mathbf{v} + \mathbf{l})]\} = \mathrm{div}\,N - \epsilon : S, \tag{28}$$

$$(\rho c)_t + \mathrm{div}(\rho c\mathbf{v}) = -\mathrm{div}\,\mathbf{l}. \tag{29}$$

$$\eta_t + \mathrm{div}\left(\eta\mathbf{v} + \frac{\mathbf{q}}{\theta}\right) = \frac{R}{\theta}. \tag{30}$$

We have verified that constitutive laws

$$S = 2\eta_1 B_s^d + 2\eta_2 B_a + 2\eta_3 \mathrm{tr}\,B\cdot I, \tag{31}$$

IOP Conf. Series: Journal of Physics: Conf. Series **894** (2017) 012134 doi:10.1088/1742-6596/894/1/012134

$$N = 2\beta_1 A_s^d + 2\beta_2 \mathrm{tr}\, A \cdot I + 2\beta_3 A_a - \alpha_3 \epsilon : \mathbf{l} - \alpha_4 \epsilon : (\mathbf{l} \times \mathbf{b}), \tag{32}$$

$$\mathbf{l} = -S_b^{-1} \langle \gamma_3 \nabla c + \gamma_1 \nabla p + \gamma_2 \nabla \theta - \alpha_3 \mathrm{rot}\, \mathbf{w} - \alpha_4 \mathrm{rot}\, \mathbf{w} \times \mathbf{b} \rangle \tag{33}$$

agree with the inequality $R \geq 0$ if

$$\eta_i \geq 0, \quad \beta_i \geq 0, \quad \alpha_1 > 0.$$

Remark 1. The micropolar effect dissapears if $c = 0$ this is why $\eta_2|_{c=0} = 0$. The concentration flux $\mathbf{l}$ also disappears if $c = 0$ and $c = 1$. Hence α_3, α_4 and k_p should be equal to zero if $c = 0$ and $c = 1$.

To complete the model (26)-(33), we should add the thermodynamic relations

$$dE_0 = \theta d\eta + \mu d\rho + z d(\rho c), \quad \theta = \frac{\partial E_0}{\partial \eta}, \quad \mu = \frac{\partial E_0}{\partial \rho}, \quad z = \frac{\partial E_0}{\partial (\rho c)}. \tag{34}$$

Pressure is defined by (12). The heat flux $\mathbf{q}$ and the rate of entropy production R are defined by Fourier law $\mathbf{q} = -\varkappa \nabla \theta$ and the formula 25 respectively.

Remark 2. In the theory of micro-polar fluids, the couple stress tensor N depends on the rate of strain tensors B and A. In the case of variable concentration, N depends on $\mathbf{l}$ also. It is proved in Appendix, that the variables $\mathbf{l}$, $\mathbf{b}$, and N are objective. Observe, that objectivity of ∇c results from objectivity of the scalar function c.

3. One-dimensional flows

We consider the flow between two parallel plains in the x direction under the hypothesis that $p_x = \mathrm{const} < 0$, $p_y = p_z = 0$. Under the assumption of incompressibility, $\rho = \mathrm{const}$, the mass conservation law takes the form of div $= 0$ and is identically satisfied in the one-dimensional case. The velocity vector takes the form $\mathbf{v} = (v(y,t), 0, 0)^T$, where y is the transverse coordinate $(-H < y < H)$ and t is the time variable. Here and in what follows, $\mathbf{a}^T$ is the transposition of the vector $\mathbf{a}$. Gyration vector is written as $\mathbf{w} = (0, 0, w(y,t))^T$ and concentration flux is written as $\mathbf{l} = (l_1(y,t), l_2(y,t), 0)^T$. In this case we have

$$\mathrm{rot}\, \mathbf{v} = (0, 0, -v_y)^T, \quad \mathrm{rot}\, \mathbf{w} = (w_y, 0, 0)^T, \quad \mathbf{b} = (0, 0, b)^T, \quad b = w + v_y/2,$$

$$\mathbf{b} \times \mathbf{l} = (-b l_2, b l_1, 0)^T,$$

$$B = \begin{pmatrix} 0 & v_y + w & 0 \\ -w & 0 & 0 \\ 0 & 0 & 0 \end{pmatrix}, \quad A = \begin{pmatrix} 0 & 0 & 0 \\ 0 & 0 & 0 \\ 0 & w_y & 0 \end{pmatrix}.$$

We consider flows without heating. By projecting equation (24) on x and y axes we get

$$\alpha_1^{-1} l_1 + b\alpha_2 l_2 = -\gamma_1 p_x + \alpha_3 w_y,$$

$$-\alpha_2 b l_1 + \alpha_1^{-1} l_2 = -\gamma_3 c_y - \alpha_4 b w_y.$$

or equivalently

$$l_1 = \frac{(-\gamma_1 p_x + \alpha_3 w_y)/\alpha_1 + b\alpha_2(\gamma_3 c_y + \alpha_4 b w_y)}{\alpha_1^{-2} + b^2 \alpha_2^2}, \tag{35}$$

$$l_2 = -\frac{-b\alpha_2(-\gamma_1 p_x + \alpha_3 w_y) + (\gamma_3 c_y + \alpha_4 b w_y)/\alpha_1}{\alpha_1^{-2} + b^2 \alpha_2^2}.$$

In one-dimensional case, the stress tensor and the couple stress tensor take the form

$$S = \begin{pmatrix} 0 & S_{12} & 0 \\ S_{21} & 0 & 0 \\ 0 & 0 & 0 \end{pmatrix}, \quad N = \begin{pmatrix} 0 & 0 & N_{13} \\ 0 & 0 & N_{23} \\ N_{31} & N_{32} & 0 \end{pmatrix},$$

where

$$S_{12} = (\eta_1 + \eta_2)v_y + 2\eta_2 w, \quad S_{21} = (\eta_1 - \eta_2)v_y - 2\eta_2 w,$$

$$N_{13} = -\alpha_3 l_2 + \alpha_4 b l_1, \quad N_{31} = \alpha_3 l_2 - \alpha_4 b l_1,$$

$$N_{23} = (\beta_1 - \beta_3)w_y + \alpha_3 l_1 + \alpha_4 b l_2, \quad N_{32} = (\beta_1 + \beta_3)w_y - \alpha_3 l_1 - \alpha_4 b l_2.$$

Thus equations (27)-(29) can be reduced to the system

$$\rho c_t = -\frac{\partial}{\partial y} l_2, \quad \rho v_t = -p_x + \frac{\partial}{\partial y} S_{12},$$

$$J\left(\rho c w_t + l_2 w_y\right) = \frac{\partial}{\partial y} N_{32} + S_{21} - S_{12}.$$

Let us consider steady flows. In this case $l_2 = \text{const}$. Channel flows are symmetrical relative to the x-axis, therefore the functions $v(y)$ and $c(y)$ are even, and the functions $w(y)$ and $b(y)$ are odd. It follows that $l_2 = 0$ and equation $(35)_2$ can be written as

$$\gamma_3 \frac{d}{dy} c = b[\alpha_1 \alpha_2(-\gamma_1 p_x + \alpha_3 w_y) - \alpha_4 w_y], \quad b = w + v_y/2. \tag{36}$$

In the case of steady flows, the equations for v and w can be written in the following form:

$$\frac{d}{dy}\left[(\eta_1 + \eta_2)v_y + 2\eta_2 w\right] = p_x, \tag{37}$$

$$\frac{d}{dy}\left[(\beta_1 + \beta_3 - \alpha_1 \alpha_3^2)w_y + \alpha_1 \alpha_3 \gamma_1 p_x\right] = 4\eta_2 b. \tag{38}$$

We integrate equation (37) taking into account that the function $b(y)$ is odd:

$$(\eta_1 + \eta_2)v_y + 2\eta_2 w = y p_x.$$

Hence, we arrive at the equation

$$b = \frac{\eta_1}{\eta_1 + \eta_2} w + \frac{y p_x}{2(\eta_1 + \eta_2)}.$$

In general, the viscosities β_i, η_i and the diffusion coefficients α_i depend on the concentration $c(y)$. To simplify our calculations we set $\beta_1 = \text{const}$ and $\beta_3 = \text{const}$. We denote $\eta_1^0 = \eta_1|_{c=0}$ and introduce dimensionless variables and functions:

$$y' = \frac{y}{H}, \quad w' = \frac{w}{w_0}, \quad \eta_1'(c) = \frac{\eta_1}{\eta_1^0}, \quad \varepsilon(c) = \frac{\eta_2}{\eta_1 + \eta_2}, \quad s_0 = \frac{4\eta_1^0 H^2}{\beta_1 + \beta_3},$$

$$w_0 = \frac{H|p_x|}{2\eta_1^0}, \quad n(c) = \frac{H^2|p_x|^2 \alpha_1 \alpha_2 k_p}{2\eta_1}, \quad m(c) = \frac{H^2|p_x|^2 \alpha_1(\alpha_4 - \alpha_1 \alpha_2 \alpha_3)}{4\eta_1^0 \eta_1 \rho D}.$$

With agreement with Remark 1, we use the next hypotheses:

$$\varepsilon = \varepsilon_0 c, \quad \eta_1' = 1 + \nu_0 c, \quad \alpha_3 = \alpha_3^0 c(1 - c),$$

$$k_p = k_p^0 c(1-c), \quad n = n_0 c(1-c)k_p^0, \quad m = m_0 c(1-c)(\alpha_4^0 - \alpha_1\alpha_2\alpha_3^0).$$

Let us introduce dimensionless functions:

$$l(c) = \frac{\alpha_3^0 \rho D k_p^0}{H^2} c(1-3c+2c^2) = l_0 \alpha_3^0 c(1-3c+2c^2),$$

$$q(c) = \frac{\alpha_1(\alpha_3^0)^2}{2H^2\eta_1^0} c(1-3c+2c^2) = q_0(\alpha_3^0)^2 c(1-3c+2c^2),$$

$$r(c) = r_0[\beta_1 + \beta_3 - \alpha_1(\alpha_3^0)^2 c^2(1-c)^2], \quad r_0 = \frac{1}{4H^2\eta_1^0}.$$

In the dimensionless form, equations (36), (38) become

$$\frac{d}{dy'}c = [1-\varepsilon(c)][\eta_1'(c)w' - y']\left(n(c) - m(c)\frac{d}{dy'}w'\right) \tag{39}$$

$$r(c)\frac{d^2}{dy'^2}w' - [l(c) + q(c)\frac{d}{dy'}w']\frac{d}{dy'}c = \varepsilon(c)[\eta_1'(c)w' - y']. \tag{40}$$

Given $c(y')$ and $w'(y')$, the dimensionless velocity

$$v' = \frac{v}{V}, \quad V = w_0 H,$$

can be found from following equation

$$\frac{d}{dy'}v' + 2\varepsilon(c)w' = -\frac{2y'(1-\varepsilon(c))}{\eta_1'(c)}.$$

4. Numerical results

Because of symmetry, we perform calculations only in the domain $0 \le y' \le 1$. As in [9], we use the no-slip conditions at the boundary $y = 1$ and the symmetry condition at the center of the channel:

$$w(1) = v(1) = 0, \quad w(0) = 0 \tag{41}$$

For the concentration, we formulate the condition

$$\int_{-1}^{1} c(y')dy' = 2\int_{0}^{1} c(y')dy' = C_0,$$

where C_0 is the total concentration of micropolar particles.

We use the following data for the calculations: $\varepsilon_0 = 0.5, n_0 = 3, m_0 = 100, s_0 = 0.7, \nu_0 = 0.03, C_0 = 0.6, q = 1, r_0 = 0.2, l_0 = 1, q_0 = 1, \alpha_1 = 1, \alpha_2 = 0.5$.

Let us consider the case $\alpha_3^0 = \alpha_4^0 = 2$. Fig. 1 depicts profiles of the velocity $v'(y')$, the concentration $c(y')$ and the gyration $w'(y')$ in the domain $0 \le y' \le 1$. Observe that concentration attains its maximal value somewhere between the center and the boundary of the channel, in agree with the Segre-Silberberg effect [12].

Fig. 2 depicts concentration for different parameters α_3^0 and α_4^0. In the case $\alpha_3^0 = \alpha_4^0 = 0$, the diffusion flux l doesn't depend on gyration. As a result, concentration attains its maximum in the center $y' = 0$, *Fig.* 2, (*a*). In the case $\alpha_4 \neq 0$ (Fig.2c,2d), concentration attains its maximum somewhere between the center and the boundary of the channel. Moreover, the greater α_3 the greater the shift of the point of maximum to the boundary. Fig. 2b concerns the case when the diffusion flux depends on gyration but only linearly and in this case the Segre-Silberberg effect does not occurs. We remind that α_3 is the weight of linear contribution of gyration and α_4 is the weight if nonlinear contribution of gyration into the diffusion flux. Thus, our calculations reveal that the Segre-Silberberg effect is genuinely nonlinear one.

IOP Conf. Series: Journal of Physics: Conf. Series **894** (2017) 012134 doi:10.1088/1742-6596/894/1/012134

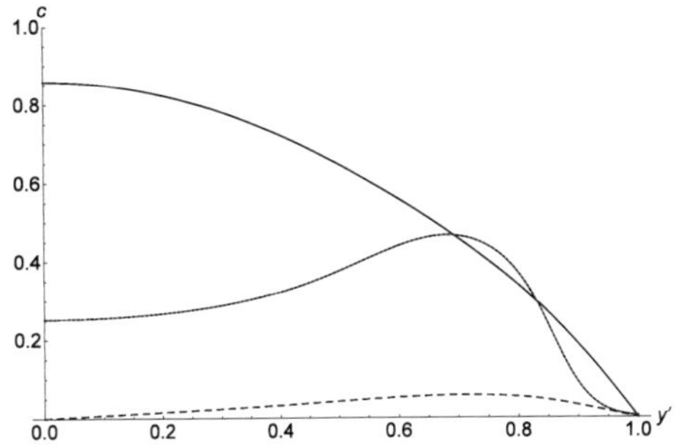

Figure 1. From the top down: velocity, concentration and gyration of the particles.

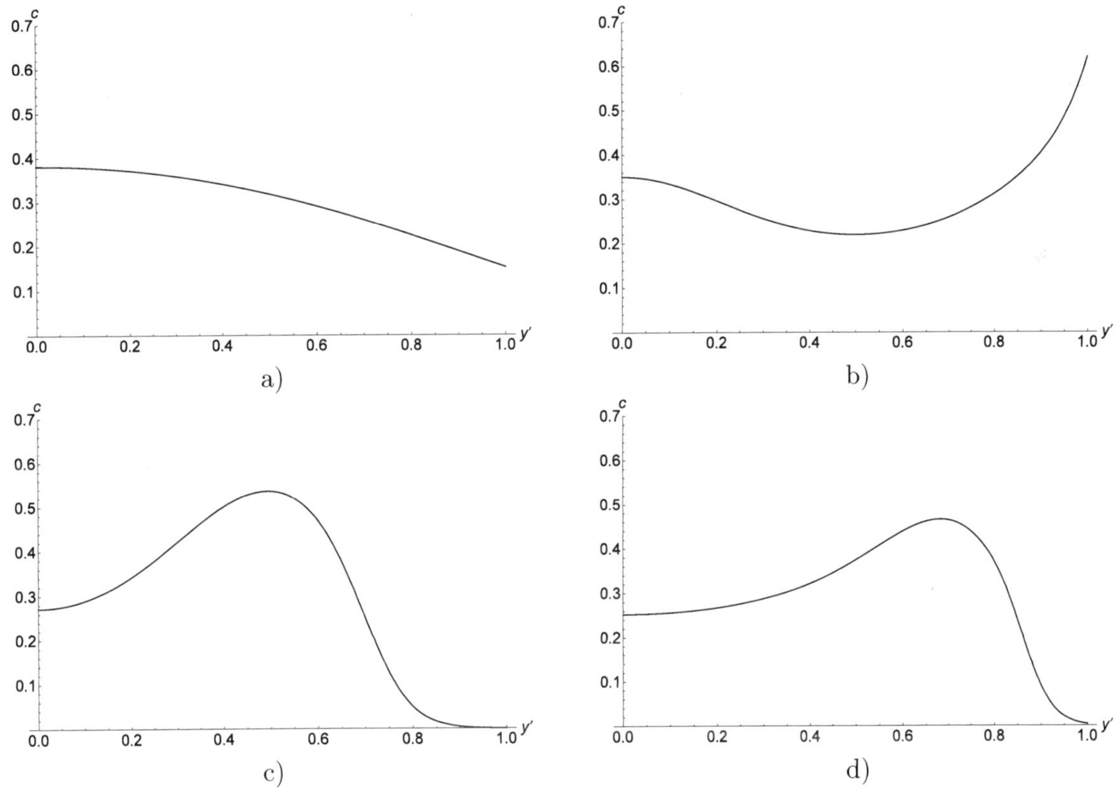

Figure 2. Concentration for different values of parameters α_3^0 and α_4^0: a) $\alpha_3^0 = \alpha_4^0 = 0$, b) $\alpha_3^0 = 2$, $\alpha_4^0 = 0$, c) $\alpha_3^0 = 0$, $\alpha_4^0 = 2$, d) $\alpha_3^0 = \alpha_4^0 = 2$

5. Conclusions

Conservation laws and constitutive equations for micropolar fluid with variable concentration of micropolar particles were derived. We should notice that mass concentration in this model obeys generalized Fick law and concentration flux depends on gyration of particles. We considered Poiseuille-like flow between two parallel planes and confirm Segre-Silberberg effect for such flow. Also it was shown that the Segre-Silberberg effect occurs if the concentration flux depends on gyration nonlinearly.

Acknowledgments
The paper is supported by Russian Science Fund (Project No. 15-11-20019).

Appendix A.
Here, we discuss questions of objectivity, i.e. frame independence [4].

Claim 1. Given an orthogonal basis $\{\mathbf{e}_i\}_1^3$ and an orthogonal matrix Q, we have the equality

$$\epsilon_{sij} \equiv \mathbf{e}_s \cdot (\mathbf{e}_i \times \mathbf{e}_j) = Q\langle \mathbf{e}_s \rangle \cdot (Q\langle \mathbf{e}_i \rangle \times Q\langle \mathbf{e}_j \rangle) \equiv \alpha_{sij}. \tag{A.1}$$

Proof. Let $Q_i^{\varphi_i}$ be the rotation around the x_i-axis through the angle φ_i. Clearly, $Q_i^{\varphi_i}\langle \mathbf{e}_i \rangle = \mathbf{e}_i$ and any orthogonal matrix Q can be represented as a composition $Q = Q_1^{\varphi_1} Q_2^{\varphi_2} Q_3^{\varphi_3}$. We have

$$Q_3^{\varphi}\langle \mathbf{e}_1 \rangle \cdot (Q_3^{\varphi}\langle \mathbf{e}_2 \rangle \times Q_3^{\varphi}\langle \mathbf{e}_3 \rangle) = Q_3^{\varphi}\langle \mathbf{e}_3 \rangle \cdot (Q_3^{\varphi}\langle \mathbf{e}_1 \rangle \times Q_3^{\varphi}\langle \mathbf{e}_2 \rangle) =$$

$$= Q_3^{\varphi}\langle \mathbf{e}_3 \rangle \cdot \mathbf{e}_3 = \mathbf{e}_1 \cdot (\mathbf{e}_2 \times \mathbf{e}_3).$$

Similarly, one can prove that equality (A.1) holds for Q_3^{φ}. Next, one can verify that

$$Q_1^{\varphi_1} Q_2^{\varphi_2} Q_3^{\varphi_3}\langle \mathbf{e}_1 \rangle \cdot (Q_1^{\varphi_1} Q_2^{\varphi_2} Q_3^{\varphi}\langle \mathbf{e}_2 \rangle \times Q_1^{\varphi_1} Q_2^{\varphi_2} Q_3^{\varphi}\langle \mathbf{e}_3 \rangle) =$$

$$Q_1^{\varphi_1} Q_2^{\varphi_2}\langle \mathbf{e}_1 \rangle \cdot (Q_1^{\varphi_1} Q_2^{\varphi_2}\langle \mathbf{e}_2 \rangle \times Q_1^{\varphi_1} Q_2^{\varphi_2}\langle \mathbf{e}_3 \rangle) =$$

$$Q_1^{\varphi_1}\langle \mathbf{e}_1 \rangle \cdot (Q_1^{\varphi_1}\langle \mathbf{e}_2 \rangle \times Q_1^{\varphi_1}\langle \mathbf{e}_3 \rangle) = \mathbf{e}_1 \cdot (\mathbf{e}_2 \times \mathbf{e}_3).$$

Hence, the claim (A.1) is proved.

Claim 2. Given an orthogonal matrix Q, we have the equality

$$Q\langle \mathbf{x} \rangle \times Q\langle \mathbf{y} \rangle = Q\langle \mathbf{x} \times \mathbf{y} \rangle, \qquad \forall \;\; \mathbf{x}, \mathbf{y}. \tag{A.2}$$

Proof. Due to the identity

$$Q\langle \mathbf{e}_i \rangle \times Q\langle \mathbf{e}_j \rangle = \alpha_{sij} Q\langle \mathbf{e}_s \rangle$$

and the equality (A.1), the claim (A.2) results from the following equalities:

$$Q\langle \mathbf{x} \rangle \times Q\langle \mathbf{y} \rangle = x_i y_j Q\langle \mathbf{e}_i \rangle \times Q\langle \mathbf{e}_j \rangle = x_i y_j \epsilon_{sij} Q\langle \mathbf{e}_s \rangle = Q\langle x_i y_j \epsilon_{sij} \mathbf{e}_s \rangle = Q\langle \mathbf{x} \times \mathbf{y} \rangle.$$

Claim 3. Given an objective vector $\mathbf{l}$, the matrix $\epsilon : \mathbf{l}$ is also objective.

Proof. Objectivity of $\mathbf{l}$ implies that [4] $\mathbf{l}' = Q^*\langle \mathbf{l} \rangle$, where $\mathbf{l}'$ is the value of $\mathbf{l}$ resulting from the passage to a new frame of references associated with the rotation Q. Given a vector $\mathbf{a}$, we have

$$\epsilon : \mathbf{l}'\langle \mathbf{a} \rangle = Q^*\langle \mathbf{l} \rangle \times \mathbf{a} = Q^*\langle \mathbf{l} \rangle \times Q^* Q\langle \mathbf{a} \rangle = Q^*\langle \mathbf{l} \times Q\langle \mathbf{a} \rangle \rangle = Q^* \cdot (\epsilon : \mathbf{l}) \cdot Q\langle \mathbf{a} \rangle.$$

Hence,

$$\epsilon : \mathbf{l}' = Q^* \cdot (\epsilon : \mathbf{l}) \cdot Q,$$

which implies objectivity of the matrix $\epsilon : \mathbf{l}$ [4].

Claim 4. Given an objective matrix B, the vector $\mathbf{b} = \mathbf{e}_i \times B\langle \mathbf{e}_i \rangle$ is also objective.

Proof. First of all, we observe that the vector $\mathbf{e}_i \times B\langle \mathbf{e}_i \rangle$ does not depend on the basis $\{\mathbf{e}_i\}_1^3$. Due to equality (A.2), we have

$$\mathbf{b}' \equiv \mathbf{e}_i \times B'\langle \mathbf{e}_i \rangle = \mathbf{e}_i \times Q^* B Q\langle \mathbf{e}_i \rangle = Q^* Q\langle \mathbf{e}_i \rangle \times Q^* B Q\langle \mathbf{e}_i \rangle =$$

$$= Q^*\langle Q\langle \mathbf{e}_i \rangle \times B Q\langle \mathbf{e}_i \rangle \rangle = Q^*\langle \mathbf{e}_i \times B\langle \mathbf{e}_i \rangle \rangle = Q^*\langle \mathbf{b} \rangle.$$

The claim is proved.

Claim 5. Given the objective vectors $\mathbf{l}$ and $\mathbf{b}$, the vector product $\mathbf{l} \times \mathbf{b}$ is also objective.

Proof. The claim follows from the equality (A.1). Indeed, we have

$$\mathbf{l}' \times \mathbf{b}' = Q^*\langle \mathbf{l} \rangle \times Q^*\langle \mathbf{b} \rangle = Q^*\langle \mathbf{l} \times \mathbf{b} \rangle.$$

IOP Conf. Series: Journal of Physics: Conf. Series **894** (2017) 012134 doi:10.1088/1742-6596/894/1/012134

References

[1] Shelukhin and Neverov V V 2014 Flows of micropolar and viscoplastic fluids within a Hele-Show cell *J. Appl. Mech. Tech. Phys.* **55** (6) 905-916

[2] Condiff D W and Dahler J S 1964 Fluid mechanical aspects of antisymmetric stress *Phys. Fluids* **7** (6) 842-854

[3] Aero E L, Bulygin A N and Kuvshinski E V 1965 Asymmetric hydromechanics *Appl. Math. Mech.* **29** (2) 297-308

[4] Eringen A C 1980 Theory of anisotropic micropolar fluids *Int. J. Engng. Sci.* **18** 5-17

[5] Eringen A C 1999 *Microcontinuum Field Theories* I,II (New York: Springer-Verlag)

[6] Lukaszewicz G 1999 *Microploar Fluids. Theory and Applications* (Boston MA: Birkhäuser Inc.)

[7] Shelukhin V V and Chemetov N V 2015 Global solvability of the one-dimensional Cosserat-Bingham fluid equations *J. Math. Fluid Mech.* **17** (3) 495-511

[8] Shelukhin V V and Khristenko U A 2013 On one slip condition for the equations of a viscous fluid *J. Appl. Mech. Tech. Phys.* **54** (5) 773-780

[9] Shelukhin V V and Růžička M 2013 On Cosserat-Bingham fluids *Z. Angew. Math. Mech.* **93** (1) 57-72

[10] Shelukhin V V and Neverov V V 2016 Thermodynamics of micropolar Bingham fluids *J. Non-Newtonian Fluid Mech.* **236** 83-90

[11] Popel A S, Regirer S A and P.I. Usick P I 1974 A continuum model of blood flow *Biorheology* **11** 427-437

[12] Segré G, A Silberberg A 1961 Radial Poiseuille flow of suspensions *Nature* **189** 209-210

[13] Leontovich M F 1985 *Introduction to Thermodynamics. Statistical Physics* (Moscow: Nauka)

[14] Shelukhin V V 2014 A poroelastic medium saturated by a two-phase capillary fluid *Continuum Mech. Therm.* **26** (5) 619-638

[15] Landau L D and Lifshits E M 1987 *Fluid mechabics. Course of theoretical physics* 2nd ed. (Oxford, New York, Beijing, Frankfurt, São Paolo, Sydney, Tokio, Toronto: Pergamon Press)

IOP Conf. Series: Journal of Physics: Conf. Series **894** (2017) 012135 doi:10.1088/1742-6596/894/1/012135

Numerical models for calculating hydrologic processes in river and lake-river systems

V S Nikiforovskaya and A F Voevodin

Lavrentyev Institute of Hydrodynamics SB RAS, Novosibirsk, Russia

E-mail: voevodin@hydro.nsc.ru

Abstract. We use one-dimensional (1D) and two-dimensional (2D) longitudinal-vertical mathematical models and their 2D+1D combination as well as numerical methods to study unsteady processes in the complex open channel systems under the influence of water management measures. The analysis shows the economic feasibility and efficiency of using the developed mathematical models to study hydrological process in water bodies. The study of the physical processes in complex water body, consisting of significantly different components, based on the use of only one chosen mathematical model, is uneconomical and inefficient from the viewpoint of computational expense.

1. Introduction

In recent decades, in our country, as well as all over the world, there has been a general trend in changing climatic conditions, leading to an increase in the number of natural disasters associated with flooding and elevated groundwater levels as a result of catastrophic floods, as well as drying up of watercourses due to lack of precipitation. In addition to natural factors, significant role in the formation of hydrothermal and environmental regimes is plaid by anthropogenic factors, such as the construction of hydropower plants on the rivers, various hydrotechnical engineering structures on the watercourses of lake-river systems, the construction of the water utilization systems in the territories adjacent to the waterways, etc. No less complexity is observed in relation to the hydrological regime in estuaries of the Siberian rivers, where due to overregulation or inadequate rainfall along with the shortage of river runoff and the effect of wind-induced surge, we can observe not only salinization of water in the estuary ducts, but also the increase in the penetration range of salty sea water into fresh river water.

To study the hydrological processes in water bodies under the impact to different natural and anthropogenic factors, one can use the set of mathematical models of different dimensionality developed at the Institute of Hydrodynamics SB RAS (IH SB RAS). These models are based on one- and two-dimensional lateral-vertical shallow water equations (Saint-Venant equations) and their combinations. Models are designed for investigation of heat and mass transfer processes, which are taking place at water bodies under the effects of different anthropogenic and natural factors.

It is assumed that the water body may represent a water utilization system, including in the general case the objects strongly differing from each other in terms of morphometric and hydraulic characteristics (separate watercourses, ponds, lakes, and their associated systems, e.g., lake-river system with tributaries or intake structures, etc.). The possible diversity of such components in the studied water utilization system or its fragment is the main problem not only in terms of choosing relevant mathematical models appropriately describing investigated physical processes, but also

Content from this work may be used under the terms of the Creative Commons Attribution 3.0 licence. Any further distribution of this work must maintain attribution to the author(s) and the title of the work, journal citation and DOI.
Published under licence by IOP Publishing Ltd

mathematical simulation of the above mentioned processes according to a certain selected mathematical model.

The set of computer programs used in the current research is a combination of mathematical models different in terms of setting up the problems (one-dimensional, two-dimensional longitudinal-vertical, etc.) with different capabilities of taking into account the effects of the various physical or any other external factors on water body [1-4]. Flexibility in the use of a set of programs is specified by provided set of corresponding correlations (linking and conjugation conditions) between individual parts of the studied water body.

2. The description of mathematical models

The numerical modeling of hydrological processes in water bodies was done using two types of mathematical models different in terms of their dimensionality.

2.1. *One -dimensional (1D) model of hydrologic processes in watercourses and their systems.* This model is based on Saint-Venant equations averaged over channel cross-section. The system allows describing dynamics of hydrological processes in water body channels with adequate accuracy [1, 2]

$$B\frac{\partial Z}{\partial t}+\frac{\partial Q}{\partial x}=q, \tag{1}$$

$$\frac{\partial Q}{\partial t}+\frac{\partial}{\partial x}\left(\frac{Q^2}{\omega}\right)=-g\omega\left(\frac{\partial Z}{\partial x}+\frac{Q|Q|}{K^2}\right)+R_w, \tag{2}$$

where t — is time; x — is the axial channel coordinate; $Z(x,t)$— is free surface level; $Q(x,t)$— is the water discharge rate; $q(x,t)$— is the lateral inflow distributed along the channel length; $B(h,x)$, $\omega(h,x)$, $K(h,x)$ — are the width of free surface, the cross-section area of channel, and the rate of discharge at the depth of h, respectively; R_w— are the external factors (for example, wind-induced surge (negative surge), or the barometric pressure gradient); g — is the acceleration gravity.

This mathematical model, being a kind of classic in the field of hydraulics, describes a wide class of problems on unsteady water flows in open channels. In this case, any of the following factors may become the source of the unsteady water flow. This can be abundant rainfall or snowmelt, the most relevant for the North-Eastern regions of Siberia and the Far East, or wind-induced surges (storms) at lakes, reservoirs, and estuarine areas of the rivers in the Far North. Moreover, this can be uneven operation mode of hydrotechnical structures, industrial enterprises, and agricultural plants; accidental discharges of industrial and domestic wastewater, as well as partial or complete destruction of hydraulic structures (hydro-electric plants, pumping stations), etc.

The solutions to the above mentioned different problems have a common goal consisting in the definition of discharge rates $Q(x,t)$ and water levels $Z(x,t)$ along the length of the watercourse (reservoir) at different points in time.

The given model can help to obtain the quantitative estimate of water resources availability in a water body. This is either a direct quantification, as for example in the problem of water resources efficient distribution and if necessary their internal territorial redistribution, or additional assessment, as for example in the problem of determining damage from flooding and elevated groundwater levels at possible destruction of hydraulic works. Such universality of generalized Saint-Venant model provides the opportunity to use it in the development of mathematical models to describe a very wide range of applied (engineering) hydro-economic problems, such as calculation of water flow in channels with floodplains [5], or for example, the below-mentioned study of salt wedge task.

Task. The sea-water penetration into river-mouth sections [2]. The feature of unsteady flow calculations on estuary sections of rivers is associated not only with necessity of taking into account the river water flow dynamics, but also to reflect the direction and strength of wind, atmospheric pressure and density inhomogeneities. In this regard, to study the salt water distribution process in the estuarine sections of rivers, we can use a mathematical model (1-2), represented as:

$$B\frac{\partial z}{\partial t}+\frac{\partial Q}{\partial x}=q$$

$$\frac{\partial Q}{\partial t}+\frac{\partial}{\partial x}(\upsilon Q)+g\omega\frac{\partial z}{\partial x}=-g\omega\frac{Q^2}{K^2}-\frac{\omega}{\rho}\frac{\partial p_a}{\partial x}+\xi BW_l|W|. \quad (3)$$

The equation of the state [6]:

$$\rho(x,t)=f(T,S). \quad (4)$$

Water temperature $T(x,t)$— and salinity $S(x,t)$— are determined from balance dispersion equation (the conservative additive and heat balance equation):

$$\frac{\partial}{\partial t}(\omega T)+\frac{\partial}{\partial x}(QT)=\frac{\partial}{\partial x}\left(\omega E\frac{\partial T}{\partial x}\right)-K_T B(T-T_e)+T_q, \quad (5)$$

$$\frac{\partial}{\partial t}(\omega S)+\frac{\partial}{\partial x}(QS)=\frac{\partial}{\partial x}\left(\omega E\frac{\partial S}{\partial x}\right)+S_q. \quad (6)$$

Here t — is time; x— is the axial channel coordinate; $p_a(x,t)$ — is the atmospheric pressure; $W_l(x,t)$ — is the wind velocity component along the channel axis; $|W(x,t)|$ — is the wind velocity modulus; $\rho(x,t)$ — is water density; ξ — is wind pressure coefficient; $T_q(x,t),S_q(x,t)$ — are the heat and salt inflows coming together with lateral inflow $q(x,t)$; k_T — is the heat transfer coefficient; T_0 — is the so called equilibrium ambient temperature; $E(h,x)$ — is the longitudinal dispersion coefficient, which is calculated by Harleman formula [7]:

$$E=20.2\cdot\sqrt{g}\cdot|v|\cdot R/C,$$

where R — is the hydraulic radius, C — is Chezy coefficient. The required functions in this model are the next ones: the free surface level $Z(x,t)$; the rate of discharge $Q(x,t)$; the density of water $\rho(x,t)$; the temperature of water $T(x,t)$ and the concentration of impurities (substances) $S(x,t)$ contained in the water flow.

The mathematical model (1D) used in the research, is a part of "Complex hydraulic models" designed at the IH SB RAS to address a wide range of open channel hydraulics problems. After thorough testing and successful adjustment to the real conditions, the latter was included to the State Fund of Algorithms and Programs (USSR State Fund of Algorithms and Programs) [8].

2.2. The two-dimensional (longitudinal-vertical) model (2D) of temperature-stratified flows in deep elongated reservoirs. The given model is based on two-dimensional equations obtained from three-dimensional equations of hydrodynamics, and averaged over channel width or watercourse, as well as assumptions about the hydrostatic law of pressure [3, 4]:

$$\frac{\partial u}{\partial t}+u\frac{\partial u}{\partial x}+w\frac{\partial u}{\partial z}=-g\frac{\partial}{\partial x}\left(\varsigma+\frac{1}{\rho_0}\int_{z_d}^{\varsigma}\rho dz\right)+\frac{1}{b}\frac{\partial}{\partial z}\left(b\nu_t\frac{\partial u}{\partial z}\right)-\frac{k}{b}\tau, \quad (7)$$

$$\frac{\partial(bu)}{\partial x}+\frac{\partial(bw)}{\partial z}=0, \tag{8}$$

where

$$k=\sum_{1}^{2}\left[1+\left(\frac{\partial b_i}{\partial x}\right)^2+\left(\frac{\partial b_i}{\partial z}\right)^2\right]^{1/2},\ \tau=\frac{\lambda}{8}|u|u,\ \rho=\rho(T,S) \tag{9}$$

Here t — is time; x, z — are horizontal and vertical coordinates; $u(x,z,t)$, $w(x,z,t)$ — are horizontal and vertical velocity components; $\zeta(x,t)$ — is free surface level; $b(x,z)$ — is channel width; $\rho(T,S)$ — is liquid density; ρ_0 — is the characteristic density value; ν_t — is the coefficient equal to the sum of molecular and turbulent viscosities; τ — is the friction resistance (shear stress) on lateral surfaces; λ — is the friction coefficient.

Heat and salinity transfer equations were obtained by averaging the relevant three-dimensional equations:

$$\frac{\partial T}{\partial t}+u\frac{\partial T}{\partial x}+w\frac{\partial T}{\partial z}=\frac{1}{b}\frac{\partial}{\partial z}\left(bK_T\frac{\partial T}{\partial z}\right)+\frac{1}{\rho_0 c_0}\left(Q_T-\frac{k}{b}q_n\right), \tag{10}$$

$$\frac{\partial S}{\partial t}+u\frac{\partial S}{\partial x}+w\frac{\partial S}{\partial z}=\frac{1}{b}\frac{\partial}{\partial z}\left(bK_S\frac{\partial S}{\partial z}\right), \tag{11}$$

where $T(x,z,t)$, $S(x,z,t)$ — are liquid temperature and salinity; K_T, K_S — are turbulent thermal diffusivity and diffusion coefficients; $\rho_0 c_0$ — is the product of the average density and heat capacity coefficient; q_n — is the heat flux over lateral surface; Q_T — is heat supply from a source or volumetric heat influx. Coefficients K_T, K_S were defined according to Prandtl theory of turbulent flows [9]. For the density calculation the state equation $\rho=\rho(T,S)$ was used.

According to this model, the following functions are unknown: the time-distance distribution of water discharge rates and levels (continuity and dynamic equations are described in variables of level and discharge rate), vertical and horizontal velocity fields, as well as temperature and salinity distribution fields (in terms of space and time) and dependent density fields. The mathematical model was developed for the study of hydro-thermal and mass transfer processes in low flow density-stratified narrow deep reservoirs.

2.3. Comprehensive 2D-1D mathematical model for calculation of hydrological processes in open channel systems. This mathematical model is developed on the basis of models being produced above, and represents the combination of one-dimensional and two-dimensional (longitudinal-vertical) models [2, 3, 4, 10].

This mathematical model can be used to study the dynamics of wave processes occurring in the water bodies complicated in structure, having components greatly differing from each other in terms of their geometric, hydraulic, and morphometric characteristics. When building one-dimensional and two-dimensional models, the real morphometric and hydraulic characteristics of the channels and adjacent floodplains were taken into account, as well as their interaction, and, if necessary, the impact of meteorological factors (wind, atmospheric pressure) on the wave processes.

Numerical modeling was carried out step by step according to the technique developed for the above mathematical models. It is described in detail in [10]. So, when modeling, the original physical object is represented schematically by relevant (single, multiple) physical model (topological diagram), i.e. the analogue of a real object. Then the selected physical model is associated with a corresponding mathematical model (task). Compliance of both models can be achieved using sets of boundary and matching conditions of different type provided by the computer programs [1-4].

Consider the example of some applied hydrodynamic problem, which was addressed using set of computer programs, created on the basis of the aforementioned mathematical models, numerical methods and algorithms, designed at the IH SB RAS [8].

3. Example. The calculation of the unsteady flow in flow-through water system of Lama Lake – Lama River [4].

The main objective of the calculations was to determine the possibility of using the developed comprehensive (2D+1D) mathematical model to study wave processes in the system of "river-lake-river" type. To this end, we considered a fragment of a lake-river system in the Putorana Lake Province located in the North-West of the Central Siberian plateau in Krasnoyarsk Territory of the Russian Federation [11]. The considered system fragment includes Lake Lama, 80 *km* long, extending to East-West, and rivers flowing in and out of the lake (Figure 1).

Lake Lama represents a narrow deep reservoir 3.7 *km* wide in its middle part with the depth of 254 *m*, and 10–13 *km* wide at the west edge (with the depth of 2.5–9 *m*).

The river flowing into the lake played a role of the cumulative inflow with elevation of water level $Z = Z_{pritok}(t)$; the river flowing out of the lake was considered as a channel with arbitrary cross-section (20 *km* long with maximum width up to 500 *m* and depth of 4–7 *m*, Figure 1).

a) Lake-river system (Putoran province)

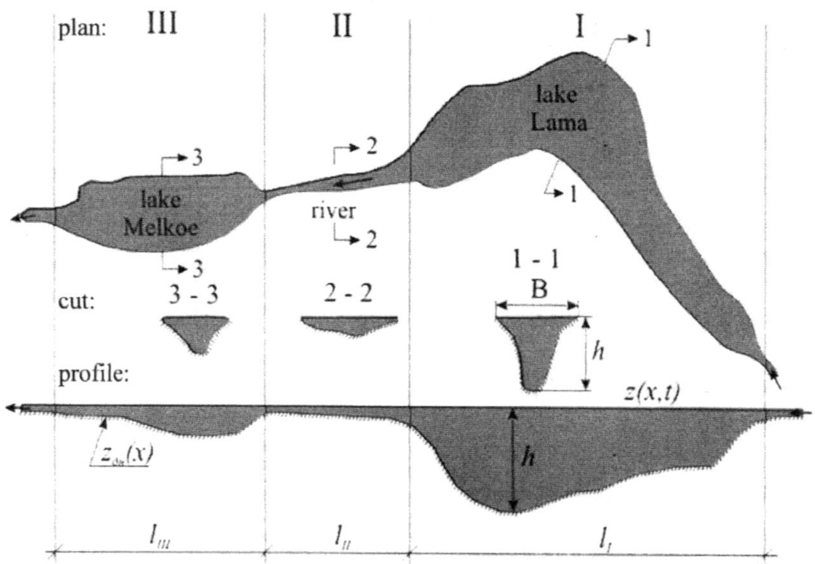

Figure 1. The lake-river system (the Putorana Lake Province).

For numerical modeling, we have built two different topological diagrams of lake-river system fragment. According to the *first* diagram, the lake-river system (both lake and river) were considered as two interrelated two-dimensional domains extended to L_1 =80 *km* and L_2 =20 *km,* respectively. The numerical calculations according to the first scheme were carried out based on two-dimensional (2*D*) mathematical model. According to the *second* approach, the lake-river system was considered as connected to each other two domains: 2D domain of lake (L_1 =80 *km*) and 1D domain of river

(L_2 =20 *km*). The calculations according to the second scheme were produced on the basis of combined (*2D*+1*D*) mathematical model. In both cases the lake and the river were described by corresponding hydraulic and morphometric characteristics. In this case, the same roughness coefficient equal to 0.030 was used in both calculating domains. Numerical calculations were based on the average annual flood, which reflects real flood waves and may well agree the possible flood waves in the territory under consideration.

The boundary conditions, when carried out calculations according to both schemes were as follows:

- At the left boundary (entering $x^{(1)}=0$) we have set condition $Z=Z(t)$, corresponding to a rise of the water of the river flowing into the lake, based on the descriptions of repeatedly observed of the water raisings and recession of flood in the spring and summer periods (from late May to early August) during the 70 days. According to the observations, water rise above the low water level in the lake was at the maximum of 4.5 m (in mid-June), which then was gradually decreased (to early or mid-August) down to the original level. Besides, at this boundary (in both cases), vertical distribution of longitudinal velocity component (elliptic law) was set [12].
- At the right boundary (the closing Lama River cross section $x^{(2)}=L_2=20$ *km*) we set the water stage discharge curve $Q=Q(z)$.
- At the joint of calculation domains (cross section of the river flowing out the lake $x^{(1)}=L_1=x_0^{(2)}$) the balance of discharges and equality of water levels were assigned.

The initial conditions were taken from numerical calculations aimed at determination of steady condition, where the rates of water discharges and levels corresponded to hydraulic regime ($Z^0=53\,m=const,\ Q_{L_2}=Q(z)$)

As noted above, the numerical simulations were aimed at comparing the relevance of both mathematical models in terms of describing the flood wave propagation process in the real lake-river system, since surveillance data is not available. In passing, we investigated the problem of choosing the method of determining the vertical distribution of the horizontal velocity $u(x_0^{(1)},z,t)$ on the left boundary of the two-dimensional domain. The study showed that setting it according to elliptic law does not violate the general flow pattern in the lake in both stationary and non-stationary conditions. The velocity set at the border, even at the first computational step ($x=\Delta x$) starting from the border, adjusts to the velocity field of the lake in both flow regimes (Figure 2).

Possibility to assign a constant value for the vertical longitudinal velocity component, when solving the two-dimensional problem, was also noted in [13]. The calculations of the unsteady flow propagation in the system according to both models were performed with a constant time step equal to 30 s, while the total duration of the calculations of the physical process of passing flood in the lake-river system was 70 days.

In the absence of observational data, the analysis was made to compare the results of calculations performed according to both mathematical models. The analysis has shown that:

1) the water levels in the lake in both steady and unsteady motion practically coincide (max divergence was equal to 0.14%); for the river, the divergence reaches a maximum of 2.03-2.5%;
2) when calculating water discharge values, divergence reaches the maximum of 9.0-9.5% at the closing cross section of river-lake system;
3) the computing cost for the calculation of 70-days flood according to two-dimensional (2D) model is almost four times higher than that of the same flood calculated based on integrated (2D+1D) model, other conditions being equal.

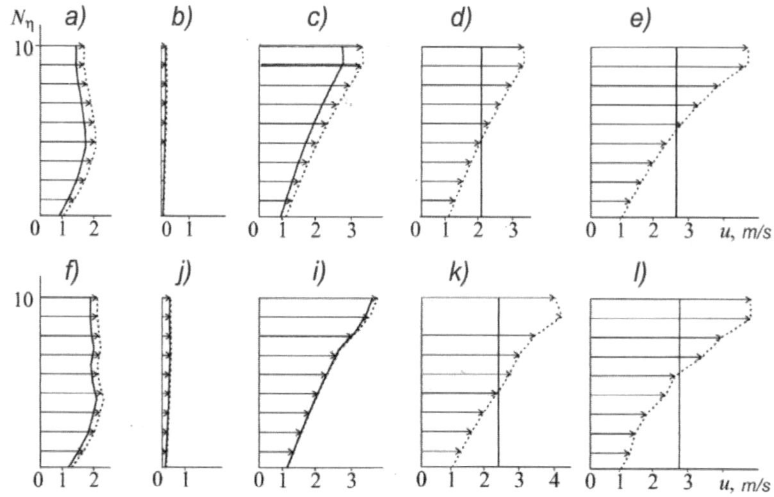

Figure 2. Horizontal velocity profiles at time points $t = 12 days$ ($(a-d)-$steady state regime), and $t = 42 days$ ($(e-k)-$ transient regime) calculated by 2D and 2D+1D models (dashed and solid lines, respectively).

4. Conclusion

Experience of solving similar tasks using the set of computer programs and techniques developed at the IH SB RAS, has shown that [4, 10, 14-16]:

a) the choice of the appropriate mathematical model to investigate hydrological processes occurring in a water body, as a rule, is dictated not only by the goals of the conducted study and structure of the research object, but also the opportunity to fulfill the most economical (in terms of computational costs) and exact numerical calculations without loss of relevancy of models to real object.

b) the developed one-dimensional and two-dimensional mathematical models, as well as numerical techniques are efficient and therefore useful for solving a broad range of problems of applied hydromechanics. Particularly, their advantage over three-dimensional models is evident in cases where the water bodies, which represent systems of quite extended real channels and reservoirs, need consideration of the specifics of the morphometry, operation hydraulic structures, local resistances, lateral inflow, etc.

References

[1] Voevodin A F and Shygrin S M 1993 *Solution methods for an one-dimensional evolution systems* Novosibirsk Nauka Sibirian Branch

[2] Voevodin A F, Nikiforovskaya V S and Ovcharova A S 1983 *Trudi AANII* S-Ptbr., Gigrometeoizdat **378** 23-34

[3] Vasiliev O F Voevodin A F and Nikiforovskaya V S 2005 *Vichislit. Technol.* **10** 29-38

[4] Voevodin A F and Nikiforovskaya V S 2012 *Meteorology and hydrology* **4** 64-74

[5] Nikiforovskaya V S 1978 *Dinamika sploshnoi sredi* **35** 89-95

[6] Mamaev O I 1964 *Izv. AN SSSR, ser. Geophiz.* **2** 309-311

[7] Hydrodynamics coastal zone and estuaries. 1970 Under red. A T Ippen, *L: Gidrometeoizdat.*

[8] Complex of hydraulic models for solution of water resources efficient management problems 1989 *Vneshtorgizdat* manufacture № 6120 MV.

[9] Blumberg A F 1977 *J. Hydraylic Division ASCE* **103** 295-310

[10] Nikiforovskaja V S 2013 *Meteorology and hydrology* **4** 79-88

[11] Popov E A and Paramurzin Yu P 1975 Water regime of rivers and lakes of the Noril region. - *Trudi of Limnology Institut*: 'Putoran lake province', Novosibirsk: Nauka (Siberian Branches), **20 (40)** pp. 32-56.
[12] Karaushev A V 1960 Problems of dynamics for the natural flows. - L.: Gidrometeoizdat, 392 p.
[13] Novicki M A 2006 Meteorology and hydrology, 2006, № 4, pp. 93-99.
[14] Nikiforovskaya V S 1985 *Hydro physics processes in rivers and water-bodies.* M: Nauka pp 211-216.
[15] Nikiforovskaya V S 2009 *Meteorology and hydrology* № 12 pp 54-64.
[16] Voevodin A F Nikiforovskaya V S and Vinogradova T A 2009 *Vestnik S-Ptbrg. Univers., ser. 7* issue 3 pp 139-144.

IOP Conf. Series: Journal of Physics: Conf. Series **894** (2017) 012136 doi:10.1088/1742-6596/894/1/012136

Duration of the zone of high electrical conductivity at the detonation of RDX of different densities

N P Satonkina

Lavrentyev Institute of Hydrodynamics, Novosibirsk, Russia,
Novosibirsk State University, Novosibirsk, Russia

E-mail: snp@hydro.nsc.ru

Abstract. This paper presents the results of measuring the duration of the zone of high electrical conductivity during detonation of RDX of different density. It is shown that the density of RDX has a weak effect on the duration of this zone.

1. Introduction

The detonation wave in condensed explosives consists of the shock wave, the reaction zone, the Chapman–Jouguet (CJ) plane (a CJ point in the one-dimensional case), and the Taylor rarefaction wave. In Zeldovich–Neumann–Doring theory [1, 2, 3], the CJ point separating the regions of subsonic and supersonic flows corresponds to the end of the chemical reaction zone. It has been suggested [1] that the reaction zone correlates with the high-pressure region and that the specific volume and pressure in the reaction zone are related by the equation

$$p - p_0 = -\frac{D^2}{v_0^2}(v - v_0),$$

where p is the pressure, v is the specific volume, and D is the detonation velocity; the subscript "0" denotes the parameters of the initial state. The chemical spike is studied using spatial distributions of mechanical parameters such as density, mass velocity, and pressure. An overview of the data the duration of the reaction zone Δt is presented in [4]; different methods give significantly different results.

This paper presents the results of measuring the duration of the zone of high electrical conductivity which may be useful for studying the reaction zone [5] as these regions correlate with each other.

2. Experiments

The techniques of electrical conductivity measurement and data processing were considered in detail in [5, 6], and here we give only a brief description.

The explosive was placed in a copper coaxial cell (Figure 1). The diameter of the explosive charge was $b = 8$ mm. The outer electrode was compound, and its parts *1, 2* were tightly connected to each other by screw threads. The inner copper electrode *3* of the diameter $c = 2$ mm was placed in the explosive and mounted in a Plexiglas plug *4*, which was fixed by a copper

Content from this work may be used under the terms of the Creative Commons Attribution 3.0 licence. Any further distribution of this work must maintain attribution to the author(s) and the title of the work, journal citation and DOI.
Published under licence by IOP Publishing Ltd

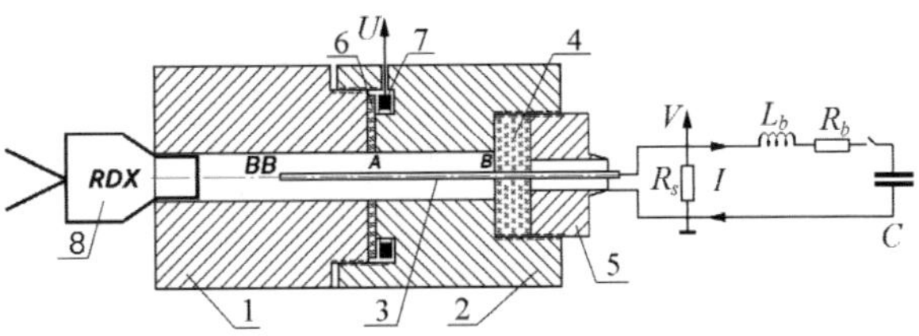

Figure 1. Electrical circuit and experimental cell.

bolt *5*. A toroidal coil *7* which served as a conductivity sensor was placed in the cavity of the outer electrode. The narrow gap between the cavity and the measuring volume of the cell was filled with Plexiglas insulator *6* $0.2 \div 0.7$ mm thick. The cavity in the coaxial cell was filled with the investigated explosive, which was initiated by RDX of density $\rho = 1.1$ g/cm^3 through an axial channel of 2.4 mm diameter in a Plexiglas plug *8*. The total length of the explosive charge was 50 mm. The critical diameter of RDX with a density close to the maximum is 2 mm [7]. Using a thick copper shell reduces it by a factor of 6 to 8 [8].

The cell was supplied with current I from a capacitor through a ballast resistor R_b; a shunt R_s was connected in parallel to the cell. The supply voltage $V(t)$ and the signal at the sensor $U(t)$ were measured. The electrical conductivity in the plane of the gap was proportional to the signal from the coil:

$$\sigma(t) = \frac{\ln(b/c)}{2\pi DM}\frac{U(t)}{V(t)}. \qquad (1)$$

Here, t is the time elapsed after the detonation front has passed the gap, D is the detonation velocity, M is the mutual inductance between the coil and the cavity circuit. The signal was corrected to eliminate the distortion of $U(t)$ due to the sensor inductance.

The spatial resolution of the circuit is estimated at one-quarter of the thickness of the insulator *6*. The error in measuring the electrical conductivity did not exceed 20%. By the time the detonation wave reached the insulator *6*, the detonation front had traveled a distance equal to more than four diameters of the explosive from the initiation point. The detonation velocity D was measured on the base $AB \approx 15$ mm. The point A corresponds to the time the detonation wave passes through the section of the outer electrode and insulator, and the point$B$ corresponds to the time the detonation wave arrives at the end of the charge.

Figure 2 shows a typical electrical conductivity profile for condensed explosives with a negative oxygen balance of about 20 %. The profile contains four regions: the increase to the maximum value σ_{max}; the decrease to the point indicated as σ_{ec}; the transition region at $0.05 > t > 0.1$ μs; the region at $t > 0.1$ μs corresponding to the rarefaction wave. In the graph, one can clearly see the triangular zone of high electrical conductivity at $0 > t > 0.05$ μs; the procedure for determining the duration of τ from the front to the intersection of the approximating lines is shown.

3. Results of the experiments

Figure 3 shows data on the durations of the zone of high electrical conductivity τ for different densities of RDX (points); there is no pronounced dependence of τ on density; a value close to 40 ns is present in the entire range of densities.

For a bulk density $\rho \approx 1.2$ g/cm ^{3}, data for different grain size are presented. Earlier [9], a decrease in the duration of the zone of high electrical conductivity τ by a factor of about two with increasing grain size of explosives was obtained for RDX, HMX, and PETN. The difference between the maximum value in Figure 3 (68 ns for a grain size of 160μm) and the minimum value

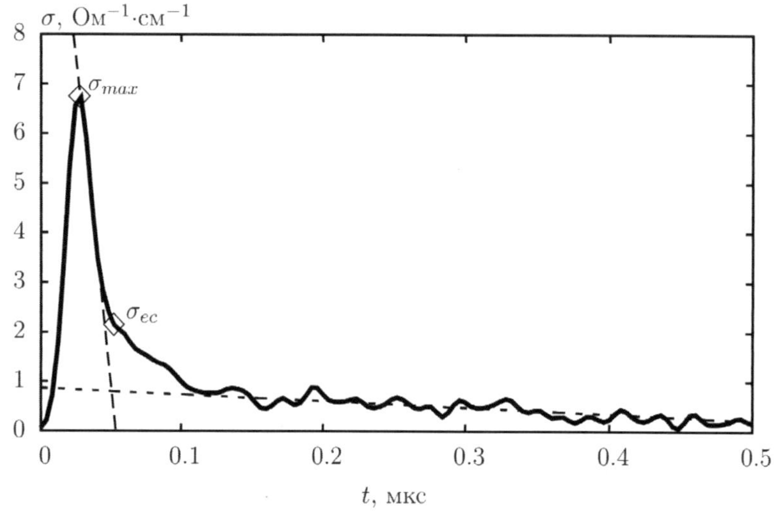

Figure 2. Electrical conductivity profile in detonating condensed explosives. The duration of the electrical conductivity peak is determined from the intersection of two approximating lines.

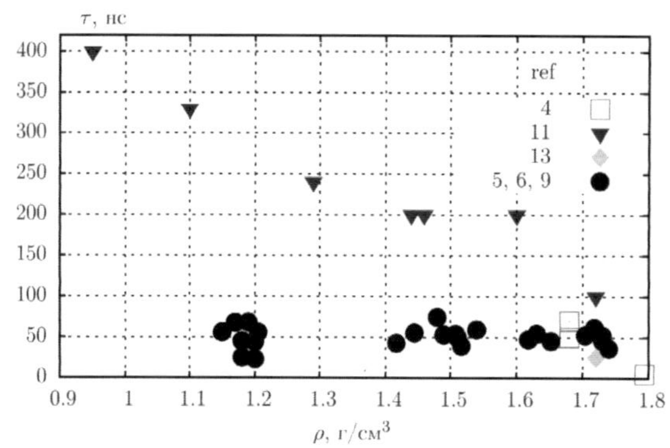

Figure 3. Duration of the chemical peak for RDX of different densities, the points are data on the electrical conductivity distribution, and the remaining symbols correspond to mass velocity and pressure profiles.

(23 ns for a grain size of 11 μm) is 45 ns. The effect of grain size on the mass velocity distribution has previously been found [10] with a similar experimental setup. This effect manifests itself in a decrease of the mass velocity and is explained by an increase in the fraction of material reacted in the front. With the preservation of the high-pressure region associated with the reaction zone, its duration changes.

Note that no effect of the grain size on the duration of the reaction zone for RDX of bulk density was found in work[11], which contradicts to the data of work[10]. The reason is in work[11] the grain size of the explosive was not small enough (150 and 450 μm).

The duration of τ for a density of 1.4–1.75 g/cm^3 is given for an average grain size of 160 μm. The same graph shows data on the duration of the reaction zone Δt obtained from the mass velocity distribution for a charge diameter ≥ 40 mm [4, 12, 13]. The values τ and Δt are similar

IOP Conf. Series: Journal of Physics: Conf. Series **894** (2017) 012136 doi:10.1088/1742-6596/894/1/012136

only for a density close to the maximum density, with the difference between them increasing with decreasing density.

It can be assumed that during detonation, the sequence of chemical reactions in the reaction zone does not depend on the charge parameters, but the geometry and density of the charge influence the pressure and mass velocity distributions in the DW. This may be responsible for the difference between τ and Δt obtained by different methods.

4. Discussion

It has been assumed [14, 15] that in the detonation of $C_aH_bN_cO_d$ type explosives, the electrical conductivity is determined by the presence of carbon in the explosive. The following model has been proposed: before the arrival of the detonation front, the explosive is a dielectric, and carbon is bound in molecules. The increase of the electrical conductivity to the maximum σ_{max} (Figure 2) corresponds to the destruction of the molecules of the original explosive with simultaneous growth of carbon grids. The decrease of the electrical conductivity from σ_{max} to σ_{ec} reflects the transition of carbon to a nonconducting form. This can be explained by carbon oxidation, as a result of which the conductive structures are thinned, and part of them is destroyed. In the region between 0.05 and 0.1 μs, $\sigma(t)$ has a lower gradient, which may be explained by a decrease of the oxidation rate due to a decrease of the concentration of the reacting components. The model described above implies that σ_{max} and σ_{ec} are characteristics of the same structure at different times, the change in electrical conductivity reflects the evolution of carbon grids, and the aggregation of individual carbon atoms into structures of different scales occurs before the electrical conductivity in the reaction zone has been reached the maximum value.

Thus, condensed carbon, as it is understood in calculations, is not released at the CJ point, but it is a residue of the structures formed in the reaction zone.

The early aggregation of carbon atoms is supported by the following observations. Anisichkin et al. [16, 17, 18, 19] assert that the aggregation of carbon atoms occurs in the reaction zone. It has been found [17] that the carbon components of highly dispersed composite explosives are fully mixed before the onset of oxidation, i.e., in the reaction zone, and oxidation reactions occur later. From energy considerations, Breusov [18] showed that the formation of carbon clusters is associated with a partial rupture of bonds in molecules and the formation and growth of a carbon skeleton. Molecular dynamics simulation [19] of the clustering of carbon upon heating of TATB, HMX, and PETN molecules has shown the formation of carbon nanostructures in the reaction zone.

The following data on the sequence of chemical reactions are reported in the literature. Calculations of Tarver [20, 21] showed that for condensed explosives, the synthesis of water occurs before the formation of CO_2. It was found [21] that the formation of water, CO_2, and N_2 and the release of condensed carbon occur simultaneously in the final stage of chemical reactions. Anisichkin [16], having analyzed the results of isotope tracer experiments, also claims that water production proceeds before the formation of CO and CO_2. Carbon is oxidized to a gaseous state after the formation of diamond and non-diamond phases. Thus, the authors of the above-mentioned papers agree that the reactions with carbon end no later than other chemical reactions.

Papers [16] and [20, 21], although disagreeing with each other in detail, nevertheless agree that carbon oxidation reactions end no earlier than others. Since the end point of chemical reactions with water and nitrogen approximately corresponds to the end of carbon oxidation, it can be assumed that the change of electrical conductivity through due to the change of the state of carbon reflects the kinetics of the reaction zone.

IOP Conf. Series: Journal of Physics: Conf. Series **894** (2017) 012136 doi:10.1088/1742-6596/894/1/012136

5. Conclusions

Experimental data on the duration of the zone of high electrical conductivity in detonating RDX are presented, showing that the density of RDX has a weak effect on the duration of this zone. It is suggested that the electrical conductivity traces the chemical state of the material during detonation. The previously proposed model of electrical conductivity and the relationship between the zone of high electrical conductivity and the reaction zone are discussed.

References

[1] Zeldovich Ya B 1940 *Theory of detonation propagation in gas systems* Zhurnal Teoreticheskoi i Eksperimental'noi Fiziki **10** 542
[2] Von Neumann J 1942 *Theory of Detonation Waves Office of Science Research and Development. Report* p 549
[3] Doering W 1943 *Uber der detonations vorgang in gasen* Analytical Physics **43**(6) 421
[4] Loboiko B G and Lyubyatinskii S N 2000 *Reaction zones of detonating solid explosives* Combustions, Explosives and Shock Waves **36**(6) 7016
[5] Ershov A P, Satonkina N P and Ivanov G M 2006 *Electroconductivity profiles in dense high explosives* Russian Journal of Physical Chemistry B **1**(6) 588
[6] Ershov A P, Satonkina N P and Ivanov G M 2004 *High-resolution conductivity profile measurements in detonating pressed explosive* Technical Physics Letters **30**(12) 1048
[7] Physics of the explosion 2002, Ed. by L.P. Orlenko, Fizmatlit, Moscow, Vol 1
[8] Kobylkin I F, Solov'ev V S and Boyko M M 1983 *Critical diameter for stationary detonation in a high-density explosives. Shell effects* Combustion, Explosives and Shock Waves **4** 484
[9] Ershov A P and Satonkina N P 2010 *Electrical conductivity distributions in detonating low-density explosives Grain size effect* Combustion and Flame **157**(5) 1022
[10] Ershov A P, et. al. 2016 *Nonideal detonation regimes in low density explosives* Journal of Applied Physics **119** 075903
[11] Dremin A N, Savrov S D, Trofimov V S and Shvedov K K 1970 *Detonation waves in condensed media* Moscow: Nauka
[12] Fedorov A V, Men'shikh A V and Yagodin N B 1999 *The structure of the detonation front in heterogeneous explosives* Khimicheskaya Fizika **18**(11) 64
[13] Utkin A V, Malyarenko S I and Kanel' G I 1989 *Decomposition of RDX in shock and detonation waves* Detonation: Proc. of the 9th All-Union. Symp. on Combustion and Explosion, Suzdal, November 19-24, Chernogolovka, p 91
[14] Satonkina N P 2015 *The dynamics of carbon nanostructures at detonation of condensed high explosives* Journal of Applied Physics **118** 245901
[15] Satonkina N P 2016 *Correlation of electrical conductivity in the detonation of condensed explosives with their carbon content* Combustion, Explosives and Shock Waves **52**(4) 488
[16] Anisichkin V F 2016 *On the Mechanism of Detonation of Organic high Explosives* Russian Journal of Physical Chemistry **35**(6) 451
[17] Anisichkin V F 2007 *Isotope studies of detonation mechanism of TNT, RDX, and HM* Combustion, Explosives and Shock Waves **43**(5) 580
[18] Breusov O N 2002 *On the Mechanism of Dynamic Diamond Synthesis from Organic Substances* Khimichaskaya Fizika **21**(11) 110
[19] Yushi Wen, Chaoyang Zhang, Xianggui Xue and Xinping Long 2015 *Cluster evolution during the early stages of heating explosives and its relationship to sensitivity: a comparative study of TATB, b-HMX and PETN by molecular reactive force field simulations* Physical Chemistry Chemical Physics **17** 12013
[20] Tarver Craig M 2001 *What is a shock wave to an explosive volecule?* in Proc. Shock Compression of Condensed Matter **CP620** 42
[21] Tarver Craig M 2012 *Effect of electric fields on the reaction rates in shock initiating and detonating solid explosives* AIP Conf. Proc. **1426** 227

Effect of sound-absorbing coatings on the disturbance evolution in a flow of a mixture of vibrationally excited gases

A I Reshetova[1, 2], **T V Poplavskaya**[1, 2], **S V Kirilovskiy**[1,2] **and I S Tsyrulnikov**[1,2]

[1]Khristianovich Institute of Theoretical and Applied Mechanics SB RAS, Novosibirsk, Russia
[2]Novosibirsk State University, Novosibirsk, Russia

Abstract. The flow around a solid plate and a plate with a sound-absorbing coating at a non-zero angle of attack in a hypersonic flow (M_∞=8.44) of a mixture of vibrationally excited carbon dioxide and nitrogen is considered. Numerical simulations are performed by solving two-dimensional unsteady Navier–Stokes equations with a two-temperature model of relaxing flows. The vibrational energy as a function of time is defined by the Landau–Teller equation. A skeleton model, which is a set of square elements arranged in a staggered order, is used for simulating the porous coating made of foamed nickel with a porosity coefficient of 95%. The distance between the elements is equal to the pore diameter of the real sound-absorbing material. Data on the evolution of disturbances on the solid plate and on the plate with the sound-absorbing coating are presented for various angles of attack and CO_2 concentrations in the mixture. The experimental and calculated data on pressure fluctuations on the plate surfaces are found to be in good agreement. The effects of various parameters of the sound-absorbing coating (depth, length, and location at the flat plate) are considered. It is shown that the sound-absorbing coating significantly reduces the intensity of pressure fluctuations on the plate surface as compared to the solid surface (up to 50% depending on the length and location of the sound-absorbing coating).

1. Introduction

The possibility of controlling the intensity of disturbances in boundary layers and the transition to turbulence is one of the important aspects of the development of perspective hypersonic aircraft. Real gas effects associated with excitation of vibrational degrees of freedom of molecules and flow nonequilibrium are manifested under real flight conditions, where high speeds and temperatures are observed. Real gas properties can significantly affect the generation and evolution of disturbances in the boundary layer and, as a consequence, the transition to turbulence. Nowadays large scientific experience has been accumulated in studying the effect of physical and chemical processes in gases on the mean flow around bodies at hypersonic velocities [1]. However, there are only a few papers on stability of such flows (e.g., [2-4]).

Disturbances formed in the shock layer are carried downstream and affect the evolution of disturbances and the laminar-turbulent transition in the hypersonic boundary layer on a model. One of the known methods for controlling the laminar-turbulent transition in moderately hypersonic flows is the method of applying a sound-absorbing coating onto the surface [5]. The method of sound-absorbing coatings has proved its effectiveness for low-temperature flows at M_∞=21 [6]. It is shown [7] that the method of sound-absorbing coatings is also effective in hypersonic flows of vibrationally excited carbon dioxide. However, the intensity of relaxation processes and their influence on the evolution of disturbances in a mixture with other gases can differ from the processes in pure CO_2.

The present research is focused on experiments conducted in the IT-302M hot-shot wind tunnel based at ITAM SB RAS [4] where the characteristics of the mean flow and pressure fluctuations on the surface of a solid plate and a plate with a porous insert made of foamed nickel (a porosity coefficient of 95%) were measured in mixtures of air and CO_2. The rectangular sound-absorbing insert was

Content from this work may be used under the terms of the Creative Commons Attribution 3.0 licence. Any further distribution of this work must maintain attribution to the author(s) and the title of the work, journal citation and DOI.
Published under licence by IOP Publishing Ltd

located on a plate of length L = 200 mm, width of 50 mm, and thickness of 25 mm at a distance of 90mm from the leading edge. The depth of the porous insert did not exceed the plate thickness.

The flow of a mixture of vibrationally excited carbon dioxide and nitrogen past a flat plate and a plate with a sound-absorbing coating at a nonzero angle of attack under the action of external acoustic disturbances was numerically simulated.

2. Computational domain and skeleton model of a porous material

The computational domain in numerical simulations of flow around the solid plate is a rectangle, and its lower boundary coincides with the plate surface, where the no-slip condition and the condition of a constant surface temperature T_w = 300 K are imposed. The left and upper boundaries are subjected to the free-stream conditions (pressure P_∞, Mach number M_∞, temperature T_∞, and vibrational energy of carbon dioxide molecules $e_{v\infty}$). The exhaustion condition is imposed on the right boundary and on the lower boundary that is not occupied by the plate surface. The computational domain height is chosen from the condition that the bow shock wave does not interact with the upper boundary. The computational grid consists of rectangular cells 0.18 mm×0.1 mm (the number of cells for the plate length of 200 mm is 322000). The computational domain and regular grid were constructed in the WORKBENCH program.

In the problem of the flow past the plate with a porous sound-absorbing coating, the computational domain is supplemented with a subdomain where the skeleton of a porous material made of foamed nickel with a porosity coefficient of 0.95 is directly simulated. The skeleton model is a set of square elements arranged in a staggered order. The distance between the elements is determined by the pore size of the real high-porosity cellular material (the pore diameter is d=2 mm). The size of the square elements (0.4 mm × 0.4 mm) was calculated in accordance with the porosity coefficient of 95% (the skeleton was only 5% of the area of the porous insert). The computational domain is covered by regular rectangular mesh inside the porous zone. The upper boundary of the porous zone is permeable. The no-slip condition and the condition of a constant surface temperature T_w = 300 K are set on other boundaries, including the skeleton walls. The total number of cells in the porous zone is 26419 for the insert with a length of 80 mm and depth of 14 mm.

3. Thermodynamic model of the flow

A system of two-dimensional unsteady Navier–Stokes equations (1)-(3), which is supplemented with the Mendeleev-Clapeyron equation (4) and with the conservation equation of vibrational energy (5) for each vibrational degree of freedom of CO_2 molecules (see below) is solved. The viscosity is calculated by Sutherland's formula; the thermal conductivity is found from the kinetic theory of gases; the heat capacity is determined as a function of temperature within the framework of the thermally perfect gas model:

$$\frac{\partial\rho}{\partial t} + \nabla(\rho\vec{u}) = 0 \tag{1}$$

$$\frac{\partial}{\partial t}(\rho\vec{u}) + \nabla(\rho\vec{u}\vec{u}) = -\nabla p + \nabla\tau_\mu \tag{2}$$

$$\frac{\partial}{\partial t}(\rho e_t) + \nabla[\vec{u}(\rho e_t + p)] == (\lambda\nabla T_{tr} + \tau_\mu\vec{u}) + \nabla\tau_\mu - \sum_{n=1}^{4} q_{tv}^{n\,(CO_2-CO_2)} - \sum_{n=1}^{4} q_{tv}^{n\,(CO_2-N_2)} \tag{3}$$

$$p = \rho R T_{tr} \sum_{i=1,2} Y_i \tag{4}$$

$$\frac{\partial}{\partial t}(\rho e_v^n) + \nabla(u\rho e_v^n) = q_{tv}^{n\,(CO_2-CO_2)} + q_{tv}^{n\,(CO_2-N_2)}, \qquad n = 1,\dots,4 \tag{5}$$

Here ρ, p, and λ are the density, pressure, and thermal conductivity, $\vec{u}$ is the velocity vector, T_{tr} is the temperature of translational-rotational degrees of freedom, n is the mode number of vibrational degrees of freedom, τ_μ is the viscous stress tensor, e_t is the gas energy, R is the universal gas constant, and Y_i is the mole fraction of the i-th component of the mixture.

The model of a thermally perfect gas is considered in this paper. The heat capacity of the gas depends on the temperature $c_p= f(T)$ due to excitation of degrees of freedom of carbon dioxide molecules. If the vibrational relaxation time τ_{vt} is much shorter than the time of convective changes in flow parameters τ_f, the gas is considered to be equilibrium. If these times are comparable ($\tau_{vt} \sim \tau_f$), then the gas is nonequilibrium, and it is necessary to take into account the time of energy exchange between the translational-rotational and vibrational degrees of freedom of molecules.

Four equations (5), which describe the conservation of vibrational energy for each vibrational degree of freedom of CO_2, are added to the Navier–Stokes equations (1)-(3) in the present paper to take into account the effect of vibrational relaxation of CO_2 molecules. The energy exchange between the vibrational and translational-rotational degrees of freedom of CO_2 molecules with a finite relaxation time is taken into account by supplementing Eqs. (5) with source terms q_{vt} (7), (8) calculated by using the Landau–Teller equation, which includes the vibrational relaxation times of CO_2 molecules in collisions with CO_2 molecules [8] and in collisions with N_2 molecules [9]:

$$q_{tv}^{n\,(CO_2-CO_2)} = \rho\left(e_v^{eq\,n} - e_v^n\right)\left(\tau_{vt}^{(CO_2-CO_2)}\right)^{-1} \quad (7)$$

$$q_{tv}^{n\,(CO_2-N_2)} = \rho\left(e_v^{eq\,n} - e_v^n\right)\left(\tau_{vt}^{(CO_2-N_2)}\right)^{-1} \quad (8)$$

Here q_{vt} is the energy flux between the translational and vibrational degrees of freedom, τ_{vt} is the vibrational relaxation time, e_v^{eq} is the equilibrium vibrational energy of the gas, and e_v is the nonequilibrium vibrational energy of gas. The source terms (7) and (8) with the opposite sign are added to the equation of translational-rotational energy (3) to avoid disturbing of the energy balance of the system of the Navier–Stokes equations.

The problem of the evolution of disturbances was solved in two stages. The first stage is the numerical simulation of a hypersonic flow past a plate without disturbances (steady problem). At the second stage, acoustic disturbances of the incoming flow are introduced into the computational domain, and an unsteady problem is solved until the unsteady solution reaches a periodic regime.

4. Numerical simulation of the flow of a mixture of vibrationally excited gases past a solid plate

A parametric study of the flow of a gas mixture of CO_2 (0.44) and N_2 (0.56) (P_∞=374.5 Pa, M_∞=8.44, $Re_1=13.6\times10^6$ m^{-1}, and T_∞=215 K) past a plate with a length L=0.4 m at various angles of attack α=5÷20° and various frequencies of external acoustic disturbances f=40÷160 kHz is performed.

Figure 1 shows the root-mean-square (RMS) pressure fluctuations on the plate (L=400 mm) aligned at the angles of attack α =5 ÷ 20°. One can see not only the disturbance growth regions, but also the regions of local maximums and minimums of the disturbance intensity.

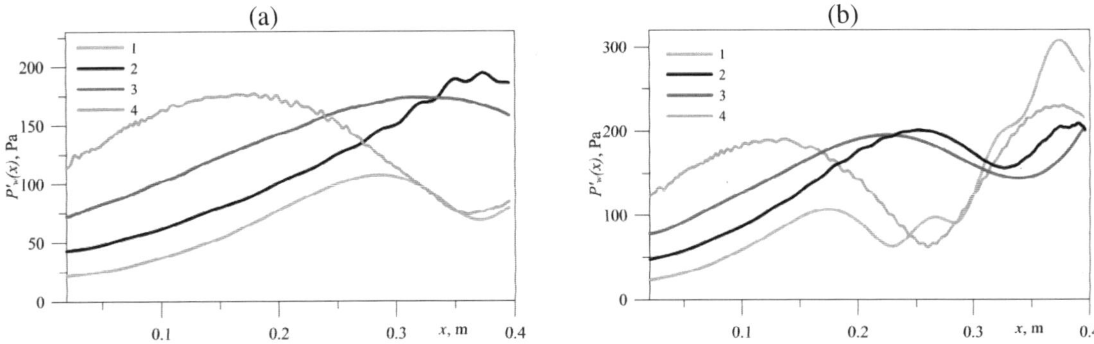

Figure 1. RMS pressure fluctuations on the surface of the solid plate under the action of a fast acoustic wave with the frequency f=120 (a), and 160 kHz (b); α=5° (1), 10.2° (2), 15° (3), and 20° (4); M_∞=8.44 and $Re_1=13.6\times10^6$ m^{-1}.

The good agreement between the experimental and numerical data of the growth rates of the disturbances on the plate surface (the ratio of the amplitudes of pressure fluctuations in two positions, near the model nose x_1=80 mm and at the plate end x_2 =180 mm, i.e. p'_2/p'_1) is demonstrated in Figure 2a. Figure 2b shows that the growth rate of disturbances as a function of the external acoustic disturbance frequency decreases with increasing angle of attack.

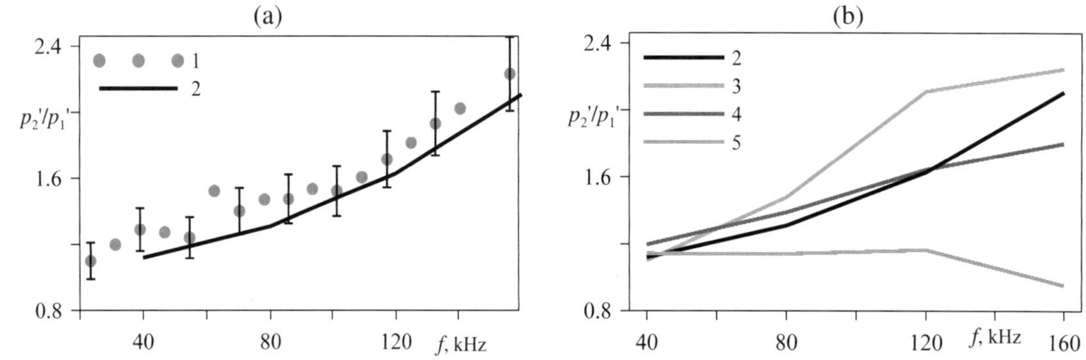

Figure 2. Growth rates of pressure fluctuations on the surface of the solid plate under the action of fast acoustic waves (M_∞=8.44 and $Re_1=13.6\times10^6$ m^{-1}). (a) – comparison with the experiment for α=10.2°: experimental data for the mixture of CO_2 and air (1); calculated data for the mixture of CO_2 and N_2 (2); (b) calculated data: for α=5° (3), 10.2° (2), α=15° (4), and α=20° (5).

The intensity of vibrational relaxation of carbon dioxide molecules depends on the CO_2 concentration in the mixture, because the equilibrium between the translational-rotational and vibrational degrees of freedom is established through collisions between gas molecules. In this paper, the effect of the CO_2 concentration in the mixture with N_2 (both gases are vibrationally excited) on the mean flow characteristics and the evolution of disturbances in the shock layer on the solid plate at the angle of attack α=10.2° is investigated.

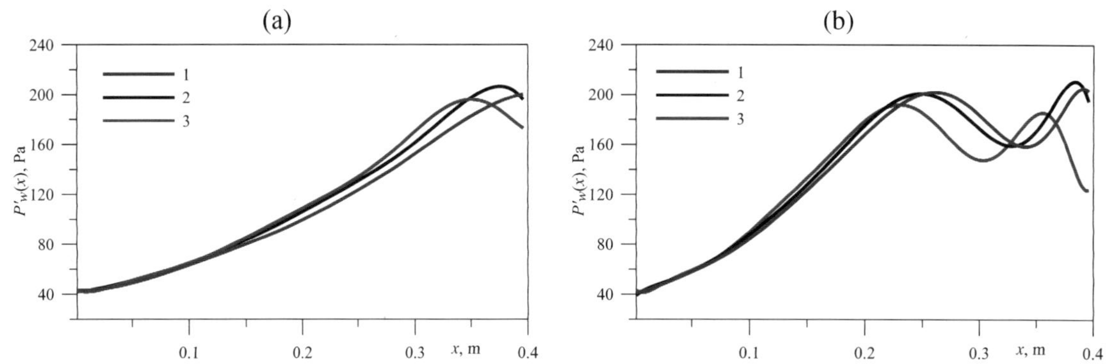

Figure 3. RMS pressure fluctuations on the surface of the solid plate under the action of a fast acoustic wave with the frequency f=120 (a) and 160 kHz (b) for different concentrations of CO_2 in the mixture: 0.22 (1), 0.44 (2), and 0.88 mol (3) (M_∞=8.44 and $Re_1=13.6\times10^6$ m^{-1}).

Figure 3 shows the distributions of the RMS pressure fluctuations on the plate surface for various concentrations of CO_2 in the mixture with N_2 under the action of a fast acoustic wave with the frequency f = 120 and 160 kHz. The excitation and nonequilibrium of the vibrational degrees of freedom of CO_2 molecules are more pronounced in the flow past a plate with a higher content of carbon dioxide in the mixture. Figure 3 shows that this leads to rapid enhancement of the disturbance amplitude along the plate; however, the maximum of the pressure fluctuation amplitude decreases in magnitude and is shifted toward the leading edge of the plate. This behavior of pressure fluctuations testifies to a damping effect of vibrational relaxation of CO_2 molecules, similar to that observed in [2].

5. Effect of the sound-absorbing coating on the mean flow characteristics and on the evolution of disturbances

Investigations of the evolution of disturbances on a solid plate are necessary for solving the problem of the effect of the sound-absorbing coatings aimed at its optimization. Sound-absorbing inserts 80 mm long and 14 mm deep are included in the computational domain at distances of 90 mm (insert I), 200 mm (insert II), and 270 mm (insert III) from the leading edge of the plate. These sound-absorbing zones are simulated by using a skeleton model (see Section 2). The calculations are performed for the mixture of CO_2 (0.44) and N_2 (0.56) (P_∞=374.5 Pa, M_∞=8.44, $Re_1=13.6\times10^6$ m^{-1}, and T_∞=215 K) for the angle of attack α=10.2°.

The results of the mean flow simulation show that the porous coatings practically do not affect the mean flow characteristics. The translational temperature profiles (x=0.25 m) are plotted in Figure 4 for the solid plate and the plate with a sound-absorbing coating. The section x=0.25 m is located downstream of the coating for the plate with insert I, in the porous region for the plate with insert II, and upstream of the porous region on the plate with insert III. It can be seen that the temperature profiles on the sound absorbing plate coincide with the profiles on the solid plate in all cases. This fact suggests that the porous coating practically does not disturb the structure of the mean flow.

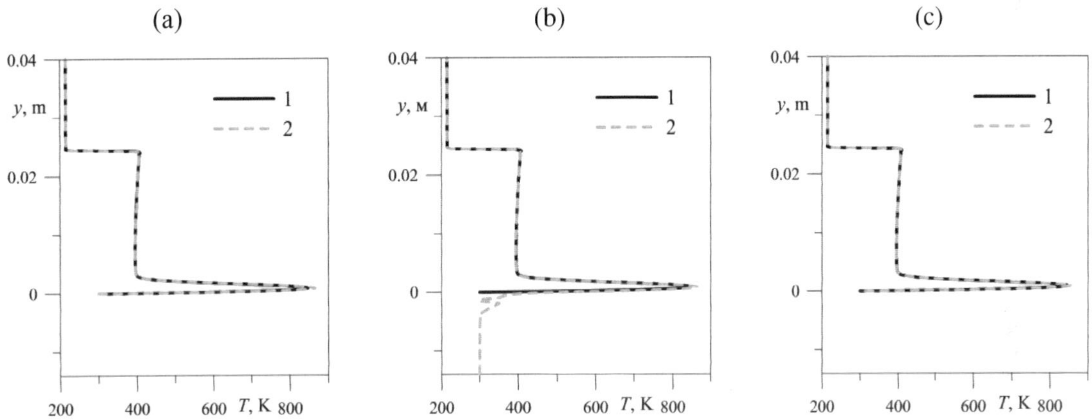

Figure 4. Computed translational temperature profiles on the solid plate (1) and on the plate with a sound-absorbing coating (2) in the section x=0.25m: (a) plate with insert I; (b) plate with insert II; (c) plate with insert III (M_∞=8.44 and $Re_1=13.6\times10^6$ m^{-1}).

The sound-absorbing inserts I and II are located in the region of the growth of disturbances for the plate at the angle of attack 10.2°, whereas insert III is located in the region of the maximum of pressure fluctuations. In all three cases, the sound-absorbing coatings effectively reduce the intensity of pressure fluctuations at frequencies higher than 80 kHz (Figure 5). However, the sound-absorbing coatings located closer to the maximum of the disturbance amplitude are more effective in reducing the intensity of pressure fluctuations.

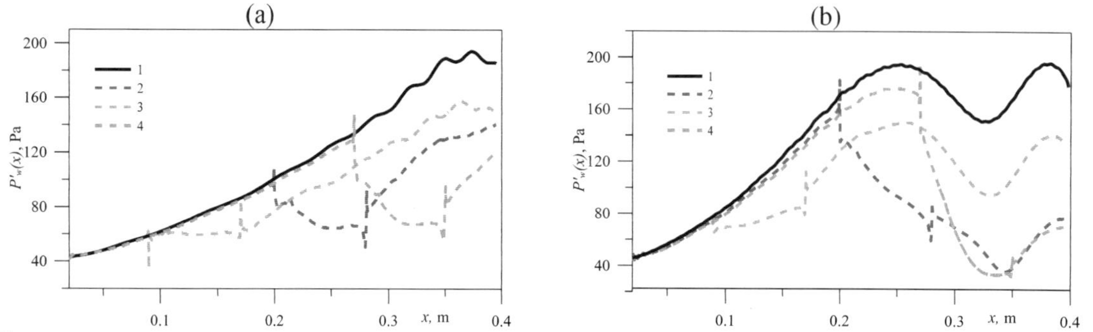

Figure 5. RMS pressure fluctuations on the plate surface under the action of a fast acoustic wave with the frequency *f*=120 (a) and 160 kHz (b); solid plate (1), plate with insert I (2), plate with insert II (3), and plate with insert III (4) (M_∞=8.44 and $Re_1=13.6\times10^6$ m^{-1}).

Two cases are considered to investigate the influence of the porous coating length: the flow past a plate with insert II 80 mm long located in the region of the maximum disturbances and with insert IV with a doubled length (160 mm). Figure 6a shows the field of instantaneous pressure fluctuations on the plate with the sound-absorbing porous coating IV. It can be seen that the intensity of pressure fluctuations is significantly reduced due to interaction with the porous zone. A quantitative comparison of the effect of the porous coating length on the evolution of disturbances is shown in Figure 6b. It is seen that the porous coating placed in the region of the local maximum of disturbances effectively reduces pressure fluctuations (in particular, up to 95% on the surface of the plate with the porous coating IV).

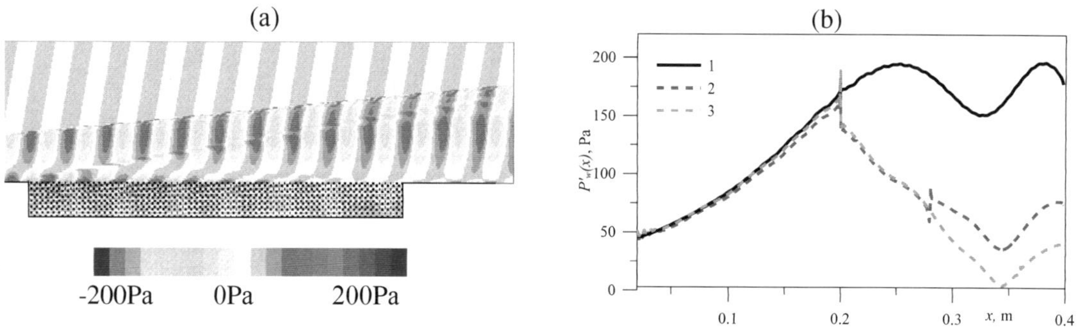

Figure 6. (a) Fields of instantaneous pressure fluctuations with the frequency *f*=160 kHz on the plate with the porous coating IV; (b) RMS pressure fluctuations on the solid plate surface (1), on the plate with the porous coating II (2), and on the plate with the porous coating IV (3) (M_∞=8.44 and $Re_1=13.6\times10^6$ m^{-1}).

6. Conclusions

The flow of a mixture of carbon dioxide and nitrogen past a solid plate and a plate with a sound-absorbing coating at a nonzero angle of attack was numerically simulated for experimental conditions in the IT-302M hot shot wind tunnel based at ITAM SB RAS. Two channels of vibrational relaxation of CO_2 molecules (in collisions with CO_2 molecules and N_2 molecules) were taken into account within the framework of the two-temperature model of relaxing flows.

Data on the evolution of disturbances on the solid plate in a hypersonic flow of a vibrationally excited mixture of carbon dioxide and nitrogen were obtained for different angles of attack, CO_2 concentrations in the mixture, and frequencies of the external acoustic disturbance. The growth rate and intensity of pressure fluctuations on the solid plate surface were demonstrated to increase with increasing angle of attack. An increase in the CO_2 concentration in the mixture was found to enhance the thermal nonequilibrium, and the intensity of pressure fluctuations on the solid plate surface decreased.

The calculated and experimental data on pressure fluctuations on the solid plate surface were found to agree well.

A skeleton model of the porous medium, which is a set of square elements arranged in a staggered order, was implemented for numerical simulations of the flow past a plate with a sound absorbing coating made of a high-porosity cellular material. The distance between the elements was equal to the diameter of the real porous sound-absorbing coating (d=2 mm).

It was shown that the sound-absorbing coating appreciably reduces the pressure fluctuations on the surface. Inserts located in regions of the local maximum or of intensely growing disturbances were found to be the most effective ones. In addition, suppression of disturbances is greatly enhanced (up to 95%) as the porous region length increases. Thus, the method of porous sound-absorbing coatings ensures effective suppression of pressure fluctuations in the flow of vibrationally excited gases.

Acknowledgments
This work was supported by the Russian Foundation for Basic Research (project No. 16-08-00674).

References

[1] Osipov A I and Uvarov A V 1992 *Physics-Uspekhi* **162**(11) 1-42
[2] Grigoryev Y N and Yershov I V 2012 *Stability of Flows of Relaxing Molecular Gases*. Novosibirsk: SB RAS
[3] Fujii K and Hornung H G 2003 *AIAA Journal* **41**(7) 1282-91
[4] Kirilovskiy S V, Maslov A A, Poplavskaya T V and Tsyrulnikov I S 2015 *Technical Physics* **60**(5) 645-55
[5] Fedorov A, Shiplyuk A, Maslov A, et al. 2003 *J. Fluid Mech.* **479** 99–130
[6] Maslov A A, Mironov S G, Poplavskaya T V, Tsyrulnikov I S and Kirilovskiy S V 2012 *J. Appl. Mech. Tech. Phys.* **53**(2) 162-72
[7] Tsyrulnikov I S, Maslov A A, Mironov S G, Poplavskaya T V and Kirilovskiy S V 2015 *J. Appl. Mech. Tech. Phys.* **56**(4) 549–57
[8] Camac M 1966 *Fundamental Phenomena in Hypersonic Flow* 195-218
[9] Losev S A 1977 *Gas-Dynamic Lasers* 1-336

Asymptotic representation for spectrum of linear problem describing flows of viscoelastic polymeric fluid

A Yegitov

Sobolev Institute of Mathematics, Novosibirsk, Russia

E-mail: eav15@mail.ru

Abstract. We study a new rheological model (a modification of a known Pokrovski – Vinogradov model). As was shown by numerical simulations, it takes into account nonlinear effects arising in flows of melts and solutions of polymers in domains with a complex boundary geometry. In the case when the main solution is a Poiseuille-type flow in an infinite plane channel (one considers a viscoelastic polymeric fluid) we obtain an asymptotic formula for the distribution of spectrum points of the linear problem. Small perturbations have such the additional property that they are periodic with respect to the variable going along the side of the channel.

1. Introduction

A new rheological model in which nonlinear effects arising in a polymer medium are taken into account with a reasonable accuracy is studied in the paper. From the physical point of view, the media under consideration is a suspension of noninteracting elastic dumbbells [1]. Each dumbbell is a couple of Brownian particles connected by an elastic force and moves in an anisotropic liquid formed by a solvent and other dumbbells.

This model is a modification of the known model by Pokrovsky and Vinogradov [2], [3]; its main element is a new rheological relation which describes the connection between kinematic characteristics of the flow and internal thermodynamical parameters. Carried out numerical experiments show that results match experimental data well [4], [5].

As the test regime we study the experimentally observed Poiseuille-type flow in an infinite plane channel for the system of the Navier–Stokes equations. The asymptotics of eigenvalues is studied for the problem linearized about a given stationary solution and possible small perturbations are harmonic functions with respect to the variable along the infinite channel with a fixed frequency.

The main result of this work continues a number of facts obtained earlier: in [6] a formal asymptotic representation of the spectrum points for the linear problem in an infinite plane channel was, in particular, found when as the class of perturbations we choose the class of functions decreasing at infinity; it follows from this formula that for the chosen model the Poiseuille-type flow is linearly unstable by Lyapunov; in [7] it is proved that there is no asymptotic linear stability by Lyapunov for Poiseuille-type flows in the class of periodic

Content from this work may be used under the terms of the Creative Commons Attribution 3.0 licence. Any further distribution of this work must maintain attribution to the author(s) and the title of the work, journal citation and DOI.
Published under licence by IOP Publishing Ltd

perturbations provided that the Fourier series contain only harmonics with a chosen limited number of frequencies.

2. Problem statement. Auxiliary information. Formulation of the main result

In [1] a new mathematical model describing flows of an incompressible viscoelastic polymeric fluid was presented. In the 2D case nonstationary flows of polymeric media are described by the following rheological model (written already in a dimensionless form):

$$u_x + v_y = 0, \tag{1}$$

$$\frac{du}{dt} + p_x = \frac{1}{Re}\{(a_{11})_x + (a_{12})_y\}, \tag{2}$$

$$\frac{dv}{dt} + p_y = \frac{1}{Re}\{(a_{12})_x + (a_{22})_y\}, \tag{3}$$

$$\frac{da_{11}}{dt} - 2A_1 u_x - 2a_{12}u_y + K_I a_{11} = -\beta(a_{11}^2 + a_{22}^2), \tag{4}$$

$$\frac{da_{12}}{dt} - A_1 v_x - A_2 u_y + \widetilde{K}_I a_{12} = 0, \tag{5}$$

$$\frac{da_{22}}{dt} - 2A_2 v_y - 2a_{12}v_x + K_I a_{22} = -\beta(a_{12}^2 + a_{22}^2). \tag{6}$$

Here t is the time, u and v are the components of the velocity vector $\mathbf{u}$ in the Cartesian coordinate system x, y, p is the hydrostatic pressure, a_{ij} is the symmetrical anisotropy tensor of the second rank; $\frac{d}{dt} = \frac{\partial}{\partial t} + (\mathbf{u}, \nabla)$ is the substantial derivative.

The remaining quantities are defined as follows: $I = a_{11} + a_{22}$ is the first invariant of the anisotropy tensor, $\bar{k} = k - \beta$, k, β are the scalar phenomenological parameters of the rheological model $(0 < \beta < 1)$, η_0, τ_0 are the initial values of the shear viscosity and the relaxation time;

$$A_1 = a_{11} + \frac{1}{W}, \quad A_2 = a_{22} + \frac{1}{W},$$

$$K_I = \frac{1}{W} + \frac{\bar{k}}{3}I, \quad \widetilde{K}_I = \frac{1}{W} + \frac{\hat{k}}{3}I = K_I + \beta I,$$

$$\hat{k} = k + 2\beta = \bar{k} + 3\beta,$$

$$Re = \frac{\rho u_H l}{\eta_0} \text{ is Reynolds number,}$$

ρ(=const) is the medium density; u_H is the characteristic velocity, l is the characteristic length; $W = \frac{\tau_0 u_H}{l}$ is the Weissenberg number.

Remark 2.1 *In formulation of the rheological model* (1)–(6) *we have the Reynolds and Weissenberg numbers and also the phenomenological parameters k, β defining the flow of the physical experiment. It follows from [8] that the relation $k = 1.2\beta$ is the most adequate to experiments with polymeric fluids.*

The system of linear equations resulting from the linearization of system (1)–(6) about a chosen stationary solution (further its components are marked with "^") in the case when the fluid is moving in an infinite plane channel was obtained in [9]. In the vector form it is formulated as follows. In the domain (see Figure 1)

$$G = \{(t, x, y)\big| t > 0,\ (x, y) \in \Pi = \{(x, y)\big|\ |x| < \infty,\ 0 < y < 1\}\}$$

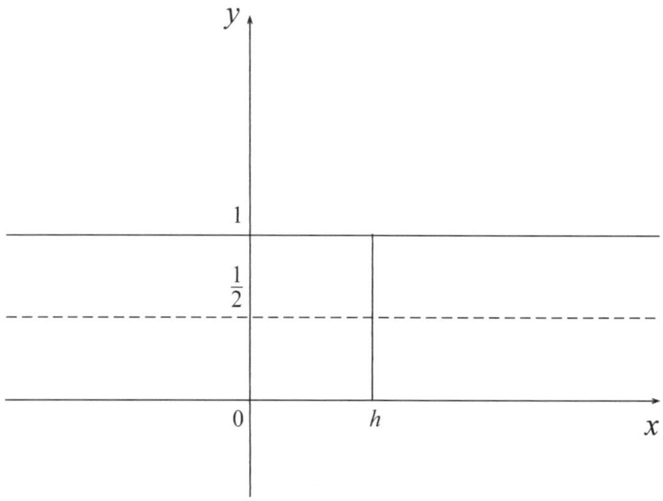

Figure 1

we need to find a solution to the following system of equations:

$$U_t + \hat{B}U_x + \hat{C}U_y + \hat{R}U + F = 0, \tag{7}$$

$$\triangle\Omega = \frac{1}{Re}\{\sigma_{xx} + 2(a_{12})_{xy}\} - 2\hat{\omega}v_x. \tag{8}$$

Here $U = \begin{pmatrix} u \\ v \\ a_{11} \\ a_{12} \\ a_{22} \end{pmatrix}$ is the unknown vector–function, $\sigma = a_{11} - a_{22}$, $\Omega = p - \frac{1}{Re}a_{22}$,

the matrices $\hat{B} = B(\hat{U})$, $\hat{C} = C(\hat{U})$, $\hat{R} = R(\hat{U})$ are written with the help of the components of the stationary solution $\hat{U}(y)$:

$$\hat{U}(y) = \begin{pmatrix} \hat{u}(y) \\ 0 \\ \hat{a}_{11}(y) \\ \hat{a}_{12}(y) \\ \hat{a}_{22}(y) \end{pmatrix}, \ \hat{B} = \begin{pmatrix} \hat{u} & 0 & -\frac{1}{Re} & 0 & 0 \\ 0 & \hat{u} & 0 & -\frac{1}{Re} & 0 \\ -2\hat{A}_1 & 0 & \hat{u} & 0 & 0 \\ 0 & -\hat{A}_1 & 0 & \hat{u} & 0 \\ 0 & -2\hat{a}_{12} & 0 & 0 & \hat{u} \end{pmatrix},$$

$$\hat{C} = \begin{pmatrix} 0 & 0 & 0 & -\frac{1}{Re} & 0 \\ 0 & 0 & 0 & 0 & -\frac{1}{Re} \\ -2\hat{a}_{12} & 0 & 0 & 0 & 0 \\ -\hat{A}_2 & 0 & 0 & 0 & 0 \\ 0 & -2\hat{A}_2 & 0 & 0 & 0 \end{pmatrix}, \ \hat{R} = \begin{pmatrix} 0 & \hat{\omega} & 0 & 0 & 0 \\ 0 & 0 & 0 & 0 & 0 \\ 0 & \hat{a}_{11}' & R_{33} & R_{34} & R_{35} \\ 0 & \hat{a}_{12}' & R_{43} & R_{44} & R_{45} \\ 0 & \hat{a}_{22}' & R_{53} & R_{54} & R_{55} \end{pmatrix}, \tag{9}$$

where $\hat{A}_1 = \hat{a}_{11} + \frac{1}{W}$, $\hat{A}_2 = \hat{a}_{22} + \frac{1}{W}$,

$$R_{33} = \frac{1}{W} + \frac{\bar{k}}{3}\hat{I} + \frac{k + 5\beta}{3}\hat{a}_{11}, \ R_{34} = -2(\hat{\omega} - \beta\hat{a}_{12}), \ \hat{\omega} = \hat{u}_y, \ R_{35} = \frac{\bar{k}}{3}\hat{a}_{11},$$

IOP Conf. Series: Journal of Physics: Conf. Series **894** (2017) 012138 doi:10.1088/1742-6596/894/1/012138

$$R_{43} = \frac{\hat{k}}{3}\hat{a}_{12},\ R_{44} = \frac{1}{W} + \frac{\hat{k}}{3}\hat{I},\ R_{45} = -\hat{\omega} + \frac{\hat{k}}{3}\hat{a}_{12},$$

$$R_{53} = \frac{\bar{k}}{3}\hat{a}_{22},\ R_{54} = 2\beta\hat{a}_{12},\ R_{55} = \frac{1}{W} + \frac{\bar{k}}{3}\hat{I} + \frac{k+5\beta}{3}\hat{a}_{22},$$

$F = \begin{pmatrix} p_x \\ p_y \\ 0 \\ 0 \\ 0 \end{pmatrix}$, "$\triangle$" is the symbol of the Laplace operator.

We will assume that on the boundary of the domain G the conditions

$$u\Big|_{y=0} = v\Big|_{y=0} = u\Big|_{y=1} = v\Big|_{y=1} = 0; \tag{10}$$

$$\Omega_y = \frac{1}{Re}(a_{12})_x \quad \text{for } y = 0, 1 \tag{11}$$

hold the initial data

$$U\Big|_{t=0} = U_0(x, y) \tag{12}$$

are prescribed. Moreover, the initial data satisfy the boundary conditions (10).

Remark 2.2 *As the basic solution we can consider, for example, a solution similar to the Poiseuille solution of the Navier-Stokes equations (see [4], [10], [11]) which is symmetric with respect to the channel's axis $y = \frac{1}{2}$ (and $\hat{p}(x,y) = \frac{1}{Re}\hat{a}_{22}(y) + \hat{p}_0 - \hat{A}x$, $\hat{p}_0$ is the pressure on the channel's axis, $\hat{A}$ is the parameter connected with the dimensionless pressure difference on the segment h).*

In [9] for $k = \beta$ the components $\hat{u}(y)$, $\hat{a}_{11}(y)$, $\hat{a}_{12}(y)$, $\hat{a}_{22}(y)$ of this solution are found explicitly, and for $k \neq \beta$ they are obtained numerically with the help of the proposed iterative process. Figure 2 shows profiles of the function $\hat{u}(y)$ for the following values of the parameters $k, \beta, \hat{D}$ $(= Re\hat{A})$ and W: $\bar{k} = 0$, $\hat{D} = 2$, $W = 1$. The graphs numbered as 1,2,3,4 correspond to the values $\beta = 0.1, 0.2, 0.3, 0.4$.

Remark 2.3 *As was noted in [4], [10] (see also [9]), the profile of the function $\hat{u}(y)$ is different from the parabolic Poiseuille profile for the viscous fluid. The maximum value of the velocity $\hat{u}_{max}$ is found by using the following formula:*

$$\hat{u}_{max} = \hat{u}(\frac{1}{2}) = \frac{\hat{D}}{8} + \int_0^{\frac{1}{2}} \frac{\frac{\bar{k}}{3}\hat{I} - \hat{a}_{22}}{\hat{A}_2}(\xi)\hat{a}_{12}(\xi)d\xi.$$

Remark 2.4 *In [9] it is proved that system (7) with the given pressure $p(t,x,y)$ t is hyperbolic (see [12]) if $\hat{A}_1 > 0$, $\hat{A}_2 > 0$ and $\hat{A}_1\hat{A}_2 - \hat{a}_{12}^2 > 0$ (see the representation of the matrices $\hat{B}$ and $\hat{C}$). These inequalities are valid when as the basic solution we chose, for example, the Poiseuille solution (this is checked explicitly for $k = \beta$ numerically for $k \neq \beta$). The information about the roots of the characteristic equation plays a significant role in the formulation of initial boundary value problems for t-hyperbolic systems.*

We will search a solution of the linear problem (7)–(12) in the class of vector-functions which have the following representation:

$$M(t,x,y) = \begin{pmatrix} U(t,x,y) \\ \Omega(t,x,y) \\ \frac{\partial\Omega}{\partial y}(= L) \end{pmatrix} = e^{-i\xi x}\begin{pmatrix} \bar{U}(t,y) \\ \bar{\Omega}(t,y) \\ \bar{L}(t,y) \end{pmatrix},\ \xi \in R \text{ is fixed parameter.} \tag{13}$$

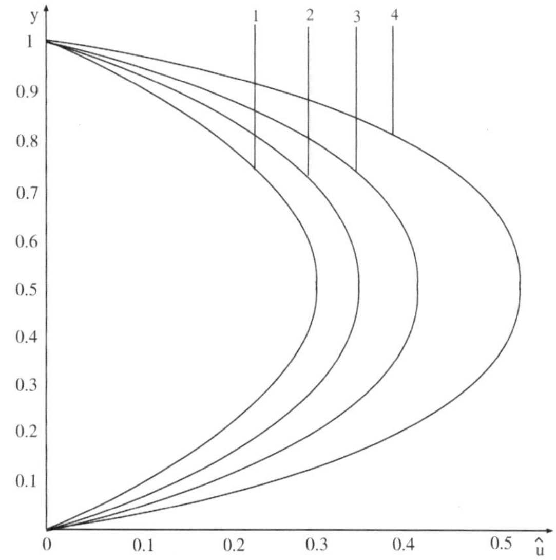

Figure 2

Applying representation (13) and using formally the Laplace transform in time, we obtain the model problem with the parameters ω and λ:

$$\hat{C}U_y^L + (\lambda I - i\xi\hat{B} + \hat{R})U^L + \tilde{F} = U_0^*(y), \tag{14}$$

$$\Omega_y^L = L^L, \tag{15}$$

$$L_y^L - \xi^2\Omega^L = -\frac{2i\xi}{Re}(a_{12}^L)_y - \frac{1}{Re}(a_{11}^L - a_{22}^L)\xi^2 + 2i\xi\hat{\omega}v^L, \tag{16}$$

where L denotes the Laplace transform of a function or a component of a vector-function, λ is the dual variable for t with respect to the Laplace transform:

$$\tilde{F} = \begin{pmatrix} -i\xi(\Omega^L + \frac{1}{Re}a_{22}^L) \\ L + \frac{1}{Re}(a_{22})_y^L \\ 0 \\ 0 \\ 0 \\ 0 \end{pmatrix}, \quad U_0(x,y) = e^{i\xi x}U_0^*(y). \tag{17}$$

The boundary conditions (10), (11) are transformed as

$$u^L|_{y=0} = v^L|_{y=0} = u^L|_{y=1} = v^L|_{y=1} = 0, \tag{18}$$

$$L^L = -i\xi a_{12}^L \quad \text{for } y = 0, 1. \tag{19}$$

Let us assume that the following inequality holds:

$$\hat{A}_2(y) \neq Re, \quad y \in [0,1]. \tag{20}$$

We are now in a position to formulate the main result of the paper.

Theorem 2.1 *If condition (20) is satisfied, then the eigenvalues of the boundary value problem (14)–(19) have the following asymptotic representation:*

$$\lambda_k = -\frac{1}{2\int_0^1 \sqrt{\frac{Re}{\hat{A}_2}}d\eta}\Big[\int_0^1 (d_{11} - d_{22})d\xi + (2k+1)i\pi\Big] + O(\frac{1}{|k|}),$$

where k is an integer number, $|k| \to \infty$, the functions d_{ii}, $i = 1, 2$ are defined as follows

$$d_{11} = \sqrt{\frac{Re}{\hat{A}_2}}\Big(- i\xi\hat{u} + R_{44} + i\xi\frac{\hat{a}_{12}}{\sqrt{Re\hat{A}_2}} + \frac{\hat{a}'_{12}R_{43}}{\hat{A}_2} + \frac{1}{2\sqrt{Re\hat{A}_2}}(\frac{1}{\sqrt{Re\hat{A}_2}})'\Big),$$

$$d_{22} = -\sqrt{\frac{Re}{\hat{A}_2}}\Big(\frac{\sqrt{Re\hat{A}_2}}{2}(\frac{-i\xi\hat{u}}{\sqrt{Re\hat{A}_2}} - \frac{2\hat{a}_{12}i\xi}{Re\hat{A}_2}) + \frac{1}{2}(2\frac{\hat{a}_{12}}{\hat{A}_2}R_{43} - i\xi\hat{u} + R_{44}) + \frac{1}{2}\sqrt{Re\hat{A}_2}(\frac{1}{\sqrt{Re\hat{A}_2}})'\Big),$$

and ξ is a fixed parameter.

Remark 2.5 *The initial boundary value problem (7)–(12) thus allows the Laplace transform with respect to the time.*

Acknowledgments
This work is partially supported by RSBR grant 17-01-00791.

References
[1] Altukhov Yu A, Golovicheva E and Pyshnograi G V 2000 Molecular approach in linear polymers dynamics: Theory and numerical experiment *Fluid Dynamics* **35** 1–9
[2] Pyshnograi G V, Pokrovski V N, Yanovski Yu G, Obraztsov I F and Karnet Yu N 1991 Defining equation for nonlinear viscoelastic (polymeric) mediums in zero approxmation by parameters of molecular theory and results for shear and stretch *DAN SSSR* **355** 612–5
[3] Volkov V S and Vinogradov G V 1984 Molecular theories of nonlinear viscoelasticity of polymers *Rheol. Acata* **23** 231–7
[4] Altukhov Yu A, Gusev A S, Makarova M A and Pyshnograi G V 2007 Generalization of the poiseuille law for plane-parallel flow of viscoelastic media *Journal on Composite Mechanics and Design* **13** 581–90
[5] Altukhov Yu A and Pyshnograi G V 2001 Entrance flows of fluid lineare polymers in 4:1 canal *Journal on Composite Mechanics and Design* **7** 16–23
[6] Blokhin A M, Yegitov A V and Tkachev D L 2015 Lineaer instability of solutions in a mathematical model describing polymer flows in an infinite channel *Comput. Math. and Math. Phys.* **55** 848–73
[7] Blokhin A M and Tkachev D L 2014 Linear asymptotic instability of a stationary flow of a polymeric medium in a plane channel in the case of periodic perturbations *J. Appl. Ind. Math.* **8** 467–78
[8] Altukhov Yu A, Gusev A S and Pyshnograi G V 2012 *Introduction into Mesoscopic Theory of Flowing Polymeric Systems* (Barnaul': Alt SPU)
[9] Bambaeva N V and Blokhin A M 2014 Stationary equations of incompressible viscoelastic polymer liquid *Comput. Math. and Math. Phys.* **54** 874–99
[10] Wassner E, Schmidt M and Munstedt H 1999 Entry flow of a low-density-polyethylene melt into a slit die: An experimental study by laser – Doppler velocimetry *J. Rheol.* **43** 1339–53
[11] Loitsyanski L G 1979 *Gas and Fluid Mechanics* (Moscow: Nauka)
[12] Godunov S K 1979 *Equations of Mathematical Physics* (Moscow: Nauka)

Cumulative jet generation in a plane parallel potential flow of the perfect incompressible fluid

N D Baykov[1] and A G Petrov[2]

[1] Moscow State University, Moscow, Russia
[2] Institute for Problems in Mechanics of the Russian Academy of Sciences, Moscow, Russia

E-mail: baikov_nd@rambler.ru, petrovipmech@gmail.com

Abstract. This manuscript deals with the plane parallel unsteady potential flow of the perfect incompressible fluid limited by a free boundary. As the result of the free boundary deformation, cumulative jets may form. A cumulative jet generation in a problem of cylindrical cavity deformation is considered in this manuscript. A numerical algorithm for this problem is proposed. Based on the boundary elements method with no saturation, the algorithm has a high order of approximation and reasonable computation time. The distribution of points along the length of the cavity boundary is required to be constant at computation time. This requirement provides the numerical stability of the algorithm. To verify the correctness of the proposed algorithm, conservation laws are used.

1. Introduction

Cumulative jet generation is a particular case of a problem with a free boundary. Its numerical modelling is one of areas in computational methods and fluid mechanics. This manuscript studies numerically cumulative jets that are formed within cylindrical cavities under different potential flows. Only plane parallel flows are considered. Fluid is perfect and incompressible.

There are various semi-analytical and numerical methods for the problem. The most notable results have been obtained by Voinov in paper [1]. It proposes a numerical algorithm for one of the simplest cases of the problem, i.e. deformation of a circular cylindrical cavity with a unit radius in a plane parallel flow. Potential and stream functions for the problem at initial time are given by the equations

$$\Phi = \frac{x}{x^2+y^2}, \quad \Psi = -\frac{y}{x^2+y^2}. \tag{1}$$

(the initial condition for the problem of flow past the circular cylindrical cavity moving with a unit velocity). Calculation results have shown that a cumulative is formed under the flow and the cavity splits into two symmetrical parts.

This manuscript proposes a numerical algorithm for the problem. It is more stable rather than the one proposed in paper [1] and allows us to improve the previous calculation results.

Cumulative jets generation under asymmetrical plane parallel potential flows with circulation is also considered in this manuscript.

To verify the correctness of the proposed algorithm, conservation laws are used.

Content from this work may be used under the terms of the Creative Commons Attribution 3.0 licence. Any further distribution of this work must maintain attribution to the author(s) and the title of the work, journal citation and DOI.
Published under licence by IOP Publishing Ltd

2. General equations

Since the proposed algorithm is based on the boundary elements method, all of the calculations are made on the cavity boundary ∂S. The basic equation is written in integral form and expresses the relationship between partial derivative of potential function Φ with respect to spatial coordinate and stream function Ψ

$$-2\pi\Psi = A\frac{\partial\Phi}{\partial s} + B\Psi, \tag{2}$$

where

$$\left[A\frac{\partial\Phi}{\partial s}\right](M,t) := \oint\limits_{\partial S} G(M,M')\frac{\partial\Phi}{\partial s'}(M',t)\, ds', \tag{3}$$

$$[B\Psi](M,t) := \oint\limits_{\partial S} \frac{\partial G}{\partial n'}(M,M')(\Psi(M',t) - \Psi(M,t))\, ds', \tag{4}$$

$$G(M,M') := \ln(r(M,M')),$$
$$r^2(M,M') := (x-x')^2 + (y-y')^2,$$
$$M(x,y),\ M'(x',y') \in \partial S.$$

Cylindrical cavity boundary in a plane parallel flow at any moment of time t is one-dimensional and can be parameterized by a single variable. Let us denote it by ζ. The following parameterization is used in the proposed algorithm:

$$ds = l(t)f(\zeta)d\zeta, \quad 0 \le \zeta \le 1, \qquad \int\limits_0^1 f(\zeta)\, d\zeta = 1, \tag{5}$$

where $l(t)$ is the overall length of the boundary and $f(\zeta)$ is a positive function that is used to manage the distribution of points along the cavity boundary. Let us call it the density function. The reason for this parameterization is numerical stability. Points with fixed values of ζ will be used to approximate cavity boundary, therefore the distribution of points along the cavity boundary will be constant at computation time. Due to the choice of density function, we can arbitrary change the number of points that are used to approximate different segments of the boundary.

The normal component of the boundary velocity is equal to the fluid velocity and can be calculated as a partial derivative of the stream function Ψ with respect to spatial coordinate s

$$V(\zeta,t) = \frac{\partial\Psi}{\partial s}(\zeta,t) = \frac{1}{l(t)f(\zeta)}\cdot\frac{\partial\Psi}{\partial\zeta}(\zeta,t). \tag{6}$$

Tangential and normal vectors are required for further calculations. They can be derived from the following equations:

$$\tau(\zeta,t) = \left(\frac{\partial x}{\partial s}, \frac{\partial y}{\partial s}\right) = \frac{1}{l(t)f(\zeta)}\left(\frac{\partial x}{\partial\zeta}, \frac{\partial y}{\partial\zeta}\right), \tag{7}$$

$$n(\zeta,t) = \left(\frac{\partial y}{\partial s}, -\frac{\partial x}{\partial s}\right) = \frac{1}{l(t)f(\zeta)}\left(\frac{\partial y}{\partial\zeta}, -\frac{\partial x}{\partial\zeta}\right). \tag{8}$$

Using the equations above, we can calculate spatial coordinates x and y as follows:

$$\frac{\partial x}{\partial t}(\zeta,t) = \frac{1}{l(t)f(\zeta)}\left(U\frac{\partial x}{\partial\zeta} + V\frac{\partial y}{\partial\zeta}\right), \tag{9}$$

$$\frac{\partial y}{\partial t}(\zeta,t) = \frac{1}{l(t)f(\zeta)}\left(U\frac{\partial y}{\partial \zeta} - V\frac{\partial x}{\partial \zeta}\right), \tag{10}$$

where U is the tangential component of velocity. It is chosen to keep the distribution of points along the cavity boundary unchanged at computation time.

Finally, we will use Bernoulli's equation:

$$\left.\frac{\partial \Phi}{\partial t}\right|_{x,y} + \frac{1}{2}\left(V^2 + \left(\frac{\partial \Phi}{\partial s}\right)^2\right) + p = 0, \tag{11}$$

where $p(t)$ is an unknown pressure function. We can express the derivative of Φ with respect to t at constant ζ using the derivative of Φ with respect to t at constant x, y by the following formula:

$$\left.\frac{\partial \Phi}{\partial t}\right|_{\zeta} = \left.\frac{\partial \Phi}{\partial t}\right|_{x,y} + U\frac{\partial \Phi}{\partial s} + V^2. \tag{12}$$

Therefore, we have

$$\left.\frac{\partial \Phi}{\partial t}\right|_{\zeta} + p = \frac{1}{2}V^2 + \frac{U}{l(t)f(\zeta)}\frac{\partial \Phi}{\partial \zeta} - \frac{1}{2}\left(\frac{1}{l(t)f(\zeta)}\frac{\partial \Phi}{\partial \zeta}\right)^2. \tag{13}$$

Note that we need only partial derivatives of Φ with respect to spatial coordinates in equation (2), therefore we can replace Φ with the following function:

$$\tilde{\Phi}(\zeta,t) := \Phi(\zeta,t) + \int_0^t p(t')dt'. \tag{14}$$

The final form of equation (11) is

$$\left.\frac{\partial \tilde{\Phi}}{\partial t}\right|_{\zeta} = \frac{1}{2}V^2 + \frac{U}{l(t)f(\zeta)}\frac{\partial \tilde{\Phi}}{\partial \zeta} - \frac{1}{2}\left(\frac{1}{l(t)f(\zeta)}\frac{\partial \tilde{\Phi}}{\partial \zeta}\right)^2. \tag{15}$$

3. Approximation

Let us choose N points along the length of the cavity boundary for numerical computations. Each point corresponds to the fixed value $\zeta_i = \frac{i}{N}$, where $i = 1, 2, ..., N$. In this case distribution of points will be constant at computation time. This requirement is used to calculate tangent components of velocity U_i. To satisfy it numerically, we require that the distances between adjacent points change in proportion to the length of the boundary. It can be expressed by the following system of equations:

$$\Delta s_i(t+\Delta t) = \frac{l(t+\Delta t)}{N} f\left(\frac{\zeta_{i-1}+\zeta_i}{2}\right), \tag{16}$$

where $i = 1, 2, \ldots, N$ and $\Delta s_i(t + \Delta t)$ is the distance between points with indices i and $i + 1$. Expressing $\Delta s_i(t + \Delta t)$ in terms of spatial coordinates x_i, y_i and velocities V_i, U_i at the moment of t, we derive a system of nonlinear equations for U_i. Note that we still can arbitrary change one of the tangent components U_i because a position of initial point ($\zeta = 0$) is not fixed. Iterative process is used to solve the system of equations.

The curvature of the boundary segment where cumulative jet is formed usually increases in course of time. To provide high accuracy of computations on the forward end of cumulative jet, a high number of points should be used to approximate it. We can guarantee this because of the density function $f(\zeta)$ choice.

To approximate equation (2), we should replace integral terms with quadrature formulas. Integral terms are represented by A and B operators. Both of them have a point of singularity and should be approximated very accurate. The advantage of the proposed algorithm is that quadrature formulas with no saturation are used. These quadratures have been proposed in paper [2]. The idea behind them is to construct quadrature formulas that yield the exact result for the first $\frac{N}{2}$ partial sums of Fourier series. Since the error of approximation by Fourier series of infinitely differentiable periodic function decaying faster than $O(\frac{1}{N^n})$ for all $n > 0$, high order of approximation is granted by the quadrature formulas. Quadrature formulas for the problem have the form

$$\int_0^1 G(\zeta_i, \zeta') \frac{\partial \Phi}{\partial \zeta'}(\zeta', t)\, d\zeta' \approx \frac{1}{N} \sum_{j=1}^{N} (\, \beta(|i-j|) + G_{ij}) \frac{\partial \Phi}{\partial \zeta'}(\zeta_j, t), \tag{17}$$

$$\beta(0) = \alpha(0), \quad \beta(m) = -\ln\left|\sin\frac{\pi m}{N}\right| + \alpha(m),$$

$$\alpha(m) = -\left(\ln 2 + \frac{(-1)^m}{N} + \sum_{k=1}^{\frac{N}{2}-1} \frac{1}{k} \cos\frac{2\pi k m}{N}\right),$$

$$G_{ij} = G(\zeta_i, \zeta_j) \quad i \neq j,$$

$$G_{ii} = \lim_{\zeta \to \zeta_i} (G(\zeta_i, \zeta) - \ln|\sin(\pi(\zeta_i - \zeta))|),$$

$$\int_0^1 \frac{\partial G}{\partial n'}(\zeta, \zeta')(\Psi(\zeta', t) - \Psi(\zeta, t))\, d\zeta' \approx \frac{1}{N} \sum_{j=1}^{N} G^n_{ij}(\Psi(\zeta_j, t) - \Psi(\zeta_i, t)), \tag{18}$$

$$G^n_{ij} = \frac{x_j - x_i}{r^2_{ij}} \cdot \frac{\partial y}{\partial \zeta}(\zeta_j, t) - \frac{y_j - y_i}{r^2_{ij}} \cdot \frac{\partial x}{\partial \zeta}(\zeta_j, t) \quad i \neq j,$$

$$G^n_{ii} = 0.$$

Most of the differential equations contain partial derivatives of Φ, Ψ, x and y with respect to spatial coordinates. These derivatives can be approximated sufficiently accurate using a cubic spline for periodic functions.

To calculate the potential function and spatial coordinates functions on the next time step, the Runge–Kutta method is applied to equations (9), (10) and (15).

4. Calculation results

The proposed algorithm was implemented and tested using the problem stated in [1], i.e. evolution of a circular cylindrical cavity with a unit radius in a plane parallel potential flow. 128 points were used to approximate the cavity boundary. Near the right edge of the cavity a cumulative jet is formed. The forward end of the jet has a high curvature, therefore a high number of points is required to approximate it. Density function

$$f(\zeta) = 1 - 0.99\cos(2\pi\zeta). \tag{19}$$

was used by the algorithm. The distribution is symmetrical about the x axis.

Cavity boundaries for eight different moments of time are shown on Figure 1, i.e. $t = 0$, 0.17, 0.31, 0.43, 0.57, 0.71, 0.86 and 0.93. All of them except the last one have been chosen to

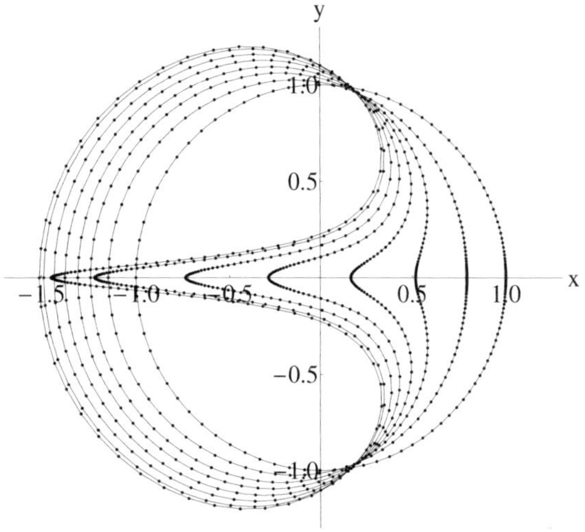

Figure 1. Circular cavity.

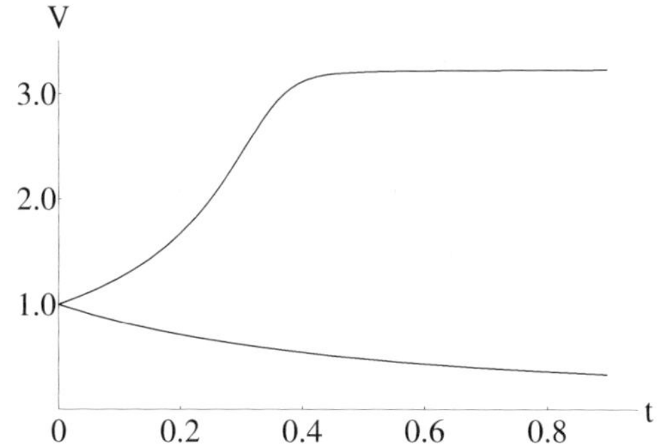

Figure 2. Circular cavity.

demonstrate results in paper [1]. The proposed algorithm has shown similar results with those obtained in paper [1], but the cumulative jet calculated by the algorithm was closer to hitting the opposite side. The left-most point of the cumulative jet and the point on the opposite side velocities are shown on Figure 2. Note that cumulative jet velocity turns stable in course of time.

To verify correctness of the proposed algorithm, conservation of energy, momentum and area were checked. The following equations were used to calculate energy and momentum:

$$E_c = -\frac{1}{2}\int_0^1 \Phi \frac{\partial \Psi}{\partial \zeta} d\zeta, \tag{20}$$

$$I_x = -\int_0^1 \Phi \frac{\partial y}{\partial \zeta} d\zeta. \tag{21}$$

When approximating the cavity boundary with 128 points, overall errors for them did not exceed $3 \cdot 10^{-4}$.

Let us consider more complex example viz. cumulative jet generation under asymmetric plane parallel potential flow with circulation.

In this case potential function at initial time is given by the following equation:

$$\Phi = \frac{\cos\varphi}{r} + \frac{\Gamma}{2\pi}\varphi, \tag{22}$$

where φ and r are angular and radial coordinates of the point on the cavity boundary and Γ is a circulation. Three different cases of $\Gamma = 1, 5, 10$ were considered. Calculation results are shown on Figures 3, 4, 5.

The results show that cumulative jet shifts from horizontal axis under the flow with circulation, therefore additional efforts are required to keep a high density of points on the jet at computation time. This difficulty was overcome by shifting points along the cavity boundary. Tangent component of velocity U_N was chosen in such a way to keep corresponding point on the

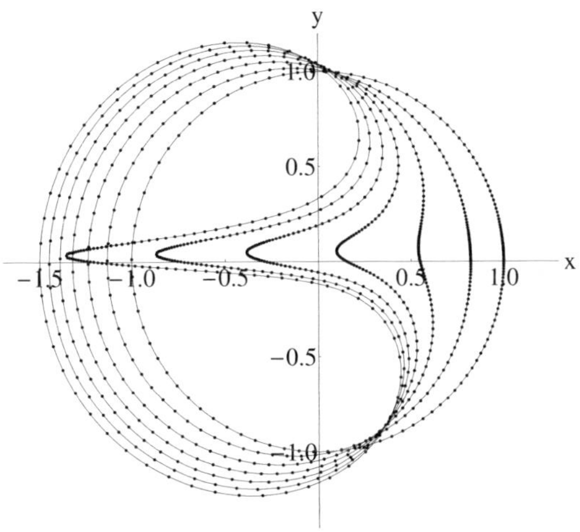

Figure 3. Flow with circulaion ($\Gamma = 1$).

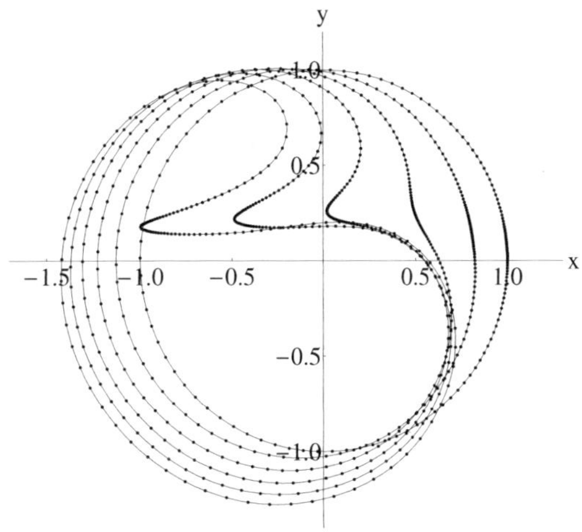

Figure 4. Flow with circulaion ($\Gamma = 5$).

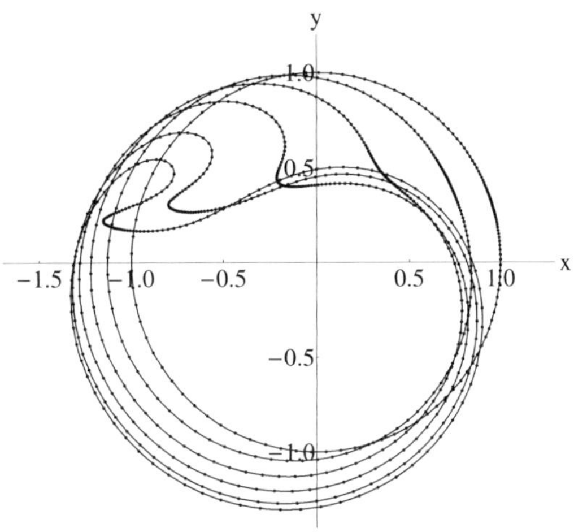

Figure 5. Flow with circulaion ($\Gamma = 10$).

forward end of the jet. Experimentally it was obtained that U_N can be set automatically using tangent component of the fluid velocity in this point. Tangent component of fluid velocity can be found as a partial derivative of potential function Φ with respect to spatial coordinate.

Note that Φ is a multivalued function because of circulation. It must be taken into account when deriving conservation laws. Conservation of linear momentum can be derived from equation

$$\frac{d}{dt}\int_S F\mathbf{n}ds = \int_S \left(\frac{\partial F}{\partial t}\mathbf{n} + v_n \nabla F\right) ds \qquad (23)$$

that is valid for all continuously differentiable single-valued functions F. Applying this equation to single-valued function $F = \Phi - \frac{\Gamma}{2\pi}\varphi$ and transforming it, we can derive the following equations:

$$\frac{d}{dt}\int\limits_S \left(\Phi - \frac{\Gamma}{2\pi}\varphi\right) n_x ds + \frac{\Gamma}{2\pi}\int\limits_S v_n \frac{\partial \varphi}{\partial x} ds = 0, \tag{24}$$

$$\frac{d}{dt}\int\limits_S \left(\Phi - \frac{\Gamma}{2\pi}\varphi\right) n_y ds + \frac{\Gamma}{2\pi}\int\limits_S v_n \frac{\partial \varphi}{\partial y} ds = 0. \tag{25}$$

Using the equations above, we can derive:

$$I_x = \int\limits_0^1 \left(-\frac{\partial \Phi}{\partial \zeta} y + \frac{\Gamma}{2\pi}\left(\frac{\partial \varphi}{\partial \zeta} y + \ln r \frac{\partial x}{\partial \zeta}\right)\right) d\zeta, \tag{26}$$

$$I_y = \int\limits_0^1 \left(\frac{\partial \Phi}{\partial \zeta} x + \frac{\Gamma}{2\pi}\left(-\frac{\partial \varphi}{\partial \zeta} x + \ln r \frac{\partial y}{\partial \zeta}\right)\right) d\zeta. \tag{27}$$

Conservation laws (26) and (27) were checked at computation time. Their overall error did not exceed $1 \cdot 10^{-4}$.

This work was supported by the Russian Science Foundation (project No 14-19-01633).

References

[1] Voinov O V and Voinov V V 1975 Numerical method of calculating nonstationary motions of an ideal incompressible liquid with free surfaces *Sov. Phys. Dokl.* **20** 179–82

[2] Petrov A G 2008 Quadrature formulas for periodic functions and their application to the boundary element method *Comp. Math. Math. Phys.* **48** 1266–83

IOP Conf. Series: Journal of Physics: Conf. Series **894** (2017) 012140 doi:10.1088/1742-6596/894/1/012140

Numerical analysis of corrugated shells rotation

S S Rassokha, S V Ladov and A V Babkin

Bauman University, Moscow, Russia

Email: rassokha@gmail.com

Abstract. In the paper, we've made an attempt to numerically analyse a rotational motion of the fluted shells collapsing under dynamic loading of expanding gases. It was shown that the primary cause of the rotation is a gas oblique shock wave interaction with the fluted surface of a shell. Also, we evaluate the influence of a shell material properties and cross-section dimensions on jet angular velocity. Obtained results are in a good agreement with known experimental data.

1. Introduction

There is a well-known effect of corrugated conical shell rotation under expanding gas dynamic loading [1-10]. Under gas pressure, a liner is collapsing and producing a rotating jet. Direction and magnitude of angular velocity depend on a corrugated cone geometry. All corrugated liners types are divided into five groups by the type and presence of the rifling on a shell two lateral surfaces. The most interesting type of a corrugated shell is the one which has rifling on both internal and external surfaces. Its cross-section can be described by five design parameters (Figure 1): a – corrugation depth; R – radius of the external lateral surface; T – wall thickness; δ – angle of indexing (angular of overlap between rifling on external and internal cone surfaces); ψ – angle between radial direction and the shortest wave side; n – total number of corrugation waves in circumferential direction.

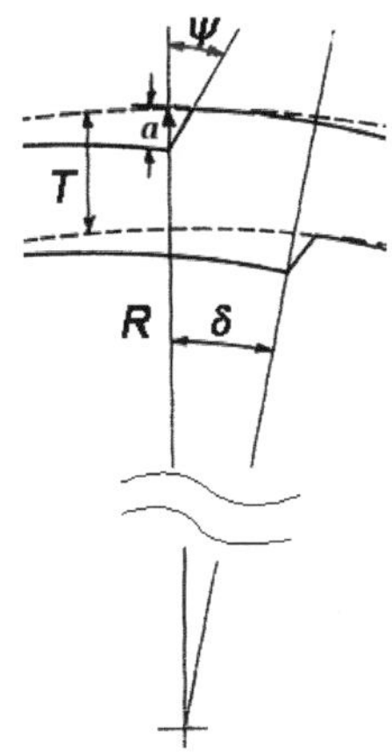

Figure 1. Corrugated shell cross-section [3].

Content from this work may be used under the terms of the Creative Commons Attribution 3.0 licence. Any further distribution of this work must maintain attribution to the author(s) and the title of the work, journal citation and DOI.
Published under licence by IOP Publishing Ltd

Such fluted shells demonstrate several non-trivial effects. High-speed jet, which is produced from these shells can change its spinning direction. For example, if we monotonically increase index angle δ then, surprisingly, jet angular velocity can increase/decrease or even change the sign to the opposite one. The same non-monotonic behavior is observed when number of flutes n is changed.

There were several ideas explaining these effects [3-9]. The most prominent one is a combination of two mechanisms – "thick-thin" and "transport" [3]. These hypotheses are mainly qualitative and cannot help formulate any quantitate relations about jet angular velocity. Besides that, explanation of the angular velocity sign change with these ideas is a bit cloudy. It is unclear how exactly "thick-thin" and "transport" mechanisms can reverse angular velocity. These ideas limitations were the reason to study fluted liners rotational movement in a more detailed manner based on the numerical calculations technique.

2. Numerical calculations

Numerical modelling was performed in ANSYS LS-DYNA. We used 2D Euler multi-material method with plain symmetry to study dynamics of a corrugated shell cross-section motion. Cross-section dimensions were chosen similar to the paper [3]: a= 0.25 mm; T= 1.0 mm; ψ = 30°; n= 16; R= 15.0 mm, δ = 6°.

Multi-material Eulerian mesh region was a square with dimensions 84 x 84 mm filled by void. Total number of elements is 705,600 (size of an element 0.1 x 0.1 mm). On the region boundaries, free flow condition was applied. The detonation was initiated at the outer circular boundary of the explosive.

Two group of material considered in simulations – explosives, with JWL equation of state and elastic perfectly plastic medium, which is modeled copper in the corrugated shell and aluminum in a case. Geometrical model of the simulation shown in Figure 2.

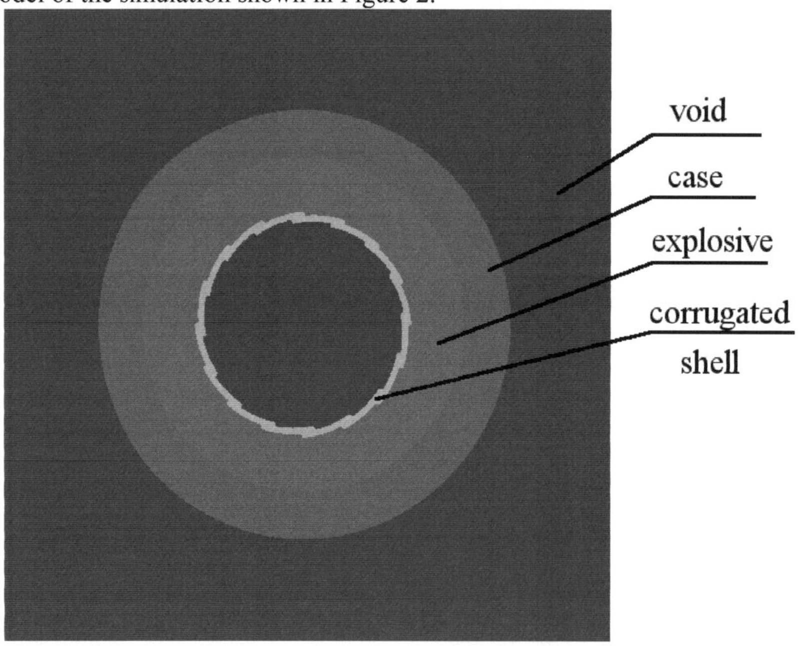

Figure 2. Geometrical model of the cross-section, which is used in simulations (blue – void; gray – case; green – explosive; yellow – corrugated shell).

First of all, we tried to understand jet angular velocity sign change phenomenon. Figure 3 represents angular velocity sign distribution in the corrugated liner cross-section. As we can see from Figure 3a, there are two group of regions in the shell – rotating counterclockwise and clockwise. This is due to the presence of the rifling on the shell outer surface. Expanding gas pushes each side of every corrugation wave to the direction perpendicular its surface. So, in Fig. 3a, each wave left side is accelerating clockwise and right side – counterclockwise.

After collapsing process is almost complete (Figure 3b) some parts of the shell turn out to be in the vicinity of the symmetry axis. If they have predominantly counterclockwise rotation (Figure 3b) then a jet will also rotate in a counterclockwise direction and vice versa.

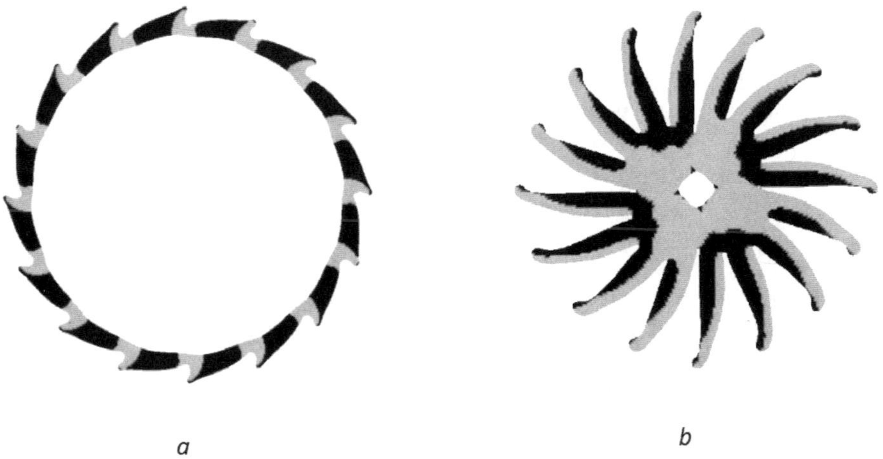

Figure 3. Angular velocity sign distribution in the corrugated shell cross-section: a) beginning of the collapsing process; b) ending of the collapsing process (gray regions – counterclockwise rotation, black regions – clockwise rotation).

Which part of the liner will be near the symmetry axis and produce a jet? It depends on their radial velocity during collapsing process. Parts with maximum radial velocity will be the first near the axis of symmetry, produce a jet and, consequently, define its angular velocity. According to the Gurney equations collapsing velocity increasing if a liner thickness decreasing [1]. So, parts of a liner with minimal thickness will have the maximum radial velocity and, consequently, produce a jet. This simple analysis is confirmed by Figure 3. and can be used to explain the non-monotonic dependence of a jet angular velocity on index angle.

If we calculate angular velocity distribution and define shell regions which are producing a jet for two value of index angle: 6 and 18 degrees (Figure 4) then we can clearly see the legitimacy of the performed analysis.

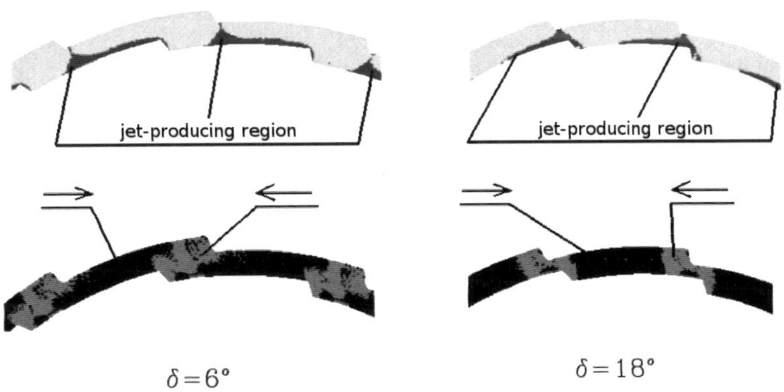

Figure 4. Correspondence between angular velocity distribution in the corrugated liner cross-section and jet-producing regions for different index angle values.

The left side of Figure 4 corresponds to δ = 6°, right side corresponds to δ = 18°. If we impose bottom row of Figure 4 on an upper row then we can clearly see that different values of δ make different regions of a shell thinner, so they move faster and produce jet with opposite rotation direction.

After obtaining a qualitative understanding of corrugated shell rotation we tried to calculate dependencies of a shell angular velocity from every key parameter (shell cross-section dimensions, properties of a shell materials) and build general dimensionless formula.

Corrugation depth a is the first parameter which was analyzed. It can be shown on a simple analytical model that a shell angular velocity must be proportional to a. The same result was obtained from numerical solution and it is in a good agreement with data from [3].

Dependencies from a radius of the external lateral surface R and wall thickness T are trickier and can't be obtained from the simple analysis. This happens because of additional factors come into a play, such as dependence of n from R etc. Numerical calculation and experimental results [3] are shown that shell angular velocity is proportional to R and inversely proportional to T.

Influence of an angle between radial direction and the shortest corrugation side ψ on a shell angular velocity can be derived from oblique shock waves theory [13]. For the values of ψ up to 30 degrees there is no significant influence on shell rotation. This result is also confirmed by [3] and numerical calculations.

The most interesting design parameters are an angle of indexing δ and a total number of corrugation waves n. As we mentioned above, shell angular velocity depends on them in a non-monotonic way which is quite difficult to describe in a simple analytic manner. That is why numerical calculation results were approximated by constructed dependencies and implement in a general formula.

Shell material was modeled by an elastic perfectly plastic model which is using three parameters – material density ρ, Young's modulus E and yield strength σ_t. Numerical calculations show that influence of material density ρ on a fluted shell angular velocity is small and proportional to the $\rho^{-1/4}$. Young's modulus is a more extreme case. It has no effect on shell spinning. Yield strength σ_t influence is more or less predictable. With an increase of σ_t shell angular velocity decreasing in such a manner that at σ_t = 1200 MPa shell angular velocity is equal to 0 rad/s. In the report [3] there is no data about an influence of material parameters on fluted shell rotation.

Summing up all numerical calculation results, we made an attempt to create a general dimensionless formula which describes which angular velocity can be compensated by a fluted shell:

$$\left(\omega R \Big/ \sqrt{\sigma_t/\rho}\right) = \left(\frac{114}{4\cdot\pi}\right)\cdot\left(\frac{\Delta p(\alpha,\beta)}{\rho\cdot V^2}\right)\cdot\left(V\Big/\sqrt{\sigma_t/\rho}\right)\cdot\left(\frac{a}{R}\right)\cdot\left(\frac{R}{T}\right)\cdot\delta_m^{0.1}\times$$
$$\times\left(0.25 - j_1(21.7\cdot\delta)\right)\cdot\left(j_1\left(0.25\cdot n^{0.89}\right)\right)\cdot\left(\exp\left(-295\cdot\left(\frac{\sigma_t}{\Delta p(\alpha,\beta)}\right)^2\right)\right), \quad (1)$$

where ω – angular velocity of a corrugated shell, which can be compensated; $\Delta p(\alpha,\beta)$ – pressure difference between adjacent sides of a wave (α and β are a wave sides inclination angles to the circumferential direction); V – shell velocity in a radial direction according to the Gurney formula; j_1 – spherical Bessel function of the first order, $j_1(x) = \frac{\sin(x)}{x^2} - \frac{\cos(x)}{x}$; δ_m – jet relative mass.

General formula is valid in the following ranges of dimensionless quantities:

$\frac{|\Delta p(\alpha,\beta)|}{\rho\cdot V^2} = 0...0.66$; $V\Big/\sqrt{\sigma_t/\rho} = 47...\infty$; $\frac{a}{R} = 0.017...0.05$; $\frac{R}{T} = 1.67...30$; $\delta_m = 0...1$; $n = 5..80$; $\delta = 0..2\cdot\pi/n$; $\frac{\sigma_t}{\Delta p(\alpha,\beta)} = 0...3$.

3. Conclusion

Numerical analysis of a corrugated shell rotation under loading of expanding gases was performed. We investigate the influence of different design parameters and shell material properties on shell rotation. Non-monotonic dependencies of shell angular velocity on index angle δ and number of corrugation waves n were explained. Based on the results of this analysis general dimensionless formula of an angular velocity which can be compensated by a fluted liner was obtained.

References

[1] Walters W P and Zukas J A 1989 *Fundamentals of Shaped Charges* (New York: Wiley–Interscience)
[2] Koch A, Jaggy P, Jaun W and Haller F 2001 *Proc. 19th Int. Symp. on Ballistics* (Interlaken) pp 1501–08
[3] August W T *et al.* 1959 *Ballistic Research Laboratories Report № 905* (Maryland: Aberdeen Proving Ground) p 347
[4] Simon J, DiPersio R and Eichelberger R J 1954 *Ballistic Research Laboratories Memorandum Report № 1231* (Maryland: Aberdeen Proving Ground) p 38
[5] Cox C M 1964 *Ballistic Research Laboratories Report № DA-33-019-ORD-3697* (Maryland: Aberdeen Proving Ground) p 147
[6] Kipp M E, Martinez R R, Hertel E S, Baker E L, Fuchs B E and Chin C L 1999 *Proc. 18th Int. Symp. on Ballistics* (San-Antonio) pp 499–506
[7] Babkin A V, Rassokha S S, Ladov S V and Odintsov V.A. 2013 Fluted liners and spin-compensation. Part 1 *Izvestia RARAN* **77** 113-25
[8] Babkin A V, Rassokha S S, Ladov S V and Odintsov V A 2013 Fluted liners and spin-compensation. Part 2 *Izvestia RARAN* **78** 87-92
[9] Smelikov V G 2016 *Stability of confined cumulation* (Moscow: Etnika Studio)
[10] Rassokha S S, Ladov S V, Kubyshkina G A and Babkin A V 2013 Performance Calculation of Shaped Charges with Shear-Formed Liners *J. Appl. Mech* **80** 031703
[11] Livermore Software Technology Corporation 2015 *LS-DYNA Keyword User's Manual* (Livermore: Livermore Software Technology Corporation)
[12] Livermore Software Technology Corporation 2015 *LS-DYNA Theory Manual* (Livermore: Livermore Software Technology Corporation)
[13] Selivanov V V, Kobylkin I F and Novikov S A 2014 *Explosive technologies* (Moscow: Bauman Press).

IOP Conf. Series: Journal of Physics: Conf. Series **894** (2017) 012141 doi:10.1088/1742-6596/894/1/012141

State dynamics of bubbly cavitation in a vicinity of quasi-empty rupture at its collapse

E S Bol'shakova and V K Kedrinskiy

Lavrentyev Institute of Hydrodynamics, Novosibirsk, Russia

E-mail: kedr@hydro.nsc.ru

Abstract. The presentation deals with one of the experimental models of a quasi-empty rupture which is formed in the liquid layer of a distilled cavitating fluid under shock loading. It is shown that the rupture is shaped as a spherical segment, which retains its topology during the entire process of its evolution and collapsing. The dynamic behavior of the quasi-empty rupture and of the structure of cavitating flow on its surface are analyzed. It is shown that rupture implosion is accompanied by the SW radiation and a transformation of the bubble boundary layer to a cavitating cluster, which takes the form of a ring-shaped vortex floating upward to the free surface of the liquid layer.

1. Introduction

Explosive volcanic processes have been intensely investigated in both experiments and numerical multi-phase models [see, for example, 1, 2]. It's well known that magma melt is saturated by a high-pressure gas. The formation of rarefaction waves induced by conduit opening violates the thermodynamic equilibrium state: diffusion processes behind the fronts of these waves arise, and intense development of bubble cavitation is observed, which can lead to magma melt rupture. An experimental model of rupture formation in a liquid layer was proposed in [3]. Those experiments were performed in electromagnetic hydrodynamic shock tubes (EM HSTs) with the diapason of storage energies up to 1.2 kJ [3]. It was shown that the shock wave generated in the liquid layer by a membrane under the action of a pulsed magnetic field initiates the development of intense cavitation processes in the liquid layer after SW reflection from its free surface. The pulsed motion of the membrane leads to the formation of a quasi-empty rupture already (in 10 μs) on the motionless membrane. The dynamics and structure of this rupture can be considered as a qualitative model of the above-mentioned process. Distilled water is a multiphase medium with very small gas bubbles, solid micro-particles, and their combinations with sizes of about 1.5 μm and densities of the order of 10^6 cm^{-3} [4]. New experimental data on the dynamics of the state of the bubble boundary layer on the rupture interface and on its transformation to a cavitating cluster at the rupture implosion "point" are reported below.

2. Statement of experiments

The test section of the small experimental setup (EM HST) includes a transparent cuvette 4 cm in diameter, which is filled with distilled water [4]. The bottom of the cuvette is a conducting Duralumin membrane, which is placed directly onto a flat helical coil connected via an electronic switch to a high-voltage capacitor bank. When the switch is closed, a pulsed magnetic field is generated on the coil,

Content from this work may be used under the terms of the Creative Commons Attribution 3.0 licence. Any further distribution of this work must maintain attribution to the author(s) and the title of the work, journal citation and DOI.
Published under licence by IOP Publishing Ltd

which pushes the membrane due to the skin effect, thus, generating a powerful SW in the liquid (the SW amplitude is of the order of 10 MPa). The process is recorded by a high-speed video camera with a frequency up to 10^5 frames/s.

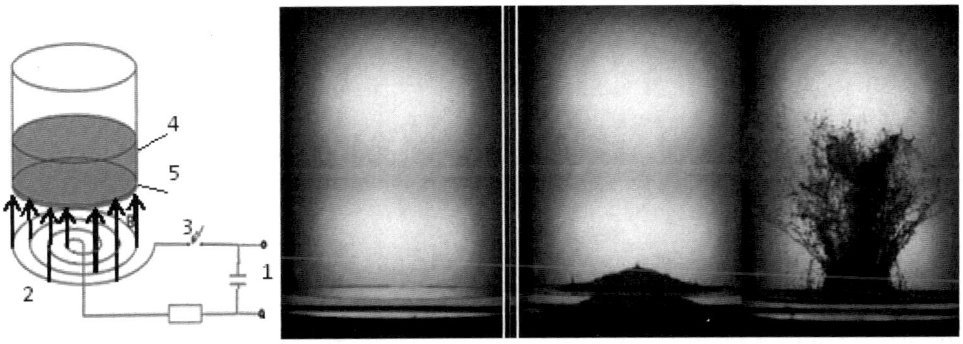

Figure 1. Principle scheme of EM HST: (4) – cuvette, high-voltage capacity bank - (1), (3) – electronic switch, (2) - flat helical coil, (5) – liquid layer. Characteristic dynamics of eruption (experimental model).

3. Experimental results.

When the electronic switch is closed, an SW is generated in the liquid layer. This SW is reflected from the free surface of the layer in the form of a rarefaction wave and leads to the formation of an intense cavitation region (**Fig. 2 a**). In 0.5 ms, a maximum size of rupture is formed on the membrane (**Fig. 2, b**). A bubble boundary layer on the surface, which covers the entire rupture interface, is clearly visible. The next two frames (**Figs. 2, c and 2, d**) demonstrate the dynamics of the collapsing rupture and SW radiation.

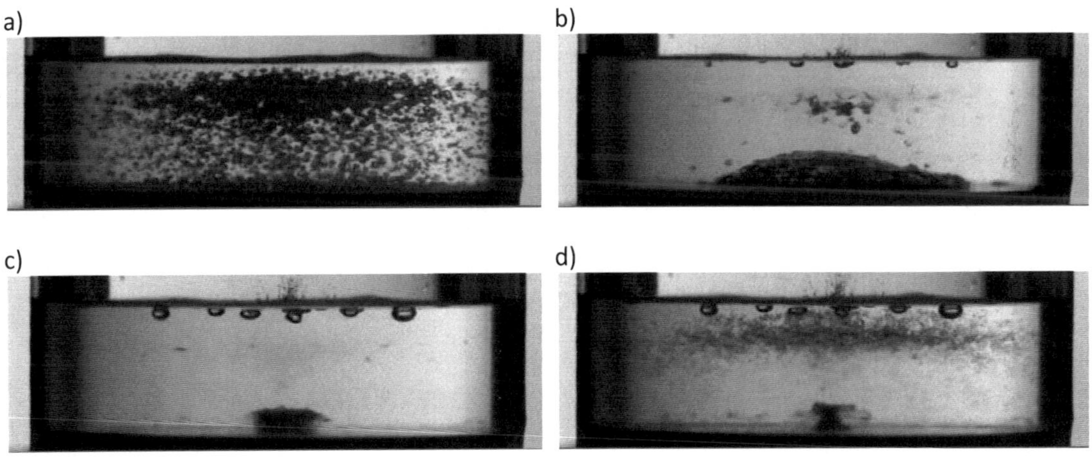

Figure2. Rupture dynamics. The loading energy 32 J, at t = a) 0.1; b) 0.5;c) 1.04;d)1.1ms.

The analysis of the rupture state dynamics shows that it acquires a shape close to a spherical segment. It is known that an implosion of an empty cavity in a homogeneous liquid near a solid wall is accompanied by the formation of a vertical cumulative jet. In the present case, however, such a situation is not observed: the rupture implodes without changing its topology. It is of interest that the liquid flow on the membrane in the course of the rupture implosion forms a ring-shaped flow of a homogeneous liquid, which is closed under the rupture with a high velocity (**Fig. 2, c**). At the end of the rupture

IOP Conf. Series: Journal of Physics: Conf. Series **894** (2017) 012141 doi:10.1088/1742-6596/894/1/012141

implosion, a dense core is formed, and the bubble boundary layer transforms to a bubble cluster surrounding the rupture. The state of the rupture surrounded by a dense bubble cluster at the instant of the implosion can be defined as a state with a high intenal energy. This is confirmed by generation of a sufficiently powerful secondary SW after the rupture implosion (**Fig.2. d**). Propagation of this SW and its interaction with the free surface leads to the formation of the secondary cavitation region (**Fig. 2d**). It was demonstrated earlier that the rupture retains its topology, i.e., a shape similar to a spherical segment. Based on this fact, we use the dimensional method, assuming that the basic parameters of the liquid state are the hydrostatic pressure p_0 and density ρ of the liquid, the storage energy E, and the coefficient **α** characterizing the ratio of the maximum potential energy (**see Fig.2b**) of the rupture to the loading energy **α** = Q/E. Let us find a combination of parameters that yields the time characteristics of the examined process:

$$T = \alpha^{1/3} E^{1/3} \rho^{1/2} . p_0{}^{-5/6} \quad (1)$$

The formulas of the rupture dynamics have the form: for its radius $r \approx 13.4 + 6.4 \cdot t - 19 \cdot t^{\ 2}$, and for its height $h \approx 0.3 + 14 \cdot t - 1t^2$

The value of **α** predicted by Eq. (1) allows to determine the value for the maximum potential energy: α = 0.0032 or Q≈ 0.1 J (at loading E= 32 J) and α = 0.02 or Q ≈ 25 J (at loading E= 1.25 kJ). It is of interest that the experimental results obtained for identical storage energy, but for different heights of the water layer in the cavity are close to each other. Thus, it is confirmed that the dynamics of cluster formation depends to a greater extent on the loading energy and is almost independent of the water layer height. **Fig 3** illustrates the cluster dynamics after the rupture implosion and generation of the secondary SW. **Fig 3a** show the secondary cavitation zone near free surface of liquid layer as result of SW- reflection.

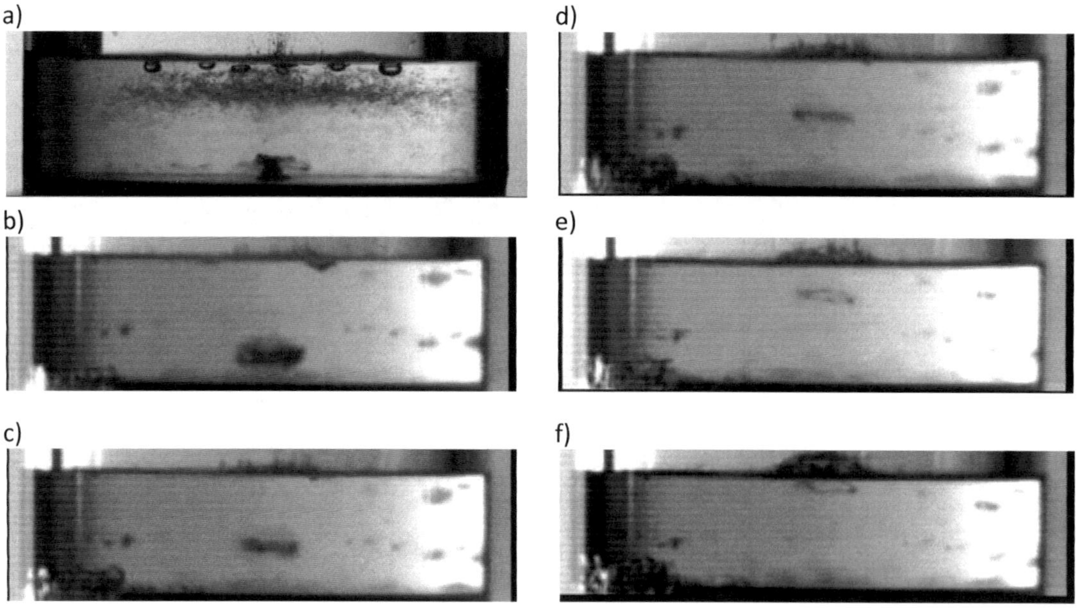

Figure 3. Dynamics of the bubble cluster. The loading energy is 32 J; the time instants are 1.1 (a), 1.3 (b), 1.7 (c), 2.1 (d), 2.7 (e), and 3.1 ms (f).

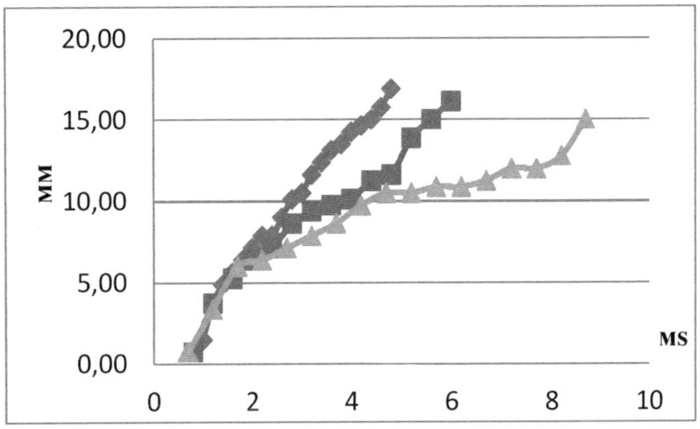

Figure 4. Dynamics of bubble cluster emergence for different loading energies. The water layer height is 1.5 cm. The loading energy is 32 (blue), 24.5 (red), and 18 J (green).

It is seen that the cluster emerges to the water layer surface and possesses a sufficient amount of the initially stored energy to move to the liquid layer surface almost uniformly (1. 5 cm in 2 ms). The character of the bubble cluster motion and the fact that a ring-shaped flow along the membrane is formed due to the rupture implosion allow us to assume that the bubble boundary layer transforms after the closure of the quasi-empty rupture to a ring-shaped cluster (vortex).

The results of the experimental analysis for the cuvette (membrane) diameter of 4 cm, water layer height of 1.5 cm, and loading energy of 18, 24.5, and 32 J are shown in **Fig.4**. It is seen that the plot of motion is an almost straight line for the greatest energy examined in this study. As the energy decreases, the plot becomes curved, and the emergence velocity becomes smaller. The bubble cluster velocity is approximately equal to 7.5 m/s for the storage energy of 32 J.

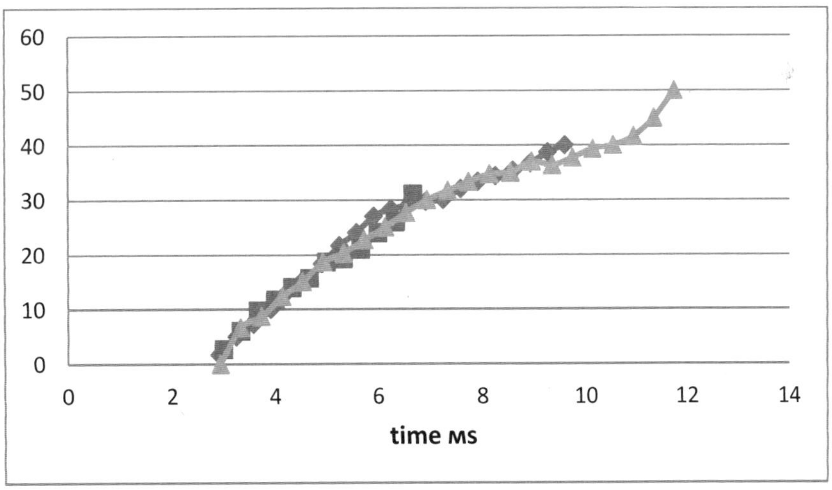

Figure 5. Dynamics of vertical coordinate of center cluster (**mm**). Diameter of cuvette (membrane) - 12.5 cm, energy of loading (permanent) - 0,8 kJ. Height of water layer - red 3 cm, blue – 4 cm, green – 5 cm.

IOP Conf. Series: Journal of Physics: Conf. Series **894** (2017) 012141 doi:10.1088/1742-6596/894/1/012141

4. Conclusions

The analysis of the experimental data shows that shock wave loading of the liquid layer leads to the rupture development on the bottom of the cuvette. The rupture shape is close to a spherical segment, and there is a bubble boundary layer on the rupture interface. After the implosion, this boundary layer transforms to a bubble cluster. The rupture implosion generates the secondary shock wave, which is confirmed by the formation of the secondary cavitation region and which forms the cluster structure. After shock wave generation, the bubble cluster emerges to the liquid layer surface. The character of the cluster dynamics indicates the formation of a ring-shaped vortex moving upward to the liquid surface. It is shown that the cluster motion is almost independent of the liquid layer height and is mainly determined by the loading energy.

Acknowledgments

This work was supported by the Russian Foundation for Basic Research (Grant No. 15-05-03336a) and partly by Siberian Branch of the Russian Academy of Sciences (Project No. III.22.3.1).

References

[1] Woods W 1995 *Reviews of geophysics* **33** (4) 495–530

[2] Gonnermann H M and Manga M 2007 *Annu. Rev. Fluid Mech.* **39** 321–56

[3] Bolshakova E S and Kedrinskiy V K 2016 *Jornal of Physics: Conference Series* **754** 042003

[4] Besov A S, Kedrinskiy V K and Palchikov E I 1984 *Letters in JTPh.* **10** (4) *67–70*

[5] Kedrinskiy V K and Bolshakova E S 2015 *The Journal of the Acoustical Society of America* **138**(3) *1829*

Institute of Physics
Dirac House, Temple Back
Bristol BS1 6BE UK

ISSN: 1742-6588
ISBN 978-1-5108-5034-7

9 781510 850347

CURRAN ASSOCIATES INC.
proceedings
.com

Annual Conference of the American Forage and Grassland Council (AFGC 2020)

The More You Know, The Smarter You Grow

Greenville, South Carolina, USA
5 – 8 January 2020

American Forage & Grassland Council